Botany

Jones & Bartlett Learning Titles in Biological Science

AIDS: Science and Society, Sixth Edition
Hung Fan, Ross F. Conner, & Luis P. Villarreal

AIDS: The Biological Basis, Fifth Edition
Benjamin S. Weeks & I. Edward Alcamo

Alcamo's Fundamentals of Microbiology, Ninth Edition
Jeffrey C. Pommerville

Alcamo's Fundamentals of Microbiology, Body Systems Edition, Second Edition
Jeffrey C. Pommerville

Alcamo's Microbes and Society, Third Edition
Benjamin S. Weeks

Aquatic Entomology
W. Patrick McCafferty & Arwin V. Provonsha

Bioethics: An Introduction to the History, Methods, and Practice, Third Edition
Nancy S. Jecker, Albert R. Jonsen, & Robert A. Pearlman

Bioimaging: Current Concepts in Light and Electron Microscopy
Douglas E. Chandler & Robert W. Roberson

Biomedical Graduate School: A Planning Guide to the Admissions Process
David J. McKean & Ted R. Johnson

Biomedical Informatics: A Data User's Guide
Jules J. Berman

Case Studies for Understanding the Human Body, Second Edition
Stanton Braude, Deena Goran, & Alexander Miceli

Clinical Information Systems: Overcoming Adverse Consequences
Dean F. Sittig & Joan S. Ash

Defending Evolution: A Guide to the Evolution/Creation Controversy
Brian J. Alters

The Ecology of Agroecosystems
John H. Vandermeer

Electron Microscopy, Second Edition
John J. Bozzola & Lonnie D. Russell

Encounters in Microbiology, Volume 1, Second Edition
Jeffrey C. Pommerville

Encounters in Microbiology, Volume 2
Jeffrey C. Pommerville

Essential Genetics: A Genomics Perspective, Sixth Edition
Daniel L. Hartl

Essentials of Molecular Biology, Fourth Edition
George M. Malacinski

Evolution: Principles and Processes
Brian K. Hall

Exploring Bioinformatics: A Project-Based Approach
Caroline St. Clair & Jonathan E. Visick

Exploring the Way Life Works: The Science of Biology
Mahlon Hoagland, Bert Dodson, & Judy Hauck

Genetics: Analysis of Genes and Genomes, Eighth Edition
Daniel L. Hartl & Maryellen Ruvolo

Genetics of Populations, Fourth Edition
Philip W. Hedrick

Guide to Infectious Diseases by Body System, Second Edition
Jeffrey C. Pommerville

Human Biology
Daniel D. Chiras

Human Biology Laboratory Manual
Charles Welsh

Human Embryonic Stem Cells, Second Edition
Ann A. Kiessling & Scott C. Anderson

Introduction to the Biology of Marine Life, Tenth Edition
John F. Morrissey & James L. Sumich

Laboratory and Field Investigations in Marine Life, Tenth Edition
Gordon H. Dudley, James L. Sumich, & Virginia L. Cass-Dudley

Laboratory Fundamentals of Microbiology, Ninth Edition
Jeffrey C. Pommerville

Laboratory Investigations in Molecular Biology
Steven A. Williams, Barton E. Slatko, & John R. McCarrey

Laboratory Textbook of Anatomy and Physiology: Cat Version, Ninth Edition
Anne B. Donnersberger

Lewin's CELLS, Second Edition
Lynne Cassimeris, Vishwanath R. Lingappa, & George Plopper

Lewin's Essential GENES, Third Edition
Jocelyn E. Krebs, Elliott S. Goldstein, & Stephen T. Kilpatrick

Lewin's GENES X
Jocelyn E. Krebs, Elliott S. Goldstein, & Stephen T. Kilpatrick

Mammalogy, Fifth Edition
Terry A. Vaughan, James M. Ryan, & Nicholas J. Czaplewski

The Microbial Challenge: Science, Disease, and Public Health, Second Edition
Robert I. Krasner

Microbial Genetics, Second Edition
Stanley R. Maloy, John E. Cronan, Jr., & David Freifelder

Microbiology Pearls of Wisdom, Second Edition
S. James Booth

Molecular Biology: Genes to Proteins, Fourth Edition
Burton E. Tropp

Neoplasms: Principles of Development and Diversity
Jules J. Berman

Perl Programming for Medicine and Biology
Jules J. Berman

Plant Biochemistry
Florence K. Gleason with Raymond Chollet

Plant Cell Biology
Brian E. S. Gunning & Martin W. Steer

Plants, Genes, and Crop Biotechnology, Second Edition
Maarten J. Chrispeels & David E. Sadava

Plants & People
James D. Mauseth

Plant Structure: A Color Guide, Second Edition
Bryan G. Bowes & James D. Mauseth

Precancer: The Beginning and the End of Cancer
Jules J. Berman

Principles of Modern Microbiology
Mark Wheelis

Protein Microarrays
Mark Schena, ed.

Python for Bioinformatics
Jason Kinser

R for Medicine and Biology
Paul D. Lewis

Restoration Ecology
Sigurdur Greipsson

Ruby Programming for Medicine and Biology
Jules J. Berman

Strickberger's Evolution, Fourth Edition
Brian K. Hall & Benedikt Hallgrímsson

Symbolic Systems Biology: Theory and Methods
M. Sriram Iyengar

Tropical Forests
Bernard A. Marcus

20th Century Microbe Hunters
Robert I. Krasner

Understanding Viruses, Second Edition
Teri Shors

Botany

Fifth Edition

An Introduction to Plant Biology

James D. Mauseth
University of Texas at Austin

JONES & BARTLETT
LEARNING

World Headquarters
Jones & Bartlett Learning
5 Wall Street
Burlington, MA 01803
978-443-5000
info@jblearning.com
www.jblearning.com

Jones & Bartlett Learning books and products are available through most bookstores and online booksellers. To contact Jones & Bartlett Learning directly, call 800-832-0034, fax 978-443-8000, or visit our website, www.jblearning.com.

Substantial discounts on bulk quantities of Jones & Bartlett Learning publications are available to corporations, professional associations, and other qualified organizations. For details and specific discount information, contact the special sales department at Jones & Bartlett Learning via the above contact information or send an email to specialsales@jblearning.com.

Production Credits
Chief Executive Officer: Ty Field
President: James Homer
SVP, Editor-in-Chief: Michael Johnson
SVP, Chief Marketing Officer: Alison M. Pendergast
Publisher: Cathleen Sether
Senior Acquisitions Editor: Erin O'Connor
Editorial Assistant: Rachel Isaacs
Editorial Assistant: Michelle Bradbury
Production Manager: Louis C. Bruno, Jr.
Senior Marketing Manager: Andrea DeFronzo
V.P., Manufacturing and Inventory Control: Therese Connell
Composition: Circle Graphics, Inc.
Cover Design: Michael O'Donnell
Rights & Photo Research Associate: Lauren Miller
Cover Image: © Tiburon Studios/ShutterStock, Inc.
Title Page Image: Courtesy of Kate Rolland
Printing and Binding: Courier Kendallville
Cover Printing: Courier Kendallville

To order this product, use ISBN: 978-1-4496-6580-7

Library of Congress Cataloging-in-Publication Data
Mauseth, James D.
  Botany : an introduction to plant biology / James Mauseth. — 5th ed.
    p. cm.
  Includes index.
  ISBN 978-1-4496-4884-8 (alk. paper)
  1. Botany—Textbooks. I. Title.
  QK47.M38 2013
  580—dc23
                    2012008432

6048

Printed in the United States of America
16 15 14 13 12   10 9 8 7 6 5 4 3 2 1

BRIEF CONTENTS

v

CONTENTS

This fifth edition of *Botany: An Introduction to Plant Biology* differs from previous editions by the addition of more coverage and emphasis on the ways in which our own biology—the biology of us humans—has an impact on plant biology. All editions of this book have discussed the importance of plants to people: plants supply us with oxygen, food, clothing, building materials, drugs, and so on. While researching and writing a new book, *Plants & People*, I became more aware that many aspects of plant biology are now affected by us and our biology. All of our ancestors for millions of years have eaten plants and breathed plant-produced oxygen, so there has never, ever been a human biology that was not one-hundred-percent dependent on plant biology. But now we people are so numerous (there are seven billion of us) that there is no longer any plant biology that is not influenced by us. Our respiration produces carbon dioxide, which plants need for photosynthesis, so this part of our biology helps plants. But we also burn coal, oil, gasoline, and firewood at such a rapid rate that we are adding much more carbon dioxide to the air than ever before—so much that we are altering the atmosphere. The extra carbon dioxide allows plants to photosynthesize and grow faster, but it also traps solar heat (this is called the greenhouse effect), and it is, thus, causing global climate change. Parts of Earth are becoming warmer, others cooler; some are drier, others have more rain, snow, and fog. Certain rivers flood more often; others are drying up. We are changing habitats throughout the entire world; we are affecting all plants no matter where they live.

We people need to eat; we like to eat. All our food comes from plants, whether we are eating fruits and vegetables or meat from an animal that was fed with corn, oats, or grass. Agriculture started at several widely separated places around the world between 9 and 11 thousand years ago, just after several great ice sheets melted away, and people have been cutting down forests and clearing land for farms, pastures, and cities ever since. We dam rivers and divert water to irrigation systems. We people have been altering habitats on a gigantic scale for thousands of years. We also produce waste on a gigantic scale, waste from our toilets, from factories, from processing plants, from fertilizer that runs off from fields and lawns during rain, from stockyards, and from chicken houses. All these send waste into our rivers or ground water, altering the biology of plants and algae that live in rivers, wetlands, and oceans. For agriculture or just for gardening, we move plants from their natural homes to new areas, thus introducing foreign species into new habitats, where some grow vigorously and crowd out native plants.

I do not want this Preface to be all doom and gloom. We can take pride in that we now recognize that plant and human biology are actually one biology; we recognize the harm we are doing, and many of us are already working to minimize our negative impacts. Some people eat less meat or none at all, thus reducing the amount of land needed for agriculture. Some of us recycle paper, aluminum, and plastics to reduce the amount of trash that goes into landfills. Many of us drive smaller cars or drive less; we walk or bicycle or take the bus. We can make a difference; we are already making a difference.

This fifth edition of *Botany* was written and illustrated to alert you to the solutions as well as to the problems. At this point in human history, we must look at plants in a new way, we must analyze the interactions of plants and people more carefully than ever before. I hope that this book helps you do both, and I hope that you help discover new ways for people to live more gently with the other organisms that share this planet.

New Features of the Fifth Edition

Most of the changes in this fifth edition were made to add emphasis to the interactions of plants and people. Box 11-5 Plants and People: Respiration discusses the ways in which we pollute rivers and ground water with nutrients that encourage bacterial growth resulting in bacterial respiration that depletes oxygen in the water and contributes to "dead zones" in oceans. Box 14-2 Plants and People: Environmental Stimuli and Global Climate Change points out that as climate changes, growing seasons become longer. Plants whose dormancy is controlled only by temperature can respond to this and grow for a longer period. Those plants whose dormancy is controlled by day length, however, cannot take advantage of the extra days of growing season. Consequently, one type of plant may out-compete the other. Box 23-2 Plants and People: Vegetarianism: Alternatives to Eating Meat discusses the many profound ways that our food choices affect animals and ecology, and it gives students information to think about the ethics and ecology of eating foods derived from animals. Other new boxes are:

1-3 Plants and People: Algae and Global Warming (discusses greenhouse gases)

2-2 Botany and Beyond: Lipids: Oils, Fats, *Trans*-Fats, and Human Health (points out that *trans*-fats are not only unnatural but also dangerous, unlike the essential fatty acids we obtain from food plants)

5-2 Plants and People: Parenchyma, Sclerenchyma, and Food (discusses why we eat certain plant parts but not others)

13-4 Plants and People: Fertilizers, Pollution, and Limiting Factors (discusses why we use low-phosphate detergents to reduce some of the harmful effects of water pollution)

Several parts of the text have been rewritten to emphasize issues that are now critically important: the greenhouse effect, global climate change, consequences of pollution, deforestation, and habitat loss, for example. In addition, new photos have been added to illustrate various aspects of agriculture; for example, crops such as wheat, cherries, potatoes, fern fiddleheads, and mushrooms, or agricultural phenomena such as harvesting wheat and sugar cane, letting fields lie fallow, selecting for wheat that will not "lodge."

Features of This Edition

The following features are used throughout *Botany* to help students learn.

Part Openers   Each of the book's four parts is introduced by a brief summary of all the chapters in that part. The opener ties together the main themes and shows how botany is a unified science, not just a body of facts to memorize.

Concepts   Each chapter opens with a section on concepts that will set the stage for the main topics and themes covered in the chapter. It provides an overview and outline of what is to follow.

Alternatives Boxes   The objective of Alternatives boxes is to show students they should think expansively. While the text describes the most common, typical aspects of plant biology, there are alternative types that are more advantageous in certain conditions.

Chapter 3   Unusual cells
Chapter 4   Rates of Growth
Chapter 5   Familiar Plants and Some Confusing Look-Alikes
Chapter 5   Simple Plants
Chapter 6   Photosynthesis Without Leaves
Chapter 10   Photosynthesis in Bacteria and Cyanobacteria
Chapter 11   Respiration in Prokaryotes
Chapter 11   Heterotrophic Plants
Chapter 12   Desert Plant Biology
Chapter 14   Simple Bodies and Simple Development in Algae
Chapter 16   Genetics of Haploid Plants

Plants Do Things Differently Boxes   These boxes compare plant biology with human biology. All of us—students, instructors, and lay people—are familiar with how our own bodies work. We understand our blood circulation, we are aware that our skeleton can grow despite being strong enough to support our weight, and we know that good nutrition requires a balanced diet of carbohydrates, fats, proteins, vitamins, and minerals. The same phenomena in plants—circulation, support, and nutrition—can appear to be completely different from the processes occurring within our bodies: rather than circulate, water moves one way in xylem, and the same is true of sugar transport in phloem. Plant nutrition mostly involves having enough dirt and water, not proteins and vitamins. In many cases, as students study a botany textbook, they might assume they are misunderstanding what they are reading about plants—plant biology just seems too exotic, too different from the students' own metabolism. The new boxes reassure students that they are understanding their reading correctly—plants really are doing things very differently from the way we do them—and the boxes go on to show students that they should not expect plants and humans to carry out these processes in the same ways. But the boxes also explain that the plant-type of biology is functional and adaptive for plants whereas our animal type of biology would not be. As an example, in Chapter 3, the box Calcium: Strong Bones, Strong Teeth, but Not Strong Plants helps students understand the plant support/skeletal system of lignified cell walls by comparing them with the biology of an animal system of calcium strengthened bones.

Plants and People Boxes   These boxes discuss ways in which plants and people influence each other. Some plants influence people by producing poisonous or irritating compounds; others produce food, medicine, and beauty. In the other direction, human activities influence plants either directly by habitat destruction and the farming of "wastelands" or by producing acid rain and global climate change. Less obviously, plants have affected the way we humans think by being the organisms in which we discovered cells, mineral nutrition, genes, genetics, artificial selection, and so on.

Botany and Beyond   Botany and Beyond boxes elaborate on subjects that, while not essential to the study of botany, help make the material more relevant and accessible.

Illustrations   The botanical world is full of color, and so is this text. Figure illustrations have been chosen to illuminate points made in the text and to show many of the plants under discussion. Many of the drawings have been redone and are both beautiful and botanically accurate. Photographs have been added where our adopters or reviewers have indicated a need for greater variety or multiple views. For this edition, many new photographs have been added to provide superior visual aids. Extensive labels have been added to many photographs to clarify the features or structures shown. Many of the micrographs are introduced with either diagrams or low-magnification micrographs to help the reader understand the orientation of the tissues in the high-magnification illustration. Selected light and electron micrographs are accompanied by interpretive line drawings to make the photographs more understandable to students.

Additional Features

Summary   The summary at each end of the chapter consists of numbered points that succinctly list the major topics and concepts discussed.

Important Terms   This feature gives a list of terms that should be understood after completing study of the chapter. Students can now review these terms online at BotanyLinks.

Review Questions   The study questions at the end of each chapter have been designed to act as a study guide, to lead students to the most important points, and to focus students' efforts on mastering the most significant concepts. "Fill-in-the-blank" questions are sentences taken verbatim from the text. The objective is for students to go back into the chapter and find the sentence: The words that correspond to the blanks are key words involving a critical concept they should pay attention to. Other questions are long, with multiple parts designed to take the students through a complex process step by step. These questions have many hints to let the students know where there are concepts that are easily misunderstood or whose importance is easily overlooked. Many of these questions draw on the students' knowledge of their own bodies.

Glossary   A comprehensive glossary defines major botanical and general biological terms. Each definition is keyed to the chapter where the principal discussion occurs.

Web Exercises   Included at the end of most chapters, original web-integrated activities invite students to explore topics in greater depth on the Web, while encouraging the use of their critical thinking skills. Jones & Bartlett Learning's extensive botany home page at **http://biology.jbpub.com/botany5e** provides the links to help them in their research along with brief introductions to place the links in context. Students can hand-in or email answers to their instructor at the instructor's discretion. Jones & Bartlett Learning monitors the links regularly to ensure that there will always be a working and an appropriate site.

Ancillaries

http://biology.jbpub.com/botany/5e   In addition to the **Web Exercises**, the website developed exclusively for *Botany* and prepared by Tharindu Weeraratne of the University of Texas at Austin, contains a variety of resources to assist in the study of Botany. The **OnLine Glossary** and the **Animated Flashcards** provide students with help for definitions of important botany terms. The flashcards correspond to the Important Terms found at the end of most chapters of the text. **BotanyLinks** point to websites that have relevance to a particular topic in each chapter. **Botany and Beyond** links go to sites that elaborate on topics not vital to understanding botany but are interesting and noteworthy. Information found here will deepen your understanding of botany by providing details on subjects such as leaf structure, acid rain, and classifying unknown plants. **Plants and People** links go to sites with information on human and plant biology comparisons. Exploring the differences between human and plant biology will give you a firmer grasp of plant structure and function. **Alternatives** links point to sites that detail strategies that plants exhibit in certain conditions. The information provided about structures and processes will explain how plants act under various conditions. **Plants Do Things Differently** links go to sites that highlight the differences between plant biology and human biology. Topics covered at these sites will show the similarities and differences between the two. Finally, **Study Quizzes** provide short true/false and multiple-choice questions for each chapter so that students can test their knowledge of the content.

Instructor's Media CD

The Instructor's Media CD, compatible with Windows and Macintosh platforms, provides adopters with the following traditional ancillaries:

- The *PowerPoint Image Bank* provides the illustrations, photograph, and tables (to which Jones & Bartlett Learning holds the copyright or has permission to reprint digitally) inserted into PowerPoint slides. With the Microsoft PowerPoint program, you can quickly and easily copy individual image slides into your existing lecture slides. If you do not own a copy of Microsoft PowerPoint or a compatible software program, a Microsoft PowerPoint Viewer is included on the CD-ROM.
- The *PowerPoint Lecture Outlines*, prepared by Dr. Stephanie Harvey of Georgia Southern State University, provides lecture notes and images for each chapter of *Botany: An Introduction to Plant Biology*. Instructors with the Microsoft PowerPoint software can customize the outlines, art, and order of presentation.
- The *Instructor's Manual*, also prepared by Tharindu Weeraratne of University of Texas at Austin, contains key chapter outlines, lecture suggestions, and laboratory exercises. It also has a wealth of supplemental information that can be included in lectures to provide extra motivation for students.
- A *Test Bank*, also prepared by Tharindu Weeraratne of University of Texas at Austin, of multiple-choice, true/false, matching, and short essay questions.
- *Unlabeled Art* consists of approximately 39 illustrations. These can be used as part of a labeling exercise in an exam or can be photocopied and given to students for taking notes. Instructors can add their own labels to customize them to their course.
- Solutions to Review Questions. The files for *Solutions to Review Questions* contain answers to all of the end-of-chapter review questions in the text.

Acknowledgments

This and previous editions have benefited from the generous, conscientious thoughts of many reviewers. They provided numerous suggestions for improving clarity of presentation, or identified illustrative examples that would improve the student's understanding and interest. It has been a pleasure to work with them. I thank them all:

Vernon Ahmadjian, Clark University
Bonnie Amos, Angelo State University
John Beebe, Calvin College
Curtis Clark, California State Polytechnic University, Pomona
Billy G. Cumbie, University of Missouri, Columbia
Jerry Davis, University of Wisconsin, La Crosse
Cynthia J. Denbow, Virginia Polytechnic Institute and State University
Nicole Donofrio, University of Delaware
Rebecca McBride-DiLiddo, Suffolk University
John Dubois, Middle Tennessee State University
Donald S. Emmeluth, Fulton-Montgomery Community College
Nisse Goldberg , Jacksonville University
Howard Grimes, Washington State University
Stephanie G. Harvey, Georgia Southwestern State University
James Haynes, State University College at Buffalo
James C. Hull, Towson University
Shelley Jansky, University of Wisconsin, Stevens Point
Roger M. Knutson, Luther College
John C. Krenetsky, Metropolitan State College of Denver
Lillian Miller, Florida Community College at Jacksonville, South Campus

Louis V. Mingrone, Bloomsburg University
Rory O'Neil, University of Houston, Downtown
John Olsen, Rhodes College
Jerry Pickering, Indiana University of Pennsylvania
Mary Ann Polasek, Cardinal Stritch College
Barbara Rafaill, Georgetown College
Michael Renfroe, James Madison University
Michael D. Rourke, Bakersfield College
Sangha Saha, Harold Washington College
James L. Seago, Jr., State University of New York at Oswego
Bruce B. Smith, York College of Pennsylvania
Garland Upchurch, Southwest Texas State University
Jack Waber, West Chester University
James W. Wallace, Western Carolina University
Katherine Warpeha, University of Illinois at Chicago
Peter Webster, University of Massachusetts, Amherst
Paula S. Williamson, Southwest Texas State University
Ernest Wilson, Virginia State University
Mark Wilson, Oregon State University
Stephen Wuerz, Highland Community College

Just like the initial production of a textbook, the preparation of a new edition is not by any means the sole effort of the author. I am fortunate to have benefited from the many contributions of numerous talented individuals through the various editions. The current editorial staff at Jones & Bartlett Learning is one of the best and most skillful. I especially thank Anne Spencer, Rachel Isaacs, Lauren Miller, Louis Bruno, Michelle Bradbury, and Molly Steinbach for their intelligent, creative solutions to many problems that had to be solved in preparing the fifth edition. I worked especially closely with Lou Bruno on all aspects of this book as well as *Plants & People*; he has a remarkable ability to keep track of dozens of details, as well as keeping everything organized and on track. I also thank my partner Tommy Navarre for his never-ending (28 years so far) support, encouragement, and confidence.

James D. Mauseth
Austin, Texas

Introduction to Plants and Botany

Concepts

Earth is becoming hotter, flooding is more frequent, and weather is more violent because we burn coal, oil, and other fossil fuels, which releases carbon dioxide into the atmosphere. Carbon dioxide is one of several greenhouse gases that allow visible sunlight to pass through the atmosphere and strike Earth's surface, heating it. The warmer rocks, soil, and water give off infrared radiation back out toward space, but greenhouse gases absorb infrared light and heat the atmosphere. It is a simple relationship: the higher the concentration of greenhouse gases in the atmosphere, the hotter the climate. As Earth becomes hotter, more water evaporates out of the oceans into the air, where it then falls as heavy rains, causing flooding throughout the world. The warmer air also causes snow and ice on mountain tops to melt faster, increasing flooding. Every summer brings more mudslides in California, larger floods on the Mississippi

Chapter Opener Image: People and all other animals must have oxygen to live. Without oxygen, we would die of asphyxiation. The plants here are photosynthesizing in the sunshine, producing the oxygen we need. The roots of these trees hold the soil in place, even on steep slopes. Without the trees, rain would wash all the soil away leaving just bare rock and this river would flood after every rain or be almost dry when there is a dry period. The lighter green rectangular area is where all the trees were cut down (called clear-cutting), probably to obtain fibers to make paper or lumber for construction. Although new plants are growing there again, most of the native plants and animals of this area were killed or frightened away during the clear-cutting; that area will not grow back to be a natural forest.

and other rivers, and more violent tornadoes and hurricanes. This is global warming, also known as global climate change.

What does this have to do with botany? Everything. Plants in the sun photosynthesize; that is, they take carbon dioxide out of the air and use it to make the chemicals that compose their bodies. Most of the weight of leaves, stems, roots, flowers, fruits, and seeds is carbon that was carbon dioxide in the air before plants captured it. As plants photosynthesize, they remove carbon dioxide from the air and lock it into their bodies, helping to keep Earth cool and counteracting the warming we are causing. An important question now is "Can plants remove enough carbon dioxide from the air to counteract the damage we are doing?" The answer is "probably not."

The balance between the addition of atmospheric carbon dioxide by us and its removal by plants is affected by several factors that are easy to understand. All animals, fungi, bacteria, and other nonphotosynthetic organisms produce carbon dioxide just by being alive. As our bodies "burn" our food (technically, as they respire it), carbon dioxide is produced; therefore, animals (including humans) have been adding carbon dioxide to the atmosphere for billions of years. The real problem began when our ancestors discovered fire: We then began burning wood and coal and, more recently, petroleum, natural gas, and other fossil fuels, adding carbon dioxide to the air in huge quantities. Until our mastery of fire, plants were actually taking carbon dioxide out of the air faster than our respiration was adding it.

Photosynthesis originated 2.8 billion years ago, and the amount of carbon dioxide in the air has decreased and Earth has cooled, until the start of the Industrial Revolution when we began burning massive amounts of fuel. If photosynthesis has been removing carbon dioxide from the atmosphere, does that mean there was more carbon dioxide in the air in the past? And, if so, was Earth warmer in the past? The answer to both questions is yes. Earth formerly had much more carbon dioxide in its atmosphere and consequently was much hotter. Earth's climate is changing now, but it has always been changing and has never stayed the same for long periods of time.

In addition to respiration and burning, carbon dioxide is added to the air as volcanoes erupt and as magma (molten rock) comes upward at mid-ocean ridges between the giant tectonic plates that carry the continents on Earth's surface. Carbon dioxide is also removed as certain algae build shells of calcium carbonate: All limestone rock on Earth is composed of vast numbers of microscopic shells of certain algae, clams, and other marine animals. At times in the past volcanoes were very active and added carbon dioxide faster than photosynthesis could remove it, causing Earth to heat up. At other times they were inactive and photosynthesis outpaced volcanism and Earth cooled.

Neither heating nor cooling has ever been severe enough to risk killing all life on Earth. Instead, when Earth was warm, rains were also heavy (because of the warm oceans), so plants grew faster and more abundantly, absorbing more carbon dioxide. When plants take most of the carbon dioxide out of the air, Earth cools and dries, and plant growth slows.

Today, we are at an unusually cool period in Earth's history. Plants have taken so much carbon dioxide out of the air there is almost none to trap the sun's heat. We are actually in an ice age right now, known as the Holocene Ice Age, but we are in its warm period, known as an interglacial period (cold periods of ice ages are called glacials because glaciers are then common on almost all mountains). Is it bad that Earth is unusually cool? Should we be burning even more petroleum and coal to heat it up?

The current coolness is exceptional, but it is the climate in which we evolved and became distinct from the other great apes. It is also the climate in which most of our food plants evolved: Wheat, rye, barley, and corn are grasses that flourish under cool, dry conditions. Grasses grow on open, treeless plains, but when Earth is warm most of its surface is covered by forests, and grasses do not grow well in the shade below trees. You may know that a significant step in the evolution of us modern humans is that, unlike our ancestors who were adapted to living in trees, we gradually evolved to walk upright on the ground, freeing our forelimbs (our arms) such that we could use our hands for holding and manipulating tools. We did not come down out of the trees until open grasslands finally appeared on Earth, and those came about in the last 30 million years as Earth became cooler, drier, and the forests receded, all due to plant photosynthesis.

More recently, we people began to cultivate our own food. Agriculture is new, having separate, independent origins in Europe, Asia, Africa, and the Americas less than about 11,000 years ago. That is a significant number: The current interglacial period we are living in now began only 14,000 years ago. Snow and ice began to melt away, people spread across more of the land, and some humans made the journey from Asia to the Americas at that time. In the very short time of just a few thousand years between 14,000 years ago and 11,000 years ago, humans progressed from being wandering hunter-gatherers to starting the first farms, then establishing villages and towns, and then civilization began with art, writing, religion, and science.

Let's go back to our original question: "What does this have to do with botany?" Again, the answer is everything. Plants changed the climate of Earth such that we can now live on it. Plants also produce the oxygen we breathe and the food we eat. We get cloth, paper, lumber, and chemicals from plants, and plants are important to us spiritually because of their beauty.

As you study the following pages, think of the many ways in which plant biology affects our own biology. And think of other organisms; we share Earth's surface not only with plants but also with all other animals, fungi, and microbes. All our biologies affect those of all other organisms, as we are all interconnected and interdependent.

Plants

Botany is the scientific study of plants. This definition requires an understanding of the concepts "plants" and "scientific study." It may surprise you to learn that it is difficult to define precisely what a plant is. Plants have so many types and variations that a simple definition has many exceptions, and a definition that includes all plants and excludes all nonplants may be too complicated to be useful. Also, biologists do not agree about whether certain organisms—particularly algae—are indeed plants. Rather than memorizing a terse definition, more is gained by understanding what plants are, what the exceptional or exotic cases are, and why botanists disagree about certain organisms.

Figure 1-1 This evening primrose (*Oenothera*) is obviously a flowering plant. It has a short stem and numerous simple leaves; its extensive root system is not visible here.

Figure 1-3 Ferns have several features in common with flowering plants; they have leaves, stems, and roots; however, they never produce seeds, and they have neither flowers nor wood.

Your present concept of plants is probably quite accurate: Most plants have green leaves, stems, roots, and flowers (**Figure 1-1**), but you can think of exceptions immediately. Conifers such as pine, spruce, and fir have cones rather than flowers (**Figure 1-2**), and many cacti and succulents do not appear to have leaves. Both conifers and succulents, however, are obviously plants because they closely resemble organisms that unquestionably are plants. Similarly, ferns and mosses (**Figures 1-3** and **1-4**) are easily recognized as plants. Fungi, such as mushrooms (**Figure 1-5**) and puffballs, were included in the plant kingdom because they are immobile and produce spores, which function somewhat like seeds; however, biologists no longer consider fungi to be plants because recent observations show that fungi differ from plants in many basic biochemical and genetic respects.

Algae are more problematical. One group, the green algae (**Figure 1-6**), is similar to plants in biochemistry and cell structure, but it also has many significant differences. Some botanists conclude that it is more useful to include green algae with plants; others exclude them, pointing out that some green algae have more in common with the seaweeds known as red algae and brown algae (**Figure 1-7**). Arbitrarily declaring that green algae are or are not plants solves nothing; the important thing is to understand the concepts involved and why disagreement exists (**Table 1-1**).

Scientific Method

The concept of a scientific study can be understood by examining earlier approaches to studying nature. Until the 15th century, several methods for analyzing and explaining the universe and its phenomena were used, with religion and speculative philosophy being especially important. In religious methods, the universe is assumed to either be created by or contain deities. The important feature is that the actions of gods cannot be studied: They are either hidden or capricious, changing from day to day and altering natural phenomena. Agricultural studies would be useless because some years crops might flourish or fail because of weather or

Figure 1-2 Conifers, like this spruce (*Picea*), produce seeds in cones; the conifers, together with the flowering plants and a few other groups, are known as seed plants.

Figure 1-4 Of all terrestrial plants, mosses have the least in common with flowering plants. They have structures called "leaves" and "stems," but these are not the same as in flowering plants. They have no roots at all.

(a)

(b)

Figure 1-5 Fungi such as (a) mushrooms and (b) brackets are not considered to be plants. They are never green and cannot obtain their energy from sunlight. Also, their tissues and physiology are quite different from those of plants. Fungi are important to plants, however, because many fungi break down dead material in the soil such as fallen leaves and rotting tree trunks; as the fungi cause these materials to rot, they release minerals and enrich the soil.

disease, but in other years, crop failure might be due to a god's intervention (a miracle) to reward or punish people. There would be no reason to expect consistent results from experiments. In a religious system, much of the knowledge of the world comes as a revelation from the deity rather than by observation and study of the world. A fundamental principle of all religions is faith: People must believe in the god without physical proof of its existence or actions.

Speculative philosophy reached its greatest development with the ancient Greek philosophers. Basically, their method of analyzing the world involved thinking about it logically. They sought to develop logical explanations for simple observations and then followed the logic as far as possible. An example is the philosophical postulation of atoms by Democritus around 400 BCE (before the common era). From the observation that all objects could be cut or broken into two smaller objects, it follows logically that the two pieces can each be subdivided again into two more, and so on. Finally, some size must be reached at which further subdivision is not possible; objects of that size are atoms.

Speculative philosophy did not involve verification; philosophical predictions were made, but no actual experiment or observation was performed to see if they were correct. A speculation is a statement that cannot be proved or disproved (e.g., "If Elvis were still alive, he would still be performing in Las Vegas."). A problem with this method is that often several alternative conclusions are equally plausible logically; only experimentation reveals which is actually true.

Starting before the 1400s, a new method, called the **scientific method**, began to develop slowly. Several fundamental tenets were established:

1. *Source of information. All accepted information can be derived only from carefully documented and controlled observations or experiments.* Claims emanating from priests or prophets—or scientists—cannot be accepted automatically; they must be subjected to verification and proof. For example, for hundreds of years, medicine was taught using a text called *Materia Medica* written by Galen, a Roman physician who lived

Figure 1-6 These green algae do not look much like plants, but many aspects of their biochemistry and cellular organization are very similar to those of plants. Some green algae were the ancestors of land plants; although not considered to be true plants, they are obviously closely related to plants.

Figure 1-7 These brown algae (*Fucus*), commonly called kelp, have very plant-like bodies as a result of convergent evolution: they are not true plants. Their biochemistry, genetics, anatomy, and reproduction differ greatly from those of plants.

TABLE 1-1	The Three Domains of Organisms

Prokaryotes
    Domain Archaea
    Domain Bacteria (including cyanobacteria)
Eukaryotes
    Domain Eukarya
        Protista: single-cell organisms (protozoans, algae); multicellular algae
        Kingdom Myceteae: fungi such as mushrooms, puffballs, bread mold
        Kingdom Animalia: animals
        Kingdom Plantae:* plants
            Division Bryophyta: mosses
            Division Pteridophyta: ferns
            Division Coniferophyta: conifers
            Division Magnoliophyta: flowering plants

*Within kingdom Plantae, many botanists recognize about 17 divisions; only the four most familiar are listed here. Many botanists conclude that algae should be included in kingdom Plantae.

in the second century C.E. (common era = A.D.). In the early 1500s, Andreas Vesalius began dissecting human corpses and noticed that in many cases Galen had been mistaken. Vesalius promoted the idea that observation of the world itself was more accurate than accepting undocumented claims, even if the claims had been made by an extremely famous, respected person.

2. *Phenomena that can be studied. Only tangible phenomena and objects are studied,* such as heat, plants, minerals, and weather. We cannot see or feel magnetism or neutrons, but we can construct instruments that detect them reliably. In contrast, we do not see or feel ghosts, and no instrument has ever detected ghosts reliably: If ghosts do exist, they must be intangible and cannot be studied by the scientific method. Anything that cannot be observed cannot be studied.

3. *Constancy and universality. Physical forces that control the world are constant through time and are the same everywhere.* Water has always been and always will be composed of hydrogen and oxygen; gravity is the same now as it has been in the past. The world itself changes—mountains erode, rivers change course, plants evolve—but the forces remain the same. Experiments done at one time and place should give the same results if they are carefully repeated at a different time and place. Constancy and universality allow us to plan future experiments and predict what the outcome should be: If we do the experiment and do not get the predicted outcome, it must be that our theory was incorrect, not that the fundamental forces of the world have suddenly changed. This prevents people from explaining things as miracles or the intervention of evil spirits. For example, if someone claims that a new drug cures a particular disease, we can check that by testing the same drug against that disease. If it does not work, the first person may have (1) made an innocent mistake, (2) tested the drug on people who would have gotten better anyway, or (3) been committing fraud; however, we do not have to worry that the difference in the two experiments is due to the fundamental laws of chemistry and physics having changed or that the first experiment's outcome was altered by benevolent spirits and the second by evil spirits.

4. *Basis. The fundamental basis of the scientific method is skepticism,* the principle of never being certain of a conclusion, of always being willing to consider new evidence. No matter how much evidence there is for or against a theory, it does no harm to keep a bit of doubt in our minds and to be willing to consider more evidence. For example, there is a tremendous amount of evidence supporting the theory that all plants are composed of cells, and there is no known evidence against it. All of our research, all of our teaching assumes that plants indeed are composed of cells, but the concept of skepticism requires that if new, contrary evidence is presented, we must be willing to change our minds. As a further example, consider people who have been convicted of crimes and then later—often years later—DNA-based evidence indicates that they are innocent: Skepticism is the willingness to consider new evidence.

Scientific studies take many forms, but basically, they begin with a series of observations, followed by a period of experimentation mixed with further observation and analysis. At some point, a hypothesis, or model, is constructed to account for the observations: A **hypothesis** (unlike a speculation) must make predictions that can be tested. For example, scientists in the Middle Ages observed that plants never occur in dark caves and grow poorly indoors where light is dim. They hypothesized that plants need light to grow. This can be formally stated as a pair of simple alternative hypotheses: (1) Plants need light to grow, and (2) plants do not need light to grow. The experimental testing may involve the comparison of several plants outdoors, some in light and others heavily shaded, or it may involve several plants indoors, some in the normal gloom and others illuminated by a window or a skylight. Such experiments give results consistent with hypothesis 1; hypothesis 2 would be rejected.

A hypothesis must be tested in various ways. It must be consistent with further observations and experiments, and it must be able to predict the results of future experiments: One of the greatest values of a hypothesis or theory is its power as a predictive model. If its predictions are accurate, they support the hypothesis; if its predictions are inaccurate, they prove that the hypothesis is incorrect. In this case, the hypothesis predicts that environments with little or no light will have few or no plants. Observations are consistent with these predictions. In a heavy forest, shade is dense at ground level, and few plants grow there (Figure 1-8). Similarly, as light penetrates the ocean, it is absorbed by water until at great depth all light has been absorbed; no plants or algae grow below that depth.

If a hypothesis continues to match observations, we have greater confidence that it is correct, and it may come to be called a **theory**. Occasionally, a hypothesis does not match an observation; that may mean either that the hypothesis must be altered somewhat or that the entire hypothesis has been wrong. For instance, plants such as Indian pipe or *Conopholis* (Figure 1-9) grow the same with or without light; they do not need light for growth. These are parasitic plants that obtain their energy by drawing nutrients from host plants. Thus, our hypothesis needs only minor modification: All plants except parasitic ones need sunlight for growth. It remains a reasonably accurate predictive model.

(a)

(b)

Figure 1-8 (a) This aspen forest in Michigan does not have a dense canopy, but it intercepts so much light that few plants survive in the shade. The herb is the bracken fern *Pteridium aquilinum.* (b) Near the aspen forest is an open area with more light; herb growth, in this case a sedge, is much more abundant.

Note the four principles of the scientific method here. First, the hypothesis is based on observations and can be tested with experiments; we do not accept it simply because some famous scientist declared it to be true. Second, sunlight and plant growth are tangible phenomena that we can either see directly or measure with instruments. Third, if we repeat the experiment anytime or anywhere, we expect to get the same results. Fourth, we interpret the evidence as supporting the hypothesis, but we keep an open mind and are willing to consider new data or a new hypothesis.

The concept of intelligent design has recently been proposed to explain many complex phenomena. Its fundamental concept is that many structures and metabolisms are too complicated to

Figure 1-9 The yellowish flowers pushing out of the pine needle litter constitute almost the entire plant body of this parasitic plant, *Conopholis mexicana.* It is attached to the roots of nearby trees and draws nutrients from them. Like fungi, it cannot obtain its energy from sunlight, but so many other aspects of its anatomy and physiology are like those of ordinary plants that we have no difficulty in recognizing that this is a true plant, not a fungus.

have resulted from evolution and natural selection. Instead, they must have been created by some sort of intelligent force or being. This may or may not be true, but this does not help us to analyze and understand the world; instead, it is used as an answer in itself that prevents further study. Photosynthesis is certainly complex, and it may have been designed by some intelligent being; however, believing that does not help us to understand photosynthesis at all, and it does not help us to plan future experiments. In contrast, the scientific method is a means through which we are discovering even the most subtle details of photosynthesis.

Areas Where the Scientific Method Is Inappropriate

Certain concepts exist for which the scientific method is inappropriate. We all believe that it is not right to wantonly kill each other, that racism and sexism are bad, and that things such as morality and ethics exist; however, both morality and ethics have no chemical composition, no mass, no electromagnetic spectrum—they are not tangible and thus cannot be studied by the scientific method. Science can study, measure, analyze, and describe the factors that cause people to kill each other or to be racist or sexist, and it can predict the outcome of these actions. Science, however, cannot say whether such actions are right or wrong, moral or immoral. Consider euthanasia: Many types of incurable cancer cause terrible pain and suffering in their final stages, which may last for months. We have drugs that can arrest breathing so that a person dies painlessly and peacefully. Science developed the drugs and can tell us the metabolic effects of using them, but it cannot tell us whether it is right to use them to help a person die and avoid pain. Biological advances have made us capable of surrogate motherhood, of detecting fetal birth defects early enough to allow a medically safe abortion, and of producing insecticides that protect crops but pollute the environment. These advances have made it more important than ever for us to have well-developed ethical philosophy for assessing the appropriateness of various actions.

Using Concepts to Understand Plants

The growth, reproduction, and death of plants—indeed, all aspects of their lives—are governed by a small number of basic principles. Each chapter in this text opens with a section called "Concepts," which discusses the principles most relevant to the topics in that particular chapter. Here in this chapter and at the beginning of your study of botany and plants, I want to introduce you briefly to some of these principles and to encourage you to use them as you read and think about plants. These concepts will make plant biology more easily understood—the numerous facts, figures, names, and data will be less overwhelming when you realize that they all fit into the patterns governed by a few fundamental concepts.

1. *Plant metabolism is based on the principles of chemistry and physics.* Weeds may seem to appear from nothing as if by magic; however, that is never true—they grow from seeds. All the principles you learn in your chemistry or physics classes are completely valid for plants.

2. *Plants must have a means of storing and using information.* After a seed germinates, it grows and develops into a plant, becoming larger and more complex; then it reproduces. The plant is taking in energy and chemical compounds and transforming them into the organic chemical compounds it uses to build more of itself. This requires a complex, carefully controlled metabolism, and there must be a mechanism for storing and using the information that regulates that metabolism. As you may already know, genes are the primary means of storing this information.

3. *Plants reproduce, passing their genes and information on to their descendants.* Because an individual obtains its genes from its parents, the information it uses to control its metabolism is similar to the information its parent had used; thus, offspring and parents resemble each other. For example, a bean seed contains genes whose information guides the seed's metabolism into constructing a new bean plant, but a tomato seed grows into a tomato plant because it received different genes and information from its parents (Figure 1-10).

4. *Genes, and the information they contain, change.* As plants make copies of their genes during reproduction, accidental changes (mutations) occasionally occur, and this causes the affected gene and its information to change. This is quite rare, and most genes (and information) are passed unaltered from parents to offspring; however, as mutations occur and change a gene's information, they basically generate new information such that the plant that grows and develops under the control of the mutated gene may be slightly (or significantly) different from its parents. Thus, over time, a gradual evolution occurs in the genes, information, and biology of plants. Consequently, in a large population of many individuals of a species, some variation exists; the individuals are not identical (Figure 1-11).

5. *Plants must survive in their own environment.* They must be adapted to the conditions in the area where they live. If they are not adapted to that area's conditions, they grow and reproduce poorly or die prematurely. Other plants whose genes result in characters that make those plants more suited to live

Figure 1-10 This bean seed is developing into a bean plant, guided by genetic information it inherited from its parents.

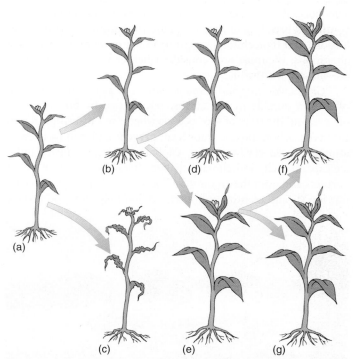

Figure 1-11 (a) A plant produces numerous offspring, many of which resemble it strongly (b). Mutations may occur that cause, for instance, leaves to be malformed and poorly shaped for photosynthesis (c); most or all these mutants die and do not reproduce. The normal plants continue to reproduce (b and d), but another mutation may occur that causes the leaves to be larger and more efficient at photosynthesis (e). These may grow and reproduce so well that they crowd out the original parental types, and the plant population finally contains only the type with large leaves (f and g).

Plants and People

Box 1-1 Plants and People, Including Students

Plants and people affect each other. The most obvious perhaps are the ways that people benefit from plants: They are the sources of our food, wood, paper, fibers, and medicines. It is difficult to excite students by listing the world production of wheat and lumber in metric tons, but just consider what your life would be like without chocolate, coffee, tea, sugar, vanilla, cinnamon, pepper, strawberries, mahogany, ebony, cotton, linen, roses, orchids, or the paper that examinations are written on. The oxygen we breathe comes entirely from plants. Plants affect each of us every day, not simply by keeping us alive but also by providing wonderful sights, textures, and fragrances that enrich our existence.

However, plants and people affect each other in ways that are not readily apparent in our day-to-day lives. Listed here are a few important topics you should be aware of. Think about their importance and how you—as an actual biological organism—interact with the other organisms on this planet.

Biotechnology is a set of laboratory techniques that allow us to alter plants and animals, giving them new traits and characteristics. Farmers have done this for thousands of years with controlled breeding of the best plants and animals (artificial selection), but biotechnology permits much more rapid, extensive alterations. We must now consider whether such manipulations are safe and worthwhile.

Global warming and *climate change* are caused by a buildup of carbon dioxide in our atmosphere caused by burning coal, oil, gas, and the trees of forests everywhere (not just tropical rain forests). Carbon dioxide traps heat, preventing Earth from radiating excess energy into space. Global warming is causing polar ice caps to melt, and climate change alters circulation of ocean currents and even the amount and pattern of rainfall. Preserving our forests and planting more trees might help stop and reverse global warming, but the possibility exists that global warming is preventing the occurrence of another ice age.

Desertification is the conversion of ordinary forest or grassland to desert. Accurate measurements are difficult, but it appears that deserts may be spreading as people cut shrubs and trees for firewood and allow goats to eat remaining vegetation. Once an area has been converted to desert, its soil is rapidly eroded, making recovery difficult. Something as simple as cheap solar cookers might prevent the Sahara desert from spreading farther across Africa.

Habitat loss results when an area is changed so much that a particular species can no longer survive in the area. Significant causes are the construction of highways, housing subdivisions, and shopping malls with enormous parking lots; these eliminate almost all species from an area, but habitats are also lost by logging, farming, mining, damming rivers, and spilling toxic chemicals. As habitat is lost, plants or animals must try to survive on the smaller remaining habitat. Once too little habitat is left, species usually become extinct.

Introduced exotics are organisms native to one part of the world but which have been brought to another part, where they thrive. Examples of introduced exotic animals are fire ants in the southern United States and zebra mussels in the Great Lakes region. Water hyacinth, purple loosestrife, and kudzu (a vine) were introduced to the United States and are now proliferating and reproducing so vigorously that they are crowding out many plants that normally grow here.

It is not realistic to believe that we humans will stop all activities that have negative impacts on our environment and on the other species with which we share this planet, but we can search for ways to minimize the harm we cause by recycling, conserving resources, and avoiding products that require pollution-causing manufacturing techniques.

(a)

(b)

Figure B1-1 Habitat loss is caused by many types of human activity. (a) This church parking lot covers acres of land previously used for grazing. Now it is used only 2 or 3 hours 1 day a week. Other than the few trees that were spared, it has no plants or animals, it prevents rain from soaking into the ground, and the asphalt leaches harmful chemicals into nearby creeks. No other business is nearby that could use this parking lot on weekdays or at night, which would at least provide additional benefit to offset the ecological damage it causes. (b) Even the construction of beautiful parks is habitat destruction.

in that area grow and reproduce more successfully and produce more offspring. Also, plants do not exist in isolation: A significant aspect of a plant's environment is the presence of other organisms. Some neighboring organisms may be helpful to the plant; others may be harmful, and most perhaps have little effect on it. This concept can be important when trying to understand a plant's structure and metabolism: One type of photosynthetic metabolism and leaf structure may function well if a particular plant always grows in the shade of taller neighbors, whereas a different type of photosynthetic metabolism and leaf structure may be necessary for a plant that grows nearby but in an unshaded area.

6. *Plants are highly integrated organisms.* The structure and metabolism of one part have some impact on the rest of the plant. When studying the biology of leaves, consider how the structure and metabolism of stems, roots, epidermis, and other parts might affect the function of those leaves. Large leaves absorb more sunlight and energy than small leaves; however, if a plant has large leaves, it may need to have a large root system to absorb water and minerals for the leaves, and it may need wide stems to conduct enough water and minerals from the roots to the leaves. Keep in mind that structure and metabolism must be integrated: The structure of a cell, tissue, or organ must be compatible with the metabolic function of that same cell, tissue, or organ. For example, if a leaf is fibrous and tough, insects may find it unpalatable and may avoid eating it; however, if the leaf is too fibrous, the fibers may block the absorption of sunlight. Such a structure would be incompatible with photosynthesis.

7. *An individual plant is the temporary result of the interaction of genes and environment.* Be careful to consider differences between an individual plant and that plant's species (the group made up of all similar plants). Consider something like a sunflower: An individual plant exists because its parents underwent reproduction and one of their seeds landed in a suitable environment, where the information in the seed's genes interacted with the environment by way of the seed's structure and metabolism. There are two concepts of "sunflower" here: (1) the actual plant that we can observe, measure, cultivate, and enjoy and that interacts with its environment, absorbs resources, responds to changes, attracts pollinators, and resists pathogens and (2) the genetic information that guides all of this and that has existed for thousands of years, evolving gradually as it has been passed down through all the ancestors of this particular individual sunflower. This information does not exist only in this one individual but rather in all of the currently living sunflower individuals. It will continue to exist in future individuals long after this generation has passed away.

8. *Plants do not have purpose or decision-making capacity.* It is easy to speak and write as if plants were capable of thinking and planning. We might say, "Plants produce roots in order to absorb water"; however, this suggests that the plants are capable of analyzing what they need and deciding what they are going to do. Assuming that plants have human characters such as thought and decision-making capacity is called **anthropomorphism**, and it should be avoided. Similarly, assuming that processes or structures have a purpose is called

teleology, and it too is inaccurate. Consider an alternative way of phrasing the sentence: "Plant roots absorb water," leaving out the phrase "in order to." The reality of the situation is that some of the information in the plant's genes causes the plant to produce roots, which have a structure and a metabolism that result in water absorption. Plants have roots because they inherited root genes from their ancestors, not in order to absorb water. Absorbing water is a beneficial result that aids in the survival of the plant, but it is not the result of a decision (anthropomorphism) or purpose (teleology).

Origin and Evolution of Plants

Life on Earth began about 3.5 billion years ago. At first, living organisms were simple, like present-day bacteria, in both their metabolism and structure; however, over thousands of millions of years cells gradually increased in complexity through evolution by natural selection. The process is easy to understand: As organisms reproduce, their offspring differ slightly from each other in their features—they are not identical. Offspring with features that make them poorly adapted to the habitat probably do not grow well and reproduce poorly if they live long enough to become mature (Figure 1-11 and Figure 1-12). Other offspring with features that cause them to be well-adapted grow well and reproduce abundantly, passing on the beneficial features to their own offspring. This is called **natural selection**. New features come about periodically by mutations, and natural selection determines which new features are eliminated and which are passed on to future generations. Evolution by natural selection is a model consistent with observations of natural organisms, experiments, and theoretical considerations.

HERMAN®

"Another one of nature's mistakes."

5-14

© 1985 Universal Press Syndicate

Figure 1-12 Mutations that produce disadvantageous features usually contribute to the death, sooner or later, of an organism. If the individual cannot undergo reproduction (because it is dead), it cannot pass the mutation on to offspring and the deleterious mutation is eliminated. (Herman, Copyright 1985, Universal Press Syndicate. Reprinted with permission. All rights reserved.)

As early organisms became more complex, major advances occurred. One was the evolution of the type of photosynthesis that produces oxygen and carbohydrates. This photosynthesis is present in all green plants, but it first arose about 2.8 billion years ago in a bacterium-like organism called a cyanobacterium. Later, cell structure became more efficient as subcellular components evolved. These components, called organelles, each provide a unique structure and chemistry specialized to a specific function. Division of labor and specialization had come about.

A particularly significant evolutionary step occurred when DNA, the molecule that stores hereditary information, became located in its own organelle—the cell nucleus. Because this step was so important and occurred with so many other fundamental changes in cell metabolism, we classify all cells as prokaryotes if they do not have nuclei (bacteria, cyanobacteria, and archaeans) or as eukaryotes if they do have nuclei (all plants, animals, fungi, and algae) (Table 1-1).

By the time nuclei became established, evolution had produced thousands of species of prokaryotes. The newly evolved eukaryotes also diversified. Some acquired an energy-transforming organelle, the mitochondrion, and some acquired chloroplasts, which carry out photosynthesis and convert the energy in sunlight to the chemical energy of carbohydrates. Those with chloroplasts evolved into algae and plants; those without evolved into protozoans, fungi, and animals (Figure 1-13).

All organisms are classified into three large groups called **domains**: domain Bacteria, domain Archaea, and domain Eukarya; within Eukarya are kingdom Plantae, kingdom Animalia, kingdom Myceteae (fungi), and protists (eukaryotes that do not fit easily into the other three eukaryotic kingdoms). Some protists are closely related to Plantae because some green algae became adapted to living on land and gradually evolved into true plants. As a consequence, early plants resembled those green algae, but as more mutations occurred and natural selection eliminated less adaptive ones, plants lost algal characteristics and gained more features suited to surviving on land. Thousands of species arose, but most became extinct, as those that were more fit grew more rapidly, survived longer, and produced more offspring. Species that did not become extinct evolved into more species and so on. The living plants that surround us are the current result of the continuous process of evolution.

Not all organisms evolve at the same rate; some early species were actually so well adapted that they competed successfully against newer species. Algae are so well suited to life in oceans, lakes, and streams that they still thrive even though most features present in modern, living algae must be more or less identical to those present in the ancestral algae that lived more than 1 billion years ago. Features that seem relatively unchanged are **relictual features** (technically known as **plesiomorphic features**; formerly called primitive features). Like the algae, ferns are well-adapted to

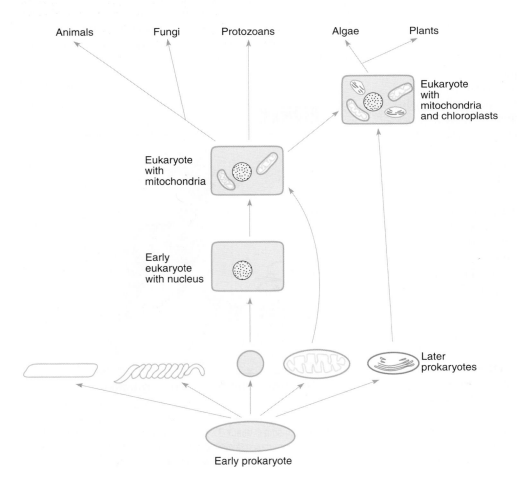

Figure 1-13 The earliest cells were simple, but over hundreds of millions of years they became more complex metabolically. Later, some evolved into simple eukaryotic cells with a true nucleus, which later developed mitochondria. Some of these evolved into protozoans, animals, and fungi. Others developed chloroplasts and evolved into plants. Protozoans and algae are protists.

Figure 1-14 This is *Gazania splendens* in the aster family. It has many derived features, especially the bitter, toxic chemicals (iridoid compounds) in its leaves; lettuce is in the same family, but plant geneticists have eliminated the genes that produce the antiherbivore features. As a result, lettuce is sweet both to us and to insects, rabbits, and deer.

certain habitats and have not changed much in 250 million years; they too have many relictual features. Modern conifers are similar to early ones that arose about 320 million years ago. The most recently evolved group consists of the flowering plants, which originated about 100 to 120 million years ago with the evolution of several features: flowers; broad, flat, simple leaves; and wood that conducts water with little friction. The members of the aster family (sunflowers, daisies, and dandelions) (**Figure 1-14**) have many features that evolved recently from features present in ancestral flowering plants. These are **derived features** (technically known as **apomorphic features**; formerly called advanced features) (i.e., they have been derived evolutionarily from ancestral features). One recent (highly derived) feature in the asters is a group of chemical compounds that discourage herbivores from eating the plants. The terms "primitive" and "advanced" are avoided in that they imply inferior and superior.

In the same way that different groups evolve at different rates, various features also evolve at different rates. For instance, the asters are a mixture of the recently derived antiherbivore compounds, less recently derived flowers, and still less recently derived wood. In addition, their bodies are covered with a waxy waterproof layer, the cuticle, that has not changed much since land plants first evolved about 420 million years ago, and their leaves contain chloroplasts and nuclei that are much like those of green algae and are extremely relictual.

Diversity of Plant Adaptations

More than 288,000 species of plants exist today. An unknown number of species, perhaps also several hundred thousand, existed at one time but have become extinct. Virtually all of this diversity came about through evolution by natural selection—survival of the fittest; however, the existence of 288,000 types of living plants means that there must be at least 288,000 ways of being fit on today's Earth. For any particular aspect of the environment, many

types of adaptation are possible. There is not one single, exclusive, perfect adaptation. Consider plants growing in a climate with freezing winters. Frozen soil is physiologically dry because roots cannot extract water from it. If the winter air is dry and windy, shoots lose water, and plants are in danger of dying from dehydration. How can plants adapt to this? Mutations that cause leaves to fall off in autumn and mutations that cause bark to develop on all exposed parts of the stem reduce the surface area through which water is lost; such mutations are advantageous in this environment, and many species have these adaptations. In other species, the entire shoot system above ground dies, but subterranean bulbs, corms, or tubers persist and produce a new shoot the next spring. Another adaptation is for the plant itself to die but its seeds to live through the harsh conditions. Finally, evergreen species retain both stem and leaves, but the leaves have extra thick cuticles and other modifications that minimize water loss.

This diversity is extremely important, and you must be careful to think in terms of alternative adaptations, alternative methods of coping with the environment. The physical and biological world is made up of gradients: It is not simply either hot or cold but rather ranges continuously from hot in some areas at some times to warm, cool, and cold in the same or other areas at other times. Roots face a range of water availability from flooded, waterlogged soil to moist, somewhat dry, and even arid soil (**Figure 1-15**). Studying and understanding plants and their survival require that we place the plant in the context of its habitat: What are the significant environmental factors? What predators and pathogens must it protect itself from? What physical stresses must it survive? What are the advantageous, helpful aspects of its habitat? When you think in terms of gradients and ranges of habitat factors, you appreciate the range of responses and adaptations that exists.

Figure 1-15 A mountain provides many types of environments. Higher altitudes are always covered with snow and ice; no plants can survive there. In the lower, warmer altitudes, the steep slopes allow rain to run off rapidly, so the soil is dry. Conifers are well-adapted to these conditions. In the flat valley bottoms, water accumulates as marshes; the roots of sedges and rushes can tolerate the constant moisture, but the roots of conifers would drown. Each type of plant is adapted to specific conditions, even though they grow almost side by side.

Plants Versus the Study of Plants

Mathematicians, physicists, and chemists study the world and derive interpretations that are considered universally valid. The relationships of numbers to each other are the same everywhere in the Universe; if intelligent beings exist elsewhere, their mathematicians will discover exactly the same mathematical relationships. The same is true for physics and chemistry.

Some aspects of biological knowledge also are universally valid: All metabolic processes, either on Earth or elsewhere in the Universe, can be predicted by mathematics, physics, and chemistry. Organisms everywhere must take in energy and matter and convert it metabolically into the substance of their own bodies. This cannot be done with perfect efficiency, and thus, all organisms produce waste heat and waste matter. Furthermore, all organisms must be capable of reproduction and must have a system of heredity.

However, biologists study organisms that exist only on Earth, and we are reasonably certain that if life exists on some other planet, it is not exactly like life here (Box 1-2). Much of what we study does not have universal truth; therefore, knowing about certain plants in certain habitats does not let us predict precisely the nature of other plants in other habitats. Much of our knowledge is applicable only to a particular set of plants or a certain metabolism. In the biological sciences, the fundamental principles have universal validity, but many details are peculiar to the organisms being studied.

As biologists study organisms, we attempt to create a model of an unknown world. Plants are part of reality, and the study of plants attempts to create a model of that reality and build a vocabulary to exchange ideas about the reality. Our observations and interpretations constitute a body of knowledge that is both incomplete and inaccurate. Because we do not know everything about all organisms, our knowledge is an incomplete reflection of reality, and at least some of our hypotheses are wrong. In college courses, even introductory botany, you study areas where we are still gathering information and our knowledge is incomplete. Often, not enough data are present to form coherent hypotheses in which we can have great confidence. You must think carefully about the things you read and hear and analyze whether they seem reasonable and logical and have been verified.

As you read, you will deal with two types of information: observations and interpretations. Most **observations** are reasonably accurate and trustworthy; we usually have to consider only whether the botanist was careful, observed correctly and without error, and reported truthfully. **Interpretations** are more difficult; they are entirely human constructs based on observations, intuition, previous experience, calculations, and expectations. How can we judge whether an interpretation has any relation to reality? A correct scientific interpretation must make an accurate prediction about the outcome of a future observation or experiment.

Your study of the material in this book, as well as your studies of plants themselves, will be easier if you keep in mind two questions that can be asked about any biological phenomenon:

1. *Are there alternatives to this phenomenon?* Do other structures exist that could perform that same function? Could a different metabolic pathway also occur? If no alternative is possible, why not? If alternatives are possible, do they exist? Did they evolve and then become extinct?

For example, consider photosynthesis: Do plants have alternative sources of energy other than sunlight? We concluded earlier in the chapter that parasitic plants do not need sunlight; they can grow in dark areas. Also, most seeds germinate underground, in complete darkness. Therefore, it seems at first glance that there are alternatives to photosynthesis; however, parasitic plants depend on host plants that do carry out photosynthesis, and germinating seeds rely on stored nutrients they obtained from their parent plant. All plants must, therefore, receive adequate sunlight, at least indirectly. On a theoretical basis, it seems possible that an insect-trapping plant like a Venus flytrap might become so efficient that it could live solely by catching animals, but none is known to be that efficient: They catch only enough insects to provide a little extra nitrogen, not enough to provide the plant with all of the energy it needs.

2. *What are the consequences?* What are the consequences of a particular feature as opposed to an alternative feature or the absence of the feature? Every feature, structure, or metabolism has consequences for the plant, making it either better or less well adapted. Some may have dramatic, highly significant consequences; others may be close to neutral. When you consider the consequences of a particular feature, you must consider the biology of the plant in its natural habitat as it faces competition, predators, pathogens, and stresses.

Consider photosynthesis again. A consequence of depending on photosynthesis is that plants with many large leaves can harvest more light than plants with few, small leaves; however, a consequence of having large leaves is that they lose more water, they are more easily seen by hungry leaf-eating animals, they are good landing sites for disease-causing fungal spores, and so on. A further consequence of obtaining energy from sunlight is that the light does not need to be hunted the way that animals must hunt for their food. Because the sun rises every morning, plants can just sit in one spot and spread their leaves (Figure 1-16). They do not need eyes, brains, muscles, or digestive systems to locate,

Figure 1-16 This tree obtains its energy from sunlight, which is always located in the sky; the plant does not have to hunt for it. Sophisticated sense organs and the power of movement are completely unnecessary; having them would not make the plant more adapted.

Plants and People

Box 1-2 The Characteristics of Life

Botany is a subdivision of biology, the study of life. Despite the importance to biology of defining life, no satisfactory definition exists. As we study metabolism, structure, and ecology more closely, we understand many life processes in chemical/physical terms. It is difficult to distinguish between biology and chemistry or between living and nonliving, but the lack of a definition for life does not bother biologists; very few short definitions are accurate, and life is such a complex and important subject that a full understanding gained through extensive experience is more useful than a definition.

Although we cannot define life, it is critically important for us to recognize it and to know when it is absent. Many hospitals use artificial ventilators, blood pumps, and drugs to maintain the bodies of victims of accidents or illness. The person's cells are alive, but is the person alive? On a less dramatic scale, how does one recognize whether seeds are alive or dead? A farmer about to spend $100,000 on seed corn wants to be certain that the seed is alive. How do we recognize that coral is alive? It looks like rock but grows slowly—but stalactites are rock and they also grow.

The ability to recognize life or its absence is important in space exploration also. The surface of Mars is dry, but water may exist within the soil; many bacteria on Earth live below ground, obtaining energy from chemicals in rock. Europa, a moon of Jupiter, has an ocean below a layer of permanent ice; on Earth, worms, clams, and bacteria live in complete, icy darkness near vents on the ocean floor, obtaining enough energy from volcanic gases to thrive, not just survive. When we explore Mars, Europa, and other parts of the solar system, how will we search for life? How will we know whether we have found it?

All living beings have all of the following characteristics; if even one is missing, the material is not alive:

1. *Metabolism involving the exchange of energy and matter with the environment.* Organisms absorb energy and matter, convert some of it into their own bodies, and excrete the rest. Many nonliving systems also do this: Rivers absorb water from creeks, mix it with mud and boulders, and then "excrete" it into oceans.

Figure B1-2a Lichens grow extremely slowly and remain dormant for months; almost no sign of life can be detected during their dormant period.

Figure B1-2b These seeds of corn (*Zea mays*) are alive and healthy, but inactive metabolically. They will germinate and become obviously alive, but only if given the proper conditions.

2. *Nonrandom organization.* All organisms are highly structured, and decay is the process of its molecules returning to a random arrangement; however, many nonliving systems also have this feature: Crystals have an orderly arrangement as do many cloud patterns, weather patterns, and ripple patterns in flowing streams.

3. *Growth.* All organisms increase in size from the time they are formed: Fertilized eggs grow into seeds or embryos, and each in turn grows into an adult. At some point, growth may cease—we stop getting taller at about 25 years of age. This too is not sufficient to distinguish living from nonliving: Mountains and crystals also grow.

4. *A system of heredity and reproduction.* An organism must produce offspring very similar to itself such that when the individual dies life persists within its progeny. Fires reproduce but are not alive.

5. *A capacity to respond to the environment such that metabolism is not adversely affected.* When conditions become dry, an organism can respond by becoming dormant, conserving water, or obtaining water more effectively; however, as mountains are raised by geological forces, erosion wears them down, and the faster that the mountains are pushed up, the faster erosion works.

In addition to these absolute requirements of life, two features are almost certainly associated with all forms of life: (1) Organisms develop, such that young individuals and old ones have distinctive features, and (2) organisms evolve, changing with time as the environment changes.

Although these various features are always present in living creatures, no one characteristic is sufficient to be certain that something is living versus inanimate. We have no difficulty being certain that rivers, fire, and crystals are not living, but when we search for life on other planets or even in some exotic habitats here on Earth, deciding whether we have actually discovered life might be quite problematical.

Plants and People

Box 1-3 Algae and Global Warming

Photosynthesis removes carbon dioxide from the atmosphere and helps keep Earth cool enough for us to live. Various organisms other than plants are photosynthetic and help reduce the amount of carbon dioxide in our air. Two types of bacteria, called "purple bacteria" and "green bacteria," have an unusual type of photosynthesis that differs from that of plants but allows those bacteria to absorb carbon dioxide from the environment and keep it out of the air. Purple bacteria and green bacteria are both rather rare, so they do not remove very much carbon dioxide. Other bacteria, called "cyanobacteria," are extremely common and carry out a type of photosynthesis that is very similar to that of plants. Cyanobacteria grow in many types of soil as well, in ponds and streams, and especially abundantly in sewage. The total amount of cyanobacteria in the world is great enough that they remove significant amounts of carbon dioxide from the air. Unfortunately, all bacteria are tiny and their bodies are rather delicate, so they break down quickly after they die and the carbon atoms in their bodies are converted back into carbon dioxide. But if the bacteria are eaten by an animal, then the carbon-containing molecules in their bodies become part of the bodies of the animals, and if those animals are then eaten by others, the carbon continues to be locked up in a body rather than going back into the air. However, this only keeps carbon dioxide out of the atmosphere for just a few years at most, because no animal lives for more than a few decades. Even long-lived animals shed hair, skin, and other parts and defecate: All these decompose, and their carbon is converted to carbon dioxide.

Algae, the close relatives of plants, are a group of organisms that are important allies in our attempts to combat global warming. There are many types of algae (most are named by the color of their pigments, such as red algae, green algae, brown algae, and so on), and all carry out photosynthesis that is almost identical to that of plants. Furthermore, algae are abundant in oceans, lakes, and rivers as well as in moist soils, rocks, and even tree bark. Just like plants, algae absorb carbon dioxide as they photosynthesize and lock it into the molecules of their bodies. Like bacteria, many algae are so delicate they decompose quickly after they die, so they do not keep carbon dioxide out of the air for long.

However, microscopic algae, called coccolithophorids (or just coccoliths), make a shell of calcium carbonate (Figure B1-3). This shell is so dense that as soon as the coccoliths die, their bodies sink to the bottom of the oceans. The cold temperatures there slow decay to such a degree that the shells do not break down for thousands, even millions of years. Consequently, as these microscopic algae grow and then die, carbon dioxide is removed from the atmosphere for very long times. One of the possibilities for combating global warming is to make these algae grow faster and more abundantly by fertilizing the oceans in areas where coccoliths already live. The hypothesis is that by adding just enough of the right kinds of nutrients, coccoliths will grow rapidly enough to start removing large amounts of carbon dioxide that will offset much of the damage we do by burning petroleum and coal. So far, experimental trials have been encouraging.

Figure B1-3 Limestone is calcium carbonate composed of the shells of sea animals such as clams, mollusks, and certain algae. As these creatures formed their shells, they removed carbon dioxide from the atmosphere, causing the climate to become cooler. This area along the Pecos River is part of a gigantic region of limestone that extends for hundreds of square miles.

catch, and consume their source of energy. Plants can be very simple and survive, whereas most animals must be complex.

Finally, no matter where you are, plants are readily available for direct observation. You can figure out a great deal by observing a plant and thinking about it (as opposed to just observing it). You will be surprised at how much you already know about plants. As you read, think about how the principles discussed apply to plants you are familiar with. It can be boring to memorize names and terms, but if you think about the material, analyze it, understand it, and see where it is not valid for all plants or all situations, you will find both botany and plants to be enjoyable.

Summary

1. It is difficult to define a plant. It is more important to develop a familiarity with plants and understand how they differ from animals, fungi, protists, and prokaryotes. The differences are presented in later chapters.
2. The scientific method requires that all information be gathered through documented, repeatable observations and experiments. It rejects any concept that can never be examined, and it requires that all hypotheses be tested and be consistent with all relevant observations. It is based on skepticism.
3. Science and religion address completely different kinds of problems. Science cannot solve moral problems; religion cannot explain physical processes.
4. Living organisms have evolved by natural selection. As organisms reproduce, mutations cause some offspring to be less fit, some to be more fit. Those whose features are best suited for the environment grow and reproduce best and leave more offspring than do those that are poorly adapted.
5. For any particular environment, several types of adaptation can be successful.
6. Our knowledge of the world is incomplete and inaccurate; as scientific studies continue, incompleteness diminishes and inaccuracies are corrected.
7. Two simple questions are powerful analytical tools: (1) What are the alternatives, and (2) what are the consequences?

Important Terms

anthropomorphism
apomorphic features
botany
derived features
domains

hypothesis
interpretations
natural selection
observations
plesiomorphic features

relictual features
scientific method
teleology
theory

Review Questions

1. Your present concept of plants is probably quite accurate. Most have roots, stems, leaves, and flowers. Can you name two plants that have cones rather than flowers? Can you name a plant that appears to not have leaves?
2. Name two types of fungi. Why were fungi originally included in the plant kingdom? Biologists no longer consider fungi to be plants because they differ in many basic _____ and _____ aspects.
3. How would you distinguish between plants and animals? What characters are important? Be careful to consider unusual plants and animals. Can all animals move? Do they all eat?
4. What are three methods for analyzing nature? Name some advantages and disadvantages of each.
5. Is it always easy to recognize that something is a living being rather than an inanimate object? Europa is a moon of Jupiter (see Box 1-2), and it is so far from the sun that there is not enough light for photosynthesis. The bottom of its ocean must be completely dark and icy cold. Are there locations like this on Earth that support life and that therefore let us hypothesize that life might also exist on Europa?
6. In the scientific method, all accepted information can be derived only from documented and controlled _____

_____ or _____. If someone claimed to have a new treatment of a disease or a new type of eye surgery, would you want some sort of documentation and proof before you let them give you drugs or operate on you?
7. The scientific method deals only with _____ phenomena and objects.
8. Physical forces that control the world are _____ through time and are the _____ everywhere.
9. The fundamental basis of the scientific method is _____. Describe this concept.
10. What is a hypothesis? A theory? Why is it important that each be able to predict the outcome of a future experiment? How do these differ from a speculation?
11. If a hypothesis makes predictions that are not accurate and do not help explain future observations, do those inaccurate predictions prove the hypothesis is not a good model of reality? On the other hand, if the hypothesis does make accurate predictions, does that accuracy prove the hypothesis is correct?
12. Do any concepts exist for which the scientific method is inappropriate? Some people suffer terribly from certain incurable diseases. Can scientific methods be used to develop

drugs that could end a person's life? Can scientific methods be used to decide whether it is right or wrong to use such drugs? Do you think the concept of right and wrong actually exists? Can you prove it with scientific methods? Are right and wrong concepts that can be measured, weighed, dissected? Do they have a chemical composition?

13. List the eight concepts that can be used to understand plants.

14. The first concept used to understand plants is that plant metabolism is based on the principles of _____ and _____. If this is true, do you think that praying for good harvests or for rain is effective?

15. The fifth concept used to understand plants is that plants must survive in their own _____. Imagine a plant adapted to a desert and one adapted to a rain forest. Do you think that the leaves of one might be different from the leaves of the other? That one might have enlarged roots that can store water and the other would not need these? Would it be easier to understand a plant's anatomy and physiology, all its biology, if we also know the type of habitat to which it is adapted?

16. What is the eighth concept used for studying plants? It is difficult to avoid using the phrase "in order to" when referring to plants. Change the following sentences to be more accurate. The first one is done as an example:
    a. Plants have leaves in order to photosynthesize. "Plants have leaves that photosynthesize," or "photosynthesis in plants occurs in leaves."
    b. Plants have flowers in order to reproduce.
    c. Plant cells divide in order to make more cells.
    d. Wind-pollinated plants have their flowers located high above the ground in order to be exposed to stronger wind.
    e. Some plants have red flowers to attract hummingbirds.
    f. After I eat dinner, my stomach secretes acid in order to help digest the food.

17. Life on Earth began about _____ _____ years ago. At first, organisms were simple, like present-day _____.

18. When organisms reproduce, some of their offspring are poorly adapted, and they do not grow and reproduce as well as the offspring that are well adapted. This is called _____ _____.

19. Look at Figure 1-12. What is the fifth concept used for understanding plants (in the section Using Concepts to Understand Plants)? We do not know for certain what the environment of this deer is, but would you guess that the deer is well adapted or poorly adapted? Is it likely to survive long enough to pass on its genes for this particular skin pattern? Now imagine

that the deer's environment changes. All hunting is outlawed. There are not even any poachers—no hunters at all try to kill deer. In this new environment, does this skin pattern matter any more? Is the deer still poorly adapted?

20. Name the three domains, and describe the types of organisms in each. Which are prokaryotes? Which are eukaryotes?

21. What are relictual (plesiomorphic) features? Derived (apomorphic) features? Which organisms seem to have a larger percentage of relictual features—prokaryotes or eukaryotes? Algae or flowering plants? Amoebas or humans?

22. It is important to distinguish between plants and the study of plants—that is, the reality of the plant world and the model of it presented in this book and in lecture. Do we already know everything there is to be known about plants? Have we discovered every plant species that exists? Are all of our hypotheses about plants correct?

23. Plants are part of _____. The study of plants attempts to create a _____ of that and build a vocabulary to exchange _____ about the reality. Can our model ever be incorrect? Can reality be incorrect? If our model—our hypotheses and theories—does not predict accurately, which is wrong—reality or the model?

24. As you read, you will deal with two types of information: _____ and _____. Which of these are reasonably accurate and trustworthy? The other is more difficult. Why? If the second type does not make any kind of prediction about the outcome of a future observation or experiment, would there be any way to determine whether it is accurate?

25. This chapter closes by suggesting that you keep in mind two questions as you study this book and plants themselves. What are the two questions?

▌ BotanyLinks

http://biology.jbpub.com/botany/5e
BotanyLinks contains a variety of resources and review material designed to assist in your study of botany.

Visit the Web Exercises area of BotanyLinks to complete these questions:

1. To study plants and their interactions with the environment, natural areas must be preserved. Where are the natural areas still available? Go to the BotanyLinks home page to complete this exercise.

2. There are many important issues affecting our air, food, water, and environment, but getting information can be difficult. Go to the BotanyLinks home page to find out how to make your opinion matter.

Introduction to the Principles of Chemistry

Concepts

P lant metabolism, like that of all other organisms, is based on the fundamental principles of physics and chemistry that govern inanimate matter. All living creatures grow, respond to stimuli, and reproduce; and these processes, however complex, involve physical and chemical reactions that can be modeled by hypotheses and verified by experimentation and observation. The bodies of plants and other organisms are made of atoms drawn from soil, air, or water, and the energy that drives their metabolism is produced by ordinary chemical reactions.

It is easy to assume that living organisms must possess special properties because they seem too elaborate to be the result of ordinary chemical and physical processes. This was the predominant view until the 1800s, when three major discoveries were made. It was found that biological compounds could be synthesized in the laboratory using inorganic chemicals and ordinary chemical processes. Next, enzymes were extracted from yeast cells, and some steps of fermentation were carried out in vitro (Table 2-1) without the presence of living cells. Finally, Louis Pasteur proved definitively that

Chapter Opener Image: These are plants of peyote, a hallucinogenic cactus (*Lophophora williamsii*). When any plant photosynthesizes, it makes a simple sugar that is then transformed into all the organic compounds the plant needs for its metabolism. Peyotes and many other plants have evolved to make chemicals that are toxic to animals; by being poisonous, the plants have some protection against being eaten by animals or attacked by fungi and bacteria. Peyotes have chemicals called alkaloids that affect our nervous systems and can cause hallucinations. Because so many people want to use them as a recreational drug, peyotes have been hunted and gathered relentlessly, and they are almost extinct in the United States.

TABLE 2-1 In Vitro and in Vivo

in vitro: in glass. This refers to studies performed in test tubes, flasks, Petri dishes, and similar containers in which some aspect of metabolism is manipulated in laboratory conditions and **the metabolic system has been removed from the organism**. Photosynthesis might be studied in vitro by breaking cells open, extracting their pigments and enzymes, and examining how they work by supplying or withholding substances such as carbon dioxide and oxygen or by changing the acidity, temperature, or light.

in vivo: in life. This refers to studies carried out with **intact cells or whole organisms**, whether in natural settings or in a laboratory. Photosynthesis might be studied in vivo by growing whole plants in bright or dim light, low or high levels of carbon dioxide, and other pairs of variables.

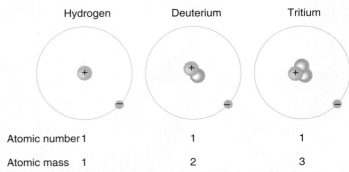

Figure 2-1 The three isotopes of hydrogen differ in the number of neutrons they contain. Because they all have the same number of protons, they have the same atomic number; they are the same chemical element and have identical properties. Neutrons only cause atoms to move more slowly.

spontaneous generation does not occur and that there is no such thing as "vital force." These advances, along with the discovery of evolution by natural selection, revolutionized the study of metabolism and showed that analytical techniques and laws of physics, chemistry, and mathematics are sufficient to understand metabolism.

Atoms and Molecules

Chemical Bonds

There are 92 natural elements, each differing from the others by the number of protons in the nuclei of its atoms (Table 2-2). The lightest, hydrogen (H), has only one; the heaviest, uranium (U), has 92. Atomic nuclei also contain neutrons in numbers roughly equal to protons (Figure 2-1); neutrons affect only the weight of the

TABLE 2-2 Essential Elements

Element	Symbol	Atomic Number	Atomic Weight	Amount Needed in Tissues Relative to Molybdenum
Hydrogen	H	1	1	60,000,000
Boron	B	5	10.8	2,000
Carbon	C	6	12	35,000,000
Nitrogen	N	7	14	1,000,000
Oxygen	O	8	16	30,000,000
Sodium	Na	11	23	
Magnesium	Mg	12	24.3	80,000
Phosphorus	P	15	31	60,000
Sulfur	S	16	32	30,000
Chlorine	Cl	17	35.4	3,000
Potassium	K	19	39.1	250,000
Calcium	Ca	20	40.1	125,000
Manganese	Mn	25	54.9	1,000
Iron	Fe	26	55.8	2,000
Cobalt	Co	27	58.9	N/A
Copper	Cu	29	63.5	100
Zinc	Zn	30	65.4	300
Molybdenum	Mo	42	95.9	1

These elements are essential for plant life; each carries out one or more vital roles. If any one is missing, a plant cannot survive. Sodium is essential for animals but not for plants. Atomic number corresponds to the number of protons in the atomic nucleus of each; atomic weight is the number of protons plus neutrons in each nucleus.

atom, not its chemical properties. Around each nucleus are electrons with a negative charge, and thus, each neutralizes the positive charge of a proton. Because of their opposite charges, protons attract electrons, and atoms with equal numbers of protons and electrons are electrically neutral.

Electrical attraction is not the only important factor in determining the number of electrons associated with a nucleus. Electrons fit only into specific orbitals and energy levels around the nucleus, and some arrangements are more stable than others. Helium, neon, and argon, for example, have their outermost energy levels exactly filled with electrons; this is the most stable arrangement possible and also results in equal numbers of electrons and protons, so the atoms are electrically neutral. If one of these atoms loses an electron, it has both an electrical charge (+1) and an unfilled energy level, which is an unstable arrangement. If an atom gains an electron, it again has an unbalanced charge (−1), and the electron is located alone in an orbital of a higher energy level, which is also unstable. Helium, neon, and argon have virtually no tendency to gain or lose electrons or to react with anything; they are called noble gases.

Chlorine, however, even when it is electrically neutral with 17 electrons matching its 17 protons, is not stable. Its outermost energy level (the third level) has its four orbitals full except for one, which contains a single unpaired electron. Each atom has an extremely strong tendency to absorb one more electron and complete this energy level; this gives the atom a negative charge (Cl^-), which tends to make it slightly unstable, but the new arrangement of 18 electrons is so much more stable that it more than compensates. In plants and animals, chlorine almost always has an extra electron, making it the chloride ion Cl^-. An atom or molecule that carries a charge is an ion; a negative ion is an **anion**.

Sodium is just the opposite; when electrically neutral, it has one electron in an orbital by itself. Although losing the electron causes the atom to become the sodium ion Na^+, the remaining electrons are in such stable orbitals with energy levels 1 and 2 exactly filled that this arrangement is favored. A positive ion is a **cation**.

This tendency of electrons to move to the most stable possible configuration, which has been named electronegativity, is the driving force behind chemical reactions. The element with the greatest affinity for electrons, fluorine, has an electronegativity of 4.0; the noble gases, with no affinity for extra electrons, have electronegativities of 0.0. Only electrons in the highest, partially

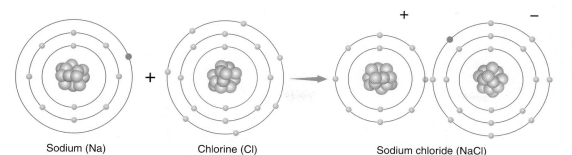

Figure 2-2 Transferring its single unpaired valence electron gives sodium a stable arrangement and gives chlorine the electron it needs to have all its energy levels full.

Sodium (Na)   Chlorine (Cl)   Sodium chloride (NaCl)

filled energy level, called **valence electrons**, are involved; they are responsible for forming chemical bonds.

Although an atom such as sodium may have a strong tendency to lose an electron, electrons virtually never fly off into space; instead, they move from one atom to another during a chemical reaction. Imagine a reaction between an atom with low electronegativity, such as sodium, and one with high electronegativity, such as chlorine. Both elements become more stable by the transfer of a valence electron from sodium to chlorine (**Figure 2-2**). It is important to emphasize that **by more stable we mean that the atoms have less energy**. This is always the case: A particle is more stable when it has less energy.

When sodium reacts with chlorine, both partners have less energy after the reaction, and thus, energy is liberated to the surroundings. Such an energy-releasing reaction is **exergonic**; if energy is released as heat, the reaction is also **exothermic**. Another exergonic reaction is the burning of hydrogen with oxygen; two hydrogen atoms transfer one electron each to an oxygen atom, resulting in water:

$$2H + \tfrac{1}{2}O_2 \rightarrow H_2O + energy$$

This reaction can be forced to run backward by adding electrical energy, as in the electrolysis of water to form hydrogen gas and oxygen gas. Because the two products of the reverse reaction have more energy than water (they absorb it from the electrical apparatus), the reverse reaction is **endergonic** (**endothermic** if energy is absorbed in the form of heat). Any endergonic reaction needs a source of energy—an exergonic reaction occurring somewhere. For water electrolysis, the exergonic reaction is the burning of coal at the electricity-generating station. Another endergonic

reaction is the formation of carbohydrates in leaves during photosynthesis; the exergonic reactions for this are the thermonuclear reactions in the sun—even though it is 93 million miles away.

In the sodium/chlorine reaction, a valence electron is actually transferred from one atom to another, converting each into an ion; in a molecule of salt, NaCl, the ions are held together by **ionic bonds**; however, in the hydrogen/oxygen reaction, the transfer is not complete. Instead, the valence electrons are shared between the nuclei (**Figure 2-3a**). Electron sharing has an extra benefit: All electrons are in the most stable orbitals, and no buildup of electrical charge occurs. Instead of two H^+ and one O^-, the result is just H_2O. A bond in which electrons are shared is described as a **covalent bond**. Whether an ionic bond or a covalent bond forms depends to a large degree on the difference in electronegativities of the two reactants.

■ Water

Electrons are often not shared equally; in water, electrons spend more time near the oxygen and less time near the hydrogen; therefore, the oxygen in water has a partial negative charge ($\delta-$; this is the Greek letter delta) and each hydrogen has a partial positive charge ($\delta+$). The orbitals are oriented so that both hydrogens and their partial positive charges are on one side of the molecule. The molecule is thus a **polar molecule**, with a slightly negative end and a slightly positive end.

Hydrogen Bonding When two water molecules come close to each other, the positive charge of one slightly attracts the negative end of the other. This is called **hydrogen bonding**. It is not nearly enough force to pull electrons from one to the other, but it is sufficient to cause them to adhere slightly (Figure 2-3b), and, thus, water is a

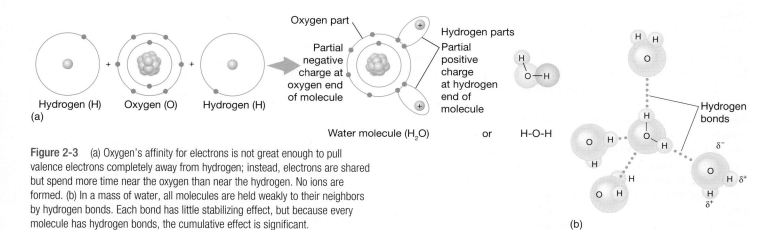

Figure 2-3 (a) Oxygen's affinity for electrons is not great enough to pull valence electrons completely away from hydrogen; instead, electrons are shared but spend more time near the oxygen than near the hydrogen. No ions are formed. (b) In a mass of water, all molecules are held weakly to their neighbors by hydrogen bonds. Each bond has little stabilizing effect, but because every molecule has hydrogen bonds, the cumulative effect is significant.

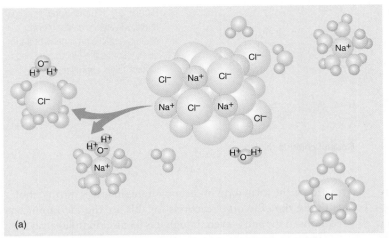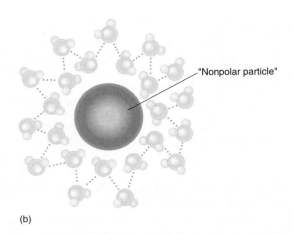

"Nonpolar particle"

Figure 2-4 (a) When a polar material like ordinary table salt, NaCl, is placed in water, the negative ends of water molecules are attracted to the sodium ions and the positive ends are attracted to the chloride ions, forming weak bonds with them. The new bonding is about the same strength as the water–water hydrogen bonding, and thus, the NaCl molecule can break apart easily. (b) If water molecules were occupying the volume of the nonpolar particle, many (at least 11 in this diagram) new hydrogen bonds could form, release energy, and stabilize the water.

rather sticky, viscous substance. It absorbs a great deal of energy without warming rapidly and requires a large amount of energy to convert to vapor. Without the stickiness caused by hydrogen bonding, water molecules could not be lifted from roots to leaves.

Substances that carry no unbalanced electrical charge, not even a partial one, are nonpolar substances, and they do not undergo hydrogen bonding. Nonpolar substances move easily past each other and flow with little viscosity. When energy is supplied their speed quickly increases, raising their temperature. They boil and turn to a gas even at a low temperature, with little energy needed. Examples are methane (see Figure 2-5) and acetylene.

Water Solubility and Lipid Solubility Many other molecules containing hydrogen bonded to oxygen also have partial charges and undergo hydrogen bonding. When they are placed in water, water molecules surround them and form hydrogen bonds with them (**Figure 2-4a**). This permits individual molecules to move into the water: The substance dissolves and is **water soluble**. When a nonpolar substance, which cannot form hydrogen bonds, is placed in water, no interaction occurs between the two types of molecules, and the substance does not dissolve. In fact, if a molecule of the substance diffuses into the water, it disrupts the water's own hydrogen bonds—this would be an energy-consuming, endergonic reaction that destabilizes the water (Figure 2-4b). If the molecule diffuses out of the water, the water molecules can form new hydrogen bonds and release energy, becoming more stable. If the substance is placed in water and agitated violently, it may mix temporarily, but it gradually separates; oil in water is an example. Nonpolar substances dissolve in other nonpolar chemicals, such as lipids (e.g., fats and oils), and are **lipid soluble**.

Acids and Bases When hydrogen combines with oxygen and forms water, the new bonding orbitals have so much less energy than the nonbonding orbitals that they are extraordinarily stable. For hydrogen to break away from water requires the input of a large amount of energy, enough to raise electrons out of the low-energy, stable bonding orbitals. In pure water, this is so rare that only about one molecule in 10 million breaks down into H^+ and OH^-, a proton and a hydroxyl ion, and when they encounter each other, they recombine and form water immediately.

The concentration of H^+ is known as the **acidity** of a solution; it is measured as pH, which is the negative logarithm of the H^+ concentration (**Table 2-3**). In many substances, hydrogen ions are held less tightly than in water molecules, and they give off protons rather easily; any substance that increases the concentration of free protons is an **acid**. Hydrochloric acid, HCl, dissolves in water to give H^+ and Cl^-; the extra H^+ donated to the solution means that HCl is an acid.

A **base** is anything that decreases the concentration of free protons; this is usually accomplished by giving off hydroxyl ions that combine with protons and form water, effectively removing free protons. NaOH breaks down into Na^+ and OH^-; the OH^- indicates that it is a base.

The protons and hydroxyl ions given off by acids and bases move onto and off of other compounds present in a solution. Be-

TABLE 2-3	Acids and Bases		
		H^+ Concentration	pH
Strong acid	Most molecules dissociate to an anion (−) and H^+. Hydrochloric acid, nitric acid, sulfuric acid	High: $1/10^3$	3
Weak acid	Few molecules dissociate to an anion and H^+. Acetic acid, citric acid, malic acid	Moderate: $1/10^5$	5
Weak base	Few molecules dissociate to a cation (+) and OH^-; these combine with H^+ from water, lowering H^+ concentration. Asparagine, glutamine, urea	Low: $1/10^9$	9
Strong base	Most molecules dissociate to a cation and OH^-. Ammonium hydroxide, potassium hydroxide, sodium hydroxide.	Very low: $1/10^{11}$	11

Strong acids and bases are those in which almost all molecules break down and liberate either a proton or a hydroxyl when dissolved in water. In weak acids and bases, many molecules do not break down.

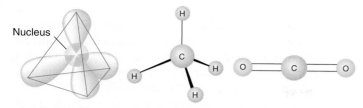

(a) Carbon (C)   (b) Methane (CH₄)   (c) Carbon dioxide (CO₂)

Figure 2-5   When an atom shares electrons with another atom, the bonding orbitals have shapes and orientations different from those of isolated atoms. (a and b) If carbon bonds to four other atoms, the bonding orbitals point to the four corners of a tetrahedron, which separates each orbital as much as possible from the others. (c) If carbon bonds to only two other atoms, two double-bond orbitals point exactly away from each other.

(a)

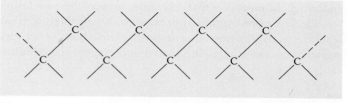

(b)

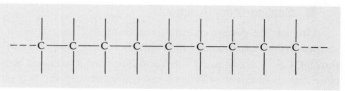

(c)

Figure 2-6   (a) A series of carbon atoms bonded to other carbons by single bonds would look something like this, with the blue spheres representing carbon atoms and the red spheres representing other atoms attached to the carbons. Because the bonds are arranged as a tetrahedron, the carbon backbone has a zigzag shape. (b and c) are simpler ways of showing the structure in (a).

cause they carry a positive or negative charge, they affect the charge on the molecule they attach to. If a **nonpolar**, water-insoluble **molecule** picks up a proton because an acid is present, the non-polar molecule becomes positively charged and water soluble.

■ Carbon Compounds

The concept of life is almost synonymous with the chemistry of carbon. With six protons in its nucleus and an electronegativity of 2.5, carbon has properties essential for life. Carbon can easily and stably exist as a neutral atom with six electrons, or it can form covalent bonds by sharing the valence electrons (usually four) of other atoms. For instance, in methane (CH_4), one carbon atom shares one electron with each of four hydrogen atoms (**Figure 2-5**), and in carbon dioxide (CO_2), it shares two electrons with each of two oxygen atoms. In fatty acids, most carbon atoms share two electrons with each of two hydrogens and each of two more carbons (see Figure 2-18). Because carbon reacts so readily with more carbon, it can form long chains and complex ring structures.

Carbon forms three types of covalent bonds. If there are four other atoms (methane, fatty acids), the carbon is linked by **single bonds**, each bonding orbital containing one electron from the carbon and one from the other atom. These bonding orbitals are arranged in a tetrahedron (Figure 2-5). In a chain of carbons, the carbon backbone is zigzag, not straight (**Figure 2-6**).

Carbon forms a **double bond** by sharing two of its electrons with one other atom that also contributes two electrons. The other two valence electrons of carbon may be in a second double bond or in two single

bonds. If two sets of double bonds are present, as in carbon dioxide, the two sets extend in opposite directions, producing a straight molecule (Figure 2-5). If one double bond and two single bonds are present, the molecule is flat and shaped like a Y (**Figure 2-7**). The double bond is extremely rigid, and the arms cannot rotate around the carbon. Many organic molecules have carbon–carbon double

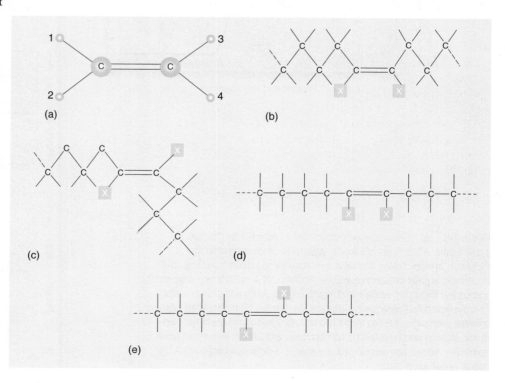

Figure 2-7   (a) The carbon–carbon double-bond system is flat: All atoms in this structure are in the same plane. Furthermore, the double bond cannot rotate, and thus, atom 1 cannot change position with atom 2; if the carbon–carbon bond were a single bond instead, such a rotation could occur thousands of times every second. (b–e) are simplified ways of presenting carbon chains with one double bond. In (b and d), the Xs are in the *cis* position. In (c and e), they are in the *trans* position.

bonds, and because the double bond cannot rotate, there are two possible forms of

$$-CX = CX-$$

With both Xs on the same side, they are in the *cis position* and when on opposite sides in the *trans position*. The cis form has physical and metabolic properties that differ from those of the trans form.

Rarely, carbon forms a triple bond by sharing three of its valence electrons with one other atom. The triple bond is not very stable (its electrons still have a great deal of energy) and is broken easily.

Mechanisms of Chemical Reactions

Second-Order Reactions

One of the easiest chemical reactions to understand is a **second-order reaction**, in which two molecules react to form a third: $A + B \rightarrow AB$. In order for A and B to react, they must be so close that their valence electrons can move between the two sets of orbitals, but as A approaches B, the two sets of electrons repel each other. A and B cannot simply drift together; they must collide violently enough that electron-cloud repulsion is overcome. The motion of molecules can be speeded by heating them; the faster an atom or molecule moves, the greater is its kinetic energy.

The energy (speed) needed to overcome electron-cloud repulsion and permit chemical reaction is called the **activation energy**. Figure 2-8a shows a reaction diagram of the potential energy of the reactants and products during a reaction. On the left, as A and B approach, they have a certain amount of potential energy related not to their speed but to the stability of their electron clouds. As they approach and begin to repel each other, they slow down, and part of the kinetic energy is converted to potential energy. If they have enough kinetic energy (if they are moving rapidly enough), they push very close together. Their potential energy is very high, and the electrons rearrange into more stable bonding orbitals, forming either ionic or covalent bonds. If this is an exergonic (energy-liberating) reaction, the valence electrons are more stable (have less energy) in the new bonding orbitals than they did in the old nonbonding orbitals. The liberation of energy lowers the potential energy level. The two electron clouds do not repel each other because of the rearrangement of the valence electrons into bonding orbitals. If A and B had not had enough speed to overcome the activation–energy barrier, they would not have shared electrons, the stabilizing bonding orbitals would not have formed, and the electron clouds of A and B would have repelled each other, flying off in opposite directions (Figure 2-8b).

The amount of electrical repulsion and therefore the energy of activation needed vary from reaction to reaction. If little repulsion occurs, even slowly moving molecules have enough kinetic

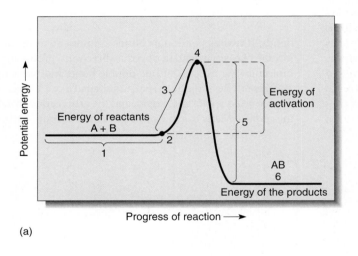

(a)

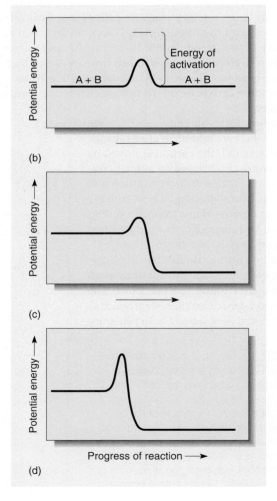

(b)

(c)

(d)

Figure 2-8 (a) A potential energy diagram of a second-order reaction. At 1, two atoms, A and B, are still too far apart to influence each other, but at 2, they begin to repel. During interval 3, they continue to approach because their momentum is great enough to overcome repulsion. At 4, orbitals rearrange from nonbonding to bonding orbitals, and energy is given up (5) and radiated away from the molecule. Consequently, after the reaction (6), the molecule AB has less potential energy than the two atoms had at 1. (b) If two atoms are moving too slowly, electron cloud repulsion is not overcome, and the two atoms bounce off each other. No reaction occurs. (c) A low energy-of-activation barrier. (d) A high energy-of-activation barrier.

Plants Do Things Differently

Box 2-1 Vitamins: Plants and Human Health

The many types of organic compounds described in this chapter actually constitute only a small fraction of the large number of diverse organic compounds that plants produce. In addition, organic compounds act as flower pigments, aromas, plant hormones, and defensive compounds that are either irritating or distinctly poisonous. A remarkable aspect of plant metabolism is that during photosynthesis plants use CO_2 and water to make a single simple organic compound, 3-phosphoglyceraldehyde. Then by altering it chemically with various enzymes, plants use 3-phosphoglyceraldehyde to construct numerous types of carbohydrates and lipids. By adding simple compounds of nitrogen, phosphorus, sulfur, and so on, all derived from the soil, plants use the 3-phosphoglyceraldehyde and its derivatives to make amino acids and proteins, nucleotides and nucleic acids, and every other compound in themselves. To build its entire body and carry out its metabolism, a green plant needs only sunlight, carbon dioxide, water, and a few mineral elements.

In contrast to a plant's well-balanced diet (light, air, water, and soil), our well-balanced diet is much more complex, requiring vitamins, essential amino acids, essential fatty acids, and a variety of other compounds. By analyzing the underlying causes of this difference, we can begin to understand certain critical concepts in biology. First, animals cannot photosynthesize because they lack the necessary pigments and enzymes; instead, animals eat plants or other animals as food and then use much of the food's carbohydrate, fat, and protein components for energy production: They respire—oxidize—it and generate energy carriers such as ATP and electron carriers such as $NADH + H^+$. After animals eat the carbohydrates, fats, and proteins, they digest them to their constituent monomers—the simple sugars, fatty acids, amino acids, and so on. Although it is possible for animals to break these down even further, that would be wasteful. It would require energy and special metabolism to break down the monomers, which would then have to be rebuilt. All plants and animals use the same amino acids, the same nucleotides, and mostly the same fatty acids and sugars. Similarly, as an animal eats a plant, it obtains all of the vitamins the plant has produced and uses them in its own metabolism. A mutation in an animal that prevents it from synthesizing a particular amino acid or nucleotide is not deleterious to the animal because it probably gets all it needs from its food. In fact, it may be slightly advantageous selectively: That animal is not using up resources to make something it gets in its diet, and thus, those extra resources can be used for other needs such as reproduction, defense, growth, and so on.

Over many generations, we expect natural selection to eliminate synthetic pathways that produce materials the animal already receives reliably in its diet. Consider humans: We have lost the ability to synthesize our own vitamins, several amino acids, and several fatty acids. Even a rather poor diet contains enough of these for humans to lead reasonably healthy lives. In fact, the existence of vitamins was discovered only when long sea voyages required sailors to eat a diet of just dry bread and beef, without any fruits or vegetables (which rotted too quickly). The lack of vitamin C (ascorbic acid) resulted in the deficiency disease scurvy, with symptoms such as bleeding gums, weight loss, and swollen joints. Captain James Cook was one of the first to notice that the disease was cured shortly after the sailors reached land and obtained a diet that included vegetables. He experimented with sauerkraut (the only way to preserve vitamin-rich cabbage) and discovered that it prevented the disease. Later studies showed that citrus fruits were even more effective, and the science of experimental nutrition was initiated. More recently, pellagra was a common deficiency disease in the southern United States. It is caused by a lack of niacin in the diet, and it affected mostly southern blacks who were so poor that they relied almost exclusively on cornmeal for the bulk of their diet; corn is remarkably low in niacin.

The rarity of deficiency diseases shows that a diet must be quite poor to cause noticeable harm to an animal. However, do animals need compounds that plants do not make? Yes, with an important example being cholesterol. This is a lipid that plants do not synthesize but that is important to animals. Consequently, all herbivorous animals must be able to make cholesterol, whereas carnivorous animals can obtain it in their food.

The rates at which different organisms have lost the ability to synthesize organic compounds vary greatly. We humans still have the capacity to make almost all the carbohydrates and fats we need, as well as most of the amino acids, but other animals have lost many of their synthetic pathways. Good examples are tapeworms and similar parasites. In their proper habitat (the digestive tract of a mammal), tapeworms receive almost all of the nutrients they need. They can synthesize only a few compounds themselves, and almost all of their metabolic resources are directed toward growth and reproduction, which occur rapidly. On the other hand, this extreme reliance on the diet means that tapeworms can survive in only a few places.

In summary, as an organism becomes more dependent on its diet, it has a greater potential for growth and reproduction when the environment provides the correct diet but a greater number of deficiency diseases when the environment provides a poor diet.

energy and the reaction occurs at low temperature (Figure 2-8c). If the energy-of-activation barrier is very great, the molecules must be moving very rapidly, and the reaction occurs only if the substance is heated to a high temperature (Figure 2-8d). It has nothing to do with how much energy can be liberated during the reaction. Wood can combine with oxygen to produce carbon dioxide and water; this is strongly exergonic and huge amounts of energy are liberated; however, the energy-of-activation barrier is so high that the wood must be heated to several hundred degrees before the reaction can occur.

Catalysts

Although there is no way to change the energy of activation of a reaction, it is possible to change the mechanism of a reaction. The reaction of A and B to form AB may be difficult because of a high energy of activation, but an alternative set of reactions may be possible, such as

$$A + D \rightarrow AD$$

$$AD + B \rightarrow AB + D$$

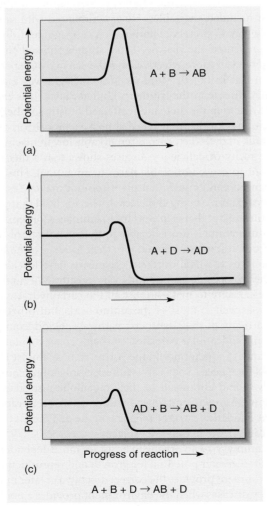

(a)

(b)

(c)

$$A + B + D \rightarrow AB + D$$

Figure 2-9   (a) The activation-energy barrier of this reaction cannot be changed, but a catalyst, D, may exist which provides an alternative reaction mechanism that has low activation-energy barriers in all steps (b and c).

in which the energies of activation for $A + D$ and for $AD + B$ are both low (Figure 2-9). If so, a mixture of $A + B + D$ results in the production of AB even at low temperature, whereas a mixture of just $A + B$ does not. A substance such as D that permits a reaction to occur rapidly at low temperature is a **catalyst**; in living organisms, virtually all catalysts are proteins called enzymes. Catalysts, enzymes included, emerge from the reaction in the same condition they entered it; the catalyst itself is not permanently altered by the reaction.

First-Order Reactions

A **first-order reaction** involves only one molecule, not the collision of two. An example is $AB \rightarrow A + B$, in which a compound breaks down into two parts. First-order reactions also have energies of activation that are overcome as the two parts of the molecule vibrate. They can be accelerated by heating or by catalysts.

Reaction Equilibria

For every reaction, an opposite back reaction can also occur; thus, not only can A and B combine to form AB, but AB can break down to form A and B. In any mixture of A and B, both reactions occur. If $A + B \rightarrow AB$ is strongly exergonic and liberates a large amount of energy, then $AB \rightarrow A + B$ is strongly endergonic and requires a large input of energy. In such a set of reactions, almost all material is ultimately converted to AB because AB is formed more rapidly than it breaks down and is more stable than A and B. If $A + B \rightarrow AB$ is only mildly exergonic, then $AB \rightarrow A + B$ is only mildly endergonic; at equilibrium, A, B, and AB are present in approximately equal quantities, and both reactions occur at about the same rate.

Organic Molecules and Polymeric Construction

Functional Groups

Millions of carbon compounds are possible, varying not only in the number of carbons they contain but also in the types and numbers of noncarbon atoms, types of bonds, and other factors. This great diversity actually consists of a small number of families of compounds whose members have similar properties. This is because the properties of a compound are due mostly to the chemical groups, known as **functional groups**, attached to the carbon atoms (Table 2-4). Because carbon compounds can be large, each may have many functional groups of various types, being simultaneously both acidic and basic, or lipid soluble in some regions and water soluble in others.

Polymeric Construction

A **polymer** is a large compound composed of a number of more or less identical subunits (**monomers**). The simplest example is construction using bricks. The bricks (monomers) are virtually identical but can be used to construct many different things: houses, lecture rooms, and sidewalks. Simple sugars such as glucose are monomers that can be polymerized into starch, cellulose, mucilage, and many other polymers.

TABLE 2-4 Functional Groups

	Name	Characteristics
—H	Hydrogen	Low reactivity; lipid solubility
—OH	Hydroxyl	Alcohol group; hydrogen bondings; water solubility
—CH₃	Methyl	Low reactivity; lipid solubility
—C(=O)OH	Carboxyl	Highly reactive; acidic; in water it acts like an acid, giving off a proton: $-C(=O)O^- + H^+$
—NH₂	Amino	Highly reactive; basic; in water it acts like a base by absorbing protons: $-NH_3^+$
—C(=O)H	Aldehyde	Moderately reactive; water solubility
—C(=O)—	Ketone	Moderately reactive; similar to an aldehyde, but the oxygen is located on an internal carbon rather than a terminal one.
—O—P(O⁻)(=O)—OH	Phosphate	Highly reactive; when transferred to a compound, that compound usually becomes much more reactive. In water, the hydrogen comes off as a proton, leaving its electron and giving this group a negative charge. It is highly soluble in water.
—SH	Sulfhydryl	Can stabilize the structure of proteins

Polymeric construction is essential to life for several reasons. First, it reduces the difficulty of construction; to build a brick building, it is necessary to know only how to make bricks and how to assemble them. If a different type of building is needed, only the assembly information needs to be changed, not the mechanism for making bricks. With sugars, for the plant to make starch, it bonds glucose together with a certain type of bond; to make cellulose, it uses a different type of bond, but the mechanism for making glucose does not change.

Polymeric construction allows an organism to have a simple basic metabolism that produces only a few types of monomers. As the physiology changes, only the assembly of monomers changes, not the basic metabolism. For example, as a plant grows, it assembles amino acids into proteins necessary for leaves and stems. At the right time, it assembles the same types of amino acids into different proteins—those for flowers. Even more dramatic, algae have the same sugars and amino acids as flowering plants, and during 400 million years of evolution, only the types of proteins formed have changed: The basic mechanism for producing amino acids is about the same.

Polymeric construction permits recycling and conservation of resources; after a polymer is no longer needed, it is depolymerized back to its monomers, which are then used in the construction of a new polymer. All of the energy expended in their construction is conserved.

Finally, polymeric construction allows various parts of an organism to work together in construction. As a fertilized egg grows into an embryo in a developing seed, surrounding tissues supply it with sugars, amino acids, and fats. The embryo can quickly assemble these into organelles and grow rapidly. If the tissues supplied the fertilized egg with only carbon, oxygen, nitrogen, and other elements, growth would be much slower.

Carbohydrates

Carbohydrates are defined by two criteria. First, they usually contain only carbon, hydrogen, and oxygen, although a few carbohydrates contain atoms such as nitrogen or sulfur. Second, the ratio of hydrogen to oxygen is close to 2:1, the same ratio as in water; the generalized chemical formula for carbohydrates is $(CH_2O)_n$.

Monosaccharides

The simplest carbohydrates are the **monosaccharides**, or **simple sugars**, glucose being a familiar example. Monosaccharides are small molecules classified by the number of carbon atoms each contains: four- (4C; tetrose), five- (pentose), six- (hexose), seven-carbon sugars, and so on. Pentoses and hexoses are the most abundant and important (Table 2-5 and Figure 2-10).

Various sugars in a class may have the same chemical formula but differ in their atomic arrangements; such molecules are called isomers. Glucose and fructose are isomers, both having the formula $C_6H_{12}O_6$ but having different chemical structures. Because the functional groups of glucose are arranged differently from those of fructose, the two molecules have distinct shapes and different chemical properties. The differences in chemistry are actually quite slight, but the differences in shape are extremely important. In order for an enzyme to catalyze a reaction, the substrate molecules must fit physically into a very precisely shaped active site. Glucose can enter the active site of certain enzymes, but fructose and other hexoses cannot. Enzymes easily distinguish between isomers by their unique shapes.

Monosaccharides are flexible because all of their carbon–carbon bonds are single bonds, not flat, rigid double bonds. When dissolved in water, the molecules flex and rotate around each carbon atom, changing shape thousands of times every second. When one end of a molecule comes close enough to the other end, the two may react, forming a closed ring (Figure 2-10). In glucose, the number 1 carbon reacts with the –OH group on the number 5 carbon, forming a six-member ring that has oxygen as part of the ring and –CH₂OH as a side group. Ring formation

TABLE 2-5 Common Monosaccharides

4C	Erythrose
5C	Arabinose
	Deoxyribose
	Ribose
	Ribulose
	Xylulose
6C	Fructose
	Galactose
	Glucose
	Mannose
7C	Sedoheptulose

Glucose

$$
\begin{array}{c}
H \\
| \\
\overset{1}{C}=O \\
| \\
H-\overset{2}{C}-OH \\
| \\
HO-\overset{3}{C}-H \\
| \\
H-\overset{4}{C}-OH \\
| \\
H-\overset{5}{C}-OH \\
| \\
H-\overset{6}{C}-OH \\
| \\
H
\end{array}
$$

Ribose

$$
\begin{array}{c}
H \\
| \\
C=O \\
| \\
H-C-OH \\
| \\
H-C-OH \\
| \\
H-C-OH \\
| \\
H
\end{array}
$$

Fructose

$$
\begin{array}{c}
H \\
| \\
H-C-OH \\
| \\
C=O \\
| \\
HO-C-H \\
| \\
H-C-OH \\
| \\
H-C-OH \\
| \\
H-C-OH \\
| \\
H
\end{array}
$$

(a)

(b) $C_6H_{12}O_6$ $C_5H_{10}O_5$ $C_6H_{12}O_6$

Figure 2-10 (a) Ribose is a 5C sugar, a pentose; glucose and fructose are both hexoses (6C). (b) Each sugar can also exist in ring form, but ring formation involves the aldehyde or ketone functional group, which thus does not exist in the ring-form sugars.

releases energy (it is exergonic), and the ring form is the more stable, more common form for a hexose dissolved in the water of a cell.

Because of ring formation, monosaccharides tend to be rather unreactive, relatively inert molecules, which is ideal for physiological functions such as construction, transport, and energy storage. Monosaccharides can be used to build structures that must be inert, stable, and long lasting. They can be transported from region to region without causing damage by reacting with structures they encounter. Monosaccharides can store energy: They are synthesized in leaves from carbon dioxide and water in an extremely endergonic process—photosynthesis. After formed, the energy-rich, stable simple sugars can be moved to sites where energy is needed, such as flowers or roots; then they are metabolized, releasing energy at the new location. Animals and fungi use the monosaccharide glucose for transport, but most plants use **sucrose**, a **disaccharide** composed of one glucose molecule plus one fructose molecule.

■ Polysaccharides

Monosaccharides can act as monomers, reacting with other monosaccharides to form polymers called **polysaccharides**. Extremely short polysaccharides, less then about 10 monosaccharides long, are called **oligosaccharides** and are named by the number of sugars they contain: disaccharides such as sucrose, trisaccharides, and so on.

Theoretically, many types of oligosaccharides and polysaccharides are possible. Consider polysaccharides that consist of just glucose: These could be disaccharides, trisaccharides, tetrasaccharides, and so on, each longer and heavier than the previous one. The bond between two monosaccharides results from the interaction of an –OH group on each (**Figure 2-11**), and, thus, there are numerous possible disaccharides of glucose, each based on the position of the linking bond. During bond formation, an entire –OH is removed from one carbon. A hydrogen is removed from the other –OH group, and water is formed; this is a **dehydration reaction**. Reversal of this, breaking the bond by adding water back to it, is **hydrolysis**.

Despite the large number of polysaccharides that are theoretically possible, very few types actually exist. The dehydration reaction is endergonic and must be coupled with an energy-producing reaction. If the cell does not provide an exergonic reaction to power the polymerization, that particular polysaccharide is not formed to any significant extent; the reaction equilibrium does not favor it. Also, the energy-of-activation barrier of polymerization is high; therefore, the reaction must be catalyzed by an enzyme. This gives organisms great metabolic control; natural selection favors evolution of enzymes that mediate production of useful polysaccharides. A mutation resulting in an enzyme that produces a harmful or useless polysaccharide would be disadvantageous and would probably become extinct. Of the few polysaccharides actually formed by plants, three are especially important.

Starch   Starch, technically known as **amylose** and **amylopectin**, is a long polysaccharide composed only of glucose residues (**Figure 2-12**). The enzyme responsible for polymerization (**starch synthetase**) recognizes only glucose molecules; these fit onto the enzyme such that the only two functional groups that ever form bonds are the —OH groups of the number 4 carbon of one glucose and the number 1 carbon of the other. Because of the way the glucoses are oriented (both face the same direction), the bond is called an **alpha-1, 4-glycosidic bond**.

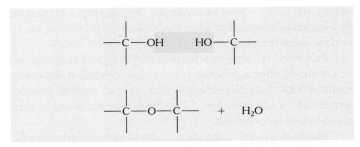

Figure 2-11 Because –OH is such a common functional group, many monomers combine by means of dehydration reactions.

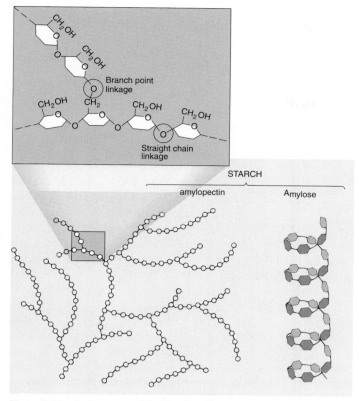

Figure 2-12 If glucoses are linked to other glucoses by alpha-1, 4-glycosidic bonds, the result is an unbranched, coiled chain called amylose, the main component of starch. A second enzyme occasionally makes a bond with the number 6 carbon, resulting in an alpha-1, 6-branched-chain amylopectin, also a component of starch.

After two glucoses have been polymerized, the polymer still has a reactive end that can be recognized by starch synthetase, so the polymer continues to grow as more glucose is added. This is controlled simply by the presence of the enzyme and glucose; control from the cell nucleus is not necessary at each step. Starch molecules become about 1000 glucose residues long, but great variation exists. The length is not important because after about 20 glucose residues, all starch molecules have about the same chemical properties and are treated the same by the cell. Glucose molecules in a polysaccharide are not really complete glucoses because of the loss of water during bond formation; and therefore, the term *glucose residue* is used.

Cells have another enzyme that polymerizes glucose units, but it adds them onto a number 6 carbon rather than the number 4 (Figure 2-12). This creates a short branch that then grows as though it were a simple starch molecule, with more glucose residues being added to its number 4 carbon. The second enzyme is not as active; therefore, branching occurs only at approximately every 20th glucose, and a cell increases or decreases the amount of branching by changing the ratio of the two enzymes. A starch molecule with almost no branching is amylose, whereas a highly branched molecule is amylopectin. The two differ slightly, and the ratio of amylose to amylopectin in starch grains varies from species to species: Potato starch is 78% amylopectin, but starch of wrinkled peas is almost entirely amylose.

Starch serves as a long-term storage chemical for energy. Sugars are excellent for storing energy because they are not very reactive and, when needed, can be metabolized and their energy liberated. Cells cannot store large amounts of sugars, however, because they absorb and hold water, causing cells to swell; starch does not do this.

Cellulose A set of enzymes called **cellulose synthases** polymerizes glucose molecules into the polymer **cellulose**. The glucose residues have an alternating orientation, and the resulting bond is a **beta-1, 4-glycosidic bond**. This is critically important to the nature of the polymer: Although both amylose and cellulose are unbranched chains of glucose joined at carbons 1 and 4, they are totally distinct chemically and biologically. Cellulose molecules form large numbers of hydrogen bonds with other cellulose molecules, crystallizing into rigid aggregates that are extremely strong. On the cell surface, cellulose molecules hydrogen bond to other polysaccharides and become cross-linked into a complex meshwork known as the cell wall).

Few organisms have enzymes capable of hydrolyzing the beta-1, 4-glycosidic bond, and thus, cellulose is remarkably inert. Wood-rotting fungi and bacteria have cellulose-degrading enzymes, but even termites, cockroaches, and cattle can live on cellulose only because their digestive tracts contain microorganisms with the proper enzymes.

Oligosaccharides Many organisms, especially animals, but also some plants, attach short chains of sugars onto proteins. These oligosaccharides are most often present on proteins near or at the cell surface and seem to be involved in cell recognition.

Amino Sugars In fungi, some sugars contain nitrogen in addition to carbon, hydrogen, and oxygen. These **amino sugars** are polymerized into long chains known as chitin that can be highly branched and cross-linked. Chitin occurs in cell walls of fungi.

Amino Acids and Proteins

Proteins are unbranched polymers of **amino acids**; they tend to be about 100 to 200 amino acid residues long. Very short proteins, with fewer than about 50 amino acids, are frequently called polypeptides. Twenty amino acids are used for protein synthesis (**Figure 2-13**); each consists of one carbon that carries four side groups: (1) –COOH, the carboxyl group that causes it to be an acid; (2) –NH$_2$, the amino group; (3) –H; and (4) a fourth group "R" that differs from one amino acid to another. The R groups are not involved in polymerization; instead, they protrude to the sides of the protein backbone, and their properties determine the property of the protein.

The R groups cause amino acids to differ structurally, chemically, and biologically. Some R groups are acidic, and others are basic. Nonpolar, hydrophobic R groups give an amino acid a tendency to interact with other nonpolar molecules such as fats and lipids. Polar, hydrophilic R groups cause their amino acids to interact with other polar molecules, avoiding nonpolar ones. Cysteine is unique in having a sulfhydryl in its R group; two sulfhydryls can interact, forming an –S—S– (disulfide) covalent link between two separate cysteines. If the two cysteines are in the same protein, the disulfide link holds the two regions together; if they are in separate proteins, the disulfide links the two proteins.

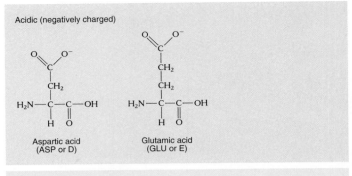

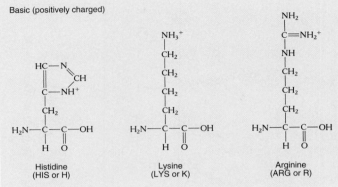

Figure 2-13 These are the twenty amino acids that occur in proteins. Two systems of abbreviations are used for the names of amino acids—an older system that involves three letters for each, and a newer system that uses just one letter each.

During protein synthesis, the carboxyl group of one amino acid reacts with the amino group of the next. Water is removed, and a peptide bond is formed (**Figure 2-14**). This sounds similar to the polymerization of monosaccharides but is actually very different: Linking of amino acids is mediated by organelles called ribosomes and involves large numbers of enzymes and intermediates. This is necessary because proteins are not as simple as polysaccharides; instead, the sequence and types of amino acids incorporated must be controlled with great precision.

Levels of Organization in Protein Structure

Primary Structure The amino acid sequence is the protein's **primary structure**. Because each amino acid has a unique R group, the sequence of amino acids produces a particular sequence of R groups. If all R groups were similar chemically, like the side groups in monosaccharides, this sequence would not be very important. However, R groups are chemically diverse, and thus, a protein has extremely complex properties. Proteins are flexible, and some regions interact with other regions of the same protein, causing the entire molecule to have a characteristic shape (**Figure 2-15**).

Secondary Structure If the R groups of an entire region are of the correct type, the protein forms a helical structure known as an **alpha helix**, one type of **secondary structure**. Other regions might form a folded, flat area called a **beta pleated sheet**. Both

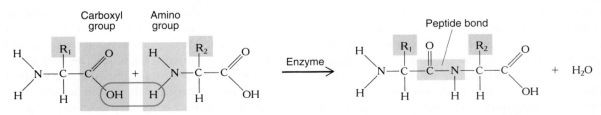

Figure 2-14 The peptide bond between two amino acid residues in a protein is formed by sharing electrons between a carboxyl carbon and an amino nitrogen. During bond formation, water is removed, and thus, this is a dehydration reaction.

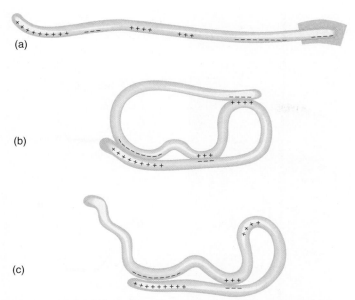

Figure 2-15 (a) The charged amino acids (acidic and basic) of a hypothetical protein. With a different primary structure, the distribution of charges would be different. (b) Various regions of a protein interact, forming a three-dimensional shape, its tertiary structure. (c) If the primary structure lacked the last four negative amino acids (box in [a]), the tertiary structure would be different.

of these secondary structures are short and affect just portions of a single protein, and it is common for certain proteins in membranes to have several alpha helices. On the other hand, many proteins have no regions of secondary structure.

Tertiary Structure The physical shape of a protein in its functional mode is its **tertiary structure**, determined largely by primary structure: Positively charged regions attract and bind to negatively charged regions (Figure 2-15), and hydrophobic R groups interact and form water-free pockets inside the folded protein. Cysteines may link two portions together with disulfide bonds. The protein's overall shape may be globular or fibrous, but the shape and nature of specific sites are usually much more important. Because most plant proteins are enzymes, the active sites must recognize substrates by shape, electrical charge, or hydrophobic properties. For example, the tertiary structure of the enzyme starch synthetase must have a site into which glucose and a growing molecule of amylose can fit. Because glucose is water soluble, the R groups near this active site should be hydrophilic. Without the correct primary structure, the protein does not fold into the proper tertiary structure, and a functional active site is not formed.

Tertiary structure is also affected by small molecules and ions whose concentrations in the protoplasm affect the shape and therefore the activity of many enzymes. Magnesium (Mg^{2+}) and calcium (Ca^{2+}) interact with negatively charged R groups and alter the tertiary structure; with high levels of magnesium or calcium, two negatively charged regions may become "glued" together, whereas without these two cations, the negatively charged regions repel.

Tertiary structure is affected by pH and heat. Protons and hydroxyl ions released by acids and bases interact with charged R groups, changing the way various regions of the protein attract or repel each other. As proteins are heated, the atoms and molecules vibrate more rapidly and unfold. If heated enough, as in cooking an egg, proteins unfold and are said to be **denatured**. With enzymes, even dilute acids or bases or just mild heating causes enough of a change to distort the active site, and the enzyme cannot function.

Quaternary Structure **Quaternary structure** refers to the interaction between two or more separate polypeptides. Quaternary structure is maintained by hydrogen bonding, the interaction of hydrophobic regions, or disulfide bridges. As with secondary structure, not all proteins have this level of organization, but for many, it is critical for proper functioning. Many enzymes consist of two or four polypeptides that must be associated to work properly; only when all are aggregated are the active sites completely formed and functional. For example, an enzyme called RuBP carboxylase is a giant enzymatic complex consisting of eight small proteins and eight large ones; the complex is functional and carries out photosynthesis only when all 16 proteins associate with the proper quaternary structure.

Quaternary structure allows **self-assembly** of certain structures; after the individual protein monomers are formed, they automatically associate into the proper structure such as a microtubule or an enzyme complex. No special constructing apparatus or metabolism is needed; however, a cell easily controls self-assembly by altering conditions that affect aggregation of the individual proteins. With a change in the concentration of calcium, magnesium, or protons, the charge of some R groups is altered, and the protein takes on a new shape, either promoting or inhibiting self-assembly into the quaternary structure.

Nucleic Acids

Nucleic acids are polymers composed of monomers called **nucleotides**. Each nucleotide is formed by the bonding of (1) a phosphate group, (2) a five-carbon sugar, and (3) a complex ring molecule that contains nitrogen and acts like a base (**Figures 2-16 and 2-17**). Nucleic acids contain only five nitrogenous bases that fall into two groups: **Pyrimidines** are molecules composed of a single ring, whereas **purines** consist of two rings. The pyrimidines are cytosine, uracil, and thymine, and the purines are adenine and guanine.

Only two types of five-carbon sugar are ever joined to these nitrogenous bases: **ribose** and **deoxyribose**. The enzymes involved in joining the sugars to the bases have such precisely shaped active sites that they accurately distinguish ribose from deoxyribose. Only four **ribonucleotides** (ribose-containing nucleotides) occur because thymine is never attached to ribose. Similarly, uracil is never attached to deoxyribose, so only four **deoxyribonucleotides** occur.

Details of nucleic acid polymerization are discussed further elsewhere, but a few points can be mentioned here. Deoxyribonucleotides and ribonucleotides are never mixed together in the same polymer, so nucleic acid polymers are of just two fundamental types: **deoxyribonucleic acid** (the famous **DNA**) and **ribonucleic acid** (**RNA**). DNA is found in the nucleus as the main component of chromosomes as well as in plastids and mitochondria. As in proteins, the sequence of monomers is critically important, and any change of sequence can have serious effects. In fact, the **sequence**

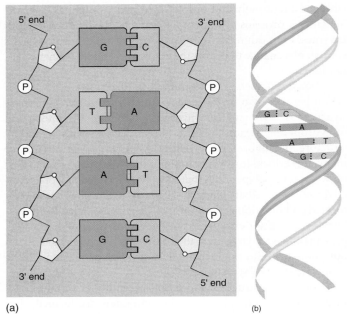

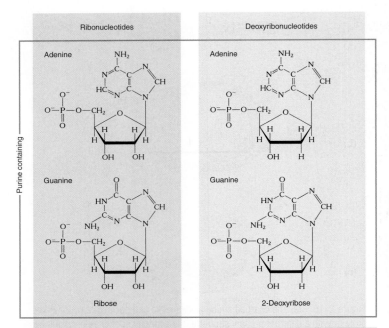

Figure 2-16 (a) As nucleotides polymerize into nucleic acid, the sugars are bound to each other by phosphate groups, making a long chain with the bases projecting from the side. In DNA, the bases of one nucleic acid molecule complement those of another nucleic acid, and the two form thousands of hydrogen bonds and adhere to each other, making double-stranded DNA. (b) The two nucleic acids are not straight but rather spiral around a common axis, forming a double helix.

of deoxyribonucleotides is the genetic information stored in the nucleus. The sequence is used indirectly to guide the polymerization of amino acids into proteins. The primary structure of DNA thus determines the primary structure of proteins and therefore also their secondary, tertiary, and quaternary structures, as well as their function.

Polymers of ribonucleotides, RNA, serve several functions. Some RNA molecules, called messenger RNAs, carry copies of genetic information from the nuclear DNA to ribosomes, the sites of protein synthesis in the protoplasm. Ribosomes are large complexes of enzymes and a second type of RNA, ribosomal RNA. A third class of RNA, transfer RNA, carries amino acids to ribosomes.

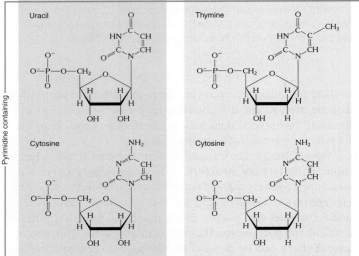

Figure 2-17 Purines (adenine and guanine, in the orange rectangle) contain two rings fused together, and pyrimidines (cytosine, uracil, and thymine, in the green rectangle) contain a single ring. The sugar ribose is found attached only to adenine, guanine, uracil, and cytosine (blue panel), whereas deoxyribose is attached only to adenine, guanine, thymine, and cytosine (tan panel). Pyrimidines are pie shaped, and a pie can be CUT (cytosine, uracil, and thymine).

Lipids

Lipids are fats and oily substances that are extremely hydrophobic and water insoluble. Like carbohydrates, they lack nitrogen and sulfur and consist mainly of carbon, hydrogen, and oxygen. Unlike carbohydrates, they have much more hydrogen than oxygen, often having only two oxygen atoms at one end of a long molecule.

The basic units of many lipids are **fatty acids**. These are long chains containing up to 26 carbon atoms with a carboxyl group at one end. If every carbon atom except the carboxyl carbon carries two hydrogens, the fatty acid is **saturated**; that is, it can hold no more hydrogen (**Figure 2-18** and **Table 2-6**). All carbon–carbon bonds are single, and all parts of the molecule can rotate;

in groups, the molecules tend to straighten and crystallize, being stabilized by interactions with closely packed adjacent fatty acids. This stability makes it difficult to melt saturated fatty acids and they tend to be solid at room temperature; examples are butter and grease.

If some carbons are double bonded to adjacent carbons, the fatty acid is **unsaturated**; carbon–carbon double bonds are rigid, and the molecule has a kink. With several double bonds (polyunsaturated), the molecules are irregular in shape and cannot align well, they have little tendency to crystallize, and they melt at low temperature; examples are olive oil and corn oil.

Botany and Beyond

Box 2-2 Lipids: Oils, Fats, *Trans*-Fats, and Human Health

The term "lipid" covers both fats and oils and several other compounds. Fats, by definition, are solid at room temperature, whereas oils have a lower melting point and, thus, are liquid at room temperature. If oils are cooled sufficiently, they solidify, and if fats are heated, as when we use them for frying, they liquefy. In a saturated fatty acid each carbon is attached to two others by what is called a carbon-carbon single bond; each carbon also has two hydrogen atoms attached to it. In an unsaturated fatty acid, at some point in the backbone, two adjacent carbon atoms are attached to each other by a carbon-carbon double bond, and each of these two carbons has only a single hydrogen attached to it. The fatty acid is not saturated with hydrogens. An unsaturated fatty acid with one double bond is monounsaturated, and an unsaturated fatty acid with two or more double bonds is polyunsaturated.

Saturated fatty acids tend to align easily with each other and have an orderly, stable packing. They have to be heated to disrupt their orderliness and make them move around as a liquid, so saturated fatty acids make up the solid lipids, the fats. The double bond of an unsaturated fatty acid causes a kink in the backbone, so neighboring fatty acids cannot align well. Instead, they make a jumble and move around even when cool; unsaturated fatty acids are the oils.

Within the bodies of plants and animals the relative proportions of saturated and unsaturated fatty acids determine whether a membrane or a lipid droplet is solid, soft, or liquid. The proportions can be changed by the cells as the surrounding temperatures change. In plants and cold-blooded animals, body temperature is similar to environmental temperature and changes with the seasons. Membranes must remain fluid at all times, so these organisms often add oily unsaturated fatty acid in winter and then replace them with saturated fatty acids in summer because the saturated fatty acids will not become too fluid (too "runny") in the heat.

Two types of double bond are possible in unsaturated fatty acids. If the two parts of the backbone lie on the same side of the double bond, this is a **cis-unsaturated fatty acid**, but if the two parts lie on opposite sides, it is a **trans-unsaturated fatty acid**, usually just called a **trans-fat** (see Figure 2-18). A polyunsaturated fatty acid can have both types of double bond. Neither plants nor animals ever make *trans*-fats naturally in their bodies. Natural dietary fats are generally beneficial; however, *trans*-fats are not essential and are never healthful in our diets. They increase the risk of heart disease; they raise the level of "bad" LDL cholesterol and reduce our "good" HDL cholesterol.

The only source of *trans*-fats in our diets is through food processing: We synthesize them artificially in factories. Unsaturated fatty acids, being oils, often cannot be heated enough to use for frying because they tend to smoke, scorch, and develop a "burnt" flavor. Also, baked goods, like cakes, cookies, health food bars, and so on, taste better if they are not "oily" at room temperature. In contrast, many fats can be heated to high temperatures; they melt into a liquid at about the temperature needed for cooking. And in baked goods they give the food a good texture. Furthermore, unsaturated fatty acids (oils) and foods made with them do not have a shelf-life as long as that of saturated fatty acids (fats) because oxygen reacts with double-bonded carbons, and we perceive this as the fat becoming rancid and inedible. If unsaturated oils are used in foods that need to be stored (such as most commercially prepared, packaged foods), then artificial antioxidants such as BHA and BHT must be added (you see them in many labels, near the end of the ingredients).

Although finding adequate supplies of natural fats needed for baking (such as lard) is possible, starting with plant oils and hydrogenating them (chemically forcing hydrogen onto their double bonds) is easier and cheaper. If the oils were hydrogenated to saturation, they would become extremely hard, too hard. Instead, the oils are partially hydrogenated, which merely saturates some double bonds and leaves others. This results in fats that have a longer shelf-life than the original oil and a better texture for baking and frying. The problem is that partial hydrogenation converts some of the natural *cis*-fatty acids into unnatural *trans*-fatty acids. The danger of *trans*-fats has only recently become known, and most food processors have redesigned their products so they no longer use *trans*-fats. This is not true, however, of all processors or of all processed foods, and it is always best to check the label and avoid anything with ingredients listed as "*trans*-fats" or "partially hydrogenated."

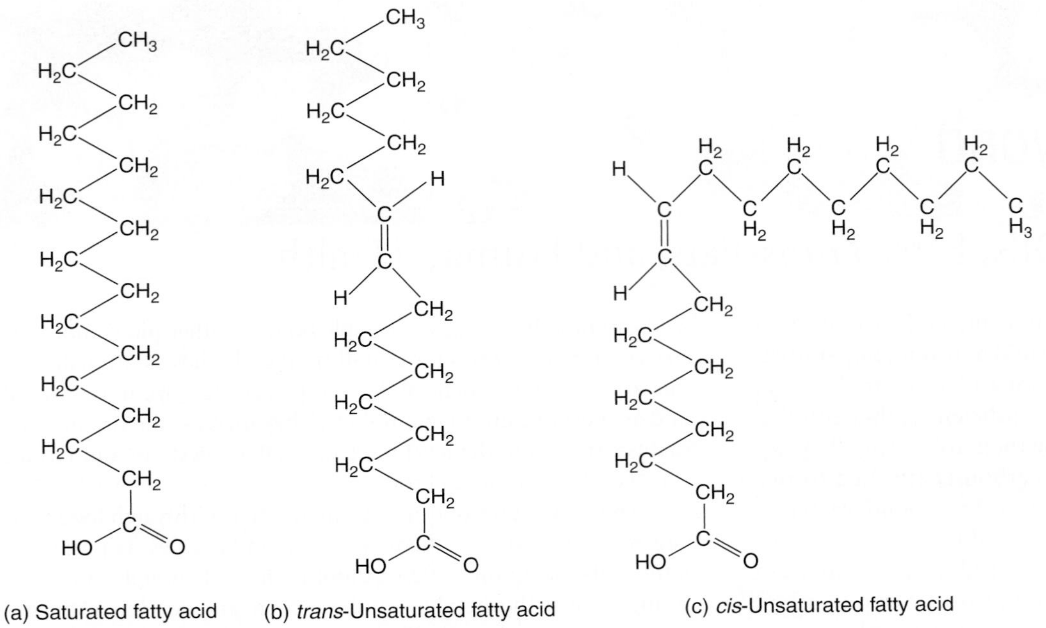

Figure 2-18 The various types of fatty acids. (a) Saturated, (b) *cis*-unsaturated, and (c) *trans*-unsaturated.

(a) Saturated fatty acid   (b) *trans*-Unsaturated fatty acid   (c) *cis*-Unsaturated fatty acid

Polymers of Fatty Acids

Cutins and Waxes Fatty acids tend to polymerize readily with each other, especially when exposed to oxygen. Plants secrete fatty acids through the outer wall of their epidermal cells, and these polymerize when they come into contact with oxygen. Short fatty acids polymerize into **cutin** and longer fatty acids form **wax**. Cutin and wax are not orderly, linear, well-defined polymers like proteins and nucleic acids. Instead, cross-linking occurs almost anywhere, and complex three-dimensional tangles result. Furthermore, the fatty acids involved are mixtures of long-chain and short-chain, saturated and unsaturated, fatty acids. The resulting cutin or wax can be extremely heterogeneous. Both cutin and wax are waterproof and reduce water loss from the plant. They also prevent fungi from invading epidermal cells.

Triglycerides **Triglycerides** are composed of three fatty acids combined with one molecule of glycerol (**Figure 2-19a**). The three fatty acids within a single triglyceride vary in length and degree of saturation and thus affect the nature of the triglyceride.

Phospholipids **Phospholipids** contain glycerol, two fatty acids, and a phosphate group (Figure 2-19b). This composition is critically important: In triglycerides, all parts of the molecules are hydrophobic, so if mixed with water, they coalesce into spherical droplets (see Figure 2-4b). The phosphate group of phospholipids, however, is extremely hydrophilic; therefore, these molecules have one end that tends to dissolve in water and one end that repels water. When mixed with water, they form a layer one molecule thick across the top of the water; this shape allows both maximum interaction of phosphates with water and minimum interaction of the hydrophobic portion with water. In cells, phospholipids form two-layered membranes, the hydrophobic layer of one contacting the hydrophobic layer of the other. This dual nature of both repelling and attracting water is exactly the property needed to build biological membranes. Once formed, the membranes stabilize by interacting with proteins that have shapes (tertiary structures) with regions of positive charges that interact with the phosphate groups.

TABLE 2-6	Common Fatty Acids
Number of Carbon Atoms: Number of Double Bonds	Name
Saturated fatty acids	
4:0	Butyric acid
6:0	Caproic acid
12:0	Lauric acid
14:0	Myristic acid
16:0	Palmitic acid
18:0	Stearic acid
26:0	Cerotic acid
Unsaturated fatty acids	
Monounsaturated	
16:1	Palmitoleic acid
18:1	Oleic acid
Polyunsaturated	
18:2	Linoleic acid*
18:3	Linolenic acid
20:4	Arachidonic acid*

*These are essential fatty acids that are necessary for human growth and development, but they are not synthesized by our bodies. These fatty acids must be present in our diet.

Cofactors and Carriers

Many enzymes by themselves convert reactants into products; however, in many cases, small molecules or ions must be present for a reaction to occur. Without ionic charge to help establish the proper tertiary structure, an enzyme may not have a properly formed active site, and no catalysis occurs. Such ions are **cofactors** for the enzyme; examples are ions of magnesium (Mg^{2+}) and iron (Fe^{3+}). Similarly, small organic molecules called **coenzymes** carry energy, electrons, or functional groups into the reaction.

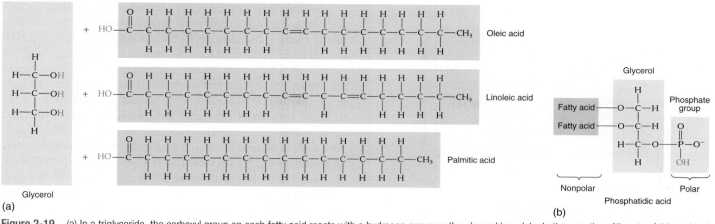

Figure 2-19   (a) In a triglyceride, the carboxyl group on each fatty acid reacts with a hydrogen group on the glycerol in a dehydration reaction. All parts of this molecule are nonpolar. These are extremely hydrophobic: In order for even one of these molecules to diffuse into water, hundreds of water–water hydrogen bonds would have to be broken. (b) A phospholipid is nonpolar on one end and polar on the other. The polar end can dissolve into water; the nonpolar end can dissolve into fats.

■ Energy-Carrying Coenzymes

The most common energy carrier is **adenosine triphosphate** (**ATP**) (**Figure 2-20**). The last two phosphate groups of ATP are attached by **high-energy phosphate bonds** whose bonding orbitals are not very stable. If a phosphate bond is broken and releases a phosphate group (P_i) and **adenosine diphosphate** (**ADP**), large amounts of energy are released as electrons rearrange into nonbonding orbitals. ADP can lose a second phosphate group to become adenosine monophosphate (AMP), which also releases large amounts of energy (**Figure 2-21**). The phosphate of AMP is attached by a stable, low-energy bond and cannot be used to drive an endergonic reaction. Because the conversions of ATP → ADP + P_i and ADP → AMP + P_i are both highly exergonic, they can be coupled with endergonic reactions and force them to proceed. ATP can force the reaction $A + B \rightarrow AB$ to proceed by first transferring a phosphate to one of the reactants in an enzyme-mediated phosphorylation: $A + ATP \rightarrow A–P + ADP$. The phosphorylated A has more energy than the nonphosphorylated A, and the energy-of-activation barrier for the reaction $A–P + B \rightarrow AB + P_i$ is lower than the energy-of-activation barrier for $A + B \rightarrow AB$ (**Figure 2-22**).

ATP is a versatile energy carrier: Almost all endergonic reactions are forced to proceed by using it; however, ATP is both highly reactive and unstable. It cannot be stored or moved from one cell to another; each cell must make its own ATP as it is needed. The majority of the ATP used in daily metabolism is generated by mitochondria, using energy derived from the exergonic breakdown of glucose or lipids. The ATP necessary to build glucose and lipids in the first place is generated in chloroplasts using the energy of sunlight.

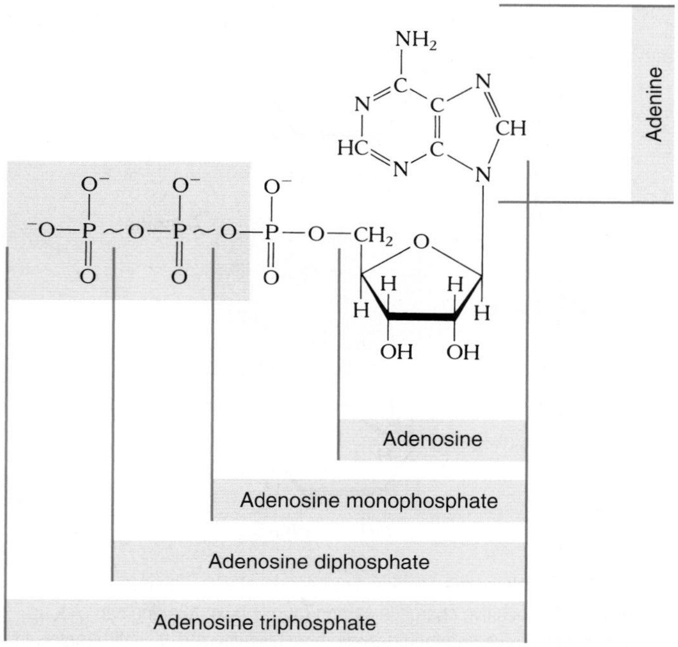

Figure 2-20   Adenosine triphosphate. Breaking off the last phosphate to produce ADP results in a more stable set of electron orbitals, and energy is given off. The same is true of removing the second phosphate, but not the third.

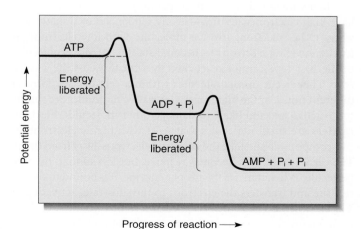

Figure 2-21   Dephosphorylation of ATP and ADP is highly exergonic; the liberated energy may be converted to heat or used to drive an endergonic reaction.

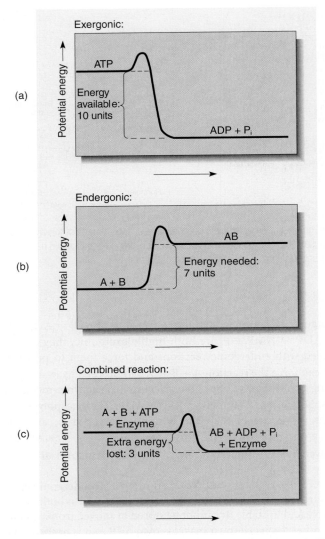

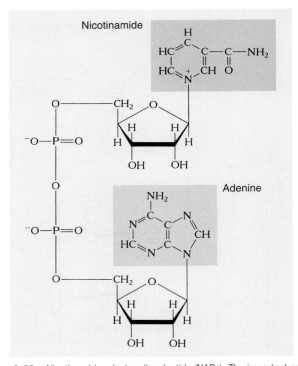

Several electron carriers exist, each with a characteristic electronegativity. Three have only moderate electronegativities: nicotinamide adenine dinucleotide (NAD⁺) (**Figure 2-23**), flavin adenine dinucleotide (FAD), and flavin mononucleotide (FMN). NAD⁺ is capable of transferring electrons between many different metabolic reactions. It picks up two electrons at once, the first of which neutralizes NAD⁺ to NAD and the second of which allows NAD to bond with an H⁺, forming reduced nicotinamide-adenine dinucleotide (NADH). (The reactions that form NADH are complex, and, to be strictly correct chemically, NADH should always be considered to be associated with a second proton: NADH + H⁺. Many textbooks do this, but others often simply write either NADH or NADH₂. These last two symbols are certainly convenient and make it easier to concentrate on other aspects of the reactions, but to balance chemical equations completely, the term "NADH + H⁺" must be used.)

Many reactions occurring at many sites produce either ATP or NADH + H⁺; these diffuse throughout the organelle and cell until by chance they run into enzymes that bind and use them. Later, the ADP and NAD⁺ are released from the enzyme and again diffuse at random until they return to the recharge sites. The cell has a pool of thousands of molecules of ATP, ADP, NADH, and NAD⁺, each of which recycles continuously.

Nicotinamide adenine dinucleotide phosphate (NADP⁺) is almost identical to NAD⁺, differing only in having an extra phosphate group. The highly energetic reactions of photosynthesis are used to force electrons onto NADP⁺, converting it to reduced nicotinamide adenine dinucleotide phosphate (NADPH + H⁺),

Figure 2-22 An exergonic reaction, such as the dephosphorylation of ATP (a), can supply power to an endergonic reaction (b) if it provides enough extra energy and if there is a mechanism, usually an enzyme, that can substitute a new common reaction mechanism for the two old separate mechanisms (c).

■ Electron Carriers

Many metabolic reactions in plants generate molecules that have high electronegativities—strong tendencies to donate electrons. Other reactions involve molecules with low electronegativities, needing electrons to become stabilized. Most molecules of the first type cannot react directly with those of the second type. They occur in separate sites within a cell, or they form at different times, or the highly electronegative molecules are simply too reactive and too likely to damage other molecules. **Electron carriers** are small molecules that react with highly electronegative compounds and take the energetic electrons away from them. The carrier itself then becomes reactive, but less so than the first molecule. When it meets the proper enzyme, it is bound to the active site and transfers the electrons to the substrate, activating it. The electron carrier is then released and returns to the site where high-energy electrons are being produced. Both ATP and electron carriers are "recharged" and reused repeatedly; they are not used up or destroyed in reactions.

Figure 2-23 Nicotinamide adenine dinucleotide (NAD⁺). The important area is the site where the positive charge is carried (purple box); this ring can pick up two additional electrons and then deposit them elsewhere. This entire molecule is the "container" for carrying two electrons around a cell. The adenine (green box) is far removed from the site where the electrons are carried, but if the adenine is altered, NAD⁺ cannot enter into the proper reactions.

which is used almost exclusively in the chloroplast to make sugars from carbon dioxide and water.

Enzymes

Enzymes are protein catalysts that accelerate certain chemical reactions by providing alternative mechanisms in which all energy-of-activation barriers are lower than in the original reaction mechanism (see Figure 2-9). Several important aspects of enzymes can be examined now.

Substrate Specificity

The atoms or molecules that an enzyme interacts with are its **substrates**, and these must fit into and be bound by the enzyme's active site if a reaction is to occur (**Figure 2-24**). The active site's size, shape, electrical charge, and hydrophobic/hydrophilic nature are determined by the protein's tertiary structure and, if present, its quaternary structure. Many enzymes have a high **substrate specificity**; that is, the active site is so distinctive that only one or two specific substrates fit into it. For example, the enzyme PEP carboxylase binds only to carbon dioxide and to PEP (phosphoenolpyruvate); therefore, the only reaction it can accelerate is

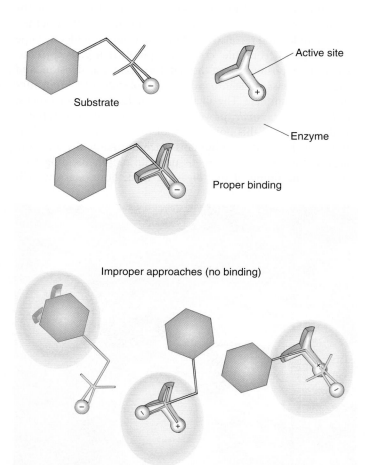

Figure 2-24 Usually only part of a substrate fits into the active site, and this part must be oriented properly. The chances are great that when a substrate molecule strikes an enzyme, it either misses the active site or is improperly positioned. The substrate may diffuse against the active site hundreds of times before it actually enters and is bound.

the addition of carbon dioxide to PEP, forming oxaloacetate. Other enzymes have low substrate specificity, binding any of several similar substrates. Often, this is because only a small part of a substrate fits into the enzyme's active site: As a molecule of amylose grows longer, only the terminal, reactive glucose residue and a new molecule of glucose fit into the active site. The great majority of the polysaccharide chain is not associated with the active site, the enzyme cannot distinguish an amylose molecule of 50 glucose residues from one of 500 residues.

Some active sites show specificity not for particular molecules but for specific functional groups, being able to bind to carboxyl groups, for example, or phosphate groups, with little or no regard for the nature of the rest of the molecule. Such enzymes carry out a particular type of reaction on many types of molecules, generally having a widespread effect on the metabolism of the cell or organelle.

Rate of Enzyme Action

The binding of a substrate to its enzyme is a second-order reaction whose rate is proportional to the concentrations of the substrates and enzyme. With a given concentration of enzyme, the reaction proceeds faster in a solution with a high concentration of substrate than in one with a low concentration. For many reactions, the enzyme must bind two substrates as well as some coenzyme or carrier such as ATP or NADH; however, these bind one at a time to the enzyme, each as a second-order reaction, and the slower one sets the rate for the entire reaction.

Binding to the enzyme is usually complex. For small, symmetrical substances such as magnesium or methane, the ion or molecule fits no matter how it enters the active site, but for a large, complexly shaped molecule, such as a phospholipid or monosaccharide, the molecule may be improperly oriented when it diffuses against the active site. If the wrong end or group enters (Figure 2-24), it does not bind to the enzyme but diffuses out again. The requirement for precise binding is greater for enzymes with high substrate specificity; consequently, the rate of binding tends to be slower because only a small fraction of substrate/enzyme collisions have the proper orientation.

Control of Enzyme Activity

Cells control the activities of their enzymes by a variety of methods. One is simply producing or not producing the enzyme. Many enzymes necessary for certain specialized types of metabolism are produced only in the cells that need them; for example, enzymes that mediate the production of chlorophyll are produced in leaf cells but not in wood or bark cells.

After an enzyme is present, its activity can usually be regulated further. Because its tertiary shape is so critical, cellular pH, temperature, and the presence of magnesium and calcium ions are also critical. Slight changes in these factors may alter all proteins within a cell, thereby exerting very broad regulatory power. More precise regulation of specific enzymes is possible by the use of activators and inhibitors, small molecules that bind to only one specific enzyme and either increase or decrease its activity. An activator may bind to some site, usually not the active site, and thus change the tertiary structure, either causing formation of the active site or improving its characteristics.

Plants and People

Box 2-3 Toxic Compounds

Consider two facts: (1) Plants live among herbivores, and (2) the world is full of uneaten plants. What is the cause? Many plants produce toxic compounds: poison ivy, death camas, poison hemlock, and deadly nightshade are familiar examples, but the full list of toxic plants is very long. Many plants we think of as being non-poisonous are only harmless because we stop eating things that are bitter or that burn our mouths: few people would eat enough chili peppers to get a lethal dose. The following are some toxins in plants we live with. Fascinating and well-illustrated accounts of many are given by *Handbook of Poisonous and Injurious Plants* (2007) (Nelson LS, Shih RD, and Bolick, MJ).

Anticholinergic (antimuscarinic) poisons. Many nerve cells have receptors that detect the presence of a small molecule called acetylcholine, which causes a reaction in the nerve cells when it binds to the receptors. Plants such as *Atropa* (deadly nightshade) and *Datura* (jimsonweed) produce chemicals that bind to one of these receptors (called the muscarinic receptor), preventing acetylcholine from stimulating proper nerve transmission.

Nicotine-like alkaloids. Alkaloids are small molecules that contain nitrogen and somewhat resemble amino acids. Alkaloids of *Conium* (poison hemlock), *Nicotiana* (tobacco), and *Sophora* (mescal bean) block another acetylcholine receptor (this one called the nicotinic subtype). Symptoms occur in the parasympathetic and sympathetic parts of the nervous system, the brain, and the junctions where nerves attach to muscles.

Convulsant poisons. The small tree *Strychnos* produces strychnine, which causes motor nerve cells to be hyperexcitable and results in convulsions. Several other plants produce convulsants.

Capsaicin. Chili peppers such as habaneros, jalapeños, and tabascos in the genus *Capsicum* contain capsaicin, which induces some of our sensory nerve cells to release a chemical called substance P, which specifically stimulates other nerve cells that normally respond when we are being burned. Artificial selection by humans has resulted in chili peppers that range from having almost no capsaicin (bell peppers) to some that are dangerous (cayenne, chiltepin, and Thai). Do you need help remembering the difference between "hydrophilic" and "hydrophobic"? Capsaicin is hydrophobic, and thus, drinking water will not cool your mouth; however, the fat in cheese, butter, and milk will absorb it and give some relief.

Cardioactive compounds. Digitoxin and digoxin in *Digitalis* (foxglove) inhibit a set of proteins that use the power of ATP to pump sodium and potassium across a membrane. When this pump is blocked in heart cells, it leads to an increase in the concentration of calcium within the cells. Although foxglove leaves are injurious if eaten improperly, these chemicals are used medicinally to slow and strengthen heartbeat.

Cyanogenic glycosides. These are nontoxic in the plant, but enzymes in our digestive system cleave them into two parts: a sugar and cyanide. Cyanide inhibits the final step in aerobic respiration, preventing generation of ATP. These toxins are found in some of our most common fruits—*Malus* (apples) and *Prunus* ("stone fruits" such as cherries, peaches, apricots)—but in parts that we do not eat such as apple seeds or peach pits. Poisoning most often occurs when apricot seeds are used as components of herbal medicines.

Mitotic inhibitors. α and β tubulin are proteins whose tertiary structure allows them to be assembled temporarily into microtubules (a quaternary structure) in certain areas of the cell then disassembled and reconstructed into new microtubules in some other area. Microtubules create the spindle that pulls chromosomes to opposite ends of the cell during nuclear division, and then they disassemble into individual molecules of tubulin after division is complete. The alkaloid colchicine in autumn crocus, *Colchicum autumnale,* prevents the assembly of microtubules. Cells exposed to this poison cannot form a spindle, and division fails. Even after we humans reach our adult size and stop growing, we still have many tissues with rapidly dividing cells, such as the lining of our digestive tract, bone marrow cells

Figure B2-3 Each leaf of poison ivy has three leaflets and the defensive compound urushiol.

that produce red blood cells, and seminiferous tubules in men. *Catharanthus roseus* (also called *Vinca rosea,* periwinkle) produces the alkaloids vinblastine and vincristine that also block cell division; these are used in chemotherapy to stop cancer cells from dividing.

Toxalbumins. These proteins bind to ribosomes and prevent them from translating messenger RNA into protein. *Ricinus* (castor bean) produces the toxalbumin ricin, one of the most toxic substances known. Even though present in only low concentrations, as few as eight beans can be fatal. *Jatropha* species (coral bean) has a higher concentration of a related toxalbumin, and a single seed can cause severe poisoning.

Urushiol. Many plants produce chemicals that are not actually toxic but that sensitize the immune system of some people. Later exposure to the same chemical results in an allergic reaction, often severe. About half of all people are sensitized to urushiol, an oily substance in species of *Toxicodendron* (poison ivy, poison oak, and poison sumac). All parts of these plants can trigger an allergic reaction.

Remember that these plants do not have toxic compounds **in order to** protect themselves, or with the **purpose** of keeping herbivores away. It is just that toxic or injurious plants are more likely to be left uneaten compared with nontoxic plants and therefore are more likely to produce more seeds and more progeny.

An inhibitor impairs the active site when it binds to the enzyme. A **competitive inhibitor** does this by occupying the active site and preventing a substrate from entering: It competes with the substrate for the same site. A **noncompetitive inhibitor** attaches to the enzyme at an area other than the active site; binding alters the enzyme such that (1) the substrate cannot bind, (2) the reaction cannot proceed, or (3) the product cannot be released. In many cases,

the inhibitor molecule is either the product of the enzyme itself or is the end product of the metabolic pathway involving the enzyme. This is **end-product inhibition**. As an enzyme or metabolic pathway functions, a product accumulates and gradually starts to inhibit the enzymes involved in its formation, preventing accumulation of excessive concentrations. If it is used up or secreted and its concentration drops, inhibition is lowered and synthesis begins again.

Summary

1. The metabolism of all organisms is based on universal principles of physics, chemistry, and mathematics.
2. Electronegativity is a measure of the capacity of an element to accept or donate electrons.
3. Electrons, orbitals, and entire atoms are more stable if they have less energy. If a reaction allows valence electrons to move into bonding orbitals that are more stable than the nonbonding orbitals, the product is more stable than the reactants. Energy is given off in this exergonic reaction.
4. In water, hydrogen ions are covalently bonded to oxygen, but water is a polar molecule with partially positive and partially negative ends. This permits hydrogen bonding.
5. Acids increase the concentration of protons in a solution, usually by liberating protons. Bases decrease the proton concentration, usually by liberating hydroxyl ions that bond with protons.
6. Carbon can form single, double, or triple covalent bonds. The single bonds are the most stable, whereas the bonding electrons in double or triple bonds still have considerable energy and can enter chemical reactions relatively easily.
7. If the activation–energy barrier is low, bonding orbitals can be rearranged easily, and even molecules with little energy (low temperature) can react. If orbital rearrangement is difficult, the barrier is high, and only high-energy (high-temperature) molecules can react.
8. Catalysts such as enzymes do not change the activation energy of a reaction; rather, they provide an alternative mechanism by which orbitals are more easily rearranged—all activation–energy barriers are lower. Catalysts do not change the total amount of energy liberated or absorbed during a reaction.
9. In all reactions, a reaction equilibrium is established in which there is a mixture of some reactants and some products, some forward reaction and some back reaction.
10. The physical and chemical properties of organic molecules are most strongly determined by the functional groups present; the backbone often is not so important.
11. Polymeric construction simplifies construction techniques and reduces the amount of information necessary for construction.
12. Three important carbohydrate polymers are starch, cellulose, and oligosaccharides. All are composed of monosaccharides.

The type of bond and the presence or absence of branching are important.

13. The primary structure—the amino acid sequence—of proteins determines how the protein folds (secondary and tertiary structure) and how separate proteins aggregate (quaternary structure).

14. Nucleic acids are of two types: deoxyribonucleic acid, a polymer of deoxyribonucleotides, and ribonucleic acid, a polymer of ribonucleotides.

15. Many lipids contain fatty acids, either saturated or unsaturated. A triglyceride consists of glycerol bonded to three fatty acids; a phospholipid consists of a glycerol bonded to two fatty acids and a phosphate group.

16. Many small molecules cycle repeatedly between two states and serve as carriers. ATP carries energy; NADH, NADPH, FAD, and FMN carry electrons.

Important Terms

acid	deoxyribonucleic acid (DNA)	nucleic acids
adenosine diphosphate (ADP)	endergonic	nucleotides
adenosine triphosphate (ATP)	endothermic	phospholipids
amino acids	enzymes	polar molecule
amylose	exergonic	polymer
base	exothermic	polysaccharides
catalyst	hydrogen bonding	proteins
cellulose	hydrolysis	ribonucleic acid (RNA)
coenzymes	lipids	self-assembly
cofactors	monomers	substrate specificity
covalent bond	monosaccharides	
dehydration reaction	nonpolar molecule	

Review Questions

1. Do plants or animals have any metabolic reactions that cannot be modeled by hypotheses and verified by experimentation and observation? What was "vital force" thought to be?

2. In the Concepts section of this chapter, three major discoveries of the 1800s are described. One concerned the synthesis of biological compounds; one involved enzymes, and the third was a discovery made by Louis Pasteur. List these discoveries, and explain why they are important.

3. Some elements are more stable as ions than as neutral atoms. Why? Give examples of biologically important positive ions.

4. What does it mean to say that after sodium reacts with chlorine, the resulting compound NaCl is more stable than the two reactants? If they are more stable, did they give off energy? If so, did they become warmer during the reaction? If so, where did the energy of that heat go?

5. If a reaction gives off energy (liberates it to the environment), is that reaction exergonic or endergonic? If the energy is given off as heat, what is the reaction called?

6. The reaction for the formation of water from hydrogen and oxygen and can be made to run in the opposite direction (water is broken down into hydrogen and oxygen) if energy is supplied. This reverse reaction occurs in plants in the process of photosynthesis. What do you think is the source of the energy? Because energy is put into the reaction (rather than being liberated), what type of reaction is this?

7. What type of bond holds sodium to chlorine in sodium chloride? What type of bond holds hydrogen to oxygen in water?

8. Examine Figure 2-3b, which shows hydrogen bonding in water. Within any plant or animal cell there are millions of water molecules that are hydrogen bonded together. This allows a plant to pull water upward through its body (water is pulled in plants and pushed in animals). Imagine how water is pulled. What would happen if there were no hydrogen bonds?

9. If a solution has a high concentration of protons (H^+), is it acidic or basic? What about if it has a high concentration of hydroxyl ions (OH^-)? When HCl is mixed with water, it breaks down to H^+ ions and Cl^- ions. Because of the H^+ ions it produces, is HCl an acid or a base?

10. The bonding between carbon atoms is important because virtually all biological molecules in plants (and our own bodies) contain carbon atoms bonded to other atoms, often to other carbon atoms. What is the shape of a molecule in which the carbon has only single bonds? What is the shape of two carbons bonded by a double bond? What about a triple bond?

11. Why is it necessary for two reactants to collide vigorously for a reaction to occur? What happens to the shape of valence orbitals if a reaction occurs? What happens if no reaction occurs?

12. Although there is no way to change the energy of activation of a reaction, it is possible to change the _____ _____. There are two names for substances that do this: _____ (is a general term) and _____ _____ (is used when this is a protein in a living organism).

13. An endergonic reaction tends to proceed slowly because it absorbs energy. How do plants force endergonic reactions to occur rapidly? Is ATP involved? Why is ATP such a versatile molecule? Because its breakdown to ADP and P_i is highly exergonic, is its synthesis endergonic?

14. Name five functional groups and give the chemical formula for each. What properties does each functional group give to the molecule to which it is attached?

15. What are carbohydrates, pentoses, and hexoses? Name several hexoses, and describe how they differ from each other. How can enzymes distinguish between them?

16. Various sugars in a class may have the same chemical formula but differ in their atomic arrangements; such molecules are called _____.

17. Examine Figure 2-10. Monosaccharides are a bit unusual in that they are just long enough that as they vibrate one end often accidentally comes close to the other end of the same molecule and the two can react. Therefore, they can exist as ring-shaped molecules or as open, chain-shaped ones. Do the two forms of the same molecule have the same chemical and physical properties?

18. When we eat plants, we are eating mostly polysaccharides. What are the two starches that we eat? One is unbranched: _____; the other is branched: _____. (By the way, if we eat more starch than we need, we store it in our bodies as the polysaccharide glycogen.) Another polysaccharide that we eat is in the cell walls; it is _____.

19. What is the technical name of the bond in starch? What is the technical name of the bond in cellulose? They both involve the same carbon atoms of the glucose molecules (the #1 and the #4 carbons). Starch and cellulose are virtually identical, except that all of the glucoses face the same way in starch, whereas in cellulose alternating glucoses face in the opposite direction. Does this matter: Can you digest starch? Can you digest cellulose?

20. Proteins are unbranched polymers composed of _____ _____ _____. The bond that holds the monomers together in proteins is called a _____ _____ bond.

21. What are the three groups found in every amino acid? What are R groups, and how do they differ from one amino acid to another? Would you agree that because R groups are not involved in forming a peptide bond they are not really very important?

22. What is the primary structure of a protein? Are proteins flexible or rigid? Are any of the bonds in the backbone a double or triple bond, or are all of them single bonds?

23. If proteins had double bonds in the backbone, they would be rigid molecules. Do you think they would then have a secondary, tertiary, or quaternary structure? Why is the tertiary structure of the hypothetical protein in Figure 2-15b different from that of Figure 2-15c?

24. The tertiary structure of a protein is affected by pH (acidity) and heat. Can you give examples of that? Our stomachs secrete acid. What do you think is its effect on our ability to digest protein? Have you ever fried or boiled an egg? What is the change caused by the heat?

25. What do the initials DNA and RNA stand for? What are the four nucleotides that occur in DNA? What are the four that occur in RNA? Which nucleotide, unique to DNA, is never found in RNA? Which nucleotide is unique to RNA?

26. What sugar occurs in DNA? What sugar occurs in RNA?

27. Fats and oils are substances known as _____. These are always hydrophobic and insoluble in water. If one of these molecules encountered a protein that has a region of hydrophobic amino acids and a region of hydrophilic ones, which region would the molecule associate with?

28. The basic units of many lipids are fatty acids. What is a saturated fatty acid? Do they tend to be straight or kinked (see Figure 2-18)?

29. What is an unsaturated fatty acid? Do they tend to be straight or kinked?

30. If you place both a stick of butter and a cup of oil in the refrigerator, the butter becomes hard while the oil remains liquid. Which is composed of saturated fats, and which is mostly composed of unsaturated ones? It is very important for plants (and animals) to have fats in their cell membranes that remain liquid—otherwise the plant would die. Do you think plants of cold climates have more unsaturated fats in their cell membranes than do plants of hot climates?

31. Fatty acids tend to polymerize with each other, especially when exposed to oxygen. If the fatty acids are relatively short, the polymer is called _____. If they are relatively long, it is called _____.

32. How does a triglyceride differ from a phospholipid? How do these two differ in their ability to dissolve in water? How does this make one especially suitable for the construction of membranes?

33. Cofactors are essential to the activity of some enzymes. Name two cofactors.

34. Each coenzyme carries one of three things into a reaction. What are the three things?

35. What is the full name of ATP? The last two phosphate groups of ATP are attached by _____-_____ _____ bonds.

36. When a reaction needs to have energy put into it, forcing it to proceed, what is the molecule that usually participates? After the reaction, what has the energy-carrying coenzyme been converted to?

37. Examine Figure 2-22. The breakdown of ATP to ADP and P_i gives off energy. The reaction $A + B \rightarrow AB$ absorbs energy. How does combining these two reactions allow the second one to proceed? Would combining the two reactions force the second one to proceed if the breakdown of ATP gave off only 2 units of energy?

38. Many reactions in plants generate molecules that have a strong tendency to donate electrons. Other reactions need

electrons to proceed. Name three electron carriers that transport electrons from one type of reaction to the other.

39. The atoms or molecules that an enzyme interacts with are its _____, and these must fit into and be bound by the enzyme's _____ _____ if a reaction is to occur.

40. Some enzymes will bind to only one or two substrates. Is their substrate specificity high or low? Other enzymes will bind to various substrates as long as they are at least somewhat similar. Is their substrate specificity high or low?

41. Cells control the activities of their enzymes by a variety of methods. Name at least five factors that affect enzyme activity.

BotanyLinks

http://biology.jbpub.com/botany/5e
BotanyLinks contains a variety of resources and review material designed to assist in your study of botany.

Visit the Web Exercises area of BotanyLinks to complete this question:

1. The bodies of living organisms contain many chemical compounds, but how many chemical elements? Go to the BotanyLinks home page to investigate.

Plant Structure

Plant structure is the physical, material body of a plant, composed of carbohydrates, proteins, lipids, minerals, water, and other components. In terms of both material and energy, it is expensive for an organism to build a body—all of the energy and resources used by a tree to construct its body could have been used instead for reproduction, increasing its number of offspring. The theory of evolution by natural selection predicts that organisms must receive enough benefit from their structure to offset the cost of building it: It must be more advantageous selectively to have a body than not to have one. The benefit is that the structure is the framework in which metabolism occurs and the means by which metabolism interacts with the environment. This can be more easily understood by considering the alternative—organisms with very little structure.

Several organisms have almost no structure whatsoever: viroids, viruses, and mycoplasmas (see Chapters 15 and 25). All are parasites that live, grow, develop, and reproduce only if immersed within the body of a host organism. The host provides a benign environment with enough nutrients for the parasite while also being free of harmful chemicals. The temperature of the host's body is ideal for the parasite, and adequate water is available. Parasites cannot live outside the host-supplied environment. With minimal structure, they have minimal capacity to interact well with the heat, cold, drought, and scarcity of nutrients in a natural environment.

Organisms whose bodies are more structured and complex are able to resist temporary adverse environmental conditions as well as exploit optimal ones. A simple bacterium can absorb mineral nutrients from certain environments; but, by constructing a root system, a plant can spread its mineral-absorbing metabolism throughout an extensive, deep volume, growing past dry, rocky, mineral-poor soil and tapping moist, rich, fertile soil. The investment for the plant

Part Opener Image: Like most plants, this tree in winter does not have all its structures at the same time. Its leaves are absent, having been shed in autumn. Likewise, flowers and fruits it had last year have been shed; and the leaves, flowers, and fruits it will have next year have not been formed yet. An animal typically has all its organs at once, but most plants produce new sets each year—the entire plant body rarely exists all at the same time.

is great, but sufficient reward is achieved in having an adequate, secure source of water and minerals.

The structures discussed in this part of the text are important to the plants themselves, but they are also important to us because we rely on them for food and many products, and as we cultivate them we affect the lives of many other organisms. Most of the food we eat is some part of a plant body. For example, we eat the leaves of lettuce, spinach, and cabbage; roots of carrots, beets, and radishes; stems of asparagus and potatoes; and fruits and seeds such as apples, oranges, peas, beans, and almonds. Some of our foods are so highly processed we may not immediately realize they were once part of a plant body: Bread, doughnuts, cakes, and pasta are made with flour (the ground seeds of wheat); beer is brewed from barley (also seeds) and hops (flowers); and wine is fermented grape juice. Our very lives depend on these plant-based foods because they supply us with energy, vitamins, and minerals. Look at your hand: Every atom and molecule, every bit of your body was at one time part of some plant. Vanilla, chocolate, cinnamon, sage, black pepper, and chili peppers are spices that do not really provide us much in the way of nutrition, but they make eating fun and few people would want to give them up. The cotton in our clothes is a plant structure, as are the fibers in paper and burlap bags and the lumber used for construction or for making guitars and pianos.

Our reliance on plant bodies goes beyond these fundamental needs. The grass lawns we play or relax on are the leaves and stems of plants, and trees provide us with shade and protection from weather. We love the beauty of flowers, and almost certainly there will be some point in your life when something wonderful or tragic will have happened and you just won't know what to say. At that point you will probably give someone flowers to express emotions you cannot put into words. We celebrate weddings, and birthdays with flowers, we give flowers to friends when they are ill or just for friendship, and we rely on flowers to express sympathy at funerals.

Because we rely on plants for so many things, we need to cultivate them on farms, orchards, and tree plantations, and this raises important issues with the rest of the world. Agriculture began as early as 9,000 years ago and has been affecting the world ever since. The first step of agriculture then as now is to clear a piece of land of its natural vegetation and animals. If we apply fertilizers or pesticides to crops, rain may wash these chemicals from the fields into nearby streams, polluting them and harming fish, frogs, and other aquatic life. Crops must be harvested, processed, and transported to markets, all of which require roads, fuel, and machinery and result in pollution. Our use of plant structures for so many things in our lives results in negative impacts on the ecology and well-being of many organisms.

Plant structures, however, may also provide solutions to these problems. Scientists are conducting research to find plants that provide more abundant or more nutritious food, better fibers, or natural resistance to insect pests. Many of our crop plants now produce much greater harvests than did the crops available just a few decades ago. For example, a fast-growing plant called kenaf can produce many more tons of fibers per acre than do pine trees, and thus less land is needed to make the same amount of paper: It may even be possible to convert some farm land back to a more natural condition suitable for wildlife.

Most plants around us are angiosperms, commonly called the "flowering plants." Most plants discussed in this text are angio-

sperms, and to a lesser extent we discuss ferns (which reproduce with spores) and conifers (which have cones, such as pines and spruces). One of the most noteworthy features of angiosperms is the diversity of bodies that occur in this group. Some angiosperms are ordinary herbs or shrubs or trees, but some are desert succulents (cacti), vines (grapes), parasites (mistletoe), bulbs (onions), and tubers (potatoes). Ferns and conifers are not so diverse, and this text would have been much shorter if it had been written 135 million years ago before angiosperms had come into existence.

Angiosperms are one of the newest groups of plants. The first true plants originated about 420 million years ago, but the oldest fossils of angiosperms are only about 135 million years old. At the time angiosperms originated the world was a very different place: The air and land were much warmer, as were the oceans. Consequently, much more water evaporated from the oceans into the air and then fell as rain on land; warm, moist conditions allowed the growth of lush vegetation that consisted mostly of ferns and conifers. The dominant land animals were dinosaurs. During the last 100 million years Earth warmed even more as volcanoes and lava released large amounts of the greenhouse gas carbon dioxide into the air. However, gradually plants took carbon dioxide out of the air as they used it in photosynthesis and built their bodies with it. As plants removed carbon dioxide from the air, less heat was trapped and Earth began to cool, until about 25 million years ago it entered the Holocene Ice Age. It was during this time of changing temperatures and climates that the angiosperms originated and evolved to have such diverse bodies.

Continental drift is another factor that affected the evolution and diversification of the flowering plants and the structures that make up their bodies. Continents are constantly changing their position on Earth's surface as hot, fluid rock flows in Earth's mantle below its crust (we live on the top of the crust). As angiosperms were diversifying, North America broke away from Europe and drifted westward, whereas South America separated from Africa and Antarctica and migrated west and northward. Today, North and South America are colliding, pushing up a range of mountains we call Central America. As the continents change their position, they move from warm areas near the equator to cold areas near the poles or vice versa, and they block circulation of ocean currents. These currents contain millions of tons of water that is either warm, if it is flowing from the equator, or cold, if coming from the poles; as continents shift they redirect the flow of that water and that heat. For example, Alaska is not as cold as it would be if warm water did not flow from the equator, past Japan and up to Alaska's southern coast. Thus, the changing climate of Earth and the changing position of the continents created an extremely varied and variable environments to which plants had to become adapted; those that did not adapt became extinct.

As you study the material in this part, think about how the structures discussed facilitate the metabolism that occurs inside them: How do the structures make the metabolism more efficient? What are the selective advantages of the structures? Are alternative structures possible for a particular metabolism? If so, what are the consequences of each? The structures that make up the body of a plant must work together to make a functional plant and they must evolve; they evolved over long periods of times and in conditions that were very different from those we live in now. If the past climate of Earth had been different, the structures covered in this text would also have been different.

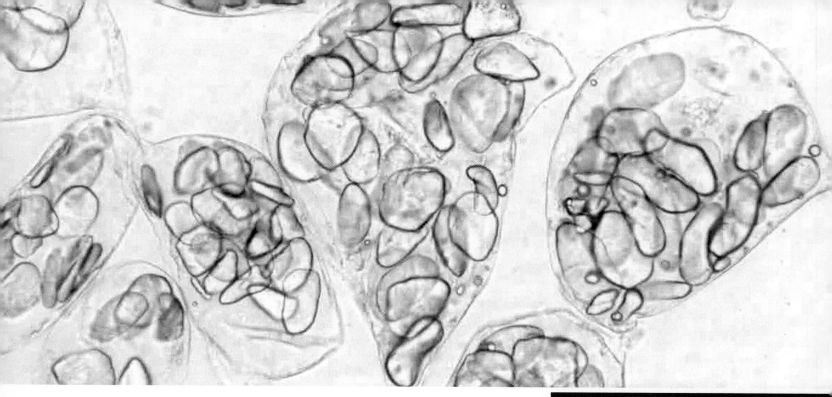

Cell Structure

chapter 3

Concepts

All organisms are composed of small structures called cells. In plants, each cell consists of a box-like cell wall surrounding a mass of protoplasm, which in turn contains its own smaller parts, the **organelles**, such as nuclei, mitochondria, and chloroplasts (**Figures 3-1** to **3-3** and **Table 3-1**). Cells are also the physical framework within which a plant's metabolism occurs. Water and salts are absorbed from soil by root cells, they are transported throughout the plant by cells of the vascular tissues, and the energy of sunlight is used in leaf cells to convert carbon dioxide and water to carbohydrates. Plant reproduction is also based on cells and cell biology: Some cells in flowers produce pigments or nectar that attracts insects that carry pollen between flowers, allowing sperm cells to contact egg cells.

Considering the large number of living organisms and the numerous types of metabolism that must be carried out, one might suspect that there are hundreds of types of cells, but actually, just a small number of cell types exist. Most differences between organisms

Chapter Opener Image: The large, irregularly shaped objects here are cells from a ripe banana fruit (*Musa acuminata*), the part we eat. Each cell is filled with many small starch grains. Most cells adhere to each other tightly, but as fruits like bananas ripen, the glue that binds their cells together dissolves, which softens the fruit and makes it more appealing for an animal to eat (the peel does not soften). This image was prepared by merely smearing a tiny bit of banana on a slide, causing the cells to separate from each other. The cell wall is so thin we see through it and observe the starch grains. Many other cell organelles are present but not visible because they are either too small or too transparent.

Outline

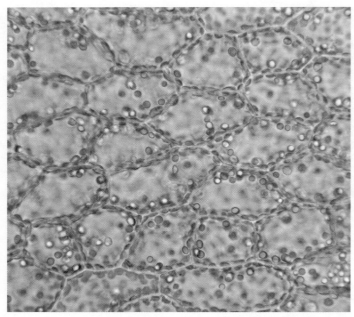

Figure 3-1 Light micrograph showing several basic features of plant cells. The walls and chloroplasts are easily visible. Other organelles occur but are difficult or impossible to see by light microscopy (×100).

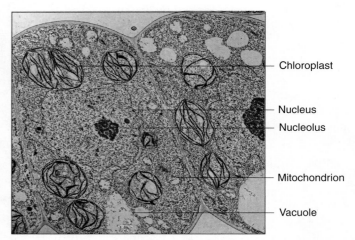

— Chloroplast

— Nucleus
— Nucleolus

— Mitochondrion

— Vacuole

Figure 3-3 A micrograph made with a transmission electron microscope of leaf cells. Especially important are the numerous membranes (×17,000).

TABLE 3-1	Examples of Plant Cell Shapes and Sizes	
Cell	Shape	Dimensions
Dividing cell in shoot or root	Cube	12 µm × 12 µm × 12 µm
Epidermal cell of lily (*Lilium*)	Flat, paving stone	45 µm × 143 µm × 15 µm
Photosynthetic cell in leaf of pear (*Pyrus*)	Short cylinder	7.4 µm diam × 55 µm
Water-conducting vessel cell in oak (*Quercus*)	Short cylinder	270 µm diam × 225 µm
Fiber cell in hemp (*Cannabis*)	Long cylinder	20 µm diam × 60,000 µm

are due to differences in associations of their cells, not in the cells themselves. Regardless of whether a root, stem, leaf, or flower is being constructed, the same basic units—cells—are required (**Figure 3-4**). Only the cell associations and minor modifications of the cells themselves change from tissue to tissue or organ to organ.

Although only a few types of cells exist, their differences are important. Any organism composed of more than one cell (a **multicellular organism** rather than a **unicellular** one) always has several types, each specialized for different tasks (Figures 3-2 and 3-4 and **Table 3-2**). As a plant develops, the cells in various parts become especially adapted for specific tasks. This **division of labor** allows the entire organism to become more efficient. Unicellular organization has a significant consequence: It does not allow division of labor or specialization. Each cell must perform all tasks: sensing the environment, gathering nutrients, excreting wastes, defense, movement, and reproduction. Because each cell

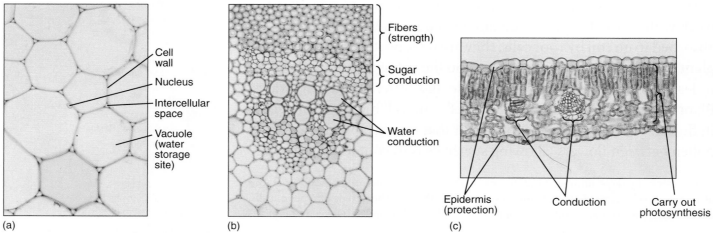

(a)

Cell wall
Nucleus
Intercellular space
Vacuole (water storage site)

(b)

Fibers (strength)
Sugar conduction
Water conduction

(c)

Epidermis (protection)　Conduction　Carry out photosynthesis

Figure 3-2 Plant metabolism, development, and survival depend on numerous cells working together in a coordinated, integrated fashion. (a) These cells store water in the center of a sunflower stem; they are relatively large and filled mostly with water. The cell walls and a nucleus are visible (×250). (b) Part of the system that conducts water and nutrients in a sunflower stem. Numerous types of cells occur in specific arrangements that permit efficient conduction. The large red cells in the center conduct water; the small grey cells above them conduct sugars. The cells at the top have thick red walls and provide strength to the stem (×60). (c) In this transverse section through a leaf of *Ligustrum,* you can see a variety of cells; those in the center carry out photosynthesis (×150).

Note: The micrograph in Figure 3.2b is 6 mm tall and has a magnification of ×60. If we were to shrink the micrograph down to life size (×1), the micrograph would be 1.0 mm tall and you would be able to see some of the cells with a magnifying glass.

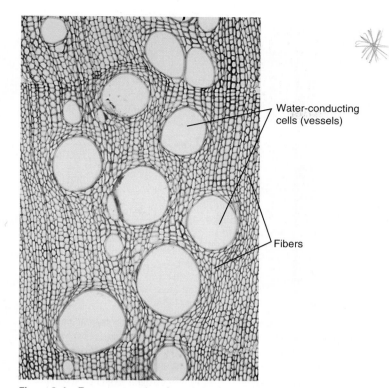

Figure 3-4 Transverse section of wood showing several cell types. Fibers with thick walls provide strength, and large open cells conduct water. These cells act as building blocks, and more wood is easily constructed by adding more cells. Stronger wood is produced by adding more fiber cells; more conductive wood results from adding more open cells (×100).

TABLE 3-2	Examples of Plant Cell Types, Specializations, and Division of Labor
Cell Type	Specialization
Cells of shoot/root tips	Cell division; produce new protoplasm
Epidermis	Water retention; cutin and wax are barriers against fungi and insects
Epidermal gland cells	Protection: produce poisons that inhibit animals from harming plants
Green leaf cells	Collect solar energy by photosynthesis
Root epidermal cells	Collect water and minerals
Vascular cells	Transport water, minerals, and organic molecules
Flower cells	Petal cells: pigments that attract pollinators
	Scent cells: fragrances that attract pollinators
	Nectary cells: sugars that attract pollinators
	Stamen cells: indirectly involved in producing sperm cells
	Carpel cells: indirectly involved in producing egg cells
	Fruit cells: produce sugars, aromas, flavorful compounds that attract fruit-eating/seed-dispersing animals

organism may result in the death of all cells, even those not initially damaged, whereas in a unicellular organism, a cell dies only if it is damaged directly. Which is more advantageous selectively: a unicellular organism composed of one generalized cell or a multicellular organism composed of specialized ones? The answer is not so obvious: Both types have existed for hundreds of millions of years, and thus, both must be considered highly adaptive and successful. Also, selective advantage depends on environment: In certain habitats, unicellular organisms are better adapted; in others, multicellular organisms survive better.

Like whole individuals, cells have a life span. During their life cycle (cell cycle), cell size, shape, and metabolic activities can change dramatically. A cell is "born" as a twin when its **mother cell** divides, producing two **daughter cells**. Each daughter cell is smaller than the mother cell, and except for unusual cases, each grows until it becomes as large as the mother cell was. During this time, the cell absorbs water, sugars, amino acids, and other nutrients and assembles them into new, living protoplasm. After the cell has grown to the proper size, its metabolism shifts as it either prepares to divide or matures and differentiates into a specialized cell.

must perform all tasks, it cannot do any one very well. Mutations that make a cell well adapted for protection make it less adapted for other functions and therefore are selectively disadvantageous (**Figure 3-5**). The same is true for modifications that improve photosynthesis, reproduction, and so on.

Multicellularity and division of labor result in a more efficient organism, but they have negative consequences as well: As each cell becomes more specialized, it depends more on the others. If a cell evolves toward having thick walls and offering maximum protection, it must rely on other cells of the organism for photosynthesis, mineral absorption, and reproduction. Damage to one part of the

Figure 3-5 (a) Cells located at the growing tip of *Pinus* (pine) are specialized for cell growth and division. Their thin cell walls allow sugar, water, and other nutrients to enter the cells easily; however, these cells are too soft and weak to be useful as bark or the shell of a nut (×400). (b) These "stone cells" provide strength in a coconut shell (*Cocos nucifera*) and therefore protect the seed. Almost the entire cell volume is wall. Protoplasm was present when the cells were young, growing and synthesizing their walls; however, after the walls were completed, cell metabolism was not necessary, and the protoplasts died (remnants are present in the tiny black hole in the center of each cell). Such modifications make it impossible for a stone cell to be a dividing, growing cell as in (a) (×200).

Cell wall

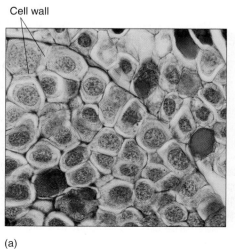

(a)

Cell wall

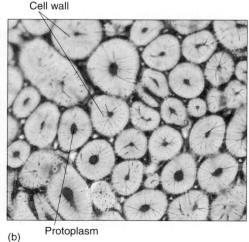

Protoplasm

(b)

Both growth and development require a complex and dynamic set of interactions involving all cell parts. That cell metabolism and structure should be complex would not be surprising, but actually, they are rather simple and logical. Even the most complex cell has only a small number of parts, each responsible for a discrete, well-defined aspect of cell life.

Membranes

All cells contain at least some membranes, and cells of eukaryotes (plants, animals, fungi, and protists) contain numerous organelles composed of membranes. Membranes perform many important tasks in cell metabolism: They regulate the passage of molecules into and out of cells and organelles; they divide the cell into numerous compartments, each with its own specialized metabolism, and they act as surfaces that hold enzymes. Without membranes, life would be impossible; indeed, many things that cause death in both humans and plants—heat, cold, many poisons, alcohol—kill because they disrupt membranes.

Composition of Membranes

All biological membranes are composed of proteins and two layers of phospholipid molecules (**Figure 3-6**). If phospholipids are poured carefully onto the surface of calm water, they automatically form a single layer (a **monolayer**) on the water's surface, with their hydrophilic ends forming hydrogen bonds to water

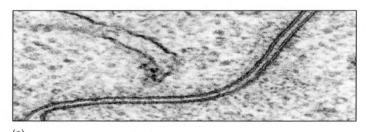

(a)

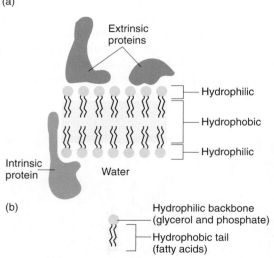

Figure 3-6 (a) In electron micrographs, membranes appear as two dark lines (proteins) separated by a light region (composed mostly of lipids) (×600,000). (b) Pictures such as (a) led scientists to believe that membranes consist of two layers of lipids with proteins located only on the surface; however, many proteins fit partly into the lipid bilayer or pass all the way through it. (c) A phospholipid has three parts: (1) two fatty acids, (2) a backbone of glycerol, and (3) a phosphate group.

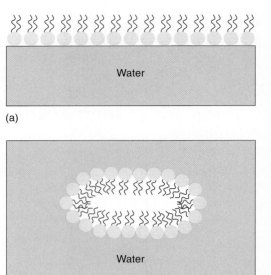

Figure 3-7 (a) In quiet water, phospholipids form a monolayer at the surface; by projecting away from the water, the fatty acids do not disrupt the hydrogen bonding between water molecules. (b) With agitation phospholipids form a bilayer; this arrangement allows the greatest number of hydrogen bonds to form between phosphates and water while simultaneously minimizing the disruption of water–water hydrogen bonds by the fatty acids.

molecules and their hydrophobic fatty acids projecting out of the water (**Figure 3-7**). If the water is agitated, the lipid layer doubles over and makes a **bilayer** in which all fatty acids are away from water and all phosphate groups are in full contact with it. Any break or tear in a bilayer membrane exposes hydrophobic fatty acids to water, and thus, membranes always reseal themselves after a rupture.

The lipid bilayer is a very thin solution. If it contains several types of lipid, they diffuse laterally throughout the membrane because they are not bonded to each other, but they cannot diffuse vertically from the membrane into the surrounding solution.

All biological membranes contain proteins as well as lipids, usually in a ratio of 60% proteins and 40% lipid (Figure 3-6b). Most proteins have large hydrophilic regions, and thus, they associate mostly with the phospholipid phosphates and with water. Many also have large hydrophobic regions that allow them to sink into the membrane and associate with the fatty acids (**Figure 3-8**). Variations in hydrophobic and hydrophilic regions mean that various proteins sit entirely on the membrane surface, are partly immersed in it, or span it entirely, with either end projecting out of opposite sides. This allows the protein to have its active site on either or both sides of the membrane or within it. Likewise, a protein may act as a hydrophilic channel that permits small hydrophilic molecules to pass through the membrane. Proteins that are even partially immersed in the lipid bilayer are said to be **intrinsic proteins**. Others, **extrinsic proteins** (also called **peripheral proteins**), are located outside the membrane and merely lie next to it.

Some intrinsic proteins contribute to the membrane's fluid nature and, like the lipids, can diffuse laterally. Other proteins interact with adjacent proteins, forming complexes or **domains** (small discrete regions) different from surrounding regions of the membrane. If all membrane components could freely diffuse

Alternatives

Box 3-1 Unusual Cells

Many plant cells are like those described in this chapter, but there are numerous types of unusual cells. Each example described here demonstrates an alternative way of organizing a cell, and of course, we can ask, "What are the consequences of each alternative?"

Dead cells. Many plant cells die as a necessary part of their development: If they remained alive they could not function as well, if at all. Cork cells in bark deposit a water-proof, enzyme-resistant chemical called suberin in their cell walls, and then they die. Suberin keeps water in the plant and resists attack by microbes, but what is the consequence of dying? By breaking down their protoplasm (called **programmed cell death**), cork cells become much less nutritious than if they remained alive. It does an animal no good to eat cork because there are no proteins, starches, lipids, nucleic acids, vitamins, or anything else of value. An animal would waste its energy chewing and trying to digest cork. Our own epidermis cells too are dead and are more protective than they would be if they were alive and nutritious. Water-conducing cells in plants, called tracheids and vessel elements, grow to their proper size and shape then undergo programmed cell death. When fully mature, each cell consists of just a wall and an empty space where protoplasm had been. Water is pulled upward through these empty cells and if protoplasm were still present, it would impede water movement. Water-conducting cells cannot function while living—they must be dead. Other cells that function after dying are those of many plant hairs, shells around seeds, and thorns and spines.

Unusual nuclei. Cells that develop into pollen grains in flowers are surrounded by a layer of cells called the tapetum. In many species, the nucleus of each tapetum cell divides once or twice more than the cells do, so at maturity, each cell has two or four nuclei. The consequence of this seems to be that each cell then has more capacity to control a very active, dynamic cytoplasm. In contrast, cells that transport sugars throughout a plant's body (sieve elements) destroy their nuclei during development. Unlike water-conducting cells, sugar-conducting cells cannot conduct if they are not alive. What, however, is the consequence of destroying the nucleus? We do not know why this is necessary, but we do know they become reliant on adjacent cells to control their metabolism.

Flagella. We think of plants as stationary and animals as mobile, but sperm cells of plants such as mosses, ferns, and even some seed plants (cycads) have flagella and swim. If sperm cells lacked flagella, they could not get to egg cells. There is, however, another alternative in most seed plants: Their sperm cells lack flagella but are carried passively to egg cells by elongate cells called pollen tubes, each of which grows out of a pollen grain.

Giant cells. Plant cells tend to be larger than animal cells, but both types are usually microscopic. Fortunately, several types of giant plant cells are extremely common and are so large you can see them with the naked eye. Cotton cloth and thread are made of long, slender cells, each up to 4 cm (1.5 inches) long; you can see individual cells by teasing cotton thread apart. These long cells grow from the surface of cotton seeds and then die and protect the seeds by forming a fluffy, soft layer. Most paper is made from water-conducting tracheids of wood, each tracheid being several millimeters long. These cells are easy to see by simply moistening a piece of paper and pulling it in two: the tiny filaments protruding from each edge are single cells. For these cells, a beneficial consequence of their extraordinary length is that it is easier to pull water through a few long cells as opposed to many short cells (if both types are the same width). Very unusual giant cells occur in some of the plants that ooze milky white liquid when poked (milkweeds in the family Asclepiadaceae and spurges in Euphorbiaceae). The milky sap is always at least mildly toxic, often extremely toxic, and is produced by cells called laticifers (**Figure B3-1**). If quite a bit of liquid drips out, it is probably produced by long, tubular cells. Tubular laticifers begin development as ordinary, small cells, but they continue to grow even after surrounding cells have reached their full size and stop enlarging. The laticifers elongate, pushing into spaces between other cells, becoming ever longer, and often branching and growing in multiple directions. Unlike cotton cells and tracheids, these must remain alive at maturity, and have hundreds of nuclei directing the synthesis of milky sap. One consequence of being giant, branched tubular cells is that even if an animal chews into just one small part of a plant, toxic sap from a wide area of the plant will flow to that site and deter the animal.

Unusual cells are mentioned here to show that even though most plant cells are rather similar to each other, there are alternatives, and with the alternatives come consequences. Most of this book focuses on the most typical, most ordinary cells, tissues, and metabolisms, but always think about the possibility of alternatives and how those might affect the biology of the plant.

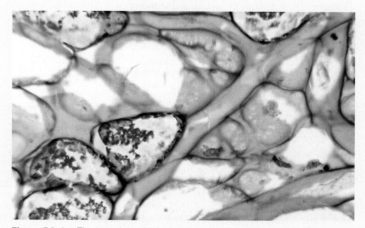

Figure B3-1 The long, branched tubes here are laticifers; each has branched many times (most branches are out of view of this high magnification), and all tips elongate and intrude between other cells. Individual cells are very long and have hundreds of nuclei (×200).

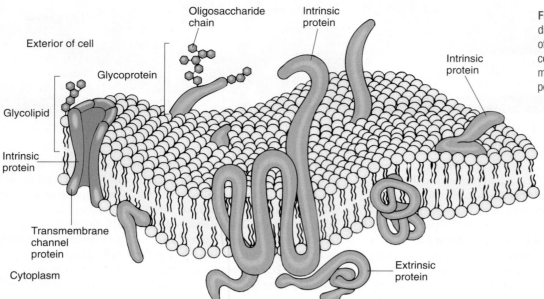

Oligosaccharide chain

Intrinsic protein

Exterior of cell

Glycoprotein

Glycolipid

Intrinsic protein

Intrinsic protein

Transmembrane channel protein

Cytoplasm

Extrinsic protein

Figure 3-8 Intrinsic membrane proteins differ in their size, shape, and location of hydrophobic, lipid-soluble regions; consequently, some sink deep into the membrane. Others span it, and some penetrate only one lipid layer.

laterally, the membrane would become homogeneous and no differentiation could occur; however, because at least some proteins are bound to their neighbors, membranes are heterogeneous and patchy, and differentiation does take place. Because the membrane is a heterogeneous liquid, it is said to be a **fluid mosaic membrane**.

Some membranes contain a small amount of sugar, usually less than 8%. The sugars occur as short-chain oligosaccharides, each with about 4 to 15 sugar residues. These oligosaccharides are bound to certain intrinsic proteins, converting them into **glycoproteins**; rarely, sugars are attached to membrane lipids (**glycolipids**). Currently, we believe these polysaccharides make the membrane more distinctive and easy to recognize, an especially important feature in animals that have an immune system.

Specialized protective cells attack and destroy anything that they do not recognize as a part of the animal's own body. The glycoproteins and glycolipids occur almost exclusively on the outer surface of the membrane that covers the cell, a position that allows maximum exposure for these polysaccharides. Glycoproteins and glycolipids may be less important in plants.

■ Properties of Membranes

Membranes have several important properties. First, they can grow. Membranes are formed molecule by molecule in certain regions of the cell; then entire pieces of membrane are moved as small bubbles or **vesicles** to different sites in the cell. When the vesicle of preformed membrane arrives at the growing membrane, the two fuse (**Figure 3-9**).

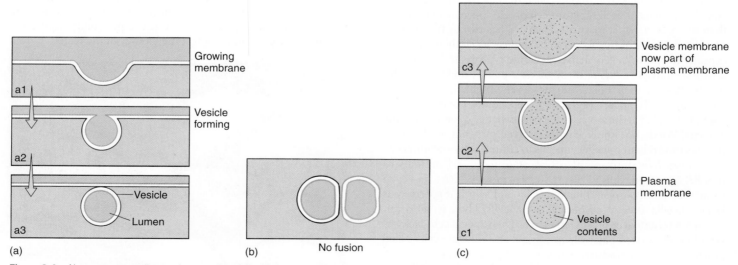

Growing membrane

Vesicle forming

a1

a2

Vesicle

Lumen

a3

(a)

No fusion

(b)

Vesicle membrane now part of plasma membrane

c3

c2

Plasma membrane

c1

Vesicle contents

(c)

Figure 3-9 Numerous organelles produce small vesicles (a) that move through the cell, either remaining discrete for long periods or fusing with other organelles or vesicles. In order to fuse, two membranes must be quite similar in composition. Here, a single membrane bilayer is shown as a double line, one orange and the other black. (b) Membrane dissimilarity prevents two compartments from fusing; for example, the nucleus does not fuse with mitochondria or chloroplasts. (c) When a vesicle fuses with the membrane at the cell surface, its contents are deposited outside the cell and the vesicle membrane becomes part of the cell surface. The vesicle membrane and the plasma membrane have been drawn with the same colors to indicate their similar composition.

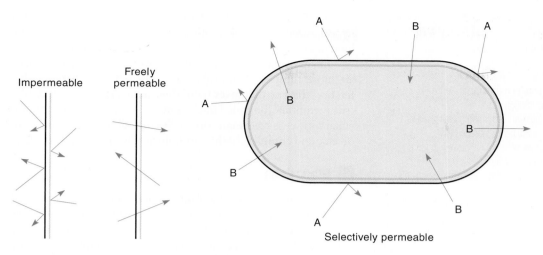

Impermeable | Freely permeable

Selectively permeable

Figure 3-10 An impermeable membrane allows nothing to pass through, whereas a freely permeable membrane stops nothing. A selectively permeable membrane allows certain materials to pass through more readily than others, with the result that chemical concentrations on the two sides differ. This membrane is impermeable to substance A but permeable to B.

In addition to permitting movement of membrane pieces, membrane fusion allows the transport of material. The volume inside the vesicle (the vesicle **lumen**) may be filled with substances that must be accumulated, broken down, or otherwise metabolized at the vesicle's destination. Vesicle movement may also release material to the outside of the cell (Figure 3-9). This **exocytosis** is a means to excrete almost anything: wastes, debris, mucilage, proteins, and polysaccharides. For example, roots slide through the soil by secreting a slippery, lubricating mucilage formed within root cells, packaged into vesicles that migrate to and fuse with the cell's surface membrane; this mucilage is then released to the exterior. In many flowers, nectar is secreted from glands by exocytosis. **Endocytosis** is basically the opposite process: A small invagination forms in the outer membrane then pinches shut, creating a new vesicle that contains extracellular material. Endocytosis is especially common in algae and other microscopic organisms that take in food particles this way, but it is not known how common it is in plants.

Permeability is an important property of membranes. All biological membranes are **selectively permeable** (also called **differentially permeable**), meaning that certain substances cross the membrane more easily and rapidly than others (Figure 3-10). Because large regions of a membrane are mostly lipid, membranes are more permeable to hydrophobic substances than to anything that carries an electric charge; however, if charged compounds such as inorganic salts, sugars, and amino acids could not enter cells at all or if they could enter only slowly, cells would starve.

Movement of charged substances is assisted by large intrinsic proteins that span the membrane and act as hydrophilic channels through it; this is **facilitated diffusion**. Other proteins, called **molecular pumps**, actually bind to a molecule on one side of the membrane, and then by using energy, the protein changes shape and releases the molecule on the other side. By using this active pumping, called **active transport**, cells accumulate substances until the interior concentration of solute far exceeds the exterior concentration. A pump working in the opposite way can actively transport materials out of the cell (Table 3-3). Except for the pumps, the membrane must be impermeable to the molecule; it would do no good to have a molecular pump for a type of molecule that could easily leak back through.

All life depends on the principle of **compartmentalization**, the formation of many compartments, each specialized for a particular process such as producing a particular substance or using a particular precursor. For example, our blood is compartmentalized away from our digestive system: The two must work together, but they must not be mixed together. If all the monomers, polymers, sugars, salts, enzymes, and vitamins of an organism were mixed together, some reactions would occur at random but would not be orderly enough to be considered life. Instead, cells are filled with many compartments—the organelles—each surrounded by its own unique selectively permeable membrane. Because the protein channels and pumps are made under the guidance of the nucleus, the cell can control the numbers and types of channels and pumps it makes and inserts into each type of organelle membrane. Organelles thus have membranes with different permeabilities or pumping capacities, depending on the instructions generated by the nucleus.

A **freely permeable membrane**, which allows everything to pass through quickly, would be rather useless for a cell, as would an **impermeable membrane**, one that does not allow anything through at all (Figure 3-10 and Table 3-4).

The last important property of membranes is that they are dynamic, constantly changing. Cell membranes are not simply established and left unchanged throughout the life of the cell. Instead, new components are constantly being inserted and old ones removed. If the function of the cell is always the same, the old and new components are similar, but if the cell must change its function, newly inserted components are different from the retracted ones. As the nature of the membrane changes, the nature of the cell changes also.

TABLE 3-3	Active Transport in Roots of Corn (*Zea*)		
	External Concentration	Internal Concentration	Increase
Potassium	1 mM	22 mM	2200%
Calcium	0.2 mM	5 mM	2500%
Chloride	1.2 mM	27 mM	2250%

Mineral nutrients are often present as a dilute solution in soil and must be accumulated by active transport carried out by root cells (the endodermis cells). The table shows the ability of endodermis cells to pump nutrients across their membranes from the exterior of the plant to the conducting tissues.

TABLE 3-4	Summary of Transmembrane Movement

Impermeable membrane: Nothing passes through; no biological membrane is impermeable to everything.

Freely permeable membrane: Virtually anything can pass through.

Selectively permeable (differentially permeable) membrane: Certain substances pass through rapidly, others pass through slowly.

Facilitated diffusion: The presence of large intrinsic membrane proteins allows hydrophilic, charged molecules to diffuse through the membrane.

Active transport: Large intrinsic membrane proteins bind a molecule and force it through the membrane, consuming energy in the process.

Exocytosis: The fusion of a vesicle with the cell membrane, releasing the vesicle's contents to the cell exterior.

Endocytosis: The invagination of the cell membrane, forming a vesicle that pinches off and carries external material into the cell.

Basic Cell Types

At the most basic level, both cells and organisms can be classified as either **prokaryotic** or **eukaryotic**. Prokaryotic cells are simpler than eukaryotic ones and are found only in domains Bac-

teria (bacteria and cyanobacteria) and Archaea (archaeans). It is hypothesized that prokaryotic cells represent the most archaic lines of evolution and that eukaryotic cells evolved from them.

Eukaryotic cells, found in plants, animals, fungi, and protists, are more complex than prokaryotic cells. The most striking difference, the one that gives them their name, is the presence of a true membrane-bounded nucleus in eukaryotic cells. In addition, eukaryotic cells have many organelles that allow them to be more diverse and complex, both morphologically and physiologically.

Plant Cells

Although various parts of a plant—roots, wood, bark, leaves, and flower parts—appear to be quite diverse, virtually all of their cells have all of the following organelles; exceptions are rare. As each type of cell develops, certain organelles may become modified and more or less abundant, but usually none is lost completely (Figure 3-11).

Protoplasm

All cells, either prokaryotic or eukaryotic, are made of a substance called **protoplasm** (Figure 3-5) (the protoplasm of a single

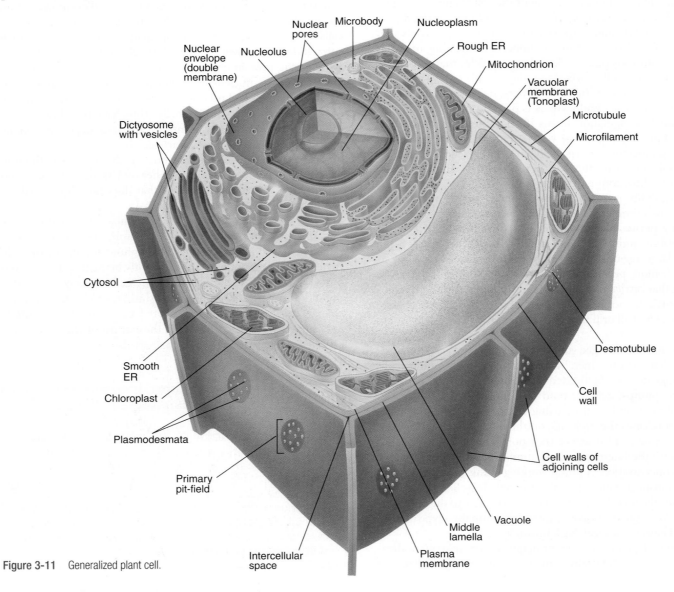

Figure 3-11 Generalized plant cell.

cell is called its protoplast). This name was given early, when it was thought that protoplasm was a distinct substance such as water, oxygen, or iron and that one of its properties was life itself. We now know that protoplasm is a mass of proteins, lipids, nucleic acids, and water within a cell; except for the wall, everything in the cell is protoplasm, composed of the following organelles.

■ Plasma Membrane

The **plasma membrane** (less frequently called the **plasmalemma**) is the membrane that completely covers the surface of the protoplasm (**Figure 3-12**). Because it is the outermost surface of the protoplast, it must be impermeable to harmful materials and permeable to beneficial ones; therefore, it is selectively permeable. Molecular pumps in the plasma membrane actively transport needed molecules inward and pump others outward for secretion. Because one side of the plasma membrane faces the external environment and the other side faces the cell, the two sides are quite different, especially in the types of protein they contain. We know very little about the proteins and lipids that make up the plasma membrane because the membrane does not separate cleanly from the protoplast. Almost all attempts to isolate the plasma membrane for study have experienced considerable contamination from other parts of the cell.

■ Nucleus

The **nucleus** (plural, nuclei) (Figure 3-11) serves as an archive, or permanent storage place, for the organism's genetic information. This text is just a brief introduction to plant biology. No one can imagine the number of pages required to store all of the information necessary for building and maintaining a single cell; however, all of that information must be stored in the DNA inside every nucleus, and the storage must be safe and permanent. Information is useless unless it, or copies of it, can be retrieved and used. The nucleus carries out information retrieval by making copies of specific parts of the DNA whenever that information is needed. The copy is not DNA, but a type of ribonucleic acid called messenger RNA. An exciting area of research involves understanding how the nucleus, in response to signals from the rest of the cell, searches its DNA for the needed information and then makes copies of it without accidentally copying other similar information.

The nucleus of a eukaryotic cell is always surrounded by a **nuclear envelope** composed of an **outer membrane** and an **inner membrane**. The nuclear envelope separates nuclear material from the rest of the cell, and it contains numerous small holes, **nuclear pores** (**Figure 3-13**), involved in the transport of material between the nucleus and the rest of the protoplasm. Nuclear pores have a complex structure and exert control over the movement of materials. If a nucleus is extracted from a cell and placed into water, it swells; this can happen only if the pores prevent material from oozing out as the nucleus absorbs water. Prokaryotes have

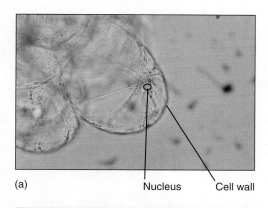

(a)

Nucleus Cell wall

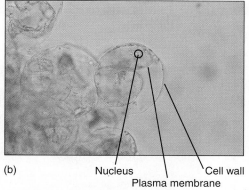

(b) Nucleus Cell wall
Plasma membrane

Figure 3-12 (a) A healthy, growing cell. Its protoplast is pressed firmly against the cell wall, and thus, its plasma membrane is not visible (×160). (b) These cells have been treated with salt to draw water out of the cell, causing the protoplast to shrink. As a result, the plasma membrane is visible (×160).

Nuclear
envelope
surface Nuclear pore

Cytoplasm

Figure 3-13 A preparation called a freeze-fracture. The cell is frozen and then tapped to cause it to break. The lipids of the membranes are weak when frozen, and thus, the fracturing often follows the membranes, separating one lipid layer from the other. Here the fracture passed irregularly through the cytoplasm and then entered the nuclear envelope. The nuclear envelope consists of the outer and inner membranes, with a space separating them; the two membranes fuse together at the nuclear pores. Nuclear pores may be distributed uniformly over the nuclear surface or may occur in bands or patches in some species (×60,000).

Box 3-2 Calcium: Strong Bones, Strong Teeth, but Not Strong Plants

Most plants and animals need hard parts. Wood is strong enough to support the weight of a tree, and bones play a similar role in animals. Seeds are often protected by resistant shells such as those of walnuts and almonds, and animal shells protect clams and oysters. Our teeth are so tough that they can chew through almost anything. Although plants and animals use hard parts for similar roles, plants rely on thick, tough cell walls, whereas animals use calcium salts. The wood of a giant redwood tree supports much more weight than does the skeleton of an elephant, but the tree must be close to 100% wood to obtain enough strength whereas the elephant's bones are only a small fraction of its body. Calcium salts are much stronger than wood.

Would it be possible for plants to use bone-like material? We can analyze this as a set of alternatives and their consequences. The present alternative—wood—consists of cellulose and a chemical called lignin. Both are carbohydrates that a plant itself makes through photosynthesis, and thus, they are readily available. And both are remarkably inert, having little impact on other aspects of the plant's metabolism. In contrast, calcium and its salts participate in many metabolic pathways, and building or resorbing shells, bones, or teeth has a broad impact on cell physiology. Shells consist of calcium carbonate, and as animals use carbonate ion (CO_3^{2-}) to build a shell, the acidity of the protoplasm is altered. Furthermore, animals can digest part of their shells if they need the calcium elsewhere, and this liberation of carbonate will again affect the pH. This is tolerable for marine organisms because they use carbonate from the surrounding seawater rather than from their own protoplasm so their pH is not affected. If the shell is resorbed later, the liberated carbonate is likewise dumped outside the animal into the seawater.

Animals like us—with an internal skeleton—use calcium phosphate in our bones and teeth. Calcium carbonate's tendency to alter pH is too dangerous for us, and our skeleton cannot use seawater as a carbonate reservoir. The phosphate ion (PO_4^{2-}) that we use has little effect on a cell's acidity, and, furthermore, if we resorb bone for some reason, the liberated phosphate is a resource for many other metabolic reactions. It is not a liability at all.

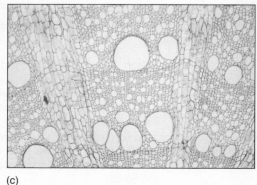

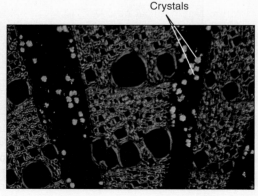

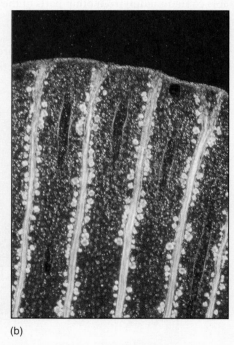

Figure B3-2 (a) A leaf clearing of maidenhair tree (*Ginkgo*), showing several red-stained leaf veins that conduct sugars out of the leaf. Such veins are the targets of aphids and other sucking insects (×15). (b) The same tissue, in polarized light (×15). (c) A cross-section of *Aristolochia* wood; crystals are present in the two bands of tissue with blue-stained walls (see [d]). This is a soft tissue in wood and is the site where sugars and other nutrients are stored (×50). (d) The same tissue as (c), but with polarized light (×50).

Plants too must be careful with calcium and its salts. Calcium carbonate's disturbance of cell acidity is just as dangerous for plants as it is for us, but plants do not use calcium phosphate either, perhaps because it requires too much phosphate. When we fertilize plants, one of the nutrients we provide is phosphate because most soils have just marginal amounts and not quite enough to let plants grow optimally. If plant cells had walls made of bone rather than cellulose and lignin, they would need a tremendous amount of phosphate and might suffer from deficiency under natural conditions. Instead, plants are self-reliant and build their walls with molecules that they make themselves.

Like animals, plants must carefully control the concentration of dissolved calcium within their cells. The concentration must be kept at extremely low levels; otherwise, it would interfere with metabolism. Roots block the entry of calcium, but so much leaks in accidentally that the plant must get rid of the excess somehow. Plants crystallize the calcium by producing oxalic acid. Calcium and oxalic acid react to form calcium oxalate, which is insoluble

and forms large crystals. Crystallized calcium is more or less inert and cannot interfere with the plant's metabolism.

Calcium oxalate crystals are hard enough to be used for strength, but plants do not use it to create cell walls as hard as bone. Calcium oxalate forms only tiny individual crystals rather than big bone-like structures. Putting calcium oxalate crystals in its cell walls would not give a plant the equivalent of an animal's strong skeleton. Instead, the plant strengthens its walls with lignin, and calcium oxalate is used only as a means of removing excess calcium from the protoplasm. By using oxalate for this, the plant benefits by not having to use its valuable phosphate nor run the risk of dealing with carbonate, but it loses the benefit of strength.

Wood with lignin is not as strong as bones or teeth, but its synthesis does not affect cell acidity. Plants can make all of the wood they need without relying on rare minerals from the soil. Plants and animals may do things differently, but there are sound biological principles underlying the differences.

no nuclear envelope; instead, DNA is simply mixed with the rest of the cell contents.

Within the nucleus is a substance called **nucleoplasm**. Like "protoplasm," this name was given before the composition was understood; we now know that nucleoplasm is a complex association of (1) DNA; (2) enzymes and other factors necessary to maintain, repair, and read DNA; (3) histone proteins that support and interact with DNA; (4) several types of RNA; and (5) water and numerous other substances that are necessary for nuclear metabolism. Nuclear DNA is always closely associated with histones, and this complex of the two is known as chromatin. As a cell ages or its metabolic activities change, so do the nucleus and nucleoplasm. In cells undergoing rapid cell division (e.g., cells in root tips, shoot tips, young leaves, and flower buds), the DNA, histones, and duplicating enzymes may dominate the nucleoplasm; however, in mature cells that are not dividing, messenger molecules and reading enzymes may be more abundant.

Inside every nucleus is one, two, or rarely several bodies called nucleoli (singular, **nucleolus**) (Figure 3-14), areas where the components of ribosomes are synthesized and partially assembled. Each ribosome contains a large amount of ribosomal RNA copied from ribosomal genes in the chromatin.

Nuclei are large, complex organelles, and frequently, they occupy a major fraction of the cell volume (up to 50%) (Table 3-5). In certain conducting cells (sieve tube members in phloem), the nucleus breaks down during cell differentiation, and the mature cell functions for several weeks or months without a nucleus. Our

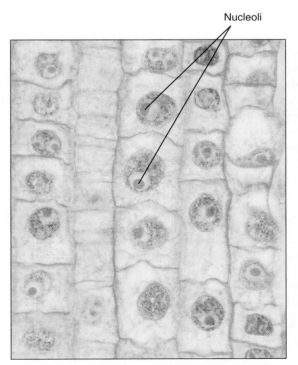

Nucleoli

Figure 3-14   Nucleoli usually stain more intensely than the rest of the nucleus; occasionally, they stain a distinct color. Root tip of hyacinth (×250).

TABLE 3-5	The Relative Volumes of Organelles in Plant Cells	
Organelle	Young Cell	Mature Cell
Nucleus	32.4	0.23
Vacuole	4.93	83.3
Mitochondria	5.35	1.01
Plastids	3.72 (proplastids)	3.16 (chloroplasts)
Dictyosomes	0.40	0.04
Hyaloplasm	52.9	12.1
Lipid	0.23	0.00

Young cells tend to be small with large nuclei; as they grow and differentiate, the cell volume increases and certain organelles become more prominent, depending on the specialization of the cell. The relative volumes of organelles (the volume of the organelle expressed as a percentage of the volume of the whole cell) are given here for a young cell from a shoot tip and for a mature, photosynthetic cell from the outer part of the stem. The size of the nucleus in both cells is the same, but the mature cell is so large that the nucleus constitutes only a small part of it, whereas the young cell is so small that the nucleus occupies one third of its volume. These values are for a cactus, *Echinocereus*.

Figure 3-15 Azalea flower buds open overnight, the petals and stamens expanding rapidly as each cell fills its central vacuole with water. The flower maintains its shape by hydrostatic pressure: central vacuoles pressing cytoplasm firmly against cell walls.

red blood cells are also enucleate (without a nucleus) at maturity. Both of these cell types are exceptional; they are enucleate only while performing a very limited type of metabolism, and they die shortly after losing the nucleus. Several types of plant cells are multinucleate.

Central Vacuole

Within young, small cells are organelles, **vacuoles**, that have just a single membrane, the vacuole membrane, also called the **tonoplast**. Vacuoles often appear to be empty (Figure 3-11 and Table 3-6) because they store mostly water and salts that cannot be preserved for microscopy; however, they sometimes contain visible crystals, starch, protein bodies, and various types of granules or fibrous materials in addition to water and salts.

As a cell grows and enlarges, vacuoles expand and merge until there is just one large central vacuole. Because it contains primarily water and salts, the central vacuole can expand rapidly, forcing the cell to grow rapidly as well (Figure 3-15). Animal cells must synthesize complete protoplasm to grow, but plant cells need only increase the amount of vacuolar water. For example, as flower buds open, petals expand from tiny to full-sized in just a few hours by simply enlarging their vacuoles, not by making new cells. Over a long period, plants must produce additional proteins, membranes, and organelles or else they would become almost pure water.

In addition to cell growth, the central vacuole functions in storage of both nutrient reserves and waste products. In seed cells, vacuoles may be filled with starch or protein that will be used when the seed germinates, perhaps 10 to 50 years after the material was deposited in the vacuole. Calcium regulates the activity of many enzymes, and plant cells keep protoplasmic calcium concentrations at the proper level by moving calcium into the vacuole, where it reacts with oxalic acid and crystallizes into an inert form. Other nutrients such as potassium may move in and out of the vacuole on a daily basis. The water-soluble pigments in many flowers, fruits, and red beets occur in vacuoles as well (Figure 3-16).

A system to excrete wastes never evolved in plants; instead, metabolic waste products are pumped across the vacuole membrane and stored permanently in the central vacuole. The tonoplast is otherwise impermeable to these wastes; therefore, they cannot leak back into the cytoplasm where they would be harmful. Holding waste inside forever does not sound like an optimal situation, but it actually is selectively advantageous: Because most of these compounds are noxious and bitter, they deter animals from eating the plants. Mutations that result in excretion might make the cells taste good, which would be selectively disadvantageous.

TABLE 3-6	Concentrations of Nutrients in Vacuoles Isolated from Barley (*Hordeum*) Leaf Cells	
Nutrient	Concentration (mM)	In Our Blood
Cl^-	56.1	101 to 111
NO_3^-	42.9	not typically measured
HPO_4^-*	64.5	0.25 to 0.43
SO_4^{2-}	25.1	not typically measured
K^+	142.2	3.7 to 5.2
Na^+	42.2	136 to 144

*Often abbreviated Pi

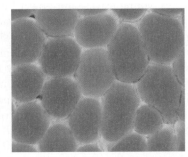

Figure 3-16 The central vacuoles of these petal cells contain dissolved pigments, making the vacuoles visible. The pale area between each oval of pigment is intercellular space, cell walls, nuclei, and all of the protoplasm other than vacuole ($\times$300).

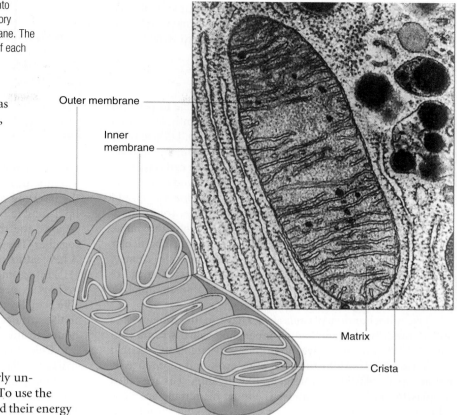

Figure 3-17 The inner mitochondrial membrane is folded into plate-like cristae, giving it a large surface area; many respiratory enzymes are intrinsic proteins embedded in the crista membrane. The large surface area makes it possible to contain many copies of each enzyme, and more reactions can occur simultaneously.

Outer membrane

Inner membrane

Matrix

Crista

The central vacuole is a digestive organelle as well. As other organelles age and become impaired, they fuse with the tonoplast and are transported into the central vacuole, where digestive enzymes break them down. The liberated monomers are transported back into the rest of the cell, where they are used again. In animal cells, which do not have central vacuoles, this task is carried out by small vacuoles called lysosomes.

■ Cytoplasm

If the nucleus and vacuole are excluded from the protoplasm, the remaining material is referred to as **cytoplasm** and contains the following structures.

■ Mitochondria

Cells store energy as highly energetic but fairly un-reactive compounds, such as sugars and starches. To use the energy, such compounds must be broken down and their energy must be used to synthesize new compounds that are both highly energetic and very reactive, the most common being adenosine triphosphate (ATP). Because these reactions involve high-energy, reactive compounds, they occur more safely within a discrete organelle than freely in the cytoplasm, where the intermediates might accidentally react with other cell components. **Mitochondria** (singular, mitochondrion) are the organelles that carry out this cell respiration (**Figure 3-17**).

Many steps of respiration are mediated by enzymes bound to mitochondrial membranes. The enzymes are located adjacent to each other so that the product of one reaction (which becomes the reactant in the next reaction) is passed directly to the next enzyme; this controls the highly reactive intermediates (**Figure 3-18**). Mitochondrial membranes are folded, forming large sheets or tubes known as **cristae** (singular, crista). This folding provides room for large numbers of enzymes. Reactions that do not involve highly reactive intermediates take place in the liquid **matrix** between the cristae. Around this complex of cristae and matrix is a second membrane, the **outer mitochondrial membrane**, which probably just gives shape and a little rigidity to the mitochondrion. The outer membrane is rather freely permeable, but the **inner mitochondrial membrane**, which forms the cristae, is selectively permeable and has numerous pumps and channels.

Mitochondria have their own DNA and ribosomes, both of which are different from those of the rest of the cell. Mitochondrial DNA is a circular molecule and lacks histones; the ribosomes are small and resemble those found in prokaryotes.

Mitochondria are dynamic organelles; they can grow larger, either as the cell does or as the cell's need for respiration increases. They are often about 1 μm in diameter but up to 5 μm in length and are even much longer in some cases. Mitochondria can also divide into two, increasing the number of mitochondria per cell. Under rare conditions, the mitochondria in some species fuse together, forming a single giant mitochondrion in each cell. Mitochondria occupy between 1% and 25% of the cell volume

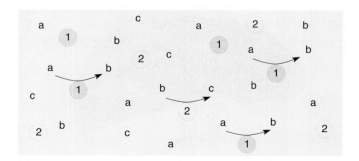

(a)

(b)

Figure 3-18 Many enzymes must work together. The product b of enzyme 1 is the reactant of enzyme 2. (a) If the enzymes are in solution, they may not be close together and b must diffuse at random until it encounters enzyme 2; in the meantime, it may accidentally react with something else. (b) If they are located side by side on a membrane, enzyme 1 can pass b directly to enzyme 2, not only speeding up the overall reaction but also eliminating the chance of accidental reactions.

(Table 3-5); the higher values are similar to those for animal cells. Because of their ability to divide and fuse, the actual number of mitochondria per cell (usually in the range of 100 to 10,000) may not be as important as their volume.

Plastids

Plastids are a group of dynamic organelles able to perform many functions. One prominent activity is photosynthesis, carried out by the green plastids, chloroplasts (Figure 3-1). Diverse types of metabolism occur in other classes of plastid: synthesis, storage, and export of specialized lipid molecules; storage of carbohydrates and iron; and formation of colors in some flowers and fruits. Plastids are the site of synthesis of amino acids: isoleucine, valine, and those that contain aromatic rings (phenylalanine, tryptophan, and tyrosine); or are derived from aspartate (lysine, threonine, and methionine). Each metabolism is associated with a particular type of plastid; as an organ changes, its plastids may also change, extensively altering their membranes and proteins. Plastids are found in all plants and algae but never occur in animals, fungi, or prokaryotes.

Like mitochondria, plastids always have an **inner membrane** and an **outer membrane** and an inner fluid called stroma (Figure 3-19). Plastids also have ribosomes and circular DNA that is not associated with histones. Plastids grow and reproduce by pulling apart. Plastids of young, rapidly dividing cells are called **proplastids** and are very simple. The inner membrane has a few folds but little surface area. When exposed to light, proplastids develop into **chloroplasts**, which are green owing to the presence of the photosynthetic pigment **chlorophyll** (Table 3-7). Because many intermediates of photosynthesis are highly reactive, energetic compounds, the controlling enzymes must be incorporated into the membranes, just as in mitochondria. This requires the inner membrane to become more extensive and elaborately

TABLE 3-7	Types of Plastids
Amyoplasts	Store starch; considered to be leucoplasts
Chloroplasts	Carry out photosynthesis
Chromoplasts	Contain abundant colored lipids; in flowers and fruits
Etioplasts	A specific stage in the transformation of proplastids to chloroplasts; occur when tissues are grown without light
Leucoplasts	Colorless plastids; synthesize lipids and other materials
Proplastids	Small, undifferentiated plastids

folded. Membrane sheets, **thylakoids**, project into the stroma. This increase in membrane area also provides more space for the insertion of the photosynthetic pigments: Chlorophyll has a lipid-soluble tail and, thus, is part of the membrane. In certain regions, the thylakoids form small bag-like vesicles that stack together. The stack of vesicles is called a **granum** (plural, grana). A key feature of photosynthesis is the active transport of protons (H^+) into a small space to build up an electrical charge; the grana vesicles are needed to accumulate these protons from the stroma. The actual conversion of carbon dioxide to carbohydrate occurs in the stroma, catalyzed by enzymes free in solution rather than bound to any of the membranes.

Chloroplasts are larger than mitochondria, approximately 4 to 6 µm in diameter. An individual leaf cell may contain as many as 50 chloroplasts. When chloroplasts photosynthesize rapidly, they produce sugar faster than the cell can use it, and thus, it is temporarily polymerized into starch grains inside the chloroplasts.

In plant tissues that cannot photosynthesize (roots, bark, wood), proplastids develop into **amyloplasts**, which accumulate sugar and store it as starch for months or years as in starchy seeds like wheat, rice, corn, or vegetables such as potatoes and yams (Figure 3-20). Each amyloplast produces large starch grains that virtually fill the stroma; few internal membranes are present. In

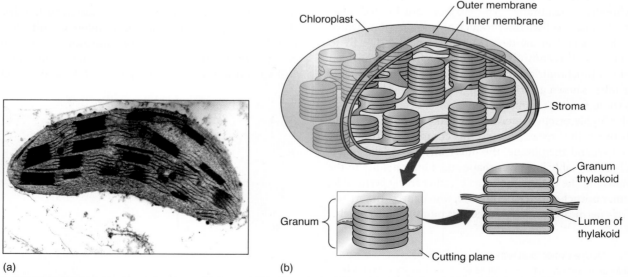

(a)   (b)

Figure 3-19 (a) All plastids have an outer and an inner membrane, but in chloroplasts, the inner membrane is extensive and highly folded. The single membranes are thylakoids, and the multiple membranes are actually stacks of flattened thylakoid vesicles; the stacks are called grana. Like mitochondrial cristae, the extra surface area of the inner membrane provides room for many copies of each enzyme. The photosynthetic pigment chlorophyll is part of the membrane lipid layer (approximately ×30,000). (b) The grana are interconnected by thylakoid membranes, and the liquid stroma surrounds the grana. Because the thylakoid membranes are selectively permeable, the concentrations of chemicals inside the thylakoid space differ from those in the stroma. This is essential for photosynthesis.

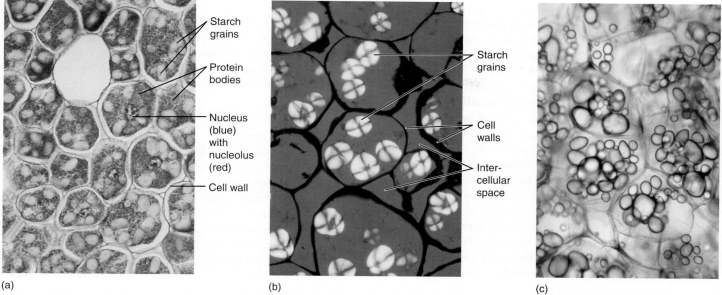

(a)   (b)   (c)

Figure 3-20   (a) Cells from a developing bean seed. The large pink bodies are starch grains; the small bluish red ones are protein bodies. These colors are not natural. The material was stained to make the colorless starch and protein visible. The starch grains occur in plastids called amyloplasts; as the bean germinates, the starch is used as a source of energy. When the starch decreases, the amyloplasts convert to chloroplasts and turn green (×150). (b) Starch grains that have been photographed in polarized light rather than being stained. Substances that have an orderly, crystalline structure shine brightly in polarized light (×200). (c) Free-hand section of fresh potato; each is filled with numerous oval starch grains. The thin walls of these cells are just barely visible (×200).

starchy vegetables, amyloplasts constitute the bulk of the tissue. If exposed to light, amyloplasts can be converted to chloroplasts. Nonphotosynthetic parasitic plants such as *Conopholis* lack chloroplasts but do have amyloplasts; no plant is ever without some form of plastid.

In some flowers and fruits—for example, tomatoes and yellow squash—bright red, yellow, or orange lipids accumulate in plastids as they differentiate into highly colored **chromoplasts** (**Figure 3-21**). An extensive, undulate system of membranes is present, but no grana, and the pigments may be present either as part of the membrane or as discrete droplets, **plastoglobuli** (singular, plastoglobulus). As fruits ripen, chloroplasts often synthesize large amounts of lipid pigments, alter their thylakoids, and convert to chromoplasts, as when apples and tomatoes change

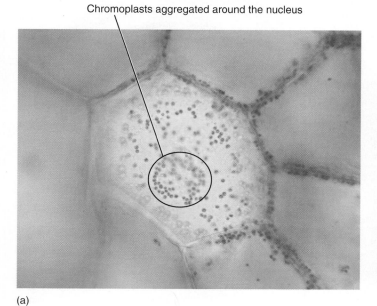

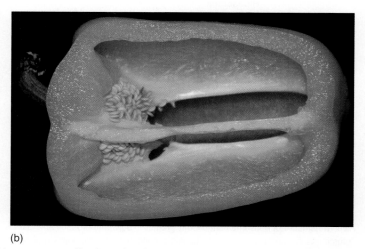

(a)   (b)

Figure 3-21   (a) Each tiny orange dot is a chromoplast; this is from an intensely bright orange pepper (b)—just a few chromoplasts provide brilliant color. Chromoplasts are located in the periphery of the cell because the central vacuole occupies most of the cell center. The largest cell appears to have chromoplasts everywhere, but we are just viewing the nearest wall, not the center of the cell. Many chromoplasts are aggregated around the nucleus.

from green to red. Lipid pigments are present in low amounts as small plastoglobuli even in leaf chloroplasts, but they are masked by the abundant green chlorophyll; they can be seen in autumn when cold weather causes chlorophyll, but not lipids, to be broken down and the leaves turn red or yellow.

Many cells have large, unpigmented plastids that have neither chlorophyll nor lipid pigments. These **leucoplasts** may be involved in various types of synthesis: Many types of fats and lipids are synthesized only in plastids and then transported to other organelles and inserted into their membranes. "Leucoplast" is a purely descriptive term—any colorless plastids, including proplastids and amyloplasts, can be considered leucoplasts.

Iron is an essential nutrient for plants and animals and is stored attached to a large protein. Although almost identical to human ferritin, the plant protein is called **phytoferritin**. Ferritin is found throughout animal cytoplasm and nuclei, but phytoferritin is stored almost exclusively in plastids. As much as 80% of the iron in leaves may be found in chloroplasts, and the leucoplasts and amyloplasts of seeds often have especially large amounts.

■ Ribosomes

Immersed in the protoplasm are **ribosomes**, particles responsible for protein synthesis (**Figure 3-22**). They are complex aggregates of three molecules of RNA (ribosomal RNA) and approximately 50 types of protein that associate and form two subunits. Compared with animal cells such as those in the liver and pancreas, most plant cells synthesize little protein and have few ribosomes; however, some do produce large amounts of proteins and are rich in ribosomes, such as the protein-rich seeds of legumes including peas and beans, and the cells that secrete the digestive enzymes of insectivorous plants. Each molecule of messenger RNA is long enough for 6 to 10 ribosomes to attach to it and read it simultaneously. The ribosomes are thus bound together by the messenger RNA, forming a cluster called a **polysome**.

■ Endoplasmic Reticulum

A typical plant cell is so small that diffusion, the random movement caused by molecular motion, carries a molecule to many different parts of the cell every second. Diffusion may be the only means by which small molecules like monosaccharides and cofactors move around the cell, but some large molecules such as proteins are carried by the **endoplasmic reticulum (ER)**, a system of narrow tubes and sheets of membrane that form a network throughout the cytoplasm (Figure 3-22).

A large proportion of a cell's ribosomes are attached to the ER, giving it a rough appearance; consequently, this ER is called **rough ER (RER)**. As an attached ribosome synthesizes a protein, it passes through the membrane and collects in the lumen. If the protein is a storage product, as in seeds of legumes, it merely remains in the ER, which may become quite swollen; however, if the protein is to be secreted (digestive enzymes, mucilages, adhesive proteins, and certain nectars), then its accumulation causes regions of the ER to form vesicles. These detach, move to the plasma membrane, and fuse with it, releasing their contents to the cell's exterior by exocytosis (Figure 3-9). In many cases, the protein must be modified before export; if so, the ER pinches off only very small vesicles

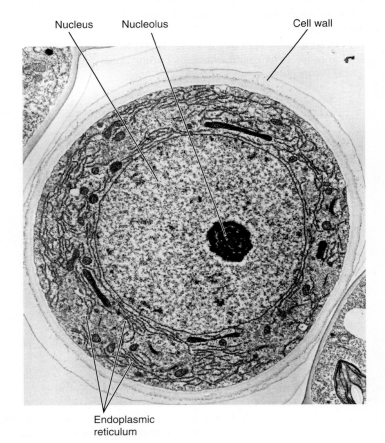

Figure 3-22 (a) Ribosomes only rarely occur free in the cytoplasm; instead, they are usually attached to membranes such as the type shown here, called ER. The amount of ribosomes attached is greater in cells that produce abundant protein (such as protein-rich seeds) or less in cells that synthesize little protein (such as the water-storage cells of Figure 3-2a). (b) Electron micrograph of a cross-section of a hair cell that secretes protein-rich mucilage. The mucilage is synthesized in the ER, which is unusually abundant for a plant cell (×3000).

Figure 3-23 (a) With rapid dictyosome activity, as in these mucilage-secreting cells, the cytoplasm may almost fill with dictyosome vesicles, and fusion of vesicles with the plasma membrane is so abundant that the plasma membrane appears scalloped (×12,000). (b) Face view of dictyosome isolated from a plant cell; the network of peripheral tubules is visible (×60,000).

Vesicles fusing with plasma membrane
Dictyosomes

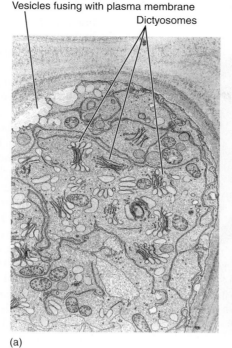

(a)

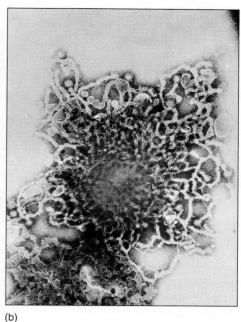

(b)

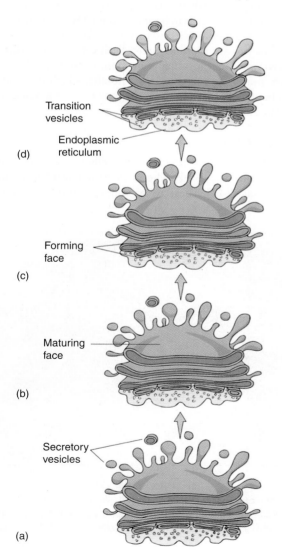

Transition vesicles

Endoplasmic reticulum

(d)

Forming face

(c)

Maturing face

(b)

Secretory vesicles

(a)

from regions where the ER is close to another organelle, the dictyosome, which carries out the protein modification.

ER that lacks ribosomes is **smooth ER** (**SER**), and it is involved in lipid synthesis and membrane assembly. The lipids range from simple to extremely complex; as they are produced, they are inserted into the membrane, and then vesicles form and pinch off, carrying the new membrane to other parts of the cell. After the ER-derived vesicles reach the correct organelle, they fuse with it, and the vesicles become a new patch of membrane in the organelle. SER is abundant only in cells that produce large amounts of fatty acids (cutin and wax on epidermal cells), oils (palm oil, coconut oil, and safflower oil), and fragrances of many flowers.

■ Dictyosomes

Much of the material secreted by a cell must first be modified by a **dictyosome**, a stack of thin vesicles held together in a flat or curved array (**Figure 3-23**). ER vesicles accumulate on one side of the dictyosome and then fuse together and form a wide, thin vesicle called a **cisterna** (plural, cisternae) that becomes attached to the dictyosome (**Figure 3-24**). Soon more ER vesicles gather next to this one and form a new cisterna. The first cisterna becomes embedded more deeply in the dictyosome as more vesicles accumulate on that side, which for obvious reasons is the **forming face**. At the other side, the **maturing face**, vesicles are being released; their contents have been processed. After separation, vesicles can move to the plasma membrane and release their contents (**Figure 3-25**). The outer edges of dictyosomes form an interconnected network of curving tubes, and these may absorb

Figure 3-24 (a) Vesicles derived from ER migrate a short distance to a dictyosome-forming face. (b) At the forming face, the vesicles fuse into a new dictyosome vesicle, called either a vesicle or a cisterna. (c) The cisterna "moves through" the dictyosome as more vesicles form on one side while other vesicles are released from the maturing face (d).

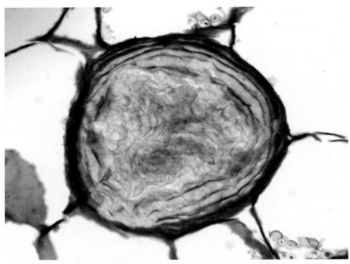

Figure 3-25 This is a mucilage cell. Mucilage-filled dictyosome vesicles secreted mucilage (stained purple) outside the protoplast (see Figures 3-9c and 3-24a). Mucilage could not penetrate the cell wall, so as it accumulated, the protoplast had to decrease in volume by removing water from the central vacuole. The cell died (probably by programmed cell death), and now the cell consists only of cell wall and mucilage (×250).

the contents from the center of the dictyosome cisterna and then detach and move away. It is not known why some dictyosomes concentrate material in the central cisternae, whereas others use the peripheral ones.

Dictyosomes can form large, complex associations. In animal cells that secrete very large amounts of protein, hundreds of dictyosomes associate side by side and form a cup-shaped structure called a **Golgi body** or **Golgi apparatus**. In the Golgi body, the dictyosomes' maturing faces are on the inner side of the cup, whereas the forming faces and associated ER are on the outside. Dictyosomes only rarely aggregate into Golgi bodies in plants, one example being in root hairs: All of the dictyosomes are part of a giant Golgi body located at the tip of the hair where growth and cell wall formation occur.

Different types of processing may occur within a dictyosome: modification of the vesicle's membrane or modification of its contents. If the vesicle is to fuse with the plasma membrane after release, the vesicle membrane must be made similar to the plasma membrane. If the vesicle is to remain separate from all other organelles, acting as a storage vesicle, its membrane must be made unique and incapable of fusion.

Alteration of the vesicle's contents involves addition of sugars to proteins, forming glycoproteins. Sugar-containing proteins occur in the plasma membrane, the cell wall, and as storage products in seeds. Strong evidence is accumulating that dictyosomes also polymerize sugars into polysaccharides used in cell wall construction.

Movement of vesicles from ER to dictyosomes and then to other sites was hypothesized on the basis of the presence of numerous vesicles located between ER and dictyosomes and the similarity of the contents of dictyosome cisternae to those of vesicles found fusing with the plasma membrane. This hypothesis predicts that if a secretory cell is given a very brief dose of radioactive sugar or protein precursors, much of the radioactivity is soon found in the ER

and later in the dictyosomes. Even later, both are nonradioactive, but vesicles at the plasma membrane and external material are radioactive. Experiments have verified both this prediction and the related hypothesis that new membrane is synthesized in the ER and then transported by vesicles to growing organelles. Although organelles appear to be distinct entities when viewed by light or electron microscopy, they are actually highly interrelated by this membrane flow. All membranes of the cell, except the inner membranes of mitochondria and plastids, actually constitute just one extensive system, the **endomembrane system**.

In a few cases, it has been possible to measure the rate of membrane flow. The insectivorous plant *Drosophyllum* has leaves covered with glands that produce a sticky secretion. The fluid contains digestive enzymes that are processed by dictyosomes in the gland's cells. After an insect has been caught, each gland produces a visible drop of digestive fluid at a rate of 1.3 μm^3 per minute. Dictyosome vesicles are about 0.15 μm in radius; thus, each has a volume of 0.014 μm^3, and approximately 100 must fuse with the plasma membrane every minute to deliver the 1.3 μm^3 of fluid and enzymes. Each vesicle has a surface area of 0.27 μm^2, so 27 μm^2 of vesicle membrane fuses with the plasma membrane every minute during secretion. From calculations of the average cell volume and surface area of the plasma membrane, these secretory cells could double their plasma membrane every 20 seconds. It is important for these cells to be able to retract membrane material from the plasma membrane and recycle it. The mucilage-secreting cells of root tips are more leisurely: Each cell has approximately 800 dictyosomes that together contribute 14 to 26 μm^2 of membrane per minute to the plasma membrane, which has a surface area of about 1,000 μm^2. Each dictyosome receives a new cisterna every 20 to 40 minutes.

■ Microbodies

Viewed by electron microscopy, cells are seen to contain numerous small, spherical bodies approximately 0.5 to 1.5 μm in diameter (**Figure 3-26**). These are so nondescript that it is not easy

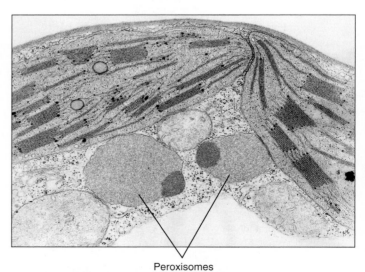

Peroxisomes

Figure 3-26 Electron micrograph showing two chloroplasts with two peroxisome-type microbodies next to them. In many plants, photosynthesis is accompanied by a process called photorespiration, in which peroxisomes produce the amino acid glycine (approximately ×66,000).

Botany and Beyond

Box 3-3 The Metric System and Geometric Aspects of Cells

The Metric System

In 1791 the French Academy proposed that a new system of measurement be adopted to simplify and regularize weights and measures. This metric system is used not only for all scientific measurements but also for engineering, commercial, and ordinary purposes in every country of the world except the United States (see the conversion tables at the end of this box). The fundamental unit is the meter; however, for long distances, the kilometer is used, and for small objects, centimeters, millimeters, and so on are more convenient. Volumes are also based on the length measurements, and weights are based on the gram, the weight of 1 cm³ of water. The metric system has two advantages: (1) Measurements and objects made in one country can be understood or used immediately in any other country, except the United States, and (2) measurements in one unit can be converted easily to other units: 35,645 meters is 35.645 kilometers, but 35,645 feet is how many yards or miles?

1 kilometer (km) = 1,000 meters (m)
1 meter = 1,000 millimeters (mm)
1 millimeter = 1,000 micrometers (μm)
1 micrometer = 1,000 nanometers (nm)
10^9 nm = 10^6 μm = 10^3 mm = 10^0 m = 10^{-3} km

These are based on multiples of 1,000. The centimeter (cm) is only 1/100 meter or 10 mm, but it is such a convenient size for measuring plants and animals that it is used even though it disrupts the otherwise perfect regularity of increments of 1,000.

Geometry of Cells

A quantitative understanding of plants provides extremely valuable insight into their biology. The following formulas are simple but will allow you to think about plants more precisely:

area of a rectangle = length × width
area of a triangle = 1/2 length × width
volume of a cubic space = length × width × height
diameter of a circle = 2 × radius
circumference of a circle = 2π × radius
area of a circle = π × radius²
surface area of a sphere = 4π × radius²
volume of a sphere = 4π/3 × radius³
volume of a cylinder
    = area of base × height
    = π × radius² × height
surface area of a cylinder
    = circumference of circle × height + ends
    = π × diameter × height + ends

Many plant cells are almost cube shaped and have edges about 20 μm long; the volume of such a cell is the volume of a cube

= L × W × H
= 20 μm × 20 μm × 20 μm
= 8,000 μm³

Do not forget to multiply the units to get μm³ (see **Figures B3-3a–d**).

Now consider the nucleus of the cell. Many plant nuclei have a diameter of about 8 μm, and thus, their radius is 4 μm. The volume of a nucleus is V = (4)(3.141)/3 × (4 μm)³ = 4.188 × 64 μm³ = 268 μm³. The nucleus, although an important part of a cell, may not be a large part. In our average cell of 8,000 μm³ volume, the proportion that is nucleus is (volume of nucleus)/(volume of cell) × 100% = 268 μm³/8,000 μm³ × 100% = 3.3%, a very small part of the cell volume.

Mitochondria are about 1 μm in diameter and are quite variable in length; let us assume that we have measured many and found the mean length to be 5 μm. Mitochondria are almost cylinders, so we can use that formula to calculate their volume:

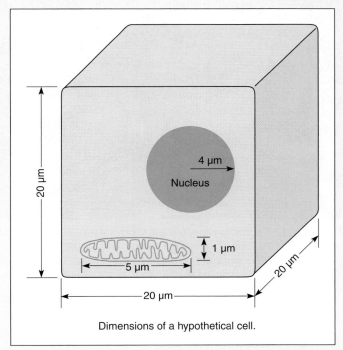

Dimensions of a hypothetical cell.

Figure B3-3a Dimensions of a hypothetical cell. Notice that the mitochondrion is not drawn to scale.

Botany and Beyond

Figure B3-3b Most of us have a set of standard metric weights readily available. A dollar bill (of any denomination) weighs exactly 1 gram (g), 2 pennies weigh 5 grams, 15 quarters and 3 pennies weigh almost exactly 100 grams, and a small Post-it Note® weighs one-tenth (0.1) of a gram.

Figure B3-3c Many rulers have centimeters (cm) along one side and inches along the other. Here, we can see that this book, chosen entirely at random, is 193 mm (about 7.5 in.) wide.

Figure B3-3d The metric system measures volumes in liters (L) and milliliters (mL). A tablespoon (foreground) is 14.7 mL, and a teaspoon (the next one back) is exactly 5 mL. One cup is slightly less than 250 mL. Drops of milk are irregular, but 20 drops of water equal 1.0 mL. The measuring cup and the beaker use the abbreviation "ml" but the correct abbreviation is "mL."

volume of a cylinder
= area of a circular cross-section × height
= (3.141) × (0.5 μm)² × (5 μm)
= 0.7852 μm² × (5 μm)
= 3.93 μm³

Mitochondria often constitute about 7.5% of a cell's volume; for our typical cell, that would be 8,000 μm³ × 0.075 = 600 μm³. How many actual mitochondria would there be in each cell? If each mitochondrion has a volume of 3.93 μm³, the number is (total volume)/(volume of each mitochondrion) = 600 μm³/3.39 μm³ = 177 mitochondria in each cell.

The surface of a cell or organelle is the space through which material enters and leaves the object by diffusion, facilitated diffusion, or active transport. Objects with large surface areas can absorb or lose material more rapidly than those with smaller surface areas. Also, surface area is a measure of the room available for placing membrane-bound enzymes and other intrinsic membrane proteins. In our hypothetical cubical cell, each surface has an area of 20 μm × 20 μm = 400 μm². A cube has six sides, and thus, the total surface area of the plasma membrane is 6 × 400 μm² = 2,400 μm². That is about 1/30 of the surface area of the period at the end of this sentence. The nucleus of the cell has a surface area of 4π × (radius)² = 12.56 × 16 μm² = 201 μm², less than 1/10 that of the cell. The cylindrical mitochondria each have a surface area of their outer membrane of 3.141 × 1.0 μm × 5 μm = 15.7 μm² (we are ignoring the ends). What, however, of their inner membrane? Assume that measurements reveal that each crista is a cylindrical tube 0.2 μm wide and 0.9 μm long and each mitochondrion contains an average of 45 cristae; then the surface area of each is 3.141 × 0.2 μm × 0.9 μm = 0.565 μm². The 45 in each mitochondrion have a total

surface area of $45 \times 0.565\ \mu m^2 = 25.4\ \mu m^2$ available for respiratory enzymes. The 177 mitochondria of the cell would have a total inner surface area of 4,496 μm^2, greater than that of the plasma membrane area (2,400 μm^2). Thus, folding the inner membrane into cristae provides a large amount of extra surface area for enzymes.

Length Conversion

1 in = 2.5 cm	1 mm = 0.04 in
1 ft = 30 cm	1 cm = 0.4 in
1 yd = 0.9 m	1 m = 40 in
1 mi = 1.6 km	1 m = 1.1 yd
	1 km = 0.6 mi

Weight Conversion

1 oz = 28 g	1 g = 0.035 oz
1 lb = 0.45 kg	1 kg = 2.2 lb

Volume Conversion

1 tsp = 5 mL	1 mL = 0.03 fl oz
1 tbsp = 15 mL	1 L = 2.1 pt
1 fl oz = 30 mL	1 L = 1.06 qt
1 cup = 0.24 L	1 L = 0.26 gal
1 pt = 0.47 L	
1 qt = 0.95 L	
1 gal = 3.8 L	

Temperature Conversion

$$°C = \frac{(°F - 32) \times 5}{9}$$

$$°F = \frac{(°C \times 9) + 32}{5}$$

Interval Equivalents

°C	°F
1°	= 1.8°
5°	= 9°
10°	= 18°

to tell exactly what they are: ER vesicles, dictyosome vesicles, or distinct organelles; however, by using special chemical reactions and stains, it has been possible to determine the contents and even the types of reactions in these bodies. Some are organelles now called **microbodies**, and there are two classes: **peroxisomes** and **glyoxysomes**. Both types isolate reactions that either produce or use the dangerous compound peroxide, H_2O_2. If peroxide were to escape through the microbody membrane, it would damage almost anything it encountered; however, both types of microbody contain the enzyme catalase, which detoxifies peroxide by converting it to water and oxygen.

Peroxisomes are involved in detoxifying certain by-products of photosynthesis and are found closely associated with chloroplasts; in animals, peroxisomes are abundant in liver and kidney cells, where they break down foreign compounds that contaminate our food. Glyoxysomes, which occur only in plants, are involved in converting stored fats into sugars. They are important during the germination of fat-rich, oily seeds such as peanut, sunflower, and coconut.

All the previously mentioned, except ribosomes, are organelles composed of membranes; the following are nonmembranous organelles.

■ Cytosol

Most of the volume of cytoplasm is a clear substance called **cytosol** or hyaloplasm. It is mostly water, enzymes, and numerous chemical precursors, intermediates, and products of enzymatic reactions (Table 3-8). Within cytosol are free ribosomes (not attached to RER), as well as skeletal structures—the microtubules and microfilaments.

■ Microtubules

Microtubules (Figure 3-27) are the most abundant and easily studied of the structural elements of a cell, and they have many functions. They act as a "cytoskeleton," holding certain regions of the cell surface back while other parts expand. Without them, cells would be just spheres, but by reinforcing specific areas, cell growth and expansion are directed to weaker areas. In other cases, microtubules assemble into arrays like an antenna which either catch vesicles and guide them to specific sites or cover a

TABLE 3-8	Subunits of the Cell
Whole cell	
Cell wall	
Protoplasm	
Nucleus	
Vacuole	
Cytoplasm	
All remaining organelles	
Cytosol	

(a)

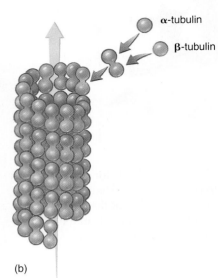

α-tubulin

β-tubulin

Figure 3-27 (a) Microtubules that are part of a dividing nucleus and are involved in pulling chromosomes to the ends of the cells. (b) Alpha- and beta-tubulin associate into a dimer called tubulin, and dimers aggregate into a microtubule. When no longer needed, microtubules depolymerize back to the monomers, which are recycled to build other microtubules.

(b)

region, thereby excluding the vesicles (**Figure 3-28**). Finally, microtubules are the means of motility for both organelles and whole cells. The framework that moves chromosomes during division of the nucleus is composed of microtubules, and microtubules can attach to and move whole nuclei, mitochondria, and other organelles.

Microtubules are composed of two types of protein with a globular tertiary structure: **alpha-tubulin** and **beta-tubulin**. These associate as dimers called **tubulin** that further crystallize into a straight tubule with a diameter of 20 to 25 nm. This is a reversible process; when a microtubule is no longer needed, it depolymerizes back into its component monomers, which disperse into the cytosol until the cell needs to assemble a new microtubule. Control of polymerization and depolymerization is not well understood. When microtubules occur as individuals or small clusters, new tubulin dimers are added to or removed from one end automatically. In other instances, microtubules occur in large clusters, often in a highly ordered

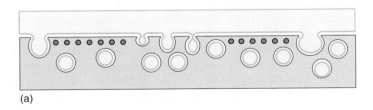

(a)

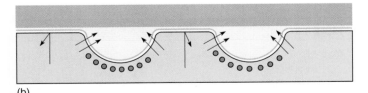

(b)

Figure 3-28 Microtubules (depicted here in purple) are often located next to the plasma membrane. (a) They may act as a screen that keeps vesicles away from the plasma membrane, or (b) they may pull the plasma membrane away from the cell wall, allowing material to accumulate there.

arrangement, and a small body is usually associated with the orderly production of microtubules. For example, when the nucleus undergoes division, an array of microtubules called the spindle is formed in the middle of the cell, and spindle microtubules push and pull the chromosomes to their proper positions.

In all animals and in some fungi and algae, a pair of organelles called centrioles is associated with the formation of the spindle. A **centriole** is made up of nine sets of three short microtubules (**Figure 3-29**); the nine triplets are held together by fine protein spokes. Centrioles were assumed to be responsible for the organization and polymerization of the spindle microtubules even though plants never have centrioles.

Much more elaborate sets of precisely arranged microtubules occur in **cilia** (singular, cilium) and **flagella** (singular, flagellum) (**Figures 3-30** and **3-31**), which appear to be identical except that cilia are short (about 2 μm) and occur in groups, whereas flagella tend to be much longer (up to several micrometers) and usually occur either singly or in sets of two or four. Both are present on many types of algae and motile fungi, but in plants, only the sperm cells have flagella. In flowering plants and conifers, no cells, not even sperm cells, ever have cilia or flagella.

In cross-section, each cilium or flagellum has a "9 + 2" arrangement: Nine pairs of fused microtubules surround two individual microtubules. The outer doublets each have two arms composed of the protein dynein, and each doublet is connected to the central pair of microtubules by protein spokes. The dynein arms convert the chemical energy of ATP into kinetic energy and bend, "walking" along the adjacent microtubule doublet. One set of doublets slides relative to the adjacent set, causing that side of the cilium or flagellum to become shorter and bend the structure. Then, as another set of microtubules slides, the cilium or flagellum bends in a different direction. The result is a powerful beating motion. If the cilia or flagella are on small organisms such as algae, fungi, or protozoans, the organism swims rapidly and gracefully.

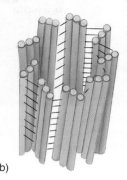

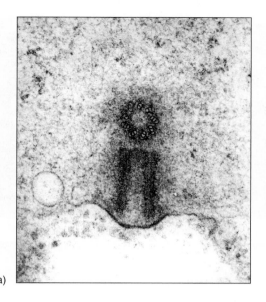

(a)

(b)

Figure 3-29 (a) The two members of a pair of centrioles are always located perpendicular to each other. Here, one is seen in transverse section, the other in longitudinal section (×100,000). (b) Each part of a centriole contains a circle of nine sets of three microtubules attached to each other by fine filaments along their sides.

Cilia and flagella can polymerize and depolymerize rapidly, but they never form autonomously; each is always associated with a **basal body**. Basal bodies appear to be identical to centrioles by electron microscopy. It was assumed that basal bodies organize the formation of flagellar microtubules, but recent studies have shown that as a flagellum grows, new monomers of tubulin and dynein are added to the tip, not the base where the basal body is located. The exact relationship between flagella and basal bodies is still not known.

■ Microfilaments

Like microtubules, **microfilaments** are constructed by the assembly of globular proteins—in this case, just one type, **actin**. Microfilaments are narrower than microtubules (only 3 to 6 nm in diameter), and they have been implicated in different types of structure and movement.

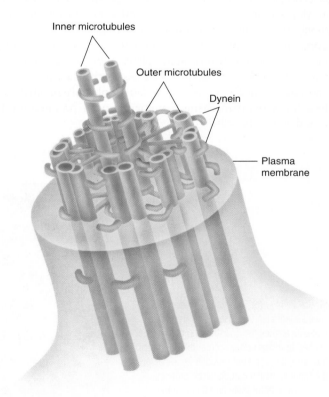

Inner microtubules

Outer microtubules

Dynein

Plasma membrane

Figure 3-31 Cilia and flagella are composed of two central single microtubules surrounded by nine sets of two microtubules. The doublets are not merely attached to each other; they actually share tubulin monomers. The outer member of each doublet has two short arms.

Figure 3-30 Flagella occur on many algal and fungal cells, especially the unicellular organisms. This is the alga *Dunalliela*. Most cells of multicellular organisms are not flagellated, but their sperm cells are (×500).

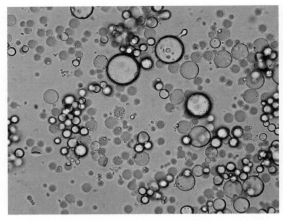

Figure 3-32 Avocado contains very large numbers of oil bodies; this view was prepared merely by smearing a small amount of ripe avocado fruit on a microscope slide. Virtually every visible thing is an oil body; other organelles are much less abundant (×200).

Figure 3-33 Calcium oxalate crystals. When calcium is part of a crystal, it is physiologically inactive and cannot affect membranes or the tertiary structure of proteins. These are cubic crystals in wood (×1000).

Storage Products

Many cells exist in an environment in which resources alternate between abundance and scarcity. To survive times of scarcity, cells must accumulate and store extra nutrients. Most often, the reserves consist of sugars that have been polymerized into starch in amyloplasts or converted into lipids and stored as large oil droplets in oily material like peanuts and sunflower seeds (Figure 3-32); however, with many other storage products, the function and advantages are not so obvious. Many plants store crystals of calcium oxalate or calcium carbonate (Figure 3-33); others accumulate large amounts of silica, tannins (Figure 3-34), or phenols. Because plants have no excretory mechanism, numerous waste products must be stored within the cells.

Cell Wall

Almost all plant cells have a **cell wall**; only sperm cells of some seed plants lack one. It is often regarded as simply an inert secretion, providing only strength and protection to the protoplasm inside; however, considerable metabolism occurs in the wall, and it should be considered a dynamic, active organelle.

The cell wall contains a considerable amount of the polysaccharide cellulose (Figure 3-5b and Figure 3-35a). Adjacent, parallel cellulose molecules crystallize into an extremely strong **microfibril** 10 to 25 nm wide. Numerous microfibrils are wound around the cell, completely covering the plasma membrane. Each cellulose polymer grows only at one end, where a complex of enzymes adds new glucose residues, one molecule at a time (Figure 3-35b). The

Figure 3-34 (a) This type of preparation is called a leaf clearing. A leaf was treated to make its tissues transparent, and then it was stained to reveal certain contents. The dark red, irregularly shaped vacuolar contents are tannins, which can denature proteins. They have a bitter taste and damage the mouth and stomach proteins of insects that try to eat them. The band running vertically through the micrograph is a leaf vein (×200). (b) The bark of a cactus. As its cells aged, they accumulated irregular masses of tannin that have stained red (×80).

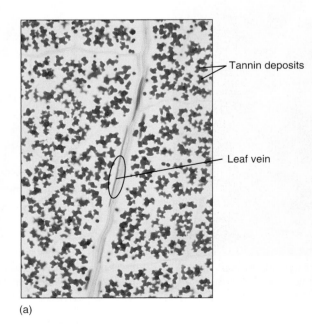

Tannin deposits

Leaf vein

(a)

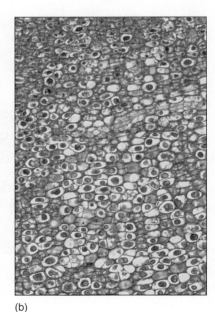

(b)

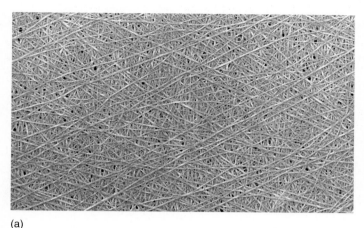

(a)

Figure 3-35 (a) Several layers of cellulose microfibrils are visible; other layers are present deeper in the wall. Each layer provides strength in the direction parallel to the microfibrils. Other wall components have been dissolved away to reveal the cellulose ($\times$7000). (b) Cellulose-synthesizing enzymes are large proteins embedded in the plasma membrane; several proteins form a cluster called a rosette. Sugar monomers are absorbed by the enzymes of the membrane inner face and then passed across the membrane and polymerized into cellulose on the outer face. The growing microfibril extends out from the rosette and lies against microfibrils already in place. Because the microfibril cannot push into the wall, the rosette is pushed through the membrane in the direction of the arrows as the microfibril grows.

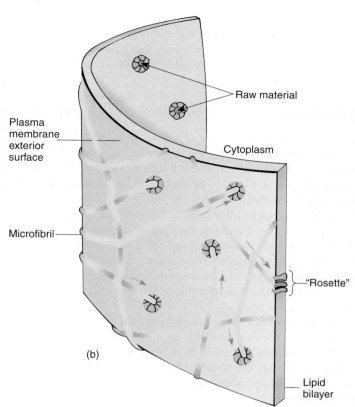

(b)

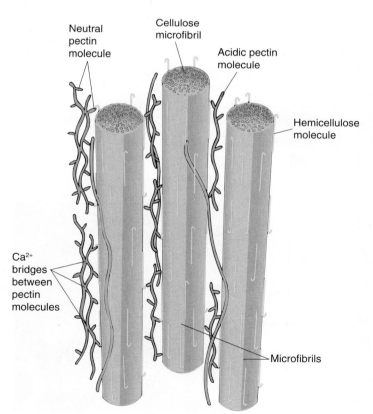

Figure 3-36 Although cellulose microfibrils lie close together, they do not bond with each other, and thus, a wall of pure cellulose is weak. Hemicelluloses and pectins are short, branched molecules that interact with several adjacent microfibrils, linking them together and forming a solid three-dimensional mesh. The relative amounts of these components can vary, resulting in some walls that are flexible and others that are rigid; some strong, and others weak.

enzymes are intrinsic proteins of the plasma membrane, and as they add new sugars to the chain, the enzymes float forward in the membrane (the chain is too heavy to be pushed backward). New cellulose molecules can be added only on the inner side of the wall, adjacent to the plasma membrane.

Cellulose microfibrils are bound together by other polysaccharides called **hemicelluloses**, which are produced in dictyosomes and brought to the wall by dictyosome vesicles. Hemicelluloses are deposited between the cellulose microfibrils and bind to cellulose with hydrogen bonds, producing a solid structure that resembles reinforced concrete (**Figure 3-36**). In multicellular plants, the wall of one cell is glued to the walls of adjacent cells by an adhesive layer, the **middle lamella**, composed of a third class of polysaccharides, **pectic substances**.

All plant cells have a thin wall called the **primary cell wall** (see Figure 3-5a). In certain cells that must be unusually strong, the protoplast deposits a **secondary cell wall** between the primary wall and the plasma membrane (see Figure 3-5b). The secondary wall is usually much thicker than the primary wall and is almost always impregnated with the compound lignin, which makes the wall even stronger than hemicelluloses alone can make it. Lignin resists chemical, fungal, and bacterial attack. Both primary and secondary cell walls are permanent; once deposited, they are almost never degraded or depolymerized, as can be done with microtubules and microfilaments.

Fungal Cells

Cells of fungi are similar to plant cells, with two important differences: (1) They do not contain plastids of any type, and (2) their walls contain **chitin**, not cellulose. Chitin is physically similar to cellulose, being tough, inflexible, and insoluble in water, but it contains nitrogen and is synthesized by a different mechanism than that used for cellulose. Whereas plant cells tend to be box shaped or fiber-like, fungal cells are often extremely narrow, long tubes with many, tiny nuclei.

Associations of Cells

In unicellular organisms, such as simple algae, protozoans, and most prokaryotes, each cell is a complete organism and does not interact directly with other cells. Such cells can communicate indirectly, at least to a small extent, by releasing chemicals that affect the other cells, but such interactions are rare, usually occurring only during the attraction of other cells during sexual reproduction.

In multicellular organisms each cell automatically, unavoidably interacts with its neighboring cells; they must share the same sources of photosynthate, oxygen, carbon dioxide, salts, and water. Whereas a unicellular organism can merely excrete its wastes across its plasma membrane, cells of a multicellular organism are not so free to do so.

Just as important are cellular interactions by which cells not only sense that they are part of a larger organism but also identify which part they are and how they should differentiate. In a developing embryo, the proper cells must be instructed to begin forming an embryonic shoot while other cells are induced to form the embryonic root, vascular tissues, seed leaves, and so on. This requires extensive, sophisticated intercellular communication.

One communication method is like that used by unicellular organisms: A cell secretes specific compounds that "inform" the surrounding cells of what it is doing metabolically and developmentally. A second method is connections between the cells. Direct physical contact between cells, as exists in animals, cannot occur in plants because the two primary walls and middle lamella are located between any two adjacent protoplasts; however, plant cells are interconnected by fine holes (**plasmodesmata**; singular, plasmodesma) in the walls (**Figure 3-37**). A plasmodesma is only about 40 nm in diameter, but the plasma membrane of one cell passes through it and is continuous with the plasma membrane of the adjacent cell. A small channel of cytosol also passes through, as does a short section of specialized ER.

The abundance of plasmodesmata in a wall is quite variable; they can occur singly or in clusters of 10 to 20 or more (**Table 3-9**). In regions of clustered plasmodesmata, the two primary walls are often particularly thin, and the area is called a **primary pit field** (Figure 3-37b and 3-38). In tissues in which considerable transport occurs (into glandular or conducting cells), many primary pit fields and plasmodesmata are present, but in regions where little movement of material happens, few plasmodesmata are found.

Because plasmodesmata are actually cytoplasmic channels from one protoplast to another, the individuality of the cells is diminished; all of the protoplasm within a single plant is part of one interconnected mass, the **symplast**. In some of the nutrient-

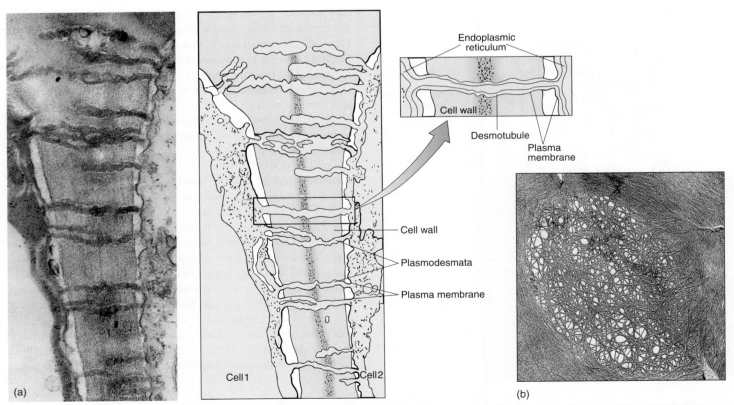

Figure 3-37   (a) Plasmodesmata are complexes that consist of fine holes in the primary walls; they also contain plasma membrane, liquid, and a tubule (called a desmotubule) attached to the ER ($\times$68,000). (b) Face view of a cell wall, showing an oval-shaped thin area, a primary pit field. The small open areas outlined by single cellulose microfibrils are holes where plasmodesmata were located. Protoplasm was removed during specimen preparation ($\times$45,000).

TABLE 3-9	Frequency of Plasmodesmata	
Tissue		Plasmodesmata/ μm²
Area of expected high rates of transport		
Salt bush (*Tamarix*)	Gland cell	17
Russian thistle (*Salsola*)	Sugar-loading cell	14
Pine (*Pinus*)	Sugar-loading cell	20–25
Abutilon	Nectary cell	12.5
Area of expected low rates of transport		
Oat (*Avena*)	Leaf cell	3.6
Onion (*Allium*)	Root cell	1.5
Tobacco (*Nicotiana*)	Outer stem tissue	0.2

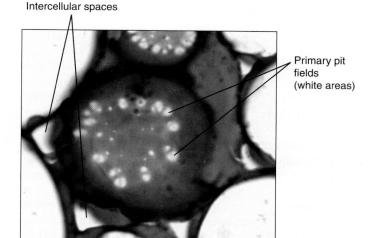

Figure 3-38 Eight cells are visible here, but a ninth cell was coming up out of the image toward us—it was cut away when the slide was made; areas 2 and 8 are contact faces where the ninth cell pressed against cell 2 and 8. The white areas appear to be holes, but they are primary pit fields, areas where the walls are especially thin and plasmodesmata (not visible) are numerous. If this same region were viewed with an electron microscope, each small white dot would look like Figure 3-37b. The "noncontact face" is the region of cell wall that faces intercellular space rather than being pressed against an adjacent cell (×450).

conducting tissues (sieve tube members and companion cells) and also in some of the reproductive tissue (microspore mother cells and endosperm), plasmodesmata are so large and numerous that the cells act more like one single large cell than like individual cells. In most tissues, however, any two adjacent cells act rather independently; the communication through the plasmodesmata is not enough to prevent individuality.

The fact that walls keep protoplasts physically separated from each other has another consequence. Walls act as a second, nonliving compartment inside plants. Water moves through cell walls by capillary action, just as it moves through tissue paper. In addition, many cells do not abut each other tightly but instead have **intercellular spaces** between them, at least at their corners (**Figure 3-38**). In some tissues, especially leaves, cells are so loosely connected that most tissue volume is intercellular space; less than half is actually symplast. These spaces plus the cell wall constitute the **apoplast**; the apoplast and the symplast together make up the entire plant. The apoplast acts as a series of channels and

spaces that permit the rapid diffusion of gases, which is necessary because plants do not have lungs. Diffusion through a gas-filled space is approximately 10,000 times faster than through a liquid-filled space. Without an apoplast, large, bulky plant tissues such as tree trunks, tubers, and fruits would be impossible; the interior tissues would suffocate.

Summary

1. All organisms are composed of cells, the fundamental units of life. Most cells contain the same types of organelles, and there are only a few basic types of cells.
2. In multicellular organisms, cells become specialized for specific tasks, making the entire organism more efficient; however, each cell then becomes dependent on the other cells of the organism.
3. Selectively permeable membranes and active transport are critically important for life; they make compartmentalization possible, in which an enclosed compartment has a chemical composition and metabolism distinct from those outside the compartment.
4. The outermost membrane of a cell is the plasma membrane; the protoplasm it surrounds is a protoplast.
5. Plant cells contain the following membrane-bounded organelles: nucleus—heredity and control of metabolism; vacuole—temporary or long-term storage, growth; mitochondria—respiration; plastids—photosynthesis (chloroplasts); ER—intracellular transport, lipid synthesis; dictyosomes—material processing; and microbodies—detoxification or lipid metabolism.
6. Plant cells contain these nonmembranous organelles: ribosomes—protein synthesis; cytosol—the liquid matrix surrounding all other organelles; microtubules and microfilaments—skeletal structure, movement of organelles; cell wall—structure, cell shape; and various storage products such as crystals and lipid droplets.
7. Animal cells do not have a cell wall, plastids, glyoxysomes, or a central vacuole; fungal cells do have a central vacuole, but they have no plastids and their wall is made of chitin, not cellulose.
8. A plant is composed of both its protoplasm (symplast) and its intercellular spaces and walls (apoplast). The apoplast can be important for diffusion of air and liquids.

Important Terms

apoplast	extrinsic proteins	molecular pumps
bilayer	flagella	nucleolus
cell wall	fluid mosaic membrane	nucleus
centriole	freely permeable membrane	organelles
chloroplasts	glycoproteins	plasmodesmata
chromoplasts	Golgi apparatus	plastids
cilia	Golgi body	primary pit field
compartmentalization	impermeable membrane	prokaryotic cell
cytoplasm	intercellular spaces	ribosomes
cytosol	intrinsic proteins	selectively permeable membrane
dictyosome	lumen	symplast
endocytosis	microbodies	tonoplast
endoplasmic reticulum	microfilaments	tubulin
(RER and SER)	microtubules	vacuoles
eukaryotic cell	middle lamella	
exocytosis	mitochondria	

Review Questions

1. In plants, each cell consists of a box-like _____ _____ surrounding a mass of _____, which in turn contains its own smaller parts, the _____ _____.

2. In Table 3-1, the size of water-conducting vessel cells in oak are_____ _ _____. Each millimeter contains 1,000 µm. Approximately how many of these cells could be laid end to end (or side by side) in 1 mm? Do you think such a cell would be visible if it were placed in your hand? How long are the fiber cells in hemp? How long is this in millimeters? How long is it in centimeters (1 cm = 10 mm)? Could you see one of these cells if it were in your hand?

3. Figure 3-1 is a light micrograph of a leaf cell, and it is typical of what you might see if you were to examine a moss leaf. What two cell structures can you see?

4. The division of labor among various types of cells in plants is important. Examine Table 3-2, and name the cells responsible for the following tasks:
   a. Collecting solar energy
   b. Transporting water
   c. Cell division and the production of new protoplasm
   d. Water retention
   e. Having pigments that attract pollinators
   f. Producing poisons that inhibit animals from harming the plants

5. Stone cells in coconut shells are part of the division of labor in plants. What modifications do they have that make it impossible for them to be dividing and growing cells (hint: see Figure 3-5)?

6. All biological membranes are composed of _____ and two layers of _____ molecules. One of these components occurs as a thin solution only two molecules thick. A layer only two molecules thick is a _____.

7. Some of the proteins that are part of a cell's membranes are actually immersed in the membrane. Others are located outside the membrane and merely lie next to it. What are the names of each type?

8. Cell membranes are described as being fluid mosaics. What does that mean? Are all membrane components free to diffuse laterally anywhere within the membrane?

9. What are the two basic components of membranes? What are three ways that material can move from one side of a membrane to the other? Which method requires the plant to use energy?

10. Pieces of membranes frequently move from one area of a cell to another, but never as just flat patches, only as vesicles. Study Figure 3-9, and then draw examples of endocytosis and exocytosis using different colors for the two sides of the membranes. Notice that the black side of the membrane bilayer always touches the blue part of the cell in Figure 3-9. Can you come up with any type of vesicle fusion that would result in the black side of the membrane touching the green part of the cell?

11. Define each of these terms: a freely permeable membrane, an impermeable membrane, and a selectively permeable membrane. During your next exam, you may begin to sweat a solution of sodium, chloride, and water, but not proteins, sugars, or nucleotides. Are the membranes of your sweat gland cells freely permeable, impermeable, or selectively permeable?

12. Circle the correct answer: Plants, animals, fungi, and protists have (prokaryotic, eukaryotic) cells. What is a key feature of eukaryotic cells?

13. The _____ _____ (less frequently called the _____) is the membrane that completely covers the surface of the protoplasm. It is (circle one: freely permeable, impermeable, selectively permeable), and it (circle one: does, does not) contain molecular pumps.

14. The nucleus serves as an archive for the organism's _____ _____, all of which is stored as __ __ __ inside every nucleus.

15. Which organelle functions as storage of both nutrient reserves and waste products? What is the name of its membrane?

16. Which organelle carries out cell respiration? Each of these organelles has two membranes—a liquid portion and folded, sheet-like membrane structures. What are the names of these four components?

17. Plastids can grow and develop into many different types of structure. What is the name for the plastids that do each of the following:
    a. Carry out photosynthesis
    b. Synthesize lipids and other materials
    c. Store starch
    d. Contain abundant colored lipids
    e. Occur when tissues are grown without light

18. Which organelle carries out protein synthesis?

19. What is the difference between RER and SER? SER is not very common in plants, but several examples are given in the text. What are they?

20. What is the role of dictyosomes in cell metabolism? What is their forming face? Their maturing face?

21. Draw a "typical" plant cell and include all of the organelles mentioned in this chapter. Most cells have specialized functions and increased proportions of certain organelles. How would you change your drawing if the cell is involved in photosynthesis? Is part of a yellow or orange petal? Secretes protein-rich digestive fluid? Needs a great deal of ATP? Stores starch? Has just grown a great amount?

22. Which of the organelles in Question 21 are composed primarily of membranes? What are some of the ways that the membranes participate in the metabolic activities of those organelles?

23. Describe how material is brought to a dictyosome. What happens to material while it is in a dictyosome, and how is it released? What are some of the things that might happen to it after it is released?

24. What are alpha-tubulin and beta-tubulin? When they aggregate, what do they form? What structure in dividing cells is made with this substance (Figure 3-27)? Algal, fungal, and animal cells move with structures made of this. What is the name of the structure?

25. Many animals and algae have cells with flagella, but do any plants have flagella? For example, do fern cells or moss cells have flagella?

26. How do microtubules act as a cytoskeleton? Are there times when they are particularly abundant? What are centrioles, basal bodies, and flagella?

27. Examine Figures 3-32 to 3-34. What types of substances occur as storage products in plant cells?

28. What chemical do clams use to make strong shells? What do we use for our bones? Do plants use either of these chemicals in the construction of strong cell walls? What problem would plants have if they used either the shell material or the bone material?

29. In cell walls, adjacent, parallel cellulose molecules crystallize into an extremely strong _____ 10 to 25 nm wide. These are bound together by polysaccharides called _____, which are produced in dictyosomes.

30. What are the most abundant components of a plant cell wall? How do the components interact, and how are they arranged with respect to each other? What are the symplast and the apoplast in plants?

31. What is the type of cell wall present in all plant cells? What is the type present in certain cells that must be unusually strong?

32. What are the extremely fine holes that interconnect plant cells? What is the name of the especially thin areas of walls where these holes are particularly common (hint: Figure 3-37)?

33. What is an intercellular space? Diffusion through a gas-filled space is approximately _____ times faster than through a liquid-filled space. Living cells deep inside a bulky object such as an apple need oxygen. Do you think the oxygen could diffuse to those cells fast enough if there were no intercellular spaces? Would you guess (you will have to guess, as the answer is not in the text) that all or most of the intercellular spaces are interconnected such that oxygen could diffuse deep into an apple without ever having to dissolve into protoplasm?

BotanyLinks

http://biology.jbpub.com/botany/5e

BotanyLinks contains a variety of resources and review material designed to assist in your study of botany.

Visit the Web Exercises area of BotanyLinks to complete this question:

1. Plant cells differ from those of animals. How do they differ? Go to the BotanyLinks home page to investigate.

Growth and Division of the Cell

Concepts

The life cycle of individual organisms includes stages of initiation, growth, and death. Individual cells also have a life cycle, the **cell cycle**. Cells are initiated by division of a mother cell, grow for a period, and usually cease to exist by dividing and producing two daughter cells (**Figure 4-1**). This is not a real death because the substance of the mother cell continues to exist in the daughter cells.

As a multicellular organism grows, this type of cell cycle is common (about 25 million cell divisions occur every second in your own body), but as parts of a plant reach their final form, most cells stop dividing (**cell cycle arrest**) and enter an extended period of growth, during which they differentiate and mature (**Figure 4-2**). For example, leaf cells stop dividing when the leaf is only a few millimeters long; they continue to grow as the leaf expands and then function for the rest of the leaf's life. Leaf cells cannot normally be stimulated to divide again; if the leaf is damaged, little regeneration is possible. In contrast, cortex cells in the stem or root also stop dividing before the organ is mature, but if damaged, the remaining cells resume division and remain active until the damaged portion has been protected with a layer of cork.

Chapter Opener Image: Plants grow longer by having groups (meristems) of dividing cells located at the tips of twigs and roots. The meristem at the tip of this fig tree (*Ficus carica*) twig had been dormant throughout winter, but now in springtime it is active and its cells divide rapidly, producing cells for new leaves and new portions of stem. While dividing, plant cells are usually extremely small (like most animal cells) but then enlarge greatly as their central vacuoles absorb water. In the larger leaf visible here, cell division has probably already stopped, yet this leaf will continue to enlarge because its cells are becoming larger, not because it is producing more cells. The smaller leaf, barely visible, might still have some dividing cells. If we would dissect the bud, we would find microscopic leaves in which most cells are still dividing.

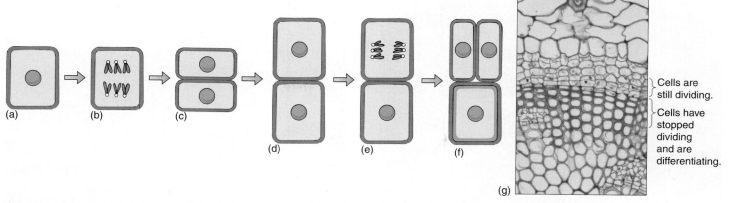

Figure 4-1 The cell cycle consists of division and growth phases. (a) The cell has grown and is ready for division; (b) first the nucleus divides, (c) and then the cytoplasm is divided by the formation of a new wall. (d) The new cells enter a new cell cycle by growing. (e) Then one divides again, but the other may begin to differentiate; in this example, its wall (red) thickens as it matures into a fiber. (f) The upper cell has finished dividing, and both daughter cells may enter new cell cycles. The lower cell continues differentiation; in some fiber cells the nucleus may divide once or twice, but usually once fiber differentiation begins, the nucleus never divides again. (g) A cross-section of a set of cells similar to those depicted in (a) to (f); the red cells have stopped dividing and are differentiating into fiber cells. The blue cells with thin walls are still capable of growth and cell division (×250).

Figure 4-2 Palms are extreme examples of plants that have only a few dividing cells; in this entire palm shoot, cell division only occurs at (1) the extreme apex, (2) in developing leaves, and (3) in developing flowers. Of all the millions of cells in the visible trunk here, none is dividing; however, most cells are still alive. Cells at the base of the trunk were produced when the plant was just a seedling and have lived for about a century.

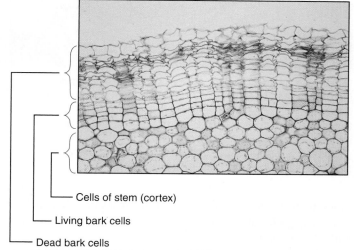

Figure 4-3 Cells in this geranium bark divide only a few times and then differentiate by placing antimicrobial compounds in their walls along with waterproofing compounds. The cells then die and degenerate to cell walls without protoplasts. Lack of a protoplast is selectively advantageous because if a fungus or insect does penetrate the wall, it encounters no nutritious protoplasm and starves (×120).

Some cells live for many years, even hundreds of years, but others die shortly after they mature. Bark cells are more protective if dead (Figure 4-3); many flower parts die only a few days after the flower first opens, and gland cells often die after a brief period of secretion.

Some cells never stop dividing. Cells in the growing points at the tips of roots and shoots are constantly cycling (Figures 4-4 and 4-5), as are those that form wood and bark (the cambium). When a cambial cell divides, one of the new daughters becomes a wood or bark cell, and the other remains part of the dividing layer. The wood or bark cell divides a few times and then differen-

Figure 4-4 In contrast to palms, this conifer has many groups of dividing cells. Each bud (eight are visible on just this twig) has a group of dividing cells (an apical meristem) at its apex. The needle-shaped leaves about to emerge probably formed most of their cells while enclosed in the bud and will now just undergo cell enlargement and maturation. Inside the woody brown stem is a layer of dividing cells (a vascular cambium) that will add more wood and bark cells, making this part of the stem wider but not longer.

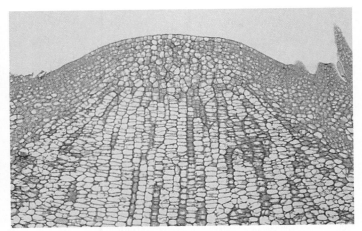

Figure 4-5 The growing shoot tip of a cactus (*Cleistocactus*). All of its cells are alive and undergo repeated cell division. One of the basic roles of shoot tips is to convert nutrients into new cells that can later differentiate into shoot tissues. At the base, cells are lightly stained because their large vacuoles do not absorb the stain; these cells cycle more slowly than the smaller, less vacuolate cells at the tip (×80).

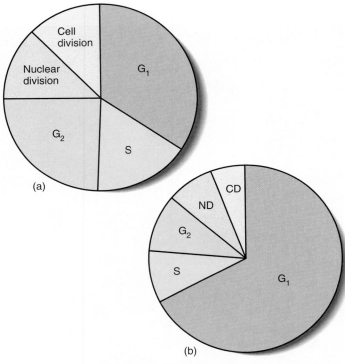

Figure 4-6 Comparison of cell cycles of different tissues or species. In a shoot tip (a), cell division may occur as soon as the cell grows to its proper size, and G_1 is only a small part of the cell cycle. Below the shoot tip (b), cells still divide but spend more of their time growing and carrying out nondivision metabolism; G_1 is a greater proportion of the cell cycle. Both (a) and (b) give only the relative amount of time spent in each phase, not the absolute time in hours or days. The cell cycle in cells like (b) is often much longer than in shoot tips (a).

tiates, matures, and usually does not divide again. The daughter cell that stays in the dividing layer grows back to its original size and divides again.

The cell cycle can be divided into a growth phase and a division phase.

Growth Phase of the Cell Cycle

In the 1800s, when the cell cycle was first being studied intensively, researchers gave the greatest attention to the division activities because many events could be identified. They assumed that between divisions cells were "resting," so the growth phase was called the **resting phase**, or **interphase** (Figure 4-6). We now know that the cell is active during interphase, and it is possible to detect three distinct phases within interphase: G_1, S, and G_2.

G_1 Phase

In G_1 (or **gap 1**), the first stage after division, the cell is recovering from division and conducting most of its normal metabolism (Figure 4-6). One important process is the synthesis of nucleotides used for the next round of DNA replication. The length of a cell cycle varies tremendously, depending on the type of cell, the type of plant, its health, its age, the temperature,

and many other factors. In single-celled organisms such as some algae, the cell cycle may be as brief as several hours; short cycle times also occur in rapidly growing embryos and roots. On the other hand, cell cycle times of 2 to 3 days or even months are not unusual in tissues or plants that grow slowly. During winter dormancy, the cell cycle may last from autumn until spring; when

TABLE 4-1 Cell Cycle Phase Durations in Root Tip Cells and Cell Cultures

Species	Total	Cycle Phase Duration (Hours)			
		G₁	S	G₂	M
Daucus carota (carrot)					
Root tip	7.5	1.3	2.7	2.9	0.6
Percent of cell cycle	100%	17%	36%	39%	8%
Daucus carota					
Cell culture	51.2	39.6	3.0	6.2	2.4
Percent of cell cycle	100%	77%	6%	12%	5%
Haplopappus gracilis					
Root tip	10.5	3.5	4.0	1.4	1.6
Zea mays (corn)					
Root tip	9.9	1.7	5.0	2.1	1.1
Allium cepa (onion)	17.5	1.5	10	4.0	2.0

Note that when carrot cells are grown in cell culture, the cell cycle becomes much longer, mostly because of G₁, which goes from being only 17% of the cell cycle in roots to 77% in cell culture.

dormancy is broken and metabolism accelerates, the cell cycle may last only a few hours. The actual length of time that any cell spends in G_1 is similarly variable. Despite exceptions, in general, G_1 is the longest part of the cell cycle (**Table 4-1**). After a cell undergoes cell cycle arrest, stops dividing, and begins to differentiate and mature, it may enter a state similar to G_1 and remain in it for life.

■ S Phase

During **S** (or synthesis) **phase**, the genes in the nucleus are replicated. A gene is a polymer of nucleotides, and each gene has a

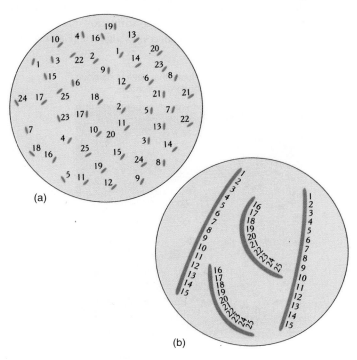

Figure 4-7 (a) If each gene were a distinct piece of DNA, ensuring replication of each would be difficult, and ensuring that each daughter nucleus received one copy of each would be even more difficult. (b) Grouping the genes into chromosomes makes them easier to manipulate. Here only two types of chromosomes are shown: long ones with genes 1 to 15 and short ones with genes 16 to 25.

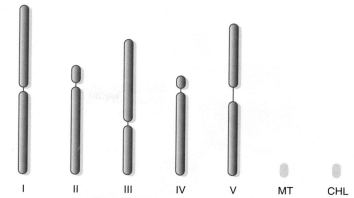

Figure 4-8 Each nucleus in the cells of *Arabidopsis thaliana* has two of each of these five types of chromosomes, one set inherited from the paternal and one from the maternal parent. The slender area in each represents the centromere. The lines labeled MT and CHL represent the DNA circles of mitochondria and plastids. The entire genome has been sequenced: We know which DNA nucleotide occurs at each site in each chromosome.

unique sequence of nucleotides. Although many genes may have similar sizes, different genes never have exactly the same sequence of nucleotides. Many higher plants and animals need about 30,000 types of genes to store the information required to make the proper enzymes, structural proteins, and hormones necessary for the organism's life. The entire complex of genes for an organism is its **genome**.

Obviously, if each individual gene floated around the cell or nucleus as an independent piece of DNA, then during division, it would be extraordinarily difficult to find each of those 60,000 genes (30,000 originals and 30,000 copies) and make certain that each daughter cell received one of each (**Figure 4-7**). This is not a problem, however, because all genes are attached to other genes by short pieces of linking DNA. Thousands of genes are attached together in a linear sequence; the entire structure is a **chromosome** (**Figure 4-8**). It might seem reasonable for all genes to link together in a single chromosome, but such a long chromosome would probably become hopelessly tangled. Only a few plants have as few as 2 chromosomes for their genome; most have between 5 and 30 chromosomes (**Table 4-2**).

TABLE 4-2 Number of Chromosomes in a Haploid Set

Species	Common Name	Number of Chromosomes
Machaeranthera gracilis	None: plant of U.S. deserts	2
Arabidopsis thaliana	None: related to mustard	5
Vicia faba	bean	6
Pisum sativum	pea	7
Allium cepa	onion	8
Zea mays	corn	10
Lilium pyrenaicum	lily	12
Oryza sativa	rice	12
Triticum aestivum	wheat	21
Homo sapiens	humans	23
Solanum tuberosum	potato	24

A haploid set is a single set of chromosomes, as carried by sperm cells and egg cells. A fertilized egg has two sets and is diploid.

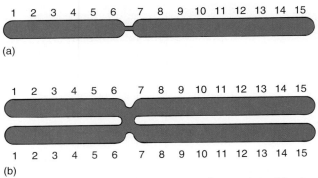

(a)

(b)

Figure 4-11 (a) Before S phase, each chromosome has one chromatid and one copy of each gene. (b) After replication in S phase, each chromosome has two chromatids and two copies of each gene. The constriction represents the centromere.

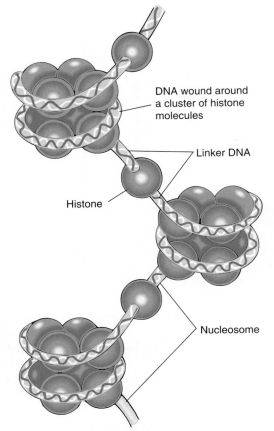

DNA wound around a cluster of histone molecules

Linker DNA

Histone

Nucleosome

Figure 4-9 Histone proteins associate into short cylinders and then DNA winds around each cylinder. The DNA-histone particles then associate into a compact arrangement.

Even with 20 different chromosomes, an organism with 30,000 genes would have an average of 1,500 genes on each chromosome, resulting in long pieces of DNA that might break if unprotected. In onion, the DNA in *each* nucleus is 10.5 meters long when all DNA molecules are measured; in lilies, it is 21.8 meters long. In eukaryotes, a special class of proteins called **histones** complexes with DNA and gives it both protection and structure (**Figure 4-9**). Chromosomes also have another structural feature, the **centromere**, which is usually located near the center of the chromosome (**Figure 4-10**). Each end is capped by a **telomere**.

During S phase, linking pieces of DNA as well as genes are replicated, and new histone molecules complex with the new DNA. Thus, entire chromosomes, not just DNA, are replicated (**Figure 4-11**). After replication is finished, the duplicate DNA molecules remain attached at the centromere, connected by a protein called cohesin. Although it would be justifiable to call this a "double chromosome" or a "pair of chromosomes," it is also called a chromosome. Now that we know how a chromosome changes during S phase, we call each half of the doubled chromosome a **chromatid**. After S phase, each chromosome has two identical chromatids; before S phase, each chromosome has just one chromatid. A chromosome after S phase is twice as large as it was before S phase.

Although many cells stop in G_1 when they cease dividing and begin to mature, most plant cells enter S phase and replicate their DNA before they begin to differentiate. This may involve just a single cycle of DNA replication, resulting in a nucleus twice as large as would be expected, or it may continue for many rounds of DNA synthesis and the nucleus becomes gigantic (**Figure 4-12**).

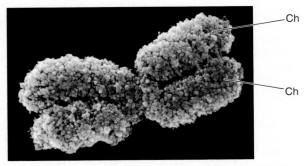

Ch

Ch

Figure 4-10 This chromosome was taken from a cell during division. It had gone through S phase of the cell cycle, so it has two chromatids (Ch), not just one. The chromatin has coiled tightly (condensed) and is visible as loops. Condensation has made the chromosome short and thick, and it can be moved around the cell easily. The pinched region in the center is the centromere.

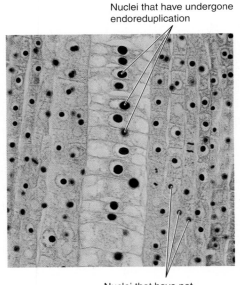

Nuclei that have undergone endoreduplication

Nuclei that have not undergone endoreduplication

Figure 4-12 In this root of *Scirpus* (bulrush, sedge), cells in the central column have enormous nuclei that have undergone endoreduplication. Each nucleus now has many copies of each gene so they can produce messenger RNA very rapidly. Consequently, the cytoplasmic ribosomes produce protein quickly as well (×200).

TABLE 4-3	Examples of Endoreduplication and Gene Amplification	
Species	Tissue	Degree*
Carex hirta	Tapetum (anther)	8 C
Viola declinata	Elaiosome	16 C
Cucumis sativus	Anther hairs	64 C
Cymbidium hybridum	Protocorm	128 C
Papaver rhoeas	Antipodal cells	128 C
Triticum aestivum	Antipodal cells	196 C
Urtica caudata	Stinging hairs	256 C
Corydalis cava	Elaiosome	512 C
Geranium phaeum	Integument	512 C
Phaseolus vulgaris	Suspensor (embryo)	2,048 C
Echinocystis lobata	Endosperm	2,072 C
Scilla bifolia	Elaiosome	4,096 C
Phaseolus coccineus	Suspensor (embryo)	8,192 C
Arum maculatum	Endosperm haustorium	24,576 C

*"C" is a symbol that represents the amount of DNA present in a nucleus; a normal body cell has 2 C before S phase and 4 C after S phase. Values that are a power of 2 represent endoreduplication; those that are not represent gene amplification.

This process is **endoreduplication**, estimated to occur in 80% of all maturing plant cells (**Table 4-3**). The resulting nucleus has many copies of each gene—as many as 8,192 copies in some cells of kidney beans (*Phaseolus vulgaris*). Endoreduplication occurs most often in hairs, glandular cells, and other cells that must have an extremely rapid, intense metabolism. The normal complement of two copies of each gene does not seem to make messenger RNA rapidly enough for such active cells. Such cells are also rich in ribosomes and produce large amounts of protein.

Gene amplification is similar to endoreduplication, but only some genes are repeatedly replicated. The amplified genes are those needed for the specialized metabolism of the mature cell (**Figure 4-13**). For example, as a protein-rich seed develops,

its cells need large quantities of the messenger RNA that codes for the storage protein, and those genes are amplified; however, it would be a waste of energy and resources to replicate all of the other genes as well. The most extreme case of gene amplification known is in *Arum maculatum* (a relative of *Philodendron*): Each nucleus has enough DNA for 24,576 copies of every gene. We do not know how many genes are being amplified and how many are present as only two copies per nucleus, and thus, we cannot compute the actual number of copies of each amplified gene; however, it is certainly more than 24,576.

■ G_2 Phase

After S phase, the cell progresses into G_2 (**gap 2**) **phase**, during which cells prepare for division. This phase lasts only about 3 to 5 hours. The α- and β-tubulins necessary for spindle microtubules are synthesized, and the cell produces proteins necessary for processing chromosomes and breaking down the nuclear envelope. In cultured animal tissue, if a cell whose nucleus is dividing is forced to fuse with one in G_1, the second cell's nucleus also begins the first steps of nuclear division. This is evidence that during G_2 the first cell produced factors necessary to start nuclear division and that these factors are located in the cytoplasm.

G_1, S, and G_2 constitute the interphase portion of the cell cycle. After G_2, division can occur.

Division Phase of the Cell Cycle

The actual division involves two processes: (1) division of the nucleus, called **karyokinesis**, and (2) division of the cytoplasm, called **cytokinesis**. There are two types of karyokinesis: mitosis and meiosis.

■ Mitosis

Mitosis is **duplication division**. It is the more common type of karyokinesis, the method any multicellular organism uses as its body grows and the number of its cells increases. It is also used by eukaryotic unicellular organisms when they are not undergoing sexual reproduction. Mitosis is called duplication division because the nuclear genes are first copied; then one set of genes is separated from the other, and each is packed into its own nucleus (**Figure 4-14**). Each daughter nucleus is basically a duplicate of the original mother nucleus and a twin of the other. Mitosis produces nuclei that are more or less exact copies of the original nucleus, except for occasional errors.

As mentioned, thousands of genes are linked together into just a few chromosomes, making separation of gene sets much easier: It is necessary to transport only one of each type of chromosome to each daughter nucleus. This is made even easier because the two new chromatids remain attached together by their shared centromere as the chromosome is replicated in S phase. It is necessary only to make certain that one half of the doubled, large chromosome goes to one end of the cell and the other half to the other end. If that happens with each chromosome, each end of the cell automatically receives one full set of genes. The mechanism that ensures this orderly separation of chromatids consists of the following four phases:

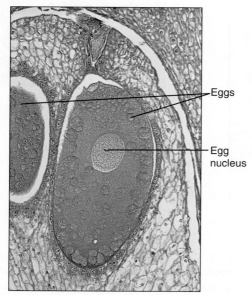

Figure 4-13   Egg cells of pines and other conifers become enormous; their nuclei are gigantic due to having large amounts of RNA. Like any egg cell nucleus, this must fuse with a sperm cell nucleus before it can divide (×20).

(labels: Eggs; Egg nucleus)

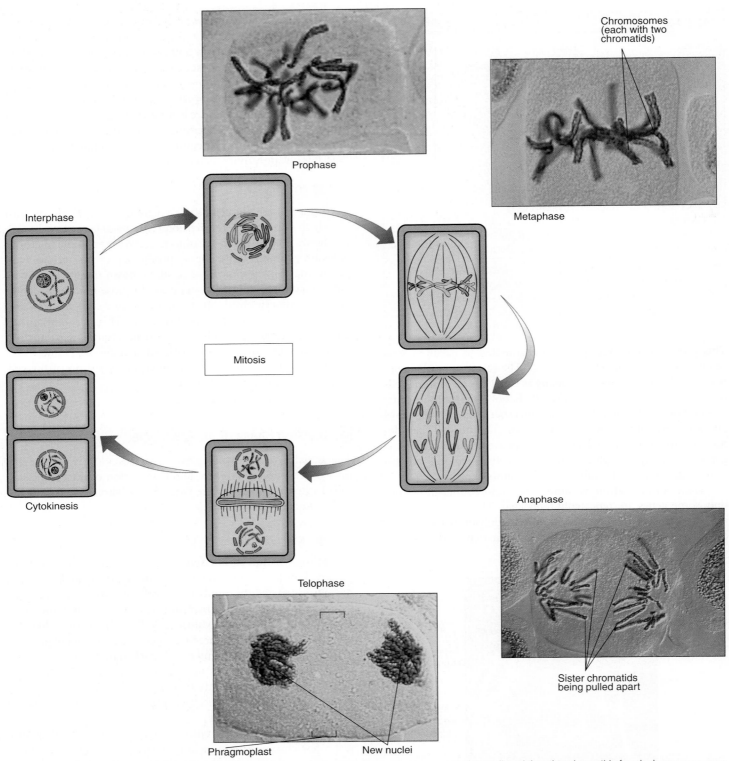

Figure 4-14 During mitotic nuclear division, one chromatid from each chromosome is pulled to one end of the cell, and the other chromatid of each chromosome goes to the other end. Details are given in the text.

Prophase During interphase, the DNA of a chromosome exists as a long, extended double helix associated with histone protein (Figure 4-9). This open configuration allows enzyme complexes to find specific genes that must be read for the information they contain, but in this condition, a chromosome may be several centimeters long. It could be wrapped around the cell hundreds of times, making it impossible to pull one chromatid from the other. During **prophase**, however, the chromosomes **condense**—they undergo a change in the way the histones associate. Chromosomes begin to coil repeatedly (Figure 4-10), becoming shorter and thicker each time. In addition, a protein framework develops to which DNA binds. **Chromosome condensation** continues

Box 4-1 Controlled Growth Versus Cancerous Growth

The actual steps of karyokinesis and cytokinesis must be controlled if cells are to divide properly, but other aspects must be regulated as well. First, the rate and frequency of cell division are important in determining how rapidly or slowly one cell produces a mass of progeny cells. Second, orientation of both cell division and cell growth affects the shape of the growing mass of cells: If all cells divide with their new walls parallel to each other, the mass grows as a column, but if new cell walls occur in any plane, the mass grows in all three dimensions. The mass grows as a sheet if new cell walls can occur in two planes but not in the third. Finally, it is important to control which cells divide: If only some cells undergo cell division, they may produce a lump or outgrowth while the rest of the mass of cells remains unchanged.

These factors are controlled accurately in plants. In a young embryonic plant, all cells divide but later, cells at the tips of roots and shoots become centers of cell division and growth while the rest of the stem and root tissues matures and carries out their functions. By controlling rate, orientation, and location of cell division, plants produce cylindrical stems and roots, thin flat leaves and petals, and massive, three-dimensional fruits.

Plants impose quiescence on certain cells and then reactivate them to growth and division later: Plants produce buds that are forced to remain quiescent for a long time, even years, but can then be stimulated by the plant to grow out as a branch or a flower. In young stems, epidermal cells grow rapidly enough to keep the stem covered, but then they mature and remain mitotically quiescent as they protect the plant. In many species, these can be reactivated years later, undergo cell division, and produce bark cells.

Cell division and growth must be controlled in animals as well. During early stages of fetal development, all cells of human embryos divide and grow. Later, some cells divide more rapidly than others, but basically, most of our cells are mitotically active during much of our growth before birth. Then cells in certain tissues and organs undergo cell cycle arrest; they mature and never divide again such as the cells of our eyes and our brains. Other cells never stop dividing and are active until we die, such as the layer of cells that produces our skin and hair and the bone marrow that generates most of our blood cells. Just as in plants, some of our cells enter a prolonged state of quiescence and later are activated to division. For example, surgical removal of part of the liver causes cells of the remaining portion to divide and restore the organ to an adequate size. Of course, this is not true of most of our organs.

Some of our cells release themselves from cell cycle arrest and begin growing uncontrolled by the rest of the body. This is cancerous growth, and its severity depends on which types of cells and organs are involved, how rapidly the cells divide, and whether the cells can migrate from their original site and invade surrounding tissues. It is well known that certain environmental factors act as carcinogens—agents that cause cancer by interfering with cell cycle arrest. Cigarette smoke is known to cause cancer of the lung and throat, and ultraviolet light triggers skin cancer.

Whereas uncontrolled cancerous growth in humans may be fatal, it does not seem to be a problem in plants. Irregular lumps and growths, called galls, may occur, but these are often caused by insects or microbes, not by the plant's own cells undergoing a spontaneous, self-induced release from cell cycle arrest. It may be that plants do form cancerous growths but that they are not a serious problem for several reasons. First, cells cannot migrate through a plant body the way that our cells migrate through our bodies; consequently, any uncontrolled growth is localized, not invasive. Second, whereas we have many organs that are each critical to our life and which occur singly (heart, brain) or in pairs (kidneys, lungs), plants have many leaves, roots, and flowers, and no single one is indispensable. Damage to one part of the plant may have little effect on the rest of the plant. By lacking the highly differentiated, complex, and tightly integrated body of humans and many animals, plants are not so threatened by diseases involving control of nuclear and cellular division.

---

until chromosomes are only 2 to 5 μm long; in this form, they can be moved around the cell more easily. In this condensed form, we can actually see the chromosomes; the uncoiled molecules of interphase are too narrow to be resolved by light microscopy.

As chromosomes condense and become visible during prophase, other events occur as well (Figure 4-14). The nucleolus becomes less distinct and usually completely disappears by the end of prophase. The nuclear envelope breaks into numerous vesicles, apparently because of the action of enzymes synthesized late in G_2. In those organisms that have a set of centrioles (mostly algae, fungi, and animals), the daughter sets that were duplicated during the previous interphase now migrate to opposite **poles** (sides) of the cell. It had been thought that they simultaneously acted as microtubule-organizing centers and produced a long set of microtubules, the **spindle**, between themselves, but even if centrioles are destroyed by laser microbeams in animals, the spindle still forms. It could be that the centrioles themselves are separated by spindle formation, and the association is a mechanism that ensures that each daughter cell receives a centriole.

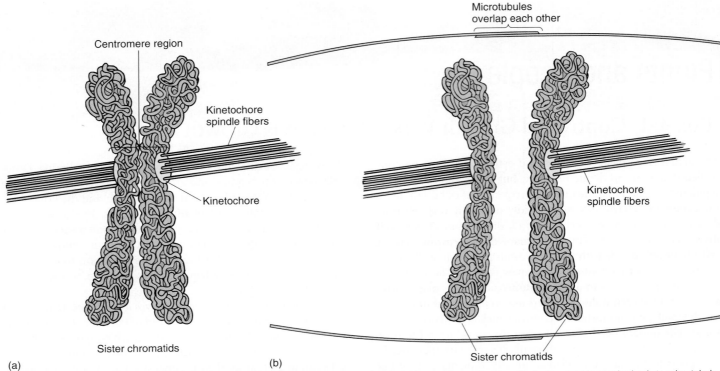

Centromere region

Kinetochore spindle fibers

Kinetochore

Sister chromatids

(a)

Microtubules overlap each other

Kinetochore spindle fibers

Sister chromatids

(b)

Figure 4-15   (a) Early in mitosis, **spindle microtubules** attach to chromosomes at the kinetochore of the centromere. Each chromatid is attached only to microtubules coming from one end of the cell. (b) Other microtubules pass from pole to pole or overlap the ends of other microtubules at the center of the spindle. As the kinetochore-attached microtubules shorten, the two chromatids of each chromosome are pulled in opposite directions. The long microtubules that do not attach to chromatids act as a framework such that chromatids are pulled apart rather than having the spindle poles pulled into the center of the cell.

The spindle is composed of hundreds or thousands of microtubules. Some microtubules extend from one pole to the center of the cell where their ends overlap the ends of other microtubules that extend from the opposite pole. The two sets together, overlapping in the center, form a large framework (**Figures 4-15** and **4-16**). Other microtubules run from a pole to a centromere. The point of attachment is a **kinetochore**; a structure consisting of two layers of proteins, one layer bound tightly to centromere DNA and the other attached to spindle microtubules. Each centromere has two kinetochore faces, one attached by approximately 15 to 35 microtubules to one end of the spindle and the other face attached by a similar number to the other end of the spindle (Figure 4-15).

Metaphase After spindle microtubules attach to centromeres, they push and pull on the chromosomes and gradually move them to the cell center; their arrangement there is called a **metaphase plate** (Figure 4-14). Viewed from the side, the chromosomes appear to be aligned in the very center, but viewed from the end, they are seen to be distributed throughout the central plane of the cell.

No distinct boundary occurs between prophase and **metaphase**; chromosomes gradually become visible and gradually move to the metaphase plate. Fragments of nucleolus and nuclear envelope may persist well into metaphase; however, a distinct boundary is evident between metaphase and the following anaphase. At the end of metaphase, a protein-degrading enzyme called separase is activated, and it digests cohesin: The two chromatids of each chromosome are suddenly free of each other. In this step, the number of chromosomes is doubled, but the size of each chromosome is halved. Each chromosome is like it was in

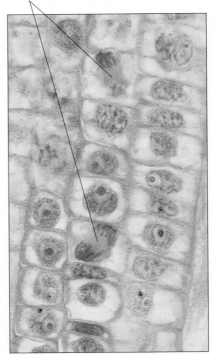

Spindle microtubules

Figure 4-16   Anaphase in hyacinth root tips. The gray material between the red chromosomes consists of masses of spindle microtubules. These roots grow quickly and cell cycle times are short, and thus, at any particular time, many cells are in mitosis rather than just in interphase (×250).

G_1, before S phase. Although the number of chromosomes in the cell is suddenly doubled, the doubling is transitory because the cell soon divides.

Anaphase **Anaphase** begins just after cohesin releases the centromeres, freeing the two kinetochore faces from each other. Spindle microtubules that run to the kinetochores shorten, depolymerizing at the end near the spindle pole, pulling each daughter chromosome away from its twin. Agents that inhibit microtubule depolymerization, such as deuterium oxide, also inhibit chromosome movement, whereas those that speed depolymerization, such as low levels of colchicine, speed movement. The energy necessary to move a chromosome from the metaphase plate to the end of the spindle is small: just 20 ATP molecules. Long chromosomes may tangle somewhat, but microtubules exert sufficient pull to untangle them and drag them to the ends of the spindle. Because the spindle is shaped like a football, as chromosomes on each side get closer to the end, they are pulled together into a compact space.

Telophase As chromosomes approach the ends of the spindle, fragments of nuclear envelope appear near them, connect with each other, and form complete nuclear envelopes at each end of the cell. The total surface area of the two new nuclei is larger than that of the envelope of the original mother nucleus; extra membrane may be derived from endoplasmic reticulum (ER). Chromosomes become less distinct because they start to uncoil. Gradually, new nucleoli appear as the ribosomal genes become active and produce ribosome subunits. The spindle depolymerizes completely and disappears. Most events in **telophase** are reversals of those in prophase.

To summarize mitosis: After G_2 is completed at the end of interphase, each chromosome has been replicated and consists of twin chromatids. Spindle microtubules from opposite poles attach to the centromeres then pull the twin chromatids of each chromosome away from each other. Two new nuclei form, each containing a full set of chromosomes and each chromosome having one chromatid. The new nuclei are identical to each other and to the original nucleus that began mitosis. The nuclei can then enter G_1 of interphase. The steps of mitosis (prophase, meta-phase, anaphase, and telophase) are part of a continuous process and intergrade with each other. Similarly, prophase and telophase intergrade with interphase.

■ Cytokinesis

The division of the protoplast is much simpler than the division of the nucleus. Although it is necessary for each daughter cell to receive some of each type of organelle, random distribution of organelles in the mother cell usually ensures this. No matter how the cell divides, each half typically contains some mitochondria, some plastids, some ER, some vacuoles, and so on. It is not necessary for each daughter cell to get exactly half of each. A single mitochondrion can divide, or a fragment of ER can grow until the cell has an adequate amount. The same is not true for genes: If one daughter cell is missing a gene or chromosome, the other genes cannot regenerate the missing information.

During prophase, a set of microtubules and actin filaments aggregates into a band running around the cell, just interior to the plasma membrane. This **preprophase band** identifies the plane of division, and it marks the region where the new cell wall will attach to the existing wall. If the preprophase band forms around the middle of a long cell, the cell will divide into two equal-sized cells, but if it forms at one end, then division will produce a large and a small cell. The preprophase band is transitory, and its microtubules quickly disassemble and are recycled into the mitotic spindle. If nuclear division occurs without cell division, a preprophase band is not formed.

Cytokinesis in plants involves formation of a **phragmoplast**, a set of short microtubules aligned parallel to the spindle microtubules. Actin filaments are also present. The phragmoplast forms in the cell center where the metaphase plate had been (**Figures 4-17 and 4-18**). Phragmoplast microtubules trap dictyosome vesicles that then fuse into a large, flat, plate-like vesicle in which two new primary walls and a middle lamella begin to form. The phragmoplast grows outward toward the walls of the original cell. It may take half a day to reach the walls of large cells. It is not known whether phragmoplast microtubules actually migrate toward the

Phragmoplast

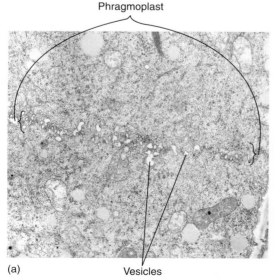

(a)

Vesicles

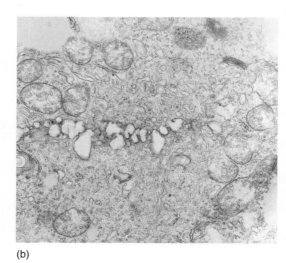

(b)

Figure 4-17 (a) Late telophase in onion. The row of small vesicles is the beginning of a cell plate. At the edges are short segments of microtubules, the phragmoplast. (b) Cell division has been interrupted by the herbicide DCPA (dimethyl tetrachloroterephalate) (both ×20,000).

Alternatives

Box 4-2 Rates of Growth

An organism, or a part of an organism, can produce more cells in a variety of ways. The two ways that are most important for you to understand are called arithmetic and geometric increase.

In **arithmetic increase**, only one cell is allowed to divide. Of the two resulting progeny cells, one continues to divide, but the other undergoes cell cycle arrest and begins to develop, differentiate, and mature. After each round of cell division, only a single cell remains capable of division, and one new body cell exists. For example, starting with a single cell, after round 1 of cell division, there is one dividing cell and one body cell. After round 2, there are two body cells, and after round 3, there are three, and so on. To obtain a plant body containing 1 million cells (which would be a very small body), the plant's single dividing cell would undergo 1 million rounds of nuclear and cellular division. If each round requires 1 day, arithmetic increase would require 1 million days, or 2,739.7 years. As you can imagine, arithmetic increase is too slow to ever produce an entire plant or animal; however, it is capable of producing the small number of cells present in very small parts of plants. For example, the hairs on many leaves and stems consist of just a single row of cells produced by the division of the basal cell, the cell at the bottom of the hair next to the other epidermal cells. The hair may contain 5 to 10 cells, and thus, all of its cells could be produced in just 5 to 10 days by divisions of the basal cell; this would be quick compared with the rest of leaf development, which might take weeks or months.

An alternative pattern of growth by cell division, **geometric increase**, results if all cells of the organism or tissue are active mitotically. Again starting with a single cell, after round 1 of cell division, there are two cells capable of cell division. After round 2, there are four, and after round 3, there are eight, and so on. The number of cells increases extremely rapidly and can be calculated as a power of 2. After round 3, there are $2^3 = 8$ cells, and after round 10, there are $2^{10} = 1,024$ cells (these calculations should be easy for you on any pocket calculator). After round 20, there are already $2^{20} = 1,048,576$ cells. If, as in our previous example, each round of cell division requires 1 day, it now takes the plant only 20 days, not thousands of years, to reach a size of 1 million cells. With geometric increase, a large plant or animal body can be produced quickly.

In fact, geometric growth is common in animals but rarely occurs in plants except when they are extremely young and small. The problem is that in plants, dividing cells are typically simple and do not carry out many of the specialized metabolic processes necessary for plant survival. They cannot be the tough, hard fiber cells that make wood strong enough to hold up a tree; they cannot be the waxy, waterproofing cells of the plant's skin that protect it from fungi, bacteria, and loss of water.

Plants grow by a combination of arithmetic and geometric increase. A young, embryonic plant grows geometrically and rapidly, with all of its cells dividing. Then cell division becomes restricted to a small group of cells at the tips of roots and shoots. After this point, growth is of the slower arithmetic type: Half of the new cells produced in each round of division remain capable of division, and the other half of the new cells develop into their mature condition and begin carrying out specialized types of metabolism. Plants are thus a mixture of older, mature cells and young, dividing cells.

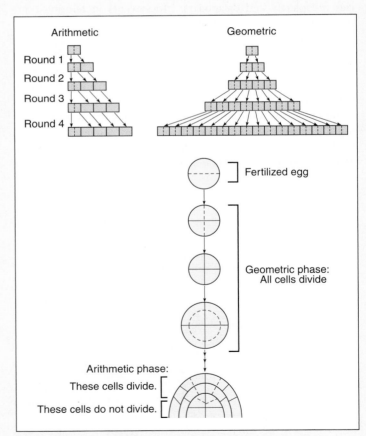

Figure B4-1 Possible types of cell division.

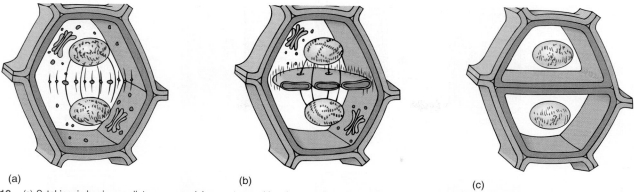

Figure 4-18 (a) Cytokinesis begins as dictyosome vesicles are trapped by phragmoplast microtubules in the space between the two new nuclei. (b) The small vesicles fuse into one large vesicle in which the new middle lamella and two primary walls will form. Plasmodesmata form at this time. (c) The cell plate enlarges toward the existing cell wall as more dictyosome vesicles fuse with the edges of the cell plate vesicle. When the cell plate reaches the existing cell walls, the vesicle membrane fuses with the plasma membrane and thus becomes plasma membrane itself. The new cell plate abuts the old cell wall, and the two become glued together with hemicelluloses and pectins.

walls. New microtubules may polymerize near the outer edge of the phragmoplast, whereas those on the inner side depolymerize. New dictyosome vesicles are trapped and added to the edge of the large vesicle; it too grows outward, following the phragmoplast. Similarly, the new walls extend outward along their edges. The phragmoplast, vesicle, and walls are called the **cell plate**. This process continues until the large vesicle meets the mother cell's plasma membrane, the two fuse, and the vesicle membrane becomes a part of the plasma membranes of the two daughter cells. Simultaneously, the new walls meet and fuse with the wall of the mother cell, completing the division of the mother cell into two daughter cells.

Many meristematic cells are small and cytoplasmic with only a few small vacuoles, but some highly vacuolate cells also divide, for example, vascular cambium cells and large cortex cells forming scar tissue over a wound. In these, the vacuole itself must be divided before either the nucleus or the cell can divide. Vacuole division is accomplished with a **phragmosome**, a set of microtubules, actin filaments, and cytoplasm. First, the microtubules and actin filaments pull the nucleus to the site where it will divide, either along the edge of the cell or often suspended in a tube of cytoplasm that extends through the vacuole. The phragmosome then enlarges as a thin plate-like sheet of cytoplasm and cytoskeleton, extending outward from the vicinity of the nucleus and pushing into the vacuole, dividing it in two. As the phragmosome advances through the vacuole, the nucleus divides, the phragmoplast and cell plate form, and then they too extend outward through the thin sheet of cytoplasm established by the phragmosome. Just after the phragmosome reaches the far side of the vacuole, the phragmoplast and cell plate pass through and join with the plasma membrane and cell wall; the cell has been divided.

■ Meiosis

In mitosis, daughter nuclei are replicates of the original mother nucleus. This is necessary for the growth of an organism but creates a problem when sexual reproduction occurs. Two **sex cells** (**gametes**) fuse together, forming a **zygote**, which then grows into a new adult. Each gamete contains one complete set of chromosomes (**Figure 4-19**). Nuclei, cells, and organisms with

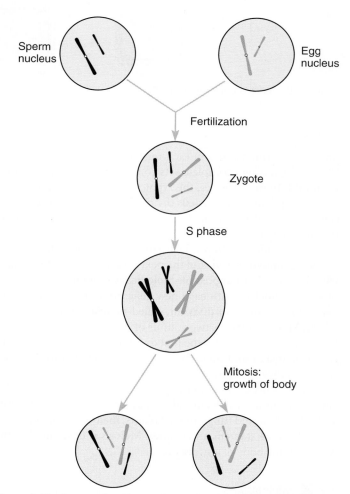

Figure 4-19 Sperm cells and egg cells are haploid, each having just one set of chromosomes; in this case, the set contains a long chromosome and a short one. After fertilization, the zygote is diploid with two complete sets: two long chromosomes and two short ones. The zygote grows into a mature plant by mitosis, so all nuclei are duplicates of the original zygote nucleus. Each is diploid, with one set each of paternal and maternal chromosomes.

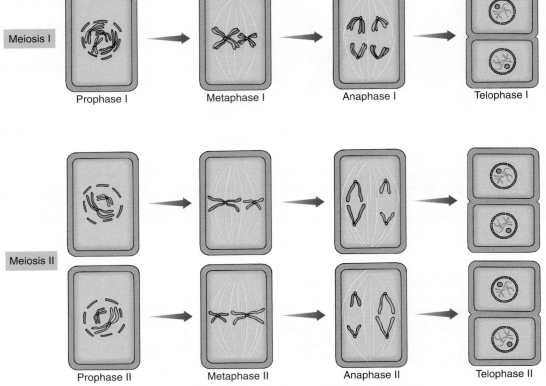

Meiosis I
Prophase I   Metaphase I   Anaphase I   Telophase I

Meiosis II
Prophase II   Metaphase II   Anaphase II   Telophase II

Figure 4-20 Meiotic nuclear division consists of two divisions without an intervening S phase. Details are given in text.

one set of chromosomes in each nucleus are said to be **haploid**. The zygote has two complete sets (one from each gamete), so it is **diploid**. In most species, the zygote grows into an adult by cell divisions in which the nucleus divides by mitosis. As a result, all cells of the adult are diploid because all nuclei are replicates of the mother nucleus.

If the adult were to produce gametes by mitosis, the gametes would be diploid as well, and the next zygote would be **tetraploid** with four sets. It is not possible biologically to double the number of sets per nucleus with each new generation; instead, it is necessary for a **reduction division**, called **meiosis**, to occur somewhere. Meiosis involves two rounds of division without allowing the S phase to occur after the first division. The two divisions are called **meiosis I** and **meiosis II**, and each contains four phases similar to those of mitosis (**Figure 4-20**).

Meiosis occurs only in the production of reproductive cells: gametes in animals and some algae and fungi and spores in plants and other algae and fungi. In seed plants, meiosis occurs in only a few cells in the stamens and ovaries. Meiosis is never used in the growth of the body of any organism.

■ Meiosis I

Prophase I All events that characterize prophase of mitosis also occur in prophase I: Nucleolus and nuclear membrane break down; centrioles (if present) separate; a spindle forms; microtubules attach to centromeres, and chromosomes condense and become visible (Figure 4-20). In addition, special interactions of

chromosomes occur that are unique to prophase I as opposed to prophase of mitosis. Because of this, prophase I is divided into five stages:

1. In **leptotene**, chromosomes begin to condense and become distinguishable, although they appear indistinct.
2. During **zygotene**, a remarkable pairing of chromosomes occurs (**Figure 4-21a**). There are two sets of chromosomes, one from the paternal and one from the maternal gamete. Because these have gone through S phase in the preceding interphase, each chromosome has two chromatids; altogether there are four sets of genes on two sets of chromosomes. During zygotene, each chromosome of one set finds and pairs with the equivalent chromosome (its **homologous chromosome**, or **homolog**) of the other set; this pairing is **synapsis**. With remarkable accuracy, the two homologous chromosomes in each pair become almost perfectly aligned from end to end. A structure, the **synaptonemal complex**, is present between the paired homologous chromosomes (Figure 4-21e). This complex is composed of a linear central protein element connected by fine transverse fibers to two lateral elements bound to the DNA of the homologous chromosomes. A synapsed pair of homologous chromosomes is called a **bivalent**.
3. As chromosomes continue to condense, they become shorter and thicker; this stage is **pachytene**. The synaptonemal complex seems to be involved in the **crossing-over** that occurs now: In several places in each chromosome, the DNA of each homolog breaks (Figure 4-21). Breaks occur in almost

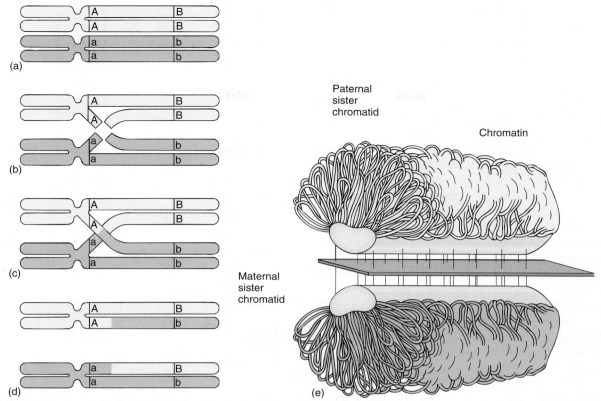

Figure 4-21 (a) During prophase I of meiosis, each paternal chromosome somehow finds and pairs with the equivalent maternal chromosome. They lie parallel to each other. (b and c) Breaks occur in similar sites on equivalent chromatids, and repair enzymes attach maternal pieces to paternal pieces, resulting in new chromatids (d). (e) The synaptonemal complex between paired homologous chromosomes. It is not known how it facilitates crossing over.

identical places in the paired homologs, and the enzymes that repair the breaks hook the "wrong" pieces together. A piece of the maternal homolog is attached to the paternal homolog, and the equivalent piece of the paternal homolog is attached to the maternal one. The full consequences of this are explained in texts on genetics and evolution. If the maternal and paternal chromosomes are absolutely identical, nothing significant has happened, but if the genes on the paternal and maternal chromosomes are slightly different, the new chromosomes that result from synapsis and crossing-over are slightly different from the original chromosomes. There are no new genes, but rather, there are new combinations of genes on each chromatid. This greatly increases the genetic diversity of all haploid cells produced: **Every sperm cell or egg cell produced by any single plant or animal (yourself included) is unique. This genetic diversity is important for evolution**.

4. After pachytene is **diplotene**. The homologous chromosomes of each bivalent begin to move away from each other but do not separate completely because they are held together at their paired centromeres and at points (**chiasmata**; singular, chiasma) where they appear to be tangled together. Some biologists believe that chiasmata are the points at which crossing-over occurred, but evidence is accumulating that chiasmata are only tangles and are not related to crossing-over at all. Under good conditions, it is possible to see all four chromatids of the paired homologous chromosomes; they are called **tetrads** at this stage.

5. In the final stage, **diakinesis**, homologs continue to separate, and chiasmata are pushed to the ends of the chromosome. The homologous chromosomes become untangled and are paired only at the centromeres.

Prophase I is the most complicated stage of meiosis; the remaining stages are simple and quite similar to the stages of mitosis.

Metaphase I Spindle microtubules move the tetrads to the center of the cell, forming a metaphase plate (Figure 4-20).

Anaphase I The homologous chromosomes separate completely from each other, moving to opposite ends of the spindle. Cohesin is not digested; thus, **centromeres do not divide, and each chromosome continues to consist of two chromatids**. Notice how this is different from the metaphase-anaphase transition of mitosis. In mitosis, separase degrades cohesin, releasing centromeres from each other, and each chromosome divides into two chromosomes, each with just one chromatid. However, in the metaphase I—anaphase I transition, homologous chromosomes separate from each other, and each still has two chromatids. One set of chromosomes is pulled away from the other set, and two new nuclei are formed. **These nuclei are now haploid** because each has only one set of chromosomes; the homolog of each chromosome in one nucleus is now in the other nucleus. Note that each chromosome in anaphase I has two chromatids; thus, each nucleus has two sets of genes, but this does not make them diploid. Any cell in G_2 has at least two sets of genes; the critical factor is to have two sets of chromosomes.

Telophase I Because the chromosomes are still doubled (each has two chromatids) and are in a G_2 state, they do not need to undergo an interphase with its G_1, S, and G_2. Also, because telophase I is basically the opposite of prophase II, some organisms go directly from anaphase I to metaphase II, skipping telophase I and prophase II. In most organisms, however, there is at least a partial telophase I in which chromosomes start to uncoil, and the nucleolus and nuclear envelope start to reappear. If the cells actually progress fully to interphase, no replication of the DNA occurs—the S phase is completely missing. This interval is called **interkinesis**. Cytokinesis may occur, but it is not unusual for this to be absent also and for both daughter nuclei (i.e., both masses of chromatin) to stay in the original, undivided mother cell.

▉ Meiosis II

If a telophase I occurs, then prophase II is necessary to prepare the nucleus for division. Prophase II is not subdivided into stages like prophase I. Metaphase II is short, and at the end of it, the centromeres divide, thereby separating each chromosome into two chromosomes, just as in metaphase of mitosis, but different from metaphase I. Anaphase II then separates each new chromosome from its replicate, and in telophase II, new nuclei are formed. Each nucleus contains just one set of chromosomes, each with a single chromatid.

To summarize, during meiosis I, each chromosome of the paternal chromosomes undergoes synapsis with the homologous chromosome of the maternal set. Crossing over results in new combinations of genes on the chromosomes. Spindle microtubules pull the paired homologs to opposite poles; therefore, at each pole, there is only one set of chromosomes, not two. The new nuclei that form temporarily after meiosis I are thus haploid. When these nuclei undergo meiosis II, the two chromatids of each chromosome are separated, as in mitosis, and the resulting nuclei each have a haploid set of chromosomes, each chromosome with only one chromatid.

In meiosis, nuclear division and cell division are often not directly linked. In some species, cytokinesis happens after both meiosis I and meiosis II, and four haploid cells result from each original diploid mother cell. In other species, no cytokinesis occurs after meiosis I, but a double cytokinesis occurs after meiosis II, again resulting in four haploid cells. For example, pollen grains can form either way, and meiosis results in four haploid cells; however, in many organisms, no cytokinesis occurs at all during the meiosis that leads to the formation of eggs. The final cell is tetranucleate and may remain that way, or three of the nuclei may degenerate and produce only a single uninucleate haploid egg cell by meiosis (**Figure 4-22**). It seems to be selectively advantageous to produce one large egg rather than four small ones.

▉ Less Common Types of Division in Plants

Cytokinesis and karyokinesis are often so closely associated that we tend to think of them as two aspects of one process. In numerous tissues, however, karyokinesis occurs without cytokinesis, and multinucleate cells are formed. The nutritive tissue of many seeds goes through a phase in which its cells are multinucleate (**Figure 4-23**), and in many algae, all cells are like this. If the cell

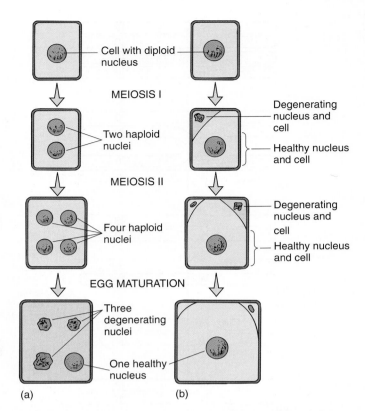

Figure 4-22 In animals and some plants, the meiosis that is part of egg production produces only one large haploid cell. (a) In some species, meiosis results in four haploid nuclei; no cytokinesis occurs and all four nuclei are temporarily in one cell until degeneration eliminates three of them. (b) In other species, after meiosis I occurs, cytokinesis places one nucleus in a small cell, and both nucleus and cell degenerate before they can undergo meiosis II. The nucleus of the larger cell does undergo meiosis II, and again, one daughter nucleus is placed in a small cell that degenerates. Only three nuclei are ever formed, and only one survives.

becomes very large and has hundreds or thousands of nuclei, it is called a **coenocyte**.

On the other hand, cell division may occur without nuclear division; this is most common in algae, fungi, and the nutritive tissues of seeds. The cells first become multinucleate and persist in that condition for some time; then cell division begins as new walls are organized around each nucleus and its accompanying cytoplasm.

In lily flowers, each haploid cell involved in producing an egg cell undergoes two rounds of mitosis without cytokinesis, resulting in a tetranucleate cell (called the first four-nucleate stage; **Figure 4-24**). Three nuclei migrate to one end of the cell, the fourth nucleus migrates to the other, then all four undergo mitosis again. The isolated nucleus divides into two nuclei as would be expected; however, as the three crowded nuclei begin to divide, their spindles fuse into a single structure. Thus, all three sets of chromatids on one side go to one pole of the spindle, and all three on the other side go to the other spindle: The result is not six haploid nuclei but instead two triploid nuclei. The cell now has two small haploid nuclei and two large triploid nuclei (called the second four-nucleate stage). The egg is always formed from one of the haploid nuclei, never one of the triploid ones.

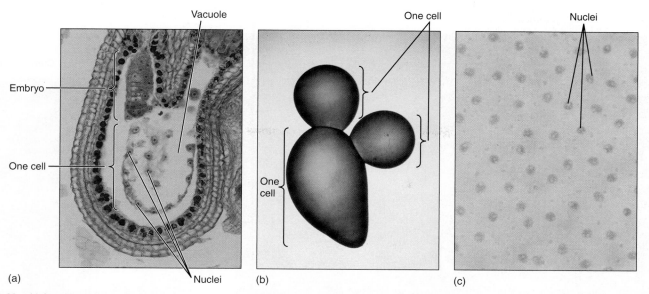

Figure 4-23 (a) A multinucleate cell in the developing seeds of *Capsella* (shepherd's purse). Although numerous nuclei have undergone mitotic divisions, the cell has not divided but has enlarged and now has a giant vacuole. The small ball of cells is the embryo, which pushes into the multinucleate coenocyte and draws nutrients from it (×80). (b) Multinucleate cells are typical of many algae. This is *Valonia*; the entire alga pictured here consists of just three giant cells (this photo is only four times life size), but each has thousands of nuclei. (c) A magnification of (b), showing some of the many nuclei, which have been stained pink (×320).

Cell Division in Algae

Nuclei

The nuclei of plants are virtually identical to those of animals in terms of structure, metabolism, mitosis, and meiosis. Apparently, the plant and animal lines of evolution diverged only after the eukaryotic nucleus had become quite sophisticated; however, nuclei are complex organelles, and mitosis and meiosis are intricate processes with many steps. Certainly plant and animal nuclei could not have arisen from prokaryotic nucleoids quickly; many intermediate steps spanning hundreds of millions of years must have been involved.

Several groups of organisms with unusual nuclear characteristics may represent lines of evolution that originated earlier in the history of eukaryotes than did plants and animals. Nuclei of many fungi have an unusual mitosis, some having both an intranuclear and an extranuclear spindle. In many fungal species, the nuclear envelope and nucleolus either do not break down at all or do so only late in mitosis. Similar behavior occurs in many groups of algae, including dinoflagellates, brown algae, euglenoids, and some green algae (**Figure 4-25a**). Gaps form in the nuclear envelope, and bundles of microtubules pass completely through the nucleus. Even more significantly, nuclei of dinoflagellates have no histones. Their chromosomes remain condensed at all times, undergoing only a slight uncoiling during interphase (Figure 4-25b). Because histone and nucleosome structure is so constant in all other eukaryotes, this unique situation in dinoflagellates is considered highly significant.

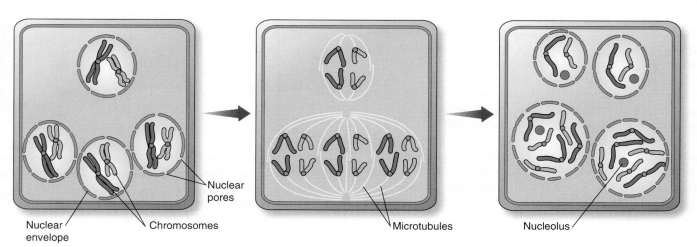

Figure 4-24 As part of egg production in lilies, certain cells go through an odd karyokinesis in which three haploid nuclei divide, but because they lie so close to each other, their spindles fuse and two new triploid nuclei are formed instead of six haploid ones.

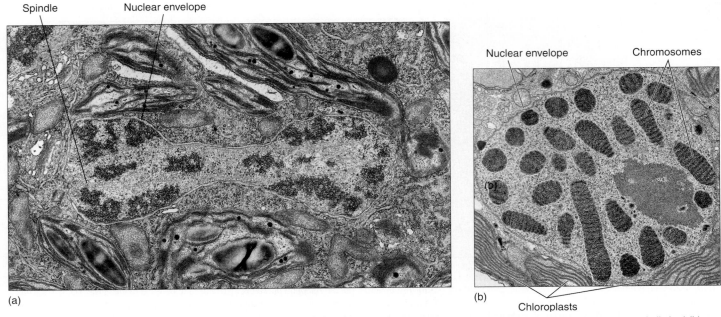

Figure 4-25 (a) Late telophase in the green alga *Cladophora*; not only is the nuclear membrane completely intact, but a portion of the intranuclear spindle is visible (×26,656). (b) An interphase nucleus of dinoflagellate. Chromosomes never completely decondense. It is not known how chemical messengers, DNA replicases, and RNA synthetases gain access to the DNA molecule (×21,000).

■ Cytokinesis

Several types of cytokinesis occur in algae. Many groups of unicellular algae have no wall, their plasma membrane being their outermost surface. During cell division, the plasma membrane pinches in two, being pulled inward as a cleavage furrow, a process remarkably similar to cytokinesis in animals (**Figure 4-26a**).

In almost all algae with walls, cell division is similar to that of plants: After telophase, a phragmoplast forms, and dictyosome vesicles establish a cell plate, which then grows radially outward—centrifugal growth—until it meets and fuses with the wall of the parental cell.

In some green algae cytokinesis occurs by a different method; the mitotic spindle depolymerizes quickly, and the two daughter nuclei lie close together. A new set of microtubules appears between them, oriented parallel to the plane where the new wall will form, which is perpendicular to the orientation of the spindle. This group of microtubules is a **phycoplast** and may be associated with division either by furrowing or by cell plate formation (Figure 4-26c and 4-26d). Phycoplasts are associated with those green algae in which the nuclear envelope does not break down during mitosis, whereas phragmoplasts are associated with the loss of the nuclear envelope. Plants arose from green algae that divide with a phragmoplast rather than a phycoplast.

In red algae cell division occurs by a phragmoplast, but the new wall grows inward from the pre-existing walls. It stops growing while still incomplete, having a large hole called a pit connection, which is filled with protein and carbohydrate material (a pit plug) (**Figure 4-27**). Otherwise, most aspects of wall formation are similar to those of other walled algae.

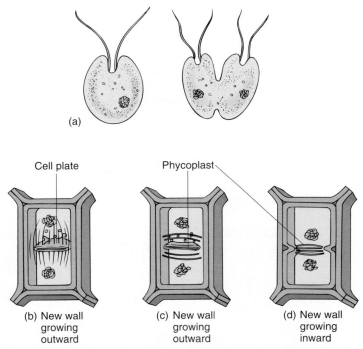

Figure 4-26 (a) In unicellular, wall-less forms, cell division occurs by infurrowing of the plasma membrane. (b) Most algae with walls divide by a phragmoplast and cell plate, just as plant cells do. (c) and (d) Some green algae have a phycoplast; the mitotic spindle disorganizes quickly, after which a new set of microtubules polymerizes perpendicular to the spindle axis. Division may then occur by cell plate growth (c) or by infurrowing (d).

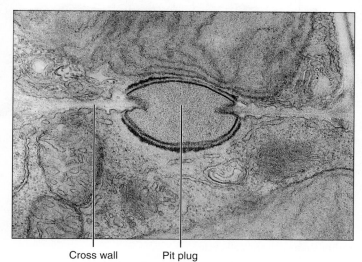

Figure 4-27 Cross walls of red algae are unusual in being incomplete, having a hole in their center. This is closed by a pit plug. *Scinaia confusa* (×15,000).

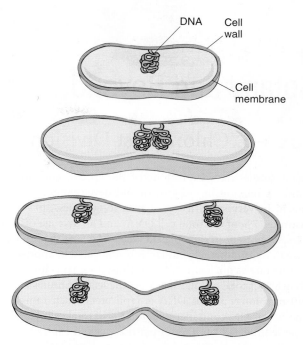

Figure 4-28 In prokaryotes, the circles of DNA are attached directly to the plasma membrane. As the cell and its membrane grow, the attached circles of DNA are separated. Cell division then occurs between the circles.

Cell Division of Prokaryotes

The events of mitosis and meiosis occur only in eukaryotes. In prokaryotes—bacteria, cyanobacteria, and archaea—mitosis and meiosis do not occur, and cytokinesis is much simpler. All genetic material occurs as one ring of DNA, which does not have any protective histones. The loop of DNA is attached to the cell's plasma membrane, and when the DNA is replicated, the two daughter loops also are both attached to the membrane (**Figure 4-28**). Because no centromere holds them together, the circles of DNA are pulled apart as the cell and its membranes grow. Several new rounds of replication may occur before cell division takes place, and thus, each cell may have as many as 20 identical circles.

Cytokinesis occurs by a process of infurrowing: The plasma membrane pulls inward and finally pinches in two. As the plasma membrane furrows inward, a new cross wall grows inward, starting from the existing wall. After completion, the cross wall splits, becoming two walls and releasing the two daughter cells; multicellular aggregates are rarely formed in bacteria and archaea. In cyanobacteria, cross walls do not break apart, and multicellular bodies are formed. The cell cycle of many bacteria can be short—only 20 minutes under ideal conditions. For many species, however, the cycle may last several days or weeks even under optimal conditions.

Division of Chloroplasts and Mitochondria

Mitochondria and plastids are constructed similarly to prokaryotes; they also contain circles of naked DNA that become separated by membrane growth. Division of the organelles is accomplished either by infurrowing or by being pulled in two (**Figure 4-29**). Because they contain DNA necessary for their

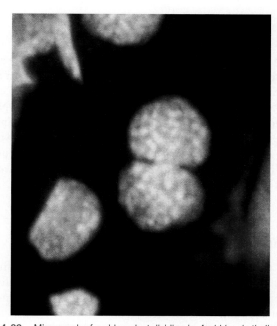

Figure 4-29 Micrograph of a chloroplast dividing in *Arabidopsis thaliana*. As the chloroplast divides, its central region is pinched inward by a contracting ring of proteins, which constricts the chloroplast into a dumbbell shape. The chloroplast shines brightly because of the way its chlorophyll reacts to the light used by the microscope: It absorbs the light and re-emits it as a different color (this is called fluorescence microscopy) (×30,000).

Botany and Beyond

Box 4-3 Chloroplast Division During Leaf Growth

Elegant studies have been done of the growth and division of plastids in relationship to growth and development of leaves in spinach. In very small leaves, 1 mm long or less, plastid DNA constitutes 7% of the total cell DNA, and an average of 76 DNA circles are present in each plastid (Table B4-3). As the leaf doubles in size to 2 mm long, plastid DNA is replicated at approximately the same rate as nuclear DNA, and thus, it remains low, approximately 8% of total cellular DNA. As the leaf continues to expand to 20 mm long, plastid DNA is replicated much more rapidly than nuclear DNA and increases to 23% of total cellular DNA. At the same time, the number of plastids per cell triples from 10 to 29, and thus, neither plastid division nor plastid DNA replication is controlled by the mechanisms that govern cell or nuclear DNA replication. At this point, each plastid has 190 DNA circles, and each cell has a total of 5,510. In the next stage of leaf growth, to 100 mm long, no synthesis of DNA occurs, and no new cells form. Instead, those already present expand; however, plastids continue to divide even though they are not making any more DNA. Con-

sequently, the number of DNA circles per plastid drops from 190 to 32, whereas the number of plastids per cell increases from 29 to 171. From these data, it is reasonable to form the hypothesis that plastid growth, DNA replication, division, and development are correlated predominantly with tissue or organ development rather than with the cell cycle.

TABLE B4-3	Development of the Plastid Genome in Spinach Leaves			
Leaf Size	1 mm	2 mm	20 mm	100 mm
Genome copies per plastid	76	150	190	32
Plastids per cell	10	10	29	171
Genome copies per cell	760	1500	5510	5470
Plastid DNA as percentage of total	7%	8%	23%	23%

Data from Scott, N. S., and J. V. Possingham, 1983. Changes in chloroplast DNA levels during growth of spinach leaves. *J. Experimental Bot.* 34:1756–67.

growth and functioning, each daughter cell must receive at least one mitochondrion and one plastid during cytokinesis; if not, the cell lacks that organellar portion of its genome and cannot produce the organelle. This happens occasionally with plastids and is often not a serious problem; the cell survives by importing sugar from neighboring cells and can grow along with the tissue. All daughter cells also lack plastids, so if this occurs in a young leaf, a white spot forms. The same phenomenon probably occurs with mitochondria but is more difficult to detect.

The metabolic stimulus that triggers the replication of nuclear DNA is not the stimulus that controls replication of organellar DNA. During the cell cycle, replication of nuclear DNA is episodic, occurring as discrete episodes that occupy a small portion of the total cell cycle, S phase; however, the "duplication" of the rest of the cell seems to be continuous: The volumes of plastids, mitochondria, cytosol, endoplasmic reticulum, and other organelles appear to increase gradually and steadily throughout interphase rather than in discrete episodes (Box 4-3).

Summary

I. Cells have a "life cycle," the cell cycle, consisting of alternating phases of growth and division. In organs that live just briefly, cells cycle only while the tissue is young and then enlarge and differentiate.

2. The interphase portion of the cell cycle consists of G_1, S, and G_2. During G_1, the nucleus controls cell metabolism. During S, the DNA is replicated; in G_2, the cell prepares for division.

3. Each DNA molecule contains thousands of regions—genes—that contain the information for proteins. A chromosome is a complex in which histone proteins bind to DNA and stabilize it. Each chromosome consists of a centromere and either one chromatid before S phase

or two chromatids after S phase. Chromatid ends are telomeres.

4. Cell division is cytokinesis, and nuclear division is karyokinesis. The words "mitosis" and "meiosis" technically refer to only karyokinesis.

5. During mitosis, spindle microtubules pull the two chromatids of each replicated chromosome to opposite ends of the cell. Because the two chromatids are virtually identical, the two new nuclei are also almost identical to each other.

6. Nuclei of any ploidy level—haploid, diploid, and higher ploidies—can undergo mitosis. During growth of the plant body, all nuclear divisions are mitotic divisions.

7. Meiosis is reduction division: diploid cells produce haploid cells. Just as important, genetic diversity is increased by meiosis.

8. During prophase I, homologous chromosomes pair (synapse) and exchange pieces of chromatid (crossing-over). At the end of prophase I, the nucleus no longer has purely paternal and purely maternal chromosomes. During anaphase I, homologous chromosomes are pulled away from each other, reducing the number of sets of chromosomes to half the original number. Haploid, triploid, and all other odd-ploid nuclei cannot undergo meiosis.

9. Meiosis occurs only in the production of reproductive cells, either gametes or spores. Meiosis is never used in the growth of the body of any organism.

10. Cytokinesis in plants and algae occurs by the formation of a large vesicle between the forming daughter nuclei. Within the vesicle, the middle lamella and two new walls form and expand until they reach existing walls.

11. Cytokinesis and karyokinesis are usually closely coordinated, but nuclear division can occur without cell division, resulting in multinucleate cells.

Important Terms

anaphase	gametes	phragmosome
arithmetic increase	genome	phycoplast
cell cycle	geometric increase	preprophase band
cell plate	haploid	prophase
centromere	histones	reduction division
chromatid	homologous chromosome	resting phase
chromosome	interphase	S phase
chromosome condensation	karyokinesis	spindle
crossing-over	kinetochore	spindle microtubules
cytokinesis	meiosis	synapsis
diploid	metaphase	telophase
duplication division	metaphase plate	zygote
G_1 phase	mitosis	
G_2 phase	phragmoplast	

Review Questions

1. In a woody plant such as a tree, which parts have cells that live only briefly and die quickly? Which parts have cells that live for several years?

2. Some cells never stop dividing. Give two examples of cells like this.

3. Interphase is also called the resting phase of the cell cycle. Why was it given that name?

4. What are the main activities of a cell while it is in G_1 phase? What kinds of organisms have short G_1 phases, and how long does a short G_1 phase last? What kinds of organisms have long G_1 phases, and how long do they last?

5. What is the main activity of the S phase of the cell cycle? What does "S" stand for?

6. Table 4-2 gives the number of chromosomes in a haploid set of chromosomes. What is the lowest number in the table? What is the highest number? Most cells in plants are diploid, having two sets of chromosomes, so the number per nucleus should be doubled. How many chromosomes are present in each diploid potato nucleus? If a potato plant has 1 million cells, each with one diploid nucleus, how many chromosomes are present in the entire plant?

7. How many chromatids does a chromosome have before S phase of the cell cycle? How many does it have after S phase?

8. Examine Table 4-1. How many hours does the cell cycle last in the root tips of corn and in onion? Which plant has cells that divide more quickly? How many cell cycles could corn and onion roots undergo in 4 weeks (the cell cycles in Table 4-1 are given in hours, not days)? If after a cell divides both daughter cells could divide and then their daughter cells could divide, how many corn cells and how many onion cells would be present at the end of 4 weeks if you started with just one cell of each?

9. What are the four phases of the cell cycle? What is the principal activity in the cell during each phase? Can any phase be eliminated or bypassed?

10. Why is mitosis called duplication division and meiosis called reduction division? What is reduced and what is duplicated: chromosomes, number of chromosomes, or number of sets of chromosomes?

11. What does it mean when chromosomes are said to condense during prophase of mitosis? How long are chromosomes after condensation is complete? How big is a typical dividing cell in a root or shoot? If a chromosome were still 20 μm long after condensation, would it be possible for division to pull half of it to one end of the cell and the other half to the other end?

12. What is the name of the set of microtubules that pull chromosomes apart? What is the name of the attachment point between microtubules and chromosomes? What is the name of the ends of the cells where the chromosomes are pulled?

13. What is a metaphase plate? What is duplicated at the end of metaphase? When this is duplicated, chromatids become free of each other. How many chromatids are there per chromosome in prophase? How many chromatids per chromosome after metaphase?

14. Imagine a nucleus that has 10 chromosomes. How many chromosomes does it have before prophase begins? How many chromosomes does each daughter nucleus have after telophase has been completed? How many chromatids are present in the one mother nucleus before prophase begins? How many chromatids are present in each daughter nucleus after telophase has been completed? How many chromatids are present if you add together all chromatids in both daughter nuclei after telophase has been completed?

15. What are the four phases of mitosis, and what is the principal activity in the nucleus during each phase?

16. Draw a single, imaginary chromosome as it would appear just as mitosis is ending. Now describe what happens to it during interphase and then during mitosis. Be especially careful to consider how many chromatids and how many copies of each gene it has at each stage.

17. How does cytokinesis occur in plants? Which organelle produces vesicles that fuse to form the cell plate? What membrane is transformed into new plasma membrane?

18. Many people consider algae to be plants even though algae do not have roots, stems, and leaves, but in many algae, cell division is different from that in true plants. Do most algae have a phragmoplast? What is the name of the structure they use?

19. Sex cells are also called _____. If a species has males and females (not all species do), males produce sex cells called _____ cells, and females produce sex cells called _____ cells.

20. How many sets of chromosomes does each sperm cell carry? How many does each egg cell have? Gametes such as sperm cells and egg cells are said to be (circle one: diploid, haploid). After a sperm and an egg have fused, the new cell is called a _____, and it is (circle one: diploid, haploid).

21. If a plant has diploid cells with 20 chromosomes in each nucleus, how many chromosomes does each of its gametes have? What is the name of the division that reduces the number of chromosomes from 20?

22. What are the five stages of prophase I, and what is the principal activity of the nucleus and chromosomes during each stage?

23. During zygotene stage of meiosis I, a remarkable pairing of chromosomes occurs. Look at the bottom part of Figure 4-19 in which "mitosis: growth of the body" has produced cells with four different types of chromosomes. The black ones came from the sperm nucleus. The red ones came from the egg nucleus. If one of these nuclei were to undergo meiosis, which chromosomes would pair during zygotene (which are homologous)—the long black one with the short black one or the long black one with the long red one?

24. What is the name of the structure that holds homologous chromosomes together after they have paired during zygotene?

25. During pachytene, crossing over occurs. Do chromosomes actually break during this process?

26. How does anaphase I of meiosis differ from anaphase of mitosis? Does the number of chromatids per chromosome change in anaphase I? Does the number of sets of chromosomes per nucleus change in anaphase I?

27. Draw all stages in the cell cycle and meiosis for a nucleus that has just one pair of homologous chromosomes, and then do the same for a nucleus that has three different types of chromosomes (six chromosomes in three sets of homologs). Draw all stages (this is not easy).

28. If a cell undergoes nuclear division but not cell division, it becomes multinucleate. If it has hundreds of nuclei, it is called a _____.

29. Do prokaryotes undergo meiosis or mitosis? How do their cells divide?

30. Describe the process of division that produces more plastids and mitochondria.

BotanyLinks

http://biology.jbpub.com/botany/5e
BotanyLinks contains a variety of resources and review material designed to assist in your study of botany.

Visit the Web Exercises area of BotanyLinks to complete this question:

1. How many chromosomes do plants have? Go to the Botany Links home page to investigate.

Tissues and the Primary Growth of Stems

chapter 5

Concepts

The body of an herb contains just three basic parts: leaves, stems, and roots (Figure 5-1). When the first land plants evolved about 420 million years ago, they were basically just algae that either washed up onto a shore or were left there as lakes and streams evaporated. They had no roots, stems, or leaves, and they just lay on the mud. As the shores gradually became crowded with such plants, some grew over others, shading them. Any plant that had a mutation allowing it to grow upright into the sunshine above the others had a selective advantage. However, being upright is not easy: Elevated cells are out of contact with the moisture of the mud; therefore, water must be transported up to them. Elevated tissues act as a sail and tend to blow down, so supportive tissue is necessary. Absorptive cells in mud are

Chapter Opener Image: Potatoes (*Solanum tuberosum*) are an unusual type of stem called a tuber. Tubers are short and broad, they stop elongating when they reach a particular length, and they have tiny, almost unrecognizable leaves because they grow underground where there is no light for photosynthesis. A potato plant has other, ordinary stems that grow upright above ground with large leaves that carry out photosynthesis. The plant transports excess sugars downward from the photosynthetic leaves to the tubers where it is stored as starch—underground where animals cannot see it. The tuber's "eyes" are buds that can germinate and establish new potato plants in the springtime, which is useful because the above-ground part of the plant usually dies in the winter. Because angiosperms have diversified so much, some use stems for one thing, others use stems for another, so we must be careful when thinking about the functions of organs.

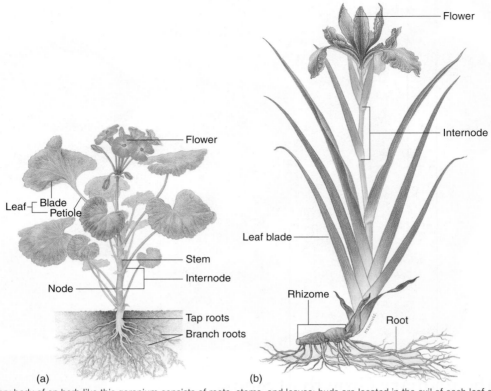

(a) (b)

Figure 5-1 (a) The primary body of an herb like this geranium consists of roots, stems, and leaves; buds are located in the axil of each leaf and may grow to be either a vegetative branch or a set of flowers. (b) This *Iris* is also an herb and never produces wood or bark. Almost all flowering plants are either broadleaf plants (eudicots) like the geranium or plants with grass-like leaves (monocots) like the *Iris*.

shaded and cannot photosynthesize; thus, sugars must be transported down to them. Shortly after plants began living on land, distinct, specialized tissues and organs began to evolve.

As early populations of land plants continued to evolve and became taller, their stems functioned primarily as transport and support structures, as they do still. But, stems of modern flowering plants have additional roles. They produce leaves and hold them in the sunlight, and during winter, they store sugars and other nutrients, such as the sugary sap of maples. Stems may also be a means of survival: Underground bulbs and corms remain alive when above-ground leaves die. Stems of many species are a means of dispersal. They spread as runners or vines, or pieces break off and are carried by animals or water to new areas where they sprout roots and grow into new plants.

Although all flowering plants possess leaves, stems, and roots, these parts have been modified so extensively in some species that they may not be recognizable without careful study. For example, cacti are often described as leafless, but they actually have small green leaves between 100 and 1000 μm long (**Figure 5-2**). Large, broad leaves would be selectively disadvantageous for these desert plants because the extensive surface area of such leaves gives up so much water to the dry air that the plant would desiccate.

Similarly, all flowering plants have stems, but in some, they are only temporary, reduced structures. Orchids such as *Campy-locentrum pachyrrhizum* and *Harrisella porrecta* consist of a mass of green photosynthetic roots connected to a tiny portion of stem; roots constitute almost the entire plant body (**Figure 5-3**). The

Figure 5-2 This prickly pear (*Opuntia*) shows that one plant can have two types of shoot: the "pad" is the main shoot, and the spine clusters are highly modified axillary branch shoots. One of the spine-bearing branches has been stimulated to develop into the first type of shoot and become a pad-like branch. The plant also has two types of leaves: (1) small fleshy green leaves on the young buds and (2) spines on the axillary branches.

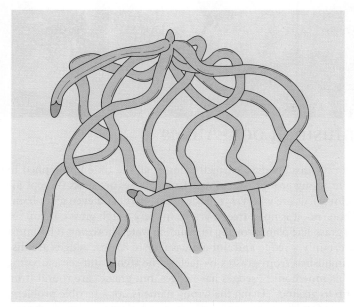

Figure 5-3 This orchid plant of the genus *Polyradicion lindenii* is composed almost completely of photosynthetic roots; only a small portion of shoot remains. Unlike most roots, these occur above ground and are green, being rich in chloroplasts.

Figure 5-4 Most epiphytic orchids resist the stresses of temporary drying because their shoots are fibrous and have a thick cuticle composed of cutin and wax.

shoot becomes active only when flowers are to be produced. One hypothesis is that this unusual body evolved because the ancestors of these species had roots that were more resistant to water stress than their stems were. Such plants could occupy drier habitats if either of two things happened: (1) Mutations were selected that caused stems and leaves to be more water conserving—this happened in most orchid species (**Figure 5-4**), or (2) mutations occurred that enhanced the root's ability to absorb carbon dioxide and carry out photosynthesis. Although thousands of plant species can withstand harsh conditions, only a handful do so by being "shootless" and having photosynthetic roots. Mutations that permit this type of body may be either rare or generally detrimental.

Some plants in the bromeliad family are nearly rootless. In the coastal deserts near Lima, Peru, fog is frequent, but rain never falls. Because the soil is always dry, roots are of little use. Plants of *Tillandsia straminea* of that region are small, herbaceous vines that lie on top of the soil (**Figure 5-5**). Plants absorb moisture through leaves made wet by fog, and they derive minerals from wind-blown dust that dissolves on moist leaf surfaces. The plants are not anchored to the soil but roll over coastal dunes as the wind blows them. Such a lifestyle would be impossible for a tree or bush because large plants could not absorb enough water or minerals without roots.

In each plant described, leaves, stems, or roots have become highly modified by natural selection, permitting survival in unusual habitats; however, in no species have any of these organs been completely lost evolutionarily. We must assume that the organ carries out some essential function. In "shootless" orchids, the residual shoot is necessary for flowering and sexual reproduction. Cactus leaves are involved in the formation of buds that produce the defensive spines. Although roots of *T. straminea* neither absorb water and minerals nor anchor the plant, they may be essential for production of critical hormones, as are roots of other plants. In all cases, analysis of many aspects of the plant's

biology reveals how the structure and metabolism of a particular organ are adaptive, how modifications affect other plant parts, and how organs have subtle functions not always obvious in more "typical" plants.

The flowering plants discussed here are formally classified as the division Magnoliophyta, but they are known informally as **angiosperms**. This group consists of approximately 235,000 species and is the largest division in the plant kingdom. Early angiosperms

Figure 5-5 These bromeliads are not rooted into the coastal sand dunes—a strong wind can blow them around. All water is absorbed from fog condensing on the leaves.

Alternatives

Box 5-1 Familiar Plants and Some Confusing Look-Alikes

Most of us have no trouble recognizing a very large number of plants, at least while they are in bloom or if certain characteristic features are present. Apples, oranges, peaches, and pears are fruits we all recognize, and if they are still on their trees, we have no trouble recognizing the whole plant as well (Figure B5-1a). Strawberries are very familiar, but you might be initially uncertain if you saw them growing on the soil surface, supported by thin, delicate stems (such stems are runners). Daisies (often called asters) are also usually easy to recognize because even though there are hundreds of species, most all look like the one in Figure B5-1b. The same is true for morning glories (Figure B5-1c), lilies, waterlilies, and many others (Figure B5-1d).

In contrast, several groups of plants can cause confusion. Many people believe any plant that is succulent and spiny is a cactus (Figure B5-1e), whether they are looking at an *Agave* (Figure B5-1f), a *Yucca* (Figure B5-1g), an *Aloe* (Figure B5-1h), or one of the succulent spurges ("spurge" may be an unfamiliar name and refers to members of the genus *Euphorbia* [Figure B5-1i)]). These are all succulent, most are spiny, and most live in deserts, but agaves and yuccas are much more closely related to lilies and irises than to cacti (their long, thick leaves are extreme versions of lily leaves). Furthermore, although many botanists also mistakenly call spurges cacti, the spurges are in a completely different family and are mostly native to Africa, whereas cacti are native to the Americas. I heard a person guess that a giant *Agave americana* must be an *Aloe vera*: He made a really good guess because the plants do look alike, even though that agave is hundreds of times larger than an aloe.

Does it really matter if people confuse these plants? Not really, because they are correct that the plant is a large, succulent spiny desert plant. If it is important to have the correct name, they can ask at a garden center or a botany department or perform a Google image search.

Other plants that are often confused with each other are roses and camellias (the flowers are quite similar; Figure B5-1j) and palm trees, cycads, and ferns (some have similar shapes and leaves; in fact, in areas where cycad are common, people often use cycad leaves instead of palm leaves on Palm Sunday; Figure B5-1k). People often describe most aquatic plants as "moss," but true mosses almost never grow in water: Long stringy ones that float are usually algae; those with stems, leaves, and flowers are flowering plants; and a few very small water plants are ferns. Many of us (even professional botanists) use the word "moss" for any very tiny plant whether it is a moss or just a small plant that resembles a moss (Figure B5-1l and Figure B5-1m).

"Grass" is used correctly almost all the time but applied to the wrong groups sometimes. Lawn grasses are grass (except for clover), as are wheat, rye, barley, rice, corn, and even giant bamboo (no, it is not a tree; it is a giant, very tough grass). If you see a grass-like plant growing in standing water, however, it is almost certainly a sedge and not a grass (Figure B5-1n). Sedges are distinguished from grasses by feeling the stems and remembering this mnemonic: "sedges have edges, but grasses are round from tip to ground." Using the wrong name is not a terrible problem; few of us know the names of very many plants, and as we discuss plants with our friends, using exactly the right name is not often necessary. It is like using the word "bug" for spiders, insects, and various other tiny animals.

There is a similar situation with the words we use for plant parts. You will often hear the word "frond" when people talk about the leaves of palms or ferns. "Frond" is just fine, but it is not a precisely defined term, it is just that for some reason, some people use it instead of the word "leaf" for palms and ferns. Similarly, "stalk" is not a precisely defined, technical term, and neither are "rind," "peel," "skin," and several others. Just keep in mind that some precise botanical terms such as "stem" and "leaf" are the same as ordinary English words, and some like "epidermis" are identical to the terms used by zoologists, even though a plant epidermis is completely different from an animal epidermis. If you have already studied a lot of zoology, be very careful as you come across botanical words that are the same as zoological or medical words: The meanings are probably very different.

Plants have common names and scientific names, and a great deal of effort is made to ensure that each species of plant or animal or any other organism has only one scientific name. For some plants the name of the plant's genus is the same as its common name; agaves are in the genus *Agave*, and citruses are in the genus *Citrus*. When we write the common name, we do not capitalize or italicize it, but we do if we are referring to the scientific name of the genus: oranges, lemons, and limes are all citruses in the genus *Citrus*. Lemons are *Citrus limon*, oranges are *Citrus sinensis*, and limes are *Citrus latifolia*. It is natural to think of "limon" as the species name of lemons, but that is not correct; the species name is *Citrus limon*. The scientific name of all species always has two words: The first is the genus name and the second is the species epithet; therefore, the species name is the genus + species epithet. When I mentioned agaves above I used the common name agave, the genus name Agave, and the scientific name of the giant agave, *Agave americana*.

Figure B5-1 (a) An orange tree. (b) Daisies. (c) Morning glories. (d) Oxblood lilies. (e) All the tall or globe-shaped plants are cacti; those with thick gray, yellow, or green leaves are agaves. (f) *Agave americana*. (g) *Yucca torreyi*. (h) Aloes. (i) A spurge (*Euphorbia*, not a cactus). (j) *Camellia* flower. (k) A cycad, often called a "sago palm." (l) *Selaginella peruviana* (despite the species epithet this plant grows in Texas, not Peru) (not a moss). (m) A liverwort (*Frullania inflata*, not a moss). (n) A sedge, not a grass.

TABLE 5-1	Types of Plant Body

Primary plant body

Derived from shoot and root apical meristems

Composed of primary tissues

Constitutes the herbaceous parts of a plant

An herb consists only of a primary plant body.

Secondary plant body

Derived from meristems other than apical meristems

Composed of secondary tissues: wood and bark

Constitutes the woody, bark-covered parts of a plant

A woody plant has primary tissues at its shoot and root tips, and a seedling consists only of primary tissues. But after a few months, wood and bark arise inside the primary tissues of stems and roots.

TABLE 5-2	Three Basic Types of Plant Cells and Tissues, Based on Cell Wall
Parenchyma:	Thin primary walls. Typically alive at maturity. Many functions.
Collenchyma:	Unevenly thickened primary walls. Typically alive at maturity. Provide plastic support.
Sclerenchyma:	Primary walls plus secondary walls. Many dead at maturity. Provide elastic support and some (tracheary elements) are involved in water transport.

diversified into several groups that are now known as basal angiosperms, eudicots, and monocots. Most **basal angiosperms** probably will not be familiar to you except for waterlilies, magnolias, and laurels. **Eudicots** are broadleaf plants such as roses, asters, maples, and others, and **monocots** are grasses, lilies, cattails, palms, philodendrons, bromeliads, and several others.

This chapter began by stating that the body of an herb contains three parts; it is necessary to explain more precisely what an herb is. Plant bodies are of two fundamental types: an herbaceous body, also called a **primary plant body**, and a woody body, known as a **secondary plant body**. An **herb** is a plant that never becomes woody and covered with bark; it often lives for less than a single year (snapdragons, petunias, beans, corn, and wheat), although many live and grow for years (irises and lilies). Its tissues are primary tissues. In woody plants such as trees and shrubs (oaks, maples, magnolias, roses, and boxwood), the wood and bark are secondary tissues (Table 5-1). A few plants are surprising: Palm trees are large, perennial, and very hard, but they do not actually have wood; thus, they are giant herbs, not woody plants.

Basic Types of Cells and Tissues

Despite the diversity of types of stems that have originated by natural selection, all share a basic, rather simple organization. The same is true for leaves and roots. Although we might suspect that a plant contains numerous types of cells, actually, all plant cells are customarily grouped into just three classes based on the nature of their walls: **parenchyma**, collenchyma, and sclerenchyma.

Parenchyma

Parenchyma cells have only primary walls that remain thin (Table 5-2). **Parenchyma tissue** is a mass of parenchyma cells. This is the most common type of cell and tissue, constituting all soft parts of a plant. Soft leaves, petals, fruits, and seeds are composed almost completely of parenchyma. Parenchyma cells are active metabolically and usually remain alive after they mature. Numerous subtypes are specialized for particular tasks (Figure 5-6).

Chlorenchyma cells are parenchyma cells involved in photosynthesis; they have numerous chloroplasts, and their thin walls allow light and carbon dioxide to pass through to the chloroplasts. Other types of pigmented cells, as in flower petals and fruits, also must be parenchyma cells with thin walls that permit the pigments in the protoplasm to be seen.

Glandular cells that secrete nectar, fragrances, mucilage, resins, and oils are also parenchyma cells; they typically contain few chloroplasts but have elevated amounts of dictyosomes and endoplasmic reticulum. They transport large quantities of sugar and minerals into themselves, transform them metabolically, and then transport the product out.

Transfer cells are parenchyma cells that mediate short-distance transport of material by means of a large, extensive plasma membrane capable of holding numerous molecular pumps. Unlike animal cells, plant cells cannot form folds or projections of their plasma membranes; instead, transfer cells increase their surface area by having extensive knobs, ridges, and other ingrowths on the inner surface of their walls (Figure 5-7). Because the plasma membrane follows the contour of all these, it is extensive and capable of large-scale molecular pumping.

Some parenchyma cells function by dying at maturity. Structures such as stamens and some fruits (such as pea pods) must open and release pollen or seeds; the opening may be formed by parenchyma cells that die and break down or are torn apart. Large spaces are often needed inside the plant body, for example, to allow gases to diffuse within leaves (Figure 5-6b and Figure 5-10e); many such spaces are formed when the middle lamella decomposes, and cells are released from their neighbors. In other cases, the space is formed by the degeneration of parenchyma cells. In a few species, such as milkweeds, as parenchyma cells die, their protoplasm is converted metabolically into mucilage or a milky latex.

Parenchyma tissue that conducts nutrients over long distances is phloem; it is discussed later in this chapter.

Parenchyma cells are relatively inexpensive to build because little glucose is expended in constructing the cellulose and hemicellulose of such thin walls. Most parenchyma cell walls are only 80 to 100 nm thick, with just 5 to 10 layers of cellulose microfibrils. Each molecule incorporated into a wall polymer cannot be used for other functions such as generating ATP or synthesizing proteins. Consequently, it is disadvantageous to use a cell with thick walls any time one with thin walls would be just as functional. Most leaves are soft, composed almost entirely of parenchyma, and are therefore not very expensive metabolically. After several weeks of photosynthesis, they replace the sugar used in their construction, and all photosynthesis after that point is net gain for the plant.

Collenchyma

Collenchyma cells have a primary wall that remains thin in some areas but becomes thickened in other areas, most often in the corners (Figure 5-8). The nature of this wall is important

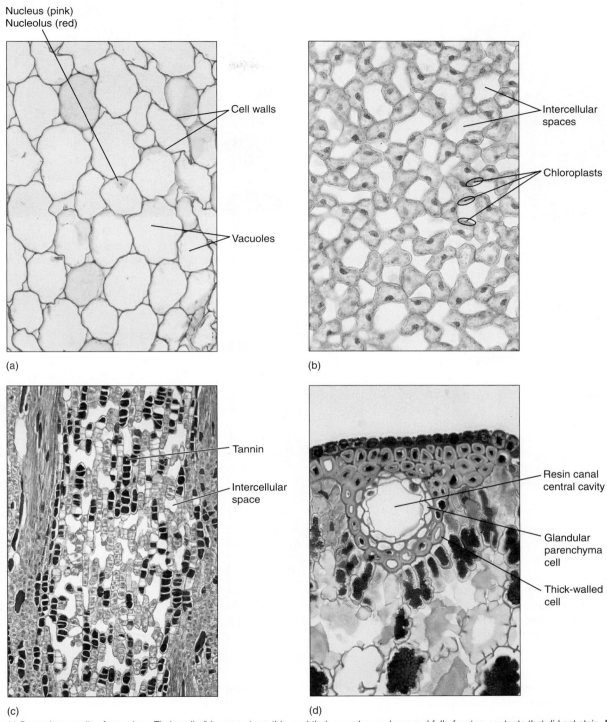

Figure 5-6 (a) Parenchyma cells of geranium. Their walls (blue green) are thin, and their vacuoles are large and full of watery contents that did not stain. Nuclei were present in all cells, but because these cells were so large and the section (slice) was cut so thin, most nuclei were cut away during the preparation of this slide. One nucleus is still present (×160). (b) Chlorenchyma cells from a leaf of privet. Because these cells are small and the section is thicker than that in (a), most cells still have nuclei (red). The structures close to the wall (blue) are chloroplasts. The large white areas are intracellular spaces where the cells have pulled away from each other, permitting carbon dioxide to diffuse rapidly throughout the leaf (×160). (c) Material taken from the center of a pine (*Pinus*) stem. Cells that have stained dark purple are filled with chemicals called tannins; these are bitter and deter insects from eating the tissue (×50). (d) A resin canal in a pine leaf. The white area is the central cavity where resin is stored, and cells that line the cavity are glandular parenchyma cells that synthesize and secrete resin. The innermost cells have thin walls, which permit movement of resin from the cells to the cavity. The outer cells have thick walls, which provide strength; cells with thick walls are not parenchyma cells (×160).

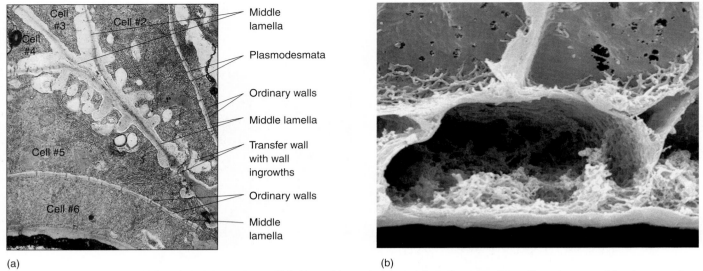

Figure 5-7 (a) Transfer cells in the salt gland of *Frankenia grandifolia*. The wall ingrowths increase the surface area of the cell membrane, providing more room for salt-pumping proteins in the membrane (×20,000). (b) These transfer cells were frozen, then broken open, and all of the protoplasm was washed away with detergent. Only the cell wall is left. All the knobby irregular areas are wall ingrowths (×5,000).

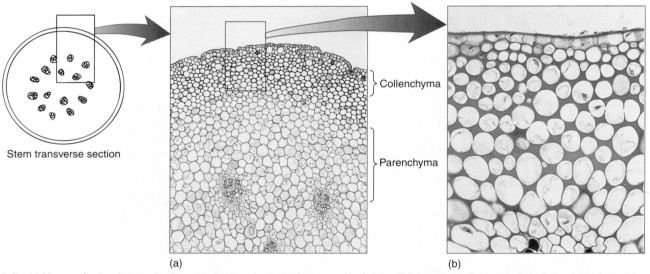

Figure 5-8 (a) Masses of collenchyma cells often occur in the outer parts of stems and leaf stalks. This is part of a *Peperomia* stem. Collenchyma forms a band about 8 to 12 cells thick. The inner part of the stem is mostly parenchyma (×50). (b) In collenchyma cells, the primary wall is thicker at the corners so the protoplast becomes rounded. No intercellular spaces are present (×150).

in understanding why it exists and how it functions in the plant. Like clay, the wall of collenchyma exhibits plasticity, the ability to be deformed by pressure or tension and to retain the new shape even if the pressure or tension ceases. Collenchyma is present in elongating shoot tips that must be long and flexible, such as those of vining plants like grapes. It is present as a layer just under the epidermis or as bands located next to vascular bundles, making the tips stronger and more resistant to breaking (**Figure 5-9**); however, the tips can still elongate because collenchyma can be stretched. In species whose shoot tips are composed only of weak parenchyma, the tips are flexible and delicate and often can be

damaged by wind; the elongating portion must be very short, or it simply buckles under its own weight.

It is important to think about the method by which collenchyma provides support. If a vine or other collenchyma-rich tissue is cut off from its water supply, it wilts and droops; the collenchyma is unable to hold up the stem. Parenchyma cells are needed in the inner tissues for support. Collenchyma and turgid parenchyma work together like air pressure and a tire: The tire or inner tube is extremely strong but is useless for support without air pressure. Similarly, air pressure is useless unless it is confined by a container. In stems, the tendency for parenchyma to expand

Figure 5-9 Shoot tips of long vines need the plastic support of collenchyma while their stems are elongating.

TABLE 5-3	Types of Sclerenchyma
Mechanical (nonconducting) sclerenchyma	
Sclereids	More or less isodiametric; often dead at maturity.
Fibers	Long; many types are dead, other types remain alive and are involved in storage.
Conducting sclerenchyma (tracheary elements)	
Tracheids	Long and narrow with tapered ends; contain no perforations. Dead at maturity. Found in all vascular plants.
Vessel elements	Short and wide with rather perpendicular end walls; must contain one or two perforations. Dead at maturity. Found almost exclusively in flowering plants. Among nonflowering plants, only a few ferns, horsetails, and gymnosperms have vessels.

is counterbalanced by the resistance of the collenchyma, and the stem becomes rigid but able to grow.

Because collenchyma cell walls are thick, they require more glucose for their production. Collenchyma is usually produced only in shoot tips and young petioles, where the need for extra strength justifies the metabolic cost. Subterranean shoots and roots do not need collenchyma because soil provides support, but aerial roots of epiphytes such as orchids and philodendrons have a thick layer of collenchyma.

Sclerenchyma

The third basic type of cell and tissue, **sclerenchyma**, has both a primary wall and a thick secondary wall that is almost always lignified (**Figure 5-10**). These walls are elastic: They can be deformed, but they return to their original size and shape when the pressure or tension is released. Sclerenchyma cells develop from parenchyma cells in mature organs after they have stopped growing and have achieved their proper size and shape. Deforming forces such as wind, animals, or snow are usually detrimental. If mature organs had collenchyma for support, they would be reshaped constantly by storms or animals, which of course would not be optimal. For example, while growing and elongating, a young leaf must be supported by collenchyma if it is to continue to grow, but after it has achieved its mature size and shape, some cells of the leaf can mature into sclerenchyma and provide elastic support that maintains the leaf's shape. Unlike collenchyma, sclerenchyma supports the plant by its strength alone; if sclerenchyma-rich stems are allowed to wilt, they remain upright and do not droop.

Parenchyma and collenchyma cells can absorb water so powerfully that they swell and stretch their walls, thereby growing; sclerenchyma cell walls are strong enough to prevent the protoplast from expanding. The rigidity of sclerenchyma makes it unusable for growing shoot tips because it would prevent further shoot elongation.

Sclerenchyma cells are of two types: conducting sclerenchyma and mechanical sclerenchyma. The latter type is subdivided into long **fibers** (Figures 5-10a to 5-10c) and short **sclereids** (Figures 5-10d to 5-10f and **Table 5-3**), both of which have elastic secondary walls. Because fibers are long, they are flexible and are most often found in areas where strength and flexibility are important. The wood of most flowering plants contains abundant fibers, and their strength supports the tree while their elasticity allows the trunk and branches to sway in the wind without breaking (usually) or becoming permanently bent (**Figure 5-11**). Our ribs are similarly strong, flexible, and elastic. Fiber-rich bark resists insects, fungi, and other pests.

Sclereids are short and more or less isodiametric (cuboidal). Because sclereids have strong walls oriented in all three dimensions, sclerenchyma tissue composed of sclereids acts brittle and inflexible. Masses of sclereids form hard, impenetrable surfaces such as shells of walnuts and coconuts or "pits" or "stones" of cherries and peaches. Flexibility there would be disadvantageous because the soft seed inside might be crushed even though the shell remained unbroken. This is similar to our fragile brain being protected by our strong, inflexible, elastic skull.

When strength or resistance is the only selective advantage of sclerenchyma, the protoplast usually dies after the secondary wall has been deposited: A mature coconut shell is protective whether its cells are alive or dead. But in some species, certain sclerenchyma cells, especially fibers, remain alive at maturity and carry out an active metabolism (**Figure 5-12**). These living sclerenchyma cells most often are involved in storing starch or crystals of calcium oxalate. Some have rather thin secondary walls, but in others, the secondary walls are just as thick as those of fibers that die at maturity and provide only support.

Like all cells, sclereids and fibers develop from cells produced by cell division; when newly formed, they are small and have only a primary wall—they are parenchyma cells. If the cell is to differentiate into a sclereid, it may expand only slightly, but if it is to develop into a fiber, it elongates greatly. When immature sclereids and fibers reach their final size, cellulose-synthesizing rosettes in the plasma membrane begin to deposit the secondary wall. The secondary wall is located interior to the primary wall; as it becomes thicker, the protoplast must shrink, usually by removing water from its central

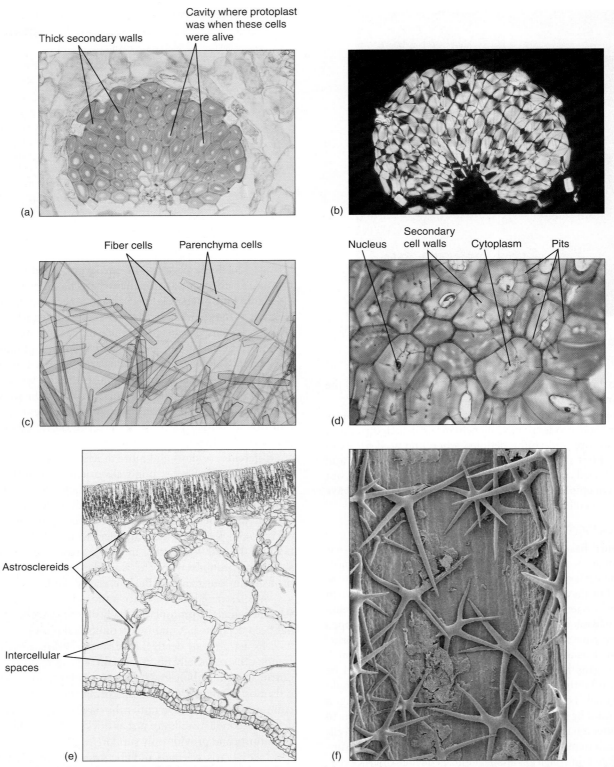

Figure 5-10 (a) A mass of fiber cells in the leaf of *Agave.* These are large, heavy, tough leaves, and fiber masses provide elastic strength. Each cell consists mostly of thick secondary cell wall; the small white space in each is an area where the protoplast had been before it died (×150). (b) The same mass of fibers as in (a) but viewed with polarized light. Secondary walls shine brightly because their cellulose molecules are packed in a tight, crystalline form, giving the wall extra strength (×150). (c) A stem of bamboo was treated with nitric and chromic acid to dissolve the middle lamellas and allow the cells to separate from each other. Fibers are long and narrow; shorter, wider cells are parenchyma (×80). (d) Sclereids are more or less cuboidal, definitely not long like fibers. These have remained alive at maturity, and nuclei and cytoplasm are visible in several. Blue-stained channels that cross the walls are pits with cytoplasm. Pits of each cell connect with those of surrounding cells so that nutrients can be transferred from cell to cell, keeping them alive (×150). (e) This portion of a water lily leaf contains large, irregularly branched cells stained red. These are astrosclereids (star-shaped sclereids). The large white spaces are giant intercellular spaces; this is an aerenchyma type of parenchyma (×40). (f) A star-shaped sclereid in a water lily leaf shown by scanning electron microscopy. These cells are located in a large intercellular space. Many have additional arms that extend back into other, smaller intercellular spaces (×200).

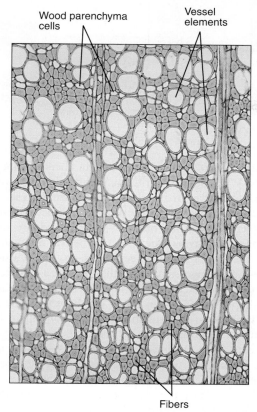

Figure 5-11 Wood is composed of several types of cells. The numerous small cells with thick walls and extremely narrow lumens are fibers, which give wood strength and flexibility. The large round cells that appear to be empty are vessel elements, discussed later in this chapter; the small cells with thin walls and large lumens are wood parenchyma cells (×60).

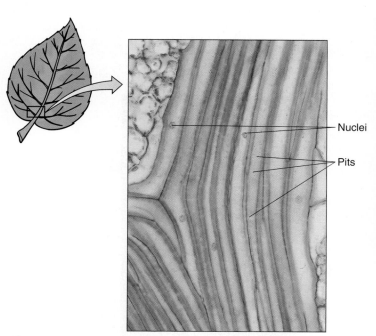

Figure 5-12 These fiber cells have nuclei: They are living cells. Their secondary walls are thick, but not so thick as those shown in Figure 5-11. The small dots in the walls are pits; these are much narrower than the pits shown in Figure 5-10d. Leaf of *Smilax* (×150).

vacuole. Secondary walls become impregnated with lignin, making them waterproof, so nutrients can enter the cell only through plasmodesmata rather than everywhere, as with parenchyma and collenchyma cells. It is important that the secondary wall not be deposited along the entire inner surface of the primary wall; small, plasmodesmata-rich areas must remain free of the secondary wall. At first, these areas are low depressions in the developing secondary wall, but as wall deposition continues, these areas become narrow **pits** in the secondary wall (see Figures 5-10d, 5-31, and 5-32a). The pits of adjacent sclerenchyma cells must meet; two pits make up a **pit-pair**. If the pits of one cell met areas of secondary wall in the neighboring cells, no water or sugars could be transferred and the protoplasts would starve.

Conducting sclerenchyma transports water and is one of the types of vascular cells; it is discussed later in this chapter.

External Organization of Stems

The terms "stem" and "shoot" are sometimes used interchangeably, but technically, the stem is an axis, whereas the shoot is the stem plus any leaves, flowers, or buds that may be present. All flowering plants have the same basic stem organization: There are **nodes** where leaves are attached, and **internodes**, the regions between nodes (see Figure 5-1). The stem area just above the point where a leaf attaches is the **leaf axil**. Within it is an **axillary bud**, a miniature shoot with a dormant apical meristem and several young leaves (**Figure 5-13**); it is either a vegetative bud if it will grow into a branch or a floral bud if it will grow into a flower or group of flowers. The bud is covered by small, corky, waxy **bud**

Figure 5-13 (a) In these buds of pistachio, leaves are modified into bud scales, just as those of prickly pear buds are modified into spines. Bud scales are waterproof and protective, but they are shed when the bud begins to grow in the spring. There are buds in the axils of the leaves and a bud at the very tip of the shoot. (b) The two axillary buds of this *Viburnum* have already begun to grow into branches, and their first young, expanding leaves are visible. These buds were formed in the spring and started growing immediately; they did not form dormant buds with bud scales.

Plants and People

Box 5-2 Parenchyma, Sclerenchyma, and Food

Most of the fruits, vegetables, and other plant parts we eat consist of almost pure parenchyma. With their thin walls, parenchyma cells are soft and easy to chew whether fresh or cooked, and they can be ground, mashed, and sliced for processed food. Whereas collenchyma and sclerenchyma are used for strength, parenchyma cells are the sites for synthesis and storage of an amazing variety of organic compounds: carbohydrates, fats, proteins, vitamins, pigments, flavors, and other nutritious materials essential to our health.

Parenchymatous foods are easy to recognize because we can bite through them easily: apples, pears, strawberries, blueberries, potatoes, lettuce, spinach, and so on. All seeds such as popcorn, beans, rice, and wheat are parenchyma as well, but in their dry, ungerminated, uncooked condition they are too hard to be edible, despite having thin cell walls. Boiling allows water to loosen hard starch grains and protein bodies and to soften the walls.

Collenchyma is not present in all plants and is never really abundant. It is almost always just a minor fraction of all the cells in leaves and stems and virtually never occurs in roots. The most familiar collenchymatous food is celery: We eat the petioles of leaves, and each ridge along the surface is a mass of collenchyma cells. Each vascular bundle inside the petiole also has a cap of collenchyma along one side. These bundles cause celery's stringiness. Other leafy vegetables with thick petioles, like rhubarb and bok choy, also have abundant collenchyma. Most of us eat these for their flavor and nutrients, not for the pleasure of chewing endlessly on collenchyma strands.

Plants can store starch and proteins in fibers, but they do so only rarely. With the low nutrition and difficulty in chewing, we never use truly fibrous material like mature bamboo shoots and wood for food. However, many vascular bundles contain fiber cells, so even when eating parenchymatous foods we still consume some sclerenchyma. The fibers may be very noticeable, as in asparagus that is a bit too old, green beans, snow peas, artichokes, pineapples, and mangoes. In others, the fibers are a bit softer, especially after being cooked, but you may still notice them in things like squash, pumpkin, and zucchini. You have probably noticed when carving jack-o-lanterns that cleaning a pumpkin is really just a matter of removing seeds and fibers.

Sclereids also occur in some of our foods. Clusters of them cause the grittiness in pears. The seed coats of beans, peas, and most other seeds are also made up of sclereids, as is the covering on seeds of corn. The pieces that get stuck in our teeth while eating corn on the cob are composed of sclereids (and the inedible cob is mostly fibers). In small seeds that are ground into flour (wheat, rye, barley), the sclereid-rich seed coats are broken down into small pieces (the bran) that are easy to eat. Sclereids also make up the brown covering of unpopped popcorn, and popping breaks up the covering. In all cases we do not digest any part of the secondary walls of sclerenchyma cells; all the walls simply pass right through us.

---

scales (modified leaves) that protect the delicate organs inside. At the extreme tip of each stem is a **terminal bud**. In winter, when all leaves have abscised, leaf scars occur where leaves were attached (**Figure 5-14**).

The arrangement of leaves on the stem, called **phyllotaxy**, is important in positioning leaves so that they do not shade each other (**Figure 5-15**). If only one leaf is present at each node, the stem has alternate phyllotaxy (the leaves alternate up the stem); two leaves per node is opposite phyllotaxy, and three or more per node is whorled. The orientation of leaves at one node with respect to those at neighboring nodes is also important. In distichous phyllotaxy, leaves are arranged in only two (di-) rows (-stichies), as in corn and irises. In decussate phyllotaxy, leaves are arranged in four rows; this occurs in only some species with opposite leaves. Finally, in spiral phyllotaxy, each leaf is located slightly to the side of the ones immediately above and below it, and leaves form a spiral up the stem. This is the most common arrangement and may involve alternate, opposite, or whorled leaves (**Table 5-4**).

All flowering plant shoots are based on this simple arrangement of nodes and internodes, and diversity and specialization

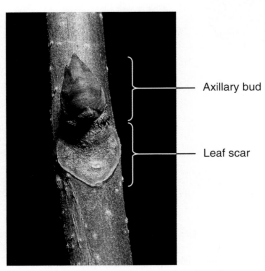

Figure 5-14 As a leaf falls from a stem, it breaks cells and creates a wound; a leaf scar is a layer of cork that seals the wound, keeping fungi and bacteria out and preventing water loss.

(a)

(b)

(c)

Figure 5-15 Phyllotaxy. (a) These young leaves of redbud (*Cercis canadensis*) are arranged with alternate phyllotaxy—one leaf per node. Each higher leaf is smaller and younger. (b) Leaves of *Salvia* show opposite phyllotaxy—two leaves per node, each pair pointing 90 degrees away from previous and subsequent pairs. (c) Irises (*Iris*) have distichous phyllotaxy—one leaf per node, arranged in just two rows. Iris leaves are flattened from side to side, not from top to bottom.

TABLE 5-4	Phyllotaxy
Alternate	Leaves one per node
Opposite	Leaves two per node
Decussate	Leaves located in four rows
Whorled	Three or more leaves per node
Spiral	Leaves not aligned with their nearest neighbors
Distichous	Leaves located in two rows only

are variations of this arrangement. In vines, internodes are especially long, whereas in lettuce, cabbage, and onions, internodes are so short that leaves are packed together (Figure 5-16). Internodes can be wide (asparagus), intermediate, or narrow (alfalfa sprouts). The diversity is not random—different types provide particular adaptive advantages in certain situations. For example, plants are physically bound to the site where they happen to germinate as a seed, which by accident may be a shady area near a more optimal sunny spot. Vines, with their elongated internodes, are a means by which a plant can "explore" its immediate surroundings, and shoots that happen to grow into a sunnier site may flourish (Figure 5-17). In some species of climbing vines, support and attachment are provided by tendrils—modified leaves or lateral branches capable of twining around small objects.

Leaves

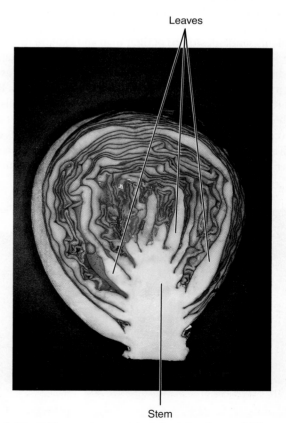

Stem

Figure 5-16 Cabbage nodes are packed closely together, and all leaves are tightly clustered. Such closely packed, large leaves with spiral phyllotaxy are poor at photosynthesis and did not evolve by natural selection; they were produced by plant geneticists.

Figure 5-17 These grape vines (*Vitis*) place their leaves in the full sun at the top of the forest canopy, even though they have not invested much sugar in building strong trunks. The pines that support these vines may ultimately die because they are so heavily shaded by the grape leaves.

Stolon   Leaves

Figure 5-18 Airplane plant (*Chlorophytum*) is a popular ornamental plant that spreads by runners in nature. Most people are familiar with it as a hanging basket plant, with its runners arching out and then drooping down and forming new plants at the tips. The new plant here has already produced roots.

The capacity to explore is even more advanced in **stolons**, also called runners (**Figure 5-18**). Their internodes are especially long and thin, and their leaves do not expand; thus, stolons extend greatly without using much of the plant's nutrient reserves. After the stolon encounters a suitable microhabitat, subsequent growth is by shorter, vertical internodes and fully expanded leaves; new roots are established, and the end of the stolon resembles a new plant. If older parts of the plant die, these vertical shoots become independent plants. After they have started growing vigorously, these plants send out stolons of their own and explore further.

In some shoots, nutrient storage is particularly important for survival; these shoots are often massive and quite fleshy, thus providing many parenchyma cells in which starch accumulates. **Bulbs** are short shoots that have thick, fleshy leaves (onions, daffodils, garlic), whereas **corms** are vertical, thick stems that have thin, papery leaves (crocus and gladiolus; **Figure 5-19**). There is no obvious selective advantage of one over the other; the type seems to depend on whether mutations affecting the stem or leaf happen to occur first.

Rhizomes are fleshy horizontal stems that allow a plant to spread underground (bamboo, irises, and canna lilies). **Tubers** are horizontal like rhizomes, but they grow for only a short period and are mainly a means of storing nutrients (potatoes). The word "tuber" is often used informally for any bulky underground plant organ.

All of these storage shoots are subterranean, which is more advantageous than an exposed surface location. Only perennial plants that go through a dormant period need storage capacity. Dormancy is the means by which perennial plants of harsh climates survive the stress of winter cold or summer heat and dryness; they often shed their leaves, reducing water loss. In order to produce new twigs, leaves, and roots in the favorable season, they must draw on stored carbohydrates. Protection of their nutrient reserves—corms, bulbs, rhizomes, and tubers—is most easily accomplished by burying them at depths that do not freeze or dry out. A subterranean location also hides them from most herbivores.

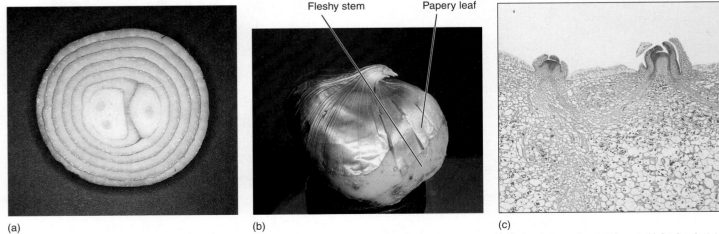

Fleshy stem   Papery leaf

(a)   (b)   (c)

Figure 5-19 (a) Onions are bulbs: They have short vertical stems and fleshy leaves. (b) Gladioluses have corms with fleshy stems and papery leaves. (c) A tuber (potato: *Solanum tuberosum*) is a short horizontal stem that grows only for a limited period of time. Its leaves are microscopic, and its axillary buds are the potato's "eyes," shown here by light microscopy (×15).

Figure 5-20 Division of labor. False Solomon's seal (*Smilacena racemosa*) consists of subterranean rhizomes (which survive harsh winters and spread through the soil) and aerial shoots (which carry out photosynthesis and flowering).

Each of these specialized tasks is accomplished by characteristic modifications in stem structure and metabolism. As with multicellularity, specialization is accompanied by division of labor: Modifications that increase a stem's ability to survive, spread, or store nutrients decrease its efficiency at other tasks. No plant consists only of rhizomes, which are not exposed to light, because most plants must photosynthesize. At least some of the rhizome's axillary buds must grow upward above ground and have green leaves (**Figure 5-20**). Similarly, a plant cannot be made up only of stolons: Modifications that allow quick, low-cost exploration are not appropriate for growth, photosynthesis, and sexual reproduction (flowering). Another example are the cactus-like spurges (*Euphorbia*) of African deserts (photograph on opening page of this chapter): Each has a thick, succulent green shoot that photosynthesizes, stores water, and grows longer each year, whereas its axillary buds grow out as slender, short red branches that become fibrous then die and function as spines. It is common for individual plants to have several types of stems and leaves, each of which contributes uniquely to the plant's survival.

Although the axil of every leaf contains a bud, only a few buds ever develop into a branch; the others remain dormant or produce flowers. Axillary buds do not become active at random; each species has a particular pattern. In shady environments, it may be advantageous for axillary buds to remain dormant so that all resources are concentrated in the growth of the vertical main shoot, the **trunk**, allowing the plant to reach brighter light in the top of the forest canopy (**Figure 5-21**). In an open environment, such dominant upward growth is not particularly advantageous; it may be better for all buds to grow and thereby maximize the rate at which new leaves are formed. On almost all

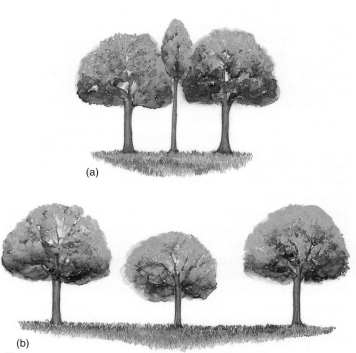

(a)

(b)

Figure 5-21 (a) In a heavily shaded forest, many axillary buds remain permanently dormant, allowing all resources to be concentrated on placing one shoot up into the light. (b) With abundant light, one main trunk is not so advantageous; instead, many axillary buds grow as branches, allowing rapid production of leaves.

Plants Do Things Differently

Box 5-3 Organs: Replace Them or Reuse Them?

Plants have three types of organs: roots, stems, and leaves. Animals have many types: heart, brain, liver, eyes, kidneys, urinary bladder, and so on. Notice that we have only one or two of each: one heart, one brain, two eyes. Plants, on the other hand, typically have many of each: dozens of stems, hundreds of roots, and thousands of leaves. Also, notice how long each lasts relative to the life of the organism. We have only one heart during our entire lifetime. If it fails, we die. Considering all the mechanical work our heart must do, it is remarkable that we do not periodically grow a new one. We have two kidneys, so if one stops functioning, we have a backup organ; however, our bodies will not grow a replacement, even if we lose both kidneys. If we are lucky enough to die of old age, the organs we have at that time are the same ones we were born with.

Plants are very different. Their biology is based on using an organ just once then throwing it away and replacing it with a new version. A tree might have thousands of leaves, but it will discard those in autumn and make an entire new set the following spring. When the tree blooms, it might have thousands of flowers, each with many petals and other parts. The petals function for just a day or two then wither and fall off. If the flower was pollinated, it will form a fruit, and then that too will fall off. After the flowering period is finished, the plant will have no sex organs whatsoever until the following year, when it makes a whole new set. Think about how different our biology is from that. Our first thought is probably relief that our sex organs do not fall off after use, even if we could grow another set next year.

What are the consequences of the plant-type alternative of depending on "one-use" organs versus the animal-type of "reusable" organs? Leaves must be exposed to sunlight if photosynthesis is to occur, so they suffer extensive exposure to ultraviolet light. Ultraviolet light causes mutations in both plants and animals, leading to skin cancer in us and to damaged, nonfunctional leaves in plants. We avoid ultraviolet light by wearing clothing (other land animals have hair, feathers, or scales) or by performing our activities in shade or indoors. But leaves must be in light. In addition, leaves become covered with dust, spores, fungi, bacteria, insect poop, and other debris. They become just as dirty as any animal, but leaves cannot groom themselves. By abscising their leaves and growing a new set, they get rid of dirty, diseased, broken-down organs and start off a new year with fresh, clean ones. If plants had to rely on the same leaves for their entire lifetime, no tree would ever make it to one hundred years old.

The same principle applies to other one-use plant organs. Some trees live to be not just hundreds of years old but thousands. Even at such an amazing age, whenever the tree undergoes sexual reproduction, it makes brand new flowers (or cones), and those are as fresh and vigorous as when the tree was only a few years old. We animals are not so lucky. When we get on in years, we may still be willing, but our reproductive organs are old. In men, sperm cells are produced by division of the same set of cells that became active when we went through puberty, and there is no reason to think that those cells will remain in better shape than those of our skin (wrinkled), eyes (failing), hair follicles (bald long ago), teeth (false). In women, cells that will mature into eggs all begin meiosis before we are even born and have been in cell cycle arrest ever since, with just one egg maturing each month. By the time we are 40 or 50 years old, so are all our potential egg cells—but not in plants, not even in old plants. All of their eggs and sperm cells are brand new made in flowers that have just been produced only a week or two before.

Even our circulatory systems follow this difference. We have just one set of blood vessels all our lives. At this point, your aorta is about 20 years old, as are the arteries that take blood up to your brain. All the blood that goes to and from your lungs is passing through veins about one fifth of a century old, and they are getting older as you read this. Not in plants, however. All woody plants make an entirely new vascular system every year. Every springtime, plants make a new layer of wood that conducts water and a new layer of bark that conducts sugars and minerals. The wood and bark from last year may still be functional, or it may be damaged by fungus; however, that does not matter. There is a set of new tissue, as healthy and functional as our human tissues were when we were babies.

Plants and animals are not totally different, actually. We threw away our baby teeth and replaced them with a set of permanent teeth when we were about 6 years old—a very plant-like thing to do. And sharks do not stop with just two sets—they constantly produce replacement teeth as long as they live. Birds replace their feathers every year, and starfish grow a new arm if one is cut off. Some lizards grow a new tail, but such capacity does not occur in us: Other than baby teeth, our organs must last us a lifetime. But any time you look at a plant, you are seeing only a tiny fraction of the plant. You are not seeing all the parts that have already existed and fallen off nor all the parts that the plant will have in the future. If all those parts could be gathered up and preserved, the total plant would be vastly greater than it ever is at any particular time.

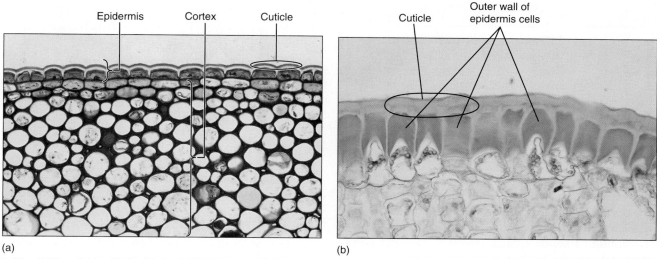

Figure 5-22 (a) The cuticle on this ivy (*Hedera helix*) has been stained dark pink; the lighter layer is wall material impregnated with cutin. Epidermal protoplasts have stained dark blue. Interior cells are photosynthetic collenchyma (×120). (b) This epidermis was taken from a desert plant, *Agave*. Its cuticle is much thicker than that of ivy in (a), and the outer walls of the epidermal cells are extremely thick. Both the wall and cuticle strengthen the leaf surface against insects and prevent water loss (×200).

plants, at least a few axillary buds remain dormant and serve as reserve growth centers. As long as the apical meristem is healthy and growing well, some axillary buds are unneeded; if the apical meristem is killed by frost, insects, or pruning, axillary buds become active and replace it, allowing the growth of the shoot to continue.

Internal Organization of Stems

Arrangement of Primary Tissues

In order to function properly, a tissue must contain the right cells in the proper arrangement. The same principle is true on a larger scale: In order to function properly, the tissues of an organ must be arranged correctly.

Epidermis The outermost surface of an herbaceous stem is the **epidermis**, a single layer of living parenchyma cells (**Figure 5-22**). This differs greatly from our human epidermis, which is much thicker, consisting of many layers of dead cells.

Also, whereas we humans have lungs and a digestive tract that absorbs certain things and expels others, all interchange of material between a plant and its environment occurs by means of its epidermis.

Because air is almost always drier than living cells, even when relative humidity is high, preventing loss of water to air is a critical function for the epidermis of land plants. Also, the epidermis is a barrier against invasion by bacteria and fungi as well as small insects. It shields delicate internal cells from abrasion by dust particles, passing animals, or leaves and stems that might rub together, and its reflectivity protects the plant from overheating in bright sunlight. Although it might seem that thick-walled sclerenchyma cells would be more protective, epidermis cells are almost always thin-walled parenchyma cells.

The outer tangential walls are encrusted with **cutin**, a fatty substance that makes the wall impermeable to water (Figure 5-22). In species that occur in mild, moist habitats, such encrustation

may be sufficient for water retention, but in most plants, cutin builds up as a more or less pure layer called the **cuticle**. Under more severe conditions, even a cuticle is not sufficient, and a layer of wax may be present outside the cuticle (**Figure 5-23** and **Table 5-5**). Cutin and wax resist digestive enzymes and provide defense against pathogens like fungi and bacteria. In some species, the cuticle is so smooth that fungal spores cannot even stick to it and are washed off by rain or shaken off by wind. Waxes are indigestible and nonnutritious, therefore, a thick layer of wax makes it difficult and unrewarding for an insect to chew into a stem.

Unfortunately, cutin and wax also inhibit the entry of carbon dioxide needed for photosynthesis—a totally impermeable epidermis would lead to a plant's starvation. However, the epidermis

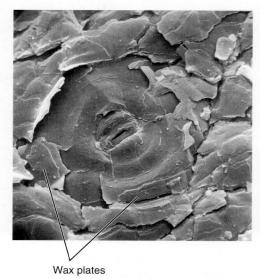

Wax plates

Figure 5-23 On this epidermis, wax is present as flat plates, but in other species it can occur as threads, beads, flakes, or a liquid (×550).

TABLE 5-5	Special Chemicals in Walls	
Cell Type	Chemical	Special Property
Collenchyma	Pectins	Plasticity (?)
Sclerenchyma	Lignin	Strength; waterproofing
Epidermis	Cutin/waxes	Waterproofing; indigestible by bacteria, fungi, animals
Endodermis (Chapter 7)	Suberin	Waterproofing
	Lignin	Waterproofing
Cork (in bark) (Chapter 8)	Suberin	Waterproofing

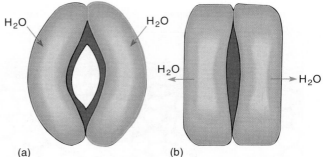

Figure 5-25 (a) When guard cells absorb water and swell, they become curved and the inner walls are pulled apart by the expansion of the back walls. (b) If guard cells lose water and shrink, they straighten, closing the stomatal pore.

contains pairs of cells (**guard cells**) with a hole (**stomatal pore**) between them (**Figure 5-24**). Guard cells and a stomatal pore together constitute a **stoma** (plural, stomata). Stomatal pores can be opened during the daytime, permitting carbon dioxide to enter the plant.

The opening of the stomatal pore is made possible by the structure of the guard cells. Guard cell walls have a special radial arrangement of cellulose microfibrils, which causes some parts to be weaker and more extensible than others. Guard cells swell by absorbing water; the walls next to the stomatal pore are rigid and do not extend, but the back walls are weak and stretch (**Figure 5-25**), causing the cells to form an arc and push apart, opening the pore and allowing entry of carbon dioxide and exit of oxygen. This unavoidably allows water vapor to escape and also increases the risk of microbes entering the plant. After sunset, when photosynthesis is impossible, it is not advantageous to keep stomata open; guard cells in most species shrink and straighten, closing the pore and preventing water loss.

In most plants, some epidermal cells elongate outward and become **trichomes,** also called **hairs** (**Figure 5-26**). Trichomes make it difficult for an animal to land on, walk on, or chew into a leaf. They shade underlying tissues by blocking some incoming sunlight, which may be too intense in summer or in high alpine environments. Trichomes can also create a layer of immobile air next to a leaf surface, which allows water molecules that diffuse out of a stoma to bounce back in rather than be swept away by air currents.

Trichomes exist in hundreds of sizes and shapes; many are unicellular, being just long, narrow epidermal cells. Multicellular ones may be a single row of cells or several cells wide, and many branch in elaborate patterns. Most trichomes die shortly after maturity, and their cell walls provide protection; but others remain alive and act as small secretory glands. Some secrete excess salt; others produce antiherbivore compounds, and those in car-

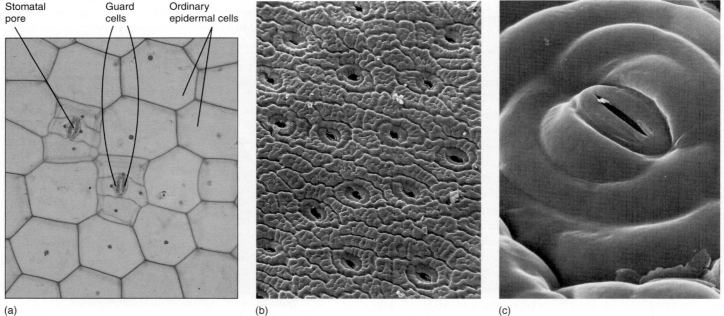

Figure 5-24 (a) Light microscope view of the epidermis of *Zebrina*; the stomatal pore, guard cells, and adjacent cells are visible, as are ordinary epidermal cells (×150). (b) Scanning electron micrograph of the epidermis of a cactus showing numerous stomata. Notice how close together stomata are; no chlorenchyma cell inside the cactus (on the other side of the epidermis) is far from a point where carbon dioxide can enter (×50). (c) The stomatal pore is surrounded by guard cells and then other epidermal cells. Notice the small particle of debris at one end of the pore; clogging by debris is a problem in dusty areas (×800).

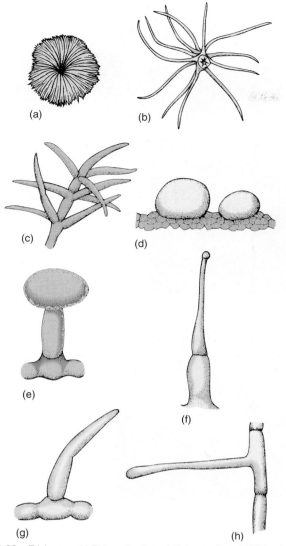

Figure 5-26 Trichomes. (a) Flat, scale-shaped trichome. (b and c) Branched trichomes. (d) Glandular trichomes—these are two large single cells. (e) Glandular trichome with a stalk cell elevating the secretory head cell. (f) Glandular trichome, capable of injecting poison. (g and h) Simple, unbranched, nonglandular trichomes.

nivorous plants secrete digestive enzymes onto trapped insects. The poisonous, irritating compounds of stinging nettle are held in trichomes.

Cortex Interior to the epidermis is the **cortex** (**Figure 5-27**). In many plants, it is quite simple and homogeneous, composed of photosynthetic parenchyma and sometimes collenchyma. In other species, it can be very complex, containing many specialized cells that secrete latex, mucilage, or pitch (resin). Some cortex cells contain large crystals of calcium oxalate or deposits of silica.

Cortex cells of most plants fit together compactly, but in fleshy stems such as tubers, corms, and succulents, cortex parenchyma is aerenchyma, an open tissue with large intercellular air spaces (see Figures 5-6 and 5-10). A few angiosperms, such as water lilies and the plants grown in aquaria, have become aquatic, living submerged in lakes or oceans. These plants have large cortical air chambers that provide buoyancy. The stems have such a strong tendency to float that no sclerenchyma is necessary for support.

Vascular Tissues For very small organisms whose bodies are either unicellular or just thin sheets of cells, diffusion is adequate for the distribution of sugars, minerals, oxygen, and carbon dioxide throughout the body (**Box 5-4**). But if any cells of the organism are separated from environmental nutrients and oxygen by just four or five cell layers, diffusion is too slow, and a vascular system is necessary. Two types of **vascular tissues** occur in plants: **xylem** (pronounced "zylem"), which conducts water and minerals, and **phloem** (pronounced either "flow em" or "flome"), which distributes sugars and minerals.

Vascular systems of plants are quite different from those of animals. The plant vascular system is not a circulatory system: Water and minerals enter xylem in the roots and are conducted upward to leaves and stems. During its passage, xylem sap travels through dead, hollow cells, not through tubes composed of living cells like our blood vessels. Once in the shoots, most water evaporates from the surfaces of stems, leaves, and flowers and is lost; the minerals and a bit of water are used by surrounding cells. Phloem cells are living; they pick up sugar from areas where it is abundant, usually leaves during summer and tubers or rhizomes in spring, and transport it to areas where sugar is needed, especially growing tips of shoots, roots, young leaves, and flowers. In

Figure 5-27 (a) The cortex of this stem of buttercup (*Ranunculus*) is the narrow band of cells between the epidermis and the vascular bundles (×20). (b) The cortex of this corn stem is even narrower, in some places being only two or three cells wide (×20).

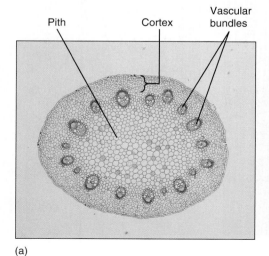

(a)

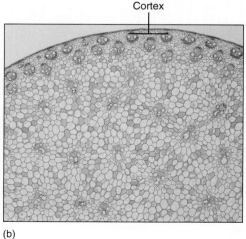

(b)

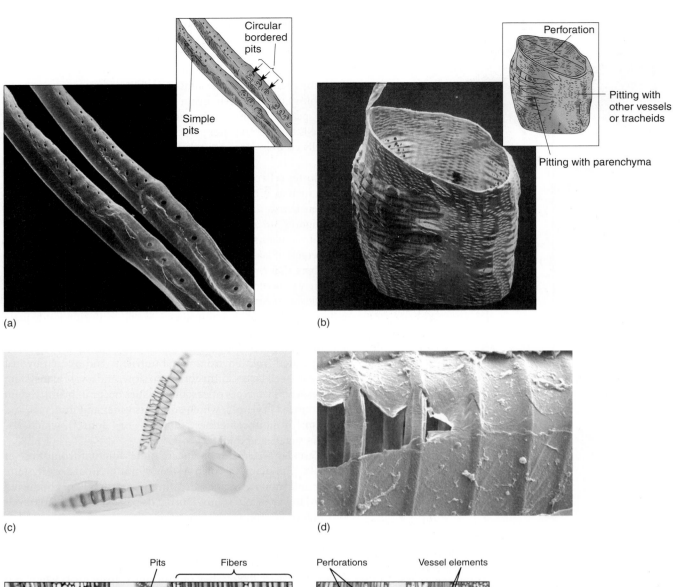

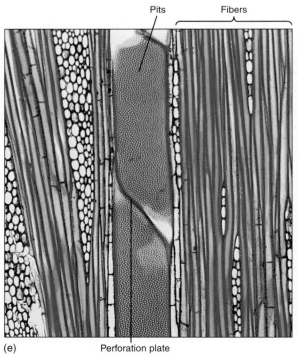

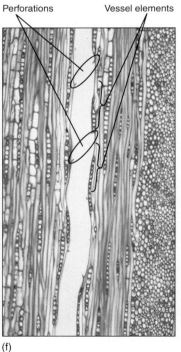

Pits   Fibers

Perforations   Vessel elements

(e)   Perforation plate

(f)

Figure 5-28 Tracheary elements. (a, c, and d) Tracheids are long cells with tapered ends. The primary wall is complete over the whole surface, and in (c), it is so thin that the inner, spiral secondary wall is visible. In (d), the primary wall tore during specimen preparation, revealing just how thin it is and permitting views of the circular bands of secondary wall which are located just interior to the primary wall and which prevent it from collapsing during water conduction. (b and e) Vessel elements tend to be wider and shorter than tracheids, but the most important feature is the perforation, the large hole at each end. (f) The perforation of one vessel element must be aligned with that of the next if water is to pass through with little friction; the stack of vessel elements is a vessel. The vessel here has been cut down the center, so the front wall and back wall are missing and the vessel appears to be just an open space. (a) ×600 (b) ×270 (c) ×100 (d) ×1000 (e) ×200 (f) ×50.

TABLE 5-6	Tracheary Elements (Conducting Sclerenchyma)	
	Tracheids	Vessel Elements
Shape	Long/narrow	Short/wide
	Ends pointed	Ends usually flat
Secondary wall	Annular	Annular
	Helical	Helical
	Scalariform	Scalariform
	Reticulate	Reticulate
	Circular bordered pits	Circular border pits
Perforations	None	Usually two: one in each end wall; terminal members with only one

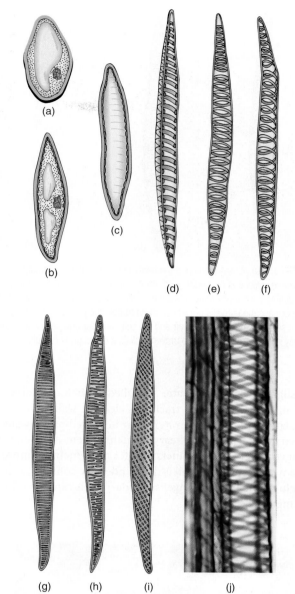

Figure 5-29 Tracheids. (a–d) The growth of a young cytoplasmic cell into a tracheid. After the cell has almost reached mature size, the secondary wall is deposited interior to the primary wall (c). After the secondary wall is finished, the protoplasm dies and degenerates, leaving only the primary and secondary walls. (d) A tracheid cut open, showing a secondary wall in the form of helical thickenings. (e) A whole cell in which the primary wall is so thin that the annular secondary walls inside are visible. (f) Whole cell with helical secondary wall. (g) Scalariform secondary wall. (h) Reticulate secondary wall. (i) Pitted secondary wall. (j) Light micrograph of a tracheid with a helical secondary wall (×500).

the later months of summer, phloem carries sugar into developing fruits and into the storage organs of perennial plants. Because sugar must be dissolved to be conducted, water is transported simultaneously in phloem.

Xylem Within xylem are two types of conducting cells: **tracheids** and **vessel elements**; both are types of sclerenchyma (**Figure 5-28** and **Table 5-6**). The term "**tracheary element**" refers to either type of cell. As a young cell matures into a tracheary element, it first must enter cell cycle arrest and stop dividing. It is initially a small parenchyma cell with only a thin primary wall, but the cell becomes long and narrow and then deposits a secondary wall that reinforces the primary wall. The cell then dies, and its protoplasm degenerates, leaving a hollow tubular wall (**Figure 5-29**).

The secondary wall is impermeable to water; thus, areas of the permeable primary wall must remain uncovered if water is to enter and leave the cell. In the simplest type of tracheary element, there is only a small amount of secondary wall, organized as a set of rings, called **annular thickenings**, on the interior face of the primary wall (Figure 5-29e). This arrangement provides a large surface area for water movement into and out of the cell, but it does not provide much strength. The primary wall must be supported because movement of the water tends to cause it to collapse inward (**Figure 5-30**). With **helical thickening**, the secondary wall exists as one to three helices interior to the primary wall (Figure 5-29f). **Scalariform thickening** provides much more strength because the secondary wall underlies most of the inner surface of the primary wall and is fairly extensive. Just as in fibers and sclereids, the area where the secondary wall is absent is called a pit (**Figure 5-31a**). In tracheary elements with **reticulate thickening**, the secondary wall is deposited in the shape of a net, as the name suggests. The most derived and strongest tracheary elements are those with **circular bordered pits** (Figure 5-31b and **Figure 5-32**). In such tracheary elements, virtually all of the primary wall is underlain by secondary wall. The pits that allow water movement are weak points in the wall, but the weakness is reduced by a **border** of extra wall material around the pit, hence the name.

Tracheary elements with annular thickenings are weak, but a large percentage of their primary wall is free of secondary wall and available for water movement (**Figure 5-33**). Pitted tracheary elements are just the opposite: They are extremely strong, but so much of the surface is underlain by lignified secondary

wall that water enters or leaves the cells slowly. Each is selectively advantageous under certain conditions and disadvantageous under others. If a species grows in perpetually wet soil and water moves easily, its cells are not in danger of collapsing, and thus, tracheary elements with annular or helical secondary walls are adaptive, not only because the large open regions of primary wall allow water to pass through quickly but also because the plant uses little sugar building secondary walls. If soil dries

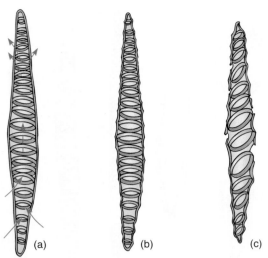

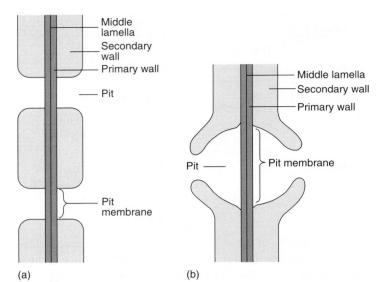

Figure 5-30 (a) As water enters a tracheary element, it passes through primary wall between regions of secondary wall. (b) Water adheres to cellulose molecules, pulling walls inward and upward as it rises in xylem; primary walls tend to collapse but secondary walls provide strength to hold the primary wall in place. (c) Under severe water stress (dry air and dry soil), annular and helical secondary walls may not be strong enough to prevent collapse of the primary wall.

Figure 5-31 (a) Side view of a pit that has been bisected. In a sclereid or fiber, the secondary wall would be thicker and the pit narrower and deeper. The diameter of the pit can vary, being broad as in Figure 5-10d or narrow as in Figure 5-12. (b) A bordered pit has a rim of extra wall material that prevents the pit from being a weak spot in the wall. With borders, pits can be broad enough to allow rapid flow of water from tracheary element to tracheary element, but not so weak as to be detrimental. Only tracheary elements have bordered pits; fibers and sclereids do not.

even slightly, however, water is pulled with more force and exerts inward traction on tracheary element walls. The stronger scalariform, reticulate, or circular bordered pitted walls are necessary to keep the tracheary elements open. Although they cannot conduct water as quickly, water movement is automatically slower because the soil is dry. Under dry conditions, using extra glucose to build stronger secondary walls can be a valuable investment.

Tracheids obtain water from other tracheids below them and pass it on to those above. They must occur in groups, lying side by side with some and having their ends overlapping the ends of others (**Figure 5-34**). Also, the pits of adjacent tracheids are aligned so that water can pass through. The aligned set of pits is a pit-pair, and the set of primary walls and middle lamella

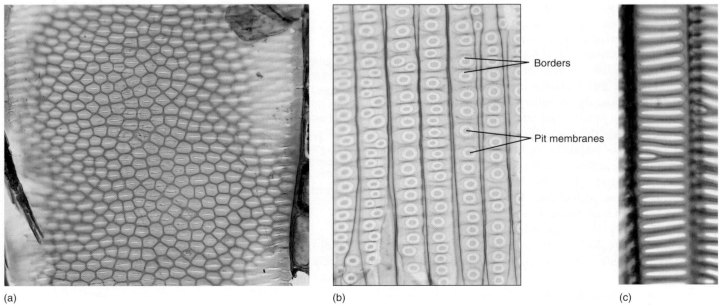

Figure 5-32 (a) Bordered pits in face view on front wall of a vessel element. The white slits are the actual openings of the pits, the pinkish area around each white region is the border, and the dark red hexagons are regions where the secondary wall lies against the primary wall (×400). (b) Bordered pits in face view on the front walls of tracheids of pine. This is the same magnification as (a); notice that tracheids are much narrower than vessel elements, and pits in pine are extremely large. (c) Scalariform bordered pits (×700).

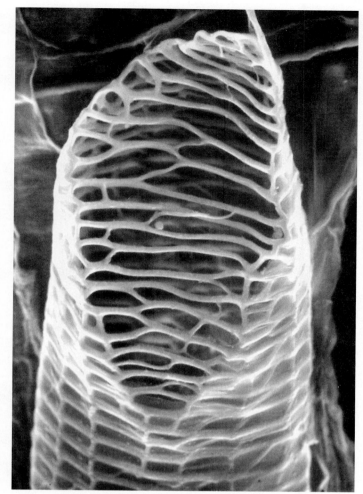

Figure 5-33 This vessel element was isolated from tissue by dissolving the middle lamella. The secondary wall (the network along the cell's end and sides) is clearly visible on the end because the primary wall was digested away before the cell died; these open holes constitute a perforation. The thin, almost transparent covering on the cell's sides is the primary wall, which was not digested. In the plant, another vessel element would have been stacked above this one, with their perforations aligned and permitting water flow (×1,000).

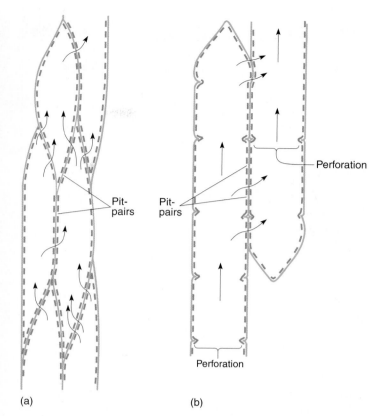

Figure 5-34 (a) Water passes between tracheids only through pit-pairs. Other than plasmodesmata (which are too tiny to be important in water conduction), no holes are present in primary walls of tracheids. (b) Water can pass between vessel elements through perforations, but it can pass from one vessel to another only through pit-pairs. Vessels are long, but not as long as a whole plant, and water must, therefore, pass from one vessel to another as it moves upward from roots to stems to leaves. Green = primary wall; brown = secondary wall.

between them constitutes a **pit membrane**, just as in fibers and sclereids (Figure 5-31). ("Pit membrane" is an old term and should not be confused with lipid/protein membranes such as the plasma membrane.) Although very permeable to water, a pit membrane—being a set of primary walls—does offer slight resistance. If each tracheid is 1 mm long, each water molecule passing from root to leaf in even a short plant only 1 meter tall (1,000 mm) must pass through 1,000 pit membranes. The friction adds up.

Vessel elements provide a way to move water with less friction than tracheids. Vessel elements, like tracheids, are individual cells that produce both primary and secondary walls before they die at maturity. In vessel elements, however, an entire region of both primary and secondary wall is missing. During the final stages of differentiation, a large hole called a **perforation** is digested through a particular site of the primary wall, often removing the entire end wall (Figures 5-28, 5-33, and 5-35).

The perforations of adjacent vessel elements must be aligned, and each element must have at least two perforations, one on each end. Whereas water must pass through pit membranes each time it enters or leaves a tracheid, it passes through perforations as it goes from one vessel element to the next. Because perforations are wide and completely lack primary walls, they cause little friction.

An entire stack of vessel elements is a **vessel**. The vessel elements on each end of a vessel have only one perforation. Whereas vessel elements, being individual cells, are only about 50 to 100 μm long, vessels can be many meters long, running all the way from a root tip to a shoot tip, although some are only a few centimeters long (Table 5-7).

Vessels must absorb water from parenchyma cells, tracheids, or other vessels, and they must pass it on. Their side walls have pits for this lateral transfer (Figure 5-34), and all of the types of wall thickening mentioned above for tracheids also occur in vessel elements. The only constant difference is that vessel elements have perforations—complete holes—in their walls, whereas tracheids only have pits (Table 5-6). Perforations greatly reduce

TABLE 5-7	Length of Vessels in Trunks of Trees	
	Vessel Length (cm)	
Species	Average	Maximum
Fagus grandifolia (American beech)	302	556
Populus tremuloides (quaking aspen)	122	132
Betula lutea (yellow birch)	119	142
Alnus rugosa (speckled alder)	105	122
Fraxinus americana (white ash)	97.1	1,829
Quercus rubra (red oak)	94.8	1,524
Ulmus americana (American elm)	94.6	853
Acer saccharum (sugar maple)	15	32
Vaccinium corymbosum (swamp blueberry)	11	120

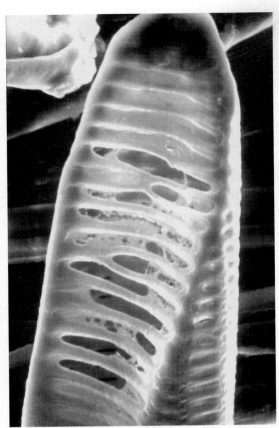

Figure 5-35 This fern (*Astrolepis*) tracheary element had digested portions of primary wall out of several pits, converting them to crude perforations. In angiosperms, the wall-digesting mechanism is more refined and removes the primary wall more completely.

friction, and water moves much more easily through a set of vessels than through a set of tracheids. Tracheids and vessel elements have plasmodesmata in their primary walls, but these are too narrow to be important in long-distance water conduction; typical diameters—plasmodesmata, 0.04 μm; pits, 1 to 2 μm; and perforations, 20 to 50 μm.

Within our human circulatory system, all arteries, capillaries, and veins share the same lumen: A water molecule or a red blood cell could circulate through every part of our blood vessel system and never encounter a barrier. In contrast, water molecules must move across pit membranes every time they enter or leave a tracheid or vessel. Some consequences are that the plant system causes more friction than does our vascular system, and solid particles cannot be transported in plants. However, a hole or leak in our system could potentially let all our blood leak out (and is often fatal), whereas a hole in a plant system affects only one tracheid or vessel, not the entire system (and is never fatal).

Tracheids evolved at least 420 million years ago, and virtually all plants with vascular tissue have tracheids. Vessels evolved more recently and occur almost exclusively in flowering plants, where they perform long-distance water conduction in roots and stems. In flowering plants, tracheids occur mostly in the fine veins of leaves but can also be found in the wood of some species. Conifers such as pines and redwoods have xylem composed exclusively of tracheids. Until recently, we thought that ferns too contained only tracheids, but they can at least partially digest their pit membranes and convert pit-pairs into sets of perforations (**Figure 5-35**).

Phloem Like xylem, phloem has two types of conducting cells, **sieve cells** and **sieve tube members**; the term "**sieve element**" refers to either (**Figure 5-36**). These are very different from tracheary elements: They are parenchyma cells, have only primary walls, and must remain alive in order to conduct. As immature sieve

elements begin to differentiate, their plasmodesmata enlarge to a diameter of more than 1 μm and are called **sieve pores**. Plasmodesmata occur in groups (primary pit fields); therefore, sieve pores also occur clustered together in groups called **sieve areas**. Sieve elements remain alive during differentiation, and the plasma membrane that lined the plasmodesma continues to line the sieve pore (**Figure 5-37**). The amount of cytoplasm within the pore also increases, and rapid, bulk movement from cell to cell becomes possible. Sieve pores of adjacent cells must be aligned if phloem sap is to pass through.

The two types of sieve elements differ in shape and placement of sieve areas (Figure 5-36). If the cell is elongate and spindle shaped (like a tracheid) and has sieve areas distributed over all its surface, it is a sieve cell (**Table 5-8**). This type evolved first and is found in older fossils and in nonangiosperm vascular plants. But because phloem conducts mostly longitudinally, it is most efficient for sieve areas on the two ends of the cell to be especially large and to have very wide sieve pores, whereas sieve areas on the sides of the cell can be rather small. Cells like this are sieve tube members; they are stacked end to end with their large sieve areas aligned, forming a **sieve tube**. These end-wall sieve areas with large sieve pores are **sieve plates**. Like vessels, sieve tubes are shorter than the plant, and the small, lateral sieve areas are important for transport of phloem sap from one sieve tube to another. Sieve

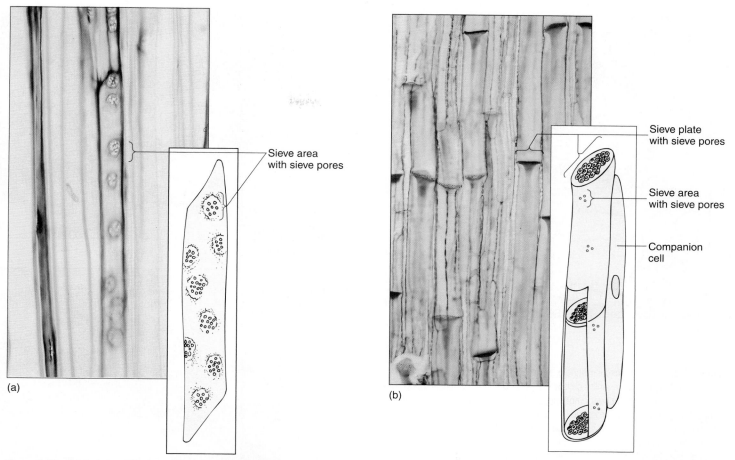

Figure 5-36 (a) A sieve cell is shaped like a tracheid and has sieve areas with sieve pores over much of its surface (×250). (b) A sieve tube member has flat end walls (sieve plates) with large sieve pores. On the side walls are only a few small sieve areas with very narrow sieve pores (×60). The micrograph of sieve cells is a high magnification; they are much narrower than the sieve tubes of (b).

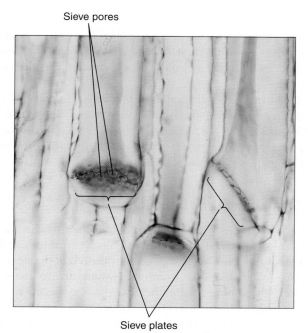

Figure 5-37 Large sieve areas present on sieve plates of sieve tube members. Phloem sap flows from one tube-like sieve tube member into the next through the wide sieve pores (×300).

tube members must have evolved at about the same time as the flower; all angiosperms have sieve tube members, but none of the nonangiosperms has them.

An unusual feature of sieve elements is that their nuclei degenerate but the cells remain alive. Cytoplasm cannot carry on complex metabolism without a nucleus, and in phloem, the necessary nuclear control is provided by cells intimately associated with conducting cells. Sieve cells are associated with **albuminous cells**, and sieve tube members are controlled by **companion cells** (**Figure 5-38**). These cells are often smaller than the accompanying conducting cell and have a prominent nucleus and dense cyto-

TABLE 5-8	Sieve Elements	
	Sieve Cells	Sieve Tube Members
Shape	Long/narrow	Short/wide
	Ends pointed	Ends usually flat
Sieve areas	Small, located over all the cell surface	On side walls: small; on end walls: very large end wall is sieve plate.
Associated cells	Albuminous cells	Companion cells
Plant division	All nonangiosperm vascular plants.	Angiosperms only
		Some relictual angiosperms

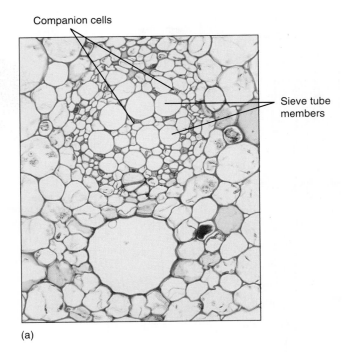

(a)

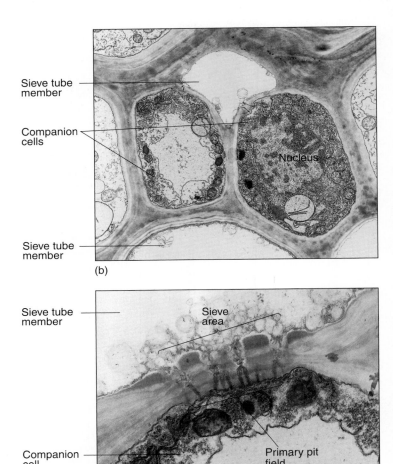

(b)

(c)

Figure 5-38 (a) In a vascular bundle, sieve tube members and companion cells occur together. Companion cells are small and rich in cytoplasm; sieve tube members are large and appear empty because most phloem sap is lost while preparing the tissue for microscopy (×150). (b) Electron micrograph showing two companion cells and two sieve tube members. (c) The connection between a companion cell and a sieve tube member is a primary pit field with plasmodesmata on the side of the companion cell (bottom), but it is a sieve area with sieve pores on the side of the sieve tube members (top) (×45,000).

plasm filled with ribosomes. Walls between conducting cells and controlling cells have many complex passages that are sieve areas on the conducting cell side and large plasmodesmata on the controlling cell side. Companion cells have another important role: that of loading sugars into and out of the sieve tube members.

Vascular Bundles Xylem and phloem occur together as **vascular bundles**, located just interior to the cortex (**Figure 5-39**). In basal angiosperms and eudicots, vascular bundles are arranged in one ring surrounding the **pith**, a region of parenchyma similar to the cortex. In monocots, vascular bundles are distributed as a complex network throughout the inner part of the stem; between the bundles are parenchyma cells. Monocot bundles are frequently described as "scattered" in the stem, suggesting that they occur at random, but they actually have precise, specific patterns that are too complex to be recognized easily.

All vascular bundles are **collateral**; that is, each contains both xylem and phloem strands running parallel to each other (Figure 5-39). The xylem of a vascular bundle is **primary xylem** because it is part of the primary plant body. In addition to conductive tracheids and vessel elements, there is usually a large proportion of xylem parenchyma and even mechanical sclerenchyma in the form of xylem fibers. The vascular bundle phloem is **primary phloem**, and mixed with sieve elements and com-

panion cells or albuminous cells may be storage parenchyma and mechanical sclerenchyma, usually as phloem fibers, although phloem sclereids also occur. The relative amounts of each cell type within a bundle vary: Storage stems have large amounts of parenchyma in the vascular tissues; vines and tendrils have extra fibers.

Cells of primary xylem are larger than cells of primary phloem, and the tracheary elements on the inner side of each bundle are much smaller than the outer ones (Figure 5-39). This is true of the primary xylem in all stems, and the reason for this becomes clear when the shoot growth is examined.

Stem Growth and Differentiation

Stems grow longer by creating new cells at their tips, in regions known as shoot **apical meristems**; cells divide by mitosis and cytokinesis, producing progenitor cells for the rest of the stem (**Figure 5-40**). When each cell divides, the two daughter cells are half the size of the mother cell, but they grow back to the original size. As they expand, they automatically push the meristem upward; the lower cells are left behind as part of the young stem. This region just below the apical meristem is the **subapical meristem**, and its cells are also dividing and growing, producing cells for the region below.

Alternatives

Box 5-4 Simple Plants

Three groups of plants have much simpler bodies than those described in this chapter. **Hornworts** (Anthocerotophyta) are a rather rare group of tiny plants whose bodies look like bits of green cellophane, often no larger than 2 to 5 mm in diameter and only a few cells thick. They have no epidermis, no stomata, no collenchyma, no sclerenchyma, no cortex, no pith, no xylem, and no phloem. Instead, they are small ribbon- or disk-shaped sheets of parenchyma cells, most of which have a single chloroplast. Each cell absorbs water and minerals, carries out its own photosynthesis, and is relatively self-reliant. When the plant reproduces, horn-like structures grow upward, break open, and release spores (there are no flowers). During this stage, nutrients are transported from chlorophyllous cells to spore-producing cells, but only by means of plasmodesmata, not phloem.

Many **liverworts** (Hepatophyta) are about as simple as hornworts. Plants of *Pallavicinia*, *Sphaerocarpos*, and *Riccia* also consist of just small green ribbons or disks of chlorophyllous cells. Bodies of *Fossombronia* and *Petallophyllum* appear leafy and more complex, but in fact, they are just ribbons whose edges are puckered. The bit of complexity present in these bodies is merely that the central portion of the ribbon is several cells thick, whereas the margins are unistratose. As in hornworts, there is no epidermis or vascular tissue.

Other liverworts such as *Asterella* and *Marchantia* have thicker, more complex bodies. The ventral side (the "belly" side facing the soil) is several layers of compact parenchyma with few chloroplasts, and the dorsal side (the "backbone" side facing the air) consists of intersecting sheets of cells that form air chambers (like the intersecting sheets of cardboard that form a six-pack carton for beer). The term "epidermis" is used for surface cells; dorsal epidermis has stoma-like air pores. Ventral epidermis has trichomes (called rhizoids) that attach the body to soil or tree bark. The most complex liverworts have "leaves" and are called leafy liverworts, but their bodies are composed of just parenchyma. As in hornworts, liverworts transfer nutrients through plasmodesmata from chlorophyllous cells to spore-producing tissues when they reproduce: Liverworts have no vascular tissues. Despite their simplicity, all liverworts grow by means of organized apical meristems.

Mosses (Bryophyta) are a bit more complex than either hornworts or liverworts. Mosses always have a stem with leaves, and in those with taller stems or larger leaves, some cells deposit a thicker wall and resemble sclerenchyma fibers. However, these walls do not contain lignin and probably have a separate evolutionary origin, so they are called stereides instead of fibers to avoid confusion. Many mosses conduct water, but along the outside of their bodies, not inside and not with xylem: Their numerous leaves are so tiny and close to the stems they form capillary spaces and act like a wick. A number of mosses do have an internal system of conducting cells in their green, leafy stems; cells called hydroids conduct water, and others called leptoids conduct carbohydrates and other nutrients. These should never be called "xylem" or "phloem" because they differ in significant ways from those two tissues.

What are some consequences of having such simple bodies? The simplicity of hornworts, liverworts, and some mosses limits their ability to survive in some habitats. Lacking conducting tissues, they must remain short, thin, or prostrate, and many quickly die in dry air. On the other hand, this same simplicity *enables* them to survive in many habitats because just a few weeks of cool moist weather is enough to allow their spores to germinate and grow into tiny, mature plants. They do not need to photosynthesize long enough to produce all the carbohydrates that a vascular plant needs for its larger, more complex body. Even deserts have rainy periods and shaded spots that stay moist for weeks; such spots will almost always have some nonvascular plants thriving there.

Figure B5-4a   Plants of hornworts (*Phaeoceros*); the bodies are just thin, green sheets of parenchyma. The columns ("horns") are reproductive structures.

Figure B5-4b   Plants of this liverwort (*Riccia fluitans*) grow in quiet streams; each shoot is less than 1 mm thick and consists of parenchyma cells. Each grows from an apical meristem that occasionally divides in two.

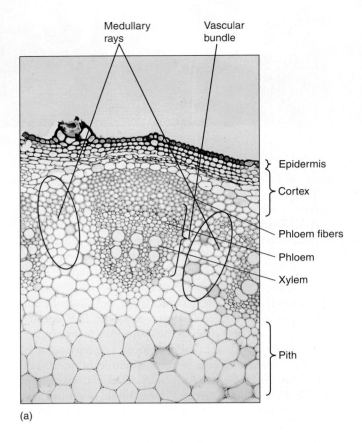

(a)

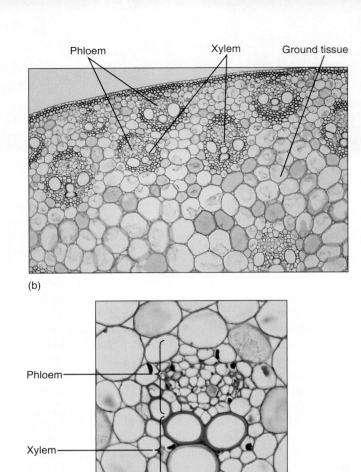

(b)

(c)

Figure 5-39 (a) In this vascular bundle of sunflower (*Helianthus*), xylem contains rows of vessels alternating with rows of xylem parenchyma. Phloem contains sieve tube members and companion cells. There is a large sclerenchyma cap of phloem fibers (×60). (b) Vascular bundles of monocots have a more complex arrangement. In this corn stem, many large bundles are located near the cortex, and fewer, smaller bundles in the ground tissue (×50). (c) High magnification of a corn bundle, showing primary xylem and phloem (×250).

In the subapical meristem, visible differentiation begins; certain cells stop dividing and start elongating and differentiating into the first tracheids or vessel elements of the vascular bundles (Figure 5-39 and **Figure 5-41**). Because they constitute the first xylem to appear, they are called **protoxylem**. Cells around them continue to grow and expand, but soon those just exterior to them also begin to differentiate; because these had expanded for a longer time, they develop into tracheary elements that are even larger than the first. This process continues until the last cells mature, and this portion of the stem stops elongating. Because these cells have had the longest time for growth before differentiation, they develop into the largest tracheary elements of all, called **metaxylem**.

Size is not the only difference between tracheary elements of protoxylem and metaxylem. Because protoxylem elements differentiate while surrounding cells are still elongating, they must be extensible and so must have either annular or helical secondary walls. Although protoxylem cells are dead at maturity, they stretch because their primary walls are glued by their middle lamellas to living, elongating cells that surround them. If protoxylem had scalariform, reticulate, or pitted secondary walls (which they never have), the cells could not be stretched; either they would

prevent the surrounding cells from growing, or the middle lamellas would break and the cells would tear away from each other. This is not a problem for metaxylem, however, because its cells differentiate only after all surrounding stem tissues have stopped elongating. Because metaxylem cells do not have to be extensible, any type of secondary wall is feasible.

A similar process occurs in the outer part of each vascular bundle. The exterior cells mature as **protophloem**, and cells closest to the metaxylem become **metaphloem**. Because the sieve elements do not have secondary walls, the walls of protophloem and metaphloem are identical; however, sieve elements are extremely sensitive to being stretched and die when stressed too much. Consequently, protophloem cells are extremely short lived, often functioning for only one day. Perhaps because they are so temporary, protophloem cells never become well differentiated; their sieve areas are small and rudimentary, and many do not have companion cells. Sieve elements of metaphloem do not differentiate until later, when all surrounding cells have stopped expanding. Metaphloem cells differentiate fully, having large sieve areas and conspicuous companion cells.

You might expect metaphloem cells to be very large, like those in metaxylem, but they are much smaller, even though they

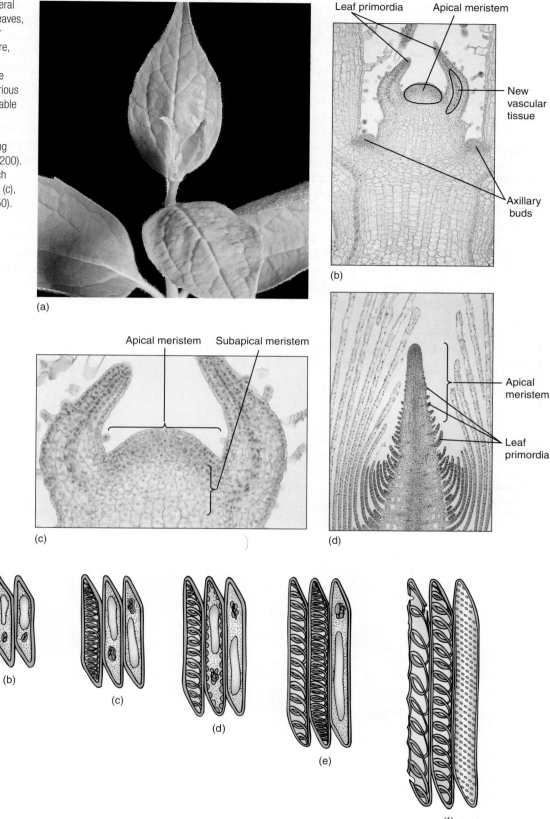

Figure 5-40 (a) This shoot tip has several fully expanded leaves, one set of small leaves, and another set of even smaller, younger leaves. Dissection would reveal even more, younger leaves. (b) Longitudinal section through a shoot tip of *Coleus* showing the apical meristem and leaf primordia of various ages. Two axillary buds are just recognizable as small outgrowths (×100). (c) Higher magnification of the *Coleus* shoot apical meristem. Vascular tissue is differentiating in the center of each leaf primordium (×200). (d) This apical meristem of *Elodea* is much taller and narrower than that of *Coleus* in (c), and it has many more leaf primordia (×50).

Figure 5-41 Development of three tracheary elements. (a) All cells are young, small, and cytoplasmic. (b) The first protoxylem element forms while all cells are still small. Notice that cell 1 has started depositing its secondary wall (red dots). (c) Cells 2 and 3 continue growing, stretching cell 1, which is dead. (d and e) Cell 2 matures while cell 3 continues to grow, stretching cells 1 and 2. Cell 1 has been stretched so much that it is on the verge of being torn open. (f) All elongation has stopped, and cell 3 differentiates as a pitted element. The primary wall of cell 1 has been torn and no longer conducts. Cell 2 is probably still capable of carrying water. In a stem, these three cells would be surrounded by numerous others.

have undergone long periods of expansion before differentiation. All cells of the region are expanding, whether they are dead functioning protoxylem, dead nonfunctioning protophloem, or living undifferentiated metaxylem and metaphloem. Cell size differs because cell division is still occurring in some cells but not in others. In xylem, all cells stop dividing, so each cell becomes larger as the tissue expands—protoxylem cells by being stretched and the soon-to-be metaxylem cells by their own growth. Protophloem cells stop dividing and differentiate; however, other phloem cells continue to divide, and they therefore remain small.

Other tissues also differentiate in the subapical region. In the epidermis, the first stage of trichome outgrowth may be visible in the youngest internodes, those closest to the apical meristem. In lower, older internodes, trichomes are more mature, and guard cells and stomatal pores may be forming. The cuticle is extremely thin at the apical meristem, but it is thicker in the subapical region and may be complete several internodes below the shoot apex. Differentiation of the pith typically involves few obvious changes: Cells enlarge somewhat; intercellular spaces expand but remain rather small, and cell walls continue to be thin and unmodified. All changes are usually completed quickly, while the cells are in or just below the subapical meristem. Similar changes occur in the cortex, except that plastids develop into chloroplasts. Protoxylem and protophloem develop quickly while the cells are still close to the apical

meristem, but metaxylem and metaphloem do not begin to differentiate until the nodes and internodes have stopped elongating.

Because the terms "epidermis," "cortex," "phloem," "xylem," and "pith" typically refer to mature cells and tissues, it is confusing to use them for the young, developing cells during differentiation. To prevent misunderstandings, a special set of terms has been agreed on: **Protoderm** refers to epidermal cells that are in the early stages of differentiation. Young cells of xylem and phloem are **provascular tissues**; the equivalent stages of pith and cortex are **ground meristem** (Figure 5-42). Thus, the subapical meristem consists of protoderm, ground meristem, provascular tissue, and more ground meristem. The rate of maturation varies greatly from species to species and also from one type of stem to another on the same plant. In stems with long internodes, such as vines, maturation is slow, and immature tissues can be found far below the shoot apical meristem. In stems with short internodes, such as bulbs and lettuce, differentiation may be completed quickly, and all tissues are mature close to the stem apex.

Now, at the end of the chapter, we can define two concepts precisely. **Primary tissues** are the tissues produced by apical meristems; **primary growth** is the growth and tissue formation that results from apical meristem activity. Secondary tissues (wood and bark) are produced by different type of meristems (vascular and cork cambia) and are discussed elsewhere.

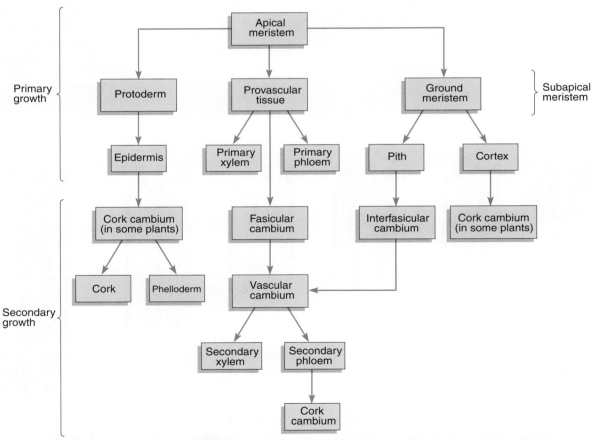

Figure 5-42 Cells of an apical meristem give rise to three types of subapical cells, which in turn give rise to particular types of mature primary tissues. Epidermis develops only from protoderm, never from provascular tissue or ground meristem; other cell lineages in the primary body are also quite simple and linear.

Box 5-5 Plants and People Grow Differently

Imagine that your parents planted a tree seed on the day you were born. That tree is probably still getting taller while your growth in height has come to an end. We achieve our maximum height when we are approximately 18 years old, but trees keep right on growing. The tips of their twigs lengthen several inches every spring and make new leaves and flowers. How tall will the tree become? If we treat them well, give them fertilizer, water, sunlight, and protect them from diseases and storms, they may never stop growing. For us, however, our height is set. We will not get to be 7 or 8 feet tall no matter how many vitamins we take, no matter how nutritious our food.

We animals have **determinate growth**. Plants have **indeterminate growth**. The maximum amount of growing an animal will do in its lifetime, and therefore, its size and the number of organs it has are determined by the genes inherited from its parents. Malnourishment may prevent us from growing properly, but even if healthy and well nourished, we cannot exceed our genetic potential. The size of most plants is not determined by their genes because even if a plant received genes for poor growth, it might still become extremely large if it lives an unusually long time.

An animal's determinate growth is related to its body's precise, invariant form with a specific number of organs. We have one head, one torso, two arms, two legs, and our spine contains 33 vertebrae and 12 pairs of ribs, and these were all present when we were born. No new ones were added later: We have **determinate organogenesis**. Plant bodies are not so precisely organized. When a plant germinates as a seedling, most of its leaves and roots are not present yet, and it has none of its branches, flowers, or fruits either. All of these will be made over the many years of its life, and there is no set number of organs. There will never be a time when it has all of its leaves because some will have already fallen off. Others will not have been made yet. Plants have **indeterminate organogenesis**.

Plants grow by special groups of cells at the tips of their shoots and roots. As these cells divide, approximately half the new cells mature and differentiate into functional cells of the stem, new leaves, and flowers and so on; however, the other cells simply grow back to their original size then divide again. Once again, about half the new cells mature, and the other half continue as cells capable of dividing. The shoot tip always has cells able to generate new cells. There is no reason for it to ever run out of cells. Animals have no set of cells like this, cells always capable of making new cells and organs for the body's growth.

Growth is not one hundred percent different in plants and animals, however. Our skin's growth is indeterminate. The innermost layer of our skin, called the stratum basale, consists of cells that never stop dividing; after each round of division, one daughter cell matures as a skin cell, and the other continues to divide. We replace our skin continuously. After a bad sunburn, some skin peels off, but the stratum basale makes more. This is the way plants grow. Our hair, fingernails, and blood also grow continuously from sets of cells that divide and provide new cells throughout our life. Becoming more muscular or fatter is different. That just involves making existing cells larger, not increasing their numbers.

Leaf growth in plants is determinate. While a leaf is small, all its cells divide repeatedly. Then, at some point, all differentiate and mature, and the leaf has no cells capable of making any new leaf cells. It expands to its full size as each cell enlarges its central vacuole. The leaf's maximum size is determined by the plant's genes. The same is true for flowers. If we give our plants good care, they may make more flowers but not larger ones.

Now for the exceptions: Tapeworms have indeterminate growth. They have a "head" that attaches to the host's intestine and a "neck" that is a set of cells that divide repeatedly. When enough new cells are present, some organize themselves into a body segment complete with organs. The dividing neck cells continue to make more cells for another segment. The tapeworm makes an indeterminate number of segments, growing to an indeterminate length. Older segments full of eggs break off and end up in the host's excrement, but the "neck" makes new replacement segments. Unlike other animals, tapeworms do not have a specific number of parts other than one head and one neck. Its plant-like indeterminate growth is based on its plant-like set of permanently dividing cells and plant-like lack of precise body pattern.

Exceptional plants are **annual plants**. These live for just 1 year and, like animals, have determinate growth. Even though they grow at their tip, for some reason those cells all stop dividing after a few weeks or months, ending the plant's organ formation. The plant then blooms and makes fruits and seeds, then it dies. **Biennial plants** live for 2 years, making many leaves in the first year, blooming and reproducing in the second and then dying. The life of annuals and biennials is programmed by their genes. No matter how much care and fertilizer we give them, we cannot keep them alive longer or get them to grow indeterminately.

Summary

1. Cells must be arranged in the proper patterns and with the proper interrelationships in order to function efficiently.
2. All angiosperms have roots, stems, and leaves, although each can be highly modified in particular environments.
3. Virtually all stems function in producing and supporting leaves and transporting water, minerals, and sugars between leaves, buds, and roots. Other stems are involved in storage, reproduction, dissemination, and surviving stress.
4. All stems consist of nodes and internodes, and they have leaves and axillary buds. Numerous types of stems exist; most individual plants have two or three types of shoots.
5. The three basic types of plant cells, based on walls, are parenchyma (responsible for most metabolism), collenchyma (plastic support), and sclerenchyma (elastic support).
6. The vascular tissues are xylem and phloem. Tracheary elements consist of vessel elements (with perforations) and tracheids (without perforations). Sieve elements consist of sieve cells (with small sieve pores) and sieve tube members (with large sieve pores on the end walls).
7. Stems of all angiosperms have epidermis, cortex, and collateral vascular bundles. In basal angiosperms and eudicots, the bundles occur as one ring surrounding pith, whereas in monocots, they have a complex distribution in conjunctive tissue.
8. Protoxylem and protophloem form while an organ is elongating and therefore must be extensible. Metaxylem and metaphloem form after elongation ceases.
9. Both primary xylem and primary phloem are complex tissues with a variety of cell types, not just conducting cells.

Important Terms

angiosperms
apical meristem
axillary bud
basal angiosperms
bulbs
circular bordered pits
collenchyma cells
companion cells
corms
cortex
cuticle
epidermis
eudicots
fibers
ground meristem
guard cells
internodes
metaphloem

metaxylem
monocots
nodes
parenchyma
perforation
phloem
phyllotaxy
pits
pit membrane
pit-pair
pith
primary growth
primary tissues
protoderm
protophloem
protoxylem
provascular tissues
rhizomes

sclereids
sclerenchyma
sieve areas
sieve cells
sieve plates
sieve pores
sieve tube
sieve tube members
stoma
tracheary element
tracheids
trichomes
tubers
vascular bundles
vessel
vessel elements
xylem

Review Questions

1. The body of an herb contains just three basic parts. What are they?
2. Stems of the first land plants functioned primarily as transport and support structures. Modern stems have these functions also, plus several more. Describe the additional functions.
3. Over the years, several different names have been used for the flowering plants. Here they are called the division _____, but they are known informally as _____. There are about _____ species of flowering plants.
4. Early angiosperms diversified into several groups that are now known as _____ _____, _____, and _____. Name three examples of plants within each line of evolution.
5. An herb has only a primary plant body. That means it has roots, stems, and leaves, but it never becomes _____ and covered with _____.
6. In general, animals have _____ growth with a fixed size and number of organs, whereas plants have _____ growth with no set size and

any number of organs. Which parts of our body grow continuously like plants? Which parts of a plant have fixed size such that no matter how well we take care of the plants those organs do not become larger? What is the name of plants that live for just 1 year? For just 2 years?

7. What are the important differences between parenchyma, collenchyma, and sclerenchyma cells?

8. Examine Figure 5-6a and 5-6b. Why do most parenchyma cells shown in Figure 5-6a lack nuclei, whereas most shown in Figure 5-6b do have nuclei? In the geranium for Figure 5-6a, did the real cells lack nuclei, or did something happen to the nuclei as the tissue was sliced to make the microscope slide?

9. What is the special name of photosynthetic parenchyma cells? Which organelle is especially abundant in these? How does wall thinness affect carbon dioxide and light?

10. It was previously stated that parenchyma cells are relatively inexpensive to build. What does that mean? How does this relate to most leaves being soft?

11. Like clay, walls of collenchyma exhibits plasticity. Does that mean it can or cannot be stretched? If a tissue is supported by collenchyma, can it still grow?

12. How does collenchyma support a tissue? Would it work if the tissue did not have parenchyma also? What happens if the tissue is not turgid (does not have enough water)?

13. Collenchyma is plastic, but sclerenchyma is elastic. What does "elastic" mean? If you stretch or deform an elastic object, will it keep its new shape or snap back to its original shape? As an organ grows, its shape changes, but after it has achieved its mature size and shape, what kind of things would deform it? Should a mature organ have plastic properties or elastic ones (e.g., if a heavy load of snow bends a branch down, should the branch stay in its bent shape when the snow melts or should it go back to the shape it grew to before the snow)?

14. What are the two types of mechanical, nonconducting sclerenchyma (hint: Table 5-3)? Which tends to be flexible and useful in wood? Which tends to be brittle and inflexible, useful in "pits" and "stones" that protect seeds?

15. Fibers and sclereids have secondary walls that are so thick and tough that the cell cannot grow. If that is true, how do fibers and sclereids grow to their mature size and shape? Do they have a secondary wall even when they are young?

16. Imagine a leaf of a palm tree. When it is fully grown, it must have sclerenchyma to support its size and weight. Do you think it has any sclerenchyma when it is tiny and just starting to grow? Would you expect that it might have collenchyma when it is medium sized and still growing?

17. Technically, the _____ is an axis, whereas the _____ is the _____ plus any leaves, flowers, or buds that might be present.

18. The point where a leaf is attached to a stem is called a _____. Just above this point is an _____ bud.

19. Figure 5-13a shows two types of buds. What are the two types and how do they differ?

20. After a leaf falls off a stem, it leaves a _____ _____ just below the axillary bud.

21. What is phyllotaxy? Corn and irises have two rows of leaves. This is known as _____ phyllotaxy.

22. When you look at a head of cabbage or lettuce, what are you seeing? Where is the stem? What about when you look at an onion? What are you seeing and where is the onion's stem?

23. Describe each of the following types of specialized shoots. Be certain to account for modifications of the leaves, internodes, and orientation of growth: stolon, rhizome, tuber, bulb, corm, and tendril. Each provides a plant with a selective advantage. What is the adaptive value of each type of specialized shoot?

24. Many scientific words are also ordinary English words. If you read that a plant has a bulb or a tuber, can you always be certain exactly what the botanical structure is?

25. What is the outermost surface of an herbaceous stem? The outer walls of this layer are encrusted with a chemical made up of fatty substance that makes the wall impermeable to water. What is the name of the substance and what is the name of the layer?

26. A stoma allows carbon dioxide to pass through the epidermis. What is the name of the two cells that control the opening and closing of a stoma? What is the name of the hole itself that carbon dioxide passes through?

27. What is the technical term for a plant hair? Plant hairs make it difficult for animals to do certain things. Name three activities that are more difficult for an animal because of a hairy leaf.

28. What is the name of the region of cells between a stem's epidermis and its set of vascular bundles? This is usually a compact parenchyma tissue, but in some aquatic angiosperms that live submerged in lakes or oceans, this region has what type of special modification? Would you hypothesize that this region is especially thick or thin in desert plants such as cacti or plants that have succulent water storage tissues (you must think about this yourself, as the answer is not in the text).

29. All flowering plants have two types of vascular tissues, _____, which conducts water and minerals, and _____, which distributes sugars and minerals. Are vascular systems of plants similar to or different from those of animals? Explain the similarity or differences.

30. As a young xylem cell matures into a tracheary element, it first must enter _____ and stop _____. It is initially a small _____ cell, but after it reaches its full size and shape, it deposits a _____ wall that reinforces the _____ wall. The xylem cell _____ and its protoplasm _____, leaving a _____ _____ wall.

31. List the five types of secondary wall deposition that can occur in tracheary elements. Which two types are most characteristic of protoxylem? What is the selective advantage of vessel elements over tracheids?

32. Consider tracheary elements with annular thickening (annular secondary walls) and those with pitted walls. Which is

weaker? Which has a large percentage of its primary wall free of secondary wall and available for water movement? Is one always more selectively advantageous than the other? Under all conditions? Explain.

33. Tracheids obtain water from _____ _____ below them and pass it on to _____ above. As water passes into or out of a tracheid, it passes through a _____-_____. The set of primary walls and middle lamella in the _____-_____ is called a _____ _____.

34. If each tracheid is 1 mm long, how many pit-pairs and pit membranes will a water molecule pass through in a plant 1 m tall? Is this a significant amount of friction?

35. What is a perforation in a vessel element? How does it reduce friction as water passes from one vessel element to another? When a water molecule arrives at the end of a vessel, how does it pass into another vessel (this is asking about vessels, not vessel elements). In Question 34, tracheids were assumed to be 1 mm long. If we assume that vessel elements are also 1 mm long (they are usually shorter, but 1 mm is an easy number to work with), how many vessel elements are there in an average vessel of American beech (Table 5-7, the lengths in the table are in centimeters)? The plant in Question 34 was said to be 1 m tall. Would an average vessel of American beech even fit into a plant this tall? If a plant 1 m tall had vessels 1 m long, how many pit membranes would a water molecule need to pass through as it traveled from root to shoot tip? Is this more or less than the answer for Question 34?

36. Which evolved more recently, tracheids or vessel elements? Do nonangiosperms such as conifers and ferns have vessel elements? Do flowering plants (angiosperms) have tracheids? Vessel elements?

37. Like xylem, phloem has two types of conducting cells, _____ and _____ _____ _____. The term "_____ _____" refers to either one. Do these die like tracheary elements or do they need to remain alive in order to conduct?

38. What is the name of the holes that interconnect conducting cells in phloem, and what are groups of these holes called?

39. Sieve tube members differ from sieve cells by having very large sieve pores on their end walls, much larger than the ones on their side walls. What is the name of these end-wall sieve areas with big sieve pores?

40. Sieve elements (both sieve cells and sieve tube members) lose their nuclei during development, but they must remain alive. What is the name of the cell associated with sieve cells? The one associated with sieve tube members?

41. During differentiation of a young cell into a sieve tube member, what are some changes that occur in the cell wall and the cytoplasm? How long do most sieve tube members live after they become mature?

42. Describe plasmodesmata, pits, perforations, and sieve pores. Which connect living cells? Which connect nonliving cells? Which occur in secondary walls?

43. All vascular bundles are _____. That is, each contains both xylem and phloem. The xylem of a vascular bundle is _____ xylem because it is part of the primary plant body, and the phloem is _____ phloem.

44. Stems grow longer by creating new cells at their tips, in regions known as _____ _____ _____. Below this region, in the subapical meristem, the very first primary xylem to appear is called _____, and cells that differentiate into xylem a little later, after they have grown larger, are called _____.

45. In the apical and subapical region, there are cells that will later give rise to epidermis and other tissues, but they are not yet mature enough to call them epidermis, and so on. What are the terms for the following cells?
    a. Epidermal cells that are still meristematic
    b. Young cells of xylem and phloem
    c. Young cells of pith and cortex

46. Describe the arrangement of tissues seen in a stem cross-section; consider monocots separately from the other angiosperms. Is the arrangement different in a stolon than in a rhizome, tuber, or corm?

47. Look at Figure 5-42. The top three rows are labeled "Primary growth" because these are found in the primary plant body, the body of an herb. The third row of the figure has five boxes with the five types of mature tissue found in the primary plant body. What are the five tissues?

48. If you live to be 100 years old, how many sets of organs will you have during your lifetime? How many hearts, stomachs, and livers? If a plant lives to be 100 years old, how many sets of leaves will it probably have? If it begins blooming when it is 10 years old, how many sets of flowers? When you are 70 years old, you will still depend on your digestive system to supply you with energy for another 30 years, but how old is that set of organs? In a 70-year-old tree, how old are the cells of the leaves that supply it with energy?

BotanyLinks

http://biology.jbpub.com/botany/5e
BotanyLinks contains a variety of resources and review material designed to assist in your study of botany.

Visit the Web Exercises area of BotanyLinks to complete this question:

1. Many plants, such as cacti, have stems with unusual shapes, but is their anatomy also unusual? Go to the BotanyLinks home page to think about this.

Leaves

chapter 6

Concepts

The term "leaf" usually calls to mind foliage leaves—the large, flat, green structures involved in photosynthesis. However, natural selection has resulted in numerous types of leaves that are selectively advantageous because they provide protection (bud scales, spines; see opening photo of this chapter), support (tendrils), storage (fleshy leaves of bulbs), and even nitrogen procurement (trapping and digesting insects). Because protecting a bud from freezing or drying is very different from photosynthesizing, leaf structures and metabolisms that are selectively advantageous for a bud scale are different from those that are selectively advantageous for a foliage leaf. This principle is true for all types of leaves: For each function, certain modifications are advantageous, whereas others are

Chapter Opener Image: This agave (probably *Agave parryi*) has survived a wildfire in West Texas. We usually think of leaves as thin, flat photosynthetic foliage leaves that are produced in the spring and drop off in autumn. But agave leaves are thick, succulent, and perennial (most live until the entire plant dies). Agave leaves do carry out photosynthesis, but they also store starch and water for these desert-adapted plants. Notice that these leaves have performed an unusual function: they protected the stem from the heat of the fire. The center of the plant—and especially the shoot apical meristem—is still healthy. Containing so much water, the leaves heat up slowly, and before the innermost leaves became hot, wind had blown the fire farther on and the stem and apical meristem were saved. The leaves even kept the ground at the base of the plant cool, protecting the roots. Leaves of most plants could not protect the shoot apical meristem like this. When we study botany, we must think about the plant in its natural habitat, the risks it faces, and how its various structures contribute to its survival.

Outline

not. Studying the various leaf types, their modifications, and the roles they play in the plant's biology leads to an understanding of how structure, metabolism, and function are related.

The shoot system, containing both stems and leaves, demonstrates both division of labor and integration of distinct plant organs. Stems and leaves must function together if the plant is to survive and reproduce, but the optimal features for each organ are quite distinct. Leaves should be flat and thin for maximum absorption of light and carbon dioxide, and most of their tissues must be alive and differentiated into chlorophyll-rich chlorenchyma to carry out photosynthesis. Stems elevate leaves and conduct material to and from them, among other activities. Maximum conduction and support with minimum expenditure of construction material require a cylindrical stem; also, much of the stem—all *of* its tracheary elements and most fibers—must die to be functional. In woody species, new cells are added to the stem xylem and phloem each year, creating a massive wood and bark and an increasingly stronger and more conductive stem. Yearly accumulation of new cells onto leaves is not feasible; leaves typically live for only several months and are rather lightly constructed; however, in open sunny habitats, elevating leaves on a tall stem is often unnecessary, and a plant may consist almost exclusively of leaves and roots. In lettuces and cabbage, leaves are short lived and delicate, but in *Agave* and *Yucca*, they are massive (to more than 2 m long), thick, and perennial (see the opening photo of this chapter and Figure 6-4a). Almost all leaves contain only primary tissues; secondary production of wood and bark in leaves is extremely rare.

The initial discussion in this chapter centers on foliage leaves because they are the most familiar. Then other leaf types are described and analyzed; by considering how these leaves differ functionally from foliage leaves, it is possible to determine the environmental factors important to leaf structure and how natural selection has produced diverse classes of leaves.

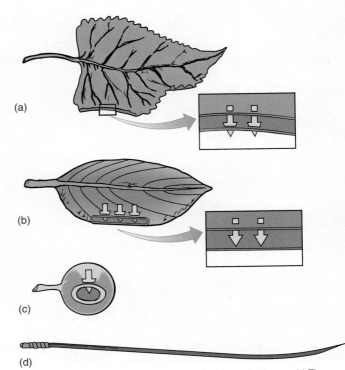

Figure 6-1 Several possible leaf shapes, all with equal volumes. (a) The common type with a thin, flat blade. All cells are exposed to light—there is a large surface area that absorbs CO_2, and none of the veins is long. (b) In a thick leaf, light cannot penetrate to the bottom of the leaf, and those cells cannot photosynthesize. Relative to the volume, little surface area is available for CO_2 absorption, but this helps conserve water. (c) A spherical leaf has maximum internal self-shading, and only a small fraction of the cells receives enough light for photosynthesis. Surface area for gas exchange is minimal. (d) A cylindrical leaf could be thin with a large surface area and no self-shading, as in (a); however, to have the same volume as (a), it would be extremely long, and conduction through the veins might be difficult.

External Structure of Foliage Leaves

The most obvious function of foliage leaves is photosynthesis, but other functions often taken for granted are just as important. Leaves must not lose excessive amounts of water. They must not allow entry of fungi, bacteria, or epifoliar algae. They must not be so nutritious and delicious to animals that they are a liability to the plant. They must not be such effective sails that the plant is blown over in a mild wind, and they must be cheap enough that the plant spends less carbohydrate building them than it recovers by their photosynthesis. If a foliage leaf fails in any of these aspects, the plant dies and the leaf's photosynthesis has been useless. Many structural and physiological aspects of leaves that make them waterproof, pathogen resistant, and so on in some way interfere with photosynthesis. Although we might assume that foliage leaves are maximally adapted for photosynthesis, these other factors affect leaf biology and must be considered as well.

During photosynthesis, leaves absorb carbon dioxide and convert it to carbohydrate by using light energy. Because sunlight comes from one direction at a time, it is best for a portion of the leaf to be flat and wide, allowing maximum exposure to light and maximum surface area for carbon dioxide absorption (**Figure 6-1**). Because chlorophyll absorbs sunlight efficiently, light penetrates only a short distance, and thus, leaves can be quite thin; in a thick

layer of chlorenchyma, the lowest cells would be in almost complete darkness, unable to photosynthesize. This flat, light-harvesting portion is the **leaf blade** (also called the **lamina**). The blade's lower side is its **dorsal surface**, and larger veins protrude like backbones (synonym: **abaxial side**); the upper side is the **ventral surface** (synonym, **adaxial**) and is usually rather smooth.

Most leaves have a **petiole** (stalk) that holds the blade out into the light (**Figure 6-2**). This prevents shading of leaf blades by those above them; self-shading would defeat the basic usefulness of the leaf. Petioles have other consequences. Long, thin, flexible petioles allow the blade to flutter in wind, cooling the leaf and bringing fresh air to its surface. If the leaf and air are still, carbon dioxide is absorbed from the vicinity of the leaf faster than diffusion brings more to it, depleting the carbon dioxide and decreasing photosynthesis (**Table 6-1**). Leaf flutter also makes it difficult for insects to land on a leaf and knocks off some that have already alighted.

If leaves are small or very long and narrow, self-shading is not a problem, and there may be no petiole; the leaf is then called a **sessile leaf** instead of **petiolate**. For example, *Aeonium* grows in arid, sunny regions, and its fleshy leaves are packed close together (**Figure 6-3**). Because the sunlight is so intense, even with self-shading

Figure 6-2 The petioles of each leaf of poplar (*Populus alba*) are long enough that the leaves do not shade each other, and all are fully exposed to light. The lower side of each leaf blade is covered with white, dead trichomes, the upper surface is hairless and green.

Leaves

Figure 6-3 *Aeonium tabuliforme* and its relatives grow in regions of intense sunlight; self-shading is not a problem, but water conservation is. Close packing minimizes water loss from stomata.

TABLE 6-1	Movement of Carbon Dioxide
Average velocity of a molecule at 30°C	381 m/s
Mean free path (distance a molecule moves before striking another molecule) at 1 atm pressure	39 nm
Number of collisions of one molecule at 1 atm pressure	9.4 billion collisions per second

As any molecule diffuses, it collides with other molecules and changes its direction of motion. Although air is almost entirely empty space, a molecule of carbon dioxide undergoes 9,400,000,000 collisions per second and moves only 0.039 mm each time.

the leaves receive enough light for adequate photosynthesis. Close packing of these leaves helps trap water molecules and prevent their escape from the plant.

In many monocots, such as grasses, irises, lilies, agaves, and yuccas, foliage leaves tend to be very long and tapered, and self-shading occurs only at the base; most of the blade is well exposed to light (**Figure 6-4** and see Figure 6-15). They also typically lack a petiole; instead, the leaf base wraps around the stem to form a **sheathing leaf base**. The blade can flex and flutter even without a petiole. Not all monocots have linear, grass-like leaves. Palms, aroids (such as *Monstera* and *Philodendron*), and bird-of-paradise plants have what appear to be petioles and laminas, but the leaves of these species are believed to have evolved from grass-like leaves, and the "petiole" may actually be a modified portion of the lamina (**Figure 6-5**).

Leaf blades

Stem

Petiole

(a)　　　(b)　　　(c)

Sheathing leaf bases

Sheathing leaf base

Figure 6-4 (a) Long, narrow leaves are common in many monocots of intensely sunny areas. Shading is no problem, although leaf bases probably do not photosynthesize at maximum efficiency. (b) These are the leaves of St. Augustine grass; the blades project horizontally and are attached to the stem by sheathing leaf bases. (c) In cattails, the sheathing leaf base can be half a meter long and provide firm but flexible attachment to the stem.

Figure 6-5 *Maranta,* a common houseplant, is a monocot that has broad leaves.

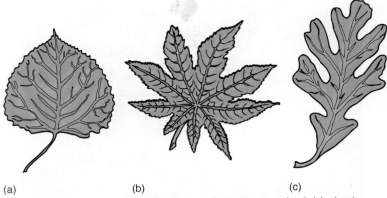

Figure 6-6 Simple leaves have numerous shapes, some so deeply lobed as to be almost compound. Poplar (a) and oak (c) have one dominant midrib from which lateral vascular bundles extend. In castor bean (b), petiolar bundles immediately diverge into several main veins.

A leaf blade may be either simple or compound. A **simple leaf** has a blade of just one part (**Figure 6-6**), whereas a **compound leaf** has a blade divided into several individual parts (**Figure 6-7** and **Figure 6-8**). Think of how this might be affected by wind: If the blade is either large or flimsy, it twists and flexes and may tear. Tearing can be prevented by making the leaf small or tough or "pretorn" (compound). A compound leaf has many small blades (**leaflets**), each attached by a **petiolule** to an extension of the petiole, the **rachis**. Leaves may be palmately compound, with all leaflets attached at the same point, or pinnately compound, with leaflets attached individually along the rachis (Figure 6-7). Even a large compound leaf has only small leaflets, all of which can flex individually without tearing. Flowing water is even denser than blowing air: Plants that grow immersed in streams typically have compound leaves with thread-like leaflets (**Figure 6-9**).

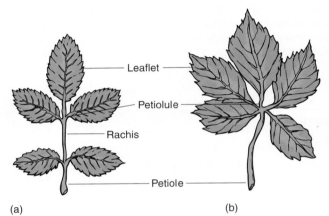

Figure 6-7 Leaflets of compound leaves have the same structure and metabolism as simple leaves. In pinnately compound leaves (rose, a) the petiolules of leaflets are attached to the rachis, but in palmately compound leaves (Virginia creeper, b) they attach to the end of the petiole. In either type, any leaflet can be damaged or can abscise without affecting the remainder of the leaf.

Figure 6-8 This is a portion of one doubly compound leaf of *Mimosa:* The central rachis (midrib) bears numerous leaflets, but each leaflet itself is pinnately compound.

Figure 6-9 *Myriophyllum heterophyllum* grows immersed in flowing streams; its leaves are very delicate and pinnately compound with thread-like leaflets (leaflets have no lamina). If the stream is quiet, some stems emerge into the air and produce thicker, tougher leaves (not present here), hence the species name "heterophyllum."

Plants and People

Box 6-1 Leaves, Food, and Death

Leaves impact our lives every day. Examples that come to mind readily are leafy foods such as artichokes (*Cynara scolymus*), cabbage (*Brassica oleracea* variety *capitata*), celery (petioles; *Apium graveolens*), lettuce (*Lactuca sativa*), onions (*Allium cepa*), and spinach (*Spinacia oleracea*). Also important are numerous herbs and spices: basil (*Ocimum basilicum*), bay leaves (*Laurus nobilis*), marjoram (*Origanum majorana*), oregano (*Origanum vulgare*), parsley (*Petroselinum crispum*), sage (*Salvia officinalis*), tarragon (*Artemisia dracunculus*), thyme (*Thymus vulgaris*), and the flavorings mint (several species of *Mentha*), spearmint (*Mentha spicata*), and peppermint (*Mentha piperita*) (**Figures B6-1a, b,** and **c**). The flavors and pungency of these are due to chemicals located within the leaves themselves or in trichomes of the leaf epidermis. Most of these chemicals probably serve as antiherbivore defensive compounds, causing animals other than humans to avoid the plants. Many classes of antiherbivore chemicals have evolved, ranging from only mildly effective ones, such as these flavors, to others that are much more powerful, such as alkaloids, many of which are toxic in small amounts and kill quickly. The alkaloids in poison hemlock and death camas are particularly effective.

An alkaloid of the leaves of one plant in particular is of interest to us—it is quite lethal but acts only slowly: nicotine in the leaves of tobacco (*Nicotiana tabacum*) (**Figure B6-1d**). If tobacco leaves are eaten, they cause vomiting, diarrhea, and even death caused by respiratory failure; however, most tobacco leaves are smoked, of course. Americans smoked an all-time high of 4,345 cigarettes per capita in 1963. Since then, consumption has declined for white males but recently has increased for other groups. Tobacco leaves contain between 0.6% and 9.0% nicotine, and an ordinary filter cigarette has 20 to 30 mg of the alkaloid, approximately 10% of which is absorbed by the lungs. Nicotine dissolves readily into our mucous membranes and passes quickly into our blood stream. Because it is transferred across the placenta, women who smoke during pregnancy may give birth to babies addicted to nicotine. Blood-borne nicotine also affects the heart, causing coronary problems: People who smoke a pack or more a day are over three times more likely than nonsmokers to die of heart disease. After nicotine is taken into the cells of the mouth and lungs of a smoker, it can cause cancer—of the lungs especially, but also of the throat, larynx, and mouth. If detected early enough, nicotine-induced lung cancer can be combated with surgery and chemotherapy, but after the cancer has spread to the lymph system, the prognosis is not good. Lung cancer causes more than 400,000 deaths per year in the United States.

(a)

(b)

(c)

(d)

Figure B6-1   (a) Basil (the dots are glands that provide its flavor). (b) Spearmint. (c) Rosemary. (d) Tobacco.

Compound leaves have other advantages. When a mild breeze blows over two leaves of equal size, one simple and the other compound, it tends to flow smoothly over the simple leaf but turbulently over the more complex surface of the compound leaf. Turbulence brings in carbon dioxide and removes excess heat. Also, an insect can crawl or a fungus can spread across the entire blade of a simple leaf rather easily, but the edges and petiolule of a leaflet act as barriers that either prevent or at least slow movement to the rest of the leaf. In some plants, if pathogens severely damage a leaf, it abscises, carrying the pathogens with it and helping to keep pests away from healthy leaves.

Is it reasonable to conclude that plants with compound leaves are better adapted than plants with simple ones? Why are there any simple leaves at all? Most very large leaves of plants are compound, apparently the only feasible architecture. The large leaves of palms and bananas are formed with a large, single lamina, but they have numerous special lines of weakness and are quickly torn to a "compound" condition by wind. The largest simple leaves, those of the Victoria water lily, float on water and thus do not need large amounts of sclerenchyma for support, and adhesion to the water's surface prevents wind damage. Among medium and small leaves, simple ones are common, perhaps because a greater percentage of a simple leaf is composed of photosynthetic cells, whereas a compound leaf has a great deal of nonphotosynthetic rachis, petiolules, and edges.

Although most pinnately compound leaves are easily recognizable as leaves, some can be mistaken for a stem with simple leaves. Close examination reveals that leaflets never bear buds in the axils of their petiolules, and the tip of the rachis never has a terminal bud. Leaflets are always arranged in two rows, never in a spiral, whorled, or decussate phyllotaxy.

Some of the tremendous range of shapes of leaves and leaflets is shown in **Figure 6-10**. There are hypotheses but no total agreement on how different leaf shapes may be adaptive; many

Figure 6-11 (a) Beans have two types of leaves: The very first leaf formed by the seedling is simple (right), but all later leaves are compound (left). (b) This *Azara lanceolata* shoot appears to have two types of leaves: The larger leaves are true leaves, whereas the smaller ones are actually enlarged stipules.

biologists believe that all function so well that none is clearly superior or strongly disadvantageous selectively. A large number of species, however, have several types of leaves (**Figure 6-11a**). In the simplest cases, the first few leaves of a seedling are distinctly different from all leaves produced later; if they were merely smaller, it could be argued that the seedling does not have enough stored energy resources to construct the large leaves that a mature plant can afford. But very often leaves of juvenile plants differ from those of adults not only in size but also shape, texture, and even simple versus compound structure. In species such as ivy or citrus, transition from juvenile foliage to adult leaves does not take place until the plant is old enough to flower, perhaps more than 10 years old. It seems reasonable to hypothesize that juvenile leaves of seedlings are adaptive in the microhabitat close to the soil surface where a seedling is located, whereas adult leaves are adapted to the more aerial microhabitat inhabited by an older, larger plant.

Other species produce two types of leaves simultaneously (Figure 6-11b), often one type on long shoots and the other on short spur shoots. In cacti, the spines are short-shoot leaves, whereas the long-shoot leaves are green, fleshy, and in most species almost microscopic. In a small percentage of species, leaf shape and size are irregularly variable even on one stem, being influenced by environmental conditions that occur at the time of leaf initiation and expansion; amounts of sunlight and minerals are perhaps the most important factors.

Despite numerous possibilities for variation within a species, leaf shape is a valuable tool for plant identification: Overall shape, including leaf base and apex, is important, as is the margin. The margin may be entire (smooth), toothed, lobed, or otherwise modified (**Figure 6-12**).

Within a leaf are **veins** or bundles of vascular tissue. These distribute water from the stem into the leaf and simultaneously collect sugars produced by photosynthesis and carry them to the stem for use or storage elsewhere. In basal angiosperms and eudicots, they occur in a netted pattern called **reticulate venation** (**Figure 6-13** and **Figure 6-14**). In monocots with long,

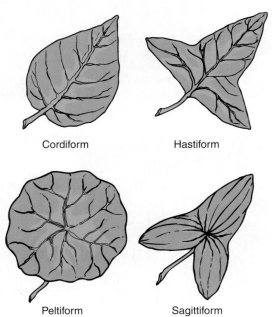

Cordiform   Hastiform

Peltiform   Sagittiform

Figure 6-10 Four common leaf shapes; numerous other types also exist.

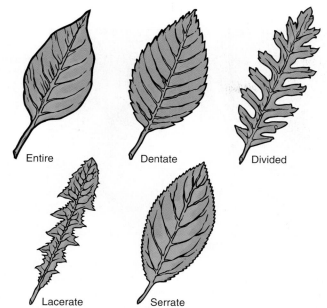

Figure 6-12 Several common types of leaf margins; in nature, hundreds of types exist, many being intermediate between two or among several other types.

strap-shaped leaves, the larger veins run side by side with few obvious interconnections: This is **parallel venation** (Figure 6-15 and Figure 6-16).

At the leaf base, usually in the petiole, is an **abscission zone** oriented perpendicular to the petiole; its cells are involved in cutting off the leaf when its useful life is over (Figure 6-17). In autumn, as deciduous leaves begin to die, abscission zone cells release enzymes that weaken their walls. As the leaf twists in the wind, cell walls break, and the leaf falls off. Adjacent undamaged

Figure 6-13 A fig leaf (*Ficus*) held up to the light reveals five prominent main veins, numerous lateral veins and the reticulate venation of all smaller veins in this eudicot leaf; no part of the leaf is without veins.

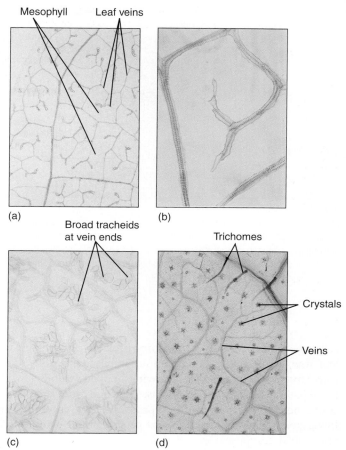

Figure 6-14 These leaves were treated (cleared) to make their cells transparent; then xylem was stained to make leaf veins visible. (a) In passionflower (*Passiflora*) leaves, veins are narrow and highly branched. At this magnification (×40), the height of the photograph represents 1 mm, and veins occur close together. Leaf veins typically occur closer together than do capillaries in our bones, tendons, and cartilage. (b) Higher magnification of privet leaf. At any point, minor veins are only two or three tracheary elements wide (×150). (c) These minor veins of crown of thorns (*Euphorbia millii*, a species that grows in deserts) are broad and have many wide tracheids at the ends (×50). (d) *Acalypha* has numerous crystals within the leaf and trichomes on the leaf surface (×80).

Figure 6-15 Cattails (*Typha latifolia*) have long, strap-like leaves common in monocots. Vascular bundles run parallel to each other from leaf base to tip.

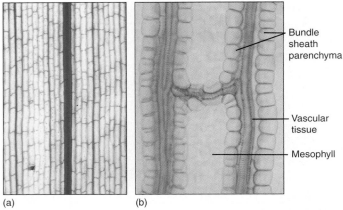

(a)   (b)

Figure 6-16 (a) Leaf clearing of corn (*Zea*), a monocot, with several sizes of longitudinal, parallel veins. If a longitudinal vein is broken by insect damage or any other problem, conduction can detour around the site by means of the fine transverse veins (×50). (b) Higher magnification showing the parenchyma cells that surround the leaf bundles in corn (×250).

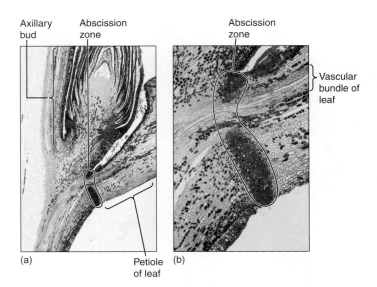

(a)   (b)

Figure 6-17 (a) Longitudinal section through a leaf axil, showing the abscission zone in the petiole and the axillary bud just above (×15). (b) Magnification of the abscission zone while the leaf is still healthy. The xylem and phloem are not interrupted until the leaf actually falls away (×80).

cells swell and become corky, forming protective scar tissue, the **leaf scar**, across the wound. Without an abscission zone, dead leaves might tear off irregularly, leaving an open wound vulnerable to pathogens. In many monocots, ferns, and cycads, an abscission zone is not formed; instead, a protective corky layer protects living stem tissues, but the dead leaf persists on the plant until it gradually decomposes (**Figure 6-18**).

Figure 6-18 As leaves of Joshua tree (*Yucca brevifolia*) age, they flex downward and do not abscise; this creates a sheath of tough, fibrous dead leaves with many sharp spines that prevent rats, mice and other small animals from climbing the plant and eating young leaves and buds.

Internal Structure of Foliage Leaves

Epidermis

Flat, thin foliage leaves (optimal for light interception) have a large surface area through which water can be lost. Water loss through the epidermis is called **transpiration** and is a serious problem if the soil is so dry that roots cannot replace lost water. The epidermis must be reasonably waterproof but simultaneously translucent, and it must allow entry of carbon dioxide. Leaf and stem epidermises are basically similar, consisting of a large percentage of flat, tabular (shaped like paving stones), ordinary epidermal cells; guard cells and trichomes (either glandular or nonglandular) may be abundant. The dorsiventral nature of leaves, however, causes their upper and lower epidermis to exist in significantly different microclimates. On a sunny day, a leaf is usually warmer than surrounding air; thus, it heats the air and convection currents rise from it. If stomata in the upper epidermis are open and losing water, the water molecules are swept away by this convection. However, air tends to be trapped on the underside of a leaf, so water loss from stomata there is not so great: Water molecules are trapped in quiet air and may diffuse back into the stomata. In most leaves, the number of stomata per square centimeter is much greater in the lower than in the upper epidermis; in many species, no stomata at all are found in the upper epidermis (**Table 6-2**).

Such unilateral distribution of stomata has other beneficial consequences. Air-borne spores of fungi and bacteria are continually landing on leaves. Rye leaves, for example, typically have 10,000 fungal spores per square centimeter, but most are on the upper surface where stomata are rare; thus, few fungi can penetrate the leaf.

TABLE 6-2	Frequency of Stomata in Leaf Upper and Lower Epidermis		
		Stomata Per Cm²	
Species	Common Name	Upper	Lower
Tilia europea	linden	—	37,000
Allium cepa	onion	17,500	17,500
Helianthus annuus	sunflower	12,000	17,500
Pinus sylvestris	pine	12,000	12,000
Zea mays	corn	9,800	10,800
Vicia faba	bean	6,500	7,500
Sedum spectabilis	stonecrop	2,800	3,500
Larix decidua	larch	1,400	1,600

(a)

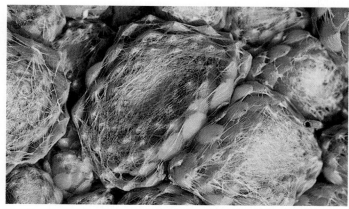

(b)

Figure 6-20   (a) Leaves of lamb's ears (*Stachys*) are so densely covered with hairs that they often appear white. The hairs reduce the amount of heat these leaves absorb from sunlight. (b) Plants of *Sempervivum* are native to high alpine regions where ultraviolet radiation is intense. The hairs of the leaf epidermis reflect part of the ultraviolet radiation away from the plant.

Leaf stomata are frequently sunken into epidermal cavities that create a small region of nonmoving air. In oleander, the lower leaf epidermis contains numerous crypts (areas where the epidermis is depressed into the leaf), and stomata and trichomes are abundant in the epidermis that lines the crypts (**Figure 6-19**). Both the recessed nature of the crypt and the presence of trichomes keep air in the crypt extremely quiet; many water molecules that diffuse out of open stomatal pores bounce around the crypt and re-enter the stomata rather than being blown away.

Leaf epidermises are often remarkably hairy, and trichomes affect leaf biology in numerous ways. They provide some shade on the upper surface of the leaf, deflecting excessive sunlight, an adaptation common in desert plants (**Figure 6-20**). On the lower surface, they prevent rapid air movement and slow water loss from stomata (Figure 6-2). In any position, trichomes make walking or chewing difficult for insects, and many glandular trichomes secrete powerful stinging compounds that prevent even large animals from eating the leaf. Insects either do not bother with the leaf or must walk so slowly on it that they become more vulnerable to their own predators. Conversely, hairs provide excellent footholds against leaf flutter for insects of the appropriate size. Poisonous glandular trichomes in stinging nettle and other species prevent mammals from eating leaf tissue but also

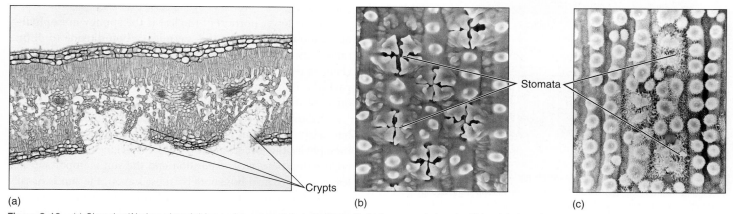

(a)   Crypts   (b)   Stomata   (c)

Figure 6-19   (a) Oleander (*Nerium oleander*) leaves have stomatal crypts filled with trichomes and stomata. This minimizes air movement near stomata, and water molecules that diffuse out of the leaf may re-enter it by random motion. Crypts occur only in the lower surface (×80). (b) Each stoma of this grass (*Elytrostrachys*) is overarched by four protrusions from surrounding cells. This reduces air movement near the stoma (×940). (c) In this bamboo (*Chusqua*), stomata are protected by protrusions and wax rods (×940).

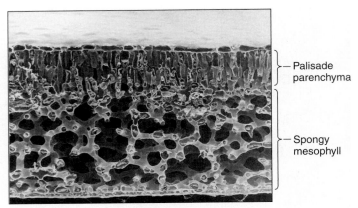

Figure 6-21   This leaf of *Laurelia* has a layer of palisade parenchyma that is two cells deep, and the spongy mesophyll is an extensive aerenchyma (×180).

protect leaf-borne insects small enough to walk, feed, and reproduce safely between the stinging trichomes.

Like stem epidermis, leaf epidermal cells contain a coating of cutin and usually also wax on their outer walls. These retain water and make digestion by fungi difficult. Their smooth, slippery surface prevents spores from sticking or allows them to be washed off by rain.

■ Mesophyll

The ground tissues interior to the leaf epidermis are collectively called **mesophyll** (Figure 6-21 and Figure 6-22). Along the upper surface of most leaves is a layer of cells, the **palisade parenchyma** (also called palisade mesophyll), which is the main photosynthetic tissue of most plants. Palisade cells are sepa-

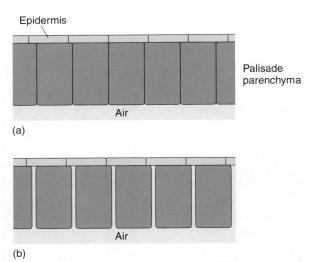

Figure 6-22   (a) If palisade cells are closely packed with no space between them, only their bottoms are exposed to the carbon dioxide in the spongy mesophyll. (b) With only slight separation of the cells, the volume of the palisade is almost unchanged, but the surface area available is increased enormously; carbon dioxide can be absorbed much more rapidly. Assume that the palisade parenchyma cells are rectangular columns $20 \times 20 \times 100$ μm long; compute the surface area exposed to CO_2 by 100 cells arranged as in (a) versus (b).

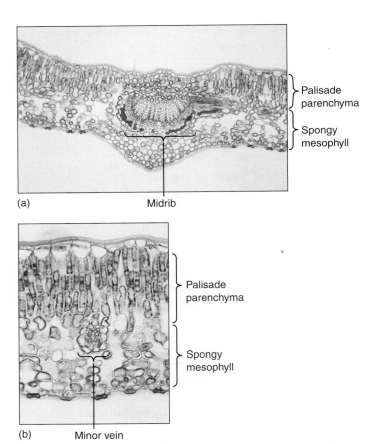

Figure 6-23   (a and b) Leaves of holly (*Ilex*) have an extremely thick palisade parenchyma consisting of at least three layers of columnar cells. Cells of the lowest palisade layer are relatively widely separated, permitting rapid diffusion of CO_2 from spongy mesophyll to upper palisade (a, ×50; b, ×150).

rated slightly so that each cell has most of its surface exposed to the intercellular spaces. Because carbon dioxide dissolves into cytoplasm slowly, the large surface gives maximum area for dissolution; tightly packed cells could not absorb enough carbon dioxide for efficient photosynthesis (Figure 6-22). Palisade parenchyma is often only one layer thick, but in regions with intense, penetrating sunlight, it may be three or four layers thick (Figure 6-23).

In the lower portion of the leaf is the **spongy mesophyll**— open, loose aerenchyma that permits carbon dioxide to diffuse rapidly away from stomata into all parts of the leaf's interior (Figures 6-21 and 6-23). If stomata were surrounded by closely packed cells, a molecule of carbon dioxide might simply bounce off a cell and escape back out the stomatal pore.

Although this arrangement is the most common, some plants have a layer of palisade parenchyma along both leaf surfaces; spongy mesophyll either occurs in the center or is lacking (Figure 6-24). The relationship between leaf position and the sun is important: For leaves that are held horizontally (most leaves), the sun is usually overhead, so having palisade parenchyma near the upper surface permits maximum absorption of light and photosynthesis. For plants that hold their leaves vertically (*Iris, Gladiolus,* and *Eucalyptus*), both sides are equally illuminated, and palisade parenchyma is equally functional on either side.

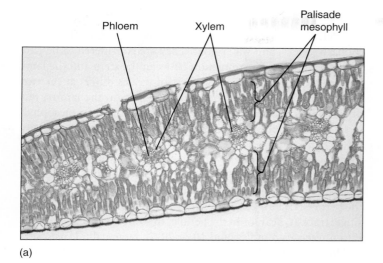

(a)

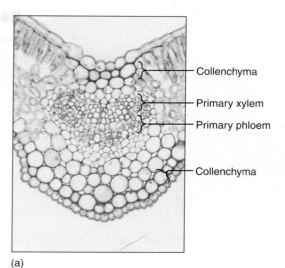

(a)

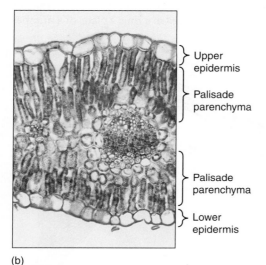

(b)

Figure 6-24 Leaves of both carnations (a) and creosote bush (b) have palisade parenchyma along both surfaces. The little spongy mesophyll present is located in the middle of the leaf (a, ×75; b, ×150).

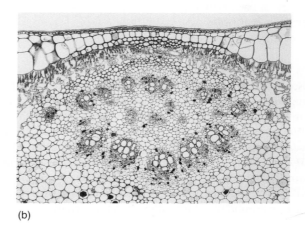

(b)

■ Vascular Tissues

Between the palisade parenchyma and spongy mesophyll are vascular tissues. A eudicot leaf usually has one large **midrib**, also called a midvein, from which **lateral veins** emerge that branch into narrow **minor veins** (see Figures 6-13 and 6-14). Minor veins are most important for releasing water from xylem and loading sugar into phloem, whereas the midrib and lateral veins are involved mostly in conduction. Vein structure changes with size: The midrib and lateral veins always contain both primary xylem on the upper side and primary phloem on the lower side (**Figure 6-25**). Because they both conduct and support the leaf blade, they may have many fibers arranged as a sheath, called a **bundle sheath**, around the vascular tissues (**Figure 6-26**). A sheath also makes it difficult for insects to chew into the vascular tissues. Many other types of nonconducting cells such as mucilage, tannin, or starch storage cells may be present in larger veins. Minor veins are the sites of material exchange with the rest of the mesophyll and must have a large surface area in contact with palisade and spongy mesophyll; they do not contain fibers or other nonconducting cells

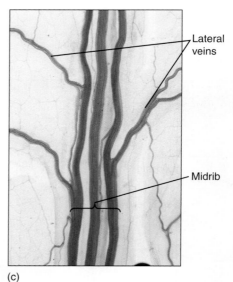

(c)

Figure 6-25 Transverse sections through leaf midribs. (a) This midrib of *Ligustrum* is rather simple, containing a single vascular bundle. Xylem is present along the top, phloem along the bottom. A small amount of collenchyma occurs on the top and bottom of the midrib (×150). (b) This midrib of rubber tree has numerous separate bundles embedded in a large mass of mesophyll (×50). (c) This is a midrib prepared as a leaf clearing (similar to Figure 6-14); the three distinct bundles have no interconnections in this region. Each transports to and from a particular portion of the blade (×15).

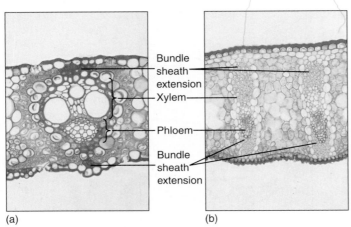

(a)   (b)

Figure 6-26 (a) This leaf bundle of sugar cane has bundle sheath extensions contacting both the upper and lower epidermis. Water diffuses from the xylem to the epidermis by capillary action, moving around the surfaces of the fibers (×100). (b) The leaves of flax have large bundle sheath extensions consisting of many layers of fibers (×45).

whose presence would interrupt this contact (**Figure 6-27**). The endings of the minor veins in some species consist of only xylem and in others only phloem, but they typically contain both.

Veins, especially larger ones, often have a mass of fibers above, below, or both—the **bundle sheath extension** (Figure 6-26). Such fibers help give rigidity to the blade and are believed to provide an additional means by which water moves from the bundle out to the mesophyll. Apparently, water moves by capillary action around the fibers rather than through them.

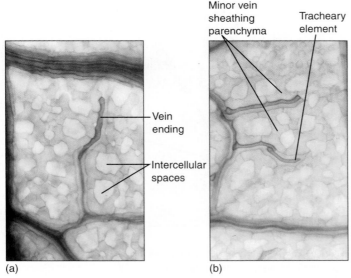

(a)   (b)

Figure 6-27 (a and b) These are minor veins, treated to reveal individual cells. Long, dark cells are tracheary elements, and phloem is out of focus, in line with the xylem. Notice how narrow these veins are: Xylem is only one element wide, so there is maximum surface area for releasing water to the leaf mesophyll. In this species, each minor vein is lined with parenchyma cells that extend outward into the spongy mesophyll (×250).

■ Petiole

Petioles may be tiny but are massive in plants like palms, rhubarb, celery, and water lilies. They are considered to be part of the leaf and are the transition between the stem and the lamina; consequently, the arrangement of tissues differs at the two ends. The epidermis may be similar to that on the lamina but often contains fewer stomata and trichomes. Petiole mesophyll is rather like cortex—somewhat compact and not especially aerenchymatous—and considerable collenchyma is present if the petiole supports a heavy lamina. Vascular tissues are the most variable; one, three, five, or more vascular bundles, called **leaf traces**, branch from stem vascular bundles and diverge toward the petiole (**Figure 6-28a**). They may remain distinct or fuse into a single trace at, near, or in the petiole. In some species, they divide into numerous bundles, and 10 or 20 (in large palm leaves, several hundred) may be found in the petiole (Figure 6-28b). Vascular bundles may either fuse with each other within the petiole or branch further; they can be arranged in a ring, a plate, or a number

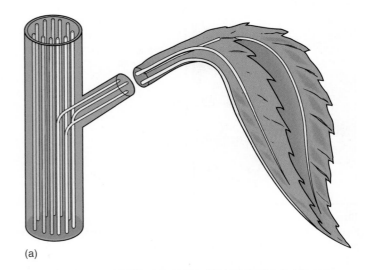

(a)

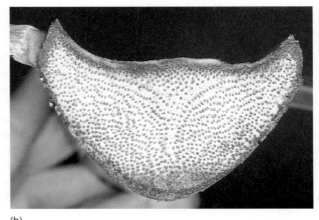

(b)

Figure 6-28 (a) This section of stem has 11 bundles, 3 of which produce leaf traces that enter the petiole. Later, the leaf traces diverge and spread throughout the blade. These traces are drawn without any branching or merging, although both are possible. (b) Cross-section of a petiole of the palm *Attalea* stained to show its many vascular bundles. Notice how numerous the bundles are compared with the leaves of Figure 6-25. This leaf, like those of most palms, is very large and must have a massive conduction system.

Botany and Beyond

Box 6-2 Botany and Beyond: Leaf Structure, Layer by Layer

The internal organization of ordinary leaves is not very complicated but it can be difficult to visualize from looking only at transverse sections. A series of sections through a leaf of privet (*Ligustrum*) is presented here. These are paradermal sections; that is, they are parallel to the epidermis. The first is at the level of the palisade parenchyma, and each successive section is deeper in the leaf. Try to imagine the appearance of the leaf from the perspective of a carbon dioxide molecule.

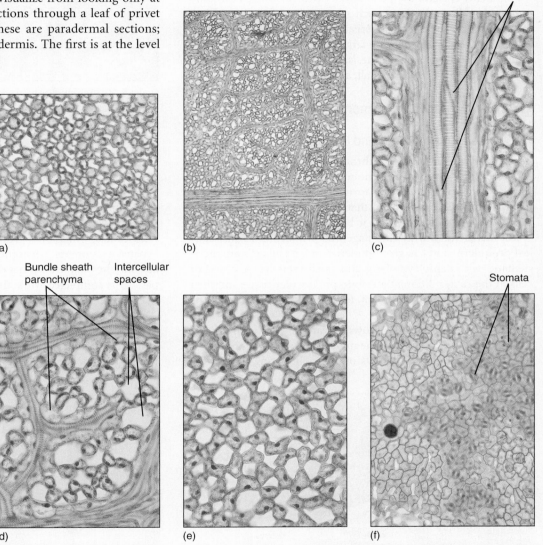

Figure B6-2 (a) Palisade parenchyma (×250). The cells are cylindrical and are not closely packed; carbon dioxide diffuses easily throughout this layer and be absorbed into the cells through their large surface area. (b) Leaf veins (×50). Just below the palisade parenchyma lies the network of leaf veins. Small lateral veins branch off of larger veins and in turn give off smaller veins. (c) High magnification of a large lateral vein (×250). This section passes through the xylem of the vein, and perforations are visible; thus, these are vessel elements, not tracheids. These have spiral thickenings: They could differentiate while the leaf was still small, and then as the leaf expanded, they could be stretched and still conduct water. (d) High magnification of minor veins (×250). All mesophyll cells are close to one vein or another. After sugar has been produced by photosynthesis, it does not diffuse very far before it is picked up by a sieve tube member and transported to the stem. (e) Spongy mesophyll (×250). In many leaves, more than half the spongy mesophyll consists of intercellular air space. This allows carbon dioxide molecules to diffuse rapidly into the leaf, away from the stomata. If spongy mesophyll were more dense, carbon dioxide molecules might be trapped near the stomata, then diffuse back out of the leaf before they could be absorbed by a chlorenchyma cell. (f) Lower epidermis (×150). The epidermis is slightly undulate, and as it was cut to make this slide, the knife passed through the center of the cells in some areas (green) and through the outer wall and cuticle in other areas (red). Stomata are abundant, but the actual stomatal pore is short and narrow; each is only a small opening through which carbon dioxide can enter.

of other patterns. If the lamina has a strong midrib, most petiole bundles fuse together and form the midvein, but other bundles may enter the lamina as small lateral veins.

We do not understand the significance of all the bundle patterns in petioles. Certain patterns may ensure the proper distribution of sugars out of different parts of the lamina, especially of large leaves, into the various bundles of the stem. However, it may be that almost any pattern functions well, and thus, mutations that cause new patterns are not selectively disadvantageous. On the other hand, future studies may show correlations between lamina, petiole, and stem.

In many species, the petiole bears two small flaps of tissue at its base called **stipules**, which serve various functions. They may protect the shoot apical meristem while the leaf is young and small. In other plants, they are large enough to contribute a significant amount of photosynthesis, but usually, when the leaf is mature, the stipules are still small, and they die early.

Initiation and Development of Leaves

Basal Angiosperms and Eudicots

Leaves are produced only through the activity of a shoot apical meristem. At the base of the meristem, cells just interior to the protoderm grow outward, forming a protrusion known as a **leaf primordium** (Figure 6-29) that extends upward as a narrow cone, growing so rapidly that it becomes taller than the shoot apical meristem. During this stage, the primordium consists of leaf protoderm and leaf ground meristem, and all cells are meristematic, with dense cytoplasm and small vacuoles. A strand of cells in the center differentiates into provascular tissue and then into primary xylem and phloem, forming a connection with the young bundles in the stem.

As the leaf primordium grows upward, it increases in thickness, establishing the bulk of the midrib. A row of cells on either edge of the primordium grows outward, initiating the lamina

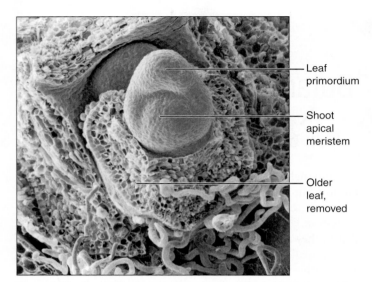

Figure 6-29 The shoot apical meristem of grape (*Vitis riparia*) can be seen in three dimensions in this scanning electron micrograph. Notice the large leaf primordium; an older leaf primordium was removed; the phyllotaxy is established at the very apex (×200).

(Figure 6-30). As a result of their activity, the young leaf consists of a midrib and two small, thin wings. All cells in the wings are meristematic, and their division and expansion enlarge the lamina rapidly (Figure 6-31). In a compound leaf, two rows of loci initiate leaflets, which then grow like simple leaves. During lamina expansion, stomata, trichomes, and vascular bundles differentiate, and the petiole becomes distinct from the midrib. The entire pattern is completed while the leaf is still very small, much less than one tenth its mature size.

In many perennial plants, leaves are initiated in the summer or autumn before they mature. After they reach the developmental state just described, they become dormant, part of a resting terminal or axillary bud. During the next growth period,

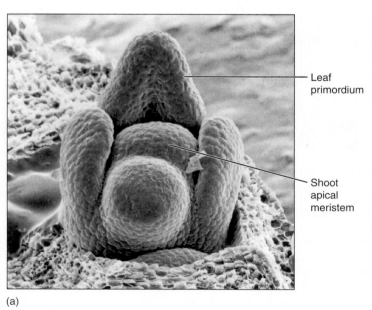

(a)

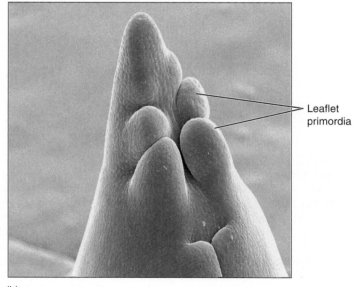

(b)

Figure 6-30 (a) The uppermost leaf primordium is the oldest, and its sides have begun growing out as the lamina of a simple leaf of grape (×270). (b) In this compound leaf, rather than forming a lamina, six leaflet primordia have been established (×150).

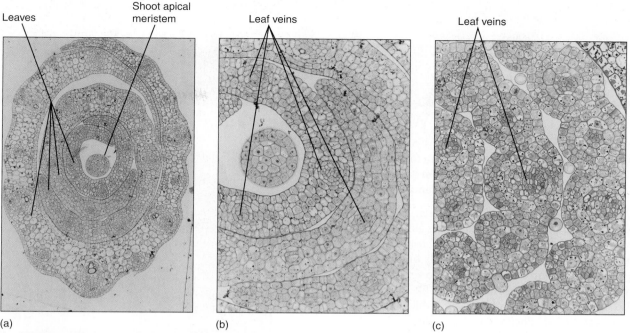

Leaves   Shoot apical meristem   Leaf veins   Leaf veins

(a)   (b)   (c)

Figure 6-31 (a) Low-magnification transverse section through young, developing leaves of the grass *Panicum effusum*. The central round structure is a section of the shoot apical meristem, and the leaves encircle it (grasses have sheathing leaf bases) (×200). (b) At higher magnification, you can see that all cells in the leaves are still small and cytoplasmic, and the central vacuoles are small. At this stage, all cells are still undergoing mitosis and cytokinesis. Veins are starting to differentiate (×500). (c) At an older stage, leaf veins are more distinct. The first protoxylem tracheary elements are mature in some veins, but most cells that will differentiate into tracheary elements are still enlarging (×500).

usually the following spring, the bud opens and the primordial leaves expand rapidly as each cell absorbs water into its vacuole and swells (**Figure 6-32**). Little or no mitosis or cell division may occur—only maturation, especially the synthesis of chlorophyll, cutin, and wax. As the leaves expand, the immature, exposed epidermis is very vulnerable, especially to insects (**Figure 6-33**). In many species of annual plants, the process is similar, with the embryo in the seed acting like the buds of perennial plants. They initiate leaves before the seed becomes dormant and dry, while it

Figure 6-32 Buds of buckeye (*Aesculus arguta*) expanding in the spring. Bud scales are being pushed back and will abscise as new foliage leaves swell. The smaller bud on the right contains only leaves, and its apical meristem may be initiating new leaf primordia that will expand a little later. The larger bud on the left has both new leaves as well as flower primordia (the small ball-like structures).

Figure 6-33 Aphids and other sucking insects typically attack young leaves and stems that have not yet developed any protective sclerenchyma around the vascular bundles.

Box 6-3 Photosynthesis Without Leaves

Photosynthesis may occur in tissues other than foliage leaf mesophyll, for example, stem cortex and bark. Carrying out photosynthesis in these alternative sites has particular consequences.

Bark. A small number of trees produce green, chlorophyllous photosynthetic bark. This is rare and occurs almost exclusively in desert-adapted trees that abscise their leaves during dry periods, after which they must rely entirely on their bark for sugar production. The trees, often called palo verde (Spanish for green stick), produce cork cells in their bark like all trees do, but they also produce cells called phelloderm, which contain chloroplasts (**Figure B6-3a**). Phelloderm is usually rather compact with few intercellular spaces, and very little of the cell walls are "free surface," that is, surface that faces intercellular space and that can absorb carbon dioxide. Consequently, carbon dioxide neither diffuses rapidly through phelloderm nor is it absorbed quickly. Furthermore, the layer of phelloderm cells is covered by several layers of cork cells, which are impermeable to gases and thus block uptake of carbon dioxide. The green bark of these trees probably carries out "recycling" photosynthesis: All living cells of the tree trunk and branches respire and give off carbon dioxide just like we animals do, and the green bark can recapture it and convert it back to carbohydrate. In trees with nonphotosynthetic bark, that carbon dioxide would simply leak out of the bark and be lost.

Cortex. Young stems of almost all plants have a chlorophyllous cortex. Most stems are too narrow to provide much surface area and thus capture little light. Whereas stomata are virtually always present in foliage leaf epidermis, they may be lacking in stem epidermis, and if so, little carbon dioxide can be absorbed. Even if stomata are present, stem cortex—like phelloderm—is typically more compact than leaf mesophyll and thus does not permit rapid diffusion or uptake of carbon dioxide.

Stem-succulent plants. Many desert plants store water in enlarged, chlorophyllous stems. Stem-succulents may produce large, thin, ordinary foliage leaves during the deserts' brief moist period, abscising these leaves when drought begins (these are "drought-deciduous" plants), or they may produce only very tiny leaves (often incorrectly said to be leafless). Either way, stem-succulent plants usually depend on their chlorophyllous cortex for much or all their photosynthesis, and their cortex typically has many leaf-like features. Their stem epidermis has a high density of stomata, sometimes having as many stomata per square centimeter as leaf epidermis. The outermost layer or two of cortex cells may be very aerenchymatous, creating a tissue similar to spongy mesophyll, allowing carbon dioxide to diffuse rapidly away from stomata and deeply into the cortex. Chlorophyllous cortex often

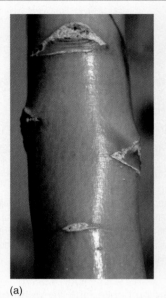 

(a)   (b)

Figure B6-3 (a) Trunks and branches of trees called "palo verde" (green stick; *Cercidium floridum*) are green because their bark contains cells with chlorophyll. No stomata are present, so no new carbon dioxide can be absorbed from the air. (b) Trunks and branches of saguaro cactus (pronounced "sa-WA-row"; there is no "g" sound at all) are green because the epidermis is still present and transparent, allowing us to see the green of the cortex cells inside. Stomata are present, and saguaros (*Carnegiea gigantea*) do bring in carbon dioxide from the air and convert it to carbohydrate. As a saguaro absorbs water after a rain, its parenchyma cells swell so the entire stem swells; the ribs flatten out and the epidermis is not torn. If it does not rain for many months, the cactus gradually loses water, despite its cuticle and wax, and shrinks; the ribs also shrink and become narrower (think of an accordion letting out air). A ribbed body like this can change volume without changing surface.

resembles palisade mesophyll, consisting of long columnar cells. Whereas palisade mesophyll is only rarely more than one or two cell layers thick, palisade cortex in stem succulents is often five to ten cell layers thick, being as much as 1.0 to 1.5 mm thick (about three to five times more than the entire thickness of a typical foliage leaf).

Any photosynthetic tissue must be well-vascularized, but cortex typically does not have its own set of vascular bundles. This vascular bottleneck can be overcome by several alternative arrangements. If the cortex is narrow, its chlorenchyma cells are close to the stem's vascular bundles. If the stem has short internodes, then leaf traces are also close to each other and provide some vascularization to the cortex as they run out to the petiole base. The cortex of cacti is exceptional in that it does have its own particular vascular system called cortical bundles. This is a network of collateral vascular bundles that branch and

anastomose forming a three-dimensional reticulate venation. Cactus cortex is well vascularized, and no matter how rapidly it photosynthesizes, phloem cells are close by and can quickly load the sugars and transport them away.

Broad succulent stems store water, which causes them a problem not usually faced by foliage leaves: how to allow volume to increase while surface area remains constant. We animals generate new skin anywhere, anytime, which is a good thing if we become overweight: our volume increases, as does our surface area, and our skin grows rather than tearing apart and revealing our insides. However, plants only generate new epidermis at shoot and root apices and in growing leaves and flowers. Stem succulents solve this problem by having an accordion-like, pleated surface (the pleats are called ribs). As water is stored and volume increases, ribs become wider; as water is lost, ribs become narrower. Volume changes dramatically but surface does not (Figure B6-3b).

In these alternative photosynthetic systems, water conservation is an important factor so none of these alternatives is free to evolve to be as effective at photosynthesis as an ordinary foliage leaf is. Most modifications that increase water conservation decrease photosynthesis and vice versa. In moist habitats, water conservation is of little importance, so modifications that increase photosynthesis will increase survival of the plants even though they cause plants to lose water more rapidly. In contrast, modifications that conserve water in dry habitats are more valuable than those that allow maximum photosynthesis. Of course, desert plants are not free to maximize water conservation to the point where they do not photosynthesize at all.

is still inside a developing fruit, and those leaves absorb water and expand rapidly during germination.

■ Monocots

Monocot leaves, like those of eudicots, are initiated by the expansion of some shoot apical meristem cells to form a leaf primordium (Figure 6-34). Apical meristem cells adjacent to the primordium grow upward along with it, becoming part of the primordium and giving it a hood-like shape. More apical meristem cells become involved until the primordium is a cylinder that completely or almost completely encircles the shoot apical meristem. This tubular portion grows upward as a sheathing leaf base, and the original conical leaf primordium, now located on one side of the top of the tube, gives rise to the lamina. The outer surface of the tube is abaxial epidermis, the inner surface is adaxial epidermis. Meanwhile, the shoot apex enlarges, forms new stem tissue, and initiates the next leaf primordium, which will develop as a tube inside the previous leaf's sheathing base. After the first leaf

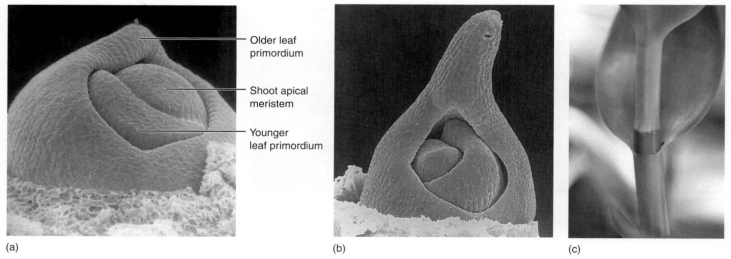

(a) —— Older leaf primordium
—— Shoot apical meristem
—— Younger leaf primordium

(b)

(c)

Figure 6-34 In a monocot, after the center of the leaf primordium has been initiated (a), neighboring regions of the shoot apical meristem become active and contribute to the primordium. This results in a leaf base that encircles the stem, forming a sheath around it (b) (a, ×150; b, ×120). (c) The leaf blade of this *Dendrobium* orchid has developed from a leaf primordium tip that resembled the tip in (b), and the sheathing base here developed from elongation growth of the small collar of leaf primordium tissue visible in (b).

has grown to its full size, the next leaf continues elongating and emerges through the opening at the top of the previous leaf's tubular sheathing base. Later, another, younger leaf will emerge and so on.

In some monocots, the lamina becomes broad and expanded like a eudicot lamina, but grasses, lilies, and many others have linear, strap-shaped leaves that grow continuously, having no predetermined size. Their lamina grows by a meristem located at its base where it attaches to the top of the sheathing leaf base. The lamina's meristematic cells remain active mitotically, producing new cells that extend the leaf. Even if a grazing animal, a range fire, or a lawnmower destroys much of the leaf, the meristem forms more lamina. This type of regeneration is not possible in most leaves.

The constant basal expansion in monocot leaves means that protoxylem and protophloem are constantly being stretched and disrupted in the basal meristem. New vessel elements and sieve tube members differentiate rapidly enough that conduction is never interrupted. Just above the basal meristem is a region where tissues differentiate: More primary xylem and phloem are initiated, as are stomata and other features. Higher above the basal meristem, tissues are mature; differentiation is similar to that for stems but is oriented upside down in monocot leaves.

Morphology and Anatomy of Other Leaf Types

Succulent Leaves

Numerous adaptations permit plants to survive in desert habitats, one of the most common being production of succulent leaves (Figure 6-35). This is characteristic of species in the families Crassulaceae (contains *Kalanchoe* and *Sedum*), Portulacaceae (contains *Portulaca* and *Lewisia*), and Aizoaceae (ice plant), among others. Succulent leaves are thick and fleshy, a shape that reduces the surface-to-volume ratio and favors water conservation. Some leaves are cylindrical or even spherical, the optimal surface-to-volume shape. The reduction in surface area, which is advantageous for water retention, has the automatic consequence of reducing the capacity for carbon dioxide uptake.

Inside the leaf, the mesophyll contains very few air spaces, reducing the internal evaporative surface area and, in turn, water loss through stomata. A lack of air spaces also makes the mesophyll more transparent, just as pure water is more transparent than soap bubbles or foam, allowing light to penetrate farther into the leaf. Photosynthesis occurs more deeply than it would in the foliage leaves described earlier. In some members of the genera *Lithops* (stone plants) and *Frithia*, leaves are so translucent they act as optical fibers; the leaves are located almost completely underground, where it is cool and relatively damp (Figure 6-36). The exposed leaf tips allow sufficient light to enter and be conducted to the subterranean chlorenchyma. Although the plants live in a harsh desert in Madagascar and southern Africa, photosynthesis actually occurs in a rather mild microclimate.

Sclerophyllous Foliage Leaves

Foliage leaves must produce more sugars by photosynthesis than are used in their own construction and metabolism, or the plant would lose energy every time it produced a leaf. This limits the amount of sclerenchyma in foliage leaves, and most leaves therefore tend to be soft, flexible, and edible.

In some species (barberry, holly, *Agave*, and *Yucca*), leaves have evolved which are perennial, existing on a plant for 2 or more years (Figure 6-37). With this extended lifetime and prolonged productivity, sclerenchymatous leaves are feasible, and their hardness makes them more resistant to animals, fungi, freezing temperatures, and ultraviolet light. Such plants are sclerophyllous and the leaves are sclerophylls. The sclerenchyma is often present as a layer just below the epidermis and in the bundle sheaths, although the epidermis itself can be composed of thick-walled cells (Figure 6-38). The cuticle is usually very thick, and waxes are abundant on leaves of many sclerophyllous species.

Leaves of Conifers

In almost all species of conifers, leaves are sclerophylls; they have a thick cuticle, and their epidermis and hypodermis cells have thick walls. Most conifer leaves contain unpalatable chemicals. Conifer leaves are always simple, never compound, and have only a few forms. Needles, either short or long, occur in all pines, firs, and spruces (Figure 6-39). Needles of longleaf pine can be 40 cm long, although most other species of pine have needles about 10 cm long.

(a)

(b)

Figure 6-35 (a) Leaves of *Senecio rotundifolia* are spherical, giving them an optimal surface-to-volume ratio for conserving water. (b) Leaves of the succulent *Dinteranthus* are hemispherical and attached to the opposite leaf, greatly reducing the exposed leaf surface. Interior tissues are mostly water-storage parenchyma. Their low levels of chlorophyll result in low levels of photosynthesis, but a lack of water is a more significant danger for these plants. Each plant consists basically of two leaves and a microscopic stem; the root system (not visible) is substantial. *Dinteranthus* and many related species have only two leaves at a time; as two new leaves expand, the existing two collapse and wither away.

Figure 6-36 *Lithops*, stone plant, is closely related to *Dinteranthus*; both are in the family Aizoaceae. *Lithops* also has only two leaves, but they are located almost entirely underground. The flat, translucent tips project above the soil and conduct light to the subterranean chlorenchyma.

Figure 6-37 Leaves of barberry (*Berberis*) are tough and hard. It is difficult for insects to bite into them or lay eggs in them; however, the plant must invest considerable glucose to make the secondary walls of sclerenchyma cells, and the leaves must therefore photosynthesize longer before they reach their break-even point.

Figure 6-39 (a) Many conifers, such as this lodgepole pine (*Pinus contorta*), have needle-shaped leaves. (b) Incense cedar (*Libocedrus*) and many other conifers have scale-shaped leaves.

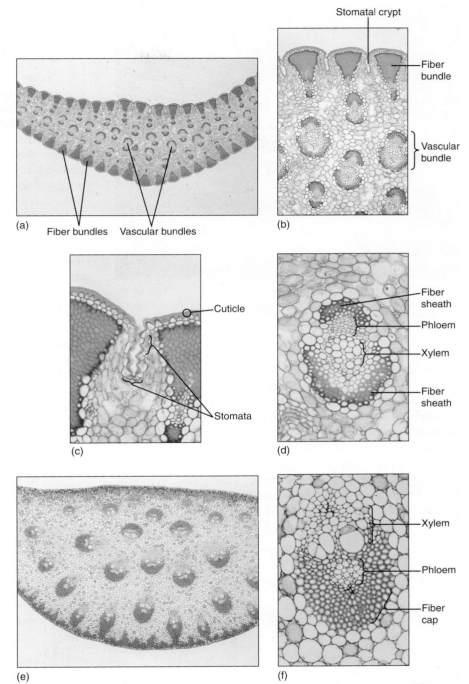

Figure 6-38 (a) *Yucca* leaves are excellent examples of sclerophylls; they are tough and fibrous, and the cuticle is thick. These features reduce the amount of light reaching the chlorenchyma, but plant growth is probably limited by scarcity of water rather than reduced photosynthesis. These leaves may function for 5 to 10 years or more (×15). (b) Bundles of fibers occur just interior to the epidermis, and the thick mesophyll surrounds many vascular bundles (×50). (c) Stomata occur in long, shallow grooves that run between fiber bundles (×150). (d) Even vascular bundles of these sclerophylls have fiber sheaths (×150). (e) Sclerophylls of *Dracaena* are very thick and have leaf veins throughout the mesophyll, unlike the single row in ordinary leaves (×15). (f) Leaf veins of *Dracaena* have extremely thick fiber sheaths (×150).

Figure 6-40 Scale-shaped leaves in monkey-puzzle tree (*Araucaria araucana*) are very large and persistent, living for many years, even as the branch enlarges and becomes woody.

Small, flat, scale-like leaves form a shield-like covering on stems of junipers, cypresses (*Cupressus*), arborvitae (*Thuja*), and others. In *Agathis, Araucaria,* and *Podocarpus,* all genera of the Southern Hemisphere, leaves are rather large, broad scales held away from the stem (Figure 6-40). Lengths up to 12.5 cm and widths of 3.5 cm have been measured in leaves of *Podocarpus wallichianus.*

Conifer leaves are mostly perennial, remaining on the stem for many years; consequently, the plants are evergreens. Needles of bristlecone pine live for at least 5 years; their vascular bundles can produce new phloem each year, but no new xylem (Figure 6-41).

The small scale leaves of juniper also persist and are photosynthetically active for many years, and leaves of *Agathis* and *Araucaria* remain even on very old trunks. Three conifers have annual leaves that are shed each autumn; larches (*Larix*), bald cypress (*Taxodium*), and dawn redwood (*Metasequoia*) are all deciduous.

■ Bud Scales

One of the most common modifications of leaves is their evolutionary conversion into **bud scales** (Figure 6-42). In perennial plants, dormant shoot apical meristems are protected from low temperatures and the drying action of wind during winter by bud scales, which form a tight layer around the stem tip. Because their role is primarily protection, not photosynthesis, bud scale structure differs from that of foliage leaves. Bud scales are small and rarely compound, so mechanical wind damage is not a risk for bud scales. Their petiole is either short or absent because they must remain close to the stem and be folded over it. To be protective, they must be tougher and waxier than regular leaves; bud scales frequently produce a thin layer of corky bark, at least on exposed portions, which provides greater protection than the simple epidermis of foliage leaves.

■ Spines

Cacti have two types of leaves. The green cactus body has microscopic green leaves, and the clusters of spines are their axillary buds: Cactus spines are modified leaves of axillary buds. The succulent, moist cactus body would be an excellent source of water for animals were it not for the protective spines (Figure 6-43). As with bud scales, spines have a distinct structure related to their function. The soft, flexible blade of a photosynthetic leaf is useless as a protective device against herbivores, but spines have no blade and are needle shaped; mutations that inhibit lamina formation have been selectively advantageous. No mesophyll parenchyma or vascular tissue is present; the mesophyll instead consists of closely packed fibers. After fibers mature, they deposit lignin in their walls, which makes them hard and resistant to decay. The cells

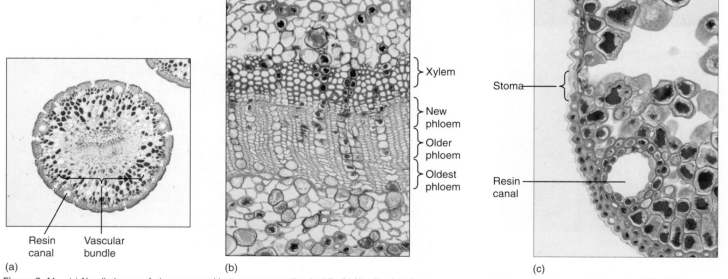

Figure 6-41 (a) Needle leaves of pine are round in transverse section (×15). (b) Needles last for many years, producing more phloem each year (×150). (c) Pine needles are sclerophylls: They have thick-walled epidermal cells with a thick cuticle. They also contain resin canals (×150).

Figure 6-42 (a) Bud scales of *Rhododendron* fit together tightly during winter, protecting the enclosed apical meristem and young leaves and flowers. (b) The bud scales of *Magnolia* have a dense covering of hair; this improves the insulation.

then die and dry out, hardening even further. Because cacti carry out photosynthesis in their stem cortex, loss of the leaf lamina was not selectively disadvantageous, but for plants that have no alternative photosynthetic tissues, mutations that cause loss of lamina are extremely disadvantageous. Spines of many other species have a similar structure.

Tendrils

The tendrils of many plants (peas, cucumbers, and squash) are another form of modified leaf (Figure 6-44). Unlike photosynthetic leaves, tendrils grow indefinitely and contain cells that are capable of sensing contact with an object. When the tendril touches something, the side facing the object stops growing but the other side

continues to elongate, causing the tendril to coil around the object and use it for support. A lamina would be detrimental, and none forms. Whereas many foliage leaves sense the direction of sunlight and reorient the lamina for maximum photosynthesis, tendrils respond by sensing solid objects and growing around them.

Leaves with Kranz Anatomy

A distinct type of leaf anatomy occurs in plants that have a special metabolism called C_4 photosynthesis. These leaves lack palisade parenchyma and spongy mesophyll but have prominent bundle sheaths composed of large chlorophyllous cells. Surrounding each sheath is a ring of mesophyll cells that appear to radiate from the vascular bundle. These plants possess a mechanism of carbon dioxide transport that requires this special **Kranz anatomy** and adapts C_4 plants to arid environments (review other tests on photosynthesis for details of C_4 photosynthesis and its relationship to Kranz anatomy).

Insect Traps

The ability to trap and digest insects has evolved in several families. Insectivory has evolved in plants that grow in habitats poor in nitrates and ammonia; by digesting insects, plants obtain the nitrogen they need for their amino acids and nucleotides. Trap leaves can be classified as either active traps that move during capture or passive traps incapable of movement. The best known passive traps are the pitcher leaves of *Nepenthes, Darlingtonia,* and *Sarracenia* (Figure 6-45a). Although the leaf appears highly modified, it is actually similar to many foliage leaves. It is thin, parenchymatous, and capable of photosynthesis. It has numerous stomata and vascular bundles as well as mesophyll containing aerenchyma and chlorenchyma. The most significant differences are that the lamina is tubular rather than flat, and it secretes a watery digestive fluid. The epidermis

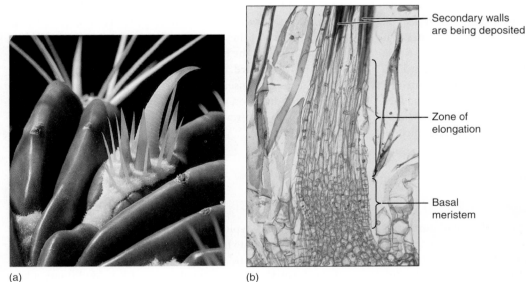

Secondary walls are being deposited

Zone of elongation

Basal meristem

(a)　　　　　(b)

Figure 6-43 (a) The spines of this barrel cactus (*Ferocactus*) are modified leaves. The cluster of spines is actually an entire axillary bud. (b) Cactus spines have a basal meristem; as new cells are formed, older cells are pushed upward. As they move out of the meristem, they fill their central vacuole with water, elongate, then deposit a thick secondary wall and differentiate into fiber cells (×50).

Figure 6-44 This tendril is a highly modified leaf: It has no leaf blade, and whereas foliage leaves stop growing after they reach a specific size, this tendril continues to grow.

(a) (b) (c)

Figure 6-45 (a) *Nepenthes* has elaborate pitcher leaves. They have an ordinary petiole and lamina, but the leaf tip is long, narrow, and pendant. Its extreme tip turns upright and develops into the hollow pitcher that contains a digestive mixture and an epidermis that can absorb nitrogen. The ultimate portion of the leaf tip is a broad flat roof that prevents rain from falling into the pitcher and diluting the digestive juices. (b) Sundew leaves have many shapes; those of *Drosera capensis* are long and narrow. Just after an insect is caught, the leaf curls and places numerous stalked glands on it. The drop at the end of each gland is both sticky and digestive. Once digestion is complete, the leaf uncurls and is ready for its next meal. (c) Each half of the blade of a Venus' flytrap leaf has three hairs; if an insect touches any two, it stimulates the trap to close.

in the digestive region must be absorptive rather than impermeable. Also, the throat of the pitcher contains numerous trichomes that point toward the liquid; it is easier for insects to walk in the direction of the trichomes, and they are thus led to their death.

Leaves of sundew (*Drosera*) (Figure 6-45b) are active traps; they too have many features in common with foliage leaves, but their upper surface is covered with glandular trichomes that secrete a sticky digestive liquid. After an insect is caught on a single trichome, adjacent trichomes are stimulated to bend toward the victim, placing their digestive drops on it as well. The entire leaf blade curls around the insect so that many trichomes come into contact with it. The modifications in these leaves have been largely metabolic rather than structural: Trichomes and lamina must be

able to sense and respond to the presence of an insect, carry on secretion and absorption, and unfold after digestion is complete.

Venus' flytraps (*Dionaea muscipula*) have leaves that are held flat, like most foliage leaves, but in *Dionaea*, this position is maintained only because motor cells along the upper side of the midrib are extremely turgid and swollen (Figure 6-45c). When an insect walks across the trap, it brushes against trigger hairs. If two of these are stimulated, midrib motor cells lose water quickly, and the trap rapidly closes as the two halves of the lamina move upward. On the margins are long interdigitating teeth that trap the insect; short glands begin to secrete digestive liquid. After digestion and absorption are complete, the midrib motor cells fill with water, swell, and force the trap open, ready for a new victim.

Summary

1. Natural selection has resulted in the evolution of numerous types of leaves involved in photosynthesis, protection, support, water and nutrient storage, and nitrogen absorption.
2. Foliage leaves must be resistant to pathogens and stresses, must produce more sugar than was used in their construction, and must not act as sails. Modifications related to these functions may prevent foliage leaves from being optimally adapted for photosynthesis.
3. Foliage leaves typically consist of a blade with or without a petiole; the blade may be simple or either pinnately or palmately compound.
4. Leaf variability may involve distinct juvenile and adult leaves or distinct leaves on long shoots and short shoots.
5. An abscission zone typically contains a separation layer in which cell walls rupture and a protective layer whose cells form a corky leaf scar.

6. Internally, most foliage leaves have an upper palisade parenchyma rich in chlorophyll and a lower spongy mesophyll that allows gas circulation. Vascular bundles composed of primary xylem and phloem are usually arranged in a netted pattern in basal angiosperms and eudicots and a parallel pattern in monocots. Minor veins load sugars and unload water.

7. Leaves are initiated only at shoot apical meristems, beginning as small conical leaf primordia. Eudicot leaves generally have diffuse growth in all parts simultaneously; most monocot leaf blades grow from a basal meristem.

Important Terms

abscission zone
bud scales
bundle sheath
compound leaf
lamina
leaf blade
leaf primordium
leaf scar

leaf traces
leaflets
mesophyll
midrib
minor veins
palisade parenchyma
parallel venation
petiole

sessile leaf
simple leaf
spongy mesophyll
stipules
rachis
reticulate venation

Review Questions

1. Leaves carry out many roles in a plant's life. Give examples of leaves that do the following:
   a. Provide protection
   b. Provide support
   c. Provide storage
   d. Obtain nitrogen
   Are leaf structures and metabolisms that are selectively advantageous for one of these functions also advantageous for all of them?

2. What is the most obvious function of foliage leaves? What are some other functions that are often taken for granted?

3. Most leaves can be quite thin. Why? What would be the condition of lower layers in a thick leaf?

4. What is the stalk of a leaf called? What is the broad, flat, thin part?

5. Box 6-1, "Plants and People: Leaves, Food, and Death," describes several ways that leaves affect animals. What is the name of the toxic chemical in leaves of poison hemlock and death camas? In leaves of tobacco? Can this poison be absorbed by the mucous membranes? Can it be passed across the placenta to an unborn baby? This chemical causes cancer of which parts of our bodies?

6. In many monocots such as grasses and irises, foliage leaves have a shape quite different from that of eudicot leaves. Their shape tends to be (circle two: long, short, tapering, wide). They lack a petiole and instead have a _____ _____ that wraps around the stem. Name three monocots that have leaves that are unusual in that they appear to have petioles and laminas.

7. If a leaf has a blade that consists of one piece of tissue, we say the leaf is _____. If the blade consists of several pieces, however, we say it is a _____ leaf. What type of leaf is shown for a rose in Figure 6-7? For a Virginia creeper?

8. How can you distinguish between a compound leaf and a twig with several simple leaves? Assuming both a simple and a compound leaf have the same texture, which is more easily eaten by an insect larva?

9. Does a plant produce only one type of leaf during its entire life or can some plants produce various types of leaf (hint: Figure 6-11)? Explain.

10. There are two types of venation in the leaves of flowering plants. Describe the type found in eudicot leaves. Describe the type in monocot leaves.

11. What does the abscission zone do in leaves? Would you guess that there are or are not abscission zones in flowers and fruits?

12. In what ways does the upper epidermis of a leaf differ from the lower epidermis? How are these structural differences adaptive?

13. This page is 22 cm wide and 27 cm tall. Calculate its surface area (remember, just like a leaf, it has an upper and a lower surface). If this page were a leaf with the same density of stomata as a leaf of sunflower (see Table 6-2), how many stomata would be on the upper side of the page and how many on the lower side?

14. The interior tissues of a leaf are called mesophyll. In most leaves, there is an upper layer of columnar cells called the _____ _____ and a lower portion of _____ _____ open, loose aerenchyma. In Box 6-2, "Botany and Beyond: Leaf Structure, Layer by Layer," Figure (a) shows the upper layer of columnar cells, but they do not look columnar. Why not? Figure (e) in the box shows the lower aerenchymatous layer. Can you tell what the shape of the intercellular spaces is from this two-dimensional micrograph (look at Figure 6-21 before you answer)?

15. Many leaves have a big, obvious vein that runs along their center, and many veins emerge from it. What is the large vein called (and can a leaf have more than one [look at Figure 6-2])? The smaller veins that emerge from it? What are the very smallest veins of a leaf called? Which veins supply water directly to mesophyll cells—the biggest veins or the smallest?

16. What is a bundle sheath in a leaf?

17. Draw a cross-section of a foliage leaf and label each part; be certain to show all the types of vascular tissues in the bundles.

18. Which parts of the leaf in Question 17 would be emphasized and which would be reduced in each of the following: tendrils, spines, bud scales, scale leaves of a bulb, and succulent leaves of a desert plant?

19. Most petioles are rather small, but they can be either massive or very long. List several examples of unusually large petiole (two are edible).

20. Shoot apical meristems make small groups of cells that protrude upward and develop into leaves. What are these small protrusions of cells called?

21. Look at the cross-sections of leaf primordia in Figure 6-31a (labeled "Leaves"). Each one wraps completely around the shoot apical meristem. Can you explain why each leaf appears to surround the meristem (hint: this is a micrograph of a grass, a monocot, and thus, it has leaves like those in Figures 6-4b and 6-4c).

22. How does the growth of eudicot and grass leaves differ? Which type would be more capable of recovering from attack by leaf-eating insects or grazing deer?

23. Dormant buds, such as the terminal and axillary buds of twigs in winter, consist of bud scales, young leaves, and leaf primordia. Which were formed by the apical meristem first and which were initiated last?

24. Figure 6-36 shows many plants of the desert plant *Lithops*. How many leaves does each plant have? (By the way, even if you grow these in ideal conditions, they never have more than this number.)

25. Sclerophyllous foliage leaves like those in holly, barberry (Figure 6-37), and *Yucca* (Figure 6-38) have many unusual features. Do they live for 1 year or longer? They are more resistant to animals that might try to eat them. Why? What type of cells do they have that most leaves do not have?

26. Leaves of conifers are perennial. Those of bristlecone pine live for at least _____ years, and their vascular bundles produce new _____ every year. Many conifers have needle-shaped leaves, but those of incense cedar are _____ shaped (see Figure 6-39b).

27. The spines of cacti are modified leaves, and as with bud scales, spines have a distinct structure related to their function. Why would the blade of a photosynthetic leaf be useless as a protective device? Cactus spines have no mesophyll parenchyma or vascular tissues. Instead, their mesophyll consists of _____ _____ _____. What chemicals do spine cells have in their walls? Are spine cells parenchyma, collenchyma, or sclerenchyma? After spine cells are mature, do they remain alive? Look at all your answers to this question: Does this modified leaf have much in common with a foliage leaf?

28. Cacti have _____ types of leaves. The green cactus body has _____ green leaves, and the clusters of spines are their _____ _____. Spines are modified leaves of axillary buds.

29. Several species have leaves that catch and digest insects, obtaining nitrogen fertilizer from their bodies. Describe how each of the following trap leaves function—the pitchers of *Nepenthes*, the flypaper leaves of sundew, the traps of Venus' flytrap. Which are active and which are passive?

BotanyLinks

http://biology.jbpub.com/botany/5e

BotanyLinks contains a variety of resources and review material designed to assist in your study of botany.

1. Leaves must survive insect attack. Go to the BotanyLinks home page for more information on this subject.

Roots

chapter 7

Concepts

Most roots have three functions: (1) anchoring the plant firmly to a substrate, (2) absorbing water and minerals, and (3) producing hormones. Firm anchoring provides stability and is therefore important for virtually all plants. Stems, leaves, flowers, and fruits then can be properly oriented to the sun, pollinators, or fruit distributors. Without proper root attachment, trees and shrubs could not remain upright, and epiphytes would be blown from their sites in the tree canopy. A highly branched rhizomatous or stoloniferous plant might resist being blown over even without roots, but their horizontal stems are usually so flexible that roots are necessary to stabilize the aerial structure.

Chapter Opener Image: These roots emerge from the basal portions of mangrove (*Rhizophora mangle*) stems in the Everglades National Park in Florida. Mangroves live in the intertidal zone, the band of seashore between high and low tide. At high tide, the roots are flooded with ocean water. This is a harsh environment because the seawater is so salty it can pull water out of most terrestrial plants and animals (that is why we can't drink seawater), and the water should make the roots drown for lack of oxygen. Mangrove roots survive these conditions because their surface tissues keep seawater away from the delicate cells inside the root, and they have an aerenchyma that lets oxygen diffuse into the root from the shoots. Few plants have adapted to the intertidal zone. Although ocean shores have existed for billions of years, mangroves became adapted to the intertidal zone only in the last several million years: There are no fossils of mangrove plants older than about 35 million years.

Outline

Although roots, like leaves, have an absorptive function, the two organs have totally different shapes. Sunlight always comes from above, but water and minerals are distributed on all sides of a root. Its cylindrical shape allows all sides to have the same absorptive capacity. Consider a leaf and a system of thin roots, both with equal volumes; the root system has a higher surface-to-volume ratio, ideal for absorption. The lower surface-to-volume ratio for leaves reduces the carbon dioxide absorption capacity, but not the absorption of light, and is actually beneficial for water conservation. Roots do not need to be adapted for light absorption, and they absorb water rather than needing to conserve it. Their cylindrical shape is also undoubtedly related to the growth of the roots through a semisolid, resistant medium. Even a light, porous soil can be most easily and thoroughly penetrated by narrow cylinders rather than thin sheets. Thus, although both leaves and roots have absorptive functions, it is selectively advantageous for them to have different shapes.

Roots are quite active in the production of several hormones; shoot growth and development depend on the hormones cytokinin and gibberellin imported from the roots. This reliance of the shoot on root-produced hormones may be a means of integrating the growth of the two systems. It is selectively advantageous for a plant to control the size of its shoot so that transpiration by its leaves does not exceed absorption by its roots, and a plant should not waste carbohydrates by constructing a larger root system than its shoot needs; the extra carbohydrate could be used for leaves or reproduction (**Figure 7-1**).

In many cases, roots have functions in addition to or instead of anchoring, absorption, and hormone production. Fleshy **taproots**, such as those of carrots, beets, and radishes, are the plant's main site of carbohydrate storage during winter. As roots of willows, sorrel,

and other plants spread horizontally, they produce shoot buds that grow out and act as new plants. This method of vegetative reproduction is quite similar to that of stoloniferous and rhizomatous plants, except that roots rather than stems are involved. In the palms *Crysophila* and *Mauritia*, roots grow out of the trunk and then harden into sharp spines. Ivy and many other vines have modified roots that act as holdfasts, clinging to rock or brick. Finally, many parasitic flowering plants (mistletoe and dodder) attack other plants and draw water and nutrients out of them through modified roots.

Distinct sets of characteristics are adaptive for different root functions. As roots specialize and become more efficient for particular tasks, they become poorly adapted for other tasks, and thus, several types of roots may occur in one plant, resulting in division of labor. For example, in addition to the large storage root, carrots and beets also have fine absorptive roots, and ivy has both holdfasts and absorptive roots. The characteristic types of structure and metabolism of each should be analyzed in terms of the function of the particular root.

External Structure of Roots

Organization of Root Systems

Roots must have an enormous absorptive surface; in order for a single unbranched root to have sufficient surface area, it would have to be hundreds of meters long, which would make conduction impossible (**Table 7-1**). Instead, plants have a highly branched root system (**Figure 7-2**). Most seed plants have a single prominent taproot that is much larger than all the rest and numerous small **lateral roots** or **branch roots** coming out of it (**Figure 7-3**). This taproot develops from the embryonic root, called the **radicle**, that was present in the seed; after germination, it grows extensively and usually becomes the largest root in the system. Carrots, beets, turnips, and other taproots sold in stores have dozens of fine lateral roots while growing, but these are removed before the products are shipped to market.

Lateral roots may also produce more lateral roots, resulting in a highly ramified set of roots analogous to the highly branched shoot system of most plants. Lateral roots can become prominently swollen like a taproot, as in sweet potatoes and the tropical vegetable manioc (cassava). If the plant is perennial and woody, roots also undergo secondary growth, producing wood and bark.

Most monocots and some eudicots have a mass of many similarly sized roots constituting a **fibrous root system** (**Figure 7-4**).

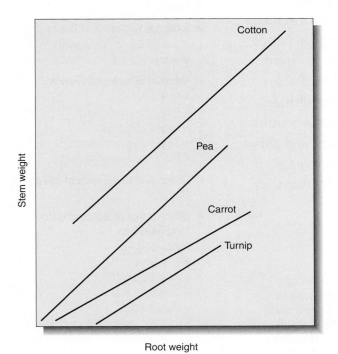

Figure 7-1 The ratio of root system to shoot system is critically important; for each species, as the root system becomes heavier, so does the shoot system by a characteristic amount. However, the relationship differs for each species: Obviously for a given amount of turnip root there is relatively little shoot, but there is much more shoot for the same amount of pea root, and even more for cotton.

TABLE 7-1	Dimensions of the Root System of Barley *(Hordeum vulgare)*, Four Weeks Old	
Number of main roots		
Seminal roots		6.5
Adventitious roots		19.2
Total length of main roots		800 cm
Number of lateral roots		2,000
Total length of lateral roots		4,900 cm
Diameter		
Main roots		0.6 mm
First-order laterals		0.2 mm
Second-order laterals		<0.1 mm

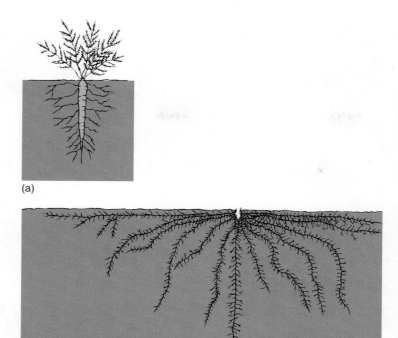

(a)

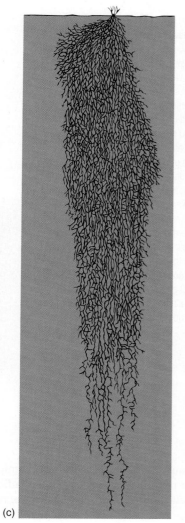

(b)

Figure 7-2 Some taproots, such as a carrot (a), become extremely swollen and are much larger than the numerous lateral roots, whereas in other species, such as sunflower (b), the taproot is about the same size as the laterals. The important criterion is that the taproot develops from the root of the embryo, the radicle. (c) A fibrous root system, as in this winter wheat, consists of many roots, none of which is the radicle. Instead, all of the main roots are adventitious roots that originated in stem tissue.

(c)

Lateral roots   Tap root

(a)   (b)

Figure 7-3 (a) The taproot of this tobacco seedling (*Nicotiana tabacum*) is definitely larger than the lateral roots, but more importantly, its development can be traced directly to the embryo. Relative size is not the critical factor; in many species, lateral roots are the ones that become enlarged. (b) The roots of radishes are obviously taproots.

Figure 7-4 Corms of *Gladiolus,* like other monocots, have a fibrous root system. The radicle died shortly after germination; no root here has developed from the radicle. As the *Gladiolus* continues to grow, it will produce more roots, each independent of the others.

External Structure of Roots   **153**

Figure 7-5 (a) A seedling of a woody plant has a few leaves and a small root system; the narrow trunk with a few vascular bundles can conduct water and nutrients between them. (b) An older woody plant has more leaves and a larger root system; the stem has more wood and bark, which increases its capacity to conduct water and sugar. Because most monocots do not undergo secondary growth, the stem of an older monocot is not wider (d) than that of a young plant (c), and it has no increased conducting capacity. Consequently, the old plant has no more leaves or roots than the young plant. (e) If a plant can produce adventitious roots, the bottleneck of the monocot stem does not matter. New roots originate near the aerial shoots and conduct directly into them, and little or no long-distance conduction occurs in the rhizome. No part of the monocot stem needs to conduct all of the water from all the roots to all the leaves and flowers, as does the trunk of a woody plant.

(a)

(b)

(c)

(d)

(e)

This arises because the radicle dies during or immediately after germination; root primordia at the base of the radicle grow out and form the first stages of the fibrous root system. As the plant ages, more root primordia are initiated in the stem tissue. Because these roots do not arise on pre-existing roots and because they are not radicles, they are known as **adventitious roots**. Adventitious roots increase the absorptive and transport capacities of the root system.

The functional significance of taproots versus fibrous root systems becomes apparent when the general growth forms of eudicots and monocots are considered. Many eudicots are perennial and undergo secondary growth, resulting in an increased quantity of healthy, functional wood (xylem) in both the trunk and roots (Figures 7-5a and 7-5b). This enlarging conduction capacity permits an increase in the number of leaves and fine, absorptive roots. This also occurs in other groups of seed plants such as conifers and cycads.

Most monocots cannot undergo secondary growth; after their stem is formed, the number of vascular bundles, tracheary elements, and sieve tubes is set, and their conducting capacity cannot be increased. Extra leaves could not be supplied with water, nor could their sugar be transported (Figures 7-5c and 7-5d). Such a shoot could not supply sugars to an ever-increasing taproot system. However, some monocots do increase their size by means of stolons or rhizomes: Their horizontal shoots branch and then produce adventitious roots (Figure 7-5e). Because these roots are initiated in the new stem tissues, they transport water directly into the new portions of the shoot, unhindered by the limited capacity of the older portions of the shoot. By this mechanism, monocot shoots can branch and grow larger, as long as they remain close enough

to the substrate to produce new adventitious roots. For them, a fibrous root system is functional, whereas a taproot system is not.

The ability to form adventitious roots is not limited to monocots; many rhizomatous and stoloniferous eudicots also grow this way naturally. Furthermore, many eudicots that never produce adventitious roots in nature do so if they are cut; this is important in the process of asexual propagation by cuttings.

▪ Structure of Individual Roots

An individual root is fairly simple; because it has no leaves or leaf scars, it has neither leaf axils nor axillary buds (Figures 7-6 and 7-7). The root tip, like that of the shoot, is the region where growth in length occurs. In roots, growth by discrete apical meristems is the only feasible type of longitudinal growth. In most animals, all parts of the body grow simultaneously (diffuse growth), whereas roots and stems elongate only at small meristematic regions (localized growth). Because the root is embedded in a solid matrix, it is impossible for all parts to extend at once; the entire root would have to slide through the soil. With apical growth, only the extreme tip pushes through soil.

Whereas the shoot apical meristem is protected by either bud scales or young, unexpanded foliage leaves, the root apical meristem is protected by a thick layer of cells, the **root cap** (Figure 7-7 and Figure 7-8). Although some soils appear soft and easily penetrable, on a microscopic scale, all contain sand grains, crystals, and other components that can damage the delicate apical meristem and root cap. Because the cap is forced through the soil ahead of the root body, it is constantly being worn away and must be renewed by cell multiplication.

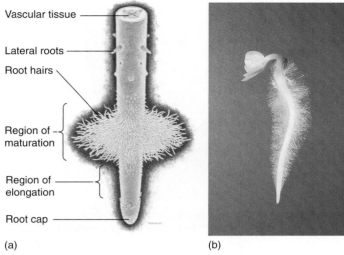

(a)   (b)

Figure 7-6 (a) The tip of a root consists of a root cap, the root apical meristem, a zone of elongation growth, and a region where root hairs are formed. (b) This radish seedling is only 3 days old and has abundant root hairs; those near the root tip are short because they are newer and younger than those near the shoot.

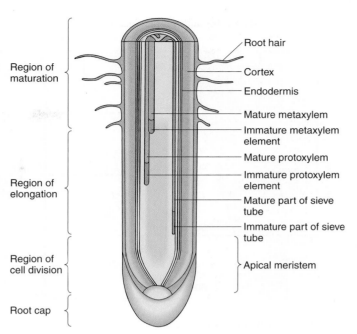

Figure 7-7 Although the exterior of a root appears rather uniform, several distinct zones of differentiation are present internally. Root hairs form only above the elongation zone, and the endodermis and first vascular tissues appear earlier than do root hairs.

Dictyosomes of root cap cells secrete a complex polysaccharide called **mucigel**, which lubricates passage of the root through the soil. It also causes soil to release its nutrient ions and permits the ions to diffuse more rapidly toward the root. Mucigel is rich in carbohydrates and amino acids, which foster rapid growth of soil bacteria around the root tip. The metabolism of these microbes is believed to help release nutrients from the soil matrix.

Just behind the root cap and root apical meristem is a **zone of elongation** only a few millimeters long within which the cells undergo division and expansion (Figure 7-7). Behind it is

the **root hair zone**, a region in which many of the epidermal cells extend out as narrow trichomes. **Root hairs** form only in a part of the root that is not elongating; otherwise, they would be shorn off.

Root hairs greatly increase the root's surface area. In a study of rye, a single plant was found to have 13 million lateral roots with 500 km of root length and a surface area of 200 m². Because of the abundant production of root hairs, however, the total surface area was doubled. Root hairs have other effects that should not be overlooked. Most pores in soil are too narrow for a root (usually at least 100 μm in diameter) to penetrate (**Figure 7-9**). Root hairs, however, being only about 10 μm in diameter, can enter any crevice and extract water and minerals from it. Furthermore, carbon dioxide given off by the respiration of root hairs combines with soil water to form carbonic acid, which helps release ions from the soil matrix. Without the acid, ions would be too firmly bound to soil particles for the root to absorb them. Root hairs are unicellular, never have thick walls, and are extremely transitory. They die and degenerate within 4 or 5 days after forming.

Behind the root hair zone is a region where new lateral roots emerge. They may occur in rows or may appear to be randomly distributed on the parent root. Outgrowth of lateral roots often depends on the soil microenvironment. If the parent root grows into a zone of rich, moist soil, numerous lateral roots form, and the pocket is fully exploited. If the soil is poor, hard, or dry, few lateral roots emerge.

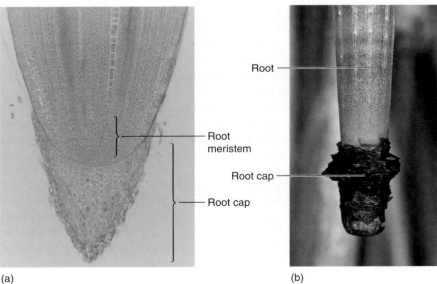

(a)   (b)

Figure 7-8 (a) The root cap is distinct from the root proper. The root apical meristem is located at the apex of the root proper but is buried under the root cap. Because no leaves, leaf traces, or branches are present, cells develop in an extremely orderly fashion in regular files. It is easy to see that these cells are derived directly from the meristem; thus, they make up primary tissues (×150). (b) Screw pine (*Pandanus*) has giant aerial roots that allow us to see the root cap more easily than with ordinary tiny roots. The root cap cells die and are sloughed off.

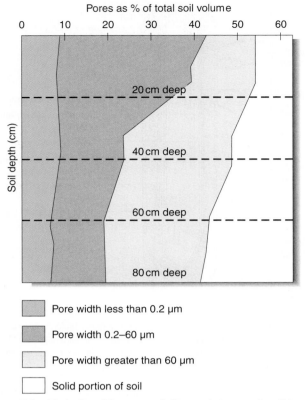

Pores as % of total soil volume

Figure 7-9 Much of a soil is composed of spaces between soil particles; in the uppermost 20 cm (uppermost dashed line) of a soil composed of sandy loam, 54% of the soil is pore space. Of this, 9% consists of extremely fine spaces less than 0.2 μm wide (pink), 33% spaces ranging between 0.2 and 6.0 μm (gray), and 12% large spaces wider than 60 μm (yellow). At deeper levels, the soil is more compacted, containing only about 42% pore space (bottom of graph); the decrease is due mostly to a compaction of intermediate-sized spaces.

Legend:
- Pore width less than 0.2 μm
- Pore width 0.2–60 μm
- Pore width greater than 60 μm
- Solid portion of soil

Internal Structure of Roots

Root Cap

To remain in place and provide effective protection for the root apical meristem, the root cap must have a specific structure and growth pattern. The cells in the layer closest to the root meristem are also meristematic, undergoing cell division with transverse walls and forming files of cells that are pushed forward (Figure 7-10). Simultaneously, cells on the edges of this group grow toward the side and proliferate. Although cells appear to extend around the sides of the root, the root is actually growing through the edges of the root cap.

Cells are small and meristematic when first formed at the base of the root cap, but as they are pushed forward, they develop dense starch grains and their endoplasmic reticulum becomes displaced to the forward end of the cell. These cells detect gravity because their dense starch grains settle to the lower side of the cell.

As cells are pushed closer to the edge of the cap, their structure and metabolism change dramatically. Endoplasmic reticulum becomes less conspicuous, starch grains are digested, and

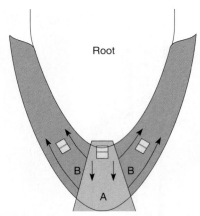

Figure 7-10 In the root cap, cells in the central portion (A) divide so that the two daughter cells are aligned with the root and the root cap grows forward. On the edges (B), cells divide and expand in such a way that cells flow radially outward.

the cell's dictyosomes secrete copious amounts of mucigel by exocytosis. Simultaneously, the middle lamella breaks down and releases cells, which are usually crushed by expansion of the root. It has been estimated that only 4 or 5 days pass from cell formation in the root cap to its sloughing off. Consequently, the cap is constantly regenerating itself. A dynamic equilibrium must be maintained between these two processes.

Root Apical Meristem

If the **root apical meristem** is examined in relationship to the root tissues it produces, regular files of cells can be seen to originate in the meristem and extend into the regions of mature root tissues (Figure 7-8). The root is more orderly than the shoot because it experiences no disruptions because of leaf primordia, leaf traces, or axillary buds. Because cell files extend almost to the center of the meristem, one might assume that cell divisions are occurring throughout it. However, the use of a radioactive precursor of DNA, such as tritiated thymidine (tritium is a radioactive form of hydrogen), can demonstrate that the central cells are not synthesizing DNA: Their nuclei do not take up the thymidine and thus do not become radioactive (Figure 7-11). This mitotically inactive central region is known as the **quiescent center**. These cells are more resistant to various types of harmful agents such as radiation and toxic chemicals, and it is now believed that they act primarily as a reserve of healthy cells. If part of the root apical meristem or root cap is damaged, quiescent center cells become active and form a new apical meristem. After the new meristem is established, its central cells become inactive, forming a new quiescent center. Such a replacement mechanism is extremely important because the root apex is probably damaged frequently by various agents—sharp objects, burrowing animals, nematodes, and pathogenic fungi.

Zone of Elongation

Just behind the root apical meristem itself is the region where cells expand greatly; some meristematic activity continues, but

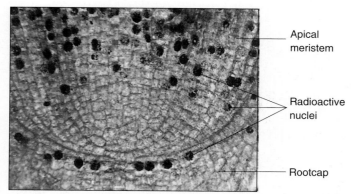

Figure 7-11 These roots were grown briefly in a solution of tritiated thymidine, a radioactive precursor of DNA. In cells that were undergoing the cell cycle, during S phase, the radioactive thymidine was incorporated into the nuclei. After a few hours, the root was killed, sliced into sections, and placed on photographic film in the dark. The radioactive nuclei caused black spots to form in the film next to the nuclei. The slide was then given a brief exposure to light so that the outlines of the cells and nuclei would be faintly visible (too much light would obscure the radioactivity-induced black spots). The quiescent center is the region where no nuclei became radioactive. Apparently no cell in the region passed through S phase while tritiated thymidine was available—these cells were in cell cycle arrest.

mostly cells are enlarging. This zone of elongation is similar to the shoot's subapical meristem region. Cells begin to differentiate into a visible pattern, although none of the cells is mature. The outermost cells are protoderm and differentiate into epidermis. In the center is provascular tissue, cells that develop into primary xylem and primary phloem. As in the stem, protoxylem and protophloem, which form earliest, are closest to the meristem. Farther from the root tip, older, larger cells develop into metaxylem and metaphloem. Between the provascular tissue and the protoderm is a ground tissue, a rather uniform parenchyma that differentiates into root cortex.

In the zone of elongation, tissues are all quite permeable. Minerals penetrate deep into the root through the apoplast simply by diffusing along the thin, fully hydrated young walls and intercellular spaces. This zone is so short that little actual absorption occurs there, and much that is absorbed is probably used directly for the root's own growth.

■ Zone of Maturation/Root Hair Zone

In the maturation zone (**Figure 7-12**), several important processes occur more or less simultaneously. Root hairs grow outward, greatly increasing absorption of water and minerals. In some electron micrographs, a thin cuticle appears to be present on root epidermal cells, but this may be just a layer of fats. The zone of elongation merges gradually with the zone of maturation; no distinct boundary exists because the differences between the two represent the gradual continued differentiation of the cells.

Although cortex cells continue to enlarge, their most significant activity is transfer of minerals from the epidermis to the vascular tissue. This can be either by diffusion through the walls and intercellular spaces (called apoplastic transport) or by absorption into the cytoplasm of a cortical cell and then transferal from cell

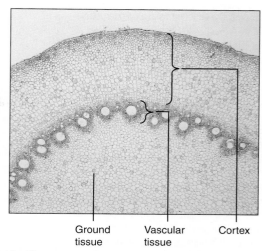

Figure 7-12 The cortex may be broad in many roots, and its interior boundary is the endodermis (×15).

to cell, probably through plasmodesmata (symplastic transport; **Figure 7-13**). Cortical intercellular spaces are also important as an aerenchyma, allowing oxygen to diffuse throughout the root from the soil or stem.

In the zone of maturation, minerals do not have free access to the vascular tissues because the innermost layer of cortical cells differentiates into a cylinder called the **endodermis** (Figure 7-13). Cells of the endodermis have tangential walls, those closest to the vascular tissue or the cortex, which are ordinary thin primary walls; however, their radial walls—the top, bottom, and side walls, that is, all the walls touching other endodermis cells—are encrusted with lignin and suberin, both of which cause the wall to be waterproof. The bands of altered walls, called **Casparian strips**, are involved in controlling the types of minerals that enter the xylem water stream. Cortex cells can not exert any control over movement of minerals within intercellular spaces; without an endodermis, minerals of any type could move from the soil into the spaces, then into the xylem, and then into all parts of the plant. Because Casparian strips are impermeable, however, minerals can cross the endodermis only if endodermal protoplasts absorb them from the intercellular spaces of the cortex apoplast or from cortical cells and then secrete them into the vascular tissues (Figure 7-13b). Many harmful minerals can be excluded by the endodermis. It is not a perfect barrier against uncontrolled apoplastic movement because in the zone of elongation, where the endodermis is not yet mature, minerals do have free access to the protoxylem, but this is only a small amount of uncontrolled movement. Many glands and secretory cavities also have Casparian strips, which prevent the glands' secretory product from seeping into the surrounding tissues.

Within the vascular tissues, many of the much larger cells of the metaxylem and metaphloem become fully differentiated and functional in the zone of maturation. The arrangement of these tissues differs from that in stems: Instead of forming collateral bundles containing xylem and phloem, the xylem of almost all plants except some monocots forms a solid mass in the center, surrounded by

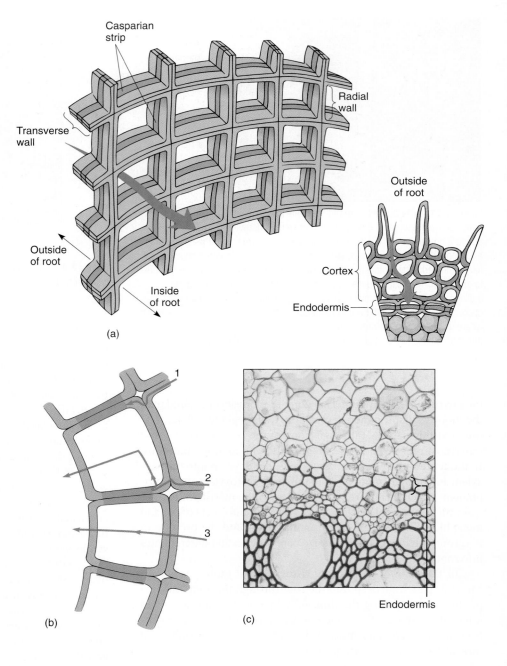

Figure 7-13 (a) The endodermis is a cylinder, one layer of cells thick, each with Casparian strips. The best analogy is a brick chimney: It is a cylinder, one layer thick. The cement is analogous to the Casparian strips. (b) No apoplastic transport occurs across the endodermis. Materials that have been moving apoplastically (1) are stopped and can proceed farther only if the endodermis plasma membrane accepts them (2). Symplastic transport (3) is not affected. Plasma membranes that are impermeable to a particular type of molecule prevent that molecule from crossing the endodermis. (c) The endodermis in this corn is relatively easy to see: In all cells, Casparian strips are visible as red lines on the radial walls. In a few cells, only Casparian strips are present, but in many, the radial walls and inner tangential walls have started to thicken; it is unusual for this development to start so early (×200).

strands of phloem; no pith is present (**Figures 7-14a** and **7-14b**). In roots of many monocots, strands of xylem and phloem are distributed in ground tissue (Figures 7-14c and 7-14d). Within the xylem, the inner wide cells are metaxylem and the outer narrow ones are protoxylem. Two to four or more groups of protoxylem may be present, the number being greater in species with larger roots with a wider mass of xylem (**Figure 7-15**); the number of phloem strands equals the number of protoxylem strands. Within phloem strands, protophloem occurs on the outer side, and metaphloem on the inner side. Other than the arrangement, the vascular tissues of the root are similar to those of the stem and leaf. Those formed first are narrowest and most extensible, often finally being torn by continued elongation and expansion of cells around them. Those formed after adjacent cells have stopped expanding are larger and, in xylem, have heavier walls, often with bordered pits.

Between the vascular tissue and the endodermis are parenchyma cells that constitute an irregular region called the **pericycle** (Figures 7-14 and 7-15). When lateral roots are produced, they are initiated in the pericycle.

■ Mature Portions of the Root

Root hairs function only for several days, after which they die and degenerate. Absorption of water and minerals in this area is then greatly reduced but does not stop entirely. Within the endodermis, cells may remain unchanged, but usually there is a continued maturation in which a layer of suberin is applied over all radial surfaces, the inner tangential face, and sometimes even the outer tangential face. This can be followed by a layer of lignin and then more suberin (Figure 7-14d). This is an irregular process, and some cells complete it earlier than others; thus, in fairly mature parts of

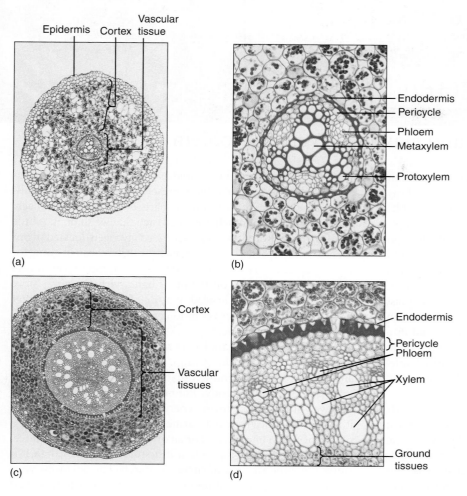

Epidermis Cortex Vascular tissue

(a)

(b)

Endodermis
Pericycle
Phloem
Metaxylem
Protoxylem

Cortex

Vascular tissues

(c)

Endodermis

Pericycle
Phloem

Xylem

Ground tissues

(d)

Figure 7-14 (a) Low-magnification view and (b) high-magnification view of a transverse section of a eudicot root (buttercup, *Ranunculus*), showing the broad cortex and the small set of vascular tissues. The xylem has three sets of protoxylem, the narrow cells at the tips of the arms (b). The central, larger xylem cells are metaxylem. Three masses of phloem are also present; this is a triarch root. The pericycle is the set of cells between the endodermis and the vascular cells (a, ×50; b, ×250). (c) Low-magnification view and (d) high-magnification view of a transverse section of a monocot root (greenbrier, *Smilax*). The vascular tissue consists of numerous large vessels and separate bundles of phloem. The endodermis is very easy to see (c, ×15; d, ×150).

the root, it is possible to find occasional cells that have only Casparian strips. These cells are called **passage cells** because they were once thought to represent passageways for the absorption of minerals; it is now suspected that they are merely slow to develop.

The result of continued endodermis maturation is the formation of a watertight sheath around the vascular tissues. In older parts of the root, it functions to keep water in. The absorption of minerals in the root hair zone causes a powerful absorption of water, and a water pressure, called **root pressure**, builds up. If it were not for the mature endodermis, root pressure would force the water to leak out into the cortex of older parts of the root instead of moving up into the shoot. This is presumably also a function of the endodermis in rhizomes and stolons; if water leaked into the stem cortex and filled intercellular spaces, oxygen diffusion would be prevented and the tissues would suffocate.

Many important events occur at the endodermis. The cortex and epidermis, aside from root hairs, may be superfluous because shortly after the root hairs die, underlying cortex and epidermis often die also and are shed from the root. The endodermis becomes the root surface until a bark can form. This happens mostly in perennial roots that persist for several years. In apple trees, the cortex is shed as early as 1 week after the root hairs die, although it may persist for as long as 4 weeks; only the root tips have a cortex. The large fibrous roots of many monocots are strictly annual; they are replaced by new adventitious roots that form on new rhizomes or stolons. In these plants, the entire root dies.

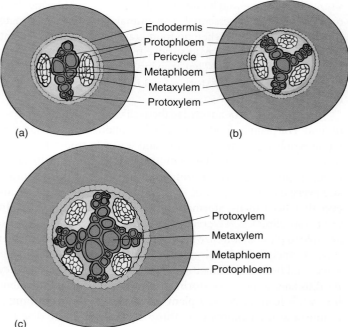

Endodermis
Protophloem
Pericycle
Metaphloem
Metaxylem
Protoxylem

(a)   (b)

Protoxylem
Metaxylem
Metaphloem
Protophloem

(c)

Figure 7-15 The number of strands of protoxylem in a root varies from one species to another and among the roots of one plant. Generally, wider, more robust roots have more protoxylem masses. Having two masses of protoxylem (and two of phloem) is diarch (a). Three is triarch (b), and four is tetrarch (c). These diagrams represent three separate roots, not young and older stages of a single root; a diarch root does not become triarch and then tetrarch as it ages.

Plants Do Things Differently

Box 7-1 Plants and People and Having a Weight Problem

Most of us probably do not spend too much time worrying about starving to death. We may start thinking about lunch midway through the morning, but for most of us, food is always available. However, throughout much of the history of civilization, people had to be careful to store up enough food to last not only through winter but through spring as well. Food supplies had to last until gardens could provide potatoes, beans, wheat, and other staples. Grocery stores are a recent luxury. How did people store food? If dry seeds like wheat and beans are kept free of moisture, they last a long time. Grapes and milk are not dry, of course, but they can be preserved by turning the first into wine and the second into cheese. Meat and fish can be dried, smoked, or salted for long-term storage. Even some animals besides ourselves do this. Squirrels stock up on nuts in autumn. Woodpeckers hide acorns; and bees store nectar by converting it to honey.

One particular method of storing food has been particularly popular with humans and is the only means available to most animals. Eat the food whenever it is available, and store it as fat inside our body. Beans can become moldy. Wine can go sour. Rats can find our supply of dried meat, but fat in our adipose tissues is safe. For many animals, finding food can be a hit or miss proposition, so storing enough energy and nutrients right inside their own body is usually essential. For animals that must hibernate through a long winter when no food is available at all, getting fat in autumn is the only way to survive from year to year. We may regret eating so much at Thanksgiving, but the pilgrims did not. Feasting was a means of storing food that might otherwise spoil during winter.

Plants too must be adapted to the availability—or the scarcity—of food. For plants, "food" is supplied by photosynthesis, and that requires only light, carbon dioxide, and water. In tropical climates where temperatures are always mild and droughts virtually never occur, plants have leaves throughout the year and photosynthesize every day, making all of the carbohydrates they need whenever they need them. Storing food reserves is not a problem. In temperate climates, evergreen trees such as pines and hollies also are able to photosynthesize most days, being inhibited only when it is extremely cold. But deciduous plants—those that drop their leaves and become dormant—are similar to animals in that they need to have a means of storing food to maintain their metabolism while leafless. When a plant abscises its leaves in autumn, it is almost as if an animal were throwing away its entire digestive tract in anticipation of growing an entire new one in spring. If a plant had no nutrient reserves inside itself when it abscised its leaves, it would not even be able to make new leaves in the following spring. It would starve to death.

Energy reserves can be stored as a variety of chemical compounds. We animals sequester our reserve energy as fats. Our adipose tissue can be located in many parts of our body—stomach, arms, and legs—but much is located such that we sit on it. A little bit of energy is stored as a polymer called glycogen, located in our liver and muscle cells, but that is only enough to keep us going for a few hours as any runner or cyclist knows. The long-term storage molecule is fat. Plants, on the other hand, virtually never store fats. They rely on starch instead. Why do plants and animals differ on such a simple feature? Let us look at the consequences of each alternative storage molecule.

An energy storage molecule must store energy, of course, but it must also do other things. It should not be too heavy, and it must be stable enough that it does not "go bad" within the plant or animal's body. Pound for pound, fat is the most lightweight means of storing energy. A pound of fat—whether it is lard, oil, butter, or margarine—stores more energy than a pound of starch or protein. For an organism that needs to move, weight will be less of a problem if it stores fat rather than if it stores the same amount of energy as starch. The next time your bathroom scale reads ten pounds more than you want, be glad we do not store starch. We would be even more overweight and have an even bigger rear end.

Plants, however, do not move around too much; thus, saving on weight is not a real necessity. For plants, the long-term stability of starch is better than the lightness of fat. Some plants save up energy for years, not just months, and starch will last that long in the plant's body. For example, many trees bloom only every other year, and their stored starch must last at least two years, maybe four. Even more dramatically, century plants bloom only once after growing for about 12 to 15 years (but not for an entire century), and fishtail palms may not bloom until they are more than 70 years old. These plants store up some starch each year, then use it all at once in a massive flowering. Fat would not last that long because it becomes rancid if exposed to oxygen, and all parts of a plant are well aerated.

Pollen and seeds are exceptional plant parts. Many flowers produce pollen with a drop of oil rather than a grain of starch, which makes it lighter and easier for wind or insects to carry it to another flower. Seeds such as peanuts, cashews, and sesame store oil and thus are lighter and smaller and more easily moved by animals. Avocados are very rich in oil, but rather than being an energy storage mechanism, it is a reward that entices an animal to eat the fruit and then spit the seed out somewhere, thereby dispersing the plant's seeds far and wide. This animal-based dispersal is important for moving seeds to new sites; otherwise seeds would fall near the base of the parent plant, germinate, and then compete with each other and with the parent for water, sunlight, and nutrients. Dispersal reduces this crowding, and it is therefore important for seeds to be light enough that an animal does not mind carrying them or does not even notice.

Also important is the site within the body where energy-storage molecules are stockpiled. If you have ever cut up chicken or steak, you already know that sometimes adipose tissue occurs as large masses. Other times it is just fine layers mixed in with muscle. In both cases, the energy-storage tissue is vascularized, and the circulatory system can deposit or withdraw fats as needed. Plants also store a little starch here and there throughout their bodies, some in cortex cells, some in pith cells, and even wood can store a bit. However, when a plant needs to store a lot of starch, it almost always relies on its roots. The enlarged roots of beets, carrots, radishes, sweet potatoes, and similar plants are filled with starch. By using roots as a storage organ, the plants are putting their reserves underground, out of sight of hungry ani-mals. Also, the soil is a stabler environment, being neither as hot during the day nor as cold at night as the air is, and similarly, it maintains a more uniform humidity, all of which contributes to the stability of the starch and the health of the storage cells. The storage tissue in roots is usually wood—a type of xylem; thus, just as in adipose tissue of animals, root storage tissues are well vascularized.

Enlarged storage roots are not confined to only familiar food plants. Many different plant families store starch in roots. Even large numbers of cacti do this, storing their water above ground in the succulent shoot, but their starches below ground in enlarged roots. It is surprising that nutrient-rich roots are not more well-protected by the plants. They are rarely poisonous, and they never have spines; therefore, if ever an animal finds the stor-age root, there is nothing to stop it from eating the entire thing.

Even though plants and animals store energy differently, the reasons are understandable when we consider the consequences of each alternative. Plants favor stability. Animals need mobility.

Origin and Development of Lateral Roots

Lateral roots are initiated by cell divisions in the pericycle. Some cells become more densely cytoplasmic with smaller vacuoles and resume mitotic activity. The activity is localized to just a few cells, creating a small root primordium that organizes itself into a root apical meristem and pushes outward. As the root primordium swells into the cortex, the endodermis may be torn or crushed or may undergo cell division and form a thin covering over the primordium. As it pushes outward, the new lateral root destroys cells of the cortex and epidermis that lie in its path, ultimately breaking the endodermis (**Figure 7-16**). By the time the lateral root emerges, it has formed a root cap, and its first protoxylem and protophloem elements have begun to differentiate, establishing a connection to the vascular tissues of the parent root.

Note how this differs from bud formation in shoots. Lateral roots are initiated deep within the root (endogenous origin), not at the surface as are axillary buds (superficial origin); they are formed in mature regions of the root, not right at the base of the apical meristem, and of course, they never develop into flowers.

Other Types of Roots and Root Modifications

The type of root just described is the most common, generalized type, comprising some or all of the roots of most plants. In other species, some roots are modified and carry out different roles in the plant's survival. Various structures and organizations are selectively advantageous for various functions.

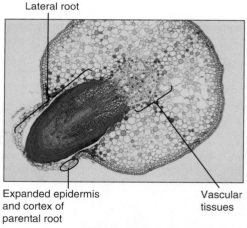

Lateral root

Expanded epidermis and cortex of parental root

Vascular tissues

Figure 7-16 This young lateral root of willow (*Salix*) was initiated in the pericycle, next to a mass of protoxylem; its growth caused considerable damage to the parent root cortex, and it broke the surface open. Entry of pathogens is prevented by formation of wound bark around the tear it caused (×50).

Storage Roots

Roots frequently provide long-term storage for carbohy-drates that accumulate during summer photosynthesis. In bien-nial species (which live for 2 years; for example, beets, carrots, and celery) and many perennials (*Datura, Phlox,* many daisies), roots are the only permanent organs: In autumn, most of the stem dies back to a few nodes located below ground at the top of the root. Carbohydrates stored in the root are used to produce a new shoot in the spring, when photosynthesis is impossible owing to the lack of leaves. It seems less economical in these species to winter-proof the shoot than to replace it using nutrients stored in the roots. Annual plants can survive without such storage capacity.

Perennial plants with permanent shoots that do not die back also store significant amounts of nutrients within themselves during winter, but roots offer certain advantages. Being subterranean, roots are less available as food than are swollen, highly nutritious, easily visible stems. Roots also have a much more stable environment, subjected to less extreme changes in temperature and humidity, which may be important for survival of storage parenchyma cells.

■ Prop Roots

The stem of a monocot can become wider, with more vascular bundles, if it can produce adventitious roots that extend to the soil. In some cases, roots are capable of extensive growth through the air: In many palms, the exposed roots can be 20 to 50 cm long (**Figure 7-17a**). In the screwpine (*Pandanus,* a monocot, not a pine), they are often 3 or 4 meters long, and the roots may grow through the air for months before reaching the ground (**Figure 7-17b**). Less dramatic examples are found at the bases of corn and many other grasses. After **prop roots** do make contact with the soil, they transport additional nutrients and water to the stem. Just as importantly, they contract slightly and place some tension on the stem, thus acting as stabilizers, much like guy wires on tall television antennas. If the roots undergo sec-

(a)

Prop roots

(b)

(c)

(d)

Figure 7-17 (a) Many monocots like this small palm (*Ptychosperma*) produce adventitious roots near the base of the stem. These provide both extra absorptive capacity and extra stability. (b and c) Screwpine (*Pandanus*) produces extremely long adventitious prop roots that not only stabilize the large, heavy trunk but also bring water and minerals into the stem. (d) In older plants of screwpine, even the branches produce adventitious roots. This water and mineral supply to the branches bypasses the trunk, where vascular bundles are already conducting at full capacity.

(a)

(b)

Figure 7-18 (a) Banyan trees (eudicots) produce adventitious roots, which, as in monocots, provide increased support and absorptive capacity. Because the giant branches are supported along their entire length, they can become much larger and more extensive than branches that are supported only at the point of attachment to the trunk. (b) Plate-like buttress roots of *Ficus subcordata*.

ondary growth and become woody, they can be extremely strong supports, permitting a branch to extend even farther from the trunk without breaking or sagging (**Figure 7-18a**). In banyan trees (eudicots of the genus *Ficus*), prop roots and branches can spread and produce massive trees many meters in diameter. Roots of certain tropical trees become tall, plate-like **buttress roots**; their upper side grows more rapidly than other parts of the root. Buttress roots brace the trunk against being blown over by wind (Figure 7-18b).

Mangroves also have prop roots, but these seem to be selectively advantageous for other reasons. The plants grow in intertidal marshes and are subjected to powerful water currents during storms and even normal tide changes; brace-like prop roots provide much more stability than a taproot system would (see opening image of this chapter). In addition, the aerial portion of the prop root is covered with numerous air chambers—lenticels—and its cortex is a wide aerenchyma. The subterranean portion of the root grows in a stagnant muck that has little or no oxygen; it is able to respire only because the aerenchyma permits rapid diffusion of oxygen from the aerial lenticels to the submerged root tissues. If mangrove roots were entirely subterranean, respiration would be extremely reduced.

■ Aerial Roots of Orchids

Many orchids are epiphytic, living attached to the branches of trees. Their roots spread along the surface of the bark and often dangle freely in the air. Although these plants live in rainforests, the orchids are actually adapted to drought conditions. In the few hours when rain does not fall, the air and bark become dry and could easily pull water out of the orchid's roots if there were no water-conserving mechanism. The root epidermis, called a velamen in these orchids, is composed of several layers of large dead cells that are white in appearance (**Figure 7-19**). Apparently, the velamen acts as a waterproof barrier, not permitting water to leave the sides of the root.

Velamen   Root tips

Figure 7-19 Aerial roots of orchids, which grow along the surface of tree bark, are covered by a white velamen that helps retain the water absorbed by the green root tip. These roots grow along the bark, not downward.

■ Contractile Roots

In *Oxalis, Gladiolus, Crinum,* and other plants (many with bulbs), roots undergo even more contraction than prop roots do. After extending through the soil and becoming firmly anchored, the uppermost portions slowly contract (**Figure 7-20**). Because the root is firmly fixed to the soil, the stem is pulled downward so that the base of the shoot is either kept at soil level or, in the case of bulbs, actually buried deeper. The contraction is caused by changes in the shape of cortex cells: They simultaneously shorten

Figure 7-20 The roots of this hyacinth have contracted, causing their surfaces to become wrinkled.

and expand radially, losing as much as one half to two thirds of their height. The vascular tissues buckle and become undulate but are able to continue conducting.

Contractile roots may be more common than is generally appreciated. Many seeds germinate at or near the soil surface; root contraction may be the means by which the shoot becomes anchored in the soil. In bulbs, corms, rhizomes, and other subterranean stems, contractile roots can be important in keeping the stems at the proper depth.

■ Mycorrhizae

Roots of most species of seed plants (at least 80%) have a symbiotic relationship with soil fungi in which both organisms benefit. The associations are known as **mycorrhizae** (singular, **mycorrhiza**), and two main types of relationships are known. In nearly all woody forest plants, an **ectomycorrhizal relationship** exists in which fungal hyphae (slender, thread-like cells) penetrate between the outermost root cortex cells but never invade the cells themselves. Herbaceous plants have an **endomycorrhizal association**, in which hyphae penetrate the root cortex as far as the endodermis; they pass through the walls of the cortex cells but cannot pass the Casparian strip (**Figure 7-21**). They invade the cell but do not break the host plasma membrane or vacuole membrane. Inside the cell, they branch repeatedly, forming a small structure called an arbuscule; this fills with granules of phosphorus that later disappear as the phosphorus is absorbed by the plant. Other hyphae fill with membranous vesicles. The plant cells lack starch grains, presumably because sugars are being transferred to the fungus. The fungus is unable to live without the sugars from the plant, and in many cases, the plant is severely

stunted if the fungus is killed. Apparently, this is a critical means of absorbing phosphorus into the roots, the mycorrhizal fungi being much more effective than root hairs.

All the modified roots mentioned to this point—storage roots, prop roots, contractile roots, aerial roots of orchids, and mycorrhizae-associated roots—have a structure virtually identical to that of the more common anchoring/absorbing roots described above. All grow by an apical meristem, and a cross-section through a mature region of each reveals epidermis, cortex, endodermis, pericycle, and vascular tissue. In nitrogen-fixing nodules and haustoria, however, the structure is highly modified.

■ Root Nodules and Nitrogen Fixation

For most plants, the scarcity of nitrogenous compounds in the soil is one of the main growth-limiting factors. Although nitrogen is abundant in air (78% of the atmosphere is N_2), plants have no enzyme systems that can use that nitrogen. Only some prokaryotes can use N_2 by incorporating it into their bodies as amino acids and nucleotides; when they die and decompose, the nitrogenous compounds are available to plants. The chemical conversion of atmospheric nitrogen into usable compounds is **nitrogen fixation**.

In a small number of plants, especially legumes, a symbiotic relationship has evolved with nitrogen-fixing bacteria of the genus *Rhizobium*. Bacteria free in the soil secrete a substance that causes root hairs to curl sharply; a bacterium then attaches to the convex side of the hair and pushes into the cell by means of a tubelike invagination of the plant cell wall. The tube is an **infection thread**, and the bacterium sits in it. The infection thread, penetrating one cell after another, extends all the way into the root's inner cortex, where adjacent cortical cells undergo mitosis and form a **root nodule** (**Figure 7-22**); the bacterium is released from the infection thread and enters the plant cell cytoplasm, where it proliferates rapidly, filling the plant cells with bacterial cells (known as bacteroids) capable of converting N_2 into nitrogenous compounds that are released to the plant cells. Energy for this process is supplied by sugars from the legume root cells, so both the *Rhizobium* and the legume benefit.

The nodule may remain rather simple or may become complex, with a meristematic region, vascular tissue, and endodermis. Such nodules function for extended lengths of time. Numerous metabolic changes have also evolved: The critical bacterial enzymes are sensitive to oxygen, being immediately poisoned by even traces of free oxygen; however, it is the plant, not the bacterium, that produces a special chemical—leghemoglobin—that binds to oxygen and protects the bacterial enzymes. Root nodules represent a sophisticated symbiosis: Plants and bacteria both benefit. The bacteria are not damaging the plants, nor are the plants parasitizing the bacteria. Without the bacteria the complex development of nodules does not occur.

This process is often described in this way: "The bacterium supplies the plant with nitrogenous compounds and in return the plant gives it sugars." This inaccurately suggests voluntary action, choice, and decision making on the part of both organisms. Instead, think in terms of natural selection: Because the bacterium receives sugars and other nutrients from the plant, any genetic mutation in the bacterium that makes the plant healthier and more vigorous is beneficial to the bacterium as well, whereas

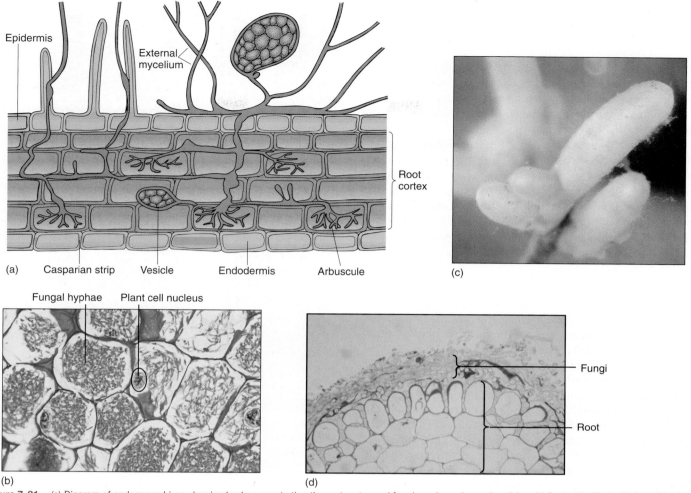

Figure 7-21 (a) Diagram of endomycorrhizae showing hyphae penetrating the root cortex and forming arbuscules and vesicles. (b) Root cells filled with fungal hyphae in the form of arbuscules. You can tell that this is a symbiotic fungus rather than a pathogenic one because the plant cells are healthy, with large normal nuclei (×220). (c) Roots of beech encased in ectomycorrhizae. The presence of the fungus stimulates the roots to be short and broad. (d) Transverse section of an ectomycorrhizal root showing that the fungal filaments are restricted to just the outermost part of the root; they do not penetrate the cortex.

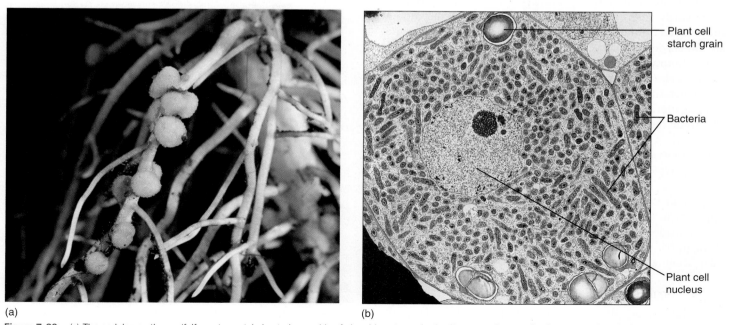

Figure 7-22 (a) The nodules on these alfalfa roots contain bacteria capable of absorbing atmospheric nitrogen and converting it to ammonia, which the plant can use to make amino acids. (b) Bacteria in an infected root nodule cell of cowpea, *Vigna unguiculata* (×5,900).

any mutation that is harmful to the plant—such as a bacterial plasma membrane that does not allow nitrogenous compounds to move to the plant—is deleterious to the bacterium. The same principle applies to the plant: Genetic mutations that aid the bacterium ultimately benefit the plant as well. We hypothesize that these organisms will continue to co-evolve and become more fully co-adapted, but this hypothesis will have to be checked by someone a few million years from now.

Working backward, we can imagine less sophisticated legumes and *Rhizobium* bacteria that did not work so well together but evolved to the present state as mutations occurred and survived by natural selection. Certain critical mutations must be rare, or perhaps several unusual mutations must occur almost simultaneously in plant and bacterium; this hypothesis is based on the observation that, although this symbiosis would probably be beneficial to virtually all plant species, only a few have it.

■ Haustorial Roots of Parasitic Flowering Plants

A number of angiosperms are parasites on other plants; because their substrate is the body of another plant, a normal root system would not penetrate the host or absorb materials from it effectively. Consequently, roots of parasitic plants have become highly modified and are known as **haustoria** (Figure 7-23). In most cases, very little root-like structure remains. Parasitism has evolved several times; thus, the structures that are termed haustoria are not all related to each other, and generalizations are difficult. Haustoria, however, typically must adhere firmly to their host either by secreting an adhesive or by growing around a small branch or root. Penetration occurs either by forcing a shaft of cells through the host's dermal system or by expanding the haustorium radially, cracking the host's epidermis. After penetration, cells of the parasite make contact with the host's xylem. In many cases, both host and parasite cells divide and proliferate into an irregular mass of parenchyma, and then a column of cells differentiates into a series of vessel elements. This results in a continuous vessel from host to parasite constructed of cells of both. Surprisingly, many parasites attack only the xylem; they do not draw sugars from the host but carry out their own photosynthesis. Others do contact both xylem and phloem and perform little or no photosynthesis. For such parasites, mutations that result in the loss of leaves or vegetative stems are not disadvantageous.

A parasitic plant named *Tristerix* spends most of its life as nothing more than a diffuse root system—a haustorium—growing inside the body of its host plant, a cactus. The body is just a network of parenchyma cells with no stem, leaves, epidermis, cortex, or vascular tissues except when *Tristerix* produces flowers.

The typical organization of a root would be nonfunctional in haustoria: The root cap, root apical meristem, and cortex would all prevent vascular tissues of the parasite from contacting the appropriate tissues of the host. Conversely, haustoria are completely inadequate for growth in soil, however rich and moist it might be. For each substrate and each microhabitat, specific adaptations are selectively advantageous.

■ Roots of Strangler Figs

While young, plants of strangler figs (several species in the genus *Ficus*) grow as epiphytes perched on a branch of a host tree. Birds eat fruits of strangler figs and then deposit the seeds on

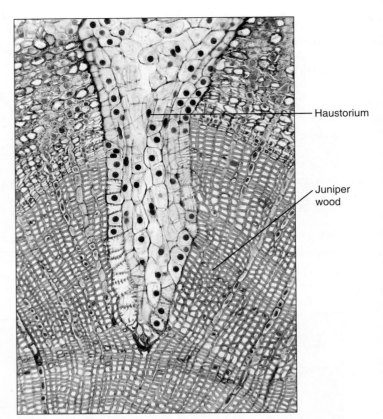

Figure 7-23   Transverse section of a branch of juniper being attacked by the haustorium (modified root) of a mistletoe (*Phoradendron*). The wood of the host juniper has been stained blue; the bark is tan. The haustorium is able to draw water and nutrients from the host vascular tissue.

Haustorium

Juniper wood

Figure 7-24   This tree in the Everglades National Park is encased in the roots of a strangler fig. The large "branch" of the fig at the upper left is actually the trunk of the fig, which is the point where the seed germinated. All portions below that are roots of the strangler fig. It would be easy to mistakenly assume the fig is a vine and the parts encasing the host tree are stems of the fig, but a close examination of the anatomy shows these are roots not branches. If this were really a vine, we would expect to see leafy branches emerging from many areas in this photo.

branches of other trees. When the strangler fig seed germinates, its roots cling to the bark of the host tree branch and then grow rapidly downward, hugging the trunk of the host tree. For some months or even years the roots have no contact with the soil, so they must absorb nutrients from rainwater that runs down the trunk of the host tree. Finally, the strangler fig's roots do reach soil and then penetrate it rapidly and branch profusely; upper portions of the roots enlarge and become woody. Strangler fig roots branch and grow at various angles, encircling the host tree's trunk, and the roots fuse to each other wherever they meet. Before

long the trunk of the host tree is encased in numerous roots of the strangler fig (Figure 7-24). Typically, the host tree finally dies and rots away, leaving the strangler fig as a self-supporting tree with its root–shoot junction high in the air.

It was believed that the tight network of strangler fig roots prevented the host tree from growing thicker, and thus the strangler fig killed the host by "strangling" it. That may be true in some cases, but it seems that more often the fig produces so many leafy branches that it shades the host tree so much it cannot photosynthesize; the host tree dies from lack of light.

Summary

1. Most roots have a variety of functions, including anchorage, absorption, and hormone production. Other roots may be specialized for nutrient storage, vegetative reproduction, or surviving harsh conditions, some even being modified into spines.

2. Roots have a root cap but nothing equivalent to leaves, nodes, internodes, or buds.

3. Adventitious roots form in organs other than other roots or the embryo; they are especially important in stoloniferous and rhizomatous plants.

4. Roots, like shoots, elongate by localized growth (apical meristems). Only the root tip and zone of elongation must slide between soil particles.

5. Root hairs greatly increase the absorptive surface area of the root system, and the carbonic acid that results from their respiration helps release minerals from soil particles.

6. An endodermis with Casparian strips prevents minerals from moving from the soil solution into the xylem. To enter the xylem, minerals must at some point cross a plasma membrane.

7. Lateral roots arise in the pericycle, deep inside the root, unlike axillary buds in stems, which arise in the outermost stem tissues.

8. Prop roots provide additional stabilization and transport for certain plants with narrow stems. Contractile roots aid in burying certain bulbs, corms, and rhizomes. Haustoria are modified roots that penetrate the tissues of host plants.

9. Most plants absorb much of their phosphorus from mycorrhizal fungi that form an extensive network both in the soil and within the root cortex.

10. A small number of plant species form symbiotic relationships with nitrogen-fixing prokaryotes. The bacteria or cyanobacteria often live in root nodules, passing nitrogen compounds to the plant and receiving carbohydrates and minerals.

Important Terms

adventitious roots
branch roots
buttress roots
Casparian strips
ectomycorrhizal relationship
endodermis
endomycorrhizal association
fibrous root system

haustoria
infection thread
lateral roots
mycorrhiza
nitrogen fixation
pericycle
prop roots
quiescent center

radicle
root apical meristem
root cap
root hairs
taproots
zone of elongation

Review Questions

1. The "Concepts" section of this chapter states that most roots have three functions. What are they? Some roots have other functions in addition to these three, or instead of them. Name an example of at least one species in which roots do the following:
   a. Store carbohydrate during winter
   b. Produce shoot buds that can act as new plants
   c. Grow out of the trunk and harden into spines

   d. Act as holdfasts
   e. Attack other plants and draw water and nutrients out of them

2. What are the two types of root systems? Give several examples of plants that have each type. Which type is associated with nutrient storage in biennial species like carrots and beets? Which is associated with rhizomes and stolons?

3. Roots must have an enormous absorptive surface. Why do plants have a highly branched root system instead of just one long root?

4. Even before a seed germinates, it already has a root; what is the name of this embryonic root? In eudicots, what does this embryonic root usually develop into? In most monocots, this embryonic root does a strange thing during or immediately after germination. What does it do?

5. What is an adventitious root? In a monocot rhizome such as that of irises and bamboo, are adventitious roots common or rare? If a bamboo rhizome grows 100 feet underground and then sends up an aerial branch, where does that branch get its water—from the roots 100 feet away or from adventitious roots right at its base (Hint: look at Figure 7-5e; you might also want to look at Question 22)?

6. What does the root cap do? Do you think this structure would have evolved if roots all grew in air like shoots do, rather than growing through dirt?

7. Roots have localized growth; the only parts of the root that become longer are the root apical meristem and the _____ of _____, a region only a few millimeters long within which the cells undergo division and expansion.

8. What would happen to root hairs if they formed in the zone of elongation? Would this happen if roots were growing in air or in water?

9. Root hairs greatly increase the root's surface area. In rye, a single plant has been found to have _____ lateral roots with _____ km of root length and a surface area of _____ m². Convert the length to miles (1 mile = 1.6 km) and the area to square feet (1 foot = 0.3 m).

10. Root hairs are narrower than roots. A root is usually at least _____ mm in diameter, but root hairs are only approximately _____ μm in diameter. Look at Figure 7-9. Could either roots or root hairs enter the pores represented by the red part of the chart? Could either or both enter the gray part? The white part? Most roots grow near the soil surface, in the upper 20 cm, where most of the soil pores are between 0.2 to 60 μm wide. If a root did not produce any root hairs at all, could the plant obtain water and minerals from these abundant, narrow pores?

11. Which part of the root detects gravity? Which organelles enable them to do this?

12. The use of a radioactive precursor of DNA can demonstrate that the central cells of the root apical meristem are not synthesizing DNA. This mitotically inactive central region is called the _____ _____. It is now believed that the central cells act primarily as a reserve of _____ _____.

13. If you could examine a transverse section through the zone of elongation in a root, would you see any mature cells? Would there be any fully differentiated epidermis cells or vascular tissues? What would the outermost cells be called, if not epidermis? In the center of the zone of elongation, what would be the tissue that would later differentiate into xylem and phloem?

14. Examine Figure 7-13. What are the four diffusion paths that a molecule might follow as it travels through the root epidermis and cortex? Why does this diagram stop at the endodermis? What happens at the endodermis?

15. Describe the shape of the endodermis. Is it a flat plane, a cylinder, a single ring of cells? What are the Casparian strips, and on which cell walls are they located?

16. Draw cross-sections of a root showing its structure at three levels: the mature region, the root hair zone, and the zone of elongation. At which level is the endodermis complete with Casparian strips?

17. Because Casparian strips are impermeable, minerals can cross the endodermis only if the endodermal _____ absorb them from the intercellular spaces or from cortical cells. Many _____ minerals can be excluded by the endodermis.

18. Examine Figure 7-14. The roots of eudicots are usually noticeably different from those of monocots in transverse section. Which has a relatively wide set of vascular tissues? Which has endodermis cells with very thick walls? Which has ground tissues in the center, and which has metaxylem in the center?

19. Which part of the root produces the primordia for lateral roots? How does the vascular tissue of the lateral root connect with that of the parent root?

20. Some perennial plants store significant amounts of nutrients in the stem during winter, but roots offer certain advantages. Describe two of these advantages.

21. Animals typically use fats in adipose tissues for long-term energy storage, whereas plants use starch in roots. How do animals benefit from using fat? How do plants benefit from using starch? Name two plants that store energy for many years. How long is long-term storage for these species? What two plant parts often use fats and why? Storage tissue in enlarged roots is vascularized. How is that important to the plant?

22. Look at the prop roots of screwpine in Figure 7.17d (screwpine is a monocot). Because the branches produced their own adventitious roots (the prop roots), can the branches obtain water and minerals without depending on the xylem in the trunk (think about Question 5)?

23. Look at the prop roots of banyan trees in Figure 7-18. These prop roots supply extra water to the branch, but how else do they help the branches?

24. What are contractile roots? They are especially common in plants with what types of shoots (e.g., in rhizomes, vines, or what)?

25. What is a mycorrhizal association? What benefit does the plant derive from the association?

26. Describe the structure of a nitrogen-fixing nodule; consider especially the relationship with the plant's vascular tissue.

27. Root nodules are _____ associations. Plants and bacteria both benefit. The bacteria are not _____ the plants, nor are the plants _____ the bacteria.

28. Many plants are parasitic on other plants. These parasitic plants have modified roots called _____. Describe some of the ways that this parasitic root penetrates the host stem or root. Do these parasitic roots have a root cap, a root apical meristem, or a cortex?

29. Except when it produces flowers, what is the body of the parasitic plant *Tristerix*? What tissues does this plant body lack, except when it produces flowers?

30. Imagine a plant that has ten roots, each 1 cm long. What is the total length of the root system? Imagine that at the tip of each root are ten lateral roots, each 1 cm long. Now what is the total length of the root system? What is the maximum distance a water molecule must travel from the farthest root tip to the base of the shoot? Distance traveled increases as a simple addition, but total absorbing capacity increases exponentially. Why is the length that is traveled important? Consider the friction of moving through tracheids and vessel elements.

BotanyLinks

http://biology.jbpub.com/botany/5e

BotanyLinks contains a variety of resources and review material designed to assist in your study of botany.

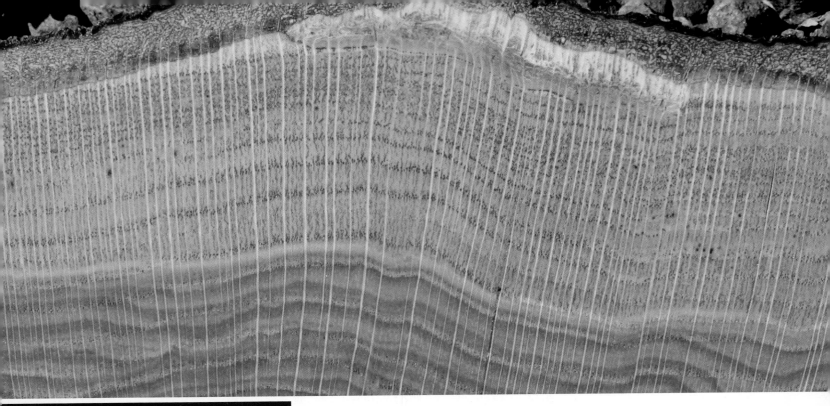

Structure of Woody Plants

Concepts

Growth by means of apical meristems has been previously described. From the meristems are derived sets of tissues: epidermis, cortex, vascular bundles, pith, and leaves. These **primary tissues** together constitute the primary plant body. In plants known technically as herbs, this is the only body that ever develops, but in woody species, additional tissues are produced in the stem and root from other meristems—the vascular cambium and the cork cambium. The new tissues themselves are wood (secondary xylem) and bark (secondary phloem and cork); they are **secondary tissues**, and they constitute the plant's secondary body. Examples of woody plants are abundant: Trees such as sycamores, chestnuts, pines, and firs are woody, as are shrubs such as roses, oleanders, and azaleas.

Chapter Opener Image: Transverse section of a water oak trunk (*Quercus nigra*). The outermost thin brown band is the bark. The broad bands of yellowish tissue and the innermost band of brown tissue are wood. The yellowish wood (sapwood) is young, and many of its cells are still alive and many of its vessels were still conducting water when the tree was cut down. The brownish wood (heartwood) is the oldest, and all its cells had died and all its vessels had cavitated. Most of the cells in wood have thick secondary walls that are heavily lignified; consequently, wood cells decompose slowly. Many trees live for hundreds of years, so the carbon dioxide they capture (sequester) and use to build wood is kept out of the atmosphere for hundreds of years. Trees also absorb unusual chemicals in the air and water and locked them into the tree rings. By carefully cutting out just a bit of a tree ring, we can analyze many aspects of the environment as it existed in the past.

The ability to undergo secondary growth and produce a woody body has many important consequences. In an herb, after a portion of stem or root is mature, its conducting capacity is set. All provascular cells have differentiated into either primary xylem or primary phloem. This capacity is correlated with the needs of leaves and roots. If the plant produces so many leaves that they lose water faster than the stem xylem can conduct, some or all the leaves die of water loss. Similarly, it would not be selectively advantageous for the plant to have so many leaves that they could produce sugar faster than the phloem could conduct it to roots, flower buds, or developing fruits.

Many herbs live for several years, however. How do they respond to the bottleneck of the limited conducting capacity of the first-year stem? In some, the first year's leaves die during winter, and in the second year, the plant produces only as many leaves as it had during the first year. In other species, adventitious roots are produced that supply conducting capacity directly to the new section of stem being formed, thus bypassing older portions of the stem. Most of these plants must remain low enough for adventitious roots to reach the soil. These are often rhizomatous, such as irises, bamboo, and ferns.

Woody plants not only become taller through growth by their apical meristems but also become wider by accumulation of wood and bark. Because wood and **bark** contain conducting tissues, their accumulation gives plants a greater capacity to move water and minerals upward and carbohydrates downward. The number of leaves and roots that the plant can support increases, as does the photosynthetic capacity.

For example, consider a tree with a trunk radius of 10.5 cm (Figure 8-1); imagine that the plant produced a layer of wood 0.5 cm thick in the previous year. As a consequence, the tree has 32.2 cm² of new wood that conducts water from roots to leaves. Assume that each leaf loses water at a rate equal to the conducting capacity of 0.1 cm² of wood; this plant can conduct enough water through its trunk to support 322 leaves. If the tree produces another 0.5 cm of wood this year, the new ring of wood will have a cross-sectional

area of 33.8 cm². It is larger than the previous ring of wood and can support conduction to a greater number of leaves—338. Even if a ring of wood could conduct for only 1 year, the plant could still produce a greater number of leaves every year, and thus, its annual photosynthetic capacity would always increase. The consequence of this ever-increasing capacity is that annual production of seeds and defensive chemicals also increases.

Only those seeds that germinate in a suitable site are able to grow into adults and reproduce. After a seed of a woody, perennial plant germinates and becomes established, it occupies its favorable site year after year. Other seeds may not be able to germinate because they do not encounter suitable sites. In springtime, all sites that had been occupied by herbs are vacant and available to the seeds of both annual herbs and woody perennials, but virtually all sites occupied by last year's perennials are still occupied by them. Many pines, oaks, and other long-lived trees hold on to the same piece of Earth for as long as 600 years and a few for up to 3,000 years; one particular tree has lived for 5,000 years. For all of that time the trees produce seeds of their own and, by their very presence, prevent the seeds of their competitors from growing at that site. Annual herbs give up their sites when they die; their seeds must compete for new sites every year.

Secondary growth also has disadvantages: A 5,000-year-old plant is 10,000 times older than an herb that germinates in April, lives 6 months, then sets seed and dies by September. It has had to battle insects, fungi, and environmental harshness 10,000 times longer, and it is a bigger, more easily discovered target for pathogens. Perennials have a greater need for defenses, both chemical and structural, than annual herbs have, and they must use a portion of their energy and nutrient resources for winterizing their bodies if they live in temperate climates. It is also expensive metabolically to construct wood and bark. The fact that wood burns so readily shows that it is energy rich. If no secondary growth occurred, this energy could be used immediately for reproduction. In fact, most woody plants do not reproduce until they are several years old; if they are killed by disease or environmental stress before they reproduce, all growth and development have been for nothing.

It must be difficult for secondary growth to arise by evolution; it has originated only three times in the 420 million years that vascular plants have been in existence, and two of those evolutionary lines later became extinct. All woody trees and shrubs alive today have descended from just one group of ancestral woody plants that arose about 370 million years ago. This group has been very successful, evolving into many species and dominating almost all regions of Earth; they include all seed plants. Within the flowering plants, herbaceousness is a new phenomenon; all early angiosperms were woody perennials, but many plants have evolved to be herbs, foregoing the woody life style. Currently, true secondary growth occurs in many eudicots, most basal angiosperms, and all gymnosperms, but not in any ferns or monocots (Table 8-1).

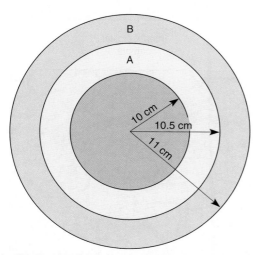

Figure 8-1 The cross-sectional area of a ring of wood is given by the formula π times the square of the outer radius minus the square of the inner radius. If each annual ring is 0.5 cm wide, when the wood has a radius of 10.5 cm, its newest ring has a cross-sectional area of 32.2 cm². Next year's ring will be larger, 33.8 cm², an increase of (33.8 − 32.2)/32.2 = 1.6/32.2 × 100% = 5% (this is not drawn to scale).

TABLE 8-1	Presence of Secondary Growth
Modern ferns	Absent in all species
Gymnosperms	Present in all species
Dicots	Present in many species, but many other species are herbs
Monocots	Ordinary type is absent in all species, but some have anomalous secondary growth

As you study this, remember that a woody plant is a combination of primary and secondary tissues: The tips of stems and roots, as well as the leaves, flowers, and fruits, are herbaceous and primary; only as portions of stems and roots become older do they begin to undergo secondary growth and become woody.

Vascular Cambium

Initiation of the Vascular Cambium

The **vascular cambium** (plural, cambia) is one of the meristems that produce the secondary plant body (Figure 8-2). In an herbaceous species, the cells located between the metaxylem and metaphloem of a vascular bundle ultimately stop dividing and differentiate into conducting tissues. But in a woody species, the cells located in this position never undergo cell cycle arrest; they continue to divide instead of maturing, and they constitute the **fascicular cambium** (Figure 8-3a; "fascicle" is an old term for bundle). In addition, some mature parenchyma cells between vascular bundles come out of cell cycle arrest and resume mitosis, forming an **interfascicular cambium** that connects on each side with the fascicular cambia. After this happens, the vascular cambium is a complete cylinder. The terms "fascicular" and "inter-

fascicular" are used only while the cambium is young; after 2 or 3 years, the two regions are usually indistinguishable, and then only the term "vascular cambium" is used.

Vascular cambia must be extended each year. The tips of roots and stems initially contain only primary tissues; however, at some time after metaxylem and metaphloem have matured, a vascular cambium arises, and that portion of the root or stem then contains both primary and secondary tissues (Figure 8-3b). During the next growing season, the apical meristem extends the axis beyond this point; a new segment of vascular cambium forms within it and joins at its base to the top of the vascular cambium formed in the previous season. The vascular cambium within a tree consists of segments of distinct ages, those near the ground being oldest and those closer to the tips of the axes being younger.

Very rarely, a vascular cambium forms in leaves that stay on a tree for many years, but just a tiny amount of secondary tissues, usually only secondary phloem without any secondary xylem, is formed in the midrib; the other veins contain only primary tissues. Vascular cambia never occur in flowers, fruits, or seeds.

Although a vascular cambium shares many features with an apical meristem, it is unique in certain aspects. It is a rather simple meristem in that it has only two types of cells, **fusiform initials** and **ray initials** (Figures 8-4 and 8-5).

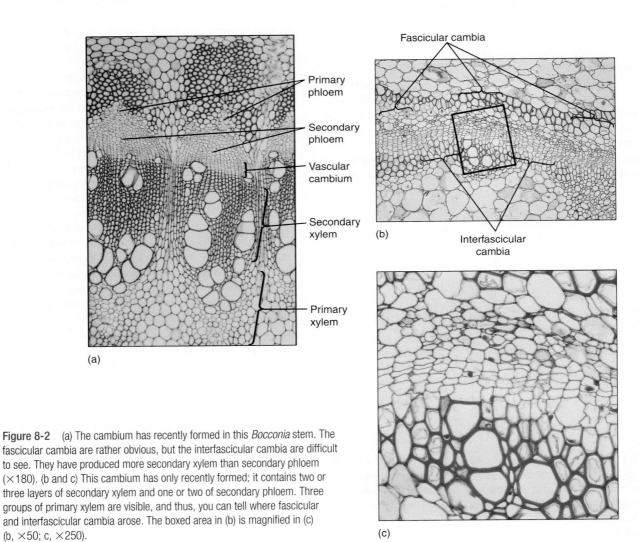

(a)

(b)

(c)

Fascicular cambia

Interfascicular cambia

Primary phloem

Secondary phloem

Vascular cambium

Secondary xylem

Primary xylem

Figure 8-2 (a) The cambium has recently formed in this *Bocconia* stem. The fascicular cambia are rather obvious, but the interfascicular cambia are difficult to see. They have produced more secondary xylem than secondary phloem (×180). (b and c) This cambium has only recently formed; it contains two or three layers of secondary xylem and one or two of secondary phloem. Three groups of primary xylem are visible, and thus, you can tell where fascicular and interfascicular cambia arose. The boxed area in (b) is magnified in (c) (b, ×50; c, ×250).

Figure 8-3 (a) Two vascular bundles and the parenchyma located between them. Some parenchyma cells have begun renewed cell division and constitute an interfascicular cambium. This zone of renewed mitotic activity is located between two fascicular cambia, so one full cambial zone will result. (b) The vascular cambium usually does not form until after a portion of shoot or root is several weeks or even many months old; so branch tips have only primary growth. After a region is old enough, the vascular cambium forms, and secondary xylem begins to accumulate to the inside of the cambium; secondary phloem accumulates to its exterior. At first, vascular bundles are still recognizable. After several years of activity, considerable amounts of secondary xylem and phloem accumulate.

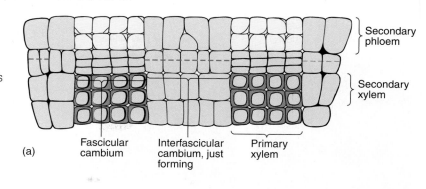

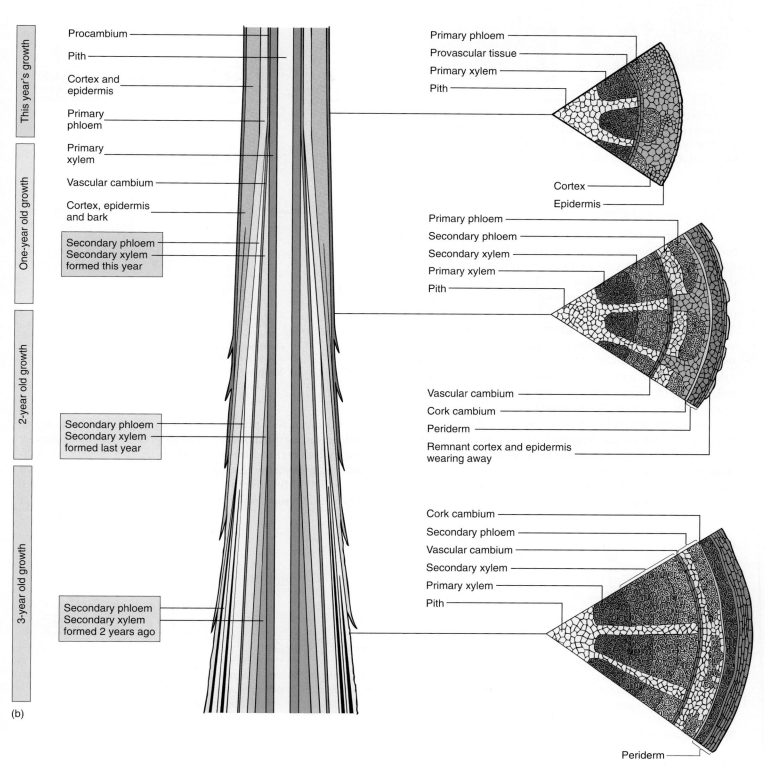

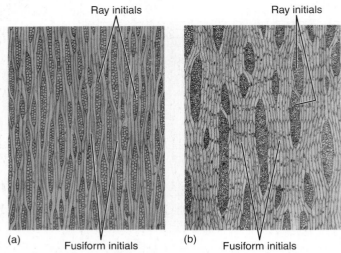

Figure 8-4 Tangential sections through a nonstoried vascular cambium of apple (*Malus sylvestris,* a) and a storied vascular cambium of black locust (*Robinia pseudoacacia,* b). The nonstoried fusiform initials are extremely long, and the ends are not aligned. In the storied cambium, fusiform initials are much shorter and occur in horizontal rows.

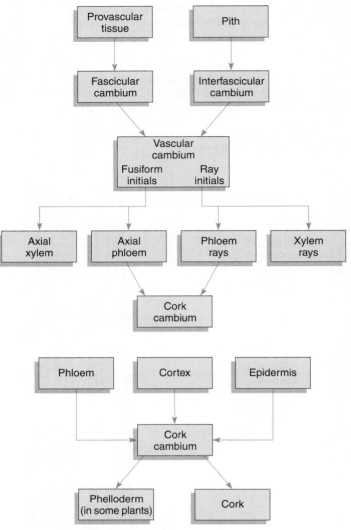

Figure 8-5 Cell lineages during secondary growth.

■ Fusiform Initials

Fusiform initials are long, tapered cells; typical lengths for fusiform initials are 140 to 462 μm in dicots and 700 to 8,700 μm (almost 1 cm) in gymnosperms. When a fusiform initial undergoes longitudinal cell division with a wall parallel to the circumference of the cambium (a **periclinal wall**), it produces two elongate cells (**Figure 8-6**). One continues to be a fusiform initial; the other differentiates into a cell of secondary xylem or secondary phloem. If the outer daughter cell remains a cambium cell, the inner cell develops into secondary xylem. But if the inner one continues as cambium, the outer cell differentiates and matures into secondary phloem. This orientation is constant: Wood never forms to the exterior of the vascular cambium, and bark never forms on the interior side. Regardless of which cell differentiates, one always remains as cambium. It is not known what factors determine which cell remains cambial and which differentiates, but within any year, both xylem and phloem are produced, almost always much more xylem than phloem.

Cambial cells produce narrow daughter cells, all of which enlarge during differentiation. Daughter cells located on the inner side, which mature into secondary xylem, increase greatly in diameter, causing the cambial cells to be pushed outward (**Figures 8-7** and **8-8**). Because the cambium is a cylinder, such outward movement results in a cylinder of larger circumference. Vascular

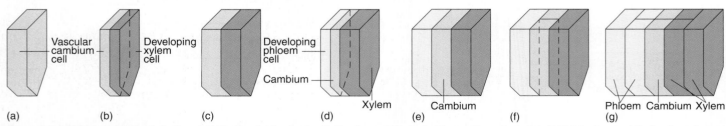

Figure 8-6 (a) The lower half of a fusiform initial before division (to simplify the diagram, the top half has not been drawn in). (b) Division by a periclinal wall results in two thin cells; the outer one remains a fusiform initial and the inner cell develops into secondary xylem. (c) Both cells enlarge to the size of the original cell. (d) The fusiform initial divides again; this time the outer cell matures as secondary phloem, whereas the inner one remains a fusiform initial. (e) The cells grow back to the original size. (f) The fusiform initial divides by an anticlinal wall, resulting in two fusiform initials. (g) Two division cycles after the radial division in (f), a new row of cells is present in the secondary xylem and phloem, produced by the new cambial cell formed in (f).

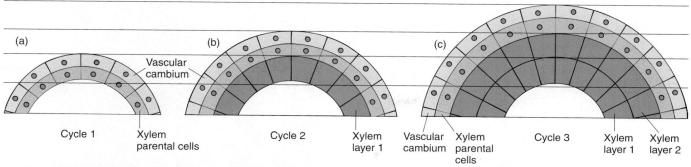

Figure 8-7 (a) After vascular cambium cells divide, progeny cells on the interior side become xylem parental cells. (b) As all cells expand to their mature size, the cambial cells are pushed outward. Cambial cells divide again, depositing a new layer of xylem parental cells exterior to the xylem that just formed (xylem layer 1). (c) The new xylem parental cells expand, pushing the vascular cambium farther outward. Cambial cells divide again. Each xylem cell is formed in place; they are not pushed inward.

cambium cells must occasionally divide longitudinally by **anti-clinal walls** (perpendicular to the cambium's surface), thereby increasing the number of cambial cells (Figure 8-6f). Without anticlinal divisions, cambial cells would be stretched wider circumferentially and finally could not function (**Figure 8-9**).

Like apical meristem cells, fusiform initials have thin primary walls, and plastids are present as proplastids. After nuclear division, a phragmoplast forms and elongates toward the ends of the cell. The phragmoplast grows about 50 to 100 μm per hour, and cell division may take as long as 10 days in species with long fusiform initials, whereas the cell cycle may be as short as 19 hours in apical meristem cells of the same plant.

■ Ray Initials

Ray initials are similar to fusiform initials except that they are short and more or less cuboidal. They too undergo periclinal cell divisions, with one of the daughters remaining a cambial ray initial and the other differentiating into either xylem parenchyma if it is the inner cell, or phloem parenchyma if it is the outer cell. One of the most significant differences between fusiform and ray initials is that the elongate fusiform initials produce the elongate cells of wood (tracheids, vessel elements, and fibers) and phloem (sieve cells, sieve tube members, companion cells, and fibers). Ray initials produce short cells, mostly just storage parenchyma and, in gymnosperms, albuminous cells.

■ Arrangement of Cambial Cells

Ray and fusiform initials are organized in specific patterns. Ray initials are typically grouped together in short vertical rows only one cell wide (uniseriate), two cells wide (biseriate), or many cells wide

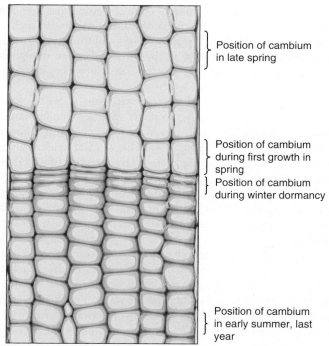

Figure 8-8 Six rows of wood cells are visible in this transverse section of a pine trunk. All cells of each row were produced by a single vascular cambium cell. Thus, all of the wood visible in this micrograph was produced by just six cambium cells. The cells at the bottom of the picture were produced when the cambium cells were located at that site; similarly, the cells in the middle of the picture were produced when the cambium existed there, having been pushed outward by the production of the cells at the bottom of the picture (×150).

Position of cambium in late spring

Position of cambium during first growth in spring

Position of cambium during winter dormancy

Position of cambium in early summer, last year

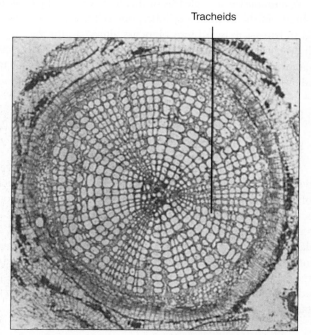

Tracheids

Figure 8-9 Many extinct arthrophytes were tree-like, having a wood of large tracheids, but their fusiform initials could not undergo longitudinal radial divisions, so no new fusiform initials could be formed and secondary growth was limited (×100).

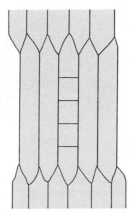

Figure 8-10 A fusiform initial can divide transversely and become a row of ray initials, after which all its derivatives differentiate as ray cells.

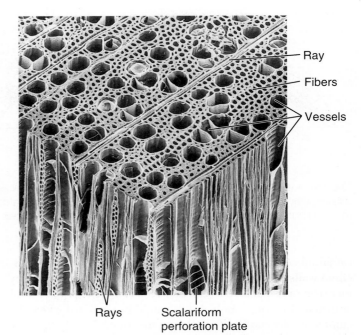

Figure 8-11 Scanning electron micrograph of a cube of wood from a tulip tree (*Liriodendron tulipifera*). Examine the relationship between the various components, especially the large vessels and the ray parenchyma (×110).

(multiseriate; see Figure 8-4). Fusiform initials may occur in regular horizontal rows (a **storied cambium**) or irregularly, without any horizontal pattern (a **nonstoried cambium**). Storied cambia have evolved more recently than nonstoried cambia and occur in only a few advanced eudicot species, for example, redbud and persimmon. The fusiform initials of storied cambia tend to be short, less than 200 μm long. The selective advantage of storied tissues is unknown.

Typically, a vascular cambium never has large regions of just fusiform initials or just ray initials. If anticlinal divisions result in many fusiform initials side by side, a central one may undergo transverse divisions and be transformed into a set of ray initials (**Figure 8-10**). Likewise, if a group of ray initials becomes unusually broad, one or several central ones may elongate and be converted to a fusiform initial. The ratio of fusiform initials to ray initials within a species is quite constant and apparently under precise developmental control.

To summarize, a single tree or shrub has only a single vascular cambium, in the form of a cylinder, one cell thick, extending upward in trunks, outward into branches almost to the very tips of each twig, and downward in roots almost to their tips. Each section of vascular cambium forms as a narrow cylinder but is pushed outward by its own production of wood cells on its inner surface.

Secondary Xylem

Types of Wood Cells

All cells formed to the interior of the vascular cambium develop into **secondary xylem**, known as **wood**. Secondary xylem contains all of the types of cells that occur in primary xylem but no new ones. Wood may contain tracheids, vessel elements, fibers, sclereids, and parenchyma. The only real differences between primary and secondary xylem are the origin and arrangement of cells. The arrangement of secondary xylem cells reflects that of the fusiform and ray initials: An **axial system** is derived from the fusiform initials, and a **radial system** develops from the ray initials (**Figure 8-11; Box 8-1**, Botany and Beyond: Wood in Three Dimensions).

The axial system always contains tracheary elements (tracheids or vessel elements or both), which carry out longitudinal conduction of water through the wood (**Figure 8-12**). In many species of woody angiosperms, the axial system also contains fibers

that give the wood strength and flexibility (**Figure 8-13** and **Table 8-2**). Most commercially important angiosperm woods contain large amounts of fibers, making them strong, tough, and useful for construction. They are called **hardwoods**, a term now used for wood of all basal angiosperms and eudicots, even those that lack fibers or are very soft, such as balsa. Woods from conifers such as pines and redwoods have few or no fibers and thus have a softer consistency. These are known as the **softwoods**, even though in many instances (such as bald-cypress, *Taxodium*) they are actually much harder than many hardwoods.

Tracheary elements and fibers are elongate, as are fusiform initials that produce them, but in many species, some immature cells undergo transverse divisions and differentiate into columns of xylem parenchyma (**Figure 8-14**). This is axial xylem parenchyma, and it is important as a temporary reservoir of water; on cloudy or humid days and at night when leaves are losing little water, wood has a temporary surplus of conducting capacity, and water is moved from roots into wood parenchyma and held there. Later, when the air is hot and dry, leaves may lose water very rapidly; water can then be drawn from wood parenchyma if the soil is too dry to supply enough water. Many desert-adapted trees have abundant xylem parenchyma. In contrast, conifers such as pine, cedar, juniper, and redwood have little or no axial parenchyma and therefore have little reserve water. For them, tough, waxy, water-conserving leaves are selectively advantageous.

The complexity of the axial system of wood varies greatly. Most gymnosperms contain only tracheids in their axial systems; fibers and parenchyma cells are sparse or absent (see Table 8-2). Some angiosperms, especially some basal angiosperms, have mostly just tracheids; however, in the majority of woody angiosperms, all possible types of cells are present, and numerous cell-cell relationships are possible. Water-storing parenchyma may be immediately adjacent to vessels, or it can be arranged such

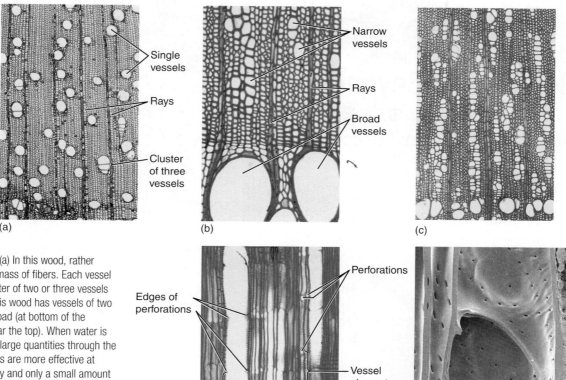

(a)   (b)   (c)

Figure 8-12   Vessels in wood. (a) In this wood, rather medium-size vessels occur in a mass of fibers. Each vessel occurs by itself or in a small cluster of two or three vessels (transverse section; ×40). (b) This wood has vessels of two very different sizes: extremely broad (at bottom of the micrograph) and very narrow (near the top). When water is plentiful, it can be transported in large quantities through the broad vessels. The narrow vessels are more effective at carrying water when the soil is dry and only a small amount of water is available to be transported (*Cotinus americana*, transverse section; ×150). (c) Vessels of *Pistasia mexicana* are arranged in long radial groups. Because most vessels touch two or three other vessels, they can share water through pits on their side walls (transverse section; ×50). (d) This longitudinal section passes through the lumen of one of the vessels. The front and back walls of the vessel elements were cut away during specimen preparation, and thus, the vessel appears empty. The perforations are visible as short projections into the lumen (radial section; ×150). (e) Simple perforation between two vessel elements: The front portion of each has been cut away and we are looking at the back half of the two elements. Water would have moved from the lower cell to the upper cell through the large hole, the perforation. There are numerous small pits in the side walls (×3,000).

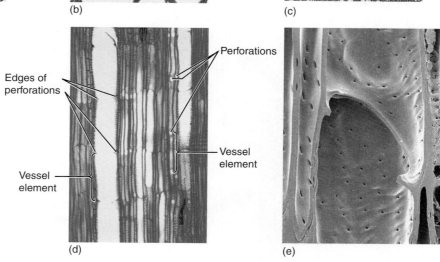

(d)   (e)

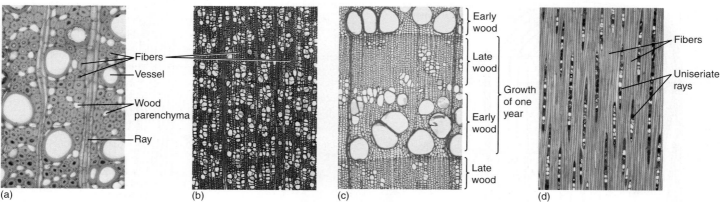

(a)   (b)   (c)   (d)

Figure 8-13   Wood fibers. (a) The fibers of this wood have extremely thick walls. Also, fibers are abundant, so this wood is quite strong (transverse section; ×150). (b) Vessels make up over half the volume of this wood, and thus, fibers are not very abundant. Furthermore, the fiber cell walls are rather thin; this would not be a strong wood (transverse section; ×40). (c) In this wood, regions that have many large vessels and few fibers (called early wood, discussed later in this chapter) alternate with regions that have abundant fibers and only a few narrow vessels (late wood). Early wood provides maximum conduction capacity, whereas late wood provides strength (transverse section; ×50). (d) A longitudinal section through fiber-rich late wood. All fiber cells are long and narrow and have thick walls (tangential section; ×50).

Box 8-1 Wood in Three Dimensions

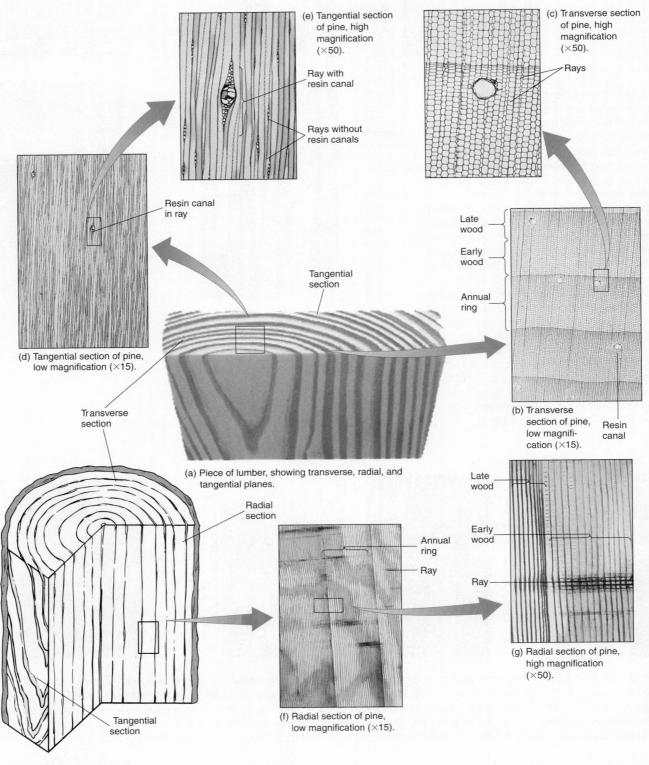

(e) Tangential section of pine, high magnification (×50).

Ray with resin canal

Rays without resin canals

(c) Transverse section of pine, high magnification (×50).

Rays

Resin canal in ray

(d) Tangential section of pine, low magnification (×15).

Late wood

Early wood

Annual ring

Tangential section

Transverse section

(b) Transverse section of pine, low magnification (×15).

Resin canal

(a) Piece of lumber, showing transverse, radial, and tangential planes.

Radial section

Annual ring

Ray

Late wood

Early wood

Ray

(g) Radial section of pine, high magnification (×50).

Tangential section

(f) Radial section of pine, low magnification (×15).

It is important to understand wood in all three dimensions, but that is not particularly easy. As you study these illustrations, think about how the cells are formed by the vascular cambium. Also try to picture how the various cells touch each other in three dimensions, which conduct water, which store water and nutrients, and so on. Pine wood is simpler because its axial system contains only tracheids, so you need to think mostly about tracheids and rays. Oak wood is more complex, with vessels, fibers, and axial parenchyma all touching the rays at some point. For some serious practice of your skill at visualizing three-dimensional objects, imagine a vascular cambium existing on one tangential face. How are the new cells formed each year?

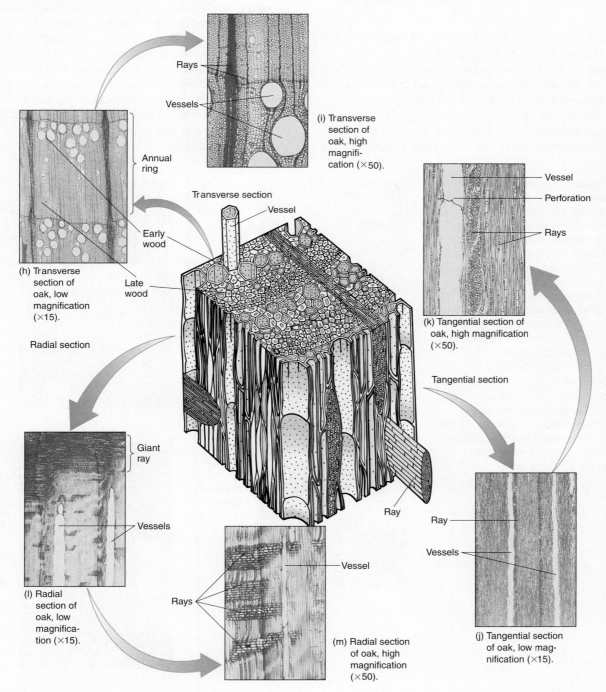

(i) Transverse section of oak, high magnification (×50).

(h) Transverse section of oak, low magnification (×15).

(k) Tangential section of oak, high magnification (×50).

(l) Radial section of oak, low magnification (×15).

(m) Radial section of oak, high magnification (×50).

(j) Tangential section of oak, low magnification (×15).

TABLE 8-2	Cell Types Present in Wood	
	Gymnosperms	Dicots*
Axial system		
Tracheids	Present	Present
Vessels	Absent (except in three groups)	Present
Fibers	Very rare	Present
Parenchyma	Very rare	Present
Radial system		
Ray parenchyma	Present	Present
Ray tracheids	Present	Absent

*Dicot wood is extremely variable; certain species lack some of these cell types, other species have them all. The relative amounts of each vary greatly.

that it never touches vessels. Fibers provide maximum strength if grouped together in masses, which is how they are usually arranged (Figure 8-14d). If fibers are located around a vessel, their secondary walls reinforce the walls of the vessel and help it resist collapse, but the presence of fibers excludes parenchyma cells.

The radial system of xylem is usually simple. In woody angiosperms, it contains only parenchyma, arranged as uniseriate, biseriate, or multiseriate masses called **rays** (Figure 8-11 and

Figure 8-15). Ray parenchyma cells store carbohydrates and other nutrients during dormant periods and conduct material over short distances radially within wood. The two basic types of ray parenchyma cells are **upright cells** and **procumbent cells** (Figure 8-15). At least in some plants, procumbent ray cells have no direct connection with axial cells, but upright cells do. The ray/axial interface can take many forms. If the upright ray parenchyma cell is adjacent to axial parenchyma, plasmodesmata occur. If the ray parenchyma is adjacent to an axial tracheid or vessel element, the tracheary element has pits in its secondary wall and the ray cell has very thin walls facing the pits. In early springtime, when trees such as maples are drawing on their nutrient reserves, the starch that had been stored in the upright cells is the first to be digested into sugar and passed into the axial tracheary elements for conduction to newly expanding buds, leaves, and flowers. Starch in procumbent cells is not digested until later and presumably must first be routed through upright cells for transfer to axial conducting cells.

In gymnosperms, xylem rays are almost exclusively uniseriate; they are multiseriate only if they contain a resin canal (**Figure 8-16a**). In addition to ray parenchyma cells, they may contain **ray tracheids**—horizontal, rectangular cells that look somewhat like parenchyma cells but have secondary walls, circular bordered pits, and protoplasts that degenerate quickly after the secondary wall is completed (Figure 8-16b).

Figure 8-14   Wood axial parenchyma. (a) To prepare wood for microscopy, it is often boiled to make it soft enough to cut into thin specimens; unfortunately, this destroys the protoplasm of any parenchyma cells. This wood was not boiled, so the starch-filled parenchyma cells are easily visible. The large cells surrounding the vessels are also parenchyma, and they can release water to the vessels in times of water shortage (temporarily dry soil) (transverse section; ×150). (b) This wood has abundant axial parenchyma—the small cells with thin walls. The small cells with thick walls are fiber cells. Because this wood was boiled, it is difficult to be certain which is parenchyma (transverse section; ×50). (c and d) The axial parenchyma in this wood forms large bands; all of the vessels occur in the bands, surrounded by parenchyma. None is in the fiber masses. This parenchyma may act as a "water jacket" around vessels, absorbing excess water when water loss from leaves is low (cool nights) and releasing it when water loss is rapid (hot days) (transverse sections; c, ×15; d, ×50).

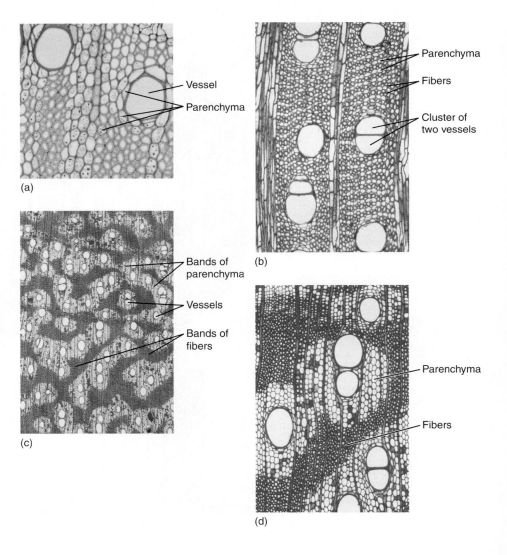

(a) Vessel / Parenchyma

(b) Parenchyma / Fibers / Cluster of two vessels

(c) Bands of parenchyma / Vessels / Bands of fibers

(d) Parenchyma / Fibers

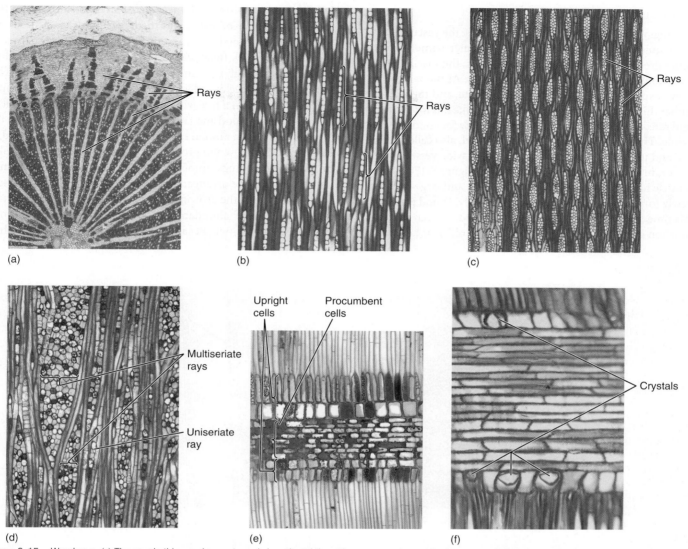

Figure 8-15 Wood rays. (a) The rays in this wood are extremely broad, and thus, they are easy to see. Each ray extends from the xylem into the phloem (×15). (b) These rays are uniseriate, only one cell wide; they are rather short, only about 10 cells tall. This is from late wood, so vessels are not present in this section (tangential section; ×50). (c) These rays are multiseriate, several cells wide. Over half the volume of this area of the wood is storage parenchyma, not conductive tracheary elements or strengthening fibers (tangential section; ×40). (d) This wood has giant multiseriate rays and small uniseriate rays (tangential section; ×50). (e) This ray has procumbent cells in the central part and upright cells along the edges (radial section; ×50). (f) Ray cells often contain crystals of stored material; four crystals are visible here in the cells along the margin. None occurs in the procumbent cells (radial section; ×150).

Figure 8-16 (a) Conifer rays often contain secretory canals that produce resin (pitch), important in preventing insects from burrowing through the wood (tangential section; ×150). (b) The central ray cells are living ray parenchyma, with large nuclei that have been stained black. The procumbent cells along the top and bottom of the ray are ray tracheids that have circular bordered pits but no protoplasm when mature (radial section; ×150).

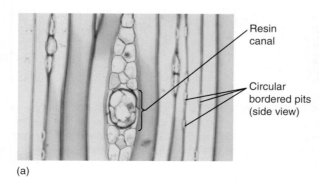

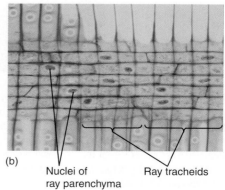

Growth Rings

In regions with strongly seasonal climates, the vascular cambium is quiescent during times of stress, either winter cold or summer drought, but when quiescence ceases, the vascular cambium becomes active and cell division begins. At the same time, the new, expanding leaves are thin and delicate, and their cuticle is neither thick nor fully polymerized. Leaves like this lose water at a rapid rate, and thus, trees need a high capacity for conduction at this time. The first wood formed is **early wood**, also called spring wood, and it must have a high proportion of wide vessels (**Figure 8-17a**) or, in gymnosperms, wide tracheids (Figure 8-17d). Later, the cuticle has thickened, transpiration is less, and large numbers of newly formed vessels are conducting rapidly. Wood produced at this time, called **late wood** or summer wood, can have a lower proportion of vessels. But the plant is a year older, is larger and

heavier, and it needs more mechanical strength to hold up the increased number of leaves and the larger branches. Late wood is stronger if it contains numerous fibers or, in gymnosperms, if it contains narrow, thick-walled tracheids. Finally, at the end of the growing season, the cambium becomes dormant again. The last cells often develop only as heavy fibers with especially thick secondary walls. In a tree with wood like that just described, it is easy to see early wood and late wood, the two together making up 1 year's growth, an **annual ring**. If a summer is unusually cold, a tree may fail to grow and produce a ring, so these are occasionally not truly annual rings. Some people prefer the term **growth ring**.

An alternative arrangement exists: In some species, vessels form throughout the growing season. Those produced in spring are neither more abundant nor obviously wider than those produced in summer. Because the wood of a growth ring has

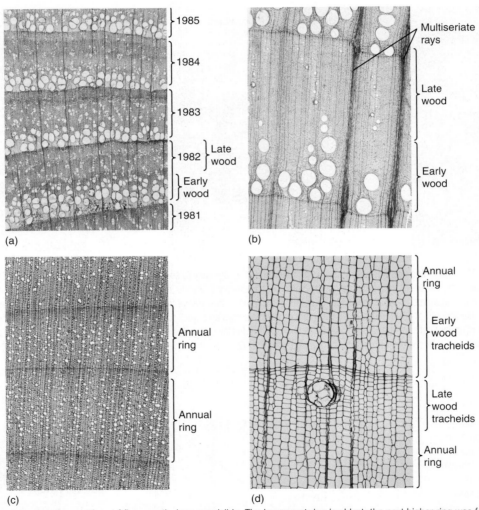

(a)

(b)

(c)

(d)

Figure 8-17   (a) In this transverse section, portions of five growth rings are visible. The lowermost ring is oldest; the next higher ring was formed 1 year after the lowest ring, and so forth. The vascular cambium is located beyond the top of the photograph. The wood is ring porous, and vessel-rich early wood is easily distinguished from fiber-rich late wood (transverse section; ×50). (b) This wood is very strongly ring porous, with large vessels formed only when the cambium first becomes active. Later, the cambium produces fibers exclusively, except for a few rare, very narrow vessels. Two large multiseriate rays are visible, but most rays are narrow and uniseriate (transverse section; ×15). (c) This wood is diffuse porous, with vessels occurring rather uniformly in both late wood as well as early wood. Growth rings are not as conspicuous as in ring porous wood (transverse section; ×40). (d) Most gymnosperm wood has only tracheids in its axial system, but it still has early wood (wide tracheids) and late wood (narrow tracheids) (transverse section; ×50).

Plants Do Things Differently

Box 8-2 Having Multiple Bodies in One Lifetime

Woody plants have two bodies. As the primary body of a woody plant ages, a vascular cambium arises inside it and produces wood and secondary phloem—an entire new body—inside the preexisting body. Think about how different the two bodies are. The primary body has leaves and axillary buds, flowers, fruits, and seeds. It has root hairs and absorbs water and nutrients. The secondary body is just wood and bark. It has no leaves, no buds, no flowers, and so on. The secondary body is, for the most part, nothing more than an ever-growing vascular/skeletal system. The two bodies look completely different and have distinct functions. Growth of the secondary body tears apart and destroys the primary phloem, cortex, and epidermis of the plant's primary body, and these dead remnants are shed as part of the plant's first bark. Shoot tips and root tips continually make more sections of primary body, but they too will be destroyed by formation of more secondary body inside them. It is dramatic for one organism to have two distinct bodies, to have one body form inside another, destroying the first. Does anything like this occur in animals? In us? Yes, and in even more dramatic fashion.

We humans undergo moderate changes in our body. When about 5 or 6 years old, we shed our baby teeth as a new set of permanent teeth forms below them. As the permanent teeth develop and enlarge, they simply push our baby teeth out and we lose them. Parts of our body—teeth with blood vessels, nerves, and living cells—just fall out much the way bark falls off a tree. Later, when we go through puberty, other changes occur. Hair follicles, especially in boys, become active and start producing thicker hair than the type children have. In girls, there is development of glandular and adipose tissue in the breasts. These and other changes, however, are really just modifications of preexisting tissues that were already present in children. There is nothing equivalent to a cambium and the formation of brand new cells.

Our puberty, however, pales in comparison to that of eels. Juvenile eels are just tiny, flat, coin-shaped marine fish that look something like a leaf. As they go through the transformation to being adults, they develop their very long, cylindrical shape and switch to being freshwater fish, migrating up rivers to spawn. Juvenile and adult eels have such different bodies that the juveniles were long considered to be a completely different type of fish. Juvenile and adult *Homo sapiens* are obviously the same species. We do not change that much during puberty.

Other animals go through more significant bodily changes. Crabs, lobsters, and beetles have an exterior exoskeleton that is so hard it cannot grow. The animal periodically produces a new, soft exoskeleton; then the animal molts—that is, it sheds its skin and old exoskeleton—and very quickly grows to a new size before its new exoskeleton hardens and prevents further growth. After some time, the animal will repeat this process so that it can grow even larger. Snakes too periodically shed an old skin, replacing it with a new one. In these examples, entire, complex tissues are being sloughed off and new body parts are formed.

Undoubtedly, the most drastic examples of individuals that have two distinct bodies are insects that go through what is called a complete metamorphosis. Their larval bodies do not look anything at all like the adult bodies. Examples are caterpillars, which metamorphose into moths or butterflies, and maggots, which metamorphose into flies. The larval body is specialized for eating and growing and has neither wings nor sex organs of any kind. In contrast, the adult body does have wings and sex organs, enabling it to fly about and find a mate and then carry out sexual reproduction and dispersal. In some cases, the adult body needs to survive only a day or two until it can mate and die (males) or lay fertilized eggs (females). Such adult bodies have either no digestive system at all or such a simple one that it can only absorb the sugar water of flower nectar. In the complete metamorphosis of caterpillars, the caterpillar spins a cocoon around itself; then its body more or less dissolves except for special sets of cells called imaginal discs. These act like meristems and produce the cells, tissues, and organs of the adult body by using the nutrients from the dissolved larval body. By the time metamorphosis is complete, the body has been completely rearranged; almost nothing exists of the preexisting larval body except that its molecules have been recycled and not wasted.

A plant's primary body differs as greatly from its secondary body as do the caterpillar body and the butterfly body of a particular species. If plants could also digest their primary bodies and rebuild them the way caterpillars do, the transformation in plants would be seen to be just as dramatic as metamorphosis in animals. However, plants do not form cocoons and do not undergo self-digestion, so the activity of a vascular cambium and a cork cambium in the production of an entire new secondary body seems unremarkable. It appears as if the plants are doing nothing more than adding a few new tissues, but the change is really much more fundamental.

Just considering ourselves, the idea of individual plants and animals having several distinct bodies may seem far fetched, but as it turns out, it is a common occurrence with each body carrying out distinct phases of the organism's life activities.

vessels located throughout it, it is said to be **diffuse porous** (Figure 8-17c), whereas species with vessels restricted mainly to early wood are **ring porous**. Examples of trees with diffuse porous wood are yellow birch, aspen, sugar maple, and American holly; trees with ring porous wood include red oak, sassafras, and honey locust. In mild tropical climates, the cambium may remain active almost continuously, and the wood of one year is difficult to distinguish from that of another; growth rings are indistinct.

Heartwood and Sapwood

The center of a log is almost always darker in color than the outer wood, and it is usually drier and more fragrant (Figure 8-18). The dark wood is **heartwood** and the lighter, moister outer region is **sapwood**. The different regions exist because vessels and tracheids do not function forever in water conduction; water columns break because of freezing, wind vibration, tension, wood-boring insects, and other factors. After the water column breaks, there is no means of pulling water upward; vessels and tracheids in which this has occurred usually never conduct water again. Although only a few water columns break at any time, after several years, all water columns in a growth ring have snapped, and that ring no longer conducts. New water-filled tracheary elements are produced by the cambium during the next year. An important problem is that a vessel is wide enough that a fungus can easily grow up through it; for vessel elements that are not conducting, a mechanism that seals them off is selectively advantageous. Wood parenchyma cells adjacent to vessels push bubbles of protoplasm through the pits into the vessel, forming a plug, called a **tylosis** (plural: tyloses), completely across it (Figure 8-19). This occurs repeatedly, and the vessel may become filled. These and other wood parenchyma cells undergo numerous metabolic changes and produce large quantities of phenolic compounds, lignin, and other dark-colored, aromatic substances that inhibit growth of bacteria and fungi. These chemicals are usually dark

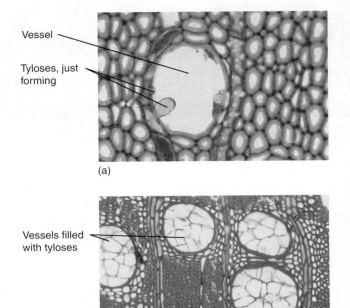

(a)

(b)

Figure 8-19 (a) Tyloses form as protoplasm from surrounding parenchyma cells pushes into a vessel. This was just beginning to form tyloses when the wood was collected for microscopy. Boiling destroyed the protoplasm (transverse section; ×250). (b) The vessels of this wood are completely occluded by tyloses; this sample came from a piece of heartwood (×50).

and aromatic, and as they accumulate, wood becomes darker and more fragrant, such as cedar wood. Ultimately, all parenchyma cells die, and conversion of sapwood to decay-resistant heartwood is complete.

A new layer of sapwood is formed each year by the vascular cambium, and on average, one annual ring is converted to heartwood each year. Thus, whereas heartwood becomes wider with age, sapwood has a more or less constant thickness (Table 8-3). Of course, this is not true of young stems and roots: Black walnut typically has wood that functions for 10 years before converting to heartwood. In a seedling or a branch only 9 years old, no heartwood is present yet.

Figure 8-18 The vascular cambium was almost perfectly circular in a cross-section in this tree trunk. The innermost annual rings, formed when the tree was a sapling, are wider than the outer rings—the tree grew more vigorously when it was younger. The dark region is heartwood, and the narrow light region is sapwood; if the tree had lived a few years longer, several more of the innermost rings of sapwood would have converted to heartwood. Bark—secondary phloem—is also present.

TABLE 8-3	Thickness of Sapwood in Dicot Trees
	Number of Xylem Rings in Sapwood
Catalpa speciosa (catalpa)	1–2
Robinia pseudoacacia (black locust)	2–3
Juglans cinerea (butternut)	5–6
Maclura pomifera (Osage orange)	5–10
Sassafras officinale (sassafras)	7–8
Aesculus glabra (Ohio buckeye)	10–12
Juglans nigra (black walnut)	10–20
Prunus serotina (wild black cherry)	10–12
Gleditsia triacanthos (honey locust)	10–12

■ Reaction Wood

In branches or trunks that are not vertical, gravity causes a lateral stress; if not counteracted, the branch would droop and become pendant. In response to such stress, most plants produce **reaction wood**. In angiosperms, this develops mostly on the upper side of the branch and is known as tension wood. In a cross-section of such a branch, growth rings are eccentric, being much wider on the top of the branch. Tension wood contains many special gelatinous fibers whose walls are rich in cellulose but have little or no lignin. These fibers exert tension on the branch, preventing it from drooping, or the tension wood may even contract, slowly lifting a branch to a more vertical orientation. Conifers form reaction wood located on the underside of the branch and is known as compression wood. It is enriched in lignin and has less cellulose; growth rings are especially wide on the lower side of the limb.

■ Secondary Phloem

Because **secondary phloem** is formed from the vascular cambium just as secondary xylem is, it too has an axial and a radial system (Figure 8-20). The axial system is responsible for conduction up and down the stem or root; it contains sieve tube members and companion cells in angiosperms, or sieve cells in gymnosperms. In both groups of plants, fibers and nonconducting parenchyma are also typically present in axial secondary phloem. In some species, there may be bands of fibers alternating with sieve tube members, but usually, these are not annual rings, and many other patterns can be seen in other species. Whereas the equivalent cells of axial secondary xylem are arranged as early and late wood and as ring porous or diffuse porous wood, no similar arrangement occurs in secondary phloem. Although the tracheary elements of secondary xylem may function for many years before being converted to heartwood, sieve tube members and sieve cells usually conduct for less than 1 year; only the innermost layer of phloem is capable of conduction.

The size, shape, and number of phloem rays match those of xylem rays because both are produced by the same ray initials (Figure 8-21). Phloem rays consist only of parenchyma cells that are used for storage, as are xylem rays, but phloem rays seem to be even more important for this (Figure 8-22). In gymnosperms, albuminous cells are ray cells.

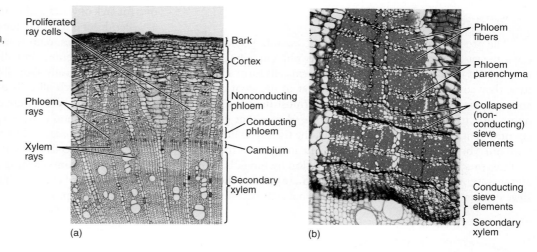

Figure 8-20 (a) This stem transverse section has secondary xylem at the bottom, a vascular cambium region, secondary phloem, and then cortex and bark at the top. Xylem rays and phloem rays meet at the cambium. Parenchyma cells of some rays have proliferated through cell division and expansion, which has prevented tearing of the bark (×40). (b) In this secondary phloem of *Artabotrys,* only the youngest axial phloem (at the bottom) contains functional sieve tube members. The youngest phloem and the vascular cambium are so soft that they are often damaged while trying to cut samples from trees; this is why they are partially crushed here. Older phloem has abundant fibers alternating with bands of phloem parenchyma and collapsed sieve tube members (×50).

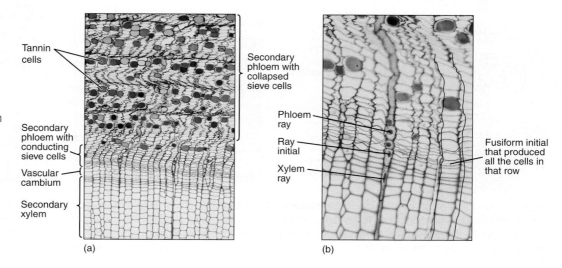

Figure 8-21 (a and b) The outermost, youngest wood and the bark of pine. As sieve cells stop functioning and collapse, phloem shrinks and becomes undulate. This causes rays to become wavy. Phloem rays meet xylem rays at the ray initials. Also, each row of sieve cells was produced by the same fusiform initial that produced the corresponding row of tracheids (transverse sections; a, ×50; b, ×150).

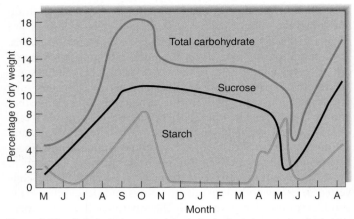

Figure 8-22 Carbohydrate accumulates in the bark of stems and roots during early summer while leaves are still present and photosynthesizing; it reaches its peak in September and October, then drops and remains steady through winter (December to April). It is released during the spring growth season, April and May. These data are for black locusts near Ottawa, Canada, where spring comes late. For plants that grow farther south, the spring release occurs earlier.

Outer Bark

Cork and the Cork Cambium

The production and differentiation of secondary xylem cells cause the vascular cambium and secondary phloem to be pushed outward. As the youngest, innermost phloem cells form and mature, they contribute to the larger diameter of the stem or root and increase pressures acting on the outermost tissues. This requires that tissues on the periphery of the plant either grow in circumference or be torn apart. Actually, the tissues do both, but both must be controlled (Figure 8-20). The integrity of the plant surfaces must be maintained against invasion by fungi, bacteria, and insects. As circumferential stretching increases and the older sieve elements die, some storage parenchyma cells become reactivated and undergo cell division. This is similar to the activation of parenchyma cells during formation of the interfascicular vascular cambium, but in secondary phloem, it results in a new cambium, the **cork cambium**, also called the **phellogen** (Figure 8-23).

Cork cambium differs greatly from vascular cambium in both structure and morphogenic activity. All of its cells are cuboidal, like ray initials. After each division, the inner cell almost always remains cork cambium, whereas the outer cell differentiates into a **cork cell** (also called a **phellem cell**). In a few species, the cork cambium may produce a cell or two to the inside that mature into a layer of parenchyma called **phelloderm**. The cork cambium, the layers of cork cells, and the phelloderm (if any) are known as **periderm**. Maturing cork cells increase slightly in volume; then their thin primary walls become encrusted with suberin, making them waterproof and chemically inert, and then they die. Cell death is probably a critical part of maturation because the protoplasm breaks down, leaving nothing digestible or nutritious for an animal to eat. In many species, some cells deposit secondary walls and mature into lignified sclereids; these usually occur in layers that alternate with cork, resulting in a periderm that is both impervious and tough (Figure 8-24). Because periderm is such an impermeable barrier, all plant material exterior to it, such as epidermis, cortex, and older secondary phloem, dies for lack of water and nutrients.

Periderm offers only temporary protection because the root or stem continues to grow interior to it, pushing it outward and stretching it circumferentially. Unlike vascular cambium, cork cambium is typically short lived; it produces cells for only a few weeks, after which all cells differentiate into cork cells and die. The layer of cork cells cannot expand much circumferentially, and after one or several years, a new cork cambium must be formed in younger secondary phloem closer to the vascular cambium. These new cork cells act as a further barrier and also block water and nutrients from reaching any secondary phloem cells located between layers of cork cambium. In this fashion, several layers of cork can build up.

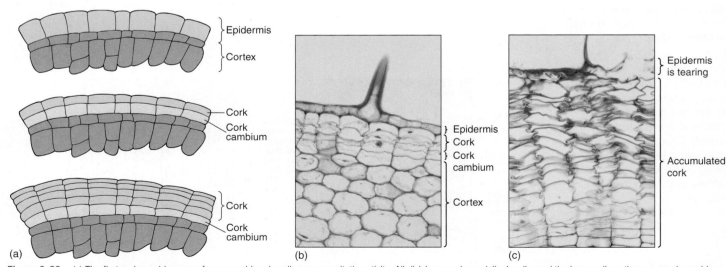

Figure 8-23 (a) The first cork cambium may form as epidermis cells resume mitotic activity. All divisions are by periclinal walls, and the inner cell continues as cork cambium while the outer cell differentiates into cork. (b) Hypodermis cells have just started to undergo cell division, resulting in formation of a cork cambium. This is a young stem of geranium; notice the base of a trichome (×150). (c) Older stem of geranium; many layers of phellem (cork cells) have accumulated owing to cork cambium activity. The phellem cells are dead and empty. A trichome was present here also, and formation of bark blocked transfer of nutrients to the trichome and other epidermal cells, killing them (×150).

Figure 8-24 (a) A thick layer of uncollapsed cork cells is present on this stem. Many parenchyma cells in the cortex and secondary phloem have converted to sclereids (b), making the bark stronger and more protective. The next cork cambium will form deep in the secondary phloem, where numerous parenchyma cells are capable of becoming mitotically active again; sclereids cannot resume cell division (transverse sections; a, ×40; b, ×150).

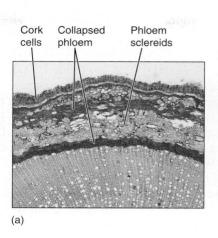

Cork cells   Collapsed phloem   Phloem sclereids

(a)

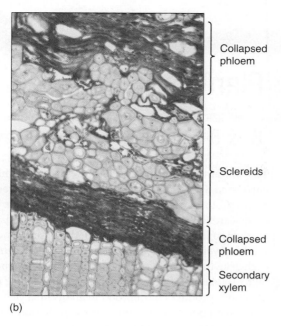

Collapsed phloem

Sclereids

Collapsed phloem

Secondary xylem

(b)

All tissues outside the innermost cork cambium comprise the **outer bark** (Figure 8-25). All secondary phloem between the vascular cambium and the innermost cork cambium is the **inner bark**. The amount of bark is quite variable from species to species; in some, only a small amount of cork is formed, so the bark is thin and consists mostly of dead secondary phloem. In others, cork is produced in large amounts and becomes 3 or 4 cm thick. Although it is usually not obvious, bark is continuously falling off the tree, but it does not accumulate at the base because wind and water carry it away.

Lenticels and Oxygen Diffusion

The impermeability of cork has negative as well as advantageous consequences. Although it keeps out pathogens and retains water, it also blocks absorption of oxygen, interfering with the respiration of the sapwood, vascular cambium, and inner bark. Bark becomes permeable to oxygen when cork cambium produces cork cells that become rounded as they mature. Because rounded cells cannot fit tightly together, intercellular spaces penetrate the cork layer, creating a diffusion pathway for oxygen. These regions of aerenchymatous cork are **lenticels** (Figure 8-26). When a new cork cambium arises interior to this one, it too forms a lenticel in the same place; the outer and inner lenticels are aligned, permitting oxygen to penetrate across all layers of the bark. Lenticel-producing regions of cork cambia are more active than adjacent regions that produce only ordinary impermeable cork; consequently, lenticels contain more layers of cells and protrude outward. In species that have smooth bark, even small lenticels are easy to identify. On plants that have thick rough bark, lenticels

Outer bark

Inner bark

(a)   (b)   (c)   (d)   (e)

Figure 8-25 (a) As this pecan trunk increases in diameter, its bark is stretched and ultimately cracks. The deepest bark is the youngest, that on the surface the oldest. (b) Bark of maple peels off in large thin sheets because numerous cork cambia form close together and each is sheetlike. (c) Cork cells of sycamore contain many chemicals; as outer patches of bark peel away, fresh patches are exposed. When their pigments oxidize, they turn grey. Each cork cambium forms as a small patch; the size and shape of the cork cambia affect the nature of the bark. (d) The bark of cork oak (*Quercus suber*) becomes extremely thick and is composed mostly of phellem, with few sclereids. Cork oaks grown in Spain and Portugal provide most of our commercial corks for bottles. When the bark has become sufficiently thick, the outer bark is peeled away; after a few years, the bark is once again thick enough to harvest. (e) This is cinnamon bark; we grind it into a powder to use as a spice. It is one of the few barks we eat.

Plants and People

Box 8-3 Dendrochronology—Tree Ring Analysis

The amount of wood produced by the vascular cambium is closely correlated with climate. Species that occur in extremely harsh climates produce very little wood, often only a single layer of vessels each year. Slow growth such as this occurs in cold regions at high altitudes or high latitudes or in hot, dry desert zones. In the tropics, however, the amount of growth each year can be large, with each growth ring being several millimeters wide and consisting of 50 to 100 layers of new cells. Even within a single plant, vascular cambium produces more wood in an optimal year than it does in a year with poor temperatures or too little rain. This can be seen easily in the varying width of tree rings.

Because growth rings reflect climate so accurately, they are used for many critical studies. Starting from the most recent, outermost ring, it is possible to count toward the center of the tree, encountering older and older rings. Not only can the age of the tree be determined, but climatic fluctuations can be inferred from the width of the rings. The oldest living trees of a region can be studied for climatic changes that occurred during the lifetime of the tree, which in some individuals is more than 5,000 years. However, the analysis of past climates can be extended beyond the age of the oldest living trees. Historical or archaeological sites may contain houses, bridges, or ships constructed with large wooden beams. Some of the outermost rings of those beams may match the patterns of the inner rings of living trees. By counting back and finding the age of that set of rings in the living wood, we can determine how old that portion of the beam is and therefore when the house, ship, or bridge was constructed. This technique has been valuable not only in dating ancient settlements but also in establishing whether people lived in times of good or poor climate (Figure B8-3).

After a tree ring sequence has been mapped out from many different sources of wood, it is important to establish an exact date to the rings. In North America and much of Europe, there are extensive forests with trees old enough to allow us to analyze tree ring sequences back for a thousand years or more. Rather recent archaeological sites yield wood samples that can be matched to the centers of living trees, extending the accurately dated sequences back even farther; then samples from older archaeological sites can be added to extend the sequence back in time, knowing the exact year each ring was produced. Even if we find a wooden post used in some ancient building or ship, we can match the ring sequence to the known sequence and determine exactly when the tree lived and when it was probably cut down to be used.

In the Middle East, in areas where many critically important civilizations flourished, such as Egypt, Babylon, and Sumer, it has only been possible to map out the tree ring sequence from the present back to CE 362 (CE is "Common Era," an alternative to AD); then we run into a lack of the necessary archaeological wood samples. By going back only to the fourth century CE, we can determine exactly when many Byzantine buildings were constructed but nothing earlier. There is a second, excellent tree ring sequence for the Middle East, but for a long time, it has been what is called a "floating" tree ring chronology. There are enough archaeological wood samples that the rings of some overlap those of others, and those overlap even others so that a well-documented sequence has been established; however, none of the samples is as recent as CE 362. None comes up to the sequence where we know exactly which ring corresponds to which year. Instead, all we know—or knew—is that the floating sequence extends over 1,503 years and that it must have begun somewhere earlier than 2200 years BCE (Before the Common Era) because we know the approximate age of some of the temples and palaces that supplied the ancient wood.

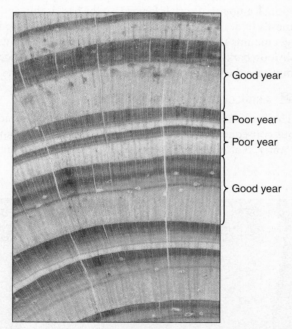

Good year

Poor year

Poor year

Good year

Figure B8-3 Notice that some growth rings in this wood are much wider than others. The innermost ring, marked "Good year," is very wide, with abundant early wood (the lighter, inner part) as well as late wood (the darker, outer part). Conditions must have been very good for the tree that year: It might have been a summer of optimal temperatures or rain; if a cultivated tree, it might have been fertilized that year; or surrounding trees that had been shading it either might have died or been cut away. The next two growth rings, both marked "Poor year," were produced when the tree was not vigorous enough to make very much wood, but then the next ring is especially broad so the tree was extremely healthy then. By simply looking at this photograph we can only tell whether the tree was growing well or not, but if we had the actual wood, we could chemically analyze traces of various chemicals in the wood, and they might indicate the exact reason for the tree's rapid or slow growth.

It may be that we can establish the absolute time of the floating sequence after all. In the North American and European sequences, a very strange ring was produced by trees in the year 1628 BCE. It is believed to have been caused by a volcanic eruption that produced so much dust that sunlight was blocked around the world, and the summer of that year was so cold that frost damage occurred in many trees, even those of warm climates. Dendrochronologists in Cornell University, the University of Heidelberg, and the University of Reading have discovered that the floating sequence for the Middle East also has an unusual ring that seems to match this one. Rather than frost damage, the ring in the floating sequence is one of extra growth. It may be that the Middle East at the time was so hot and dry that the cooling caused by the volcanic dust cloud was only enough to produce extra rain and optimal growing conditions rather than frost and poor growing conditions.

This hypothesis—that the strange ring in the floating sequence occurred in the same year as the strange ring in other, anchored sequences—makes predictions that other rings should match. If those predictions are correct, that will be evidence that the floating sequence has been dated exactly, and from it, we can then put exact dates to the buildings, statues, and ships that the various wood samples came from.

Figure 8-26   (a) Low magnification of stem of *Aristolochia,* with a large lenticel. Rapid production of cork cells in lenticels cause the epidermis to rupture. Although they cannot be seen here, small intercellular spaces allow oxygen to diffuse through the lenticel into the trunk or root (×15). (b) Lenticels of birch (*Betula*) widen as the trunk increases in circumference. Occasionally, they divide into two lenticels when the central cork cambium cells begin producing compact cork instead of aerenchymatous cork. (c) This bark of honey mesquite (*Prosopis glandulosa*) contains many small patches of cork, each produced by a cork cambium that arose from parenchyma cells in the secondary phloem, produced a few layers of cork, and then stopped functioning. Each patch of cork is surrounded by secondary phloem, and the outer patches of cork are older than the inner ones. Lenticels are not visible but would be located at the deepest part of each fissure.

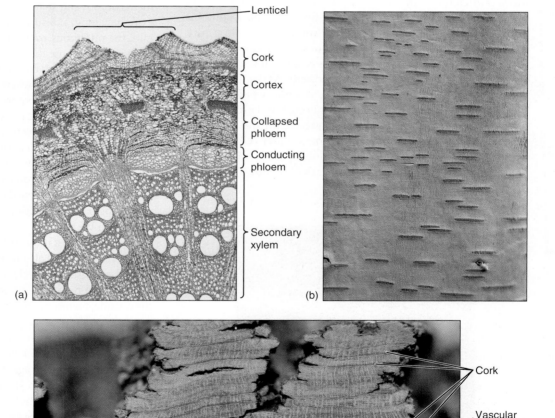

can be almost impossible to see, but generally, they are located at the bases of the cracks in the bark. When the bark of cork oak is made into bottle corks, it is necessary to cut them so that the lenticels do not run from the top of the cork to the bottom.

Initiation of Cork Cambia

The timing of initiation of the first cork cambium is far more variable than that of the vascular cambium. In some species, the first cork cambium arises before a twig or root is even 1 year old. On stems this is often detectable as the surface color changes from green to tan. In other species, the first cork cambium forms only when that region is several years old; until then, the epidermis and cortex are retained. Epidermises more than 40 years old have been reported. Delayed formation of bark is common in plants that depend on cortex chlorenchyma for much of their photosynthesis, as cacti do. The first cork cambium may arise in a number of tissues: epidermis, cortex, primary phloem, or secondary phloem. Subsequent cork cambia may form shortly afterward, sometimes in the same season, but usually a year or two later. If the growth in diameter is slow, new cork cambia may arise at intervals of as much as 10 years. These later cork cambia usually form deep in the secondary phloem.

The first bark on young stems usually differs from bark formed when the stem is older. If the first cork cambium arises by reactivation of epidermal cells, the first outer bark contains only periderm and cuticle and is very smooth. If the first cork cambium arises in the cortex, the first outer bark contains periderm, cortex, and epidermis; this too is smooth and contains any cortical secretory cells that were present. As the first bark is shed and later cork cambia arise in the secondary phloem, they produce an outer bark that contains only cork and phloem. The nature of these later barks depends greatly on the cell types present in the trapped secondary phloem: Fiber cells produce fibrous, stringy bark, sclereid-filled phloem produces hard bark, and so on. It is not unusual for the bark of a young tree to be

Figure 8-27 This trunk of *Casuarina* (red beefwood) still has fragments of its first bark, but the underlying newer bark is also visible. Young branches have only the first type of bark, older parts of the trunk only the second type.

dramatically different from the bark it will have when it is older (**Figure 8-27**).

Secondary Growth in Roots

Roots of gymnosperms and woody angiosperms undergo secondary growth, as do the stems. A vascular cambium arises just like the interfascicular cambium, when parenchyma cells located between the primary xylem and primary phloem become active mitotically, as do pericycle cells near the protoxylem (**Figure 8-28**). The new vascular cambium has the same star shape as the primary xylem, but it soon becomes round as the cambium in the sinuses of the primary xylem produces more secondary xylem than do the regions of cambium near the arms of protoxylem. Consequently, some portions of the cambium are pushed outward more rapidly than others. When a circular cambium is achieved, the unequal growth stops, and all parts grow at similar rates.

Root vascular cambium contains both ray and fusiform initials, and in many cases, wood produced in the root is quite similar to that of the shoot, having sapwood and heartwood and being ring porous or diffuse porous as the stem is. Typically, however, the wood of roots is not identical to that of stems of the same plant, and they may be totally dissimilar when the conductivity requirements of root and stem differ. For example, many cactus roots are extremely long, nonsucculent, cable-like structures. After a brief rain, the numerous root tips absorb water from a large region of surface soil. Because the soil can dry within hours after such a rain, water must be conducted rapidly and in large quantity into the succulent cactus shoot where it can be stored. Once in the shoot, however, conduction requirements are totally different; the shoot may be quite small and consist mostly of parenchyma, with very little wood and just a few narrow vessels. The water is stored so effectively that most conduction is probably by diffusion through the cytoplasm rather than by xylem. Furthermore, the narrow roots are strengthened by numerous wood fibers, whereas the small succulent shoots are strengthened only by turgor. In the shoot, the xylem and phloem rays are gigantic, providing a large region for water storage. In the roots, which do not store water, rays are small and narrow and consist mostly of sclerenchyma.

Perennial roots also form bark; the first cork cambium usually arises in the pericycle, causing the endodermis, cortex, and epidermis to be shed. The cork cambium produces cork cells to the outside, forming a protective layer, and in some species a layer or two of phelloderm as well. Lenticels also occur, being especially prominent near lateral roots. The bark on roots may be similar to that on the stem of the same plant, but in some species, there are significant differences, again related to the differing metabolisms and microhabitats of the two organs.

Several mechanisms exist by which the storage capacity of a woody root can be increased. The ray parenchyma of the secondary xylem offers considerable volume, and storage capacity is increased if the rays become larger. Axial parenchyma in the wood can also be used, and in many storage roots, wood is almost pure parenchyma (**Figure 8-29a**). For example, the orange part of a carrot root that we eat is wood consisting mostly of wood parenchyma and just a few vessels. When we peel the carrot, we remove a thin bark and all secondary phloem.

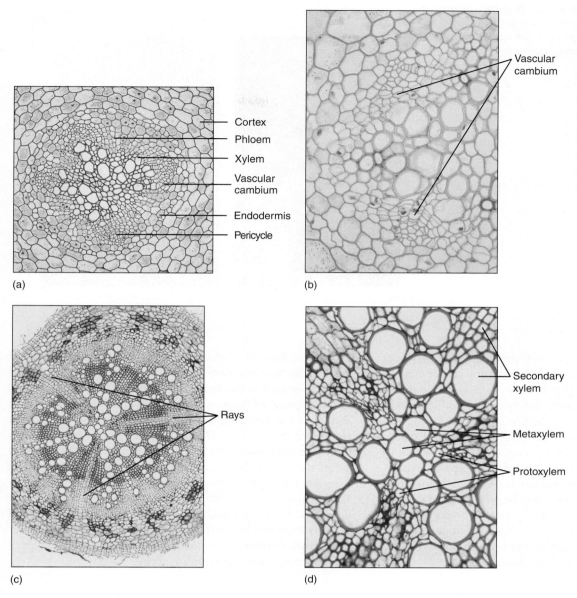

Cortex
Phloem
Xylem
Vascular cambium
Endodermis
Pericycle

(a)

Vascular cambium

(b)

Rays

(c)

Secondary xylem

Metaxylem

Protoxylem

(d)

Figure 8-28 (a and b) Young roots. The vascular cambium of a root such as this baneberry arises as an undulate cylinder located around the xylem, interior to the phloem (a, ×50; b, ×150). (c and d) Older roots. The organization of root wood is similar to that of stem wood: an axial system and a set of rays. In (c), there are three large rays; the high magnification of (d) reveals that these rays are aligned with the protoxylem of the primary xylem (c, ×40; d, ×150).

Anomalous Forms of Growth

Anomalous Secondary Growth

The development, cellular arrangement, and activity of vascular cambia in gymnosperms and most woody angiosperms are remarkably similar. However, there are alternative types of cambium structure and activity, and analysis of the consequences of various types of arrangements reveals a great deal about secondary growth. Because alternative cambia produce secondary bodies that differ from the common type, their growth is called **anomalous secondary growth**.

Roots of Sweet Potatoes In sweet potatoes (*Ipomoea batatas*), the amount of storage parenchyma is increased dramatically by an anomalous method of secondary growth. Numerous vascular cambia arise, not around the entire mass of primary xylem, but around individual vessels or groups of vessels (Figure 8-29b). The cambia act normally, except that the xylem and phloem produced

are almost purely parenchyma. New vessels may also be surrounded by another new cambium, and the process is repeated. As the sweet potato becomes quite large, it may contain hundreds of cambia of various ages; the secondary tissues are an irregular matrix of parenchyma, a few sieve tubes, some vessels, and vascular cambia.

What is the selective advantage of producing so many vascular cambia instead of just one? It may be that the rate of cell production is important. Because the root must become large very quickly, having just one cambium may be too slow. Multiple cambia all functioning simultaneously speed the production of storage capacity.

Included Phloem In several eudicots, a vascular cambium of the common type arises and produces ordinary secondary xylem and phloem. After a short period, however, the cambium cells stop dividing and differentiate into xylem: There is no longer a cambium. Cells in the outermost, oldest secondary phloem then become reactivated and differentiate into a new vascular

Figure 8-29 The part of a carrot (a) or radish (b) we eat is the wood of the storage root. It does not look like wood at first glance; it is mostly parenchyma with a few vessels and no fibers. This wood is well adapted for rapid production of long-term storage tissue (×40).

(a)   (b)

cambium that acts just like the first, producing ordinary secondary xylem and phloem, then differentiating and ceasing to exist. Notice in **Figure 8-30a** that because the second cambium arose in the outermost phloem, the xylem it produces is located exterior to the phloem of the first cambium. From interior to exterior there is first xylem, first phloem, second xylem, and second phloem. Then a new vascular cambium arises in the outermost second phloem and so on. This type of secondary phloem, located between two bands of xylem, is **included phloem**. The selective advantage may be protection of phloem from insects and other pests by one to several layers of wood. Also, tissue relationships between xylem and included phloem differ from those between xylem and ordinary phloem, but the advantages of this have not yet been studied.

Unequal Activity of the Vascular Cambium In ordinary growth, all areas of the vascular cambium have about equal activity, and the stem or root is therefore round in cross-section. However, in some species of *Bauhinia* and certain other woody vines, two sectors of the cambium are very active, while two are almost completely inactive (**Figure 8-31**). The stem grows outward in two directions but remains thin in the other two, and it soon becomes a thin, flat, woody ribbon. The selective advantage of this may be related to flexibility. As an ordinary round stem becomes wider in all directions, its conducting capacity increases, but its flexibility decreases. As the *Bauhinia* stem becomes wider, its conducting capacity increases, but its flexibility remains about the same. For many vines, flexibility is selectively advantageous, and this type of anomalous secondary growth is adaptive.

Figure 8-30 (a) This stem of *Iresine* shows bands of secondary xylem alternating with bands of included phloem. Each band of xylem and the phloem just external to it were produced by one cambium; the next set was produced by a different cambium (×40). (b) *Bougainvillea* also has included phloem, but it occurs in patches, not complete bands (×40).

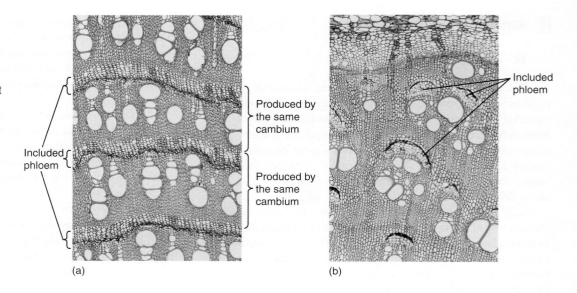

(a)   (b)

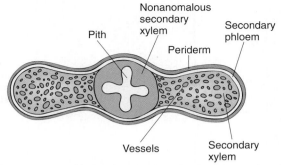

Figure 8-31 The cambium on two opposite sides of this *Bauhinia* is rapidly producing cells, many of which differentiate into wide vessel elements. The two alternate portions of cambium produce very few cells, and all mature without much expansion.

Secondary Growth in Monocots None of the monocots has secondary growth like that in gymnosperms, basal angiosperms, and eudicots, but some do become tree-like and "woody," such as Joshua trees, dragon trees, and palms. The first two groups undergo a process of anomalous secondary growth; the palms have an unusual type of primary growth.

In Joshua trees (some members of the genus *Yucca*; **Figure 8-32**) and dragon trees (in the genus *Dracaena*), a type of vascular cambium arises just outside the outermost vascular bundles (**Figure 8-33**). It originates from cortex cells in the same manner that the interfascicular vascular cambium arises in eudicots. This cambium, however, produces only parenchyma; conducting cells are completely absent. Columns of some of the parenchyma cells undergo rapid division and produce narrow cells that differentiate into **secondary vascular bundles** containing xylem and phloem. The outermost cells of each bundle develop into fibers with thick secondary walls. The parenchyma cells that do not divide like this form a secondary ground tissue, the arrangement of which is almost identical to that of primary tissues. They are "woody" because of the fibers, and they have more conducting capacity and greater strength each year, so branching is feasible.

■ Unusual Primary Growth

Palm trees are unusual in that their trunks do not taper at the tips like those of eudicot or gymnosperm trees, and they do not branch. The palm trunk is all primary tissue consisting of vascular bundles distributed throughout ground tissue; each bundle contains only primary xylem and primary phloem, all derived from the shoot apical meristem (**Figure 8-34**). A vascular cambium never develops, and true wood and secondary phloem do not

Figure 8-32 Joshua tree (*Yucca brevifolia*) is an arborescent monocot in the lily family. Because it has secondary growth, even though of an unusual type, the ability of its trunk to conduct increases, and both branching and increased numbers of leaves are feasible without adventitious roots of the type necessary for screwpine.

occur; the trunk does not grow radially. The trunk is hard and "woody" because each vascular bundle is enclosed in a sheath of strong, heavy fibers.

A palm seedling does not have a full set of leaves and a wide trunk. For the first few years of life, the palm trunk becomes wider and the number of leaves increases. This happens without secondary growth because during the seedling years, palms produce numerous adventitious roots from the base of the short stem. Each root adds extra vascular bundles, and the portion of stem above each new root can have that many more bundles than it does below the root. For example, if the stem has 100 bundles at one point and if just above this it produces five adventitious roots that each have eight bundles, then above these roots the stem can have 140 bundles. This increase in width and addition of adventitious roots in palms is called **establishment growth**, a form of primary growth. At some point, this process ceases. No new adventitious roots are established, and the conducting capacity is set for the lifetime of the plant. This same type of primary growth occurs in other monocots with extremely broad stems such as corms and bulbs.

Figure 8-33 Secondary growth in arborescent monocots. (a) The broad zone of the vascular cambium region and immature secondary ground tissues are visible. The bundles are secondary, and a secondary cortex also forms (×50). (b) The inner part of the stem, with the primary vascular bundles and ground tissue on the bottom and the secondary bundles at the top (×15). (c) Primary bundles (×50). (d) Secondary bundles (×50).

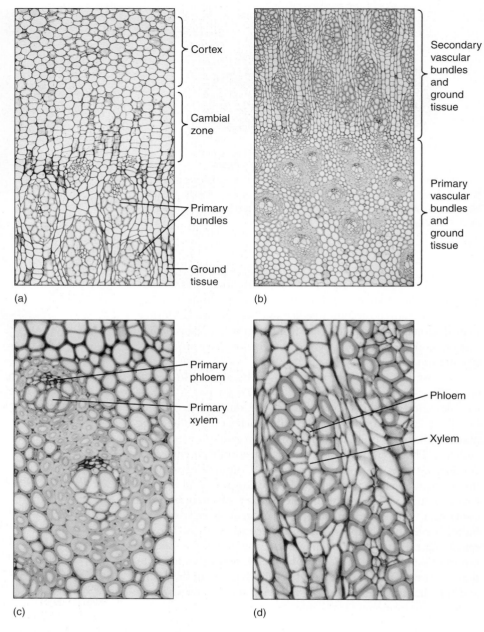

(a)

Cortex

Cambial zone

Primary bundles

Ground tissue

(b)

Secondary vascular bundles and ground tissue

Primary vascular bundles and ground tissue

(c)

Primary phloem

Primary xylem

(d)

Phloem

Xylem

Figure 8-34 Palm trees are monocots, so they have many vascular bundles rather than wood. This large palm trunk has hundreds of vascular bundles; the outermost, very dark bundles consist of just fibers, without xylem or phloem. This is shown about one-fourth life size.

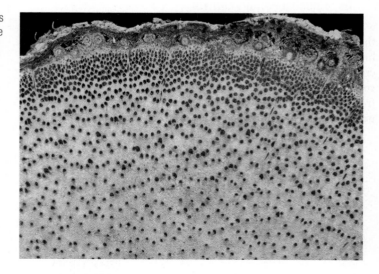

Plants Do Things Differently

Box 8-4 Thinking about the Growth of Wood

Although the growth of trees and their wood is perhaps the most complex type of growth you will ever try to understand, it really is not too difficult. Use your imagination to picture a complicated object in three dimensions and then imagine watching it grow. To start, imagine a tree without any leaves, flowers, or fruits. Now imagine removing the bark from all the woody areas, and then paint all the wood—every bit of the trunk, branches, and roots but not the twigs or root tips where there is epidermis. If you imagined this correctly, you have one single, continuous coat of paint that indicates where the vascular cambium is, and you have thousands of twig tips and root tips, each with an apical meristem. All active meristems of the tree are now visible in your imagination (the axillary buds and root primordia are inactive, but if your brain is up to it you can try to imagine those as well—just remember they are only on twigs and root tips, not in woody parts).

Hold that image in your mind, and now imagine you are making bread and have mixed yeast cells with water and flour and then kneaded them until the yeast cells are evenly distributed throughout the dough. Each yeast cell absorbs water and nutrients and then enlarges and divides; each daughter cell then repeats this process. All parts of the dough contain living cells and all are growing at the same rate. This is diffuse growth, and if we were doing this on the space station without gravity, our ball of dough would grow into a perfect sphere, becoming larger in all directions equally.

Plants have localized growth, and at the tip of every living twig and root there is a small mass of dividing cells, the shoot or root apical meristem. They have localized growth. If the meristems were to grow like the bread dough (equally in all directions), then the stems and roots would end up looking like lollipops. But all the meristem cells next to existing stem or root tissue act differently from the other meristem cells: They divide less and less and finally stop dividing, having become part of the new stem or root rather than continuing to be part of the meristem. If all meristem cells slowed their rate of division and became stem or root cells, the stem or root would stop growing, and they would not have indeterminate growth. It is important that some meristem cells become added to the growing stem or root, while at the same time some of the meristem cells continue to be meristem cells.

Fortunately, our own hair, nails, and skin have localized growth just like this. Let's consider the growth of one of our hairs, as an example. Small blood capillaries at the base of the hair follicle bring water and nutrients to the base of the hair shaft, the area where hair cells are still alive and dividing. The hair "meristem" cells absorb the nutrients, grow, divide, and then do this again. Those dividing cells that are closest to the hair shaft (and farthest from the capillaries) gradually stop dividing and instead differentiate into dead hair cells. As time passes more hair meristem cells are added to the base

of the shaft and our original cells are thus pushed upward. Each of our hairs grows at its base and is pushed upward out of the hair follicle; the outermost end of each hair is the oldest part (and thus the most likely to become a split end), and the base of the hair is the newest. Our hairs do not have growth rings like tree wood, but the outer end of each hair is old and the inner end is younger (think of the problems this causes for those who dye or bleach their hair).

When growth is localized instead of diffuse, the location of the meristem is important. In roots and stems the meristem is at the apex, the outermost part, so the apical meristem is pushed away from the plant body by the very shoot or root that it is producing. But the dividing cells of our hairs are located at the base of the hair; they are basal meristems and they remain in the same place; it is the hair itself that is forced to move as new cells are added to it. The same is true for our skin (the innermost basal layer contains the living, dividing cells, whereas the outer part that we see is dead) and nails (we only see the dead nail cells, and we can actually watch those cells being forced away from the fingernail base, the "quick," by applying fingernail polish and seeing the new, fresh, dead unpainted nail cells being pushed out of the fingernail base).

Remember that one of the themes of this text is to think about alternatives and consequences. I am going to describe an alternative way of growing that does not exist; this is a theoretical alternative that has never evolved, as far as we know. We can imagine that some tree might have two vascular cambia located back to back and shaped like a hollow cylinder (the same shape as a real vascular cambium). The inner cambium would be a "wood cambium": Its cells would divide, and those cells on the inner side closest to the existing wood would differentiate into new wood cells. We could say this wood cambium is an apical meristem because it is at the apex of the wood it is producing and it is being pushed outward as wood accumulates on its inner side. This wood cambium would be just like a shoot apical meristem except that it would be a cylinder instead of a tiny point, and, for a whole tree, it would have the shape of the paint we imagined in the first paragraph. Lying right to the outside of the wood cambium would be our hypothetical "phloem cambium": All its cells would grow and divide, and the cells closest to the mature phloem would differentiate to be new phloem cells, and the rest of the phloem cambium cells would continue to be phloem cambium. Like the cells that produce our skin, nails, and hair, the meristem cells cannot be pushed deeper into the body; instead, the new tissues must be pushed out. So our phloem cambium would be a basal meristem and the new phloem would be pushed outward, just as our hair and nails are pushed outward. In this hypothetical, nonexistent model, our tree would grow wider because of its "apical" wood cambium and its "basal" phloem cambium lying back to back.

A minor problem is that as new cells are added to the wood, the wood becomes thicker and consequently so does its circumference: Its surface area increases. Because the "wood cambium" and the "phloem cambium" are on the surface of the wood, those two cambia also must become larger. We have a familiar example for understanding this: We have to be extremely careful around a newly born baby because the five boney plates that make up its skull are not yet solid; they are still growing at the edges (sutures) where each piece of bone meets its neighbors. As the baby's brain grows, so does its skull, as the sutures between each plate add new cells to the edges of those plates. The bones in a baby's skull become larger as a "meristem" of dividing cells along the edges of each plate adds new cells to each plate. And as each plate becomes larger, its perimeter becomes larger, so that "meristem" along the edges must increase, must become longer. It is just the same for the meristems that cover the surface of a growing cylinder. The baby's skull adds enough new "meristem" cells so that all parts of the edge of each plate is completely covered in dividing cells and thus each plate grows everywhere along its edge.

Now let's go back to the tree. Instead of having two meristems back to back, there is just one single layer of cells, just one meristem, the vascular cambium. As its cells divide, sometimes cells on the inner side stop dividing and differentiate as wood, and sometimes instead it is cells on the outer side that stop dividing and become phloem. In both situations it is critically important that some cells continue to divide and act as meristem cells. Because the addition of new cells to the wood causes the wood's surface to increase in circumference and because the cambium is pressed tightly against the wood, the cambium's circumference must also increase. The cambium has two options: Its cells could occasionally divide to make both daughter cells lie side by side within the cambium and neither one of those would turn into either wood or phloem. In this case the cambium itself would grow and have more cells. By doing this often enough the cambium will always have enough cells to completely cover the wood cylinder, no matter how broad it becomes and no matter how huge its circumference. Alternatively, rather than dividing to make new cambium cells, the cambium cells could just enlarge and become bigger cells. But as the tree continues to grow the cambial cells will also need to grow, and at some point they would reach a size that is just too large for a living, dividing cell to survive. At that point cell division would have to stop, and then the tree could no longer increase in width. This type of cambium did indeed evolve in one group of plants (species of *Sigillaria*), and the trunks it produced never grew to be very wide.

To summarize, shoots and roots of woody plants become longer as new cells are added to their apices in a form of localized, apical growth. Woody shoots and roots become wider through the action of a single vascular cambium that is only one layer of cells thick, and that layer produces xylem to its inner side and phloem to its outer side *and* it also produces new cambial cells. Each year, as the trunk or branch or root becomes longer, a new section of vascular cambium must differentiate, and it attaches itself to the pre-existing cambium, and in the following year this will be repeated. At any given time the tree has only one cambium.

When trying to understand how trees grow, it is important not to make too much comparison with our bones. We refer to wood as a tree's "skeleton" or that wood supports a tree the way our bones support and strengthen our body. But keep in mind that our bones are composed of living cells, most of which are encased in a layer of calcium phosphate. This may sound like a wood cell with a lignified secondary wall, but bone cells are alive and can remodel themselves, and bones are constantly taking some cells from certain spots, adding other cells to other spots, and remodeling the calcium (especially in women). Broken bones heal if set properly because our bodies can readjust the cells and the calcium phosphate, but the wood in a broken branch never heals. Some new wood may be laid down around it if the cambium survives, but the broken splinters inside never heal. And do not compare our growth in height with that of a tree; as we grow taller, the ends of our bones are constantly being reshaped and remodeled. New cells are added to the ends, and old cells are removed from other areas. This never, ever happens in plants; new wood cells are never added to the top end or bottom end of an existing piece of wood; instead, a piece of wood only grows wider through its vascular cambium. A branch or trunk becomes longer only because an apical meristem makes a new section of twig, and at some point that develops a new section of cambium that then starts to make wood, and this new cambium is attached to the end of the old cambium.

Summary

1. Cells and tissues produced by apical meristems are primary tissues, whereas those produced by the vascular cambium and cork cambium are secondary tissues.

2. The vascular cambium contains fusiform initials that produce the elongate cells of secondary xylem and phloem, and ray initials that produce ray cells.

3. Secondary xylem contains the same types of cells as primary xylem, but the arrangement differs. Growth rings with early wood and late wood are usually present.

4. In an old trunk, branch, or root, the central xylem is heartwood; it is dry and its parenchyma has died. The outer, younger xylem is sapwood; it is involved in water conduction and its parenchyma is alive.

5. Cork cambium produces resistant cork cells; the cork and all exterior tissues are the protective outer bark. Lenticels permit oxygen diffusion into the organ.

6. Palms appear woody because their numerous vascular bundles contain many fibers. Certain other monocots have anomalous secondary growth, in which a vascular cambium produces secondary vascular bundles.

Important Terms

annual ring	included phloem	reaction wood
anticlinal walls	late wood	ring porous
axial system	lenticels	sapwood
bark	nonstoried cambium	secondary phloem
cork cambium	periclinal wall	secondary tissues
corks	periderm	secondary vascular bundles
diffuse porous	phellem cell	secondary xylem
early wood	phellogen	softwoods
establishment growth	primary tissues	storied cambium
fusiform initials	radial system	tylosis
growth ring	ray initials	vascular cambium
hardwoods	ray tracheids	wood
heartwood	rays	

Review Questions

1. Growth by apical meristems has been previously described. In woody species, additional tissues are produced in the stem and root by two other meristems, the _____ _____ _____ and the _____ _____. The new tissues themselves are _____ (_____ _____) and _____ (_____ _____ and _____). They are the secondary tissues, and they constitute the plant's secondary body.

2. Name at least three examples of woody plants that are trees and three that are shrubs.

3. Many herbs live for several years. How do they respond to the bottleneck of the limited conducting capacity of the first-year stem?

4. Which groups of plants have secondary growth? Which never do?

5. Imagine a tree that has a radius of 20 cm and that produces a new layer of wood 0.5 cm thick (the outer radius of the new wood is 20.5 cm and the inner radius is 20 cm). What is the cross-sectional area of the new layer of wood (CSA = πr^2)? If each leaf needs 0.1 cm^2 of wood to supply it with water, how many leaves can the tree have? If the tree produces a new ring of wood next year that is again 0.5 cm thick, how many leaves can the tree have next year (assume that wood conducts water for only 1 year, not 2 years)?

6. Woody plants are almost always perennial plants, often living for many years. Describe some advantages of this with regard to a plant's ability to occupy a favorable site. Describe some disadvantages with regard to how long a plant must survive pathogens and harsh conditions.

7. It must be difficult for secondary growth to arise by evolution. How many times has it evolved in the 420 million years that vascular plants have been in existence? How many of those have become extinct?

8. What is the name of the vascular cambium that arises within vascular bundles? Between vascular bundles? After the two have formed, what is the shape of the vascular cambium? Is it a series of strips that are aligned up and down the trunk or root, or is it a complete cylinder?

9. Imagine a tree that is growing in springtime. All woody parts that were present last year have a vascular cambium, but all of the twigs that are brand new have just primary growth with vascular bundles. When will a vascular cambium form in these twigs? Will it connect to the old, pre-existing vascular cambium, or will it remain separate? Describe how this occurs.

10. Figure 8-3 is complex but easy to understand. Figure 8-3b shows a tree trunk, and this year is its fourth year of growth. A transverse section at the base of Figure 8-3b would show how many layers (growth rings) of wood (circle one: 1, 2, 3, 4)? Would a transverse section through the top of Figure 8-3b show any layer of wood this year? Would it show a layer of wood at the end of next year? Would there be any transverse section that does not show the pith and primary xylem; that is, are pith and primary xylem ever lost or destroyed by the formation of wood? The orange layer is labeled "Provascular tissue" in the top right of Figure 8-3b, but what is it labeled in the middle and lower right diagrams?

11. What are the two types of cells in the vascular cambium? Can either type be converted into the other?

12. Look at Figure 8-6. Part (a) shows the lower half of a fusiform initial (actually most fusiform initials would be much taller than this), and part (b) shows the same cell after it has divided and one of the daughter cells is developing into a xylem cell. Did the fusiform initial divide with a periclinal wall or an anticlinal wall as it went from part a to part b? Part (d) shows that the same fusiform initial has now produced a second cell that is developing into a phloem cell. Was that phloem cell produced by a periclinal or an anticlinal division of the fusiform initial? Part (f) shows the fusiform initial dividing and forming a second fusiform initial. Is this division occurring by a periclinal wall or an anticlinal wall?

13. Are fusiform initials parenchyma cells, collenchyma cells, or sclerenchyma cells? Do they have chloroplasts, chromoplasts, or proplastids?

14. Look at Figure 8-8. Vascular cambium is not present; it was located far above the top of the micrograph, but you can see there are six rows of wood cells. How many fusiform initials were involved in producing the wood visible in this figure? Look at all the cells in any particular row. Were all the cells of that row made by divisions of one fusiform initial or by many different fusiform initials?

15. Are ray initials longer or shorter than fusiform initials? Can they undergo both periclinal and anticlinal divisions or just one or the other?

16. Ray initials are typically grouped together in short vertical rows only one cell wide (_____), two cells wide (_____) or many cells wide (_____).

17. True or false: Typically a vascular cambium never has large regions of only fusiform initials or ray initials.

18. What fraction of cells formed to the interior of the vascular cambium develop into secondary xylem, known as wood—less than half, exactly half, or more than half all cells?

19. In gymnosperms (conifers like Christmas trees), what types of cells occur in the axial system (the cells produced by fusiform initials)? Which cells are very rare (Hint: Table 8-2)? Wood of gymnosperms is called _____ because a certain type of cell is absent from their axial system. Which type of cell is that?

20. In basal angiosperms and eudicots, what types of cells occur in the axial system? What is the only type of cell present in the rays of eudicots? The wood of eudicots is called _____ _____ because it has the type of cell lacking in gymnosperms. Which type of cell is that?

21. What types of cells are derived from fusiform initials? What types of cells are derived from ray initials? Is it theoretically possible to have a vascular cambium without ray initials or without fusiform initials?

22. Look at the woods of Figure 8-13. In part (b), all vessels have about the same diameter and are narrow, but the wood in part (c) has some very wide vessels (in the early wood) and some narrow ones (in the late wood, both types produced in the same year). The answer is not in the text, but would you think that these two species have different water conducting needs? Is it possible that one species lives in areas with very rainy spring times and the other lives in an area that is rather dry all the time?

23. What is a growth ring? How does early wood differ from late wood? Do all species of wood show strong differences between these two phases of a growth ring?

24. Why do some people prefer the term "growth ring" rather than "annual ring"? What can sometimes happen if a summer is unusually cool? Do you think this occurs more frequently in California or in Alaska?

25. If especially wide vessels are produced early in the growing season and only narrow, sparse vessels are produced later, the wood is said to be _____ porous; however, if vessels have similar size and abundance throughout a growth ring, the wood is _____ porous. Name three examples of each type.

26. What changes occur as sapwood is converted to heartwood?

27. What is a tylosis? How does formation of tyloses slow the spread of fungi in wood? Why doesn't a tree make tyloses in vessels that are still conducting water?

28. The thickness of sapwood is some indication of how long a growth ring of wood is able to function. Look at Table 8-3. About how long does wood of catalpa function? How about wood of wild black cherry or honey locust?

29. Because secondary phloem is formed from the vascular cambium just as secondary xylem is, it too has an _____ _____ system and a _____ system. Why do the size, shape, and number of phloem rays match those of xylem rays?

30. In a cross-section of a tree, where are the oldest growth rings—in the outer region or nearer the pith? Where is the oldest secondary phloem—near the outside of the tree or near the cambium?

31. What causes the outermost tissues of a woody stem or root to become pushed outward and expanded?

32. The layer of cells that produces cork has a technical name and an ordinary name. Give each name.

33. Periderm consists of at least two types of cells. Sometimes a third is present. What are these three types of cell?

34. Why does a layer of periderm offer only temporary protection? When the plant makes a new layer of cork cambium, does it make the new layer to the outer side of the failing periderm or to the inner side of it, deeper in the secondary phloem?

35. In which tissues does the first cork cambium form? When does it usually arise? In which tissues do later cork cambia form?

36. Are geraniums herbs or woody plants? Do they ever form bark (Hint: Figure 8-23)?

37. What type of plant produces the cork used for wine bottles and cork boards—pines, oaks, maples, eucalyptus, or none of these (Hint: Figure 8-25)?

38. What is the function of lenticels? How are intercellular spaces important for this function?

39. Do roots form wood and bark, or are these secondary tissues present only in stems?

40. Describe the anomalous secondary growth in roots of sweet potatoes. What is the selective advantage of this unusual type of secondary growth?

41. Describe the formation of included phloem. How did the included phloem of *Iresine* in Figure 8-30a become surrounded by xylem?

42. Describe the secondary growth of monocots like Joshua trees and dragon trees. What are secondary vascular bundles?

43. Describe the growth of a palm seedling. For the first few years of life, the seedling becomes wider, but without secondary growth, how is it able to do this? Each adventitious root adds something. What is it? If the palm stem has 100 vascular bundles and then forms five new adventitious roots with eight bundles each, how many vascular bundles can the stem have above these new adventitious roots? Would the stem be wider or narrower above these adventitious roots?

44. Why can a monocot like an iris branch and increase its number of leaves? Is the fact that the shoot is a rhizome with adventitious roots important? Is water transported from one end of the shoot to the other?

45. What is the name of the analysis of tree rings? How is that used to study past climates? How is it used to establish the date when ancient buildings and ships were constructed?

46. In tree ring analysis, what is a floating sequence? Why has the sequence for the Middle East been floating, whereas we know the exact dates of each ring in the sequences for North America and Europe? What is so important about the volcanic eruption in 1628 BCE? Do you think it is possible that the unusual ring in North American tree sequences was produced by a volcanic eruption in North America and the unusual ring in the Middle East sequence was caused by a different phenomenon that occurred in a different year? If so, then has the floating sequence been anchored?

BotanyLinks

http://biology.jbpub.com/botany/5e

BotanyLinks contains a variety of resources and review material designed to assist in your study of botany.

Visit the Web Exercises area of BotanyLinks to complete this question:

1. How is wood related to the environment? Go to the Botany Links home page for more information on this subject.

chapter 9

Flowers and Reproduction

Concepts

Reproduction can serve two very different functions: (1) producing offspring that have identical copies of the parental genes or (2) generating new individuals that are genetically different from the parents. Under certain environmental conditions, species that are genetically diverse survive better than genetically homogeneous species; under other conditions, just the opposite is true (in a genetically diverse species, individuals differ slightly from each other, as is true of humans). A plant that has been able to survive and grow to reproductive maturity is relatively well adapted to its location, so any progeny that are genetically identical to it are at least as well adapted as it is. Any progeny that are not genetically identical to the parent may or may not be well adapted to the conditions to which the parent is adapted.

Chapter Opener Image: We talk about angiosperms as being flowering plants, but they are just as important as the group of plants that make fruits rather than cones. Some fruits are edible, like these cherries (*Prunus avium*), and many are not, such as the shell of a peanut. All fruits develop from the central part of a flower, and all contain seeds, unless scientists have altered them, as in seedless grapes and watermelons. Edible fruits like cherries are the result of an important change in evolution, the cooperation of plants and animals. This is coevolution, which results when changes in the plants—such as evolution to have edible fruits—benefits some animals (which obtain food), and then evolution in those animals benefits the plants. For example, an animal eats the fruit but does not harm the seeds, defecating the seeds at some place away from the plant and, thus, distributing the plant's seeds to many new locations. Bees, butterflies, moths, and birds obtain nectar when they visit flowers and then they carry pollen from one flower to another; in most cases this is also the result of coevolution between the plant and its pollinator.

If the environment is stable during several lifetimes, it is selectively advantageous for an organism to reproduce asexually by budding or sending out runners, thus producing new similarly adapted individuals. If the environment is not stable, however, such offspring may find themselves in conditions for which they are poorly adapted; if all are identical, all may die. Instability of the environment can result from many factors (e.g., landslides, avalanches, and road building in forests kill existing vegetation, opening up new, sunny sites that are good for quickly growing, sun-loving weedy plants). Irregular climatic events, such as unusually severe freezes, droughts, floods, or hurricanes, also disrupt plant communities. If all members of a species are equally susceptible to low temperatures, all may be killed by a rare freeze, but if the species is genetically diverse, some individuals may survive. Even though most die, the few survivors may be sufficient to repopulate the site (Table 9-1). A current concern is that we are adding so much carbon dioxide to the atmosphere that our climate is changing too rapidly for many plant and animal species to be able to adapt.

With asexual reproduction, progeny are never more fit than the parent, but during sexual reproduction, sex cells of one plant combine with those of one or several others, resulting in many new gene combinations. Sex cells are so small that many can be produced by a single plant, and many new combinations of genes can be "tested" rather inexpensively. For example, a single large tree can produce thousands of flowers and millions of pollen grains, each genetically unique, yet the tree uses only a few grams of carbohydrate, protein, and minerals (Figure 9-1). Similarly, thousands of egg cells can be produced using only a small amount of resources. The pollen from one plant can be blown or carried by pollinators to the flowers of hundreds or thousands of other plants, and one plant may receive pollen from numerous other individuals. The thousands of seeds produced by a single sexually reproducing plant represent thousands of natural genetic experiments.

TABLE 9-1	Sexual and Asexual Reproduction

Sexual reproduction

Progeny are genetically diverse.
Some are less adapted than the parent but others are more adapted.
Offspring cannot colonize a new site as rapidly because not all progeny are adapted for it, but some can colonize different sites with characteristics not suitable for parents.
Changes in habitat may adversely affect some progeny, but others may be adapted to the new conditions.
Isolated individuals cannot reproduce.

Asexual reproduction

All progeny are identical genetically to parent and to each other.
All are as adapted as parent is, but none is more adapted.
Rapid colonization of a new site is possible.
All may be adversely affected by even minor changes in the habitat.
Even isolated individuals can reproduce.

During seed and fruit maturation, those embryos with severely mismatched genes abort and use no further resources. The tree finally produces hundreds or thousands of fruits and seeds. The total reproductive effort may be a significant drain on the tree's resources, but it produces numerous embryos, many of which are at least as genetically fit as the tree is and perhaps even more fit. In both stable environments and changing ones, sexual reproduction provides enough diversity of progeny that at least some are well adapted.

As a further example, think of sexual reproduction in humans: The children produced by a particular couple are variable, not identical to each other or to either parent. Some of the children may have a particularly advantageous combination of genes and be more healthy or intelligent or athletic or creative than either parent. Others may have combinations of genes that result in congenital

(a)   (b)

Figure 9-1 (a) These apple fruits have developed from flowers, and the seeds inside the fruits developed from egg cells inside the flowers. All egg cells produced by this tree have similar genes, but sperm cells were necessary to fertilize the eggs, and bees may have brought many types of sperm cells from many different apple trees. Although every seed here has the same maternal parent (the tree in the photo), it is possible that all the seeds have different paternal parents. The seeds are not genetically identical. (b) These anthers of a lily (*Lilium*) flower are each releasing thousands of pollen grains; because they were produced by meiosis with crossing over, each pollen grain is at least slightly different genetically from all the others produced by the same plant.

problems such that the children survive only through medical help. Most children are more or less the same as the parents. The diversity is important.

Sexual reproduction also has negative aspects. Two individuals are required, and sex cells must move from one plant to another. In seed plants, pollen may be carried by wind, insects, and birds, but each results in the loss of many pollen grains or the need to produce nectar. Furthermore, potential sex partners may be widely scattered. For example, in a population of trees, those few individuals growing at the highest altitudes may have no neighbors, whereas those growing at lower altitudes have numerous close neighbors. The flowers of the highest individuals may receive no pollen and thus produce no seeds during some years. In contrast, plants that reproduce asexually can do so at any time, even when completely isolated. Some flowering plants are self-fertile and can undergo self-pollination, but they lose the benefit of receiving new genes from another plant.

Some plants reproduce both sexually and asexually. Strawberries have flowers and sexual reproduction involving genetically diverse embryos and seeds, but they also spread rapidly and asexually by runners. Bamboos are perennial grasses that flower and set seed only occasionally (in some species, only once every 80 years), but their rhizomes grow vigorously and establish many new plants asexually. Kalanchoes produce large numbers of seeds each year, but they also produce such large numbers of plantlets along their leaf margins that they can be weeds in both nature and in greenhouses (**Figure 9-2a**).

Seeds, which are produced by sexual reproduction, often have a means of long-distance dispersal: Strawberries are eaten and the seeds later defecated; bamboo fruits and seeds are carried by

(a)   (b)

(c)   (d)

Figure 9-2   (a) *Kalanchoe* plants are called maternity plants because they produce plantlets complete with stems, leaves, and roots along their leaf margins. Although hundreds of plantlets can be produced, all nuclear divisions are mitotic—duplication division—and thus, all of these plantlets are genetically identical to the parent; none is superior. (b) Chollas, species of *Opuntia*, have branches only weakly attached to the trunk. If an animal brushes against a branch, the spines stick to the animal and the branch is pulled from the plant. After the animal dislodges the branch, it roots and grows into a new plant. Long-distance distribution almost as extensive as that of seeds can occur. (c) All the trees in this photograph are a part of the same plant, each a sprout from a single root system. The plant actually covers several acres. (d) All grapes used for wine are the same species (*Vitis vinifera*), but there are hundreds of varieties, each of which must be propagated vegetatively to ensure every plant produces grapes with exactly the same flavor. Here you can see an entire field planted just with cuttings. Every plant is a clone and each is genetically identical to the others.

winds. The consequence of this is that seeds may become widely scattered and germinate in numerous diverse sites, each site differing from the others in its microclimate, soil conditions, and exposure to predators and pathogens. Of course, a seed carrying an embryo with a combination of genes selectively advantageous for growth in a dry site may land in a wet site and not survive, but with the large numbers of seeds produced, it is statistically probable that some seeds will land in sites for which they are well suited.

In contrast, new plants that are produced asexually are usually not capable of long-distance dispersal; runners, rhizomes, and plantlets result in new plants that become established in the same microhabitat as their parent. After a single plant becomes established in a suitable site, by reproducing asexually it can quickly fill the area with replicas of itself, all of which are as fit as it is.

Asexual Reproduction

Within angiosperms, numerous methods of asexual reproduction have evolved. One of the most common is **fragmentation**: A large spreading or vining plant grows to several meters in length, and individual parts become self-sufficient by establishing adventitious roots. If middle portions of the plant die, the ends become separated and act as individuals. Certain modifications improve the efficiency of fragmentation. In many cacti, branches are poorly attached to the trunk, and the plant breaks apart easily. The parts then form roots and become independent (Figure 9-2b). In some members of the saxifrage, grass, and pineapple families, plantlets are formed where flowers would be expected; these look like small bulbs and are called bulbils.

In willows and many thistles, adventitious shoot buds form on roots and then grow into plants. Adventitious buds may grow out even while the parent plant is still alive, and a small cluster

of trees may in fact consist of just a single individual. A grove of aspens that covers several acres in Utah has been discovered to be a single plant (Figure 9-2c).

Sexual Reproduction

Sexual reproduction in angiosperms involves flowers, which produce the necessary cells and structures. To understand flower structure, one must first understand the plant life cycle.

The Plant Life Cycle

The life cycle of mammals such as humans is simple: Diploid adults have sex organs that produce haploid sex cells called **gametes**, either **sperms** or **eggs**, by meiosis. Individuals that produce sperms are called males, of course, and individuals that produce eggs are females. One sperm and one egg are brought together, forming a new single diploid cell, the fertilized egg or **zygote**, which then grows to become a new individual that is diploid and resembles its parents.

In plants, the life cycle is more complex. The plants you are familiar with—trees, shrubs, and herbs—are all just one phase of the plant life cycle, called the **sporophyte phase** or **sporophyte generation**. A critical factor is that **sporophytes** are always diploid, like most adult animals, and they have sex organs (located in the flowers in angiosperms) with cells capable of undergoing meiosis. In animals, meiosis results in haploid gametes, but in plants, it results in haploid **spores** (Figure 9-3). The difference between gametes and spores is great: Gametes can fuse with other gametes in a process called **syngamy** or **fertilization**, thereby producing the diploid zygote. A gamete that does not undergo syngamy dies because it cannot live by itself and usually cannot grow into a new, haploid individual (unfertilized eggs of some insects such as bees are exceptional and develop into sterile workers).

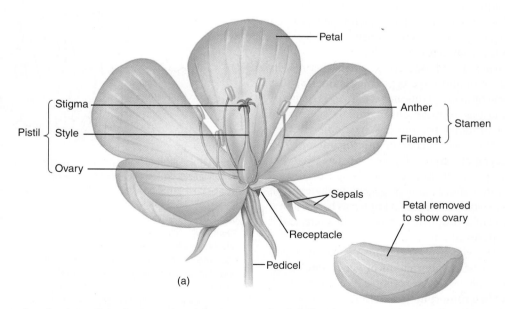

(a)

Figure 9-3 (a) Most flowers have four types of structures: sepals, petals, stamens, and a pistil. The pistil shown here is actually composed of five parts called carpels that have merged into one structure.

(*continues on next page*)

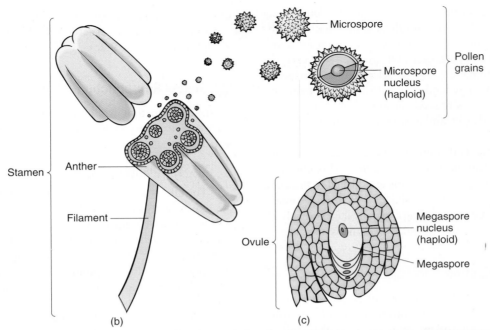

(b)   (c)

Figure 9-3 (*continued*)   (b) In the anthers of stamens, the central cells undergo meiosis and each produces four daughter cells called microspores or pollen grains.
(c) In the flower's central organs, the carpels are ovules, each containing only one cell that undergoes meiosis; often three of the daughter cells die and the one survivor becomes the megaspore.

Plant spores are just the opposite: They cannot undergo syngamy, but each undergoes mitosis and grows into an entire new haploid plant called a **gametophyte**. It is called a gametophyte because it is the plant (-phyte) that produces gametes (gameto-). During sexual reproduction, when a sporophyte reproduces, it does not produce a new diploid plant like itself but rather a haploid plant (simple plants like hornworts, liverworts, and mosses are gametophytes). Furthermore, in all vascular plants, a haploid gametophyte does not even remotely resemble a diploid sporophyte. It is a tiny mass of cells with no roots, stems, leaves, or vascular tissues, but it is an entire plant (**Figure 9-4**). Gametes are formed by the haploid plants by mitosis, not meiosis. The gametes then undergo syngamy, forming a zygote that grows into a new, diploid sporophyte, and the life cycle is complete (**Figure 9-5**).

Mammalian gametes are of two types: small sperm cells (**microgametes**) that swim and large eggs (**megagametes**) that do not. This is also true of many plants and is known as oogamy. In oogamous plants, just as in oogamous mammals, sperms are produced by one type of individual and eggs by a different type of individual; hence, there are "male" or **microgametophytes** and "female" or **megagametophytes** (Figure 9-4). The two types of gametophytes have grown from two types of spores: microgametophytes from **microspores** and megagametophytes from **megaspores**. Having two types of spores is known as heterospory. Typically, just one kind of sporophyte occurs in a life cycle, and it produces both microspores and megaspores.

A life cycle like this, with two generations—sporophyte and gametophyte—is said to be an **alternation of generations** (Figure 9-5). Because gametophytes do not resemble sporophytes at all, this is an alternation of **heteromorphic generations**. This is a complex life cycle, with at least three distinct plants (one sporophyte and two gametophytes). The human life cycle, like that of all other animals, does not have anything equivalent to the haploid generation.

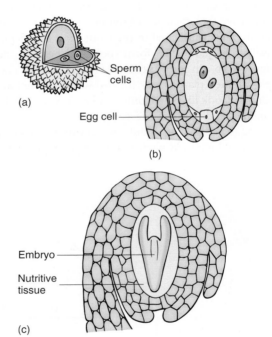

Figure 9-4   All seed plants produce two types of gametophytes. (a) Microspores (pollen grains) develop into microgametophytes (also called pollen grains). The microgametophyte body is so small that it has just three cells and fits inside the pollen cell wall. In angiosperms, pollen grains convert to microgametophytes before the anther opens and sheds the pollen. The microgametophyte produces two sperm cells located within its own protoplasm. (b) The megaspore develops into a megagametophyte. It is slightly larger than a microgametophyte and has seven cells, one of which has two nuclei. One of the cells is the egg. The megaspore and megagametophyte are never released from the flower of the parent sporophyte, so they appear to be just sporophyte tissues, not one plant growing inside another. (c) After one sperm cells fertilizes the egg cell, the new egg cell nucleus is diploid and the cell is a zygote. It develops, by mitosis, into a new sporophyte, shown here as the embryo in an immature seed. The entire ovule develops into a seed; the ovary develops into a fruit.

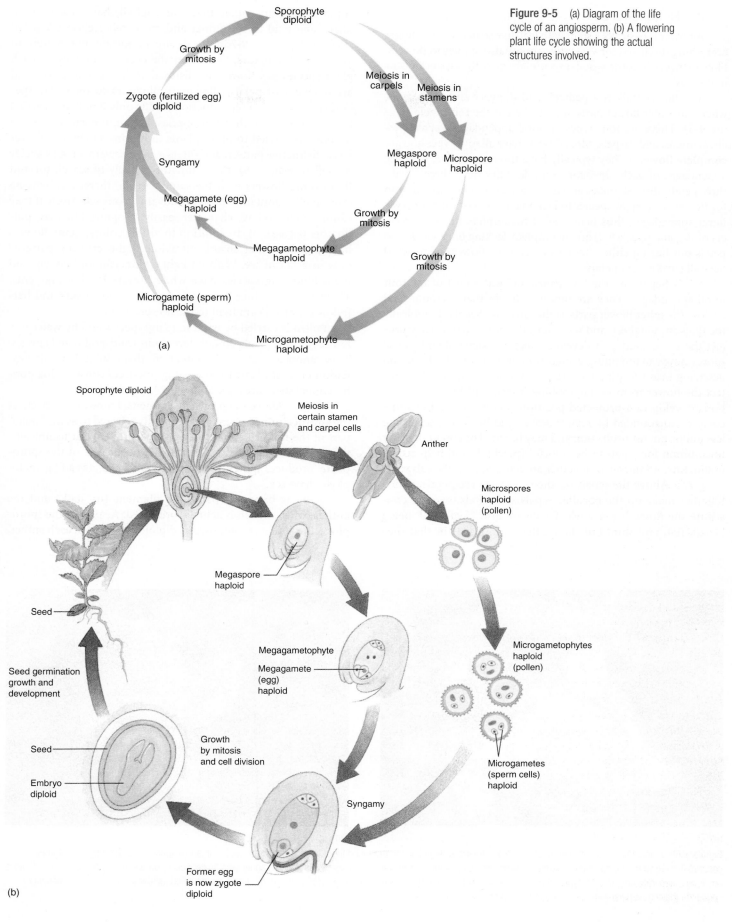

Figure 9-5 (a) Diagram of the life cycle of an angiosperm. (b) A flowering plant life cycle showing the actual structures involved.

(a)

Sporophyte diploid

Growth by mitosis

Zygote (fertilized egg) diploid

Meiosis in carpels

Meiosis in stamens

Syngamy

Megaspore haploid

Microspore haploid

Megagamete (egg) haploid

Growth by mitosis

Megagametophyte haploid

Growth by mitosis

Microgamete (sperm) haploid

Microgametophyte haploid

(b)

Sporophyte diploid

Meiosis in certain stamen and carpel cells

Anther

Microspores haploid (pollen)

Megaspore haploid

Seed

Megagametophyte

Megagamete (egg) haploid

Microgametophytes haploid (pollen)

Seed germination growth and development

Growth by mitosis and cell division

Seed

Embryo diploid

Syngamy

Microgametes (sperm cells) haploid

Former egg is now zygote diploid

Flower Structure

A flower is basically a stem with leaf-like structures, so almost everything discussed for vegetative shoots also applies to flowers. Flowers never become woody; secondary growth does not occur in flowers.

The flower stalk is a **pedicel**, and the very end of the axis, where the other flower parts are attached, is the **receptacle** (Figure 9-3). There are four types of floral appendages: sepals, petals, stamens, and carpels. Most flowers have all four types and are **complete flowers**. They typically have three, four, five, or more appendages of each type; for example, lilies have three sepals, three petals, three stamens, and three carpels. It is not uncommon for flowers of certain species to lack one or two of the four basic floral appendages, thus being called **incomplete flowers**. Flowers of *Begonia* (**Figure 9-6**) are incomplete, lacking both sepals and petals but having either stamens or carpels; flowers of pigweed have all parts except petals.

Sepals **Sepals** are the lowermost and outermost of the four floral appendages. They are modified leaves that surround and enclose the other flower parts as they mature. Sepals are typically the thickest, toughest, and waxiest of the flower parts. They protect the flower bud as it develops, keeping bacterial and fungal spores away, maintaining a high humidity inside the bud, and deterring insect feeding (**Figures 9-7a and 9-7b**). Sepals also protect the flower from nectar-robbing insects and birds. If flower buds develop in a protected position such as beneath a spiny cover or surrounded by regular leaves and branches, sepals are less important for protection and may be small or absent. It is not uncommon for sepals to be colorful (petaloid) and help attract pollinators. All the sepals together are referred to as the **calyx**.

Petals Above the sepals on the receptacle are **petals**, which together make up the **corolla**. Sepals and petals together constitute the flower's **perianth**. Petals are also "leaf-like," being broad, flat, and thin, but they differ from leaves in that they contain pigments other than chlorophyll, have fewer or no fibers, and tend to be thinner and more delicately constructed (Figures 9-7c and 9-7d). Petals are important not merely in attracting pollinators, but rather the correct pollinators. Each plant species has flowers of distinctive size, shape, color, and arrangement of petals, allowing pollinators to recognize specific species. Sexual reproduction results only if pollen is carried to other flowers of the same species; it cannot occur efficiently if pollen is carried to other plants indiscriminately. If a flower has a distinctive pattern and offers a good reward such as nectar or pollen (**Figure 9-8**), the pollinator is likely to search for and fly to other flowers with the same pattern, thereby enhancing cross-pollination; mutations are advantageous selectively if they cause flowers to have characters easily recognized by their pollinators (**Figure 9-9**). In addition to visible colors, many flowers have pigments that absorb ultraviolet light, creating patterns only insects can see. Without light, colors cannot be seen, and night-blooming species have white flowers lacking pigments. Their petals produce volatile fragrances, and insects and bats follow the aroma gradient to the flower.

Pollen is carried by wind in many species and by water in a few. Typically, petals do not develop in wind-pollinated species: They cannot attract wind so mutations that inhibit their differentiation prevent plants from wasting resources constructing nonfunctional structures (see Figure 9-27).

Stamens Above the petals are **stamens**, known collectively as the **androecium**. Stamens are frequently referred to as the "male" part of the flower because they produce pollen, but technically, they are not male because the flower, being part of the sporophyte, produces spores, not gametes. Only gametes and gametophytes have sex.

Stamens have two parts, the **filament** (its stalk) and the **anther**, where pollen is actually produced. As part of the sporophyte, the anther is composed of diploid cells, and in each anther,

(a)

(b)

Figure 9-6 In most species of flowering plants, the stamens and carpels are produced together in the same plant, but in a few species like this *Begonia,* stamens occur in flowers without carpels (a), whereas carpels occur in stamen-less flowers (b). In species in which stamen-less flowers occur on one plant and carpel-less flowers occur on a separate one, the two types of plants look so much alike that, in contrast to mammals, they usually cannot be identified as staminate (having stamens) or carpellate (having carpels) without looking at the flowers.

Petals   Sepals

Protective bracts

(a)

(b)

(c)

(d)

Figure 9-7 (a) The sepals of this rose form a tight covering over the rest of the flower as it develops, protecting the inner parts. When the microspores (pollen) and megaspores (in the ovules) are ready, the sepals bend outward and the flower opens. (b) Sunflowers and daisies are actually clusters of many small flowers that together have the appearance of a single flower. In this case, the entire cluster of flowers is protected by a set of bracts that look and function like sepals. It is not necessary for the individual flowers to have protective sepals. (c) Petals, like these of the mallow, are leaf like in being thin and flat, but in most species, they are less leaf like than sepals. (d) Leaves and petals of *Arbutus* can be compared here; these petals are highly modified, and the five petals of each flower have fused together, forming a tubular corolla.

Edible decoy stamens

Figure 9-8 Some flowers do not produce nectar; the pollinator eats part of the stamens and pollen instead. This is more expensive metabolically for the plant because pollen is rich in protein, whereas nectar is composed of carbohydrate. In these flowers of cannonball tree (*Couropita*), the lower stamens are modified for edibility; they are large and showy and produce little or no pollen. The pollen-producing stamens are small and inconspicuous, easily overlooked by pollinators.

Figure 9-9 Distinctive patterns like this are known as nectar guides; they direct the pollinator to the nectar and position the animal to pick up pollen.

Sexual Reproduction   **207**

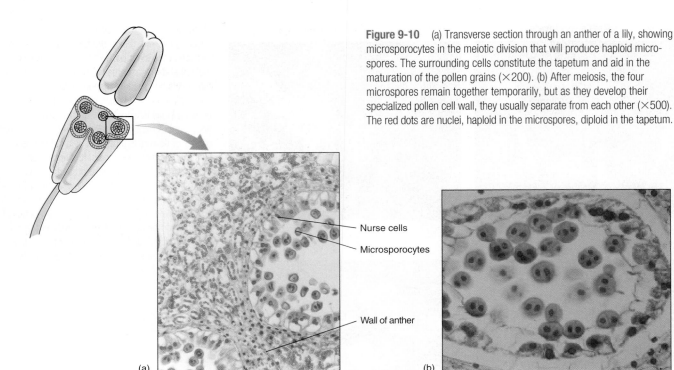

Figure 9-10   (a) Transverse section through an anther of a lily, showing microsporocytes in the meiotic division that will produce haploid microspores. The surrounding cells constitute the tapetum and aid in the maturation of the pollen grains (×200). (b) After meiosis, the four microspores remain together temporarily, but as they develop their specialized pollen cell wall, they usually separate from each other (×500). The red dots are nuclei, haploid in the microspores, diploid in the tapetum.

Nurse cells

Microsporocytes

Wall of anther

(a)

(b)

four long columns of tissue become distinct as some cells enlarge and prepare for meiosis (**Figure 9-10a**). These **microspore mother cells** or **microsporocytes** continue to enlarge and then undergo meiosis, each producing four microspores (Figure 9-10b and **Table 9-2**). Neighboring anther cells, in a layer called the **tapetum**, act as nurse cells, contributing to microspore development and maturation. Microspores initially remain together in a tetrad, but later separate, expand to a characteristic shape, and form an especially resistant wall. They are then called **pollen**. The anthers open (**dehisce**) along a line of weakness and release the pollen.

The wall of a pollen grain is a cell wall; however, it is quite complex structurally. It has an inner layer called the intine, composed of cellulose, and an outer layer called the exine that consists of the polymer sporopollenin. It has one or several weak spots, germination pores, where the pollen opens after it has been carried to the stigma of another flower. Sporopollenin is remarkably waterproof and resistant to almost all chemicals; it protects the pollen grain and keeps it from drying out as it is being carried by wind or animals. The exine can have ridges, bumps, spines, and numerous other features so characteristic that each species has its own particular pattern (**Figure 9-11**). In many cases, it is possible to examine a single pollen grain and know exactly which species of plant produced it. Because sporopollenin is so resistant, pollen grains and their characteristic patterns fossilize well. By examining samples of old soil, botanists can determine exactly which plants grew in an area at a particular time in the ancient past.

Carpels **Carpels** constitute the **gynoecium**, located at the highest level on the receptacle (**Figure 9-12a**). Carpels have three main parts: (1) a **stigma** that catches pollen grains, (2) a **style** that elevates the stigma to a useful position, and (3) an **ovary** where megaspores are produced. A flower can have zero (some imperfect flowers) to many carpels; usually they are fused together into a single compound structure, frequently called a pistil. Inside the ovary are **placentae** (singular, placenta), regions of tissue that bear small structures called **ovules**. Ovules have a short stalk, called a funiculus, that carries water and nutrients from the placenta to the ovule by means of a small vascular bundle. The ovule has a central mass of parenchyma called a **nucellus**. Around the nucellus are two thin sheets of cells (integuments) that cover almost the entire nucellus surface, leaving only a small hole (micropyle) at the top (Figure 9-12c). As in anthers, some nucellus cells, usually only one in each ovule, enlarge in preparation for meiosis; these are **megaspore mother cells** or **megasporocytes**. After meiosis, usually three of the four megaspores degenerate, and only one survives, becoming very large by absorbing the protoplasm of the other three. Megaspores differ from microspores (pollen) because the ovule and the carpel do not dehisce and the megaspore remains enclosed inside the carpel.

TABLE 9-2	Stamen and Carpel Structure

Stamen
    Filament
    Anther
        Microsporocytes
Carpel
    Stigma
    Style
    Ovary
        Ovary wall
        Placenta
        Ovule(s)
            Funiculus
            Integuments and micropyle
            Nucellus
                Megasporocyte

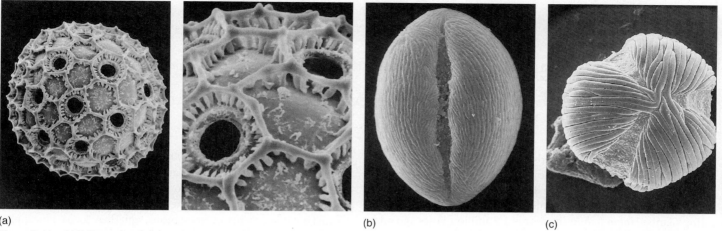

Figure 9-11 (a) Pollen grains of *Cobaea* (no common name), with the pollen wall forming hexagonal ridges. The numerous holes are germination pores where a pollen tube can emerge after the pollen lands on a stigma (×2,000). (b) Pollen of *Lycium* (boxthorn) has a single long groove from which the pollen tube emerges (×4,000). (c) Pollen of *Macrolobium* (a legume) has three germination grooves (×4,000).

An ovule develops into a seed after its egg is fertilized, and the surrounding ovary develops into a fruit. Each ovary might have either one or many placentae, each bearing one or many ovules; ovaries with just a single ovule develop into fruits with a single seed (such as avocado or peach). Many-seeded fruits (e.g., tomato, cantalope, and squash) have numerous placentae, each with many ovules. Orchid pistils always consist of three carpels,

each of which might have tens of thousands of ovules, and some orchid fruits are estimated to contain up to one million tiny seeds.

Aside from bearing ovules, the rest of the carpel is at least somewhat leaflike. Figure 9-12a shows three carpels fused together as a pistil; the outermost layer is basically leaf lower epidermis and typically has stomata. The innermost layer is upper epidermis, and all middle layers are mesophyll and vascular bundles.

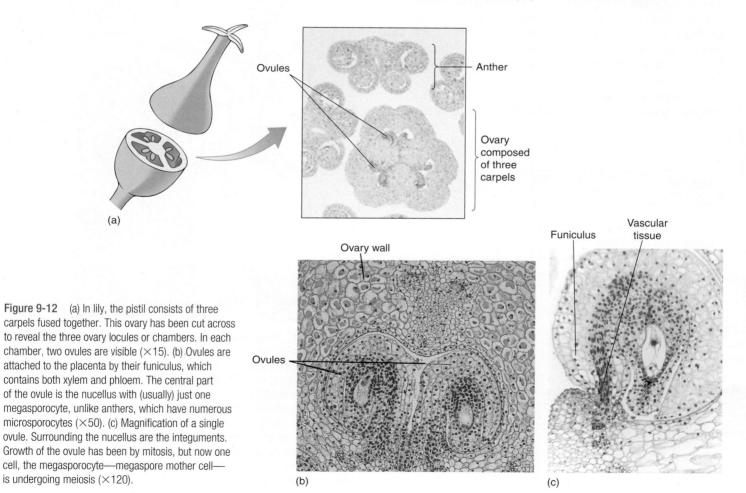

Figure 9-12 (a) In lily, the pistil consists of three carpels fused together. This ovary has been cut across to reveal the three ovary locules or chambers. In each chamber, two ovules are visible (×15). (b) Ovules are attached to the placenta by their funiculus, which contains both xylem and phloem. The central part of the ovule is the nucellus with (usually) just one megasporocyte, unlike anthers, which have numerous microsporocytes (×50). (c) Magnification of a single ovule. Surrounding the nucellus are the integuments. Growth of the ovule has been by mitosis, but now one cell, the megasporocyte—megaspore mother cell—is undergoing meiosis (×120).

Plants and People

Box 9-1 Flowers, Fruits, Seeds, and Civilization

Flowers and the fruits and seeds that result from them have always been important to us. One of the first hominids—animals on our line of evolution—was *Australopithecus africanus;* fossils of its teeth, 3 million years old, show that it was adapted to eating plants. *Homo habilis* lived only 1.8 million years ago and likewise had a diet strongly dependent on plants. Our species, *Homo sapiens,* appeared evolutionarily about 500,000 years ago and survived by hunting game and collecting fruits, seeds, and edible roots.

Approximately 11,000 years ago, a momentous change occurred. Small groups of humans began cultivating plants rather than just gathering them (Table B9-1). This happened in the Middle East, an area that is now Iran, Iraq, and Syria, also known as the Fertile Crescent. The plants were common, wild species such as wheat, barley, peas, and lentils. Farming required major changes in human society: People had to stay in one place to tend and protect crops rather than follow herds of game animals. Permanent villages were necessary, as were the rules and regulations needed when people live together in close proximity. Land, huts, and harvests are tempting targets for thieves, and thus, defense, both individual and collective, became a necessity—government had to be created.

Civilization advanced rapidly. Neolithic (New Stone Age) agricultural societies were widespread by 6000 BCE, and the ox-drawn plow was in use as early as 4500 BCE. New plants were cultivated: olives, date palms (Figure B9-1), and grapes for eating and for wine. Independently, the peoples of Southeast Asia domesticated rice and soybeans; in the New World the Incas, Mayans, and Aztecs cultivated potatoes, corn, tomatoes, beans, squashes, cocoa, pineapples, and peanuts.

The importance of fruits and seeds to the survival of both individual humans and societies is reflected in the prominent position they were given in early art. Ancient Egyptians carved likenesses of date palms, barley, and wheat as long ago as 3000 BCE, and on Crete in 1800 BCE, Minoan artists depicted, in

TABLE B9-1	Locations Where Plants Were Domesticated
Africa	
Coffee, melons, okra, sorghum, watermelon	
Central America and Mexico	
Beans, cocoa (chocolate), corn, hot peppers (chili peppers), squash, sweet potato, tomato, vanilla	
Europe	
Beets, carrots, dill, mints (peppermint, spearmint), mustard, oregano, rosemary, thyme	
Fertile Crescent (Near East)	
Date palms, barley, figs, flax, grapes, olives, onion, opium poppies, peas, wheat	
Far East	
Apricot, grapefruit, orange, peach, rice, soybean, tea	
South America	
Beans, coca (cocaine), hot peppers (chili peppers), peanut, pima cotton, pineapple, potato, yucca (cassava)	
Southeast Asia	
Banana, black pepper, cinnamon, coconut, cucumber, eggplant, mango, sugar cane	

Figure B9-1 Images like this occur frequently in ancient Sumerian art. The person is either a priest or a king, and the object he faces is a very strange plant. In earlier depictions the plant is recognizable as a date palm, but through the years this important food tree became a symbol for life itself and artists began to draw it more stylized and with mystical symbols surrounding it. The person is holding a cone-like object in one hand and a pail in the other. Date palms grow as carpellate ("female") plants and staminate ("male") plants. To ensure a good crop, pollen can be collected in a pail and then applied to the carpellate plants with a brush to increase the rate of pollination and fruit production. The scene, carved before 2000 BCE, indicates that the very earliest farming societies understood this sophisticated aspect of botany.

addition to the plants already mentioned, figs, saffron crocus, pomegranate, and lupine. They also showed several plants that must have been grown purely as ornamentals: lilies, myrtle, narcissus, and roses.

As societies become wealthier, time and resources become available for leisure and pleasurable pursuits. Gardens dedicated to beautiful, fragrant flowers, not food crops or medicinal plants, arose early. Expeditions in search of exotic ornamental plants are described in the oldest epic poem in the world, the story of Gilgamesh of Sumer, and an ancient Egyptian monument, the Palermo Stone, records a plant-gathering expedition by King Snefru in 2900 BCE. Inventories list the extensive plant-collecting trips and gardens of Ramses III.

Growing beautiful flowers and enjoying exotic fruits have never lost their popularity; after the Spanish discovered America, the initial exploitation was for gold and silver, but there soon followed expeditions dedicated solely to gathering plants, especially ornamental ones. Two of the many famous plant explorers sent to the United States are David Douglas, who explored the Pacific Northwest for the Horticultural Society of London in the 1820s, and the father and son team, André and François-André Michaux, sent by the government of France in 1785.

Even today, gardening and growing flowers for pleasure are considered by many an essential part of our civilization. As technology increases, cultivating flowers, or at least having a potted plant on a window sill, seems to maintain contact with our past.

Gametophytes

Microgametophyte Microspores develop into microgametophytes. In all angiosperms, each microgametophyte is very small and simple, consisting of at most three cells located within the original pollen cell wall (Figure 9-13). The microspore nucleus migrates to the side of the pollen grain and lies next to the wall.

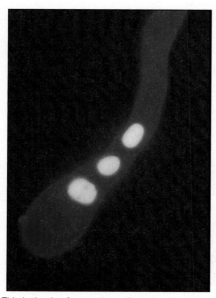

Figure 9-13 This is the tip of a growing pollen tube. It was treated to make the nuclei fluoresce (glow) so that they can be found easily despite being so small. The lowest nucleus is probably the tube nucleus, the upper two are probably sperm cell nuclei, each located within a sperm cell. The two sperm cells themselves are located within the protoplasm of the pollen tube (×5,000).

There it divides mitotically, producing a large **vegetative cell** and a small lens-shaped **generative cell**, which subsequently divides and forms two sperm cells. The entire microgametophyte consists of the vegetative cell and the two sperm cells (the microgametes). Although extremely simple, this is a full-fledged plant; see examples of gametophytes that are more complex than these of angiosperms and that are obviously complete plants.

In approximately 30% of angiosperm species, formation of sperm cells occurs even while pollen is still located within the anther. In the majority of angiosperm species, the pollen is released from the anther at about the time the generative cell has formed, and sperm cells are not produced until after the pollen has been carried to a stigma.

After a pollen grain lands on a stigma, it germinates by producing a **pollen tube** that penetrates into the loose, open tissues of the stigma (Figure 9-14). The pollen tube absorbs nutrients from the stigma and grows downward through the style toward the ovary. The microgametophyte (the pollen tube) is protected and nourished by the style tissue. As the pollen tube grows downward, it carries the sperm cells to the ovule. Almost all of the pollen cytoplasm is located at the tip of the pollen tube; the rest of the tube and the pollen grain are filled with a giant vacuole. Pollen grains are too small to store much starch, fat, or protein; if they could not absorb nutrients from the style, the pollen tubes could not grow long enough to reach the ovules.

Megagametophyte Within the ovule the surviving megaspore develops into a megagametophyte. In one type of development, the nucleus undergoes three mitotic divisions, producing two, four, and then eight haploid nuclei all in a single, undivided cell (Figure 9-15). The technical term for a multinucleate megagametophyte is **embryo sac**. The nuclei migrate through the cytoplasm,

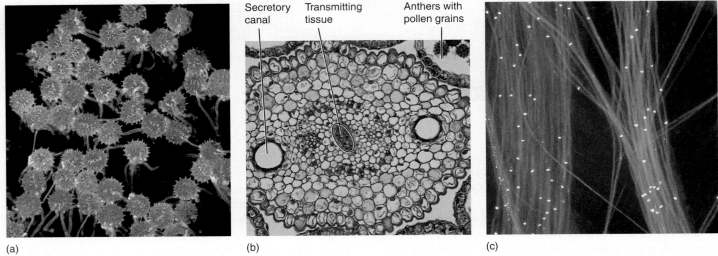

(a)   (b)   (c)

Figure 9-14   (a) Shortly after landing on a stigma of the correct species, a pollen grain may germinate by pushing out part of the wall as a pollen tube. This penetrates the loose tissues of the stigma and style. This was photographed with ultraviolet light, which causes the pollen wall and callose to fluoresce and shine (×50). (b) Tissues of the style allow the pollen tubes to grow through them easily. Some styles are hollow and lined with a rich, nutritious transmitting tissue; other styles are solid but also have transmitting tissue. Sunflower (*Helianthus*) style (×80). (c) These pollen tubes have penetrated deep into the style. The bright spots are callose plugs, which seal off the protoplasm at the tip from the empty parts of the pollen tube (×200).

pulled by microtubules, until three nuclei lie at each end and two in the center. Walls then form around the nuclei, and the large, eight-nucleate megaspore becomes a megagametophyte with seven cells, one of which is binucleate. The seven cells are one large **central cell** with two **polar nuclei**, three small **antipodal** **cells**, and an **egg apparatus** consisting of two **synergids** and an **egg** (the megagamete) (Table 9-3). Like the microgametophyte, the megagametophyte is a distinct plant. As with the pollen, the megagametophyte obtains all of its nourishment from the parent sporophyte.

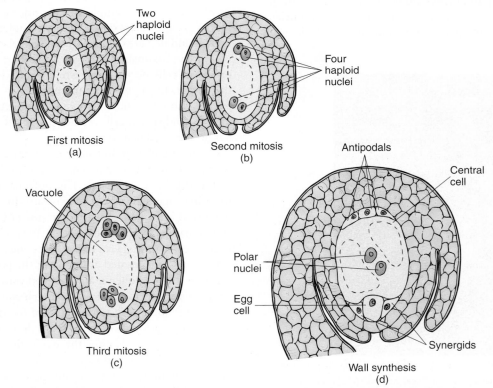

Figure 9-15   As a megaspore develops into a megagametophyte, three sets of mitosis occur (a to c) without any cytokinesis; thus, the megagametophyte is temporarily an eight-nucleate coenocyte (c). Then nuclei migrate; cytoplasm accumulates around each, and walls are established (d). Much of the egg is not covered by wall. The central cell is binucleate and is mostly vacuole.

TABLE 9-3 Gametophyte Development

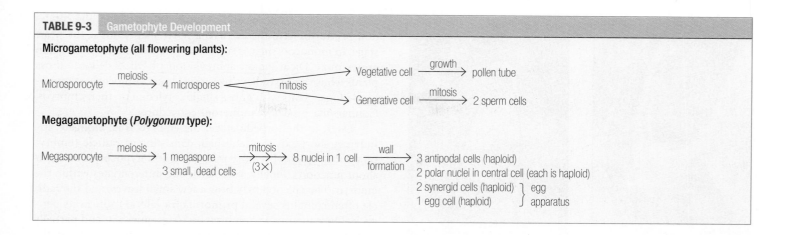

Microgametophyte (all flowering plants):

Microsporocyte →(meiosis)→ 4 microspores →(mitosis)→ Vegetative cell →(growth)→ pollen tube
                                            →(mitosis)→ Generative cell →(mitosis)→ 2 sperm cells

Megagametophyte (*Polygonum* type):

Megasporocyte →(meiosis)→ 1 megaspore / 3 small, dead cells →(mitosis)(3×)→ 8 nuclei in 1 cell →(wall formation)→ 3 antipodal cells (haploid)
2 polar nuclei in central cell (each is haploid)
2 synergid cells (haploid) } egg
1 egg cell (haploid) } apparatus

Fertilization

Syngamy of sperm and egg involves both **plasmogamy**, fusion of the protoplasts of the gametes, and **karyogamy**, fusion of the nuclei. As a pollen tube grows downward through the style toward the ovule, it is guided to the ovule's micropyle by some means. It penetrates the nucellus and reaches the egg apparatus, then enters one synergid. The pollen tube tip bursts and releases both sperm cells, one of which migrates through the synergid protoplasm toward the egg. As it does so, the sperm cell's plasma membrane breaks down, and it loses most of its protoplasm. The sperm nucleus enters the egg, then is drawn to and fuses with the egg nucleus, establishing a diploid zygote nucleus.

Because the sperm sheds its protoplasm as it passes through the synergid, it contributes only its nucleus with the set of nuclear genes during karyogamy. The sperm does not carry mitochondria or plastids into the egg, and thus, organellar genes from the pollen parent are rarely inherited by the zygote. Instead, the mitochondrial and plastid genes of the embryo are inherited only from the ovule parent. Gymnosperms are different: Their sperm cells retain their plastids while destroying their mitochondria. The zygote inherits mitochondria and a nucleus from the egg, a nucleus and plastids from the sperm cell.

In angiosperms only, the second sperm nucleus released from the pollen tube migrates from the synergid into the central cell. It undergoes karyogamy with both polar nuclei, establishing a large **endosperm nucleus** that is triploid, containing three full sets of genes. Because both sperm nuclei undergo fusions—one with the egg nucleus and the other with the polar nuclei—the process is called **double fertilization**. The endosperm nucleus is extraordinarily active and begins to divide very rapidly by mitosis, with cell cycles lasting only a few hours.

The endosperm nucleus initiates a dynamic cytoplasm and the central cell enlarges enormously, usually without cell division, into a huge cell with hundreds or thousands of nuclei (see **Figure 9-16**). Finally, nuclear division stops, and dense cytoplasm gathers around the nuclei. Walls are constructed, thus forming cells. An example of this is a coconut full of "milk." The hollow center of the coconut is one single cell, and the milk is its protoplasm. The white coconut "meat" is the region where nuclei form cells. A green, immature coconut is full of milk but has almost no meat; as it ripens, the coenocytic milk is converted to a thick layer of cellular meat. All of this tissue, both coenocytic and cellu-

lar, is called **endosperm**, and it nourishes the development of the zygote. No other megagametophyte forms as much endosperm as a coconut does. More typical examples are grains such as wheat, rice, oats and corn, in which most of the grain is endosperm. Endosperm of these also pass through a milk stage, but by the time of harvest, the endosperm has become cellular, starch-filled, and dry enough to be hard (but if corn is picked for corn on the cob, it is collected just as endosperm is converting from coenocytic to cellular).

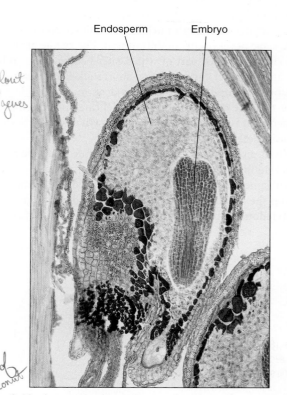

Endosperm   Embryo

Figure 9-16 In many species, endosperm development is accompanied by some cytoplasmic division, and the endosperm is therefore a mass of multinucleate "cells," each with variable amounts of protoplasm and nuclei and with irregular shapes. This is the only plant tissue in which nuclear and cytoplasmic divisions are so poorly coordinated, possibly because the endosperm is just a temporary tissue that is consumed by the embryo before or soon after seed germination (×180).

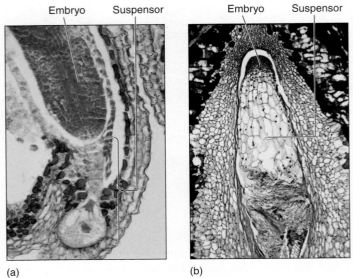

Embryo   Suspensor   Embryo   Suspensor

(a)   (b)

Figure 9-17 (a) The suspensor of shepherd's purse (*Capsella*) has one large bulbous cell and a stalk of smaller cells. The young embryo is being pushed deep into the endosperm (×600). (b) The suspensor of *Tristerix* is much more massive than that of *Capsella,* being larger than the embryo until the last stages of seed maturation. The embryo is still just a small ball of densely cytoplasmic cells (×300).

Embryo and Seed Development

As the endosperm nucleus proliferates, the zygote also begins to grow, but always by both nuclear and cellular divisions; a coenocytic stage never occurs in the embryo. The zygote grows into a small cluster of cells, part of which later becomes the embryo proper, and the other part becomes a short stalk-like structure, the **suspensor**, which pushes the embryo deep into the endosperm (**Figure 9-17**). The suspensor is usually delicate and ephemeral in angiosperms; it is crushed by the later growth of the embryo and is not easily detectable in a mature seed.

Cells at one end of the suspensor continue to divide mitotically, developing into an embryo. The cells at first are arranged as a small sphere, the globular stage. The end of the embryo farther from the suspensor initiates two primordia that grow into two **cotyledons** in basal angiosperms and eudicots (**Figures 9-18**

and **9-19**), such as beans and peanuts. While young, the cotyledon primordia give the embryo a heart shape; this is the heart stage. In monocots such as corn, only one cotyledon primordium grows out. "Dicot" is an abbreviation of "dicotyledon," those plants whose embryos have two cotyledons. Monocots are monocotyledons, plants with only a single cotyledon on their embryos. Conifers like pine have numerous cotyledons.

Later, in the torpedo stage, the embryo is an elongate cylinder: A short axis is established, consisting of **radicle** (embryonic root), **epicotyl** (embryonic stem), and **hypocotyl** (the root/ shoot junction). Finally, vascular tissue differentiates within the embryo. The epicotyl may bear a few small leaves, and the radicle often contains several primordia for lateral roots in its pericycle. Once mature, the embryo becomes quiescent and partially dehydrates, and the funiculus may break, leaving a small scar, the hilum. In green peas, the two halves of each pea are the two cotyledons, and the stalk attaching each pea to the pod (the fruit) is the funiculus.

In most basal angiosperms and eudicots, cotyledons store nutrients used during and after germination; during embryo development, the cotyledons become thick and filled with starch, oil, or protein, whereas the endosperm, which is supplying the nutrients, shrinks. When the seed is mature, the cotyledons are large, and the endosperm may be completely used up. We are eating mostly cotyledons when we eat beans, peas, peanuts, almonds, pecans, and other seeds that easily separate into two halves.

In monocots, the one cotyledon generally does not become thick and full; instead, the endosperm remains and is present in the mature seed. When eating cereals such as wheat, rice, oats, and corn, we are eating almost purely endosperm. During germination, the cotyledon acts as digestive/absorptive tissue, transferring endosperm nutrients to the embryo. Some eudicots are intermediate: Cotyledons store some starch and protein, but a considerable amount of endosperm remains in the seed, and both methods of nutrition are used during germination. A mature seed in which endosperm is rather abundant is an **albuminous seed** (Figure 9-19). If endosperm is sparse or absent at maturity, the seed is **exalbuminous**.

The amount of embryo growth and development that occurs before dormancy sets in is extremely variable. Orchids, bromeliads, and a few other species have small, dust-like seeds in which

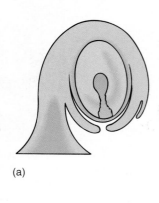

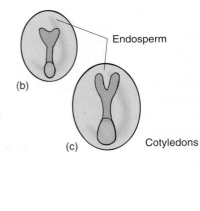

Endosperm

(b)

(c)

Cotyledons

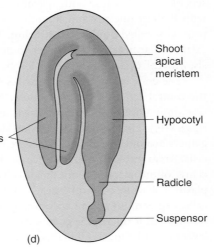

Shoot apical meristem

Hypocotyl

Radicle

Suspensor

(d)

Figure 9-18 Embryo development. (a) Globular stage. (b) Heart stage. (c) Torpedo stage. Embryo root (radicle), cotyledons, and hypocotyl are present. At this stage, the first xylem and phloem may become distinguishable in the hypocotyl. (d) Mature embryo. A shoot apical meristem is shown; in some species, even some leaf primordia and a small stem are present.

(a)

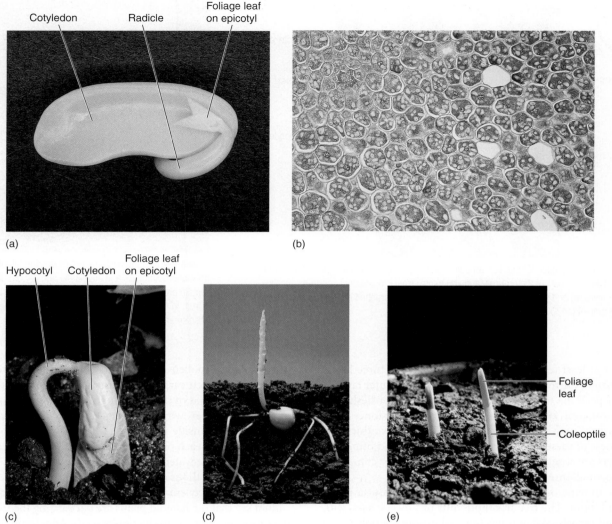

Figure 9-19 (a) This bean seed has begun germinating; the radicle has extended and will develop into a taproot. The two cotyledons (one was removed) were digesting their carbohydrates and proteins and transporting nutrients to the root and shoot apical meristems. Beans have a well-developed epicotyl with several partially expanded leaves; one is visible here. (b) These cells of bean cotyledon are full of starch (stained pink) and reserve protein (stained blue) (×100). (c) The small leaves present on the epicotyl are expanding during germination; because they were rather completely formed before the embryo became dormant, the leaves can now mature rapidly, and photosynthesis begins almost as soon as germination is complete. The hypocotyl is curved: This allows the leaves to be dragged gently up through the soil, protected by the cotyledons. (d) Most of a corn seed is the endosperm; the embryo is less than half the volume of the seed. During germination, the one cotyledon secretes digestive enzymes into the endosperm and absorbs the resulting monomers. (e) Corn seeds also produce many small leaves before becoming dormant, and they can begin photosynthesis immediately after germination. The leaves are protected from the soil by a tubular sheath called a coleoptile.

the embryo is only a small ball of cells with no cotyledons, radicle, or vascular tissue. In seeds of most angiosperm species, all of the parts are present, and the epicotyl contains two or three young leaves in addition to cotyledons; these leaves can begin photosynthesis immediately after germination (Figure 9-19). In corn, the embryo is even more advanced: It contains as many as six young leaves while in the seed. A fully mature corn plant often has only 10 or 12 leaves. Over half of the leaf production of the new sporophyte occurs while it is embedded in the parental gametophyte, which is itself embedded in the previous sporophyte.

The embryo and endosperm develop from the zygote and megagametophyte central cell, respectively, both located in the nucellus of the ovule. Soon after fertilization, or even before, synergids and antipodals break down in most species. The nucel-lus expands somewhat but later is crushed by the expansion of the embryo and endosperm and is usually not detectable in mature seeds.

The integuments that surround the nucellus expand and mature into the **seed coat** (also called the **testa**) as the rest of the ovule grows. In their last stages of maturation, they may become quite sclerenchymatous and tough (Figure 9-20). It is seed coat color that lets us distinguish between black, red, and pinto beans.

■ Fruit Development

As the ovule develops into a seed, the ovary matures into a **fruit**. Development varies with the nature of the carpels as well as the nature of the mature fruit. The stigma and style usually wither away, as do sepals, petals, and stamens, although they

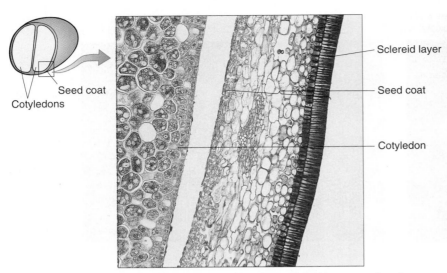

Figure 9-20 This seed coat of bean has an outer layer of sclereids and inner parenchymatous layers. The cotyledon is part of the embryo, and you might expect to see endosperm or nucellus between the cotyledon and seed coat; however, those have been crushed during the later stages of seed development (×200).

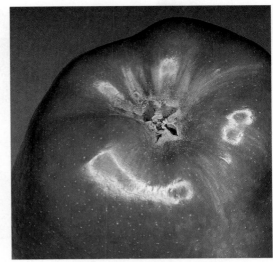

Figure 9-21 In apples, the petals die and fall off after pollination. Stamens and sepals die and dry out, but they remain present, whereas carpels and receptacle develop into a fruit. In most species, sepals, petals, and stamens all abscise after pollination.

may persist at least temporarily (**Figure 9-21**). Often three layers become distinct during growth: The **exocarp** is the outer layer—the skin or peel; the middle layer is the **mesocarp**, or flesh; and the innermost layer, **endocarp**, may be tough like the stones or pit of a cherry or it may be thin (**Figure 9-22**). The relative thickness and fleshiness of these layers vary with fruit type, and often one or two layers are absent. The entire fruit wall, whether composed of one, two, or all three layers, is the **pericarp**.

Before moving on to the next section, think about a peach tree in full bloom. The tree has thousands of flowers, each with stamens and carpels. When it has finished producing pollen and megaspores, it has technically finished its own reproduction and now has tens of thousands of offspring in the form of micro-gametophytes (pollen grains) and megagametophytes (inside each ovule in each carpel). Bees carry pollen from other peach trees to the stigmas of this one, and before long, it has thousands of pollen tubes growing inside its styles. You see one diploid peach tree, but inside it are thousands of haploid pollen tubes (which are also peach plants) and megagametophytes (also peach plants). All of the water, minerals and photosynthates needed to keep all of these thousands of plants alive are being supplied by the leaves, roots, xylem, and phloem of the diploid peach tree. We cannot say that the haploid plants are parasitizing the tree because they are essential to its production of seeds. After fertilizations have occurred, hundreds of zygotes develop into embryos, each of which is the offspring of microgametophytes and megagametophytes and

(a)

(b)

Figure 9-22 (a) A pecan nut is only the seed and innermost part (endocarp) of a fruit. The exocarp (green portion here) and mesocarp (brown) are tough, fibrous, and completely inedible. The fruit opens while on the tree, allowing the endocarp and seed to be shed. (b) Coconuts in stores are usually just the endocarp and seed; in nature, these are surrounded by a thick, fibrous, and buoyant mesocarp and a thin epicarp. Dark spot (an "eye") on endocarp is a soft spot where the seedling root and shoot will emerge during germination.

the "grand-offspring" of the peach tree itself. This tree has other "grand-offspring" as well, produced by its pollen that was carried to other peach trees. This tree is the ovule parent of its own seeds and the pollen parent of thousands of seeds in many other trees.

Flower Structure and Cross-Pollination

The production and development of spores, gametes, zygotes, and seeds are complex, elaborate processes, but they are not the only functions of flowers. Flowers are also involved in the effective dispersal of pollen and seeds. Because numerous mechanisms carry out these processes, numerous types of flowers and fruits exist.

Cross-Pollination

Cross-pollination is the pollination of a carpel by pollen from a different individual; **self-pollination** is pollination of a carpel by pollen from the same flower or another flower on the same plant. In any plant population, there is genetic diversity. Random mutations will have produced some new genes that offer improved fitness and some that are deleterious. With cross-pollination, sperm cells and egg cells from different plants unite, resulting in new combinations of genes, at least a few of which may be better adapted than either parent. But self-pollination has about the same result as asexual reproduction because all genes come from the same parent. No possibility exists of bringing in new genes that might provide more fitness than those inherited from the parent; however, if a plant is isolated by distance or lack of pollinators from potential cross-pollination partners, self-pollination allows it to set seed and propagate its genes rather than lose them when the plant dies.

Many mechanisms have evolved that decrease the probability of self-pollination and increase the chances of cross-pollination with its accompanying genetic diversity.

Stamen and Style Maturation Times

Self-fertilization in flowers that have both stamens and carpels is prevented if anthers and stigmas mature at different times (Figure 9-23). In many species, anthers release pollen while stigma tissues are immature and unreceptive; the style may not have elongated yet, and the stigmas may be near the base of the flower while the anthers are at the top, elevated by elongated filaments. Exposed pollen lives only briefly, being susceptible to desiccation in dry air and to damage to its DNA by ultraviolet light. When the stigma and style become mature, there may be no living pollen left in the flower; thus, all pollination is affected by younger flowers just opening their anthers.

This is not a very effective means of ensuring cross-pollination. On plants with many flowers that do not open simultaneously, older flowers could be self-pollinated by freshly opened flowers of the same plant. Even if pollen does come from other plants, many of the near neighbors are probably closely related because many of the seeds of a plant fall and germinate near the plant, producing a cluster of plants that are at least as closely related as first cousins.

Stigma and Pollen Incompatibility

In many species, especially many important crop species, self-pollination is inhibited by **compatibility barriers**, chemical reactions between pollen and carpels that prevent pollen growth. In one incompatibility system, as a pollen tube grows, the stigma and style test proteins on the tube's surface; if one of these proteins is produced by a gene that matches an incompatibility gene in the carpel-bearing plant, the stigma and style block any further growth of the pollen tube. In self-pollination, all pollen tube genes match those of the stigma and style, and blocking occurs (Figure 9-24).

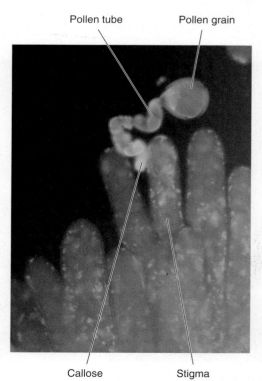

Pollen tube   Pollen grain

Callose   Stigma

Figure 9-24 This pollen grain has started to grow on an incompatible stigma, but the stigma has blocked it, causing callose to form. The callose fluoresces in ultraviolet light, so it shines brightly in this micrograph (×200).

Figure 9-23 The stamens of this cactus flower are mature and shedding pollen, but the stigmas (green) are pressed together and are unreceptive. Later, the stamens wither and the stigmas spread open, ready to receive pollen from a different flower.

In another common system, the critical proteins are deposited on the outer surface of the developing pollen grain by the anther's tapetum. Any match of proteins produced by the incompatibility genes blocks germination of the pollen grain. The interaction between a pollen tube and a style involves a haploid genome (pollen tube) and a diploid one (carpel), whereas the second system involves two diploid genomes (anther and carpel). The diploid/diploid system involves twice as many genes and therefore has a much greater probability of pollen rejection. With the diploid/diploid system, not only is self-pollination prevented, but inbreeding between close relatives is blocked as well.

Monoecious and Dioecious Species

Among incomplete flowers there is a significant difference between flowers that lack sepals or petals and those that lack stamens or carpels. The latter two organs are **essential organs** because they produce the critically important spores; if either organ is absent, sexual reproduction is dramatically affected. Flowers that lack either or both essential organs are not only incomplete but also **imperfect flowers** (see Figure 9-6). If a flower has both, it is a **perfect flower** even though it may lack either sepals or petals or both. Sepals and petals do not produce spores and are considered **nonessential organs**.

It is necessary to consider the entire plant and the entire species as well as individual flowers. Stamens produce pollen that results in sperm production, and carpels are involved in egg production; thus, a species must have both types of organs. Plants that have perfect flowers satisfy this requirement. But if some flowers of the species are imperfect, having no stamens, for instance, then other flowers must have stamens by being either perfect or by being imperfect because of a lack of carpels. A large number of combinations is possible: A species may have individuals that produce only staminate flowers and others that produce only carpellate flowers—this is **dioecy**, and the species (not the flower or the plant) is said to be **dioecious** (pronounced "dye EE cy" and "dye EE shus"). Examples of dioecious species are marijuana, dates, willows, and papaya. In dioecious species, the life cycle actually consists of four types of plants: (1) microgametophytes, (2) megagametophytes, (3) staminate sporophytes, and (4) carpellate sporophytes.

Monoecy is the condition of having staminate flowers located on the same plant as carpellate flowers; **monoecious** species include cattails and corn—ears are clusters of fertilized carpellate flowers, and tassels bear numerous staminate flowers (**Figure 9-25**). In some members of the cucumber family (including melons, squash, and pumpkins), the type of flower produced varies. Young plants and those growing in a poor environment produce staminate flowers, whereas older plants and those growing in a good environment produce perfect or carpellate flowers. If fertilization occurs, the carpels develop into large fruits, and only a healthy, robust plant can afford to do this. A young or poorly growing plant cannot supply enough carbohydrate and protein for fruit development, but it can supply enough to produce pollen.

Dioecy is an extreme adaptation that ensures cross-pollination; a plant that produces only one type of spore cannot pollinate itself. It is similar to the condition of separate genders in mammals—no individual can fertilize itself. Both conditions ensure that fertilization is by sex cells that are not identical genetically, thus increasing the genetic diversity of the offspring.

Animal-Pollinated Flowers

The evolution of animal-mediated pollination had a dramatic impact on the evolution of flowering plants. In wind-pollinated gymnosperms, things such as brightly colored petals, fragrances, and nectar are a waste of material and energy. But once insects began visiting early angiosperms, mutations that resulted in pigments, fragrances, or sugar-rich secretions became adaptive. For pollen that is carried by wind, the probability that any particular grain will actually land on a stigma is very low, whereas if it is carried by an insect that flies from flower to flower, the probability of a pollen grain reaching a stigma is much improved.

When this insect-flower association began around 120 million years ago, neither insects nor flowers were particularly sophisticated. Insects probably visited all kinds of flowers, not recognizing the different species. As a result, pollen often landed on the stigma of the wrong species, where it was useless. Mutations that increased a plant's distinctiveness, its recognizability by an insect—flower color, size, shape, fragrances, and so on—became selectively advantageous. Rewards for the insect such as sugary nectar or protein-rich pollen and stamens were advantageous. Mutations were adaptive if they increased the capacity of insects to recognize flowers that offered abundant nectar or pollen. Because it is expensive energetically for an insect to fly, the ability to recognize the most nutritious flowers from a distance while flying is advantageous. As a result, many lines of insects and flowers underwent **coevolution**, a flower

(a)   (b)

Figure 9-25 (a) Ears of corn are really large groups (inflorescences) of carpellate flowers surrounded by protective leaflike bracts (the husks). The corn "silks" are long styles. (b) Corn tassels are inflorescences of staminate flowers.

(a)

(b)

Figure 9-26 (a) Any median longitudinal section of this poppy results in two halves that are mirror images of each other; it is radially symmetrical, that is, actinomorphic. (b) These *Penstemon* flowers are bilaterally symmetrical, zygomorphic; only if cut from top to bottom are the two halves mirror images. A pollinator, also bilaterally symmetrical, fits well only if it approaches the flower properly. Anthers are at the top of each flower's throat and will put pollen only onto a bee's back.

becoming adapted for visitation by a particular insect and the insect for efficient exploitation of the flower. Coevolution has also occurred between flowers and birds and between flowers and bats.

The shape of the flower is particularly important, as a pollinator actually makes contact with it. Most flowers are radially symmetrical; that is, any longitudinal cut through the middle produces two halves that are mirror images of each other. These flowers (and stems and roots) are **actinomorphic** or **regular** (**Figure 9-26**). But all insects, birds, and bats are bilaterally symmetrical—only one longitudinal plane produces two halves that are mirror images. In many species, flowers and pollinators have coevolved in such a way that the flowers are now also bilaterally symmetrical—**zygomorphic** (Figure 9-26). When a pollinator approaches a zygomorphic flower, only one orientation is comfortable for it; any misalignment prevents the pollinator's head or body from fitting the flower's distinctive shape. As a result, as the pollinator feeds at the flower, pollen is placed on a predictable part of its body. When it visits the next flower, pollen is rubbed directly onto the stigma. Not only is pollen carried to the appropriate flower, but it is carried directly to the stigmas. In contrast, a pollinator can approach an actinomorphic flower from any direction, and pollen may be brushed onto any part of the body; when it visits the next flower, the pollen-carrying part may not come into contact with the stigmas and effect cross-pollination.

Most plant/animal relationships are a battle: Animals try to eat plants, lay eggs in them, or do other harmful things. Plants defend themselves with poisons, spines, and sclereid barriers. Mutations that permit plants to make more effective deterrents benefit the plant but not its animal pests. But both plants and animals tend to benefit from the pollination relationship, and coevolution is possible. Mutations that make a plant more recognizable or more convenient to its pollinators help both the plant and the animal, just as certain mutations in pollinators are beneficial to both organisms. But neither side should give more than it has to. Some bees now are "nectar robbers," being able to get nectar from flowers without having to carry any pollen. And a num-

ber of orchids, for example *Chiloglottis,* trick wasps into carrying pollen for free. These orchid flowers look, smell, and feel just like a female wasp, and male wasps try to mate with them. The wasps then fly off to the next orchid flower and the next, transferring pollen between each of them. The orchids get pollination without having to provide nectar, and the wasp gets nothing more than a good time.

Wind-Pollinated Flowers

In species that are wind pollinated, a totally distinct set of modifications is adaptive. Attracting pollinators is unnecessary; thus, mutations that prevent the formation of petals are selectively advantageous, and the energy saved can be used elsewhere in the plant. Sepals are also often reduced or absent, and the ovaries need no special protection; thus, the entire flower may be tiny. Zygomorphy provides no selective advantage. After pollen is released to the wind, the chance of any particular pollen grain landing on a compatible stigma is small, so huge numbers of grains must be produced. Large, feathery stigmas are adaptive by increasing the area that can catch pollen grains (**Figure 9-27**). In general, wind-pollinated individuals produce up to several thousand small flowers; although each flower is tiny, the entire plant has a large total stigmatic surface area.

Pollination is aided by the growth pattern of the plant population. Wind-pollinated species—like grasses, oaks, hickories, and all conifers—grow as dense populations in range lands or forests. Within 1 km^2 may be found thousands of plants and, more importantly, millions of stigmas. Species that occur as widely scattered, rare individuals must rely on animal pollination.

Ovary Position

The ovary and ovules must be well protected from pollinators. Paradoxically, a flower must bring a hungry animal to within millimeters of protein-rich ovules in order to effect pollination. Adaptations that maximize the separation are long styles and stamen filaments. Burying the ovaries deep within the flower provides

Figure 9-27 Many wind-pollinated trees have such tiny flowers people do not realize they are flowering plants at all. This is an entire set (an inflorescence) of flowers of poplar (*Populus*), each flower consisting of just numerous stamens.

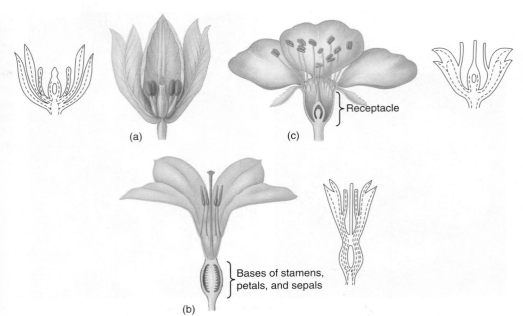

Figure 9-28 (a) In flowers with superior ovaries, carpels sit above the other organs, as would be expected because they are initiated last by the flower apical meristem. (b) This is an inferior ovary resulting from the fusion of the bases of sepals, petals, and stamens to the base of the carpel. Four appendages are involved whose vascular bundles (dashed lines) are still present and distinct. (c) In some species, after all of the appendage primordia are initiated, the receptacle tissue grows upward and surrounds the ovaries. This can be detected because the vascular bundles around the ovary are receptacle bundles, not appendage bundles.

further protection. In some species, after all flower organ primordia are initiated at the receptacle apex (all parts are microscopically small when initiated), the primordia crowd together, and the bases of the stamens, petals, and sepals fuse, creating a thick layer of protective tissues around the ovaries, which appear to be located below the other organs (see Figures 9-21 and **Figures 9-28a and 9-28b**). Two terms describe this: We can either say that the ovary is an **inferior ovary** or that the other parts are **epigynous**. Inferior ovaries also can result if receptacle tissue grows upward around the ovary (Figure 9-28c). The more common arrangement, in which no fusion to the ovary occurs and the ovary is obviously above the other flower parts, is a **superior ovary** or **hypogynous parts**. Intermediate, partially buried ovaries are **half-inferior** with **perigynous** flower parts.

Inflorescences and Pollination

The positioning of flowers on an individual plant is important; few species have plants that produce only a single flower. Instead, many flowers are produced either within a single year or over a period of many years. A large mountain ash or cherry tree produces thousands of flowers every year for well over 100 years. Many important factors determine whether a flower is seen and visited by pollinators—for example, the positions of flowers relative to other flowers, leaves, and trunk; height from the ground; and distance from an open, uncluttered flight path for pollinators (Figure 9-29).

Reproductive success is measured in terms of the number of healthy, viable seedlings that become established. One important factor for this is pollination. It might seem optimal to have large flowers that can be easily seen by pollinators; unfortunately, they can also be seen by herbivores, and because ovules and pollen are rich in protein, they are good food sources for pests. Production of the pedicel, receptacle, sepals, and petals can be thought of as packaging and advertising cost; the larger the flower is, the larger the cost. These costs can be made more acceptable (less disadvantageous selectively) by increasing the number of ovules (potential seeds) per flower. Large flowers tend to have numerous ovules and small flowers fewer. Small flowers exist because small plants cannot afford large flowers and their accompanying large seeds, which are very expensive. Also, a large flower with many ovules is a big risk—if it is found and eaten, a major reproductive investment has been lost; however, damage to a small flower with only a few ovules is less significant. Also, if many flowers are grouped together, an **inflorescence**, they give a collective visual signal to pollinators: One small flower may be overlooked, but not a hundred close together (**Figure 9-30**). Furthermore, in an inflorescence, the plant is able to control accurately the timing of the initiation, maturation, and opening of the flowers. Consequently, the plant can be in bloom and available to pollinators for several weeks even though each flower lasts only a day or two.

Large flowers can produce more nectar than small flowers, but this makes them a more tempting target for nectar robbers, animals that take nectar without carrying pollen. Furthermore, a flower should not produce enough nectar to completely satisfy even a legitimate pollinator because no incentive would remain for the pollinators to go to another flower; it is better if the pollinators get only enough to make them interested in searching for more.

(a)

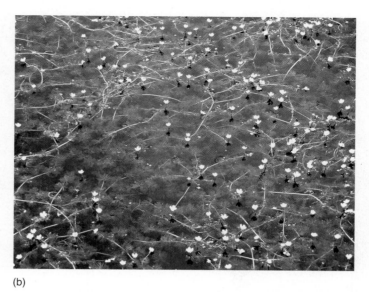
(b)

Figure 9-29 (a) Flowers of sausage tree (*Kigelia*) are pollinated by bats, which do not like to fly among the clutter of leaves because their sonar does not work well there. Long stalks allow flowers to hang in open air where bats have free access to them. (b) Angiosperms that grow under water in streams and lakes usually have flowers that project above the water's surface, being accessible to flying pollinators (*Ranunculus aquatilis*).

Many other factors affect the relative selective advantage or disadvantage of flower number and size as well as the number of ovules per flower. Considering the numerous types of plants, pollinators, and environments, the diversity of flower and inflorescence types is not surprising. In the simplest arrangement, flowers occur individually in leaf axils or as a transformation of the shoot apex. When grouped into inflorescences, two basic arrangements occur: (1) **determinate inflorescences** and (2) **indeterminate inflorescences**.

A determinate inflorescence has only a limited potential for growth because the inflorescence apex is converted to a flower, ending its possibilities for continued growth (**Figures 9-31a** and **9-31b**). Typically, but not always, the terminal flower opens first, and then lower ones open successively. In the simplest type, below the terminal flower is a bract with an axillary flower, which also may have a bract and axillary flower, and so on.

In an indeterminate inflorescence, the lowest or outermost flowers open first, and even while these flowers are open, new flowers are still being initiated at the apex. A raceme has a major inflorescence axis, and the flowers are borne on pedicels that are all approximately the same length (Figure 9-31c). Catkins are similar to racemes, differing in that the flowers are imperfect, either staminate (Figure 9-27) or carpellate, and all flowers of a single inflorescence are the same, and thus, each species must have both staminate catkins and carpellate ones. Catkins almost always contain very small, wind-pollinated flowers. A spike is similar to a raceme except that the flowers are sessile, lacking a pedicel (Figure 9-31d). A spadix (plural, spadices) is a spike-like

(a)

(b)

Figure 9-30 (a) Palm flowers are small and inconspicuous, but when grouped together, their inflorescence has even more impact than a single large flower. (b) The inflorescence of *Combretum* is even more striking than the individual flowers. Notice how the flowers mature and open sequentially.

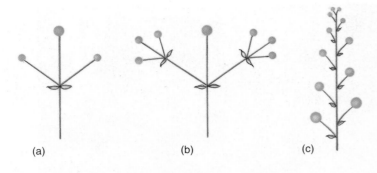

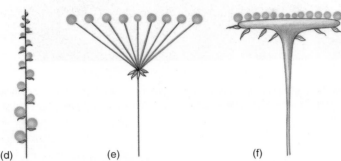

Figure 9-31   Inflorescence types: (a) Simple determinate inflorescence. (b) Compound determinate inflorescence. (c) Raceme. (d) Spike. (e) Umbel. (f) Head. Larger circles represent flowers that open earlier than those depicted as smaller circles.

TABLE 9-4	Agents of Dispersal
Agent	**Descriptive Term**
Animals	Zoochory
Attached to animal	Epizoochory
Eaten by animal	Endozoochory
Birds	Ornithochory
Mammals	Mammaliochory
Bats	Chiropterochory
Ants	Myrmecochory
Wind	Anemochory
Water	Hydrochory
Dispersed by the plant itself	Autochory

Seeds, fruits, and asexual propagules can be dispersed by many means. These are a few of the most common types.

inflorescence with imperfect flowers, but both types occur in the same inflorescence, most often with staminate flowers located in the upper portion of the inflorescence and carpellate flowers in the lower portion, although they can intermingle. The main inflorescence axis is thick and fleshy with minute flowers embedded in it; the entire inflorescence is subtended or enclosed by a petal-like bract called a spathe. A panicle is a branched raceme with several flowers per branch.

Several types of indeterminate inflorescences have no dominant main axis. In umbels, the inflorescence stalk ends in a small rounded portion from which arise numerous flowers (Figure 9-31e). Their pedicels are long and arranged so that all flowers sit at the same height, forming a flat disk. A head is similar to an umbel except that the flowers are sessile and attached to a broad expansion of the inflorescence stalk (Figure 9-31f); numerous bracts may surround the inflorescence during development. Heads are almost synonymous with the aster family, sunflowers and dandelions being easily recognizable examples. In this group, the inflorescences are so compact and highly organized that they mimic single flowers; what appear to be the petals are really entire flowers, ray flowers, in which the petals are very large and fused together. The center of the inflorescence is composed of a different type of flower, disk flowers, in which the corollas are short and inconspicuous.

Fruit Types and Seed Dispersal

Fruits are adaptations that result in the protection and distribution of seeds. Many different agents disperse fruits and the seeds they contain: Gravity, wind, water, and animals are the most

common (Table 9-4). The principles involved in fruit function are somewhat opposed to each other: Fruits that are tough and full of fibers or sclereids, such as pecans, walnuts, brazil nuts, and coconuts, offer maximum protection but are heavy and expensive metabolically. Also, the protected seed must be able to break out to make germination possible; a more fragile fruit is better for that. If animals are to disperse the seeds, part of the fruit must be edible or otherwise attractive, whereas the seed and embryo must be protected from consumption. A division of labor often occurs—some parts being protective, others attractive, and still others allowing germination.

True Fruits and Accessory Fruits

The term "pericarp" refers to the tissues of the fruit regardless of their origin. In most cases, this is the ovary wall, but in many species, especially those with inferior ovaries, the receptacle tissues or sepal, petal, and stamen tissues may also become involved in the fruit. The terms "pericarp" and "fruit" have been applied to both types of fruit; thus, now the term **true fruit** is used to refer to fruits containing only ovarian tissue, and **accessory fruit** (or false fruit) is used if any nonovarian tissue is present (Figure 9-32). Apples develop from inferior ovaries, and the bulk of the fruit is enlarged bases of sepals and petals; only the innermost part is true fruit derived from carpels (see Figure 9-21).

Fusion of carpels also affects the nature of fruits. If the fruit develops from a single ovary or the fused ovaries of one flower, it is a **simple fruit**, the most common kind. If the separate carpels of one gynoecium fuse during development, an **aggregate fruit** results, such as raspberries. If during development all of the individual fruits of an inflorescence fuse into one fruit, it is a **multiple fruit**, as in figs, mulberries, and pineapple (Figure 9-32b). These are also largely accessory fruits because in addition to the ovary tissue, the inflorescence axis, bracts, and various flower parts contribute to the mature fruit.

Classification of Fruit Types

There are several ways of grouping or classifying fruits. In one method, emphasis is placed on whether the fruit is **dry** or **fleshy**. A dry fruit is one that is not typically eaten by the natu-

(a)

(b)

Figure 9-32 (a) The red, edible flesh of strawberry is really the receptacle, not carpel tissue. It is therefore an accessory fruit. The "seeds" are true fruits, each derived from one carpel of the flower. (b) Pineapples develop from the coalescence of all the many true fruits of one inflorescence, so they are a multiple fruit. In addition, many noncarpellary tissues become involved, and are therefore accessory fruits as well. Bracts are visible here.

ral seed-distributing animals; fleshy fruits are those that are eaten during the natural seed distribution process.

A further classification of dry fruits emphasizes fruit opening: **Dehiscent fruits** break open and release the seeds, whereas **indehiscent fruits** do not (Table 9-5). For the most part, fleshy fruits are indehiscent. Animals chew or digest the fruit, opening it; if uneaten, the fruit rots and liberates the seeds. Although many dry fruits are dehiscent, others must rot, or the seed must break the fruit open itself.

Perhaps the simplest fruits are those of grasses (all cereals such as corn and wheat): Each fruit develops from one carpel containing a single ovule. During maturation, the seed fills the fruit and fuses with the fruit wall, which remains thin: "Seeds" of corn and wheat are actually both the fruit and its single seed, fused into a single structure. It has little protection and no attrac-

tion for animals. The fruit/seed falls and germinates close to the parent sporophyte. The fruits are indehiscent, and during germination, the embryo absorbs water, swells, and bursts the weak fruit. These fruits are caryopses.

Beans and peas form from a single carpel that contains several seeds. The fruits, known as legumes, are dry and inedible (unless eaten while still young and green, such as snowpeas and green beans) like caryopses but do not fuse to the seeds. At maturity, the two halves of the fruit twist and break open (dehisce) along the two specialized lines of weakness (Figure 9-33). After liberation from the legume, the seeds are protected from small animals and fungi by only the seed coat.

Fruits or seeds carried by wind must be light; they often have wings (maples) or parachutes (dandelions) that keep them aloft as long as possible. Such fruits are dry and weigh much less than

TABLE 9-5	Fruit Types

1. Fleshy fruits
   Berry: a fleshy fruit in which all three layers—endocarp, mesocarp, exocarp—are soft (grape, tomato).
   Pome: similar to a berry except that the endocarp is papery or leathery (apple).
   Drupe: similar to a berry except that the endocarp is hard, sclerenchymatous (stone fruits: peach, cherry, plum, apricot).
   Pepo: a fleshy fruit in which the exocarp is a tough, hard rind; the inner soft tissues may not be differentiated into two distinct layers (pumpkin, squash, cantelope).
   Hesperidium: exocarp is leathery (*Citrus*).

2. Dry fruits
   Indehiscent fruits
   Developing from a single carpel
   Caryopsis: simple and small, containing only one seed, and the testa (seed coat) becomes fused to the fruit wall during maturation (grasses: wheat, corn, oats).
   Achene: like a caryopsis, but the seed and fruit remain distinct. Fruit wall is thin and papery (sunflowers).
   Samara: a one-seeded fruit with winglike outgrowths of the ovary wall (maples, alder, ash).

   Developing from a compound gynoecium (a compound pistil)
   Nut: although the gynoecium originally consists of several carpels and ovules, all but one ovule degenerate during development. Pericarp is hard at maturity (walnut).
   Dehiscent fruits
   Developing from a single carpel
   Legume: fruit breaks open along both sides (beans, peas)
   Follicle: fruit breaks open on only one side (columbine, milkweeds)
   Developing from a compound gynoecium
   Capsule: opens many ways:
   Splitting along lines of fusion (*Hyperium*)
   Splitting between lines of fusion (*Iris*)
   Splitting into a top and bottom half (primrose)
   Opening by small pores (poppy)
   Schizocarp: Compound ovary breaks into individual carpels called mericarps.

3. Compound fruits
   Aggregate fruits: carpels of flower not fused, but grow together during fruit maturation (raspberry).
   Multiple fruit: all the fruits of an inflorescence grow together during fruit maturation (pineapple).

Figure 9-33 (a) At maturity, different layers in this dry fruit contract in opposite directions, causing the pod to twist and break open. (b) As these maple fruits drop, the wings cause them to rotate and float, thus remaining aloft and traveling much farther on the wind. (c) The spiral, hairy tails of these geranium fruits let them float on the wind over long distances.

(a)

(b)

(c)

one gram. Fruits and seeds that are transported by water (ocean currents, streams, and floods) can be larger and heavier, but they must be buoyant and resist mildew and rot. An excellent example is coconuts that float from one island to another.

Animals carry fruit in a variety of ways. Dry fruits with hooks or stickers catch onto animal fur or feathers; sticky, tacky fruits glue themselves onto animals, and fleshy, sweet, colorful fruits are eaten. Edible fruits require particular specialization because the fruit must be consumed but the seeds must not be damaged; they must not be crushed by teeth or gizzard or be digested. Immature fruits deter frugivores by being hard, bitter, or sour, such as unripe apples or bananas. At maturity, the fruit is soft, sweet, flavorful, and typically strikingly different in color. The enclosed seeds often have hard seed coats or endocarps (peach and cherry) that allow them to pass through an animal unharmed (see Figure 9-22). Dinosaurs may have been especially poor seed dispersers: They had no hair for hooks to tangle in, and their digestive tracts were often too long, hot, and fermentative for seeds to survive.

Grapes and tomatoes are examples of simple fleshy fruits: They are berries because all parts of the fruit are soft and edible (Table 9-5). The seed coat provides some protection against being crushed during chewing, but perhaps more important are the size of the seeds and their slippery seed coat, which allows the seeds to slip out from between our molars. Pomes (apples and pears) differ from berries in two ways. First, they develop from inferior ovaries, and thus, the inner tissue is the true fruit and the outer tissue is accessory fruit. Second, the seeds are protected because the innermost fruit tissues, the cores, are bitter and tough. The most elaborate fleshy fruits are drupes (peaches and cherries), in

Figure 9-34 This seedling is as happy as a clam at high tide: It has enough fertilizer to last for months. Many of its competitors have been digested; others are being suffocated.

which the innermost tissues, the endocarp, are extremely hard and totally inedible, full of sclerenchyma. The mesocarp becomes thick, fleshy, juicy, and sweet, and the exocarp forms the peel, whose color informs the frugivore of the fruit's ripeness. The endocarp of a drupe is known as the pit and is often mistaken for the seed, but the true seed and its seed coat are located inside. Drupes provide maximum attraction to animals with minimum danger to the seed.

Because frugivores are a major danger to seeds, it is necessary to think about how edible fruits evolved: They must offer some advantage that outweighs the danger. Animals have predictable habits and migration patterns. Ants, birds, and mammals carefully select the sites where they feed, nest, and sleep; therefore, seeds are not distributed at random but are moved through the environment to specific sites. One example is mistletoe: Its seeds are sticky and adhere to a bird's beak as it eats the fruit. The bird probably feeds and cleans itself in the same species of tree; thus, the seeds are rubbed off the beak directly onto a proper host tree. Few seeds are ever lost by dropping to the ground or being deposited in the wrong tree, as would happen with wind or water dispersal. Just as efficient is dispersal of fruits and seeds of marsh plants. The birds and animals that live in marshes typically spend little or no time in other habitats, so seeds carried by these animals are almost certainly distributed to favorable sites.

A special benefit of seed distribution by animals is that the "deposited" seed may find itself in a small (or large) mound of "organic fertilizer" (Figure 9-34). Seeds adapted to pass through an animal's digestive tract are resistant to digestive enzymes and can tolerate the high level of ammonium in fecal matter. Because many seeds of plants cannot tolerate these two factors, the adapted seeds find themselves in a microenvironment that is not only nutrient rich but also excludes some competitors.

Summary

1. Asexual reproduction results in new individuals genetically identical to the parent. The parent and all progeny are equally adapted to the habitat.

2. Sexual reproduction results in progeny that differ from each other genetically. There is a range of fitness, and some progeny may have a combination of maternal and paternal genes that causes them to be even better adapted than the parents.

3. All seed plants have a life cycle that consists of an alternation of heteromorphic generations. Microgametes and megagametes are produced by microgametophytes and megagametophytes that grow from microspores and megaspores. Most species have only one type of sporophyte, capable of producing both types of spores. Dioecious species have two types of sporophytes, one of which produces microspores and the other megaspores.

4. Flowers contain all or some of the following: nonessential organs—pedicel, receptacle, sepals, and petals; and essential organs—stamens and carpels.

5. In flowering plants, the microgametophyte consists initially of a vegetative cell and a generative cell, but the latter divides into two sperm cells. The megagametophyte consists of the egg cell, two synergid cells, a binucleate central cell, and three antipodal cells.

6. After fertilization, the zygote (fertilized egg) grows into an embryo; polar nuclei fuse with the second sperm nucleus, establishing the endosperm nucleus and endosperm, and the integuments develop into a seed coat. The entire structure is a seed. The ovary develops into a fruit.

7. Inflorescences are groups of many flowers. They may be determinate or indeterminate and may contain several types of specialized flowers.

8. Fruits protect seeds during development and may aid in their dispersal and maturation. Fruits are either dry, inedible, dehiscent, or indehiscent, or they are fleshy, edible, and indehiscent.

Important Terms

alternation of generations	gametophyte	pollen tube
anther	heteromorphic generations	radicle
carpels	hypocotyl	receptacle
coevolution	imperfect flowers	seed coat
compatibility barriers	incomplete flowers	seeds
complete flowers	inflorescence	sepals
cotyledons	karyogamy	spores
cross-pollination	ovary	sporophytes
double fertilization	ovules	stamens
embryo sac	pedicel	stigma
endosperm	perfect flower	style
epicotyl	petals	syngamy
fruit	plasmogamy	zygote
gametes	pollen	

Review Questions

1. Reproduction can serve two very different functions. What are they?

2. Under what conditions is it selectively advantageous for a plant to produce offspring identical to itself?

3. Imagine a plant that has been well adapted to a particular habitat, and it has been reproducing only asexually; thus, all of its offspring are identical to it. If the climate or other conditions change such that the plant can no longer survive in the new conditions, what will happen to all of its offspring? Why?

4. With asexual reproduction, are progeny ever more fit, more adapted than the parent?

5. Look at the apples in Figure 9-1a, all growing on a single tree. Do all of the seeds in all of the apples in the photograph have the same maternal parent? Do they all have the same paternal parent?

6. Sexual reproduction produces offspring that are not identical to each other or to either parent. Usually, some are more well adapted than the parents. Some are more poorly adapted, and most are about as adapted. What is one of the beneficial aspects of this diversity?

7. In stable populations, ones that are neither increasing nor decreasing in abundance (e.g., there are a million trees now and there will be a million trees a thousand years from now), about how many of a plant's seeds survive and grow to adulthood, being able to replace it when it dies? If during the plant's lifetime it produces 100,000 seeds, how many do not survive, do not grow, and cannot replace it when it dies. (Hint: Do not think of humans, as we are an increasing population, not a stable one. Almost all our children survive, but that is not true of any other species.)

8. Describe the life cycle of us humans. Are the tissues and organs in our bodies made up of diploid cells or haploid ones? Our reproductive organs make sperm cells in males and egg cells in females. Is this done by mitosis or meiosis? When our bodies make gametes, does every cell in our body become a sperm cell or an egg cell, or do just some of the cells in our reproductive organs do this? Can our haploid sperm and egg cells undergo mitosis and grow into new animals that look like us but are haploid instead? (Haploid eggs do grow in bees. They develop into males.) After a fertilized egg (a zygote) is formed, is it diploid or haploid? Can it immediately undergo meiosis to make four new sperm cells or egg cells, or can it only grow by mitosis into another person?

9. The plants we are familiar with are called sporophytes or the sporophyte generation. Do these plants have bodies made up of diploid cells or haploid ones? In which organs does meiosis occur (two are correct)—leaves, stems, stamens, roots, carpels, or petals? When plants undergo meiosis, do all cells of the plant undergo meiosis? Do all cells become haploid, or does meiosis occur in just a few cells in some flower parts?

10. In animals, meiosis produces gametes (sperm cells and egg cells), but that does not happen in plants. When some of the cells of a sporophyte undergo meiosis, what types of cells are produced? What do they grow into?

11. What is a gametophyte? How many different types of gametophytes are there in a plant life cycle? What do they look like (Hint: see Figure 9-5b)? Do the sporophytes and gametophytes of seed plants ever look like each other?

12. Draw and label a microgametophyte of a flowering plant. What type of gamete does it produce? Draw and label a megagametophyte of a flowering plant. What type of gamete does it produce?

13. Flowers typically have many parts, although some flowers can be missing some of the standard parts. What is the name of each of the following parts:
    a. The stalk of the flower
    b. The end of the stalk, where the other parts are attached
    c. The parts that are usually green and protect the flower bud as it develops
    d. The parts that attract pollinators
    e. The parts that produce pollen
    f. The parts that receive pollen and which contain ovules

14. If a flower has all of the parts c, d, e, and f in Question 13, they are said to be _____ flowers. If they are missing any of these parts, they are said to be_____ _____ flowers. Parts e and f are especially important because they produce the reproductive cells, the spores. If a flower has both e and f in Question 13, they are _____ flowers. If they are missing either e or f or both, they are _____ flowers.

15. In flowers that are pollinated by wind or water, which of the parts in Question 13 is often missing?

16. What is the difference between a perfect and an imperfect flower? What is the difference between a complete and an incomplete flower?

17. A stamen usually has two parts. The stalk is called a _____ _____ and an upper portion, the _____ _____, which produces the pollen. Only some of the cells in the upper part undergo meiosis and become pollen grains (microspores). Those cells are called _____ _____ mother cells or _____. Neighboring anther cells, in a layer called the _____ _____, act as nurse cells.

18. Carpels usually have three parts: a _____ that catches pollen grains, a _____ that elevates the first part, and an _____ where megaspores are produced. In this last part, there are placentae that bear small structures called _____ each with a short stalk called a funiculus and a central mass of parenchyma called the nucellus. One cell in the nucellus will be the _____ mother cell or _____.

19. The megaspore in most flowering plants grows into a megagametophyte that has seven cells and eight nuclei. Name and describe the seven cells.

20. After pollen lands on a stigma, it is far away from the ovule with the megagametophyte, which holds the egg (the megagamete). How are the two sperm cells transported from the stigma to the egg?

21. In angiosperms, as a sperm cell enters the egg, it loses both _____ and _____ such that only the sperm cell nucleus contributes any DNA to the new zygote. In gymnosperms, however, the sperm cell loses only its _____ such that the zygote inherits both plastids and a nucleus from the sperm cell.

22. After the sperm cells enter a synergid, one fertilizes the egg in a two-step process. First there is fusion of the sperm cell's protoplasm with that of the egg, a step called _____. Then the sperm cell nucleus fuses with the egg cell nucleus, the second step, called _____.

23. What happens to the second sperm nucleus, the one that does not fertilize the egg cell? What is the tissue that develops from this second "fertilization"? How is coconut related to this?

24. In most eudicot seeds, the parts of the embryo are very easy to see. Describe each of these parts:
   a. Cotyledons
   b. Radicle
   c. Hypocotyl
   d. Epicotyl
   e. Extra credit portion: Answer the following question if you have read about the radicle with regard to a taproot system and a fibrous root system: How is the radicle involved in these different types of root systems?

25. What are albuminous and exalbuminous seeds? Consider corn, peas, and beans. Which of these seeds are which?

26. After pollination and fertilization, as the ovule develops into a seed, the ovary matures into a _____. Many of these have three parts. The _____ is the skin or peel. The _____ is the flesh, and the innermost layer, the _____, may be tough like the pit of a cherry.

27. What is the difference between self-pollination and cross-pollination? If pollen is transferred from the stamens of a flower to the stigma of the very same flower, is that cross-pollination or self-pollination? If it is transferred from the stamen of one flower to the stigma of another flower on the very same plant? If it is transferred from the stamen of one flower to the stigma of another flower on a different plant, but a plant that is a clone of the first one? Animals do not have pollen, of course, but why is it that most animals never have to worry about the equivalent problem of self-fertilization? Why cannot most animals fertilize themselves?

28. Describe a species that is dioecious. Name several examples (Hint: see Figure 9-6). In order for sexual reproduction to occur in a dioecious species, how many separate plants must be involved? Name them all.

29. Describe a monoecious species. Name several examples.

30. In several pairs of plant and animal species, plants have become modified such that only its animal partner can pollinate it, and the animals have become modified such that they are especially well adapted to pollinate just their plant partner. What is the name of this type of evolution that results in two organisms becoming particularly adapted to and dependent on each other?

31. Explain the following terms: inferior ovary, superior ovary, actinomorphic flower, zygomorphic flower. How is each of these modifications selectively advantageous?

32. What is an inflorescence? The inflorescences of a sausage tree (Figure 9-29) hang far down out of the tree. How is this of benefit to bats? Why can the inflorescence of *Combretum* (Figure 9-30b) attract more pollinators than can the individual flowers?

33. After pollination and then fertilization, what usually happens to each of the following: stigma, style, carpel, ovule, integuments, and zygote?

34. Fruits are often classified as dry or fleshy. What is the difference? Which of these two are dehiscent, which are indehiscent?

35. In ordinary English, we use the word fruit to mean something sweet and juicy; however, the following things are fruits: peanut shells, pea pods, bell peppers, and chili peppers. What is the characteristic that lets us know these really are fruits even though they are not sweet? In contrast, bananas are fruits that do not have this characteristic (they are sterile and new plants must be grown from buds that sprout near the base of the plant).

36. Some things that we call fruits are not true fruits but instead are accessory (false) fruits. What is the red part of a strawberry, and what are the true strawberry fruits (Hint: see Figure 9-32a)? In an apple, what is the fleshy part that we eat, and what is the core that we throw away?

BotanyLinks

http://biology.jbpub.com/botany/5e

BotanyLinks contains a variety of resources and review material designed to assist in your study of botany.

Visit the Web Exercises area of BotanyLinks to complete these questions:

1. How does living underground affect flowering, pollination, and seed dispersal? Go to the BotanyLinks home page for more information on this subject.

2. What kinds of flower modifications would be selectively advantageous for desert plants? Go to the BotanyLinks home page for more information on this subject.

Plant Physiology and Development

When you are studying plant structure, you can often see the material directly with the naked eye, although light or electron microscopy may be necessary to see some structures. In plant physiology and development, however, the objects of our study—chemical reactions and metabolic pathways—cannot be seen at all. Instead, the results of experiments, measurements, and analyses are studied, and hypothetical reactions and pathways are set down on paper. From these, predictions are made and new observations planned to test the hypotheses. It is easy for us to study and memorize metabolic pathways, chemical formulas, and diagrams of physiological control mechanisms without appreciating that these have never been, and cannot ever be, seen directly. Every one is a theoretical model that is consistent with the majority of the available data and that is logically and internally consistent.

When experienced anatomists see an unusual structure, they may be able to recognize instantly that it is new to science or is at least a significant variation. For physiologists, it is not so easy: If an experiment on photosynthesis does not come out as expected, it may be that a new type of photosynthesis is being discovered; it may be that the current theories of photosynthesis are not completely accurate and have made an erroneous prediction, or it may be that an experimental error has occurred. It can be difficult for students to

Part Opener Image: Although the peak of O'Malley Summit in Alaska is covered in snow in winter, the summers are warm enough that some plants grow here. They must have a metabolism that allows them to grow and develop rapidly in the brief summer and to survive the long periods of subzero weather throughout winter.

appreciate the tremendous amount of careful, ingenious work that must be done just to establish that a particular theoretical metabolic pathway truly represents the reactions that occur in certain plants. Whereas it is relatively easy to determine that natural selection has resulted in the evolution of many types of leaves, stems, roots, flowers, xylem, phloem, and so on, it is much more difficult in physiology. Numerous differences in microhabitats, water availability, heat, cold, soils, pests, and plant diseases have resulted in diverse types of structures that are selectively advantageous under various conditions. It is logical to expect the same to be true of metabolism; we do know that there are several varieties of photosynthesis and respiration, and there may be others that have not yet been discovered.

As you study the material in this section, keep in mind that, just as is true for structure, organization is of fundamental importance for metabolism. The chemical and physical reactions that constitute plant metabolism are highly ordered and not at all random. This orderliness is maintained by the input of energy (see Chapters 10 and 11) acting on materials brought into plants from the environment (see Chapters 12 and 13). There are many types of order, and the information necessary to establish the proper reactions acting on the proper material is stored in the genes, both in the nucleus and in the plastids and mitochondria. The mechanisms by which the genes control the interaction of energy and matter, such that a plant of a particular species results, are discussed in Chapters 14 and 15.

Energy Metabolism: Photosynthesis

chapter 10

Concepts

Probably the most important concept concerning cells and all of life itself is that living organisms are highly ordered, highly structured systems. The universe as a whole is constantly becoming less orderly; its disorder (**entropy**) is increasing. Prokaryotes, protists, fungi, plants, and animals, however, represent phenomena in which particles become more orderly. A plant absorbs diffusely scattered molecules of carbon dioxide, water, and minerals and organizes them into organic molecules, cells, tissues, and organs. Each plant carries this out with such precision that each species of plant is easily distinguishable from others. After death, decay is the process by which an organism's molecules become more disordered and scattered—their entropy increases.

Chapter Opener Image: All organisms must obtain the energy needed to drive their metabolism. Most plants obtain energy through photosynthesis using the green pigment chlorophyll. All animals and fungi obtain their energy by either eating plants or by eating animals that ate plants. Some plants, called holoparasites, like this orange dodder (*Cuscuta*), do not carry out photosynthesis. Instead, they send specialized roots called haustoria (visible here as short pegs) into the bodies of ordinary green, photosynthetic plants and extract sugars, water, and minerals from them and convert the sugars into the organic compounds they need. This may seem like an odd type of evolution: sunlight is abundant and free, so it seems like photosynthesis would be a good way to obtain energy. But the pigments and organelles necessary for photosynthesis are expensive and complex, and stealing energy from another plant is an alternative that has evolved in many plant groups.

Because living organisms are part of the natural world described by the laws of physics and chemistry, the decrease in the entropy of living organisms must obey physical laws. This is accomplished by putting energy into the living system, the source of energy being sunlight. To be accurate, we must consider the sun and life together: Atomic reactions that generate sunlight cause greater disorder in the sun than sunlight causes order in living organisms. The entire system (sun + life) becomes more disordered. Because there is no means of putting energy into an organism's body after death, an increase in entropy cannot be prevented.

Sunlight maintains and increases the orderliness of life by two methods: (1) directly, in the process of photosynthesis, which produces complex organic compounds, and (2) indirectly, in the respiration of those organic compounds, either by the organism itself or by another organism that eats it. These two methods of supplying energy and maintaining orderliness—photosynthesis and respiration—are the basis for a major, fundamental distinction in the types of organisms. **Photoautotrophs** are organisms that gather energy directly from light and use it to assimilate small inorganic molecules into their own tissues. Photoautotrophs include all green plants, all cyanobacteria, and the few bacteria capable of photosynthesis. **Heterotrophs** are organisms that cannot do this but instead take in organic molecules and respire them, obtaining the energy available in them. Heterotrophs include all animals, all completely parasitic plants (**Figure 10-1**), all fungi, and nonphotosynthetic prokaryotes. Gathering energy by taking in organic material has the advantage that part of the material can be used as construction material instead of fuel. At least some of the amino acids, fatty acids, and sugars in food can be built into the organism's own polymers and the rest respired for energy.

Figure 10-1 Total parasites such as this broomrape (*Orobanche*) are heterotrophs like animals. Like parasitic tapeworms or blood flukes, they absorb monomers such as monosaccharides and amino acids.

TABLE 10-1	Differences Between Photoautotrophs and Heterotrophs	
	Photoautotrophs	Heterotrophs
Source of energy	Sunlight	Food: carbohydrates, proteins, fats
Source of building material	Carbon dioxide	Food
Organisms	Photosynthetic plants Algae Cyanobacteria Photosynthetic bacteria	Animals, protozoa Nonphotosynthetic parts of ordinary plants: wood, roots, bark Completely parasitic plants Fungi Most bacteria

Photoautotrophs must build all of their own molecules using just carbon dioxide, water, and various nitrates, sulfates, and other minerals.

Tremendously important consequences follow from the fact that photoautotrophs and heterotrophs differ in their sources of energy and building material (**Table 10-1**). Sunlight and carbon dioxide do not need to be stalked, hunted, and captured; thus, sensory organs, muscles, and central nervous systems like those of animals are unnecessary. Conversely, the ocean is full of microscopic bits of food, and animals such as sponges and corals can gather it the way plants gather carbon dioxide. The mode of nutrition has had overriding influence on the bodies and metabolisms of plants and animals.

Tissues and organs are also either photoautotrophic or heterotrophic. Chlorophyllous leaves and stems are photoautotrophic, whereas roots, wood, and flowers are heterotrophic and survive on carbohydrates imported through phloem. During winter, if all leaves have abscised, the entire plant may be composed of heterotrophic tissues, and it maintains its metabolism by respiring stored starch.

Tissues often change their type of metabolism; young seedlings are white and heterotrophic while germinating underground; they survive on nutrients stored in cotyledons or endosperm. Seedlings become photoautotrophic only after they emerge into sunlight. Immature fruits may be green and photosynthetic; however, in the last stages of maturation, chloroplasts are converted to chromoplasts, and metabolism then depends on imported or stored nutrients (chromoplasts are plastids that contain large amounts of pigments other than chlorophyll). Young leaf primordia are green, but they grow more rapidly than their own photosynthesis would permit; they have a mixed metabolism of photosynthesis and carbohydrate import.

Photosynthesis is a complex process by which carbon dioxide is converted to carbohydrate. This involves endergonic reactions driven by adenosine triphosphate (ATP) and requiring new bonding orbitals filled by electrons carried to the reaction by reduced nicotinamide adenine dinucleotide phosphate (NADPH). Before this can happen, ATP and NADPH themselves must be formed in highly endergonic reactions driven by light energy. In order to understand this, you must first understand the nature of light and pigments along with the concept of reducing power.

Energy and Reducing Power

Energy Carriers

Energy enters the biological world through photosynthesis, a process that converts light energy to chemical energy. The sun's light is captured by certain plant pigments that use the energy in chemical reactions. Unfortunately, the energized pigments can enter into only two chemical reactions, although plants have thousands of different reactions in their metabolism. Several theoretical ways exist of transporting energy from energized pigments into endergonic reactions:

1. Allow the pigments to enter into every reaction necessary. A problem is that the energized pigments are large molecules; therefore, they are not very mobile and never move across membranes (Figure 10-2). Furthermore, they are too energetic; they can react with almost anything and would be difficult to control.

2. Allow the energized pigments to make one or several smaller, less energetic intermediates that can be moved and controlled easily. Such a method has evolved: Photosynthetic reactions produce ATP, an extraordinary molecule. Its high-energy phosphate bonds carry enough energy to force almost any reaction to proceed, and it can enter into almost every reaction for which energy is needed. In those that it does not enter, other energy carriers, often relatives of ATP, are involved; the most frequent is **guanosine triphosphate**, which also carries high-energy phosphate bonds.

Although ATP is an essential molecule, it constitutes only a tiny fraction of the plant body. Each molecule is recycled and reused repeatedly, thousands of times per second. ATP is converted to ADP and phosphate by metabolic reactions, but the phosphate can be reattached with a high-energy bond by the reactions of either photosynthesis or respiration. Each molecule is an energy carrier, shuttling between reactions that release energy and those that consume it.

There are three methods by which adenosine diphosphate (ADP) can be phosphorylated to ATP (Table 10-2). The first, **photophosphorylation**, involves light energy in photosynthesis; animals, fungi, and nonchlorophyllous plant tissues cannot perform photophosphorylation because they lack the necessary pigments and organelles. Instead, they respire some of the high-energy compounds they have consumed as food or imported by phloem. Compounds with high-energy phosphate groups are produced, and these compounds force their phosphate onto ADP, making ATP. This is **substrate-level phosphorylation**. In the last stages of respiration, ADP is phosphorylated to ATP by **oxidative phosphorylation**. Each process occurs in a distinct site within the cell, and each captures energy from distinct types of exergonic reactions. Photophosphorylation occurs only in chloroplasts in light, but substrate-level and oxidative phosphorylation occur in all parts of the plant at all times.

Reducing Power

Earth's atmosphere is about 21% oxygen; therefore, many compounds are found in their oxidized form: carbon as carbon dioxide (CO_2), sulfur as sulfate (SO_4^{2-}), nitrogen as nitrate (NO_3^-),

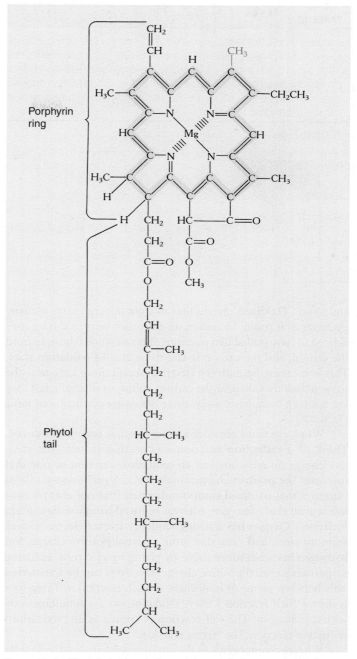

Figure 10-2 The tail of chlorophyll *a* contains only hydrogen and methyl functional groups, so it is hydrophobic and dissolves into the chloroplast's membrane lipids, immobilizing it. The porphyrin ring system of alternating double and single bonds acts as an antenna that captures light energy. The magnesium atom carries the electrons involved in photosynthesis. If the colored methyl group were an aldehyde group (—CHO), the pigment would be chlorophyll *b*.

TABLE 10-2	Methods of Synthesizing ATP	
	Energy Source	Site
Photophosphorylation	Sunlight	Chloroplasts
Substrate level phosphorylation	Reactions not involving oxygen	Cytosol
Oxidative phosphorylation	Oxidations with oxygen	Mitochondria

TABLE 10-3　Calculation of Oxidation States

Calculating the oxidation state is often simple. Always assume that H is $+1$ and O is -2; carbon, sulfur, and nitrogen are variable, because they can add or lose electrons rather easily. Carbon dioxide and carbohydrates are neutral molecules, so the oxidation state of carbon must be opposite and equal to that of all other oxidation states combined. C is $+4$ in carbon dioxide and $+0$ in carbohydrate. During photosynthesis, four electrons must be added to each carbon in carbon dioxide so that the bonding orbitals of carbohydrate can exist; this lowers its oxidation state from $+4$ to 0. In malic acid the five oxygens contribute -10 and the six hydrogens add up to $+6$, so there are still $4+$ charges to be accounted for by the four carbons: $4+$ charges divided by 4 carbons equal one $+$charge per carbon. Carbon in malic acid is at the $+1$ oxidation state and is not as reduced as carbon in carbohydrate.

CO_2 (carbon dioxide)	O -2	O -2	C $+4$	
CH_2O (carbohydrate)	O -2	H $+1$	H $+1$	C 0
COOH | H—C—H　(malic acid) | H—C—OH | COOH	$5 \times O$ $5 \times -2 = -10$	$6 \times H$ $6 \times +1 = +6$	$4 \times C$ $4 \times +1 = +4$	

and so on. "**Oxidized**" means that an atom does not carry as many electrons as it could. In carbon dioxide, each oxygen can be considered to have pulled two electrons almost completely away from the carbon, and the carbon is said to be at a $+4$ **oxidation state**. This is speaking figuratively; electrons spend more time near the oxygen than they do near the carbon, sulfur, or nitrogen, but they are not torn completely away; these bonds are covalent, not ionic (Table 10-3).

When electrons are added to an atom, it becomes **reduced**. Think of a **reduction reaction** as one that reduces the positive charge on an atom and an **oxidation reaction** as one that increases the positive charge (Table 10-4). A preliminary rule of thumb is that **oxidized compounds** often (but not always) contain a great deal of oxygen, whereas **reduced compounds** contain hydrogen. Oxygen has a strong tendency to pull electrons away from an atom and raise that atom's partial positive charge, but hydrogen becomes more stable by giving up electrons, reducing its partner's partial positive charge. Electrons can be transferred only between atoms or molecules, so each reaction in Table 10-4 is only a "half reaction." Every oxidation occurs simultaneously with a reduction. The full reaction is known as an "oxidation-reduction reaction," or "redox reaction."

Whereas compounds in the environment are predominantly in the oxidized state because of our oxygen-rich atmosphere,

TABLE 10-4　Reductions and Oxidations

	Oxidation State
Reductions	
$X + e^- \longrightarrow X^-$	$0 \longrightarrow {}^-1$
$X^{+2} + e^- \longrightarrow X^+$	$+2 \longrightarrow +1$
$X^- + e^- \longrightarrow X^{-2}$	$^-1 \longrightarrow {}^-2$
Oxidations	
$X \longrightarrow X^+ + e^-$	$0 \longrightarrow +1$
$X^+ \longrightarrow X^{+2} + e^-$	$+1 \longrightarrow +2$
$X^{-2} \longrightarrow X^- + e^-$	$^-2 \longrightarrow {}^-1$

most compounds in organisms are in the reduced state. Carbon is often in the form of carbohydrate, where its oxidation state is $+0$; nitrogen is frequently present as an amino group, NH_3 (N^{3-} H^{1+} H^{1+} H^{1+}); and sulfur is present as SH_2 (S^{2-} H^{1+} H^{1+}). Thus, in addition to needing energy, organisms also need **reducing power**, the ability to force electrons onto compounds. Reducing power is especially important to plants because they take in carbon dioxide and water—the most highly oxidized forms of carbon and hydrogen—and then convert them to carbohydrates, fats, and other compounds that are very reduced. Heterotrophs have less of a problem: When they consume plants, they get compounds that have already been reduced. They do need some reducing power, however, when they synthesize compounds that are very reduced, such as fatty acids.

Just as with energy, an optimum solution for moving and handling reducing power—electrons—is to use small molecules that are semistable and mobile. The two molecules used most often are nicotinamide adenine dinucleotide (NAD$^+$; Figure 10-3) and the closely related nicotinamide adenine dinucleotide phosphate (NADP$^+$). Both can pick up a pair of electrons and a proton, thereby becoming reduced to NADH and NADPH. When they reduce a compound by transferring their electrons to it, the proton is released and NAD$^+$ or NADP$^+$ is regenerated. Rather than having a large number of carrier molecules, the cell recycles each molecule, using it thousands of times a second as it shuttles between electron-producing and electron-consuming reactions.

Because NAD$^+$ and NADP$^+$ take electrons away from other molecules, they are **oxidizing agents**—they oxidize the material they react with. It is also possible to say that the material has reduced the NAD$^+$ or NADP$^+$. During the process, NADH and NADPH, two strong **reducing agents**, are formed. They have a powerful tendency to place electrons onto other molecules, reducing those molecules and becoming oxidized themselves (Figure 10-3b). The tendency to accept or donate electrons varies greatly and is known as a molecule's **redox potential** (Table 10-5). Cells contain a variety of electron carriers that differ in their tendency to accept or donate electrons.

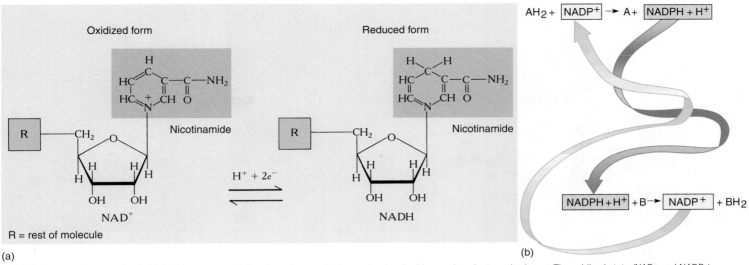

Figure 10-3 (a) The full structure of NAD⁺ is presented elsewhere; here, only the portion involved in carrying electrons is shown. The oxidized state (NAD⁺ and NADP⁺) carries a partial positive charge on the nitrogen and three double bonds in the ring. When it is reduced by two electrons, the positive charge disappears from the nitrogen, numerous bonding orbitals within the ring are changed, and the top carbon picks up a proton. No bonding orbitals are formed between NAD⁺ and the electron donor; therefore, NADH is free to diffuse away after picking up electrons. Similarly, as it donates electrons to some substrate, the ring bonding orbitals revert to the NAD⁺ form, none of which binds it to the substrate, so the NAD⁺ diffuses away. The same is true for NADP⁺. (b) In the upper reaction, two electrons are passed from AH_2 to NADP⁺: AH_2 has been oxidized to A, and NADP⁺ has been reduced to NADPH + H⁺. The NADPH is free to diffuse to another site, where it passes the two electrons onto another molecule, B, reducing it to BH_2 and becoming oxidized back to NADP⁺.

■ Other Electron Carriers

Cytochromes **Cytochromes** are small proteins that contain a cofactor, heme, which holds an iron atom (**Figure 10-4**); the iron carries electrons and cycles between the +2 and +3 oxidation states. Cytochromes are intrinsic membrane proteins; they are an integral part of the chloroplast's thylakoid membranes and

TABLE 10-5	Redox Potentials of Electron Carriers		
Oxidized	Reduced	Number of Electrons Transferred	Redox Potential (volts)
X (Fe₄S₄)	X⁻	1	−0.73
Ferredoxin (ox)	Ferredoxin (red)	1	−0.43
Q	Q⁻	1	−0.35
NADP⁺	NADPH + H⁺	2	−0.32
NAD⁺	NADH + H⁺	2	−0.32
FAD	FADH₂	2	−0.22
FMN	FMNH₂	2	−0.22
Cyt b⁺³	Cyt b⁺²	1	−0.00
Ubiquinone	Ubiquinone-H₂	2	+0.10
Cyt c₁⁺³	Cyt c₁⁺²	1	+0.22
Cyt c⁺³	Cyt c⁺²	1	+0.25
Cyt a⁺³	Cyt a⁺²	1	+0.25
Cyt f⁺³	Cyt f⁺²	1	+0.36
Plastocyanin (ox)	Plastocyanin (red)	1	+0.37
Cyt a₃⁺³	Cyt a₃⁺²	1	+0.38
1/2O₂ + 2H⁺	H₂O	2	+0.82

Compounds with a large negative redox potential (at the top of the list) tend to donate electrons and exist in the oxidized state: X, ferredoxin, NADP1, FAD. Compounds with a large positive redox potential (at the bottom of the list) tend to accept electrons and exist in the reduced form: Cyt a_3^{+2}, H_2O.

cannot be removed without destroying the membrane. Consequently, they carry electrons only between sites that are extremely close together within a membrane rather than diffusing throughout the stroma as NADPH does.

Plastoquinones **Plastoquinones**, like cytochromes, transport electrons over short distances within a membrane (Figure 10-4b). After they pick up two electrons, they also bind two protons. Their long hydrocarbon tail causes them to be hydrophobic, so they dissolve easily into the lipid component of chloroplast membranes.

Plastocyanin Like cytochromes, **plastocyanin** is a small protein that carries electrons on a metal atom—in this case copper. When oxidized, the copper ion is in the +2 oxidation state, but as it picks up the electron, it is reduced one level to the +1 oxidation state. Plastocyanin is loosely associated with chloroplast membranes; it can move a short distance along the surface, but it does not travel far.

■ Photosynthesis

As the name implies, photosynthesis is a process that uses light energy to synthesize something. The term is so general that it could be applied to many different types of reactions, but whenever a botanist uses the term "photosynthesis," the reaction being discussed is the combination of carbon dioxide with water to form carbohydrate (**Figure 10-5**). Think about why these particular compounds are part of photosynthesis. First, both carbon dioxide and water are abundant and cheap, occurring almost everywhere in large quantities. The exception is the lack of water in severe deserts like the Sahara, where very little life exists simply because water is scarce. It is important to have a metabolism based on abundant compounds. It is necessary also for raw materials to be cheap; that is, the plant must be able to obtain them

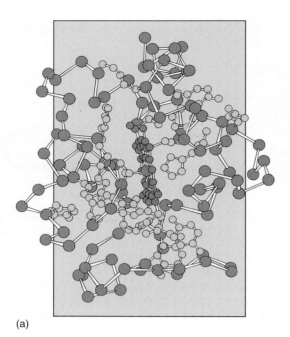

(a)

Plastoquinone A$_{45}$

Plastoquinone C

(b)

Figure 10-4 (a) In cytochromes, iron is not bonded directly to any amino acid but is held by heme, a box-like porphyrin ring similar to the portion of chlorophyll that holds the magnesium ion. (b) The distinguishing feature of the quinone class of electron carriers is that each has two ketone groups (the double-bonded oxygen) whose carbon atoms are part of a ring structure.

without expending much energy. Water and carbon dioxide are excellent because they diffuse into the plants automatically from soil, air, or water.

Another important quality of carbon dioxide and water is that they are very stable and contain little chemical energy, so it is possible to deposit a large amount of energy into them. The carbohydrates they form are a good means of storing energy because all reactions leading to carbohydrate breakdown have high energy-of-activation barriers. Despite being energy rich, carbohydrates are stable and chemically unreactive.

Finally, both the reactants and the products of photosynthesis are nontoxic; it is safe to absorb large quantities of carbon

dioxide and water and to store high concentrations of carbohydrates. Many substances that are critical for life are extraordinarily toxic if they become even slightly concentrated; chlorine, sodium, ammonium, and many vitamins are just a few examples.

During photosynthesis, the carbon of carbon dioxide is reduced and energy is supplied to it, converting it to carbohydrate (Figure 10-5). The carbon atom in carbon dioxide is at the +4 oxidation state, whereas carbon atoms in carbohydrate are, in general, at +0. Four electrons must be found and placed into new bonding orbitals around the carbon atom to reduce it. This is not easy because carbon is more stable in the oxidized state than in the reduced state: Carbohydrates such as wood and sugar can burn, releasing energy, but carbon dioxide does not. Therefore, a source of electrons and a source of energy are necessary for photosynthesis: The electron source is water, and the energy source is light. Water and light, however, do not act on carbon dioxide directly; instead, they create the intermediates ATP and NADPH by a process called the **light-dependent reactions**. In a separate set of reactions, the **stroma reactions** (formerly known as dark reactions), ATP and NADPH interact with carbon dioxide and actually produce carbohydrate (**Figure 10-6**).

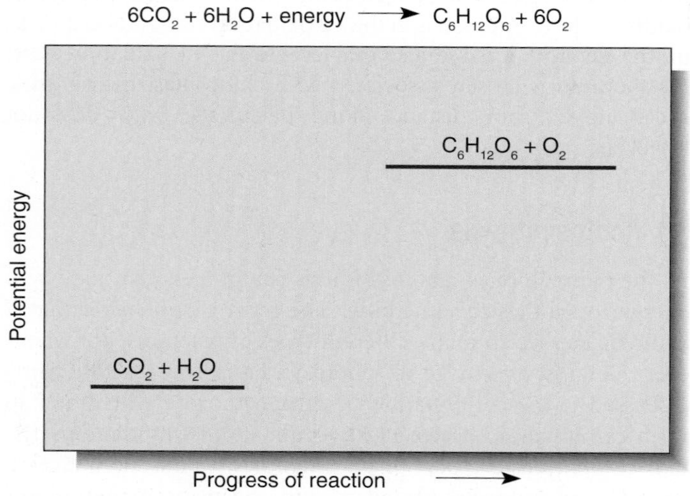

■ **The Light-Dependent Reactions**

The Nature of Light Light is one small segment of the **electromagnetic radiation spectrum**, which encompasses gamma rays, X-rays, ultraviolet light, infrared light, microwaves, and radio waves, in addition to visible light (**Figure 10-7**). Radiation can be thought of and treated physically either as a set of particles called **quanta** (singular, **quantum**), also called **photons**, or as a set of waves. The various types of radiation differ from each other only in their wavelengths and the amounts of energy each individual quantum contains. Short wavelengths (cosmic rays, gamma rays, and ultraviolet light) have relatively large amounts of energy in each quantum, whereas long wavelengths (infrared, microwaves, radar, and radio waves) have relatively little. Because

$$6CO_2 + 6H_2O + energy \longrightarrow C_6H_{12}O_6 + 6O_2$$

C$_6$H$_{12}$O$_6$ + O$_2$

Potential energy

CO$_2$ + H$_2$O

Progress of reaction

Figure 10-5 Although this chemical equation succinctly summarizes photosynthesis, it reveals virtually nothing of the reaction mechanism or the many carriers and enzymes that participate. We cannot draw a reaction diagram because photosynthesis does not occur by the direct interaction of six molecules of carbon dioxide with six of water; however, the relative potential energies can be shown, indicating that this is an endergonic process.

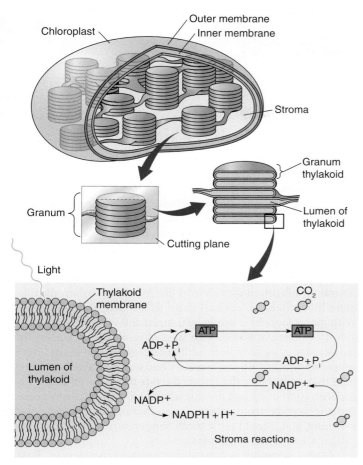

Figure 10-6 Light-dependent reactions of photosynthesis occur by means of membrane-bound carriers, but the actual formation of carbohydrate occurs in the chloroplast liquid (stroma). ATP-ADP and NADP⁺-NADPH diffuse between the two regions. No region of the chloroplast is far from a membrane, so the distances traveled are only a few hundred times the diameter of a molecule.

humans use visible light for vision (that is why it is visible), we are more sensitive to and familiar with this region of the spectrum. We cannot see other wavelengths but we can feel infrared radiation as heat, and our bodies react to ultraviolet light by becoming tanned or burned. We distinguish differences in quantum energy (wavelength) of visible light as differences in color. Most of us see all wavelengths from red (760 nm) through orange, yellow, green, blue, indigo, to violet (390 nm). Certain insects see some near ultraviolet but in general, all animals see in the range from 350 to 760 nm, which is also the radiation that plants use for photosynthesis.

The Nature of Pigments Most materials absorb certain wavelengths more than other wavelengths. If a substance absorbs all wavelengths except red, then red light either bounces off or passes through it, and the substance appears red to us. Any material that absorbs certain wavelengths specifically and therefore has distinctive color is a **pigment**, but we more often think of pigments as substances that absorb light as part of their biological function. Some pigments, such as melanin (the pigment of our skin), absorb light and thereby protect other light-sensitive substances. Other pigments, such as those in flowers, fruits, or the skins of animals, are important for the light they do not absorb, which gives the pigments their color, allowing them to be useful to the organisms in attracting mates, pollinators, or frugivores or in hiding from predators (Figure 10-7b).

Photosynthetic pigments transfer absorbed light energy to electrons that then enter chemical reactions. Because only absorbed energy can be used, a theoretically ideal photosynthetic pigment would be black: It would absorb and use all light, not letting any escape. The pigment should at least absorb high-energy radiation (ultraviolet light and gamma rays) instead of the fairly weak visible light. Rather surprisingly, the critical pigment, **chlorophyll _a_**, is not like this at all (see Figure 10-2). It absorbs only some red and some blue light, letting most of the rest pass

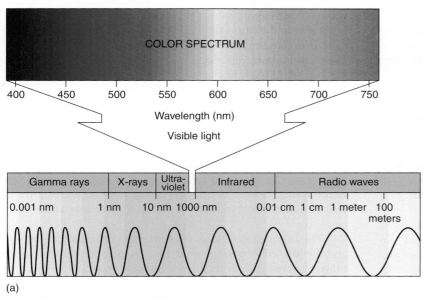

(a)

(b)

Figure 10-7 (a) On the left end of this electromagnetic spectrum are gamma rays. If treated as waves, they have extremely short wavelengths, less than 1 nm. If treated as particles, each quantum is highly energetic. Moving to the right, wavelengths become longer and quanta less energetic. The region to which our eyes respond—the visible region—is enlarged. (b) These apples have an abundance of pigments that absorb all colors except red. Earlier, while still developing, they had the pigment chlorophyll, which absorbs red and blue strongly, but reflects green.

Plants and People

Box 10-1 Photosynthesis, Global Warming, and Global Climate Change

Our atmosphere is critically important to life on Earth; ironically, its composition is the product of that very life. The free oxygen (O_2) we breathe is produced solely by oxygenic photosynthesis; there is no other source. Oxygenic photosynthesis originated 2.8 billion years ago: We know this because for millions of years the newly produced oxygen reacted with iron, forming a worldwide stratum of rust in ancient rocks. After all iron had been oxidized, free oxygen began accumulating in the atmosphere, and its concentration has been increasing ever since. Simultaneously, photosynthesis pulled carbon dioxide out of the atmosphere, converting it first to 3-phosphoglyceraldehyde and then to all of the other organic compounds that exist. Every single organic molecule started out as carbon dioxide snared by RuBP carboxylase. Most organic molecules are digested and respired by aerobic organisms, a process that returns carbon dioxide back to the atmosphere rather quickly. But millions of tons of trees have died and fallen into stagnant swamps where a lack of oxygen prevented decay: The carbon in their wood—all of the cellulose, hemicellulose, and lignin—was converted to coal and did not return to the atmosphere. Petroleum also is probably derived from photosynthetically fixed carbon dioxide. The point is that respiration does not release all carbon back to the atmosphere; therefore, photosynthesis is gradually causing carbon dioxide concentration in the air to decrease.

Three groups of organisms have had especially important impacts on atmospheric carbon dioxide: coccoliths, mollusks, and humans. Coccoliths are microscopic algae that build shells of calcium carbonate, as do mollusks. When they die, their shells and the carbon they contain sink to the bottom of the ocean and decompose only slowly. All limestone and vast carbonate deposits on the ocean floor represent millions of tons of carbon dioxide removed from the atmosphere by clams, barnacles, and unimaginable numbers of algae.

We humans were no different from any other aerobic organism until we made a fateful discovery: how to use fire. Since then, we not only oxidize food in our mitochondria, but we also oxidize wood, coal, oil, and gas, putting carbon dioxide back into the atmosphere and raising its concentration measurably.

Why does the concentration of atmospheric carbon dioxide matter? Think of carbon dioxide as a pigment; its absorption spectrum is low for visible light but high for infrared wavelengths. Visible light from the sun passes easily through the atmosphere: It is not absorbed by nitrogen, oxygen, or carbon dioxide. As it strikes Earth's surface, some is reflected immediately back out into space, and a small amount is absorbed by biological pigments such as chlorophyll in leaves or rhodopsin in eyes, where it powers photosynthesis or vision; however, most visible light has no effect other than to warm rocks, soil, and water, causing them to radiate the extra energy away as long-wavelength infrared light. Many of these infrared quanta pass directly back through the atmosphere without hitting a carbon dioxide molecule because the concentration of carbon dioxide is so low (0.03% of air), but many quanta are absorbed by atmospheric carbon dioxide molecules, causing them to become warmer. This energy is trapped in the Earth/atmosphere system and warms our world. This is called the **greenhouse effect** because the glass in greenhouses works the same way, as does the glass in a parked car. Carbon dioxide is a **greenhouse gas**.

An important balance exists between the atmospheric concentration of carbon dioxide and life: With less carbon dioxide, more heat would be lost and Earth would be frozen, like Mars. With more, more heat would be trapped and our world would be as hot as Venus, at 800°C, with lakes of molten lead. During the industrial age, we have been adding carbon dioxide to the atmosphere by burning oil, gas, and coal, and we have destroyed forest trees that can remove the carbon dioxide by photosynthesis. The concentration of carbon dioxide is increasing in the atmosphere, and the average temperature is also increasing. This is **global warming**, and it could cause mean temperatures to be 2°C or 3°C (3°F or 4°F) warmer in the next century.

Global warming is having numerous consequences. First, surface water of the oceans is becoming warmer; therefore, more water evaporates into the air. Much of our weather in North America comes as winds blow eastward across the North Pacific. The water is cold and the air picks up only enough moisture to keep the Pacific Northwest wet; by the time it moves to the Central Plains states, it has so little moisture left that only grasses, not forests, thrive. But as surface waters of the Pacific become slightly warmer, vastly more moisture will evaporate into the wind and be carried to the Mississippi drainage basin. This increased rainfall could cause much better farming conditions in the Central Plains, and catastrophic flooding in most river valleys where cities are located. El Niño years show the gigantic flooding that results from slight warming in just one area of an ocean.

Global warming is also causing rapid melting of snow and glaciers in mountains and of ice caps in the Arctic and Antarctic. It is difficult to comprehend, but Antarctica is a large continent covered by ice 1 to 2 miles thick. As the world's ice fields melt, that water is added to oceans, increasing their volume and causing sea level to rise. Sea levels are rising even more because as ocean water warms, it expands. Coastal cities will be flooded, and so will coastal wetlands where ducks, geese, and hundreds of other species live.

The additional freshwater flowing from melting ice caps and flooding rivers has other impacts. Freshwater is lighter, more buoyant than seawater. It spreads outward from a river's mouth as a cap, only gradually mixing with seawater as waves agitate the two. With more flooding, this cap becomes more extensive. Being fresh, marine algae cannot live in this layer, so they are forced into lower, darker waters where they cannot photosynthesize so well. The mixing of freshwater with seawater dilutes the nutrients in

seawater, also slowing the growth of algae, and algae are of course the basis of the entire food web in the oceans: All ocean life is ultimately dependent on algae.

Another consequence of global warming is that not all areas are being affected equally. Wind patterns are being changed such that certain areas are becoming warmer, others cooler, some wetter, and others drier. This is **global climate change**; if circulation of ocean currents like the Gulf Stream is altered, climates in vast areas of the world will change dramatically. At what point will we decide that weather patterns have truly changed and that a particular city should be abandoned because of frequent flooding? Some agricultural areas will become too dry to farm, causing at least personal misery as crops fail, perhaps causing massive starvation. New areas will become optimal for farming, but it will take a great deal of confidence to be certain that the new weather patterns are stable enough for people to risk starting again in an unknown area. Millions of lives will be (are being) disrupted, some suddenly and catastrophically in floods and others slowly and inexorably as conditions decline. In addition to the problems global climate change causes us, we are also forcing it onto all of the other organisms with which we share the planet and which bear no responsibility for the damage we are doing.

The Kyoto Protocol is a treaty designed to reduce production of greenhouse gases. Signed by 166 countries, it went into effect in 1994, but the United States—the largest producer of carbon dioxide—did not sign the treaty. Some people contend that if we do not act decisively, greenhouse gases will build up rapidly. Others point out that China and India—with combined populations of almost 2 billion people—are modernizing so rapidly that soon they will be producing more carbon dioxide than the United States is. They conclude that greenhouse gases will be produced as long as coal and oil are available and people need energy.

One human interaction with the atmosphere seems to be going well. The atmosphere protects us from harmful radiation from space by absorbing high-energy cosmic rays, X-rays, and ultraviolet light; very little of these wavelengths reach ground level. If not for water in the atmosphere, the first two would kill us outright. Ultraviolet light can induce mutations in our DNA and cause skin cancer and blindness, but atmospheric oxygen (produced by oxygenic photosynthesis) weakly blocks some ultraviolet light; fortunately, a by-product of its absorbing high-energy quanta is the conversion of oxygen to **ozone** ($3O_2 \rightarrow 2O_3$), which is even more effective at blocking ultraviolet light. Man-made chlorofluorocarbons from air conditioners and cans of hairspray and deodorant, among other products, had been escaping into the atmosphere and destroying ozone, thus increasing the amount of ultraviolet radiation reaching us. Effort has been made to eliminate production of chlorofluorocarbons worldwide, and its levels in the atmosphere are dropping, thus allowing ozone to accumulate again and continue protecting us. This one success story shows that we have the power to protect the environment, the environment that protects us.

Figure B10-1a After this tree is killed by the chain saw, it will stop photosynthesizing and will stop removing carbon dioxide from the atmosphere. Even if converted to lumber and used as part of a house, it will eventually decay or rot or be burned, being oxidized back to carbon dioxide and increasing the amount of CO_2 in the air.

Figure B10-1b Tropical rain forests are not the only forests being damaged. Clear-cutting like this is common in the western United States and Canada.

through, especially high-energy radiation. Why does chlorophyll *a* lack what seem like ideal characteristics?

First, chlorophyll, like all other biological pigments, does not use high-energy quanta because they have too much energy. Each is so powerful that it would knock electrons completely away from the pigment, disrupting bonding orbitals and causing the molecule to break apart. Notice in Figure 10-2 that all bonds in the chlorophyll ring system are double bonds that alternate with single bonds (conjugated double bonds); this bond system is excellent for absorbing quanta, but if even a single electron is knocked out, the entire structure becomes useless. It is selectively disadvantageous for a plant to produce a photosynthetic pigment that is destroyed by the light it absorbs; the molecule would break down, and all of the ATP that had been expended earlier in its construction would be wasted. Fortunately for our own molecules, the atmosphere's ozone layer protects us by absorbing high-energy radiation.

Long-wavelength radiations, such as infrared and microwaves, have so little energy per quantum that they cannot appreciably boost an electron's energy. Instead, they make the pigment molecule warmer, as in a microwave oven, but this is not especially useful for chemical synthesis. Visible light contains just the right amount of energy per quantum. When one of these quanta is absorbed by the pigment, an electron is **activated**—raised to an orbital of a higher energy level. We say that the electron and

the molecule go from the **ground state** to an **excited state** (Figure 10-8). Under the right conditions, this high-energy, excited electron can be used in chemical reactions. If it is not used, it returns to its original, stable ground orbital by emitting a new quantum of light, one with less energy and a longer wavelength than the one that it absorbed. The release of light by a pigment is called **fluorescence** (Figure 10-8c).

Two of the most useful pieces of information about a photochemical process are its action spectrum and the absorption spectrum of its pigment (**Figure 10-9**). An **absorption spectrum** is a graph that shows which wavelengths are most strongly absorbed by a pigment, whereas an **action spectrum** shows which wavelengths are most effective at powering a photochemical process. To initiate a photochemical process, light must first be absorbed; therefore, the action spectrum of a process must match the absorption spectrum of the pigments responsible. The absorption spectrum of chlorophyll *a* shows that it absorbs red light (especially 660 nm) and blue light (440 nm) very well and other wavelengths only slightly. It would be better if it could absorb a greater number of wavelengths, but it simply does not. Chlorophyll *a* is the

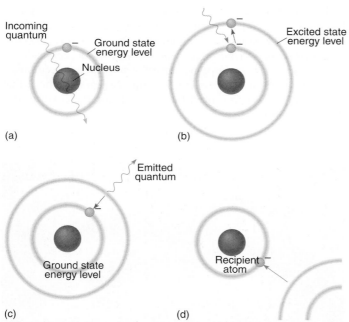

(a) Incoming quantum / Ground state energy level / Nucleus

(b) Excited state energy level

(c) Ground state energy level / Emitted quantum

(d) Recipient atom

Figure 10-8 (a) If a quantum has the wrong wavelength for a pigment, it passes through the pigment's bonding orbitals without being absorbed. Chlorophyll looks green because most green light passes through it. (b) If the quantum has the correct wavelength, the correct amount of energy, it is absorbed, and the electron must move to a new orbital whose energy level corresponds to the electron's new energy load. (c) The excited state is unstable; it may stabilize itself by having the electron emit enough energy (fluoresce) to drop back to its original ground state energy level. (d) The electron can also be stabilized by moving to a more stable orbital on an entirely different atom. This is the critical process in photosynthesis; without this step, life would not exist.

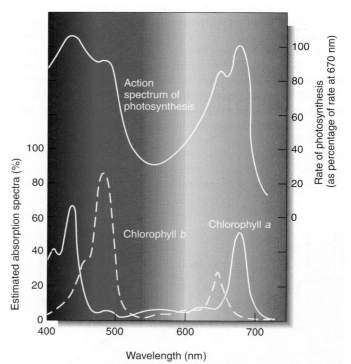

Figure 10-9 The absorption spectra of chlorophyll *a* and chlorophyll *b* and the action spectrum of photosynthesis. On the bottom axis is the wavelength of light, with short (blue) wavelengths to the left and long (red) ones to the right. The vertical axis of the absorption spectra is the amount of light absorbed by the pigment (scale on left); for the action spectrum, it is the amount of photosynthesis carried out (scale on right). Chlorophylls absorb little of the very short wavelength light at 400 nm, and little photosynthesis occurs; however, light at slightly longer wavelengths, about 425 nm, is absorbed well by chlorophyll *a*, and photosynthesis proceeds. Quanta with intermediate wavelengths pass right through the pigment, and photosynthesis is low, but in the 650 to 680 nm range (red), considerable absorption occurs. Because the absorption spectra of chlorophyll *a* and *b* differ, more wavelengths are harvested. If the two matched perfectly, chlorophyll *b* would be useless.

essential photosynthetic pigment in all plants, algae, and cyanobacteria and has existed unaltered by evolution for about 3 billion years. To put this in perspective, the entire Milky Way Galaxy takes 250 million years to make one rotation about its center, so in 12 full rotations of our galaxy, no alteration in the structure of chlorophyll *a* has been selectively advantageous.

Accessory pigments are molecules that strongly absorb wavelengths not absorbed by chlorophyll *a*. The absorbed energy is passed on to chlorophyll *a*. In effect, the accessory pigments overcome the narrow absorption of chlorophyll *a* and broaden the action spectrum of photosynthesis. We know that accessory pigments are involved because the action spectrum of photosynthesis does not perfectly match the absorption spectrum of chlorophyll *a*. The most common accessory pigments in land plants are *chlorophyll b* and the *carotenoids* (**Figure 10-10**); algae have other types. Chlorophyll *a* and chlorophyll *b* are large, flat molecules with almost identical porphyrin ring structures. Their phytol tails are hydrophobic and dissolve into the lipid portion of the thylakoid membrane. When packed tightly in a membrane by their phytol tails, their porphyrin rings lie parallel to each other, which causes the electron orbitals of one molecule to interact with those of the two adjacent molecules. Hundreds of chlorophylls act somewhat like one molecule, and the energy absorbed by one can be rapidly transferred to another in a different part of the complex. This transfer, called **resonance**, allows chlorophyll *b* to absorb wavelengths that chlorophyll *a* would miss and then to transfer the energy to chlorophyll *a* for use in chemical reactions. Carotenoids are poor at this type of resonance and transfer only approximately 10% of their energy; they seem to be more important in absorbing excessive light and thus protecting chlorophylls.

Photosystem I When light excites an electron in chlorophyll *a*, the electron is so unstable that it either reacts with almost anything or fluoresces, wasting its energy. Chlorophyll molecules and the electron carriers involved in photosynthesis must therefore be tightly bound to the thylakoid membranes of the chloroplast. Chlorophyll's electron must react only with the proper molecule, which should be close enough for the reaction to take place instantly, before fluorescence can occur. All pigments and carriers that work together are packed into a granule called a **photosynthetic unit**, and the thylakoid membranes are filled with millions

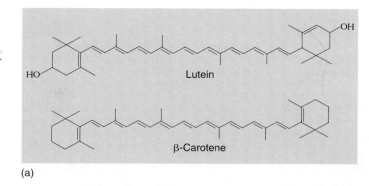

(a)

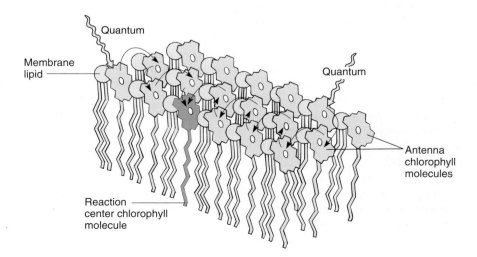
(b)

Figure 10-10   (a) There are two types of carotenoids: Carotenes lack oxygen, but xanthophylls (such as lutein) have it. Both are accessory pigments that protect the chlorophyll from excess sunlight. (b) Carotenoids are always present in leaves, but usually the abundant chlorophyll masks their presence. In autumn, chlorophyll breaks down, and we can see the carotenoids.

of these granular arrays (**Figure 10-11**). Each photosynthetic unit contains about 300 molecules of chlorophylls *a* and *b* and carotenoids. In some photosynthetic units, chlorophyll *b* is plentiful, and in others, it is less abundant. Those with little chlorophyll *b* have been named **photosystem I**; those in which chlorophyll *b* is present at levels almost equal to *a* are **photosystem II**. The photosystem I units are involved in the following reactions.

Figure 10-11   Only special chlorophyll *a* molecules—reaction centers—undergo the initial photochemical reaction of photosynthesis, but they are surrounded by other chlorophyll *a* molecules as well as accessory pigments; regardless of which pigment absorbs light, the energy is transferred to the reaction center.

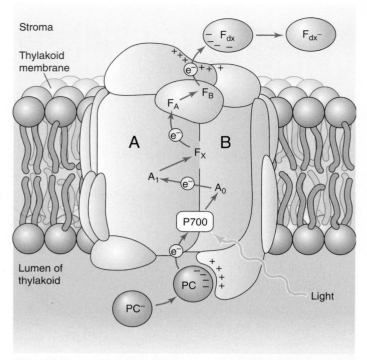

Figure 10-12 Reaction center of photosystem I. A and B are two major proteins that bind chlorophyll; many other components are clustered around these proteins. PC = plastocyanin; F_{dx} = ferredoxin; A_0, A_1, F_X, F_A, and F_B are electron carriers.

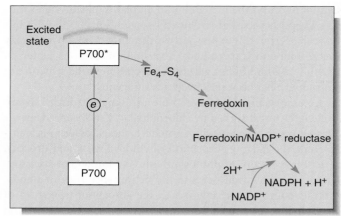

Figure 10-13 In photosystem I, energy is absorbed by a pair of P700 chlorophyll *a* molecules, raising two electrons to an excited energy level; from here, they pass onto Fe_4–S_4 ("F_X"), then onto ferredoxin, and finally onto ferredoxin-NADP reductase. After two electrons have reduced ferredoxin-NADP reductase, they are transferred simultaneously to $NADP^+$, reducing it to $NADPH + H^+$.

When light strikes any pigment of a photosystem I array, either chlorophyll *a* or an accessory pigment, the energy is transferred to the **reaction center**, a structure that contains a pair of special molecules of chlorophyll *a* whose properties differ from all other molecules of chlorophyll *a* in the unit (**Figure 10-12**). This pair of chlorophylls of the photosystem I reaction center is given the special name **P700** because they absorb red light of 700 nm most efficiently. The energy excites an electron of P700, which is then absorbed by a membrane-bound electron acceptor known as "F_X." This is a transfer of an electron; no bonding orbital is formed. The exact chemical nature of F_X is not known, but it contains iron and sulfur and is sometimes designated Fe_4S_4. When X absorbs an electron from P700, it becomes a powerful reducing agent, with a redox potential of -0.73 volts (Table 10-5). The transferred electron is still extremely unstable, and the reduced F_X immediately passes it onto **ferredoxin**, which is also located in the thylakoid membrane (**Figure 10-13**). Ferredoxin is a small protein (10,500 to 11,000 daltons; a dalton is the weight of one hydrogen atom) with an active site consisting of two iron atoms bound to two sulfur atoms. Reduced ferredoxin is also a strong reducing agent, with a redox potential of -0.43 volts. Electrons are passed from ferredoxin to an enzyme, **ferredoxin-NADP$^+$ reductase**, which then reduces $NADP^+$, converting it to NADPH, as its name indicates. Ferredoxin carries only one electron, but two are needed simultaneously to reduce $NADP^+$. Ferredoxin-NADP reductase carries two electrons, but it can be reduced one electron at a time; then it transfers those two electrons together to $NADP^+$. Although NADPH is also a strong reducing agent, it is stable enough to move away from the membrane safely without the risk of reducing things indiscriminately, as the previous electron carriers might.

Photosystem II Photosystem I efficiently produces NADPH, but the reaction center P700 chlorophyll *a* loses electrons during the process. In this oxidized state, bonding orbitals could easily rearrange, causing the molecule to break down and be destroyed. There must be a mechanism that adds electrons back to the P700, reducing it so that it can work repeatedly.

The mechanism that reduces P700 is photosystem II (**Figure 10-14**). Photosystem II can be best described by working backward from photosystem I: A molecule of **plastocyanin**, which contains copper, donates an electron to the chlorophyll *a* of the photosystem I reaction center. The plastocyanin is now oxidized, lacking an electron; it must reacquire one because it also is too expensive a molecule to donate just one electron and then never work again. It receives its new electron from a complex of cytochrome molecules, called the **cytochrome b6/f complex**, which in turn gets an electron from a molecule of plastoquinone. This receives electrons from another carrier, **Q**, a molecule of quinone, which in turn receives electrons from **phaeophytin**. Phaeophytin is actually a chlorophyll *a* molecule that does not contain a magnesium atom. Phaeophytin becomes oxidized as it donates an electron to Q, so it must obtain another electron, which it does when a chlorophyll *a* molecule absorbs light and is activated. This is a different chlorophyll *a* from the one in photosystem I; it is the reaction center of photosystem II and has the name **P680**.

We may seem to be going in circles: taking an electron from one chlorophyll *a*, P680, to pass it onto another chlorophyll *a*, P700, which then sends it to $NADP^+$, but the physical differences between the two molecules of chlorophyll *a* are crucial. The one in photosystem II gets new electrons from water, not plastocyanin. The important thing is that water is cheap enough to just throw away after the electrons are removed. Water breaks down into protons (H^+), which the plant uses, and oxygen (O_2), which it discards. Whereas all electron carriers are large, expensive molecules that the plant must construct itself, water is simply brought in. The electrons are stripped off, the protons are used, and the oxygen is discarded through stomata.

Figure 10-14 The two photosystems work together to transfer electrons from water to NADPH. The scale on the left indicates reducing power, the redox potential (see Table 10-5). The higher a molecule is in the chart, the greater its capacity to force electrons onto another molecule. Because of its shape, this diagram is called a **Z scheme**.

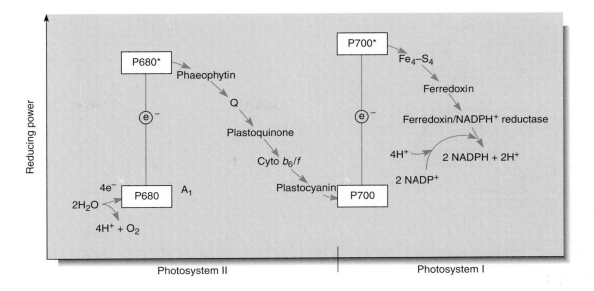

Photosystems I and II together are an efficient system. Electrons are passed from water to P680 in photosystem II, their energy is boosted by light, and then they move through an **electron transport chain**—the various electron carriers—to P700 in photosystem I. Their energy is boosted by light again, and they pass through a short second electron transport chain to $NADP^+$, reducing it to NADPH. This last step requires that protons be added to $NADP^+$; these protons are present in the water surrounding the membrane (water is always a mixture of H_2O, H^+, and OH^-). It would be simpler if photosystem I could receive electrons directly from water, but that does not happen. Besides, the electron transport chain between P680 and P700 is necessary for the production of ATP.

The Synthesis of ATP The light-dependent reactions produce the reducing agent NADPH that actually places electrons onto the carbon of carbon dioxide in the stroma reactions, but the stroma

reactions are highly endergonic and must be driven by being coupled to the exergonic splitting of ATP. The necessary ATP is also generated by the light reactions, but the process is indirect. It is photophosphorylation because light is involved, but a more specific name is often used: **chemiosmotic phosphorylation**. To understand it, we must take a closer look at the structure of chloroplasts. The inner membrane of chloroplasts folds inward, forming flattened sacs called thylakoids (**Figure 10-15**). In certain regions, these swell slightly and form rounded vesicles. All thylakoids in one region form vesicles at the same spot, so they occur in sets called **grana** (singular: granum). Thylakoids that lie between grana are **frets**. The liquid surrounding the thylakoid system is the **stroma**, but notice especially that there is another compartment, the **thylakoid lumen.**

The thylakoid lumen is a critically important compartment because some of the enzymes and electron carriers of the photosystems are embedded in the membrane layer facing the lumen, whereas other enzymes are in the membrane layer facing the stroma (**Figure 10-16**). Reactions that break down water and produce oxygen and protons are located on the lumen side of the thylakoid membrane, in the granum areas. This membrane is not permeable to protons; therefore, as light reactions run, the thylakoid interior accumulates protons, and their concentration increases. The molecules of ferredoxin-NADP reductase that generate NADPH are located on the other side of the membrane, facing the stroma. The protons they attach to $NADP^+$ are those present as a result of the natural breakdown of water: $H_2O \longrightarrow H^+ + OH^-$. As protons are absorbed, their concentration in the stroma decreases. Furthermore, during electron transport between P680 and P700, the electron carrier plastoquinone moves a proton from the stroma to the thylakoid lumen every time it carries an electron between phaeophytin and the cytochrome b6/f complex. This also contributes to the increased concentration of protons in the thylakoid lumen and to the decreased concentration of protons in the stroma.

The strong difference between the concentrations of protons inside the thylakoid lumen and exterior to it in the stroma quickly becomes so powerful that protons begin to flow out of the

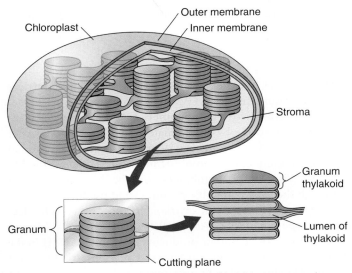

Figure 10-15 Grana are stacks of small thylakoid vesicles compressed together; frets are regions of thylakoid that connect one granum to another. The lumen of the thylakoid region is continuous with that of the fret region. The liquid surrounding all of the thylakoids is stroma.

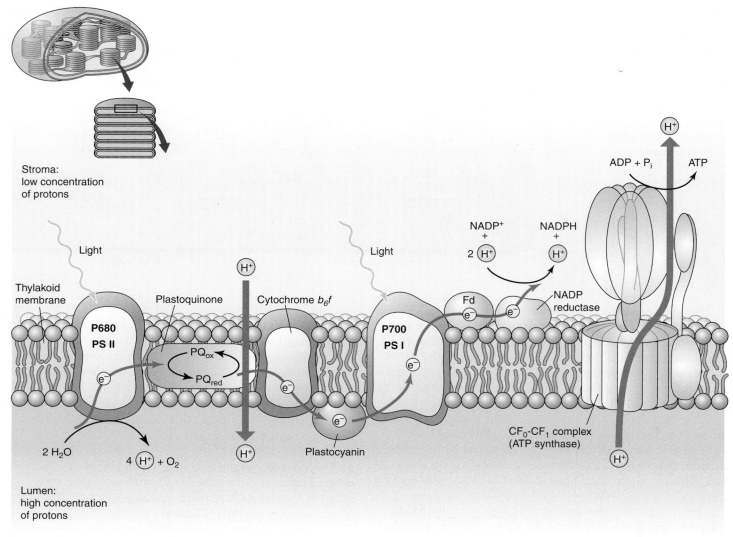

Figure 10-16 The water-splitting, proton-producing reactions of photosystem II take place on the lumen side of the thylakoid membrane. Plastoquinone is like NADP$^+$ in that when it picks up electrons it also picks up a proton. This occurs on the stroma side of the membrane, but the reduced plastoquinone must diffuse to the other side of the membrane to pass electrons on to the cytochrome b6/f complex. The proton then dissociates and is deposited in the lumen, adding to the growing pool of protons. When NADPH is formed, it picks up protons from the stroma. This and the plastoquinone pumping result in a deficiency of protons in the stroma. Protons return to the stroma by passing through ATP synthetases; their passage is exergonic and powers the phosphorylation of ADP to ATP.

lumen through special channels in the membrane. These channels are complex sets of enzymes that can synthesize ATP from ADP and phosphate; the entire complex is called **ATP synthetase** (Figure 10-16). The ATP synthetase of chloroplasts is known specifically as the **CF$_0$-CF$_1$ complex**. CF$_0$ is the portion of the enzyme spanning the membrane where the actual proton channel is located. CF$_1$ is the portion of the enzyme that phosphorylates ADP to ATP. The power required to force phosphate onto ADP and establish the high-energy bonding orbitals of ATP comes from the flow of protons through the ATP synthetase channels. In a car battery, electron flow through wires powers the starter; in chloroplasts, proton flow through ATP synthetase channels powers phosphorylation of ADP to ATP.

When electrons flow smoothly from water to NADPH, the process is called **noncyclic electron transport** (see Figure 10-14

and Figure 10-17). The chemiosmotic potential that builds up does not produce quite enough ATP for the stroma reactions: There is too little ATP relative to the amount of NADPH produced. This problem is overcome by an alternate route for electrons. After they reach ferredoxin in photosystem I, they can be transferred to the plastoquinones of photosystem II instead of being used to make NADPH. The plastoquinones carry the electrons along just as though they had gotten them from Q and use their energy to pump more protons into the thylakoid lumen. This is **cyclic electron transport**, and with it, chloroplasts make extra ATP without making extra NADPH, thus producing ATP and NADPH in the proper ratios for the stroma reactions. Cyclic electron transport is a simple light-powered proton pump, and similar types occur in bacteria. This may have been the original power system that evolved first, billions of years ago.

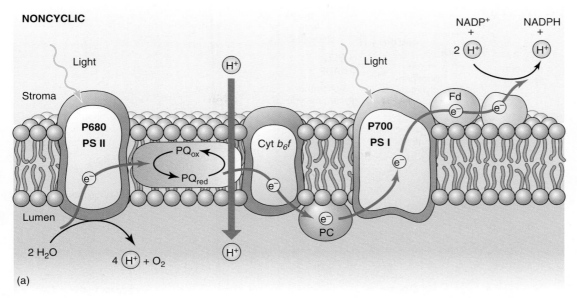

NONCYCLIC

Light

Stroma

P680 PS II

PQ$_{ox}$

PQ$_{red}$

Cyt b_6f

Light

P700 PS I

Fd

e$^-$

e$^-$

NADP$^+$ + 2 (H$^+$)

NADPH + (H$^+$)

H$^+$

e$^-$

e$^-$

e$^-$

e$^-$

PC

Lumen

2 H$_2$O

4 (H$^+$) + O$_2$

H$^+$

(a)

Figure 10-17 (a) In noncyclic electron transport, electrons flow through the Z scheme from water to NADPH. (b) Cyclic electron transport is much simpler: Electrons flow from P700 to plastoquinone, which carries a proton to the lumen and returns the electron to P700.

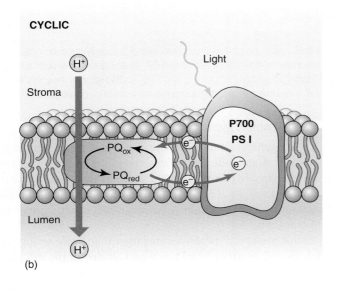

CYCLIC

H$^+$

Light

Stroma

P700 PS I

PQ$_{ox}$

PQ$_{red}$

e$^-$

e$^-$

e$^-$

e$^-$

Lumen

H$^+$

(b)

■ The Stroma Reactions

Conversion of carbon dioxide to carbohydrate occurs in the stroma reactions, also called the **Calvin/Benson cycle**, or the **C$_3$ cycle** (Figure 10-18). These reactions take place in the stroma, mediated by enzymes that are not bound to thylakoid membranes (see Figure 10-6). In the first step, an **acceptor molecule** (**ribulose-1,5-bisphosphate; RuBP**) reacts with a molecule of carbon dioxide. Because RuBP contains five carbons and one more is added from carbon dioxide, you might expect a product that contains six carbons. However, the new molecule breaks apart immediately, while still on the enzyme; stable bonding orbitals cannot be formed between all six carbon atoms while so many oxygen atoms are present and pulling electrons to themselves. Instead, orbitals rearrange and two identical molecules are formed that each contains three carbons: **3-phosphoglycerate**, hence the name C$_3$ cycle. The abbreviation PGA is often used for 3-phosphoglycerate.

The enzyme that carries out this reaction has many names; the most common is **RuBP carboxylase** (**RUBISCO**). This is one of the largest and most complex enzymes known—a giant complex of two kinds of protein subunits. There are eight copies of a small protein, each with a molecular weight of 14,000 to 15,000 daltons, and eight copies of a large protein, each with a molecular weight of 53,000 to 55,000 daltons. The entire enzyme has a molecular weight of about 480,000 daltons. Not only is the tertiary structure of each protein subunit important, but their quaternary structure as a complex is critical: When all eight subunits are properly assembled, the large subunits form the functional active site. RuBP carboxylase can constitute up to 30% of the protein in a leaf, making it the most abundant protein on Earth. Without it, there would be almost no life at all; all photosynthesis that produces oxygen is mediated by this enzyme. A few photosynthetic bacteria use an enzyme composed of just large subunits, lacking any small ones. RuBP carboxylase is crucial to the production of food; without it, heterotrophs would starve.

Like chlorophyll *a*, RuBP carboxylase is by no means ideal. Its active site recognizes and binds to carbon dioxide only poorly, and it has low substrate specificity, frequently putting oxygen rather than carbon dioxide onto RuBP. Yet this enzyme is highly conserved evolutionarily. The amino acid sequences of RuBP carboxylase from all plants are virtually identical. Apparently, all mutations that cause any change in structure, however slight, disturb the active sites and are selectively disadvantageous.

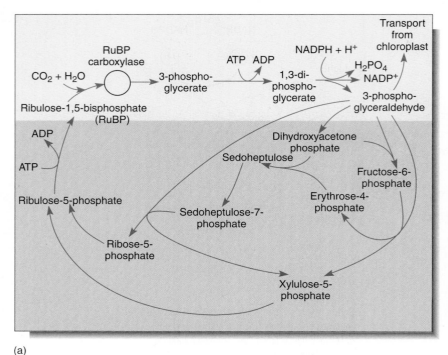

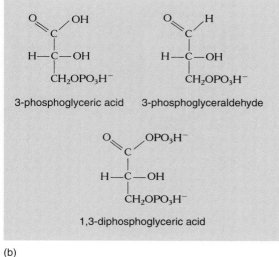

Figure 10-18 (a) In the yellow area are the first steps of the stroma reactions, also known as the C₃ cycle; the product is two molecules of 3-phosphoglyceraldehyde. Some of this is transported out of the chloroplast, and the rest undergoes reactions (blue area) that form a new molecule of the acceptor, RuBP. (b) At various times, acids such as phosphoglycerate and malate are written as phosphoglyceric acid and malic acid; the "-ic acid" ending refers to the whole acid. The "-ate" ending refers to the acid's anion, the negatively charged portion left after the proton dissociates. In protoplasm, most of the acids occur as free anions, not intact neutral acids still holding their protons.

It is important to realize that the first step of the stroma reactions is carboxylation only. Electrons and energy are added in the next two steps: ATP donates a high-energy phosphate group to the 3-phosphoglycerate, converting it to **1,3-diphosphoglycerate**, which then is reduced by NADPH to **3-phosphoglyceraldehyde** (PGAL); a phosphate comes off in this step also. The carbon is now both reduced and energized.

The rest of the stroma reactions are complex, but the important point is that as they operate some 3-phosphoglyceraldehyde can be taken out of the chloroplast and used by the cell to build sugars, fats, amino acids, nucleic acids—basically anything the plant needs. The rest of the PGAL remains in the chloroplast and undergoes several more stroma reactions, which convert it to RuBP, the original acceptor molecule. The principle involved is important: To incorporate carbon dioxide, the plant needs the acceptor RuBP, and the two react on a one-to-one basis. To assimilate large amounts of carbon dioxide, the plant either needs large amounts of RuBP or needs to use a few RuBP molecules repeatedly. Plants use the second strategy. As 3-phosphoglyceraldehyde is formed, some of it is reconverted to RuBP by the rest of the stroma reactions, and some is exported to the cytoplasm. The chloroplast does not need to import quantities of RuBP from the rest of the cell; it just recycles the small amount that it has. Imagine a chloroplast that has 1,000,000 carbon atoms inside it as the various intermediates of the stroma reactions; after three carbon dioxides have been assimilated, there are 1,000,003 carbon atoms. When one molecule of 3-phosphoglyceraldehyde is exported, the carbon pool returns to 1,000,000, and a steady state is maintained. In very young leaves with growing chloroplasts, little or no 3-phosphoglyceraldehyde is exported; it is retained, and the pools of C₃ metabolites increase in numbers of molecules.

■ Anabolic Metabolism

3-Phosphoglyceraldehyde is an amazingly versatile molecule: Using it plus water, nitrates, sulfates, and minerals, plants construct everything inside themselves. The entire fabric of the organism can be synthesized. This is also the basis of all animal metabolism because animals either eat plants or eat other animals that eat plants.

Most biological molecules are larger than 3-phosphoglyceraldehyde so it must be rearranged and altered in the cytoplasm to build up larger, more complex molecules. This constructive metabolism is called **anabolism**, and it consists of **anabolic reactions**.

Anabolic pathways are numerous, but two are especially important with regard to energy metabolism: the synthetic pathways of polysaccharides and fats, which are storage forms of energy and carbon. The NADPH and ATP produced by photosynthesis are excellent sources of energy, but they cannot be stored for even a short time. They are so reactive and unstable that they would break down. A plant cannot stockpile them to survive times when photosynthesis is impossible nor can they be transported over long distances; therefore, even if leaves had an abundant supply, roots would starve.

Several types of storage compounds have evolved that solve these problems.

1. Short-term storage: ATP and NADPH can be used within the cell and last only briefly.
2. Intermediate-term storage: The simple sugar glucose and the disaccharide sucrose are stable enough to be moved from cell to cell, either in the vascular tissue of a plant or in a blood

stream. They are also sufficiently stable to last for weeks or months. A problem with storing large quantities of monosaccharide or disaccharide is that they cause cells to absorb water by osmosis.

3. Long-term storage: Starch is a large, high-molecular-weight polymer of glucose, too large to be transported. It is even more stable than glucose, lasts for years, and does not cause

the cell to absorb water. Lipids are an even more concentrated storage form of energy that can be synthesized rapidly and stored in large quantities.

The Synthesis of Polysaccharides The anabolic synthesis of glucose is **gluconeogenesis** (Figure 10-19). In reactions similar to those of C_3 metabolism, part of the PGAL exported to the cytoplasm is

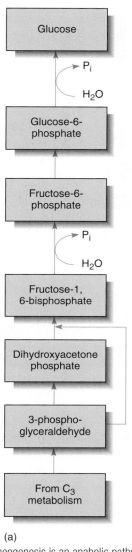

(a)

(b)

(c)

(d)

Figure 10-19   (a) Gluconeogenesis is an anabolic pathway in which large molecules are built up from small ones. This process may occur in chloroplasts, amyloplasts, or cytosol. (b) Most of the sugar (sucrose) we eat and drink comes from sugar cane, which is a type of grass with large stems more than 2 m tall. This harvester cuts the shoots off at ground level, slices the stems into pieces called "joints," and loads them into the cart behind the tractor. Pieces of leaves are so light they blow away. (c) These "joints" of sugar cane must be taken to a refinery and have their sap squeezed out and refined into raw sugar within 18 hours. Immediately after the stems are cut in the field, enzymes begin to convert the disaccharide sucrose into the monosaccharides glucose and fructose, both of which are less valuable than sucrose. (d) Sucrose is obtained by boiling cane sap to concentrate the sucrose until it becomes so concentrated it crystallizes. The crystals are centrifuged out and stored in giant piles as "raw sugar," shown here; the remaining liquid is molasses. Raw sugar is converted to refined (white) sugar by dissolving it again and reboiling it to recrystallize it, after which it is filtered. The raw sugar in this warehouse will last indefinitely if it is kept dry.

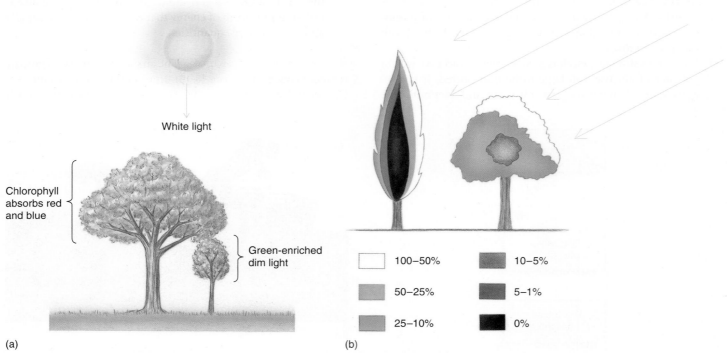

White light

Chlorophyll absorbs red and blue

Green-enriched dim light

	100–50%		10–5%
	50–25%		5–1%
	25–10%		0%

(a)   (b)

Figure 10-20 (a) Plants growing in the shade of other plants receive not only dim light, but light depleted of red and blue. (b) Not all leaves of a tree receive equal amounts of light; this self-shading is more severe in latitudes farther away from the equator. Colors indicate the percentage of full sunlight that reaches the various parts of a tree.

converted to **dihydroxyacetone phosphate**; one molecule of this condenses with one molecule of unconverted 3-phosphoglyceraldehyde to form the sugar **fructose-1,6-bisphosphate**. This loses a phosphate to become **fructose-6-phosphate**, and part of this is rearranged, converting it to **glucose-6-phosphate**. Both fructose-6-phosphate and glucose-6-phosphate are versatile, useful molecules that enter many metabolic pathways. In plants, the glucose-6-phosphate is polymerized into polysaccharides: amylose, amylopectin, or cellulose. It is similarly essential in animals.

■ Environmental and Internal Factors

A plant's photosynthesis is affected by its environment in many ways.

■ Light

From a plant's viewpoint, light has three important properties: (1) quality, (2) quantity, and (3) duration.

Quality of sunlight refers to the colors or wavelengths it contains. Sunlight is pure white because it contains the entire visible spectrum. During sunset and sunrise, sunlight passes tangentially through the atmosphere, and a large percentage of the blue light is deflected upward; consequently, light at ground level is enriched in red, which is easily visible. This period of red-enriched light lasts only a few minutes and probably has little effect on photosynthesis. At noon, sunlight passes nearly vertically through the atmosphere, more blue light is transmitted, and even though the blueness of the sky suggests that all reds, greens, and yellows have been blocked, in fact, enough of all of these wavelengths penetrate to Earth's surface to allow efficient

photosynthesis. This is true of plants in deserts, grasslands, and the top layer—the canopy—of a forest; however, herbs and shrubs that grow near soil level in a forest are understory plants, and the light that they receive has already passed through the leaves of the canopy (**Figure 10-20**). As light penetrates those leaves, red and blue are absorbed by chlorophyll, so the dim light received by understory plants is especially depleted in these critical wavelengths. It is selectively advantageous for them to have extra amounts of accessory pigments so that they can gather the wavelengths available and pass the energy on to chlorophyll *a*. Similarly, algae that grow near the surface of lakes or oceans receive complete light, but water absorbs red and violet. Algae at deep regions receive mostly green and blue light and must have special accessory pigments capable of absorbing these wavelengths efficiently (**Figure 10-21**).

Quantity of light, which refers to light intensity or brightness, is affected by several factors. More light is available for photosynthesis on a clear than on a cloudy day; understory plants receive dim light; lower branches and branches on the shaded side of a plant receive less light (Figure 10-20b). Plants growing in the shadow of mountains or in deep canyons receive much less light than plants that grow on slopes that face the sun. Plants growing near the equator receive intense light because the sun is always more or less directly overhead at noon, whereas plants near the poles receive very little light. Even during the summer the sun is low at noon, and light is scattered by the atmosphere.

Think about how intensity of sunlight varies during the day and affects photosynthesis. Examine the solid line labeled "300 ppm CO_2" in **Figure 10-22**. This was derived from many experiments in which the rate of photosynthesis was measured

Figure 10-21 Many kelps (brown algae) grow in deep ocean water and receive mostly green and blue light. Their accessory pigments absorb these wavelengths; therefore, the kelps appear yellow-brown to us.

but faster on brighter days. At point **b** in the graph, plants that received more light did not photosynthesize faster than those that received slightly less. Where the curve turns flat, there was enough light to saturate the process. In these conditions, the limitation was lack of carbon dioxide; in the experiments in which more carbon dioxide was available (blue line, 1000 ppm CO_2), photosynthesis went faster. Thus, at point **b**, carbon dioxide was the rate-limiting factor. At point **c**, light was so intense that it damaged the plant by overheating it and bleaching the pigments.

At point **a**, if the lack of light prevents photosynthesis from proceeding faster, there must be adequate amounts of carbon dioxide. As soon as ATP and NADPH are produced, they move to the stroma and are used by the waiting enzymes and carbon dioxide; then ADP and $NADP^+$ diffuse back to the thylakoid membranes and wait for another quantum. Conversely, at point **b**, where light is bright and the low concentration of carbon dioxide is rate limiting, there is so much light that as quickly as ADP and $NADP^+$ come to the thylakoids, they are reprocessed immediately into ATP and NADPH. These then move to the stroma, where they must wait for a carbon dioxide molecule.

To the left of the **light compensation point**, it appears that there was no photosynthesis, even though some dim light was provided. The problem actually lies with the technology of measuring photosynthesis. Either the amount of carbon dioxide absorbed or the amount of oxygen released must be measured, but both of these gases are involved in respiration as well as photosynthesis. To the left of the light compensation point, photosynthesis was absorbing carbon dioxide more slowly than respiration was releasing it; thus, it *appears* as though no photosynthesis occurred. The same problem arises when we try to measure respiration: Photosynthesis distorts the measurement, but then we can turn off the lights and stop photosynthesis. The light compensation point is the level of light at which photosynthesis matches respiration. Plants that are grown for a long time in conditions below the light compensation point respire faster than they photosynthesize; they gradually consume their reserve carbohydrates and fats and starve to death. For plants

for plants grown under different intensities of light, but all with 300 ppm carbon dioxide in the air. Near the left side at point **a**, those plants that received dim light absorbed little carbon dioxide, whereas those grown in brighter light absorbed more carbon dioxide. Under these normal levels of carbon dioxide, light is the limiting factor. Photosynthesis is slow on dull, overcast days

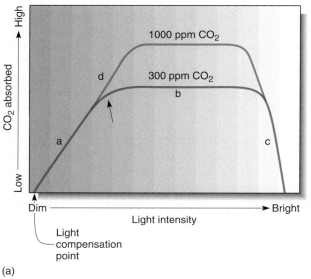

(a)

(b)

Figure 10-22 (a) Light and photosynthesis; details are given in text. The unit ppm is parts per million; for 300 ppm, in 1 million liters of air, 300 liters are carbon dioxide. (b) These grape vines have abundant sunlight, water, and fertilizer. They could probably photosynthesize more rapidly only if more carbon dioxide were present.

Figure 10-23 (a) While young, the leaves of dusty miller are completely obscured by trichomes, protecting the leaf from strong sunlight and insects. (b) These cacti (*Epithelantha*) live in environments where sunlight is extremely intense; their spines are so abundant and closely spaced that they shade the stem and prevent chlorophyll from being damaged.

(a)

(b)

grown in light brighter than the light compensation point, photosynthesis exceeds respiration, and the extra sugar can be used for growth and reproduction.

In the brightest environments, the air is so clear that sunlight is frequently too intense during summer months. Protective adaptations are necessary, and in many species, mechanisms have evolved that provide shade. A common method is the production of a thick layer of dead trichomes, plant hairs (**Figure 10-23**). A heavy coating of wax can also reflect light, and cutin is especially good at absorbing the more harmful short wavelengths. Part of the value of carotenoids and other accessory pigments is that they shade the chlorophyll, absorbing some of the most damaging wavelengths.

The intensity of light, the actual number of quanta that strike a given area per unit time (e.g., that strike 1 cm²/s), is greatest at noon in midsummer when the sun is most directly overhead and is less whenever the sun is lower—morning, evening, and winter. Light may be too intense at midday but optimal when the sun is lower; some species (iris, eucalyptus) have adapted to this by means of vertical leaves. The lamina face is exposed fully only in mornings and afternoons, but at noon, only the leaf edge is exposed. In other plants, leaves orient vertically automatically when stressed—they wilt and hang down.

Understory plants of forests are adapted to low light. If a roadway is cut into a forest and plants adjacent to the cut are suddenly exposed to full sunlight, the shock may cause them to wilt and die. The same phenomenon occurs when trees are blown down during storms, floods, or avalanches and the surrounding plants are exposed.

The **duration** of light refers to the number of hours per day that sunlight is available. At the equator, days are 12 hours long throughout the year. Farther north or south, days become longer in summer; maximum length occurs near the poles, where the day is 24 hours long in midsummer, and only night occurs in midwinter. In middle latitudes, winter days are short, and sunlight is weak because the sun is so low in the sky. Under these conditions, even evergreen plants are unable to undergo very much photosynthesis; however, because temperatures are low, the plants are growing little and have a low rate of respiration.

Even deciduous trees and biennials can survive by means of stored nutrients.

During summer, days are longer and light is brighter because the sun is higher in the sky. The amount of energy obtained by photosynthesis easily exceeds that consumed by respiration and growth. In many plants, longer days cause greater amounts of photosynthesis, but in others, chloroplasts become so full of starch that photosynthesis stops, even though light is present. At night, starch is converted to sugar, which is then transported out of chloroplasts and can be used for growth or stored in amyloplasts in tubers, corms, or other such organs. By morning, leaf chloroplasts can resume photosynthesis.

■ Leaf Structure

Leaf structure of most temperate and tropical plants is quite standard: palisade parenchyma above and spongy mesophyll below. This structure is excellent for absorbing carbon dioxide but inefficient for conserving water. If plants of hot, dry habitats had this leaf architecture, they would have to keep their stomata closed so much of the time that they would starve. Instead, their leaf cells are frequently packed together without intercellular spaces. Water loss is reduced because the small internal surface area retards water evaporation, but with so little surface area, it is difficult to dissolve carbon dioxide from the air into the cytoplasm. This slows photosynthesis, but apparently this tradeoff, slow growth versus water conservation, is selectively advantageous.

Another method of minimizing water loss while maintaining photosynthesis is to reduce external surface by means of cylindrical leaves. Water movement from interior air spaces to exterior air is minimized because so few stomata are present (**Figure 10-24**). Photosynthesis is reduced because absorption of carbon dioxide is slowed.

■ Water

The amount of water available greatly affects photosynthesis. Most plants keep their stomata open during the day, permitting entry of carbon dioxide, but water is inevitably lost. At night,

(a)

(b)

(c)

Figure 10-24   (a) These living stone plants (*Lithops*) of African deserts conserve water in several ways. They have only two leaves at a time; when two new ones form, the old two die. Not enough water is available for four leaves. The leaves are fleshy and pressed together, such that they form a cylinder with minimal surface area through which water can be lost. (b) This is a plant of *Haworthia cooperi*. Its short stem is located several centimeters underground, and it produces many cylindrical leaves that are just long enough to reach the soil surface. The tips of the leaves, visible here, have transparent epidermis, and the mesophyll is also transparent because it has no intercellular spaces. Consequently, light enters through the window-like tip and then passes deep into the rest of the underground leaf, where cells with chloroplasts are located: photosynthesis actually occurs underground. Because most of each leaf is subterranean, they stay cooler in summer, they are less visible to animals, and the air in the soil is richer in carbon dioxide than is air above ground. (c) This transverse section shows the leaf is so transparent you can read through it. It is excellent for transmitting light to parts of the leaf that are underground.

carbon dioxide cannot be used, and stomata are closed, retaining water within the plant. If the soil becomes dry and water is not readily available, a plant keeps its stomata closed even during the day, and carbon dioxide cannot enter (**Figure 10-25**). The small amount of carbon dioxide produced by the respiration of the leaves can be reused photosynthetically, but this is a minor amount. In addition to numerous structural modifications that conserve water, metabolic adaptations also exist; two of the most important are C_4 metabolism and Crassulacean acid metabolism (CAM).

■ C_4 Metabolism

An important factor for plants is the amount of water lost for each molecule of carbon dioxide absorbed. Ideally, this ratio is low. Carbon dioxide diffuses into a leaf faster if its concentration in air is higher or if its concentration inside leaf protoplasm is lower. Plants can do nothing about the carbon dioxide level in the external air, and many also have poor control over the protoplasmic concentration. RuBP carboxylase has a low affinity for carbon dioxide; as carbon dioxide concentration drops, enzyme-substrate binding slows. Even while carbon dioxide is still rather abundant in protoplasm, the enzyme is only rarely picking it up. With this relatively high concentration, carbon dioxide diffusion into the leaf is slow, whereas water loss may be high.

RuBP carboxylase occasionally binds to oxygen instead of carbon dioxide, acting as an oxygenase and producing one molecule of 3-phosphoglycerate and one of phosphoglycolate. This latter molecule is transported from the chloroplast to peroxisomes and mitochondria, where some of it is converted to the useful amino acids glycine and serine, but much of the phosphoglycolate is broken down to two molecules of carbon dioxide (**Figure 10-26**). The breakdown is **photorespiration**, an energy-wasting process. The energy and reducing power used to produce the two reduced carbons of phosphoglycolate are completely lost. Photorespiration is extremely exergonic but is not used to provide power to endergonic reactions. Because phosphoglycolate may be toxic, photorespiration protects the plant whenever RuBP carboxylase picks up oxygen; however, it is an expensive defense because up to 30% of all ATP and NADPH produced by the chloroplast can be immediately lost by photorespiration.

Apparently RuBP carboxylase cannot be significantly improved. At the time when RuBP carboxylase originated by evolution,

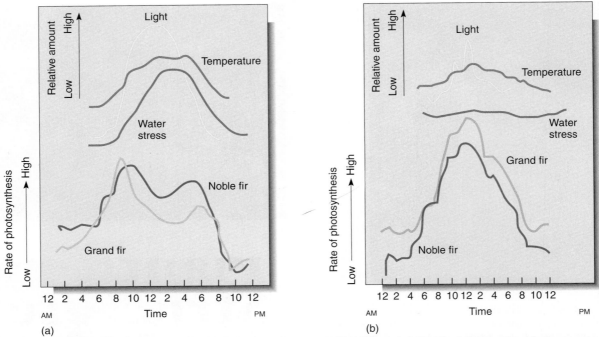

Figure 10-25 (a) Light intensity exceeded the light compensation point for these trees just after 6 AM, and photosynthesis increased rapidly; however, by early afternoon, photosynthesis dropped even though light and temperature were adequate. The problem was a lack of water (water stress), and probably stomata had begun to close around noon. (b) On a cloudy day, water stress never became a problem and stomata remained open. Even though there was less total light than in (a), there was more photosynthesis for the day.

virtually no free oxygen was present in the atmosphere, and carbon dioxide levels were high. RuBP carboxylase has existed for billions of years, and virtually no structural mutations have survived; they must not have produced superior versions of the enzyme. An alternative is to improve its working conditions. RuBP carboxylase should be compartmentalized in a site where carbon dioxide

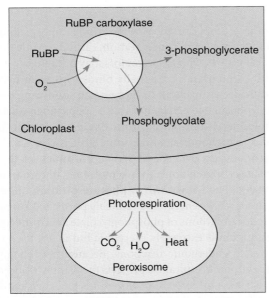

Figure 10-26 If RuBP carboxylase puts oxygen onto RuBP, one of the products is phosphoglycolate, which is transported to a peroxisome and broken down during photorespiration. All of the energy that was present in the phosphoglycolate is wasted. Mitochondria are also involved.

concentration is high and oxygen concentration is low. This has evolved in some plant groups and is known as C_4 **metabolism** or C_4 **photosynthesis**. Basically, C_4 metabolism is a mechanism by which carbon dioxide is absorbed, transported through, and concentrated in a leaf, whereas oxygen is kept away from RuBP carboxylase.

C_4 metabolism occurs in leaves with Kranz anatomy (**Figure 10-27**). In such leaves, mesophyll is not distributed as palisade and spongy parenchyma; rather, each vascular bundle has a prominent chlorophyllous sheath of cells, and around the sheath are mesophyll cells. Mesophyll cells contain the enzyme **PEP** (phosphoenolpyruvate) **carboxylase**, which has a very high affinity for carbon dioxide. Unlike RuBP carboxylase, as the carbon dioxide concentration drops lower and lower, PEP carboxylase continues binding to it rapidly and firmly. Carbon dioxide concentrations inside the leaf are kept very low, and carbon dioxide diffuses inward rapidly whenever stomata are open. The ratio of water lost to carbon dioxide absorbed is favorably low. Also, PEP carboxylase has a high specificity for carbon dioxide; it never picks up oxygen. PEP carboxylase is an ideal enzyme except that it does not perform the critical reaction that results in 3-phosphoglycerate. Despite its shortcomings, RuBP carboxylase is still the only enzyme that carries out the necessary reaction.

PEP carboxylase adds carbon dioxide to PEP, producing oxaloacetate, which has four carbons, hence the name C_4 metabolism. Oxaloacetate is reduced to malate by a molecule of NADPH, and further reactions may occur, depending on the species. Malate from throughout the mesophyll moves into the bundle sheath and breaks down into pyruvate by releasing carbon dioxide (Figure 10-27b). This reaction is powerful enough to drive the forma-

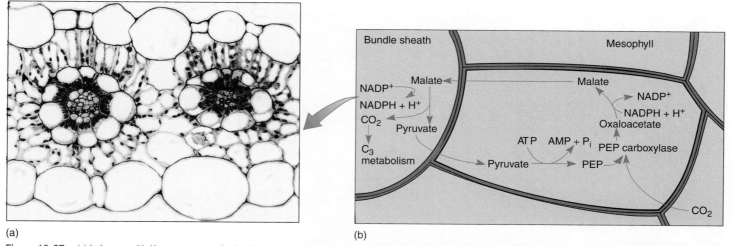

(a)　　　　　　　　　　　　　　　　　　　　　　　(b)

Figure 10-27　(a) In leaves with Kranz anatomy, the bundle sheath around all veins is prominent and rich in chloroplasts. The bundle sheath chloroplasts are located as closely as possible to the vascular tissues. The mesophyll cells are arranged in a sheath around the bundle sheath (×250). (b) Knowing the reactions of C_4 metabolism is not sufficient for understanding it; only by realizing that some reactions occur in separate compartments does it become logical.

tion of a new molecule of NADPH, so, the process results in the transport by malate of both carbon dioxide and reducing power. Because all of the malate from a large volume of mesophyll decarboxylates in the small volume of the bundle sheath, carbon dioxide concentration in the sheath is very high. Also, because NADPH is synthesized by this unusual method in bundle sheath cells, the bundle sheath chloroplasts primarily carry out cyclic electron transport, pumping protons and making ATP. Without noncyclic electron transport, there is no breakdown of water or production of oxygen. RuBP carboxylase, located exclusively in the bundle sheath chloroplasts, is in ideal conditions—a high carbon dioxide concentration with low or no oxygen—and carries out efficient C_3 stroma reactions with low production of phosphoglycolate. C_4 plants have little photorespiration (**Table 10-6**).

Like any other carrier, malate must be shuttled back to its recharge site, the mesophyll. In the bundle sheath it is converted to pyruvate by the release of carbon dioxide; pyruvate moves back to the mesophyll and receives a phosphate group from ATP, which converts it to PEP.

The selective advantage of C_4 metabolism depends on the environment. Photorespiration increases with temperature, so it is more of a problem in hot climates. Under warm, dry conditions, C_4 metabolism has a strong selective advantage over C_3 metabolism: Much less water is lost during carbon dioxide

absorption. Also, abundant light is available to generate the extra ATP needed to convert pyruvate to PEP. The ATP used to make PEP means that C_4 metabolism is not free; under cool conditions, photorespiration may be slow enough that it loses less energy than C_4 metabolism. Also, many cool habitats are also moist, so water conservation by stomatal closure is not as critical. The critical temperature above which C_4 metabolism is more advantageous selectively than C_3 metabolism varies among species but averages around 25°C.

C_4 metabolism and Kranz anatomy have evolved several times; most C_4 species are monocots of hot climates such as corn, sugarcane, sorghum, and several other grasses, but a considerable number of eudicots, also from warm, dry regions, are also C_4 species (**Table 10-7**).

Crassulacean Acid Metabolism

Crassulacean acid metabolism (**CAM**) is a second metabolic adaptation that improves conservation of water while permitting photosynthesis. It is so named because it was first discovered in those members of the family Crassulaceae that have succulent

TABLE 10-6	Types of Carbon Dioxide Processing		
	C_3 Plants	C_4 Plants	CAM Plants
Ultimate carboxylase	RuBP carboxylase	RuBP carboxylase	RuBP carboxylase
Adjunct metabolism	None	CO_2 transfer	CO_2 storage
Adjunct carboxylase	None	PEP carboxylase	PEP carboxylase
Photorespiration	High	Low	Moderate
Stomata open	Day	Day	Night

TABLE 10-7	Plant Families Having C_4 Species
Family	Familiar Example
Aizoaceae	Ice plants
Amaranthaceae	Amaranths
Asteraceae	Asters, daisies
Chenopodiaceae	Pigweeds, beets
Cyperaceae	Sedges
Euphorbiaceae	Spurges
Poaceae	Grasses
Nyctanginaceae	Four o'clocks, bougainvillea
Portulacaceae	Purslanes
Zygophyllaceae	Creosote bush

Figure 10-28 (a) *Sempervivum arachnoideum,* a member of the Crassulaceae. Like many desert succulents, it has CAM metabolism as well as other adaptations that conserve water. (b) This barrel cactus (*Ferocactus*) is a CAM plant. (c) Photosynthetic cells of CAM plants have large vacuoles, which permits the accumulation of C₄ acids. These are cortex cells, arranged in palisades, similar to leaf palisade mesophyll (×250).

leaves (**Figure 10-28** and **Table 10-8**). The metabolism is almost identical to that in C₄ plants: PEP is carboxylated, forming oxaloacetate, which is then reduced to malate or other acids. These acids are not transported but simply accumulate, in effect storing carbon dioxide. This occurs at night. These plants differ from C₃ and C₄ plants in that their stomata are closed during the hottest periods and open only at night when it is cool. Coolness reduces transpiration, and usually air is calmer at night; therefore, water molecules near stomata are not blown away immediately and may diffuse back into the plant. Opening stomata at night is effective for conserving water, but the lack of light energy creates a problem for photosynthesis: A plant cannot store ATP and NADPH during the day for use at night; they are not stable enough and each cell has too little. Hence, carbon dioxide is stored on acids until

daytime, when stomata close and the malate or other acids break down, releasing carbon dioxide for C₃ metabolism. The released carbon dioxide cannot escape because stomata close at dawn.

Crassulacean acid metabolism is not particularly efficient; the total amount of carbon dioxide is so small that it may be entirely used in C₃ metabolism after just a few hours of sunlight. Furthermore, RuBP carboxylase is protected from photorespiration only in the morning when internal carbon dioxide levels are high. Later in the day, photorespiration may be high, but the carbon dioxide released by the breakdown of phosphoglycolate is trapped and refixed.

Crassulacean acid metabolism is selectively advantageous in a hot, very dry climate where survival rather than rapid

Alternatives

Box 10-2 Photosynthesis in Bacteria and Cyanobacteria

A process as complex as photosynthesis had to evolve by numerous steps, and many early types of photosynthesis are present in bacteria and cyanobacteria. By examining them, we can understand plant photosynthesis more fully.

Cyanobacterial photosynthesis is the most similar to that of plants. This is not surprising because chloroplasts arose as endosymbiotic cyanobacteria living inside early eukaryotic cells. Cyanobacterial light reactions are almost identical to those in chloroplasts, having chlorophyll *a* but lacking chlorophyll *b*. Their accessory pigments are phycobilins, open-chain tetrapyrrole rings (Figure B10-2a), which act like carotenoids by absorbing wavelengths that chlorophyll cannot and then transferring the energy to chlorophyll and activating an electron. One class of phycobilins, phycocyanin, absorbs most strongly at approximately 620 to 640 nm and is blue; the other class, phycoerythrin, is red because it absorbs maximally at 550 nm. Both occur bound to proteins, and these aggregate into small nodules (phycobilisomes) visible by electron microscopy. Most other aspects of photosynthesis in cyanobacteria are identical to those of plants; electrons are taken from water, passed through photosystem (PS) II to PS I, and are then used to reduce $NADP^+$ to NADPH. Oxygen is liberated as a waste product, as in plants; therefore, both have **oxygenic photosynthesis**. Cyanobacteria do not have chloroplasts, but they do have extensive sheets of infolded plasma membrane that contain pigments and electron carriers; the folded membrane forms accumulation spaces for protons and the generation of a chemiosmotic gradient.

Purple bacteria and green bacteria do not contain chlorophyll, either *a* or *b*, but instead have **bacteriochlorophylls** (Figure B10-2b). Like chlorophyll, these are closed tetrapyrroles with a long tail, but they have certain side groups that the chlorophylls lack. Carotenoid accessory pigments are present, as in chloroplasts. PS I operates, but PS II is not present to put electrons back onto bacteriochlorophyll. Instead, the original electron comes back through a series of carriers, so electron flow is cyclic, similar to that in plant photosynthesis (Figure B10-2c; compare with Figure 10-17). One carrier is quinone; thus, protons are pumped across the photosynthetic membranes, creating a strong chemiosmotic gradient that causes ATP to be generated, just as in eukaryotic chloroplasts. Because there is no PS II, oxygen is not formed. This is **anoxygenic photosynthesis**. The photosynthetic apparatus of purple bacteria consists of extensive arrays of membranes connected to the plasma membrane. In green bacteria, photosynthetic membranes are cylindrical vesicles that occur in clusters surrounded by another "membrane." The enclosing membrane does not look like a typical bilayered membrane in electron micrographs and is probably different physiologically as well.

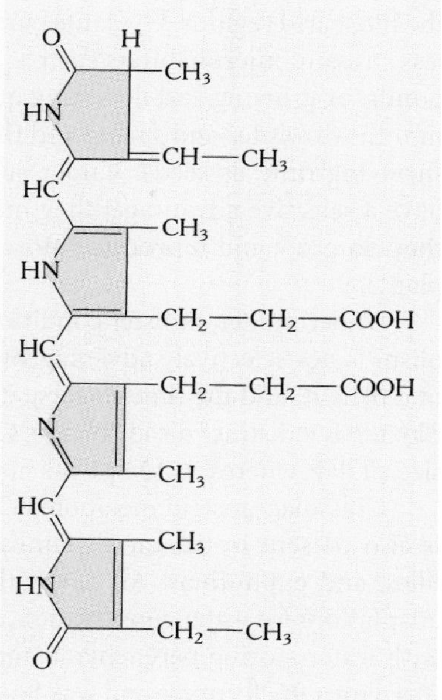

Figure B10-2a A typical phycobilin, containing four pyrrole groups—a tetrapyrrole. The two ends of the pyrrole are not attached to each other as they are in chlorophyll.

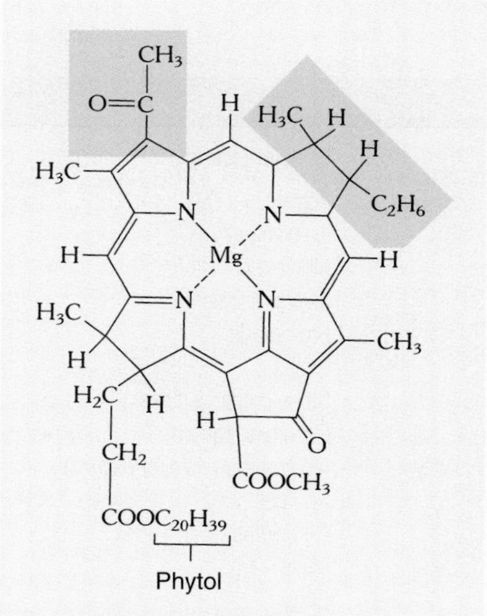

Bacteriochlorophyll *a*

Figure B10-2b Bacteriochlorophyll *a* is similar to chlorophyll *a*, differing only at the sites indicated by color. Is this similarity analogous, the product of two independent lines of evolution, or homologous, the genes for one pigment having evolved from those for the other? These two molecules are not sufficient evidence to judge, but we are almost certain that all photosynthetic systems are closely related evolutionarily.

Alternatives

Notice that bacterial photosynthesis is basically just light-powered proton pumping; it is extremely simple and not very effective. Bacteriochlorophyll does not capture enough light energy to activate an electron strongly enough for it to move onto NADP$^+$; instead, electrons must cycle repeatedly, gradually building up enough of a chemiosmotic gradient to finally produce NADPH. In contrast, chlorophyll captures enough energy to reduce NADP$^+$ in just two steps. Also, plants and cyanobacteria obtain all of the electrons they need from water, which is always present wherever organisms are alive. Purple and green bacteria recycle electrons repeatedly as they pump protons, but the electrons used to reduce NADP$^+$ must come from other substances, such as sulfur and hydrogen, which are not always reliably present.

Although plant photosynthesis is complex, think about it in terms of being composed of several parts that originated as simple pumps, cyclic electron flows, and chemiosmotic synthetic pathways. Each could originate as a simple metabolic function that then evolved to become more sophisticated. Gradually, one part was added to another, then another, and finally, chlorophyll-based oxygenic photosynthesis resulted.

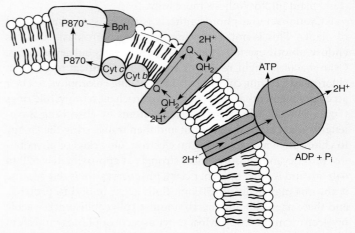

Figure B10-2c Electrons flow cyclically from activated bacteriochlorophyll *a*, pumping protons with quinone carriers. This establishes a chemiosmotic gradient that generates ATP. Bph = bacteriopheophytin.

TABLE 10-8	Plant Families Having CAM Species
Family	**Familiar Example**
Agavaceae	Agaves, yuccas
Aizoaceae	Ice plants
Asclepiadaceae	Milkweeds
Asteraceae	Asters, daisies
Bromeliaceae	Bromeliads
Cactaceae	Cacti
Crassulaceae	Stone crops, sedums
Cucurbitaceae	Cucurbits, squash
Didiereaceae	(no common, familiar example)
Euphorbiaceae	Spurges
Geraniaceae	Geraniums
Lamiaceae	Mints
Liliaceae	Lilies
Orchidaceae	Orchids
Oxalidaceae	*Oxalis*
Piperaceae	Peppers, *Peperomia*
Polypodiaceae	(a fern)
Portulaceae	Purslanes
Vitaceae	Grapes
Welwitschiaceae	(a gymnosperm)

growth is most important. In these habitats, unaided C_3 metabolism is so wasteful of water that C_3 plants cannot survive. In the most arid regions, C_4 plants barely get by, growing in slightly less stressful microhabitats such as near temporary streams and ponds, or growing and flowering quickly during the cool, moist months of winter and spring and then dying in the summer and surviving only as seeds. Under such conditions, CAM plants have a selective advantage; they may not grow luxuriantly, but they do grow and reproduce more successfully than C_3 and C_4 plants.

Under milder, moister conditions, Crassulacean acid metabolism is not selectively advantageous. Water conservation is less of a benefit, and the limited capacity to absorb and store carbon dioxide is a distinct disadvantage. C_3 and C_4 plants photosynthesize all day, whereas CAM plants may stop before noon.

Crassulacean acid metabolism has evolved several times and is also present in the cactus family, many orchids, bromeliads, lilies, and euphorbias. All have other metabolic and structural adaptations for water conservation, such as succulent bodies filled with water-storing parenchyma and covered by a tough epidermis with a thick cuticle and wax layer.

Summary

1. All physical systems have a tendency to become disordered, to increase in entropy. Living organisms have a high degree of order and regularity, maintained by the input of energy, the ultimate source of which is the sun.

2. All photosynthetic plants are autotrophs, but some parasitic plants and many plant tissues are heterotrophic.

3. When electrons are passed from one atom to another, the recipient becomes reduced, and the donor becomes oxidized. Most of a plant's raw materials are highly oxidized and must be reduced as they are assimilated.

4. Photosynthetic pigments respond to light quanta that have just enough energy to boost electrons one or two energy levels.

5. Accessory pigments have absorption spectra different from that of chlorophyll *a*, and they therefore absorb different wavelengths and transfer energy to chlorophyll *a*.

6. Photosystems I and II work together, transferring electrons from water to NADPH. The electron's energy is boosted twice, once at P680 and again at P700.

7. The light-dependent reactions result in a chemiosmotic gradient. Driven by concentration differences, protons flow from the thylakoid lumen to the stroma through ATP synthetase channels, powering the phosphorylation of ADP to ATP.

8. Cyclic electron transport permits production of extra ATP without synthesis of NADPH or production of free oxygen. Noncyclic electron transport results in production of both ATP and NADPH, but the amount of ATP is not sufficient for the stroma reactions.

9. All photosynthetic plants use the C_3 stroma reactions mediated by RuBP carboxylase: A molecule of RuBP is carboxylated, energized, and reduced, resulting in two molecules of 3-phosphoglyceraldehyde.

10. C_4 metabolism is an adjunct to C_3 metabolism, not a replacement. PEP carboxylase, which acts as the initial carboxylating enzyme, has a great affinity for carbon dioxide.

11. Crassulacean acid metabolism is similar to C_4 metabolism, except that it accumulates and stores carbon dioxide at night while stomata are open and releases it during the day while stomata are closed.

Important Terms

absorption spectrum
action spectrum
anabolic reactions
anoxygenic photosynthesis
ATP synthetase
bacteriochlorophylls
C_3 (Calvin/Benson) cycle
C_4 metabolism
chemiosmotic phosphorylation
chlorophyll
Crassulacean acid metabolism (CAM)
cyclic electron transport
cytochromes
electron transport chain
entropy
gluconeogenesis

greenhouse effect
heterotrophs
light compensation point
light-dependent reactions
noncyclic electron transport
oxidation state
oxidative phosphorylation
oxidized compounds
oxygenic photosynthesis
ozone
3-phosphoglyceraldehyde
photoautotrophs
photons
photophosphorylation
photorespiration
photosystem I
photosystem II

pigment
plastocyanin
plastoquinones
quality of sunlight
quantum
reaction center
reduced compounds
reducing power
ribulose-1,5-bisphosphate (carboxylase)
RUBISCO
stroma
stroma reactions
substrate-level phosphorylation
thylakoid lumen
Z scheme

Review Questions

1. What is the meaning of the word entropy? Does the entropy of a plant increase or decrease while it is alive? After it is dead?

2. Name several examples of photoautotrophs and several of heterotrophs. How do photoautotrophs obtain energy? Can a plant be heterotrophic while a seedling and photoautotrophic when older?

3. ATP is an important chemical involved in many of a plant and animal's metabolic reactions. Yet any plant has only a small amount of it. Can you explain this? When ATP enters a reaction and forces it to proceed, what is the ATP converted into? What then happens to that molecule?

4. Name the three methods of phosphorylation.

5. What is a reduction reaction? Why does a reduction reaction always occur simultaneously with an oxidation reaction?

6. In organic molecules, we calculate the oxidation state of carbon by assuming that each oxygen has an oxidation state of _____. Each hydrogen has an oxidation state of

_____. Calculate the oxidation state of carbon in each of the following: CO_2, CH_2O, and malic acid.

7. Two of the following are oxidizing agents and two are reducing agents. Which are which: NAD^+, $NADP^+$, NADH, and NADPH?

8. In photosynthesis, what is the ultimate source of electrons? What are the benefits of this molecule in terms of its toxicity and the cost of the plant to obtain it?

9. Describe the absorption spectrum of chlorophyll. Why does it match the action spectrum of photosynthesis?

10. Chlorophyll does not use high-energy quanta. Why not? What would happen to the chlorophyll if it did? It also does not use long wavelength radiation either. Why not?

11. The most common accessory pigments in land plants are chlorophyll _____ and the _____. Algae that live in deep water have other accessory pigments because only _____-_____ light penetrates deeply into water.

12. Name the electron carriers that transport electrons from photosystem II to photosystem I. Which ones contain metal atoms, and which do not?

13. When photosystem I produces NADPH, its reaction center P700 chlorophyll *a* loses electrons. What would happen if photosystem II did not supply new electrons to P700?

14. When electrons are removed from water, protons are liberated. Does this occur in the stroma or inside the thylakoid lumen? Can protons move directly across the membrane? Describe the chemiosmotic mechanism of ATP synthesis in chloroplasts.

15. Is ADP converted to ATP directly by the reaction center chlorophylls? Do the enzymes that synthesize ATP obtain the necessary energy by interacting directly with the reaction center chlorophylls?

16. What chemical is the acceptor of carbon dioxide in the C_3 cycle? What enzyme catalyzes the reaction, and what is the product?

17. RuBP carboxylase is by no means an ideal enzyme. Describe some of the problems with its active site and its substrate specificity. If we compare the amino acid sequences of this enzyme from many different species, they are almost identical. What is the significance of this uniformity?

18. Which chemicals are useful for energy storage on a short-term basis? Which are for intermediate term and which are for long term?

19. What is the "quality" of light? How does it differ for plants in deserts, grasslands, and the canopy of a forest versus for plants in the understory? How does it differ for algae that grow near the surface of a lake or ocean versus those that inhabit deep water far below the surface?

20. How is the quantity of light affected by a plant's location relative to the equator or the poles? On one side of a mountain or the other? On one side of a valley or the other?

21. Imagine a leaf in bright light but an atmosphere with no carbon dioxide. Would RuBP carboxylase be functioning? Would the NADP be in the reduced or oxidized form?

22. Name some of the brightest environments. Describe some protective adaptations that plants may use to shade themselves.

23. An important factor for plants is the amount of water lost for each molecule of carbon dioxide absorbed. How could the plant be harmed if it loses a lot of water for each carbon dioxide molecule, that is, if the ratio is high? Would this be more important for a plant in a rainy habitat or one in a desert?

24. In a C_4 plant, where is PEP carboxylase located? Where is RuBP carboxylase located?

25. In a CAM plant, are stomata open during the day or the night? How does this affect the amount of water the plant loses when its stomata are open?

26. As a CAM plant makes and stores acids during the night, how does this affect the plant's acidity (its pH)? Think about the acidity of your own blood. Do you think it is allowed to vary by any large amount?

27. In habitats where water conservation is not especially necessary, is CAM metabolism more or less advantageous than C_3 or C_4 metabolism? Why?

28. What is global warming? What is the main gas that causes it? What would happen if the Earth's atmosphere had a lower concentration of CO_2 than it has now? What would happen if it had more?

29. What is the Kyoto Protocol? How many countries have signed it? Name one country that has not signed it. What are two substances burned in the United States (and all other countries) that produce CO_2? Which two countries have large populations and may soon surpass the United States in production of greenhouse gases?

BotanyLinks

http://biology.jbpub.com/botany/5e

BotanyLinks contains a variety of resources and review material designed to assist in your study of botany.

Energy Metabolism: Respiration

chapter 11

Concepts

The light-dependent reactions of photosynthesis produce an excess of energy and reducing power, both of which are stored as glucose and starch; an important corollary must be the ability to recover that energy and reduced carbon. Recovery may occur later, when photosynthesis is impossible, such as night or in winter when the plant is leafless. Also, energy recovery may occur at a different site from photosynthetic capture: Glucose may be converted to sucrose, transported to apical meristems, vascular cambia, or any other heterotrophic tissue, then broken down to recover the energy (**Figure 11-1**).

Chapter Opener Image: The leaves of these quaking aspen trees (*Populus tremuloides*) turn yellow in autumn because their chlorophyll breaks down, revealing the presence of other pigments that had been masked by the abundant chlorophyll. The leaves are not able to carry out photosynthesis without chlorophyll, but they will drop off the trees soon anyway. Throughout spring and summer the leaves had been photosynthesizing and supplying the trees with energy; but, throughout late autumn, winter, and early spring, photosynthesis will be impossible while the trees are leafless. During this period, the trees obtain ATP by respiration, just as we animals do all the time. All plants use respiratory metabolisms that are almost identical to those of animals to provide the ATP necessary to power their growth, development, and survival. Even during the summer when leaves are present, all the living parts of trees that are not green (flowers, fruits, inner bark, sapwood, and roots) depend on respiration. The leaves at night have no source of ATP other than respiration.

Outline

- **Types of Respiration**
  - Anaerobic Respiration
  - Aerobic Respiration
  - Heat-Generating Respiration
  - Pentose Phosphate Pathway
  - Respiration of Lipids
  - Photorespiration

- **Environmental and Internal Factors**
  - Temperature
  - Lack of Oxygen
  - Internal Regulation

- **Total Energy Yield of Respiration**

- **Respiratory Quotient**

- **Fermentation of Alcoholic Beverages**
  - Beer
  - Wine
  - Spirits
  - Warnings

Box 11-1 Plants and People: Fungal Respiration

Box 11-2 Alternatives: Respiration in Prokaryotes

Box 11-3 Plants Do Things Differently: Plants, Babies, and Heat

Box 11-4 Alternatives: Heterotrophic Plants

Box 11-5 Plants and People: Respiration

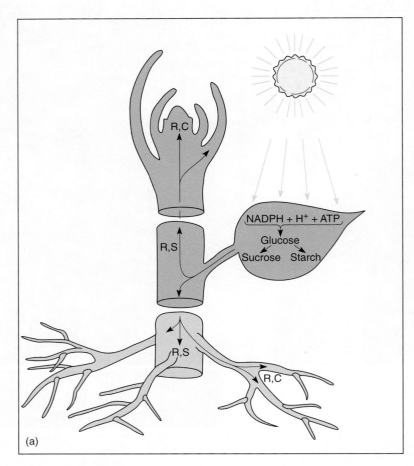

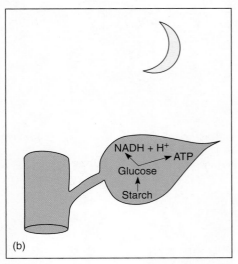

Figure 11-1 (a) When light is available, autotrophic tissues produce more ATP and NADPH than needed for cell metabolism. Glucose is produced, part of which is stored as starch in the leaf and part converted to sucrose and transported to areas that are either completely heterotrophic or are growing more rapidly than their own photosynthesis can support. Sucrose is converted back to glucose and is either respired (R) or used in the construction (C) of cell structures such as cellulose, lignin, amino acids, and nucleic acids. In roots, trunks, fruits, and seeds, some glucose may be polymerized to starch and stored (S) for months or years. (b) Even autotrophic tissues are heterotrophic in the dark, surviving on respiration of stored starch.

Respiration is the process that breaks down complex carbon compounds into simpler molecules and simultaneously generates the ATP used to power other metabolic processes (Figure 11-2). During respiration, carbon is oxidized. Its oxidation state goes from +0 to +4 as electrons are removed by NAD+, which is converted to NADH in the process. This is basically the opposite of photosynthesis, in which NADPH carries electrons to carbon, reducing it.

The NADH generated by respiration is a good reducing agent, but it is produced in much larger quantities than needed for constructive reduction reactions. Plants use mostly NADPH

from photosynthesis for their reductions, not NADH, and most of the compounds that animals and fungi consume are already reduced. Because NADH contains a great amount of energy, it is selectively advantageous for an organism to be able to oxidize it so as to generate even more ATP.

Oxidation of NADH to NAD+ requires transferring electrons from it onto something else. The ideal recipient would be abundant, cheap, and nontoxic after it is used (Figure 11-3). These are also the three ideal characteristics of the source of electrons in photosynthesis, and the chemical involved in both processes

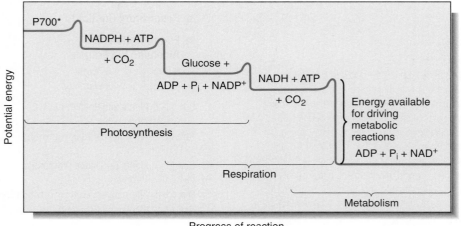

Progress of reaction

Figure 11-2 This hypothetical reaction diagram shows the relative levels of potential energy of the major compounds of photosynthesis and respiration. Energized P700 has the greatest potential energy, most of which is trapped in the formation of ATP and NADPH. When these reduce carbon dioxide and make glucose, much of the energy is conserved. Respiration transfers the energy to ATP and NADH. During other metabolic reactions, the breakdown of ATP and NADH yields large amounts of energy. Depending on the metabolic pathway, some of this energy is retained and some lost.

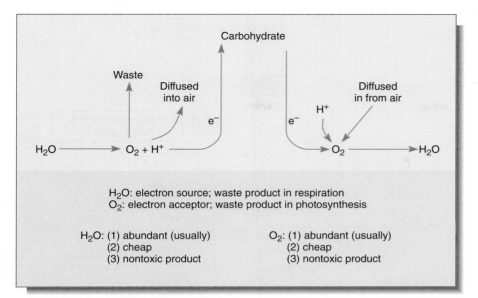

Figure 11-3 It is selectively advantageous for an organism's metabolism to be based on raw materials that are abundant and cheap and produce nontoxic waste products. Mutations that cause an individual to require rare or expensive compounds or those that break down into toxic products are selectively disadvantageous.

is water. As it turns out, the result of respiration is the reverse of photosynthesis. Electrons are transferred from carbon in carbohydrate by means of reduced NADH, which carries them to an electron transport chain, which in turn deposits them onto oxygen, reducing it. As electrons are added, protons are attracted and incorporated, converting oxygen to water. If this reverse analogy were carried one step further, the electron transport chain would have to give off light, but that would be a complete waste of energy; instead, it conserves the energy as high-energy phosphate-bonding orbitals of ATP.

Respiration is also important to the plant because numerous intermediate compounds of its many steps are useful as starting materials for several anabolic pathways. A molecule of glucose, after entering the respiratory pathway but before being broken down to carbon dioxide and water, may be picked up in the form of an intermediate by a nonrespiratory enzyme that diverts it into a pathway that produces amino acids, fats, nucleic acids, lignin, and other molecules.

Types of Respiration

Cellular respiration falls into two categories: aerobic and anaerobic. Respiration that requires oxygen as the terminal electron acceptor is **aerobic respiration**. Under certain conditions, oxygen is not available, and an alternative electron acceptor must be used. This is **anaerobic respiration**, respiration without oxygen, often called **fermentation**. Because animals and plants must have oxygen for their respiration, they are known as **obligate** or **strict aerobes** (Table 11-1). At the opposite extreme are certain bacteria called **obligate anaerobes**, which carry out anaerobic respiration exclusively; such bacteria are actually killed by oxygen. Many fungi and certain types of tissues in animals and some plants are **facultatively aerobic** (or **facultatively anaerobic**): If oxygen is present, they carry out aerobic respiration, but when oxygen is absent or insufficient, they switch to anaerobic respiration. Although many fungi, especially yeasts, can live indefinitely anaerobically, plant and animal tissues can survive this way for only a short time. They must eventually obtain oxygen and switch back to aerobic respiration or they die.

Anaerobic Respiration

Glucose is broken down during anaerobic respiration by a metabolic pathway called **glycolysis** or the **Embden-Meyerhoff pathway**. Glycolysis and gluconeogenesis are essentially the same pathway, with the reactions running in opposite directions (**Figure 11-4**). Although all intermediates are the same, as are many of the enzymes, the two processes use different enzymes at certain key steps. This allows a cell to regulate the two processes so that one is stopped while the other runs; it would be useless for both pathways to operate simultaneously within a single cell.

In glycolysis, ATP phosphorylates glucose to glucose-6-phosphate, which is then converted to fructose-6-phosphate. A second molecule of ATP then phosphorylates this to fructose-1,6-bisphosphate, which breaks down into 3-phosphoglyceraldehyde and dihydroxyacetone phosphate. The latter can be converted into a second molecule of 3-phosphoglyceraldehyde, and both can be oxidized to 1,3-diphosphoglycerate. During this oxidation step, electrons are transferred from a carbon of 3-phosphoglyceraldehyde to NAD^+, converting it to NADH (as with NADPH, to be strictly correct when balancing equations, NADH should be considered to be associated with a proton: $NADH + H^+$). The 1,3-diphosphoglycerate is energetic enough that an enzyme can

TABLE 11-1	Relationships Between Types of Organisms and Presence or Absence of Oxygen	
	Oxygen Present	Oxygen Absent
Obligate aerobes	Aerobic respiration; able to live	No respiration; death
Obligate anaerobes	Oxygen destroys certain vital metabolites; death	Fermentation; able to live
Facultative organisms	Aerobic respiration; able to live	Fermentation; able to live

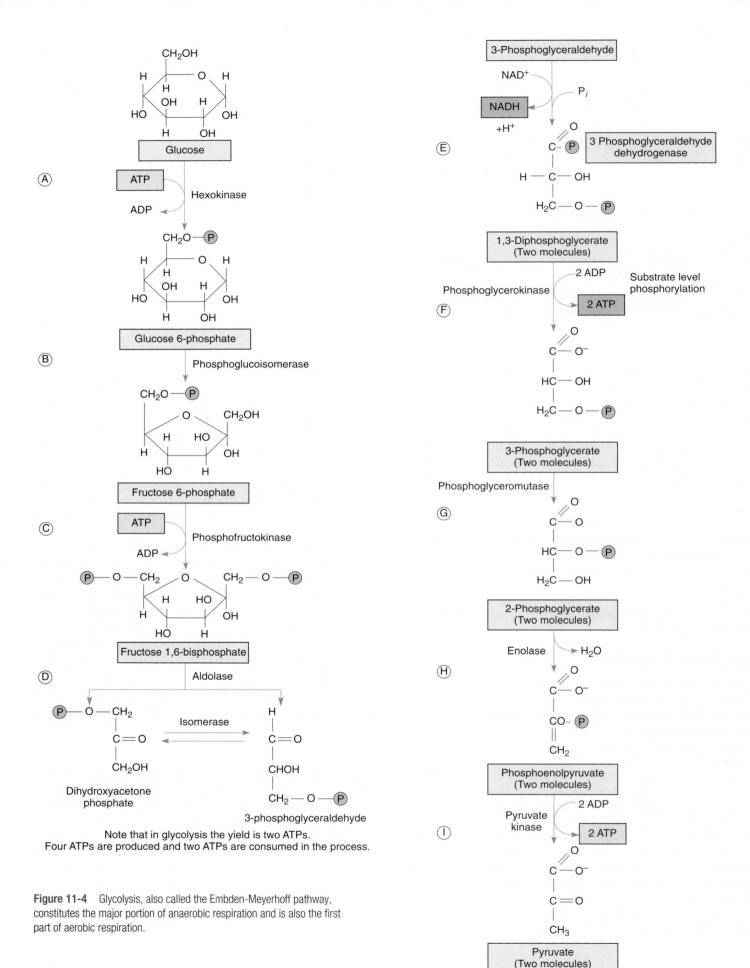

Figure 11-4 Glycolysis, also called the Embden-Meyerhoff pathway, constitutes the major portion of anaerobic respiration and is also the first part of aerobic respiration.

Plants and People

Box 11-1 Fungal Respiration—The Prehistoric Industrial Revolution

Most college students know at least a little about fermentation—you probably already know that beer and wine are fermented even though you never much worried about the Embden-Meyerhoff pathway. Perhaps you know already that aerobic respiration of the fungi known as yeast causes bread to rise and holes to form in Swiss cheese, but you may not have realized that the use of these fermentations started out as lifesaving processes several thousand years ago.

When we think about the origins of agriculture between 2000 and 3000 BCE, we often think of the domestication of wheat, rice, and corn, but the use of fruits and animal products was necessary not only for diversity in the diet but also for diversity of vitamins and essential nutrients. Think about life in 3000 BCE—no supermarkets, of course, and no canned food, frozen food, or refrigerated food. People grew their own food, or collected it, or traded for it.

Now imagine the first cold winds and rains of autumn. How do you preserve the abundant food from summer to allow your survival through winter? It will be 7 long months before grains can be planted, before the dormant buds on grape vines open and put out their new leaves and flowers. Even after spring, the first crops cannot be harvested until May or June at the earliest. Hopefully, cows will give birth and there will be milk (cows lactate only while they have calves too young to eat grass). How do you preserve food for the winter?

Grains such as wheat dry naturally just before harvest and can be stored rather easily; they are so dry that microbes cannot grow. One problem is that they must be kept dry—protected not only from winter's dampness but also from the water produced by their own slow respiration.

Fruits also can be dried, but these cannot be kept dry reliably without cellophane or plastic bags. Under ancient storage methods, dried fruit often drew enough moisture from wet winter air to allow fungi and bacteria to grow on them. One way to prevent microbial growth was discovered early: fermentation. If sugary fruits were partially respired in a sealed jar (anaerobic conditions), production of ethanol would finally kill all microbes, sterilizing the fruit. As long as the jar remained sealed, the fruit (or fruit juice) was safe, including all of its minerals, vitamins, amino acids, and so on. Beer is most often fermented barley, and barley grows best in cool, damp, northern climates where it is most difficult to keep grains dry.

Natural, unpasteurized milk becomes sour in just 2 or 3 days at room temperature (and, of course, in 3000 BCE just about all temperatures were room temperature) as bacteria grow rapidly, but if the right fungi are added, milk is only partially degraded and the resulting cheese is dry and stable enough to stop bacterial growth. Covered with a layer of wax to protect it from air-borne bacteria, it can last for years.

Using yeast to leaven bread (to make it rise and be spongy) has basically the opposite effect. If wheat is ground to flour, mixed with water, and then baked, the resulting bread can be so hard and dry that it will last forever, even in your mouth—mastication (chewing) is an ordeal. However, carbon dioxide from aerobic respiration of baker's yeast becomes trapped by the bread dough, causing the bread's texture to be lighter and more open. This bread does not preserve as well, but it is easier to eat.

---

transfer one of its phosphate groups onto an ADP, converting it to ATP and changing the 1,3-diphosphoglycerate into 3-phosphoglycerate, a process called **substrate level phosphorylation**. The enzyme is phosphoglycerate kinase; the kinases constitute a large group of enzymes that remove phosphate groups from substrates. Phosphorylases are just the opposite, adding phosphates to substrates.

Although this is basically a reversal of the stroma reactions, ribulose-1,5-bisphosphate does not occur next as one might expect. Instead, 3-phosphoglycerate is converted first to 2-phosphoglycerate and then to phosphoenolpyruvate (PEP), the same metabolite that is the carbon dioxide acceptor in C_4 metabolism. PEP is also energetic enough that an enzyme can transfer its phosphate group onto ADP to make ATP; dephosphorylation causes PEP to become pyruvate.

If no oxygen is present, anaerobic respiration has now removed all the energy possible. From each molecule of glucose, four ATPs were generated, and two were consumed; thus, there is a net production of two ATPs. Occasionally, the cell can start with glucose-6-phosphate and use one less ATP to get started. The ATP generated in anaerobic respiration is used for other metabolic reactions such as protein synthesis, nucleic acid replication, microtubule assembly, and ion transport. Indeed, these metabolic pathways are the reasons for respiration, and the ADP they generate migrates back to the sites of glycolysis and is rephosphorylated to ATP.

The reduction of NAD^+ to NADH during glycolysis is a problem. If the cell needs reducing power, the NADH can be used, and it regenerates the NAD^+ necessary to keep glycolysis running. For example, roots absorb nitrates and sulfates that must be reduced, and during the synthesis of fatty acids, large amounts of reducing power are needed. Many of these reductions can be carried out with this extra NADH (**Figure 11-5**), but usually a cell does not need as much reducing power as is produced during respiratory ATP

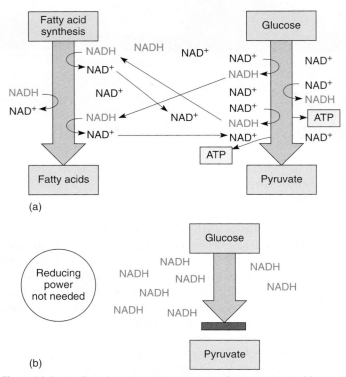

Figure 11-5 (a) If a cell needs reducing power, as for the synthesis of fatty acids, those reactions are actually electron-accepting reactions and they regenerate NAD⁺, allowing the continued production of ATP by glycolysis. (b) If reducing power is not needed, NAD⁺/NADH accumulates as NADH. Without NAD⁺, glycolysis stops: 3-phosphoglyceraldehyde cannot be oxidized to 1,3-diphosphoglycerate, and no ATP can be formed.

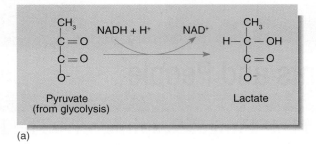

(a)

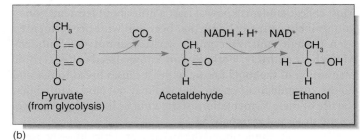

(b)

Figure 11-6 (a) When we exercise slowly enough for blood circulation to keep up, all our muscular activity is aerobic, but with rapid, intense, and prolonged activity, blood does not carry oxygen to the muscles rapidly enough and lactic acid fermentation begins. Lactate accumulation causes cramps and muscle pain. (b) Alcoholic fermentation involves the conversion of pyruvate to acetaldehyde before reduction by NADH + H⁺. Some goldfish survive low oxygen by respiring glucose to ethanol, for example, when ice on a lake surface prevents oxygen from dissolving into the water.

production; consequently, NADH accumulates; all the NAD⁺ is consumed, and glycolysis stops for lack of NAD⁺. Without further glycolysis, no ATP would form and death would result. The big problem is converting NADH back to NAD⁺ by dumping its electrons onto something. In animal tissues under anaerobic conditions, the electron acceptor is pyruvate: NADH reacts with it to form **lactate**, the anion of **lactic acid** (Figure 11-6). In plants and fungi, pyruvate is first converted to **acetaldehyde**, and then NADH reacts with that, forming **ethanol** (**ethyl alcohol**; Figure 11-7a). Anaerobic conditions occur in plants and fungi growing in mud beneath stagnant water, especially in swamps and marshes (Figure 11-7b). Rice seeds germinate and grow anaerobically until the shoots reach oxygenated water (Figure 11-8).

Ethanol and lactate are not especially good solutions to the problem of NADH accumulation. The pyruvate consumed is always present in adequate amounts, of course, because glycolysis itself produces it, but it is not really "cheap" because many of its bonding orbitals have high-energy electrons. Furthermore, pyruvate could be used as a monomer for many types of synthesis. A lack of oxygen forces a cell to use a very valuable molecule as an electron dumping ground. Even worse, the products, either lactate or ethanol, are toxic. If they accumulate in the tissues or environment, they damage or even kill the cells producing them (Table 11-2).

Considering the negative aspects of anaerobic respiration, how could natural selection have produced something so inefficient? At the time life arose and respiratory pathways were evolving, Earth's atmosphere contained reduced, hydrogen-rich compounds but no oxygen. Consequently, pyruvate and acetaldehyde had to be used as electron acceptors. Millions of years later, photosynthesis based on chlorophyll *a* evolved, and oxygen was released to the environment as a result of the water-splitting involved. After free oxygen became relatively abundant, the mutations leading to aerobic respiration began to be selectively advantageous.

Even in the presence of an oxygen-rich atmosphere and aerobic respiration, anaerobic respiration still has some selective advantage. Although the method is far from ideal, it does allow certain organisms to survive in particular environments. For example, because rice seedlings are capable of anaerobic respiration, they germinate and establish themselves during times of floods when other plants or seedlings are suffocating (Figure 11-8). In the absence of oxygen, the alternative to anaerobic respiration is death (see Table 11-1). Although this metabolism is expensive for the rice, by the time the flooding subsides, rice plants are already well rooted, have several leaves, and outcompete other species whose seeds are just beginning to germinate.

■ Aerobic Respiration

With oxygen present, the problems of using pyruvate or acetaldehyde are eliminated; oxygen is absorbed and acts as the terminal electron acceptor. Oxygen is inexpensive because it is absorbed and distributed by molecular diffusion, which requires neither active transport nor ATP consumption. Oxygen is abundant in most situations, and the product of reduction is water, which is not only nontoxic but actually beneficial. Aerobic respiration

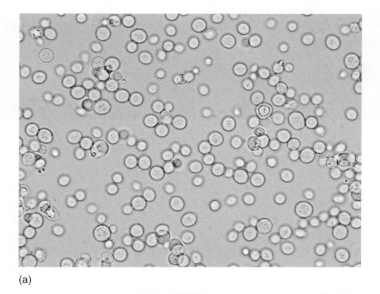

(a)

(b)

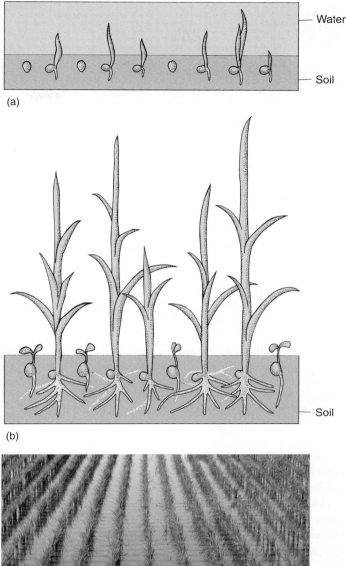

(a)

(b)

(c)

Figure 11-7 (a) In wine-making, yeast cells (*Saccharomyces*) ferment the glucose present in grapes and excrete the waste product ethanol. Naturally fermented wine has a maximum alcohol content of 14%; at that point, the waste product kills the yeast. Beer has a lower alcohol content, 5.5%, not because its yeast is more sensitive to the ethanol but because it is the legal limit. Fermentation must be stopped artificially, usually by heating the beer to kill the yeast. Alcoholic beverages with alcohol contents higher than 14% have extra alcohol added or part of their water removed (×430). (b) The leaves and upright stems of these marsh plants exist in a highly oxygenated aerobic environment, but the rhizomes and roots exist in an anaerobic muck. Aerenchyma tissues in the stems permit the diffusion of some oxygen from leaves to rhizomes and roots, but some anaerobic respiration may be occurring. The trees in the background only survive where there is adequate oxygen in the soil.

Figure 11-8 (a) Even under flooded, anaerobic conditions, rice seeds germinate and begin to grow because their embryos carry out facultative anaerobic respiration. Other seeds (round) remain dormant because they are incapable of fermentation; being dormant, they have a low rate of metabolism and the small amount of oxygen present may keep them from dying immediately. (b) When flooding subsides, oxygen is available, and nonrice seeds germinate; however, they are shaded by rice plants, and the soil is already filled with rice roots. The rice plants have a tremendous advantage. (c) Rice seedlings, still flooded.

consists of three parts: (1) glycolysis, (2) the citric acid cycle, and (3) oxidative phosphorylation in an electron transport chain.

Glycolysis The initial steps of aerobic and anaerobic respiration are identical: glycolysis by the Embden-Meyerhoff pathway to pyruvate (Figure 11-4). This produces ATP and NADH just as before, but with oxygen present, the NADH migrates to electron carriers that oxidize it back to NAD⁺, permitting glycolysis to continue. Glycolysis occurs in the cytosol.

The Citric Acid Cycle Because pyruvate is not needed as an electron acceptor, it can be used in a number of metabolic pathways. One of the main pathways takes advantage of the large amount of energy remaining in pyruvate and rather than using it for its structure, breaks it down and generates more ATP. This one pathway has three names: the **citric acid cycle**, the **Krebs**

Box 11-2 Respiration in Prokaryotes

Basic Reactions

In all plants, animals, fungi, cyanobacteria, and most bacteria and archaea, the compounds used for respiration are organic—sugars, fats, amino acids, and so on; however, certain prokaryotes (**lithotrophs**) metabolize inorganic compounds and extract energy from them, examples being hydrogen, sulfur, iron, and nitrogen.

Electron Donors

Hydrogen respiration is easy to understand. The reaction $2H_2 + O_2 \rightarrow 2H_2O$ liberates two electrons and releases so much energy that it easily forces electrons onto NAD^+, producing NADH. This then donates the electrons to an ordinary electron transport chain that produces ATP. Hydrogen sulfide (H_2S) and thiosulfate ($S_2O_3^{-2}$) can both be converted to sulfur (S^0) and then to sulfate (SO_4^{-2}). These reactions release energy and electrons that either go into an electron transport chain or react directly to produce ATP. The end product is sulfuric acid, which is toxic like the ethanol produced in plant and yeast fermentation. As long as the environment absorbs the acid and keeps it dilute, the bacteria are unharmed. *Thiobacillus thiooxidans* grows well even at pH 2. Oxidation of ferrous iron (Fe^{+2}) to ferric iron (Fe^{+3}) releases an electron that can be used to form a single molecule of ATP (**Figures B11-2a and b**). Likewise, ammonium (NH^{+4}) can be oxidized to nitrite (NO_2^-) and then to nitrate (NO_3^-), releasing six energetic electrons that can be used for ATP synthesis.

Electron Acceptors

Just as various prokaryotes use various reduced compounds as sources of electrons and energy, other prokaryotes use various highly oxidized compounds as electron acceptors. The most common is nitrate (NO_3^-), which is converted to either NO_2^- or N_2. Sulfate also acts as an electron acceptor and is converted to hydrogen sulfide. These processes are the opposite of those described previously here, in which these same compounds were used as substrates for respiration because they could be oxidized. How can they act as a source of electrons one time but as acceptors another time? The key is the presence or absence of oxygen: Oxygen has such a strong tendency to absorb electrons that in its presence the balance between H_2S and SO_4^{-2} is shifted in favor of SO_4^{-2}. Without oxygen, the balance is shifted the other way.

Certain organic compounds act as electron acceptors when oxygen is not present; the process is fermentation. Bacteria ferment diverse substrates such as numerous sugars and organic acids, resulting in products like ethanol, lactate, propionate, acetate, acetone, methanol, and carbon dioxide. Fermentations of amino acids (e.g., bacteria decomposing animal bodies), and other compounds that contain nitrogen or sulfur are often called putrefactions and produce odiferous compounds: hydrogen sulfide (aroma of rotten eggs), isobutyric acid, cadaverine (from lysine), putrescine (from ornithine), and ammonia. Researchers are experimenting with methods to increase fermentation of cellulose and lignin to ethanol or methane to be used as fuel, but as you know, these are extremely inert, nonreactive compounds.

Integrated Metabolism

Relationships between photosynthesis, respiration, and carbon are simple but important. All photosynthetic plants are autotrophs. They take in carbon dioxide, reduce it to carbohydrate

Figure B11-2a   This stream in Colorado is in a natural condition with clean water.

Figure B11-2b   This stream is located only a few miles from the other, but it is polluted with run-off from mining wastes. Iron-oxidizing bacteria use the ferrous iron as a substrate, respiring it to ferric iron and generating ATP. Virtually no other life occurs in such polluted conditions.

by photosynthesis (carbon fixation), transport it to another part of the plant, and then use most to build structures. They respire the rest to produce ATP. All animals are heterotrophs: Their food consists of reduced carbon compounds, and a small amount is used to construct animal tissues; however, the majority is respired as an energy source. All animals lack the Calvin cycle; no animal is able to fix carbon dioxide.

In prokaryotes, similar relationships hold and others also occur. Green bacteria, cyanobacteria, and some purple bacteria are photosynthetic (phototrophic) and have the Calvin cycle (they can fix CO_2: they are autotrophic). They are therefore **photoautotrophs** like plants. Most other bacteria, like animals, take in organic substances and use them both for ATP generation and for polymer construction; these are **conventional heterotrophs**. Photosynthetic bacteria, however, commonly have food and light available, and most purple bacteria and green bacteria are **photoheterotrophs**: They absorb and use organic carbon for construction and use photosynthesis almost exclusively to generate ATP. Little or no photosynthetic energy is used to reduce $NADP^+$ to NADPH because NADPH is not in great demand, because the Calvin cycle is unnecessary if carbohydrate is present in food.

Lithotrophs use carbon dioxide as their primary carbon source; they are **lithotrophic autotrophs**. Examples are the colorless (nonphotosynthetic) sulfur bacteria, hydrogen bacteria, nitrifying bacteria, and iron bacteria. As they oxidize hydrogen, sulfur, iron, and so forth, the energy goes to make ATP, part of which is used for regular metabolism and part to pump protons, creating a pH gradient that can be used to reduce $NADP^+$ to NADPH. Their energy metabolism is somewhat like bacterial photosynthesis. Neither metabolism can produce a compound as energetic as activated chlorophyll.

A very few bacteria (*Beggiatoa*, *Thiobacillus*, and some others) are **lithotrophic heterotrophs**: They get their energy by oxidizing sulfur compounds and they take in organic compounds for structural uses. They do not use carbon dioxide, and apparently they lack the Calvin cycle.

The ability to live as a heterotroph offers great advantages. By taking in carbon in its reduced form, heterotrophs use all their respiratory or fermentative energy for growth and reproduction. Autotrophs, which must use energy to reduce carbon dioxide, seem to be at a disadvantage. Their one advantage is that an autotroph has probably never starved to death; carbon dioxide is always abundant enough to sustain the life of autotrophs. The same is not true for the food supply of heterotrophs, whether that food is organic or not; famine is a common occurrence not only for animals and fungi but also for bacteria.

cycle, or the **tricarboxylic acid cycle**. These names reflect different facts about the cycle. One of the intermediates is citrate, the anion of citric acid. Much of the pioneering work on this metabolism was carried out by Hans Krebs. Finally, several of the intermediates are tricarboxylic acids—that is, each has three carboxyl (-COOH) groups.

TABLE 11-2	Electron Acceptors
Pyruvate	
1. Abundant when needed	
2. Expensive	
3. Toxic product (lactate)	
Acetaldehyde	
1. Abundant when needed	
2. Expensive	
3. Toxic product (ethanol)	
Oxygen	
1. Usually abundant when needed	
2. Cheap	
3. Nontoxic product (water)	

In the citric acid cycle, pyruvate is transported from the cytosol, where glycolysis occurs, across the mitochondrial membranes to the mitochondrial matrix. There it is oxidized and decarboxylated: As electrons are transferred to NAD^+, bonding orbitals holding the last COO^- rearrange and CO_2 is liberated. Carbon dioxide and NADH are produced, along with a two-carbon fragment called **acetyl** (Figure 11-9). The carbon dioxide and NADH remain free in the matrix solution, but the acetyl becomes attached to a carrier molecule, **coenzyme A (CoA)**, the resulting combination being **acetyl CoA**. Like pyruvate, acetyl CoA can be used in many synthetic pathways, but here we are interested in its entry into the citric acid cycle by transfer of the acetyl group to an acceptor molecule, oxaloacetate, a compound with four carbons (Figure 11-10). The oxaloacetate is converted to a six-carbon compound, **citrate**, which is then rearranged to **cis-aconitate**, which in turn is transformed to **isocitrate** (Figure 11-10, steps A, B, C, and D). In the next step, one of the carbons of isocitrate is oxidized by passing electrons onto NAD^+, creating NADH. The oxidized carbon is liberated as carbon dioxide, leaving **α-ketoglutarate**, which has only five carbons (step E). This too is oxidized by NAD^+, liberating another carbon dioxide, and the four-carbon remnant becomes

Figure 11-9 The process by which pyruvate is attached to CoA releases a carbon dioxide molecule and forms NADH. This step does not occur during anaerobic respiration; not only would it use up pyruvate needed as an electron acceptor, but it would generate even more NADH. Both of those results would be detrimental under anaerobic conditions, but both are beneficial when oxygen is present.

attached to a new molecule of CoA in the process, forming **succinyl CoA** (step F). The energy released by the breakdown into free CoA, and free succinate can power phosphorylation of ADP to ATP (step G). The succinate still contains considerable energy and is oxidized to **fumarate** as electrons and protons are passed to **flavin adenine dinucleotide** (**FAD**), reducing it to $FADH_2$ (**Figure 11-11**). Although the molecule becomes oxidized by this step, no carbon dioxide is lost; the fumarate is also a four-carbon compound. It reacts with a water molecule and becomes **malate**, which passes a final set of electrons onto NAD^+ and is transformed into the original acceptor molecule, oxaloacetate (steps H, I, and J).

It was mentioned above that the benefit of the citric acid cycle is the generation of more ATP; nevertheless, at only one step has ATP been produced. Instead, there are four steps in which more NAD^+ is reduced to NADH and one in which FAD is reduced to $FADH_2$. Excess NADH and the related deficiency of NAD^+ were seen to be problems in anaerobic respiration, and thus, the citric acid cycle at first seems to be a real contradiction. However, in the next step, the electron transport chain, the energy in NADH and $FADH_2$ drives the synthesis of ATP, and NADH is simultaneously oxidized back to NAD^+.

The Mitochondrial Electron Transport Chain: Chemiosmotic Phosphorylation The mitochondrial inner membrane, like the chloroplast inner membrane, contains sets of compounds capable of carrying electrons (**Figure 11-12**). Although many of the actual carriers differ in the two organelles, the principles of electron transport are the same. The carriers that react with NADH or oxygen are placed precisely and asymmetrically in the crista membranes, both on the matrix side only (**Figure 11-13**). The exact order of carriers in plant mitochondria is complex, and numerous types exist. Only the most well-understood are presented later here.

NADH diffuses to the membrane and passes electrons to a protein that has **flavin mononucleotide** (**FMN**) bound to it as a cofactor. The FMN is reduced, and the NADH simultaneously oxidized to NAD^+, which can migrate back to the site of the citric acid cycle. Reduced FMN ($FMNH_2$) passes the electrons to a set of iron- and sulfur-containing, proteinaceous electron carriers, which transfer the electrons to one or several quinones, one of which is **ubiquinone** (also called coenzyme Q; **Figure 11-14**). From the pool of quinones, electrons are transferred to **cytochrome b**. The next carriers in sequence are more quinones, then **cytochrome c_1, cytochrome c, cytochrome a,** and **cytochrome a_3**. Cytochromes a and a_3 are part of a large enzyme complex known as **cytochrome oxidase,** which contains several proteins and two copper ions that mediate transfer of electrons from the iron in

cytochrome to oxygen. As oxygen is reduced, it picks up two protons and becomes water.

If this electron transfer were the only thing to happen in the electron transport chain, it would be better than anaerobic respiration because the cell could avoid the waste of pyruvate and the synthesis of the toxic lactate or ethanol, but much more is accomplished. Mitochondria also perform chemiosmotic phosphorylation. As with chloroplasts, the fate of the protons released and absorbed during electron transport is important. In mitochondria, when NADH reacts with FMN, two protons are transferred as well as electrons. One proton comes from the NADH and the other from the water, both on the matrix side of the membrane (a few water molecules are always splitting spontaneously [$H_2O \rightarrow H^+ + OH^{-2}$]; water is not split specifically as in photosystem II of photosynthesis). When $FMNH_2$ reduces any quinone, the two protons are released on the other side of the membrane into the lumen of the crista. This step of electron transport acts as a proton pump; a large concentration of protons begins to build up in the lumen while there is a lack of protons in the matrix. Ubiquinone and the other quinones just before cytochrome c_1 act in a similar manner, pumping protons out every time they carry electrons. Also, as oxygen receives electrons from the electron transport chain, it forms water by picking up two protons from the matrix, decreasing the proton concentration. In a short time, a proton concentration gradient develops that is strong enough to cause the protons to migrate back into the matrix.

The flow of protons from the crista lumen to the mitochondrial matrix can be used to synthesize ATP. As in chloroplasts, ATP synthetase channels in the membrane use the flow of protons to force a phosphate group onto ADP, creating ATP. This is a chemiosmotic phosphorylation, and because of it, the NADH is an excellent if indirect source of ATP rather than a problem. The electrons that each molecule of NADH contributes to the mitochondrial electron transport chain provide enough power to create three ATPs.

$FADH_2$ also passes its electrons to the mitochondrial electron transport chain, but it reacts with ubiquinone instead of FMN; thus, the first step in proton pumping does not use these electrons. Therefore, $FADH_2$ contributes less to the proton gradient than does NADH, providing enough power for production of two ATPs rather than three.

The NADH Shuttle NADH produced by glycolysis cannot cross the mitochondrial inner membrane and donate electrons directly to the electron transport chain because the inner membrane is impermeable to such large molecules. Instead, a series of chemical reactions carries (shuttles) reducing power across the membrane

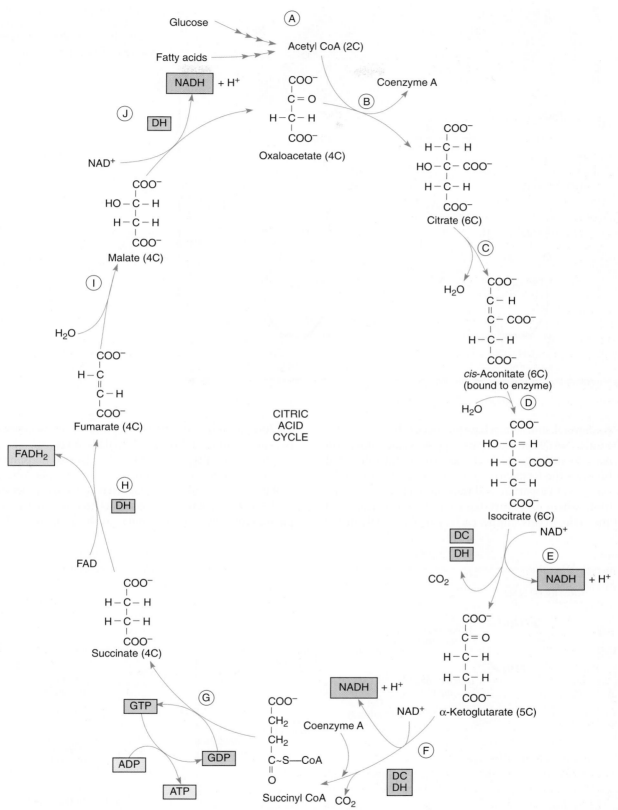

Figure 11-10   The steps of the citric acid cycle. This cycle is an important part of aerobic respiration, even though it does not consume oxygen and generates only one molecule of ATP directly. DH = dehydrogenation (oxidation); DC = decarboxylation.

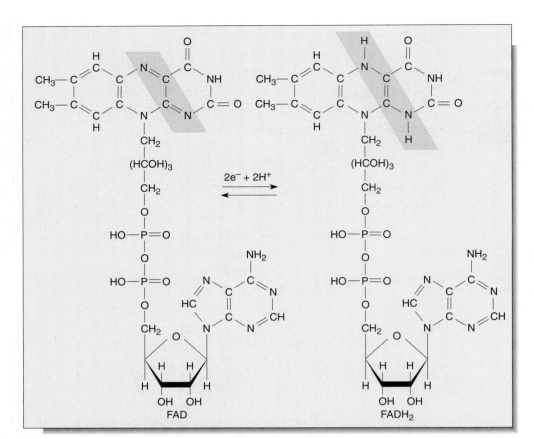

Figure 11-11 Flavin adenine dinucleotide (FAD) is a large organic electron carrier similar to NAD+. When it is reduced by two electrons, it picks up two protons, becoming FADH$_2$. Unlike NADH + H+, both protons are covalently attached to FADH$_2$ (shown in blue box).

(Figure 11-15). Several shuttle mechanisms occur. In the malate-aspartate shuttle, NADH in the cytosol powers the conversion of oxaloacetate to malate, which crosses to the mitochondrial matrix and powers the formation of a new molecule of NADH. Malate is converted to aspartate and transported back out of the mitochondrion, where it is converted to oxaloacetate again and can repeat the cycle. In this shuttle, each cytosolic NADH drives

the formation of a matrix NADH and the consequent oxidative phosphorylation of three ADPs to three ATPs.

In a second type of shuttle, the glycerol phosphate shuttle, cytosolic NADH reduces dihydroxyacetone phosphate to glycerol phosphate, which is transported across the inner membrane to the matrix. There it converts back to dihydroxyacetone phosphate, reducing FAD to FADH$_2$ in the process. Each cytosolic

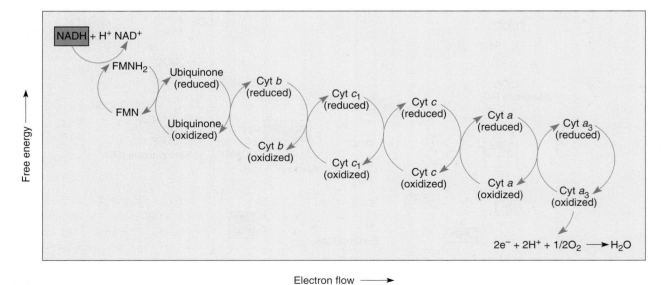

Figure 11-12 Membrane-bound electron carriers of the mitochondrial electron transport chain. As in chloroplasts, their positions and movements are important (see Figure 11-13). Electrons are brought to membrane-bound carriers by mobile carriers such as NADH and FADH$_2$, which are produced in the matrix.

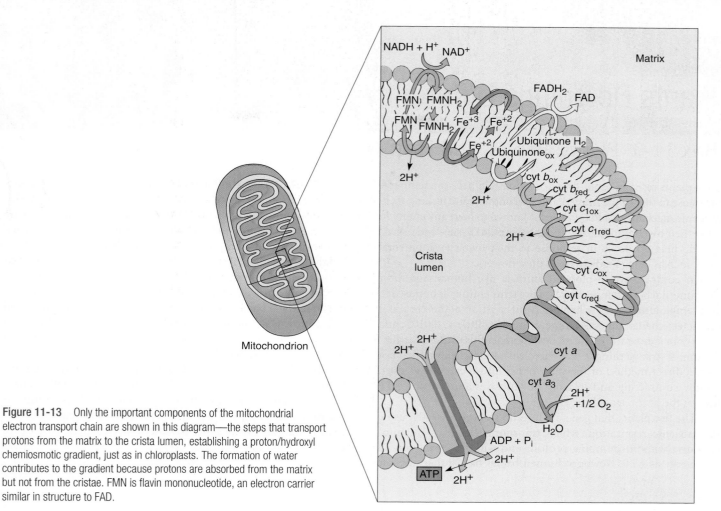

Figure 11-13 Only the important components of the mitochondrial electron transport chain are shown in this diagram—the steps that transport protons from the matrix to the crista lumen, establishing a proton/hydroxyl chemiosmotic gradient, just as in chloroplasts. The formation of water contributes to the gradient because protons are absorbed from the matrix but not from the cristae. FMN is flavin mononucleotide, an electron carrier similar in structure to FAD.

Figure 11-14 Ubiquinone (coenzyme Q) is a quinone that can carry two electrons and two protons simultaneously.

NADH results in the formation of one matrix $FADH_2$, which can drive the formation of only two ATPs; not as much energy is conserved in this shuttle.

In at least some plants, NADH can cross the outer mitochondrial membrane and react directly with ubiquinone at the outer surface of the inner membrane. A shuttle mechanism is not necessary, but the step of proton pumping by FMN is bypassed, decreasing the amount of ATP that can be generated.

■ Heat-Generating Respiration

During glycolysis, the citric acid cycle, and mitochondrial electron transport, small amounts of energy are lost in each step even though a great deal of energy is conserved by the synthesis of ATP (see Figure 11-2). The total chemical energy of all products (ATP, carbon dioxide, and water) is less than that of all reactants (gluclose-6-phosphate and oxygen); the difference is "lost" as heat and increased entropy (disorder). For example, compost piles become warm because of the heat loss during the respiration of the fungi and bacteria that decompose the compost. Heat loss is usually an inefficient aspect of respiration; in most cases, it would be selectively advantageous for a plant to produce more ATP from each molecule of glucose respired and lose less heat to the environment.

The heat "lost" during respiration by warm-blooded mammals is vitally necessary to maintain body temperature, and when

Plants Do Things Differently

Box 11-3 Plants, Babies, and Heat

Most plants resemble cold-blooded animals in having a temperature close to that of the immediate environment and lacking thermogenic mechanisms. Plants are not known to have any ability to sense their own temperature, and the few plant tissues and organs that do heat up do so in response to their own developmental state, such as being ready to open their flowers.

In contrast, warm-blooded animals like humans monitor their internal body temperature—core temperature as opposed to skin temperature—by the hypothalamus portion of their brain. It can detect changes in core temperature of as little as 0.01°C, and when core temperature drops, the hypothalamus causes heat generation. A strange thing is that we warm-blooded animals do not have a direct method of thermogenesis: All we can do is shiver, rapidly contracting and relaxing our muscles and wasting ATP just for the heat released.

The few plants that generate heat do so more directly, using thermogenic respiration, which transports electrons in mitochondria without pumping protons and without saving any of the energy as ATP. Newborn human babies have a thermogenic

mechanism rather similar to that of plants. They have a special type of fat, called brown fat, in which the cells are filled with numerous mitochondria, but these mitochondrial membranes are permeable to protons. Mitochondrial electron transport occurs, along with proton pumping; however, no chemiosmotic gradient builds up, and no ATP is synthesized. Thus, all energy is converted to heat. This is an extremely efficient method of generating heat, but for some reason, we give this up when we are several months old. While still babies, we acquire the ability to shiver and lose all of our brown fat. Mammals that hibernate, however, retain brown fat thermogenesis: Hibernating bears stay warm just like human babies, and almost like plants.

Brown fat cell and plant thermogenic respiration have similar results—the generation of heat. It is not known whether one is more suited to animals and the other to plants. Perhaps in animals the mutation of proton-permeable membranes occurred earlier than mutations of the genes for alternative carriers, whereas in plants the opposite occurred.

we are chilled, we must generate even greater amounts of heat by shivering: Our muscles contract and relax rapidly, breaking down large amounts of ATP and releasing the stored energy as heat.

Some plants also generate large amounts of heat. In the voodoo lily (*Sauromatum guttatum*), parts of the inflorescence become much warmer than the surrounding air, causing amines

and other chemicals to vaporize and diffuse away as chemical attractants for pollinators. Skunk cabbage (*Symplocarpus foetidus*) often begins floral development while covered with snow; it melts the snow cover and exposes its flowers by generating large quantities of heat (**Figure 11-16**). These and other plants produce heat much more efficiently than humans do; they have alternative electron carriers that apparently do not pump protons during electron transport in mitochondria. Consequently, there is no proton gradient and no chemiosmotic production of ATP. The energy in NADH is converted entirely to heat, and the tissues become quite warm (**Figure 11-17**).

In ordinary mitochondria, cyanide (CN^-), azide (N_3^-), and carbon monoxide (CO) interfere with the last electron carrier, cytochrome oxidase. When they are present, electrons cannot pass from cytochrome c to oxygen; without electron flow and consequent ATP generation, both animals and plants die unless capable of anaerobic respiration. In plants that generate heat, the alternative electron carriers do not interact with cyanide, azide, or carbon monoxide, and heat is generated even if these chemicals are present. Heat-generating respiration can thus be studied by poisoning normal aerobic respiration with cyanide; consequently, heat-generating respiration is usually called **cyanide-resistant respiration**. A better name is **thermogenic respiration**.

The term "cyanide-resistant respiration" is somewhat misleading because it suggests that plants can be immune to cyanide poisoning. The analogy that because anaerobic respiration allows

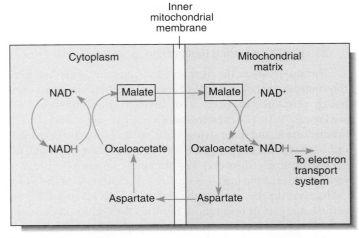

Figure 11-15 The malate–aspartate shuttle allows reducing power to be transferred into mitochondria when NADH itself cannot cross the mitochondrial membranes.

Figure 11-16 By generating heat internally, skunk cabbage (*Symplocarpus foetidus*) maintains a high enough temperature to carry out active metabolism even when covered by snow. When it is ready to emerge, it produces even more heat and melts the snow, revealing the inflorescence to pollinators.

cells to survive in anaerobic conditions, cyanide-resistant respiration must allow cells to survive in the presence of cyanide is incorrect. Plants never encounter high concentrations of cyanide in natural conditions; if they did they would be killed because aerobic ATP generation is blocked. "Thermogenic respiration" and "heat-generating respiration" are less confusing terms.

Many aspects of thermogenic respiration are still unknown. The alternative carriers appear to be present in all mitochondria, and they seem to be resistant to stressful conditions such as cold or desiccation. One hypothesis is that during stress, the ordinary, proton-pumping carriers do not function as well, causing ATP production to drop, but the alternative carriers continue to transport electrons, allowing NADH to be converted back to NAD$^+$, which can then be used to keep glycolysis running and producing ATP. If this hypothesis is correct, thermogenic respiration is a metabolism that typically functions only rarely and helps plants survive stress conditions, and it has been adapted for actually generating heat in a small number of species.

■ Pentose Phosphate Pathway

The intermediates of all respiratory pathways can be used in other pathways to make various compounds; it is not inevitable that they will be completely oxidized to carbon dioxide and water. The **pentose phosphate pathway**, which involves several intermediates that are phosphorylated five-carbon sugars (pentoses), is an important source for many fundamental compounds. It is usually included in discussions of respiration because it begins with glucose-6-phosphate, gives off carbon dioxide, and involves oxidations that produce NADPH (**Figure 11-18**); however, its importance as a source of respiratory energy is much less significant than its role as a synthetic pathway. The pentose phosphate pathway transforms glucose into four-carbon sugars (erythrose) and five-carbon sugars (ribose) that are essential monomers in many metabolic pathways. The ribose-5-phosphate produced can be shunted into nucleic acid metabolism, forming the basis of RNA (*ribo*nucleic acid) and DNA (deoxy*ribo*nucleic acid) monomers, the nucleotides. In meristematic cells, large amounts of DNA must be synthesized during the S-phase of a short cell cycle; the pentose phosphate pathway is an extremely important part of the metabolism of these cells.

The four-carbon sugar erythrose-4-phosphate is the starting material in the synthesis of many compounds. Two important types are lignin and anthocyanin pigments. Tissues such as wood, fibers, and sclereids deposit large amounts of lignin into their secondary walls during development, and erythrose-4-phosphate is in great demand. During differentiation, these cells use the pentose phosphate pathway but pull out erythrose-4-phosphate, not ribose-5-phosphate. Flower petals and brightly colored fruits divert erythrose-4-phosphate from the pentose phosphate pathway to anthocyanin production while synthesizing their pigments. The pentose phosphate pathway also occurs in plastids, where it supplies erythrose-4-phosphate for synthesis of amino acids such as tyrosine, phenylalanine, and tryptophan.

In addition to these four- and five-carbon sugars, the pentose phosphate pathway also produces NADPH. Although many anabolic reductions in the cytoplasm use NADH rather than NADPH, the reduction of nitrate (NO_3^-) to amino acids (NH_3^+) can be accomplished only by NADPH, which can be generated by the pentose phosphate pathway.

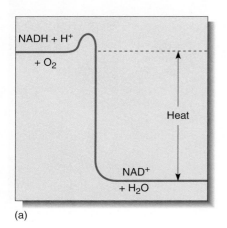

(a)

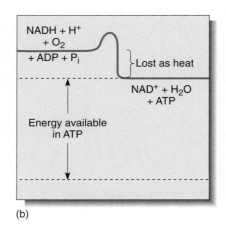

(b)

Figure 11-17 If NADH breaks down to NAD$^+$, a large amount of energy is liberated. (a) If this powerful exergonic reaction is not coupled to any endergonic reaction, all of the energy is converted to heat. (b) When coupled to the endergonic synthesis of ATP from ADP, most of the energy is conserved and little is converted to heat.

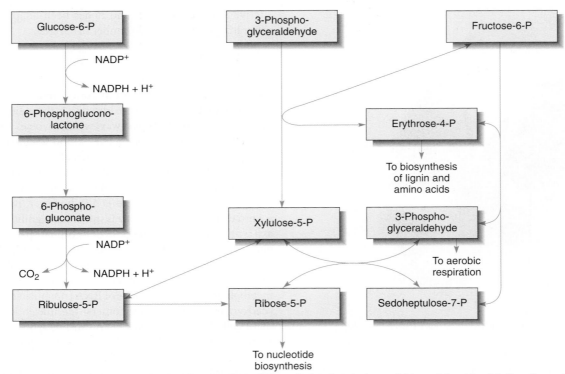

Figure 11-18 The pentose phosphate pathway, showing all intermediates. If ribose-5-phosphate is drawn off into nucleic acid metabolism, the pentose phosphate pathway is shifted in favor of ribose production. If erythrose-4-phosphate is diverted to lignin metabolism, the pentose phosphate pathway reaction equilibria are shifted toward erythrose production.

Many of the reactants in the pentose phosphate pathway are the same as those in glycolysis, and both pathways occur in the cytosol. They are best understood as interconnected and simultaneous pathways. In meristematic cells, the pentose phosphate pathway is active and shifted in favor of ribose production. In wood cells, it is also active but produces erythrose. In other cells, it may be much less active and glycolysis may dominate, producing NADH that then powers ATP production in mitochondria. Imagine a cell as it is produced in the vascular cambium (meristematic, needs nucleic acids) and then differentiates into lignified wood parenchyma (needs lignin), which after the wall is mature,

takes on the role of storing starch during the summer and releasing it in the spring (Figure 11-19). Energy metabolism is adjusted at each stage in a major way, and smaller changes in the rates of reactions may occur on a day-to-day basis (Figure 11-20).

■ Respiration of Lipids

Some tissues, especially oily seeds and dormant apical meristems, store large amounts of lipid, usually as triglycerides or phospholipids. During germination or release from dormancy, lipids undergo catabolic metabolism in which they are broken

Figure 11-19 (a) At the cambium, the pentose phosphate pathway may be producing ribose-5-phosphate, but in the differentiating vessels, it is producing erythrose-4-phosphate. In all cells, glycolysis and the rest of aerobic respiration are also occurring simultaneously with the pentose phosphate pathway (×400). (b) In petals, erythrose-4-phosphate, used in the production of pigments, is produced by the pentose phosphate pathway.

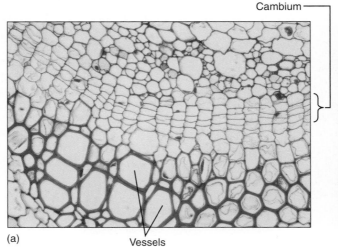

(a)

(b)

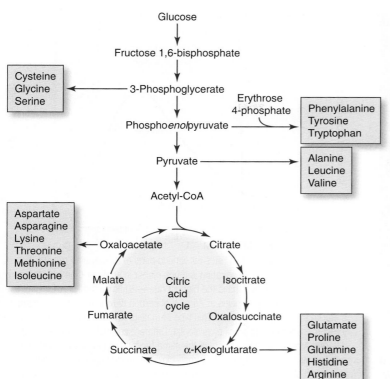

Figure 11-20   All 20 amino acids are constructed from intermediates of respiration. Although too much sugar harms our health, plants construct everything from glucose and some minerals.

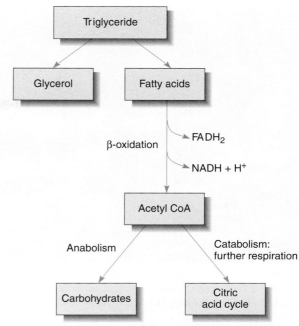

Figure 11-21   During the respiration of lipids, fatty acids are separated from glycerol and undergo β-oxidation to acetyl CoA units, producing $FADH_2$ and NADH. In germinating seeds, much of the acetyl CoA is converted to glucose and fructose by gluconeogenesis then are polymerized into sucrose. The sucrose is transported by phloem from cotyledons to embryo meristems, where it is used in construction or respired for energy.

down into glycerol and three fatty acids (triglycerides) or glycerol phosphate and two fatty acids (phospholipids). Fatty acids are then further broken down into two-carbon units—acetyl CoA—by a process called **β-oxidation** in either cytosol or microbodies called glyoxysomes. For example, an 18-carbon fatty acid would be converted into nine acetyl CoA units. As each acetyl CoA is formed, one FAD is reduced to $FADH_2$, and one NAD^+ is reduced to NADH, both of which can carry electrons to mitochondria and drive production of ATP by means of the electron transport chain (**Figure 11-21**). Acetyl CoA may be used for synthesis of carbohydrates and other compounds, or it may enter the citric acid cycle and be further respired.

Photorespiration

Photorespiration occurs only when RuBP carboxylase adds oxygen rather than carbon dioxide to ribulose-1,5-bisphosphate, resulting in one molecule of 3-phosphoglycerate and one of phosphoglycolate. Phosphoglycolate is dephosphorylated to glycolate, which is then transported to microbodies called peroxisomes. Glycolate can be converted to glycine in the peroxisomes, and the glycine may be transferred to mitochondria, where it is respired to carbon dioxide and water with no conservation of energy in either ATP or NADH; all energy is wasted. In some cases, some of the glycine can be converted to serine; both are useful amino acids. Other mechanisms also produce these amino acids, however, and direct measurements show that for many C_3 species, photorespiration wastes as much as 30% of the energy trapped by photosynthesis.

Environmental and Internal Factors

As with photosynthesis, numerous environmental factors influence the rate of respiration. It is necessary to consider the integration of respiration into the total biology of the plant as well as the metabolic pathways involved.

Temperature

Temperature greatly influences respiration in a plant growing under natural conditions. In all environments, shoots located in air are at a different temperature than roots in soil. Shoots are subjected to great changes of temperature from day to night, often more than 20°C.

In most tissues, an increase in temperature of 10°C, in the range between 5°C and 25°C, doubles the respiration rate, a magnitude of increase seen in many enzyme-mediated reactions. Below 5°C, respiration decreases greatly (**Figure 11-22**). Above 30°C, respiration still increases, but not so rapidly; at such high temperatures, oxygen probably cannot diffuse into tissues as rapidly as the tissues use it. Above 40°C, respiration, like many other processes, slows greatly, probably because of enzyme damage or disruption of organelle membranes.

Lack of Oxygen

Because plants are not as active as animals, much lower oxygen concentrations—as little as 1% to 2%—maintain full rates of plant respiration. Oxygen concentration in the atmosphere is so stable that it does not cause variations in respiration, but

Figure 11-22 Trees in cold winter weather are not necessarily dormant; sunlight warms them enough for metabolism to occur. Although respiration rates are low, it provides enough energy for critical metabolic reactions.

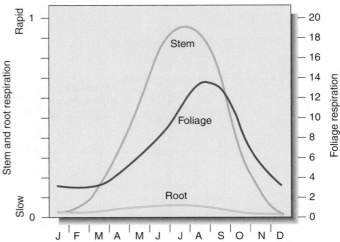

Figure 11-23 Respiration rates for various parts of loblolly pine trees in a 14-year-old plantation. Respiration is much higher in summer than in winter for all organs, and leaves have the highest rate—20 times the scale for the other organs. The rate is given as "grams of carbon dioxide released per square meter of soil surface per hour." This type of measurement is more useful to a forest manager than to a cell physiologist. Different scientists or studies need different types of measurements of the same function. A cell physiologist might express respiration as "grams of carbon dioxide released per cell per hour," "per gram of protein per hour," or "per gram of tissue per hour." Each gives a different result: Leaves have small cells with thin walls, so a gram of leaf tissue has much protein and many cells, whereas a gram of woody stem or branch is mostly inert, nonrespiring cellulose walls with few cells and little proteinaceous protoplasm. It can be difficult to select an appropriate basis for stating rates of respiration, and comparing studies that use different bases is not easy.

variations occur in a cell's access to oxygen. During daylight hours, chlorophyllous tissues and organs produce oxygen, some of which is used in respiration. At night, oxygen is not produced, but it can diffuse into the large intercellular spaces of the plant if it can penetrate the closed stomata. Because the cuticle is not absolutely impermeable to oxygen and because the atmosphere contains such a high concentration, oxygen diffuses in rather well, even through closed stomatal pores. Less oxygen is available, but the temperature is lower, and thus, demand for oxygen is reduced.

Oxygen availability is much more variable for roots; in well-drained soil, the large quantity of gas located between soil particles is depleted in oxygen owing to respiration by roots, fungi, bacteria, protists, and soil animals. During and after rain, soil air is displaced by water, and roots either have lesser amounts of oxygen (hypoxia) or none (anoxia) available. Anaerobic respiration allows roots to survive for a short period, but it is not sufficient for root growth and healthy metabolism. If continually flooded, roots of almost any species die. Those that survive do so primarily by oxygen diffusion from the stems through aerenchyma channels in the cortex or pith, such as in petioles of water lilies and cattails. Floods that submerge the bases of shoots or woody trunks are especially damaging if they last more than a few days. Rice seedlings are the only certain example of plants that can grow normally without oxygen for a prolonged period.

In thick tubers (potatoes) or bulky roots (beets, carrots), it is not well established whether respiration is completely aerobic or at least partially anaerobic. Such organs usually have a significant amount of intercellular space through which oxygen might diffuse, but the concentration of oxygen is very low. Similarly, sapwood and cambium of trees with thick bark may be hypoxic; even though lenticels are present, the diffusion path is long, and the dense inner bark, cambium, and sapwood lack intercellular spaces. Developing embryos inside large seeds and fruits are reported to respire anaerobically.

Internal Regulation

Like virtually all other processes, respiration is subject to specific metabolic controls. Cells that have an active metabolism, such as glands that secrete protein or epidermal cells that secrete waxes, have a high level of aerobic respiration, whereas meristematic cells produce ribose by means of the pentose phosphate pathway. Nearby cells, perhaps those of the spongy mesophyll or collenchyma, may have a much lower respiration rate (Figure 11-23). During fruit maturation, respiration usually remains steady or increases gradually until just before the fruit is mature, at which point a sudden burst of respiration is triggered by endogenous hormones. Conversely, within seeds, after an embryo is mature, its respiration decreases so dramatically that it is difficult to measure, and the seed becomes dormant. In seeds with a true dormant period, virtually no respiration occurs even if the embryo is surgically removed and given water, warmth, and oxygen because metabolic inhibitors suppress respiration.

Total Energy Yield of Respiration

During anaerobic glycolysis, four molecules of ATP are synthesized, whereas either one or two ATPs must be used to initiate the process, depending on whether glucose or glucose-6-phosphate is the initial substrate. The NADH + H$^+$ generated cannot be used for energy, and thus, the net result is two molecules of ATP for every molecule of glucose fermented (Table 11-3).

Alternatives

Box 11-4 Heterotrophic Plants

Are ordinary stems and foliage leaves really adaptations that improve photosynthesis? Or do plants have plant-like structure just because that is the way things are? It is easy to take really common things for granted and not even ask ourselves if there are alternatives. We can create many models and consider theoretical aspects, but we always want to verify our theories with real data, with empirical evidence. That is a principle of the scientific method. To see just how much of ordinary plant structure is the result of natural selection that improves photosynthesis, we need plants that do not photosynthesize, that are heterotrophic, and we need plants that have been heterotrophic for so long that their bodies have had many millions of years to evolve without being restricted by the need to photosynthesize. The plants we need are **parasitic plants**.

Approximately 4,000 species of parasitic plants occur in about 18 families (Table B11-4), and the ability for one plant to attack another has arisen evolutionarily about 12 times. Because there has been a dozen different starting points and because some groups had more time than others to become adapted to their hosts, parasitic plants differ widely from each other in structure, physiology, and so on. Two fundamental types of parasitic plants occur.

Hemiparasites have chlorophyll and produce all, or at least part, of their own glucose; they merely obtain water, minerals, and perhaps some organic compounds from their hosts. Hemiparasites still have leaves with stomata and rather ordinary stems. Their roots, however, penetrate the epidermis or bark of host plants rather than soil.

Holoparasitic plants have neither chlorophyll nor photosynthesis: All of their ATP is produced by aerobic respiration of glucose obtained from the host plant, and they probably need little reducing power. In all cases, after the ancestral parasitic plants became holoparasitic and lost any need for photosynthesis, mutations that curtail leaf development became beneficial: Most **holoparasites** have either small, scale-like leaves or none at all. Dodder (*Cuscuta*) is one of the few holoparasites that has any ordinary plant-like features left in its body (Figure B11-4a). It has

Figure B11-4a   The slender orange stems are *Cuscuta*, a holoparasite that still has leaves and stems, but its internodes are extremely long and slender and its leaves tiny. It is producing adventitious roots that develop as haustoria, penetrating to the vascular tissues of the host, *Justicia*.

Figure B11-4b   *Prosopanche americana* is holoparasitic, and most of its highly modified body remains underground; it has no need for photosynthesis, obtaining all its water and nutrients from the roots of its host. This flower is also highly modified: The black and white structure is a massive set of fused stamens.

(continued)

TABLE B11-4	Families that Contain Parasitic Plants

Parasitic plants occur in many families, but these listed here have some of the most common or most familiar parasites. For more information and excellent images, go to the Parasitic Plant Connection (http://www.parasiticplants.siu.edu/index.html).

Balanophoraceae
Hydnoraceae
Loranthaceae (mistletoes native to North America)
Orobanchaceae
Rafflesiaceae
Santalaceae
Viscaceae (mistletoes native to Europe)

long, slender stolon-like shoots with tiny leaves; it twines around host plants, inserting haustoria (modified roots) into them and then growing onward. Its plant-like features of nodes and internodes seem to be necessary for it to spread from branch to branch of a host or from one plant to another.

Most holoparasites are subterranean. Their seedling roots invade host roots then the parasite's embryo develops into a rhizome-like shoot with small, tough, protective, scale-like leaves. Being underground, it is invisible to herbivores, and temperatures and humidity are more stable, less variable than above ground. Many photosynthetic plants also have rhizomes, but many of their axillary buds grow upward as aerial, chlorophyllous branches; that never occurs in holoparasites. A few of their axillary buds grow above the soil surface as inflorescences, displaying flowers to pollinators (**Figure B11-4b**).

In other holoparasite species, development is even less like that of ordinary plants: The subterranean "rhizome" has no leaves at all, no axillary buds, nodes, or internodes. The growing apex consists of a mass of meristematic cells but has neither leaf primordia nor a root cap. Internally, the "rhizome's" tissues are not typical of either root or shoot, having neither a single ring of

bundles typical of a stem nor a central mass of xylem as ordinary roots do. Even the surface is just a rough, irregular proliferation of cells rather than a smooth sheet of epidermis. The term "runner" is often used to indicate that it is not an ordinary plant organ.

Several holoparasitic plants live entirely within the body of their hosts. After their seedling root penetrates the host, the rest of the embryo dies. The "roots" grow deeper into the host, penetrating its cortex, phloem, xylem and pith, and they make up 100% of the parasite's body. The parasite has no shoot at all. The term "roots" is in quotation marks because these structures have no features typical of roots. Technically this is an **endophyte**, a plant that lives within another. The endophyte of the mistletoe *Tristerix aphyllus,* which grows inside the cactus *Trichocereus chilensis,* consists only of a branching web of uniseriate filaments of parenchyma cells; after some months (?) or years (?), the filaments become multiseriate as their cells undergo longitudinal divisions (**Figure B11-4c**). At some point, a bit of phloem develops and an occasional, isolated vessel element—The entire plant body has no roots, stems, leaves, epidermis, cortex, pith, and only a few cells of xylem and phloem. Detailed studies are needed, but it appears that every surface cell of the endophyte body absorbs

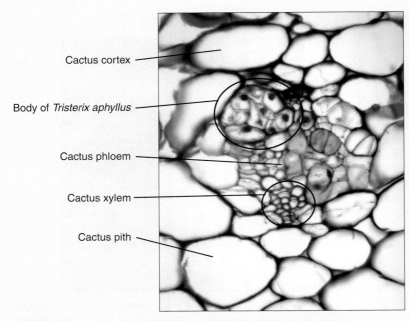

Cactus cortex

Body of *Tristerix aphyllus*

Cactus phloem

Cactus xylem

Cactus pith

Figure B11-4c The small cluster of cells with large red-stained nuclei are part of the body of *Tristerix aphyllus*, a holoparasite whose vegetative body grows entirely within the body of its host, in this case, the cactus *Trichocereus chilensis*. The body of *T. aphyllus* consists of just parenchyma, lacking leaves, stems, epidermis, and virtually all other features of an angiosperm body.

Figure B11-4d The dark blobs do not belong to the visible shoot; instead, they are the flowers of a holoparasitic plant, *Pilostyles thurberi*, growing completely inside its host, the shrub *Dalea*. Holoparasites typically must emerge from underground or from inside their hosts when they flower.

water, minerals, and organic molecules from the host's tissues. With the entire body being absorptive, there is little need for conduction. When stimulated to flower, parenchyma cells just below the host epidermis proliferate into a nodule of callus, an adventitious inflorescence apical meristem forms, and then it breaks its way through the host epidermis. Because it has an ordinary apical meristem, the inflorescence has epidermis, cortex, a ring of vascular bundles, and pith (but no chlorophyll). Endophytes of *Viscum minimum*, *Pilostyles thurberi*, and several *Rafflesia* species are similar (Figure B11-4d).

After a green photosynthetic photoautotroph has produced glucose, it then synthesizes all other organic compounds needed for its structure and metabolism (Figure 11-20). It may be that some holoparasites do the same thing, they may merely obtain glucose from their hosts and then use various types of respiration and anabolic pathways to synthesize all other organic molecules. It would not be surprising, however, to find some species of parasitic plants that have become dependent on their hosts for at least some amino acids, lipids, or vitamins. Despite being so modified, the endophytic bodies of holoparasites are healthy, thriving, and well adapted to respiration-based metabolism in a unique environment.

Holoparasitic plants show us that many features of ordinary plants must be very expensive and risky: After a plant has an alternative to photosynthesis, ordinary plant features are no longer adaptive, and mutations that prevent their development are advantageous. On the other hand, for plants that do photosynthesize, plant-like bodies are worth the trouble.

During aerobic respiration, glycolysis again yields two ATPs directly; in addition, the two NADHs can be transported to mitochondria, where their electrons power the formation of two or three more ATPs. The conversion of each pyruvate to acetyl CoA yields another NADH. Because two pyruvates are produced from each initial glucose, six more ATP per glucose are produced. Within the citric acid cycle, each original molecule of glucose yields two molecules of ATP, six of NADH, and two of FADH$_2$; the total is 24 ATPs. Aerobic respiration can produce as many as 38 molecules of ATP, making it significantly more efficient than anaerobic respiration.

The pentose phosphate pathway yields only two NADPH per glucose-6-phosphate if either ribulose-5-phosphate or erythrose-4-phosphate is drawn off for anabolic metabolism. If neither of these is removed, the various intermediates continue to cycle until all carbon of the glucose is completely oxidized to carbon dioxide and six NADPHs are produced. These may be used for cellular reductions of nitrate to amino acids, sulfates to sulfhydryls, or carbohydrates to fats. NADPH not consumed in anabolic reductions may contribute protons to the mitochondrial electron transport chain and indirectly produce two molecules of ATP.

The amount of ATP produced by fatty acid respiration depends on the length of the fatty acid and whether acetyl CoA enters the citric acid cycle; with maximum respiration, up to 40% of the total energy in a fatty acid is conserved in ATP. Thermogenic respiration and photorespiration produce no ATP. Also, keep in mind that intermediates may be diverted from all respiratory pathways and be used in synthetic reactions, so complete respiration may not occur; ATP production thus is less.

Respiratory Quotient

An action spectrum is a valuable tool for studying light-mediated phenomena such as photosynthesis. For respiration, a similar type of information is useful. A theoretical calculation can be made of the amount of oxygen consumed by each type of respiratory substrate. For example, the complete aerobic respiration of glucose should consume six molecules of oxygen and produce six molecules of carbon dioxide (Table 11-4). This ratio of carbon dioxide liberated to oxygen consumed is known as the **respiratory quotient (RQ)**; for glucose RQ = 1.0. Acids can enter the citric acid cycle and be oxidized to carbon dioxide, producing NADH for the mitochondrial electron transport chain. Because acids contain relatively large amounts of oxygen in their molecular structure, less is needed to convert them to carbon dioxide and

TABLE 11-3	Production of ATP by Various Types of Respiration		
Anaerobic respiration		2 ATP $\longrightarrow$ 2 ATP	
Aerobic respiration			
Glycolysis		2 ATP $\longrightarrow$ 2 ATP	
	2 NADH $\longrightarrow$ 4 or 6 ATP	$\longrightarrow$ 4 or 6 ATP	
Pyruvate/acetyl CoA	1 NADH $\longrightarrow$ 3 ATP ($\times$2)	$\longrightarrow$ 6 ATP	
Citric acid cycle	1 ATP ($\times$2)	$\longrightarrow$ 2 ATP	
	3 NADH $\longrightarrow$ 9 ATP ($\times$2)	$\longrightarrow$ 18 ATP	
	1 FADH$_2$ $\longrightarrow$ 2 ATP ($\times$2)	$\longrightarrow$ 4 ATP	
Pentose phosphate pathway	2 to 6 NADPH $\longrightarrow$ 0 to 12 ATP		
Heat-generating respiration	None		
Lipid respiration	Variable		
Photorespiration	None		

TABLE 11-4 Respiratory Quotients of Various Compounds

$$RQ = \frac{CO_2 \text{ liberated}}{O_2 \text{ consumed}}$$

Glucose

$$C_6H_{12}O_6 + 6O_2 \longrightarrow 6CO_2 + 6H_2O$$

$$RQ = \frac{6CO_2}{6O_2} = 1.0$$

Lipid

$$C_{18}H_{34}O_2 + 25.5O_2 \longrightarrow 18CO_2 + 17H_2O$$

$$RQ = \frac{18CO_2}{25.5O_2} = 0.7$$

Fermentation

$$C_6H_{12}O_6 \longrightarrow \text{alcohol} + CO_2$$
No O_2 consumption

Citric acid

$$C_6H_8O_7 + 4.5O_2 \longrightarrow 6CO_2 + 4H_2O$$

$$RQ = \frac{6CO_2}{4.5O_2} = 1.33$$

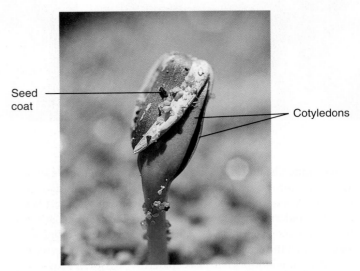

Seed coat

Cotyledons

Figure 11-24 Sunflower seeds store oils in their cotyledons; when they germinate, the oils are respired, giving the seed a low RQ value. The cotyledons of this seed, although still enclosed by the seed coat, have turned green and are carrying out photosynthesis, so both oils and sugars are available for respiration. After a few days, all the oil will be consumed; then only photosynthetically produced sugars will be available, and the seedling will have a high RQ value, near 1.0.

water, so their RQ is high, greater than 1.0. On the other hand, fatty acids contain virtually no oxygen; therefore, when they are respired, enough oxygen must be consumed to oxidize not only every carbon but every hydrogen as well; thus, the RQ is very low, often about 0.7. Of course, anaerobic respiration consumes no oxygen, although its carbon dioxide production is very high.

After these RQ values are calculated, it is relatively easy to measure the amounts of gases exchanged during respiration and thus gain information about the respiratory metabolism. For example, peanut seeds are rich in lipids and oils. As they germinate, the seedling's RQ is initially low, indicating that the lipids are being respired. If they are kept in the dark, the RQ remains low until the seedling exhausts its nutrient reserves and dies. But if they are allowed to germinate in the light, the RQ gradually increases toward 1.0 as photosynthesis begins to produce glucose needed for respiration (Figure 11-24). The increase is not immediate because much of the first products of photosynthesis are used for constructive anabolic reactions in the leaf cells, so a rise in the RQ indicates that leaf photosynthesis is meeting anabolic needs and producing extra for respiration.

Fermentation of Alcoholic Beverages

The ethanol of alcoholic beverages is always produced by the fermentation (anaerobic respiration) of glucose by yeasts. Sugars present in fruits can be fermented immediately, but starches in seeds and tubers must be depolymerized to glucose before they can be fermented. Ethanol produced by fermentation finally kills the yeast if it builds up to a concentration of about 18%. To obtain a stronger concentration of alcohol in the beverage, the fermented mixture must be distilled, a process in which it is heated to concentrate the alcohol while removing some of the water. Alternatively, extra alcohol can be added to a beverage rather than distilling it. These factors combine to give us three basic types of alcoholic beverages: beers, resulting from partially fermenting starchy seeds; wines, produced by a more complete fermentation of sugary fruits; and

spirits, in which their ethanol concentration is increased by distillation or by adding alcohol.

Beer

Beer is made by fermenting starchy cereal grains, especially barley, wheat, corn, or rice. Barley is by far the most common type of beer, and if a different grain is used, it is typically specified. For example, beers made from wheat are becoming more popular and are always called "wheat beer." Before a cereal grain can be fermented, at least part of its starch must be converted to glucose. This is done by moistening the grains and allowing them to germinate. The moistened embryo secretes enzymes that break down the starch located in endosperm. Barley has the greatest number of enzymes and converts starch to sugar more quickly than do other grains. Once the seedling root is visible, sprouting has proceeded long enough, and the grains are dried to prevent any further enzymatic reactions. Sprouted, dried barley grains are called malt (this is also the malt in a malted milkshake). To brew the beer, malt is mixed with water in a large vat; the mixture is called the mash. If malt is the main ingredient, the beer will have a strong flavor, and the quality of the sprouting and roasting of the barley is very important. But often, at least in the United States, unsprouted grains or even corn syrup or potatoes are added. These are very cheap sources of extra starch that can be acted upon by the enzymes in the malt. By adding these adjuncts the beer is less expensive and typically has less flavor.

Hops are the dried carpellate inflorescences of a tall, vining plant, *Humulus lupulus*, closely related to *Cannabis*, marijuana. Hops have a bitter taste and aroma and are added so beer will not be too sweet: the greater the amount of hops, the more bitter the beer.

Once all the ingredients are in the mash, it is warmed to 68 to 73°C and allowed to stand for several hours while the various enzymes are active. The chaff and solid materials are then strained

Plants and People

Box 11-5 Respiration

For the most part people do not affect plant respiration very much, and plant respiration does not have too much of an impact on people. Food storage is one aspect of respiration in which plants and people influence each other. Stored fruits and vegetables respire if they are alive, and if they are moist like fresh potatoes, carrots, peaches, and lettuce they respire especially rapidly. Even dry foods like beans, lentils, wheat, and rice respire, but very slowly. Frozen and canned foods are dead and do not respire.

Respiration in dry foods is important because of the water produced. Remember that the formula for respiration is $C_6H_{12}O_2 + 6O_2 \rightarrow 6CO_2 + 6H_2O$. This water is released from the mitochondria and moistens the cell's protoplasm, causing the food to become wet and thus susceptible to attack by fungi (molds and mildews). Even if not attacked by fungi, the extra moisture may cause the cells to become more active metabolically, and seeds might even germinate while in storage. To keep wheat, corn, and similar foods dry, air is blown through the storage containers to remove any moisture produced by respiration (Figure B11-5a).

In moist stored foods, the extra moisture produced by respiration is not much of a problem. A fresh watermelon has so much water already that the little bit produced by respiration is insignificant. Instead, the problem is that respiration uses up part of their carbohydrates, the very reason we cultivate and harvest most of them. Each week that these plants are stored, there is less

carbohydrate available to us when we eat the food. A fresh potato is packed with starch, whereas a potato that has been stored for months has respired away much of its starch and is much less nutritious to us. Respiration in such foods is minimized by keeping them cool. Low temperatures slow down metabolism and thus slow down respiratory loss of carbohydrates.

Sometimes we harm plant respiration indirectly, without even realizing it. We pollute rivers and lakes with fertilizers; this happens when we dump sewage, even cleaned up sewage, into rivers. The wastes we flush down our toilets are food to many types of bacteria, and such bacteria are able to grow rapidly and respire rapidly. With enough human fertilizer, the bacteria proliferate and use up most of the oxygen available in the water or in the mud on the bottom of a lake or river. Plant roots in those areas may then suffer from hypoxia (low oxygen) or anoxia (no oxygen), and if the roots die, the plants may die also. Of course, fish need even more oxygen than roots do, so hypoxia caused by excessive bacterial growth may cause fish to die. A current problem is that the Mississippi River carries huge amounts of human waste, as well as that from farm animals such as cattle, pigs, and chickens. In addition, it also is polluted with agricultural fertilizers that are washed out of fields by rain. As the Mississippi's water reaches the Gulf of Mexico it fertilizes the bacteria there; the bacteria grow and use so much oxygen that no fish or algae are able to survive. Our pollution causes a huge dead zone in the Gulf of Mexico (Figure B11-5b). Similar dead zones occur in other oceans where rivers drain fertilizer-enriched water from farms and cities.

(a)

Figure B11-5a   Grains such as wheat, rice, and corn can be stored in "elevators" like these for months or years, but only if they are kept dry. Grains are never harvested and put into storage unless they are already extremely dry (they must not be harvested during rainy periods). Even so, the very slow respiration of each kernel produces a tiny bit of water, and if it accumulates, fungi can grow and spoil the grain. The elevators have large fans that blow dry air through the grain.

(b)

Figure B11-5b   The dead zone in the Gulf of Mexico is caused by the rapid growth and respiration of bacteria, which are nourished by fertilizers that pollute the Mississippi River and its tributaries.

out and the sugar-rich liquid, called wort, is boiled to inactivate the enzymes and kill any microbes. Once it has cooled, yeast is added. For beers that will be ales, bitters, or stouts, *Saccharomyces cerevisiae* is used, but for lagers and Pilsners, *S. uvarum* is added. Most beers in the United States are lagers. The yeasts are allowed to ferment the wort from 1 week to 12 days. The beer is filtered and pasteurized to stop all further fermentation and then is aged for 2 or 3 weeks. The carbon dioxide produced by fermentation is allowed to escape during brewing, so to make beer frothy when it is poured, new carbon dioxide is added to the beer artificially just before it is bottled.

Various types of beer result from controlling the malting process, the adjuncts, hops, and the type of yeast used. *S. cerevisiae,* the yeast used for ales, floats to the top of the fermenting vats (called top fermentation), and it works best at a warm temperature. This species of yeast produces compounds called esters that add light, fruit-like flavors to the beer. Common types of ale are brown ale, pale ale, and porter (or stout, made with extra hops and more caramelized barley). *S. uvarum,* used for lager beers, sinks to the bottom of the vats (bottom fermentation) and functions at a cool temperature. Cool, bottom fermentation does not produce esters, so lager beers have a crisper taste. Most of the beer consumed in the United Stated is pale lager, with an alcohol content of only 3% to 6%. The concentration of alcohol depends on the amount of starch converted to fermentable sugar and whether fermentation is artificially stopped at a particular point. With a greater conversion of starch to sugar and a longer fermentation, the concentration of alcohol increases.

Light beers have fewer calories than ordinary beer. This can be achieved in either of two ways. One is to begin brewing with a mash that contains fewer carbohydrates. When brewing is halted, regular beer still has many starches and unfermented sugars present, but light beer has fewer of them. This gives the beer fewer calories. Many people believe that most calories in beer come from the alcohol, which may be true for light beers but not for regular beers, and certainly not for stouts. These have relatively high amounts of starches and sugars, giving the beer its body and fullness. The second way to make light beer is to convert more of its starches to sugars; this ferments to a beer with much higher alcohol content but low in starch. Water is added at the end to lower the alcohol content to the normal 3% to 6% level for beer, but it also dilutes the flavor.

Sake is fermented rice. It is usually called rice wine, but because it is a cereal, technically it should be classified as a beer. However, sake is made by a unique process that differs from brewing beer. The rice is polished to remove all fruit and seed coats and then is steamed to cook and soften its starch grains, as if it were to be eaten. Instead, it is inoculated with the fungus *Aspergillus oryzae* and allowed to set for several days. Enzymes from the *Aspergillus* depolymerize the starch to fermentable sugars. After the rice gets a distinctly moldy aroma, it is mixed with warm water, and then *S. cerevisiae* is added. No hops or adjuncts are added as they would be for beer, and instead of fermenting for only a week, sake is fermented for almost a month. During this time the *Aspergillus* enzymes continue to convert more starch to sugar, and the yeast ferments that to ethanol. At the end, the sake is separated from the solids by filtration through cloth, and extra ethanol is added to bring the final concentration up to about 20% to 22%; this makes it a fortified beer. Sake is thus much stronger than beer, and it is served hot.

■ Wine

Wines are fermented fruit juices that are rich in sugars. When the word "wine" is used by itself, it refers to fermented grapes of the species *Vitis vinifera*; if other fruits are used, then they are named, such as elderberry wine or peach wine. As anyone who has shopped for or selected wine knows, there seems to be an endless variety of wines, such as chardonnay, pinot noir, cabernet sauvignon, and merlot. It may seem that these represent numerous species of grape, but actually all are varieties of just one, *V. vinifera*. Wine was first produced about 6000 BCE, and since then farmers have selected numerous varieties (there are now 15,000 varieties), each with its own special flavors, aromas, and colors, but always cultivated from *V. vinifera*. This is a little surprising because there are 60 species of grape in the genus *Vitis*. These others are typically used for jams, jellies, juices, raisins, or for eating fresh (table grapes are *Vitis labrusca*), but they are not used for wine.

Grapes are harvested when they reach proper maturity and their sugar content is suitable. Typically, 20% to 30% of the grape's weight is sugar (mainly glucose and fructose) when harvested. They are washed and then crushed to obtain the juice. For white wines the skins (exocarps) are removed, but for red or rosé wines the skins are mixed in with the juice. The skins not only provide the color of red wine, they produce various flavors and aromas. The chemical resveratrol is produced when yeasts attack the skins; there are claims that resveratrol lowers blood pressure, prolongs life, and has anticancer activity, but none of these has been proven definitively in humans. *S. cerevisiae* is added to the juice, and the mixture is cooled slightly; white wines are fermented at 10 to 15°C (50–59°F) and red wines slightly warmer at 25 to 30°C (77–86°F).

Fermentation vats have valves that allow carbon dioxide to escape without allowing oxygen to enter. If oxygen were to enter the tank, certain bacteria would convert the ethanol to acetic acid, which would turn the wine into vinegar. After about 8 to 10 days the liquid is removed from the skins and allowed to continue fermenting for up to 1 month. During fermentation, dead yeast cells and other particles settle to the bottom of the fermentation tank, and crystals of tartaric acid (cream of tartar) also sink to the bottom. The wine must be drawn out of the tank carefully without disturbing the sediment because wine with sediment is less appealing. The tartaric acid crystals are a hallmark of grape fermentation, and archaeologists search from them in old pots to ascertain whether particular peoples were producing wine at a certain time or place.

Fermentation continues as long as sugar is present and the concentration of ethanol has not reached lethal levels (18–20%). If all the sugar is fermented, the wine is a dry wine, but if some sugar remains, it is a sweet wine. Most wines produced in the United States have an alcohol content between 12% and 14% and have some sugar left, so their fermentation must be stopped artificially. This is usually done by microfiltering the wine to remove all yeast; the alternative is to heat the wine enough to kill the yeast (to pasteurize the wine), but that damages the flavor.

At this point the wine begins an aging process. White wines are aged only briefly, between 12 and 18 months, but red wines may be aged for up to 5 years. Typically, wine is aged in large oak barrels, where chemicals in the wine interact with compounds in the heartwood. These reactions alter the flavor and aroma of the wine and even create new types of longer-chain alcohols that give the wine a more viscous, substantial feel in the mouth. When the wine master has decided the wine is ready, it is transferred to bottles and either shipped to market or set aside for more aging in the bottles.

Champagnes and sparkling wines are made by adding sugar to a bottle of wine that still has a few live yeast cells in it. The wine then continues to ferment, but all the carbon dioxide is trapped inside and builds up pressure. When the bottle is uncorked, the carbon dioxide forms bubbles and makes the wine effervesce (or, if not done properly, the wine shoots across the room). To be called "champagne" the wine must be produced in the Champagne region of France. All other effervescent wines are "sparkling wines," although those of Italy are usually referred to as prosecco. Inexpensive sparkling wines are made by simply carbonating white wine in the same way that soft drinks are carbonated.

If extra ethanol is added to a wine, it becomes a fortified wine. Examples are sherry, port, and Madeira. Such wines are then aged, either in bottles with no exposure to oxygen or in oak barrels that permit a slight oxidation and evaporation, giving the wine a richer feel in the mouth. For example, tawny port is made by stopping fermentation while half the grape sugars are still present. Distilled spirits are added to bring its alcohol level to 20%, and then the wine is aged for 10 years in oak barrels that allow a very small amount of oxygen to act on the mix, converting it from wine to port. During this time the bright red pigment of the grape skins precipitate and change color to a tawny brown.

The effects of global climate change could be devastating for vineyards and wine production. Even with irrigation and a sunny climate, weather still plays a crucial role in the quality of the grapes. Vineyards are typically some of the most costly land used for agriculture, and all the production facilities are located close to most vineyards; juice is rarely shipped more than a few miles. If an area's climate changes to one slightly warmer or wetter, if the date of the last frost of spring changes, or if rains fall at a different time, wine grapes could be rendered useless. Most existing vineyards have already been carefully chosen to be suitable for a particular variety of grape and for the production of a particular type of wine, so any change in weather or climate will almost certainly hurt the crop rather than improve it.

■ Spirits

Spirits are alcoholic beverages with an ethanol content above 20%. Once fermentation has produced enough alcohol to raise its concentration to 20%, the solution is toxic to the yeast and they die. To obtain a higher alcohol content, either strong alcohol (always ethanol) must be added or the solution must be distilled. During distillation the solution is heated, which mostly causes the alcohol to evaporate, along with some water and some of the flavoring compounds. Distillation is possible because different liquids boil at different temperatures, and ethanol boils at 78°C (173°F), which is much cooler than the boiling point of water at 100°C (212°F). By keeping the mixture above 173°F and below 212°F, ethanol and certain flavors evaporate and most water remains behind. The vapors are carried away to a pipe that is chilled so the vapors condense into a new liquid, one that has

a higher concentration of alcohol, less water, and an altered set of flavorings. This second liquid can be redistilled to raise the alcohol even higher and to achieve a different flavor. Distillation can raise the alcohol content only to 95.6%, making it 191 proof liquor; the "proof" number is twice the alcohol content.

The type of spirit that results from distillation depends on the initial fermented material. Distillation of grain-based fermentations produces whiskeys, vodka, and gin. For example, Scotch and Irish whiskies (spelled with "ie") use primarily or only malted barley, but American whiskeys (spelled with "ey") use up to 80% corn as well, and rye and wheat may also be added. Gin also begins with the fermentation of malted barley, corn, and rye, but during distillation juniper berries are added for their flavor and aroma. Vodka is basically just pure ethanol and water: The objective of distillation is to raise the concentration of ethanol and to eliminate all flavors. Consequently, the initial fermentation for vodka is made with whatever carbohydrate is cheapest: wheat, corn, potatoes, and even sugar beets. Brandy results from distillation of grape wine (if fruit wines are distilled, they produce fruit brandies). Distillation of fermented molasses creates rum, whereas tequila and mescal are distilled from fermented juices of agave plants: tequila from *Agave tequilana* and mescal from *A. americana*. Both agaves are unusual in that they store their carbohydrates not as starch (a polymer of glucose), as most plants do, but as inulin, a polymer of fructose. The agaves must be cooked to break the inulin down to simple sugars before fermentation can start. Tequila agaves are steamed, whereas those for mescal are roasted slowly (**Figure 11-25**; despite its name, mescal does not contain mescaline).

■ Warnings

Fermentation of plant material always produces ethyl alcohol (ethanol), and this is classified as a depressant. This seems counterintuitive, because light drinking causes people to be more lively and less inhibited, but even moderate drinking reduces a person's ability to focus their attention and slows their reaction speed.

Drinking ethanol has both beneficial and harmful consequences. Moderate consumption, about two drinks per day (two bottles of beer or two glasses of wine or two ounces of liquor), lowers the risk of heart disease. Red wine seems to be the most beneficial because it has resveratrol (see above). In contrast, deaths related to ethanol—including driving while under the influence of alcohol (DUI, drunk driving)—is the fifth greatest cause of death among North Americans. Also, drunk drivers don't just kill themselves; they kill and maim innocent bystanders. Ethanol is not at all addictive to many people, but it is to others and results in alcoholism. Ethanol easily passes from mother to fetus through the placenta and can cause fetal alcohol syndrome if the mother consumes too much ethanol while pregnant. The effect of ethanol on us depends on the amount we drink in proportion to our body size, which in turn determines how much ethanol passes into our blood. When the concentration of ethanol in our blood (our blood alcohol content [BAC]) reaches 0.02%, it starts to interfere with both our coordination and our reaction time, and it may cause us to have impulsive behavior. At a BAC of 0.15% people are drunk and have trouble walking or doing other simple things. A BAC of 0.4% is usually fatal. In the United States it is illegal to drive if your BAC is 0.08% or higher. As a general rule a person can metabolize the ethanol in one standard size drink in about 60 to

Figure 11-25 Mescal is made from *Agave americana*, which has long, spiny leaves but which stores its carbohydrate in its broad, short stem. The leaves must be cut off, leaving just the stem (the "heart"), which you see here. These hearts are being transferred from the truck to a stone-lined pit; burning charcoal will be added and the entire pit will be covered to allow the hearts to roast slowly, converting their carbohydrates to simple sugars that can be fermented.

90 minutes; if you drink faster than that, chances are good that you will get drunk.

Ethanol is only one of many types of alcohol. Other types we encounter commonly are isopropyl alcohol (isopropanol), used as rubbing alcohol, and methyl alcohol (methanol), used as wood alcohol. All alcohols are poisonous, but our bodies are able to detoxify ethanol if we drink only small amounts. With larger amounts we develop a hangover, and even larger amounts cause death. One to several college students die every year from drinking so much ethanol (usually high strength liquor) that it disrupts the lipids in their cell membranes (remember that biologists kill and preserve specimens of plants and animals in a mixture of ethyl alcohol and formaldehyde). Methyl alcohol is very toxic by itself, and our livers convert it into the even more poisonous substances formaldehyde and formic acid. At doses too low to kill us, methyl alcohol causes blindness. Isopropyl alcohol is converted to acetone in our bodies; you may be familiar with acetone as fingernail polish remover.

Summary

1. Respiration provides a cell with both ATP and small carbon compounds that are important in various metabolic pathways.

2. The type of respiration carried out by a cell depends on environmental factors and the state of differentiation of the cell.

3. Anaerobic respiration—fermentation—is inefficient because only two ATP molecules are produced for each glucose respired; however, under anaerobic conditions, it is selectively more advantageous than its alternative, death.

4. Some plant and animal tissues are facultatively anaerobic, but no large plant or animal can live for prolonged periods without aerobic respiration. They are all obligate aerobes.

5. Three common electron acceptors are oxygen, acetaldehyde (in plants), and pyruvate (in animals); only oxygen is abundant and cheap and results in a harmless waste product, water.

6. During the complete aerobic respiration of glucose, electrons are removed and transported, ultimately, to oxygen. As oxygen accepts electrons, it picks up protons, and water is formed.

7. Aerobic respiration consists of glycolysis, the citric acid cycle, and oxidative phosphorylation via the mitochondrial electron transport chain. Passage of electrons through the electron transport chain creates a chemiosmotic gradient of protons and hydroxyl ions.

8. Some tissues of some plants generate heat by carrying electrons on an alternative set of electron carriers that do not pump protons; no ATP is formed, so all energy of NADH is released as heat.

9. The pentose phosphate pathway is a complex interaction of metabolites that can produce erythrose (for lignin), ribose (for nucleotides), or NADPH (for reducing power).

10. Respiration increases as temperature increases, within the range of approximately 5°C to 25°C. During warm days and nights, respiration is much more rapid than during cool periods.

Important Terms

acetyl CoA	facultatively aerobic	lactate (lactic acid)
aerobic respiration	fermentation	obligate aerobes
anaerobic respiration	flavin adenine dinucleotide (FAD)	pentose phosphate pathway
citric acid cycle	flavin mononucleotide (FMN)	respiration
cyanide-resistant respiration	glycolysis	respiratory quotient (RQ)
cytochrome oxidase	hemiparasites	strict aerobes
Embden-Meyerhoff pathway	holoparasites	thermogenic respiration
ethanol (ethyl alcohol)	Krebs cycle	tricarboxylic acid cycle

Review Questions

1. Most plants store excess energy from photosynthesis. Name some times when recovery of stored energy may occur. Does energy recovery always occur in the same sites where photosynthetic capture occurred?

2. When a storage organ becomes active after dormancy and mobilizes its reserves, the starch is usually converted to sucrose for transport. Why is starch not transported?

3. The oxidation of NADH to NAD^+ requires transferring electrons from it onto something else. What are characteristics the ideal recipient would have? What is the recipient actually used? What does it turn into as electrons are added to it?

4. Cellular respiration falls into two categories. What are the two categories? When is each one used?

5. Define obligate aerobe, obligate anaerobe, and facultative aerobe. Name some type of organism that is an example of each.

6. Under what conditions does plant tissue experience lack of oxygen? How is ATP generated from glucose without oxygen?

7. Glycolysis can proceed only if _____ is available. If it cannot be regenerated from NADH, the organism will starve, despite having glucose available.

8. The reduction of NAD^+ to NADH during glycolysis is a problem. Why is this less of a problem if roots absorb nitrates and sulfates or if the plant is synthesizing fatty acids?

9. Anaerobic and aerobic forms of respiration differ in the ultimate electron acceptors for each process. What is the electron acceptor in each? What are the advantages and disadvantages of each?

10. With oxygen present, not only can NAD^+ be regenerated, allowing glycolysis to continue, but even more _____ is formed in the regeneration process.

11. During anaerobic respiration in animals, the electron acceptor is pyruvate. What is the acid (or the anion of which acid) that is formed? During hard, fast exercise, your own muscles probably synthesize this acid. What is the sensation you feel?

12. Plants do not make the acid mentioned in Question 11. What do plants make instead during anaerobic respiration? Pyruvate is converted to _____, and then NADH reacts with that, forming _____.

13. Considering the negative aspects of anaerobic respiration, how could natural selection have produced something so inefficient?

14. What are the three basic parts of aerobic respiration? Which steps occur in mitochondria and which in cytosol?

15. Are the initial steps of aerobic and anaerobic respiration similar or dissimilar?

16. For the citric acid cycle to occur, pyruvate must be transported from _____ where glycolysis occurs, across the _____ to the _____. It is transported as a two-carbon fragment called acetyl. What is the name of the carrier molecule that transports it?

17. The text mentions that the benefit in the citric acid cycle is the generation of more ATP, yet at only one step is ATP produced. How is the citric acid cycle involved in more ATP production than just that (Hint: what is the fate of NADH and $FADH_2$)?

18. Examine Figure 11-13. Where are the electron carriers of the mitochondrial electron transport chain located—in the matrix, the membrane, or in the crista lumen? As they carry electrons, some of them also deposit protons (H^+). Are the protons deposited in the matrix or in the crista lumen?

19. What is a chemiosmotic potential in mitochondria?

20. The flow of protons from the crista lumen to the mitochondrial matrix can be used to synthesize ATP. Describe how this occurs.

21. During glycolysis, is all of the energy of glucose converted to energy in ATP? Why do compost piles become warm?

22. Some plants generate large amounts of heat. Name two examples. How is the generation of heat beneficial to each of these plants? What is the name of heat-generating respiration?

23. The intermediates of all respiratory pathways can be used in _____ _____ to make various compounds. It is not _____ that they will be completely oxidized to carbon dioxide and water.

24. The pentose phosphate pathway has some intermediates that are pentoses. What is a pentose? Name the one that is shunted into nucleic acid metabolism.

25. The four-carbon sugar erythrose-4-phosphate is the starting material in the synthesis of many compounds. Name two important examples.

26. Consider a meristematic cell preparing for mitosis and a young xylem cell differentiating into a fiber or a tracheary element; each uses the pentose phosphate pathway, but for different products. Explain why.

27. In most tissues, if the temperature is increased by 10°C, how much is the rate of respiration increased?

28. The total energy yield of respiration relates the number of ATP molecules synthesized per glucose molecule respired. What is the yield for anaerobic respiration? For aerobic respiration? For heat-generating respiration?

29. What is the definition of the respiratory quotient (RQ)? The complete aerobic respiration of glucose consumes six molecules of oxygen and produces six of carbon dioxide. What is the RQ for this process? If you measured the RQ of a plant and found that RQ = 0.7, would you suspect that the plant was respiring lipids or citric acid?

Transport Processes

chapter 12

Concepts

One fundamental aspect of life itself is the ability to transport specific substances to particular sites, moving molecules against the direction in which they would diffuse if left alone. After death occurs, atoms, ions, and molecules diffuse, moving from regions of higher to lower concentration, and the organization of protoplasm decays; the disorder of the components increases. Diffusion also occurs during life but proceeds more slowly than the controlled and oriented transport processes that tend to increase the order within the plant or animal body. Transport processes consume energy, and many are driven by the exergonic breaking of ATP's high-energy phosphate-bonding orbitals.

Specific transport occurs at virtually every level of biological organization: Enzymes transport electrons, protons, and acetyl groups. Membranes transport material across themselves. Cells transport material into and out of themselves as well as circulate it within the protoplasm. Entire organisms transport water, carbohydrates, minerals, and other nutrients from one organ to another—between roots, leaves, flowers, and fruits.

Chapter Opener Image: All organisms must transport materials throughout their bodies. Just like us, the horses have a heart and circulatory system made up of arteries and veins, and blood circulates continuously. Like all vascular plants, the tree (a live oak, *Quercus virginiana*) has two systems, xylem, which transports water and minerals upward from the soil to all other parts of the plant, and phloem, which carries sugars and amino acids from one area of the tree to another. In both tissues, material is moved in just one direction; it is not circulated. The xylem and phloem are not interconnected by anything similar to our capillaries.

Plants have only a few basic types of transport processes, and the fundamental principles are easy to understand. They are grouped here into **short-distance transport**, which involves distances of a few cell diameters or less, and **long-distance transport** between cells that are not close neighbors.

Many types of short-distance transport involve transfer of basic nutrients from cells with access to the nutrients to cells that need them but are not in direct contact with them. Such transport requirements arose when early organisms evolved such that they had interior cells that were not in contact with the environment. Short-distance transport became necessary to the survival of internal cells.

Long-distance transport is not absolutely essential in the construction of a large plant. Many large algae have no long-distance transport, nor do sponges, corals, or similar animals; however, the ability to conduct over long distances is definitely adaptive, especially for land plants (Figure 12-1). Before xylem and phloem evolved, a plant's absorbing cells could not penetrate deep into soil because they would starve so far from photosynthetic cells, nor could they have transported their absorbed nutrients very far upward. Being limited to the uppermost millimeter or two of soil meant that absorptive cells could not reach more permanently moist, deep soil where there are more minerals; the uppermost layers dry quickly and free minerals are leached away by rain. With xylem and phloem, roots that penetrate deeply can be kept alive and their gathered nutrients can be carried up to the shoot.

Vascular tissues make it selectively advantageous for shoots to grow upright, elevating leaves into the sunlight above competing plants. This elevation is feasible because photosynthetically produced sugars can be transported downward to other plant parts. Such elevation of photosynthetic tissues resulted in tall plants that could also place their reproductive tissues at a high elevation, enabling spores or pollen to be distributed widely by wind. After insect-mediated pollination evolved, it was adaptive to have flowers located high, in an easily visible position. The evolution of transport processes affected all aspects of plant biology and permitted later evolution of many new types of plant organization.

Vascular tissues also act as a mechanism by which nutrients are channeled to specific sites, resulting in rapid growth and development of those sites (Figure 12-2). At certain times, nutrients are transported to apical meristems, promoting growth and leaf primordium initiation; at other times, nutrients are directed to flower buds or young fruits, and at other times, production of wood and bark is supported. The combination of short- and long-distance transport has resulted in the ability of some plants to become large and complex enough to survive temporary adverse conditions, such as drought, heat, or attack by pathogens.

Because almost everything transported by a plant (or animal) is dissolved in water, the ability of water to move throughout a plant is important. Water is an unusual liquid: It is heavy and viscous, and it adheres to cell components as well as to soil, factors that affect transport processes.

Related to transport processes are **isolation mechanisms** that inhibit movement of substances. Plants are adept at synthesizing organic polymers impermeable to a variety of substances. The epidermis with its cutin-lined walls keeps water in the shoot after it has been transported there by the xylem (Figure 12-3). The Casparian strips of the endodermis prevent diffusion of minerals from one part of a root to another. Isolation mechanisms are essential if transport is to be useful.

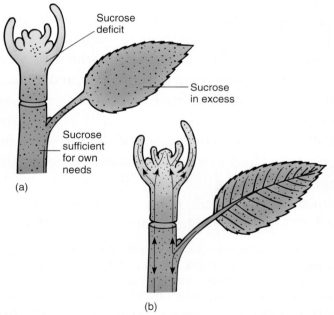

Figure 12-2 (a) Imagine an angiosperm that has no vascular tissue. Leaves would produce large amounts of glucose, but because diffusion is slow over long distances, the sugar would diffuse out of the leaf only slowly and in small quantities. The shoot apex has almost no chlorophyll; if it had to depend solely on its own photosynthesis, growth, leaf initiation, and leaf expansion would be extremely slow. (b) With vasculature, glucose can be transported from regions of excess to regions of need; apical meristems and leaf primordia can thereby grow very rapidly. Vascular tissues also make the minerals gathered by an extensive root system available to regions that need them.

Figure 12-1 This is the type of shoot whose vascular system is shown in the opening photograph of the chapter: It is an *Opuntia* cactus with flattened stems and tiny leaves that will soon abscise. Xylem and phloem are arranged in a net-like pattern (not visible here) with leaves located at every vertex of the net. Xylem is a means of long-distance transport, carrying water from the tips of this plant's long roots, up into and throughout this stem. Short-distance transport occurs as water moves out of xylem and then from cell to cell until it reaches the epidermis.

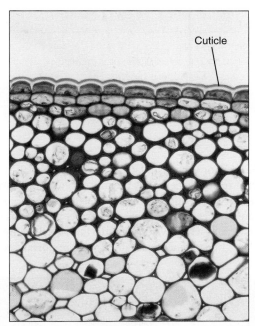

Figure 12-3 The cuticle (pink), composed of cutin, is a waterproof epidermal layer that acts as an isolation mechanism, retaining water within the plant and keeping pathogens out (×80).

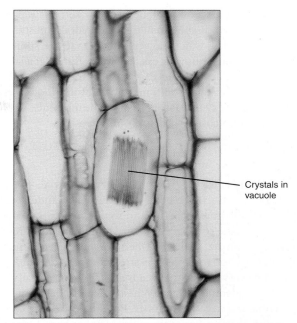

Figure 12-4 These crystals are located in a cell vacuole. Calcium and oxalic acid were transported across the vacuole membrane by molecular pumps, concentrating them until they crystallized into these needle-shaped raphide crystals (×80).

Diffusion, Osmosis, and Active Transport

The first thing to consider is the mechanism by which material moves through a solution and crosses a membrane. The simplest method is **diffusion**, in which the random movement of particles in solution causes them to move from areas where they are in relatively high concentration to areas where they are in relatively low concentration. Diffusion through a membrane is technically known as **osmosis**.

Membranes are of three types: **Freely permeable** membranes allow all solutes to diffuse through them and have little biological significance. **Completely impermeable** membranes do not allow anything to pass through and occur as isolation barriers. **Differentially** or **selectively permeable membranes** allow only certain substances to pass through; all lipid/protein cell membranes are differentially permeable. Hydrophobic molecules diffuse easily through any cell membrane, whereas many polar, hydrophilic molecules can cross differentially permeable membranes only if the membranes have special protein channels through which the molecules can diffuse. Water molecules, even though highly polar, pass through all membranes, but their movement is more rapid if the membrane has protein channels called **aquaporins**.

Most membranes also have membrane-bound **molecular pumps** that use the energy of ATP to force molecules across the membrane, even if that type of molecule is extremely concentrated on the receiving side; this is **active transport**. The molecular pump, which is a protein, binds to both the molecule and ATP; when ATP splits into ADP and phosphate, the energy is transferred to the pump, forcing it to change shape, carry the molecule across the membrane, and release it. The membrane must otherwise be extremely impermeable to the molecule or it would leak back. Proton pumping in photosynthesis and respiration are examples of active transport.

All cell membranes are important in transport processes; the plasma membrane governs movement of material into and out of the cell. Substances can move across the vacuolar membrane by either osmosis or active transport; the vacuole acts as an accumulation space for sugars, pigments, crystals, and many other compounds (**Figure 12-4**). The endoplasmic reticulum and dictyosome membranes transport material that then accumulates in vesicles. These vesicles may be relatively permanent, remaining in the cell for long periods of time, or they may be a means of **intracellular transport**, in which the vesicles migrate through the cytoplasm and fuse with another organelle. During fusion, the membranes merge and the vesicle contents are transferred into the organelle. This is a common means of moving material from the endoplasmic reticulum to dictyosomes or from either of these organelles to the cell exterior by fusion with the plasma membrane. During cell division, the new cell plate (the two primary walls and middle lamella) is formed by the coalescence of vesicles from both endoplasmic reticulum and dictyosomes.

Water Potential

Like any other chemical, water has free energy, a capacity to do work. For most chemicals, this energy is called its *chemical potential*. Because water is so important in botany, its chemical potential is usually referred to as **water potential** and has the symbol ψ (pronounced "sigh"). Water potential, the free energy of water, can be increased several ways: Water can be heated, put under pressure, or elevated. The energy of water can be decreased by cooling it, reducing pressure on it, or lowering it.

Water's capacity to do work can be changed in other ways as well. When water adheres to a substance, these water molecules form hydrogen bonds to the material and are not as free to diffuse as are other water molecules; their capacity to do work has decreased. Consider a small beaker of water; if it is pure water, it can flow, move, dissolve material, and hydrate substances, but if a sponge is added, water molecules adhere to the sponge material and can no longer flow or easily dissolve things. If a large amount of sugar is added instead of a sponge, the results are the same. Syrup is just a sugar solution, but the water molecules in syrup have less capacity to do work than do the molecules of pure water.

Water potential has three components:

$$\psi = \psi_\pi + \psi_p + \psi_m$$

In this equation, ψ_p is **pressure potential**, the effect that pressure has on water potential. If water is under pressure, the pressure potential increases and so does water potential. If pressure decreases, so do the pressure potential and water potential. Pressure can be positive (when something is compressed) or negative (when something is stretched). Most liquids cannot be stretched very much, but because water is cohesive, it actually can resist considerable tension. When water is under tension, pressure potential is a negative number. Potential is measured in units of pressure, usually in **megapascals** (**MPa**) or **bars**. One megapascal is approximately equal to 10 bars or 10 atmospheres of pressure. Pure water at one atmosphere of pressure is defined as having a water potential of zero.

ψ_π ("sigh pie") is **osmotic potential**, the effect that solutes have on water potential. In pure water, no solutes are present and osmotic potential is given the value of 0.0 MPa. Adding solutes can only decrease water's free energy because water molecules interact with solute molecules and cannot diffuse easily; therefore, osmotic potential is always negative. If water molecules do not interact with the added molecules, the substance does not dissolve.

It is important to be careful here: Adding acid to water only seems to make water more active. The solution may have more free energy than the water, but it does not have more free energy than both pure water and the original concentrated acid.

Osmotic potential is related to the *number of particles* present in solution; that is, a solution composed of 2 g of glucose in 100 mL of water has an osmotic potential twice as negative as a solution of only 1 g of glucose per 100 mL of water. This has some unexpected results: If a molecule of starch containing 1,000 glucose units is hydrolyzed to 1,000 free glucose molecules, the osmotic potential of the solution becomes much more negative because there are now 999 more particles in solution than previously. Using terms such as "increase," "decrease," "larger," and "smaller" can be confusing when dealing with negative numbers. If the osmotic potential goes from –0.01 to –0.1 MPa, is it increasing or decreasing? It is least confusing to use the terms "more negative," "less positive," and so on.

ψ_m is **matric potential**, water's adhesion to nondissolved structures such as cell walls, membranes, and soil particles. Adhesion can only decrease water's free energy, and thus, matric potential is always negative. In soils, matric potential is important because so much of the soil water is tightly bound to soil particles, but in living cells, matric potential usually is much less important than osmotic potential or pressure potential and usually is ignored entirely. The water potential equation for living cells is usually considered to be just $\psi = \psi_\pi + \psi_p$ (**Table 12-1**).

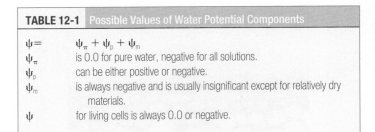

TABLE 12-1	Possible Values of Water Potential Components
$\psi =$	$\psi_\pi + \psi_p + \psi_m$
ψ_π	is 0.0 for pure water, negative for all solutions.
ψ_p	can be either positive or negative.
ψ_m	is always negative and is usually insignificant except for relatively dry materials.
ψ	for living cells is always 0.0 or negative.

Movement of water is related to water potential; substances diffuse from regions where they are more concentrated to regions where they are more dilute. This can be stated more precisely for water: Water moves from regions where water potential is relatively positive to regions where water potential is relatively more negative (**Figure 12-5**). This statement contains several important points:

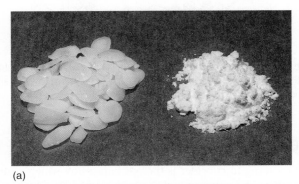

(a)

(b)

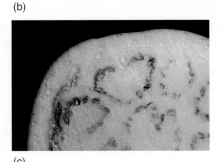

(c)

Figure 12-5 (a) The material on the left is potassium hydroxide; on the right is starch. They were photographed immediately after being exposed to air, while they were dry. (b) Photographed after 1 hour in humid air. Water has moved from where it was more concentrated (the air) to less concentrated. The potassium hydroxide holds water by forming a solution with a very negative osmotic potential. Water is held to the starch by adhering to the long polysaccharide molecules. Water is not obvious in the starch, but think of saltine crackers left unwrapped. (c) By adding salt to eggplant, water can be drawn from the tissues, making them easier to cook.

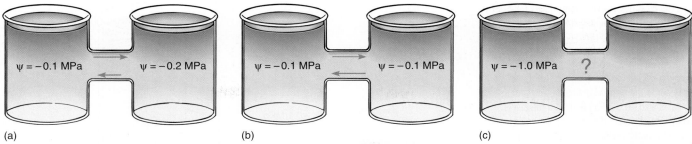

Figure 12-6 (a) If a solution whose water potential is –0.1 MPa is connected to a solution whose water potential is –0.2 MPa, the two are not in equilibrium. Water moves from a region of relatively positive water potential to a region of more negative potential, as indicated. Water molecules move in both directions, but more move to the right in any particular instant. Both beakers must have solute dissolved in them (otherwise each would have ψ = 0.0 MPa), and the –0.2 MPa solution must have twice as much solute. Because the concentration of solute molecules in the left beaker is lower, they are less able to restrict the movement of water molecules than those in the right beaker. (b) If the water potential of two solutions is identical, they are in equilibrium, and no net movement of water occurs; in each second, equal numbers of water molecules move to the right and left. (c) A ψ of –1.0 MPa is very negative and has a strong tendency to absorb water, but we cannot be certain that water will move to it in this case; the right beaker may contain a solution with a ψ of –1.1 MPa.

1. Water moves whenever there is a difference in water potential within the mass of water. All protoplasts are interconnected and most cell walls are fully hydrated; therefore, basically all of a plant body is one mass of water; water can move between regions in the plant if the water potentials of the regions are not equal (**Figure 12-6a**). As a consequence, the water potential of any particular cell may change many times a day as various parts of a plant lose or gain moisture (**Figure 12-7**).
2. If the water potentials of two regions are equal, the regions are in equilibrium, and there is no net movement of water.

Water still diffuses back and forth, but on average, equal numbers of water molecules diffuse into and out of a site (Figure 12-6b). Unless frozen, water is always in motion, always moving within a plant from areas where it is abundant or under pressure to areas where it is rare or under tension.
3. Water potentials must always be considered in pairs or groups. Because water moves from one site to another, the water potentials of the two sites are important. Knowing one single water potential does not allow us to predict whether water will move (Figure 12-6c).

Temperature is not a factor because the solutions being compared are assumed to be at the same temperature. This is not strictly correct if the water potentials of leaves and roots are compared, but the difference is not significant.

Cells and Water Movement

Some examples may help. Imagine a cell with a water potential of –0.1 MPa. It contains solutes that cause the osmotic potential to be some unknown negative number. The cell is turgid and presses against the cell wall, but the cell wall presses back equally, causing pressure on the cell, and the pressure potential is some unknown positive number. We can ignore matric potential because it is usually such a small number. The osmotic potential, whatever it is, plus the pressure potential, whatever it is, equals –0.1 MPa (**Figure 12-8a**). Now imagine the cell being placed in a beaker of solution that also has a water potential of –0.1 MPa. The two water potentials are equal, the cell and solution are in equilibrium, and no net water movement occurs—the cell neither shrinks nor swells. Water molecules do move between the cell and the solution, but approximately equal numbers move in each direction every second (Figure 12-8b).

Now imagine the same cell placed in a solution with a water potential of –0.3 MPa (Figure 12-8c). The water potentials are not equal, that of the solution being more negative (has more solutes) than that of the cell, and thus, water moves from the cell into the solution. How much water moves? As water leaves the cell, the solutes that remain in the protoplasm become more concentrated. Because more solutes are present per unit water, osmotic potential becomes more negative. As water moves out, the protoplasm volume becomes smaller, so the protoplast presses against

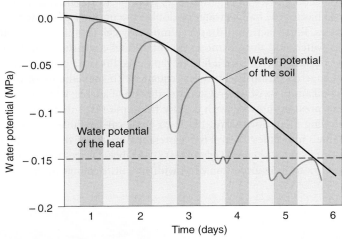

Figure 12-7 If soil is watered well and then allowed to dry over a period of days, the water potential of the soil solution gradually and smoothly becomes more negative. Every day, leaves lose water more rapidly than xylem replaces it; therefore, leaves dry slightly, and leaf water potential becomes more negative. At night, stomata close and xylem transport rehydrates the leaf tissue, but in each cycle, leaves dry more than the previous night. This is not very serious until daytime leaf water potential becomes more negative than the wilting point (dashed line). When leaves wilt, many metabolic processes are adversely affected. Wilting point varies from species to species and is much more negative for xerophytes than for mesic plants. After soil becomes extremely dry, even nighttime rehydration does not bring leaf water potential above the wilting point; the plant is at its permanent wilting point, and severe stress damage may occur. All growth stops, and leaves and developing flower buds or fruits may die and be abscised.

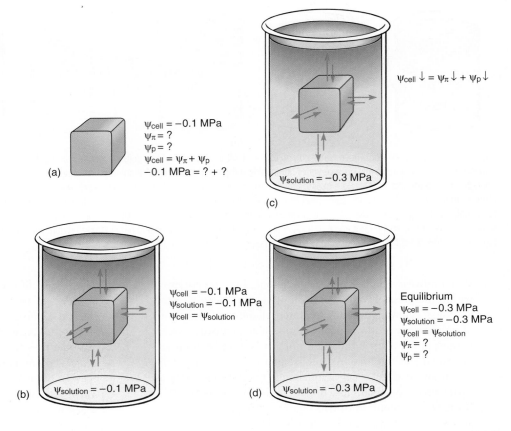

Figure 12-8 (a) A cell whose ψ is –0.1 MPa but whose values of ψ_π or ψ_p are unknown. (b) If the cell is placed in a solution with a ψ of –0.1 MPa, the cell is in equilibrium with the solution and no net movement of water occurs. (c) If the cell is placed in a solution with a ψ of –0.3 MPa, the cell loses water to the solution until the cell's water potential is also –0.3 MPa (d), even if the cell is killed by loss of water. The fact that the cell's water potential becomes more negative means that the osmotic potential or the pressure potential or both also become more negative.

(a)
$\psi_{cell} = -0.1$ MPa
$\psi_\pi = ?$
$\psi_p = ?$
$\psi_{cell} = \psi_\pi + \psi_p$
-0.1 MPa $= ? + ?$

(c)
$\psi_{cell} \downarrow = \psi_\pi \downarrow + \psi_p \downarrow$
$\psi_{solution} = -0.3$ MPa

(b)
$\psi_{cell} = -0.1$ MPa
$\psi_{solution} = -0.1$ MPa
$\psi_{cell} = \psi_{solution}$
$\psi_{solution} = -0.1$ MPa

(d)
Equilibrium
$\psi_{cell} = -0.3$ MPa
$\psi_{solution} = -0.3$ MPa
$\psi_{cell} = \psi_{solution}$
$\psi_\pi = ?$
$\psi_p = ?$
$\psi_{solution} = -0.3$ MPa

the wall with less force and the wall presses back less; therefore, pressure potential becomes less positive. Because both osmotic potential and pressure potential are decreasing (becoming more negative), so is the water potential of the cell. At some point, the cell's water potential (ψ_{cell}) reaches –0.3 MPa and is in equilibrium with the solution; then net water movement ceases (Figure 12-8d). Of course, as water moves from the cell into the experimental solution in the beaker, the solution becomes more dilute, and its osmotic potential and water potential become less negative; its pressure potential does not change because pressure cannot build up in an open beaker. Therefore, equilibrium actually occurs slightly above –0.3 MPa, but because most beakers are much larger than most cells, the amount of water that moves is much more significant to the cell than to the beaker solution.

Consider the relative importance of osmotic potential and pressure potential in this example. In order for osmotic potential to become twice as negative (e.g., from –0.15 to –0.3 MPa), the cell has to lose half its water or double the number of solute particles. Either action is drastic; almost any cell dies if it loses half its water, and thus, osmotic potential does not usually increase or decrease more than a few megapascals. Pressure potential can change enormously, however, usually with movements of only small amounts of water. Consider the top of a table: Its molecules are pressing upward with exactly the same force that gravity is pulling them downward. Placing a book on the table causes the table's molecules to exert more pressure upward as their bonding orbitals are stretched (Figure 12-9). The table changes the amount of upward pressure it can exert with very little change in shape.

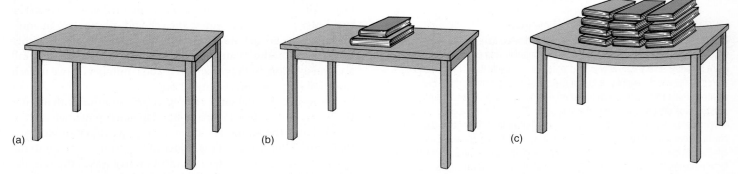

(a) (b) (c)

Figure 12-9 (a) The molecules in this table top are exerting just enough pressure upward to counteract gravitational force: There is no movement upward or downward of the table top, and thus, forces must be in equilibrium. (b) Adding several books increases gravitational force, but no net movement occurs. The table top is slightly bent, which stretches its molecules, but they resist just enough to counter the new force. (c) Even more force is perfectly balanced by the table top. If the books were removed, the stretching on the table would stop, and the molecules would go back to exerting only enough pressure to counteract its own weight—the table would not fly upward.

Box 12-1 Water and Ecology

Water is essential to life. The regions of the world that have no water have no life, whereas those areas that do have water—even just a little water—have at least a few organisms. Think about the various ways that water is available to us living creatures. We can set aside the water potential equations for a while and simply think about aspects of the world that we already know. This is an opportunity to use our two analytical questions: What are the alternative ways in which water is present? What are the consequences of each alternatives?

The Water Available in Water

Most of the world is covered with water. Oceans, lakes, marshes, ponds, rivers, snow fields, and glaciers occupy more of Earth's surface than does land. It might seem that there is plenty of water in water, but we know that water comes in a variety of forms—fresh, salty, brackish, and so on. Let's consider several alternative forms and the consequences each carries. Lakes and rivers contain fresh water, that is, water that has very few solutes dissolved in it. It starts out as rain, water that condenses from clouds and is basically distilled water, pure water. As it collects in streams, lakes, and rivers, this "water" continues to be more or less pure water with very few minerals and salts dissolved in it. Mountain streams are clear and clean because the water is too pure for algae to grow in it—Like other organisms, algae need minerals such as nitrates, potassium, magnesium, and so on. Larger rivers are less pure because they have received some of their water as runoff from fertilized fields, lawns, and golf courses, and cities dump their sewage into rivers. Sewage is always rich in dissolved minerals and organic chemicals, and as it mixes with river water, the water becomes "drier," its osmotic potential becomes more negative. River water, however, typically never becomes so mineral-enriched that it is "dry" like the salt water of an ocean; just the opposite, river water often reaches a balance of water and minerals that supports abundant life. That may sound great, but the result is that algae grow so abundantly that they degrade the river. As the algae die, their bodies sink and are decomposed by bacteria, a processes that uses up so much oxygen that fish may suffocate. This process is called **eutrophication**, but you are probably fortu-

nate enough to have never seen a eutrophied river. In the 1960s, it was realized that we could never keep all of our sewage and agricultural runoff out of rivers, but if we could at least keep phosphate levels low, we could prevent eutrophication. Algae can not grow without phosphate. If that nutrient is absent, there will be no algal growth even though all other essential elements are present. As it turns out, most phosphate pollution was due to laundry detergents, and since the 1960s, we have switched to alternative detergents that are free of phosphates.

In contrast to the fresh water of rivers, oceans contain saltwater. Rivers bring in water and dilute salts, but the water will evaporate, form clouds, and then fall as rain again and run to the oceans, carrying another load of dilute salts. This is a mechanism that moves salts from land to ocean and concentrates them. Currently, the salts in seawater are so concentrated that land plants

Figure B12-1a This small part of Yosemite National Park has many microhabitats that differ in water availability. The waterfall has abundant water when the stream is flowing, but it goes dry or freezes at times. Its spray zone provides some water, but the cliff face is wet only during a rain—it has no moisture holding capacity. Trees grow where soil retains moisture longer than on the cliff face, but some areas with soil are in full sun, whereas others are in shade; thus, rates of transpirational water loss differ.

Figure B12-1b This cactus *Armatocereus procerus* grows in a remarkably dry area of coastal Peru. It is shriveled and sunburned from lack of water. It appears that this area receives rain only during El Niño years, and those occur only about once every 10 or 15 years. These plants must routinely go for 10 years or more with no rainfall whatsoever.

(continued)

and animals cannot use ocean water to hydrate themselves: We would die if we tried to survive by drinking seawater just as plants would die if we irrigated them with it. Marine organisms—algae and animals adapted to life in salt water—have plasma membranes and vacuolar membranes able to move salts either into or out of the protoplasm as necessary to keep the cells properly hydrated. Seagulls have glands that purify seawater and throw away the salt, but we land organisms cannot do that.

It might seem as if marine organisms would have a safe, stable environment, never having to worry about a lack of water, but seawater's ratio of salt to water changes because of many factors. As rivers flow into an ocean, they deposit buoyant fresh water, which floats like a cap on the seawater (salts in seawater make it denser than fresh water). Marine algae cannot live in the cap of fresh water, and must remain in the deeper saltwater, where they do not have as much light for photosynthesis. As the fresh river water spreads out, the thickness of its layer diminishes, and at some point, wave action mixes fresh and saltwater together. The mouth of a river then is a complex gradient of fresh and saltwater, of river-borne minerals, and of temperatures (rivers are usually warmer or cooler than the ocean they flow into).

The complex mixing at the mouth of a river is a rich environment that provides numerous habitats for many types of creatures. Unfortunately, we have interfered greatly with these ecosystems. The simplest to understand are the Rio Grande as it flows into the Gulf of Mexico and the Colorado River as it flows into the Sea of Cortez: Both the United States and Mexico use so much of the river water for drinking, manufacturing, and irrigation that these two mighty rivers are nothing more than trickles that barely reach the coast. They no longer provide the millions of gallons of fresh water that would have maintained the coastal ecosystems, and without this outward flow of fresh water, the salty seawater approaches into the mouth of the rivers, seeping underground and making ground water so salty that wells have to be abandoned. Fresh water marshes have been destroyed by brackish water that floods in at high tide because freshwater plants are unable to extract water from seawater.

The concentrations of salt and water in seawater also change greatly at the North and South Poles. The Arctic Ocean near the North Pole is frozen over with a sheet of ice, and at the South Pole, the continent of Antarctica is covered with snow fields and glaciers. Ice and snow are pure water, and as they melt in the summer, they dilute the salts in seawater. It may seem like this would not matter, that these are such cold places that they must be lifeless except for a few polar bears in the north and penguins in the south, but that is not true. Despite having a temperature near freezing, polar water is so teeming with microscopic algae and animals that it is one of the most life-filled habitats in the entire

Figure B12-1c   Cacti are not the only plants that survive drought by storing water in succulent bodies. The deserts of Africa and the Middle East have many succulent plants as well; this is *Huernia recondita* in the milkweed family, and there are also succulents related to geraniums and to dandelions among others.

world. The addition of fresh water as glaciers and icebergs melt in summer is a problem that they have had to become adapted to. In the winter, there is the opposite problem as temperatures drop below freezing and ice forms. As water freezes into ice, it leaves its salts behind; therefore, as the poles become more ice bound in winter, the remaining water becomes ever more salty, another dilemma for the creatures in it. An even worse problem, however, is that the increasing saltiness causes increasing density, and the cold, salty water sinks. Fish and seals easily swim upward and remain safe, but algae and microscopic animals are carried downward into darkness where they become food for benthic creatures, those that live on the ocean floor.

The intertidal zone—the region of coast that lies between the levels of low and high tides—is an area where available water changes greatly and quickly. As the tide goes out, sea anemones, barnacles, and algae are at first wet with seawater, but as the water on their surface evaporates, the concentration of salt increases. If exposed long enough on a hot day, they may become not only dry but encrusted with salt, and it is necessary for them to try to retain water in their bodies against the osmotically dry salt on them. In contrast, a rain shower would drench them in fresh water, and they would then need to struggle to prevent a loss of minerals out of their bodies. Organisms near low tide level are only exposed briefly, but those near the high-tide mark spend most of their life just above the safety of the ocean's water.

The Water Available in Air

Air supplies water to land plants in the form of rain, fog, dew, frost, snow, hail, and even just humidity. Air also pulls water out of plants, sometimes gently, sometimes relentlessly—it is the motive

force for transpiration. Water availability is more complex than just the amount of rain that falls per year per square mile. Also important are the timing and regularity of precipitation, as well as the rate at which water is lost between rains. Some desert regions, for example, the Atacama Desert in Peru and Chile, are believed to have never received any rain at all, ever. At the other end of the scale are fog and cloud forests, regions where the air is saturated most of the time: Plants are dripping wet and soil is always moist. Far from being ideal habitats, fog forests are often populated by dwarf trees, trees stunted because the humidity prevents transpiration so that xylem cannot transport minerals up from the roots. Most land habitats are rather more ordinary, having rainy days that alternate with sunny ones, but we can immediately see that there are thousands of alternatives here—the ratio of rainy days to sunny ones, the amount of rain that falls each time, the seasons when it falls, the dryness of the air between rains, and so on. Who would guess that Seattle and Austin receive virtually identical amounts of rain? Each get about 35 inches of rain per year, but in Seattle, a bit falls every week, whereas Austin gets 0.6 to 6 inches every other month or so. Between rains, Seattle is cool and cloudy and plants transpire mildly, but in Austin, rains are separated by day after day of intense sunlight with low humidity; thus, plants transpire rapidly. To this variation must also be added differences in soil texture: Those in the Pacific Northwest are thick and organic and hold rainwater in place for days, whereas Austin's soil is often less than 0.5 inch thick and it lies over porous limestone: Rainwater drains away immediately. Another consideration is the seasonality of rainfall. California's chaparral vegetation is famous for its fires that rage through it in summer, destroying homes and threatening cities. It burns like this because it has a wet winter but a dry summer; if its rains came in the summer instead, they would suppress the vegetation's tendency to burn.

Water is water, but there are many ways in which it is available to plants, animals, and other organisms. Likewise, there are even more ways in which organisms have become adapted such that they can use this essential resource.

A similar process occurs in cell walls; imagine placing the cell, now with a water potential of –0.3 MPa, in a beaker of pure water. Some water moves inward, diluting the solutes and causing the osmotic potential to become less negative, but the change is not significant; however, the extra volume of the water causes the protoplast to swell and press against the wall with more force. The wall presses back with equal force and pressure potential rises rapidly, even though only a small amount of water moves in (Figure 12-10). How high will it rise? We cannot predict its value; however, it will go high enough that osmotic potential plus pressure potential will equal zero, and the cell will be in equilibrium with pure water.

Can a cell ever absorb so much water that it bursts? Animal cells often burst if placed in pure water, a process called **lysis**, but plant cells can never burst (Figure 12-11). Walls, either primary or secondary, are always strong enough to resist breakage by water absorption. Even the thinnest, most delicate walls of mature parenchyma cells can exert enough pressure on the protoplast to raise the pressure potential high enough to counterbalance osmotic potential, however negative it might be.

Immature, growing cells have weak, deformable walls and cannot generate enough pressure to stop water absorption. The cell grows rather than bursts. Under these conditions, the cell may increase greatly in size. With such a large influx of water, solutes in the cell may become significantly diluted; osmotic potential and water potential may become less negative, and the cell reaches hydraulic equilibrium with surrounding cells and growth stops.

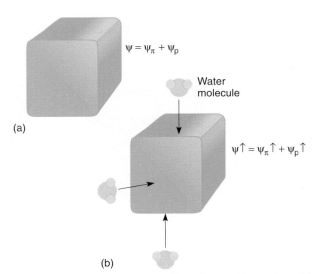

$$\psi = \psi_\pi + \psi_p$$

Water molecule

(a)

$$\psi\uparrow = \psi_\pi\uparrow + \psi_p\uparrow$$

(b)

Figure 12-10 (a) A healthy cell, turgid and full of water and protoplasm. It is swollen and firm, just like the cells of unwilted leaves. Its walls are stretched and are pressing back against the protoplasm that is pressing on them. (b) If even a small amount of water enters the cell, the slight increase in volume causes the walls to stretch and push back, and pressure builds inside the cell. Consequently, ψ_p becomes much more positive and so does the cell's water potential. Because of the slight increase in the amount of water in the cell, the salts, sugars, amino acids, and all other solutes are now slightly more dilute; therefore, ψ_π becomes slightly less negative also, but this change is insignificant compared with the change in ψ_p.

Figure 12-11 Some aspects of biotechnology processes require that the cell wall be digested away with cellulase enzymes, leaving behind a naked protoplast. The digestion mixture must contain just enough solute—usually sucrose or the sugar-alcohol mannitol—that the cells lose water and shrink away from the wall. If not, the protoplasts burst when the wall is removed. The solute concentration must be adjusted carefully; if too strong, too much water is pulled out of the cells and they die. (a) Cells in suspension culture before treatment. (b) Cells in 6% mannitol with the protoplasts just pulling back from the wall (both ×500).

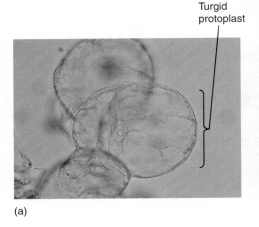

(a)

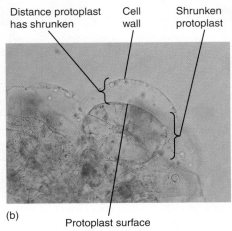

Turgid protoplast

Distance protoplast has shrunken

Cell wall

Shrunken protoplast

(b)

Protoplast surface

In growing regions, however, such as the tips of stems and roots, and in expanding leaves, cells can keep their osmotic potential and water potential very negative despite the influx of water, either by actively pumping in solutes through the plasma membrane or by hydrolyzing giant starch molecules into thousands of glucose molecules. After the proper size is reached, growth can be stopped either by strengthening the wall so that it exerts more pressure and raises the pressure potential or by stopping the import of solutes or the hydrolysis of starch, allowing the osmotic potential to rise. Either way, rising pressure potential or osmotic potential causes the cell's water potential to rise and reach equilibrium with the surrounding cells, stopping the net inflow of water.

Although plant cells cannot absorb so much water that they burst, water loss can be a serious problem. Imagine that our demonstration cell, now in pure water and with a water potential of 0.0 MPa, is placed in a strong sugar solution with a water potential of −2.0 MPa. Water moves out of the cell, osmotic potential

becomes slightly more negative, and the pressure potential drops rapidly. In such a strong solution, long before the cell reaches equilibrium it loses so much water that the protoplast shrinks in volume and no longer presses against the wall. Plants never absorb so much water that their cells burst, but they frequently lose enough water to wilt because their protoplasts do not press firmly against the cell walls. The wall is not stretched and does not exert any pressure back; therefore, the pressure potential drops to 0.0 MPa. The point at which the protoplast has lost just enough water to pull slightly away from the wall is called **incipient plasmolysis** and is quite important (**Figures 12-12 and 12-13**). Up to that point, the cell has lost very little water, so its volume change and osmotic potential change have not been great, but because the pressure potential is now zero, the water potential equation is

$$\psi = \psi_p + 0$$

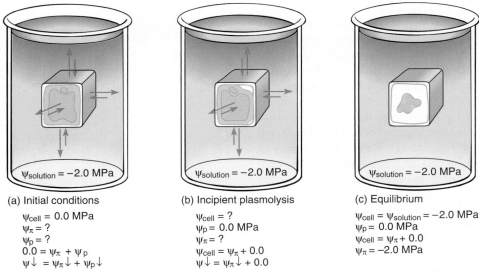

(a) Initial conditions

$\psi_{cell} = 0.0$ MPa
$\psi_\pi = ?$
$\psi_p = ?$
$0.0 = \psi_\pi + \psi_p$
$\psi\downarrow = \psi_\pi\downarrow + \psi_p\downarrow$

(b) Incipient plasmolysis

$\psi_{cell} = ?$
$\psi_p = 0.0$ MPa
$\psi_\pi = ?$
$\psi_{cell} = \psi_\pi + 0.0$
$\psi\downarrow = \psi_\pi\downarrow + 0.0$

(c) Equilibrium

$\psi_{cell} = \psi_{solution} = -2.0$ MPa
$\psi_p = 0.0$ MPa
$\psi_{cell} = \psi_\pi + 0.0$
$\psi_\pi = -2.0$ MPa

$\psi_{solution} = -2.0$ MPa (in each beaker)

Figure 12-12 (a) When placed in a solution with a strongly negative water potential, the cell loses water rapidly and cell volume drops. ψ_p becomes less positive and ψ_{cell} becomes more negative; ψ_π changes only a small amount. (b) Incipient plasmolysis is the point at which the protoplast has shrunk just enough to pull away from the wall, and thus, ψ_p is zero and ψ_{cell} equals ψ_π. (c) If the cell does not reach equilibrium at incipient plasmolysis, it continues to lose water, and ψ_{cell} continues to become more negative until it reaches −2.0 MPa. The pressure potential here cannot become a negative number; therefore, the changing water potential is due to a changing osmotic potential. During plasmolysis, the cell loses enough water to change the concentration of solutes significantly.

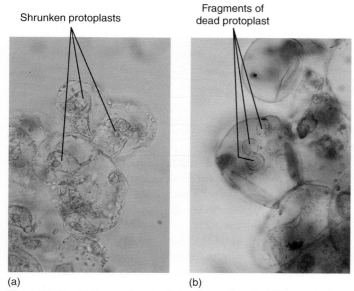

Shrunken protoplasts

Fragments of dead protoplast

(a)   (b)

Figure 12-13 (a) These cultured cells have been placed in 12% mannitol for several hours and are severely plasmolyzed (compare with Figure 12-11). (b) After a few days of severe plasmolysis, the cells have died (both ×500).

TABLE 12-2	Water Potentials of Various Tissues Under Certain Conditions
Megapascals	
0.0	Leaves at full turgor
−0.05	Fertilizer solution
−0.2	Most roots in dry soil
−0.5	Leaves of plants in well-watered soil; leaf growth good
−1.0	Leaves of plants rooted in dry soils; leaf growth slow
−2.5	Seawater
−2.7	1 molal sucrose or mannitol
−3.0	Leaves of plants rooted in very dry soil; for many plants, growth is slow or has stopped. Some plants have died.
−6.0	Leaves and twigs of desert shrubs in very dry soil
−20	Dry, viable seeds capable of germination

If the cell has not reached equilibrium at the point of incipient plasmolysis, it continues to lose water, and the protoplast pulls completely away from the wall and shrinks. The cell has become **plasmolyzed**. Water potential continues to become more negative entirely because of the osmotic potential as solutes become more concentrated. Most plants at the equilibrium point of −2.0 MPa would die of severe water loss.

Although such severe desiccation kills most cells, some can survive it easily. The embryos in most seeds are much drier, having water potentials as low as −20 MPa. Less dramatically, the leaves of desert shrubs in dry soil have water potentials as low as −2.0 to −6.0 MPa (Table 12-2). For most plants of temperate climates, a leaf water potential below −1.0 MPa stops leaf growth, although leaves can survive such desiccation for many days or weeks.

Short-Distance Intercellular Transport

Most plant cells communicate with their neighboring cells, transferring water, sugars, minerals, and hormones at least. This movement occurs by a variety of mechanisms. First, all living cells are interconnected by plasmodesmata, the fine cytoplasmic channels that pass through primary cell walls. All of the protoplasm of one plant can be considered one continuous mass, referred to as the **symplast**.

Material also is transferred from one cell to another by transport across the plasma membrane. The methods involve osmosis, molecular pumps in the plasma membrane itself, or fusion between transport vesicles and the plasma membrane. Once across a plasma membrane, a molecule initially resides in the cell wall. The wall is probably thin and permeable, and the molecule can penetrate it easily, diffusing across it to an intercellular space or laterally through it, spreading along the cell surface (Figure 12-14). Most small molecules can move easily through both the wall and the intercellular spaces; the two together are called the **apoplast** of the plant.

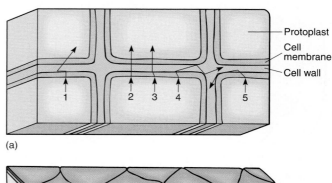

(a)

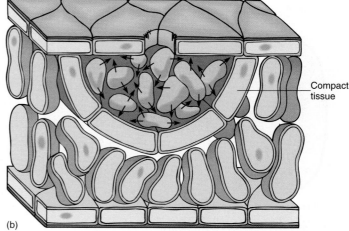

(b)

Figure 12-14 (a) After being released from a cell, a molecule diffuses in a series of random short paths as it collides with and bounces off other molecules. Probability favors its entering a neighboring cell (molecules 1, 2, and 3), but it can diffuse laterally along the wall as well (molecules 4 and 5). (b) In many glands, the apoplast is large, so movement between cells may be faster and easier than movement within cells. Such glands often have a lining of compact tissue that isolates the gland, preventing the secreted material from permeating the entire region.

In glands, the apoplast is mostly intercellular space through which molecules move easily, usually toward the surface of the gland. In nonglandular regions, the apoplast is mostly cell wall. The secreted molecule is probably absorbed by a cell neighboring the one that secreted it. In most parenchymatous tissues, primary walls are thin (less than 1 μm thick), and the contact faces between two cells are so extensive (10 to 20 μm²) that the probability is much greater that a molecule will either diffuse more or less directly into the next cell or re-enter the cell from which it came. If the molecule was originally secreted by active transport, the original cell membrane is probably impermeable to it, at least in that area, so return to the original cell is usually not possible. This is probably the most common mechanism for movement of water, sugar, and other nutrients between parenchyma cells within cortex, pith, or leaf mesophyll.

■ Guard Cells

The opening and closing of stomatal pores are based on short-distance intercellular transport. At night, when stomata are closed (except for CAM plants), guard cells are somewhat shrunken and have little internal pressure. They are in hydrau-

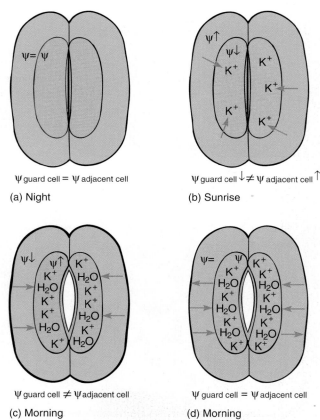

ψ guard cell = ψ adjacent cell

(a) Night

ψ guard cell $\downarrow$ ≠ ψ adjacent cell $\uparrow$

(b) Sunrise

ψ guard cell ≠ ψ adjacent cell

(c) Morning

ψ guard cell = ψ adjacent cell

(d) Morning

Figure 12-15 At night, guard cells and adjacent cells are in hydraulic equilibrium (a), but at sunrise, potassium is pumped into guard cells, increasing solute concentration (b). Osmotic potential and water potential become more negative, and water flows in (c), causing guard cells to swell and open the pore. As pressure builds, pressure potential rises, counteracting the falling osmotic potential; therefore, the guard cell's water potential rises and moves back into equilibrium with the adjacent cell, but by the time that happens, the stomatal pore is open (d). At night, when the stomata must close, all steps are reversed.

lic equilibrium with surrounding cells: Water enters and leaves guard cells at approximately the same rate; no net change occurs in the amount of water. When guard cells must open, such as just after sunrise, potassium ions (K⁺) are actively transported from surrounding cells into guard cells (Figure 12-15). Once inside the guard cells, the potassium cannot leave because the plasma membrane is impermeable to it: Potassium pumping is possible but diffusion is not. The loss of potassium causes the water potential in adjacent cells to become less negative, whereas absorption of potassium causes water potential in guard cells to become more negative. Adjacent cells and guard cells are thrown out of hydraulic equilibrium by potassium pumping, and water diffuses out of surrounding cells across their plasma membrane, across the two primary walls and middle lamella, and into guard cells across their plasma membrane. The extra water and potassium cause guard cells to swell, bend, and push apart, opening the stomatal pore. After they are open, potassium pumping stops. Water movement brings guard cells and adjacent cells into water potential equilibrium again, and net water movement stops. When guard cells must close, the process is reversed: Potassium is pumped from guard cells into surrounding cells and water follows. Guard cells and adjacent cells are in equilibrium when stomata are fully open and fully closed.

Notice also that guard cells of fully opened and fully closed stomata are both in equilibrium with surrounding cells (and thus also with each other), even though they all have different internal conditions: When open, the guard cells' abundant potassium gives them a very negative osmotic potential, which is countered by turgor pressure, giving a large pressure potential. This results in only a small negative water potential. When closed, the guard cells have less potassium; therefore, only a small negative osmotic potential. This is countered by less turgor and therefore a less positive pressure potential. These cells too have only a small negative water potential. Even though two cells might have very different internal conditions, they can have the same water potential and be in hydraulic equilibrium.

■ Motor Cells

The leaves of sensitive plant (*Mimosa pudica*), prayer plant (*Oxalis*), and many other species move slowly and reorient themselves by flexing and folding in response to a variety of stimuli (Figure 12-16). The location of flexure is either the entire midrib or the point at which the petiole attaches to the lamina or stem. The cells at these "joints," called **motor cells**, are similar to guard cells: They can either accumulate or expel potassium and thus adjust their water potential and turgidity.

In Venus' flytrap, the leaf can close rapidly, in less than a second, but it requires several hours to reopen. Motor cells are located along the midrib, and when they are shrunken, pressure in other midrib cells causes the two halves of the blade to be appressed, and the trap is closed. Trap opening occurs as potassium is slowly accumulated by motor cells, water diffuses in, and the motor cells become turgid. Closure is not caused by pumping potassium out of motor cells; that would be too slow. Instead, the membrane suddenly becomes freely permeable to potassium, and it rushes out instantly. The water balance is rapidly changed, and water too floods out, allowing the motor cells to virtually collapse; the trap then shuts quickly enough to catch insects.

Plants and People

Box 12-2 Farming "Wastelands"

It has been proposed that we develop plants that tolerate high levels of salt and use them as crop plants in arid, marginal "wastelands" where fresh water is scarce. This might be possible in two ways: We could examine desert-adapted plants to see whether any have useful properties such as nutritious seeds, medicinal drugs, or useful fibers. Alternatively, desert-adapted plants could be studied to identify which of their features make them drought resistant; then the corresponding genes could be identified, cloned, and transferred to some of our crops that are not now drought adapted. Once either of these types of plants is available, they could be grown in semidesert regions or perhaps even in deserts using a little irrigation with seawater or brackish water.

This concept has many problems. The first is that it is one thing to find plants with enough drought and salt tolerance to grow in seawater during a 1-year-long experiment, but it is another thing to irrigate an area with seawater year after year. Because they have little rain, most arid regions have only small, temporary rivers that have not been able to carve effective drainage channels: They often end in dry lakes with no outlet. After a rain, water is present for a few weeks, but it evaporates, leaving its minerals behind. The rivers do not carry minerals out to sea continuously as do rivers in the eastern United States and other moist areas. If seawater is used for irrigation in an arid area, tremen-

dous amounts of salt would be deposited as the water evaporates. This would continue year after year until the salt concentration becomes so high that a salt desert is created, such as occurs in the Great Salt Lake in Utah, the Dead Sea in the Middle East, or the Devil's Golf Course in Death Valley, where accumulated salt is so abundant it forms crystals 2 or 3 feet tall. Nothing could grow—not the new crop plants, not the genetically engineered plant, and certainly not the original plants, which would all be destroyed when the region was plowed to start the project. Even using fresh water rather than seawater causes a gradual accumulation of salt, and California's Central Valley is already facing a serious problem of salt accumulation in the soil.

Another problem with this type of project is the concept that deserts and semideserts are wasteland, that they must be "developed" and "improved." Although this type of thinking may have been popular at one time, many people now disagree with it. These areas, in their natural state, have an intrinsic worth. They are home to a great diversity of plants, animals, fungi, and other species. Even if none of these "wasteland" species is ever discovered to contain a medicinal drug or other useful feature, does that give us the right to exterminate them? Is it really necessary to bring more land into cultivation when so many Americans are overweight or eat foods that have been processed specifically to remove calories from them?

(a)

(b)

Figure B12-2 (a) This desert would be considered "wasteland" by some people, but it is where many rare and unusual plants live. This area has not only high biodiversity (there are many different species of plants, animals, fungi, and other organisms) but also peace, quiet, beauty, and a chance for people to reconnect with nature. The plants that look like upside-down green octopuses are ocotillos (pronounced oh ko TEA yoz; *Fouquieria splendens*). Fortunately, this "wasteland" is protected from "developers": it is part of Big Bend National Park (b).

(a)

(b)

Figure 12-16 (a) Each leaflet of the compound leaf of *Oxalis* is joined to the petiole by motor cells; here they were photographed in the morning when it was cool, and the sunlight was not intense. Motor cells are turgid and leaflets are held into the sunlight. (b) *Oxalis* plants in full, intense sunlight. The motor cells have lost potassium and water; thus, they are not turgid. Leaflets hang down, minimizing their exposure to light. Later in the afternoon, when sunlight is not so intense, or if the shadow of a tree moves across the plant, the motor cells will absorb potassium, then water, and raise the leaflets again.

Transfer Cells

The rate at which material can be actively transported depends on the number of molecular pumps present, which in turn depends on the surface area of the plasma membrane: The larger the membrane, the more molecular pumps it can hold. In certain specialized **transfer cells**, the walls are smooth on the outer surface but have numerous finger-like and ridge-like outgrowths on the inner surface. The plasma membrane is pressed firmly against all of the convolutions and thus has a much larger surface area than it would if the wall were flat. Consequently, room is available for many molecular pumps, and high-volume transport can occur across these transfer walls. Transfer cells are found in areas where rapid short-distance transport is expected to occur: in glands that secrete salt, in areas that pass nutrients to embryos, and in regions where sugar is loaded into or out of phloem.

Long-Distance Transport: Phloem

Although the exact mechanism by which water and nutrients are moved through phloem is not known, most evidence supports the **pressure flow hypothesis**. Membrane-bound molecular pumps and active transport are postulated to be the important driving forces.

The sites from which water and nutrients are transported are **sources**. During spring and summer, leaves are dominant sources, as their photosynthetically produced sugars are exported to the rest of the plant. At other times, such as early spring, before new leaves have been produced by deciduous trees, the sources are storage sites such as tubers, corms, wood and bark parenchyma, and fleshy taproots. Cotyledons and endosperm are sources for embryos during germination. Within sources of many species, sugars are **actively transported** into sieve elements—sieve tube members in angiosperms, sieve cells in plants other than angiosperms. In other species, phloem is loaded by the **polymer trap mechanism**: Conducting-cell plasma membranes are permeable to monosaccharides and disaccharides but not to polysaccharides. Simple sugars merely diffuse into the conducting cells and then are polymerized into polysaccharides that cannot diffuse

back out. In both loading mechanisms, cells surrounding sieve elements, both companion cells and other phloem parenchyma cells, are important in loading phloem; many of the cells are transfer cells. It is common now to think of the functional unit as consisting of both a conducting cell and one or several companion cells, and the term **STM/CC complex** is used.

As sugars accumulate in sieve elements, the sieve element protoplasm becomes more concentrated (**Figure 12-17a**). Consequently, both its osmotic potential and its water potential become more negative. This causes hydraulic disequilibrium between sieve elements and surrounding cells, and water diffuses into the sieve elements. In any other cell, the increased volume of sugars and water would cause the protoplast to expand and press against the cell wall, but sieve elements are unique, being living cells with relatively large holes in their walls, up to 14 μm wide in cucumbers and pumpkins. When pressure starts to build in these cells, "protoplasm" is squeezed through the sieve pores into the next cell. Sieve element protoplasm is not like that of most cells: The vacuolar membrane (tonoplast) disintegrates, allowing vacuolar water to mix with part of the cytoplasm, creating an extremely watery, nonviscous substance—phloem sap. The majority of the protoplasm is held firmly to the walls, probably by microtubules or microfilaments, and is not carried away with the watery central phloem sap.

In sources, phloem loading occurs along numerous vascular bundles such as the fine veins in leaves, the network of bundles in tubers and corms, and the inner bark in storage roots and stems. With this massive loading, pressure builds quickly, and a large volume of material flows from the source. Pressures as high as 2.4 MPa have been measured in some sieve tubes (human blood pressure is about 0.016 MPa). The rate of transport can be high; up to 660 cm/hr has been measured in leaves of corn (**Table 12-3**). The actual amount of sugars and other nutrients (excluding water) transported by phloem per hour is called the **mass transfer**. Vascular bundles vary not only in the speed at which their phloem translocates but also in the amount of phloem present. The number of bundles leaving a source is also important. To make comparisons easier, mass transfer can be divided by the cross-sectional area of phloem to obtain the **specific mass transfer** (**Table 12-4**).

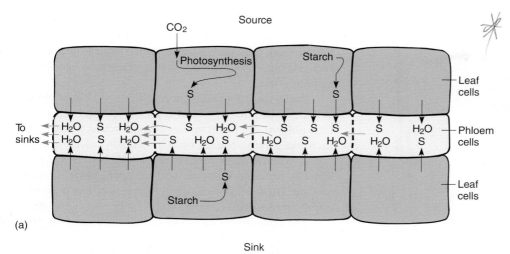

Source

CO_2

Photosynthesis

Starch

Leaf cells

Phloem cells

To sinks

Starch

Leaf cells

(a)

Sink

From source

Starch

Respiration

Anabolic construction

(b)

Figure 12-17 (a) As sucrose (S) is actively transported into sieve elements, ψ_p and ψ_{cell} become more negative, moving away from equilibrium with companion cells and other neighboring cells. Water moves into the sieve elements, squeezing phloem sap out through the sieve pores. Because the sugary water escapes through sieve pores, pressure does not build high enough to stop the influx of water. Water potentials of sieve elements and surrounding cells never reach equilibrium as long as sucrose is being pumped. (b) In sinks, sucrose is actively transported out of sieve elements, and all processes work in reverse compared with sources.

TABLE 12-3	Speed of Phloem Sap Translocation
	Maximum Speed (cm/hr)
Picea (spruce) stem	13.2
Pinus (pine) stem	48
Fraxinus (ash) stem	48
Ipomoea (morning glory) stem	72
Ulmus (elm) stem	120
Triticum (wheat) leaf	168
Heracleum (cow parsnip) stem	210
Helianthus (sunflower) stem	240
Zea (corn) leaf	660

TABLE 12-4	Specific Mass Transfer	
Leaves of	Maximum Specific Mass Transfer (g/hr/cm² of phloem)	
Digitaria (crab grass)	4.7	
Cynodon (Bermuda grass)	6.6	
Paspalum (a grass)	12.7	
Cocos (coconut)	32	
Triticum (wheat [roots])	180	
Ricinus (castor bean)	252	

Sinks are sites that receive transported phloem sap, and they are extremely diverse. Storage organs are important in perennial plants during summer, but also important are meristems, root tips, leaf primordia, growing flowers, fruits, and seeds. On even a small tree there may be thousands of sinks, each receiving nutrients. Not all sinks are active simultaneously; most plants do not produce flowers and leaves at the same time, and fruits can develop only after flowers. Within sinks, sugars are actively unloaded from sieve elements into surrounding cells. The loss of sugar causes phloem sap to be more dilute, and its osmotic potential and water potential tend to become less negative; thus, water diffuses outward into the surrounding cells (Figure 12-17b). As a result, even though phloem sap flows rapidly into a sink, the end cells of the phloem do not swell. As quickly as nutrients are loaded at sources, they are unloaded at sinks.

Because phloem sap is under pressure, the danger exists of uncontrolled "bleeding" if phloem is cut. Vascular bundles are broken open frequently, especially by chewing insects and larger animals. Two mechanisms seal broken sieve elements. The first is **P-protein** (P for phloem), found as a fine network adjacent to the plasma membrane inner surface of uninjured sieve elements. When phloem is ruptured, the phloem sap initially surges toward the break; this rapid movement sweeps P-protein into the cell center, where it becomes a tangled mass. When it is carried to a

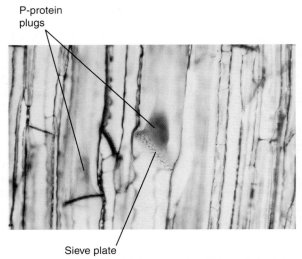

P-protein
plugs

Sieve plate

Figure 12-18 When this material of squash (*Cucurbita*) was being prepared for microscopy, it was cut open, causing the phloem sap to surge toward the cut, sweeping P-protein along and forming P-protein plugs, visible as dark brown masses. Sieve pores in the sieve plates are also visible (×500).

sieve area or sieve plate, the P-protein mass is too large to pass through and forms a **P-protein plug** (Figure 12-18).

Within uninjured phloem there is another polymer as well, **callose**. Apparently, it stays in solution only if it is under pressure; when injury causes a pressure drop, callose precipitates into a flocculent mass and is carried along with the P-protein to the nearest sieve areas. There the callose contributes to the plug, and leaking is prevented.

In monocots with long-lived stems, such as palms and Joshua trees, sieve tube members live and function for many years, even hundreds of years, but in all other plants, individual sieve elements have a lifetime of only months or even weeks. They stop transporting and are replaced by new phloem cells from the provascular tissues or vascular cambium. After they cease to function, callose deposits seal them permanently.

A further aspect of phloem transport is important to consider. As sugar is actively transported into phloem in sources, what happens to the water potential of the cells losing the sugar? Shouldn't it become less negative? No, it remains unchanged, because the sugars are being exported at the same rate they are being synthesized in leaves. Chlorenchyma cells absorb carbon dioxide molecules, but this does not cause the water potential to become more negative because the carbon dioxide is synthesized into sugar and exported. Millions of molecules of carbon dioxide may pass through a chlorenchyma cell without any long-term impact on its osmotic or water potentials. In sources such as tubers, sugar export has no impact on the storage cell water potential because as rapidly as it is exported, new sugar appears by depolymerization of starch. The same is true at sinks: Storage cells do not accumulate sugar as sucrose but instead polymerize it into starch. Thousands of molecules may be absorbed, but because they are polymerized into one molecule, no change in water potential occurs. In growing cells of sinks such as meristems and buds, the imported sugar is polymerized to cellulose, hemicellulose, and other carbohydrates. It can also be metabolized into amino acids, fatty acids, and nucleotides; these are

then polymerized into proteins, fats, and nucleic acids, and thus, again, little or no change in osmotic potential and water potential occurs. In all cells, part of the imported sugar is respired, but this converts it to carbon dioxide that is expelled; so again there is no osmotic effect.

Plants control the direction and rate of flow of phloem sap. While dormant in late winter and early spring, buds receive very little phloem sap, but after they become active, phloem transport increases greatly. Not all buds are equally affected; some grow rapidly whereas others, even though located quite close by, receive virtually nothing. While flowers are open, phloem transport is low, but after fertilization, when the ovary begins to develop into a fruit, transport increases.

The direction of transport can also change. Leaf primordia and young expanding leaves are sinks; imported sugar allows them to develop much more rapidly than they would if they had to be completely autotrophic (see Figure 12-2). In addition, they also need large amounts of nitrogen, sulfur, potassium, and other minerals. After leaves reach a critical size, they become self-supporting, able to photosynthesize rapidly enough to meet all their own needs (Figure 12-19). Shortly afterward they become sources, exporting material. Phloem transport is reversed in the leaf and petiole; molecular pumps must now load the phloem, not unload it. Plasma membranes may be altered, or one set of sieve elements may cease to function and may be replaced by an entire new set of cells. The early primary phloem often lives for less than a few weeks.

Young,
expanded
leaves

Shoot tip
with meristem
and leaf
primordia

Cotyledons

Figure 12-19 This bean contains two prominent sources, the cotyledons, which are supplying sugars, amino acids, and minerals to the rest of the seedling. The shoot tip with its meristem and leaf primordia are sinks, as are the roots. The first two leaves are expanded and are probably sources now, but they were sinks while they were developing.

Leaves near a stem tip export upward to the shoot apical meristem, while leaves farther back export toward the trunk and roots. As the shoot apex grows, leaves that had been near the apex are left behind and their transport shifts from upward to downward. If the apex produces flowers and then fruit, the direction of transport may shift again so that all leaves send sugars upward.

Long-Distance Transport: Xylem

Properties of Water

The movement of water through xylem is based on a few simple properties of water and solutions. One property is that water molecules interact strongly with other water molecules, behaving as if weakly bound together; when frozen, the molecules become strongly bound to each other. Because of this, liquid water is said to be **cohesive**, and any force acting on one molecule acts on all neighboring ones as well.

Another property of water is that its molecules interact with many other substances—it is **adhesive**. Almost all substances in plants, except lipids, interact with water: Cellulose, enzymes, DNA, sugars, and so forth have a shell of water molecules rather firmly attached to them. Occasionally, a water molecule vibrates out of one of these shells and is replaced by another water molecule, but in general, adhesion makes these water molecules less free to move around than other water molecules.

Water also adheres firmly to soil particles. When soil is quite moist, roots can absorb the free liquid water between soil particles, but as the soil dries, the remaining water adheres firmly to the soil and cannot be absorbed easily, if at all (Figure 12-20). Even though the soil may contain considerable water, it is unavailable to the plant. The same is true of seawater; water molecules interact so strongly with the salt molecules that land plants cannot pull the water away.

Another property of water is that it is heavy, and lifting it to the top of a tree requires a great deal of energy. If water were lighter, less energy would be involved.

Water Transport Through Xylem

Water movement through xylem and plants as a whole is governed by the principles of water relations just described. The **cohesion-tension hypothesis** is the most widely accepted model of the process. When stomatal pores are open, they unavoidably allow water loss. The apoplastic space of spongy mesophyll and palisade parenchyma is filled with moisture-saturated air, so water molecules have a strong tendency to diffuse from intercellular spaces to the atmosphere. Even relatively humid air has a tremendous capacity to absorb water: At 50% relative humidity, warm air can have a water potential as negative as −50.0 MPa. This water loss is called **transstomatal transpiration**. The cuticle and waxes on the epidermal surfaces are fairly efficient isolation mechanisms, being so hydrophobic that very little water passes through them; however, some water is lost directly through the cuticle by **transcuticular transpiration** (Table 12-5).

Consider a leaf in early morning: Stomata are closed, air is cool, and relative humidity is high. The air may have cooled

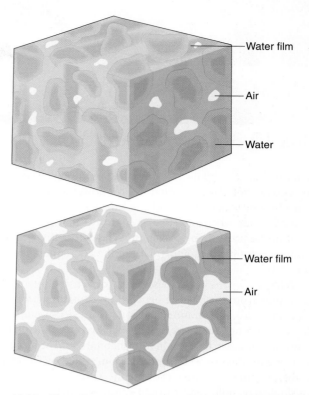

Figure 12-20 Wet soil contains water both as a film covering all surfaces of soil particles and as small masses held in capillary spaces formed where soil particles touch (a). The latter is held weakly and can be easily absorbed by roots. (b) Dry soil contains only tightly bound films of water. This adhesion is measured by ψ_m, which can be so negative that the soil's water potential is also very negative. This water cannot be absorbed by the roots of most plants.

enough during the night to allow dew to form. Cells within the leaf are turgid and probably in equilibrium with each other, all having a water potential between 0.0 and −1.0 MPa (see Figure 12-7). As the sun rises, stomata open and begin losing water; the air warms, and its relative humidity decreases. As transpiration causes epidermal cells and mesophyll cells near stomata to lose water, their water potentials become more negative, going out of equilibrium with surrounding cells. The disequilibrium

TABLE 12-5	Transstomatal and Transcuticular Transpiration*	
	Transstomatal	Transcuticular
Herbaceous plants		
Coronilla varia (crown vetch)	1,810	190
Stachys recta (mint family)	1,620	180
Oxytropis pilosa (locoweed)	1,600	100
Woody plants		
Betula pendula (European white birch)	685	95
Rhododendron ferrugineum (rhododendron)	540	60
Pinus sylvestris (Scots pine)	527	13
Picea abies (Norway spruce)	465	15
Fagus sylvatica (European beech)	330	90

*Rates are mg H_2O/dm²/hour (dm is decimeters; 1 dm = 10 cm). Surface area includes both sides of the leaf.

Alternatives

Box 12-3 Desert Plant Biology

Plants in many habitats have more than enough water; their leaves do not wilt, and they do not abort flowers or fruits because of insufficient water. Plants adapted to deserts, however, must survive periods when moisture is so scarce that more water is lost from the plant than is taken in. What alternatives make life possible in dry regions, and what are the consequences of each alternative?

What do we mean by "desert" or "dry environment"? Dryness results from the balance between precipitation (rain, snow, dew, and so on), evaporation, and soil texture. Precipitation may fall evenly throughout the year or seasonally, often as winter snow or summer thundershowers. In some habitats, rain occurs as drizzle lasting for days or weeks and is accompanied by cloudy, cool weather; in others, it comes torrentially, with 2 to 8 inches falling in a single day, followed by bright sunny weather. Periods without precipitation might last just a few weeks, but it is not unusual for rainless periods to last months in the Chihuahua, Sonora, and Mohave Deserts of the southwestern United States.

Evaporation is related to the relative humidity of air. Hot air can hold enormous amounts of water vapor; cool air can hold less. Precipitation occurs when air cools so much that its holding capacity drops below its actual moisture content: Excess water vapor condenses as fog, dew, rain, or snow; however, even cold air can be dry (think of chapped lips and dry skin in winter), and hot air can be humid (think of muggy summer days). Plants face water problems when periods of low precipitation coincide with periods of dry air.

Texture affects a soil's capacity to store precipitation. Rocky, sandy soils, especially those on slopes, hold almost no water and become dry just days after a heavy rain. Fresh lava flows in Hawaii are deserts despite receiving rain almost every day. Alternatively, fine soils with some clay and abundant humus hold large amounts of water for months: Roots have a steady supply to draw from, but even this is a subtle complexity for plants: Sandy desert soils are actually beneficial because they allow even a light rainfall to penetrate down to the root zone. If the desert had rich soil with clay and humus, light rains would be held in the uppermost layers, leaving roots dry.

Averages are not especially helpful in understanding deserts and desert-adapted plants: Variations from year to year are important. The vegetation of any desert is the result of plants surviving not merely the droughts of average years but also rare protracted droughts that last several years. For example, cacti dominate the deserts of Mexico and the American southwest because they survive exceptionally long droughts that kill off any plant that can only survive average droughts. If every year were an "average year," these plants would become so abundant that they would overgrow the cacti, shading them and ultimately killing them.

Two alternatives by which plants adapt to dry habitats are **drought avoidance** and **drought tolerance**. Most deserts have either a brief period when they are moist or they have small areas where water collects. Many drought-avoiding plants are known as desert ephemerals (ephemeral means short lived). They are small plants that complete their life cycle in just a month or two: Seeds germinate, seedlings grow, the plant flowers, produces new fruits and seeds in just a few weeks, while the soil is moist after a rain or a snow melt. By the time drought arrives, the plants have died, but their seeds are ready for the next moist season. These plants avoid dry conditions. Two consequences are that these plants (1) can live in deserts and (2) can never become large or perennial.

Drought-tolerating plants live through dry periods, losing water more rapidly than they gain it. Most accomplish this by being succulent: A high percentage of their body consists of water-storage cells. Stem succulents such as cacti, euphorbias, and stapeliads store water in pith and cortex. Leaf succulents such as agaves, yuccas, echevarias, and lithops have very thick leaves. Desert-adapted bulbs are leaf succulents in which the upper portion of each leaf extends above ground and is thin and photosynthetic, whereas the lower portion of each leaf is subterranean, thick, fleshy and able to store water and nutrients through adverse periods. Desert-adapted bulbs survive for years in a dormant state, not producing any aerial leaves until moisture is sufficient. Root succulents are less common, but they include plants like yams (*Dioscorea*). Root succulents typically have nonsucculent shoots and leaves that die back during drought, being replaced in the next moist season by new shoots that sprout from the "root crown," a bit of shoot located at the top of the root. Wood succulents produce wood with a high percentage of parenchyma cells or a type of fiber that stores water; examples are baobab trees (*Adansonia grandidieri*), boojum trees (*Idria columnaris*), and elephant trees (*Bursera microphylla*).

Each of these alternatives has particular consequences. Water stored in wood is near vessels and may be especially effective in preventing cavitations, but it must then be transported out to leaves and flowers as in nondesert plants. Succulent leaves store water near photosynthetic tissues, keeping them hydrated and photosynthesizing even during very dry periods. Stem succulents may have thin, flat ordinary foliage leaves that are ephemeral, abscising when dry seasons start, after which the stems must perform all photosynthesis. Other stem succulents, such as cacti, never have large foliage leaves and must rely solely on their stems for photosynthesis; such plants are always good at conserv-

ing water but can never take full advantage of occasional moist periods.

A completely different way of tolerating drought is the capacity to survive protoplasmic desiccation. Many desert-adapted mosses have no capacity to store water, and they are perennial, not ephemeral. As the habitat becomes drier, so does their protoplasm until they become dormant for months. The water potential of their cells becomes extremely negative, which would kill most plants, but these are merely inactive, not dead. On days with sufficient dew, fog, or light rain, the mosses rehydrate in just an hour or so and quickly resume ordinary

metabolism, including photosynthesis, growth, and development. Their activity will last as long as the environment remains sufficiently moist, but this might be just a few hours. Having a body water content that changes with habitat moisture is called **poikilohydry** (cold-blooded animals are poikilothermic)—it also occurs in some liverworts, some species of *Selaginella* ("resurrection plants"), the tiny cactus *Blossfeldia*, and a few other vascular plants. Lichens are arguably the supreme examples of poikilohydry, being capable of drying to very low water content for months without dying, even if growing on rocks exposed to full summer sun.

does not become major because water diffuses into these cells from other cells and apoplastic spaces deeper within the leaf. But this water movement out of the deeper mesophyll cells causes their water potentials to become more negative, away from equilibrium with even deeper cells (Figure 12-21).

Finally, this gradient of water potentials reaches a tracheid or vessel member. As water molecules move out of tracheary elements into mesophyll parenchyma cells, the water potential within the xylem water column becomes more negative. The loss of water from tracheary elements does not really affect the xylem osmotic potential because solutes are very dilute to begin with.

Here water's cohesive properties are more important: As a water molecule leaves the xylem, it does not leave a hole behind but instead drags other water molecules along with it. All water molecules of the plant are hydrogen bonded together, but the water molecules in the xylem can move upward most easily. That water is purest, is not bound to proteins and cellulose, is not locked into hydration shells around solutes, and so on. As water molecules diffuse out of xylem in the leaves, cohesive forces pull water upward through the xylem, all the way from the roots (Table 12-5 and Table 12-6). Think of an icicle: If the top molecules are pulled upward, the entire mass of icicle is lifted.

Figure 12-21 As water moves out of the leaf into the air, the tissues dry and a water potential gradient becomes established. Water flows from the xylem, where water potential is least negative, toward air, where water potential is most negative.

1. ψ tracheids > ψ sheath

2. ψ sheath > ψ apoplast

3. ψ sheath > ψ spongy mesophyll

4. ψ spongy mesophyll > ψ apoplast

5. ψ apoplast > ψ air
   ψ air = −50 MPa

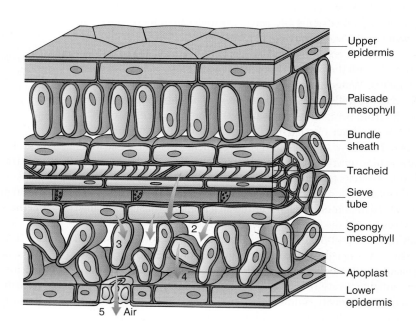

Upper epidermis

Palisade mesophyll

Bundle sheath

Tracheid

Sieve tube

Spongy mesophyll

Apoplast

Lower epidermis

5 ↓ Air

TABLE 12-6	Speed of Xylem Sap Translocation
	Maximum Speed (cm/hr)
Evergreen conifers	120
Sclerophyllous plants	150
Diffuse porous trees	600
Ring porous trees	4,400
Herbs	6,000
Vines	15,000 (0.6 mi/hr)

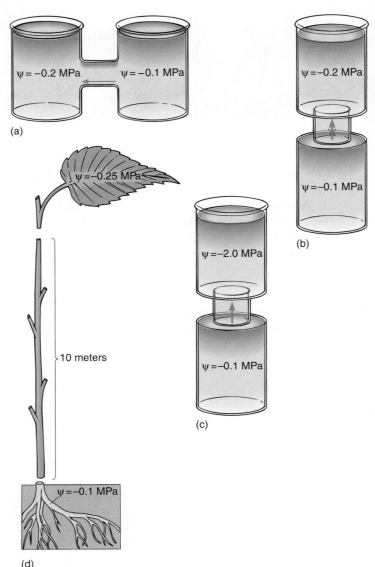

(a)

(b)

(c)

(d)

Figure 12-22 (a) If water can move laterally between two solutions, no lifting is involved. The slightest difference in water potential results in water movement. (b) If water must move upward against gravity, work must be done. A slight difference in water potential may not be enough to cause water movement; only a large difference will (c). (d) To overcome gravity and friction, the water potential of plant tissues receiving water must be at least 0.2 MPa more negative than that of roots for every 10 meters of height separating them. In the case illustrated here, with a difference of 0.15 MPa, water would not move up the stem. The water potential of the leaf would have to become 0.5 MPa more negative.

Water is heavy, and water molecules in the uppermost tracheary elements must lift the weight of the entire water column. There is tension (pull) on these molecules, and consequently, the **pressure potential is a negative number**; as water moves into the leaf mesophyll, the xylem water potential becomes more negative because of an increasingly negative pressure potential. In vertical stems, water must move directly upward in the xylem, and the water's weight is a significant consideration. In the examples discussed earlier, water could move laterally between a cell and a beaker; therefore, no lifting was involved. In vertical xylem, the weight of water counteracts its tendency to rise into areas of more negative water potential. Consequently, if leaf xylem water potential is only slightly more negative than root xylem water potential, the water does not move. For every 10 meters of height, leaf water potential must be at least 0.1 MPa more negative than root water potential (**Figure 12-22**). In trees such as elms and sycamores that are typically more than 30 meters tall, leaf water potential must be at least 0.3 MPa more negative than root water potential simply to overcome the weight of water. This is accomplished automatically: When stomata open in the morning, leaf cells lose water, and their water potentials become more negative; however, water does not begin moving upward in the xylem until drying causes the water potential of leaf cells to become sufficiently more negative. Stolons, rhizomes, and horizontal vines have no such problem; long-distance transport is horizontal and no lifting is involved, so gravity is not a factor. A few plants grow as pendant epiphytes, their stems dangling down from the branches of the host plant. Their stems and leaves are lower than their roots, and gravity assists water movement (**Figure 12-23**).

Water is extremely adhesive, and its molecules interact strongly with the polymers of the cell walls of tracheids and vessel elements. Water molecules adjacent to the walls tend to remain fixed to the walls and also tend to prevent neighboring water molecules from being drawn upward by transpiration/cohesion. This results in a layer of relatively immobile water that does not move easily. In narrow tracheids and vessel elements, this immobile water is a significant fraction of the water column. The resulting friction hinders water's movement and contributes to the tendency of water to remain stationary even when leaves have a more negative water potential than roots. Imagine lifting an icicle: You must pull against its weight and the friction of the icicle in a tube (the cell walls), but lifting the top of the icicle raises the entire water column unless it breaks. As a rough approximation, to overcome friction, leaf water potential must be at least 0.1 MPa more negative than root water potential for every 10 meters of height; therefore, considering both friction and gravity, a difference of 0.2 MPa is needed for every 10 meters of height. In plants with numerous wide vessels, friction is

less, and less than 0.2 MPa is needed; however, in plants with narrow tracheary elements, even more than 0.2 MPa is necessary. Also, in plants that have only tracheids, water molecules must be pulled through pit membranes when entering and leaving each tracheid, further contributing to friction.

Returning to the plant in our example, transpiration causes leaf cells to lose water, and their water potentials become more negative; water moves into them from tracheary elements, and tension pulls on water molecules in the xylem. When the water potential in the uppermost tracheary elements has become sufficiently more negative than that in the lower elements, friction and gravity are overcome and water moves upward. This causes

Figure 12-23 This epiphytic cactus (*Rhipsalis*) grows upside down. Its roots cling to the bark of a big branch of a rain forest tree, and its slender stems dangle straight down. Because its transpiration surfaces are lower than its roots, water flows downward from the roots, and gravity actually assists xylem conduction rather than hindering it, as in most plants.

TABLE 12-7	Water Potentials of Soils	
	Water Potential (MPa)	
Soil Type	10% Moisture	30% Moisture
Sand	−0.05	−0.001
Loam	−0.5	−0.005
Clay	−10.0	−0.1

rain has stopped. The diversity of pore sizes allows root hairs to absorb water but prevents gravity from pulling the water so deep into the soil that roots cannot reach it.

Many roots remain healthy with their water potentials as low as −2.0 MPa; they still absorb water from soils that are quite dry (Figures 12-24; see Table 12-2), but if the plant is 10 meters tall, the leaves would have to have a water potential at least slightly more negative than −2.2 MPa to overcome friction and gravity. This is also possible in some species, but typically, the leaves would either be dormant or preparing for abscission.

When both soil and air are dry, plants are greatly stressed. Even if stomata close, transpiration continues, at least through the cuticle. Leaf water potentials become more negative, but water cannot move upward easily because the soil is so dry. Tension on the water columns increases, and at some point, cohesion is overcome: Hydrogen bonding is broken over a large region, and the water column breaks. This breaking is called **cavitation**, and the water column acts just like a broken cable. Molecules above the cavitation point are drawn rapidly upward because they are now free of the weight of the water below them; those below the cavitation point rush downward because nothing supports their weight. Between the two portions is space called an **embolism** (often called an air bubble), which expands until its surface encounters a solid barrier such as a pit membrane. The water/embolism surface cannot pass through pit membranes, but it can pass through perforations because they are open holes (Figure 12-25). When an embolism forms in a tracheid, only that tracheid loses its water, but when an embolism forms in a vessel element, it may expand through perforation after perforation until the entire vessel has been emptied.

Cavitation often means that that tracheid or vessel can never conduct water again. The cohesive bonding that permits leaf transpiration to draw water upward has been disrupted. Under unusual conditions, embolisms are occasionally "healed." If all of the surrounding cells are full of water and if the night is so cool and humid that transpiration stops, enough water may seep into the embolism to fill it and re-establish a continuous water column. More typically, any water that seeps in simply flows down the side of the tracheary element but is not able to fill it.

Because cavitation destroys the usefulness of an entire vessel or tracheid and because a plant invests considerable energy and reduced carbon in making tracheary elements, features that minimize cavitation are selectively advantageous. Adhesion between water and the cell wall is just such a feature, giving the water column extra strength so that it does not cavitate easily. This is most effective in narrow elements, where the reinforcement affects all water molecules, even those in the center. In wide elements, the central molecules are freely mobile and cavitate almost as easily as pure water. To clarify, cavitation breaks hydrogen bonds in water

the lowermost xylem cells to pull water inward from the root cortex, which in turn pulls water in from the root epidermis. Water potential of the root epidermis and root hairs becomes more negative than that of the soil, and water moves automatically into the root.

Long-distance water transport occurs in this manner as long as the soil is sufficiently moist. A sandy soil that has 30% moisture has a water potential of approximately −0.001 MPa, almost equal to that of pure water (Table 12-7). Water is held in soils by cohesion and adhesion as wedges and droplets between soil particles, and in sandy soils, root hairs can easily draw water from moist soils. Gravity and evaporation also pull water away from the droplets in sandy soils, so these soils dry quickly after a rain. As the soil dries, the most mobile molecules are removed, and those tightly bound to soil particles remain (Figure 12-20). In a dry soil, not only is less water present, but it is also relatively immobile. Clay soils are composed of thin flakes with a high surface-to-volume ratio. When wet, they hold a great deal of water, but it is firmly bound as a hydration layer. No root can pull water away from clay that is even slightly dry. Loam soils consist of sand, silt, and clay and have a diversity of pore sizes. During a rainfall, loam soils absorb large amounts of water and then hold it for weeks after the

(a)

(b)

(c)

Figure 12-24 (a) Many desert shrubs withstand severe desiccation; even though their water potentials become extremely negative, the cells survive, although they may become inactive. This *Nolina parryi* is in full bloom despite its very dry habitat, Joshua Tree National Park. (b) Plants such as this floating aquatic stream plant (*Ludwigia*) do not tolerate water stress at all; if their water potential falls very far below −0.2 MPa, they die quickly. (c) During a drought lasting 8 months with no rain at all when this photograph was taken, the ground became so dry the bamboo died. However, the trees in the background were still healthy, and a few vining weeds climbing the dead bamboo were not merely surviving but actually growing.

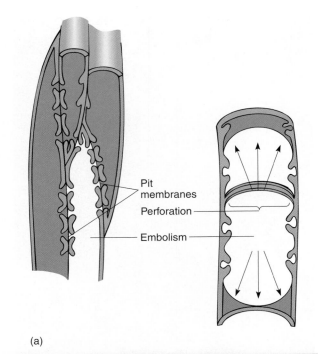

(a)

(b)

Figure 12-25 (a) Severe tension can overcome the cohesion of water molecules and cause an embolism to form and expand rapidly. The embolism can pass through holes such as perforations but is stopped by pit membranes. If an embolism occurs in a tracheid, it cannot spread beyond the tracheid, but if a vessel element cavitates, the embolism spreads throughout the entire vessel. (b) This is the trunk of a large water oak (*Quercus nigra*) that was cut down. As the chainsaw cut through the xylem vessels the water in them was cavitated, so now there is no hydrogen bonding to hold the water in the wood. The water here is pouring out of the wood, not just dripping, and the flow continued for almost a full minute. If this wood had been composed of tracheids, almost no water would have come out.

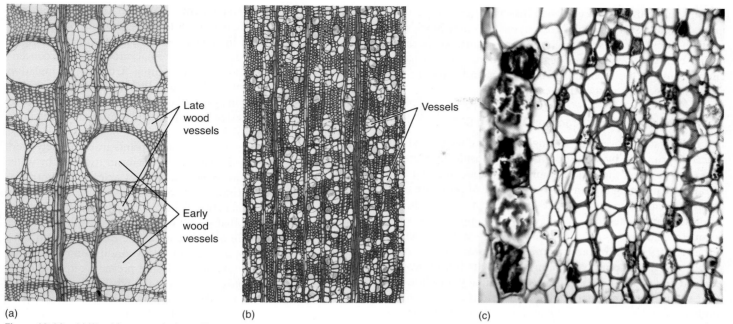

(a)   (b)   (c)

Figure 12-26 (a) Wood from a tropical tree. There are many broad vessels in the early wood, each of which can conduct water rapidly from the moist soil. Late wood has narrower vessels (×100). (b) This wood is from a tree of temperate climates; the vessels are narrow and abundant. No single vessel can conduct very much water, but if one cavitates, only a small fraction of the conducting capacity of the wood is lost (×100). (c) This is wood of a succulent relative of geranium, *Pelargonium carnosum*. Broad, open red cells are vessels. Narrow red cells are fibers (only about five are present), and all cells with thin blue primary walls are wood parenchyma cells. This plant stores water in its wood, minimizing chances of cavitation (×200).

but does not damage cell walls at all (remember that tracheary elements are dead when conducting); all tracheary elements in heartwood are cavitated, but the wood is still strong.

This wall-induced reinforcement of water columns is believed to be the feature that allows plants to reach the heights they do. Redwood trees in northern California and southern Oregon are the tallest plants known; they grow to 100 meters in height (the record is 115 m—379 feet—for a tree named Hyperion), and water is pulled upward the entire distance through their tracheary elements. This cannot be duplicated with glass or metal capillary tubes; the water columns are too fragile to support their own weight without the reinforcing that cell walls provide. One hundred meters appears to be the limit for xylem; even with reinforcement, the cohesive forces at the top of the water column cannot support the weight of 100 meters of water hanging from them.

Some remarkably exquisite types of wood have evolved that are elegantly adapted for the various conditions. In the moist tropics, water is always abundant; thus, the soil is never dry, and water always moves easily. Reinforcement is not necessary, and the wood is full of wide vessels (**Figure 12-26**). In drier temperate areas, especially rocky slopes, water is frequently scarce, and water stress common. It is selectively advantageous for plants in such an environment to have narrow vessels or even wood with tracheids only, as conifers have. In temperate areas with good rainfall, plants usually have a moist spring and produce early wood (also called spring wood) with large vessels; the summer is drier, and they then produce late wood (summer wood) with narrower vessels or only tracheids. During the summer, the wide vessels of the early wood cavitate, and conduction occurs primarily or entirely in the late wood (**Figure 12-27**).

Eventually, all vessels and tracheids cavitate. Dry conditions in summer cause many cavitations, as do freezing in winter, vibration

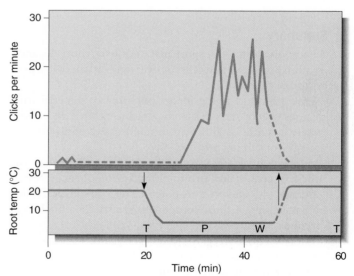

Figure 12-27 Cavitation of vessels and tracheids causes audible clicks that can be heard by sensitive microphones. This graph shows the induction of cavitation in castor bean xylem; plants were grown in water solutions that permitted water uptake to match transpiration loss—there was little tension on xylem water columns, and no cavitations occurred. Then the roots were cooled (arrow at left) to 5°C to inhibit root absorption of water, but shoots were kept warm to encourage transpiration. Water columns were stretched and immediately began breaking. After 20 minutes, embolisms were forming at almost 30 per minute, and the plants were wilting. At 45 minutes, the root solution was warmed to permit water absorption, and cavitation quickly stopped.

in wind, and damage by burrowing insects. After tracheary elements cavitate, surrounding parenchyma cells may block them off with tyloses or by secreting gums and resins. As more tracheary elements cavitate in a region of wood, adjacent parenchyma cells synthesize antimicrobial compounds then die, and the region becomes part of the heartwood.

■ Control of Water Transport by Guard Cells

Bulk water movement through xylem is influenced and powered primarily by water loss to the atmosphere. Although water loss through the cuticle is important, transstomatal transpiration is more significant whenever stomatal pores are open. Open stomata represent a trade-off between carbon dioxide absorption and water loss. Whenever water supply in the soil is adequate, water loss is actually advantageous—water movement is the primary means of carrying minerals upward from roots to shoots, and the evaporative cooling that results from transpiration can prevent heat stress in leaves and young stems; however, if the soil is too dry to supply water, transpiration represents an immediate, potentially lethal threat due to desiccation. Numerous mechanisms have evolved that control stomatal opening and closing. Each mechanism is keyed to a particular environmental factor, and their interaction results in great sensitivity to potential stresses in the habitat.

If the leaf has an adequate moisture content, light and carbon dioxide are the normal controlling factors. For most healthy, turgid plants, light most often controls guard cell water relations. Blue light is the important, triggering wavelength, and the action spectrum of opening closely matches the absorption spectrum of a flavin or flavoprotein pigment. It is not yet known how absorption of light by the pigment leads to potassium pumping.

The presence of light also leads to photosynthetic fixing of carbon dioxide; the decrease in internal carbon dioxide concentration may also lead to stomatal opening. Artificial manipulation of the amount of carbon dioxide available can stimulate guard cells to open or close in light or dark. At night, with no photosynthesis, carbon dioxide levels are high and presumably contribute to stomatal closing.

All of these mechanisms in healthy plants are completely overridden by a much more powerful mechanism triggered by water stress. As leaves begin to dehydrate, they release the hormone abscisic acid. This hormone immediately causes guard cells to close the stomatal pore even in blue light and low concentrations of carbon dioxide, factors that would otherwise favor opening. Water stress-induced closure often occurs in the early afternoon on a warm, dry day if root uptake and xylem conduction cannot keep up with transpiration. Stomatal closure prevents carbon dioxide uptake and stops photosynthesis even though light is available.

In plants with Crassulacean acid metabolism, stomata open at night and close in the morning. Temperature is particularly important for these plants; if night temperatures are too high, stomata may remain closed for days or weeks at a time. The low night temperatures typical of their desert habitats are essential for stomatal opening and carbon dioxide absorption. Under conditions of mild temperatures and abundant moisture in the tissues, such as after a spring rainfall, CAM plants convert to C_3 metabolism, opening their stomata in the morning and picking up carbon dioxide with RuBP carboxylase directly. When the soil dries after several days, they revert to CAM and night opening of stomata.

■ Summary

1. Living organisms transport materials over short distances (within organelles and cells) or over long distances (between nonadjacent cells).

2. Active transport is the forced pumping of material from regions where it is relatively unconcentrated to regions where it is more highly concentrated; active transport is an energy-consuming process.

3. Water potential (ψ) measures the capacity of the water to do work; in cells, it has two important components: osmotic potential (ψ_π) and pressure potential (ψ_p). In soils and rather dry materials, a third component, matric potential (ψ_m), becomes important.

4. Water moves from regions where it is relatively concentrated (similar to pure water, ψ near zero) to regions where it is less concentrated (dissimilar to pure water, ψ more negative).

5. A cell's water potential can be made more negative by pumping solutes such as K^+ and sucrose into it or by depolymerizing polymers to monomers, especially starch to glucose. Reversing these processes increases a cell's tendency to lose water.

6. A cell grows if its wall is too weak to counteract the tendency of water to enter the cell. Pressure potential cannot rise high enough to raise the cell's water potential and bring the cell into equilibrium with its environment.

7. Incipient plasmolysis is the point at which the protoplast has lost just enough water that it no longer presses against the wall, and ψ_p equals zero. If the cell continues to lose water, it becomes plasmolyzed.

8. The water potentials of guard cells, motor cells, and sieve elements become more negative as these cells are forcibly loaded with solutes by active transport. Water flows into the cells, causing guard cells and motor cells to swell but the phloem sap to be squeezed out of sieve elements through sieve pores.

9. Water begins to flow from roots to shoots when the shoots lose water to the air and their water potential becomes negative enough to draw water from the roots and overcome the effects of gravity and friction.

10. As both air and soil become dry, tension on water columns in xylem increases; cohesion may be overcome and some water columns cavitate, forming embolisms.

11. Water's adhesion to the sides of tracheary elements helps prevent embolisms; the presence of tracheids as opposed to vessels limits the amount of damage done if an embolism does form.

12. For non-CAM plants, light and low carbon dioxide stimulate stomata to open if moisture is adequate. Water stress overrides all other controls and causes stomata to close.

Important Terms

active transport	eutrophication	P-protein plug
adhesive	incipient plasmolysis	pressure flow hypothesis
apoplast	long-distance transport	pressure potential
aquaporins	matric potential	selectively permeable membranes
callose	molecular pumps	short-distance transport
cavitation	osmosis	sinks
cohesion-tension hypothesis	osmotic potential	sources
cohesive	plasmolyzed	STM/CC complex
diffusion	poikilohydry	symplast
embolism	P-protein	water potential

Review Questions

1. Plants have both short-distance and long-distance transport. How long are the distances involved in each?

2. Plants have several tissues that act as isolation mechanisms. Name two.

3. Define diffusion. How does this differ from osmosis?

4. What is a freely permeable membrane? How does it differ from a completely impermeable membrane or a differentially permeable membrane?

5. Are the membranes in plant cells freely permeable, completely impermeable, or differentially permeable?

6. Would active transport be possible if the molecular pumps were located in a freely permeable membrane?

7. Like any other chemical, water has a free energy measured by its water potential. Name three simple ways water potential can be increased.

8. What are the three components of water potential? Which of these potentials measures water's interaction with dissolved material?

9. The pressure potential of water measures the effect of pressure on water. If a cell had no cell wall, would it have a pressure potential?

10. Unless frozen, water is always in _____, always _____ within a plant from areas where it is abundant or under pressure to areas where it is rare or under tension.

11. Imagine that you have two solutions of glucose in water. One solution consists of 1 g of glucose in 100 mL of water. The other consists of 10 g of glucose in 100 mL. Which solution has a more negative osmotic potential?

12. In a beaker of pure water, what is the water potential? Does water potential become more positive or more negative as you add solute to it? Put pressure on it? Add it to dry clay? Add acid to it? In each case, which water potential component is changing?

13. By adding salt to eggplant, water can be drawn from the tissues. Which has a more negative water potential—eggplant or salt crystals?

14. What can you say about the water potentials of two solutions (or of a solution and a cell, or of two cells) when they are in equilibrium? At equilibrium, is there any net movement of water?

15. In each of the following pairs, circle the one that would probably have the more negative water potential: cell of a wilted leaf—cell of a turgid leaf; guard cells of an opening stoma—regular epidermal cells; guard cells of a closing stoma—regular epidermal cells; root cortex cell—moist soil; phloem cell being loaded with sucrose—leaf chlorenchyma cell; phloem cell unloading sucrose—tuber cell storing starch; clay with 10% moisture—silt with 10% moisture.

16. Imagine a cell with a water potential of –0.1 MPa being placed in a beaker of solution that also has a water potential of –0.1 MPa. Are the two water potentials in equilibrium? Would any water molecules be moving between the cell and the solution? Would there be a net movement of water? Now imagine a root in moist soil, and imagine that the root cortex cells have a water potential of –0.1 MPa and that the soil solution also has a water potential of –0.1 MPa. Would there be any net movement of water into the root?

17. Now imagine the same root as in Question 16 being placed in a dry soil in which the little soil water present has a water potential of –1.0 MPa. Would water move from the soil into the root or would it move from the root into the soil?

18. Circle the correct word of each pair.
    a. If a cell [absorbs, loses] water, it will become turgid.
    b. If a cell [absorbs, loses] water, it will become plasmolyzed.
    c. If a cell [absorbs, loses] water and it has a very [weak, strong] wall, it will grow.

19. Plants never absorb so much water that their cells _____, but they frequently lose enough water to _____ because their protoplasts do not press firmly against the _____ _____.

20. All of the protoplasm of one plant can be considered to be one continuous mass, called the _____ _____. Walls and intercellular spaces of a plant are called the _____ of the plant.

21. In glands, the apoplast consists mostly of _____. In nonglandular regions, the apoplast is mostly _____ _____.

22. Are opening and closing of stomatal pores based on short-distance or long-distance transport? Is osmosis or active

transport involved in opening and closing? What ion is especially important?

23. What is a motor cell? Name two plants that have them. Describe how motor cells adjust the position of a leaf.

24. In phloem transport, the sites from which water and nutrients are transported are known as _____ _____ are sites that receive transported phloem sap.

25. During spring and summer, which organs are the dominant sources of sugar? When are tubers or fleshy taproots likely to be important sources?

26. As sugars are pumped into sieve elements, water follows. What happens? Do sieve elements merely become turgid?

27. Phloem sap is under pressure. What is the danger associated with this? How are P-protein and callose involved in counteracting this danger?

28. As sugar is actively transported into phloem in sources, what happens to the water potential of the cells losing the sugar? Does it become less negative?

29. Consider the pressure flow model of phloem transport. How do sugars and water enter the phloem from the source? How do sugars and water move from one phloem cell to another?

30. Can the direction of phloem transport change? Does phloem ever transport material into a leaf?

31. Water is both cohesive and adhesive. What do these words mean, and how do they affect water's movement in a plant?

32. What is a leaf like early in the morning with respect to conditions that affect water movement? Describe what happens as the sun rises.

33. What is the speed of xylem sap translocation in ring porous trees?

34. In xylem, the pressure potential is a negative number. Why? What does the weight of the water have to do with this? Sto-

lons, rhizomes, and horizontal vines do not have a problem with the weight of water. Why? In the epiphytic cactus *Rhipsalis,* the weight of water actually makes conduction easier. Why?

35. Many roots remain healthy with their water potentials as low as _____.

36. Imagine lifting an icicle. You must pull against its _____ and the _____ of the icicle in a tube (the cell walls), but lifting the top of the icicle raises the _____ icicle, unless it breaks.

37. Describe the cohesion-tension model of water movement through xylem. Would the weight of water be more of a problem in an upright tree or in a stolon? Why?

38. The breaking of a water column is called _____. What breaks, the hydrogen bonds of the water or the cell walls of the tracheary elements?

39. When a water column cavitates, an air bubble is formed. The technical name is an _____.

40. In which habitat would you expect more cavitations—moist tropics or drier temperate areas? In which would you expect wood to have wide vessels, in which would you expect wood to have narrow vessels or just tracheids?

41. If a vessel cavitates and fills with air, can it ever be refilled with water?

BotanyLinks

http://biology.jbpub.com/botany/5e

BotanyLinks contains a variety of resources and review material designed to assist in your study of botany.

Soils and Mineral Nutrition

chapter 13

Concepts

Ajo ll organisms need elements such as nitrogen, phosphorus, calcium, magnesium, and sulfur. Plants must absorb these from soil and then use them and the glyceraldehyde-3-phosphate from chloroplasts to build all of their chemical components, however complex. This is an important concept: Plant metabolism is based on sunlight and chemicals present in water, air, and soil. No animal is able to survive on just minerals and one simple carbohydrate; they must obtain minerals and complex organic compounds in their food.

Most of the elements that are essential for plant growth and development are present in the crystal matrix of minerals. The elements become available to roots as rocks weather and break down, creating soil. During soil formation, rocks are converted gradually into dissolved ions and inorganic compounds. Because they are derived from the rock minerals, their role in plant nutrition is called **mineral nutrition**.

The term "mineral nutrition" covers a variety of types of plant metabolism. For some elements, after the mineral is absorbed from the soil, it can be used immediately as it

Chapter Opener Image: This field has been out of production for a year; no crop is cultivated on it, and we say that it is a "fallow" field. Being fallow helps the soil in many ways. Soil particles break down a bit and release more nutrient ions; the straw from last year's crop decomposes and releases nutrients that had been bound in the walls and protoplasm; many disease organisms die if there are no living plants for them to attack; and cyanobacteria take nitrogen from the air and convert it into a form that can be used by living organisms. We often think of soil as being just a matrix of minerals, but each handful contains millions of bacteria, archaeans, fungi, and microscopic animals and plants that affect the soil's quality. In the year or two that this field is fallow, it is tilled several times to prevent weed growth.

TABLE 13-1	Number of Organisms in Soil
Insects	670,000,000 per hectare*
Arthropods	1,880,000,000 per hectare
Bacteria	1,000,000,000 per gram of soil
Algae	100,000–800,000 per gram of soil
Earthworms	1,800,000 per hectare
Weight of worms	990 kg
Worm casts	396,800 kg
*1 hectare = 10,000 m² = 2.45 acres.	

is. An example is potassium, which is used by cells such as guard cells to adjust their turgor and water relations. Simple potassium ions are sufficient. Mineral elements such as iron and magnesium are more complex because they must be incorporated into compounds such as cytochromes or chlorophyll molecules before they are useful. Nitrogen is even more complicated: Like carbon, its oxidation state is important. Consequently, it must be reduced, and elaborate electron transport chains are necessary to convert it to useful forms.

The term "soil" covers a wide variety of substances. The various soils are important to plants not only in supplying minerals and harboring nitrogen-fixing bacteria, but also in holding water, supplying air to roots, and acting as a matrix that stabilizes plants, preventing them from blowing over. Critical aspects of soil are its chemical nature, which determines which mineral elements are present; its physical nature, which reflects its porosity, texture, and density; and its microflora and microfauna—the small animals, fungi, protists, and prokaryotes that live, respire, and gather food within the soil.

The microflora and microfauna of soil deserve special mention. It is easy to think of soil in terms of its chemical and physical properties only, but that would be an incomplete concept of soil. Most soils contain large amounts of microbes and tiny animals that are extremely important to plants (Table 13-1). Although microscopic to us, they are about the same size as roots and root hairs, and they interact extensively with root systems. For example, many soil microbes supply plants with nitrogen: Nitrogen is not found in rock matrixes, so soil formation does not make nitrogen available to plants. Instead, the primary source of nitrogen is molecular nitrogen (N_2) of the atmosphere, but only certain bacteria and cyanobacteria have enzymes that convert molecular nitrogen into forms useful for metabolism. When these microbes die and decay, their organic nitrogen compounds are released to the soil and become available to plants.

Just as the foods of animals vary, soils also vary in the quantities of minerals present, the texture of the soil, and the organisms present. Also, plants vary with regard to the amounts of minerals they require and their capacity for absorbing and processing minerals. All of these factors affect a plant's health and, to a large extent, determine the types of plants that exist in a particular area.

Essential Elements

Much research in mineral nutrition involves experiments in which a single element is supplied to a plant in excessive quantities or withheld from it. This is accomplished by growing the plant in a **hydroponic solution** in which the chemical composition is carefully controlled. Hydroponic experiments were for-

Figure 13-1 In a hydroponics experiment such as this one at Penn State University, the test solution is often just a water solution in a bottle of boron-free glass. Air must be bubbled through the liquid to permit root respiration (clear plastic tubes). The bottles are wrapped in foil or painted black to exclude light, more closely resemble a soil environment, and prevent the growth of algae.

malized and used extensively by Julius von Sachs in 1860. In such experiments, a solution is developed that supports plant growth. At first, this must be by trial and error, with numerous chemicals being added to the solution, and plants are tested to see whether they survive. Many tests are necessary because some of the chemicals may be toxic, even in low concentration. Other chemicals needed by plants are toxic if present at too high a concentration. Also, the form in which a chemical is present makes a difference: Nitrogen is important, but in addition to nitrate and ammonia, which plants use, other nitrogen compounds exist that plants do not use. Also, when numerous compounds are added to the same solution, unsuspected reactions may occur that create toxic compounds or convert useful compounds to useless ones.

After a solution of known composition is found which supports plant growth, an identical solution can be prepared except that one component is left out (Figure 13-1). If that component is not necessary for plant growth—that is, if it is not **essential**—plants grow normally, but if the excluded element is essential, the plant cannot grow correctly. For example, an experimental solution might contain nickel; if nickel is left out of the second solution, plants would grow well, perhaps even better than in the first solution; therefore, nickel is not an essential element. The second solution then becomes the main test solution. If a third solution is prepared similar to the second but lacking potassium, for example, no plant could survive; therefore, potassium is an essential element and must always be included (Figure 13-2). Using these hydroponic techniques, Sachs established a minimal nutrient solution that would support plant growth; it contained calcium nitrate, potassium nitrate, potassium phosphate, and magnesium sulfate. Currently, several solutions are known to support reasonable growth of most plants; two are Hoagland's solution and Evan's modified Shive's solution. Many experiments begin with one of these.

It may seem simpler just to grind up a plant and then extract and measure all of the chemical elements present, but plants actually absorb many elements they do not need—the endodermis

Figure 13-2 These plants are being grown in a hydroponic solution that contains all known essential elements except one—magnesium. Even though nitrogen, sulfur, and the other elements are abundant, they are of little use to plants if one essential element is missing. Growth and reproduction are governed by the least abundant factor, not the most abundant.

TABLE 13-2	Elements Essential to Most Plants
Macronutrients	
Carbon	Organic compounds
Oxygen	Organic compounds
Hydrogen	Organic compounds
Nitrogen	Amino acids; nucleic acids; chlorophyll
Potassium	Amino acids; osmotic balance; enzyme activator; movement of guard cells and motor cells
Calcium	Controls activity of many enzymes; component of middle lamella; affects membrane properties
Phosphorus	ATP; phosphorylated sugars in metabolism; nucleic acids; phospholipids; coenzymes
Magnesium	Chlorophyll; activates many enzymes
Sulfur	CoA; some amino acids
Micronutrients	
Iron	Cytochromes; nitrogenase; chlorophyll synthesis
Chlorine	Unknown; possibly involved in photosynthetic reactions that liberate oxygen
Copper	Plastocyanin
Manganese	Chlorophyll synthesis; necessary for the activity of many enzymes
Zinc	Activates many enzymes
Molybdenum	Nitrogen reduction
Boron	Unknown

Except for boron, all these elements are also essential for us humans. But unlike plants, our diet must also provide us with fluorine, iodine, cobalt, selenium, chromium, and sodium. We obtain fluorine by adding it to our drinking water (fluoridation) and iodine is obtained by adding it to salt or by eating a large amount of seafood. Our lives depend on sodium and if we lose too much by sweating, we can quickly die. Most plants have no need for sodium at all; the exceptions are C4 plants and CAM plants, which need trace amounts.

simply cannot exclude them completely. A living plant usually contains at least trace quantities of every element present in the soil, whether essential to the plant or not.

The **essential elements** discovered by Sachs are called the **major** or **macro essential elements** because they are needed in large quantities by plants (**Table 13-2**). If dry plant material is analyzed, calcium, nitrogen, potassium, phosphorus, magnesium, and sulfur are present at concentrations of between 0.1% and 3.0% of the plant's dry weight. Even as the first hydroponic experiments were being performed, botanists realized that the available chemicals were quite impure and that trace quantities of other elements were present. Despite their best efforts, they could not prepare solutions that had absolutely no copper, zinc, or many other elements. Thus, it was not possible to conclude that any particular element was completely nonessential; it was possible to determine only that relatively large amounts of most chemical elements were not essential.

As chemical methods improved, purer compounds were produced, and mineral nutrition experiments could be repeated with more certainty that the element tested had been almost completely excluded. With such improved chemicals, it was soon discovered that there exists a group of **minor** or **micro essential elements**, also called **trace elements**. Iron, boron, chlorine, copper, manganese, molybdenum, and zinc are required in extremely low concentrations by plants. Iron is the exception, being needed in amounts intermediate between those of the major and minor elements.

Although our reagents today are much purer than those of 100 years ago, it is still impossible to create a solution that contains absolutely only the nine major and seven minor elements. Our purest water has traces of many contaminants, the very best glass containers release silicon and boron into the solution, and all forms of chemical reagents have some trace contaminants. Chlorine was only recently discovered to be essential, at least for some plants; it remained undetected until it was realized that despite the purity of the test solutions, experimental plants receive adequate chlorine if scientists touch the plant because the skin of one fingertip has sufficient chlorine for an entire plant. Typical hydroponic experi-

ments had shown that plants could grow in chlorine-free solutions, but when the studies were repeated with all experimenters wearing plastic gloves, the plants did not survive. Because such small quantities of chlorine are sufficient, we must suspect that other elements may also be necessary in similarly minute quantities; testing will continue whenever purer reagents become available.

■ Criteria for Essentiality

An element must meet three basic criteria to be considered essential.

1. The element must be necessary for complete, normal plant development through a full life cycle. If a particular element is required for any aspect of a plant's growth, differentiation, reproduction, or survival, that element is essential. This logical and necessary criterion can be difficult to test. In a hydroponic experiment, the solution and plant must be carefully protected from contamination by dust and insects, both of which are mineral rich. Experiments must occur in laboratories, growth chambers, or specially controlled greenhouses. Laboratory conditions are not the same as the natural environment, however, and part of a plant's ability to survive depends on its response to stress: cold, heat, drought, and pathogens. It may be that certain elements needed only under unusual conditions are not anticipated in a growth chamber experiment.

It is not feasible to study large trees in greenhouses or growth chambers, both because of their size and because it takes so many years to complete a life cycle. Because it is virtually impossible to test such species, the practical assumption is made that the major elements and most minor ones are essential for all plants, having been tested and found essential for many small herbs. This is a safe assumption if the role of the element is known. For example, nitrogen is present in amino acids and nucleic acids; because nothing can live without these, nitrogen is essential. The same is true for iron (cytochromes), calcium (middle lamella, enzyme control), and others. Chlorine is now known to be involved in the water-splitting reactions of photosynthesis, but the role of boron is still not known for certain, so we cannot assume with confidence that all plants require it. Furthermore, some desert plants have large amounts of silica in their epidermal cell walls; silicon is an essential element for them, but most plants appear not to need it at all.

2. The element itself must be necessary, and no substitute can be effective. This criterion is relatively straightforward; in most instances, if one essential element is absent, the presence of a chemically related element does not keep the plant alive. In some cases, elements can be substituted in specific enzymatic reactions or transport processes when studied in vitro, but even so, the entire plant cannot live with only the substitute. Some plants that require chlorine can survive if given large amounts of bromine, but such elevated levels of bromine do not occur in nature; therefore, the substitution works only in the laboratory.

3. The element must be acting within the plant, not outside it. The complexity of test solutions makes it difficult to analyze results in many cases. Iron is an essential element, but it is soluble in only a very limited range of acidity. Although iron may be added in adequate quantities, it often reacts with other chemical components and forms an insoluble precipitate. A test chemical can cause the precipitated iron compound to break down and form a soluble iron compound. With the new availability of iron, the plant grows well, even if the test element is not needed or absorbed. Without careful analysis, the test element would appear to be essential even though it is not used inside the plant.

This criterion has interesting ramifications. Most plants absorb phosphorus only poorly and depend on an interaction with soil fungi that absorb phosphorus and transfer it to the plant. If soil fungi are killed, plants grow poorly. If an element is essential to the fungus, is it then also essential to the plant even if the plant does not use it directly? This is one of those points that can be debated endlessly without being resolved; it is better to understand the biology than to dispute definitions.

Mineral Deficiency Diseases

Causes of Deficiency Diseases

Virtually all types of soil contain at least small amounts of all essential elements; under natural conditions, it is rare to encounter plants whose growth and development are seriously disrupted

Figure 13-3   As water evaporates from the soil, more water moves upward, carrying dissolved minerals that crystallize as the water evaporates. When soil water is present, its osmotic potential is extremely negative, as is its water potential (its pressure potential is zero). Roots cannot pull water out of such osmotically dry soil; they would die before their water potential became sufficiently negative. At the edge of a salt flat, salt concentrations are lower, and plants such as *Atriplex* (saltbush), *Sueda* (seepweed), and many grasses occur.

by a scarcity or an excess of mineral elements. It is especially uncommon to find plants suffering from an overabundance of a particular mineral. In many cases, the unnecessary ions are not even absorbed by the roots. Certain types of excess minerals, if absorbed, can be precipitated in vacuoles as crystals; although a cell may contain large amounts of the mineral, only the ions actually in solution have any significant effect on metabolism.

Desert soils often have excessive amounts of all available minerals because ground water moves upward, carrying dissolved minerals with it. These can reach such strong concentrations that the water potential of the soil solution is extremely negative and roots are unable to extract water from it. Plants are unable to grow, not because of mineral toxicity but because of osmotic drought (**Figure 13-3**). Some species are adapted to less severely salty regions; saltbush (*Atriplex*) absorbs both water and salt but passes salt directly through the body and secretes it from salt glands located on leaves. This produces a coating of salt crystals that is thought to be selectively advantageous (**Figure 13-4**). The salt reflects away some of the excessive sunlight. It also makes the bush an unacceptable food because desert-dwelling animals must avoid salty foods in order to balance their salt/water intake.

One of the most widespread examples of toxicity caused by elevated levels of single minerals is aluminum toxicity in acid soils. It is one of the main limitations on crop production on acid soils, which often occur in tropical areas with abundant rain and decaying plant matter. Mine tailings—the piles of discarded dirt extracted from mines after ore has been removed—often have such high levels of heavy metals that few or no plants grow on them.

Mineral-deficiency disease does not seem common in natural populations. Some soils may have such low concentrations of certain essential elements that some species are unable to thrive on them. For example, a type of soil called serpentine soil is extremely deficient in calcium, and few plants grow on it. Some species are more sensitive than others to low concentrations of

Figure 13-4 (a) Certain mangroves (*Avicennia*) of tidal marshes secrete salt through special glands. Manipulation and transport of such large amounts of salt require tremendous energy use, but salt excretion not only permits growth in saline habitats but also provides protection against herbivores. (b) Most plants neutralize excess salts by precipitating them as crystals such as these.

(a)   (b)

essential elements (**Figure 13-5**). Because of competition with other plants and pathogens, the sensitive plants probably weaken and die early and do not reproduce successfully. Consequently, species especially sensitive to a particular deficiency typically do not occur in the plant community growing on soil deficient in that element.

Deficiency diseases are most commonly encountered in non-native crop plants or ornamentals. Crop plants especially have

Figure 13-5 Beans are especially sensitive to deficiency of zinc. On soils low in zinc, many plants grow well and are healthy but leaves of beans develop chlorosis and brown spots where the cells die.

undergone human artificial selection (not natural selection) for traits such as rapid growth and high fruit/seed yield that require large amounts of nitrogen and mineral nutrients. Without fertilization, these plants often have poor growth and symptoms of deficiency diseases. As shown in **Table 13-3**, entire forests can be fertilized. The data show the results of fertilizing forests in Tennessee with 335 kg/hectare of fixed nitrogen, either with or without phosphorus. The addition of nitrogen was always beneficial, as virtually all soils are deficient in fixed nitrogen, but the addition of phosphorus often did not improve tree growth more than nitrogen alone. Soils naturally contain adequate phosphorus for many forest species.

The very act of harvesting crops leads to soil depletion: Fruits, seeds, tubers, and storage roots often have the greatest concentration of minerals in a plant. Harvesting them removes those minerals from the area; the rest of the plant body is relatively mineral poor and does not contribute much to re-enriching the soil even if it is plowed back in. With human populations, consumption of food may occur thousands of miles from the site of plant growth; even worse, waste products are dumped into rivers and carried away, never being returned to the soils. Under natural conditions, minerals are returned in the form of manure, which is typically deposited in the general region where feeding occurs. The most extensive crop removal is harvesting a forest for timber (**Table 13-4**), but several methods of harvesting are possible:

TABLE 13-3	Fertilizing Forests	
	Increase in Growth (%)	
	Nitrogen Alone	Nitrogen + Phosphorus
Hickory	261	172
Northern red oak	167	100
Cucumber tree	107	207
Chestnut oak	61	70
Black cherry	52	71
Yellow poplar	48	69
Dogwood	36	173

TABLE 13-4	Mineral Loss Through Crop Removal				
	Minerals Removed from Forest (kg/hectare/year)				
Harvest Method	N	P	K	Ca	Mg
Complete removal, with roots	17.6	2.3	12.6	12.8	3.6
Complete removal of shoot, leaving roots	16.1	1.9	10.3	11.7	2.9
Wood only, leaving bark, narrow branches, and roots	4.6	0.8	3.8	4.3	1.3

Plants Do Things Differently

Box 13-1 Plants Eat Dirt; Animals Eat Protoplasm

At its most fundamental level, plant nutrition is almost identical to that of animals, virtually indistinguishable. All cells depend on the same amino acids, nucleic acids, sugars, and with a few exceptions, the same lipids (plants never use cholesterol). Small molecules such as ATP and vitamins such as thiamin, riboflavin, and folic acid perform exactly the same functions in both types of organisms. At the same time, however, the two types of organisms could hardly differ more. No organism can synthesize mineral elements, of course, so plants and animals share that obvious similarity, but differences abound if we consider how an organism obtains organic molecules. Plants can be described first because they are so easy. They themselves make absolutely everything organic within their own bodies. It might be a bit difficult for zoologists, medical students, and dietitians to truly grasp this point. Every plant itself makes every organic molecule found within its body. A balanced diet for a plant is dirt, dirt, and more dirt, with carbon dioxide and water, morning, noon, and night. Photosynthesis converts carbon dioxide and water into glyceraldehyde-3-phosphate, and starting with just this simple small molecule and some minerals, a plant constructs everything it uses in its life—absolutely everything.

Animals lack many of these synthetic pathways and must obtain many organic compounds in their diet. We humans, like all other organisms, use a universal set of twenty amino acids in our proteins, but we cannot make nine of these ourselves. We must obtain these nine in the food we eat or we become ill and could even die. Several fatty acids cannot be synthesized by any tissue, cell, or organelle of our bodies. The list of essential nutrients is especially dramatic when it comes to vitamins, the organic molecules so fundamentally important in such small amounts that they were the first chemicals to be discovered as being essential dietary factors. Thirteen molecules have received this designation so far, and no one would be surprised if others are added to the list with further research. Every plant makes all of its own vitamins;

we must get most of ours from our food. If an organic molecule is always reliably present in an animal's food, then mutations that prevent the synthesis of the molecule are actually beneficial. The animal saves energy by not synthesizing compounds it will get in its diet anyway, and that energy can be used to carry out other life activities. If the vitamin is truly always available in the diet, then it is redundant for the animal to synthesize it as well.

The differences in nutritional resources used by plants versus animals are also great. Plants obtain nutrients in the form of elements or as simple compounds present in the environment, such as CO_2, H_2O, K^+, Mg^{2+}, SO_4^{2-}, and so on. An animal begins with food in the mouth, but the nutrients occur as monomers in complex polymers, which in turn are parts of cell structure. Minerals must be digested away from organic molecules; for example, iron must be digested out of hemoglobin and myoglobin before it is absorbed into the blood stream. Although animals save energy by not needing to synthesize many molecules, they must go through much more effort to obtain their food and convert it to forms that can be absorbed. And their food usually also contains indigestible fur, feathers, bones, teeth, and dirt. Plants never take in such debris.

Plants are not completely self-sufficient nutritionally. Most rely on bacteria for converting atmospheric nitrogen gas (N_2) into a chemical form such as nitrate (NO_4^{2-}) or ammonium (NH_4^+) that plants and animals can use. Some plants have gone so far as to actually cultivate these bacteria within their own bodies, within nitrogen-fixing nodules on roots of alfalfa, for example, or within special chambers in liverworts. Although plants can take up phosphorus from the soil on their own, they usually obtain it more efficiently by entering into a symbiotic relationship—called a mycorrhizal association—with certain soil fungi that are more effective at scavenging phosphorus. Other plants have decided that animals have the right idea; the plants either are parasitic on other plants, or they capture and consume animals.

removal of every part of the tree versus trimming and debarking, which permit the small branches, leaves, and bark to remain in the ecosystem.

■ Symptoms of Deficiency Diseases

The particular symptoms of a mineral deficiency are more closely related to the particular element that is lacking than to the plant species; usually, all plants that suffer from a scarcity of a particular essential element show the same symptoms. One common symptom is **chlorosis**; leaves lack chlorophyll, tend to be

yellowish, and are often brittle and papery. Deficiencies of either nitrogen or phosphorus cause another common symptom, the accumulation of anthocyanin pigments that give the leaves either a dark color or a purple hue. A lack of certain elements causes **necrosis**, the death of patches of tissue (necrosis can also be caused by bacterial, viral, and fungal infections). The location of the necrotic spots depends on the particular element: Potassium deficiency causes leaf tips and margins to die, whereas manganese deficiency causes the leaf tissues between veins to die even though all of the veins themselves remain alive and green.

(a)

(b)

Figure 13-6 (a) This rose leaf is suffering from iron deficiency. Because the element is immobile, the little iron present in the plant cannot be transferred from older leaves to younger ones; therefore, young leaves show the disease symptoms. Cells near the veins have chlorophyll; those farther away are chlorotic. (b) These mature tomato leaves are suffering from a deficiency of phosphorus. Because the element is mobile, the plant can transfer atoms of phosphorus from older leaves to newer ones.

Mobile and Immobile Elements

An important diagnostic aspect is whether symptoms appear in young leaves or older leaves. This is related to the **mobility** of the essential element. Boron, calcium, and iron are **immobile elements**; after they have been incorporated into plant tissue, they remain in place. They do not return to the phloem and cannot be moved to younger parts of the plant. A plant that grows in a soil deficient in boron, calcium, or iron is probably able to grow relatively well until the few available ions have been absorbed. Growth is normal until the soil is exhausted; further growth suffers mineral deficiency, and the newly formed tissues are affected (**Figure 13-6a**).

The elements chlorine, magnesium, nitrogen, phosphorus, potassium, and sulfur are **mobile elements**; even after they have been incorporated into a tissue, they can be translocated to younger tissue. After the soil becomes exhausted of one of these elements, older leaves are sacrificed by the plant. The mobile elements are salvaged and moved to growing regions (Figure 13-6b). The adaptive value of this is easy to understand: A leaf photosynthesizes most efficiently right after it has first expanded and less efficiently as it ages. The plant increases its overall photosynthetic rate by sacrificing old, inefficient leaves and using the minerals to construct new, efficient leaves.

The immobility of certain ions is not understood; boron, calcium, and iron are initially moved upward from roots into shoots, flowers, and fruits, so transport mechanisms do exist for them. Mutations that would result in the degradation of cytochromes in old leaves and the recovery of iron should be selectively advantageous. Animals have trouble with mineral recovery as well. For example, humans do not recycle the large quantities of iron in dead red blood cells; it is simply discarded, even if this results in anemia.

Soils and Mineral Availability

Soils are derived from rock by processes of **weathering**. The initial rock may be volcanic (granite, basalt), metamorphosed (marble, slate), sedimentary (sandstone, limestone), or other types, but two things are important: Rock has a crystalline structure, and trapped within the structure are numerous types of contaminating ions and elements (**Figure 13-7**). Its crystal matrix prevents rock from being a suitable substrate for plant growth; it may contain most essential elements, but as long as they are part of the matrix or trapped by it, they cannot be absorbed and used by plants. Also, any water held by the rock as part of the crystal structure is unavailable to plants.

Two fundamental processes of weathering convert rock to soil: **physical weathering** and **chemical weathering**. As its name implies, physical weathering is the breakdown of rock by physical forces such as wind, water movement, and temperature changes. Ice is an important agent. During winter days, water from rain or melted snow seeps into capillary spaces within rock; at night,

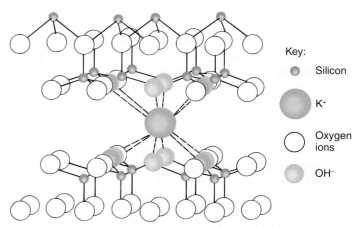

Figure 13-7 Rock is a complex, highly contaminated crystal; as it weathers, it gradually breaks down into soil. Within the crystal matrix of rock, numerous atoms of many different elements occur. As rock breaks apart, atoms at the surface are liberated into the soil solution. This is potassium trapped in vermiculite, a type of clay.

Key:
- Silicon
- K^+
- Oxygen ions
- OH^-

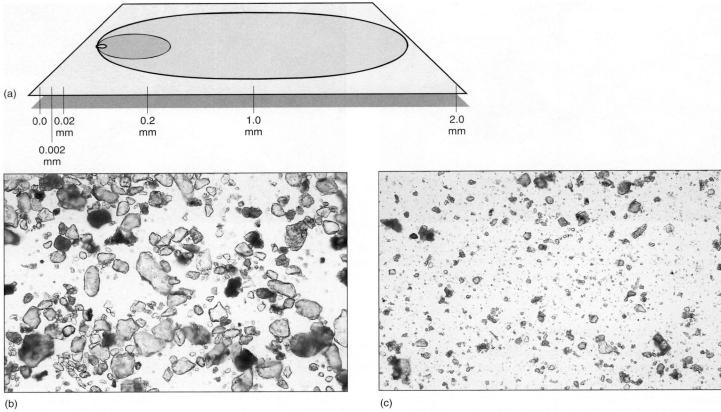

(a)

(b)

(c)

Figure 13-8 (a) Relative sizes of soil particles: From 0.2 to 2.0 mm is coarse sand; the next smaller is fine sand, and so on. It is easy to calculate the approximate amount of surface area, volume of rock, and volume of air/water space in a soil composed of only one type of particle. Assume you have 1 cm³ of pure coarse sand; the number of particles present is calculated by dividing 1 cm³ by the volume of a cube measuring 2.0 mm on each side (the volume of the particle plus its surrounding air/water space). Next, multiply the number of particles by the volume of a sphere with a radius of 1.0 mm (each sphere has a diameter of 2.0 mm). This gives the total volume of rock in 1 cm³ of soil. Subtract this from 1 cm³ to obtain the volume of air or water that can be held by the soil. Finally, calculate the surface area of each sphere and multiply that by the number of particles to obtain the total surface area that is releasing minerals into the 1 cm³ of soil. Now do this for fine sand, silt, and clay. Which soil has the most surface area? The greatest amount of air and water? (b) Particles of fine sand, viewed by light microscopy (×60). (c) Particles of silt and micelles of clay (×60).

when temperatures fall below freezing, the expansion of water as it becomes ice causes cracks to widen. Portions of rock, ranging from small flakes to large pieces, are broken off. This is a slow process, but gradually, the average particle size of the rock is reduced.

Runoff from rainstorms, avalanches, and similar forces wash rock fragments and pebbles into streams and rivers, where physical weathering accelerates. In rapidly moving streams, the rocks are scoured by suspended sand grains; with high flow rates, rocks and boulders are carried downstream, grinding against each other and the stream bed. Wind-blown sand is a powerful erosive force, as are glaciers, which are extensive during the periodic ice ages.

Physical weathering produces a variety of sizes of soil particles; the largest ones that are technically important to soil are grains of **coarse sand**, with a size range of 2.0 to 0.2 mm. Particles only one tenth this large (0.2 to 0.02 mm) are **fine sand**, and those one tenth of this (0.02 to 0.002 mm) are **silt**. The finest particles, smaller than 0.002 mm in diameter, are **clay particles**, technically known as **micelles** (Figure 13-8).

The various particle sizes affect soil texture and porosity. In sands, particles are large and fit together poorly, so a great deal of space remains between particles. The spaces permit rapid

gas diffusion, and roots in sandy soils typically are never starved for oxygen. During rain, the spaces fill with water, but typically, they cannot hold it against gravity because the spaces are too broad to act as capillary tubes. After rain stops, most of the water percolates downward and enters aquifers, flowing underground to wells, springs, and streams. Water that remains in the soil is held by capillary adhesion/cohesion (Figure 13-9) and is said to be the **field capacity** of the soil. Much of this water is available to roots.

Chemical weathering involves chemical reactions, and the most important agents are acids produced by decaying bodies, especially those of plants and fungi. In addition, many organisms secrete acids while alive, and the carbon dioxide produced during respiration can combine with water, forming carbonic acid. When an acid dissolves in water, it dissociates into a proton and an anion, both of which interact chemically with rock's crystal matrix and with embedded contaminant elements. In regions with a great deal of warmth, moisture, and abundant decaying vegetation, such as tropical regions, chemical weathering can be extremely rapid, and thick soils accumulate in a short time. In drier regions with long, cold winters and less vegetation, such as temperate mountains and the prairies of the plains states

(a)

(b)

Figure 13-9   (a) This sandy soil has such large spaces between its particles that it cannot hold water well; it dries quickly after a rain. It has a low field capacity. (b) As water is pulled away from soil by gravity, it percolates downslope into valleys. Soil at the bottom of ravines has water available longer than soil on the sides of a slope; consequently, more vegetation grows at the bottom of a valley than on the sides.

and Canada, fewer acids are available and chemical weathering is slower. It may take as long as 10 million years for just 1 cm of soil to form. Chemical weathering is greatly increased if rock has already been reduced to sands and silts by physical weathering; these have a large surface area for the acids to attack.

Chemical weathering decreases soil particle size, but more importantly, it alters soil chemistry. As the crystal matrix dissolves, matrix elements become available to the plant, and trapped elements are liberated. As the matrix breaks down, positively charged cations are freed; thus, the residual undissolved

Figure 13-10   As rock weathers, the negatively charged components are most resistant, so the rock fragments have a negative surface charge. Because of this, the cations released by weathering do not completely leave the rock, but are held by very weak electrical attraction.

matrix has a negative charge (**Figure 13-10**). With coarse particles such as sand and silt, the surface-to-volume ratio is so small that the charge is not important, but with clay micelles, the total amount of surface per unit of soil volume is great. Because the particles have a negative charge, cations such as K^+, Ca^{2+}, Mg^{2+}, and Cu^{2-} are held near the particles' surfaces. The bonding is much weaker than that of the crystal matrix, therefore, roots can absorb the cations. This attraction to the micelle surface is beneficial; without it, many important cations would be washed deep into the soil by passing rainwater.

■ Cation Exchange

Because cations are loosely bound to micelle surfaces due to their charge, roots cannot absorb them directly. Instead, the cations must first be freely dissolved in the soil solution; this is done by **cation exchange**. Roots and root hairs respire, giving off carbon dioxide. As this dissolves in the soil solution, some reacts chemically with water, forming carbonic acid, H_2CO_3. This breaks down into a proton and a bicarbonate ion (**Figure 13-11**), which can further dissociate into a second proton and a carbonate ion. The presence of protons acidifies the soil solution adjacent to roots and root hairs; as the protons diffuse, they bounce close to a bound cation at a micelle surface. The presence of the proton's positive charge disrupts the electrical attraction of the cation, liberating it and trapping the proton. Because the proton was derived from waste carbon dioxide, its loss does not hurt the plant. The liberated cation may diffuse in the direction of the root and be absorbed and transported upward by xylem, or it may diffuse away from the root or strike another micelle, liberating either a proton or another cation. Over time, however, large quantities of cations are absorbed. Acidity caused by secretion of acids by bacteria and fungi, by the decomposition of humus, and by acid rain also result in the liberation of cations.

Soil particles are not the only structures that hold cations. Decaying organic matter also forms negatively charged matrixes. Cellulose crystals of cell walls in mulch and humus are especially

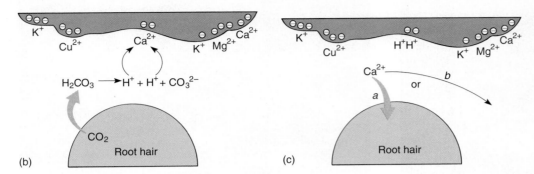

(a)

(b)

(c)

Figure 13-11 (a) The reaction of water and carbon dioxide results in carbonic acid (H_2CO_3), most of which dissociates into a proton and a bicarbonate anion. Some of the bicarbonate dissociates further, releasing another proton and a carbonate anion. (b) Protons from carbonic acid may diffuse close enough to a cation to disrupt its attraction to a soil micelle, liberating it (c). As the cation then diffuses through the soil, it may encounter a root (arrow a) or it may not (arrow b).

valuable not only for holding cations liberated from rock weathering but also for retaining the essential elements released by decaying protoplasm. Such organic matter also holds water and greatly improves the quality of any type of soil.

▪ Soil Acidity

Soil pH, the concentration of free protons in the soil solution, is important for cation exchange and the retention of cations in the soil during heavy rain. As acidity increases (pH becomes lower), the greater concentration of protons causes more cations to be released from soil micelles; these may be absorbed by roots or washed away in ground water. An extremely acid soil (pH of 4.0 to 5.0) tends to lose cations too rapidly and becomes a relatively poor soil. On the other hand, highly alkaline soils (pH of 9.0 to 10.0), which are frequent in dry climates, have too few protons to allow cation release, and concentrations of minerals can become excessively high.

Soil pH affects the chemical form of certain elements, causing them to change solubility. In acidic soils, aluminum and manganese can become so soluble as to reach toxic levels. In alkaline soils, iron and zinc become quite insoluble and unavailable to plants, but molybdenum is more soluble at a high pH. In general, a pH between 6.5 and 7.0 is best for many elements, especially iron, zinc, and phosphorus.

Many factors affect soil acidity, such as the chemical nature of the original rock, but probably the most important factor is rainfall. With high rainfall, there tends to be an abundance of vegetation that produces acids by means of respiration, excretion, and decay. With low rainfall, not only is there little vegetation, but there may not be any washing out of the soil. Cations build up, increasing soil alkalinity by increasing the concentration of hydroxyl ions. Just as dissociation of an acid produces a proton and an anion, dissociation of a base produces a hydroxyl and a cation:

$$NaOH \longrightarrow Na^+ + OH^-$$

Because some soils are more acidic than others, plants have adapted to differences in the availability of essential elements (Table 13-5). Natural selection favors mutations that allow desert plants to cope with alkaline soils, whereas plants of wet areas must become adapted to acid soils. For example, azaleas, camellias, gardenias, and rhododendrons absolutely must have acidic soil; if planted into alkaline soils, they show stress symptoms immediately, often suffering from lack of iron as well as general poor health (Figure 13-12). Alfalfa, apples, broccoli, and hydrangeas require an alkaline soil, whereas most plants do best if the soil pH is near 6.0 to 6.5. Some plants are able to tolerate a wide range of soil acidity, but typically, plants show best health and vigor only in soil with the optimal pH.

TABLE 13-5	Species Adapted to Acidic, Neutral, and Alkaline Soils	
Acidic (pH 4.5–5.5)	Neutral (pH 5.5–6.5)	Alkaline (pH 6.5–7.5)
azalea	carrot	apple
blueberry	*Chrysanthemum*	*Asparagus*
Camellia	corn	beet
cranberry	cucumber	cabbage
fennel	pea	cauliflower
Gardenia	*Poinsettia*	lettuce
potato	radish	onion
Rhododendron	strawberry	soybean
sweet potato	tomato	spinach

Figure 13-12 This azalea leaf is from a plant growing in alkaline soil. Azaleas require acid soils and suffer iron deficiency in alkaline soils.

Botany and Beyond

Box 13-2 Acid Rain

Acid rain is a silent killer, destroying forests, streams, and lakes throughout the world. As we burn fuels rich in sulfur, such as much of the coal used to generate electricity, the sulfur burns to sulfur dioxide and is emitted through the smokestack of the generating plant. In the air, sulfur dioxide reacts with water to form sulfuric acid, which then dissolves into the water droplets of clouds. As the drops fall as rain, snow, or sleet, they carry the sulfuric acid with them as **acid rain**, also called acid precipitation.

Acid precipitation damages plants in many ways. Because the cuticle is not absolutely impermeable, some acid slowly moves directly into the plant tissues and damages leaves, flowers, fruits, and cones. Perhaps more significantly, most of the acid enters the soil and accelerates cation exchange, causing positively charged ions to be released from the soil particles and to be washed away. Soil is left depleted of nutrients, and plants suffer from mineral deficiency. Downwind of the most heavily polluting industrial centers of Germany, entire forests are dying or are dead. Pollution from the United States and Canada is causing extensive damage to North American forests.

As acid rain accelerates cation exchange, minerals are washed from the soil and enter streams, in effect fertilizing them and causing rapid growth of algae. In small quantities, this provides more food for fish, turtles, and other aquatic animals, but in many cases, the algal growth is so abundant that it forms a massive, impenetrable layer across the top of a lake or slowly moving river. As the algae die, their bodies sink and are attacked by decompos-

ers, mostly bacteria. Bacterial decomposition consumes oxygen, and before long, the bottom of a lake or quiet river is an anaerobic dead zone with too little oxygen to support any animal life. This process is called **eutrophication**, and the result is a lake or river that is basically dead, having little life other than a mat of algae at its surface.

Figure B13-2 These conifers have been damaged by acid rain. The acid rain does not always destroy a forest but instead may alter the species composition and diversity.

The Endodermis and Selective Absorption of Substances

Elements in the soil solution, whether essential or not, can enter roots either by crossing a plasma membrane and entering the symplastic protoplasm phase of the plant or by diffusing along cell walls and intercellular spaces in the apoplastic phase. In the first method, the selective permeability of the plasma membrane and the presence or absence of molecular pumps control entry of ions and molecules—certain substances can be excluded and others can be actively transported in; however, a substance can penetrate the root epidermis and cortex simply by moving through the water in cell walls and intercellular spaces. No metabolic control exists, and even harmful substances enter. If all substances could enter the xylem transpiration stream, they would have access to all cells of the plant body.

The endodermis prevents uncontrolled, apoplastic diffusion in roots. The Casparian strips on all radial walls are imperme-

able to water and water-borne solutes; nothing can cross it simply by diffusion (Figure 13-13). For a substance to penetrate beyond the root cortex, it must first enter the endodermal cell protoplasm by being accepted across the plasma membrane of a cell in the root epidermis, cortex, or endodermis. Its highly selective permeability allows the endodermis to control which elements enter the transpiration stream.

Mycorrhizae and the Absorption of Phosphorus

The roots of most plants form a symbiotic association with soil fungi, and this relationship is called a **mycorrhiza**; the symbiosis permits plants to absorb phosphorus efficiently. In the most common type, vesicular/arbuscular mycorrhizae, some fungal filaments penetrate root cortex cells and then branch profusely, forming a small tree-shaped arbuscule inside the cell; other filaments swell into balloon-like vesicles. The fungus collects phosphorus from the soil and transports it into arbuscules,

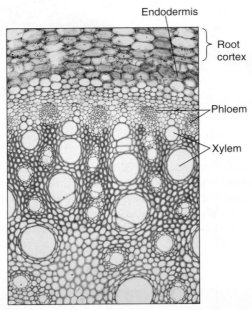

Endodermis
Root cortex
Phloem
Xylem

Figure 13-13   The endodermis in roots prevents uncontrolled diffusion of minerals into the xylem (×50).

where it accumulates as granules. After the arbuscules fill with phosphorus, the granules gradually disappear as phosphorus is transported into the root cell protoplasm. After the transfer is complete, the arbuscule collapses and the root cell returns to normal. This mycorrhizal symbiosis is essential to most plants; plants in sterilized soil grow poorly and show signs of phosphorus deficiency even if the soil contains adequate amounts of phosphorus. In soils with very high levels of available phosphorus, mycorrhizae may be less important.

Nitrogen Metabolism

Nitrogen does not occur as a component of rock matrixes nor as a contaminant in rock; the most abundant source of nitrogen is the atmospheric gas N_2. This nitrogen is relatively inert chemically and is useless to almost all organisms; it must be converted to chemically active forms. This process, called **nitrogen metabolism**, consists of (1) nitrogen fixation, (2) nitrogen reduction, and (3) nitrogen assimilation.

Nitrogen Fixation

Nitrogen fixation is the conversion of N_2 gas into nitrate, nitrite, or ammonium, all forms of nitrogen that are substrates for a variety of enzymes. One means of nitrogen fixation is human manufacturing; the fertilizer industry synthesizes either nitrate or ammonium from atmospheric nitrogen, but this is an extremely expensive and energy-intensive process. About 25 million tons of nitrogen fertilizer are produced annually.

Natural processes fix over 150 million tons of nitrogen annually. Lightning is important; the energy of a lightning strike passing through air converts elemental nitrogen to a useful form that dissolves in rain and falls to the earth. However, nitrogen-fixing bacteria and cyanobacteria are by far the most important means

of fixing atmospheric nitrogen, annually converting 130 million tons of nitrogen to forms that plants and animals can use. These organisms have **nitrogenase**, an enzyme that use N_2 as a substrate. It forces electrons and protons onto nitrogen, reducing it from the +0 to the −3 oxidation state (Table 13-6). Ammonia, NH_3, is the product; it immediately dissolves in the cell's water and picks up a proton, becoming the ammonium ion, NH_4^+ (ammonia is the nonionized form and ammonium is the dissolved, ionized form; in both, nitrogen is in the −3 oxidation state). Nitrogenase is a giant enzyme complex composed of two distinct enzymes (dinitrogenase composed of four proteins, and dinitrogenase reductase composed of two proteins); it has a molecular weight of 300,000 Daltons and contains numerous atoms of iron, molybdenum, and sometimes vanadium (depending on the species). It is extremely sensitive to oxygen and functions only if oxygen is completely excluded from it.

Some nitrogen-fixing microorganisms are free living in the soil; examples are *Azotobacter, Clostridium, Klebsiella,* and *Nostoc* (Figure 13-14). The nitrogen they fix is used in their own metabolism and becomes available to plants and fungi only when they die and their bodies decay. Other nitrogen-fixing organisms live symbiotically, growing inside tissues of host ferns and seed plants (Table 13-7). The best known examples are root nodules on legumes such as alfalfa; the nodules are growths of root tissue whose cells contain bacteria of the genus *Rhizobium*. Plants such as alders (*Alnus*), bog myrtle (*Myrica gale*), and *Casuarina equisetifolia* are pioneer plants that are the first to grow in poor, nitrogen-deficient soils such as bogs, sand dunes, and glacial rubble. These species obtain their nitrogen from a symbiosis with the prokaryote *Frankia*. The symbiotic bacterial cells use part of the fixed nitrogen for their own growth and reproduction, but they also permit large amounts to leak out into the protoplasm of the surrounding root cells. Symbiotic nitrogen fixers usually produce fixed nitrogen at a much greater rate than free-living microorganisms, perhaps because they have the energy resources of the plant at their disposal.

The rate at which plant/prokaryote symbioses fix nitrogen is strongly influenced by the stage of development of the plant: When soybeans begin to produce their protein-rich seeds (40% protein, the richest seeds known), nitrogen fixation in the roots increases greatly. As much as 90% of all nitrogen fixation occurs during the phase of seed development, whereas only 10% occurs during all the vegetative growth that precedes it. Furthermore, if legume crop plants are given high levels of nitrogen fertilizer,

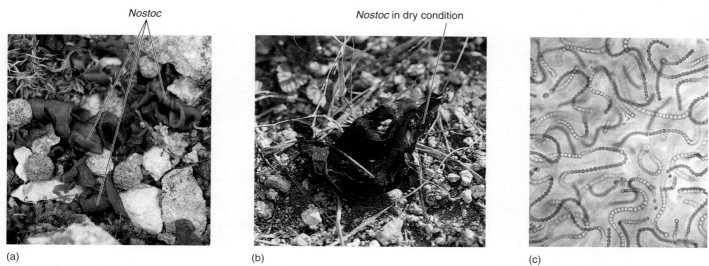

Nostoc

Nostoc in dry condition

(a)   (b)   (c)

Figure 13-14 Cyanobacteria are common components of most soils, although usually they are quite inconspicuous. These are colonies of the cyanobacterium *Nostoc*. The colony swells and fixes nitrogen rapidly when wet (a), but it becomes dormant and crisp when dry (b). Even though extremely desiccated with an extraordinarily negative water potential, the cells are alive and revive within seconds of receiving water. (c) The largest of these *Nostoc* cells are heterocysts, which carry out nitrogen fixation ($\times$300).

plants that have not yet formed nodules do not produce them, and those that already have bacteroid-filled nodules decrease the amount of nitrogen fixed and even allow the nodules to senesce. With adequate nitrogen available in the environment, it is selectively advantageous not to pass glucose on to bacteria.

▪ Nitrogen Reduction

Nitrogen reduction is the process of reducing nitrogen in the nitrate ion, NO_3^-, from an oxidation state of +5 to the −3 oxidation state of ammonium, which is also the oxidation state of nitrogen in amino acids, nucleic acids, and many other biological compounds (see Table 13-6). Nitrogenase automatically reduces nitrogen during the fixation process, and if a plant can absorb that form of nitrogen, no further reduction is necessary. Also, as organic matter decays in the soil, ammonium is released and becomes available. Unfortunately, for plants, ammonium is an extremely energetic compound that numerous species of soil bacteria use as "food," oxidizing it to produce ATP. In the process, ammonium is converted to nitrate. Such soil bacteria are so common that ammonium lasts only a short time in soil, and the predominant form of nitrogen available to roots is nitrate.

Reducing nitrate back to ammonium requires eight electrons for each nitrogen atom and a great deal of energy. In the first step, two electrons are added, reducing nitrogen from +5 to +3 and forming nitrite, NO_2^-. The enzyme is nitrate reductase, and it carries electrons by means of a molybdenum atom (Figure 13-15). Just like electron carriers in photosynthesis or respiration, when nitrate reductase reduces nitrate, it becomes oxidized and must pick up more electrons. It gets these from reduced flavin adenine dinucleotide ($FADH_2$) and reduced nicotinamide adenine dinucleotide (NADH); the ultimate source of energy and electrons is respiration.

In the second step, nitrite reductase adds six electrons to nitrite, reducing it to ammonium. The process is not well understood, but extremely strong reducing agents are needed. Apparently, nitrite reduction in leaves is powered by reduced ferredoxin from the light reactions of chloroplasts; however, roots are a more important site of nitrite reduction, and of course, they have no light reactions. Instead, they use reduced nicotinamide adenine dinucleotide phosphate (NADPH) produced by the pentose phosphate pathway. Although ATP is not consumed during nitrogen reduction, the process is expensive energetically because the NADH, NADPH, and ferredoxin used are no longer available for ATP synthesis in the mitochondrial electron transport chain.

In leaves, nitrate and nitrite are present as a result of breakdown of amino acids, nucleic acids, and other nitrogenous compounds during normal metabolism; nitrogen reduction recycles the nitrogen and conserves it within the plant.

▪ Nitrogen Assimilation

Nitrogen assimilation is the actual incorporation of ammonium into organic molecules in the plant body. The process is similar to that of an electron transport chain: Reduced nitrogen passes through a series of carriers that function repeatedly but in the long run are not changed.

The acceptor molecule is glutamate (the ionized form of glutamic acid); it reacts with ammonium and ATP, producing glutamine and ADP (Figures 13-16 and 13-17). Glutamine transfers the ammonium, now referred to as an **amino group**, to α-ketoglutarate, transforming both molecules into glutamate. The original acceptor has been regenerated, and an extra molecule of glutamate has been produced which can in turn transfer its amino group to another molecule. If the next molecule to receive the amino group is oxaloacetate, the amino acid aspartate is produced. If pyruvate receives the amino group, the amino acid alanine is produced.

These glutamate-mediated transfers of amino groups are the basis for incorporating nitrogen into the plant's metabolism and synthesizing all of the amino acids, nucleotides, chlorophyll, and

TABLE 13-7	Plants that Form Associations with Nitrogen-Fixing Prokaryotes

Prokaryote	Plant
Cyanobacteria	
Anabaena	Ferns: *Azolla*
Nostoc	Cycads: all genera examined
Actinomycetes	
Frankia	Angiosperms
	Betulaceae (birch and alder family)
	Casuarinaceae (beefwood family)
	Eleagnaceae (oleaster, Russian olive family)
	Myricaceae (bayberry family)
	Rhamnaceae (buckthorn family)
	Rosaceae (rose family)
Eubacteria	
Rhizobium	Angiosperms
	Fabaceae (legume family)
	Ulmaceae (elm, hackberry family)

These plants of alfalfa (*Medicago sativa*) will be used to feed cattle, horses, and other livestock. Like many species in the legume family (Fabaceae), this has root nodules that contain nitrogen-fixing *Rhizobium* bacteria. Consequently, alfalfa plants can grow well in poor soil and produce protein-rich leaves, which are especially nutritious as animal feed.

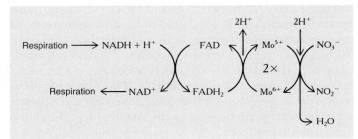

Figure 13-15 The electrons that reduce nitrate to nitrite are brought to it by a short electron transport chain. FAD and molybdenum are actually bound to the nitrate reductase enzyme, but NADH and NAD⁺ diffuse between the enzyme and sites of respiration.

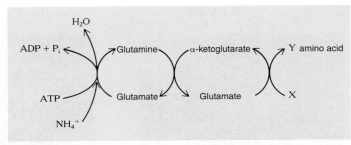

Figure 13-16 Nitrogen assimilation occurs by means of an "amino transport chain." The acceptor molecule is glutamate, which is already an amino acid, having one amino group. When it picks up ammonia, it becomes glutamine, an amine, having two amino groups. One is passed on to α-ketoglutarate, regenerating the acceptor and producing a new carrier. The next step is variable: By using various acids at X, the cell produces different amino acids at Y. If oxaloacetate is used, the amino acid aspartate results.

many more compounds. The transfer of an amino group from one molecule to another is **transamination**.

Nitrogen assimilation usually occurs in roots, the site of either absorption of nitrate or ammonium or transfer of ammonium from symbiotic prokaryotes. Much of the assimilated nitrogen must be transported to the shoot through the vascular tissues. Several nitrogen-rich compounds are common: Asparagine and glutamine both contain two amino groups rather than just one, as in the corresponding amino acids (Figure 13-17). In legumes, allantoic acid and allantoin are common transport forms, and in alders, nitrogen is carried from roots as citrulline.

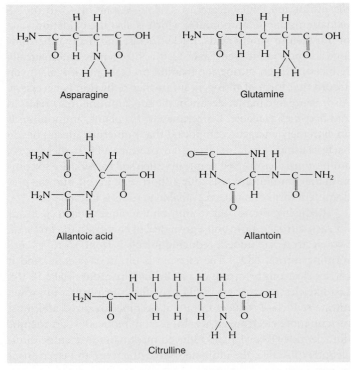

Figure 13-17 These chemicals each have a high nitrogen-to-carbon ratio and are transport forms of reduced nitrogen. Once moved through the phloem to a nitrogen sink, the compounds are catabolized and the amino group is used in the synthesis of amino acids, nucleic acids, and other compounds.

Plants and People

Box 13-3   From Fertility Gods to Fertilizers

The field of mineral nutrition is excellent for studying the development of biological thought. It is easy to assume that people have always understood the importance of nitrogen, potassium, phosphorus, and other elements, even if they did not study it explicitly. After all, humans have been farming for thousands of years, but the way we think of things now is different from other approaches to understanding the world. The first agricultural societies, those of Sumer, Egypt, India, and China, thought in terms of fertility gods; plants were believed to grow well or poorly according to the whims of divine intervention.

The earliest attempts to understand plant growth in a non-mystical way were formulated by ancient Greeks. Their thoughts are usually summarized by the phrase "Plants eat dirt." This is not as simplistic as it sounds. The Greeks believed that the universe contained only four "elements": earth, water, fire, and air. All things, plants and animals included, were constituted of various combinations of those elements. In such a world, the idea of plants as transmuted earth makes sense. Plants do not grow unless their roots are in earth, they must have water, and of course, they can burn, so the element fire must be contained as well.

This concept remained basically unchanged until the 1600s. By then, it was known that there are many elements, although the concept of "element" was not perfectly clear. The periodic chart had not been developed yet by Mendeleev, and the belief that lead could be changed into gold was still held. In 1644, J. B. van Helmont published the results of an important experiment. He had made a cutting of a willow and allowed it to form roots. When it weighed 5 pounds, he planted it in a container holding exactly 200 pounds of soil. He watered the willow and allowed it to grow for 5 years, then removed it and cleaned off the roots, being careful not to lose any soil. The plant had increased in weight to 169 pounds, but the soil had lost almost nothing—it still weighed 199 pounds, 14 ounces. Clearly the plant was not composed primarily of transformed earth; van Helmont concluded that water was the important transformed element (carbon dioxide was not yet known to exist).

This was a tremendous advance for two reasons. If water could be transmuted to vegetable matter, then it must be a compound and not an element. The second reason lay in the concept of experimentation. Greek science had been based on observations combined with thought, reasoning, and logic, but without experimentation. Van Helmont's results showed not only that the Greek conclusion was incorrect, but that their method of study without experimentation was inadequate.

Van Helmont's work was followed by that of John Woodward, who tested various types of water without soil: rainwater, stream water, and water from mud puddles. He found that rainwater was least effective in promoting plant growth, even though it was known to be the purest type of water. Chemical methods still were not advanced enough for him to be able to analyze why water from puddles was best. We now know that it is richest in the minerals necessary for plant growth, whereas rainwater is almost completely lacking in them.

A problem that impeded study of plants and animals was the belief that living creatures contain a vital force, something that was not chemical or physical and was assumed to be beyond study. It was believed that when an organism died and decayed, its vital force passed into the soil and made it fertile. Try to think like someone in 1700, only 50 years after the Pilgrims landed at Plymouth Rock. You would know that soil could be fertilized with bonemeal, manure, fish scraps, or compost (Table 13-8). All of these things had been living and could presumably add vital force to the soil.

At this time, two scientists, Nehemiah Grew and Marcello Malpighi, were making the first studies of the microscopic structure of plants and animals. Unfortunately, some of the early observations were misinterpreted and reinforced the concept of the existence of vital force. For example, the presence of microscopic holes in various types of cells caused people to think in terms of filtration: It was postulated that roots contained fine pores that allowed water and vital humours (liquids) to pass into the plant while non-nutritive soil sap was excluded. After partially purified by root filtration, the vital humours from the soil were thought to be transported upward, being further filtered by the pits and perforations of the xylem cells. This filtration presumably purified the vital force, allowing it to be transmuted into plant cells. This incorrect hypothesis misled scientists, and little thought was given to the role of minerals, which seemed too simple to be very important for life.

TABLE 13-8	Natural Sources of Organic Nitrogen
	Percent Nitrogen
Dried blood	12
Peruvian guano	12
Dried fish meal	10
Peanut meal	7
Cottonseed meal	7
Sludge from sewage treatment plant	6
Poultry manure	5
Bone meal	4
Cattle manure	2

These organic fertilizers are rich in nitrogen, but their usefulness was discovered by trial and error and by assuming that they contained vital force, not by chemical analysis.

(continued)

Plants and People

Modern analysis of plant and animal mineral nutrition was further hindered by another incorrect concept—that of spontaneous generation. It was believed that the vital force of dead plants and animals, if not absorbed by plant roots, would transform itself directly into new plants and animals. When vital force became sufficiently concentrated in soil, it was thought to cause worms, ants, mushrooms, and ferns to come into existence spontaneously. For example, the vital force of a dead animal was believed to cause the spontaneous generation of maggots and that of dead trees to produce mushrooms. Microscope studies would not reveal that maggots develop from fly eggs and mushrooms are composed of fungal filaments until years later.

In the 1800s, several important discoveries laid the foundation of modern biological studies based on chemical and physical principles. The first breakthrough came with the work of N. T. de Saussure in 1804; by this time, it was possible to prepare or purchase many chemical compounds in rather pure and well-defined condition. De Saussure designed and carried out the first well-planned hydroponics experiment and proved that nitrogen is essential for plant growth and development. The work of Julius Sachs followed, establishing the basic concepts of mineral nutrition in plants. It is difficult for us to appreciate how important these botanical discoveries were, but imagine the impact on 19th century scientists: Plants could be grown to maturity in the laboratory using no soil or other natural products. Containers, solutions, even the newly discovered electric lights, were all man made and humanly controlled. Hydroponic experiments established that life, as exemplified by plant life, is described by the laws of physics and chemistry, not metaphysics. Furthermore, contemporaneous with Sach's research, the theory had just been formulated that cells come only from preexisting cells, and in the late 1800s, Louis Pasteur proved that no vital force exists and that spontaneous generation does not occur. At approximately the same time, the first artificial synthesis of a biological compound, urea, proved that vital force was unnecessary in the construction of the material of protoplasm. The 1800s were exciting times for the philosophy of science and the concept of our relationship with nature. The study of metabolism became a science based completely on chemistry, physics, and mathematics; metaphysics and mysticism were eliminated.

■ Other Aspects of Prokaryotes and Nitrogen

Whereas certain bacteria reduce and fix nitrogen, others oxidize it, which adversely affects plants. When organisms die and decay, their nitrogenous compounds become available to the roots of living plants; however, because these compounds are highly reduced, they can be used as an energy source if oxygen is available. Certain soil bacteria are **nitrifying bacteria**: They oxidize ammonium to nitrite (*Nitrosomonas, Nitrosococcus*), and others oxidize nitrite to nitrate (*Nitrobacter, Nitrococcus*). The entire process is called **nitrification**. Both types of nitrifying bacteria are so common that nitrite never builds up in the soil. Nitrification has important ecological consequences; although plants can absorb and use nitrate, it is more readily washed from soil because it is negatively charged and remains in solution, whereas ammonium is positively charged and is bound to soil particles. Also, when ammonium is absorbed by a plant, the nitrogen is reduced already; however, when nitrate is absorbed, large amounts of ATP must be used to reduce it before it can be incorporated into amino acids, nucleotides, and so on.

Denitrification is a process in which certain bacteria (*Hyphomicrobium, Pseudomonas*) reduce nitrate to gaseous nitrogen, N_2. Nitrate or nitrite is used as an electron acceptor during energy production. Whereas nitrification results in nitrate that plants cannot use as easily as ammonium, denitrification results in nitrogen gas, which plants cannot use at all.

■ Obtaining Nitrogen From Animals

Soils in bogs and swamps typically have very little nitrogen available because of nitrifying and **denitrifying bacteria**. Many bog-adapted plants obtain a significant fraction of their reduced nitrogen by catching animals; these are **carnivorous plants**. Carnivory has originated many times, in many different groups of plants; thus, there are numerous mechanisms that trap and digest animals. It is important to point out that trap leaves photosynthesize as well as catch, digest, and absorb insects. Carnivorous plants obtain their energy through photosynthesis, not from the fats, carbohydrates, or proteins of the animals they consume.

Ant-plants are flowering plants and ferns that also obtain reduced nitrogen from animals; examples are *Myrmecodia, Hydnophytum* and *Solanopteris* (**Figure 13-18**). They are epiphytic, with their roots attached to tree trunks or branches, and bark is a nitrogen-poor "soil." Ant-plant stems are swollen and have hollow chambers that ants use as living spaces, as bathrooms and as graveyards for ant corpses. As wastes and dead ants decompose,

Figure 13-18 The holes and chambers (not visible) in this hypocotyl of *Hydnophytum formicarium* form naturally. Ants live inside and supply the plant with nitrogenous waste.

ant-plants absorb nitrogenous compounds either by means of absorptive areas on the chamber walls themselves or by means of adventitious roots that grow into the waste-filled chambers. Rather than trapping and harming the ants, the plants provide living space; therefore, this relationship is mutually beneficial and is said to be **mutualistic**.

Storage of Minerals Within Plants

Neither plants nor animals can be certain that essential minerals will be available in appropriate quantities throughout their lives, so storage mechanisms are necessary. Animals tend to have large reservoirs of at least some elements: The copious amounts of protein in muscle can be broken down to amino acids that then donate their amino groups to supply nitrogen. Bones and teeth store both calcium and phosphate. For some reason, plants rarely store minerals or nitrogen as crystallized or polymerized forms the way animals do, even though a plant's mineral requirements change greatly. During formation of flower buds and then especially during maturation of fruits and seeds, the need for nitrogen, phosphorus, potassium, and other essential elements increases greatly, whereas wood synthesis may require huge amounts of carbohydrate but almost no mineral resources. Trees such as cherry, which blossom in early spring and then produce both fruits and leaves at the same time, have a tremendous demand for essential elements early in the year but require very little during summer.

Apparently, all parts of a plant except for seeds store minerals in soluble form in the central vacuoles of cells. Nitrogen can be concentrated a little by being converted to compounds with multiple amino groups such as asparagine, citrulline, and so on, the same compounds that are used in nitrogen transport. Phosphates, sulfates, and other mineral nutrients apparently are simply sequestered in the central vacuole in the same forms in which they are used metabolically by the cell. This method does not permit storage of large amounts of minerals because high concentrations would be toxic, but plant cells differ from our cells by having very large vacuoles and just a tiny amount of cytoplasm squeezed between the vacuole membrane and the cell membrane. Pound for pound, plant tissues need only very small amounts of minerals compared with animals.

In contrast, seeds do store minerals. Seeds must be both lightweight but also packed with enough resources to get a seedling established quickly. Amino acids are stored as particles of protein, protein packed so tightly that it typically crystallizes into a structure called a protein body. In many seeds, protein bodies themselves contain inclusions of another crystalline form, crystals of a substance called phytin (technically, myo-inositol-hexaphosphate). Myo-inositol-hexaphosphate is a six-carbon sugar that has six hydroxyl groups, each carrying a phosphate bound to it. Being an acid, myo-inositol-hexaphosphate ionizes and loses protons, H^+, when dissolved in water, but when the developing seed concentrates it and precipitates it in the protein body, it puts cations on it rather than returning the protons, and the cations used are Mg^{2+}, Ca^{2+}, Zn^{2+} and K^+. This mineral-holding form is phytin, and it permits the protein body to not only store amino acids and phosphate but all of these other essential elements as well.

Plants and People

Box 13-4 Fertilizers, Pollution, and Limiting Factors

Plants in nature usually do not grow as vigorously as they potentially could. For example, desert plants typically grow more rapidly if given extra water, plants in shady areas grow better if given a bit more light, and prairie grasses, which already have enough water and light, benefit from extra nitrogen fertilizer. Any plant grows at a particular speed and vigor because it is limited by some factor, such as too little water or light or nitrogen fertilizer. An important concept is that there is only one single **limiting factor** at a time for any plant. Desert plants given more light or fertilizer will not grow faster—it is water that is limiting them. But if we do give them extra water, their growth rate will increase until some other factor becomes limiting, perhaps lack of nitrogen. Although growing slowly, the plants could get nitrogen quickly enough, but now that they are growing faster, their ability to obtain nitrogen from the soil may be limiting. If we give the plants both water and nitrogen fertilizer, they may grow even more rapidly, but some other factor will become limiting. On many farms plants are irrigated and fertilized and planted far enough apart that they do not shade each other and their roots do not interfere with each other. Such plants grow much more rapidly than they would in nature, but they are still limited, in this case, by their own genetics, their own innate metabolic capacity. There is always a limiting factor.

The concept of a limiting factor is important in understanding techniques for reducing the damage caused by pollution. Under natural conditions the water in rivers and lakes has so few nutrients that algae grow slowly, and they are so sparse the water is blue. In the mid-20th century pollution from farms and cities fertilized rivers and lakes and allowed algae to grow more vigorously. As rain drained from fields, lawns, gardens, and golf courses, it carried much of the fertilizers that had been applied to stimulate the growth of crops, flowers, and grass. Also, most household waste that is flushed down toilets is an excellent organic fertilizer. With all these extra inputs of nutrients, populations of algae became so dense that "pond scum" floated near the surface of rivers and lakes, and the water was green because of the abundance of microscopic algae (Figure B13-4a).

It is difficult to stop all pollution, but it is only necessary to control one single pollutant to create a limiting factor. Phosphate was chosen as the target. Phosphorus is an essential element, and it is naturally low in pure water. A large amount of the phosphate pollution in rivers comes from laundry detergent and dishwashing soap. With a little effort phosphate-free detergents were invented, and now they are used almost universally, so there is much less phosphate pollution. The concentration of phosphate in rivers dropped so low that algae could no longer thrive, and their populations fell to more normal levels. Even though the water is still heavily polluted with nitrates, sulfates, and other nutrients, the algae cannot use them as long as phosphate is kept low enough to limit their growth. If we could reduce the phosphate runoff from farms and lawns, the levels of algae would drop even more and rivers and lakes would be even cleaner (Figure B13-4b). It would be better to control and reduce all types of pollutants, but by keeping one at limiting levels we can at least minimize some of the damage caused by pollution.

(a)

(b)

Figure B13-4   (a) This river runs through extensive agricultural areas in east Texas and is polluted by fertilizers from fields and manure from livestock; because the water is so rich in nutrients, algae and bacteria grow well, giving the water a brown color. (b) This stream in North Carolina runs through a forested area without agriculture, so it does not receive as much pollution as the river in (a). There are many towns and homes along this river, but because modern detergents do not contain phosphates, the wastewater dumped into this river is almost phosphate-free, so algae and other microbes cannot grow in it and the river has clear, blue water.

Summary

1. Plants synthesize all their body's materials using only carbohydrate and a few essential elements.

2. Three criteria must be met for an element to be considered essential: (1) it is necessary for the completion of a full life cycle; (2) it cannot be replaced by a chemically similar element; (3) it must act inside the plant.

3. Macro or major essential elements are needed in relatively large amounts, whereas micro or minor or trace elements are required in only extremely small amounts.

4. Deficiency diseases are rare in nature, probably because susceptible plants are outcompeted by deficiency-tolerant plants; however, deficiency symptoms are frequent in crops and horticultural plants if proper fertilizing is not maintained.

5. Deficiency of an immobile element produces symptoms in young leaves and buds, whereas deficiency of a mobile element results in symptoms in older organs.

6. Cations are released from the surface of soil particles if protons, which result from the respiration of roots and soil organisms, disrupt their electrical attraction.

7. Mycorrhizae are the principal means by which most plants absorb phosphorus.

8. Nitrogen metabolism consists of fixation, reduction, and assimilation.

9. Nitrogen must be obtained from the air, but only certain prokaryotes—some free living, others symbiotic—have the necessary enzyme, nitrogenase.

10. Most plants absorb nitrate, which must be reduced by the plants, using large quantities of NADH, NADPH, and ferredoxin.

11. Once formed within plant cells, ammonium is assimilated by transamination, being passed from glutamate to various amino acids.

Important Terms

acid rain
amino group
carnivorous plants
cation exchange
chlorosis
denitrifying bacteria
essential elements
eutrophication

field capacity
hydroponic solution
immobile elements
major (macro) essential elements
micelles
mineral nutrition
minor (micro) essential elements
mobile elements

necrosis
nitrifying bacteria
nitrogen assimilation
nitrogen fixation
nitrogen reduction
trace elements
transamination
weathering

Review Questions

1. Most elements that are essential for plant growth are present in the _____ _____ _____ _____. How do the elements become available?

2. What is meant by the microflora and microfauna of the soil? Why are they important?

3. How many bacteria typically occur in one gram of soil?

4. What is a hydroponic solution? After a solution of known composition which supports plant growth is found, how would you go about testing whether all of the chemicals in the solution are necessary for plant growth?

5. It is simple to grind up a plant and then extract and measure all of the chemical elements present, but this does not tell us which elements are essential. Why not?

6. List the major or macro essential elements. Why are they called that?

7. List the minor or micro (trace) essential elements. Why are they called that?

8. Are our modern reagents so pure that we can prepare a hydroponic solution that contains only the nine major and seven minor essential elements and nothing else? Can we be certain we have discovered all the minor essential elements?

9. What are the three criteria an element must meet to be considered essential?

10. Do you think anyone has performed a mineral nutrition experiment that examined the entire life cycle of a giant redwood? A giant cactus? A mistletoe that is parasitizing an oak tree? What would be the problems involved?

11. To be essential, an element must be acting within the plant, not outside it. What does this mean? Discuss the roles of iron and phosphorus in determining which elements are essential for a plant.

12. We humans, like all other organisms, use a universal set of twenty amino acids in our proteins, but we cannot make _____ of these ourselves. How many are plants unable to make? How many do plants need to take up through their roots?

13. The differences in nutritional resources used by plants versus animals is also great. Plants obtain nutrients in the form of _____ or as simple _____ present in the environment. An animal begins with food in the mouth, but the nutrients occur as _____ in complex _____ which in turn are parts of _____ _____.

14. Name the essential element involved in each process: changes the osmotic potential in guard cells as they swell or shrink, present in all nucleotides and amino acids, carries electrons in chlorophyll, present in cytochromes, present in plastocyanins, involved in nitrogen metabolism, present in the water-splitting enzyme of photosystem II, present in ATP, present in the middle lamella, and important in regulating the activity of many enzymes.

15. Under natural conditions, is it rare or common to encounter plants whose growth is disrupted by a scarcity or an excess of mineral elements? Describe two habitats that often have excessive amounts of mineral elements (Hint: do not forget to study the figures as well as the text).

16. Question 15 specified natural conditions. What about non-native crop plants and ornamentals? Do they often encounter mineral deficiencies?

17. Name and describe two common symptoms of deficiency diseases.

18. What is the difference between a mobile and an immobile element? Where are the first deficiency symptoms for each? Which elements are immobile?

19. Do we understand why certain minerals are immobile in plants? What is one essential element that we humans do not recycle well in our own bodies?

20. What are the two fundamental processes of weathering that convert rock to soil?

21. What are size ranges of particles of coarse sand, fine sand, silt, and clay particles (also called clay micelles)? Coarse sand can absorb a great deal of water during a rain, but what happens to the water soon after the rain stops? What would happen instead if the soil consisted of clay micelles?

22. Chemical weathering dissolves the surface of the crystal matrix of rock. As the matrix breaks down, what is released—positively charged cations or negatively charged anions? As a result, the residual crystal matrix has what kind of charge?

23. Many factors affect soil acidity, but probably the most important factor of all is _____. Describe how this factor exerts its effect.

24. Most soils are _____ to _____ acidic. Desert soils, which have very little leaf litter or other organic matter, are usually _____ (_____).

25. Describe cation exchange. How would decomposition of mulch and humus affect cation exchange? Why is acid rain so damaging?

26. The roots of most plants form a symbiotic association with soil fungi. Name the association and the element that it supplies to the plant.

27. Which essential element is neither a component of rock matrixes nor a contaminant in rock? Although this element is abundant in air, no plant or animal can absorb it from the air and use it. Why not?

28. Name the three steps in the conversion of N_2 into organic nitrogen that is part of a plant. Which type of organism is capable of performing each step?

29. What is nitrogenase? What chemical does it use as a substrate? It is extremely sensitive to _____ and can function only if _____ is completely excluded from it.

30. The electrons that reduce nitrate to nitrite are brought to it by a short electron transport chain involving $FADH_2$ and NADH (see Figure 13-15). In which other reactions have you seen these two electron carriers?

31. How does nitrogen reduction affect carbon fixation in the stroma reactions? How many molecules of NADPH must be diverted from one pathway to the other? How would respiration be affected?

32. What is nitrogen assimilation? What is the acceptor molecule? What does it react with? What is the product?

33. What is transamination? If an amino group is passed onto oxaloacetate, which amino acid is produced? If an amino group is transaminated onto pyruvate, which amino acid is produced?

34. Figure 13-17 shows several chemicals that carry nitrogen up from the roots to the rest of the plant. What is the unusual feature of these chemicals that make them good carriers?

BotanyLinks

http://biology.jbpub.com/botany/5e

BotanyLinks contains a variety of resources and review material designed to assist in your study of botany.

Visit the Web Exercises area of BotanyLinks to complete this question:

1. How important is acid rain? Is it an extensive problem? Go to the BotanyLinks home page to begin your search for more information on this subject.

Development and Morphogenesis

chapter 14

Concepts

All plants begin as a zygote—a single cell—that undergoes enlargement and cell division as it grows and develops into an individual plant. It undergoes **morphogenesis** (generation of the shape of the plant and its various organs) and **differentiation** (an increase in complexity as some cells become different from each other). Often the zygote divides into a basal cell and an apical cell: This establishes **polarity**, the formation of the root/shoot axis. Also, the zygote's spherical **symmetry** changes to the embryo's radial symmetry. Radial symmetry dominates roots and stems, and typically, only leaves, sepals, and other flower parts have bilateral symmetry.

Cell differentiation occurs with the first cell division. The basal cell develops a size, shape, and metabolism that differ from those that the apical cell develops. Later, surface cells differentiate into epidermis cells, including guard cells and trichomes; inner cells

Chapter Opener Image: A seed has just germinated, and the seedling is beginning to grow and develop into a new plant. It does not yet have any shoot at all (the region below the cotyledons is the hypocotyl); its microscopic shoot apical meristem is probably just starting to make its first leaf primordia and the first cells of the nodes and internodes. At this point, it is impossible to tell what species this is as many dicot seedlings look just like this. But as new cells are formed, each differentiates in a particular way, guided by its genes and hormones such that the leaves have characteristic features, as do the nodes, internodes, axillary buds, and later on the flowers and fruits. Ultimately, the actions of all the genes guide the development of a plant body that is easy to recognize as being a particular species. During this process, some new cells develop into epidermis, others into xylem, some into phloem and so on, even though all have the same genes. This is the result of differential gene expression, as explained in this chapter.

Outline

differentiate into cortex, phloem, xylem, and pith. Such differentiation requires that each cell be capable of identifying its position relative to other cells and also of informing other cells of the types of developmental and morphogenic changes it is undergoing. For example, cells that will differentiate into epidermis cells must first detect that they are on the plant's surface. Also, surface cells in roots must develop into epidermis tissue that consists only of pavement cells and root hairs, whereas surface cells of stems of the same plant develop into an epidermis with pavement cells, trichomes, and guard cells; those of petals might develop into pavement cells only, but filled with pigments. To develop properly, a pre-epidermis cell must not only detect that it is a surface cell, but also which organ it is a part of.

Consider also the differentiation of guard cells. They must occur in pairs (due to being sister cells in some species but derived from separate mother cells in other species), and they must coordinate where the stomatal pore forms; it would do no good if each were stimulated to develop as a guard cell but then each tried to form a pore on one of the sides that does not face the other guard cell. Furthermore, pairs of guard cells do not occur at random on stems and leaves, nor are they clustered together: There is a **pattern establishment mechanism** that informs each cell of its location relative to other epidermis cells. It appears that after one cell begins to differentiate as a guard mother cell, it produces a substance that moves into surrounding cells and inhibits them from also differentiating as guard mother cells. As the inhibitor spreads, it becomes more dilute until it no longer prevents distant cells from acting as guard mother cells. If subsurface cells are differentiating into fibers, they too inhibit adjacent surface cells from undergoing guard cell morphogenesis.

Inner cells of young regions of shoots also must detect their position and differentiate accordingly. Typically, the first cells below the apical meristem to become visibly differentiated are sieve tube elements of protophloem. They are located part way between the epidermis and the center of the stem, and each develops several micrometers away from other newly forming sieve tube elements: Like guard mother cells, each region that begins to differentiate as phloem must inhibit adjacent regions from also becoming phloem, thus establishing a pattern of one ring of bundles in dicots, numerous bundles in monocots. Vascular bundles always contain a strand of xylem located just interior to the phloem; thus, the first sieve tube elements may be capable of stimulating another pattern: xylem formation. This pattern, however, does not affect all cells equally; only cells located between the phloem and pith are stimulated to undergo xylogenesis, whereas cells located between the phloem and epidermis are unresponsive.

Whereas guard cell morphogenesis requires coordinated differentiation of two cells, morphogenesis of vessels requires coordination of thousands of cells. As each new vessel element differentiates from a young subapical parenchyma cell, it must recruit the next cell above itself to also develop into a vessel element, thus extending the vessel upward. And each element must make perforations aligned with the perforation of the element below it (which is slightly older and more mature) and the one above it (which is slightly younger). Each newly recruited cell must then recruit the next cell to be part of the vessel and so on. Differentiation of sieve tubes requires not only the progressive recruitment of sieve tube members upward in the shoot's newly developing vascular bundle, but also the coordinated differentiation of a companion cell for each sieve tube member.

In most cases, we assume that cells inhibit or promote activities in nearby cells by secreting chemicals, but it may be that the enlargement of some cells—or the lack of enlargement—causes physical stresses in tissues. Compression or stretching may also be important in establishing patterns of cells and tissues.

Development and morphogenesis also occur at the level of organs and the whole plant. While still small, a seed plant embryo establishes a shoot apical meristem and a root apical meristem. The shoot apical meristem generates new cells for both itself and for the elongating stem. It also establishes leaf primordia in characteristic phyllotactic patterns, and leaf primordia in turn establish axillary buds. Although each axillary bud develops its own apical meristem, a chemical messenger (a hormone) from the stem's apical meristem inhibits most of the stem's axillary buds, keeping them quiescent and thus controlling the pattern of branching and the shape of the plant. Vascular bundles differentiate upward toward the shoot apex and branch at the proper place such that leaf and bud traces interconnect the stem's vascular system with those of leaves, flowers, and branches.

Plant bodies tend to be much simpler than those of animals, but even so, plant growth and development are not trivial. A great deal of information is required to establish polarity and symmetry, then patterns of cells, tissues, and organs. And the differentiation and morphogenesis that converts each pattern into actual cells, tissues, and organs require even more information. Virtually all information needed for plant development is stored as genes located in nuclei, plastids, and mitochondria. As an embryo develops and later as a seed germinates, it develops into a complete plant relying entirely on its own self-contained information and the intercommunication between its parts.

But plants live in habitats with environmental factors such as light, gravity, water, temperature, seasons, and other organisms. All plants have evolved to have the capacity to detect at least some environmental factors and to alter their development and morphogenesis in adaptive ways. This is an important concept: Plants respond to certain environmental cues in ways that make the plants more likely to survive and reproduce successfully in the habitats they find themselves in. For example, all living stem and leaf epidermis cells synthesize some cuticle: If they are alive, they make cutin. But many plants also can detect environmental moisture and respond adaptively by producing a thicker, more effective cuticle if conditions are dry and a thinner, less expensive cuticle if it is humid. Similarly, seeds germinate and grow if given water and warm temperatures, but if buried under several centimeters of soil when they germinate, they detect gravity and respond adaptively: Shoots grow upward, eventually reaching light, whereas roots grow downward into deep soil that may be more moist. Gravity is not necessary for growth, but it acts as a source of information about where the seedling is located with respect to light and moist soil.

In these and many other cases, the environment provides information that allows plants to alter aspects of development and morphogenesis that would occur anyway. In other cases, environmental cues are needed for the plant to alter its development radically and to complete its life cycle. Many plants detect season (spring, summer, and autumn) by measuring the length of nights, and this stimulates the adaptive response of producing flowers at the right time of year or restricting growth and initiating preparations for winter. Without this environmental information, the plants would continue to grow vegetatively without ever forming flowers or without producing cold-resistant dormant buds, and this would be disastrous for the plant. Note that here too all information needed to actually produce flowers or cold resistance is present in the plant's genes—it is just that the plant itself cannot activate those genes until triggered by environmental signals.

It is important to distinguish between adaptive responses and mere consequences of environmental stress. When fire kills a plant, the plant's death is a result of overly adverse conditions; it is not adaptive to the plant. Similarly, a late frost in spring may cause developing leaves to become malformed. The resulting shape is not adaptive; it is merely that frost disturbed leaf morphogenesis destructively.

This chapter describes several environmental cues that plants detect, the chemical signals used by plant parts to communicate with each other, and several widespread types of development and morphogenesis. The following chapter discusses genes and the mechanisms by which their information is accessed and used.

Environmental Complexity

If all plants were extremely small and lived in completely uniform, nonvarying environments, most would probably be simple and would experience little selective pressure for the evolution of complex shapes, tissues, organs, and metabolism. The most uniform, constant conditions occur in oceans and large lakes, where water buffers rapid changes in temperature, acidity, oxygen concentration, and other factors. Under such stable conditions, small organisms such as algae, protozoans, and sponges are simple. But most organisms exist in a heterogeneous environment: Gravity comes from only one direction. The sun is either to the side or overhead but never below. Temperatures are lower on the shady side of a plant. Moisture depends on depth below or height above the soil surface. This mosaic of conditions changes over minutes, days, seasons, or longer periods of time. It is selectively advantageous for plants to sense these differences and changes and to respond to them.

Most plants are so large that their bodies exist in several different microenvironments. Consider a small tree: Its roots are in soil, which is usually moister, cooler, and darker than air; the highest branches are in open air, exposed to full sunlight and the full force of wind, storms, rain, and snow (Figure 14-1). The trunk base and lower branches are in an intermediate environment—less stable than soil and less variable and severe than open air. A vertical tree trunk is oriented to best resist gravitational attraction (its own weight), whereas horizontal branches are highly stressed unilaterally by gravity.

In springtime, the shoot can become warm enough for active metabolism even though the soil remains cold or frozen. The plant parts must communicate with each other, or shoot buds would become active and expand before roots were capable of transporting water to new leaves. In autumn, increasing night length and declining air temperatures signal impending winter

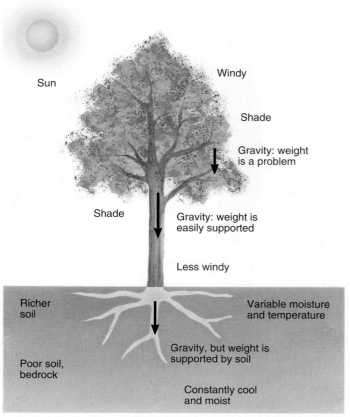

Figure 14-1 All plants, especially large ones, live in a complex mosaic of microenvironments that differ from each other, often dramatically. Each factor also changes throughout a day or a season.

and the need for dormancy; roots are informed about changing seasons by chemical signals from the shoot.

The need for intercommunication and coordination also exists within a limited region of the body. For example, leaf parts act in a coordinated fashion during development such that the petiole has enough xylem and phloem to facilitate the transport needs of the blade. Too little conductive tissue causes the blade to suffer water stress or an inability to export sugars; too much conductive tissue is a waste of energy and material. Even on the intracellular level, organelles must communicate with each other because their metabolisms are interrelated.

All levels of communication have in common a basic mechanism. Information about the environment or the metabolic status of the organ must be **perceived**. The plant must sense environmental cues such as changes in temperature, moisture, or day length, or the nucleus must receive chemical signals if conditions in the surrounding cytoplasm change. Next, information must be **transduced**, or changed to a form that can be either acted upon or transported. Finally, there must be a **response**: The plant must enter dormancy, produce flowers, change the type of leaf production, and so on (Figure 14-2). If any of these steps is missing, the plant cannot respond to the environment. These principles apply to all organisms, and of course, higher animals have the most elaborate mechanisms. Their sophisticated sense organs for sight, taste, touch, hearing, and smell perceive external conditions. These sense organs transduce the perceived information to

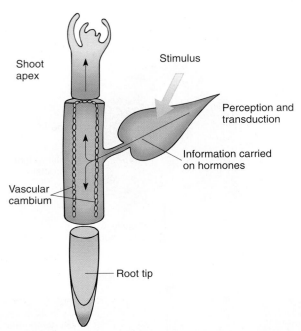

Figure 14-2 The signal for dormancy and preparation for winter is short days (long nights) in autumn, perceived by leaves. The chemical messenger transported from the leaves causes the shoot tip to produce bud scales instead of leaves; the vascular cambium fills with many small vacuoles and stops mitosis and cytokinesis. Young xylem mother cells differentiate to a predetermined stopping point and then become dormant; roots slow their growth greatly but usually do not stop completely. Roots and the cambial region cannot perceive the approach of winter themselves but depend on leaves as sites of perception. Although the shoot apices could perceive it themselves, the entire plant is integrated as a whole by being cued by the leaves.

a transportable form such as nerve impulses or hormones that are secreted into the blood stream. Animal responses are also typically highly elaborate, involving precisely controlled movement and increased activity of organs.

Sensory systems and response mechanisms of plants tend to be simpler, and their signal transport is usually slow, involving movement through cortex parenchyma or phloem. This does not mean that animal systems are superior to those of plants. "Superior" has no meaning. The proper question is this: "Which is more adaptive, which is more advantageous selectively?" In terms of evolution and natural selection, animal systems are successful adaptations for organisms that must detect food or predators and must move in order to capture that food or avoid being captured. Any animal with a plant-type mechanism of perception, transduction, and response would soon starve or be eaten.

On the other hand, rapid response mechanisms would not be adaptive in plants. They are extremely expensive to build and maintain and are unnecessary because most environmental conditions important to plants change only slowly. Eyes, nerves, muscles, or adrenal glands are not needed by plants for absorption of carbon dioxide, water, minerals, or light or for perception of autumn and preparation for winter dormancy. For sexual reproduction, animal-pollinated plants do need sophisticated perception and response mechanisms, but basically, they simply "rent" those of their pollinators, paying with nectar or other rewards. Finally, the very sophistication of animal sensory/response systems makes animals more vulnerable, more easily injured, and

more dependent on avoiding dangers such as fires, floods, freezes, and predators. Plants are typically much more resilient than animals, being able to survive burning because of thick bark or resprouting from rhizomes and bulbs; many withstand flood by being tough or flexible. Predators can consume most of a plant's leaves, wood, or roots without actually killing the plant.

Light

Besides energy for photosynthesis, light also provides two important types of information about the environment: (1) the **direction** or, more precisely, the gradient of light. This allows a plant to grow or orient its leaves toward a region of bright light, which increases the light available for photosynthesis. (2) The **duration** of light (length of the day) provides information about the time of year. Air temperature is unsuitable because cool autumn temperatures may be followed so quickly by severe cold that plants do not have enough time to become dormant. But day length is an infallible indicator of season.

Gravity

It is selectively advantageous for many plants to orient themselves or their parts with respect to the direction of gravity. In some cases, gravity itself is important because it causes weight stress. A vertical stem supports more weight than does a similar stem growing at random. Whenever a plant is bent or tilted because of flooding or the slipping of a hillside, the plant must change its growth back to an upright direction; if it continued to grow at an angle, it would need to produce many more fibers to support its weight. In some situations, the direction of gravity is instead a guide to other important factors. Roots that grow downward are more likely to encounter water and minerals. Shoots that grow upward grow above other plants and encounter better conditions for photosynthesis, pollination, and seed distribution. Normally, shoots do this by growing toward the brightest light, the open sky, but shoots of seeds that germinate deep in the soil must determine which way is up while in the dark. Direction of gravity is their only reliable guide. Most bilaterally symmetrical flowers must be aligned with the body symmetry and flight pattern of their pollinator. Such flowers must be bilaterally symmetrical vertically and project horizontally because insects and birds do not fly upside down or sideways. The flower must orient itself along the same environmental gradient that the pollinator uses—the gravitational gradient of up versus down (**Figure 14-3**).

Although gravity does not change with time, the force it exerts on a particular organ does change as the weight supported by the organ changes. The pedicel of an apple flower supports almost no weight, but the same pedicel must later support the weight of a fully grown fruit (**Figure 14-4**). The extra fibers are not produced until needed. Similarly, a young branch must be strong enough to support a small amount of weight, whereas a larger branch must support more.

Touch

Although plants do not move around like animals, their parts frequently grow against objects and respond to this contact. Certain types of contact are detrimental, for example, when a root grows against a stone or a branch rubs against another branch.

Figure 14-3   These *Penstemon* flowers have detected and responded to gravity in two ways. First, they all project horizontally, not at random and not toward the sun. Second, they are bilaterally symmetrical (their right side is a mirror image of their left side), and the plane of symmetry is vertical. Thus, flower symmetry matches the symmetry of bees, and flower projection allows bees to land easily.

In these cases, a thick bark is adaptive as a protective layer, and it is produced only where needed, only in the area being touched. Other types of contact are beneficial: After a tendril touches an object, it grows around the object and uses it as a support. When a fly touches sensitive trigger hairs on a Venus' flytrap, the trap closes,

Figure 14-4   Weight is a source of information to the plant about the amount of collenchyma or sclerenchyma needed to counteract the gravitational attraction on the plant. Initially, the flower stalk was strong enough only to support the weight of a flower; by sensing and responding to gravity, it is now strong enough to support the weight of an apple.

catching and holding the insect during digestion. In some cases, the contact is between two growing primordia and is a normal developmental feature. Many flowers that have fused petals or carpels start with separate primordia that grow together and fuse, acting as a single unit during development (Figure 14-5). In each case, the physical action of touching is similar, but each organ responds in a distinct way that is adaptive for the plant; the response of each would be inappropriate if it occurred in the others.

Figure 14-5   (a) The carpels of *Catharanthus roseus* are initiated separately and consist of protoderm and ground meristem, but they soon crowd into each other and fuse into one syncarpelous gynoecium (×70). (b) At the points of contact (arrows), the protoderm changes into mesophyll rather than epidermis (×6,000).

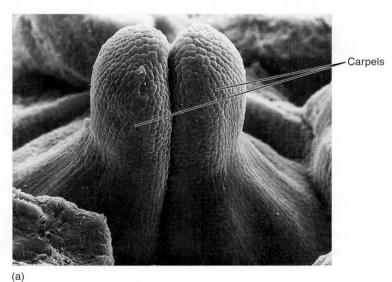

Carpels

(a)

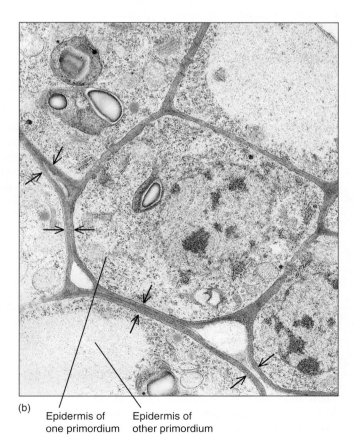

(b)

Epidermis of one primordium

Epidermis of other primordium

Temperature

Temperature fluctuates in a predictable pattern on both a daily and a yearly basis. Changing temperatures can induce many specific types of plant development. Although most plants appear to be quiescent and virtually lifeless in winter, a considerable amount of critically important metabolism is occurring. This metabolism usually does not proceed at temperatures above 1°C to 7°C.

Cold temperatures are required for the normal flowering of biennial and many perennial plants. Species of perennial trees that are adapted to habitats with cold winters, such as apples, typically require near-freezing temperatures to break the dormancy of their flower buds, which they had formed in the previous summer. If the trees are grown in areas with warm winters, they form flowers that never open. In contrast, biennial plants spend their first year in a vegetative phase and cannot be induced to form flowers. Their first cold winter **vernalizes** them, it causes them to switch to a state in which they can sense and respond to a stimulus that induces flower formation (Figure 14-6). In the year after the vernalizing winter, the plants resume vegetative growth, but when exposed to the proper stimulus (usually the short nights of early summer), they respond by producing flower buds. In their first year (prevernalization), summer has no effect; in their second year (postvernalization), it induces flowering. If never vernalized, they never produce flowers.

Low temperatures are required to induce deep dormancy in temperate trees and shrubs. Short days of autumn induce plants to initiate preparations for winter and enter a mild state of dormancy, but the most resistant stages are not entered until the plant actually experiences a week or two of cool temperatures. In contrast, cool temperatures are required to break the dormancy of many seeds, allowing them to germinate when their habitat becomes warm and moist.

Water

Although water is an absolute prerequisite for life, its presence probably does not act like a signal in the way other factors do. If enough water is available, plants grow; if not, plants wilt and perhaps even die. Although roots often appear to grow toward water, they actually grow in all directions, and those which, by chance, grow toward water grow more rapidly because they are in a favorable environment. Roots that grow away from water grow slowly, but only because they enter an environment too dry to permit growth. Roots do not turn and grow toward water in the way they turn and grow toward gravity.

Water scarcity triggers specific adaptive responses. One of the first is production of the hormone abscisic acid, which causes guard cells to lose potassium and close stomatal pores. This occurs in most plants even while cells have enough water to carry on basic metabolism. If water stress continues or becomes more severe, new responses are triggered that may inhibit production of new leaves, increase the cuticle on existing leaves, or even initiate abscission of leaves.

Responding to Environmental Stimuli

Plant responses to the diverse types of information present in the environment can be grouped into four simple classes (Table 14-1).

(a)

(b)

Figure 14-6   (a) In its first year a biennial plant has only a very short stem, with all its leaves attached close together, and is unable to flower. (b) In its second year a biennial plant produces a large group of flowers, almost always on a tall stalk. After the fruits and seeds mature, the entire plant dies. This is mullein, *Verbascum thapsus*.

Tropic Responses

A **tropic response** is a growth response oriented with regard to the stimulus. For example, growth toward a bright light is a phototropic response (Figure 14-7 and Table 14-2). A **positive** tropism is growth toward the stimulus; a **negative** one is growth away from the stimulus, and **plagiotropism** is growth at an angle (Figure 14-8). Most tap roots are positively gravitropic, growing downward in response to gravity, whereas shoots are negatively gravitropic,

TABLE 14-1	Types of Plant Responses to Stimuli	
Response	Mechanism	Oriented
Tropic	Growth	Yes
Nastic	Turgor changes	No
Morphogenic	Changes in basic metabolism	No
Taxic	Swimming	Yes

TABLE 14-2	Prefixes for Stimuli
Stimulus	Prefix
light	photo-
gravity	gravi- (formerly geo-)
touch	thigmo-
chemical	chemo-

growing upward in response to gravity. Branches and secondary roots grow horizontally or at an angle, plagiogravitropically.

When touch is the stimulus, the response is **thigmotropism**. Positive thigmotropism occurs when a tendril touches an object and, by growing toward it, wraps around it. Pea tendrils are extremely sensitive to touch: After a brief rubbing with a light-

weight thread, tendrils gradually coil toward the stimulated side, even if a physical object is no longer present.

Pollen tubes of flowering plants are suspected of displaying positive **chemotropism**, growing along the style to the ovary by following a gradient of chemical released from the ovule, probably from the synergids. However, this is still uncertain: The chemical responsible is not known, and pollen tubes would have to be sensitive to extremely slight variations in its concentration.

Tropic responses often involve a change in direction of growth. Tendrils grow in a slow spiral, called circumnutation, until they touch an object, and then they change to oriented growth. Changing direction involves **differential growth**: one side of the tendril must grow more than the other. This can occur by (1) growth on the contact side slowing, (2) growth on the opposite side accelerating, or (3) both. After a tendril starts differential growth, it continues growing differentially until it is mature a few hours or days later. In contrast, if a plant is tilted, its shoots begin differential growth until they are again oriented properly, and then they grow straight ahead with equal growth on all sided, perhaps for years.

(a)

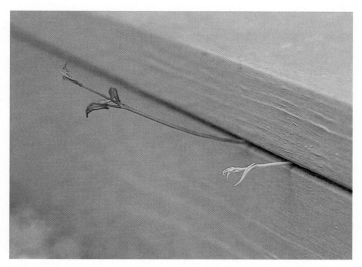

(b)

Figure 14-7 (a) Construction of porch steps trapped these plants, but by growing toward light (positive phototropism), their shoot tips found spaces between boards and emerged from a suboptimal dark environment into a more suitable sunny environment. (b) Sunflowers (*Helianthus annuus*) have an unusual type of phototropism: Their "flower" (it is really a set of many small flowers) follows the sun all day long, so they turn from east to west every day for several weeks until finally the stalk becomes too tough and woody to move. Notice that every sunflower is facing the same direction, almost directly toward the sun.

Figure 14-8 The central shoot of this tree grows vertically, probably due to both positive phototropism and negative gravitropism, but the branches grow at an angle to vertical, plagiotropically.

Nastic Responses

A **nastic response** is a stereotyped nongrowth response that is not oriented with regard to the stimulus. For example, the trapleaf of a Venus' flytrap has six large, sensitive trichomes. If a fly or other insect touches any two of these, the trap closes. It does not matter if the fly was moving north or south or up or down, the trap always closes in the same manner in this thigmonastic response. Furthermore, the trap does not grow shut; it closes as motor cells on the midrib upper side suddenly lose turgor. "Positive" and "negative" are not used because the response is not oriented with regard to the stimulus.

Many pollinators are active only at night or during the day, and the flowers they pollinate are open only at the appropriate time. Sepals and petals spread open when the sun rises in **diurnal** species (active during daylight) and as it sets in **nocturnal** ones (active at night). Although presence of the pollinator is the critical factor for pollination, the cue that stimulates flower opening is the presence or absence of light, not the presence or absence of pollinators. The opening and closing always happen in the same manner, even if light is given artificially from the west, north, or south, above, or below. This response is **photonastic**.

Nastic movements are based on changes of turgor pressure rather than growth; thus, movements can be repeated. This occurs in many leaves that undergo **sleep movements** with the blade elevated in the day, lowered at night. Such leaves often have an enlarged area, a pulvinus (plural, pulvini) at their petiole base, and it contains motor cells that cycle between being turgid and flaccid.

Morphogenic Responses

A **morphogenic response**, sometimes called a **morphogenetic response**, causes a change in the "quality" of the plant; that is, a fundamental change occurs in the metabolism of a tissue or even the entire plant. Because day length is such an excellent indicator of season, photomorphogenic responses are numerous: the induction to form flowers (which later open photonastically; **Figure 14-9**),

Figure 14-9 Conversion from the vegetative to the floral condition is a common photomorphogenic response. Day length is controlled in these commercial greenhouses to ensure that all the poinsettias bloom simultaneously at Christmas. The growers could just as easily make them bloom on the 4th of July by controlling day and night length.

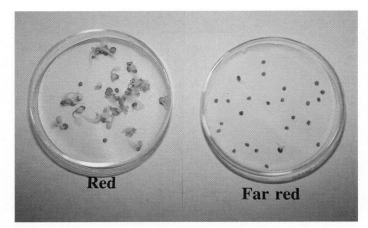

Figure 14-10 In these seeds and many others, a two-part mechanism ends dormancy: Cold winter temperatures and rain destroy or wash out an inhibitory chemical, and then light triggers germination. The light-dependent mechanism ensures that the seeds do not germinate while deeply buried under leaf litter and soil. Red light induces germination, but far-red light (infrared) blocks germination.

the induction of dormant seeds to germinate (**Figure 14-10**), the induction of buds to become dormant. An example of a gravimorphogenic response is the formation of fibrous wood when a stem or branch is tilted and becomes stressed by gravity. Thigmomorphogenic responses include formation of extra bark where branches rub against an object and formation of a suture when petal or carpel primordia grow against each other (see Figure 14-5).

Taxis

Taxis is a response in which a cell swims toward (positive taxis) or away from (negative taxis) a stimulus. Even in plants like mosses, ferns, cycads, and maidenhair tree (*Ginkgo*), sperm cells swim to egg cells by following a chemical gradient (chemotaxis). In algae, chemotaxis is similarly important for reproduction, and in many species, phototaxis allows them to swim toward light for photosynthesis or away from light that is too intense.

Communication Within the Plant

Perception and Transduction

Many, possibly most, responses occur in tissues or organs different from those that sense the stimuli. The site of perception is not the site of response, so a form of communication must exist. In plants, most sites of perception and response are not specialized for those functions but seem to be rather ordinary cells. Day length is probably perceived by all living leaf cells; no specialized region of cells has been discovered. Low temperatures for **vernalization** appear to be detected by buds, which do not contain a particular group of cells specialized just for temperature perception. In root caps, certain cells called **statocytes** do have large starch granules, **statoliths**, that sink in response to gravity; statoliths are too dense to float in cytoplasm and always settle to the bottom of the cell, thereby distinguishing "down" from "up" (**Figure 14-11**). This is our best example of a set of specialized perceptive cells. The trigger hairs on Venus' flytrap leaves are also a

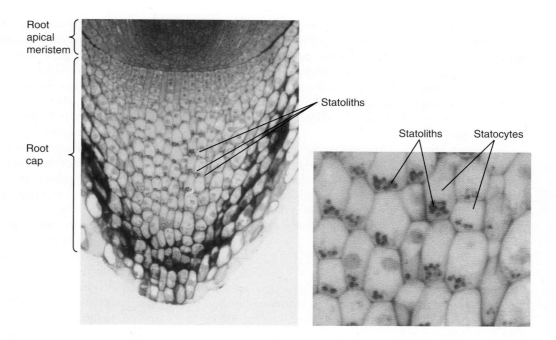

Figure 14-11 Cells located centrally in this root cap are statocytes, and their starch grains are statoliths. Regardless of vertical or horizontal position, the statoliths are located at the gravitationally lower side. It is necessary to distinguish between the gravitational and the morphological bottom in gravity-sensing systems.

Root apical meristem

Root cap

Statoliths

Statoliths   Statocytes

discrete perceptive mechanism, but it is not known which cells within the hairs are responsible.

The site of perception is tentatively assumed to be the site of transduction, where the stimulus is converted into a form that can be transmitted and can trigger a reaction at a response site (see Figure 14-2). Transduction is still a complete mystery in almost all plant responses; we do not know how changes in temperature, light, weight, or humidity are converted into chemical signals.

Two factors are important in perception and transduction: presentation time and threshold. **Presentation time** is the length of time the stimulus must be present for the perceptive cells to react and complete transduction. Presentation time for root gravitropism is easy to understand: A root must lie on its side long enough for statoliths to sink to the new bottom of the cell. If the root is returned to vertical before they can settle, no perception occurs. In many tropic responses, only a brief touch or unilateral lighting is sufficient to cause curvature; presentation times are often only a few seconds.

After the stimulus has acted long enough to fulfill the presentation time, a response occurs even if the stimulus is removed. For example, the vernalization of many biennial plants has a presentation time of only one or a few days; after this, the plants still flower at the proper time even if kept in warm, nonvernalizing conditions. Tendrils of peas do not bend thigmotropically in the dark, but if they are rubbed for several seconds—their presentation time—in the dark, they bend when placed in light even though they are no longer being touched.

Threshold refers to the level of stimulus that must be present during the presentation time to cause perception and transduction. In phototropism, plants are extremely sensitive to very dim unilateral light if they are in an otherwise dark environment; the threshold for curvature is low. In bright conditions, the threshold is higher, and the unilateral light must be much stronger to trigger curvature. In Venus' flytraps, the threshold for stimulating leaf closure is moderate: The trigger hairs must be firmly bent. This is advantageous in preventing wind or rain from triggering

trap closure; the moderate threshold almost guarantees that the trap contains an insect every time it closes (**Figure 14-12**).

Related to threshold is the level of **response** relative to the level of stimulation; the alternatives are all-or-none responses and dosage-dependent responses. In an **all-or-none response**, after the threshold and presentation time requirements are met, the stimulus is no longer important; the response is now completely internal. Individuals respond identically whether they received strong

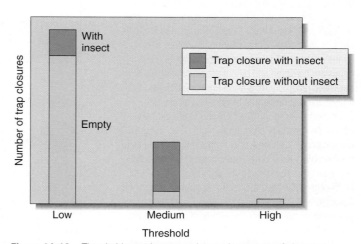

Figure 14-12 Threshold must be appropriate to the amount of change a stimulus can cause. If Venus' flytrap hairs had an extremely low (sensitive) threshold, they would be capable of detecting and catching every insect, but wind and rain would cause so many useless closings that the leaf would be inefficient because it would miss insects whenever it was closed unnecessarily. With a medium threshold, it captures more insects because it is open and ready much of the time. Small insects may escape because they cannot bend the trigger hairs enough to meet the threshold. If the threshold were too high, no insects would be caught because none could bend the hairs. The trap would close only when larger animals brushed against it, but these animals are too big to be enclosed in the trap; thus, all closures would be unproductive. We can hypothesize that natural selection results in a threshold appropriate for the most abundant size of insects.

or weak stimuli, regardless of whether the stimulus was brief or long lasting. For example, many species are induced to flower by environmental conditions; after the minimal threshold and presentation time requirements are met, the plants flower fully, limited only by their general health, vigor, and nutrient reserves. Until they receive the proper stimulus, they produce no flowers; their flowering is all or none. Examples are poinsettia, chrysanthemum, *Hibiscus syriacus,* and oats.

In **dosage-dependent responses**, the amount or duration of the stimulus affects the amount or duration of the response. In species of this type, individuals that receive only minimum stimulation flower poorly, even if the plant is quite healthy. Those that receive longer or stronger stimulation produce more flowers. Examples are turnip, marijuana (*Cannabis sativa*), and some varieties of cotton and potato.

■ Chemical Messengers

Almost all plant communication is by a slow mechanism: transport of hormones through the plant. **Hormones** are organic chemicals produced in one part of a plant and then transported to other parts, where they initiate a response. A critical aspect is that hormones act at very low concentrations. Hormones are synthesized or stored in regions of transduction and are released for transport through either phloem or mesophyll and cortex cells when the

appropriate stimulus occurs. At the site of response, hormones bind to receptor molecules, usually located in the plasma membrane, and thereby trigger a response. Hormones appear to be released into general circulation and are not carried specifically to the target. Many regions that are not target regions are exposed to the hormone but do not respond because they do not have the proper receptor molecules. In some instances, a plant hormone acts directly on the cells that produce it.

At one time, plant hormones were believed to carry in their structure much of the information necessary for the response. We now know that plant hormones are quite simple in structure. The receptor cell and its nucleus contain almost all of the information necessary for proper response, and hormones serve only to activate the response. An analogy is a computer, its programs, and commands. The computer is capable of carrying out numerous functions and processes, but only if properly controlled; the same is true of cells, tissues, organs, and entire plants. Computer programs contain the information needed to run the computer, just as the nuclear, plastid, and mitochondrial genes contain the information needed to run cells. Both computer programs and genomes contain numerous sets of information. On a computer, commands such as OPEN, PRINT, SAVE, and COPY select subroutines or programs. Hormones are thought to act as commands that activate programs within the target cells (**Figures 14-13a** and **14-13b**).

Figure 14-13 A cell responds to a hormone only if it has receptors for that hormone. (a) Evidence suggests that some receptors (R) are in the plasma membrane and that others are in membranes such as the endoplasmic reticulum. Once bound, the hormone-receptor complex (h-R) may cause a metabolic change immediately, or the complex may migrate to another site, such as the nucleus. (b) In many responses, some nuclear genes are activated and others are repressed. Cells may have receptors for several hormones (R_A and R_B); if hormone A is present, it binds and activates (or represses) program A. Other programs are unchanged. One of the results of program A might be to withdraw the receptors of either A or B from the membrane or to add receptors for C or D, thus changing the sensitivity and type of response possible. A second cell (the cell on the right) may have a different program (A program 2) activated by hormone A; the response is cell specific, not just hormone specific. (c) The effects of a hormone are often quite different when the hormone is applied alone or with or after a second hormone. In this hypothetical cell, development occurs only if hormones A and B are applied simultaneously or if hormone D is applied after hormone C.

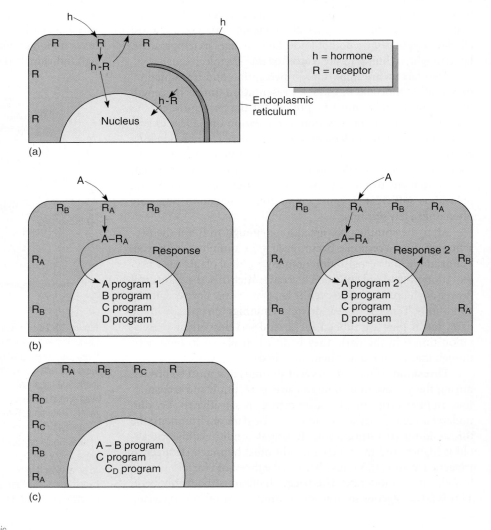

In higher animals, because so many systems and responses must be activated, many distinct hormones are necessary. Because plants are much simpler, their responses can be controlled by fewer hormones; even so, the handful of known plant hormones seems inadequate. It is likely that many plant hormones are still unknown to us. Also, many responses are activated not by one hormone but by a combination or a sequence of several hormones (Figure 14-13c), or a particular hormone elicits different responses when present at different concentrations. The following are examples of the most well-studied hormones.

Auxins   The first plant hormone discovered was **auxin**. In 1926, it was identified as the chemical messenger involved in positive phototropism in oat seedlings. Identifying it chemically at that time was impossible because it is present in such low concentrations. Experiments had to be performed by allowing auxin to diffuse out of a seedling leaf tip into a small block of agar, which was then used as if it were a small dose of auxin (see Figures 14-25g and 14-25h). The auxin was later identified as being **indoleacetic acid** (**IAA**), which could be synthesized artificially and applied to plants under various conditions to find other responses that IAA might either mediate or inhibit (**Figure 14-14**). The search was successful—dozens of responses were found (**Table 14-3**). Many compounds chemically related to IAA were also found to be as effective or even more so. It was hypothesized that IAA was only one of many natural auxins, each with its own effect and role, but further findings were not consistent with that hypothesis. Analysis of IAA metabolism showed that the compounds were converted to IAA by the plant's enzymes.

TABLE 14-3	Examples of Responses Involving Hormones
Hormone	**Response**
Auxin	Abscission suppression; apical dominance; cell elongation; formation of roots in cuttings; fruit maturation; tropisms; xylem differentiation
Cytokinin	Bud activation; cell division; fruit and embryo development; mimics the effects of phytochrome and red light in several cases; prevents leaf senescence
Gibberellin	Converts some plants from juvenile to adult condition; converts other species from adult back to juvenile condition; involved in flowering; releases some seeds and buds from dormancy; stem elongation; stimulates pollen tube growth
Abscisic acid	Initiation of dormancy; resistance to stress conditions; stimulation of growth at very low doses; stomatal closure; probably not involved in abscission
Ethylene	Formation of aerenchyma in submerged roots and stems; fruit ripening and abscission; initiation of root hairs; latex production

Many synthetic compounds mimic the effect of auxin or the other hormones; for clarity, only natural products are called hormones. The term "**plant growth substance**" is used for any hormone-like compound, natural or artificial (Figure 14-14). Naphthaleneacetic acid, an artificial compound, produces effects in plants that are for the most part indistinguishable from those of IAA. 2,4-Dichlorophenoxyacetic acid (2,4-D) is auxin like but so powerful that it disrupts most normal growth and development even in low concentrations, making it valuable as an herbicide.

IAA is related to the amino acid tryptophan; so far, at least four separate metabolic pathways are known that convert tryptophan to IAA. Some plants, corn, for example, synthesize IAA by different pathways at different stages of development, and each pathway has its own characteristic set of controls. The most active centers of auxin synthesis are shoot apical meristems, young leaves, and fruits. IAA is present in root tips, but is believed to be transported there from the shoot rather than being synthesized there.

The concentration of substances as powerful as hormones can be controlled not only by synthesis but also by destruction and by conversion to an inert storage form. Two pathways for IAA destruction are known: removal of the side group and oxidation of the five-member ring. IAA is converted to an inactive form by **conjugating** (attaching) it to various compounds. In the conjugated form, IAA is safe from destruction; it can be stored indefinitely in seeds, and it can be transported from cotyledons to the epicotyl during germination. Conjugation allows rapid regulation of the level of free IAA. Up to 80% of the IAA in oat seeds is conjugated, but this can be deconjugated, releasing free IAA during germination more quickly than synthesis could.

In addition to hormone transport by phloem, a second mechanism exists for auxin only: **polar transport**. In shoots and leaves, auxin moves basipetally—from the apex to the base of the plant, and in roots it moves acropetally toward the root apex. Movement is about 11 mm/hr regardless of whether the tissue is in a vertical, horizontal, or upside-down orientation. By means of the polar transport system, auxin movement through the plant can be maintained independently of the variation in phloem transport caused by changing sinks and sources for carbohydrates and minerals.

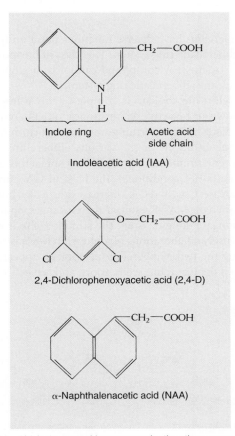

Figure 14-14   IAA is the natural hormone auxin; the others are synthetic. They are called plant growth substances.

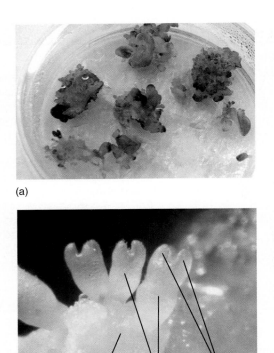

(a)

(b)
Callus Embryos Cotyledons

Figure 14-15 Cells of most dicots, such as this tobacco, can be grown in culture if provided with auxin, cytokinin, some vitamins, minerals, and sugar. The ratio of auxin to cytokinin is important, as are the absolute concentrations. (a) At one ratio, the cells proliferate as a callus composed of parenchyma. (b) At another ratio of auxin to cytokinin, buds form in the callus and then grow into embryos.

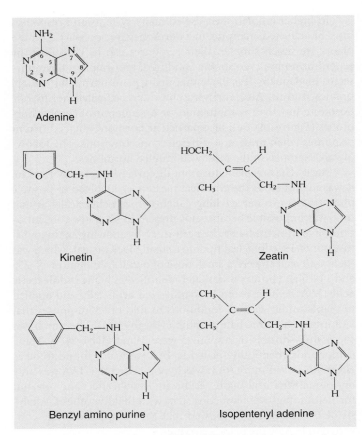

Adenine

Kinetin   Zeatin

Benzyl amino purine   Isopentenyl adenine

Figure 14-16 Both natural and artificial cytokinins are related chemically to adenine. The size and chemical nature of the group on C_6 is critical.

Cytokinins **Cytokinins** were named for the fact that their addition to a tissue culture medium containing auxin and sugar stimulates cell division—cytokinesis (**Figure 14-15**). The first one discovered, kinetin, is an artificial cytokinin; however, two natural ones, zeatin and isopentenyl adenine, have been found, and more are suspected to exist (**Figure 14-16**). Cytokinins are purines, related to adenine; extensive testing of adenine analogues has been done to determine which aspects of its chemical structure are critical to the molecule's ability to act as a cytokinin. The most active compounds have a side group containing four to six carbon atoms attached to C_6. If this side group is longer or if complex groups occur at other areas, the molecule does not act like a cytokinin, apparently lacking the proper shape and charge to bind with the cytokinin receptor molecule.

Like auxin, cytokinins are involved in dozens of responses in all parts of the plant (Table 14-3). One important response is root-shoot coordination. As roots begin to grow actively in the spring, they produce large amounts of cytokinins that are transported to the shoot, where they cause dormant buds to become active and expand. The richest concentrations of cytokinins generally occur in endosperm and are apparently involved in controlling the development and morphogenesis of the embryo and seed.

Gibberellins At least 125 **gibberellins** are known, and rather than being named, they are just numbered: GA_1, GA_2, . . . GA_{125} (**Figure 14-17**). GA_3 has the name gibberellic acid, and at first it received the greatest amount of study because it could be obtained easily from fungi and used in experiments; however, it now appears that GA_1 and GA_{19} are more active and important in

plants. Gibberellins have diverse functions but a unifying structure, the gibberellane ring system. This class of hormones is thus defined by structure: A compound cannot be a gibberellin if it does not have the gibberellane ring system.

Gibberellin metabolism is complex. Only a few gibberellins are known to be active as hormones; others are precursors or intermediates in transforming one active form into another. Relative concentrations of the various gibberellins change in response to environmental signals. When spinach is exposed to long days (summer conditions), the level of GA_{19} undergoes a fivefold decrease; GA_{20} and GA_{29} increase drastically, but GA_{17} and GA_{44} do not change. As a result of these changes, spinach stems begin to elongate: long days are the stimulus; shoot elongation is the response; and alterations of gibberellin levels is the mechanism linking the two. Gibberellin metabolism appears to occur in all parts of the plant, but seeds, roots, and leaves are especially

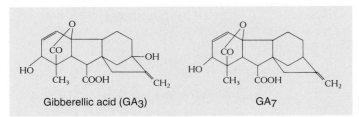

Gibberellic acid (GA3)   GA7

Figure 14-17 All gibberellins are based on gibberellane; the most common forms are GA_3 and GA_7. Not all of the 125 gibberellins occur in plants; at least 15 have been found only in fungi.

important. Gibberellin transport is studied by giving plants synthetic, radioactive gibberellins and then tracking the movement of the radioactivity. Both xylem and phloem sap become radioactive, and it is believed that the large amounts of gibberellins synthesized by active root tips are loaded into xylem. Gibberellin movement from leaves and seeds appears to correlate more closely to phloem sap dynamics. Like all other hormones, gibberellins are involved in numerous responses (**Figure 14-18**).

Abscisic Acid This class apparently contains the single compound **abscisic acid (ABA)** (**Figure 14-19**). As its name suggests, it was thought to play a role in the abscission of fruits, leaves, and flowers, but currently, we do not know whether that is true of many species or only sycamore, the species in which it was discovered. ABA is widely regarded as a growth inhibitor, possibly involved in inducing dormancy in buds and seeds; however, dormancy is not just inhibition of growth, but a complicated set of changes that prepare the plant or seed for adverse conditions.

ABA appears to be especially involved in many types of stress resistance. Heating leaves, waterlogging roots, chilling, and high salinity have all been found to cause sudden increases in ABA. If healthy plants are pretreated with ABA, they become much more resistant to stressful conditions. When plants begin to wilt, the concentration of ABA in leaf cells increases dramatically from about 20 μg/kg fresh weight to 500 μg/kg, and guard cells close stomatal pores. Wilt-induced production of ABA overrides all

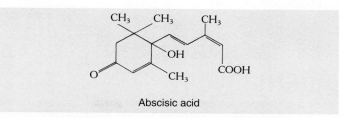

Abscisic acid

Figure 14-19 ABA is transported rapidly between cells and through phloem; therefore, its presence in a tissue is not proof that it was produced there. ABA can be synthesized from mevalonic acid in roots, stems, leaves, fruits, and seeds of various species.

other stomatal controls; other mechanisms that normally cause stomata to open are ineffective when ABA is present. The stimulus for ABA production appears to be soil dryness, which is detected by roots. When soil dries to some critical level, ABA production is dramatically increased and the ABA moves upward through the xylem to the leaves, causing stomatal closure.

ABA can be removed by being converted to phaseic acid, which has no known hormonal activity.

Ethylene **Ethylene** is the only gaseous plant hormone, and it has the simplest structure (**Figure 14-20**). Its most commonly studied effects occur during fruit development. Fruits such as apple,

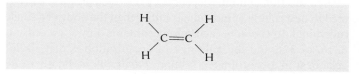

(a)

(b)

Figure 14-20 (a) Ethylene is a simple, small molecule. Many of its effects are blocked by carbon dioxide, whose size and shape are similar enough to those of ethylene that it can bind to ethylene's receptor and block normal response. (b) Bananas produce large amounts of ethylene as they ripen. This diffuses away and stimulates other types of fruits (called climacteric fruits) to ripen rapidly. People often place a single banana, or even just a banana peel, into a plastic bag with other fruits: The plastic traps the ethylene, which then causes the other fruits to ripen faster and more fully. Similarly, a bruised apple produces large amounts of ethylene, which can cause other apples to over-ripen, thus, the old saying "One bad apple spoils the barrel."

Figure 14-18 Many biennial plants grow as a rosette in their first year and then elongate rapidly (bolt) in their second year, producing a tall shoot that bears numerous flowers. The bolting is controlled by gibberellic acid; if it is absent, the plants remain short. The plant shown here, *Arabidopsis thaliana,* completes this cycle in just a few months, not two years: Notice that all its large leaves are arranged in a rosette, then after it was several months old, it bolted and produced the flowering stalks with long internodes.

Box 14-1   Simple Bodies and Simple Development in Algae

Angiosperms have complex bodies, which require them to control many aspects of development, but there are alternatives. Green algae have much simpler bodies. **Unicellular algae** such as *Chlamydomonas* have bodies consisting of just a single cell. No middle lamella is formed during cell division; therefore, nothing holds the two daughter cells together and they just swim away from each other (they have flagella). Each individual has a body (the single cell) but no tissues or organs. They develop polarity with a front end that faces forward as they swim, and a rear end.

Colonial green algae are slightly more complex. These cells also lack a middle lamella but instead produce a gelatinous material that holds cells near each other: As a zygote divides and multiplies, all daughter cells remain loosely bound together. There is so little interconnection and interdependency that each group is considered to be a colony of individuals rather than the body of a single individual. In *Gonium,* each colony contains only a few cells (4, 8, 16, or 32), and the only sign of organization is that all flagella beat in a coordinated fashion. *Pandorina* is slightly more derived because it shows a trace of differentiation; the colony swims in one direction, and anterior cells are slightly different from posterior ones. *Volvox* colonies contain up to 50,000 *Chlamydomonas*-like cells and are easily visible without a microscope (**Figure B14-1a**). Slight differentiation exists in that up to 50 cells in the posterior half of a colony are specialized for reproduction only.

In green algae that do produce a middle lamella, daughter cells remain bound together and form a single integrated individual with a multicellular body. After the zygote of *Ulothrix* divides,

one of the two daughter cells becomes a colorless adhesive "holdfast" cell that glues itself to a stone or seashell. The other daughter cell produces no adhesive, and its plastids develop into chloroplasts: Already the body has polarity, differentiation of two cell types, and can detect which cell is in contact with a solid substrate. The chlorophyllous cell divides repeatedly but always with crosswalls parallel to each other, and thus, the body grows as a radially symmetrical **filamentous body** one cell thick but many cells long. This requires morphogenetic mechanisms that precisely control the orientation of cell division. In some filamentous species, a few cells undergo a longitudinal division, and then one of the two daughter cells grows out as a filamentous "branch." **Laminar bodies**, such as those of *Ulva* (sea lettuce; **Figure B14-1b**), are slightly more complex: A cell settles down and begins to grow into a *Ulothrix*-like filament, but its cells divide in two directions and thus form a sheet. Then all cells divide once in a third direction; as a result, the sheet becomes two layers thick. Both filamentous and laminar bodies remain simple with regard to cell differentiation, having only holdfast cells and chlorophyllous cells. Also, their bodies are soft and flexible, being supported and moved by water, so they do not have phototropism or gravitropism.

Several groups of green algae have **parenchymatous bodies**. *Chara* has a stem-like body up to 1 m long, divided into "nodes" and "internodes," with whorls of branches arising in a precise pattern at internodes. The body is several cells thick, with large multinucleate cells in the center of each internode and smaller uninucleate cells surrounding them as well as composing the nodes and branches. All constitute a true parenchyma tissue

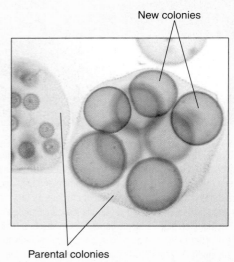

Figure B14-1a   Colonies of *Volvox*; all cells are almost identical; a few are specialized for reproduction. Little differentiation occurs.

Figure B14-1b   Individuals of the laminar alga *Ulva* are two cells thick. The holdfast is not visible here.

because they are derived from an apical meristem that contains a prominent apical cell that divides in all three planes. Reproduction in *Chara* is significant. Its sperm cells are produced in a truly multicellular structure (a gametangium) whose outer cells are sterile; only the inner cells convert to sperm cells (**Figure B14-1c**). At maturity, the outer cells separate slightly and the motile sperm cells swim away. The egg is formed as the terminal cell of a short filament three cells long, but the subterminal cell subdivides, and those cells grow upward and surround the egg, differentiating into a variety of specialized cells. After fertilization, the sterile cells surrounding the fertilized egg deposit thickenings on their inner walls, those adjacent to the zygote. The resting structure thus consists not only of a thick-walled zygote but also protective sterile cells. Both sperms and eggs are produced in gametangia consisting of two types of tissue (fertile and sterile): Reproduction requires the integrated functioning of several types of cells. This is an organ-level of differentiation. Furthermore, germination of the zygote is reported to be promoted by cold temperatures and by red light; development responds to environmental cues.

Although all algae are much simpler than true plants, all undergo at least some development and morphogenesis, and *Chara* and its relatives have numerous sophisticated developmental metabolisms.

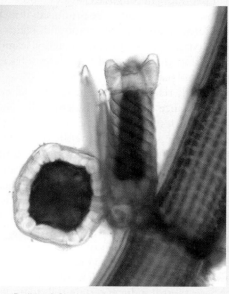

Figure B14-1c Bodies of *Chara* are several cells thick and are true parenchyma, formed by an apical meristem. Bright orange round structures are about to release sperm cells, yellow, elongate structures contain eggs; each of these gametangia consists of some fertile cells surrounded by several types of sterile cells. Individuals of *Chara* undergo considerable differentiation.

avocado, banana, mango, and tomato are **climacteric fruits**: They ripen slowly as they mature, but in the final stages, numerous developmental changes occur rapidly. Starches are converted to sugars. Cell walls break down and soften. Flavors and aromas develop, and color changes. Ethylene stimulates these changes, but at first, so little ethylene is present that the changes occur slowly; however, one effect of ethylene in these fruits is the production of more ethylene, which constitutes a **positive feedback system**: The concentration increases exponentially and rapidly. The sudden burst of ethylene and the rapid completion of maturation of the fruit are known as a **climacteric**; at this time, ethylene production can be as high as 320 nL/g/hr (one nL = one nanoliter = one billionth of a liter). In nonclimacteric fruits, such as cherry, lemon, and orange, ethylene does not stimulate its own production; therefore, ethylene levels remain stable and no sudden change occurs just before maturity.

Because ethylene controls the ripening of most of our important food fruits, it is important commercially. It is used in harvesting cherries, cotton, and walnuts by causing their uniform abscission; it also synchronizes flowering and fruiting in pineapple, making harvesting easier. Ethylene can be drawn out of

unripe fruits by storing or transporting them in a partial vacuum. When they reach market, air pressure is returned to normal; ethylene accumulates, and ripening occurs. Fruits may also be treated with 2-chloro-ethylphosphonic acid (commercial trade name Ethrel), which breaks down and releases ethylene.

Being a gas, ethylene moves rapidly through tissues by diffusion rather than by specific transport mechanisms. In many cases, it acts as a final effector for auxin. Arrival of auxin at a target site often causes that site to produce ethylene, which diffuses rapidly and triggers responses in the adjacent area more quickly than the auxin itself could. Being a gas also allows ethylene to escape from plants through stomata and lenticels, which makes it a means of detecting flooding. When rivers flood, ethylene builds up in submerged stems and induces the formation of aerenchyma in certain species.

Other Hormones Several other compounds show hormone-like activity but are either not present in all plants or are much less well studied.

Brassinosteroids are complex chemicals involved in leaf morphogenesis, root and stem growth, and vascular differentiation. If plants of *Arabidopsis thaliana* are artificially mutated to disrupt brassinosteroid synthesis, the plants develop abnormally but can be rescued by applying small amounts of brassinosteroids.

Jasmonic acid is involved in defense against animals and fungi. When animals chew into plants, membrane lipids are released, and one of them, linolenic acid, is converted to jasmonic acid. This travels to healthy cells and induces them to activate the genes necessary to synthesize alkaloids and chemicals that inhibit protein-digesting enzymes; both of these interfere with an animal's ability to eat and digest. When fungi attack plants, they secrete small peptides and oligosaccharides, and several of these somehow cause jasmonic acid to be produced; it then stimulates plant cells to form an antifungus compound. Plants can attach a methyl group to jasmonic acid, creating methyl jasmonate, which diffuses away from the plant and stimulates adjacent plants to become resistant to pathogens.

Salicylic acid, related to aspirin (acetyl salicylic acid), is involved in resistance to pathogens, especially viruses. Artificial inoculation of leaves with viruses causes salicylic acid levels to rise, and these in turn activate disease resistance genes. The resistance mediated by salicylic acid is not confined to just the site where the viruses were applied; instead, all parts of the plant become resistant. This is known as systemic acquired resistance. As with methyl jasmonate, methyl salicylate is volatile and is carried by wind to other plants, alerting them to the presence of pathogens in the area. Both chemicals are part of a method of plants communicating with each other. Also, staminate flowers of voodoo lilies (*Sauromatum*) produce salicylic acid, which then travels downward through the thick inflorescence axis until it reaches the base; there it induces certain cells to undergo thermogenic respiration, heat up, and give off foul-smelling fragrances that attract pollinators.

Activation and Inhibition of Shoots by Auxin

Auxin is often described as a growth hormone, whereas ABA is considered an inhibitor; unfortunately, such characterizations are confusing. Hormones simply carry information about the status of a particular region, nothing more; whether the elicited response is inhibition or activation depends on the site of response. An example of the complexity is provided by shoot tips.

As shoot apical meristems grow and initiate the new cells of shoots and leaf primordia, they also produce the auxin, IAA. Young leaves are also a rich source of this hormone. No external signal must be perceived to initiate auxin production; instead, this is a means of integrating the plant during ordinary growth. Large quantities of auxin indicate to cells that shoots are elongating and producing new leaves. Although neither signal perception nor transduction occurs, transport takes place. In stems auxin undergoes basipetal, polar transport at a speed of 5 to 20 mm per hour, perhaps by means of molecular pumps in plasma membranes. This downward flow of auxin surrounds all stem cells, and at least three cell types are set to respond to it, each response unique to the particular cell type.

Cell Elongation

In cells of the young internodes just below the apical meristem, auxin triggers cell elongation. When IAA contacts these

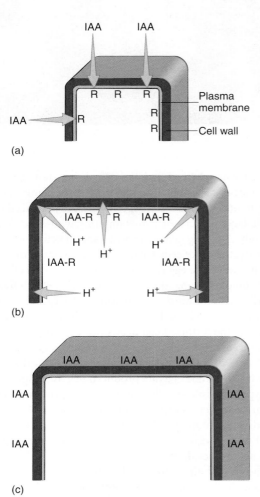

(a)

(b)

(c)

Figure 14-21   Cells in the subapical region have auxin receptors. If the apex is growing and producing auxin, IAA is present to bind to the receptors (a). After stimulation by auxin binding, the plasma membrane pumps protons from the cytoplasm into the wall (b), weakening it and allowing turgor pressure to stretch it. (c) Cell elongation stops once the maximum cell size is reached, and adding more auxin does not cause any more elongation; perhaps the receptors have been removed from the membrane.

responsive cells, which are prepared for growth, it binds to a receptor, thought to be a small protein called *ABP1* (*Auxin Binding Protein 1*). The cells begin to transport protons actively out across the plasma membrane (**Figure 14-21**). This has the effect of acidifying the cell wall. The protons break some of the chemical bonds that hold one cellulose microfibril to another and activate enzymes that weaken other bonds so that the wall becomes weaker. If the protoplast is turgid and pressing against the wall, it exerts enough pressure to stretch the weakened wall and growth results. Immature cells neither excrete protons nor grow if auxin is lacking. At lower internodes, fully grown, mature cells apparently lack the proper auxin receptors because auxin does not cause them to extrude protons or grow.

Apical Dominance

The second site of response to apically produced auxin is the buds located in leaf axils; their response is not cell elongation

(a)

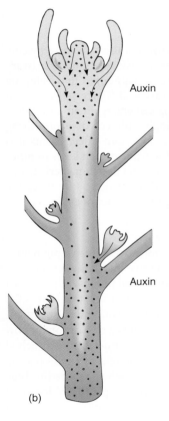

(b)

Figure 14-22 (a) This twig of viburnum was pruned 2 months before the photo was taken. The axillary buds had been completely inhibited by the shoot apical meristem at the time of pruning, but then both became active once pruning stopped the flow of auxin from the shoot apex. (b) The concentration of auxin is greatest at the shoot apex and less at lower levels; at sites where it drops below the threshold level, it can no longer suppress axillary buds, and one or several become active. Other factors must be involved because it is not always the lowest bud that becomes active first, even though it should be the first to encounter sufficiently low auxin concentrations.

but rather inhibition of growth. Apically produced auxin induces dormancy in these axillary buds, the result being that each shoot tip has only one active apical meristem, a phenomenon called **apical dominance** (Figure 14-22). This is a threshold response: As the terminal shoot apical meristem grows away, the concentration of auxin around an axillary bud gradually decreases until at some point it drops below the threshold. Inhibition cannot be maintained, and the axillary bud becomes active and grows out as a branch or flower. As the axillary bud grows, it produces auxin but does not inhibit itself, although it does inhibit all its own newly formed axillary buds.

Differentiation of Vascular Tissues

The third site of response to auxin produced in shoot tips is the vascular cambium; the response is cell division and morphogenesis. In springtime, as air temperatures rise and buds become active, their auxin moves basipetally, activating the dormant vascular cambium. Auxin not only stimulates cambial cells to begin mitosis and cytokinesis but also causes new daughter cells to differentiate into xylem cells. If an apical meristem is destroyed, by insects or a late frost for example, the basipetal flow of auxin stops, vascular differentiation is interrupted, internode elongation ceases, and apical dominance is broken. Some axillary buds, now free of apical dominance, becomes active and re-establishes a flow of auxin that maintains the vascular cambium and any other cells that depend on it.

Three separate target tissues give three distinct responses, not because there are three separate chemical messengers—there is only auxin—but because part of their previous differentiation was preparation to respond to auxin in a particular way. It is important to realize that auxin carries no information except that the shoot apex is healthy and active. Each target site must have receptor molecules that interact with IAA and therefore detect its presence or absence. It is not known whether all three targets have the same receptors or whether each has a unique type, nor is it known how the interaction of IAA with the receptor triggers the response.

Interactions of Hormones in Shoots

In some species, apical dominance may involve only the presence or absence of auxin; in others, there is an interplay of two or three hormones. Active roots synthesize cytokinins that are transported to the shoot and stimulate axillary buds. Whether buds become active or remain dormant depends on the relative amounts of the two hormones. If a plant is growing vigorously, its roots are active and cytokinin levels are high; many buds at a distance from a shoot apical meristem have a low auxin/high cytokinin ratio and become active. Such a mechanism is adaptive because if a plant is growing well, activating more dormant buds increases the rate of new leaf production. The role of ABA in apical dominance is uncertain. It is present in quiescent buds but does not decrease either just before or as buds are becoming active and growing out.

Apical dominance in prickly pear cacti (*Opuntia polyacantha*) is more elaborate. The spine clusters are short shoots, and the spines are highly modified leaves. If the spine cluster is excised and placed in a tissue culture with cytokinin, the dormant short shoot apical meristem grows out as a long shoot—a new "pad" similar to a normal branch. If the culture medium contains gibberellins instead of cytokinin, the short shoot apical meristem produces more spines— it acts as a rejuvenated short shoot.

A second interaction of hormones occurs in the vascular cambium. Auxin alone activates the cambium and elicits differentiation of xylem, but gibberellin is also present in a healthy stem and causes some of the new cells to differentiate as phloem. Without the interaction of both auxin and gibberellin, a normal, functional vascular system would not develop.

Hormones as Signals of Environmental Factors

Leaf Abscission

Whereas normal growth of shoots and roots results in large flows of auxin and cytokinin, respectively, environmental factors also influence hormone concentrations. Hormones communicate to various parts of a plant the information that a particular part has encountered an environmental change. Export of ABA by wilted leaves has been mentioned, and another example involves abscission of leaves and fruits. A young leaf produces large amounts of auxin, but production falls to a low but steady level in a mature leaf. As long as auxin flows out through the petiole, activity in the abscission zone is inhibited (**Figure 14-23**). If the leaf is damaged by animal feeding or water stress, auxin production drops to such a low level that its flow through the petiole does not keep the abscission zone quiescent. Perception and transduction in this case may be simply that insect or wilt damage makes it impossible for the impaired cells to produce enough auxin to inhibit the abscission. Old age of the leaf may also result in the lack of sufficient auxin, but evidence suggests that autumn conditions stimulate production of ethylene, which then suppresses auxin production and transport in time for abscission before winter.

Fruits are prevented from abscising prematurely by the presence and export of sufficient amounts of auxin through the pedi-

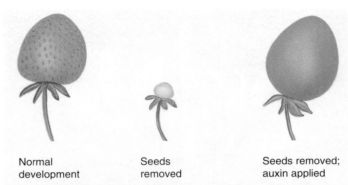

Figure 14-24 Auxin's effects on fruit ripening are often studied in the strawberry because the tiny true fruits (achenes) are located on the outside of the large red false fruit. The achenes are sources of auxin necessary for development of the false fruit. If they are removed, no development occurs unless auxin is applied experimentally.

cel. Fruit ripening is under the control of both auxin and ethylene, at least in edible, fleshy fruits. Initial transformation of the ovary wall into a fruit is a response to auxin synthesized in developing embryos and transported to the ovary wall (**Figure 14-24**). Auxin stimulates many changes, including cell enlargement and differentiation; there is usually surprisingly little cell division during the formation of fruits, even large ones. One effect of auxin is release of ethylene by the developing fruits, which leads to other aspects of ripening in both climacteric and nonclimacteric fruits. At maturity, the high concentration of ethylene stimulates the pedicel abscission zone, overriding the presence of auxin.

Tropisms

Light is the stimulus in phototropism; therefore, a pigment is needed, and its absorption spectrum must match the action spectrum (just as the spectra of chlorophyll and photosynthesis must match each other). The action spectrum indicates that blue is the most effective wavelength for phototropism, and the pigment whose absorption spectrum matches that is a small protein, now called phototropin, with two flavin mononucleotides attached as chromophores. Phototropin is a component of a cell's plasma membrane and becomes phosphorylated when exposed to blue light. Blue light stimulates other aspects of plant development also (it inhibits hypocotyl elongation once a seedling has emerged from underground; it stimulates chlorophyll and carotenoid synthesis and causes opening of stomata at dawn), and together these are called **blue-light responses**.

Oat seedlings are often used to study phototropism. They are grown in the dark (surviving on their endosperm), and if illuminated briefly with blue light from one side, they bend toward it. Plants grown in darkness do not have chlorophyll so that pigment is not present to interfere with studies of absorption and action spectra or studies of presentation time and threshold amount of light needed. Oat seedlings have an outermost protective leaf called a **coleoptile** (pronounced coal ee OP tile—the p is pronounced), which shows a strong positive phototropic response; oat coleoptiles are the organ studied most often.

The site of perception of blue light is the tip of the coleoptile, and if it is covered with an opaque hood, light direction is

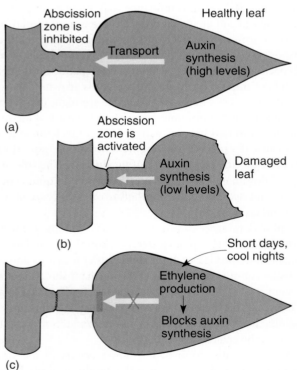

Figure 14-23 (a) A healthy leaf produces and transports enough auxin to suppress activity in the abscission zone. (b) A damaged leaf produces less auxin, insufficient to prevent abscission. (c) Autumn stimuli may cause the production of ethylene, which then suppresses auxin synthesis and transport. An alternative hypothesis postulates that autumn conditions directly suppress auxin production and ethylene is not necessary.

not detected. Similarly, if the coleoptile tip is cut off before unilateral illumination, the rest of the coleoptile does not perceive the light. We do not know the steps of transduction, but the result is an asymmetric redistribution of auxin: Auxin is transported from one side of the coleoptile to the other. Neither synthesis nor destruction of auxin is affected. After the differential redistribution of auxin is established, the darker side of the coleoptile, or the stem in other plants, receives extra auxin, and thus it grows more rapidly and bends toward the light. When the coleoptile or stem points directly at the light, neither side is brighter nor darker; as a result, differential auxin transport stops, and the organ grows straight ahead. The site of response, where differential growth causes bending, is approximately 5 mm below the coleoptile tip (**Figure 14-25**).

In positive gravitropism in roots, the root cap acts as the organ of perception (see Figure 14-11). After the lower side of the root is detected, a growth inhibitor is transported to that lower side of the root cap and then into the root, where it slows growth on the lower side. We know almost nothing about how the

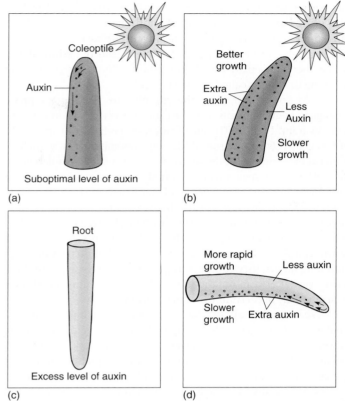

Figure 14-26   (a and b) Shoots may contain slightly too little auxin for fastest growth. Thus, redistribution in a coleoptile apex caused by unilateral light causes the side with more auxin to have a level that is more nearly optimal, whereas the other side, which receives less, has even poorer growth than before. (c and d) Roots may contain slightly too much auxin for fastest growth. In root positive gravitropism, the horizontal position of a root cap causes statoliths to fall to the side of statocytes, causing a downward redistribution of auxin. The lower side receives extra auxin and becomes even more inhibited, whereas the side that receives none has a more optimal amount and faster growth.

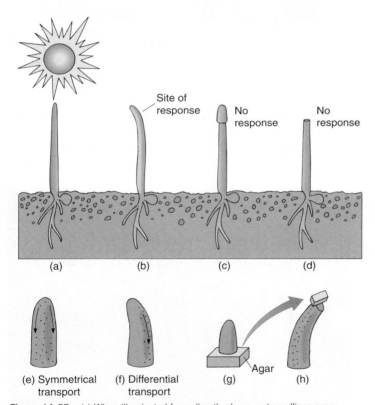

Figure 14-25   (a) When illuminated from directly above, oat seedlings grow upward. (b) When a young oat seedling is exposed to light from one side, its outermost sheathing leaf, the coleoptile, bends and grows toward the light. If the coleoptile apex is covered (c) or cut away (d), no response occurs to unilateral illumination, and thus, the tip is the site of perception. If the site of response is covered, bending occurs and the site of response is not involved in perception at all. (e) In dark conditions or with overhead lighting, auxin is transported symmetrically down the coleoptile, causing equal amounts of growth everywhere. With unilateral illumination (f), auxin is redistributed, with the darker side transporting more auxin than the lighted side, so the darker side grows faster, resulting in curvature. (g and h) Auxin can be collected in small blocks of agar or other absorptive material and then placed asymmetrically on a decapitated coleoptile; the side receiving auxin grows, but the other side does not.

position of the statoliths is transduced, nor are we certain of the growth inhibitor. Much circumstantial evidence indicates that it is auxin. How can auxin be a growth stimulator in coleoptiles and stems but an inhibitor in roots? It depends on the concentration and the tissue's sensitivity. Small amounts of auxin stimulate growth, and larger amounts inhibit it (**Figure 14-26**). Coleoptiles and stems typically have suboptimal amounts, so redistribution causes the side with more auxin to have a more nearly optimal level and that side grows faster. Roots are more sensitive to auxin and ordinarily have optimal or excess amounts; therefore, redistribution causes the side with more auxin to have such an excess it is inhibitory.

Flowering

Ripeness to Flower

Almost all plants must reach a certain age before they can be induced to flower; only a few species (*Chenopodium rubrum*, *Pharbitis nil*) can be induced as seedlings. Annual plants need to be only several weeks old before they become competent to respond to a floral stimulus, but many perennials must be 5 or 10 years old; beech trees (*Fagus sylvatica*) must be approximately

40 years old. Before this time, conditions that should induce flowering have no effect. Virtually nothing is known about the metabolic difference between the **juvenile phase**, when plants are incapable of being induced to flower, and the **adult phase**, when they are sensitive to floral stimuli. The conversion from juvenile to adult is called **phase change**.

Cold temperatures are the stimulus responsible for phase change in biennial plants, and the process is vernalization (see Figures 14-6 and 14-18a). The site of perception is the shoot apex itself; if it is cooled while the rest of the plant remains warm, vernalization occurs; however, if the rest of the plant is cooled while a small heater keeps the apex warm, no vernalization occurs. Presentation time is as short as 1 day in some plants. The transduction process is known to require oxygen, carbohydrates as an energy source, and an optimal temperature just above freezing, between 1°C and 7°C. Vernalization results in a stable change. If the plants are returned to warm conditions but are not given the floral stimulus of short days and long nights, they continue to grow vegetatively, without flowering, year after year, but they retain their vernalization and flower whenever the floral stimulus is finally given.

We know little about what stimulates phase change in perennial plants. Often a tree that is capable of flowering looks the same as one that cannot flower: Leaves, twigs, phyllotaxy, and bark all seem similar. Ivy (*Hedera helix*) is a common plant with distinct juvenile and adult morphology: Juvenile ivy has palmately lobed leaves and climbs (it is the form you see on walls), whereas adult ivy has entire, ovate leaves and upright stems that do not climb. Ivy can persist in the juvenile condition for years, growing extensively over walls and fences, and then individual branches may undergo phase change and start to flower while adjacent branches continue in the nonflowering juvenile phase.

Several genera of cacti have a much more pronounced difference between juvenile and adult phases. Juveniles of all species of *Melocactus* and *Backebergia* grow as cacti with ordinary bodies for many years, but when they undergo phase change, the adult shoot has a narrower pith and cortex, the cortex is not green and not photosynthetic, a different kind of wood is produced, phyllotaxy changes, epidermis is immediately replaced by bark, and the size, shape, and abundance of spines are different. The adult phase differs so strongly from the juvenile phase that a single plant appears to be two entirely different plants grafted together (**Figure 14-27**). In *Espostoa* and *Facheiroa*, when a phase change occurs, it affects only one narrow strip of the shoot: As the shoot apical meristem continues to function, one side of the shoot it produces is adult (with characters like that of adult *Melocactus*) and able to produce flowers, and the other sides of the same shoot are juvenile and cannot flower (**Figure 14-28**).

■ Photoperiodic Induction to Flower

The conversion of an adult plant from the vegetative to the flowering condition may be the most complex of all morphogenic processes. This is not one process—different mechanisms exist in different species. In certain annual species, size appears to be the only important factor: Peas and corn initiate flowers automatically after a particular number of leaves has been produced, regardless of environmental conditions; flowering is controlled by internal mechanisms. In many species, perhaps most, transition to the flowering condition is triggered by **photoperiod**—day

Figure 14-27   The green base of this *Melocactus* is its juvenile phase, and the red top is its adult phase; the two are parts of a single shoot produced by just one apical meristem. While young, the plant grew for years as a juvenile, with a body resembling that of many cacti, but it was unable to flower. Once old enough, it underwent phase change and the apical meristem switched to producing the narrower, very spiny adult phase (called a cephalium), which can flower. The adult portion cannot photosynthesize because of the dense covering of spines. For the rest of its life, this plant will continue to grow as an adult, with the cephalium becoming taller and the juvenile base remaining as it is— no new photosynthetic tissues will be formed.

Figure 14-28   The green parts of this *Espostoa* are its juvenile body, and the brown strips are its adult phase. While young, it grew as just a juvenile that had about 20 branches, but as each branch became old enough, each underwent a phase change. Only a strip of each branch became adult, however; the rest of each branch continued to develop with juvenile characters. Flowers are produced only from the brown adult regions. Unlike *Melocactus*, new photosynthetic tissues are produced every year in *Espostoa*.

TABLE 14-4	Photoperiodic Species*	
Long-night Plants (Short-day Plants)	**Short-night Plants (Long-day Plants)**	**Day-neutral Plants**
chrysanthemum (*Chrysanthemum*)	barley (*Hordeum vulgare*)	balsam (*Impatiens balsamina*)
cockleburr (*Xanthium strumarium*)	cabbage (*Brassica* sp.)	bean (*Phaseolus*)
Japanese morning glory (*Pharbitis nil*)	clover (*Trifolium pratense*)	corn (*Zea mays*)
kalanchoe (*Kalanchoe blossfeldiana*)	hibiscus (*Hibiscus syriacus*)	cotton (*Gossypium hirsutum*)
poinsettia (*Euphorbia pulcherrima*)	petunia (*Petunia* sp.)	holly (*Illex aquifolium*)
strawberry (*Fragaria chiloensis*)	radish (*Raphanus sativus*)	rhododendron (*Rhododendron* sp.)
violet (*Viola papilionacea*)	wheat (*Triticum aestivum*)	tomato (*Lycopersicon esculentum*)

*All plants fall into one of these three photoperiod categories; this table lists only a few familiar examples.

length—which acts as a season indicator (Table 14-4). One subclass of these plants blooms when days are short (spring or fall) and are **short-day plants**. Another subclass, **long-day plants**, is induced to bloom when days are long, in summer. Plants that do not respond to day length are **day-neutral plants**.

We know much about how plants measure day length. First, the pigment **phytochrome** detects the presence or absence of light (Table 14-5 and Figure 14-29). Phytochrome has a light-absorbing portion (a chromophore) attached to a small protein of about 125,000 daltons. Phytochrome occurs as dimers in cells. When phytochrome absorbs red light with a wavelength of approximately 660 nm, the chromophore changes shape, and this forces the protein to change its folding. This affects many of its properties, one of which is its hydrophobicity; in the refolded state, it is more hydrophobic and binds to membranes. A second altered property is its absorption spectrum; it now absorbs not at 660 nm but in the far-red (almost infrared) region of 730 nm. When this form absorbs far-red light, however, it refolds back to the red-absorbing form and releases from the membrane. The two forms are called P_r (red absorbing) and P_{fr} (far red absorbing). Also, P_{fr} reverts to P_r in darkness. Apparently P_r is inactive metabolically but becomes morphogenically active and exerts its effect when it absorbs red light and is converted to P_{fr}: P_{fr} is the active form and brings about metabolic responses. P_{fr} is transported into the nucleus and binds to one or more proteins, and the combination of the two apparently activates particular genes. After a plant is given red light, converting phytochrome to the active P_{fr} form, exposure to far-red light converts phytochrome back to the inactive P_r form. If far-red light is given quickly enough after red light, phytochrome does not have enough time to affect cell metabolism, and no effect is seen; however, if far-red comes long enough after red for the presentation time to be fulfilled, P_{fr} is able to complete the transduction process and far-red light can no longer cancel the red light stimulation (see Figure 14-10).

A plant experiencing a short day in nature automatically receives a long night (Figure 14-30). Similarly, long days are always accompanied by short nights. Night length is actually the critical factor. A long-day plant is in reality a short-night plant. It can be placed in a growth chamber and artificially given both long days and long nights—for example, 16 hours of light and 16 hours of dark in a 32-hour "day." If day length is the important factor, the plant should

TABLE 14-5	Photomorphogenic Responses for Which Phytochrome Is the Photoreceptor

Inhibition of internode elongation
Development of proper leaf shape
Increase in number of stomata per leaf
Increase in amount of chlorophyll
Decrease in apical dominance; formation of a more highly branched, bushy plant
Increased accumulation of carotenoid pigments in tomato fruits
Initiation of formation of storage organs such as tubers and bulbs
Initiation of dormancy in woody plants
Initiation of vegetative reproduction such as the formation of runners in strawberry or the formation of plantlets in *Kalanchoe*

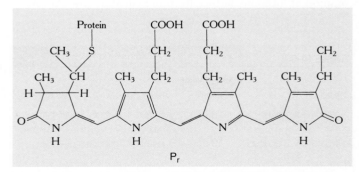

Figure 14-29 The chromophore or light-detecting portion of one particular phytochrome. The rest of the molecule is a protein. The two portions are joined by a sulfur atom. We now know that there is a family of at least five closely related phytochromes; some have overlapping functions, some have distinct functions. Many details are being investigated now.

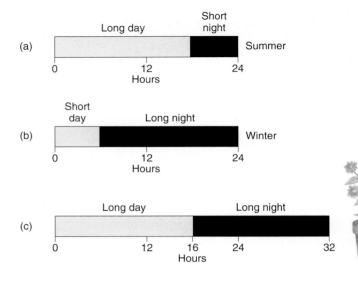

(a) Long day / Short night — Summer
0 — 12 Hours — 24

(b) Short day / Long night — Winter
0 — 12 Hours — 24

(c) Long day / Long night
0 — 12 — 16 — 24 — 32
Hours

Short-day
(long-night)
plant

Long-day
(short-night)
plant

Figure 14-30 (a and b) Natural 24-hour light/dark cycles must have short nights any time the days are long (summer) and long nights when days are short (winter); thus, plants could detect season by measuring either day or night. When photoperiodism was first discovered, it was assumed that day length was important; therefore, all our terminology was based on that. (c) With electric lights, it is possible to create a long photoperiod—32 hours long here—with both long days and long nights. Under these conditions, short-day (long-night) plants bloom, but long-day (short-night) plants do not.

flower because it has long days; if night is the critical factor, then it should not flower because it does not have short nights. When the experiment is done, the long-day plant does not flower, indicating that night length is critical, not day length. Similar experiments have shown that short-day plants really are long-night plants: If given a 16-hour cycle (8 hours light/8 hours dark—both day and night are short), they do not flower, but if given long nights, even accompanied by artificially long days, the plants are induced to flower.

Each species has its own particular requirements for long or short nights; that is, not all "long nights" have to be the same length. Instead each species has a **critical night length**; if a short-night plant receives nights shorter than this critical length, it flowers, whereas a long-night plant must receive nights longer than its own critical night length. Because critical night length varies from species to species, it is possible for a long-night plant and a short-night plant to bloom under the same conditions if the critical night length for the short-night plant happens to be longer than that for the long-night plant (**Table 14-6**).

Day length/night length control of flowering and other processes such as the initiation or breaking of dormancy is more common at locations farther from the equator. Away from the

equator, nights become progressively shorter from winter to summer and then progressively longer from summer to winter. The greater the distance from the equator, the greater the length of the longest winter night and the shorter the length of the shortest summer night. Thus, if two species are to bloom just after the beginning of May, a species in the southern United States or Mexico must have a critical night length shorter than that of a species in the northern United States or Canada.

Presentation time varies considerably; in some morning glories, one photoperiod of the proper length induces flowering, whereas at least 1 or 2 weeks of proper photoperiods are necessary for other species. The accuracy with which night lengths can be measured varies, but the most accurate species known is the long-day (short-night) plant henbane (*Hyoscyamus niger*); it must have nights shorter than 13 hours, 40 minutes. If the nights are even 20 minutes too long, 14 hours long, for example, it does not flower.

When phytochrome was discovered to be responsible for measuring night length, it was hypothesized that most phytochrome was converted to P_{fr} by the end of a day and then reverted slowly back to P_r during the night. It was postulated that if it could completely revert in the dark, before the next sunrise, metabolic changes would be triggered; however, it is now known that virtually all P_{fr} converts back to P_r within 3 or 4 hours, a time far too short to be a night-measuring clock by itself. It must instead interact with another type of clock that occurs in plant cells (a circadian rhythm, discussed later here).

The sites of perception for night length are young leaves. It is possible to stimulate one leaf with a spotlight and induce the plant to flower, even if the apical meristem, the site of response, is not illuminated (**Figure 14-31**). Because the site of perception is not the site of response, a chemical messenger must be transmitted between the two. If a leaf is photo induced and then immediately cut off the plant, no flowering occurs; if it is allowed to remain attached for several hours, the flowering stimulus is synthesized and transported out of the leaf. If the leaf is then removed, the plant still flowers. An obvious experiment is to induce a leaf by

TABLE 14-6	Concurrent Flowering of Long-day and Short-day Plants	
Species	Critical Night Length	Response
Xanthium strumarium	8.4 hr	Flowers if nights are longer than 8.4 hr
Hyoscyamus niger niger	13.6 hr	Flowers if nights are shorter than 13.6 hr

Although *Xanthium strumarium* is a short-day plant and *Hyoscyamus niger* is a long-day plant, both flower if given days 14 to 12 hr long; those days have nights 10 to 12 hr long, longer than the 8.4-hr critical night length of *Xanthium* but shorter than the 13.6-hr critical night length of *Hyoscyamus*.

(a)   (b)

Figure 14-31 If a short-night plant is given long nights, it does not flower, but it is possible to cause flowering by illumination with 15 minutes of dim red light; the plant acts as though it has received 2 short nights separated by a 15-minute day. (a) The red light "night break" does not have to be given to the entire plant; if a narrow beam of red light shines on a single leaf while all of the rest of the plant remains in darkness, the plant flowers (b).

giving it the proper night length, then collect the sap that is transported through the petiole and assay it for the hormone that acts on the apical meristem. This has been done hundreds of times by many people, without any repeatable success; the process is not as simple as we had at first thought.

An extremely interesting set of results has been obtained by grafting together plants with different photoperiod requirements. Individuals of the tobacco species *Nicotiana silvestris* are long-day plants. The species *N. tabacum* has two types of individuals: those of the cultivar *N. tabacum* cv. Trabezond are day-neutral plants, and those of *N. tabacum* cv. Maryland Mammoth are short-day plants. When long-day and day-neutral plants are grafted together and given long days, the day-neutral plant is also induced to flower, presumably by a floral stimulus that passes through the graft union. Similarly, day-neutral plants are induced to flower if grafted to short-day plants and given short days. Grafting one day-neutral plant to another does not increase flowering, so the grafting by itself has no effect.

When the combination of short-day plants grafted to day-neutral plants was given long days, the short-day plants did not flower as expected, and the day-neutral partners flowered about the time they would have if not grafted to anything. When long-day plants grafted to day-neutral plants were exposed to noninductive short days, however, not only did the long-day plants not flower, but the day-neutral partners were also prevented from flowering. Apparently, under noninductive conditions, long-day plants actually produce an inhibitor of flowering. Flowering in these long-day tobacco plants seems to be controlled by a switch from inhibitor production to promoter production, whereas in the short-day tobacco plants, flowering is controlled only by the presence or absence of a promoter. Neither the promoter nor the inhibitor has been isolated, but gibberellins are suspected to be involved in the synthesis or activation of the promoter.

A new advance in our understanding of the control of flowering is the discovery that when leaves are given the proper day length, a gene called *FT* is activated. This gene codes for a protein called FT that is carried through phloem to the apical meristems. This is an exciting breakthrough, but control of flowering is complex and

we must anticipate that there are many other steps to be discovered with many variations involving numerous genes other than just *FT*.

■ Endogenous Rhythms and Flowering

Plants contain **endogenous rhythms**; that is, certain aspects of their metabolism cycle repeatedly between two states, and the cycle is controlled by internal factors, by an internal clock. The most obvious example of this is in the "sleep movements" of the leaves of plants like prayer plant (*Oxalis*). In the evening, leaflets drop down, and in the morning, they raise themselves to the horizontal position as motor cells increase their turgor. It is easy to assume that this is a photonastic response, but if the plants are placed in continuous darkness, the leaflet position continues to change: They return to the up position about every 24 hours and move down about 12 hours later, controlled not by dawn or dusk but by their own internal clock. In many flowers, the production of nectar and fragrance is also controlled by an endogenous rhythm and occurs periodically even in uniform, extended dark conditions.

The underlying mechanism that constitutes the clock is poorly understood but is known to involve a **negative feedback loop**. Imagine just two proteins, *A* and *B*, with *A* activating *B*, and with *B* inhibiting *A*. As *A* works, *B* becomes more active and thus inhibits *A*, but as *A* becomes inhibited, it can no longer activate *B*. Consequently, *B*'s activity lowers, which allows *A*'s activity to increase. The system can cycle rhythmically indefinitely (compare with the positive feedback of ethylene in climacteric fruits). In *Arabidopsis thaliana*, the negative feedback loop involves at least three genes rather than just two: *TOC1* (and light) activate *LHY* and *CCA1*, but one of the effects of these latter two is to inhibit *TOC1*.

When placed in uniform conditions (constant temperature and either constant darkness or constant light), the endogenous cycle typically differs slightly from 24 hours, being either somewhat longer or shorter (**Figure 14-32**); however, in nature, the rhythm is exactly 24 hours long because light is able to **entrain** (reset) the rhythm. The pigment responsible for detecting the light for entrainment is phytochrome. Each morning, the red

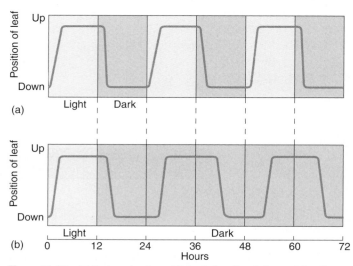

Figure 14-32 (a) Under natural conditions, a circadian rhythm matches the cycle of light and dark, being exactly 24 hours long. (b) In continuous darkness, most circadian rhythms have periods slightly longer than 24 hours. In nature, sunrise resets the clock by acting on phytochrome every morning.

TABLE 14-7	Metabolic Processes that Undergo Endogenous Rhythms in Plants
Accumulation of metabolites for CAM	
Changes in leaf position	
Growth rate of stems and roots	
Mitosis	
Opening and closing of stomata	
Rates of enzyme activity	
Respiration and other metabolisms	
Root absorption of minerals	

light of sunrise resets the rhythm so it can never get out of synchronization with exogenous light/dark cycles.

Many types of endogenous rhythms have a period that is not 24 hours long. Cytoplasmic streaming and the spiraling motion of elongating stem tips have periods of only a few minutes to a few hours; these are ultradian rhythms. If a period is approximately 24 hours long, it is a **circadian rhythm**, the most common kind (Table 14-7). The release of gametes in brown algae is controlled by a 28-day lunar rhythm. Some seeds have an annual rhythm of germinability: If stored in uniform conditions and periodically provided with moisture and warmth, they germinate only at times of the rhythm that correspond to springtime.

The involvement of endogenous circadian rhythms in flowering was discovered during dark interruption experiments: A short-day (long-night) plant can be prevented from flowering by interrupting long nights with a brief (15 minutes or less) exposure of red light. This is detected by phytochrome, and the

plant acts as though it has received two short nights separated by a 15-minute day. Short-day plants given a very long night—continuous darkness—have an endogenous rhythm of sensitivity to light breaks. If the light break is given at 6 hours into the dark period, or at 30 hours (24 + 6), 54 hours (24 + 24 + 6), and so on, the light break prevents flowering (Figure 14-33). These times correspond to darkness in a normal environment. But if the light break is given at a time when the endogenous rhythm would be "expecting" normal daylight conditions, such as at 16, 40 (24 + 16), or 64 (24 + 24 + 16) hours after the beginning of the dark treatment, the light break does not stop flowering. Plants kept in uniform, dark conditions undergo an endogenous cyclic sensitivity and insensitivity to red light interruption of the critical night length. Just how the endogenous rhythm and the critical night length work together to stimulate flowering is not known.

◼ ABC Model of Flower Organization

All flowers share the same basic developmental organization. Most have four types of "appendages": Outermost are the sepals, then petals, stamens, and finally carpels occur in the center. There are many variations: Wind-pollinated flowers often lack sepals and petals, and in many species certain flowers lack carpels and others lack stamens (in corn, for example, the flowers in the cob have carpels and develop into the fruit we eat, whereas the tassels at the top of the corn plant have stamens but no carpels). Many of the plants we cultivate in flower gardens are "double" flowers, which means they have extra petals but no stamens (Figure 14-34). It is typical for eudicot flowers to have five of each appendage: five sepals, five petals, five stamens, and five carpels (and monocots

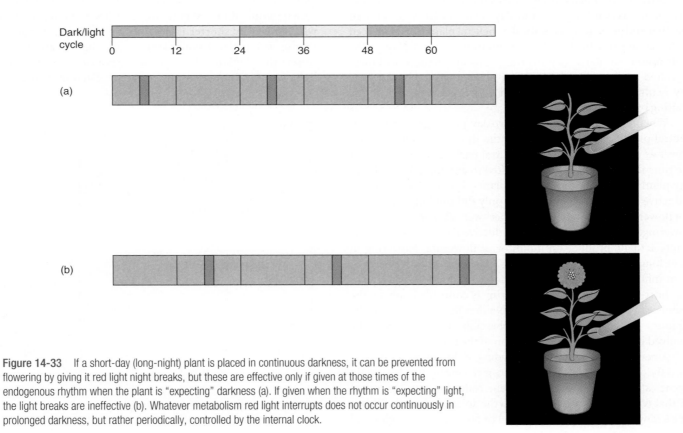

Figure 14-33 If a short-day (long-night) plant is placed in continuous darkness, it can be prevented from flowering by giving it red light night breaks, but these are effective only if given at those times of the endogenous rhythm when the plant is "expecting" darkness (a). If given when the rhythm is "expecting" light, the light breaks are ineffective (b). Whatever metabolism red light interrupts does not occur continuously in prolonged darkness, but rather periodically, controlled by the internal clock.

(a)

(b)

Figure 14-34 (a) This wild rose has only a few petals and is a "single" flower. (b) Many cultivated roses have been bred to be "double" flowers, with a mutation that causes them to produce extra petals in the place of stamens.

typically have three of each), but some flowers have dozens of petals, 100 or more stamens, and numerous carpels (**Figure 14-35**).

Despite the diversity it appears as if the basic organization of most flowers can be explained by a hypothesis called the ABC model. This model postulates that three fundamental genes, A, B, and C, interact to control the basic aspects of flower organization. Gene A affects the outermost, lowest regions of a flower shoot apical meristem, gene C controls the uppermost, central region, and gene B partially overlaps the other two (**Figure 14-36a**). In the outermost, lowest, doughnut-shaped region where only gene A is expressed, the primordia (similar to leaf primordia on a vegetative shoot apical meristem) develop into a whorl of sepals (flower appendages usually occur in whorls, not in spirals like leaves). Primordia in the next region upward and inward are influenced by both genes A and B, and they develop as petals. The combination of genes B and C produces

stamens, and in the very center of the flower where only gene C is expressed, primordia develop as carpels.

This model was proposed and is supported by many types of aberrant flowers. For example, if gene B is mutated and nonfunctional, sepals are produced in the outer part of the flower and carpels occupy the inner parts. There are no petals or stamens. More importantly, when these flowers are examined closely, they have an outer whorl of sepals, as is normal, and then an inner whorl of sepals where we would expect petals. Then there is a whorl of carpels where we would expect stamens. Finally, there is the innermost set of normal carpels. Although such a flower has just two types of appendages, they are still arranged in four whorls. The important thing is that normal appendages are produced in the "wrong" regions (sepals in the petal zone and carpels in the stamen zone; Figure 14-36b). A mutation that causes an organ to develop in an unexpected site

(a)

(b)

Figure 14-35 (a) This *Graptopetalum* is the type of flower often thought of as being a typical eudicot flower: It has five sepals, five petals, five stamens (this actually has 10, which is quite common), and five carpels. (b) This flower of *Epiphyllum* (in the cactus family) differs from that of a typical eudicot by having dozens of petals (30 are visible here), almost 100 stamens (many are too short to be seen here), and 15 carpels. Whereas each floral apical meristem of most species produces only a few primordia, those of cacti produce hundreds.

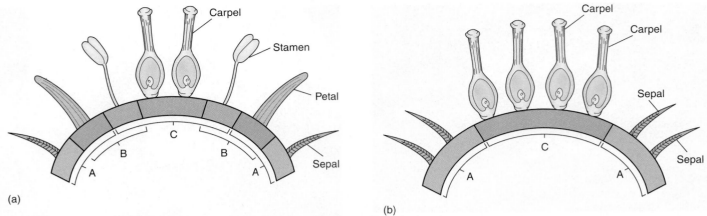

(a)   (b)

Figure 14-36   (a) The ABC model proposes that the outer part of a floral apical meristem is controlled by gene *A*, the inner part by gene *C*, and gene *B* overlaps the other two. Notice the outermost area is controlled only by gene *A* and produces sepals; the innermost is controlled only by gene *C* and produces carpels. Petals are produced by the zone where both *A* and *B* are active, and stamens occur where both *B* and *C* are active. (b) If a mutation inactivates gene *B*, then the floral apex has only two zones; the one controlled by gene *A* produces two whorls of sepals, and the one controlled by gene *C* produces two whorls of carpels. If gene *B* is inactivated, then the floral apex does not produce either petals or stamens.

is called a **homeotic mutation**. Mutations in gene *A* result in carpels in whorl 1 (where sepals would be expected) and stamens in whorl 2 (where petals would be expected).

The ABC model has been so successful that it is now the basis for numerous experiments. Many plant geneticists induce mutations in well-studied species such as *Arabidopsis thaliana* and *Antirrhinum majus* (snapdragon), searching for more details

about the ways in which these genes are controlled and how they work together. Because there are so many types of flowers, we expect to discover not only many versions of *A*, *B*, and *C* but also many other genes that subtly or dramatically influence *A*, *B*, and *C*. An example of modified flowers whose development still is unknown is those that consist of just stamens or carpels without any other appendages (**Figure 14-37**).

(a)   (b)

Figure 14-37   (a) Each cluster on this slender, pendant stalk is an oak (*Quercus*) flower, consisting of just stamens (you are most likely to notice these staminate flowers in springtime, after they have been shed and are lying on sidewalks, streets, and cars). Elsewhere on the oak tree are flowers that consist of just carpels. (b) Cattails (*Typha*) produce two types of extremely reduced flowers: At the top are thousands of flowers that consist of just stamens, and at the bottom are flowers that consist of just carpels. Both types of flower are extremely simple, consisting mostly of either stamens or carpels without sepals or petals. A few hairs may be present. In this photograph the carpellate flowers have been pollinated and have developed into fruits; you are unlikely to ever see carpellate flowers of cattails unless you dissect them from the plant before they emerge above the leaves.

Plants and People

Box 14-2 Environmental Stimuli and Global Climate Change

Global climate change is causing our world to rapidly become warmer and wetter, but plant mechanisms for detecting and responding to environmental stimuli are changing more slowly, if at all. As we burn oil, coal, and natural gas and as we convert forests into pasture for cattle, we increase the amount of greenhouse gasses in the atmosphere, causing the air, soil, lakes, and oceans to become warmer. As ocean temperatures rise their surface waters evaporate faster, making the atmosphere more humid and increasing the amount of rain and snow that later fall on land. Temperatures do not increase uniformly everywhere; instead, circulation patterns in the atmosphere and oceans are affected, so some areas become warmer, others cooler, some wetter, others drier.

Changing climate will have profound effects on all plants, not only on those that respond to temperatures but also on those controlled by day length. Increasing temperatures affect two critically important events for temperate plants: The date of the last frost in spring occurs earlier, and the time of the first frost in autumn comes later. The frost-free growing season in many areas starts earlier and ends later: Plants have a longer growing season.

Plants that germinate or bud out solely based on temperatures can take advantage of this longer growing period, and many seem to be thriving. But for plants controlled by photoperiod, their critical night length does not change, they germinate or bud out at the same time in spring as they have for centuries, and they go dormant at their typical time in autumn. They are not able to take advantage of the extra days of warmth in spring and autumn; instead, they are dormant when they could be photosynthesizing, growing, and reproducing. And, just as bad, their respiration during dormancy is higher than before because it is controlled by environmental temperature: Not only are the plants not producing sugars photosynthetically as long as they could, they are now respiring away their carbohydrates faster. They will have less reserve nutrients available when they resume growth in springtime.

Now consider the interaction of photoperiodic plants and temperature-controlled plants. They occur together in the same habitat and compete with each other for water, minerals, room for their roots, and so on. As warm temperatures occur earlier in spring, the temperature-controlled plants get a head start over the photoperiod-controlled ones, and the same is true in autumn. It is likely that the photoperiod-controlled plants will suffer in this competition, and the ratio of the two types of plants in the ecosystem will change.

As temperature in general increases, the snow-free habitable zone in alpine areas gradually rises to higher elevations. Similarly, habitable zones are expanding northward in the Northern Hemisphere, southward in the Southern. Areas near the North and South Poles are more hospitable. Again, temperature-controlled plants may benefit from this: If their seeds happen to occur in the newly warmer areas, they should be able to grow and reproduce. But the same is only partially true for photoperiod-controlled plants. These should be able to grow higher on any mountain on which they exist already: The critical night length is the same up and down the mountain. But close to the poles, a few days at the beginning of summer (June 20 or 21) have sunlight for 24 hours: There is no night at all for a few days. And at the beginning of winter (December 20 or 21), several days have no sunlight. From early winter to early summer, day length increases from 0.0 to 24 hours. At the equator daylight always lasts for 12 hours and night is also 12 hours, all year long. Between these two extremes, days get slightly longer each day in lower latitudes and much longer each day in high latitudes. For plants that need very long days to bloom (for example, 17 hours, with 7 hours of night), that occurs in May or June in the northern part of the United States and southern Canada, but it occurs in March in northern Alaska and Canada. If a longer growing season would allow that species to grow that far north, its critical night length would occur too early (March) while the plant is still a seedling: It could grow in the new habitat but not reproduce there.

It is important to remember that plants and their control mechanisms do evolve. Hundreds of different plant species differ in their critical night length, and this variation came about through evolution by natural selection. The important question is whether these mechanisms will evolve rapidly enough to allow plants to adapt to the changing climate. We do not know the answer, but, in general, such evolution is slow and we are causing the climate to change rapidly.

Summary

1. It is selectively advantageous for organisms to be able to sense and respond to significant aspects and changes in their environments.

2. Communication between body parts of an organism is essential to the integration and coordination of the organism's metabolism and development. Certain parts may need to respond to environmental or metabolic changes that they cannot sense themselves.

3. Plants must perceive important environmental information, transduce it to a communicable form, and respond to the transduced information.

4. Four ways in which plants respond to stimuli are tropic responses (oriented growth), nastic responses (stereotyped turgor changes), morphogenic responses (changes in quality), and taxis (oriented swimming).

5. In general, sites of perception are different from sites of response and must be linked by a means of communication. Presentation time and threshold are important elements in perception. Most communication appears to be by hormones.

6. Currently, the known classes of plant hormones are auxins, cytokinins, gibberellins, abscisic acid, ethylene, jasmonic acid, and several others.

7. The response to a hormone depends on which hormone is acting, the preparation of the responding cell, and the simultaneous or sequential presence of other hormones.

8. Flowering may involve the following steps: competence ("ripeness") to be induced, occurrence of inductive conditions, sufficient health to produce flowers, and later stimuli to induce flowers to open.

9. Flowering and many other seasonal responses are controlled by night length in many species. Phytochrome is involved in measuring night length.

10. Plants contain endogenous rhythms, cyclic changes in their metabolism. The rhythms most frequently are circadian, having a period of approximately 24 hours. These rhythms affect numerous aspects of plant metabolism.

Important Terms

ABC model of flowers
abscisic acid (ABA)
adult phase
all-or-none response
apical dominance
auxin
blue-light responses
circadian rhythm
climacteric
critical night length
cytokinins
day-neutral plants
differential growth
differentiation
dosage-dependent responses

endogenous rhythms
ethylene
gibberellins
homeotic mutation
hormones
indoleacetic acids (IAA)
juvenile phase
long-day plants
morphogenic response
nastic response
negative feedback loop
pattern establishment mechanism
perception of a stimulus
phase change
photoperiod

phytochrome
polar transport of auxin
polarity
positive feedback system
presentation time
response to a stimulus
short-day plants
statocytes
statoliths
symmetry
taxis
threshold for a stimulus
transduction of a stimulus
tropic response
vernalization

Review Questions

1. Most plants are so large their bodies exist in several different environments. What does this mean? Give examples for a single plant.

2. For a stimulus to cause a plant to change its metabolism or development, there must be three actions: _____ _____, _____, and _____.

3. Before a plant can respond to environmental factors, they must be perceived. Give examples of several environmental cues that plants perceive.

4. After a plant perceives a stimulus and before it can respond, there is another step called _____. What happens in this step?

5. Name five environmental factors that plants detect as stimuli. In each case, what types of information are provided to the plant?

6. Different prefixes are used for various stimuli. What is the prefix for each of the following:
   a. Light
   b. Gravity
   c. Touch
   d. Chemical

7. How does a positive tropic response differ from a negative tropic response? What is a diatropic response?

8. How does a nastic response differ from a tropic response? What is a morphogenic response?

9. In the perception of a stimulus, what are presentation time and threshold?

10. Responses can be either of two types: _____-_____-_____ and _____-_____.

11. What is an all-or-none response? Give some examples of plants that flower with an all-or-none response.

12. What is a dosage-dependent response? Name some plants that flower this way.

13. What are the four ways that a plant can respond to a stimulus? Define and give examples of each. Why are some responses classified as "positive," "negative," or "plagio-," whereas others are not?

14. Almost all plant communication is by a slow mechanism. What is the mechanism?

15. Name the known classes of plant hormones. What are the characteristics of plant hormones?

16. Name one natural auxin: _____.

17. The concentration of substances as powerful as hormones can be controlled by three methods in plants. What are the three methods?

18. Auxin, like many substances, can be transported through the phloem; however, there is a second mechanism that exists just for auxin transport. What is it?

19. Like auxin, cytokinins are involved in dozens of responses in all parts of the plant. Name one important response.

20. About how many gibberellins are known to exist? Are all of them active hormones?

21. What is the only gaseous hormone? How is it involved in fruit ripening?

22. What is apical dominance? How is auxin involved?

23. Describe the mechanism by which auxin appears to control leaf abscission.

24. Bending of plant parts toward or away from stimuli requires _____ _____.

25. Studies of phototropism often use which part of an oat seedling: _____?

26. True or false: Despite years of intensive research, we still do not have even an outline of the mechanism that actually causes plants to begin producing flowers.

27. Almost all plants must reach a certain age before they can be induced to flower. Before that point, they are said to be in the _____ stage. Afterward, when they can be induced, they are in the _____ stage.

28. Light is often the most important environmental factor for inducing a plant to flower, but it is not the intensity of light. What it the important aspect?

29. What are long-day plants? Short-day plants? Day-neutral plants? What is the critical night length?

30. Which pigment is responsible for detecting presence or absence of light? What are its two forms?

31. We know that short-day plants do not really need short days, but actually long nights. Describe the type of treatment that proves this. Is some kind of light given during the night or some period of darkness given during the day?

32. What is the difference between endogenous and exogenous rhythms? Give several examples of each type, and be certain to include several that have either short periods or long ones.

33. List eight metabolic processes that undergo endogenous rhythms in plants.

34. If an endogenous rhythm has a period of about 24 hours long, we say it is a _____ rhythm.

35. Imagine a short-day plant (really a long-night plant) being given continuous darkness. If it is given a flash of red light at a time when its endogenous rhythm would be "expecting" daylight, will that red light inhibit flowering? What about if the red light is given at a time when the plant's endogenous rhythm is "expecting" a long night?

BotanyLinks

http://biology.jbpub.com/botany/5e

BotanyLinks contains a variety of resources and review material designed to assist in your study of botany.

Visit the Web Exercises area of BotanyLinks to complete this question:

1. How can tissue culture be used to study plant growth and development? Go to the BotanyLinks home page for more information on this subject.

chapter 15

Genes and the Genetic Basis of Metabolism and Development

Concepts

Plants are composed of numerous types of cells. Each cell type is unique because it has a distinct metabolism, based largely on proteins such as enzymes, microtubules, and membrane proteins. Although all cells carry out a fundamental metabolism involving respiration, amino acid synthesis, and so on, some of the reactions in each cell type differ from those in other cell types. Each may have one or more characteristic types of enzymes or other proteins. For example, enzymes involved in synthesis of flower color pigments are present in petal cells but not in cells of roots, wood, and bark (**Figure 15-1a**). Also, sclerenchyma cells contain all enzymes necessary for producing and lignifying secondary walls, but these enzymes and metabolic pathways are not present in parenchyma cells (Figures 15-1b, 15-1c, and 15-1d).

Cells also differ in shape, again largely owing to differences in their proteins. All tracheids and vessel elements probably have the same enzymes and metabolism for secondary wall deposition, but the pattern of wall deposition varies, guided by a pattern of protein microtubules in the protoplasm (Figure 15-1e). Similarly, cell divisions occur in precise

Chapter Opener Image: This gardenia flower is a cultivated form that has "double" flowers. A mutation causes it to produce petals at places in the flower where we would expect stamens, consequently a double flower has at least twice the number of petals of ordinary flowers and are, thus, more showy. This character is controlled by genes in the nuclei of the plant's cells, and, by controlling which versions of the genes are present, plant scientists control which flower form is present.

(a)

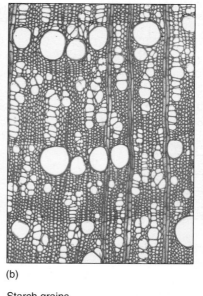
(b)

Intercellular spaces   Chlorenchyma cells

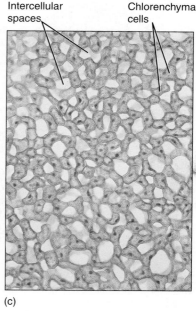
(c)

Starch grains

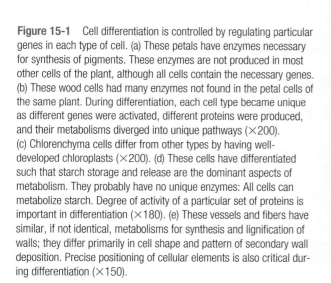
(d)

Fibers   Vessel

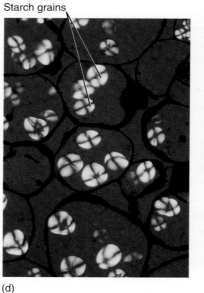
(e)

Figure 15-1 Cell differentiation is controlled by regulating particular genes in each type of cell. (a) These petals have enzymes necessary for synthesis of pigments. These enzymes are not produced in most other cells of the plant, although all cells contain the necessary genes. (b) These wood cells had many enzymes not found in the petal cells of the same plant. During differentiation, each cell type became unique as different genes were activated, different proteins were produced, and their metabolisms diverged into unique pathways (×200). (c) Chlorenchyma cells differ from other types by having well-developed chloroplasts (×200). (d) These cells have differentiated such that starch storage and release are the dominant aspects of metabolism. They probably have no unique enzymes: All cells can metabolize starch. Degree of activity of a particular set of proteins is important in differentiation (×180). (e) These vessels and fibers have similar, if not identical, metabolisms for synthesis and lignification of walls; they differ primarily in cell shape and pattern of secondary wall deposition. Precise positioning of cellular elements is also critical during differentiation (×150).

patterns, and as a result, there must be an underlying pattern in the cell that causes the mitotic spindle to have the proper alignment.

The information needed to construct each type of protein is stored in genes, but because an organism grows by mitosis—duplication division—all of its cells have identical sets of genes. As each cell differentiates and develops a unique suite of proteins, the underlying developmental process is the **differential activation of genes**. In the maturing epidermis, genes that code for cutin-synthesizing enzymes are turned on, whereas in other cells they remain turned off (**Figure 15-2**). On the other hand, genes for P-protein remain quiescent in all except phloem cells. Studies of development and morphogenesis examine the mechanisms by which some genes are activated and others are repressed.

During protein synthesis, the correct amino acids must be incorporated in the proper sequence because this determines both the structure and all other properties of the protein. The cell must contain a source of information that holds the sequence informa-

tion for all its proteins; this information archive is **DNA**, **deoxyribonucleic acid**. DNA is a linear, unbranched polymer composed of four types of deoxynucleotide monomers, usually abbreviated A, T, G, and C (**Table 15-1**). Once actually polymerized into DNA, the base portion of each nucleotide monomer protrudes as a side group. It is the sequence of nucleotide side groups that is the information needed to synthesize proteins correctly. A **gene** is each region of DNA that is responsible for coding the amino acid sequence in a particular protein. Each type of protein has its own gene.

Both environment and protoplasm also contain information vitally important for plant growth, morphogenesis, and survival. The environment provides informative cues about season, moisture availability, time for seed germination, time for flowering, and direction of gravity. These environmental and metabolic signals must be converted into chemical messengers that enter the nucleus and interact with genes. If the signals indicate that the cell is to differentiate into a vessel element, all genes that produce

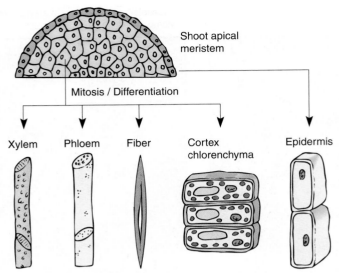

Shoot apical meristem

Mitosis / Differentiation

Xylem   Phloem   Fiber   Cortex chlorenchyma   Epidermis

Figure 15-2 Because all body cells are produced by mitosis, they all have the same genes. Certain basic metabolism genes ("housekeeping genes") are probably active in all cells, but specific pathways become active during differentiation, probably because the genes that code for the enzymes of those pathways become active. Once fully mature, the cells may go back to basal metabolism. Epidermal cells often produce cutin only when differentiating, not after maturity. Sieve elements lose their nuclei as part of maturation, and tracheary elements digest away all of their protoplasm; they have no metabolism at all when mature.

enzymes necessary for synthesis and lignification of a secondary wall must be located and activated. Genes that code for proteins that guide a particular pattern of wall deposition also must be turned on. Conversely, the cell must be inhibited from undergoing any further cell division; genes involved in mitosis and cytokinesis must be repressed.

Several techniques permit botanists to locate the genes for many proteins; the genes can then be isolated in vitro, duplicated and their nucleotide sequences revealed. Currently, our knowledge is still limited, but these techniques of **DNA sequence analysis** are so powerful that progress is extremely rapid. Similar techniques make it possible to alter the DNA sequence and then insert the gene back into a plant cell. As the cell grows, divides, and differentiates, the altered DNA either produces an altered protein if the coding region was changed or produces the protein at an unusual

TABLE 15-1	Nucleotides			
Symbol	Base	Nucleoside	Ribonucleotide	Deoxyribonucleotide
T	thymine	thymidine	—	dTTP
A	adenine	adenosine	ATP	dATP
G	guanine	guanosine	GTP	dGTP
C	cytosine	cytidine	CTP	dCTP
U	uracil	uridine	UTP	—

Nucleic acids contain five bases, abbreviated T, A, G, C, and U. One base plus a five-carbon sugar is a nucleoside, and with a phosphate attached, a nucleoside becomes a nucleotide:

$$base + sugar = nucleoside$$
$$base + sugar + phosphate = nucleotide$$

When DNA is synthesized, deoxyribonucleotides—those that contain the sugar deoxyribose—are used; when RNA is made, ribonucleotides—containing ribose—are used. Uracil does not occur in DNA, and thymine is absent from RNA.

time or place if its control site was changed. These **recombinant DNA techniques**, sometimes called **genetic engineering**, are helping us understand the processes that occur between the perception of a stimulus and the plant's response to that stimulus. In addition, recombinant DNA techniques permit us to change features of plants—for example, making them more resistant to insects or having seeds and fruits that are more nutritious for us.

Storing Genetic Information

Protecting the Genes

It is critically important that the information in DNA be stored accurately for a long time; if storage is not safe, the information produced by the DNA will be inaccurate and probably useless or even harmful. There are several ways in which DNA is kept relatively inert and safely stored.

1. *DNA does not participate directly in protein synthesis.* Instead, DNA produces a messenger molecule, **messenger RNA (mRNA)**, which carries information from DNA to the site of protein synthesis. The mRNA, not DNA, is exposed to the numerous enzymes, substrates, activators, and controlling factors of protein synthesis (**Figure 15-3**). If mRNA is dam-

Figure 15-3 When a gene is active, its sequence of nucleotides guides the synthesis of "heterogeneous nuclear RNA," which is modified into messenger RNA and transported to the cytoplasm. mRNA binds to ribosomes that translate (read) its nucleotide sequence and polymerize amino acids in the proper order, thus creating a protein. Ribosomes on the left have just started; thus, their proteins are still short; ribosomes on the right have read almost all the RNA, so their proteins are longer, almost complete.

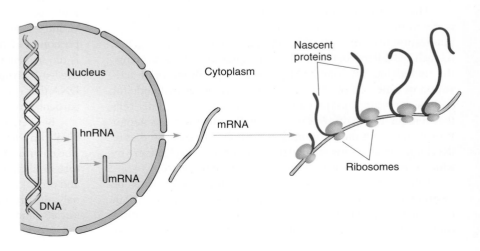

Nucleus   Cytoplasm   Nascent proteins

hnRNA   mRNA   Ribosomes

mRNA   DNA

aged, it can be replaced with more copies of mRNA. Within a single cell, thousands of individual molecules of a particular enzyme may be needed; if the DNA itself had to direct the synthesis of each protein molecule, it would probably be damaged long before enough protein had been synthesized. Instead, however, DNA directs the production of several copies of mRNA, each of which directs the production of hundreds of protein molecules.

2. *Most DNA is stored in the nucleus, protected from the cytoplasm by the nuclear envelope.* During interphase, the nuclear envelope forms the outer boundary of the nucleus, keeping most cytoplasmic components out and the proper nuclear substances in. The DNA of plastids and mitochondria is protected from cytosol enzymes by being located within plastids and mitochondria themselves. Nuclear genes in plants store information for approximately 20,000 to 40,000 types of proteins; plastid DNA encodes only about 50 to 100 genes, and mitochondrial DNA specifies less than 40.
3. *Histone proteins hold most nuclear DNA in an inert, resistant form.* Histones are a special class of proteins found in all organisms that have nuclei (plants, animals, fungi, algae, and protozoans). There are five types—H1, H2A, H2B, H3, and H4. The last four are among the most highly conserved proteins known; that is, the sequence of amino acids in the histones of one organism is virtually identical to the sequence in any other organism. For example, the H4 histone contains 103 amino acids, and its sequence in higher animals, such as cows, differs from its sequence in higher plants, such as peas, at only two sites (**Figure 15-4**). Histone proteins are so essential that virtually any change in their amino acid sequence causes the organism to die or at least not reproduce.

Histones form aggregates and DNA wraps around them, forming a spherical structure called a **nucleosome**. Histone H1 then binds nucleosomes into a tightly coiled configuration. In this mode, the DNA/protein structure—**chromatin**—is so dense that enzymes cannot penetrate it, and DNA is relatively inert. Even if it is exposed directly to DNA-digesting enzymes, called **DNases** (also written as **DNAases**), histone-bound DNA is not extensively damaged. However, chromatin is still sensitive to regulatory molecules and can be unpacked in preparation for synthesis of mRNA or for replication of DNA during the S phase of the cell cycle.

Plastids, mitochondria, and prokaryotes have no histones or nucleosomes; their DNA is naked.

```
                                                                     -200
AGCACCCTCCCACCTCATCCCACCCTTCTGATCTCAATCCAACGTCGCATCTCCACCGTCTCG
GCCACAGCTCTGACCCAGCCCGCCAAACCACCGGTCCAATCTCTCGGCCACGTCACCGATCCG

                                                 -150
CGGATCGACCCAGCGAAGTCCCTCCCGCCCCCAAAGTCCCCCAAATCTTGCAGTTCCCTCCTA
CGGCATCTCTCCCCCGGATCGCCGTCTCGACCGTCCACTCCATCCGCATCCAACGGCAGCCAC

                                     -100
AATCCTCCCCATATAAACCAACCCCCCGCCCTCAGATCCCTAATCCCATCGCAAGCATCAGAC
ACGCCTCCTCCAACCTCTCGACCCCTTTAAGACGCCCTTCGCCCCACCCAGCAAATCACAGCA

                  -50
TCCCTCCAAAGCAGGCAGCAGCTCCTCTTCTTCCTAATCACACTATCTCGGAGAGGAGCGGCC
CCAGACGCCACCCACCACCGTTCCTCCCATCCCACACTCGCTCGCAGCTCGAGATCGTCGGCC

+1
ATG  TCT  GGG  CGC  GAC  AAG  GGC  GGC  AAG  GGG  CTG  GGC  AAG  GGC  GGC  GCC
---  --C  ---  ---  -G-  ---  --A  ---  ---  --C  --A  ---  ---  ---  ---  ---
     Ser  Gly  Arg  Asp  Lys  Gly  Gly  Lys  Gly  Leu  Gly  Lys  Gly  Gly  Ala
                    Gly

+50
AAG  CGG  CAC  CGG  AAG  GTC  CTC  CGC  GAC  AAC  ATC  CAG  GGC  ATC  ACC  AAG
---  --C  ---  ---  ---  ---  ---  ---  --T  ---  ---  ---  ---  ---  ---  ---
Lys  Arg  His  Arg  Lys  Val  Leu  Arg  Asp  Asn  Ile  Gln  Gly  Ile  Thr  Lys

    +100
CCG  GCG  ATC  CGG  AGG  CTG  GCC  AGG  AGG  GGC  GGC  GTG  AAG  CGC  ATC  TCC
---  ---  ---  ---  C--  ---  --G  C--  C--  ---  ---  ---  ---  ---  ---  --G
Pro  Ala  Ile  Arg  Arg  Leu  Ala  Arg  Arg  Gly  Gly  Val  Lys  Arg  Ile  Ser

      +150
GGC  CTC  ATC  TAC  GAG  GAG  ACC  CGC  GGC  GTC  CTC  AAG  ATC  TTC  CTC  GAG
--G  ---  ---  ---  ---  ---  ---  ---  ---  --G  ---  ---  ---  ---  ---  ---
Gly  Leu  Ile  Tyr  Glu  Glu  Thr  Arg  Gly  Val  Leu  Lys  Ile  Phe  Leu  Glu

          +200
AAC  GTC  ATC  CGC  GAC  GCC  GTC  ACC  TAC  ACC  GAG  CAC  GCC  CGC  CGC  AAA
---  ---  ---  ---  --T  ---  ---  ---  ---  ---  ---  ---  ---  ---  ---  --G
Asn  Val  Ile  Arg  Asp  Ala  Val  Thr  Tyr  Thr  Glu  His  Ala  Arg  Arg  Lys

               +250
ACC  GTC  ACC  GCC  ATG  GAC  GTC  GTC  TAC  GCG  CTC  AAG  CGC  CAG  GGC  CGC
---  ---  ---  ---  ---  ---  ---  ---  ---  ---  ---  ---  ---  ---  ---  ---
Thr  Val  Thr  Ala  Met  Asp  Val  Val  Tyr  Ala  Leu  Lys  Arg  Gln  Gly  Arg

               +300
ACC  CTC  TAC  GGC  TTC  GGA  GGC  TAG  ATTTGTGTGGTGAAGCAACTTCCTCGTTTGC
---  ---  ---  ---  ---  --C  ---  --A  GGGCCGGCCGGCCGACGGGAGTCACTCTTTG
Thr  Leu  Tyr  Gly  Phe  Gly  Gly

     +350                                                   +400
TCTGTGATCTGTGCTGTCGTAGATGAGATTTACTGATTTGGCGTGCGCCGGTTGTATTCTGTC
TCGCCGCCTGCAGATTCCAGAAGCCTGATGAAGCCCCGACTTGTTTAGTTCGCTATTTCCTCT
```

Figure 15-4 The nucleotide sequence for the gene for histone H4 in wheat is presented in the top row of each set of lines. In the second row is the sequence for the same gene for a different type of wheat. Where the two genes are identical, only a dash appears for the second gene. The portion of the gene that codes for protein begins in the fifth line (labeled +1), and the amino acid sequence—the primary structure of the protein—is given in the third row of each set of lines. Wherever a mutation has caused the second gene to code for an amino acid different from the first gene, that amino acid is given in the fourth row. Although 18 mutations have occurred, the resulting amino acid sequence is unchanged except at one site (most of these mutations have no effect because the genetic code is redundant—discussed later in the chapter). The noncoding regions of the gene—from the beginning to +1, and from +309 to the end—are not highly conserved, and the two genes differ greatly in these sites. (Sequence data obtained by T. Tabata and M. Iwabuchi.)

The Genetic Code

Twenty types of amino acids are used in synthesizing proteins, but only four different nucleotides are present in DNA or mRNA; consequently, it is not possible for one nucleotide alone to specify one amino acid because 16 amino acids would be left without nucleotides to code for them. Similarly, nucleotides cannot be used simply in pairs of two, such as AU for isoleucine or CC for proline, because there are only 16 possible pairs. It is necessary for nucleotides to be read and used in groups of three; 64 possible triplets, known as **codons**, can be made using four nucleotides. **Table 15-2** shows which amino acid is coded by each codon. Notice that codon refers to triplets in mRNA, not in DNA.

With 64 possible triplets, a surplus of 44 codons remains after each amino acid is paired with a codon. Three codons—UAA, UAG, and UGA—are **stop codons**; they signal that the ribosome should stop protein synthesis. AUG is the **start codon** that signals the point in mRNA where protein synthesis should begin. The extra 40 codons also code for amino acids, so most amino acids have two or more codons. For example, both UUU and UUC code for phenylalanine, and CAU and CAC code for

TABLE 15-2 The Codons of mRNA

First Base	Second Base				Third Base
	U	**C**	**A**	**G**	
U	UUU Phenylalanine	UCU Serine	UAU Tyrosine	UGU Cysteine	U
	UUC Phenylalanine	UCC Serine	UAC Tyrosine	UGC Cysteine	C
	UUA Leucine	UCA Serine	UAA *STOP*	UGA *STOP*	A
	UUG Leucine	UCG Serine	UAG *STOP*	UGG Tryptophan	G
C	CUU Leucine	CCU Proline	CAU Histidine	CGU Arginine	U
	CUC Leucine	CCC Proline	CAC Histidine	CGC Arginine	C
	CUA Leucine	CCA Proline	CAA Glutamine	CGA Arginine	A
	CUG Leucine	CCG Proline	CAG Glutamine	CGG Arginine	G
A	AUU Isoleucine	ACU Threonine	AAU Asparagine	AGU Serine	U
	AUC Isoleucine	ACC Threonine	AAC Asparagine	AGC Serine	C
	AUA Isoleucine	ACA Threonine	AAA Lysine	AGA Arginine	A
	AUG (*START*) Methionine	ACG Threonine	AAG Lysine	AGG Arginine	G
G	GUU Valine	GCU Alanine	GAU Aspartic acid	GGU Glycine	U
	GUC Valine	GCC Alanine	GAC Aspartic acid	GGC Glycine	C
	GUA Valine	GCA Alanine	GAA Glutamic acid	GGA Glycine	A
	GUG Valine	GCG Alanine	GAG Glutamic acid	GGG Glycine	G

The triplets in mRNA are codons; the DNA triplets in the gene itself are their complements. The first base of the codon is listed on the left and the second base at the top. Within each large box are the four possible third bases and the amino acid specified by each.

histidine. Because multiple codons exist for most amino acids, the genetic code is said to be **degenerate**. Degeneracy further protects DNA: A mutation in DNA might change a codon in mRNA from UUU to UUC, for example, but because both code for phenylalanine, the same protein is produced before and after this particular mutation.

The genetic code is almost perfectly universal; all organisms and genetic systems but one share the genetic code shown in Table 15-2. Viruses, prokaryotes, fungi, animals, and plants all use the same codons to specify particular amino acids, and the same is true for plastid DNA. Only in mitochondria are several codons changed. This almost universal commonality of the genetic code is one of the strongest pieces of evidence that life arose only once on Earth and that all living organisms have evolved from one ancestral organism.

The Structure of Genes

Most genes, up to 90% in any cell, are quiescent most of the time and are activated and read only when the cell needs the particular enzymes they code for. Each gene must have a structure that allows controlling substances to recognize the gene, bind to it, and activate it at the proper time.

Genes are composed of a **structural region** that actually codes for the amino acid sequence, and a **promoter**, a controlling region involved in regulating the synthesis of mRNA from the structural region (**Figure 15-5**). The promoter is located "upstream" from the structural region, that is, to the 5′ side. It varies in length from gene to gene but can be several hundred nucleotides long. Certain regions are particularly important; one, called the **TATA box**, is a short sequence about six to eight base pairs long rich in A and T. If the TATA box is damaged by either mutation or experimental

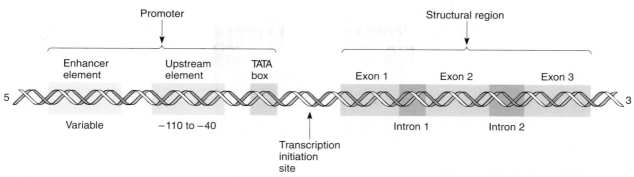

Figure 15-5 Carbons in ribose and deoxyribose are numbered from 1′ to 5′, with phosphate attached to the 5′ carbon and the hydroxyl used in polymerization attached to the 3′ carbon. Any nucleic acid has a 5′ end and a 3′ end. A gene is always written with the 5′ end, and thus the promoter, on the left. Nucleotides are numbered beginning at the left boundary of the structural region; nucleotides to the left of this are given negative numbers and are said to be *upstream*. Because DNA is double stranded, its length is measured as the number of nucleotide pairs or base pairs, whereas RNA, being single stranded, is measured as the number of nucleotides or bases.

treatment, the RNA-synthesizing enzyme **RNA polymerase II** does not bind well. Most eukaryotic genes have other promoter sequences called **enhancer elements** located even farther upstream, as many as several hundred base pairs away from the structural region of the gene. When a hormone alters cell metabolism, it does so by producing intracellular chemical messengers that activate genes either by binding directly with the promoter region or by binding with proteins that then interact with the promoter. After activating agents have bound to the promoter, RNA polymerase II can attach.

After RNA polymerase II binds to the promoter, it migrates downstream (toward the 3′ end of the DNA strand) toward the structural region; however, it does not create any RNA until it is approximately 20 to 30 nucleotides below the TATA box. The RNA polymerase might be expected to search for the DNA equivalent of the AUG start codon of mRNA, but that is not the case.

If some of the DNA is artificially removed between the TATA box and the normal start site, the RNA polymerase begins synthesizing RNA farther downstream than normal.

Like those of prokaryotes, genes of plastids and mitochondria often have multiple promoters, not just one. Plastids use two types of RNA polymerase, one imported from the nucleus and one they make themselves and which is remarkably similar to that of the bacterium *Escherichia coli*.

The structural portion of genes contains two distinct types of regions: exons and introns. **Exons** are sequences of nucleotides whose codons are eventually expressed (**ex**on, **ex**pressed) as sequences of amino acids in proteins, and **introns** are sequences of nucleotides that are not expressed, but instead **in**tervene between exons (Figure 15-5 and **Figure 15-6**). Several plant genes have just two or three introns: the gene for RuBP carboxylase and the genes for the storage proteins glycinin and phaseolin of

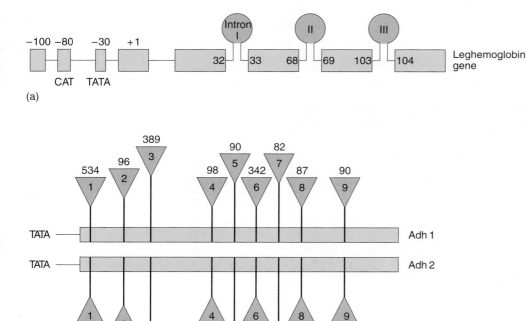

Figure 15-6 Two ways of illustrating introns and exons in maps of genes. (a) The gene for leghemoglobin in legumes has three introns: Intron I occurs between bases 32 and 33 of the finished mRNA; intron II between bases 68 and 69, and so on. (b) Two genes in corn produce two similar enzymes, both called alcohol dehydrogenase and distinguished as Adh1 and Adh2. The numbers above and below the triangles indicate the number of bases present in each intron. The two genes are similar, having nine introns located at the same positions; corresponding exons in Adh1 and Adh2 are the same length, but corresponding introns may be quite different. Intron 1 is 534 bases long in Adh1 but only 95 bases long in Adh2.

legume cotyledons. The gene that codes for the protein portion of phytochrome has five introns, one of which is 1,500 base pairs long. Genes with 25 introns occur, as do genes with no introns. It is very common for exons to make up much less than half the structural part of a gene.

Transcription of Genes

After RNA polymerase binds and encounters the start signal, it begins actually creating RNA, a process called **transcription**. The two strands of DNA separate from each other over a short distance, and free ribonucleotides diffuse to the region (**Figure 15-7**). If a ribonucleotide containing cytosine approaches a DNA nucleotide that contains guanine, the two form three hydrogen bonds

and remain together, at least temporarily. Wherever DNA contains T, a free A can form two hydrogen bonds to it; similarly, free U forms two hydrogen bonds with A in DNA, and so on with G and C. RNA polymerase binds the free ribonucleotide, holds it, and catalyzes the formation of a covalent bond, forming RNA. As each covalent bond is formed, two high-energy phosphate-bonding orbitals are broken. Polymerization is thus a highly exergonic process that cannot be easily reversed. Free ribonucleotides diffuse to and pair with both strands of DNA, but RNA polymerase is located only on the strand that has the gene; the complementary strand is not read.

Transcription proceeds rapidly, incorporating about 30 ribonucleotides per second, with the DNA double helix unwinding ahead of the moving enzyme. After RNA polymerase moves off the promoter/initiation site, a new molecule of RNA polymerase binds and begins synthesizing another molecule of RNA (**Figure 15-8**). Whereas two molecules of DNA wrap around each other into a double helix, RNA/DNA duplexes do not; the RNA polymer that emerges from RNA polymerase releases from the DNA.

RNA polymerase continues to act until it encounters a transcription stop signal in the DNA. The stop signal of several genes consists of two parts. The first is a short series of DNA nucleotides that are self-complementary; that is, the RNA transcribed from them can double back on itself and hydrogen bond to another part of itself. This results in a small kink, a **hairpin loop** that is believed to affect RNA polymerase. Just downstream of this region of DNA is a long series of adenines. Various protein factors also are involved; in their absence, RNA polymerase sometimes continues transcribing, reading the next region of DNA as if it were also part of the gene.

RNA polymerase transcribes both introns and exons into a large molecule of **hnRNA** (**heterogeneous nuclear RNA**) that is rapidly modified by nuclear enzymes. Introns are recognized, cut out, and degraded back to free ribonucleotides. Exons are spliced together, resulting in an RNA molecule, all of which codes for

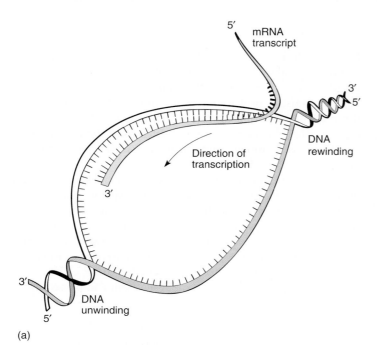

(a)

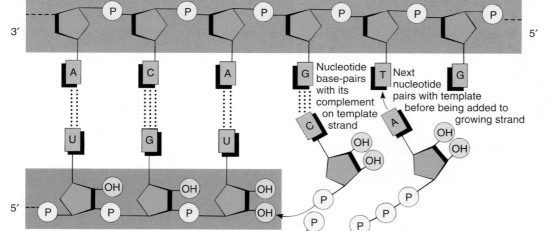

(b)

Figure 15-7 When a gene is turned on, the DNA double helix separates over a short region (a); free ribonucleotides diffuse in and pair with the region of temporarily single-stranded DNA (b). The formation of two or three hydrogen bonds between the DNA deoxyribonucleotides and the free ribonucleotides allows the DNA sequence to control the sequence of the RNA being formed.

Figure 15-8 DNA can be carefully extracted from a nucleus, allowed to spread out, and then prepared for examination in an electron microscope (a). These are ribosomal genes being transcribed into long RNAs that will later be cut into three separate rRNAs. The diagram (b) explains each type of line.

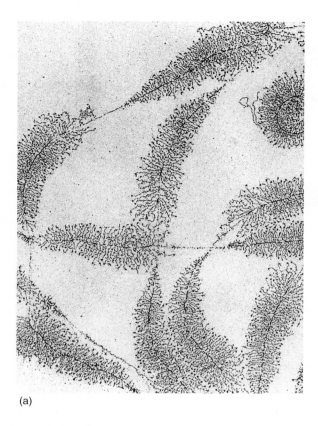

(a)

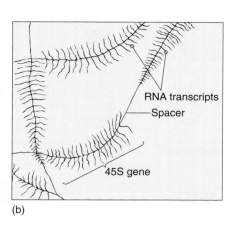

RNA transcripts

Spacer

45S gene

(b)

amino acids. All RNA destined to become mRNA is somehow recognized by an enzyme that binds to it and attaches a series of adenosine ribonucleotides on its 3′ end, forming a **poly(A) tail** approximately 200 bases long, the only exception being the mRNAs for histone proteins (**Figure 15-9**). Another step in message processing involves changing the first nucleotide into 7-methyl guanosine. Ultimately, a completed mRNA is produced and transported from nucleus to cytoplasm.

Plastid and mitochondrial genomes have prokaryotic organization. Their genes are not separated by long stretches of spacer DNA, but instead genes occur in sets, one following another immediately. One set of promoters causes RNA polymerase to transcribe not just a single gene but actually several before it releases. This long, multigene transcript is then processed differently than is hnRNA.

Protein Synthesis

In the process of protein synthesis, ribosomes bind to mRNA and "read" its codons. Guided by the information in the nucleotide sequence of the mRNA, the ribosomes catalyze the polymerization of amino acids in the order specified by the gene from which the mRNA was transcribed.

Ribosomes

Ribosomes are small particles that "read" the genetic message in mRNA and construct proteins guided by that information. Each is composed of two subunits, one larger than the other, and each is made up of both proteins and **ribosomal RNA rRNA(1)**

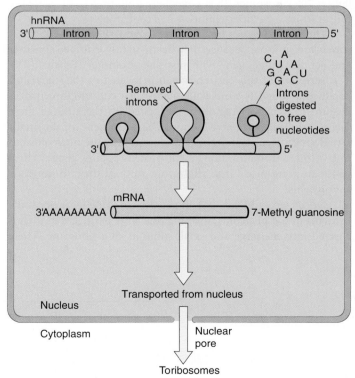

Figure 15-9 The primary transcript, hnRNA, has its introns cleaved out and the exons spliced together; the introns are depolymerized back to free nucleotides. An enzyme adds up to 200 adenosine ribonucleotides to the end of the RNA; these are not coded by thymidines in the DNA. The first nucleotide at the 5′ end is converted to 7-methyl guanosine.

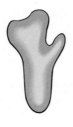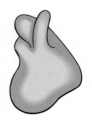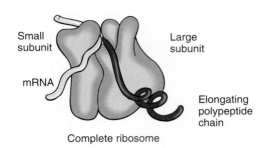

Small subunit

Large subunit

Small subunit

mRNA

Large subunit

Complete ribosome

Elongating polypeptide chain

Figure 15-10 Ribosomes have two subunits, one large and one small. When they fit together there is a groove for mRNA to pass through, a channel through which the growing protein emerges, and a channel into which amino acid carriers enter.

TABLE 15-3	Components of 80S Cytoplasmic Ribosomes	
	Size of RNA Molecules	Number of Proteins
Small subunit	18S	approx. 33 ?
Large subunit	28S	approx. 49 ?
	5.8S	
	5S	
	4 molecules	approx. 84

(Figure 15-10 and Table 15-3). The small subunit contains one molecule of rRNA, the large subunit one molecule each of three types of rRNA. The number of proteins present in eukaryotic ribosomes is known to be greater than 80, but the exact number is still uncertain. Ribosomes found in the cytoplasm of eukaryotes are designated **80S**, meaning that they are relatively large and dense; ribosomes of plastids, mitochondria, and prokaryotes are smaller, lighter **70S ribosomes**. (S is a Svedberg unit, used to measure the rate at which a particle sediments in a centrifuge.)

The nuclear genes for the three largest rRNAs of 80S ribosomes are unusual because they occur tightly grouped together and act as a single gene with just one promoter (**Figure 15-11**). Transcription of this cluster, which is located in the nucleolus, produces a long RNA molecule that is then cut into three pieces (Figure 15-8). Short regions are digested off the ends of these pieces, resulting in three rRNA molecules. The gene for the smallest rRNA molecule (5S RNA) is not located in the nucleolus but out in the chromatin along with protein-coding genes. These genes are transcribed into rRNA, not mRNA; they do not code for proteins.

Once transcribed, rRNA molecules combine with ribosome proteins in the nucleolus, forming one large particle that then is cleaved into the large and small subunits of a ribosome. These subunits are then transported out of the nucleolus through the nucleus to the cytoplasm.

All cells need large numbers of ribosomes. Most genes are present in a diploid nucleus as only two copies, but just two copies would be inadequate for producing the hundreds of thousands of rRNAs needed. Instead, rRNA genes are highly amplified; that is, many copies of each are present. In flax plants, each nucleus may have up to 120,000 copies of the gene for the small rRNA and 2,700 copies of the gene cluster for the three large rRNA molecules.

■ tRNA

During protein synthesis, amino acids are carried to ribosomes by ribonucleic acids called **transfer RNA (tRNA)**. tRNAs are necessary because a codon cannot interact directly with an amino acid; the genetic code can be read only by a ribonucleic acid that has a three-nucleotide sequence, called an **anticodon**, that is complementary to and hydrogen bonds to the codon. For example, UUU and UUC are both codons for phenylalanine; the tRNAs that carry phenylalanine have the complementary anticodons of either AAA or GAA. GAA looks backward for something that must bind to UUC; by convention, nucleotide sequences are written 5′ to 3′, which is GAA. An alternative is to write the numbers to show it is being written backward for easier comparison with the codon: 3′-AAG-5′.

There are as many types of tRNA as there are codons that specify amino acids; stop codons do not have tRNAs. All tRNAs have the same parts, an anticodon and an **amino acid attachment site** at its 3′ end consisting of the sequence CCA (**Figure 15-12**). A special class of enzymes recognizes each tRNA and attaches the correct amino acid to it. This step, called **amino acid activation**, must be precise. If the wrong amino acid is placed onto a tRNA, it becomes incorporated into the protein as if it were the correct amino acid, causing the protein to have an erroneous structure.

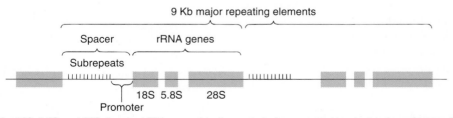

9 Kb major repeating elements

Spacer

rRNA genes

Subrepeats

18S 5.8S 28S

Promoter

Figure 15-11 The genes for the 18S, 5.8S, and 28S ribosomal RNAs occur together as a cluster, separated by short regions of spacer DNA. All three genes are transcribed as just one molecule of RNA, spacers included; spacer RNA is cut out, resulting in three individual rRNA molecules. Upstream from the 18S gene is the promoter and then a region made up of 10 to 15 copies of a short sequence about 135 base pairs long, called subrepeats; they are not transcribed. On either side of the rRNA gene cluster are more rRNA clusters, as many as 2,700, all occurring end to end as a gene "family." Kb is kilobases; 1 Kb = 1,000 bases.

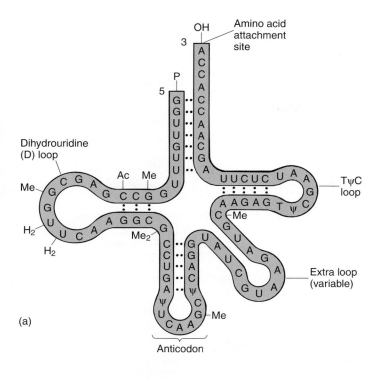

(a)

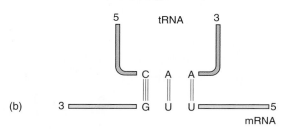

(b)

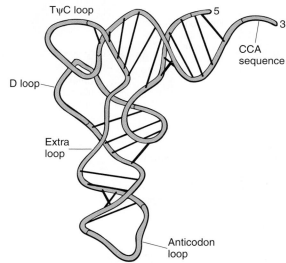

(c)

Figure 15-12 (a and b) All tRNAs have this general shape, with at least three loops caused by self-complementary base pairing. The small fourth loop is present only in some. The anticodon is at the middle loop and the amino acid is attached to the CCA at the 3′ end. This is one of the six tRNAs that carry leucine; it has the anticodon CAA, which recognizes the codon 3′-GUU-5′, normally written UUG. The unusual symbols (Ac, Me, ψ) indicate bases that are chemically modified into unusual forms. (c) tRNA folds into an L shape rather than lying flat.

Transfer RNAs contain bases not found in other nucleotides. tRNAs are transcribed from tRNA genes, and like all ribonucleic acids, they contain the common bases A, U, G, and C; however, after these are polymerized into tRNA, enzymes modify some bases to unusual forms (Figure 15-12a).

Each tRNA carries an amino acid only briefly; after it is activated, it rapidly encounters a ribosome that is "reading" the codon complementary to its anticodon. It gives up its amino acid and shuttles back to the cytosol and is reactivated. Even though each tRNA can cycle like this many times per second, millions of tRNAs are needed in every cell, and hundreds or even thousands of copies of each tRNA gene may be present in each nucleus.

The nucleus contains genes for all 61 types of cytoplasmic tRNA, but plant mitochondria have genes for only a few tRNAs and must import the rest from the cytoplasm. Plastids have genes for only about 30 types of tRNA, but seem to not need the rest.

■ mRNA Translation

Initiation of Translation The synthesis of a protein molecule by ribosomes under the guidance of mRNA is called **translation**. Protein synthesis begins with a complex **initiation** process involving the start codon AUG. This codes for the amino acid methionine, but two types of tRNA actually carry methionine, and they have different properties. One, called initiator tRNA, binds to the ribosome small subunit even before any mRNA is present (**Figure 15-13**). Several initiation factors, mostly proteins called **eukaryotic initiation factors** (**eIFs**), also bind to the small subunit. In this condition, the complex is competent to bind to mRNA, after which the large subunit of the ribosome binds to the complex and the initiation factors are released. Just how mRNA is recognized is not known, but the initiator tRNA is important for finding the AUG start codon and positioning the small subunit.

It is critically important that the ribosome be properly aligned on mRNA because the sequence of nucleotides can be read in any set of three. The sequence CUUGCACAG can be read as CUU GCA CAG and would code for leucine alanine glutamine (**Figure 15-14a**), but if the ribosome binds incorrectly, shifted downstream by one nucleotide, mRNA is read as _ _ C UUG CAC AG _, the codons for phenylalanine histidine and either serine or arginine, depending on the next nucleotide. Reading nucleotides in the wrong sets of three is a **frameshift error**; because virtually all codons are misread, frameshift errors typically result in completely useless proteins.

Elongation of the Protein Chain mRNA lies in a channel between the two ribosome subunits. Extending outward from the mRNA channel are two grooves, each wide enough for a tRNA to fit into it such that the tRNA anticodon can touch an mRNA codon. When both channels contain activated tRNAs, adjacent codons are being read (Figure 15-13).

At the time of initial binding, the small subunit already contains initiator tRNA in the **P channel** (P for protein). The adjacent **A channel** (A for amino acid) is empty, but numerous molecules enter it at random. Some are activated tRNAs, but if their anticodons do not complement the exposed codon at the bottom of the A channel, they diffuse out. If a tRNA with the proper, complementary anticodon enters, hydrogen bonds form between the codon and the anticodon. This holds the tRNA in place long enough for an even more stable binding to occur. Enzymes located

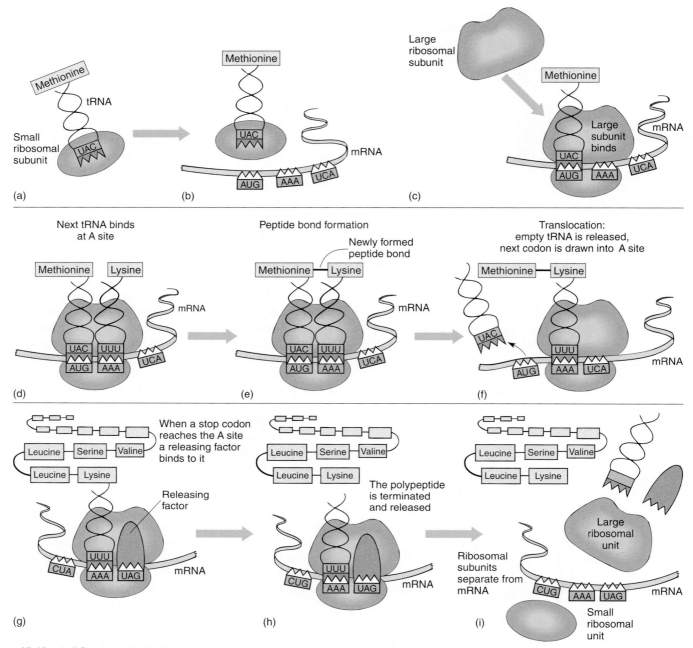

Figure 15-13 (a–i) Protein synthesis; the steps are explained in the text.

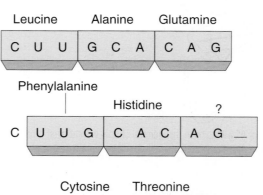

Figure 15-14 Because there are no spacers between codons, the ribonucleotides can be read in three possible ways. Each results in completely different proteins, only one of which has the proper primary structure.

Box 15-1 Genetic Engineering and Evolution

An important area of genetic engineering is the production of herbicide-resistant crop plants. The concept is to engineer plants that are not harmed by an herbicide, such that fields of the plant can be sprayed with the herbicide to kill weeds without harming the crop itself. There are arguments for and against the basic idea of using herbicides rather than using organic gardening, but for the moment, let us consider just the genetic engineering aspect.

An extremely effective herbicide called **glyphosphate** became available in the 1970s (sold with the trade name Roundup). Glyphosphate has many favorable features. First, it inhibits an enzyme necessary for the synthesis of three aromatic amino acids (tyrosine, tryptophan, and phenylalanine); this alone would make it deadly by blocking protein synthesis due to missing amino acids. In addition, however, these amino acids themselves are used in other essential plant pathways, and thus, glyphosphate is especially lethal. Second, the enzyme that it inhibits occurs only in plants; animals do not have this enzyme (animals must obtain these amino acids in their diet), so being exposed to glyphosphate does not harm animals. Third, glyphosphate rarely pollutes water because it binds so strongly to soil that it does not wash into ground water or streams. In fact, it is applied by spraying it onto leaves rather than mixing it with soil. Finally, it breaks down quickly into harmless products.

The one drawback of glyphosphate is that it kills all plants. It can be applied to fields only before the crop seeds germinate, killing weeds that germinated before the crop did, or if crop plants are tall and weeds are short, it can be sprayed below the crop's leaves and onto the weeds' leaves.

Genetic engineering entered the picture when it was discovered that bacteria have a gene called *CP4* that synthesizes the three amino acids but is immune to glyphosphate. Plant scientists isolated the gene and genetically engineered soybeans to use this gene in addition to the natural plant gene. These soybeans would not die if sprayed with glyphosphate. An entire field of these soybeans could be sprayed at any time to kill weeds without harming the crop. Currently, glyphosphate-resistant alfalfa, canola, cotton, corn, and sugar beets have been genetically engineered (such plants are often called **genetically modified [GM]** plants). This is so effective that many farmers have stopped using other herbicides and rely on glyphosphate. An added benefit is that they do not have to plow their fields before planting; plowing inhibits weeds by burying their seeds deeply; however, it loosens soil so much that it greatly increases erosion, and it requires large amounts of fuel.

The extensive use of a potent herbicide like this creates a strong selection pressure for resistance among other plants: Any weed that has a mutation that makes it resistant to glyphosphate will thrive in a field with fertilizer, irrigation, pesticides, and no competition from other weeds. People thought it would be almost impossible for resistance to evolve because the plant enzyme is so fundamental. How could an entire new enzyme or whole new pathway evolve in a short period of time? But glyphosphate-resistant weeds have appeared already in many parts of the world, and not because they have a new enzyme. Instead, plants are variable in their capacity to transport glyphosphate through phloem. Most transport glyphosphate throughout their body, and it accumulates in shoot and root tips, stopping growth. But a small number of plants have something different about their phloem (we do not know what) that causes them to transport glyphosphate to leaf tips—It harms the leaf tips but does not hurt the rest of the plant. In natural environments, this unusual transport may not help the plant at all, but it is extremely beneficial in a field being sprayed with glyphosphate. Glyphosphate did not cause glyphosphate resistance to come into existence; this feature would have been present already in a naturally variable population of plants. It is just that these plants suddenly are more adapted because glyphosphate is being used in their environment; they will become a greater part of their population as susceptible plants are killed.

Other types of resistance should be expected. It may be that plants already exist that have degrading enzymes that can break glyphosphate down or that have membranes that are impermeable to it. The continued use of glyphosphate will give these plants a selective advantage, and their numbers will increase. It is a simple, clear-cut case of evolution by natural selection.

This experience should remind us that plants and evolution can respond in numerous ways, many of which we might not anticipate. Other areas in which GM crops are being produced and where we might want to be cautious are drought resistance and insect resistance. Plants have already been genetically engineered to require less water for their survival. They require less irrigation, so less water can be diverted from rivers and lakes, which would then retain more of their natural condition. Would seeds of drought-resistant crop plants then be able to invade natural, dry areas and become invasive pests on their own, crowding out the natural plants? A potent insect-killing protein is produced by the bacterium *Bacillus thuringiensis*; it is easy to transfer that gene into plants, creating plants that resist insects. The gene, however, is also expressed in pollen of genetically modified plants, and poisonous pollen blows away, especially from corn fields, landing on the leaves of other plants, coating them with a poisonous "dust" that could harm any kind of insect.

As is so often the case, advances in technology bring with them both benefits and risks. By analyzing each opportunity for its merits and dangers, we should be able to increase our quality of life without harming the environment, but we must always remember that we are dealing with organisms that can evolve and respond to the altered conditions we are creating.

in the large ribosomal subunit break the bond between the methionine and its tRNA in the P channel, simultaneously attaching the methionine to the amino group of the amino acid on the tRNA in the A channel. This reaction needs no outside source of power: The broken bond is a high-energy bond, whereas the newly formed peptide bond is only a low-energy bond.

The empty tRNA, freed of its methionine, is released from the P channel, and the ribosome pulls itself along the mRNA for a distance of three nucleotides. This movement is powered by GTP and seems to be performed by proteins on the large subunit. As the ribosome slides along the mRNA, the A channel slides to the next codon, and the tRNA with the two amino acids attached becomes surrounded by the P channel, not the A channel. When the proper activated tRNA diffuses into the A channel and hydrogen bonding between codon and anticodon occurs, the large subunit enzymes again release the short protein chain (now two amino acids long) from the tRNA in the P channel and attach it to the amino acid on the tRNA in the A channel, creating a protein three amino acids long. This process repeats until the ribosome reaches a stop codon.

Termination of Translation When a stop codon is pulled into the A channel, normal elongation cannot occur; no tRNA is present with an anticodon complementary to a stop codon. Instead, a release factor enters the channel and stimulates the large subunit enzymes to initiate the normal reactions. The high-energy bond holding the protein to the P-channel tRNA is broken, and a new bond is formed with water, releasing the protein from both the tRNA and the ribosome. The tRNA is released, and the small subunit disassociates from the large subunit and releases the mRNA. All components diffuse away, but as soon as the small subunit encounters an initiator tRNA and other initiating factors, it binds to another mRNA and the process begins again.

In summary, protein synthesis involves the following steps: RNA polymerase transcribes hnRNA, being guided by the sequence of DNA nucleotides in a gene. The hnRNA is processed into mRNA as introns are cut out and exons are spliced together. After moving from nucleus to cytoplasm, mRNA binds with ribosomes, and tRNAs carry amino acids to the ribosome-mRNA complex. The ribosome translates the mRNA codon by codon, and tRNAs fit into the ribosome only when their anticodon is complementary to the codon in the A channel. After the protein is completed, it is released and the ribosome subunits detach from the mRNA. All components diffuse away from each other, but the ribosome subunits can translate the next mRNA they happen to meet. The mRNA can be translated again by other ribosomes.

Control of Protein Levels

As cells undergo differentiation and morphogenesis, their metabolism and structure become different from those of other cells because of the presence of proteins, especially enzymes, unique to that cell type. A central question in developmental biology is the mechanism by which distinct cell types control the activities of genes so that they undergo the proper differentiation and obtain the proper set of proteins. There are several points at which protein synthesis and activity theoretically could be controlled: making a gene physically available for transcription; nature of the promoter region; processing of hnRNA into mRNA; transport of mRNA from nucleus to cytoplasm; binding of mRNA to the ribosome small subunit; rate of translation; processing of protein, and activation or inactivation of the protein.

Many enzymes and structural proteins are present in a cell in an inactive form. Tubulin is an excellent example: A cell may contain a large pool of tubulin monomers and then aggregate them into microtubules at a specific time. Microtubules can appear rapidly without the need for gene activation or protein synthesis. Similarly, enzymes are often completely inactive until phosphate groups are added by a class of enzymes, called kinases, that are themselves activated by the arrival of hormones. Hormones arrive at the plasma membrane and bind to receptors, which then synthesize second messengers that enter the cell. These activate the phosphorylating enzymes that in turn activate the dormant enzymes, leading to a significant change in the cell's metabolism. This is mostly an activating rather than a differentiating mechanism; these cells are already prepared for highly specific responses.

It would be extremely inefficient if a cell transcribed all its genes and synthesized all of its possible proteins when only a few are needed. For maximum efficiency, a cell should not even synthesize the mRNA for proteins it does not need. We expect that the most fundamental level of control of morphogenesis would occur at the level of transcription.

We know very little about control of transcription in eukaryotes, especially plants, but our knowledge is increasing rapidly. In many cases, gene activity is controlled by **transcription factors**, proteins that bind to promoter or enhancer regions and activate genes. Many transcription factors have a sequence of amino acids that fits into the large groove of a DNA double helix, recognizing a specific sequence of nucleotide bases. Interestingly, most transcription factors must act as dimers: Two similar factors first recognize each other, and only then can they bind to the proper promoter. By acting as dimers, a moderate number of transcription factors can control a very large number of genes; for example, three proteins A, B, and C can form six dimers: AA, AB, AC, BB, BC, and CC. The binding of transcription factors to DNA creates a structure to which RNA polymerase can bind. Many transcription factors can be classified together based on their overall structure: "Helix-turn-helix" transcription factors have two regions of α helix (part of their secondary structure) joined by a looping section of protein; in "basic leucine zipper" transcription factors, the two parts of a dimer are held together by interactions of leucine amino acids on their surfaces, and so on.

Because transcription factors come from somewhere else and bind to DNA, they are said to be **trans-acting factors**; promoters, enhancers, and TATA boxes are part of the gene itself and thus are **cis-acting factors**.

Gene expression is also controlled by a family of short RNA molecules called **micro-RNAs**. Many types of micro-RNAs are now known, and there is a large number of names as well; however, each name ends in -RNA, so they are easy to recognize when you read about them. These short pieces of RNA act in many ways, some recognizing sites on DNA itself and binding there, and others recognizing sites on hnRNA or mRNA and attaching to those molecules. In many cases, binding by micro-RNAs inhibits gene expression or mRNA translation. Because RNA consists of nucleotides, it has the potential to recognize sequences in DNA and RNA very precisely and thus exert precise control over gene action.

Analysis of Genes and Recombinant DNA Techniques

Nucleic Acid Hybridization

The two halves of a DNA double helix can be separated by heating them just enough to break the hydrogen bonds between complementary bases. This separation, which produces a solution of single-stranded DNA molecules, is called both **DNA melting** and **DNA denaturation.** If the solution of single-stranded DNAs is cooled slowly to a temperature low enough for hydrogen bonds to reform, two molecules with complementary sequences form hydrogen bonds and stick together whenever they collide. If the two halves of one piece encounter each other, all of their bases pair and adhere firmly even at a relatively high temperature (**Figure 15-15a, b, and c**), but just by chance, many pieces have short sequences that are complementary, whereas most of their sequences are not. If these encounter each other, they form only a small number

Figure 15-15 (a and b) Mild heating of DNA disrupts hydrogen bonding, allowing the two halves to separate into single-stranded molecules. (b and c) Cooling allows them to reanneal. (b and d) In the millions of nucleotides in a single DNA molecule, short sequences occur many times simply by chance, so almost any two pieces have a few short complementary sequences; however, much of the sequences are mismatched and do not adhere. The poorly bonded DNA duplexes fall apart if the solution is still warm, giving each piece another chance to encounter its true complementary partner. (e) If two identical pieces of DNA are broken at random, each produces pieces perfectly complementary to those produced by the other, but because the pieces are unequal in length, a large amount of single-stranded DNA remains after allowing them to reanneal.

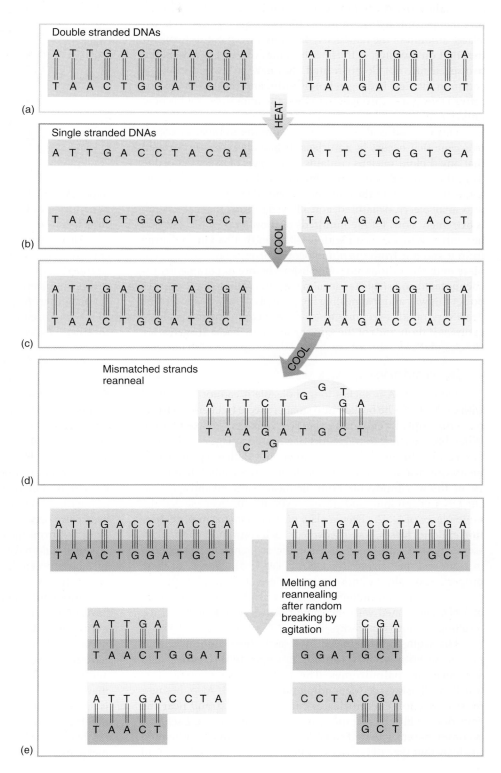

of hydrogen bonds and probably fall apart again if the mixture is cooling only slowly (Figure 15-5a, b, and d). As the degree of complementarity decreases, the stability of pairing decreases.

The reformation of double-stranded DNA by cooling a solution of single-stranded DNAs, called both **DNA hybridization** and **reannealing**, is used to determine the relatedness of two types of DNA. For example, DNA can be extracted from two organisms, cut into small pieces, and melted. The sample from one is attached to a filter, and the other sample, which has been made radioactive (or fluorescent), is poured over it at a temperature that permits hybridization. If the two organisms are related, a large amount of the radioactive DNA hydrogen bonds firmly to the DNA on the filter, so the filter becomes very radioactive. If the two species are not closely related, they have so few sequences in common that few hydrogen bonds form, and most of the radioactive DNA pours through; therefore, the filter does not become radioactive.

This method is also used to measure the number of copies of a gene that occurs in a nucleus. It had always been assumed that any diploid nucleus contains two copies of every gene, one from the paternal parent and one on the homologous chromosome inherited from the maternal parent. If such DNA is broken into small pieces, denatured, and allowed to cool, the reannealing is extremely slow. Of the hundreds of thousands of pieces, each has only two that can pair with it—its own that had melted away from it and one from the homologous chromosome. Thus, each piece undergoes thousands of collisions before its complement is encountered and stable reannealing occurs. Although most genes do act this way and reannealing is extremely slow, a portion reanneals very rapidly, seeming to encounter complementary pieces easily, as if thousands of the same gene occur. These are genes for rRNA and tRNA, which have thousands of copies in each nucleus.

■ Restriction Endonucleases

Natural DNA is such a long molecule it cannot be worked with easily. It can be broken into smaller pieces by chemical treatment or simply by agitating a solution violently (Figure 15-15e). When breaking it into fragments of a more manageable size, it is often critical to cut it at specific known sites so that repeat experiments can yield the same pieces as the first experiment. Before the 1970s, this was impossible; both chemical treatment and agitation cut DNA at random, and no two experiments ever yielded the same pieces of DNA. Then a class of bacterial enzymes, **restriction endonucleases**, was discovered. Each restriction endonuclease recognizes and binds to a specific sequence of nucleotides in DNA and then cleaves the DNA (Table 15-4). Because of these properties, we always know exactly where DNA will be cut by a particular restriction endonuclease, and when two identical batches of DNA are treated with the same restriction endonuclease, the resulting fragments are always the same.

The sequence recognized by a restriction endonuclease is present in both strands, running in opposite directions; such sequences are **palindromes**. The sequence can be read "forward" or "backward," depending on the strand. Also, because most restriction endonucleases cut each DNA strand near the ends of the palindrome, the two cuts are not aligned. Each end is complementary to and capable of pairing with any other end made by that type of restriction endonuclease; the ends are said

TABLE 15-4	Sites Recognized by Restriction Endonucleases
Endonuclease ↓	Recognition Site
EcoR1	5'-GAATTC-3' 3'-CTTAAG-5' ↓ ↑
BamH1	5'-GGATCC-3' 3'-CCTAGG-5' ↓ ↑
HindIII	5'-AAGCTT-3' 3'-TTCGAA-5' ↓ ↑
Hpal	5'-GTTAAC-3' 3'-CAATTG-5' ↑ ↓
PstI	5'-CTGCAG-3' 3'-GACGTC-5' ↑

to be "sticky" (**Figure 15-16**). Currently, more than 100 restriction endonucleases have been discovered, giving us a large choice of cleavage sites.

All fragments produced by a particular class of restriction endonucleases have exactly the same sequence in their single-stranded ends. Thus, if fragments made from the DNA of one organism are mixed with those of another organism, they adhere to each other. A DNA repair enzyme, **DNA ligase**, can be added to the mixture to repair the cuts so that the two fragments join together. DNA prepared by this method is **recombinant DNA**.

■ Identifying DNA Fragments

Evolutionary Studies After restriction endonucleases have acted, the DNA fragments can be identified and used. Most simply, fragments are used directly to study the evolution of DNA. For example, plastid DNA is a small molecule containing only 60,000 to 80,000 base pairs. Treatment with the Pst I restriction endonuclease produces about 10 to 20 fragments. These can be separated by gel electrophoresis and then made visible by staining, producing a **restriction map** of the plastid DNA (Figure 15-17). The number of fragments reveals the number of Pst I sites and the number of base pairs between them in the plastid DNA. If plastid DNA is extracted from the chloroplasts of a second, closely related species, it should have the same number of Pst I sites and the same spacing between them as in the first species. If two species are not closely related, their fragment profiles differ: There is a **restriction fragment length polymorphism** (RFLP). A mutation may have altered one of the Pst I sites so that Pst I neither recognizes nor binds to it. If so, one less fragment is present, and one of the remaining fragments is extra long. Conversely, a mutation may have altered an ordinary sequence, converting it to a new Pst I site and thus producing an extra fragment. In addition, two species may have equal numbers of fragments, but with a fragment that is longer in one species than the corresponding fragment in the other: A mutation has added extra base pairs to one species or removed some from the other. RFLP analysis is easy, quick, and inexpensive, and it can be done with various restriction endo-

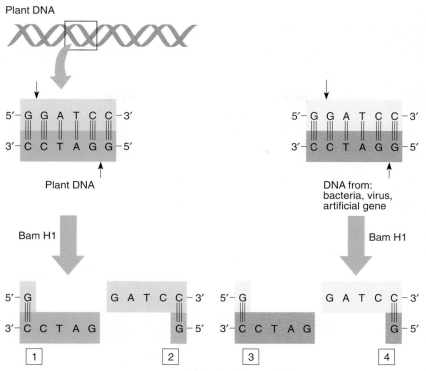

Figure 15-16 Most restriction endonucleases cut near the ends of palindromes, resulting in pieces with sticky ends. DNA pieces from different sources can be mixed together and tend to adhere, making it much simpler to work with them. Both pieces 1 and 3, even though from different organisms, have the same sequence at their ends because Bam H1, like all restriction endonucleases, binds to and cuts the DNA only if it finds a particular sequence. The same is true of pieces 2 and 4.

nucleases, each providing its own information about how much change has occurred in the DNA.

For further analysis, some or all fragments can be sequenced; that is, the identity of every nucleotide can be revealed in order by techniques described below. Similar analyses can be performed with mitochondrial DNA, but mitochondria are used more often in studies of animals and fungi. For plant studies, plastid RFLP analysis is more convenient. Nuclear DNA cannot be used easily because there is too much of it; it produces thousands of fragments, and the addition or loss of a few cannot be detected.

Physiological Studies In many cases, the objective of the experiment is to locate and isolate one particular fragment, such as the

Figure 15-17 (a) Plastid DNA from species A can be cut with a particular restriction endonuclease, resulting in several pieces that can be identified by *gel electrophoresis.* The enzyme digest is placed on a gel slab and a voltage is applied. Phosphate groups on the nucleotides cause the pieces to move in the electrical field, and the shortest DNA pieces slip quickly through the gel matrix, thus moving the farthest. The largest DNA pieces can barely move through the gel; therefore, they stay at the top. After staining, the gel reveals the number and length of pieces. Here, we have indicated which band on the gel corresponds to each position in the plastid DNA circle, but you would not know that from this type of experiment. (b) In a closely related hypothetical species B, a mutation has changed a nucleotide between a and b so that that region does not have the proper sequence for the enzyme, so no cutting occurs. Electrophoresis shows that the two species have three bands in common and that species B has one band as long as the combined length of the missing two.

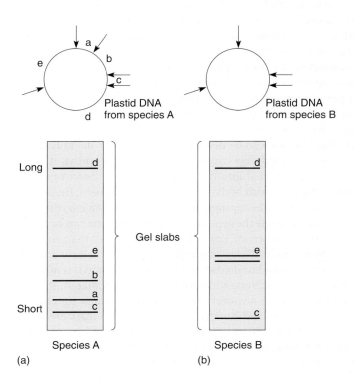

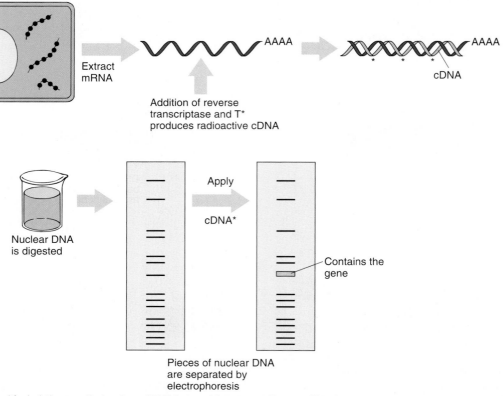

Figure 15-18 The method for isolating a particular piece of DNA that contains a gene of interest. T* indicates radioactive thymidine; the gel slab contains fragments made from restriction endonuclease treatment of nuclear DNA. The method is described in the text.

one that contains a specific gene. The following method can be used if any cell forms large quantities of the protein; for example, developing cotyledons of legumes such as beans and soybeans produce large amounts of the storage proteins phaseolin and glycinin, respectively. While the cells are actively synthesizing the protein, ribosomes can be extracted from them and mRNA obtained. The mRNA can be mixed with **reverse transcriptase**, a virus enzyme that synthesizes DNA using RNA as a template (**Figure 15-18**). This **complementary DNA** (**cDNA**), is complementary to the exons of the gene. It is synthesized using radioactive nucleotides; it can then be placed on the various DNA fragments produced by the restriction endonuclease treatment of nuclear chromatin. The cDNA forms hydrogen bonds to whichever fragment contains the gene, and this bonding can be detected by assaying for the radioactivity. The gene itself may contain a restriction site so that it is cut in two and separated into two fragments; the radioactive cDNA probe hybridizes with both fragments. If this happens, the experiment may have to be repeated with other restriction enzymes that do not cleave inside the gene itself. The best enzyme can be found only by trial and error. Few cells in plants, other than those in cotyledons, produce large amounts of just one protein; as a result, most cells contain hundreds of types of mRNAs, and it is not possible to know which one codes for the protein of interest. Reserve proteins of seeds, however, are some of our most significant, important foods, and hundreds of billions of dollars are spent every year buying storage proteins in the form of beans, wheat, corn, and soybeans. These significant proteins are the objects of intensive study and genetic manipulation.

A second method can be used if mRNA is difficult to obtain, but something is known about the amino acid sequence in the protein, even just a little of it. If the sequence of 5 to 10 amino acids is known, the genetic code in Table 15-2 can be used to guess the probable sequence of nucleotides in the corresponding exon. For example, if the known protein sequence contains phenylalanine, the exon may contain either UUU or UUC. After two or three probable nucleotide sequences have been chosen, they can be synthesized by machines that produce DNA in the form of a radioactive probe. This is used to detect which fragment contains the complementary sequence.

If neither of these approaches is possible, all of the restriction endonuclease fragments from the plant are modified and mixed with bacteria. The bacteria take up the fragments; some take up many; others absorb one, and some take up none. In many cases, the plant DNA is digested by the bacterium's own restriction endonucleases, but often it is incorporated into the bacterium's DNA and replicated. The bacteria are spread over dozens of Petri dishes, and every bacterium grows into a small colony. Each colony is tested to see if it is making the protein of interest. Any colony that is producing the protein must be replicating the gene as well as transcribing and translating it. That colony can be transferred to a new Petri dish and grown. Then DNA is isolated from some of its members and is treated with the same restriction endonuclease that was used to obtain the original plant fragments. At least one fragment of the bacterial digest should match in size one fragment of the plant digest; that fragment contains the gene.

It is not possible to compare each of the dozens of bacterial fragments with the tens of thousands of fragments from the original plant material, but that is no longer a problem. Genetically modified bacteria are used, and we know every type of fragment that the bacterial DNA will be digested into when treated with any restriction endonuclease if it has not picked up foreign DNA from our experiment. Any fragment that does not match the maps of the bacterial DNA must correspond to the plant fragment that was introduced.

Expression profiling uses cDNAs to examine gene expression during development or to compare development in one species with that in another. For example, all mRNA can be extracted from a young leaf primordium and converted to cDNA; then each type of cDNA is separated from the others. Each type is attached to many glass microscope slides in an orderly matrix, a **DNA microarray**: Each slide will have thousands of microscopic dots, each dot being a different cDNA and each slide being identical to all other slides (**Figure 15-19**). Then mRNA is extracted from an older leaf primordium and placed onto one of the slides such that each mRNA can find and bind to the spot that has its cDNA. Excess mRNA is washed off, and chemicals are added to detect DNA/RNA pairs; then the slide is examined to see which spots have picked up mRNA. Spots that have no mRNA bound to them correspond to genes that became inactive as the leaf developed from the younger to the older stage. Similarly, mRNA from flower petal primordia might be added to another slide to see which genes are active in both leaf and flower primordia. This does not tell us what the genes are or even their sequence, but by examining microarrays, we can study general gene expression patterns.

DNA Cloning

The method of placing DNA fragments into bacteria, as just described, is an extremely useful technique of **DNA cloning**. The colony that contains the important fragment can be subcultured and grown easily, and each time a bacterium divides, a new copy of the experimental DNA fragment is made. The bacteria can be cooled, induced to become dormant, and stored for years. When the gene is needed again, the bacteria are revived and cultured. As long as the proper restriction endonuclease is used, the gene can be obtained easily.

Placing the original plant DNA fragments into bacteria is much easier than it may seem. Fragments are not simply mixed with bacteria, but are typically combined with plasmids or virus DNA. A **plasmid** is a short, circular piece of DNA that occurs in bacteria and acts like a tiny bacterial chromosome. Plasmids can move from one bacterium to another and are readily taken up from solution by bacteria. Recall that many restriction endonucleases cut DNA near the ends of a palindrome, resulting in sticky ends. If the plasmid or virus DNA is cut with the same enzyme used to isolate the DNA fragment from the plant, then when the two batches of DNA are mixed, some of the plant DNA adheres to plasmid DNA by means of their complementary ends. DNA ligase is then added to bond the pieces covalently into one continuous double helix (**Figure 15-20**).

Several plasmids, such as pBR322, have been genetically engineered to be ideal DNA fragment **vectors** (carriers). In their current highly modified form, they have factors that allow them

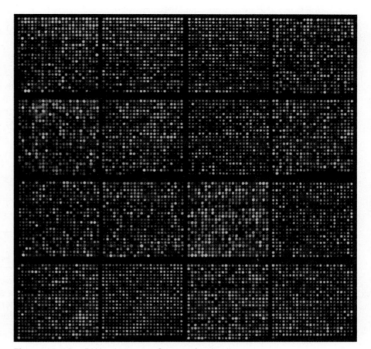

Figure 15-19 DNA microarray. See text for details.

to enter bacteria readily and be replicated. They have antibiotic resistance genes, and they have single sites for several restriction endonucleases so that fragments can be incorporated in precisely known positions. After the fragment is added, the plasmid is mixed with bacteria and absorbed. After a few hours, all bacteria are treated with the appropriate antibiotics; those that did not absorb a plasmid die. The survivors have the plasmid and therefore DNA fragments from the original plant digest.

Several viruses are being used like plasmids; they contain a short fragment of DNA and can infect bacteria and then replicate. As with plasmids, certain viruses have been modified into highly efficient vectors for genetic engineering.

Genetically modified bacteria are easy to culture and handle in the laboratory, but because their genomes are short circles, they cannot accept long fragments of DNA for cloning. Most plant chromosomes are hundreds of times larger than a typical bacterium, and thus, even a short chromosome from a plant nucleus would have to be cut into thousands of fragments for cloning in bacteria. **Yeast artificial chromosomes** (**YACs**) can be used instead. A YAC contains the essential parts of a yeast chromosome (telomeres, a centromere, replication start sites) plus a gene for drug resistance and parts of plasmid that can accept experimental DNA cut with the proper restriction endonuclease. Such a YAC is tiny and thus has room for a very large amount of DNA—up to 1 million base pairs—to be cloned. After the experimental DNA is inserted, the YAC is placed into a yeast cell, which is much larger than a bacterial cell and can hold more DNA. The yeast cell is then cultured and allowed to grow and multiply. Every time the yeast cells replicate their own chromosomes, they also replicate the YAC. The yeast can be cultured until billions of cells (and billions of YACs) are present, and then some can be used to obtain the cloned DNA; the rest can be frozen indefinitely for use later.

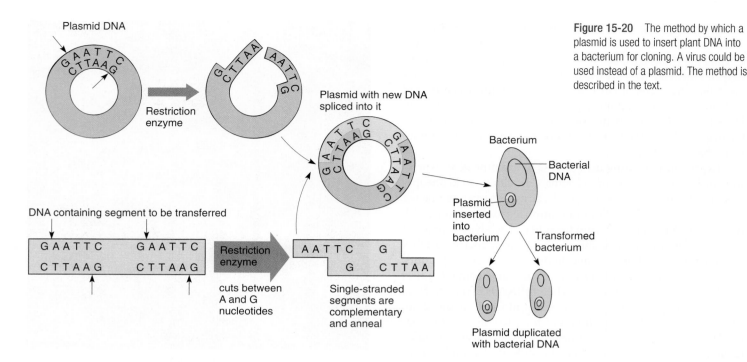

Plasmid DNA

Restriction enzyme

Plasmid with new DNA spliced into it

Bacterium

Bacterial DNA

Plasmid inserted into bacterium

Transformed bacterium

DNA containing segment to be transferred

Restriction enzyme cuts between A and G nucleotides

Single-stranded segments are complementary and anneal

Plasmid duplicated with bacterial DNA

Figure 15-20 The method by which a plasmid is used to insert plant DNA into a bacterium for cloning. A virus could be used instead of a plasmid. The method is described in the text.

The Polymerase Chain Reaction A powerful technique for DNA cloning is the **polymerase chain reaction** (**PCR**), in which only enzymes, not living bacteria, are used. The sequence to be amplified is heated to separate the two strands of the helix; after cooling, DNA polymerase is added, and it replicates both strands of DNA. This enzyme can add nucleotides only to a preexisting nucleic acid; therefore, before the replication can begin, it is necessary to add two types of primer DNA, each complementary to a short region at either end of the sequence to be cloned. End sequences are known if closely related genes are being studied, but often no part of the nucleotide order is known. In that case, artificial DNA—usually a short sequence such as AAAAAA—is added chemically to both ends; then the primer TTTTTT will be effective.

After the primer has hydrogen bonded to the sequence, a molecule of DNA polymerase attaches and begins working toward the other end. After replication is completed, the mixture is heated temporarily to separate the two strands; then the process is repeated. A heat-stable enzyme, Taq polymerase extracted from hot springs bacteria, is used because it is not denatured by the heating.

After each heating/replication cycle, there are twice as many copies of the sequence being amplified. Consequently, extremely small amounts of DNA can be cloned very rapidly, and PCR is used for rare copies of DNA. Examples are small amounts of cDNA made from the mRNA of just a few cells, as well as the DNA present in the early stages of viral infection (HIV, which causes AIDS, is detected very early with PCR) or the DNA present in a hair, a drop of blood, or a bit of skin at the scene of a crime.

DNA Sequencing

Two methods are currently used to sequence DNA. In the **chain termination method**, DNA to be sequenced is first cloned to obtain a large sample and is then divided into four batches. To each batch are added all the enzymes and free nucleotides neces-

sary to carry out DNA duplication. To one tube a small amount of dideoxyadenosine is also added. A *dideoxynucleotide* can be added to a growing DNA; however, it cannot react any further, and the growth of the DNA stops. Nucleotides cannot be added to it. In this tube with dideoxyadenosine, the DNA acts as a template and replication begins; when a T is reached in the template molecules, a few incorporate a dideoxyadenosine and stop. Most growing DNA strands incorporate a normal A and keep on growing; then a few stop at the next T, a few more stop at the next, and so on. When the reaction is complete, the test tube contains thousands of DNA molecules of hundreds of sizes, but each size corresponds to the point where a T occurred in the template DNA being analyzed. Similarly, the second test tube contains a small amount of dideoxy T, the third dideoxy C, and the fourth dideoxy G (**Figure 15-21**).

Until the last few years, the method of analyzing the results was as follows. When all four batches are finished, they are loaded into separate lanes in a gel electrophoresis apparatus and allowed to separate, as in Figure 15-17 and **Figure 15-22**. Each lane contains bands corresponding to the sizes of the DNA molecule. For example, if the DNA contained T as the third, seventh, eighth, and fifteenth base, the lane containing dideoxyadenosine has bands corresponding to DNAs that are 3, 7, 8, and 15 nucleotides long. The DNA sequence can be read immediately (Figure 15-22). At present, gels are almost never used; instead, all processing of DNA samples is now done by machines. Sequences are read and transferred directly to databases.

In the **pyrosequencing method**, DNA is added to a solution with all enzymes for replication. In addition, there are other enzymes that release light when pyrophosphate is present (pyrophosphate is the set of two phosphate groups given off when a nucleotide is added to growing DNA; it is represented as two yellow circles in Figure 15-7b). Then a single nucleotide is added; if it is the correct nucleotide, DNA polymerase incorporates it

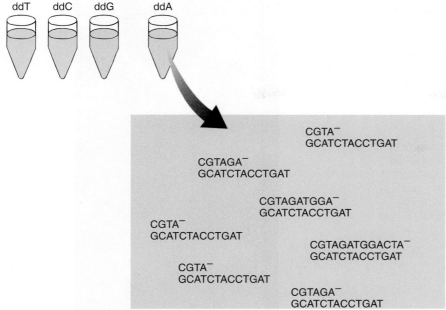

Figure 15-21 Chain termination method of sequencing. Into four test tubes, identical pieces of DNA have been placed; all received everything necessary, but each received a small amount of chain terminator. The one depicted here received dideoxyadenosine. If too much of the chain terminator is added, almost all stop at the first T; if too little is added, almost no chains stop at any T, and most finish the entire template. The correct amount produces a good number of copies of every possible length. After incubation, the contents of each test tube would be placed into a lane of a gel electrophoresis machine and separated by length, as shown in Figure 15-22.

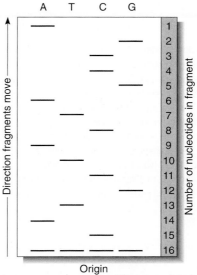

Figure 15-22 Diagram of a gel for a DNA fragment. Each lane marked A, T, C, or G corresponds to one of the test tubes in Figure 15-21. You can read the nucleotide sequence, starting at the top with the shortest, fastest nucleotide. A real gel would be much longer and have hundreds of bands.

and releases pyrophosphate, which causes a flash of light. The sequencing machine records the nucleotide that caused the flash and then introduces another nucleotide to continue the process. When the wrong nucleotide is introduced, there is no flash, and the sequencer notes that, then washes it away and introduces another. The process is repeated thousands of times as the template DNA is replicated.

Computer programs analyze the sequences, searching for regions that might be promoters or enhancers, TATA boxes, AUG start sites, and areas that might be boundaries between exons and introns. If a region is found that appears to have these gene-like features, it is referred to as an **open-reading frame** (**ORF**). To determine whether it actually is a gene, databases are searched to see if the ORF matches the sequence of any known genes. If there is no matching sequence, it continues to be called an ORF. Often there will be some sequence similarity with a known gene in the same or another organism; if the similarity is high (90% or more bases are the same), this ORF is probably a gene with the same function as the known gene, but if the similarity is lower, we are less certain—This may be a real gene, a mutated gene, an ancient duplicated gene that is evolving into junk DNA or simply a random sequence that resembles a gene.

■ Sequencing Entire Genomes

The sequencing described above is only effective for fragments less than several hundred bases long. To sequence a plastid or mitochondrial genome, organelles are extracted from a cell.

Their circles of DNA are isolated and then cloned and divided into several batches. The DNA is cut into fragments, each batch being cut with a different restriction endonuclease, and then each fragment in each batch is sequenced. In the first batch, we might have several hundred fragments, all perfectly sequenced, but we do not know the order of the fragments. The same is true of the second batch, but each fragment here will have some portion that matches one or several portions of fragments in the first batch. By aligning the equivalent regions in the two batches, we should be able to align all fragments in proper order. Sequences from the third batch act as a verification (**Figure 15-23**). To sequence the immense genome inside nuclei, each chromosome is isolated from the others and sequenced individually by the same method.

■ Genetic Engineering of Plants

Recombinant DNA techniques have made it possible for botanists to identify, isolate, and study the structure and activity of many genes. They also allow botanists to insert genes into plants that do not normally contain those genes. For example, cotton plants have received a gene from the bacterium *Bacillus thuringiensis*, which codes for a protein toxic to caterpillars but not to other insects or mammals (including humans). If the gene protects the plant, it could prevent as much as $100 million in crop losses annually. Botanists have transferred genes from desert petunias into normal petunias, resulting in plants that require 40% less water. If

■ Endonuclease 1
■ Endonuclease 2
■ Endonuclease 3

Figure 15-23 Very long pieces of DNA are sequenced by cloning the DNA, dividing it into several identical batches and then cleaving each batch with a separate restriction endonuclease. Each fragment is then sequenced. If we had only used endonuclease 1, we would have gotten fragments a–f, but we would not know their proper order. By comparison with the fragments from a second endonuclease, we would see that the left end of fragments a, g, and l are the same. Fragment h matches the right end of a, all of b, and the left end of c, and fragment j matches parts of c and d and so on.

such drought-resistance genes can be transferred to crop plants, the amount of water needed for irrigation will be greatly reduced. Tomatoes have been engineered such that the enzymes which cause them to become mushy are inhibited without inhibiting the other enzymes involved in developing flavor and aroma. The tomatoes can be allowed to ripen fully on the vine but are still firm enough to ship to market.

Obtaining some genes is not very difficult, but at present, inserting them into nuclear DNA properly so that they can be transcribed and translated is not easy. A gene may code for an improved, more nutritious storage protein and should be expressed in cotyledons during seed formation, but if the gene inserts into a region of nuclear DNA that codes for root, stem, or wood characteristics, the engineered gene either may not be activated or may become active in the wrong place. Research on the nature of promoter sites and their interaction with chemical messengers is especially intensive. Many genes have been located that are expressed only in particular tissues such as cotyledons, wood, chlorenchyma, or epidermis. Their promoters are valuable because a cotyledon promoter can be attached to our foreign gene before insertion, increasing the likelihood the gene will respond to the appropriate chemical messenger, regardless of where it inserts into the DNA (**Figure 15-24**).

After a gene and promoter have been prepared, they are attached to an insertion vector, usually the **ti plasmid** from the bacterium *Agrobacterium tumefasciens* (**Figure 15-25**). The plasmid has been modified by botanists by adding genes for antibiotic resistance. The completed plasmid-promoter-gene is mixed with plant cells in tissue culture; they are allowed to grow and then are treated with an antibiotic (often, hygromycin). Those cells that did not take up the plasmid, and those that took it up but are not expressing it are killed by the herbicide. Those cells that have taken up the plasmid and are transcribing the gene and translating the mRNA into protein are resistant to the herbicide and survive. They can be cultured further. The surviving cells are induced to form new plants, which in many cases not only carry the gene and express it but pass it on to their progeny when they undergo sexual reproduction.

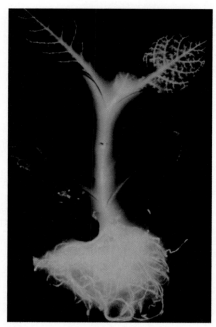

Figure 15-24 The gene for a luminescent protein from fireflies was inserted into a plant cell in tissue culture; then the cell was induced to grow into a full plant. Apparently, this gene had a promoter that is normally activated in every cell because all cells are glowing. By attaching different promoters to this gene before the genetic engineering, it is possible to study when and where the promoter is normally active. Biologists hope to find promoters that, when attached to this gene, cause only specific cells or tissues to luminesce; then those promoters can be used whenever we want to affect specifically those types of cell or tissue.

■ Viruses

In the past, much of the emphasis in studying viruses in plants was on preventing the spread of virus diseases in crops. Now many viruses are especially interesting as possible vectors for plant genetic engineering. Also, because most are RNA viruses (described below), they contain many unusual enzymes that are useful for the experimental manipulation of nucleic acids.

■ Virus Structure

Viruses are extremely small particles that usually contain only protein and nucleic acid. They were originally discovered in 1892 as factors that cause disease but were so small they could not be seen with a microscope and could pass easily through filters with pores fine enough to trap even bacteria. They were never actually seen until the development of the electron microscope (**Figure 15-26**). They were originally thought to be very small cells, but now we know that is not true: They have no protoplasm, no organelles, and no membranes. They cannot carry out their own metabolism independently—they must always parasitize cells of some sort. The great majority of viruses, especially those that attack plants, consist of only one or a few types of proteins and a small amount of nucleic acid (**Figure 15-27**).

Plant viruses always have a simple morphology, either long or short rods or even round particles. Tobacco mosaic viruses are

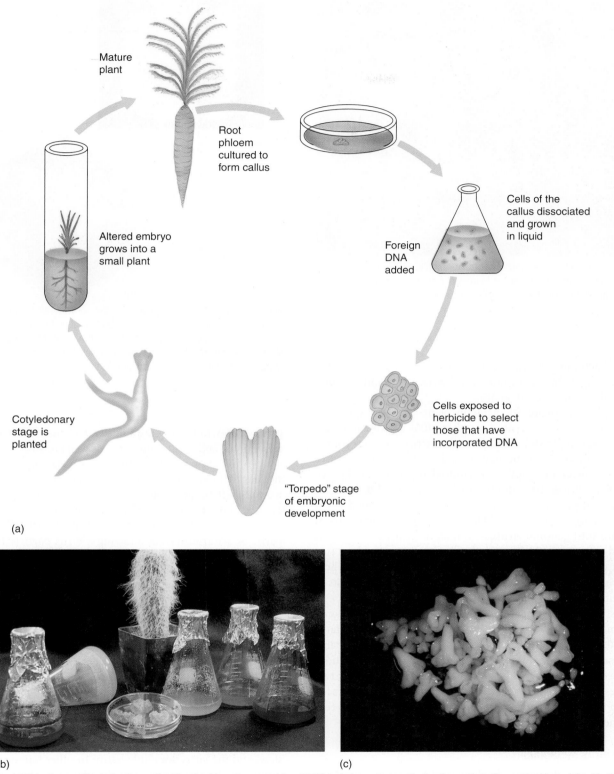

(a)

(b)

(c)

Figure 15-25   (a) This diagram illustrates the method involved in using recombinant DNA techniques to genetically alter a plant. The method is explained in the text. (b) Several components of a tissue culture experiment: the original plant (the cactus *Cephalocereus senilis*), callus culture on agar in foreground, liquid cultures in flasks. The red liquid cultures have been experimentally induced to form pigments. The flask tipped on its side shows that these cells can grow so extensively that they consume all the liquid medium and form a solid mass of cells. (c) Tissue culture experiments can also produce embryos. If they have been genetically engineered, they grow into altered plants.

Labels in diagram (a):
- Mature plant
- Root phloem cultured to form callus
- Cells of the callus dissociated and grown in liquid
- Foreign DNA added
- Cells exposed to herbicide to select those that have incorporated DNA
- "Torpedo" stage of embryonic development
- Cotyledonary stage is planted
- Altered embryo grows into a small plant

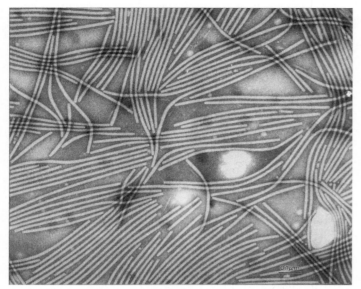

Figure 15-26 Turnip mosaic virus particles are long, narrow filaments. In this preparation, viruses have been isolated from plant cells, washed, and treated with heavy metals to make them visible in an electron microscope (×180,000).

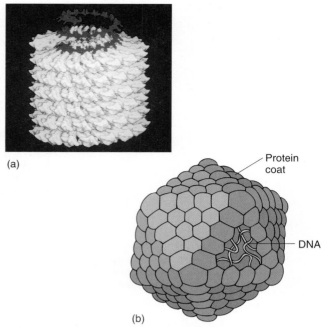

Figure 15-27 Most viruses, but especially plant viruses, are extremely simple. (a) Tobacco mosaic virus contains only one type of protein, which binds to the DNA, forming the long rod. In this form, the DNA is well protected. (b) In the virus shown here, the protein does not bind so closely to the nucleic acid but rather forms a coat, often called a capsid, around the nucleic acid.

rods 15 × 300 nm, whereas citrus tristeza viruses are rods up to 2,000 nm long. Spherical or polyhedral virus are common, often having a diameter of about 60 nm.

The diversity of nucleic acids in plant viruses is great. Of the known viruses, the greatest number (400) are **retroviruses**, which contain single-stranded RNA. Retroviruses are our source of reverse transcriptases, needed to make cDNA. Many viruses that cause cancer in humans are animal retroviruses. Twelve known plant viruses have double-stranded DNA, and there are two groups of unusual types: 10 contain double-stranded RNA, and 15 contain single-stranded DNA. Furthermore, many plant viruses are **split genome viruses**—not all of their nucleic acid is packaged as one particle. Instead, the virus consists of at least two different particles, and both must be transmitted to a new host cell for viral metabolism to occur. Alfalfa mosaic virus consists of four distinct particles.

Most plant viruses have enough nucleic acid to code for only a very small number of proteins. Tobacco mosaic virus contains only 6,400 nucleotides; its one and only coat protein contains 158 amino acids; thus, a minimum of $158 \times 3 = 474$ nucleotides are needed just for that. Three other genes exist in this virus; two large ones code for replicase enzymes, and a smaller one codes for a protein thought to mediate spread of the virus from cell to cell. Most other plant viruses are similarly small.

■ Virus Metabolism

Viruses must always invade a living cell in order to reproduce, and all known types of organisms are attacked by viruses: plants, animals, fungi, protozoans, algae, and prokaryotes. Viruses that attack bacteria are called **bacteriophages** or phages, but these are viruses just like the others. Viruses that attack bacteria and animals are usually extremely specific: A particular type of virus can attack only a certain species of host. The same is true of many plant viruses, but others are able to attack many related plant species. Fungal and algal viruses are less well known.

To invade a plant, viruses rely on damage to living cells, such as through the action of aphids, chewing insects, or open wounds left by pruning or breaking. If an insect has chewed or sucked on an infected plant, the virus particles adhere to its mouth parts and feet. If these penetrate a host cell, some virus particles are transferred, allowing a virus particle to enter the protoplasm directly, completely bypassing the cell wall. During experimental studies of virus activity, healthy plants are infected by first abrading the epidermis with sandpaper to create fine breaks in the cells and then rubbing into the wound a solution of virus particles extracted from an infected plant; alternatively, freshly crushed, infected leaves are rubbed into the abrasion.

After the virus is inside a suitable host cell, protein molecules of the coat fall away from each other, releasing virus nucleic acid. If the virus is a DNA virus, its liberated DNA is recognized by the plant's own RNA polymerases, and transcription of viral genes into mRNA begins. These mRNAs are translated by the plant's ribosomes, and within minutes of infection, new viral protein appears in the cell. Transcription and translation of some viruses are rather slow, and the virus does little harm. In other types, viral mRNAs dominate the ribosomes, and the plant's protein-synthesizing apparatus is completely taken over. At the same time that RNA polymerases are transcribing viral DNA, DNA-replicating enzymes bind to and copy the DNA, making new viral double helixes. As the number of viral DNA molecules increases, the rate of viral transcription and translation also increases. Finally, all plant cell metabolism is redirected to viral metabolism.

Retroviruses, which have RNA but no DNA, may act as an mRNA and be picked up by a plant ribosome and translated. Usually one of the first proteins to result is reverse transcriptase, which binds to the RNA and synthesizes a complementary molecule of DNA. Another enzyme then synthesizes the other half of the DNA duplex, and the rest of the steps are the same as for a DNA virus. Some retroviruses bring molecules of reverse transcriptase with them as part of their coat; therefore, reverse transcription is the first step.

As virus components become more abundant, viral nucleic acid moves into surrounding cells by means of plasmodesmata, and perhaps by passing through the wall as well. Viral nucleic acid that enters phloem spreads rapidly; the only tissues that are somewhat safe are root and shoot apical meristems, which do not contain phloem. Even in a heavily infected plant, the shoot apex itself may remain virus free.

The rate at which viruses take over host cell metabolism is variable; in many plants, it is slow and viruses never become abundant. Such plants show few disease symptoms, and the presence of virus is detected only by electron microscopy. This type of virus is common in many crop plants. If the virus multiplies quickly and dominates the cell, symptoms such as chlorosis, necrosis, and leaf curling may appear (**Figures 15-28** and **15-29**). Such viruses are **virulent**.

Some viruses, once freed of their coat protein, insert viral DNA into the host DNA. Little is known about this process, but apparently, viral DNA lies next to host DNA, and then breaks are made in both molecules. The ends join and the cuts are healed by DNA ligase. Such viral DNA usually lies dormant, perhaps simply because so little of the eukaryotic DNA is expressed in any particular cell. However, if this insertion happens in young plant tissues, viral DNA is replicated along with plant DNA during the S phase of every cell cycle, and both daughter cells receive infected DNA during mitosis. Such viruses are termed **temperate** and may produce no symptoms whatsoever; they may remain hidden forever and die when the plant dies. In some, however, they become active and reproduce just like virulent viruses. The factors that induce conversion from temperate to virulent are not known. Perhaps the viral DNA inserted itself downstream from the promoter region for a gene that is now being activated as part of the cell's differentiation. After the promoter region allows RNA polymerase to bind and transcribe, the viral DNA is transcribed as well. This hypothesis has not yet been proven, and other factors may be important.

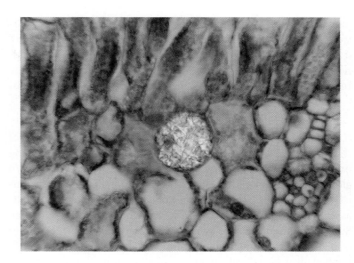

Figure 15-28 These cells of tobacco leaf are infected with tobacco mosaic virus. Although the cytoplasm is almost filled with virus particles, the surrounding cells appear healthy, and the vacuoles are virus free (×400).

■ Formation of New Virus Particles

At some point, viral components assemble into new particles (Figure 15-27). For many viruses, such as tobacco mosaic virus, this is a self-assembly process. Viral coat protein has a tertiary structure that causes it to bind to viral DNA. This binding then permits it to attract and adhere to more viral protein, and a new protein/DNA viral particle is quickly assembled. The protein monomers by themselves do not adhere to each other; viral DNA must be present. No other components are necessary, and even in a test tube, the two self-assemble into infectious virus particles.

In many types of animal viruses and bacteriophages, one of the last proteins made is an enzyme that destroys the host cell, causing it to burst (lyse) and release virus particles into the environment. This

(a)

(b)

Figure 15-29 Squash mosaic virus affects not only squash, but related plants as well, such as watermelon. Symptoms occur in both leaves (a) and fruits (b).

does not happen with plant viruses; perhaps the cell wall is too massive to be digested. Instead, virus particles remain in the cell until it is broken open by insects or larger animals. If an infected leaf is never eaten but instead is abscised in autumn, virus particles are released as the leaf decomposes; these too can be spread by animals.

Origin of Viruses

Many, if not most, viruses are actually portions of genes of the host species or a species closely related to the host. As living organisms are damaged or die and decay, their nuclei break down along with the rest of the cell material. It is possible that occasionally a fragment of a chromosome codes for proteins that can self-assemble into a crude coat and have some infectious potential. Such a fragment has the possibility of acting as the forerunner of a virus; the fragment does not have to be as efficient as a full-fledged virus because it can evolve just like any other genetic system. With a long span of time, natural selection favors mutations in this DNA fragment that improve its survival rate.

An alternative hypothesis postulates that viruses are highly evolved, extremely efficient parasites. Whenever an organism can obtain required compounds from its environment, it is selectively advantageous not to waste energy synthesizing them, but even the most reduced, simplified parasites are vastly more complex than viruses. Could viruses be reduced parasitic bacteria instead? This seems doubtful for most viruses because the metabolism of plant and animal viruses should then resemble bacterial metabolism, but instead, it matches plant or animal metabolism.

Plant Diseases Caused by Viruses

Plants suffer from at least a thousand different virus-caused diseases. Because symptoms are similar to those of mineral deficiency or other environmentally caused problems, detection can be difficult (**Figure 15-29**). Few effective treatments exist for plants infected with viruses. We do not try to cure entire crops of viral disease; rather, we try to maintain healthy, uninfected breeding stock for virus-free seeds or cuttings. Heat treatment inactivates some viruses; entire plants can be kept in hot (35°C to 40°C) growth chambers or greenhouses for several weeks to several months, after which they may be free of virus. Even these techniques are ineffective in most plants and against most virus diseases; the best policy is to use virus-free plants and protect them from infection. Virus-free plants are obtained by shoot meristem propagation or sometimes by normal sexual production of seeds, if stamens and carpels can grow and set seed more rapidly than virus particles can infect them.

Summary

1. All information required to specify protein primary structure—the sequence of amino acids—is stored as the sequence of deoxyribonucleotides in DNA.
2. Cell differentiation is based largely on differential activation of genes and control of the processing of heterogeneous nuclear RNA into messenger RNA.
3. The exact details of the mechanism by which a plant hormone induces differential activation of either nuclear or organellar genes are not known. Binding of a hormone to its receptor results in the formation of transcription factors that bind to DNA promoter regions.
4. The genetic code consists of triplets of nucleotides, each triplet coding for only one amino acid, or for STOP or START. The code is degenerate, each amino acid being coded by several codons.
5. Genes consist of a promoter region that contains enhancer elements and a structural region that usually contains both exons and introns.
6. In transcription, RNA polymerase attaches to the promoter region, moves to a start site, and then polymerizes RNA, being guided by base pairing in a short region of single-stranded DNA. Both introns and exons are transcribed.
7. Heterogeneous nuclear RNA is processed to mRNA and then transported to the cytoplasm where it binds to ribosomes. Each ribosome has a large and a small subunit, four molecules of rRNA, and approximately 80 proteins.
8. Amino acids are carried to ribosomes as part of an activated tRNA, each of which has an anticodon complementary to the codon for the amino acid it carries. All tRNAs have similar structures.
9. Restriction endonucleases cut DNA at specific sequences; the resulting pieces can be melted to the single-stranded state. Single-stranded nucleic acids from different sources can be mixed and allowed to hybridize, either as a measure of their relatedness or as part of the construction of a new molecule of DNA.
10. Specific sequences of DNA can be synthesized artificially by incorporating one copy into a vector and then inserting the vector into a bacterium. As the bacterium reproduces, the sequence of DNA is reproduced as well.
11. Most viruses are short pieces of DNA or RNA that contain a few genes closely related to normal host genes. Most plant viruses have RNA, not DNA, and a coat of just one type of protein.
12. Viruses infect plants through wounds and then divert the plant's nucleic acid and protein-synthesizing metabolism to the synthesis of more virus molecules, which then self-assemble into complete virus particles.

Important Terms

anticodon
bacteriophages
chromatin
codons
complementary DNA (cDNA)
differential activation of genes
DNA cloning
DNA denaturation
DNA hybridization
DNA ligase
DNA microarry

eukaryotic initiation factors (eIFs)
exon
expression profiling
gene
introns
messenger RNA (mRNA)
palindromes
polymerase chain reaction (PCR)
promoter region of a gene
recombinant DNA
restriction endonucleases

restriction map
retroviruses
reverse transcriptase
ribosomal RNA (rRNA)
ribosomes
start codon
stop codons
structural region of a gene
transcription
transfer RNA (tRNA)
yeast artificial chromosomes (YAC)

Review Questions

1. Plants are composed of numerous types of cells that are all unique because they have distinct metabolisms. What are these metabolisms based on?

2. The information needed to construct each type of protein is stored in _____.

3. Because an organism grows by duplication division, all its cells have (choose one: identical, unique) genes.

4. What is meant by the differential activation of genes? Explain how this affects the synthesis of cutin and P-protein.

5. Cutin, lignin, and chlorophyll are not proteins. How is it possible for genes to control the synthesis of these polymers?

6. A gene is made up of (choose one: RNA, protein, DNA, carbohydrate).

7. For each of these symbols, write out the full name of the base and of the nucleoside. Indicate which are components of DNA and which are components of RNA:
   a. T
   b. A
   c. G
   d. C
   e. U

8. The text describes three important ways that cells protect their genes. List these methods.

9. DNA does not participate directly in the synthesis of proteins. What molecule transfers the information in DNA to the site of protein synthesis?

10. Histones are proteins in the nucleus, and DNA wraps around them forming a spherical structure. What is the name of the structure?

11. What is the name of the enzyme that digests DNA?

12. What is a codon? How many exist? Why is the genetic code described as being degenerate?

13. Examine Table 15-2. Which amino acids are coded by the following codons? (Hint: for the first base [letter] in each of the codons listed below, look at the leftmost column of the table labeled "First Base." For the second base in each codon, look at the top row of bases, and for the third base, look at the column on the right-hand side of the page).

   a. UUU
   b. UCU
   c. UAU
   d. UGC
   e. CUU
   f. CGA
   g. CGG
   h. AAA
   i. AAG
   j. GAG
   k. GGU
   l. GGC
   m. GGA
   n. GGG

14. What is unusual about the answers for Questions 13k, 13l, 13m, and 13n? Because the multiple codons exist for most amino acids, the genetic code is said to be _____.

15. Examine Table 15-2. What do the following codons specify:
   a. UAA
   b. UGA
   c. UAG
   What happens when a ribosome comes to one of these codons?

16. Examine Table 15-2. The codon AUG codes for the amino acid methionine and also for "start." What does "start" mean?

17. What are the two portions of a gene? Which portion actually codes for the amino acid sequence?

18. Describe the promoter and structural region of a nuclear gene. What is a TATA box? What are exons and introns?

19. The structural portion of genes contains two distinct types of regions—exons and introns. Which consists of codons that are eventually translated into the amino acid sequence of a protein, and which consists of codons that are not expressed?

20. After RNA polymerase binds to DNA, it begins making mRNA. What is the name of this process?

21. What is the name of the cytoplasmic particles that translate mRNA into proteins? What type of RNA brings amino acids

to this particle? This RNA has a region called an anticodon. What does the anticodon do?

22. One of the codons in mRNA that specifies the amino acid phenylalanine is UUC. What is the anticodon on the tRNA that carries phenylalanine?

23. What is a frameshift error? Does the binding of methionine tRNA to the small subunit help reduce frameshift errors? If so, how?

24. If a molecule of mRNA had the following sequence, what would be the sequence of the gene that coded the mRNA? Which amino acids would be incorporated into the protein?

gene: 3'-_____-5'
mRNA: 5'-UCGAACAAUUGUCCCGGCCUC-3'
protein:

25. Describe the steps of polypeptide elongation by ribosomes.

26. When a stop codon is pulled into the A channel of a ribosome, what happens? Is there any tRNA with an anticodon that is complementary to the stop codon?

27. There are three main classes of RNA involved in protein synthesis: _____ RNA, _____ RNA, and _____ RNA.

28. Describe DNA melting and DNA reannealing.

29. What is a restriction endonuclease?

30. What are palindromes, and how are they related to restriction endonucleases? Why are they useful for inserting one piece of DNA into another?

31. DNA fragments can be identified. For example, if plastid DNA is treated with Pst I restriction endonuclease, it cuts the DNA into about 10 to 20 fragments. If these are separated on an electrophoresis gel and then stained, this is a _____ _____ of the plastid DNA.

32. Imagine that plastid DNA of species A is treated with a restriction endonuclease in order to make a restriction map. Now imagine that species B is not very closely related to species A, but species C is. If plastid DNA of species B and species C are treated with the same restriction endonuclease that was used for species A, will the restriction map of species B or species C be more similar to that of species A?

33. If mRNA is extracted from a plant and mixed with reverse transcriptase, what is the name of the DNA that is produced? If it was synthesized with radioactive nucleotides, what can be done with it?

34. Briefly describe how a piece of DNA can be cloned using bacteria. Why would it be important to use the same type of restriction endonuclease in the harvest phase as in the preparation phase? After the DNA had been cloned, how would you go about sequencing it?

35. Briefly describe how a piece of DNA can be cloned using PCR.

36. In the genetic engineering of plants, after a gene and promoter have been prepared, they must be inserted into the plant with an insertion vector. A very important vector comes from *Agrobacterium tumefasciens*. What is the vector's name? After the vector and the gene-plus-promoter have been added to many plants, the plants are treated with a herbicide. Why? Which plants will survive?

37. Corn has been genetically engineered to produce the anti-insect poison of a bacterium *Bacillus thuringiensis*. What is one benefit of this? What are some of the risks?

38. Viruses are extremely small particles that usually contain only _____ and _____ _____.

39. The diversity of nucleic acids in plant viruses is great. Which class of viruses contain single-stranded RNA? These viruses are the source of which important enzyme (Hint: this is the enzyme needed to make cDNA)?

40. There are three hypotheses about the origin of viruses. Briefly describe each hypothesis.

41. About how many plant diseases are caused by viruses?

BotanyLinks

http://biology.jbpub.com/botany/5e

BotanyLinks contains a variety of resources and review material designed to assist in your study of botany.

Visit the Web Exercises area of BotanyLinks to complete these questions:

1. How many viruses attack plants? Do they attack organisms like algae, fungi, or other viruses? Go to the BotanyLinks home page to begin researching this subject.

2. Does being a scientist have moral and ethical implications? Now that we really can genetically engineer our food plants and clone animals (even humans?), what ethical responsibilities do scientists have? Go to the BotanyLinks home page for discussion on this subject.

PART 3

Genetics and Evolution

The focus of this part is the mechanism by which DNA and its information are passed from parent to progeny; how that DNA changes over time, resulting in new types of organisms; and how the environment and natural selection have interacted to produce hundreds of thousands of species of prokaryotes, algae, fungi, animals, and plants.

DNA serves as an archive of information that, interacting with environmental information and resources, produces an organism's body—its structure, metabolism, and biology. DNA also carries the information needed to make more DNA so that during reproduction—either of the cell or of the entire organism—each progeny receives copies of the DNA and the information it contains. Errors occur occasionally, and molecules of DNA are not copied perfectly; each version may code for slightly different enzymes or in some other way produce an organism slightly different from parents or siblings. If the world were absolutely uniform, we might expect that whenever a mutation resulted in a new type of individual, either the new type or the original would be favored by natural selection until it was the only type left; a more advantageous metabolism or structure would be more advantageous everywhere.

However, our world is complex, containing aquatic and terrestrial environments, areas with mild temperatures and others with temperature extremes, and a diversity of rainfall, soil types, pathogens, pollinators, and so on. Consequently, alternative biologies may each be selectively more advantageous in different conditions. The diversity and richness of the environment result in diversity and richness of organisms.

Part Opener Image: The wheat being harvested here has been artificially bred to have many desirable traits. One is that the plants stand upright so that the machine can harvest them easily. Another is that the grains have proteins that are both nutritious and have good qualities for making bread or pasta. The science of genetics analyzes and modifies the traits of the organisms. Evolution is the process in which the traits of a species change from generation to generation.

Chapter Outline

chapter 16

Genetics

Concepts

Genetics is the science of inheritance. The chemical basis of genetic inheritance is the gene, the sequence of DNA nucleotides that guides the construction of RNA and proteins and also controls construction of more copies of the genes themselves. If all plants of a species had exactly identical nucleotide sequences in their DNA, then all those plants would be identical physically, but virtually all genes occur in multiple forms known as **alleles**, the alleles of a particular gene differing from each other in their sequence of nucleotides.

Alleles arise by mutation; if the nucleotide sequence is changed (mutated) in any way, the new sequence is a new allele. Mutations can occur in any gene in any individual, and thus, gradually a population of separate plants comes to have a variety of alleles. The types of alleles that a single individual has are called its **genotype**, and the expression of those alleles in the individual's size, shape, or metabolism is **phenotype** (Figure 16-1). As a result of mutations, the population of individuals has varying genotypes and phenotypes. They are not identical, as is apparent from considering humans.

Chapter Opener Image: The red poppies are all plants of the same species, *Papaver rhoeas*, and, therefore, they are able to interbreed. As insects go from one flower to another, they transfer pollen between plants. Each pollen grain is a haploid male gametophyte that contains two sperm cells, and if pollen lands on the stigma of the correct flower, it grows as a pollen tube, carrying the sperm cells toward the egg cells in the ovules in the ovary. Poppy flowers are perfect (they have both stamens and carpels), so each flower can be the paternal parent (the "father") to many seeds in other plants as well as the maternal parent of the seeds in its own fruits. This sexual reproduction allows alleles present in some plants to be brought together with alleles in others, creating genetic diversity and possibly resulting in offspring that are more well adapted to the environment than any of the plants shown.

Figure 16-1 Flower color in four o'clocks (*Mirabilis jalapa*) is controlled in part by a gene that has two alleles. The DNA sequence of one results in a protein whose primary and tertiary structures cause it to synthesize red pigment. The DNA sequence of the other allele codes for a protein whose tertiary structure is misformed; it has no enzymatic activity, and no pigment is produced. Therefore, flowers are white. If only a small amount of pigment is produced, the flower is pink. Flower color is the plant's phenotype; the type of alleles present is its genotype.

An important concept of inheritance is the selective advantage of reproduction, which may be either sexual or asexual. In asexual reproduction, each offspring is identical to its parent and siblings, having exactly the same DNA and thus the same alleles. Although this might seem like the safest, most efficient mechanism for producing large numbers of offspring to carry the parent's genes into future generations, the progeny are never more fit than the parent, and no progeny is adapted to any environment other than the parent's environment. During a drought, plants with poor capacity to withstand water stress are outcompeted by plants whose alleles confer superior water stress resistance. During an infestation of pathogenic fungi, individuals without alleles that confer immunity are destroyed. There may be no individual resistant to both stresses (**Figure 16-2**).

Sexual reproduction is a mechanism by which an organism combines its alleles with those of other, possibly better adapted individuals, thereby increasing the probability that copies of its own alleles survive. If a plant is susceptible to drought, sexual reproduction is advantageous because some of its sex cells may fuse with those of a drought-resistant plant. If so, the new zygote should grow into an individual resistant to water stress, thus adding some protection to the alleles derived from the susceptible parent. Although the original plant may die during a drought, copies of its alleles survive in its progeny, protected by alleles from the resistant parent (Figure 16-2). The second parent should benefit as well, as it is almost certainly not superior in all attributes. The drought-sensitive parent may carry alleles for resistance to fungal attack, which, during times of fungal attack, protect the alleles of the drought-resistant parent in the offspring, making the progeny of this sexual reproduction resistant to both fungi and water stress.

Sexual reproduction involves a large degree of chance and risk. Fitness and survival capacity are not governed by just one or two genes, but by almost all genes, including those responsible for the proper construction of membranes, the functioning of organelles, the production and transport of hormones, and so on. One cannot say that any aspect is trivial and does not matter; therefore, the presence of one or two particular alleles in a sex cell

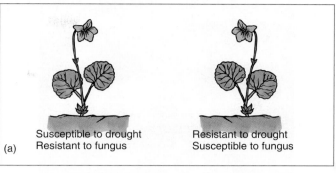

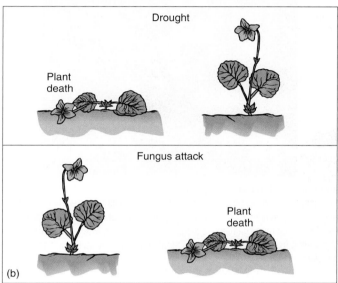

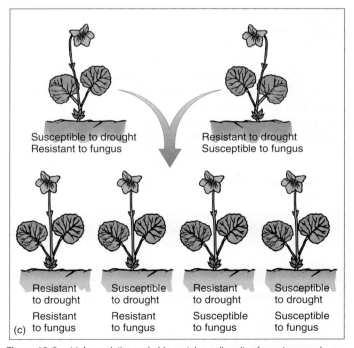

Figure 16-2 (a) A population probably contains a diversity of genotypes and phenotypes. (b) Without sexual reproduction, selectively advantageous alleles of one individual cannot be combined with those of another, and multiple stresses might kill all individuals. (c) With sexual reproduction, traits of one individual are combined with those of another, often producing individuals more fit than either parent.

Figure 16-3 Being diploid like most plants, this lupine (*Lupinus*) has one set of chromosomes from its paternal parent and one from its maternal parent, but the meiosis that precedes formation of each pollen grain and each ovule ensures that none of its many gametes is the same as any other. Insects will carry pollen to this plant from many other lupines, some of which may be better adapted than this plant is.

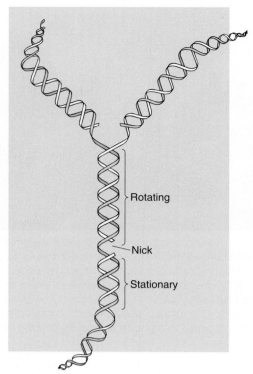

Figure 16-4 In order for DNA to unwind, one strand must be cut; otherwise, the entire unreplicated part would have to spin rapidly, which is impossible. With a single cut, only the region between the replicating enzymes and the cut must rotate.

is not the key feature, but rather the combination of all the alleles. Because of synapsis and crossing-over in meiosis, each plant produces thousands of types of sex cells, each with a unique genotype, which then fuse with the sex cells of many other individuals. For example, when plants bloom, they produce thousands or millions of pollen grains that are carried away and fertilize ovules on many other plants; simultaneously, its own ovules are receiving pollen from numerous plants (Figure 16-3). As a result, thousands of fertilizations may occur that involve this one plant's alleles. Many may produce poor combinations of alleles that have little survival value, but at least a small percentage should have the best attributes of both parents and should produce healthy, genetically sound plants with a high capacity to survive all stresses. It is not necessary or even advantageous for all potential progeny to survive, only the fittest. The plants and animals around us today are the successful survivors of evolutionary experimentation.

Replication of DNA

Before a cell undergoes nuclear division, either mitosis or meiosis, DNA is replicated during S phase of the cell cycle. Replication doubles the amount of DNA, and each gene exists in at least two copies, one on each of the two chromatids, one of which goes to each daughter nucleus during anaphase.

As DNA replication begins, chromatin first becomes less compact, opening sufficiently to allow entry of the necessary replicating factors. DNA does not release from histones; instead, nucleosome structure remains intact. Next, one strand of the

DNA double helix is cut, and the two strands separate from each other in a short region, forming a small "bubble" called a **replicon** (Figures 16-4 and 16-5). With the double helix open, free nucleotides diffuse to regions of single-stranded DNA and pair with its bases along both strands. These are ribonucleotides, not deoxyribonucleotides, and they are polymerized into short pieces of **primer RNA** approximately 10 nucleotides long. The primer RNAs then act as substrates for the DNA-synthesizing enzyme, **DNA polymerase**. It now enters and adds deoxyribonucleotides onto the end of the primer RNA using the open DNA as a guide. This method of replication, in which each strand of DNA acts as the template for making the complementary strand, is **semiconservative replication** because each resulting double helix contains one new molecule and has conserved one old one.

DNA polymerase adds deoxyribonucleotides only to the 3′ end of the growing nucleic acid, and thus, one strand of the open DNA is copied in one direction and the other strand in the opposite direction (Figure 16-5: top strand toward the left, bottom strand toward the right; both are actually 5′ ⟶ 3′). As DNA continues to unwind and open in both directions, new pieces of primer RNA form and new pieces of DNA grow from them toward the existing fragments (called Okazaki fragments). The original primer RNAs are depolymerized, and new Okazaki fragments are joined to the first ones. On the other end of each strand, replication is continuous, one long chain forming from the original primer RNA.

Each chromosome contains only a single DNA double helix, and in each chromosome, hundreds or thousands of sites occur

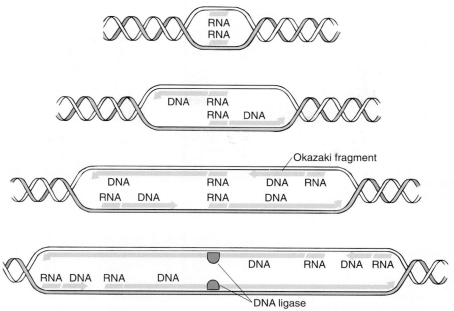

Figure 16-5   After a replicon opens, two pieces of primer RNA are formed; then DNA polymerase adds new DNA to the 3′ end of the primers, and both strands elongate as the replicon unwinds at both ends. As the replicon continues to open, Okazaki fragments are initiated periodically; they grow to the downstream fragments, and the enzyme DNA ligase joins them into single molecules.

where replication is initiated. As DNA uncoils and then separates, it has a forked appearance; this region is a **replication fork**. As DNA continues to uncoil and open, replication forks at each end of each replicon travel along the DNA. Finally, each replication fork runs into one of the adjacent replicons; then all new pieces of DNA are **ligated** (attached to each other with covalent bonds) into two new, complete molecules. The large number of initiation sites allows numerous DNA polymerases to work simultaneously. Each is able to incorporate approximately 1,000 nucleotides every second into growing DNA fragments, and replicons extend at the rate of about 0.5 μm/min. Each replicon is about 45,000 to 180,000 base pairs long (45 to 180 kbp) and can be replicated in about 1 to 3 hours. Because S phase of the cell cycle is 3 to 10 hours long, only about one third to one tenth of the replicons are active at any particular moment. If only one DNA polymerase could act on each chromosome, S phase would require 40 to 60 days (**Table 16-1**).

TABLE 16-1	Amount of DNA per Haploid Set of Chromosomes		
		Nucleotide Pairs (in billions)	Meters of DNA
Arabidopsis thaliana	relative of mustard	0.12	0.043
Beta vulgaris	beet	1.2	0.41
Zea mays	corn	2.3	0.78
Pisum sativum	pea	4.7	1.62
Vicia faba	bean	14	4.8
Allium cepa	onion	16.3	5.62
Lilium	lily	31.7	10.9

The values are known for only a few species at present, but more are being analyzed. Diploid nuclei have twice as much as shown in this table.

As replication forks advance, DNA partially dissociates from the histone octamers, but as all enzymes migrate forward and the two new DNA double helixes are complete, they immediately reassociate with histones into complete nucleosomes. Some of the old histone octamers go to one double helix and some to the other, and new histone octamers are added to both. Any particular segment of DNA remains unpackaged for only a few moments.

Mutations

A **mutation** is any change, however large or small, in DNA. The smallest mutation, affecting the least amount of DNA, is a **point mutation** in which a single base is converted to another base by any of various methods (**Figure 16-6**). If a piece of DNA is lost, the mutation is a **deletion**; the addition of extra DNA is an **insertion**. Under some conditions, a piece of DNA becomes tangled and breaks, and during repair, it is put in backward as an **inversion**.

Causes of Mutations

A **mutagen** is something that causes mutations. Several that are important are certain chemicals, ultraviolet light, X-rays, and radiation from radioactive substances. Many mutagenic chemicals are man made and are increasing in our environment. One mutagen, nitrous acid, reacts with cytosine and converts it to uracil, giving the DNA a G-U base pair in place of a G-C base pair (**Figure 16-7**). Numerous classes of DNA repair enzymes exist, one of which recognizes this G-U pair as abnormal and changes it back to G-C, but in meristematic cells that are replicating their DNA, DNA polymerase may arrive before the repair enzyme; if so, one new strand is formed with A complementing the U. In later replication, the A guides the incorporation of thymine, and the original G-C base pair becomes A-T.

Figure 16-6 (a) Consider this initial sequence of DNA; the box is not important. (b) A point mutation consists of the change in just a single base pair—here G-C is converted to A-T. If this sequence is part of a gene rather than just spacer DNA, this change in sequence has resulted in a new form of the gene, a new allele. (c) The six base pairs in the yellow box of the initial sequence have been deleted. Deletions often remove hundreds of base pairs. On the upper strand, the right boundary of the deletion has removed a C and an A and left a T, changing CAT to AAT. The AUG mRNA start codon is coded by the DNA triplet CAT. Therefore, this may be an important mutation, destroying a start codon, but this may not be a triplet; the real triplets may have been GCA TAA or TGC ATA. The diagram provides too little information to predict the significance of the mutation. (d) An insertion is the addition of one or more base pairs. (e) In this inversion, the DNA has broken at two points and then flipped over and been reinserted; the original top strand of the insert is now the bottom strand.

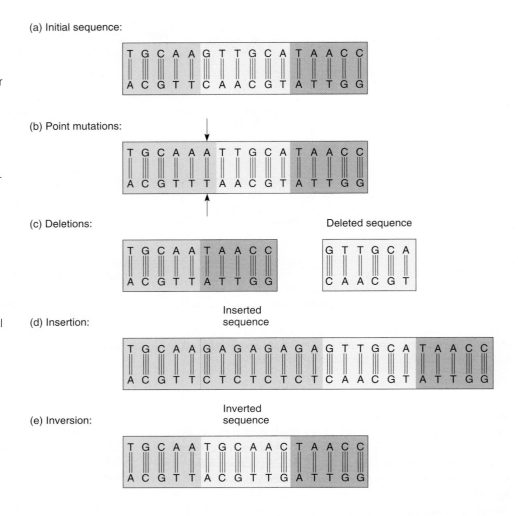

Figure 16-7 Nitrous acid converts cytosine to uracil. This strand is mutated, and all strands derived from it will be mutated, having A-T base pairs. The guanine is not affected and produces the original allele. One new allele has come into existence.

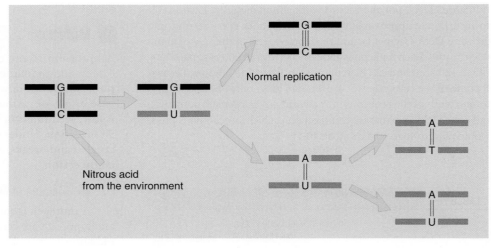

Deletions can be caused in several ways; one method is due to short regions of a self-complementary sequence. As DNA unwinds ahead of the DNA polymerase, one strand with a self-complementary sequence may form a small loop. DNA polymerase may pass by this loop without reading any of the bases in it; the bases complementary to those in the loop are left out of the new DNA molecule, so it is shorter than it should be.

Insertion mutations can be caused by many methods because a variety of enzymes cut and rejoin DNA as part of repair pro-

cesses. If a small piece of foreign DNA is present after cutting, it may accidentally be incorporated into the chromosomal DNA by DNA ligase.

Transposable elements are one of the most interesting and useful causes of insertion and deletion mutations. These are pieces of DNA that readily change their positions from one chromosome to another. Transposable elements have two basic forms—insertion sequences and transposons. **Insertion sequences** are only a few thousand base pairs long and contain the genes that code for the enzymes actually involved in cutting the insertion sequence out and splicing it into DNA somewhere else. A **transposon** is like an insertion sequence except that it may be much longer and carries genes that code for proteins not associated with transposition. The deletion and insertion mutations caused by transposable elements can vary in severity. If the element inserts into spacer DNA, the effect is not very important, but if it inserts into a gene, it totally disrupts either the promoter or the structural region.

Mutations caused by transposable elements may be one of our most powerful tools for genetic analysis and engineering. Often, a metabolic pathway is studied by exposing thousands of plants to mutagens such as X-rays or nitrous acid, then examining the plants to find any that have mutations disrupting the metabolic pathway of interest. These are then studied further to determine how the pathway was affected, but we never actually know which gene was disrupted. The gene is given a name (**Table 16-2**), but we do not know its DNA sequence, its location on a chromosome, or the nature of its promoter or structural regions. However, transposons are now being sequenced and engineered to contain markers such that the transposons can be applied to plants or cultured cells, where they cause insertion mutations. The mutants of interest are located (**Figure 16-8**), and DNA from the mutants is cut into pieces by restriction endonucleases, denatured, and combined with radioactive DNA complementary

TABLE 16-2	Names of Genes in Pea Plants	
Gene	Symbol	Phenotype
Chlorophyll synthesis	*alb, alt, au, auv, ch₁, ch₂, cov, cvit, lum, pa, py, vac, xa₁, xat, yg*	Each affects chlorophyll synthesis, changing the color from green to dark bluish green
Leaf structure		
Aeromaculata	f_1	Few airspaces between epidermis and palisade parenchyma
Stomata	sa_1, sa_2, sa_3	Leaves have fewer stomata than in the wild type
Leaf development		
Afila	*af*	Leaflets develop as tendrils
Clavicula	t_1	Tendrils develop as leaflets
Unipetiole	*up*	Leaves have only one pair of leaflets
Unifoliata	*uni*	Leaflets are converted to just a single leaf
	cri, crif, cris	Folded, crinkled leaves
	cont, lat, st, x	Leaves of unusual size

Although there are thousands of alleles in plants, only a few ever become obvious and only a small number of those are ever studied and given names. These are just a few of the named alleles in pea (*Pisum*).

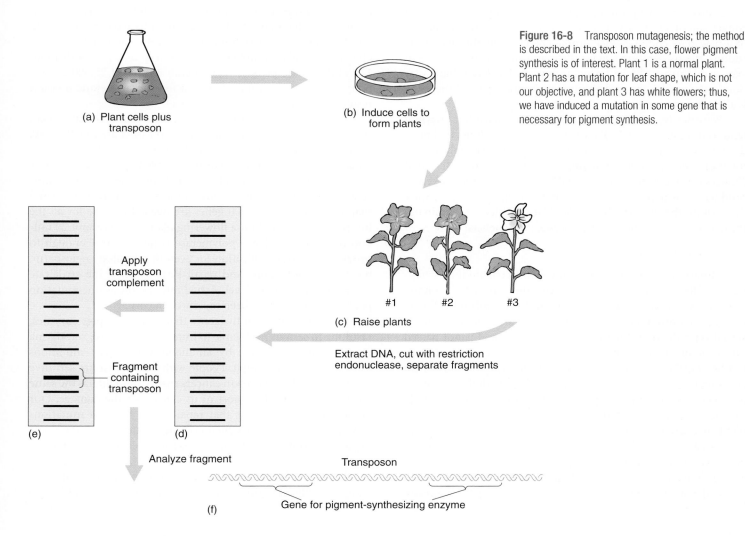

(a) Plant cells plus transposon

(b) Induce cells to form plants

Figure 16-8 Transposon mutagenesis; the method is described in the text. In this case, flower pigment synthesis is of interest. Plant 1 is a normal plant. Plant 2 has a mutation for leaf shape, which is not our objective, and plant 3 has white flowers; thus, we have induced a mutation in some gene that is necessary for pigment synthesis.

#1   #2   #3

(c) Raise plants

Extract DNA, cut with restriction endonuclease, separate fragments

Apply transposon complement

Fragment containing transposon

(e)   (d)

Analyze fragment

Transposon

Gene for pigment-synthesizing enzyme

(f)

to the transposon. After the piece is found, it can be sequenced. Because we know the transposon's sequence already, the DNA on either side of the transposon must be the gene of interest. Transposon mutagenesis offers the possibility of quickly correlating metabolic pathways, proteins, genes, and even promoters for virtually any aspect of an organism.

■ Effects of Mutations

The effect and significance of a mutation depend on its nature, its position, and its extent. If it occurs in spacer DNA between two genes, it may have no effect whatsoever. Also, point mutations and small insertions and deletions in introns appear to be unimportant generally—they change a portion of hnRNA that will not be incorporated into mRNA. Within exons, a small mutation may not be important if it only changes a codon into another codon that specifies the same or a similar amino acid. Changing codons to ones that specify very different amino acids may not matter if they are located in a part that is not critical to the protein's functioning.

On the other hand, a mutation in an exon may cause the gene to code for a protein whose active site is disrupted and the protein cannot function (Figure 16-6). An insertion mutation may cause the gene to code for protein so long that it cannot fold properly. Also, mutations in promoter regions can completely inactivate a gene or cause it to be active at the wrong time or place. Even point mutations can have profound effects; for example, nitrous acid conversion of a G-C base pair to A-T might cause the formation of a new start codon. Other simple changes can eliminate start codons, the recognition site for distinguishing the boundaries of exons and introns, TATA boxes, and so on. The larger the mutation is, the greater the probability that a critical part of the DNA is affected.

Statistically, mutations are almost always harmful. Enzymes tend to be approximately 300 to 400 amino acids long, and hundreds of trillions of proteins could possibly exist. Yet only a small fraction would be useful in living organisms. Any mutation that changes the structure of proteins, rRNA, or tRNA is more likely to produce a less useful than a more useful form. The majority are deleterious, the minority beneficial. Natural selection eliminates the deleterious mutations and preserves the beneficial ones.

■ Somatic Mutations

Mutations can occur at any time in any cell, but if they happen in cells that never lead to sex cells, they are called **somatic mutations**. For example, a gene in a leaf primordium cell may undergo a mutation, but because leaves are not involved in sexual reproduction, the mutation is somatic and is not passed on to the plant's offspring, regardless of whether the mutation is advantageous or disadvantageous. When the leaf falls off in autumn, the mutated gene decomposes along with the rest of the leaf. The same is true of any mutation in roots, wood, or bark. In general, somatic mutations are not very important for most plants because they affect such a small portion of the plant and are not passed on to the offspring. A somatic mutation might not ever result in an altered phenotype. The mutation may occur in a leaf cell nucleus, but in a gene that is inactive in leaves, such as a gene that affects root hair growth or bark formation. In species

that undergo extensive vegetative reproduction, such as blackberries, alder, and prickly pear cactus, somatic mutations can be important if they affect a part of a plant that gives rise to a vegetative offshoot. As the offshoot grows and reproduces vegetatively, the patch of mutated cells may finally include an axillary bud and then affect a flower and its sex cells. The mutation then becomes a sexual one.

■ DNA Repair Processes

Because most mutations are deleterious and occur frequently enough to be a significant problem, it is selectively advantageous for organisms to have DNA repair mechanisms that recognize and remove mutations. Certain mechanisms recognize base mismatches, loops, or other problems; other enzymes minimize the number of errors that occur in the first place.

The DNA repair rate must be neither too efficient nor too ineffective. In organisms that have very short genomes, such as prokaryotes and perhaps algae and protozoans, DNA polymerases can replicate a full genome without errors. Perfect replication does not occur every time, but nonmutated replication can sometimes occur. With the larger genomes of more derived animals and plants, one set of chromosomes can virtually never be replicated without mutations if no repair mechanisms are present. DNA polymerase may make an error only once in every 1 million nucleotides, but if the genome is several billion base pairs long, every replication produces numerous errors. Most angiosperms have about 10 to 100 billion base pairs per haploid set of chromosomes (we humans have only 3 billion base pairs). One of the smallest numbers known is in *Arabidopsis,* with about 120 million base pairs (containing an estimated 25,500 genes); every round of replication before a mitotic or meiotic division results in about 120 errors (Table 16-1).

As a zygote grows into an adult, every cell cycle introduces new mutations, and in the adult, no two nuclei are exactly alike; after several cell cycles, DNA would be useless if there were no repair mechanisms. It would be impossible to produce sperms or eggs that were not extensively mutated, and under such circumstances, complicated organisms could not exist. It is estimated that a minimum of 15,000 genes are needed to code for all the information required for a flowering plant. This is about 20 million base pairs, so DNA polymerase is not accurate enough to provide error-free replication for even the simplest angiosperm, but DNA proofreading and repair systems bring the error rate down to an acceptable level. Actual rates as low as one mutation per 500,000 genes have been measured in corn.

Mutations occur in the genes that code for repair enzymes, resulting in serious problems. In humans, the disease xeroderma pigmentosum is caused by an inability to repair mutations caused by ultraviolet light. People with this disease are sensitive to sunlight and develop skin cancers easily. Sunlight has the same effect on all of us, but most of us can repair the damage. Our bodies have mechanisms that repair other types of mutations as well, but our modern chemical society may be contaminating our environment with mutagens for which we have no repair mechanisms.

DNA repair mechanisms must not be perfect or even extremely efficient. If every replication were perfect and if no mutations ever arose, all cells of an individual would have absolutely identical nuclei, as would all sperm and egg cells. Sexual reproduction would be useless because all eggs and sperms of a species would carry identical genes. Mutations must occur at a low enough rate that they do not endanger every individual but rapidly enough that a species evolves as its environment changes. With no variation, there would be no differences for natural selection to act on; if all are identical, none has a selective advantage.

Monohybrid Crosses

Sexual reproduction between two individuals is called a **cross**. The meiotic divisions that precede a cross reduce the number of sets of chromosomes per cell from the diploid number to the haploid number. Consequently, each sex cell—that is, each sperm cell and egg cell—contains one complete set of genes. Furthermore, each sperm cell contains all the genes necessary to construct a new plant; the same is true of each egg cell. The zygote (the fertilized egg) has two complete sets of genes.

Within a population, mutations produce new alleles, and the genotypes of individuals within the population differ. Of the plants that grow in an area and that can interact sexually, many may have the same allele of a particular gene, but other individuals may have other alleles, other versions of the gene. Consequently, the alleles carried by a particular sperm cell may or may not be identical to the homologous alleles of the egg it fertilizes.

Monohybrid Crosses with Incomplete Dominance

In a **monohybrid cross**, only a single character is analyzed and studied; the inheritance of other traits is not considered. For instance, a plant with red flowers might be crossed with one that produces white flowers, and only the inheritance of that flower color trait is studied. Characters involving flower shape, leaf structure, and photosynthetic efficiency are also being inherited simultaneously, but in a monohybrid cross, only one is studied. This makes the analysis and understanding of the results much simpler. After basic principles of a particular trait are known, then its interaction with another factor (dihybrid cross) or two other factors (trihybrid cross) can be studied. Mendel gave his attention to monohybrid crosses, which are easy to understand; before him, people tried to analyze many characteristics at a time and became hopelessly confused.

Consider the cross just mentioned: A plant with red flowers is bred to one with white flowers. It does not matter which flower produces pollen and which produces ovules. After the cross is made, seeds and fruit develop; the seeds are planted, and when mature, the new plants are allowed to flower so that their flower color can be examined. The flowers of all plants in this new generation are pink, resembling each parent somewhat, but not exactly like either (**Figure 16-9**). The parents are called the **parental generation**; the offspring of their crossbreeding are the **F₁** or **first filial generation**, and if these interbreed, their offspring are the **F₂** generation.

The molecular biology of this monohybrid, flower color cross is easy to understand. Each parent is diploid and thus has two copies of the gene involved. In the red-flowered parent,

both alleles produce mRNA that is translated into functional enzymes involved in synthesis of red pigment. In a white-flowered parent, both alleles are defective. It may be that each produces an mRNA that when translated results in a protein unable to perform the necessary reaction, or the promoter region may be mutated and can no longer interact with a chemical messenger. Whatever the cause, there is no pigment, and the flower is white.

Using genetic symbols, the red-flowered parent is *RR* and the white-flowered one is *rr* (Figure 16-9). Each parent is said to be **homozygous** because each has two identical alleles for this gene. The pink-flowered F_1 has received an *R* allele from one parent and an *r* allele from the other, so its genotype is *Rr*; it is **heterozygous** because it has two different alleles for this gene. We use these symbols even though we have neither isolated the gene nor analyzed its nucleotide sequence; *R* and *r* are simply labels. Most genes are known only by their phenotypes and the labels given to them by geneticists (Table 16-2). With a genotype of *Rr*, the plant produces mRNA, half of which carries the defect; thus, only half the normal amount of enzyme is produced. This results in less pigment being formed, only enough to make the flower look pink, not red. Neither parental trait dominates the other, so this pair of alleles shows **incomplete dominance**: The heterozygous phenotype differs from both homozygous phenotypes.

When analyzing the possible outcomes of crosses and breeding, one must understand the types and quantities of gametes involved. All of the plants we are considering are diploid and form haploid spores by meiosis. The *RR* parent has chromosomes as shown in Figure 16-9b, and regardless of how chromosomes separate during meiosis, all spores receive an *R* allele. In the *rr* parent, all spores receive an *r* allele. Spores are not sex cells, but as they develop into gametophytes, they divide by mitosis—duplication division—so all cells of the microgametophyte (pollen) have the same allele, as do all cells of the megagametophyte (inside the ovule). Because the *RR* parent produces only *R* spores, it also produces only *R* sex cells, both sperms and eggs in typical bisexual flowers. Similarly, the *rr* parent produces, indirectly, only *r* gametes. When the two plants are interbred, an *R* gamete unites with an *r* gamete, establishing a heterozygous (*Rr*) zygote that grows into a heterozygous adult by means of mitotic cell divisions. No other outcome is possible; it does not matter which gamete is which, an *R* sperm and an *r* egg result in the same type of zygote as an *r* sperm and an *R* egg.

When the heterozygote matures and flowers, each spore mother cell produces two types of spores (**Figure 16-10**), not just one as is true of a homozygote. Because each cell has one *R* allele and one *r* allele, during the first meiotic division, one daughter cell receives *R*, and the other receives *r*. The second division of meiosis results in two *R* spores and two *r* spores. If this is occurring in the anther, half the pollen grains and hence half the sperms have *R*, and the other half have *r*. In the nucellus of many plants, only one megaspore survives. If this gene has no effect on spore metabolism (and a gene for flower color is probably inactive in spores), then half the time *R* cells survive and half the time *r* cells live. In heterozygote parents, the two types of sperms and eggs are produced in equal numbers.

Figure 16-9 (a) A monohybrid cross analyzing the trait of flower color. Details are explained in the text. (b) The phenotypes of parents, gametes, and progeny.

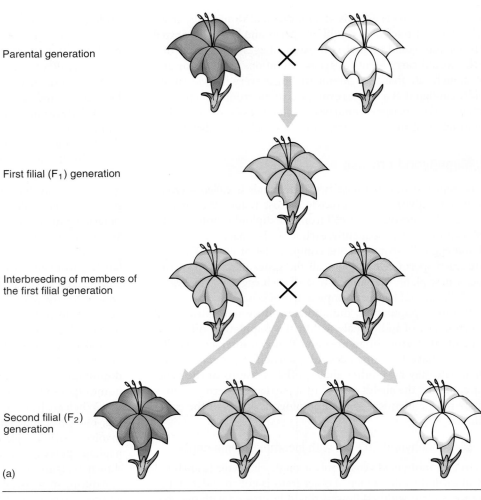

Parental generation

First filial (F$_1$) generation

Interbreeding of members of the first filial generation

Second filial (F$_2$) generation

(a)

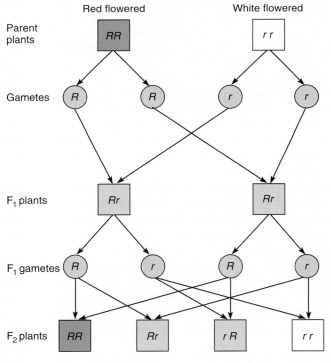

Red flowered

White flowered

Parent plants RR rr

Gametes R R r r

F$_1$ plants Rr Rr

F$_1$ gametes R r R r

F$_2$ plants RR Rr rR rr

Genotype ratio: 1 RR:2 Rr:1 rr
Phenotype ratio: 1 Red:2 pink:1 white

(b)

Key: R = red flower, r = white flower

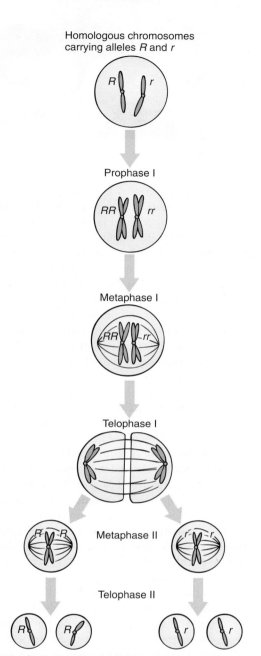

Homologous chromosomes carrying alleles *R* and *r*

Prophase I

Metaphase I

Telophase I

Metaphase II

Telophase II

Figure 16-10 Production of haploid cells by meiosis in a heterozygote. Homologous chromosomes pair (synapse) during metaphase I, then the paternal chromosome is separated from the maternal chromosome during anaphase I: One type of allele is separated from the other. As sister chromatids separate after metaphase II, each daughter nucleus receives identical alleles.

■ Crossing Heterozygotes with Themselves

When a plant's own pollen is used to fertilize its own eggs, the cross is a **selfing**. A plant can also be selfed by being crossed with another plant with exactly the same genotype. Selfing heterozygotes has interesting, instructive consequences; 50% (on average) of all sperms and eggs contain the *R* allele, and 50% have *r* (Figure 16-10). Not all zygotes are identical genotypically: Some are *RR*, having resulted from an *R* sperm and an *R* egg; others are *rr* (*r* sperm and *r* egg), and some are *Rr* (either *R* sperm and *r*

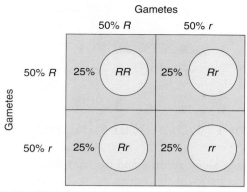

Gametes

Figure 16-11 A Punnett square makes it easy to analyze and understand the results of a cross. Gametes from one parent are listed across the top; those of the other parent are listed on the left side.

egg or *r* sperm and *R* egg). Selfing a heterozygote produces three types of F₁s, some of which (*Rr*) resemble the parents and others (*RR* and *rr*) the grandparents. Again, to analyze the results, we must wait for the zygote to develop into an embryo, then plant the seeds and wait until the new plants are old enough to flower. Because each genotype produces a distinct phenotype, the genotype of each plant is known simply by looking at the flowers. If a large number of heterozygotes (pink-flowered plants) are selfed and large numbers of F₁ plants grown, approximately one fourth are red (*RR*), one half pink (*Rr*), and one fourth white (*rr*). This is an important ratio, typically represented as 1:2:1, and should be memorized immediately.

The reason for the proportion of these genotypes is explained in Figure 16-11. A **Punnett square** can be set up in which all types of one gamete, say, the egg, are arranged along the top of the square, and all types of the other gamete are arranged on the left side. The boxes are then filled in with the allele symbol above it and to the left. Because the gametes are produced in a ratio of 1 *R* : 1 *r*, listing them as in Figure 16-11 automatically represents their relative numbers in nature.

The Punnett square does not represent the outcome of any one cross; if a single heterozygous flower is selfed, it may produce only one or two seeds. If you plant just one seed, there is one chance in four that it will have red flowers, two chances in four pink, and one chance in four white.

The 1:2:1 ratio was one of the great discoveries of Gregor Mendel (1822-1884), an Austrian monk who performed experiments with pea plants. His cross-breeding experiments became the basis for modern genetics. Mendel discovered that in a selfing of this type, the recovery of the parental types means that genetic material must be composed of particles, such that the *R* genetic material can be separated from the *r* genetic material in the *Rr* heterozygote. Before Mendel's work, it was thought that the genetic material was a fluid. Two fluids cannot mix and then separate again perfectly. The constancy of the 1:2:1 ratio made it more logical to think in terms of discrete particles, genes, that never lost their identity regardless of the crosses in which they participated.

Furthermore, the 1:2:1 ratio can be realistically interpreted only in terms of each individual plant having two copies—being diploid—and each sex cell having one copy—being haploid. If

Box 16-1 Botanical Philosophy and Popular Culture

When thinking about genetics and the contributions of Gregor Mendel, it is important to keep in mind the state of scientific knowledge at that time. Some of the greatest scientific geniuses of all time were discovering and documenting the fundamental concepts of biology. Only 27 years earlier than Mendel's work, M. J. Schleiden and T. Schwann had proposed the cell theory, the concept that all organisms are composed of cells with nuclei and that all existing cells come from preexisting cells. We take this for granted today, but it means that all living organisms are fundamentally the same, that life may have arisen only once, and that all cells are descended from those first cells. In 1849, Wilhelm Hofmeister proved that during plant reproduction the new embryo develops primarily from the egg after receiving some "influence" from the pollen tube. This was revolutionary because until that time everyone firmly believed that in both plants and animals, the new embryo developed from the sperm cell—the female parent might provide nourishment and protection, but the new generation was really a continuation of the paternal parent (could this belief have been based not on observation of nature but on observation of society and history dominated by males?). Hofmeister's conclusion that the female parent also contributes to the new offspring was both shocking and also very important for Mendel's later observation that genes are passed equally from both parents.

In 1859, Charles Darwin published *Origin of Species by Natural Selection,* which postulated that organisms change through time, that they are not constant (A. R. Wallace came to the same conclusion at the same time). The discovery of evolution by natural selection caused a great deal of trouble for Mendel: Natural selection requires diversity and change, whereas Mendel's theories of genetic inheritance showed that differences in phenotype were due only to the mixing and separating of unchanged, constant genes. The idea of evolution had aroused great controversy and serious philosophical battles between proponents of evolution and of biblical creation. This battle had been raging for 6 years when the paper by Mendel (a monk) appeared. Had he published 10 years earlier, he might have found instant acceptance, and Darwin would have had a more difficult time.

A further difficulty for genetics, however, was its postulation of genes as particles, not as fluids. This was at a time when fluids were considered the basis for most of biology, with sieve pores in the phloem and pits and perforations in the xylem considered to be effective by acting as filters that separated the various fluids that were thought to exist. Fortunately, in 1879, H. Fol showed that during angiosperm reproduction, a sperm nucleus (a particle) entered the egg, and in 1888, E. Strasburger described meiosis, giving a firm foundation to the concept of genes being particulate. If genes are on the chromosomes, then suddenly it is possible to see how Mendel's ratios and independent assortment could occur. This was summarized in 1915 by T. H. Morgan in *The Mechanism of Mendelian Heredity.* Finally, in 1927 and 1928, H. J. Muller and L. J. Stadler showed that genes could be artificially changed (mutated) by the use of X-rays. Thus, genes could be stable enough to permit Mendelian genetics to operate and yet variable enough to allow natural selection to operate as well. It is important to keep in mind that these were not the only biologists working at the time, nor were these the only discoveries. A huge volume of information was being generated, much of which concerned unusual, exceptional, or aberrant cases and which thus made the fundamental discoveries more difficult to recognize as being important. Also, we must always try to appreciate the profound influence of the general philosophy of the time—a male-dominated society produced male-dominated theories of biology. It is likely that our own sincerely held beliefs are causing us to improperly evaluate some of our own biological observations.

---

each plant had only one copy, pink-flowered heterozygotes would be impossible, whereas if each had three, phenotypes such as dark pink (*RRr*) and light pink (*Rrr*) should also be present. Keep in mind the state of scientific knowledge in 1865 when Mendel was working. The concept that all organisms are composed of cells with nuclei had only recently been proposed; mitosis and cell division were very poorly understood, and meiosis would not be discovered for another 23 years. The concepts of chromosomes and homologous pairs would not be well established until the 20th century, and the existence of mRNA was not confirmed until the mid 1960s (**Box 16-1**).

■ Monohybrid Crosses with Complete Dominance

The situation in which only half as much product of an enzyme, such as the red pigment discussed above, is produced in a heterozygote is not universal. In certain species or with other traits, cytoplasmic control mechanisms may cause the enzyme to function until a specific amount of product is synthesized. The enzyme may have to work faster or longer, but the final amount of product is the same whether the plant has two functional alleles or only one. In other situations, the amount of enzyme might be monitored such that the nonmutant mRNA of the heterozygote is translated more frequently or rapidly.

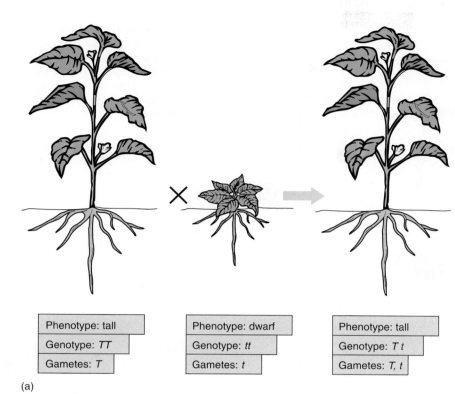

| Phenotype: tall |
| Genotype: *TT* |
| Gametes: *T* |

| Phenotype: dwarf |
| Genotype: *tt* |
| Gametes: *t* |

| Phenotype: tall |
| Genotype: *T t* |
| Gametes: *T, t* |

(a)

Figure 16-12 (a) In this monohybrid cross, the trait "tall" shows complete dominance over the trait "dwarf." In a *Tt* nucleus, the *T* allele may be transcribed twice as much as each *T* allele in a *TT* nucleus. The resulting mRNA may be translated twice as much, or the protein may work twice as long or twice as fast. A *TT* plant may contain a level of *T* protein far above the threshold for responsiveness. (b) Most of the wheat plants in this field have blown over and are lying on the ground (we say the wheat has "lodged"). This is a tall variety of wheat: Its stems are too tall and slender to be strong enough to hold themselves upright. The harvesting machine (a "combine") will have to cut the wheat so low to the ground that dirt and rocks may be brought into the machine, damaging it. Also, if it rains before harvest, the stems lying close to the soil may rot. (c) All the wheat plants in this field are still upright and will be easy to harvest; none of these plants has lodged. This is a short variety of wheat: Plant geneticists obtained the seeds for this crop by crossing plants that all had short, strong stems. This field can be harvested easily, and there is little chance any of the wheat will rot, even if it rains before harvest time.

(b)

(c)

In either case, the phenotype of the heterozygote is like that of the parent with two effective alleles. That trait is said to be **dominant** over the other version of the trait, which is **recessive**. An example is height. A tall plant with a *TT* genotype produces sex cells that are all *T*. A short plant, genotype *tt*, produces only *t* sex cells (**Figure 16-12**). When the tall plant is crossed with the short one, all F₁ progeny have the *Tt* genotype, but they all have the phenotype of being tall, indistinguishable from their *TT*, homozygous dominant parent. We do not know what protein is produced by the *T* allele, but even with only one functional allele, enough product is made to permit normal growth. *Tt* plants are tall, and the tall character completely dominates the "dwarf" character.

Knowing the molecular biology of genetic systems, you can predict that when heterozygotes are selfed, two types of sperm cells (*T* and *t*) and two types of eggs (*T* and *t*) are produced. The Punnett square for the cross is as in **Figure 16-13**, and a genotype ratio of 1 *TT*:2 *Tt*:1 *tt* is expected. The prediction is correct, but what will the phenotype ratio be? One out of four plants will be tall due to a *TT* genotype, and two out of four will be tall due to a *Tt* genotype. Thus, three fourths have the tall phenotype, and one fourth have the short phenotype. Whenever a cross is made and a phenotype ratio of 3:1 is seen, we should suspect that the parents are heterozygous and the character shows complete dominance.

Test Crosses

Remember that we can only see an organism's phenotype; to discover its genotype, we must perform crosses, and the test cross is one of the most useful. When traits with incomplete dominance

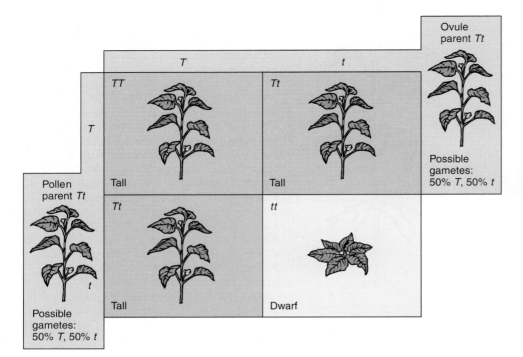

Figure 16-13 In setting up the Punnett square for a selfing of *Tt* plants, first establish the genotypes of the two parents. Then determine what types of gametes are produced and in what proportions, and fill in the squares with the genotypes. From the genotypes, the phenotypes in each square can be determined.

are studied, the genotype of any plant is easy to determine from its phenotype. If the trait has **complete dominance**, it is difficult to know what the genotype of any particular plant is unless the plant shows the recessive trait. Imagine that you are studying the inheritance of tallness; you have selfed some heterozygotes and planted the resulting seeds. In your greenhouse or garden, there are now hundreds of plants, approximately 75% of which are tall and 25% short. You know the short ones are *tt*, and if you need a plant with the *tt* genotype for experimentation, you know automatically to choose short plants. But if you need to experiment on plants with the *TT* genotype, how can you tell which they are? A tall plant picked at random is more likely to be a *Tt* plant because there are twice as many of them as *TT* tall plants.

The genotype can be revealed by a **test cross**, a cross involving the plant in question and one that is homozygous recessive for the trait being studied. All gametes produced by a homozygous recessive parent carry the recessive allele, which is unable to mask the homologous allele in the resulting F₁ zygote. If the plant being tested is actually homozygous dominant (*TT*), 100% of the progeny will be heterozygous (*Tt*) and tall (**Figure 16-14**). If the tall plant is heterozygous, half its progeny in the test cross will be tall (*Tt*) and half short (*tt*). If the cross produced thousands of seeds, you might not want to plant all of the seeds, just some, and if a large enough number of seeds are grown, the chance of accidentally not choosing any of the *Tt* seeds is very low, and if even one of the F₁ plants is short, the test parent must have been *Tt*.

After the actual genotypes of plants are known, those that are homozygous dominant can be gathered and planted in special areas, kept free of all natural pollinators, and allowed to breed among only themselves. All of their progeny will be homozygous dominant and can be used in breeding experiments. The homozygous recessives can also be kept as a special line, being selfed and kept pure. Such groups are **pure-bred lines** and are both useful and valuable. It is not possible to maintain the heterozygotes

like this because they do not breed true; that is, their progeny are not exactly like them. One fourth are homozygous recessive; one fourth are homozygous dominant and cannot be distinguished visually from the one half that are heterozygotes.

So far, we have considered only one trait, height, but for plants that are important crop or horticultural species, dozens or even hundreds of traits are cataloged. Seed companies maintain hundreds of different lines of corn, for instance, in which the genotype of many characters is known for each line (**Figure 16-15**). Many times, plants are collected from the wild, so nothing is known about their genotype unless it is immediately obvious from the phenotype. From looking at a few collected plants, however, it is not possible to tell which characters are dominant and which are recessive; carefully controlled and recorded crosses must be made.

When test crosses must be made on annual plants, the results are usually not known until after the plants have died. Their genotypes are then known, but the plants cannot be used for experimentation or breeding. In such cases, it usually is necessary to do both test crosses and experimental crosses simultaneously, not knowing which plants have the correct genotype for the experiment being performed. After the test cross results are complete, the parents that had the proper genotype can be identified, and the experimental crosses that involved them can be analyzed.

■ Multiple Alleles

Each gene may have many alleles, not just two as in the examples discussed so far (*T* and *t*, *R* and *r*). A protein of average size consists of about 300 amino acids, so the coding portion of its mRNA must have about 300 codons, each containing three nucleotides. The gene is therefore at least 900 nucleotides long, not counting introns and promoters. At least 900 sites exist at which point mutations can occur, and of course, any mutation may involve several nucleotides. Consequently, the gene may exist in many forms, called **multiple alleles**. When genes are polymorphic,

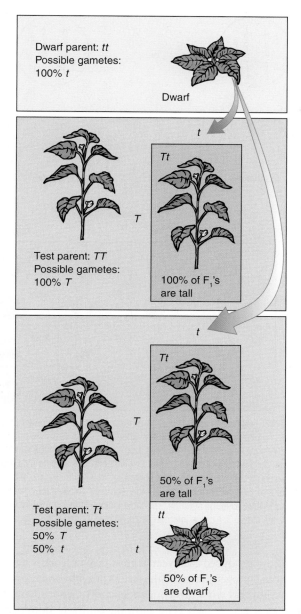

Figure 16-14 Test crosses to determine whether a plant with the dominant phenotype is heterozygous or homozygous. Of the data in this diagram, you would not know the genotypes of the test parents before the test cross; those data are what you are trying to discover. If you want to find only a single homozygote or heterozygote, you would need to do test crosses on only one or a few plants. Imagine, however, that you had done an experimental cross for a trait that you assumed would give you a 3:1 ratio; it would be necessary to do test crosses on a large number of progeny plants with the dominant phenotype just to confirm your experiment.

having multiple alleles, numbers, such as X_1, X_2, X_3, X_4, and so on, are used rather than capital and lowercase letters. Certain mutations still result in the production of a protein with the normal sequence, but most lead to altered protein structure. Some of these proteins are similar to the original protein, perhaps having similar or even identical enzymatic activity. However, many carry out the proper reaction more slowly or are not accurately controlled by regulatory mechanisms in the cytoplasm; a normal, wild-type phenotype may not be produced. With multiple alleles, the con-

Figure 16-15 Maintaining pure-bred lines is relatively easy for perennial plants; after the plant is growing, it can produce pollen or ovules for experimentation for years. Annual plants are much more difficult because pollination, seed gathering, planting, and record keeping must be done annually. Here, wheat seeds are being frozen in liquid nitrogen; after several years, some will be thawed and planted. Their seeds will be frozen for further storage.

cept of dominance is more complex; one allele may produce the proper quantities of the functional protein, whereas a different allele produces more, another produces less, and a fourth produces one with altered activity. Many different phenotypes are possible.

Within a population of plants, as many types of gametes can be produced as there are different types of alleles. A heterozygous X_1X_2 plant can be crossed with a heterozygous X_3X_4 one, resulting in progeny such as in **Figure 16-16**: X_1X_3, X_1X_4, X_2X_3, and X_2X_4. Four distinct types of F_1 plant are produced, none of which has the genotype of either parent.

Dihybrid Crosses

A **dihybrid cross** is one in which two genes are studied and analyzed simultaneously, rather than just one, as in a monohybrid cross. Every cross involves all the genes in the organism, but the terms "monohybrid" and "dihybrid" refer only to the number being analyzed.

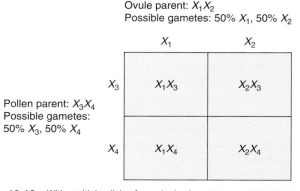

Figure 16-16 With multiple alleles for a single character, numerous types of crosses become possible; however, we still determine, from the parental genotype, all possible gametes and then construct a Punnett square in the usual fashion.

When two genes are studied, the results of the crosses depend on the positions of the genes on the chromosomes. If they are on different chromosomes, the alleles for one gene move independently of the alleles for the other gene, but when two genes are close together on the same chromosome, the alleles for one gene are chemically bound to the alleles for the other gene and move together. The situation in which the genes are on separate chromosomes is easier to understand and is explained first.

■ Genes on Separate Chromosomes: Independent Assortment

Consider a plant heterozygous for two traits, for instance seed coat texture and color, with a smooth seed coat (*S*) showing complete dominance over a wrinkled seed coat (*s*), and a yellow seed coat (*Y*) being dominant over a green seed coat (*y*). The plant's genotype is *SsYy*, and its phenotype is smooth yellow seeds. We know that if we consider only the gene for color, the plant produces two types of gametes in approximately equal numbers, some carrying *Y* and some carrying *y*. Also, if we consider only texture, half carry *S* and half *s*. How do the alleles of the two genes relate to each other? As in monohybrid crosses, knowing the types of gametes that can be formed is the key to understanding the patterns of inheritance.

If the two genes are on separate chromosomes, the alleles of one gene move independently of the alleles of the other gene; this is called **independent assortment**. The chromosomes have duplicated during S phase and each has two copies of each allele (**Figure 16-17a**); all chromosomes align on the metaphase plate during metaphase I, where homologous chromosomes also pair with each other. There are two *Y* alleles, two *y* alleles, two *S* alleles, and two *s* alleles at the metaphase plate. During anaphase I, homologous chromosomes separate from each other, and both *Y* alleles move to one pole because they are on the two chromatids of one chromosome, still held together by its centromere. Both *y* alleles, located on the two chromatids of the homologous chromosome, move to the other pole. There is no way to predict which pole will receive which type of allele. Similarly, the *S* alleles separate from the two *s* alleles and move randomly to the poles. In some cells, the pole that receives the *Y* alleles at telophase I also receives the *S* alleles, but in other cells, the *Y* and *s* alleles end up together. During meiosis II, the two chromatids of each chromosome separate from each other, resulting in four types of haploid cells in equal numbers: *SY*, *sY*, *Sy*, and *sy*. Any single microspore or megaspore mother cell produces only two types of haploid cell: the set *SY* and *sy* or the set *sY* and *Sy*. All four types are produced by a single plant because some mother cells produce one set and some produce the other set.

After the possible types of gametes are known, the Punnett square can be set up, as in Figure 16-17b. Any single fertilization results from the syngamy of one sperm cell and one egg cell and produces only one of the 16 possible zygote genotypes shown in Figure 16-17b. All 16 types of zygote occur only if we study many fertilizations; to have the ratios come out accurately, we have to analyze hundreds of progeny. In plants, it is usually easy to obtain large numbers of fertilizations because pollen and ovules are produced in large amounts. After pollinated, most plants produce dozens or even thousands of seeds, enough progeny that all 16 zygote genotypes occur in about the expected ratios. But in large animals such as mammals, reproduction may be infrequent, and only one or two

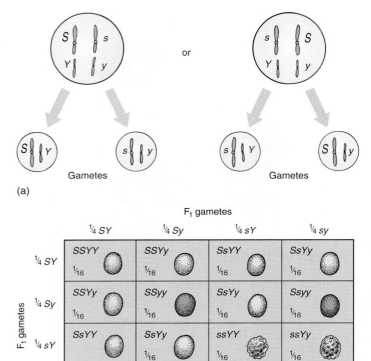

(a)

(b)

Figure 16-17 (a) During anaphase I, chromosomes move independently of each other, so one pole receives *S* and the other receives *s*; likewise, one gets *Y* and the other *y*. But in about half the cells, both *S* and *Y* move to the same spindle pole by chance, whereas in other dividing cells, one spindle pole receives *S* and *y*, again by chance. (b) A Punnett square for a dihybrid cross is set up just like one for a monohybrid cross; establish the types and relative abundance of gametes, and then fill in the squares. The table looks a little formidable, but it really consists of two 3:1 ratios intermingled.

offspring are produced each year; a great deal of work is necessary to get enough progeny to verify the results of a dihybrid cross.

In a dihybrid cross involving independent assortment of two heterozygous genes, each gene showing complete dominance, a characteristic phenotype ratio occurs, just as is true of the 3:1 ratio in a monohybrid cross. The ratio is 9:3:3:1, with 9/16 of the plants having the dominant phenotype for both traits (in our example smooth yellow seed coats), 3/16 with the dominant phenotype of the first trait and the recessive phenotype of the second (smooth, green), 3/16 with the first trait recessive and the second dominant (wrinkled, yellow), and 1/16 in which the plants have the recessive phenotype of both traits (wrinkled, green) (**Table 16-3**). The 9:3:3:1 ratio results only if all four types of gametes are produced in equal numbers and have equal opportunity to participate in reproduction. The alleles *Y* and *y* must be independent of *S* and *s* during meiosis I; this automatically happens if they are on different chromosomes.

Notice that if only one trait is considered, it behaves as in a monohybrid cross: Plants with smooth seeds outnumber those with wrinkled seeds by 3:1, and those with yellow seeds are three times more abundant than those with green seeds. Similarly, the monohybrid genotype ratios are also present—1 *SS*:2 *Ss*:1 *ss* and 1 *YY*:2 *Yy*:1 *yy*. Considering two genes simultaneously does not affect their inheritance at all.

TABLE 16-3 Genotype and Phenotype Ratios of a Dihybrid Cross*

Phenotype	Ratio	Genotype	
Smooth, yellow	$\frac{9}{16}$	$\frac{1}{16}$ *SSYY*: $\frac{2}{16}$ *SSYy*: $\frac{2}{16}$ *SsYY*: $\frac{4}{16}$ *SsYy*	or *S__Y__* †
Smooth, green	$\frac{3}{16}$	$\frac{1}{16}$ *SSyy*: $\frac{2}{16}$ *Ssyy*	or *S__yy*
Wrinkled, yellow	$\frac{3}{16}$	$\frac{2}{16}$ *ssYy*: $\frac{1}{16}$ *ssYY*	or *ssY__*
Wrinkled, green	$\frac{1}{16}$	$\frac{1}{16}$ *ssyy*	must be *ssyy*

*The two individuals are heterozygous for two characters showing complete dominance and independent assortment.
†The notation ____ indicates that either allele may be present; e.g., *S__yy* represents either *SSyy* or *Ssyy*, both of which produce a smooth green phenotype.

Crossing-Over

Independent assortment can also occur if two genes are located far apart on the same chromosome such that crossing-over occurs between them during prophase I, after homologous chromosomes have paired and a synaptonemal complex is formed. Because no preferential sites for crossing-over seem to exist, the farther apart two genes are, the greater the possibility that crossing-over will occur between them. Most plant chromosomes are so long that crossing-over occurs several times within each chromosome during each prophase I. Consequently, the two ends act like separate entities; if the gene for seed coat color were at one end of the chromosome and the gene for seed coat texture were at the other, they would still undergo independent assortment. If the two genes are close together on a chromosome, however, crossing-over may not occur and the two alleles may move together during meiosis I, as described below.

Genes on the Same Chromosome: Linkage

If two genes occur close together on a chromosome, they usually do not undergo independent assortment during meiosis I; instead, the two **genes** are said to be **linked**. Consider a plant heterozygous for the two traits of the seed coat again, but now imagine the genes occurring close together on one chromosome. During meiosis, haploid cells that are either *S* or *s*, *Y* or *y* are formed, but a new complexity arises. Because the genes are linked, several types of heterozygote are possible: an *SsYy* individual may have the alleles *S* and *Y* on one chromosome and the alleles *s* and *y* on the homologous chromosome, but an *SsYy* individual may have *s* and *Y* linked together and *S* and *y* linked (Figure 16-18a). We must consider the types of gametes that can be formed. If crossing-over is ignored, the first individual produces haploid cells with the genotypes *SY* and *sy* only, whereas the second plant would produce *sY* and *Sy* gametes only. The result of selfing is not a 9:3:3:1 ratio, but something drastically different.

The most instructive cross is a test cross using a double homozygous recessive parent: *ssyy*. If we continue to exclude crossing-over (imagine that the two genes are extremely close together), the gametes from the first type of heterozygote will be *SY* and *sy* and the gametes from the homozygous recessive parent will be all *sy*. Two types of F₁ will be produced, as shown in Figure 16-18b, smooth yellow and wrinkled green, and they occur in a ratio of 1 to 1. The two phenotypes are like those of the parents.

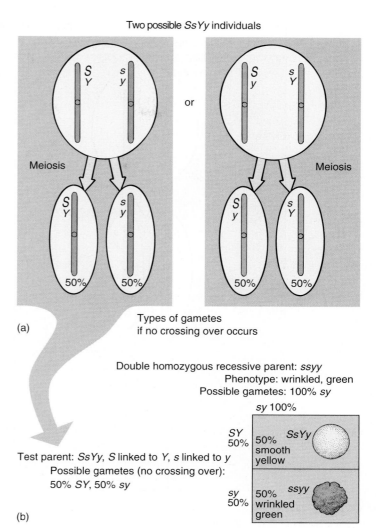

(a)

Types of gametes if no crossing over occurs

Double homozygous recessive parent: *ssyy*
Phenotype: wrinkled, green
Possible gametes: 100% *sy*

Test parent: *SsYy*, *S* linked to *Y*, *s* linked to *y*
Possible gametes (no crossing over):
50% *SY*, 50% *sy*

(b)

Figure 16-18 (a) Two types of *SsYy* individuals are possible, but in most instances, their phenotypes are identical. Only genetic tests can distinguish which is which, based on the unique gametes produced by each. (b) A test cross, with the plant on the left in (a) being the test parent. Try setting up the Punnett square for a test cross with the plant on the right in (a). A test cross of two closely linked genes produces results very different from those of a test cross involving nonlinked genes that assort independently. What would the Punnett square be like if *S* and *Y*, *s*, and *y* were not linked?

If the genes are not extremely close together, crossing-over may occur; for example, in 10% of the cells undergoing meiosis, a crossing-over might happen; therefore, the plant would produce four types of gametes, but not in equal numbers. Of the gametes for the first type of heterozygote, 47.5% would be *SY*, 47.5% *sy*, 2.5% *sY*, and 2.5% *Sy* (Figure 16-19a). The last two are **recombinant chromosomes** formed from a crossing-over of the homologous chromosomes and recombination of alleles. The first two are **parental type chromosomes**. In the majority of the cells, *Y* is still linked to *S* and *y* is linked to *s*, but in a minority, *Y* has become linked to *s* and *y* to *S*. A Punnett square alone is not sufficient to show the relative proportions of progeny, so the percentages of gamete must be added. A test cross now would result in 47.5% of the F₁ progeny being smooth and yellow, 47.5% wrinkled and green, 2.5% wrinkled yellow, and 2.5% smooth green (Figure 16-19b). The first two types are like the parents, but the last

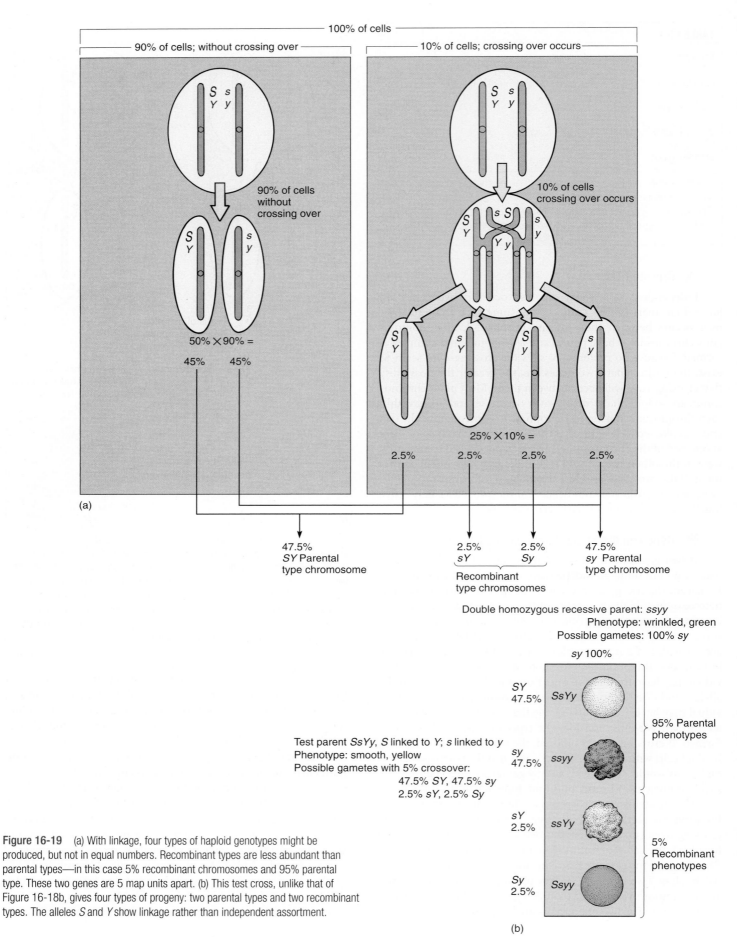

Figure 16-19 (a) With linkage, four types of haploid genotypes might be produced, but not in equal numbers. Recombinant types are less abundant than parental types—in this case 5% recombinant chromosomes and 95% parental type. These two genes are 5 map units apart. (b) This test cross, unlike that of Figure 16-18b, gives four types of progeny: two parental types and two recombinant types. The alleles *S* and *Y* show linkage rather than independent assortment.

Botany and Beyond

Box 16-2 Whose Genes Do You Have?

You received half your chromosomes from your father and the other half from your mother. So half of the DNA you had when you were a zygote—a single cell—was your father's and half was your mother's. But beyond that? One fourth of your genes came from your paternal grandfather (your father's father), one fourth from your paternal grandmother, and so on for your maternal grandparents. But, can we say that one fourth of your DNA came from each of your grandparents? That is not so easy. Of the thousands of cells in your father's testes, one in particular underwent meiosis and produced four sperm cells, one of which gave rise to you. During prophase I of that meiosis, the chromosomes your father got from his father paired (synapsed) with the ones he got from his mother (your grandmother), and they underwent crossing-over. If each homologous chromosome were to break in exactly the same place, then when they are repaired and meiosis continued, each would have exactly the same amount of DNA. Often, however, homologous chromosomes do not break at quite the same spot; after repair one might have a bit more

DNA, the other a bit less, one might be mostly paternal, the other mostly maternal. As long as they are not too unbalanced, the chromosomes will be functional. But because of this, we cannot say that exactly one fourth of your DNA came from each of your grandparents: Some might have given you a bit more, and others a bit less.

With each generation we go back, the number of ancestors doubles and the number of genes we received from each drops by half. By the time we get to the 15th generation we have 32,768 great-great-etc.-grandparents (2^{15}), and each contributed 1 in 32,768 of our genes. We believe, however, humans only have somewhere between 30,000 and 35,000 genes. If we assume each generation is about 20 years (some people have children earlier, some later), then this 15th generation lived about 300 years ago, about 1700 CE, not long after the first microscopes were invented. Of course, each of us is descended from many more generations before that, but even so, we may not have actually gotten any of their DNA.

two are recombinant types produced at the same percentage as crossing-over occurs. We cannot actually "see" crossing-over with a microscope, but we can infer that because 5% of the F_1s have a recombinant phenotype, then 5% of the chromatids must have undergone crossing-over.

The rate of crossing-over is directly proportional to the physical spacing between the genes on a chromosome. Two genes that produce 6% recombinant F_1s are closer together than two genes that produce 10% recombinant F_1s. By analyzing as many mutant alleles as possible, we can measure "space" between them in recombination percentages, each 1% being called one **map**

unit or one **centimorgan**; one map unit, on average, is approximately one million base pairs. By this means, a genetic map can be constructed. If enough genes can be mapped, we may find that A is linked to B which is linked to C; even if A and C are so far apart that they assort independently, by knowing that both are linked to a common gene, B, we know that all three are on the same chromosome. On the other hand, if A, B, and C cannot be shown to be linked to another gene F, that does not mean that F is on a different chromosome; it might be that intervening genes have not yet been mapped (**Figure 16-20**). A set of genes known to be linked is called a linkage group; when all genes are mapped,

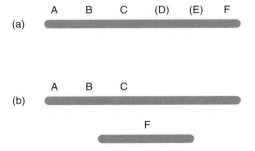

Figure 16-20   (a) *A* and *C* might be far enough apart for crossing-over to occur in virtually all nuclei, so they undergo independent assortment. If each can be shown to be linked to *B*, they must be on the same chromosome. If *A*, *B*, and *C* are not known to be linked to *F*, it might be because they are on separate chromosomes (b) or because they are on the same chromosome but intervening genes have not yet been mapped, as shown in (a). (c) Because of incomplete mapping, *ABC*, *JK*, and *STUV* are three linkage groups even though there is only one chromosome.

there will be exactly as many linkage groups as there are chromosomes, but if only a few genes are mapped, there may be more linkage groups than chromosomes.

Multiple Genes for One Character: Quantitative Trait Loci

Individual phenotypic traits are the result of complex metabolic processes involving numerous enzymes; therefore, many separate genes may affect any single trait. The gene R was described as affecting flower color by producing an enzyme that synthesized a red pigment, but that enzyme requires the proper substrate, which is present only if it is synthesized by a different enzyme controlled by a distinct gene. If this gene is present as an ineffective allele, there will be no substrate and thus no pigment, regardless of whether the first flower color gene is present as allele R or allele r. In *Arabidopsis thaliana*, abscisic acid (ABA) causes developing seeds to become dormant at maturity. In several mutants (Table 16-4), mature embryos continue to grow as if germinating and do not enter dormancy. One mutant, *aba*, is unable to produce enough abscisic acid. A second mutant, *abi-3*, produces plenty of abscisic acid but does not respond to it—perhaps its receptor is the affected protein. Although *aba* and *abi-3* involve different genes, they both result in the same phenotype. They are multiple genes for one trait even though they are not part of one metabolic pathway.

Most synthetic pathways involve at least four or five intermediates, and their four or five genes all affect the same trait. Complex traits such as leaf shape, cold hardiness, and general vigor can be the results of dozens of genes and rather than having phenotypes that are simply dominant and recessive, the trait may show a continuous gradient of values. Having multiple genes for each trait is referred to as **epistasis**. Some intermediates may be produced by several pathways, each with its own enzymes; a mutation in one is not necessarily particularly severe because the alternate pathway may increase its activity and produce sufficient amounts of the intermediate. If this happens, the mutation is masked and is not reflected in the phenotype, even if it is present in the homozygous condition. Extremely complex crosses, involving hundreds or thousands of progeny, may have to be performed to determine what fraction of a particular phenotype is correlated with a particular gene; the genes or other portions of DNA associated with such traits are **quantitative trait loci**. Conversely, when an intermediate is part of several metabolic pathways and is produced by only one enzyme, a mutation in that enzyme's gene affects all of the pathways and alters several different traits. Multiple phenotype effects of one mutation are called **pleiotropic effects**. For example, any mutation that affects the protein portion of phytochrome affects all developmental processes controlled by phytochrome, and mutations that alter the level of pyruvate affect the citric acid cycle, amino acid synthesis, and C_4 metabolism.

Other Aspects of Inheritance

Maternal Inheritance

In the crosses described above, the alleles of both parents are transmitted equally to the progeny, a situation called **biparental inheritance**. All genes in the nucleus undergo biparental inheritance. An unexpected feature of both plants and animals is that during fertilization, the sperm cell loses most of its cytoplasm and only the sperm nucleus enters the egg. Consequently, the zygote obtains all its plastid and mitochondrion genomes from the maternal parent; this is **uniparental inheritance**, more specifically **maternal inheritance**.

Mitochondrion genetics are difficult to study for several reasons. First, each cell has many mitochondria, up to hundreds, and each mitochondrion has several circles of DNA. Each cell has multiple copies of each mitochondrion gene and may have received diverse types of alleles from its maternal parent. Simple Mendelian ratios are not produced. Second, the presence of many of the mitochondrion DNA mutants does not result in easily recognizable phenotypes. The plants usually simply grow somewhat more slowly, or if the mutation has a severe effect, they may die as embryos. Consequently, the presence of mitochondrion mutants may not be detectable by examination.

Plastid genetics are somewhat easier if alleles affecting chlorophyll synthesis are studied. Consider a plastid whose DNA has mutated so that the plastid can no longer produce any chlorophyll; it may be red or orange because carotenoids are no longer masked by chlorophyll, or it may be white if no pigment at all is produced. If a zygote receives only plastids like these, the embryo and seedling will be red, orange, or white. Such seedlings are rare, but in large commercial nurseries that grow millions of seedlings, they appear from time to time. The plants can be kept alive artificially if they are grafted onto a normal plant that supplies them with carbohydrates (Figure 16-21). Grafted plants can then be used for genetic experiments.

When an achlorophyllous plant is crossed with a green, chlorophyll-bearing plant, the outcome depends on which plant is the **pollen parent** and which is the **ovule parent**. If the achlorophyllous parent provides the pollen, 100% of the zygotes will be green because they receive all of their plastids from the normal parent by way of the egg; all mutant plastids are destroyed during sperm differentiation and syngamy. But if the achlorophyllous plant is the ovule parent, all progeny also have mutant plastids and are achlorophyllous.

Plastid inheritance is also responsible for certain types of **variegation** in plants, the presence of spots or sectors that are white, red, or orange on a plant that is otherwise green (Figure 16-22). Because plant cells have many plastids, only rarely do all the plastids of an egg have the same alleles unless the parent has only one type, as in the cactus of Figure 16-21. More often, each cell has a mixture of plastid types. During cell division, some plastids move into each end of the dividing cell; usually random chance alone

TABLE 16-4	Multiple Genes for One Character	
Allele	ABA*
wild type	169
aba	10
abi-3	279

*ng/g fresh weight of seed tissue.

Figure 16-21 Plastid mutations in the cactus genus *Gymnocalycium* seem to be common; quite often totally red or orange seedlings appear in cactus nurseries. These lack chlorophyll but can be kept alive by grafting them onto a green cactus that supplies them with sugar. Many plants like *Coleus* and maples have bright red leaves; they do have chlorophyll, but it is masked by large amounts of other pigments.

(Figure 16-22a). If it occurs in a very young leaf, the cells grow quite a lot and produce a large spot; if it happens in an older leaf that has reached almost full size, the patch is small.

When variegation affects stems, the achlorophyllous patch may include an axillary bud. If so, the entire bud and the branch that grows from it are achlorophyllous. Any flowers it produces can be used for genetics experiments, providing either pollen or ovules. Experiments on both corn and morning glory have shown the achlorophyllous character to be maternally inherited; thus, it is a feature of the plastid DNA as just described. Recombinant DNA technology is now greatly accelerating our efforts to understand plastid and mitochondrion genetics. In both organelles, the DNA is a circular double helix, one of its ends being attached to the other, and no histones are present. Plastids contain many circles, 200 in each spinach plastid and 1,000 in each of those of wheat, so rather than being haploid or diploid, wheat plastids are 1,000 ploid. Plastid DNA may constitute up to 21% of the total DNA content of a cell. In angiosperms, plastid DNA tends to be about 150 kbp long. In plants, mitochondrion DNA circles are very large, well over 200 kbp (up to 2,400 kbp), more than ten times larger than mitochondrion DNA in animals.

Genomes of plastids contain genes for numerous enzymes as well as a complete set of plastid ribosomal RNAs and transfer RNAs. Currently, it is estimated that 80 to 100 enzymes and membrane proteins are coded by plastid DNA, and many of these are essential: the large subunit of RuBP carboxylase, components of the proton channel/ATP synthetase complex, cytochromes, and proteins that bind chlorophyll *a* (**Figure 16-23**). Mitochondrion DNA also contains genes for rRNA and tRNA, as well as those that code for proteins involved in ATP synthetase complexes, electron

results in each daughter cell getting some of both types of plastid (Figure 16-22b); however, occasionally, one daughter cell by chance receives only mutant plastids. It is white, red, or orange and is able to survive because surrounding normal cells supply it with sugar. If this occurs in a leaf primordium or a young internode, the cell continues to grow and divide along with the surrounding cells, resulting in a patch of cells that lack chlorophyll

Figure 16-22 (a) Variegated plants of *Syngonium podophyllum*. Each green cell probably has a mixture of normal chloroplasts and plastids with a mutation that prevents chlorophyll synthesis. Cells in the white area have only mutant plastids. (b and c) Cells usually have many plastids and mitochondria, so random distribution of the organelles is usually sufficient to ensure that both cells receive some (b). If mutant plastids are present, then occasionally when cytokinesis occurs one end of the cell may by chance have only mutant plastids (c). Most cells have only two sets of (nuclear) chromosomes; therefore, all of the elaborate mechanisms of mitosis are necessary if each daughter cell is to receive one complete set.

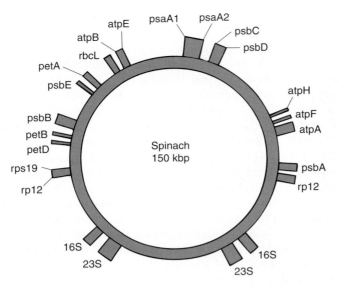

Figure 16-23 A map of the plastid DNA circle from spinach; the letters are the abbreviations for the names of the genes. For many species, the entire plastid circle has been sequenced so that we know the exact position of every nucleotide.

transport, and NADH dehydrogenase complex. In many cases, organelle genes must work in coordination with nuclear genes. The interaction of these three genomes within one cell is now an area of very active research.

As mentioned in the beginning of this chapter, recombination of genes during meiosis and sexual reproduction creates greater genetic diversity, which is the raw material of natural selection: Sexual reproduction results in more rapid evolution. But plastids and mitochondria, being derived from endosymbiotic prokaryotes, have no sex, no genetic recombination. However, there are mechanisms that transfer genes from the organelles into nuclei where they become nuclear genes. We think this happens very rarely (once every several million years), but gradually, most of the prokaryotic ancestral genes of plastids and mitochondria are now located in the nucleus (they are recognizable in DNA sequencing projects because they lack introns and other features of eukaryotic genes), and the DNA circles within the organelles themselves have only a few genes left. The organelle-derived genes that are now located in the nucleus have all of the benefits of nuclear genes: They are not destroyed as sperm cells mature; syngamy of sperm and egg brings two sets of genes (possibly with different alleles) together into one nucleus, and then during meiosis, homologous chromosomes pair and undergo crossing-over, generating genetic diversity. These genes can evolve rapidly, whereas those still located in the organelles evolve more slowly.

Lethal Alleles

The phenotypic result of a mutation can vary in severity from almost undetectable to **lethal**; that is, its presence can kill the plant. Genetically inherited lethal mutations are most often recessive and are fatal only if present in the homozygous condition; a heterozygous plant has enough normal protein to survive. The reason for the recessive nature of most lethals is simple: A

dominant lethal would kill the plant, even in the heterozygous condition; therefore, it would rarely be passed on to any offspring.

The presence of a **lethal allele** can be difficult to detect if its effect occurs early. If it affects basic metabolic functions such as respiration, DNA replication, cellulose synthesis, or the structure of histones, plants homozygous for it probably die while still very young; in some instances, the gametes die even before fertilization. The only seeds that complete development are heterozygotes or dominant homozygotes. This results in unusual genotype and phenotype ratios. A selfing of a heterozygote for a lethal condition, Ll, produces zygotes in the typical 1 LL:2 Ll:1 ll genotype ratio, but the ll zygotes die. The seeds and plants that live are 1 LL:2 Ll genotypically; such a ratio should always lead one to suspect a recessive lethal allele.

Plants have an alternation of heteromorphic generations, the diploid sporophytes alternating with the haploid gametophytes. In gametophytes, each gene is present only as a single allele. No homologous chromosomes occur, and recessive and lethal alleles are not masked by the presence of a dominant homologous allele. In animals, this haploid phase is restricted to the sperm and egg, which are simple and have many inactive genes. As a result, numerous genes with deleterious mutations are not active during the brief haploid condition; they are turned on only during growth and differentiation from a zygote to an adult, when their effect is mitigated by their diploid condition. In plants, however, the haploid phase consists of at least a pollen grain with three cells and a megagametophyte with seven cells. This phase is still quite simple and most genes are inactive; however, in both gametophytes, many types of central metabolism are occurring, and any highly deleterious allele causes that gametophyte to develop poorly or not at all. If the gametophyte dies without producing a sperm or an egg, the allele is eliminated.

Multiple Sets of Chromosomes and Gene Families

Whereas most animals have diploid body cells and haploid gametes, plants are much more diverse. Rarely, a spore is formed without undergoing meiosis, resulting in diploid gametophytes and diploid gametes. If these fertilize a normal haploid gamete, a triploid zygote results. Although this would be instantly lethal in almost all animals, plants tolerate it well, and a triploid sporophyte develops from the zygote. Triploids, however, are sterile because they cannot undergo pairing of homologous chromosomes during prophase I; chromosomes come together in threes, and the rest of meiosis is aberrant.

Occasionally, cells fail to undergo mitosis after S phase DNA replication, and the cell becomes tetraploid (or diploid if it happens in a gametophyte). Tetraploid cells are usually perfectly healthy; if they produce a part of the plant that initiates flowers, diploid pollen and eggs are produced, again raising the possibility of triploid zygotes if fertilized by haploid gametes. These processes can occur in virtually every conceivable combination, and plants that are 3n (three sets of chromosomes, triploid), 4n, 5n, 6n, . . . up to several hundred n are common. All plants with more than two sets of chromosomes are **polyploid**. All plants with an even ploidy level can undergo meiosis and are fertile; all odd ploidy levels are sterile like triploids.

A change in chromosome number per nucleus in plants can also come about by **nondisjunction**: During the second division of meiosis, the two chromatids may remain together (fail to disjoin), so one daughter cell receives both copies of the chromosome (the two chromatids), whereas the other receives none. The cell with the extra chromosome probably survives quite well and forms a functional sperm or egg cell. If this gamete, which is diploid for one chromosome, is involved in fertilization, the new zygote will be triploid for that chromosome. It received a normal, haploid chromosome set from one parent and a haploid set plus the extra chromosome from the parent in which nondisjunction occurred. In plants, this condition is tolerated often, and only by studying the karyotype can it be discovered. In animals, this is almost invariably fatal. The human congenital disease Down syndrome is caused by nondisjunction of the very small chromosome 21. Down syndrome individuals are triploid for just a few genes yet have severely disrupted metabolism. Fetuses triploid for longer chromosomes abort spontaneously.

Plants with even ploidy levels grow and reproduce successfully, often more vigorously than diploid individuals of their species. Initially, these have complex genetics. For example, a tetraploid can have any of the following genotypes for red/white flower color: *RRRR, RRRr, RRrr, Rrrr,* or *rrrr.* Its gametes can be *RR, Rr,* or *rr.* Notice that polyploids have more copies than needed for every gene. In all diploid plants, even heterozygotes with only one functional allele are usually healthy. For nonessential genes, a complete absence of functional alleles may not be harmful, such as when *rr* produces white flower color. Consequently, the two or three extra copies of each gene in a tetraploid plant are surplus DNA. Mutations that completely incapacitate one of the four alleles of a tetraploid usually are not deleterious, as one, two, or three functional alleles remain in every nucleus. For polyploid plants, deleterious mutations tend to have little effect on phenotype; therefore, they do not affect survival as much as they would in a haploid cell or a diploid plant. As a result, deleterious mutations are not eliminated quickly by natural selection, and the extra copies of a gene may rapidly become nonfunctional; as long as the genes are still so similar to the wild-type allele that they can be recognized as having originated as duplicates of it, they are called **paralogs**. Before many generations have passed, so many genes have changed that polyploids act like diploids, their "extra" copies having mutated. Almost half of all flowering plant species are actually polyploids; those that now appear to be perfectly normal diploids probably underwent conversion to the polyploid condition so long ago that all their extra alleles have mutated extensively and can no longer be considered alleles of those that are still functional. Nondisjunction, insertions, and deletions may have changed the numbers and sizes of the chromosomes so much that the four originally homologous chromosomes now appear as just two different pairs. Conversion back to the diploid state is an uneven process, and some species act like diploids for certain genes but tetraploids for others.

What do the extra alleles mutate into? At first, they may simply be alleles that allow the cell to produce a particular protein more rapidly or produce slightly different versions of the same protein. For example, the five histone proteins that make up nucleosomes are needed in such abundance that each nucleus may have up to 600 paralogs of the genes, not just two as for

Figure 16-24 As corn seeds mature, they fill with the storage protein zein. Each nucleus has two families of zein genes, each with about 25 copies of the gene, and thus, zein mRNA can be produced very rapidly. It is not essential that all zein molecules have identical amino acid sequence (they will merely be digested by the embryo when the seed germinates); therefore, zein paralogs evolve independently of each other, but still, each produces a zein protein 90% identical to those produced by any other member of the gene family.

most traits; these multiple copies constitute a **gene family** (Figure 16-24). Each paralog will evolve independently of the others, and some may ultimately produce defective proteins (but some must continue to produce the proper protein). Mutations in the promoter regions may prevent them from ever being activated, so they turn into "junk," acting as nothing more than spacer DNA. Deletions, inversions, and movements by transposons may break them up. DNA sequencing has now revealed that there are large families of genes in which one gene codes for a useful protein, whereas the other genes have nucleotide sequences that are obviously related to the useful gene but are inactive and apparently code for nothing.

This extra DNA is actually extremely valuable—it is the raw material for the evolution of new genes. Some may mutate into genes that produce enzymes almost identical to those being coded by the original form of the allele. The original enzyme may work best at low temperature, whereas the new form may have a higher optimum temperature. The plant can now produce two types of enzymes and function well in both warm and cool days, or the species may be able to extend its range, the new enzyme allowing it to survive in hot deserts and the original enzyme allowing it to live in its original habitat (Figure 16-25). Mutations in the promoter region instead may allow the structural portion to produce the same protein as the original gene but at a different time or place in response to a different chemical messenger.

In other cases, mutations in the extra DNA may result in totally new genes that produce proteins whose function is not at all related to that of the original gene. Mosses, ferns, conifers, and flowering plants have evolved from green algae. Because algae have no genes for flower color, lignin synthesis, and so on, the evolution of these species has involved the evolution of entire new metabolic pathways. All of these genes had to arise by the gradual mutation of surplus alleles into nucleotide sequences that code for useful proteins.

Box 16-3 Genetics of Haploid Plants

This chapter discusses diploid zygotes formed by syngamy, the fusion of a haploid sperm with a haploid egg. Each zygote has two chances of obtaining at least one functional allele for every gene in its nucleus. If one gamete happens to supply a nonfunctional allele for a particular gene, the zygote will probably survive if the other gamete provides a functional allele: The zygote will be heterozygous for that gene, but often a single functional copy is enough. Almost all plants familiar to us benefit from the safety net of having two sets of genes in each nucleus.

Haploid plants also exist, however. Each has only a single parent, not two, and thus, each nucleus has only a single set of chromosomes, not two. Mosses are a familiar example: Each moss plant is initiated when a haploid spore germinates and grows. Spores differ from gametes in being able to undergo mitosis and growth but being unable to fuse with another cell. A gamete, on the other hand, undergoes syngamy with another gamete but can not grow or divide (unfertilized eggs of bees are exceptional and develop into haploid workers). After a haploid moss spore germinates, it grows as a branching network of green parenchyma cells, and later, some cells take on the shape of a shoot apical cell and produce upright green, leafy little moss plants. Every nucleus in every moss plant is haploid.

How does this affect the genetics of these plants? Imagine a gene that has two alleles, *A* and *a*. Some plants in the moss population might be *A* and others *a*, but none is ever *AA, Aa,* or *aa.* If *a* produces a nonfunctional enzyme, then all of the *a* mosses will be in danger and probably die before they can reproduce. When they die and decompose, the nuclei and all of the genes—including the *a* allele—break down; consequently, this allele will be very rare in the moss population. In diploid plants and animals, non-functional alleles can persist for many generations in the heterozygous condition: It is usually when they occur in homozygotes, *aa,* that natural selection eliminates them. Because mosses are haploid, nonfunctional alleles cannot be protected by heterozygosis and natural selection eliminates them quickly. Moss populations tend to have "healthy" genomes in which almost all alleles are functional.

Is this good or bad, or more accurately, what are the consequences of quickly eliminating mutant alleles? Evolution by natural selection operates only if there is genetic diversity. Mutations produce new versions of genes, versions that often are less functional than the original versions but which may actually be better in a different habitat, one that is drier or wetter, warmer or cooler, or which has different animals and fungi. If the plant population has both the original and the new allele, it may be able to adapt to a wider range of habitats, or as the climate changes, a more diverse set of alleles for many genes may allow the population to adapt rapidly enough to survive the new climatic conditions.

Imagine a new allele arising in some diploid plant. If it occurs in a somatic cell, it cannot be passed on to another individual, but if the mutation occurs in a sex cell, then a copy of the new allele will be incorporated into some of the sperms and eggs and then into zygotes. Almost certainly, the diploid zygote has gotten a functional allele from the other parent, and thus, the zygote will be heterozygous and will survive—and so will the new allele. Every time this plant reproduces, more copies of the allele will be duplicated and passed on. At some point, a seed carrying this allele might land in a microhabitat where this allele is adaptive, not useless, and the allele will then help the population adapt to a broader range of habitats.

Now imagine a similar allele being formed by mutation in a haploid plant. Again, if it forms in a somatic cell, it does not matter; however, if it forms in a spore, then it will soon find itself as the sole allele for that gene in a haploid plant, and almost certainly the plant will suffer from not having a functional wild-type allele. Under very lucky circumstances, the spore with this new allele might land in a habitat where the allele is adaptive; if so, this may be beneficial, but the chances are low that a spore with the very first copy of new, mutant allele would land in a suitable area. It is more likely that the spore will land near the parent, in a habitat similar to the parent's habitat and the spore with the mutant allele will develop poorly or die, eliminating the allele within a few days or weeks after it formed. The gene pool of the moss population will return to being the same as it was before the mutation and will have no new genetic resources for helping it adapt to new or changing conditions.

Figure B16-3 All of the short plants beside this stream are mosses, and all are haploid. Each grew from a haploid spore, not from a diploid embryo in a seed.

Figure 16-25 After a plant becomes tetraploid, mutations in the extra copies of a gene are tolerable as long as (1) at least one allele is left functional, and (2) the extra copies do not mutate into a lethal allele. In this example, two chromosomes have changed morphology, becoming shorter; they could have become longer by picking up pieces of DNA by unequal crossing-over. Several copies have mutated into spacers and some into junk that still has much sequence homology with functional alleles. Others have evolved into new alleles of the original genes (d_3 to d_5; a_2 to a_3; f_3 to f_4), and some have evolved into entirely new genes (a_1 to g_1; d_4 to h_1).

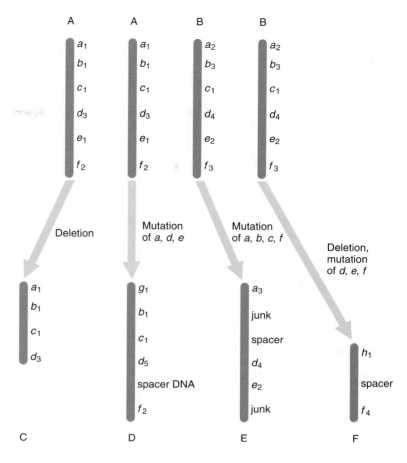

Summary

1. Most genes occur as slightly different forms, alleles, in which the DNA sequence varies. Alleles arise by mutations.
2. An organism's physical and physiological features constitute its phenotype, whereas its suite of alleles is its genotype.
3. During sexual reproduction, two sets of alleles, maternal and paternal, are brought together into one cell. During prophase I of meiosis, alleles of homologous chromosomes are rearranged into new chromosomes.
4. Mutations are any changes in the sequence of nucleotides in DNA. Larger mutations are often more significant than small ones, but any type can potentially affect the phenotype. The great majority of mutations are selectively disadvantageous.
5. DNA proofreading and repair enzymes keep the mutation rate low enough to allow complex organisms to live, but they are imperfect enough to allow a low rate of mutation, permitting evolution by natural selection.
6. If a diploid organism has two identical alleles for a gene, the organism is homozygous for the gene; if it has two distinct alleles, the organism is heterozygous for that gene. If the phenotype of heterozygotes is indistinguishable from that of one of the homozygotes, the trait shows complete dominance. If the heterozygote's phenotype differs from that of both homozygotes, the trait shows incomplete dominance.
7. When heterozygotes with complete dominance are selfed, the F_1 generation should have a phenotype ratio of three dominant to one recessive, and a genotype ratio of one

homozygous dominant to two heterozygotes to one homozygous recessive.
8. A test cross is one in which one parent is known to be homozygous recessive for the trait of interest; this allows the phenotype of the other parent to be expressed in every progeny, thus revealing the genotype of the parent.
9. Almost any gene has multiple alleles; that is, there are many mutant forms of the gene. Any particular phenotype trait is probably the result of the action of many enzymes or other proteins; the alleles for all these genes may affect the expression of the phenotype.
10. A dihybrid cross is one in which two traits are studied simultaneously; if the responsible genes are on separate chromosomes or if they are widely separated on the same chromosome, they show independent assortment.
11. Two genes that occur close together on a chromosome are linked. Crossing-over during prophase I breaks up linkage groups.
12. In sperm cells of both plants and animals, organelles are destroyed, so sperms do not contribute mitochondrion or plastid genes. All organelle genes are provided in the egg and show maternal rather than biparental inheritance.
13. Lethal genes cause such a severe disruption of metabolism that the individual dies. Many kill the organism before it reaches reproductive maturity, so they can be passed on only rarely, masked in the heterozygous condition.

Important Terms

alleles	incomplete dominance	polyploid
biparental inheritance	independent assortment	Punnett square
complete dominance	lethal allele	pure-bred lines
cross fertilization	linked genes	quantitative trait loci
dihybrid cross	map unit	recessive allele
DNA polymerase	maternal (uniparental) inheritance	recombinant chromosomes
dominant allele	monohybrid cross	self-fertilization (selfing)
F_1 (first filial) generation	mutagen	semiconservative replication
gene family	mutation	somatic mutations
genotype	paralogs	test cross
heterozygous	parental generation	transposon
homozygous	phenotype	

Review Questions

1. What is an allele? Do all genes come in only one form such that every plant that has that gene has exactly the same nucleotide sequence? Look at the gene in **Figure 16-26**. How do the various alleles differ from each other?

AGCACCCTCCCACCTCATCCCACCCTTCTGATCTCAATCCAACGTCGCATCTCCACCGTCTCG
GCCACAGCTCTGACCCAGCCCGCCAAACCACCGGTCCAATCTCTCGGCCACGTCACCGATCCG

 −150
CGGATCGACCCAGCGAAGTCCCTCCCGCCCCCAAAGTCCCCCAAATCTTGCAGTTCCCTCCTA
CGGCATCTCTCCCCCGGATCGCCGTCTCGACCGTCCACTCCATCCGCATCCAACGGCAGCCAC

 −100
AATCCTCCCCATATAAACCAACCCCCCGCCCTCAGATCCCTAATCCCATCGCAAGCATCAGCA
ACGCCTCCTCCAACCTCTCGACCCCTTTAAGACGCCCTTCGCCCCACCCAGCAAATCACAGCA

 −50
TCCCTCCAAAGCAGGCAGCAGCTCCTCTTCTTCCTAATCACACTATCTCGGAGAGGAGCGGCC
CCAGACGCCACCCACCACCGTTCCTCCCATCCCACACTCGCTCGCAGCTCGAGATCGTCGGCC

+1
ATG	TCT	GGG	CGC	GAC	AAG	GGC	GGC	AAG	GGG	CTG	GGC	AAG	GGC	GGC	GCC
---	--C	---	---	-G-	---	--A	---	---	--C	--A	---	---	---	---	---
	Ser	Gly	Arg	Asp	Lys	Gly	Gly	Lys	Gly	Leu	Gly	Lys	Gly	Gly	Ala
				Gly											

+50
AAG	CGG	CAC	CGG	AAG	GTC	CTC	CGC	GAC	AAC	ATC	CAG	GGC	ATC	ACC	AAG
---	--C	---	---	---	---	---	---	--T	---	---	---	---	---	---	---
Lys	Arg	His	Arg	Lys	Val	Leu	Arg	Asp	Asn	Ile	Gln	Gly	Ile	Thr	Lys

+100
CCG	GCG	ATC	CGG	AGG	CTG	GCC	AGG	AGG	GGC	GGC	GTG	AAG	CGC	ATC	TCC
---	---	---	---	C--	---	--G	C--	C--	---	---	---	---	---	---	--G
Pro	Ala	Ile	Arg	Arg	Leu	Ala	Arg	Arg	Gly	Gly	Val	Lys	Arg	Ile	Ser

+150
GGC	CTC	ATC	TAC	GAG	GAG	ACC	CGC	GGC	GTC	CTC	AAG	ATC	TTC	CTC	GAG
--G	---	---	---	---	---	---	---	--G	---	---	---	---	---	---	---
Gly	Leu	Ile	Tyr	Glu	Glu	Thr	Arg	Gly	Val	Leu	Lys	Ile	Phe	Leu	Glu

+200
AAC	GTC	ATC	CGC	GAC	GCC	GTC	ACC	TAC	ACC	GAG	CAC	GCC	CGC	CGC	AAA
---	---	---	---	--T	---	---	---	---	---	---	---	---	---	---	--G
Asn	Val	Ile	Arg	Asp	Ala	Val	Thr	Tyr	Thr	Glu	His	Ala	Arg	Arg	Lys

+250
ACC	GTC	ACC	GCC	ATG	GAC	GTC	GTC	TAC	GCG	CTC	AAG	CGC	CAG	GGC	CGC
Thr	Val	Thr	Ala	Met	Asp	Val	Val	Tyr	Ala	Leu	Lys	Arg	Gln	Gly	Arg

+300
ACC	CTC	TAC	GGC	TTC	GGA	GGC	TAG	ATTTGTGTGGTGAAGCAACTTCCTCGTTTGC
---	---	---	---	---	--C	---	--A	GGGCCGGCCGGCCGACGGGAGTCACTCTTTG
Thr	Leu	Tyr	Gly	Phe	Gly	Gly		

+350   +400
TCTGTGATCTGTGCTGTCGTAGATGAGATTTACTGATTTGGCGTGCGCCGGTTGTATTCTGTC
TCGCCGCCTGCAGATTCCAGAAGCCTGATGAAGCCCCGACTTGTTTAGTTCGCTATTTCCTCT

Figure 16-26

2. What is the phenotype of a plant? Can you see it? Measure it? Describe it. (Hint: Could you describe the phenotype of a geranium well enough to distinguish it from a maple tree?) What is the genotype of a plant?

3. As DNA replication begins, chromatin becomes less compact. Why?

4. When DNA is read by RNA synthase, mRNA is made. What is the name of the enzyme that reads DNA and makes more DNA?

5. Describe DNA replication. What are Okazaki fragments? Why does each chromosome have thousands of replication start sites instead of just one?

6. Name the types of mutations that may occur and describe how some of them happen. With regard to UV-induced mutations, think about the fact that most leaves last for just a few months and then are replaced by new leaves the next spring. The next time you are getting a suntan, remember that your dermis must last as long as you live.

7. Statistically, are mutations almost always beneficial or harmful? Why?

8. What is the name of a mutation that occurs in a cell that never leads to sex cells? Why is this type of mutation less important than one that affects gametes?

9. Imagine that as a leaf is growing, one of its cells is hit by ultraviolet light, which causes a mutation in a gene that is involved in producing flower pigments. Will that be an important mutation? Will it ever actually affect a flower's color? When the leaf is abscised in autumn, will the gene have ever been transcribed to make mRNA?

10. Examine the value for lily in Table 16-1. How many billion pairs of nucleotides are there in a diploid nucleus? If DNA synthase is very accurate and if DNA repair mechanisms are so good that, on average, there is a mistake made once in every ten million nucleotides, how many mistakes are made on average every time a lily nucleus replicates? If a meter is 39 inches long, how many inches of DNA are there in every diploid nucleus of every cell in the body of a lily?

11. There are pieces of DNA that readily change their position from one chromosome to another. These are called _____ _____. When

one of these moves to a new position, it might move into a gene and cause a mutation. Describe why this may be one of our most powerful tools for genetic analysis and engineering.

12. What are monohybrid crosses? Is it possible to make a cross in which only one single character is actually involved, or is it just that one single character is being analyzed and the others are being ignored?

13. Imagine that a plant with red flowers is bred to one with white flowers and when they make fruits, the seeds are collected and grown into new plants. The new plants, when old enough, have only pink flowers. Which plants are the parental generation? Which are the first filial generation? If the pink-flowered plants are interbred with each other and the resulting seeds planted, what would that generation be called?

14. Assume the plants in Question 13 are like those in Figure 16-9. What is the genotype of the plants with red flowers? Those with white flowers? Those with pink flowers? When a plant with red flowers makes pollen grains, will they all carry the *R* allele? Will any carry the *r* allele? Or will half have *R* and half have *r*? Will any of the pollen grains be *Rr*? Explain your answer.

15. In plants like those of Questions 13 and 14, with red, pink, and white flowers, does flower color show complete or incomplete dominance?

16. What do homozygous and heterozygous mean? In the plants of Questions 13 and 14, which plants are which?

17. Imagine you are crossing two plants that have these two genotypes—*Tt* and *tt*. What are the two genotypes of the gametes that the *Tt* plant can make? What is the one genotype of the gametes that the *tt* plant can make? Set up a Punnett square to analyze the cross. Does the Punnett square need to have four boxes or just two? If you were crossing a *TT* plant with a *tt* plant, how many boxes would be necessary in a Punnett square?

18. Imagine you are given two tall plants, but you do not know their genotype (see Figure 16-12). There are three possibilities—they might both be *TT* or both *Tt* or one *TT* and the other *Tt*. Imagine you cross the two plants, wait for them to produce seeds, and then plant the seeds. If all resulting plants are tall, can you then determine what the genotypes of the two parents are? If approximately half the resulting plants are tall and about half are short, can you then figure out the genotypes of the parents?

19. In a field you find three types of the same plant: some with long leaves, some with short leaves, and some of intermediate length. What would you suspect to be the genotypes of each? You measure hundreds of plants and find that about half have leaves of intermediate length, whereas one fourth have long leaves and one fourth have short leaves. Does this support your estimate of the genotypes? What kinds of crosses would you do to test it further? In each cross, what types of gametes would each parent produce and in what ratios?

20. What is a test cross? Why do we have to use it? When we just look at a plant, can we see its genotype? Can we see its phenotype?

21. When traits show complete dominance, the genotype can be revealed by a test cross. In the example of tall and dwarf plants (Figure 16-14) why is it necessary to use the dwarf parent? In general, why is it necessary to use a parent that is homozygous recessive?

22. What does the term "multiple alleles" mean? Think about flowers such as roses that have many different colors. Do you think that flower color genes in roses might have multiple alleles? Now think about a protein like RuBP carboxylase in which all RuBP carboxylase molecules have the same amino acid sequence. We can extract this protein from almost any plant, and the structure will always be the same. Do you think that there are multiple alleles for RuBP carboxylase?

23. We say that genes that are close together on the same chromosome are linked. What does that mean? Imagine that in a diploid nucleus, one chromosome has the allele *R* and right next to it on the same chromosome is the allele *T* for a different gene. Are these linked? Imagine that *r* is right next to *t* on the homologous chromosome. If this cell never undergoes meiosis, does this matter at all? If it does undergo meiosis, what are the two genotypes that the gametes are most likely to have? Look at Figure 16-17 for help if you need it.

24. In Question 23, if one chromosome has *R* and *T* and the homologous chromosome also has *R* and *T*, does linkage matter? What is the single type of genotype that this plant can produce in its gametes?

25. The answer to Question 23 will be different if *R* and *T* are very far apart on their chromosome (and if that is true, then automatically *r* and *t* will also be very far apart on their chromosome). If it undergoes meiosis, there are four genotypes that can be produced, not just two. Why? What are the four genotypes? Which represent recombinant chromosomes? Which represent parental chromosomes? Look at Figure 16-18 for help if you need it.

26. Think about the same two genes that were mentioned in Question 23—*R* and *T*, present in the heterozygous condition as *R* and *r*, *T* and *t*. Now imagine that the gene for flower color (the alleles *R* and *r*) is not on the same chromosome as the gene for plant height (the alleles *T* and *t*). Because the two genes are on separate chromosomes, they show independent assortment. What does that mean? When this nucleus undergoes meiosis, it can make gametes with four different genotypes. Why? A very important aspect is that the four different genotypes will be formed in almost exactly equal numbers. Why?

27. Figure 16-17 shows a dihybrid cross but it can be analyzed as a monohybrid cross of just smooth seed coat (*S*) and wrinkled seed coat (*s*) with complete dominance that should show a 1:2:1 ratio. Examine Figure 16-17 and see if you can find these results. Likewise, it can be analyzed as a monohybrid cross of yellow seed coat versus green. Can you find these in the expected ratio?

28. What is maternal inheritance? Which characters show this type of inheritance?

29. Why is it easier to study plastid inheritance than mitochondrial inheritance? What is unusual about both? Do plastids and mitochondria undergo sexual reproduction? Syngamy?

30. Imagine that you have a plant with no chlorophyll. It is completely white because of a mutation in a plastid gene. Imagine that you cross this plant with an ordinary green plant. In one cross, you transfer pollen from the white plant onto the stigma of the green plant. After seeds form and you plant them, what color do you expect the seedlings to be—white, green, or pale green? At the same time as the first cross, you also transfer pollen from the green plant onto the stigma of the white plant. After you obtain these seeds and plant them, what color do you expect the seedlings to be? Will they be the same as in the first cross? Explain the results.

31. What is a lethal allele? Would you expect most lethal alleles to be recessive (that is, only *ll* is fatal), or would you expect many of them to be dominant (*LL* and *Ll* are both fatal)? Do you think a plant carrying a dominant lethal allele would survive very long? Do you think such a plant could become abundant in the population?

32. If a cell has one set of chromosomes, it is said to be haploid. If it has two sets, it is diploid. What term is used if it has three sets? Multiple sets?

33. Plants with multiple sets of chromosomes (polyploid plants) have more sets than they need. They are often healthy and vigorous. What might happen to their extra chromosomes and genes over long periods of time?

BotanyLinks

http://biology.jbpub.com/botany/5e

BotanyLinks contains a variety of resources and review material designed to assist in your study of botany.

Visit the Web Exercises area of BotanyLinks to complete these questions:

1. To study the genetics of inheritance, we need to know which genes are on which chromosomes. How are genes mapped? Go to the BotanyLinks home page for information on this subject.

2. Genetics and molecular genetics are becoming increasingly important, but how do they affect your life? Go to the BotanyLinks home page to begin researching this subject.

Population Genetics and Evolution

chapter 17

Concepts

Evolution is the gradual conversion of one species into one or, in some cases, several new species. It occurs for the most part by natural selection: Mutations cause new alleles or new genes to arise that affect the fitness of the individual, making it less or more adapted to the environment than other individuals without the new allele or gene. If the mutation is deleterious, the individual may grow or reproduce slowly or even die early without reproducing. Either way, the new allele has less chance of increasing in the population and is more likely to be eliminated as the individuals that carry it die out. If the mutation is beneficial, the individual should grow and reproduce better than other individuals, producing a greater number of gametes and ultimately a greater number of seeds than

Chapter Opener Image: Take a look at the many organisms in this photo taken in Zion National Park. The gray sagebrushes all look more or less the same, as do all the yellow grasses and the green juniper trees. But we would have no trouble telling the people apart. Every day, we recognize our friends and colleagues from among the hundreds of other people we see. Even though all people are members of the same species, *Homo sapiens*, we have many genetic differences, many differences in our alleles. As people reproduce sexually, half of the alleles of one are combined with those of the other, and no two children will be identical (except identical twins). Even though all the sagebrushes (*Artemisia tridentata*) look alike to us, each one too differs from the others in at least half of its alleles, so as they interbreed, their offspring too are variable. In just this photograph, there are hundreds of sagebrush plants, but in the southwestern United States there are millions, and all can theoretically interbreed. All make up a population that can share alleles.

Outline

Figure 17-1 (a) The population in this diagram originally consisted of 29 individuals, 20 of which carried allele 1 and 9 allele 2. Allele 2 produces individuals that are more vigorous, and thus, allele 2 has increased in the next generation (b) from 9 of 29 (31%) to 30 of 38 (78.9%). Allele 2 produces a phenotype (triangle) distinct from allele 1 (round); therefore, as the allele frequency of the population changes, so does the phenotype. In this diagram, the extra vigor of allele 2 is allowing the population as a whole to enlarge, from 29 to 38 individuals, perhaps by crowding out other species or by entering new habitats.

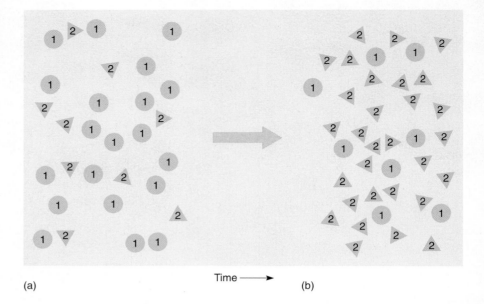

(a)　　　Time ⟶　　　(b)

the other individuals. The abundance of the new allele increases relative to the original alleles (**Figure 17-1**). As evolution by natural selection continues, the types and abundances of alleles present in the species change, and consequently, the phenotype also changes.

Evolution is an extremely slow process that may require thousands of generations and millions of years to produce obvious changes in a species. Because it is so slow compared with the length of a human lifetime, it is easy to understand that it went undetected until recently, just as continental drift, mountain building, and valley formation by erosion went unnoticed. A human lifetime is too short to perceive changes in these processes; even the full history of accurate observation and detailed record keeping is not long enough to detect the fact that, since the times of the ancient Chinese, Sumerians, and Egyptians, continents have moved closer together, mountains have risen higher, and species of plants and animals have changed. It is not surprising that people thought in terms of a constant, unchanging world composed of specific mountain ranges, rivers, lakes, continents, plants, and animals.

Figure 17-2 As the science of geology developed, it was soon realized that river deltas, such as this delta of the Mississippi, contained so much sediment that they must represent the transformation of landscapes, that mountain ranges must have eroded and no longer exist. But this means that the world is not constant, as God had supposedly created it, but changing and evolving.

During the explorations of Africa, the Americas, and Australia near the time of Columbus, so many plants and animals were discovered that people could not help but notice remarkable similarities between many species. Various types of roses resembled each other more than they resembled other plants, just as did many new types of lilies, orchids, and so on. People began to realize that giant basins of sediment, such as the deltas at the mouth of the Nile, the Ganges, and the Mississippi, held enough eroded material to account for entire mountain ranges; the concept of a changing landscape—geological evolution—began to develop (**Figure 17-2**). An important corollary was the realization that in order for so much sediment to have formed, Earth must be millions (actually billions) of years old, not merely 6,000 years old, as had been calculated from the genealogy of the Book of Genesis in the Old Testament (see the timeline diagram on the inside of the back cover). This is one of those discoveries that is easy for us to underappreciate; the quantity of sediments proved that either Earth had a very long history or that God had created sediments. This gave theologians a difficult dilemma and caused many people to be even more skeptical of priestly interpretations of natural phenomena. One of the areas of doubt was the idea that all species had been created at once. Everyone was vividly aware that the Church already had made a grave error by censuring Galileo, insisting that Earth was the center of the universe and that the sun, planets, and stars all revolved around it (**Figure 17-3**).

As careful observations of nature became common after the Renaissance and as the scientific method proved to be an accurate method for analysis, the concept of the evolution of species became widespread. Scientists searched for an understanding of the mechanism by which it occurred. Finally, Alfred Russel Wallace and Charles Darwin independently discovered the basis—natural selection—in the mid-1800s, the critical explanation being given in Darwin's *Origin of Species* published in 1859. The concept of natural selection—survival of the fittest—had an electrifying effect on all biologists and led to a revolution in all aspects of thinking. Suddenly, many observations had a rational explanation; every character could be interpreted in terms of whether

Figure 17-3 *Galileo Before the Inquisition.* Kepler, Newton, and Galileo proved that the sun is the center of the solar system, but church doctrine considered this heresy; as the home of the one organism created in the image of God, the Earth should be the center of the Universe. The Inquisition forced Galileo to recant his belief in a sun-centered solar system and placed him under house arrest for the rest of his life. Attempts were made to confiscate and burn all copies of his book; only a few survived. Less famous scientists were frequently punished more severely. Misinterpretation of the physical world by religious doctrine is not a fundamental aspect of religion, but rather the rejection of observation and experimentation in favor of speculation without confirmation.

it made a species less or more adapted to its environment. And because so many environments exist, it is logical to expect that many types of plants and animals should also exist, each having genotypes and phenotypes that make it particularly adapted to its own environment. Before the discovery of natural selection, flowers were thought to have been created by God for the delight of the human eye and nose; that philosophy did little to explain bilaterally symmetrical flowers, inferior ovaries, or wind pollination. Science had been greatly hampered by trying to interpret the world in terms of the mind of God and what had been created to feed, clothe, and house the descendants of Adam and Eve.

Fortunately, the discovery of evolution by natural selection happened at about the same time as other critical discoveries— the discovery of genes and chromosomes, cell theory, proving that spontaneous generation does not occur, sophisticated hydroponic cultures of plants, the discovery of enzymes and the carrying out of some metabolic steps in a test tube without living cells, and the artificial synthesis of biological compounds. All of these combined to move biology firmly out of the realm of metaphysics/theology and into that of scientific analysis and interpretation.

For those of you who believe in divine creation rather than evolution by natural selection, keep in mind that all of our experience has shown the scientific method to be greatly superior to theological interpretation of scriptures for understanding and predicting the nature of the physical/biological world. Equally important, only ethical/philosophical systems allow us to understand and solve moral problems; science cannot do that. It is not logical or consistent for a person to accept certain discoveries of science (photosynthesis, DNA, respiration, vessels, sieve tubes) while rejecting others (evolution) if one type of discovery is just as well documented as the other.

Population Genetics

In this chapter we discuss the genetics of groups of individuals of one species. **Population genetics** deals with the abundance of different alleles within a population and the manner in which the abundance of a particular allele increases, decreases, or remains the same with time. The genetic recombination that occurs during sexual reproduction is important only if the two sexual partners have differing genotypes. A cross between two plants that are $A_1A_1B_1B_1 \times A_2A_2B_2B_2$ produces offspring that have the genotype $A_1A_2B_1B_2$, which is different from that of either parent. Crossing-over also increases genetic diversity in populations.

The total number of alleles in all the sex cells of all individuals of a population constitutes the **gene pool** of the population. Imagine that gene A has four alleles: A_1, A_2, A_3, and A_4. If the population consists of 1 million individual plants, each with 100 flowers and each flower producing on average 100 sex cells (sperm cells and egg cells), the gene pool contains 10 billion haploid sex cells. The alleles are probably not present in equal numbers; for instance, 60% of all gametes may be A_1, 20% A_2, 15% A_3, and 5% A_4. Will this ratio be the same for the population next year or in the next generation? Early in the 20th century, two biologists, G. H. Hardy and G. Weinberg, demonstrated mathematically that if only sexual reproduction is considered, the ratio remains constant over time; even if an increase or decrease occurs in the total number of individuals or gametes, the ratios do not change. Sexual reproduction alone does not change the gene pool of a population; if no other factors were involved, the gene pool would remain constant forever.

Factors That Cause the Gene Pool to Change

Although it is possible theoretically for a gene pool to remain constant, in reality, changing allele frequencies are the rule because populations are always affected by factors other than sexual reproduction.

Mutation All genomes are subjected to mutagenic factors, and mutations occur continually. Because of mutation, existing alleles decrease in frequency, and new alleles increase. Whether or not mutation is significant depends in part on the population's size. In the population described earlier, there were 10 billion copies of gene A, and 60% were the allele A_1. Therefore, there were 10,000,000,000 $\times$ 0.6 = 6,000,000,000 copies of A_1; if one mutates to become a new allele, A_5, then the frequency of A_1 drops to 5,999,999,999, which is still pretty close to 60%. A_5 exists as just one copy out of 10 billion, and thus, its frequency is extremely small; however, the presence of one copy is an infinite increase over zero copies, and the existence of even a single copy should not be ignored.

Accidents Accidents are events that an organism cannot adapt to, such as the collision of a large meteorite with Earth. When a meteorite of sufficient size strikes Earth, a large region of Earth's surface is destroyed, killing all life in the area. All organisms, along with all their alleles, are eliminated. If by chance the area of impact had included the single individual that had just obtained A_5 by mutation, then A_5 would no longer exist. If the population in the impact area has the same gene frequencies as the general population, the alleles are eliminated in the same proportions as they exist generally, and no change in allele frequency occurs. If the impact area had an unusually high number of a particular allele,

(a)

(b)

Figure 17-4 Severe drought and heat across all of Texas in 2011 resulted in many wildfires in all parts of the state. (a) A fire in a pine forest near the East Texas town of Bastrop killed almost all trees, none of which appears to have been adapted to survive fire. Because all trees were killed, this fire was not a selective force; there was no differential survival. (b) A fire near the West Texas town of Fort Davis damaged many trees, shrubs, and succulents but seems to have killed only a few. These plants with long leaves are sotols (*Dasylirion leiophyllum*), and although all were burned severely, many actually survived; however, others were killed completely. If the ones that survived differed genetically from those that died (perhaps they had genes for thicker bark or more heat-resistant organs), then the fire was a selective force: The gene pool after the fire will be different from what it was before the fire.

for instance A_3, then A_3 is affected more than the other alleles; therefore, the allele frequency of the gene pool of the survivors is altered. This change of allele frequency, however, does not make the species any more able to survive another hit by a meteorite.

Many phenomena qualify as accidents. A volcanic eruption produces poisonous gases and molten rock that destroy everything within a limited area. Infrequent floods, hailstorms, or droughts can act as accidents for plants too small and delicate to become adapted to those events; all individuals in the affected area are killed. For other species that consist of plants with larger bodies, such natural phenomena act as selective forces, removing the weaker, less well-adapted individuals but not affecting the more well-adapted members (**Figure 17-4**). The continental drift of Antarctica southward from a temperate region to the South Pole was an accident for all the plants living on it. While located in the temperate latitudes, Antarctica had a rich flora with abundant plant life; as it drifted southward, it entered a region too cold and severe for any plant life to survive. All individuals and their alleles were eliminated (**Figure 17-5**).

Accidents can be small events as well as large ones. After an allele is formed, A_5 for instance, its numbers may increase as it is used in sexual reproduction; after a few years, there may be 10 individuals with the A_5 allele. The individuals are closely related and probably located close together because most seeds do not travel very far. It is possible for a local disturbance to eliminate all 10: an avalanche, a herd of grazing animals, or the construction of a highway.

Artificial Selection **Artificial selection** is the process in which humans purposefully change the allele frequency of a gene pool. The most obvious examples are the selective breeding of crop plants and domestic animals (**Figure 17-6**). Plant breeders continually examine both wild populations and fields of cultivated plants, searching for individuals that have desirable qualities such as resistance to disease, increased protein content in seeds, and the ability to survive with less water or fertilizer. When plants with beneficial qualities are found, they are collected and used in breeding

programs to produce seed for future crops. Consider just wheat, rice, and corn: Almost the entire world populations of these three species consist of cultivated plants; very few of the natural ancestors still exist in the wild. Consequently, the gene pool for each is made up almost entirely of alleles that have been artificially selected for thousands of years.

Figure 17-5 In earlier times, Antarctica was located in temperate latitudes and supported many plants, such as this *Glossopteris*. Continental drift moved Antarctica to the South Pole, into conditions for which no alleles are able to produce an adapted phenotype. All plants perished.

(a)

(b)

(c)

(d)

Figure 17-6 Artificial selection by crop breeders has resulted in increased frequency of certain alleles and elimination of others. The allele frequency of the population changes dramatically from year to year, depending on whether farmers decide to grow one variety or the other. Cabbage (a), broccoli (b), and cauliflower (c) are all the same species (as are brussels sprouts as well), and all evolved from the same ancestor; however, their evolution was controlled by artificial selection for certain traits in each variety. Recently, artificial selection has produced the variety Violet Queen that stores large amounts of pigment (d).

Artificial selection is also used to produce ornamental plants that flower more abundantly or for a longer time. Artificial selection has also been used to alter flower color and size and to make the plants hardy in regions where they otherwise could not grow. The trees cultivated for lumber and paper are also subjected to artificial selection.

Artificial selection is often carried out in conjunction with artificial mutation. Plants are exposed to mutagens such as acridine dyes or irradiation with ultraviolet light or gamma rays to increase the number of new alleles that come into existence. The plants are allowed to grow to see how the new alleles affect the phenotype, and those plants with the desired phenotypic traits are used in selective breeding programs.

Natural Selection **Natural selection**, which is the most significant factor causing gene pool changes, is usually described as survival of the fittest: Those individuals that are most adapted to an environment survive, whereas those less adapted do not. However, natural selection is such an important factor in evolution that it must be given careful attention. Two conditions must be met before natural selection can occur:

1. The population must produce more offspring than can possibly grow and survive to maturity in that habitat. This condition is almost always valid for plants anywhere on Earth. Most

plants produce hundreds of seeds, which often germinate near the parent plant (**Figure 17-7**). Even in species with wind-dispersed seeds, such as maples or milkweeds, most seeds do not travel far. Consequently, the ground can be covered with hundreds of seedlings crowded closely together, and there simply is not enough room to accommodate the physical bulk of so many plants as they grow.

Besides limited resources, the number of individuals that can survive in a particular habitat is affected by predators, pathogens, and competitors. All plants are faced with attack by herbivorous animals, ranging from almost microscopic mites and nematodes to much larger beetles, reptiles, birds, and mammals. Animals not only eat plants but may also lay eggs in them, bore into tree trunks for nesting sites, walk on them, and rob nectar without carrying out pollination. Pathogenic fungi and bacteria are similarly harmful. Competitors are other organisms that use the same resources the plant needs to survive. When root systems grow together, the two plants compete for the same water and nutrients. If two species are pollinated by the same species of insect or bird, they must compete for the attention of the pollinators. In a forest, plants compete for light: Those that can grow tallest receive the most light; those that have shorter trunks receive only dim light. All of these activities adversely affect

Figure 17-8   The summer of 1988 produced one of the most severe droughts on record in the American midwest, our primary corn-growing area. Under natural conditions, such a drought would have resulted in the natural selection of drought-resistant plants; all corn plants of 1989 would have been the progeny of the resistant plants that survived, and all would have carried alleles for drought resistance. Natural selection did not occur, however, because there was no genetic variability among the corn plants; corn is one of our most highly inbred crops, and all seed corn is produced by careful crosses of absolutely uniform parents. Consequently, of the trillions of plants affected by the drought, none was more or less resistant than any other. Any that survived did so not by genetics, but by accident: receiving a little extra rain or being in a valley where extra moisture collected.

Figure 17-7   Even if animals carry away many of the fruits of this red elderberry (*Sambucus racemosa*), most fall and germinate in its immediate neighborhood, and there are too many seedlings to fit into the space physically. Also, the tree will produce seed abundantly for many years.

the plant's ability to reproduce and may cause the plant's death.

2. The second condition necessary for natural selection is that the progeny must differ from each other in their types of alleles (Figure 17-8). If they are all identical, all are affected by adversity in the same way and to the same degree. Under crowded conditions, probably all are stunted similarly, all grow poorly, and finally, none reaches reproductive maturity. If all individuals of a species are equally susceptible to a pathogenic fungus, no increase in survivability, and fitness occurs as the result of a fungal attack. Even if some survive and reproduce, they are identical genetically to those that died, so no change occurs in allele frequency; natural selection has not occurred.

When genetic diversity exists among individuals, differential survival can occur. If some members of the population have an allele that gives them increased resistance to fungi, those plants should fare much better during an outbreak of fungus in the population than those lacking the allele. If the fungus is so virulent that it often kills the plants it attacks, the allele frequency of the population is changed radically after infection—the resistance allele constitutes a much greater percentage of the gene pool. Natural selection operates even if the fungus only weakens plants but does not kill them outright; the weakened plants should produce fewer seeds than do the resistant, healthy plants.

Competition for water and nutrients among crowded root systems also acts as a selective force; alleles that allow roots to absorb water and nutrients more effectively have an advantage. If most roots can no longer absorb water when the soil has a water potential of −0.8 MPa, an allele that alters root metabolism such that it can extract water even at −1.0 MPa enables the plant to grow when others cannot. The mutation has a selective advantage—the plant is more fit or adapted to conditions of crowding.

Natural selection refers to the differential survival among organisms that have different phenotypes. Natural selection can act only on preexisting alleles; it does not cause the mutations. The presence of a fungus does not cause plants to become resistant; if none had been resistant before, all individuals would be adversely affected. The advantageous allele must exist first. But if an allele for resistance does exist, natural selection can cause the *population* to become resistant by the preferential survival of resistant individuals, even though it cannot cause an *individual* to become resistant. A population or species evolves, but an individual does not. Similarly, a population may become more adapted but a single individual does not change; it cannot become more adapted during its lifetime. An attack by a fungus may stimulate a plant to activate genes that produce antifungal chemicals, but this is a case of causing a plant to express genes it already has; it is not creating new genes or alleles.

Natural selection does not always result from the action of an agent outside the organism. Although many selection pressures are external—pathogenic fungi and dry, hot climate—any factor that causes one plant to produce more progeny than other plants is a selective factor (Figure 17-9). If an allele causes chloroplasts to photosynthesize more efficiently, plants with that allele can pro-

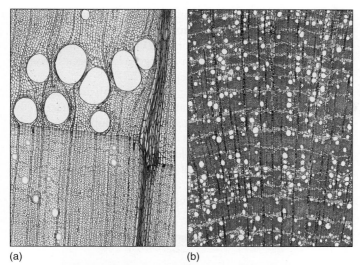

(a)   (b)

Figure 17-9 (a) This oak wood contains wide vessels that can conduct water with relatively little friction but which cavitate easily (×80). (b) This *Casuarina* wood has vessels so narrow that the water in them is strongly reinforced and cavitation is almost impossible under natural conditions. Even though vessels are completely internal, they still affect the plant's survival relative to its environment (×80).

duce carbohydrates more rapidly than plants that lack the allele; the former plants grow faster and produce more seeds, at least half of which carry the advantageous allele.

Factors That Are Not Part of Natural Selection Natural selection can be understood more clearly and accurately if you realize that certain factors are not part of natural selection at all: Natural selection does not include **purpose**, **intention**, **planning**, or **voluntary decision making**. Whenever we say that "plants do something in order to . . . ," we are suggesting that the plants can plan their activities and have purpose, which is not true. It is not correct to say that certain plants have disease-resistance genes in order to protect themselves against fungi. The resistance of a population of plants to a fungus is nothing more than the result of the preferential survival of the plants' ancestors because they had an allele for resistance, whereas their competitors did not. Although plants that have this allele in the presence of the fungus have a selective advantage, the plants do not have the allele in order to protect themselves. Similarly, plants do not produce nectar in order to attract pollinators; those plants in the past that did produce nectar happened to be pollinated more often than others that did not produce nectar, so the alleles for nectar production were increased in the population. Currently, those plants that secrete nectar are visited by pollinators, but there is no purpose, intent, or planning by the plant. Only humans and other primates act with intent and purpose, and this applies only to our voluntary actions. Even humans do not digest food in order to have an energy supply; rather, our autonomic nervous system and cell metabolism have automatic responses to the presence of food in the small intestine that cause the secretion of digestive juices, the absorption of monomers, their distribution through the hepatic portal system, and their respiration by cytoplasm and mitochondria. We have no control over this result of our evolutionary history. This may seem to be just a trivial problem of semantics because everyone knows what we mean, but statements should be accurate, not sloppy. If we are not meticulous in how we express our ideas, we

will not be meticulous in how we think, and important details will be lost.

Situations in Which Natural Selection Does Not Operate

Further understanding of natural selection can be gained by considering several cases in which it does not operate. As already mentioned, it cannot operate if all individuals of a population are identical genetically or if it is impossible to become adapted to a certain condition. Competition does not occur in a habitat that can support the growth and reproduction of all individuals; if survival is universal, natural selection does not occur. Situations like this occur in newly opened habitats such as a plowed field. All seeds present may germinate and grow vigorously, even the ones not well adapted for competition. Because all survive, no natural selection has occurred. Other examples are the sides of a road cut, a recently burned area, or a recently flooded plain covered with rich sediments. We must be careful here; if the road cut passes through a heavy, dark forest, the newly exposed sides may be too bright and exposed for seedlings from plants adapted to the forest shade. In this case, the environment favors those plants that can tolerate full sunlight and suppresses those that require shade; many selection pressures, but not all, have been eliminated, and some natural selection can still occur.

Multiple Selection Pressures

In many cases, the loss of individuals and reduced reproduction are not caused by a single factor, such as a pathogenic fungus. Instead, the plants are also affected by insect attack, drought, cold, need for pollinators and need for a mechanism to disperse their seeds, as well as the efficiency of their own metabolism, such as the ability of their membranes to pump ions, the capacity to reduce nitrogen, or the efficiency of producing just enough P-protein in the phloem without a wasteful, nonuseful excess.

A mutation that produces an allele that would result in improved fitness is potentially advantageous selectively, but it may never have the opportunity to improve the fitness of the plant or the species. A mutation that results in improved cold hardiness may be eliminated from the gene pool if the plant carrying the new allele is killed by fungus or drought or cannot reproduce because of poor competition for pollinators. Such a loss of this allele is simply an accident.

However, if the new allele for cold hardiness does survive it may be able to improve the species. If cold winters are common, this allele greatly improves fitness, and its frequency may increase rapidly. If cold winters are infrequent, they do not exert a strong selection pressure, the allele does not improve fitness very much, and its frequency may remain low for years until a harsh winter does occur. Until that time, the allele's frequency is determined by several factors. It may be tightly linked to an allele that is strongly advantageous for an important condition. For instance, the cold hardiness allele may be on the same chromosome as the allele for resistance to fungi. If the two genes are so close on the same chromosome that crossing-over virtually never occurs between them in prophase I, the presence of fungi selects not only for the antifungal allele but also for the cold hardiness allele, just by coincidence.

The cold hardiness allele may affect the plant in various ways besides the ability to withstand cold; that is, pleiotropic effects may operate. If these are also advantageous under common conditions, the allele increases in frequency, but often many of the "side effects" are disadvantageous. Many improvements to phenotype have some negative aspects, at least in terms of cost. Increased cold hardiness may be due to thicker, more sclerified bud scales with a thick layer of wax. These require the input of increased amounts of nutrients and energy that could have been used to produce more seeds. Thus, whereas this allele may be strongly advantageous in an environment with frequent cold winters, it may be disadvantageous in an environment where winters are always mild. Whether a particular allele is beneficial or not depends entirely on the habitat, which may change with time. As the environment changes, the selection pressures change and certain features become more or less advantageous.

Rates of Evolution

From the examples given, it seems that the allelic composition of a population could change rapidly, within a few generations, but that is not typically the case. Most populations are relatively well adapted to their habitat, or they would not exist. Very few mutations produce a new phenotype so superior that it immediately outcompetes all other members of the population. At the extremes, there are many species of seedless plants (lycopods, *Equisetum*, ferns) that have persisted for tens of millions of years without diverging into new species (**Figure 17-10**). In contrast, very rapid speciation is occurring in a group of asters in Hawaii:

Shortly after Kauai formed 5.2 million years ago, an aster seed arrived, thrived, reproduced, and spread rapidly in the unpopulated island. Its descendants have diversified into three distinct genera, each with many species, and on average, a new species arises in this group once every 500,000 years.

It is difficult to identify the presence of particular alleles in a population unless they result in an easily identifiable effect on the phenotype. Consequently, most studies of evolution concern the changes in gross structures such as flowers, leaves, fruits, shoots, and trichomes. But these complex structures are the product of the developmental interaction of many genes. Any new mutation results in a more adaptive structure only if the effects of the new allele fit into the already existing highly integrated mechanism

(a)

(b)

Figure 17-11　(a) The ancestors of the cactus family were large, woody trees with rather ordinary dicot leaves. The cactus genus *Pereskia* still contains members quite similar to the ancestors, as shown here. Apparently few genes had been modified by the time *Pereskia* appeared. (b) This *Gymnocalycium* is also a cactus, but its phenotype is significantly different from the ancestral condition; apparently all critical genes involved in leaf production have mutated so much that they are nonfunctional or absent. Genes involved in stem elongation now produce short stems, and this species contains genes for succulence that were not present in the ancestral *Pereskia*-like species; these new genes probably are highly mutated forms of "extra" genes from a tetraploid ancestor or arose by other methods of gene duplication.

Figure 17-10　Plants of *Equisetum* have jointed stems, often with whorls of branches at each node; they are easily recognizable, even as fossils. This group originated millions of years ago and had many species at one time, most of which are extinct. There are now only 15 species of living plants, and even though many are widespread and well-adapted to many areas, they do not appear to be evolving into new species.

of morphogenesis without causing serious disruptive effects. As systems become more intricate, the probability decreases that any random change is beneficial.

Evolutionary changes that result in the loss of a structure or metabolism can come about quickly, however, and for the same reason: complexity. If a feature becomes selectively disadvantageous, many of the mutations that disrupt its development become selectively advantageous. Because disruptive mutations outnumber constructive mutations, loss can occur relatively rapidly. For example, the ancestors of cacti lived in a habitat that became progressively drier; large, thin leaves were advantageous because they carried out photosynthesis but were disadvantageous because too much water was lost by transpiration (Figure 17-11). Mutations that disrupted formation of the lamina were advantageous, and cacti lost their leaves in perhaps as little as 10 million years, whereas the evolutionary formation of leaves in seed plants had required over 200 million years. Leaves could not be lost too quickly, however, because the plants would be left with virtually no photosynthetic surface area. Mutations that caused the complete absence of leaves could not be selectively advantageous until other mutations had occurred that permitted the stem to remain green and photosynthetic, that prevented the early formation of an opaque bark, and that slowed the metabolism of the plant to a level compatible with reduced photosynthesis. The loss of leaves could occur only simultaneously with or after these modifications of the stem.

Speciation

As natural selection operates on a population for many generations, the frequencies of various alleles and consequently the phenotype of the population change. At some point, so much change has occurred that the current population must be considered a new species, distinct from the species that existed at the beginning. Natural selection has caused a new species to evolve, a process called **speciation**.

At what point can we conclude that a new species exists? It is not possible to give an exact definition of species that is always valid, but generally, two organisms are considered to be members of distinct species if they do not produce fertile offspring when crossed. Many exceptions exist, however. The individuals of maples in the western United States cannot naturally reproduce sexually with those in the eastern United States, but that is only because the pollen does not travel that far; if an eastern plant is brought close to a western one artificially, they can cross-fertilize and are, therefore, considered the same species. On the other hand, many orchid species grow together in nature without producing hybrids, but artificial cross-breeding may produce healthy, fertile hybrids. Nevertheless, they are considered distinct species because they look different from each other. So, if two plants freely interbreed in nature, they are members of the same species; if they do not interbreed even when manually cross-pollinated, they are separate species, and if they do not interbreed naturally but they do artificially, it is a judgment call on the part of the scientists who work with the plants and are most familiar with them.

Speciation can occur in two fundamental ways: (1) **phyletic speciation**, in which one species gradually becomes so changed that it must be considered a new species (Figures 17-12a and b), and (2) **divergent speciation**, in which some populations of a species evolve into a new, second species while other populations either continue relatively unchanged as the original, parental species or evolve into a new, third species (Figures 17-12c and d).

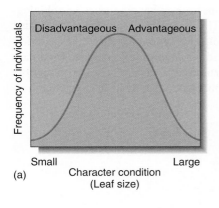

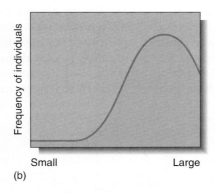

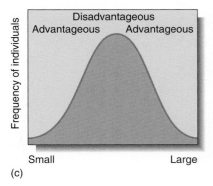

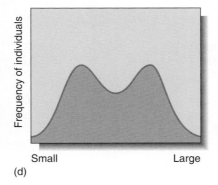

Figure 17-12 (a and b) In phyletic speciation, all of a species gradually changes because one particular aspect of a character is advantageous for all individuals. Here, all leaves become larger; perhaps the climate is becoming more humid or herbivorous insects are less of a problem. (c and d) In this scenario of divergent evolution, both extremes of the condition are more advantageous than intermediate values; the climate may become drier in some areas, favoring the smaller leaves, but moister in other areas, favoring plants with larger leaves.

■ Phyletic Speciation

Millions of years are often required for a species to evolve into a new one. The critical feature is that as new beneficial alleles arise and are selected for, they become spread throughout the entire population. This movement of alleles physically through space, called **gene flow**, occurs in many ways, such as by pollen transfer, seed dispersal, and vegetative propagation.

Pollen Transfer   Pollen grains each carry one full haploid genome, and all alleles of a plant are present in its pollen grains. Wind-distributed pollen, such as that of ragweed, grasses, and conifers, can travel great distances. If a new allele is carried by some of the pollen grains, it can move to very distant plants; if the pollen grain's sperm cells fertilize an egg, a new seed is formed whose embryo contains the new allele.

Animal-mediated pollination also contributes to gene flow; both birds and insects tend to spend most of their time in a small area; therefore, although birds can fly long distances, they usually spread pollen through a smaller area. Nevertheless, allele movement can be rapid.

Seed Dispersal   The fruits and seeds of some plants fall close to the parent, but many species have long-distance dispersal mechanisms. Seeds and fruits can be carried by wind, floods, and stream flow. They can be carried to islands by rafting, in which they are trapped above water on a tangled mat of floating debris. Seeds or fruits that are spiny or gummy stick to the fur or feathers of animals; migratory animals can be especially important in dispersing seeds (**Figure 17-13**). Most birds reduce their weight before migration by preening themselves of all adhesive seeds, but if just one seed is overlooked, its alleles are transferred; as the seed germinates, grows into a new plant, and reproduces, the new alleles can be spread throughout the new site.

Vegetative Propagation   If a species produces small, mobile pieces that reproduce vegetatively, these too contribute to gene flow.

If these various mechanisms are sufficient to enable alleles that arise in one part of the species' range to travel to all other parts, the species remains relatively homogeneous, even as the entire species evolves into a new species. Alleles that arise at various geographic sites ultimately come together by gene flow; then

Figure 17-13   After the hooks on this cocklebur fruit tangle in fur or feathers, the fruits will be carried far from the parent plant. When the seeds inside germinate and grow into new plants, their genes can combine with the genes of other cockleburs that are also in the new location.

meiosis, crossing-over, and genetic recombination rearrange them into thousands of combinations (**Figure 17-14**).

■ Divergent Speciation

If gene flow does not keep the species homogeneous throughout its entire range, divergent speciation may occur; if alleles that arise in one part of the range do not reach individuals in another part, the two regions are **reproductively isolated**.

Reproductive isolation can occur in many ways, but the two fundamental causes are abiological and biological reproductive barriers.

Abiological Reproductive Barriers   Any physical, nonliving feature that prevents two populations from exchanging genes is an **abiological reproductive barrier**. The original species is physically divided into two or more populations that cannot interbreed; if speciation results, it is called **allopatric** or **geographic speciation**. Mountain ranges are frequently reproductive barriers because pollinators do not fly across entire mountain ranges while feeding or gather-

Figure 17-14   Gene flow. This range is almost divided in two by a series of mountains over which pollen and seeds cannot travel; this slows down gene flow and homogenization of the population. The numbers indicate the generation in which the allele flows: 1, gene flow in the first generation; 2, gene flow from plants that grew from seeds produced by the gene flow of the first generation; and so on. Here, three generations were necessary for the new allele to become distributed to the lower side of the mountains, seven generations to become widespread throughout the range. In a species of annual plants, each generation is 1 year long; for biennials, a generation is 2 years long, and gene flow is slower. In trees that do not produce their first flowers, pollen, or seeds until they are many years old, gene flow can be very slow.

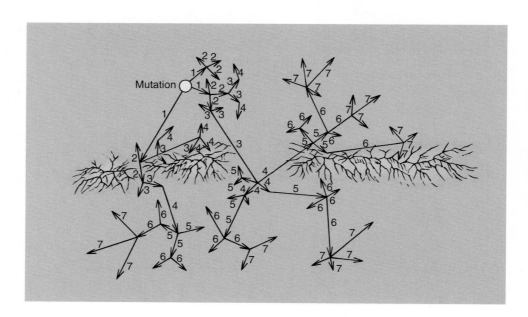

(a)

(b)

Figure 17-15 (a) These small flowering herbs (*Caltha*) and shrubs (willow, *Salix*) grow in high alpine habitats near snow. A mountain range, such as the Tetons in Wyoming (b), provides many suitable patches of habitat near their summits, but each patch is separated from others by valleys that are warm or by peaks that are too dry. Seeds are only rarely carried from one patch to another, so gene flow is restricted: The *Caltha* in this photo is part of a small population that can interbreed but is relatively isolated from other populations on other peaks of this same mountain range.

ing pollen. Seeds may occasionally be carried across mountains by birds or mammals, but probably too rarely to be significant. Rivers are often good barriers for small animals, but they rarely prevent plant gene flow by means of seed dispersal. Deserts and oceans are effective barriers. Plants adapted to the harsh conditions of mountain tops are reproductively isolated from plants on adjacent mountain tops by the intervening valleys because pollinators and seed dispersers do not often travel from mountain to mountain (**Figures 17-15a** and **b**). Any seeds dropped into the valley would not be able to compete with lowland plants and would die. Ultraviolet (UV) light and dry air are barriers to very long-distance wind dispersal of pollen; during the ride on wind, pollen is damaged by UV light and dry air and often dies before it travels far.

Biological Reproductive Barriers Any biological phenomenon that prevents successful gene flow is a **biological reproductive barrier**. Differences in flower color, shape, or fragrance can be effective barriers if the species is pollinated by a discriminating pollinator. A mutation that inhibits pigment synthesis and causes a normally colored plant to have white flowers might prevent the pollinator from recognizing the flower. The flower is not visited, and gene flow no longer occurs between the mutant plant and the rest of the species even though the individuals grow together. Timing of flowering can be important: If some flowers open in the evening, they probably do not interbreed with those that open in the morning. Differences in flowering date are critical because of the brief viability of pollen after it leaves the anthers. When two groups become reproductively isolated even though they grow together, the result is **sympatric speciation**.

For example, one species of monkey flower, *Mimulus lewisii*, has wide pink flowers and a small amount of concentrated nectar, which are ideal for bumblebees. The closely related *M. cardenalis* flowers are narrow and dark red and have large amounts of dilute nectar, a hummingbird pollination syndrome. The two species grow together but do not hybridize because bumblebees do not visit hummingbird flowers (they cannot even fit into them) and vice versa, but the two species can be crossed artificially and produce healthy hybrids. Apparently just a few floral genes have major effects

in *Mimulus*, so a few mutations resulted in strikingly different flowers, completely different pollinators and thus biological isolation.

Evolutionary changes in pollinators can also act as reproductive barriers for plants. If a plant population covers a large area, some parts of the range probably have characteristics different from those of other parts, such as elevation, temperature, and humidity. These variations may be important to pollinators and seed distributors if not to the plant. If so, the animals may diverge evolutionarily into distinct species, each limited to a small part of the plant's range. Little or no flow of the plant's genes occurs because of the restricted movements of pollinators and seed dispersers. These examples of biological reproductive barriers prevent pollen from moving from one plant to another, and thus, neither pollination nor fertilization occurs. Consequently, these are called **prezygotic isolation mechanisms**: They act even before a zygote can be formed.

The environmental diversity of a large geographic range can lead directly to divergence of the plants themselves. Although a plant species may occupy an extensive, heterogeneous range, mutations will gradually arise that are particularly adaptive for certain aspects of specific regions of the range. When these new alleles arrive at that part of the range by gene flow, their frequency there is increased by natural selection. When they arrive at other parts of the range, their frequency remains low or they are eliminated there if they are neutral or selectively disadvantageous for conditions in these sites (**Figure 17-16**). Even with active gene flow and interbreeding, different subpopulations of the plant species emerge, each adapted to its own particular portion of the total range. As this process continues, each subpopulation becomes progressively more distinct and may be recognized and named as a subspecies. Finally, because of the large number of unique, characteristic alleles, each may become sufficiently distinct phenotypically that the two subpopulations can no longer interbreed: Their genomes are too different. At this point, **postzygotic internal isolation barriers** are in place, and the two subpopulations must be considered separate species.

One of the earliest postzygotic barriers to arise is **hybrid sterility**: The two populations occasionally interbreed or are

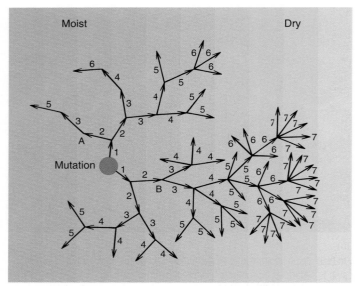

Figure 17-16 A mutation has produced an allele highly adaptive in dry parts of the range but neutral or deleterious in moist parts. Because it originated in a less than optimal environment, in generation 1, only two offspring survive. In generation 2, each of these produces two more that survive, one in a moister habitat (A) and one in a drier region (B). The allele is deleterious at A, and only a single progeny survives in generation 3; at B, two survive, both in drier habitats, and in generation 4, each produces three surviving progeny. As the alleles flow, by chance, into regions that are more optimal (drier), reproductive success is greater, and the allele frequency increases. As they flow, by chance, into unsuitable habitats, reproductive success is low and the allele never becomes abundant.

Figure 17-17 One of the pioneer plants that arrived and colonized the Hawaiian Islands was an aster, related to dandelions. Because of the low levels of competition and because so few individuals existed initially, each mutation became a significant part of the gene pool even while only one or a few copies existed; therefore, genetic drift was rapid. *Argyroxyphium,* silverswords, are basically giant, perennial dandelions. Their leaves grow in a rosette, and their many "flowers" are like dandelion "flowers," actually being groups of tiny flowers. Silverswords grow only in Haleakala National Park in Hawaii. Goats ate them almost to extinction before the park was established.

artificially cross-pollinated and produce viable seed, but the seed grows into a sterile plant. Spore mother cells in anthers and ovules in the sterile hybrid are unable to complete meiosis because of a failure of synapsis: "Homologous" chromosomes are no longer truly homologous, and they fail to form pairs and synaptonemal complexes during prophase I. Without fertile pollen or ovules, no seed is formed, and the mixture of the two genomes ends when the hybrid plant dies. The two populations continue to diverge and become even more distinctly separate species. If cross-pollination occurs, alleles from one parent may code for proteins incompatible with those coded by alleles from the other, and not even a sterile hybrid can result; instead, the zygote or embryo dies early in development. This is called **hybrid inviability**.

After internal isolation barriers are established, evolutionary divergence should be even more rapid because new alleles cannot be shared; gene flow between the two populations comes to a stop. Each species becomes more easily recognizable as new characters evolve in each, characters which are not present in its sibling species. The two species are initially so similar that they are obviously related; such closely related species are grouped together into genera by taxonomists.

Divergent evolution may result in numerous types of new species; in some cases, one subpopulation changes into a new species, while the remaining part of the population continues relatively unchanged as the original species. In other cases, both subpopulations change so much that two new species emerge and the original species no longer exists. The original species has become extinct, although it has numerous progeny that form the members of two new species.

Adaptive Radiation **Adaptive radiation** is a special case of divergent evolution in which a species rapidly diverges into many new species over an extremely short time, just a few million years. This usually occurs when the species enters a new habitat where little or no competition or environmental stress exists. The best examples are the colonization of newly formed oceanic islands such as the Hawaiian or Galapagos Islands. After being formed by volcanic activity, they are initially devoid of all plant and animal life, but eventually a seed arrives, carried by a bird, wind, or ocean currents. After the seed germinates and begins to grow, it is free of danger from herbivores, fungi, bacteria, or competition from other plants. It must be relatively adapted to the soil, rainfall pattern, and heat/cold fluctuations, but otherwise, its life is remarkably free of dangers. If this plant is self-fertile or if it reproduces well vegetatively, it successfully colonizes the area.

All offspring greatly resemble the first, **founder individual(s)** because the initial gene pool is extremely small; if just one seed is the **founder**, the original gene pool consists of its two sets of alleles. This homogeneity may last only briefly because, with the lack of competition, pathogens, and predators, fewer forces act as selective agents. Consequently, new alleles build up in the population much more rapidly than can occur in the parental population on the mainland (Figure 17-17). Although the population is small, it is more subject to accidents, and thus, the gene pool can change rapidly and erratically; this is **genetic drift**, and the island population soon becomes heterogeneous. Natural selection for adaptation to soil types, drainage, climate, and metabolism is still operative; as a result, divergence is based largely on physical factors in the environment.

If the island is very far from any seed source (the mainland or other islands), like Hawaii is, hundreds of years may pass before the arrival of more seeds of either the same or another species. Even if it is relatively close to the mainland, like the Galapagos, seeds arrive only rarely. For thousands of years the only animals might be birds

(a)

(b)

Figure 17-18 Cacti and euphorbias have undergone extensive convergent evolution: Many early euphorbias were leafy trees (a) like early cacti (see Figure 17-11a), but some have evolved to be succulent (b) like many cacti. Both (a) *Sapium sebiferum* and (b) *Euphorbia gymnocalycioides* are members of the *Euphorbia* family, Euphorbiaceae.

Euphorbia   *Viscum*
(host)   (parasite)

Figure 17-19 Both cacti and euphorbias are parasitized by distinct types of mistletoe that have also undergone convergent evolution. Flower buds of *Viscum minimum* are shown here emerging from the body of *Euphorbia lactea*. *Tristerix aphyllus* infects a cactus called *Trichocereus chilensis*. Both mistletoes are dwarf plants that spend most of their life buried deep inside the host's succulent body.

or airborne insects, typically spiders; land reptiles (lizards) and mammals (mice) arrive by rafting, but this can be extremely rare.

Adaptive radiation can also occur in mainland populations if the environment changes suddenly and eliminates the dominant species of a region. With the absence of these species, competition changes and other species that had not been able to compete well before can now occupy the new areas; while few in number, they undergo genetic drift, rapidly producing many unusual genotypes. Within a short time, many new species are recognizable, and adaptive radiation has occurred.

■ Convergent Evolution

If two distinct, unrelated species occupy the same or similar habitats, natural selection may favor the same phenotypes in each. As a consequence, the two may evolve to the point that they resemble each other strongly and are said to have undergone **convergent evolution**. The most striking example is the evolutionary convergence of cacti and euphorbias (see Figure 17-11, **Figures 17-18**, and **17-19**). Cacti evolved from leafy trees in the Americas; as deserts formed, mutations that prevented leaf formation were advantageous because they reduced transpiration. Other selectively advantageous mutations increased water storage capacity (succulent trunk) and defenses against water-seeking animals (spines). In Africa, the formation of deserts also favored a similar phenotype, and the euphorbia family became adapted by these means. As a result, the succulent euphorbias are remarkably similar to the succulent cacti, even though the ancestral euphorbias are quite distinct from the ancestral cacti. Two groups cannot converge to the point of producing the same species; only the phenotypes converge, not the genotypes. For example, cactus spines are modified leaves, whereas euphorbia spines are modified shoots.

Box 17-1 Zoos, Botanical Gardens, and Genetic Drift

Genetic drift, adaptive radiation in small populations, and artificial selection are important factors in conserving endangered species. With only a small number of individuals maintained in zoos, botanical gardens, and national parks, the gene pools of the captive plants and animals are so small that rapid fluctuation and drift can cause them to evolve rapidly. An endangered species protected in captivity may be lost because it evolves into one or several species adapted to highly artificial, humanly maintained conditions. Even if it still looks like the ancestral species that was placed into the zoo, it differs in its resistance to infections, parasites, and predators and its ability to understand and perform mating interactions. Most zoologists now believe that habitat preservation is the only true means of maintaining an endangered animal species. Zoos cannot accommodate the numbers of individuals necessary and cannot provide realistic habitats that act as agents of natural rather than artificial selection. Many of the progeny born in zoos would be poorly adapted for survival in the wild; if they are kept alive by antibiotics and special diets, the presence of nonbeneficial alleles is maintained in the population. Natural populations have a tremendous death rate among juvenile animals, a factor necessary for natural selection; with this eliminated in zoos, the gene pool of the captive population rapidly diverges from that of the natural population, being greatly enriched in deleterious mutations. The zoo population may become incapable of surviving without human care. The same is true of plant species maintained in botanical gardens, where often the most unusual varieties are given special attention, increasing the presence of exotic alleles that would be selectively disadvantageous in natural conditions. Artificial and natural selection result in very different gene pools. This is not a condemnation of zoos or botanical gardens: They have provided an essential service in saving many species from extinction, which at present would be the only alternative for them, but it should be remembered that a species in the wild consists not only of its individuals but of its genetic diversity interacting with its environment.

In order to provide truly safe sanctuary for endangered species, each park or habitat preserve may have to be expanded to encompass a population large enough to be stable against genetic drift. Even Yellowstone National Park is not large enough to maintain a genetically stable population of grizzly bears, although the buffalo seem to be doing well. Many individuals of a plant species may survive in a small area, such that 1 km^2 may hold a large enough number for their genetic stability on a theoretical basis. An important aspect, however, is to preserve enough habitat to maintain the pollinators and seed dispersers as well. The habitat must also be diverse enough to contain realistic selection pressures, including pathogens, herbivores, and so on. If these are eliminated owing to insufficient habitat preservation, the plant species experiences unnatural selection. Recent forest fires are a good example. Fire is a natural part of many ecosystems and is an agent of natural selection for many features; if fires are suppressed, nonresistant individuals survive, and the allele frequency of the population is altered.

Although artificial selection improves our cultivated plants, it can have disastrous effects on natural populations. For many species, individual specimen plants are valuable commercially, such as orchids, bromeliads, and cacti with particularly beautiful shapes or flowers. In many cases, it is easier and cheaper to collect these plants from natural populations than to cultivate them in nurseries. Extensive plant collecting actually threatens some species with extinction, but more often it has the very serious impact of removing most of the healthy plants from a population. Because the plants that are left are those that are weak, unhealthy, or misshapen, the remaining population has an increased frequency of alleles that produce unhealthy, poorly adapted plants. Even though the collector does not wipe out an entire population, he or she may do significant damage to the population's genetic resources. We inflict the same condition on parrots, toucans, and tropical fish.

Evolution and the Origin of Life

The species present today have evolved from those that existed in the past, which evolved from those that existed before them. It is appropriate now to consider how life arose and what it was like initially.

The most seriously considered hypothesis about the origin of life on Earth is that of **chemosynthesis**. Basically, the chemosynthetic hypothesis attempts to model the origin of life using only known chemical and physical processes, rejecting all traces of divine intervention. It was first proposed by the Russian scientist A. Oparin in 1924 and then by J. B. S. Haldane in England. They postulated that before the origin of life, the surface of Earth was different from the way it is now, and the chemicals present then could react spontaneously, producing more complex chemicals that could in turn continue to react. Over millions of years, reactions might produce all the molecules necessary for life, and these might aggregate into primitive protocells. From the protocells, natural selection would guide the evolution of true, living cells. Four conditions would have been necessary for the chemosynthetic origin of life: The primitive Earth would have to have had (1) the right inorganic chemicals, (2) appropriate energy sources, (3) a great deal of time, and (4) an absence of oxygen in its destructive molecular form, O_2.

■ Conditions on Earth Before the Origin of Life

Chemicals Present in the Atmosphere Earth condensed from gases and dust about 4.6 billion years ago; it was initially hot and rocky and had an atmosphere composed mostly of hydrogen. Because hydrogen is such a light gas, most of this first atmosphere was lost into space. It was replaced by a **second atmosphere** produced by release of gases from the rock matrix composing Earth and from heavy bombardment by meteorites. Both sources would have provided gases such as hydrogen sulfide (H_2S), ammonia (NH_3), methane (CH_4), and water. All of these are found in volcanic gases and in meteorites that still strike Earth. Molecular oxygen was absent; it had already combined with other elements, resulting in compounds such as water and silicates. The early second atmosphere was a **reducing atmosphere** due to the lack of molecular oxygen and the presence of powerful reducing agents (Figure 17-20).

Figure 17-20 The outer planets, except for Pluto, are so massive that their gravity has retained their original hydrogen atmosphere. In addition, meteor bombardment and other activities have added ammonia, methane, and other components that make these atmospheres similar to Earth's early second atmosphere. Reactions in the atmosphere may be producing organic compounds.

Energy Sources There must have been a complex chemistry in the early second atmosphere because it was exposed to powerful sources of energy. Foremost was intense UV and gamma radiation from the sun. These radiations have energetic quanta that knock electrons from atoms, creating highly reactive free radicals. Part of the ammonia would have decomposed to hydrogen and nitrogen, and some of the methane would have converted to carbon monoxide and carbon dioxide, increasing the complexity of the atmosphere.

Heat was another source of energy available to power reactions. One heat source was the coalescence of gas and dust to form Earth: As the particles fell toward the center of gravity, they accelerated and then collided, converting kinetic energy to heat. A second heat source was radioactive decay of heavy elements like uranium and radium; this decay was extremely intense 4.5 billion years ago. Even today, enough radioactive decay remains to keep Earth's core molten. The chemicals present in the early second atmosphere were also dissolved in the ocean's water, and whenever they came into contact with hot or molten rock, endergonic reactions could proceed, resulting in the formation of more complex chemical compounds.

Electricity was abundant on a gigantic scale; much of Earth's water was initially suspended in the atmosphere because of the high temperature of the air and the planet's surface. When sufficient planetary cooling occurred, rains fell, but the rainwater evaporated upon hitting the hot surface. Rainstorms must have been immense, lasting for thousands of years and generating tremendous amounts of lightning. As each lightning strike flashed through the atmosphere, it triggered more chemical reactions, and the resulting products were washed downward by rain: On the hot surface, they would have boiled, producing even more chemical reactions. After further planetary cooling, the surface temperature dropped to less than 100°C, and liquid water began to accumulate as streams, lakes, and oceans.

Volcanoes also produce lightning around their throats as they erupt (Figure 17-21). Considering the immense amount of volcanism that occurred, this lightning would have supplied a

Figure 17-21 Rapid movement of gases during a volcanic eruption generates the electrical potential necessary for lightning. Electrical discharges through the volcanic gases produce organic compounds.

significant amount of energy. Also, volcanic lightning occurs through the clouds of venting gases where hydrogen sulfide, methane, ammonia, and water are most concentrated.

Time Available for the Origin of Life   The time available for the chemosynthetic origin of life basically had no limits, simply because of the lack of free molecular oxygen. Without oxygen, no agent was present to cause the breakdown and decomposition of the chemicals being created. If molecular oxygen had been present, the chemicals either would have not formed or would have oxidized soon after formation; without oxygen, they could accumulate for millions of years. The ocean of that time has been called a "dilute soup" or a "primordial soup" containing water, salts, and numerous organic compounds that became increasingly complex as time went on. As much as 1.1 billion years may have elapsed between the time Earth solidified and life arose.

Chemicals Produced Chemosynthetically

After the writings of Oparin and Haldane, the first experimental tests of the chemosynthetic hypothesis were performed in 1953 by a graduate student, S. Miller, at the University of Chicago. He constructed a container that had boiling water in the bottom and a reducing atmosphere in the top; electrodes discharged sparks into the gases, simulating lightning: As the water boiled, steam rose, mixed with the atmosphere and was acted on by the electrical sparks and then condensed and fell back into the water to be cycled again (**Figure 17-22**). Miller let his first experiment cycle for a week and noticed that the solution had become dark from the accumulation of complex organic compounds that had formed. When he analyzed their composition, he found that many different substances were present, including amino acids. Since then, this type of experiment has been performed numerous times, testing the effects of varying atmospheric composi-

tions, using several types of energy sources, or including metal ions in the water. Virtually all of the small molecules essential for life can be formed this way: amino acids, sugars, lipids, nitrogen bases, and so on.

These experiments tell us what is theoretically possible; direct analysis of meteorites and lunar samples reveals what has actually happened in nonliving environments. Rock samples brought back from the moon by the Apollo astronauts contain various organic compounds, including amino acids. The interiors of meteorites, uncontaminated by the fall through the atmosphere or contact with soils, have contained alcohols, sugars, amino acids, and the nitrogenous bases that occur in nucleic acids. With regard to the formation of monomers, the chemosynthetic hypothesis represents a plausible model.

Formation of Polymers

Monomers present in the early ocean had to polymerize if life were to arise, but polymerization required high concentrations of monomers. Given enough time, the oceans would have changed from a dilute soup to a concentrated one, but that probably was not necessary. Numerous mechanisms would have produced pools of highly concentrated reactants.

An obvious method of concentration is formation of seaside pools at high tide that evaporate after the tide goes out. With intense sunlight the pools would have been warm, perhaps even hot, and polymerization reactions could occur. With the return of high tide, the polymers would be washed into the sea and accumulate. Monomers could also have accumulated when ponds and seaside pools froze; ice is relatively pure water and the monomers become increasingly concentrated in whatever water has not yet frozen. This might produce a class of polymers distinct from those formed by evaporation at high temperature.

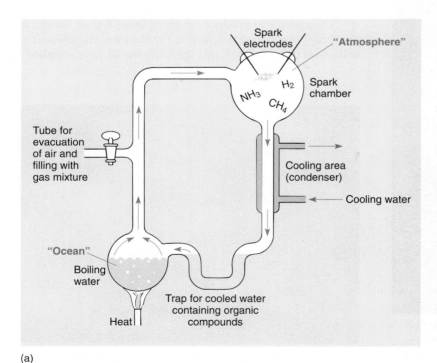

(a)   (b)

Figure 17-22   (a) Diagram and (b) photograph of apparatus used by Miller to show that the first steps in the chemosynthetic origin of life were possible.

Absorption by clay particles could have concentrated monomers, and clays are receiving great attention now. Because clay particles are tiny fragments of rock, they have a regular, crystalline surface; organic molecules adhere to them in a particular orientation, not simply at random. Thus, binding to a clay particle is similar to binding to an enzyme. Furthermore, the crystalline matrix of clay contains contaminating ions of iron, magnesium, calcium, phosphate, and other charged groups that are typically present at the active sites of enzymes. Considerable experimentation is being done to determine whether clays might have both concentrated monomers and acted as the first primitive catalysts.

■ Aggregation and Organization

The next step in the possible chemical evolution of life would have been aggregation of chemical components into masses that had some organization and metabolism. Fatty, hydrophobic material would have accumulated automatically as oil slicks in quiet water or as droplets in agitated water. Fatty acids would have occupied the outermost layer, accompanied by other molecules such as proteins that had a hydrophobic portion and a hydrophilic one. The interior may have been mostly hydrophobic, but proteins that had a hydrophobic exterior and hydrophilic interior would have added complexity. Aggregation of certain types of proteins would have resulted in large regions of hydrophilic sites.

These first aggregates would have formed basically at random, controlled only by relative solubility. If some of the proteins had some enzymatic activity by chance, the aggregate would have had some simple "metabolism"—perhaps the conversion of some molecule, absorbed from the sea, into another molecule; the aggregates would have been heterotrophs completely.

These aggregates are not postulated to have been alive or even to have been early stages of life because at that point no means of storing genetic information existed. Without genetics, natural selection cannot occur. Countless numbers of these aggregates might have formed and disassociated throughout all of the oceans over millions of years. Their existence may have had a significant effect on the chemistry of the oceans. Some may have provided appropriate conditions for certain proteins to be very active enzymes that might have produced quite complex products. However, others may have been able to degrade such products, and an equilibrium may have been established between formation and destruction of ever more elaborate molecules and polymers.

At some point, presumably an aggregate formed that did have a heritable information molecule able to direct the synthesis of products useful to the aggregate. By "useful" we mean that the product helped the aggregate persist longer, without being broken up by wave action or dissipating by diffusion. It may have helped the aggregate grow or even helped the information molecule replicate. With the presence of heredity, everything changed; any mutation that caused the production of a more efficient, more advantageous enzyme or structural protein provided a strong selective advantage and could be passed on to progeny aggregates. Currently, attention is focusing on RNA as the first heritable information molecule. As the aggregate increased in size, perhaps largely by absorbing material from the ocean, its information molecule would replicate. If the aggregate divided into two by either wave action or surface tension and if each half contained an information molecule, reproduction occurred.

■ Early Metabolism

The aggregates would have been complete heterotrophs, absorbing all material from the ocean and modifying only a few molecules. As aggregates continued to consume certain nutrients, however, scarcity occurred; any aggregate that could produce an enzyme capable of synthesizing the scarce molecule from an abundant one still available in the ocean would have had a strong selective advantage. With this, there would have been a metabolic pathway two steps long involving two enzymes (**Figure 17-23**). Being able to synthesize a valuable, scarce molecule from an abundant, free one would have given that aggregate great advantage over the others; it would have had a more rapid metabolism and would have grown and reproduced more rapidly.

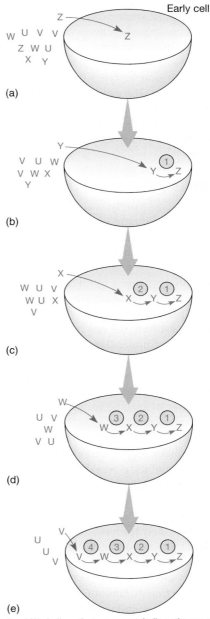

Figure 17-23 (a–e) We believe that many metabolic pathways evolved backward as raw materials in the environment became scarce. Currently, the pathway for photosynthesis has gone as far as it can, requiring only water and carbon dioxide.

As its abundance increased, it and its descendants would finally have become so numerous as to use the second chemical faster than it could be formed chemosynthetically, and scarcity would have occurred again. Natural selection would again favor any genetic system that could extend the metabolic pathway another step to include an abundant precursor. Recall that in photosynthesis in modern plants, the only two precursor molecules that need to be drawn from the environment are water and carbon dioxide—both are abundant and cheap. From them, metabolic pathways extend and ramify such that all carbon-containing compounds are derived from them. No preformed organic molecules are necessary.

Energy metabolism must have been important also, and glycolysis must have evolved early because it is present in virtually all organisms. The aggregates and first cells may have absorbed some ATP and generated more by fermentation with glycolysis for millions of years. At some point, electron transport systems and hydrogen ion pumping systems would have evolved. The first would have been powered by light, and an early form of photosynthesis would have evolved. Higher plant photosynthesis is elaborate, but much simpler types occur even today in certain photosynthetic bacteria. The evolution of photosynthesis would not have been as complex or as difficult as chloroplasts would lead us to believe. Oxidative electron transport, as occurs in mitochondria, could not have evolved because free molecular oxygen still did not exist.

■ Oxygen

The evolution of chlorophyll *a* and photosynthesis that liberates oxygen had two profound consequences: (1) it allowed the world to rust, and (2) it created conditions that selected for the evolution of aerobic respiration. Until that time, photosynthesis had involved the bacterial pigment bacteriochlorophyll, which liberates sulfur from hydrogen sulfide rather than oxygen from water. Earth retained its reducing atmosphere, and chemosynthesis of complex precursor molecules could continue. However, after chlorophyll *a* evolved, the raw material for photosynthesis became water, and free molecular oxygen, O_2, was released as a waste product. We know this evolutionary step occurred 2.8 billion years ago because the oxygen rapidly combined with iron and formed ferric oxide—rust. Sedimentary rocks of this age contain a thick red layer of rust, indicating when oxygen-liberating photosynthesis arrived and how long the oxygen/iron reactions continued. Only after all of the free iron in Earth's oceans had oxidized did oxygen finally begin to accumulate in the atmosphere. The atmosphere present today was derived from the early second atmosphere by this addition of oxygen from photosynthesis. It is an **oxidizing atmosphere**.

The period of rusting was critically important for all life because it kept the concentration of free oxygen very low. It can be difficult for us humans—obligate aerobic animals—to appreciate just how toxic free oxygen is; we need it to live, but our bodies have numerous mechanisms to keep it under control and to prevent it from reacting at random in our bodies. The same is true of plants, fungi, algae, and many bacteria. Had iron not been present, the concentration of free oxygen in water and air might have increased rapidly and killed everything; possibly there would have been no survivors. With iron, however, oxygen concentration remained low for millions of years; it was an environmental danger but was not instantly, universally lethal. Any mutation that produced a mechanism that could detoxify oxygen had great selective advantage. Of course, the best way is to add two electrons to it and let it pick up two protons and turn into water, which is exactly what cytochrome oxidase does. This is the last electron carrier in the mitochondrion transport chain, and with its evolution, aerobic respiration became possible. Oxygen was transformed from a dangerous pollutant to a valuable resource. The evolution of aerobic respiration did not occur in mitochondria but rather in bacteria, and some species seem to retain very relictual, early versions.

The buildup of atmospheric oxygen had other important effects; under the influence of UV light, oxygen is transformed into ozone, O_3. Ozone is extremely opaque to UV light and thus prevents most of it from penetrating deep into the atmosphere and reaching Earth's surface. This immediately removed UV light as an energy source and must have greatly slowed the chemosynthetic formation of organic molecules. Biological photosynthesis became the main means of bringing energy into the living world. The lack of UV radiation made the surface a safer place to live; UV radiation damages nucleic acids, and while the atmosphere lacked ozone, probably nothing lived near the top of the oceans or on land. With the ozone shield, organisms could move higher in the oceans and onto the mud of seashores. The transition to terrestrial life became possible.

■ The Presence of Life

The chemosynthetic theory postulates a long series of slow, gradual transitions from completely inorganic compounds to living bacteria. At which stage can we say that life came into being? Can the aggregates be considered alive? This is a difficult question because the theory delineates no absolute demarcation between living and nonliving objects (**Figure 17-24**). The chemistry of living creatures is more complex than that of nonliving objects, but it does not possess any unique properties. The physics of living and nonliving systems is identical. As so often in the living world, we are dealing with a continuum, not simply two mutually exclusive alternatives. To look for a dividing line would be simplistic; it is more important for us to understand the processes in their complexity.

Figure 17-24   The fossil remains of an early prokaryote that lived approximately 3.5 billion years ago.

Summary

1. When considering the reproductive effort of an entire population, it is important to consider all possible alleles for each gene, their relative frequency in the gene pool, and their effects on reproductive success.

2. Sexual reproduction alone does not cause relative abundance of alleles in the gene pool to change. Mutations bring about new alleles but do not greatly affect allele frequency.

3. Both artificial and natural selection increase the fitness of a population to its habitat by preventing certain individuals from breeding and passing their alleles on to the next generation. Artificial selection often increases the plant's fitness for artificial habitats.

4. In order for natural selection to operate, a population must produce more offspring than can possibly survive in the environment, and offspring must be genetically diverse. These two conditions occur almost everywhere in nature at all times.

5. Natural selection operates only on the phenotypes of preexisting alleles.

6. Plants and most other organisms never do anything with purpose or intent. Only humans and higher primates have purpose, intent, and will and then only with regard to voluntary actions.

7. Organisms face multiple selection pressures and must be adapted to all aspects of their environment. Certain aspects are more important than others, and it may be less dangerous to be poorly adapted to one factor than to another.

8. Most adaptations have negative consequences as well as advantageous ones; retention or loss of these alleles from the gene pool is related to the relative benefit versus cost of the adaptation. As an environment changes, the selective value of the phenotype may also change.

9. In phyletic speciation, all of a species evolves into one new species. In divergent speciation, part of a species evolves into a new species while the rest of the population continues as the original species or evolves into a different new species.

10. Gene flow is inhibited or completely prevented by reproductive barriers, either abiological aspects of the environment or biological aspects of the organisms themselves or of their pollinators and seed dispersers.

11. Life is believed to have originated on Earth by the process of chemosynthesis. This hypothesis postulates that reactions of inorganic compounds in Earth's early second atmosphere resulted in formation of organic compounds that coalesced into simple aggregates with rudimentary metabolism. After a system of heredity developed, evolution by natural selection made it possible for truly living cells to come into existence.

Important Terms

abiological reproductive barrier
adaptive radiation
allopatric speciation
artificial selection
biological reproductive barrier
chemosynthesis
convergent evolution
divergent speciation
founder
gene flow
gene pool
genetic drift
hybrid inviability
hybrid sterility
natural selection
phyletic speciation
population genetics
reducing atmosphere
speciation
sympatric speciation

Review Questions

1. Evolution is the gradual conversion of one species into one or several new species. It occurs for the most part by _____ _____. Mutations cause new _____ or new _____ to arise, which affect the _____ of the individual.

2. Relative to a human lifetime, is evolution a process too slow or too fast for us to see it easily?

3. As the idea of geological evolution developed, it had an impact on biological thinking. Describe the impact. What was the importance of river deltas?

4. The discovery of evolution by natural selection happened at about the same time as other critical discoveries. Name some of these discoveries. Can you describe how they might have affected biological thinking?

5. Genes and inheritance have an impact on the lives of parents and their progeny, the F_1 and F_2 generations. Population genetics also deals with genetics, but in a different way. What is the definition of population genetics? How can the abundance of an allele be different in a population as compared with an individual of the population? (Hint: How can the phenotype of a population differ from that of an

individual? Can a population have red flowers, pink flowers, and white flowers? Can one individual have all three types of flowers?)

6. What is the gene pool of a species? Why are some alleles more common than others?

7. Many mutations can be corrected by DNA proofreading/repair enzymes. What would happen to the further evolution of a species if a perfect repair mechanism evolved such that no mutation ever went uncorrected?

8. How does an accident differ from natural selection? Would an unusually severe drought with no rainfall at all for 50 years be a selective or a nonselective destructive force?

9. How does artificial selection differ from natural selection? Do you think the ancestors of lettuce had such soft, non-bitter leaves? Could today's lettuce plants survive in natural conditions?

10. One of the conditions necessary before natural selection can occur is that the population must produce more offspring than can possibly grow and survive to maturity in that habitat. Do you think this is true for most plants or most animals? (Do not think about humans. We are an exceptional case.) Think about some tree that you see on campus or near your home. Estimate the number of seeds it makes every year and the number of years it might live. How many seeds (offspring) will it produce in its lifetime? Do you think it would be possible for them all to germinate, grow, and reproduce?

11. Another condition necessary for natural selection to occur is that progeny must differ from each other in their types of alleles. Do you think this is true for most plants or animals? Be careful for some familiar crop plants. Bananas have no seeds. They must be propagated by taking cuttings from buds at the base of the plant. All of the bananas in a store are clones—they are all the same plant. Can natural selection affect these bananas? Most apple varieties are also clones. Are they being affected by natural selection?

12. Many people assume that natural selection depends on something outside the plant in the environment affecting the plant, but is that true? Wood is always inside the plant. It is never in contact with the environment. Can the phenotype of wood affect a plant's fitness? Its survival (Hint: Figure 17-9)? If a mutation caused a plant to not be able to make either vessels or tracheids, just fibers, in its wood, do you think that would affect the tree's fitness compared with all of the other trees in the population?

13. Name several factors that are not part of natural selection.

14. Change each of the following sentences into ones that are true: (a) Plants produce root hairs in order to absorb more water. (b) Plants close their stomata at night to conserve water. (c) DNA produces mRNA in order to keep ribosomes and potentially harmful high-energy enzymes away from itself. (d) Stems and roots produce bark to keep pathogens out. (e) Cacti and euphorbias have succulent bodies in order to store water. (f) Certain plants have C_4 metabolism in order to reduce photorespiration. (g) Some plants time their flow-

ering by night length in order to avoid damage caused by blooming too early in an unusually mild spring.

15. Are plants subjected to just one type of selection at a time or are there multiple selection pressures affecting a plant? The text gives an example of a mutation that improves cold hardiness but that might be eliminated from the gene pool despite being beneficial. Why might it be eliminated?

16. Why is it that evolutionary changes that result in the loss of a structure or metabolism can come about quickly?

17. Modern cacti evolved from ancestors that were leafy trees (Figure 17-11), and in the early stages of cactus evolution, leaves were lost; however, the capacity to produce leaves was not lost quickly. Why not?

18. Look at the two cacti in Figure 17-11. Both the leafy *Pereskia* in part a and the very desert-adapted *Gymnocalycium* in part b evolved from the same ancestors—leafy trees that were not themselves cacti. In which genus, *Pereskia* or *Gymnocalycium,* has natural selection favored more of the new alleles produced by mutations? In which genus has natural selection retained more of the original alleles and eliminated almost all of the new alleles produced by mutation?

19. What is phyletic speciation? What is divergent speciation?

20. What is gene flow? How does it occur?

21. The Hawaiian Islands were formed as underwater volcanoes that became so tall that they protruded above the ocean's surface. Being formed from molten rock, they were absolutely sterile at first, but now they are home to many types of plants and animals. How did the ancestral plants and animals get to Hawaii? Was this a type of gene flow?

22. What is an abiological reproductive barrier? Name several. Is something that is an abiological reproductive barrier for a plant necessarily also one for an animal? Would you guess that small herbs have more abiological reproductive barriers than do birds?

23. Differences in flower color, shape, or fragrance are listed as being potential biological reproductive barriers. Do you think they would matter at all in a wind pollinated species? If a plant species has several flower colors (for example, red, white, and pink), could a mutation in the plant's pollinator—but not in the plant species itself—be a biological reproductive barrier? For example, a mutation that makes it unable to see red?

24. One of the earliest postzygotic barriers to arise is hybrid sterility. Two plants can be crossed, but the resulting seeds grow into sterile plants that cannot reproduce. Is this as important in plants as it is in animals (Hint: How many types of hybrid flowers and fruits are there, compared to hybrid animals)?

25. What is genetic drift? What is a founder? Are these important in large populations or in small ones? How are these phenomena important when we create national parks and preserves to protect endangered species?

26. What is convergent evolution? Euphorbias and cacti are often used as examples of convergent evolution. Have their genotypes converged? Have their phenotypes converged?

Both have spines, but if we look carefully we can tell that those of cacti evolved from _____ and those of euphorbias evolved from _____. They have converged, but not so much that we cannot tell them apart.

27. What is the theory of chemosynthesis?

28. Describe the conditions present on Earth at the time of its early second atmosphere. What chemicals and energy sources were present? What chemical in particular was absent? Can you speculate about what conditions would be like on Earth today if meiosis, crossing-over, and sexual reproduction had never evolved?

29. Life originated on Earth so long ago that erosion and weathering and metamorphosis of rock have destroyed most traces of the processes that occurred, but oxygen is an interesting exception. What is the source of all the oxygen in our atmosphere? What happened to the first oxygen to be produced? Why was this important to the survival of all other life forms?

BotanyLinks

http://biology.jbpub.com/botany/5e

BotanyLinks contains a variety of resources and review material designed to assist in your study of botany.

Classification and Systematics

Concepts

Starting approximately 400 million years ago, a long series of mutations and natural selection gave rise to the plants that are alive today. During the intervening eons since then, that evolutionary line has progressed and diversified, branching into more and more lines of evolution. Thousands of these have become extinct and are known only by fossils; thousands of other evolutionary lines are represented by the approximately 280,000 species of living plants. The hereditary relationships of any group of organisms constitute its **phylogeny**, basically the evolutionary history of each member of the group. Some phylogenetic studies attempt to create models of the evolutionary relationships of very large groups, for example, the steps that were involved as plants, animals, and fungi evolved from

Chapter Opener Image: You probably had no trouble at all recognizing this as a cactus even though it is unlikely you have ever seen it before; it is *Oreocereus trollii*, native to the high deserts of Argentina and Bolivia. It is recognizable as a cactus because it has so many features we associate with cacti: its stems are succulent, it has obvious spines, and it has no visible leaves. Most cacti do indeed have these characters because these characters arose early in the evolution of cacti and have been inherited for generation after generation ever since. But habitats are variable and plants diversify, so some cacti now have lost their spines and others are not as succulent as this one. Features such as succulence and spines have evolved in other families; for example, many spurges resemble cacti very greatly. When we attempt to classify organisms, we try to group together those that have similar features because they are closely related, and we try not to be fooled by characters that have evolved two or three times in different groups.

some early ancestor. Other studies focus on the relationships of smaller groups, such as the species of the legume family (e.g., peas, beans, and lentils). The goal of modern plant **systematics** is to understand each of these evolutionary lines and to have a system of names—**nomenclature**—that reflects their relationships accurately.

In the past, plants were given names merely so that people could communicate about medicinal plants, crops, poisonous plants, and so on. The task of taxonomists was to discover and identify new species and give them unique names, but with the discovery of evolution by natural selection, the basis of naming plants suddenly changed: Natural selection showed that all organisms are related to each other genetically, all are part of a large phylogeny. Taxonomists (also called systematists), scientists who specialize in classification and naming, immediately realized that the most scientifically valid system of assigning names to species would be one that reflected evolutionary relationships. At the end of the 19th century, taxonomists adopted the goals of (1) developing a **natural system of classification**, a system in which closely related organisms are classified together, and (2) assigning plant names on the basis of phylogenetic relationships. The nomenclature would reflect the natural system of classification. Because organisms range from closely to distantly related, taxonomists devised a classification and a nomenclature with numerous levels. In this system, closely related species are placed into a genus, and closely related genera are grouped together into a family, and so on.

Today, as in the past, there are many regions of Earth that have not been adequately explored. These probably contain thousands of plant species that have never been collected, named, or studied. Plant systematists still have the goals of searching for new species and using them to construct more accurate models of plant evolution. It is important to remember that the pattern of evolutionary diversification—the phylogeny—is a reality. Our classification systems are hypotheses, models that attempt to map that evolution. Because our knowledge is incomplete and imperfect, the current classification systems are only approximations. Most are probably close approximations, and we do not expect major changes; however, we do expect numerous smaller changes to be made periodically for many years.

Levels of Taxonomic Categories

Plants have varying degrees of relatedness, and a natural classification system reflects this in its numerous levels.

The most fundamental level of classification is the **species**, which ideally and theoretically is a set of individuals closely related by descent from a common ancestor. (The word "species" is both singular and plural; the word "specie" refers to money.) Members of a species can interbreed with each other successfully but cannot interbreed with individuals of any other species (Table 18-1). Some species are not so predictable; they may not interbreed well with closely related species, but an occasional cross-pollination results in a viable seed that grows into a fertile adult. If this occurs frequently, the two plant groups may best be considered subspecies of a single species. The two subspecies must be so similar genetically that the chromosomes of the two parents can function in the same nucleus, and when the plant is

TABLE 18-1	The Taxonomic Categories	
	Examples	
	Dicot	Monocot
Kingdom	Plantae	Plantae
Division	Magnoliophyta	Magnoliophyta
Class	Magnoliopsida	Liliopsida
Order	Fabales	Liliales
Family	Fabaceae (Leguminosae)	Liliaceae
Genus	*Lupinus*	*Hymenocaulis*
Species	*Lupinus texensis*	*Hymenocaulis caribaea*
Common name	Texas bluebonnet	Spider lily

The scientific name of a species is always a binomial, consisting of the genus and species. The species epithet is often descriptive: *Lupinus texensis* occurs in Texas; other common species epithets are "longifolia," "grandiflora," "vulgare" (common), "acaulis" (rosette, without a stem), and so on.

mature, spore mother cells can undergo meiosis successfully. For all of this to happen, the organisms must be extremely closely related, having undergone divergent evolution from a common ancestor only recently so that few mutations have accumulated since they diverged. But as mutations continue to accumulate in each subspecies, finally either no viable seed results when the two are crossed or the seed grows into a sterile plant. The two then must be considered separate species.

Closely related species are grouped together into **genera** (singular, **genus**). Deciding whether several species are closely related enough to be placed together in the same genus is difficult. No objective criteria exist; the decision is entirely subjective and is often the cause of great dispute. Some taxonomists, generally referred to as "lumpers," believe that even relatively distantly related species should be grouped together in large genera. Other taxonomists, called "splitters," prefer to have many small genera, each containing only a few species that are extremely closely related. For example, some taxonomists believe that cranberries and blueberries are so similar that they should go into the same genus, *Vaccinium*; others think that cranberries are distinct enough that segregating them into their own genus, *Oxycoccus*, more accurately reflects evolutionary reality. Both groups of taxonomists agree that the two sets of species are closely related, but they have different opinions as to how much evolution has occurred since the time of the most recent common ancestor. The critical concern is that the genera are natural, that all of the species included in the genus are related to each other by a common ancestor, and that all descendants of that common ancestor are in the same genus. Such a group is **monophyletic**. In an unnatural, **polyphyletic group**, members have evolved from different ancestors and may resemble each other only as a result of convergent evolution. Whenever systematists discover that they have accidentally classified unrelated plants together in an unnatural, polyphyletic group, they immediately search for a more accurate classification composed only of monophyletic groups.

The level above genus is **family**, each family being composed of one, several, or often many genera. Most families are well defined, with widespread agreement as to which species and genera belong in a particular family. As examples, consider how easy it is to recognize the following families: cacti, orchids, daisies, palms,

Box 18-1 Development of Concepts of Evolution and Classification

Our modern classifications have their origins in ancient Greece. Theophrastus, a student of both Plato and Aristotle, wrote extensively about plants around 300 BCE and established many important concepts. He distinguished flowering and nonflowering plants, recognized sexuality in plants, and understood that fruits develop from carpels. Theophrastus described almost 500 species, and our genus names *Asparagus, Narcissus,* and *Daucus* (carrot) can be traced directly back to him. Pliny the Elder (Caius Plinius Secundus; CE 23-79), a Roman lawyer and natural historian, wrote voluminously on almost every subject. His largest work, *Natural History,* was an attempt to describe everything in the world. Despite its inclusion of many fanciful creatures based only on folk tales, it served as the definitive, authoritative source of information on most subjects for more than 1,000 years. The most important book on plant classification from the ancient world is *Materia Medica* by Dioscorides. Written in the first century CE, it describes 600 plant species and how they can be used to treat disease. It was the best, often the only, source of information about preparing herbal medicines. *Materia Medica* was at last superseded by more accurate work during the Renaissance, but it is still published today for its immense historical value as a direct link to early Greek science.

Between the time of Dioscorides and the Renaissance, almost no works of great value in natural history were written. Changes began to occur in the 15th century, and Europe entered an age of exploration. Prince Henry of Portugal sent ships on expeditions down the west coast of Africa, and the explorers returned with new plants and animals and knowledge of new lands and peoples. Simultaneously, exploration of physics, chemistry, astronomy, and geology began. An important result of these explorations was the discovery that Pliny and other ancient authorities had been wrong in many subjects. Until then, people thought the ancient Greeks and Romans had represented a Golden Age and were basically infallible. The realization that the ancients had made errors meant that the answers to questions about the world had to be sought in the world itself, not in ancient books. Exploration—not only geographical but also scientific, philosophical, and religious exploration—became an obsession.

In botany, this new, independent thinking first became apparent in the publication of medically oriented plant books, called herbals. By the middle of the 16th century, they began to contain careful, precise descriptions of plants based on first-hand observation of plants, not by paraphrasing *Materia Medica.* Most herbals contained fewer than 1,000 plants, so a reasonably good botanist could become familiar with all of them; however, as exploration continued, especially after the discovery of the Amer-

icas, the number of plant species became too large for this type of familiar treatment. It became necessary to develop a classification system so that a person would have some means of identifying an unfamiliar specimen and finding it in the ever-larger herbals.

Several important ideas developed at this time. One was the concept of a genus as a group of similar species, established by Gaspard Bauhin (1560-1624). Although this may seem like an obvious notion, it was a profound breakthrough at the time. The concept of species had been easy—God was believed to have created all of the types of organisms—the species. But this concept did not explain why there should be groupings of species, as the concept of genus implies. Why would God create several types of roses, several types of mints? This was not a trivial question then. Until the theory of evolution by natural selection, the world was viewed as a reflection of the mind of God, who presumably had created it. If there were thousands of species of grasses but only three of cattails, this would reveal something about how God thinks. It was obvious to anyone familiar with the increasing number of plant and animal species that some species resembled each other very closely. But why?

Our system of nomenclature, of **scientific names**, can be traced directly to Carolus Linnaeus, a professor of natural history at the University of Uppsala in Sweden during the middle and latter part of the 18th century. He adopted the genus system of Bauhin, created a large numbers of genera, and placed every species into one genus or another. Every species had both a genus name and a species epithet, the basis of our present **binomial system of nomenclature**. Before Linnaeus, rather than names, plants had paragraph-long descriptions of their features, which was obviously cumbersome; after Linnaeus, each plant had a name of just two words.

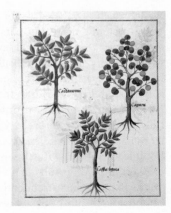

Figure B18-1a Prior to the Renaissance, most European botanists considered ancient books to be a better source of knowledge than plants themselves, so they merely copied drawings from pre-existing books, and each drawing became less accurate and less useful. Left: cardamon; right: *Capparis;* cassia lignea.

In 1753, Linnaeus published *Species Plantarum,* a treatment of all plant species known in the world at that time. This book made botanical studies of all types much easier because specimens could be identified quickly; however, the system often placed unrelated species into the same genus on the basis of homoplasic (analogous) characters.

The idea of the evolution of species developed slowly over many years; it did not begin with Darwin and Wallace by any means. Bauhin recognized resemblance early, and an expansion of the idea of similarity occurred at the Jardin des Plantes in Paris. Antoine-Laurent de Jussieu wanted to organize the garden in a logical way so that plants resembling each other were planted together (the garden still exists, you can visit it). The accompanying catalog was *Genera Plantarum* (1789): It was the first major attempt at a natural system of classification, and people could not help but wonder whether some species within each genus had not evolved from others. Perhaps similar genera had changed, evolved. But how? Science was still dominated by the idea of divinely created types.

Unfortunately, J. B. P. de Lamarck at this time presented his theory of evolution by **inheritance of acquired characteristics**. This was the incorrect idea that all cells of the body produced fluids that diffused to the genitalia, where the fluids were concentrated and formed into sperm cells or egg cells. As an individual changed, the secreted fluids and therefore the characteristics carried by the gametes would change. The evolution of giraffes was explained as follows: Ancestors to giraffes had been born with short necks; however, as they stretched their necks to reach leaves in high trees, their necks lengthened—long necks in the adults would be the acquired characteristic—and subsequently, the genetic fluid passed to their gametes carried information about long necks.

Consequently, their offspring had longer necks. Mendel showed that this could not be right because genes are nonvarying particles, not variable fluids. Evolution by natural selection explains giraffes by postulating that variability of neck size occurred among the earliest ancestors of giraffes; alleles in some animals produced short necks and in others slightly longer necks. Those with longer necks could reach more leaves; therefore, they had better survival and reproductive success. Over millions of years, various mutations produced new alleles that resulted in even longer necks, and these new alleles were increased by natural selection. The theory of inheritance of acquired characteristics was never widely accepted, and it gave evolution a bad reputation for many years.

In 1859, Charles Darwin and Alfred Wallace each propounded the theory of evolution by natural selection, which gave natural systems of classification an immediate validity. Taxonomists quickly understood why some species with a genus resemble each other so strongly: They are descendants of a common ancestor and are related by evolution. Similarly, closely related genera constitute a family.

The theory of evolution by natural selection became accepted while two German botanists, A. Engler and K. Prantl, were working on a monumental classification of all the world's plants, *Die Natürlichen Pflanzenfamilien,* published in 1915. It organized species on a phylogenetic, natural basis, as understood by Engler and Prantl. In those early days of evolutionary studies, organisms were assumed to have evolved from the simple to the complex (amoebae to humans). Thus, in flowering plants, the wind-pollinated species that have no sepals, and petals were placed first; the sunflowers, with their complex floral structures, were considered more advanced. It was quickly discovered that simplicity may also be the result of evolutionary simplification. Because no automatic correlation exists between simple and primitive or between complex and advanced, the Engler and Prantl classification has many errors, although as a compendium of information it is still an unmatched vital resource. Other major classifications were produced by C. E. Bessey (early 1900s) of the University of Nebraska, A. Takhtajan in Russia, and A. Cronquist (1980s) at the New York Botanical Garden.

Starting in the mid 20th century, systematics began incorporating many more types of characters. In addition to morphology, they used anatomy, physiology, enzyme differences, transmission and scanning electron microscopy, pollen features, and most recently DNA sequences. Also, a new analytical technique called **cladistics** was developed, which formally addresses the science of grouping items together on the basis of similarity.

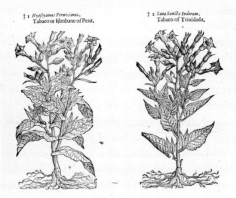

Figure B18-1b During and after the Renaissance, people realized they could make accurate drawings and studies of plants only if they examined the plants themselves. Much more realistic images were produced. Tobacco plants.

and grasses. The reason for this probably is the age of most families. A line of evolution must be very old—usually at least several tens of millions of years—to diversify into several genera and many species. Even closely related families have been separate evolutionary lines for at least 20 or 30 million and often 50 million years. The common ancestor and many of the earliest species probably have become extinct. Therefore, although two families have many characters in common, enough features are unique to each family to make them recognizable as distinct from each other.

The levels above family are **order**, **class**, **division**, and **kingdom**. Although you might expect universal agreement at the level of kingdom, even the boundaries of the plant kingdom, kingdom Plantae, are disputed. Some believe that green algae should be included because they are almost identical biochemically to vascular plants. Others believe that, despite the biochemical similarity, green algae should be excluded because they are so different morphologically and anatomically.

Except for kingdom, genus, and species, the names must have a certain ending to indicate the classification level. Division names end in *-phyta,* for example Chlorophyta (green algae), Coniferophyta (conifers), and Magnoliophyta (flowering plants) (Table 18-1). Class names end in *-opsida,* order names in *-ales,* and family names in *-aceae* (pronounced as if you were spelling the word "ace"). Genera and species names do not have standard endings. The scientific name of a species is its genus and species designations used together and either underlined or italicized; for example, tomato is *Lycopersicon esculentum.* Note that the species name is not "esculentum" but is *Lycopersicon esculentum.* Esculentum is the **species epithet**, the word that distinguishes this species only from the other species of the genus *Lycopersicon.* We cannot refer to tomatoes as "esculentum" because that species epithet is used for many different species in different genera: for example, buckwheat, *Fagopyrum esculentum,* and taro, *Colocasia esculenta.*

Some families have two names; the above rules were adopted to regularize family names so that each family is named after one of its genera, using the *-aceae* suffix. The rose family was named Rosaceae, based on the genus *Rosa.* This worked well for most families, but for some, the old name was so well known and so familiar that it was kept as well. Examples are Asteraceae (Compositae: daisies, asters, sunflowers), Fabaceae (Leguminosae, legumes: beans, peas, peanuts), Aracaceae (Palmae, palms), Poaceae (Gramineae, grasses: wheat, rice, corn), Brassicaceae (Cruciferae, mustards), and Apiaceae (Umbelliferae, umbels: celery, dill).

The word "**taxon**" (plural, taxa) is used to refer to any of the above groups in a general way. For example, some systematists study smaller taxa such as species and genera; others are concerned with higher taxa such as orders and divisions.

Cladistics

The study of phylogeny centers on examining the similarity of one species to others. A species evolves into two species as different populations accumulate distinct alleles; even when enough divergence has occurred to create separate species, the two still resemble each other strongly. As they continue to evolve, each acquires its own mutations, and because they cannot interbreed,

(a)   (b)

Figure 18-1 Members of the family Araceae (the aroids, Jack-in-the-pulpit) all have a spathe (the collar-like sheath) and spadix (the central column), as in this *Amorphophallus* (a) and *Anthurium* (b). In all extant species, the genes that control the formation of these structures are descendants of the genes of the original, ancestral species. The genes that exist today have resulted from thousands of rounds of DNA replication during plant growth, crossing-over, syngamy, and then more rounds of DNA replication as the zygote grows. This has been repeated millions of times since spathes and spadices first evolved millions of years ago. During all this, mutations and natural selection have led to obvious varieties of spathes and spadices, but the basic relatedness, the homology, is still obvious.

they cannot share the new alleles. Thus, they differ from each other more as time passes. Distantly related plants have been on separate lines of evolution for millions of years, time enough for so many mutations to accumulate that they resemble each other only slightly.

Phylogenetic studies are complicated by the fact that plants can resemble each other for two distinct reasons: (1) they have descended from a common ancestor, or (2) they have undergone convergent evolution. Features similar to each other because they have descended from a common ancestral feature are **synapomorphies (homologous features)**. For instance, almost all members of the anthurium family are easily recognizable because of their spathe and spadix inflorescence (**Figure 18-1**). All members have these structures because all have inherited their inflorescence genes from a common ancestor. The spathe and spadix are homologous in all *Anthurium* species. Homologous features are the ones critically important for making taxonomic comparisons and the only ones that can be used to conclude that species are related.

The second cause of resemblance, convergent evolution, results when two distinct evolutionary lines of plants respond to similar environments and selection pressures. Under these conditions, natural selection may favor mutations in each line that result in similar phenotypes. Features like this are **homoplasies (analogous features)** and should never be used to conclude that plants are closely related. A striking example is the convergent evolution of cacti and the succulent euphorbias. Both occur in

Figure 18-2 (a) Cactus spines are modified bud scales, which makes them modified leaves. They occur in clusters as all bud scales do, and they never branch. *Discocactus.* (b) *Euphorbia* spines are modified branches. They bear small leaves that may themselves have axillary buds; therefore, the spine can branch, as in this *E. meloformis.* Although the spines in cacti and euphorbia are often called homoplasies, they are so different it is difficult to confuse the two. (c) In contrast to homoplasy, another problem arises when a structure evolves so much as to be almost unrecognizable. It took a great deal of study to be certain that cactus spines are modified leaves, and they continue to evolve. These flat, thin structures are spines of *Tephrocactus papyracanthus*; they protect the cactus by disguising it. This species grows among grasses and is almost impossible to find unless it is flowering. (d) The spines of this *Tephrocactus strobiliformis* have evolved to be glands, extrafloral nectaries: Each droplet is a secretion of dilute sugar water that attracts ants to the plant. The ants protect the plant as they defend their source of food (plants b, c, and d were cultivated by B. Barth).

deserts where a succulent water-storing body is advantageous, and spines are selectively advantageous by deterring animals from eating the plants to get the water. The two families are not considered closely related on the basis of spines and succulence, however, because these features do not share a common ancestry; they converged evolutionarily because of similar selection pressure.

Determining whether a similarity is due to homology (common ancestry) or analogy (convergent evolution) can be extremely difficult. In the case of spines, it is easy: Cactus spines are modified leaves that are always smooth and never branched, occurring in clusters on an extremely short shoot (**Figure 18-2**). *Euphorbia* spines are modified shoots that branch and have small, scale-like leaves. Being modified shoots, they occur singly in the axil of a leaf, never in clusters. If we look beyond the analogous similarities, the dissimilar features are dramatic; flowers of cacti are large and have many parts, whereas those of euphorbias are small and unisexual, either staminate or carpellate. The wood of cacti is parenchymatous and has few fibers, whereas that of euphorbias is fibrous and hard. Most important, the plants that most closely resemble the succulent euphorbias in important homologous features are the nonsucculent euphorbias of Africa. Succulent cacti resemble a completely different set of plants, nonsucculent cacti of the tropical regions of the Americas. Thus, the evolution of spines and succulence occurred independently in each group.

A corollary to the assumption that similar plants are closely related is the assumption that dissimilar plants must not be closely related. Studying lack of similarity can also be difficult because in some cases a small genetic change results in dramatic phenotypic changes (**Figure 18-3**). Mutations that affect production, distribution, and sensitivity to hormones result in large changes of the phenotype between two closely related species. Also, mutations that affect early stages of development such as the embryo or bud meristems can cause closely related species to look deceptively dissimilar. In many evolutionary lines, species of large woody trees have evolved into herbs; the main transition involved is the ability to flower early and then not survive a harsh winter. It must be a simple evolutionary process involving few genes because it has occurred many times. For example, nearly half the species of the Violaceae are herbs in the genus *Viola*—violets and pansies—but the early members of the family were woody trees, as are the members of the relictual, tropical genus *Rinorea*. In the cabbage family (Brassicaceae), artificial selection by humans has resulted in one species that has dramatically different forms: Broccoli, brussels sprouts, cabbage, and cauliflower are all subspecies of a single species, *Brassica oleracea.*

For taxonomic studies, a mosaic of evidence is available—some valid and useful, some misleading. Systematists attempt to look at as many features as possible on the assumption that misleading evidence will be outweighed by valid characters. Also, certain features are considered more significant; those that result

(a)　(b)

Figure 18-3 This small, leafless shrub (a, *Acacia aphylla*) may not appear at first glance to have much in common with this larger, leafy tree (b, *Acacia drepanolobium*), but careful examination reveals homologous similarities in many critical features of the flowers and stems, which allow assignment to the same genus, *Acacia*. In this case, genes controlling flower and stem morphogenesis have changed little during evolution, whereas genes controlling elongation growth of stems, formation of leaf primordia, and formation of bark have all changed dramatically.

from numerous metabolic interactions and the influence of many genes tend to change more slowly than those with a simple metabolism controlled by just one or two genes. Loss of features, such as loss of sepals and petals in wind-pollinated flowers, is often rapid because deleterious mutations occur commonly; without the natural selection of insect pollination to eliminate these mutations, they are retained and cause a loss of perianth. After those genes have become so mutated that a flower lacks sepals and petals, the evolution of new ones would be slow and difficult.

Currently, taxonomists study virtually every aspect of plants using a wide variety of tools. Simple observation of major parts is still important, but scanning electron microscopy to study hairs, stomata, cuticle, and waxes is also used. Internal structure is studied with both light and transmission electron microscopy; the nature of plastids in the phloem has been found to be an extremely valuable character. Various aspects of metabolism are important, ranging from the types of pigments in flowers to the presence of Crassulacean acid metabolism (CAM) and C_4 metabolism and specialized defensive, antipredator compounds.

DNA sequencing is a new tool for analyzing evolutionary relationships. Two plants are considered separate species only if they differ in significant, heritable ways; therefore, the sequence of nucleotides in the DNA of each species must differ from that of all other species. When plant phenotypes are studied, the actual objective is to determine differences in genotype. In the past, this could be done only by examining phenotypic features. Now DNA can be examined directly and mutations can be identified, even if they do not cause any detectable change in phenotype (Table 18-2).

Like traditional taxonomic data, DNA sequencing data have their own ambiguities and limitations. Any two species, even those closely related, must differ at many points (nucleotide pairs) of their DNA. Simply counting the number of sequence differences is not enough, because even individuals of a very uniform, homogeneous species may have hidden mutations that do not affect the phenotype. Furthermore, any two species in the same family probably share the great majority of their DNA sequences in identical form, with only a few genes mutated. How many differences are

needed to conclude that two genomes represent two distinct species? There is no simple answer.

DNA sequencing permits a new type of study: the evolution of DNA itself. Sequencing allows us to investigate the evolution not only of taxa but of centromeres, telomeres, DNA replication start sites, enhancer elements, boundaries between introns and exons, and so forth.

■ Understanding Cladograms

A **cladogram** is a diagram that shows evolutionary patterns by means of a series of branches. Each point at which a cladogram branches, called a **node**, represents the divergence of one taxon into two, and all of the branches that extend from any particular point represent the descendants of the original group. That ancestor is their **common ancestor** (Figure 18-4). Any ancestor (any node) and all of the branches that lead from it constitute a **clade** (thus, this is a cladogram). Because each node represents a taxon dividing into two, it also represents some detectable change that creates a new group: After the divergence, one of the taxa differs from the other. For example, a plant in a species with red flowers might undergo a mutation that gives it white flowers. As long as the two can still interbreed, this is not especially significant; however, the change in flower color may cause different pollinators to visit it, and it might become reproductively isolated from its red-flowered ancestors. If the white-flowered individuals thrive and become numerous—but cannot interbreed with the red-flowered ones—this is a new species, and the divergence should be represented as a node with two branches coming from it. In this case, white flowers are a derived condition, an **apomorphy**, and red flowers are the **ancestral condition**.

Both populations might continue to evolve and diverge, giving rise to a phylogeny like that shown in Figure 18-5. Node 1 here represents the same taxon as node 1 in Figure 18-4, but now with more time, both its lineages have diverged: At node 5, some that had continued to have red flowers have diverged into two groups, one continuing to have red flowers and the other mutating to have white flowers. White flowers have now originated twice. Synthesis

TABLE 18-2 DNA Sequences Used for Phylogenetic Studies

Plastid DNA

All plants have plastids, and it is usually easy to grind leaf tissue and obtain large numbers of chloroplasts, each of which has many circles of DNA. Plastid genomes are small, each containing less than one hundred genes; therefore, any gene to be studied automatically constitutes a significant fraction of the DNA sample. Plastids are inherited maternally, through the egg; sperm cells do not have plastids. There is no meiotic recombination of plastid genes.

rbcL	This encodes the large subunit of the photosynthetic enzyme RUBISCO. It is universally present in plants (except for some nonphotosynthetic parasitic plants) and is rather long (1,428 base pairs), so it has many sites to study. Certain regions are highly conserved, thus sequences from various species can be aligned using these regions. The more variable regions are then compared. Because it is so stable and certain regions change so slowly, *rbcL* is useful for studying large groups of families or the relationships of angiosperms to other plant groups.
ndhF and *matK*	These encode the F subunit of NADP dehydrogenase (*ndhF*) and a maturase gene (*matK*; its protein helps remove introns properly). Both are shorter than *rbcL* and evolve more rapidly. They are useful for studying relationships within families.
trnL and *trnF*	These are spacers and do not code for any protein or functional RNA. There is little selection pressure against mutations, so these are extremely variable and can be used to study closely related species.

Mitochondrial genes

Although these are commonly used in studying animal phylogeny, they are rarely used in plants. For some reason, plant mitochondrial DNA circles rearrange easily, and the order of genes therefore changes rapidly. Sometimes even a single cell contains various versions of mitochondrial DNA. These are too variable to be useful for most phylogenetic studies.

Nuclear genes

It is easy to grind up tissue, extract thousands of nuclei, then burst them and obtain large amounts of DNA. But each diploid nucleus contains billions of base pairs and thousands of genes, yet only two copies of most genes. The gene of interest may be extremely rare and difficult to find. However, ribosomal genes are present as thousands of copies in each nucleus, grouped together in arrays. Genes within each array are separated by internal transcribed spacers (ITS), and sets are separated by intergenic spacers (IGS).

18S and 26S ribosomal genes	These are long (1,800 and 2,600 base pairs, respectively), stable genes with conserved and variable portions present in all eukaryotes. Conserved regions allow alignment. Variable regions change slowly enough that these are used to study phylogeny of large groups within the plant kingdom, as well as comparing plants with algae, fungi, and animals.
ITS	The internal transcribed spacers in nuclear ribosomal genes are used at the genus and species level.
Protein-encoding genes	Several genes that encode proteins are being used more commonly, despite their difficulty. Phytochrome genes have been especially revealing. Most angiosperms have a family of four phytochrome genes, *A*, *B*, *C*, and *E*, all of which have descended from a single ancestral gene that was duplicated in an ancestor of seed plants. One of the duplicate genes then diverged into *A* and *C*; the other became *B* and *E*. (Phytochrome *D* occurs only in the mustard relatives, having recently evolved from *B*.) It is necessary to be certain which version of phytochrome gene is being sequenced, but this set provides a great deal of valuable data.
Whole genomes	By 2009, the entire genome of four species of angiosperms had been sequenced: *Arabidopsis thaliana*, cottonwood, grape, and rice. Such a task is extremely difficult but provides vast amounts of information for a tremendous variety of studies. At present, full genomes of one or two plant species are being completed every year. In 2008, the "1000 Plant Genomes Project" was started; its goal is not to sequence the entire genome of 1000 species but to sequence all the genes that code for proteins (these genes produce mRNA, so they are easy to find). The concept is that these genes are especially important because they code for enzymes and structural proteins and because their sequences will provide information for cladistic studies.

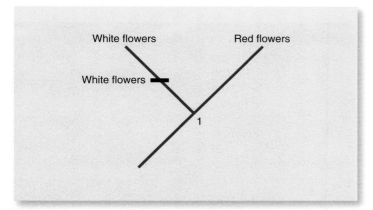

Figure 18-4 A simple cladogram showing the phylogenetic relationships between an ancestral group (at node 1) and its two descendants. The ancestral condition was to have red flowers. The group diverged into two; now one group has white flowers (an apomorphy), and the other still has red flowers. Be careful that this is treating only a very limited amount of information. It does not indicate whether we are dealing with two species and their ancestral species or whether we are comparing two genera or two families.

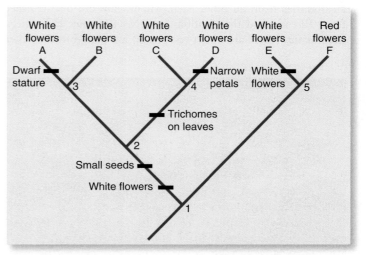

Figure 18-5 A more complex cladogram with more taxa and characters. We have data for the following characters: flowers (white or red), seeds (large or small), leaves (with or without trichomes), plant height (tall or dwarf), and petals (wide or narrow). In all taxa, the fruits are berries. Imagine that the ancestral group had red flowers, large seeds, leaves without trichomes, tall shoots, and broad petals (we probably would not know this before hand: the cladogram is helping us determine what the ancestors were like). See the text for details.

Box 18-2 Cacti as Examples of Evolutionary Diversification

The cactus family (Cactaceae) is a large group that has undergone extensive evolutionary diversification, with the modification of many characters. Many of the changes are easy to see and understand, so this is a good group to use as an example. First, several morphological and DNA features unite the entire family, the most obvious is that all have spines that evolved from modified bud scales. They are present in almost all members (*Blossfeldia* no longer makes spines), and no other family produces spines this way.

The first divergence at node 1 (see Figure B18-2o) produced one clade labeled *Pereskia* and another clade that leads to all the other cacti. The *Pereskia* clade actually branches into several species (not shown here), all of which lack many derived characters: These are large trees with ordinary broad, thin photosynthetic leaves, hard woody trunks, and no succulent tissue. These features are what we would expect in an ordinary eudicot tree; therefore, we think that many of these are ancestral characters, not derived ones. Pereskias are so similar to each other and so different from other cacti that they are grouped together as a subfamily, Pereskioideae.

Nodes 2 and 3 lead to *Maihuenia* and *Opuntia*. Plants in these clades are very similar to pereskias, but each has several derived characters. Plants in both are small, woody shrubs with small, cylindrical leaves that appear to consist of just a midrib without any lamina. Their cortex is thicker, with more water-storage parenchyma than that of pereskias. Opuntias differ from maihuenias and all other cacti in having hundreds of tiny bristle-like spines (called glochids) that are a synapomorphy distinguishing all opuntias from all other cacti. Maihuenias are distinguished from *Pereskia* and the opuntias by DNA sequence differences. Maihuenias (there are only two species) are placed in their own

subfamily Maihuenioideae, and the opuntias (there are many genera and species) are in subfamily Opuntioideae (prickly pear cacti and chollas).

Node 4 is a real surprise. One branch leads to *Blossfeldia*, a monotypic genus (there is only one species, *B. liliputana*) of plants that have many derived characters. They are tiny dwarfs only about 1 cm tall when mature. Their foliage leaves are reduced to just leaf primordia. They have stomata in their stem epidermis, and their wood has an unusual type of tracheid but no xylary fibers. Until recently, *B. liliputana* was placed up by node 10 among several other genera that have similar, highly derived features, but DNA sequencing by several people indicates that it is a sister to all of the cacti from node 5.

Figure B18-2b *Pereskia sacharosa,* spines are modified bud scales. Notice the large green foliage leaf.

Figure B18-2a Plant of *Pereskia sacharosa.*

Figure B18-2c *Maihuenia poeppigii.* This is a cushion plant, a short, extremely compact tree approximately 20 cm tall; this is a common adaptation to cold, windy alpine habitats. Yellow structures are fruits.

Cacti from node 5 are extremely variable, but all are more "cactus-like" than pereskias and maihuenias. These really look like cacti: They are succulent, have water-conserving and water-storing adaptations. Those at node 8 (*Leptocereus, Pachycereus,* and their relatives) still have many ancestral features, especially their hard, fibrous wood that has few adaptations for desert life. Taxa from node 6 are extremely derived with fewer ancestral characters: They tend to have spherical or short cylindrical bodies (which have a low surface-to-volume ratio that helps reduce water

Figure B18-2f *Blossfeldia liliputana.* This is a mature, full-grown plant, approximately 1 cm tall.

Figure B18-2d Tiny but obvious foliage leaves of *Maihuenia poeppigii.*

Figure B18-2g *Mammillaria lasiacantha* and many of its relatives are small, desert-adapted plants with many derived features that occur only in cacti.

Figure B18-2e *Austro-cylindropuntia schaferi* has leaves without a lamina; its wood is hard and fibrous, and it has little succulent tissue.

Figure B18-2h *Echinocactus horizonthalonius* is one of many species of barrel cacti in North America.

loss by transpiration). Many have specialized spines that shade the body, disguise it, or secrete nectar. Also, this clade occurs only in North America, not South America, and geographic distribution is often an important character (closely related species tend to be found in the same area, not scattered all across the world).

All clades from node 9 are predominantly South American; only a few individual species have spread to the Caribbean, Florida, Central America, or Mexico. *Hatiora* and its relatives are epiphytic, living only in rain forests, not a habitat we typically associate with cacti. Of the remaining clades, *Parodia, Gymnocalycium,* and their relatives are small, globular South American cacti that greatly resemble the North American cacti of node 6:

Figure B18-2j *Pachycereus pringlei* has a very thick cortex and is capable of storing large amounts of water. Its wood, however, is like that of *Pereskia*—fibrous and strong enough to support tremendous weight.

Figure B18-2i *Leptocereus weingartianus* is the shrubby, brushy cactus sprawling out of the undergrowth; its stems are narrow, with little succulent tissue, and it is highly branched, like an ordinary noncactus bush. The tall, upright cacti are *Pilosocereus polygonus.*

Figure B18-2k *Hatiora salicornioides* is an epiphytic cactus with tiny, narrow stems; it never becomes a large plant.

In fact, early taxonomists often classified them together. One of the original reasons for considering them distinct is that the two clades (node 6 and node 10) occur so far from each other, separated by thousands of miles of ocean and Amazon rain forest (both are significant isolation barriers). Later, certain flower, fruit, and seed differences were noticed, and more recently, DNA sequencing also supports the idea that North American small cacti and South American small cacti are remarkable cases of convergent evolution within the same family. The most recent common ancestor of node 6 and node 10 is node 5, and plants at that node must have lacked many derived features while having many relictual ones (shrubby, hard wood, little succulence).

This family has tremendous diversity. It contains trees, shrubs, and dwarfs (but no herbs—all still make at least a tiny bit of wood); succulents and nonsucculents; plants adapted to deserts, rain forests, grasslands, and alpine regions (some grow next to snowbanks); flowers pollinated by insects, bats, and hummingbirds. The monophyly of the family is not in doubt, however. All are universally accepted as being members of the same family-level clade.

Figure B18-2n *Oreocereus celsianus* are adapted to high, cold alpine conditions on the border between Bolivia and Argentina.

Figure B18-2l *Parodia maassii* are small globular South American cacti whose body size and shape are similar to those of *Mammillaria* because of convergent evolution.

Figure B18-2m *Gymnocalycium horstii* has obviously diverged greatly from its *Pereskia*-like ancestors.

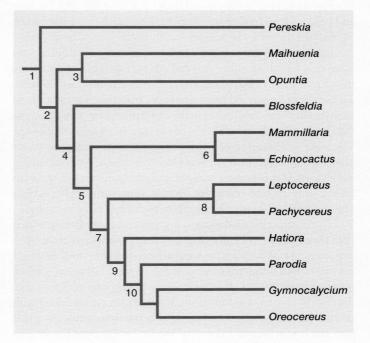

Figure B18-2o Simplified cladogram of several clades in Cactaceae.

of pigments requires many steps and many enzymes, and mutations in any of those genes could cause the same phenotype. The two sets of white flowers almost certainly have different genotypes. This is homoplasy. The original white-flowered lineage has diverged twice: After node 2, one group has trichomes on its leaves, and the other continues to have smooth leaves. At node 3, dwarfism originates in one line; at node 4 one group mutates to have narrow petals.

We can speak of many clades here: All of the species depicted, plus their common ancestor at node 1, form a clade. Species E and F plus node 5 are a smaller clade, as are the four species (A to D) that have descended from node 2. It is especially important to notice what is not a clade: There are now five species with white flowers (A, B, C, D, E), but we cannot classify E with the others—they do not form a clade. The **most recent common ancestor** of all white-flowered species is node 1, and a natural, monophyletic clade includes the ancestor and all of its descendants. If we would try to classify A–E as a group but leave out F, we would be creating a **paraphyletic group**, one that does not contain all the descendants of the most recent common ancestor (Table 18-3).

Think about what a systematist is faced with. There are six species, some with white flowers, some with trichomes on their leaves, some with a dwarf stature, and some with narrow petals. In constructing a cladogram, we search for shared derived characters, synapomorphies. Species C and D both have trichomes on their leaves, but no other species does; therefore, this is a synapomorphy that shows these two are closely related. Clades A–D share the synapomorphy of having small seeds. Species A is the only one that is dwarf: No other species has this feature, so it does not help us understand the phylogeny—it is **uninformative**. All species still produce fruits that are berries. This has not changed anywhere; this shared ancestral condition is a **symplesiomorphy**, and it too is uninformative for arranging these species.

Given these six species and these characters, a systematist could create any number of cladograms. Between node 1 and species A we could imagine many steps in which flower color changed from white to blue to yellow and then back to white, with the blue and yellow forms having gone extinct. This might have happened: We can see the six living species and their characters; however, we cannot see any of the ancestors (they are all long since dead), and there may have been other descendants that died out and thus are missing from our cladogram. Given the information we do have, we use a principle called **parsimony**: We prefer the simplest possible hypothesis, and we do not make a hypothesis any more complicated than it needs to be. Parsimony is used in all fields of biology. The most widely accepted theory of water movement in vascular plants is that water moves upward through xylem. We could imagine all sorts of much more complicated mechanisms, but we focus on the simplest one and move on to a more complex hypothesis only if data indicate that the simple hypothesis is not accurate.

In constructing cladograms, a great deal more data are used than those shown in Figure 18-5. For living species, we have DNA sequences in which each base pair is a data point, and thus, we might have dozens or hundreds of informative sequence differences when comparing six species. However, when a computer program uses the data to construct a cladogram, it almost always comes up with several that are equally simple—**equally parsimonious**—but that have the taxa arranged differently. When this happens, the various cladograms are analyzed to see if any clade appears repeatedly; if so,

TABLE 18-3 Cladistics

The diagrams in Figs. 18-5 and 18-6 are cladograms, branching trees that show the evolutionary lines and relationships of the existing species or genera being studied. The forking points in cladograms represent times when one group evolved into two distinct groups, usually with a new feature being present in one of the forks, the ancestral feature being present in the other fork. Many, perhaps most, studies of evolution now use cladograms, and there is a set of important terms that describe characters:

Plesiomorphies are characters that were present in a group's ancestors. They are relictual or primitive characters. For example, some cacti still have leaves like their ancestors had. The presence of leaves is a plesiomorphy in cacti.

Symplesiomorphies are ancestral characters shared by two or more modern groups. These are often called shared primitive traits. Symplesiomorphies are not usually helpful in analyzing evolutionary relationships. Most dicots have simple leaves, and the ancestor of flowering plants probably did too. The fact that many dicots now still have this symplesiomorphy does not indicate that all of them are closely related to each other.

Apomorphies are derived characters, new characters that were not present in the ancestors.

Synapomorphies are shared derived characters, that is, they are derived characters that occur in two or more modern groups because those groups are closely related. They share a recent ancestor that had this character, and they have inherited it. Synapomorphies are important characters we search for. The spathe and spadix inflorescence of aroids (Fig. 18-1) is a synapomorphy that indicates that all members in this family are closely related and form a natural group.

Autoapomorphies are unique derived characters. If a derived character occurs in only one group, it indicates that the group should be considered as a unit, but it does not help us understand what it is related to. The big ears of corn that we eat are autoapomorphies. They occur only in corn and indicate that we would be making a mistake if we combined corn (*Zea mays*) with some other species. Corn is a grass with many relatives, and if one other species had ears also, we would assume the two species are closely related, but no other species has this. The ears of corn are unique.

Homoplasies are the result of convergent evolution. They are characters that appear to be the same in two or more groups, but in fact they did not evolve from the same ancestral character. Plants called spurges in the genus Euphorbia resemble cacti but are not closely related to them. The similarities that fool us when we examine cacti and spurges are homoplasies. Trying to distinguish homoplasies from synapomorphies is always a concern in evolutionary studies. If we accidentally classify two unrelated groups together because we mistake a homoplasy for a synapomorphy, we have created an unnatural, polyphyletic group.

we have confidence that that particular clade might be an accurate depiction of the phylogeny. Sometimes there is simply not enough data, and then usually the various species are shown as arising from the same node. This is an **unresolved polychotomy**. These become the focus of more intense study.

As you study cladograms, keep in mind that they are hypotheses. Each is a model supported by some evidence, but we do not have all possible information about all species either living or extinct. DNA sequencing projects are progressing very rapidly and providing large amounts of data. These can be combined with morphological, anatomical, and physiological data to refine the cladogram and to understand not only how taxa are interrelated but also how the characters themselves have evolved.

Cladograms and Taxonomic Categories

How are names or taxonomic levels assigned to a cladogram? The only taxonomic unit with an objective definition is species: Individuals that can interbreed. But there is no such objective definition for genus or family or so on. It has been proposed that every node and every clade in cladograms be names, but that would result in an unacceptably large number of names. Even worse, most cladograms are still not stable: They change as new species and data are added. Many systematists have decided to continue using the old names for genera, families, and orders unless they are shown to be definitely polyphyletic or paraphyletic. This has resulted in several well-known families being combined into one or divided into two. Informal names are temporarily being used for certain groups until we have more confidence that we truly understand their phylogeny. An especially important example for us involves angiosperms. Until recently, we believed that all angiosperms were either monocots or dicots, that the "monocot/dicot" divergence occurred extremely early (**Figure 18-6**); however, now there is strong evidence that early angiosperms diverged into several clades before that happened, and the living descendants of those early-diverging clades are no longer called dicots. They are just called the **basal angiosperms** and have not been given a formal taxonomic name. The rest of the dicots are called **eudicots**, and this too is an informal name. The old word "dicot" has to be abandoned because the group it referred to is now known to be paraphyletic (it included basal angiosperms and eudicots but left out monocots, which share a common ancestor with eudicots).

Other Types of Classification Systems

Artificial Systems of Classification

The natural system of classification, which attempts to follow the evolutionary history of the organisms being categorized, is only one of several types of classification system. Another fundamental type is an **artificial classification system**, in which several key characters, often very easy to observe, are chosen as the basis of classification. Good examples are roadside floras and picture guides to plants, birds, and mammals of national and state parks. The botanical classifications in these are often based primarily on flower color: All plants with white flowers are grouped together, as are all those with red flowers, and so on. Within each category, the next classification category might be the plant's habit: Trees are grouped together, as are shrubs, herbs, vines, and so on. These systems are described as artificial because many plants in a category are not closely related to each other by descent from a common ancestor; furthermore, they are separated from their close natural relatives in other categories simply because they differ in flower color or habit.

Artificial systems typically have the goal of easy plant identification by means of obvious characters such as flower color and plant habit. Alternatively, an artificial system may be designed to group together plants with economically or scientifically important features. From a practical standpoint, carpenters and woodworkers are more interested in color, texture, grain, and hardness of a wood than its phylogeny. Gardeners might classify plants according to their ability to tolerate shade, full sun, frost, or alkaline soils.

Artificial classifications are only used as adjuncts to natural systems. As physiologists examine the phylogeny of all species with C_4 metabolism, they find that C_4 metabolism has evolved several times; therefore, all C_4 metabolisms are not expected to be identical. Similarly, someone studying the metabolism of petal pigmentation might classify flowers artificially according to color, deciding to investigate the synthesis of red pigment first. But there are two very different types of red pigment: anthocyanins in most angiosperms and betalains in the order Caryophyllales (beets, cacti, portulacas). If some scientists were working with red rose flowers and others were studying red beets, they would get inexplicably conflicting results about the enzymes and pathways involved in pigment synthesis.

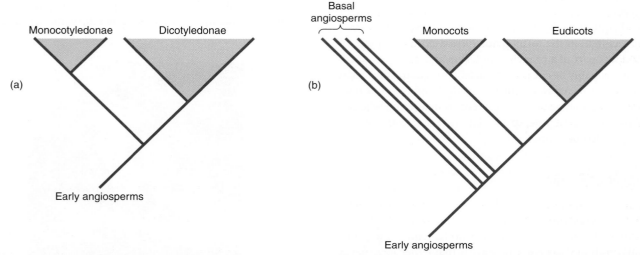

Figure 18-6 (a) A cladogram that represents the historical view that all angiosperms were either monocots or dicots. These had the formal taxonomic rank of class: Monocotyledonae and Dicotyledonae in some books, Liliopsida and Magnoliopsida in others (Table 18-1). (b) This very simplified cladogram summarizes a current hypothesis of evolutionary relationships within angiosperms. The "basal angiosperms" diverged from early angiosperms before the monocot/dicot split. All basal angiosperms had been known as dicots, as had all eudicots, but that concept of dicot would be a paraphyletic group, as it would leave out some descendants (the monocots) of the early angiosperms. We do not use paraphyletic clades, and therefore, we need a new name for the basal angiosperms (they are usually just called basal angiosperms) and for the dicots to the right of monocots (they are now called eudicots).

Box 18-3 Genealogy Versus Clades: Your Family History Is the Opposite of a Clade

It is easy to confuse cladistics with genealogy. When tracing your family tree, you consider your parents, their parents, the next set of parents, and so on into your past. You are trying to locate your past relatives, but the real point of a genealogy is to map out all the people who might have contributed alleles to your own genome. You received all your alleles from your two parents, but those alleles came from your four grandparents, your eight great-grandparents, and your 4,096 great-, great-, great-, great-, great-, great-, great-, great-, great-, great-grandparents, and so on. At each generation you are considering the potential input of new alleles, and also you double the number of ancestors you are considering. Your family tree becomes broader as you go farther back.

A cladogram is just the opposite. We are interested in one single set of ancestors or founder individuals and how their progeny change with time. The **ancestral group** is genetically isolated from all other groups (remember, only members of the same species can interbreed), so their only source of new alleles is mutations within the descendants. A cladogram attempts to map out the origins of new alleles or of important new combinations of alleles that lead to detectable changes in phenotype.

A genealogy maps the genetic convergence of characters from many ancestors into one descendant over a few generations. A phylogeny maps the evolutionary divergence of characters and progeny over thousands or millions of generations.

◼ Classification Systems for Fossils

A third type of classification, used for fossil organisms, combines features of both artificial and natural systems. The goal is to understand the evolution of the fossil and to identify both its ancestors and its relatives that might have later evolved into other species. This requires a natural system. Typically, we do not know much about the fossil; thus, superficially similar ones are grouped together for convenience—a basically artificial system. The groupings are **form genera**: All fossils with the same basic form or structure are classified together. For example, a piece of fossil wood similar to the wood of modern pines, spruces, and larches is classified in the form genus *Pityoxylon*. If the piece of fossil wood was part of a branch with leaves and cones attached, there would probably be enough characters present to allow us to assign the wood to a natural genus, but if the fossil contains only wood, not enough characters are present to determine whether it came from an ancient pine, spruce, larch, or some other group that has since become extinct.

◼ Taxonomic Studies

People have been hunting for and discovering new plant types since before recorded history to find not only new food plants but also ornamental plants and beautiful woods. Exploration continues today as biologists enter new areas that either have never been explored or have been visited only once or twice. Tropical rain forests of Brazil, Central America, Africa, and Southeast Asia are so poorly explored that any serious expedition returns with previously unknown species of plants and animals. Some plants are less than 1 cm tall when fully grown and have an extremely limited range; it is easy to walk past them without noticing them (**Figure 18-7**). On the other hand, some large plants are too big to overlook, but they grow in only one small area and can be missed easily by explorers. Finally, a species may resemble others so closely that a scientist may ignore it, thinking it is already well known; the fact that it is a new species might be revealed only by careful study of certain details.

Botanical exploration is also concerned with discovery of new facts about already recognized species. No species is made up of absolutely identical individuals; therefore, numerous samples

Figure 18-7   This *Lipanthes* orchid is fully grown; even in full flower, it could be overlooked easily. The photo is twice life size.

must be collected to gather information about the variability of its features as well as its geographical and ecological range. Exploration may be dedicated to gathering either seeds or live material so that the plant can be propagated for research, horticulture, or food. Now that recombinant DNA techniques make it possible to transfer genes from one species to another, live preservation of as many species as possible is particularly important.

Preliminary studies of newly collected plants include many diverse activities. Specimens must first be identified using diagnostic keys and personal knowledge. An experienced taxonomist often recognizes the family of an unfamiliar species almost immediately; even the genus may be obvious. If so, further identification is usually simple. In an unfamiliar region in which many species are unknown, identification can be more difficult, and numerous characters must be studied and compared with those in published descriptions of the plants of the area.

If the plant can be identified, it may be studied to see if it provides new information about the species. Even if it turns out to be absolutely typical in every way, that too is important. If the specimen cannot be identified (**Box 18-4**) or if it does not match any description, the taxonomist may suspect that it is a new species. Typically, the specimen is then sent to a specialist experienced with the species most closely related to the specimen.

Declaring a plant to be a new species is easy; proving it and describing it properly are difficult. Anyone can declare a plant to be a new species, name it, and write about it, even if it is obviously not a new species; this has caused countless problems. To overcome this, taxonomists from all over the world have established an *International Code of Botanical Nomenclature* that describes precisely the steps necessary for naming a new species. A valid name, one never previously used, must be declared and must be accompanied by a detailed description of the species in Latin and usually also in English, French, German, Spanish, or Russian. The name and description must be published in a widely circulated journal, a step that prevents many problems. The journal's editors send the description to at least two independent specialists to verify that it is a previously unknown species and that the name has never been used before. The description must also include the designation of a **type specimen**; this is a single preserved plant that truly carries the name.

When new species are named, very little is known about them; as more research is carried out in the following years, enough variation may be discovered to warrant the recognition of a second species. For example, the range of leaf sizes, types of trichomes, and types of nectaries may warrant classification into two or three species. The original type specimen determines which type of leaf size, trichome, and nectary goes with the first name. When doubt arises as to which type of plant goes with which name, the type specimen must be checked. Type specimens are often kept in special fireproof cabinets and are not allowed out of the herbarium. During World War II, Allied bombing destroyed the herbarium in the Berlin Botanic Garden, and thousands of type specimens burned to ashes. Germans had been very active in plant exploration during the first part of that century and much of the 1800s, and the loss of those type specimens has caused tremendous problems. To prevent a recurrence of this disaster, other specimens, as similar as possible to the type specimen, are sent to many herbaria around the world; these are **isotypes**.

The Major Lines of Evolution

All organisms are grouped into three domains: Bacteria (with cyanobacteria), Archaea, and Eukarya (**Figure 18-8**). The most significant event in evolution was the origin of life itself, probably about 3.5 billion years ago. The first organisms were simple, consisting of a cell membrane, protoplasm, and some means of inheritance, probably RNA; they almost certainly had no distinctive nucleus or other membrane-bounded organelles. Such organisms, either living or extinct, are prokaryotes. This line of evolution diversified rapidly into numerous clades, as evidenced by the presence of thousands of living species of bacteria, cyanobacteria, and archaeans. A significant step in prokaryote evolution was the development of a type of photosynthesis that liberates oxygen and is based on chlorophyll *a*. We do not know exactly which species of bacterium this occurred in, but their descendants are known as cyanobacteria, a clade within Bacteria, and one was the ancestor of chloroplasts.

The next major evolutionary event was the conversion of a prokaryote into a eukaryote, having a membrane-bounded nucleus. This must have been an extremely gradual procedure with many intermediates because many living species still have characteristics intermediate between prokaryotes and eukaryotes. A significant aspect was the origin of mitochondria; this was a gradual process in which a bacterium capable of aerobic respiration began living inside the protoplasm of an early eukaryote whose own capacity for aerobic respiration was less sophisticated. Both organisms would have benefited from their association; thus, this was a symbiosis, not a parasitism, and because one lived inside the other, it was an endosymbiosis. This is the hypothesis of the endosymbiotic origin of mitochondria, and there is a great deal of evidence to support it: Mitochondrial DNA, genes, and enzymes are more similar to those of certain bacteria than they are to the nuclei of the cells in which they are located. The most parsimonious (simplest) cladograms of mitochondrial genes are those that include bacterial genes as the early ancestral genes;

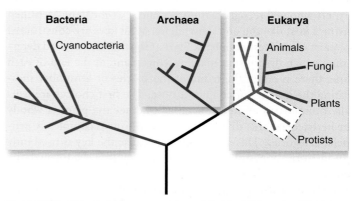

Figure 18-8 This cladogram represents our latest hypotheses about the interrelationships of the three domains: Bacteria, Archaea, and Eukarya.

trying to include eukaryotic genes into a mitochondrial gene tree requires many more nodes and steps and is much more complicated (less parsimonious).

Organisms with an organization similar to that of early eukaryotes were formerly classified in a "kingdom Protista," but that classification was never satisfactory because it included members that were extremely diverse and obviously not closely related to each other. The criterion for classifying organisms in kingdom Protista was simply that they were eukaryotes but lacked the derived characters of plants, animals, or fungi; however, classification is based on shared derived characters (synapomorphies), not on a lack of them. Actually, the old classification of protista was a **grade classification** (as opposed to a clade): "Protistans" were placed together because they had a low level (low grade) of evolutionary advancement. Similarly, ferns and other plants that reproduce without seeds used to be classified together as the grade "Cryptogamae." We no longer use classifications based on lack of characters, but the old names are often convenient; therefore, we use them informally, never capitalized, and occasionally in quotation marks to emphasize that "protists" do not form a clade.

After eukaryotes evolved to the level of having mitochondria, endoplasmic reticulum, and a true nucleus, numerous evolutionary lines emerged. One clade contained organisms that would later diverge into animals and fungi. Surprisingly, DNA studies are consistent with the hypothesis that fungi are more closely related to animals than they are to plants, despite superficial similarities such as lack of movement and reproduction by spores. The other early clade established an endosymbiosis with cyanobacteria, which then evolved into chloroplasts, producing the first algal cells. Just as with mitochondrial genes, phylogenies of plastid genes are most parsimonious if they go back to cyanobacterial genes instead of trying to include nuclear genes. The early algae continued to diversify, and approximately 400 million years ago, some became adapted to living on land, establishing the clade of true plants, **kingdom Plantae**. From these early pioneers, the major evolutionary lines that diverged were simple plants with neither seeds nor vascular tissues (mosses, liverworts, and hornworts), plants that do not produce seeds but do have xylem and phloem (ferns and similar plants), and the seed-bearing vascular plants (gymnosperms and angiosperms). The plants around us today represent thousands of evolutionary lines descended from early land plants, which in turn were derived from algae. Our classification system does not yet have every step identified, and not all relationships are clearly understood; however, the major features appear to be consistent with the data, and we expect that only relatively minor changes will be made as our knowledge increases.

Plants and People

Box 18-4 Identifying Unknown Plants

When a plant must be identified, it can either be sent to an expert who will recognize it immediately, or it can be identified using a **key**, like the one shown here. Most keys are constructed of pairs of choices (couplets). Start at the first choice, and determine which description matches your plant; at the end of each line, there is either the name of the species or a number indicating the next set of couplets to use. The first choices are usually quite broad and easy, such as whether the plant is woody or herbaceous; this divides the plants into two completely artificial categories. This is acceptable because the key is used only for identification, not classification. As you proceed through the choices, you finally come to a name, meaning that you have tentatively identified your plant. The key should be accompanied by descriptions of all the species and, if possible, drawings or photos. After arriving at the name in the key, you should turn to the description to see whether it matches.

In this key, I have incorporated only the most common plants found in grocery stores, so you may try to identify a plant that is not in this key. If the plant does not resemble any plant in this key, you will soon run into a couplet in which neither description seems appropriate. In real life, this could mean that you have discovered a new species or a known species growing in an area where it had not been known to exist. Unfortunately, it most often means that you have made a mistake somewhere in the keying process.

Keys in floras or monographs usually are constructed in stages; a preliminary, small key identifies the plant to family or genus, and subsequent keys identify it further to species. As you go to each new key, you can check the description of the family or genus to see whether your keying has been correct to that point. With some experience and knowledge of the plants of an area, most people can recognize the family immediately and per-

haps even the genus; therefore, they can skip the first keys and go directly to the appropriate level. The key shown here is really three keys: The first distinguishes between fruits and vegetables, and the other two are separate keys for fruits and vegetables; most of you should be able to identify your specimen as a fruit or a vegetable without having to go through a key. On the other hand, mushrooms present a problem. You do not know how I would treat a mushroom—it is reproductive like a fruit but has not developed from an ovary or associated structure; you have to use the first key for peculiar specimens like this.

As you use the key, keep in mind that you may have selected a plant that is not available in my store, so it will not be in the key.

Key to Fruits and Vegetables Commonly Found in Grocery Stores

13. Leaves are tubular, base of plant is white with numerous roots . . . . onion
Allium cepa

13. Leaves are flat, not tubular; without roots when sold. 14

    14. Leaves are thin and flat, forming either a loose or a semicompact mass . . . . .lettuce
Lactuca sativa

    14. Leaves are thicker, fleshy, forming a very compact mass, never loose . . . . .cabbage
Brassica oleracea var. *capitata*

15. Vegetable is composed mostly of stem, with or without small scales. 16

15. Vegetable is composed mostly of petioles 17

    16. Stem is green, with small scales, grows above ground . . . . asparagus
Asparagus officinalis

    16. Stem is brown or tan or reddish, subterranean. . . . . .potato
(if orange with orange flesh, this is a root, yam)
Solanum tuberosum

    17. Vegetable is almost entirely thick green petioles with just a small amount of
highly dissected lamina . . . . celery
Apium graveolens

    17. Vegetable is mostly white petioles with some green; lamina entire . . . . .bok choy
Brassica chinensis

Fruits, Including False Fruits

18. Fruit is considered fruit even by most lay people, mostly sweet and edible raw; some sour, acidic 19

18. Fruit is not sweet, commonly considered vegetables by lay people 34

    19. Fruit is sour, very tart 20

    19. Fruit is edible even when fresh, but also used in cooking; many are very sweet 22

        20. Fruit is a compound false fruit, yellow-brown with many scales; usually sold with leaves attached to top. . . . pineapple
Ananas comosus

        20. Fruit is simple, not compound, not false fruit. 21

            21. Fruit is yellow . . . . lemon
Citrus limon

            21. Fruit is green . . . . .lime
Citrus aurantifolia

    22. Fruit has an inedible rind or peel that must be removed before the endocarp is eaten 23

    22. Entire fruit is often eaten, although some people may peel some of these. 29

        23. Fruit is small, each one is usually consumed by one person 24

        23. Fruit is larger, not customarily eaten by one person 26

            24. Fruit is longer than wide, a curved yellow cylinder . . . . banana
Musa paradisiaca

            24. Fruit is spherical or slightly flattened 25

                25. Fruit is orange. . . . . .orange
Citrus sinensis

                25. Fruit has short brown hairs. . . . . kiwi fruit
Actinidia chinensis

        26. Rind is extremely hard, like rock . . . . coconut
Cocos nucifera

        26. Rind is not so hard, can be cut by knife. 27

        27. Fruit is very large, longer than wide; flesh is red with black seeds . . . . watermelon
Citrullus vulgaris

27. Fruit is smaller, more or less spherical, flesh never red 28
    28. Rind is smooth, flesh green . . . . honeydew melon
                        Cucumis melo
    28. Rind is netted, flesh orange . . . . cantaloupe
                        Cucumis melo var. *cantalupensis*
29. Fruits are sold as clusters of fruits attached to infructescence stalks . . . . grapes
                        Vitis vinifera
29. Fruits are not attached to each other at time of sale 30
30. Fruits are blue, small, round . . . . blueberries
                        Vaccinium corymbosum
30. Fruits are not as above 31
    31. Fruits are red false fruits with "seeds" on outer surface . . . . strawberry
                        Fragaria chiloensis
    31. Fruits are not as above 32
        32. Fruit is red, soft, juicy, not sweet, with many seeds . . . . tomato
                        Lycopersicon esculentum
        32. Fruit is not as above 33
            33. Fruit is variable, typically yellow/brown and pear-shaped but may be greenish and round; always with small nests of stone cells that give the flesh a gritty texture . . . . pear
                        Pyrus communis
            33. Fruit is variable, often similar to above, but not pear-shaped and never with nests of stone cells; apex of fruit (opposite stalk) typically retains five sepal tips and often has withered stamens also . . . . apple
                        Malus sylvestris
34. Fruit is not much longer than wide, spherical or oval 35
34. Fruit is more than three times longer than wide 38
    35. Fruit is hollow . . . . bell pepper
                        Capsicum annuum
    35. Fruit is not hollow 36
        36. Fruit is green, solid, oily with one large brown seed . . . . avocado
                        Persea americana
        36. Fruit is not as above, is red or purple 37
            37. Fruit is red . . . . tomato
                        Lycopersicon esculentum
            37. Fruit is purple . . . . eggplant
                        Solanum melangena var. *esculentum*
38. Fruit has many prominent seeds and a cob . . . . corn
                        Zea mays
38. Fruit is without a cob 39
    39. Fruit is narrow, flattened, less than 1 cm wide . . . . string bean
                        Phaseolus vulgaris
    39. Fruit is much wider than 1 cm 40
        40. Fruit is yellow . . . . summer squash
                        Cucurbita pepo var. *melopepo*
        40. Fruit is not yellow 41
            41. Fruit has small white, raised points; stalk less than 0.5 cm broad . . . . cucumber
                        Cucumis sativus
            41. Fruit has smooth green and yellow skin, without raised points, stalk at least 1 cm broad . . . . . . zucchini
                        Cucurbita pepo var. *medullosa*

Summary

1. One of the goals of classification and nomenclature is to give each species a single unique name.
2. The classification system being developed at present is a natural one, attempting to reflect the actual evolutionary relationships—the phylogeny—of all species.
3. Artificial classification systems are now used only to identify plants or to categorize useful features but are never the basis for naming species.
4. Closely related species are classified in one genus, related genera are grouped into families, and so on through orders, classes, divisions, and kingdoms.
5. Each taxonomic category should be monophyletic, all of the organisms in it having evolved from the same ancestral group. Taxonomic groups discovered to have had two or more distinct origins are the result of convergent evolution; they are polyphyletic and the category is not a natural one; its members must be reclassified into monophyletic groups.
6. Homologous characters (synapomorphies) are those that have evolved from the same ancestral character, whether they now are similar to each other or not; these can be used as guides to phylogeny and the construction of a natural classification system.
7. Analogous characters (homoplasies) are those that resemble each other but have arisen independently; they are a part of convergent evolution and often cause confusion and misclassification because they appear to be homologous characters.
8. When a new species is declared, it must be given a name never before used, and it must be accompanied by a type specimen and a complete Latin description published in a widely circulated journal. The internationally recognized rules are occasionally not followed, leading to an invalid name.
9. All organisms are classified as Bacteria, Archaea, or Eukarya.

Important Terms

analogous features
ancestral condition
apomorphy
artificial classification system
basal angiosperms
binomial system of nomenclature
clade
cladistics
cladogram
class
common ancestor
division
equally parsimonious
eudicots
family

form
genus
grade classification
homologous features
homoplasies
inheritance of acquired
    characteristics
isotypes
key
kingdom
monophyletic
natural system of classification
node
nomenclature
paraphyletic group

parsimony
phylogeny
polyphyletic group
scientific names
species
species epithet
symplesiomorphy
synapomorphies
systematics
taxon
type specimen
uninformative
unresolved polychotomy

Review Questions

1. What is the name of the science of giving things names?
2. One of the critical goals of biological nomenclature is to provide each species with a _____ name. After the discovery of evolution by natural selection, taxonomists realized that the most scientifically valid name for each species would be one that reflected _____ _____.
3. Imagine that two completely unrelated species in very different families were accidentally given the same name. What kind of confusion would this cause? What if both were separately the subject of several scientific studies, and results were published in papers that did not mention the family name. Would it be possible to know which study concerned which species?
4. At the end of the 19th century, taxonomists adopted the goal of developing a natural system of classification. In a natural system, which kinds of organisms are classified together?
5. Originally, the main task of taxonomists was to discover new species and give them unique names. In a natural system of classification, they must also figure out the evolutionary relationships of the species they discover. Which do you think is more difficult: the original task or the modern one? Describe some of the reasons why it is more difficult.
6. Imagine trying to identify plants in your region by using a roadside flower guide. This type of a classification system is not a natural system. Instead, it is an _____ classification system.
7. For fossils, we often must use a mix of artificial and natural classification. What is a form genus? If you found a specimen of fossil wood—but only wood, with no leaves or cones or bark—and it looked like wood of modern pines, what genus would you classify it in?

8. Theophrastus is often called the father of botany. When did he live? He had two very famous teachers. Who were they?

9. What is the book *Materia Medica*? Who wrote it and when? What kind of information does it contain?

10. In the 14th and 15th centuries, explorers brought to Europe information and specimens from Africa, the Americas and Asia, and this led to the realization that Pliny and other ancient Greek and Roman authorities were incorrect in many of their writings. Why was this significant? What did this realization lead to?

11. Linnaeus is well known for inventing the binomial system of nomenclature. In this system, each species has a _____ name and a _____ name.

12. Of all taxonomic categories, only species has an objective definition. What is it?

13. Closely related species are grouped together into _____ (singular, _____). By the way, what is the singular of "species"?

14. Study Table 18-1 and then, without looking at it, write all of the categories in order. Then write all the group names for Texas bluebonnet and for spider lily. Another bluebonnet, in the same genus as Texas bluebonnet, is Chisos bluebonnet, *Lupinus harvardii*. Write all of the group names for this species. Only one single word differs between it and Texas bluebonnet.

15. A critical concern of systematics is that the genera are _____, that all the species included in the genus are related to each other by a common ancestor, and that (circle one: all, most, at least half) of the descendants of that common ancestor are in the same genus. Such a genus is (circle one: polyphyletic, monophyletic).

16. In an unnatural, _____ group, members have evolved from different ancestors and may resemble each other only as a result of convergent evolution.

17. Give definitions for synapomorphy, symplesiomorphy, and homoplasy.

18. Sometimes two species appear similar because of convergent evolution not because they are closely related. Usually, it is safe to assume that if two plants do not resemble each other, they are not closely related. Is that always safe? Can two closely related plants be very dissimilar (Hint: Figure 18-3)?

19. Think about the apples we eat. They are in the genus *Malus* which is a member of the rose family, Rosaceae, along with roses, peaches, cherries, and even strawberries plus many other familiar plants. Almost all members of this family are trees or shrubs, and we are certain that the ancient ancestor was also a tree. If we are correct, is woodiness an apomorphy or plesiomorphy for the rose family? Being herbaceous is something new and rare in this family. Is it a plesiomorphy or an apomorphy? Apples, strawberries, and most other members of Rosaceae do not have a pit, but peaches and cherries do. We think this is a new character that originated just once, not in the ancestor of all the rose family but just in the ancestor of peach, cherries, apricots, and plums. If we are correct, is having a pit a symplesiomorphy, a synapomorphy, or a homoplasy? Rose bushes have spines and so do cacti, but the two families are not closely related. The spines of roses did not evolve from the same ancestral structure that cactus spines evolved from. The spines in these two families are a _____ and do not indicate that the families are closely related.

20. Have you tried using the Plants and People Box 18-4 to key out plants? Couplet 6 is used to distinguish between beets and turnips and relies on a character that is very easy to assess. What character is it? Couplet 4 distinguishes between carrots, beets, and turnips, even though only carrots are mentioned in couplet 4. If the plant being identified is a beet, what do you need to do at couplet 4 and in the next steps? Onions appear twice—first in couplet 11 and then in couplet 13. Why might onions be confusing enough to need to appear in two different places?

21. What are type specimens? Why are they important?

22. What is a monograph?

BotanyLinks

http://biology.jbpub.com/botany/5e

BotanyLinks contains a variety of resources and review material designed to assist in your study of botany.

Algae and the Origin of Eukaryotic Cells

Concepts

Life originated over three billion years ago as simple prokaryotic organisms (**Figure 19-1 and see the timeline on the inside of the back cover**). Some time later, perhaps less than a billion years after that, eukaryotic cells had come into existence. At first, these too were extremely simple, probably having the rudiments of a membrane-bounded nucleus, endoplasmic reticulum, microtubules and so on, but they did not have mitochondria or plastids, at least not initially. Mitochondria arose first, as an early eukaryotic cell engulfed but did not digest a bacterium capable of aerobic respiration. The two organisms lived together, one inside the other, and both benefited: This is an **endosymbiosis**. Over hundreds of millions of years, this gradually evolved into a more sophisticated form of eukaryotic cell with real mitochondria, and as it did so, it diverged into many species. Some of these species died out rather quickly; others were more successful and persisted, also diversifying into many forms. Currently, there are numerous groups of organisms, often called **protozoans** and **algae** (**Figure 19-2**)

Chapter Opener Image: What does a whale have to do with algae? Whales eat microscopic animals called krill, and krill eat algae. Without algae, there would be no whales, and for that matter, there would be no other marine animals: algae are the basis of the food chain in the oceans, as well as in most rivers and lakes. The bodies of algae are organized differently than those of plants, so algae are considered to be close relatives of plants but not actually plants. Like plants, almost all algae are photosynthetic (there are parasitic algae) and they produce a significant amount of the oxygen we breathe. Algae are the ultimate food of all marine animals. Many algae are unicellular or have very simple multicellular bodies that have not changed much in millions of years, so they are often considered to be "primitive." But true plants have existed for more that 420 million years, and none has ever become as well adapted to living in the oceans as algae are.

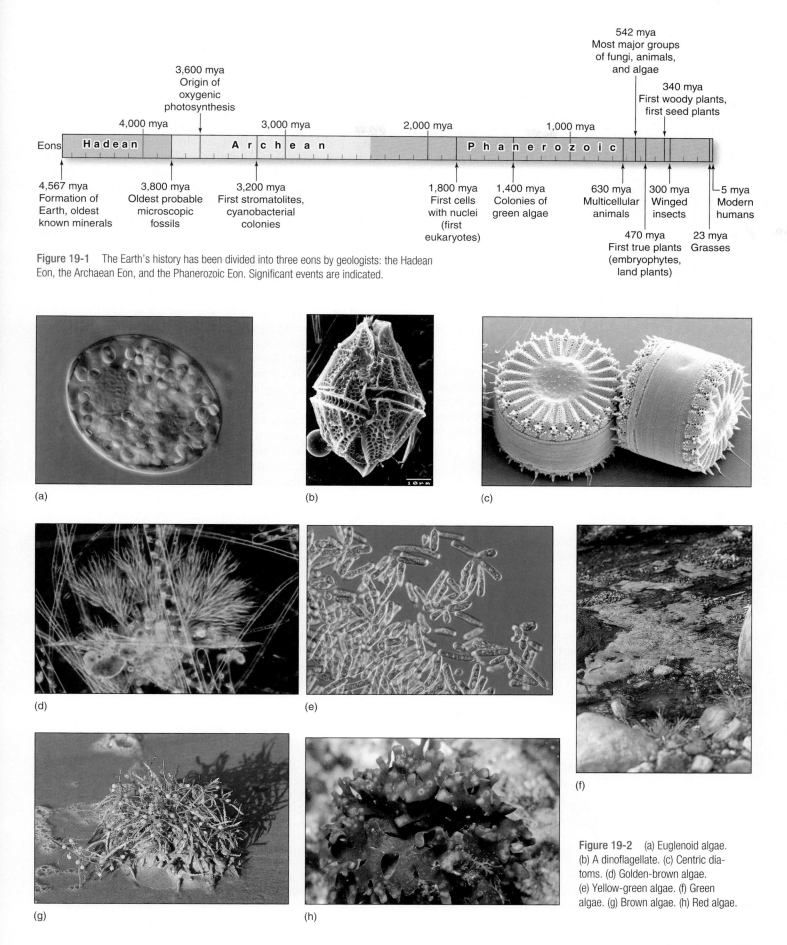

Figure 19-1 The Earth's history has been divided into three eons by geologists: the Hadean Eon, the Archaean Eon, and the Phanerozoic Eon. Significant events are indicated.

Figure 19-2 (a) Euglenoid algae. (b) A dinoflagellate. (c) Centric diatoms. (d) Golden-brown algae. (e) Yellow-green algae. (f) Green algae. (g) Brown algae. (h) Red algae.

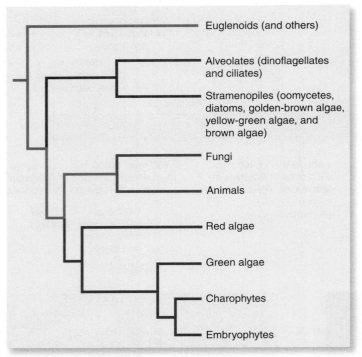

Figure 19-3 One recently proposed cladogram of the phylogeny of algae. The organisms that we call algae are parts of numerous clades, but they do not make up one single monophyletic clade. The ancestor that is common to all algae is also common to animals, fungi, and plants: It is the ancestor of all eukaryotes. Consequently, we use "algae" as an informal name.

that are the living descendants of clades that originated from those early steps in eukaryote evolution. Eventually, one of those would establish a clade that much later diverged into animals and fungi—multicellular organisms that have nuclei and mitochondria but not plastids (**Figure 19-3**).

Another line of early eukaryotes, ones that did have mitochondria, entered another endosymbiosis, this time with a photosynthetic cyanobacterium, which later evolved into a chloroplast. This line gave rise to algae, including green algae which in turn produced true plants—**embryophytes**. As you can imagine, numerous modifications are needed to convert an undigested

cyanobacterium into the chloroplasts in angiosperm leaves. Fortunately, during this long process, several clades originated that still have some extant members whose plastids have numerous prokaryote characters; chloroplasts of red algae especially resemble cyanobacteria.

We are fortunate that many organisms still exist that have many ancestral characters, characters that we believe were present in early eukaryotes. These are almost always unicellular and in the past were grouped together in a kingdom called "Protista," but this was a grade level of classification: They were grouped together because they were considered to represent a low grade, a low level of evolution: They lacked derived characters of plants, animals, and fungi. We no longer group organisms on the basis of what they lack; therefore, "Protista," the formal name, has become "protista," the informal name. More importantly, however, DNA sequencing studies, corroborated by electron microscopy and other techniques, are revealing the phylogenetic relationships of these organisms (Figure 19-3).

This chapter focuses on the origin and diversification of photosynthetic eukaryotic cells. This includes the various types of algae, especially the green algae, which are closely related to the ancestors of true plants, the embryophytes. Reproductive structures are the critical factors in distinguishing algae from plants. Algae are defined as photosynthetic individuals whose reproductive structures are completely converted to spores or gametes that, when released, leave nothing but empty walls (**Figure 19-4**). In plants, the reproductive structures are always complex and multicellular, and only some of the inner cells become reproductive. Outer cells constitute a protective layer and persist after the reproductive cells are released. This may seem trivial, but the algal method of reproduction is a cellular process, reflecting the simple, cellular level of organization most species have throughout their bodies. Reproductive structures of plants involve two tissues—reproductive and sterile—functioning together as an organ; this reflects a more complex organ level of integration in the entire body.

Origin of Eukaryotic Cells

The differences between prokaryotes and eukaryotes are deep seated, involving many of the most basic metabolic processes and cellular organization (**Table 19-1**).

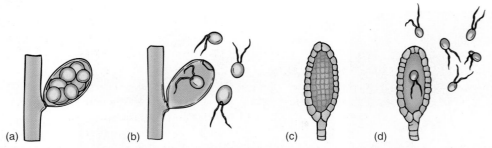

Figure 19-4 Algae are distinguished from plants (embryophytes) primarily by the nature of the reproductive structures. (a and b) In algae, the sporangium or gametangium may be either one large cell, as shown here, or multicellular, but every cell is converted into a reproductive cell. When reproductive cells are released (b), only the wall of the original cell is left. (c and d) Sporangia and gametangia of plants are always multicellular and only the inner cells differentiate into spores or gametes; after they are liberated (d), a residual layer of sterile cells remains.

TABLE 19-1	Differences Between Prokaryotes and Eukaryotes	
Character	Prokaryotes	Eukaryotes
DNA		
form	small circles	chromosomes
introns	rare	common
histones	absent	present
Nucleus		
nuclear envelope	none	present
nucleolus	none	present
true nucleus	none	present
Mitosis		
spindle	none	present but variable
centromeres	none	present
Meiosis	none	present in most
Gene exchange	conjugation or none	sex in most
Membranous organelles	none or lamellae in some	plastids, mitochondria, endoplasmic reticulum, dictyosomes, vacuoles
Ribosomes	70S	80S cytoplasmic 70S in organelles
Flagella	not 9 + 2	9 + 2

DNA Structure

In prokaryotes, DNA is "naked," not complexed with proteins, its numerous negative charges being neutralized by calcium ions instead. Each genome is just a short circle of DNA, often containing only approximately 3,000 genes. Prokaryote DNA contains few introns, and mRNA processing is relatively simple.

In eukaryotes, DNA is more elaborately packaged, being complexed with nucleosome histones and forming chromosomes. Each DNA molecule is long, capable of carrying thousands of genes. Nevertheless, no genome ever consists of just a single DNA molecule; at least two occur per haploid set, and 10 to 20 are more common. Many eukaryotic genes contain introns, and large segments of DNA do not code for any type of RNA.

Nuclear Structure and Division

The DNA of prokaryotes lies directly in the cytoplasm; no nucleus is present. As a prokaryotic cell grows, its plasma membrane expands; the DNA circles attached to it are separated from each other by cell growth. The cell then pinches in two.

The nuclei of plants are virtually identical to those of animals and fungi in structure, metabolism, mitosis, and meiosis. Apparently these three clades diverged only after eukaryotes and their nuclei had become quite sophisticated; however, nuclei are complex organelles, and mitosis and meiosis are intricate processes with many steps. Certainly plant and animal nuclei could not have arisen from prokaryotic nucleoids quickly; many intermediate steps spanning hundreds of millions of years must have been involved.

Most of the DNA of eukaryotes is located within a nucleus, separated from the cytoplasm by two nuclear membranes with nuclear pores; this is the source of their name (eukaryote = true nucleus). A nucleolus is present as well. Eukaryotic nuclei are typically haploid or diploid, having only one or two copies of each of their various chromosomes, and thus, the rather haphazard prokaryotic method of simple membrane expansion would not work reliably. Instead, mitosis ensures that each progeny cell receives one of each type of chromosome. In almost all eukaryotes, meiotic nuclear division occurs as part of sexual reproduction. The critical phenomena are pairing of maternal and paternal homologous chromosomes followed by crossing-over and genetic recombination. These complex procedures result in an interchange of alleles, such that a great diversity of spores or gametes is produced.

Several groups of organisms with unusual nuclear characteristics may represent lines of evolution that originated earlier in the history of eukaryotes than did plants and animals. Some have both an intranuclear and an extranuclear spindle, and in some, the nuclear envelope and nucleolus either do not break down at all or do so only late in mitosis. Gaps may form in the nuclear envelope, and bundles of microtubules pass completely through the nucleus.

Nuclei of dinoflagellates have no histones. They do have what appear to be chromosomes, but these remain condensed at all times, undergoing only a slight uncoiling during interphase.

Organelles

Prokaryotes lack membrane-bounded organelles: Their cytoplasm is rather homogeneous, containing only ribosomes and storage granules. A few species have gas vacuoles, and photosynthetic species have folded membranes that project into the cell. All eukaryotes have nuclei and mitochondria, and plants and algae also have plastids. In addition, dictyosomes, endoplasmic reticulum, vacuoles, and vesicles are components of the eukaryotic endomembrane system. These membranes result in the division of the cell into numerous compartments, permitting several types of highly specialized metabolism to occur within the same cell. Ribosomes of prokaryotes are 70S, being smaller and denser than the 80S ribosomes in the cytoplasm of eukaryotes. Plastids and mitochondria also have 70S ribosomes instead of 80S ribosomes.

Flagella and cilia are remarkably uniform in eukaryotes, consisting of a 9 + 2 arrangement of microtubules. Only a few prokaryotes have flagella, and these have a totally different type of construction, never a 9 + 2 arrangement. They are not composed of microtubules or tubulin.

Origin of Mitochondria and Plastids: The Endosymbiont Theory

Prokaryotes represent the types of organisms that arose first. Until the early 1970s, it was assumed that some had given rise to eukaryotes by gradually becoming more complex and developing an endomembrane system, parts of which eventually specialized so much they became mitochondria and plastids. This was the **autogenous theory**.

In the 1960s, the **endosymbiont theory** was revived. In 1905, K. C. Mereschkowsky had speculated that plastids and mitochondria might be prokaryotes living inside eukaryotic cells. At that time, there was no way to test the concept, but by the late 1960s, plastids and mitochondria were discovered to have their own DNA and ribosomes, both having prokaryotic features. Furthermore, these organelles divide by a type of cleavage similar to that of prokaryotes, and they also lack microtubules. Before long, other prokaryotic traits were discovered: Their DNA is a small closed circle of naked DNA, and organelle genes are organized like prokaryotic genes. Their ribosomes are sensitive to the same antibiotics that interfere with prokaryotic ribosomes.

Origin of Mitochondria The endosymbiont theory postulates that a prokaryotic cell had evolved to the point of having some eukaryotic features such as 80S ribosomes and a nuclear envelope (**Figure 19-5**). It would have been heterotrophic, living by engulfing and digesting other cells. One of these early eukaryotic cells engulfed a type of bacterium but then did not digest it, perhaps because the larger cell had inefficient digestive enzymes or because the engulfed cell had a more resistant membrane. Such an association might be advantageous to both: The engulfed cell could be efficient at aerobic respiration but poor at obtaining nutrients. If it could obtain sugars from the larger cell and "leak"

ATP back to it, both cells might be better off. Each species might have survived better when in an endosymbiosis than when each partner was living independently. Such an endosymbiosis would still have many obstacles to overcome—increasing the compatibility between the two cells and coordinating growth, development, and reproduction—but the descendants of the engulfed organism had the potential to evolve into a mitochondrion. DNA sequence data indicate that mitochondria are most closely related to bacteria known as proteobacteria and that the endosymbiosis occurred approximately two billion years ago. All known extant eukaryotes have mitochondria—no living eukaryote is primitively amitochondrial.

Origin of Plastids Plastids could arise similarly if the engulfed partner were photosynthetic (Figure 19-5). Because all chloroplasts contain chlorophyll *a* and produce oxygen but none has bacteriochlorophyll, cyanobacteria were suspected to be the ancestors. Two groups called the green and purple bacteria are unlikely candidates because they have very different pigments and do not produce oxygen when they photosynthesize. Prochlorophytes are a type of cyanobacteria that arouse great interest because they have both chlorophyll *a* and *b* and some lack phycobilin pigments. The first prochlorophyte discovered, *Prochloron*, not only resembles chloroplasts biochemically but also exists as an obligate symbiont inside marine invertebrates, called didemnid ascidians, supplying them with carbohydrates whenever the ascidians lie in sunny areas. A second prochlorophyte, *Prochlorothrix*, has been discovered growing freely in lakes in Holland. 16S rRNA sequencing has now shown that chloroplasts, *Prochlorothrix*, and *Prochloron* are all definitely closely related, but neither cyanobacterium is the immediate ancestor of chloroplasts. Instead, all three share a common ancestor.

Chloroplasts have originated several times, but not in the way we had anticipated. Chloroplasts in green, red, brown, and other types of algae differ from each other in so many ways that we had expected that various types of cyanobacteria and early eukaryotes had formed protochloroplast endosymbioses. However, accumulating data from DNA sequences, electron microscopy, and biochemistry are consistent with a very different hypothesis: There was a single primary endosymbiosis and then multiple secondary endosymbioses. The **primary endosymbiosis** gave rise to a clade containing red algae, green algae, and a small group called glaucophytes (Figure 19-3). One feature of glaucophyte chloroplasts is that they still actually produce a thin film of cyanobacterial wall between themselves and the cell. Red algal chloroplasts have chlorophyll *a* (but not *b*) and the cyanobacterial pigment phycobilin, organized into particles called phycobilisomes (**Figure 19-6**). Green algal chloroplasts have no trace of bacterial wall or phycobilin, but instead have chlorophylls *a* and *b* and carotenoid accessory pigments, all of which are similar to chloroplasts in true plants (embryophytes). Glaucophytes are just a small group of algae. Red algae are a larger and more diverse group. Green algae are extremely numerous, and some are the sister group of embryophytes. Land plants are highly modified descendants of green algae that are in turn derived from the eukaryotes of the original primary chloroplast endosymbiosis.

Several events of **secondary endosymbiosis** produced other lines of algae. Euglenoids arose after an early eukaryote engulfed

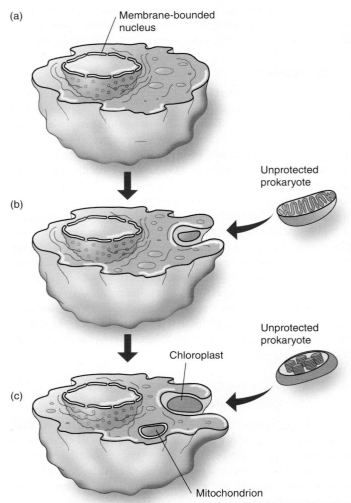

Figure 19-5 The endosymbiont theory postulates that a prokaryote (a) evolved to the level of having a membrane-bounded nucleus; this must have been an early step because all eukaryotes have a nuclear envelope and all are rather uniform in structure. (b) The early eukaryote engulfed but did not digest a prokaryote, which then evolved into a mitochondrion. (c) Finally, some descendants of that early eukaryote would engulf but not digest a prochlorophyte type of cyanobacterium, which would evolve into a plastid.

Labels in figure:
(a) Membrane-bounded nucleus
(b) Unprotected prokaryote
(c) Unprotected prokaryote
Chloroplast
Mitochondrion

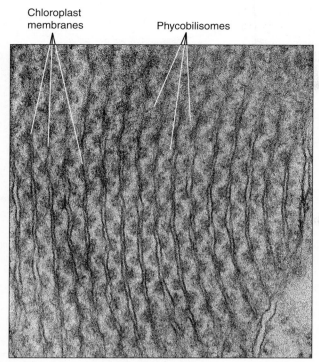

Figure 19-6 Internal membranes of red alga chloroplasts do not form grana stacks like those of green plants. Their phycobilisomes are visible, attached to the membrane. *Porphyridium* (×115,000).

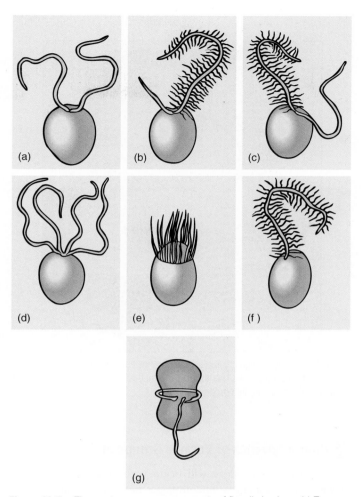

Figure 19-7 The most common arrangements of flagella in algae. (a) Two equal whiplash. (b) A short whiplash and a long tinsel. (c) Two equally long flagella, one of each type. (d) Four whiplash. (e) Numerous flagella located in a ring (rare). (f) One tinsel flagellum (mostly just in some brown algae). (g) One flagellum trailing, one in a transverse groove (dinoflagellates).

an entire green alga: A eukaryote engulfed another eukaryote. Since then, the cell of the green alga has become so reduced that mostly just the green algal chloroplast remains.

Another group of early eukaryotes, called **heterokonts** (or stramenopiles), were involved in one or several endosymbioses with entire cells of red algae, and these gave rise to brown algae, golden-brown algae, yellow-green algae, and diatoms. Heterokonts are named for the fact that all have two flagella of different types: one flagellum is smooth; the other has hairs (**Figure 19-7**). All heterokonts have this flagellum heteromorphy; thus, they seem to form a monophyletic clade, but not all have chloroplasts. We think that heterokonts diversified somewhat, and then some entered into secondary endosymbioses and became photosynthetic, whereas others did not. A lack of chloroplasts in these heterokont protists is an ancestral condition. We are not yet certain whether there was more than one secondary endosymbiosis that produced pigmented heterokonts; there is a diversity of structure and biochemistry in the extant chloroplasts, as the red alga symbiont became modified in different ways in different lines (**Figure 19-8**). Most are similar in having chlorophylls *a* and *c* (not *b* and not phycobilins), and most have four chloroplast membranes, not just the two (outer and inner) of red algae, green algae, glaucophytes, and true plants. Of the four chloroplast membranes, the two innermost membranes correspond to the original inner and outer membranes of the red alga chloroplast, a third membrane that encompasses these and corresponds to the red alga plasma membrane, and an outermost fourth membrane that corresponds to part of the heterokont cell's endoplasmic reticulum (called its cER for chloroplast ER; **Figure 19-9**). No heterokont chloroplast still has any remnant of the red alga cell

wall, but some have a tiny bit of red alga nucleus, called a nucleomorph, which still has a nuclear envelope and a few genes. These cells have four types of DNA: heterokont eukaryotic nucleus, red alga eukaryotic nucleomorph, chloroplast prokaryotic DNA circles, and mitochondrion prokaryotic DNA circles.

Types of Cytokinesis Several types of cytokinesis occur in algae. Many groups of unicellular algae have no wall, their plasma membrane being their outermost surface. During cell division, the plasma membrane pinches in two, being pulled inward as a cleavage furrow, a process remarkably similar to cytokinesis in animals.

In almost all algae with walls, cell division is similar to that of plants. It involves a phragmoplast, consisting of short microtubules oriented parallel to the spindle microtubules. In some green algae, cytokinesis occurs by a **phycoplast**, with microtubules oriented parallel to the plane where the new wall will form, which is perpendicular to the orientation of the spindle. A phycoplast may be associated with division either by furrowing or by cell plate formation. Embryophytes arose from green algae that divide with a phragmoplast rather than a phycoplast.

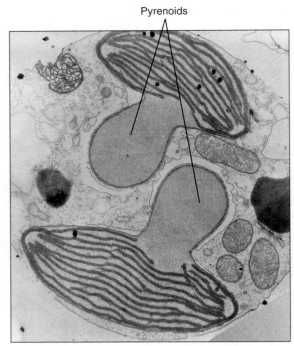

Pyrenoids

Figure 19-8 Brown alga chloroplast membranes associate into stacks of three or, less often, stacks of two. The large clear structures are pyrenoids, involved in polymerization of sugars into polysaccharides.

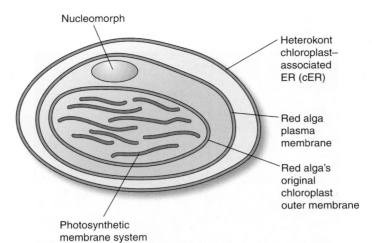

Nucleomorph

Heterokont chloroplast–associated ER (cER)

Red alga plasma membrane

Red alga's original chloroplast outer membrane

Photosynthetic membrane system

Figure 19-9 Many heterokont chloroplasts have four membranes: an outermost cER membrane, an inner membrane that corresponds to the plasma membrane of the red alga endosymbiont, and two innermost membranes corresponding to those of the chloroplast of the red alga. The innermost, photosynthetic membrane also corresponds to that of the original cyanobacterium. This entire structure would be located within the cytoplasm of a heterokont algal cell.

Characteristics of Various Groups of Algae

The descriptions of algae will begin with green algae because they are in the clade that encompasses not only the ancestors of embryophytes but also the primary endosymbiosis that originally produced chloroplasts. Red algae are described after that, and then attention is turned to the lines that resulted from the secondary endosymbiosis of plastids.

Green Algae

From an evolutionary standpoint, the **green algae** (Table 19-2) constitute an extremely important group: Not only could some early green algae organize complex differentiated multicellular bodies, but some moved onto land—the ancestors of embryophytes, true plants. Perhaps because most of the animals and plants with which we are familiar are multicellular and terrestrial, we tend not to appreciate just how complex these phenomena are. Consider that organisms spent as many as 1 billion years at the level of unicellular organization (Figure 19-1); then among algae true multicellularity evolved only a few times. The transition to having a complex photosynthetic body capable of living on land was so difficult that it appears to have occurred only once.

Green algae have remarkable developmental and metabolic plasticity: They are resilient and survive many types of disturbances and changes. Mutations that cause cells to adhere after cytokinesis have not been lethal, nor were many mutations that affected swimming, orientation of cell division, or coordination of karyokinesis and cytokinesis. Bodies composed of several types of specialized cells have not evolved in most other algal

TABLE 19-2	Divisions of Algae

Division Euglenophyta—euglenoids

There are many nonpigmented euglenoids, and this group is classified by zoologists as a member of their phylum Protozoa. The photosynthetic species have pigments similar to those of green algae. Euglenoids seem to be extremely ancient, having a cell organization similar to that which evolved hundreds of millions of years ago (see Fig. 19-2a).

Division Pyrrhophyta—dinoflagellates

These lack histones and have intranuclear mitosis with chromosomes that never decondense during interphase. Many botanists now consider them to be an isolated line that originated during the very first stages of the evolution of eukaryotic cells (see Fig. 19-2b).

Division Chrysophyta—diatoms, golden-brown algae, yellow-green algae

This is a diverse group, sometimes divided into several separate divisions, sometimes included with the brown algae. They and the brown algae have similar biochemistry, especially photosynthetic pigments (chlorophylls *a* and *c*) and unusual storage products (see Fig. 19-2c, d, and e).

Division Chlorophyta—green algae

Green algae are extremely diverse structurally but very homogeneous and well-defined biochemically, being almost identical to true plants in terms of basic metabolism. They are universally considered to be the ancestors of true plants (see Fig. 19-2f).

Division Phaeophyta—brown algae

A large group of species that often have large, complex bodies and that are common along rocky coasts. Some kelps have well-defined tissues, one of which strongly resembles phloem and is involved in long-distance transport of organic molecules (see Fig. 19-2g).

Division Rhodophyta—red algae

Many species of this large group have simple filamentous bodies, but the filaments may be aggregated and resemble parenchyma. Unlike all other eukaryotic groups, the red algae never have flagella at any time. They are also unique in having a strongly prokaryotic organization to their chloroplasts: There are phycobilisomes as in cyanobacteria. However, nuclear structure and mitosis do not appear to be unusual (see Fig. 19-2h).

groups. Although brown algae and red algae do show considerable sophistication in certain types of multicellular bodies, they are metabolically intolerant of ecological changes; few can live in fresh water, soil, air, or inside animals as many green algae do.

The diversity in green algae is tremendous. Examining it will give you a basis for understanding the diversity possible among living organisms and appreciating that the metabolisms and organizations of humans and flowering plants are by no means the only solutions to biological problems.

■ Body Construction in Green Algae

The most relictual and simple body is probably the motile single cell. Numerous green algae are unicellular. From this, many evolutionary possibilities exist:

1. **Motile colonies** (**Figure 19-10a** and Figure 19-18): If cells adhere loosely, the resulting structure is a **colony**, not an individual organism. All cells are similar, and none is particularly specialized; there may be some differentiation into two or three somewhat distinct cell types.
2. **Nonmotile colonies** occur if the cells lose their flagella or never develop them (Figure 19-10b). Although a nonmotile cell is simpler than one with flagella, the flagellated form is believed to be ancestral because all motile algae have the same type of flagella, the 9 + 2 arrangement of microtubules.
3. A **filamentous body** results if cells are held tightly by a middle lamella and if all cells divide transversely (**Figure 19-11a** and b and Figure 19-19). If occasional cells undergo longitudinal division, the filament branches. Often some portions of the body differ from others; for example, one end may serve as a means to attach the filament to a rock (a **holdfast**), or other parts may produce spores or gametes.

4. A **membranous body** results if the orientation of cell divisions is controlled precisely such that all new walls occur in only two planes (Figure 19-11c). The result is a sheet of cells that can become more extensive but remains thin. A membranous body is more strongly affected by currents and wave action and more likely to be torn than is a filamentous body.
5. If cell division occurs regularly in all three planes, a bulky, three-dimensional **parenchymatous body** results (Figure 19-11d and Figure 19-22). All cells are interconnected by plasmodesmata, and a true parenchyma tissue is formed.
6. A **coenocytic** or **siphonous body** results if karyokinesis occurs without cytokinesis, and giant multinucleate cells result (**Figure 19-12** and Figure 19-21). The cells can grow to several centimeters in diameter, but they usually remain fairly simple despite their size. Some coenocytes, such as *Acetabularia,* can be surprisingly complex (**Figure 19-13**).

Evolutionary development of control over the orientation of cell division and adhesion must have been difficult. The fossil

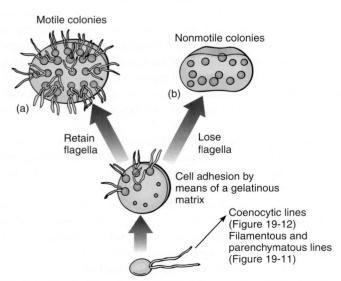

Figure 19-10   Evolution of colonial body type in green algae. The association of cells is controlled, not random. Depending on the species, each colony has a specific number of cells held in a particular shape such as a sphere, a flat plate, or a curved plate. No cell is free to divide at random.

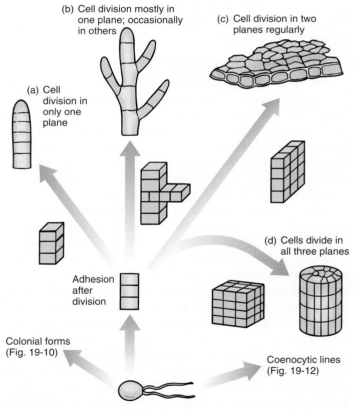

Figure 19-11   Evolution of filamentous, membranous, and parenchymatous body types in green algae. The formation of a multicellular body requires control over timing, position, and orientation of cell division, along with cell-cell adhesion mediated by a middle lamella. (a) All cells divide in only one plane, resulting in a uniseriate filament. (b) If cells occasionally divide in a different plane, the filament branches. (c) A two-dimensional sheet of cells results if all cells divide in two dimensions regularly. (d) A three-dimensional body is produced if cells divide regularly in all three planes. All cells share plasmodesmata with all adjacent cells; this is a true parenchyma tissue.

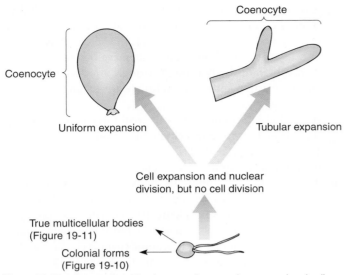

Figure 19-12 Coenocytic bodies in green algae may be more or less isodiametric (spherical) or long, branched, or unbranched tubes. Tubular forms are called siphonous forms.

Figure 19-13 Individuals of *Acetabularia* are unusual for being giant single cells, but even more remarkable, they are uninucleate. Each cap is composed of hollow tubes of cytoplasm and vacuole, and at times of reproduction, the sole nucleus begins rapid, repeated mitosis, resulting in hundreds or thousands of nuclei. Cytoplasmic streaming carries them upward into the cap segments; when each segment has several hundred nuclei, it either becomes a dormant cyst and is released or produces hundreds of gametes.

record indicates that the ability to grow as a siphonous form came into existence 230 million years earlier than the ability to organize even the simplest filament.

Life Cycles of Green Algae

The angiosperm type of life cycle, an alternation of heteromorphic generations, can be traced to green algae, and the theory of life cycles can be discussed here. In unicellular algae that have no sexual reproduction, such as euglenoids, the life cycle is really just the cell cycle; mitosis and cytokinesis constitute reproduction. However, with the evolution of sex, two processes are critical: meiosis, which segregates out haploid sets of chromosomes, and syngamy, which brings two sets back together.

After sexual reproduction evolved, while all organisms were still unicellular (no multicellular organism is relictually asexual), the simplest life cycle may have been the following: A diploid cell undergoes meiosis, and each cell then exists as a haploid individual, able to undergo mitosis (**Figure 19-14**). Because this is a unicellular organism, cell division is both "growth" and asexual reproduction. Some of these cells can also act as gametes and fuse, producing a diploid zygote that also is able to reproduce asexually—and simultaneously grow—by mitosis. Finally, at some time, some of these undergo meiosis again. Little difference exists between gametes, zygotes, and individuals, none of which is very specialized. Both haploid and diploid cells are capable of growth, division, and reproduction. Such species are **dibiontic**; that is, there is an alternation of generations between haploid and diploid.

In **monobiontic species**, specialization occurs in that only one free-living generation exists. In some monobiontic species, the haploid phase represents the individual, and the only diploid cell is the zygote, which is capable only of meiosis, not growth or mitosis (Figure 19-14a and b). The haploid body, either unicellular or multicellular, carries out photosynthesis and growth.

In other monobiontic species, the diploid phase represents the individual, vegetative growth phase; the only haploid cells are the gametes, which can undergo only syngamy (Figure 19-14c and d).

In a dibiontic species in which both stages are multicellular, the gametophyte (haploid phase) and sporophyte (diploid phase) may resemble each other strongly, and **alternation of isomorphic generations** occurs (**Figure 19-15a**), or the two may be very different in appearance and construction (**alternation of heteromorphic generations**), which allows them to exploit different ecological niches almost as if they were two species; gametophytes do not compete directly with sporophytes (Figure 19-15b).

All sporophytes, both in algae and embryophytes, produce spores by meiosis, but many algal sporophytes also produce spores by mitosis; these spores are diploid and grow into a new sporophyte in a form of asexual reproduction. Some algal gametophytes produce spores by mitosis; these are haploid and develop into new gametophytes, also a form of asexual reproduction. In both cases, this is basically similar to asexual reproduction in angiosperms in which some plants produce plantlets along their leaf edges, offset with bulbs or when pieces of cacti break off, root, and act as new plants.

During the earliest stages of the evolution of sexual reproduction, gametes were **isogamous** (identical), but **anisogamy** (slight differences in gametes) and **oogamy** (pronounced either OH-oh-gam-ee or oh-AH-gam-ee) later evolved (**Figure 19-16**). Gametes are produced in **gametangia**. Sperm cells or microgametes form in microgametangia, and egg cells or megagametes form in megagametangia. Spores are formed in **sporangia**, either megasporangia or microsporangia, depending on the size of the spore. A critical feature that distinguishes algae from embryophytes is that algal gametangia and sporangia are organized on a cellular level: Individual cells develop into gametes or spores, and when these are released, the gametangium or sporangium is nothing but empty cell walls. In contrast, embryophyte gametangia and

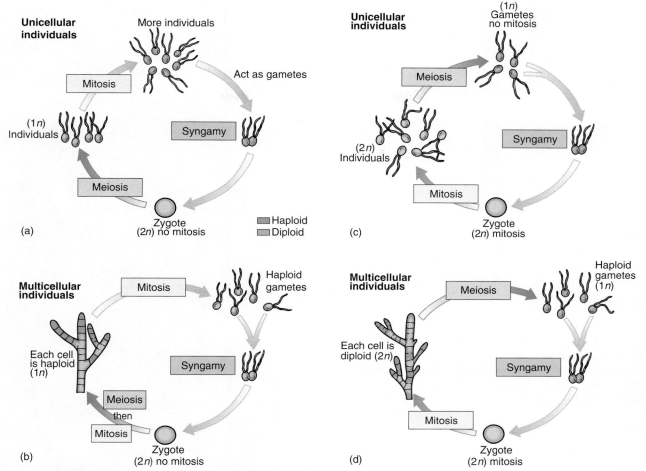

Figure 19-14 In monobiontic life cycles, only one generation—one phase—is capable of undergoing mitosis. In most algae (a and b), the haploid phase is dominant, undergoing mitosis that results in either more unicellular individuals or a multicellular individual. Only one diploid cell occurs—the zygote—and it cannot undergo mitosis; it cannot produce more diploid cells. (c and d) In a few algae and in all animals, the monobiontic life cycle is dominated by the diploid phase, the only haploid cells being the gametes.

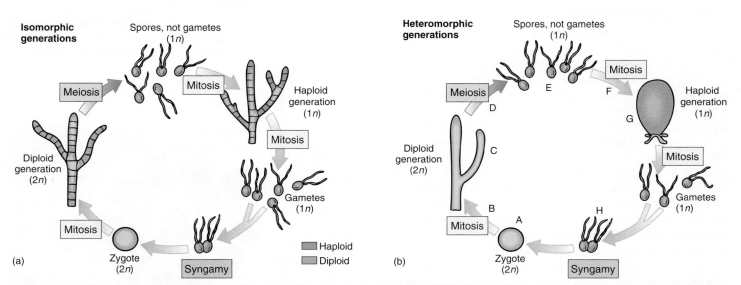

Figure 19-15 Dibiontic life cycles involve an alternation of either isomorphic (a) or heteromorphic (b) generations. With isomorphic generations, the similarity of body types can be so great that it is very difficult to know if an individual is a gametophyte or a sporophyte. (b) In dibiontic algae with an alternation of heteromorphic generations, the two types of individuals could never be confused. In flowering plants, the mitosis at B would be the growth of the embryo into the sporophyte plant body, such as an oak tree or a lily, and the meiosis at D would occur only in the sporogenous cells of anthers and ovules. The mitosis at F would be the formation of the tube nucleus, generative nucleus, and sperm cells in pollen and the growth of the megaspore into the antipodals, central cell, synergids, and egg cell of the megagametophyte. The syngamy at H would be fusion of one sperm nucleus with the egg. In flowering plants, the zygote begins growth immediately without a resting period, and flagella are not formed at any stage.

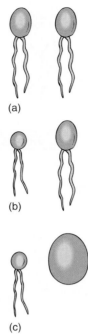

(a)

(b)

(c)

Figure 19-16 (a) Isogametes are identical gametes. Theoretically, they could be either motile or nonmotile, but in reality, all known isogametes do have flagella. (b) If gametes have any visible differences but are still similar, they are anisogametes. (c) Oogametes are obviously different in size, and the megagamete is virtually always nonmotile. In most algae and plants, the microgamete has flagella. In flowering plants, conifers, and some algae, microgametes are also nonmotile.

sporangia always have one or more outer layers of protective cells, cells that do not turn into gametes or spores; when the reproductive cells are released, one or several layers of complete cells remains.

■ Representative Genera of Green Algae

Unicellular Species *Chlamydomonas* is one of the simplest chlorophytes. It is unicellular and, like all green algae, has chlorophyll *a* and *b*, carotenoids, and xanthophylls; its starch is formed in chloroplasts just like that of true plants (**Figure 19-17**). It has two anterior flagella that resemble each other; therefore, it is not heterokont like brown algae and their relatives. Like most motile green algae, it has normal mitosis, meiosis, and syngamy. Its life cycle is simple: A haploid cell resorbs its flagella, divides mitotically, and forms 2, 4, 8, or 16 new cells that grow new flagella, each of which swims until it encounters a compatible cell. The cells recognize each other by reactions at the tips of their flagella. They undergo plasmogamy and karyogamy, forming a large zygote that sheds the four flagella and sinks to the bottom in a quiescent state. Germination is by meiosis to form four biflagellate haploid individuals; the zygote is the only diploid cell.

Motile Colonial Species In the motile colonial line of evolution, cells that greatly resemble *Chlamydomonas* are produced when the zygote divides, but the progeny cells are held together by a gelatinous matrix (**Figure 19-18**). In *Gonium,* each colony contains only a few cells (4, 8, 16, or 32), and the only sign of organization is that all flagella beat in a coordinated fashion. *Pandorina* is about the same size as *Gonium* but is slightly more derived because it shows a trace of differentiation; the colony swims in one direction,

Chloroplast   Flagella   Nucleus

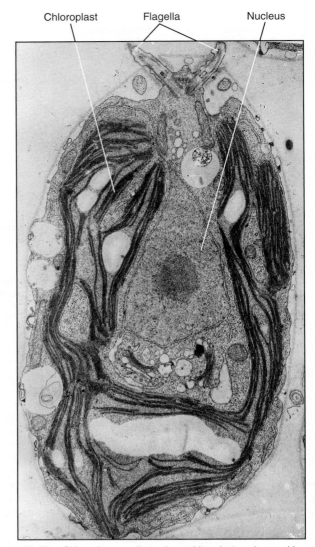

Figure 19-17 *Chlamydomonas.* Its nucleus, chloroplast, and pyrenoid are visible, but only the bases of the two flagella can be seen. The flagella are at the anterior end; that is, they pull the cell forward rather than pushing it. Several starch grains are present.

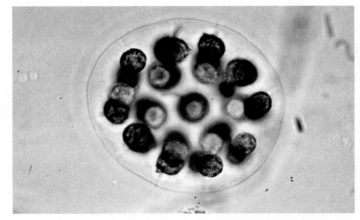

Figure 19-18 *Eudorina elegans* is a small motile colony of green algae. In each species, the number of cells and their spatial arrangements are constant. The numbers and orientations of cell divisions are controlled.

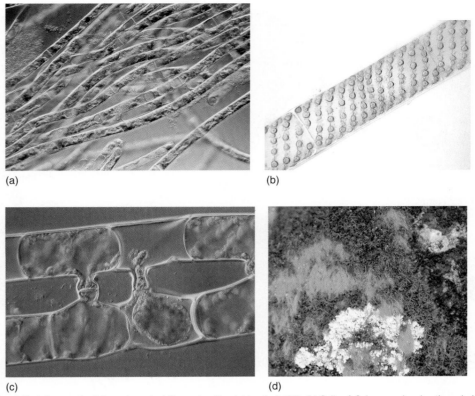

(a) (b)

(c) (d)

Figure 19-19 (a) Individuals of *Ulothrix* are uniseriate, unbranched filaments of haploid cells (×80). (b) Cells of *Spirogyra,* showing the spiral, band-shaped chloroplasts. (c) Compatible filaments of *Spirogyra* have been brought together in culture and are undergoing conjugation. This type of plasmogamy is unusual in algae. (d) The yellow material is a green alga, *Trentopohlia,* growing with lichens and mosses on the surface of a rock in Austin. A rock face in a dry area is not a typical habitat for algae, but *Trentopohlia* is so well adapted that it not only thrives, it grows large enough to be easily visible. Its green chlorophylls are masked by abundant yellow pigments.

and anterior cells are slightly different from posterior ones. *Volvox* is the stunning conclusion of this line: Its colonies contain up to 50,000 *Chlamydomonas*-like cells and are easily visible without a microscope. Differentiation exists in that up to 50 cells in the posterior half of a colony are specialized for reproduction only.

Filamentous Species Members of the genus *Ulothrix* are simple types of filamentous green algae (**Figure 19-19a**). Their life cycle is monobiontic; it involves only one free-living multicellular generation, and it is haploid. It consists of one row of cells that are all rather similar except for the basal cell, which is modified into a holdfast. Filament cells may produce asexual zoospores mitotically; these have four flagella, swim briefly, and then settle and grow into new filaments. During sexual reproduction, some cells produce gametes, which can be identified because they have only two flagella, not four. *Ulothrix* is isogamous, and the gametes, which all look like cells of *Chlamydomonas,* pair and fuse. The zygote germinates by meiosis, producing four haploid **zoospores**, each of which swims for a period, then loses its flagella, attaches to a substrate, and grows into a new filament.

Spirogyra is an extremely common fresh-water filamentous green alga (Figure 19-19) found in streams and ponds throughout North America. Its cells have beautiful spiral band-shaped chloroplasts that wind around the cell just below the plasma membrane. Swimming gametes are not formed; instead, filaments undergo **conjugation**. Each filament is haploid and is either a + or − mating type. If compatible filaments drift against each other, a conjugation tube forms between cells, and the − protoplasts migrate

through the tube and fuse with the + protoplasts. Karyogamy follows, although as many as 30 days may pass before nuclei fuse. The cell becomes dormant, thick walled, and resistant; it later germinates and grows into a new filament. Meiosis occurs immediately after karyogamy, and thus, the spore is haploid.

Laminar Species *Ulva* is slightly more complex than *Ulothrix,* but many of its stages are almost identical. A quadriflagellate, haploid zoospore settles down and grows into a *Ulothrix*-like filament. The cells divide in two directions and thus form a sheet; then all cells divide once in a third direction so that the sheet becomes two layers thick. *Ulva* has a dibiontic life cycle with an alternation of isomorphic generations (see Figure 19-15a and **Figure 19-20**). During sexual reproduction, gametophyte cells produce biflagellate anisogametes, the smaller gametes being produced on one gametophyte and the larger ones on a different gametophyte. Having two different types of individuals in one generation is dioecy, just as in flowering plants. The zygote grows into a filament and then into a double-layered sheet just like the gametophyte.

Coenocytic Species A dibiontic life cycle with an alternation of heteromorphic generations is illustrated by *Derbesia,* the organism in which it was discovered (see Figure 19-15b and **Figure 19-21**). Before going into the details of the life cycle, think for a second about how organisms are studied. Often a biologist goes on a field trip; collects plants, insects, birds, or whatever; and preserves them by drying or fixing them in formaldehyde. It is difficult to maintain things in a living condition. Now imagine trying to do this 100 years ago when transportation was slow, laboratories did

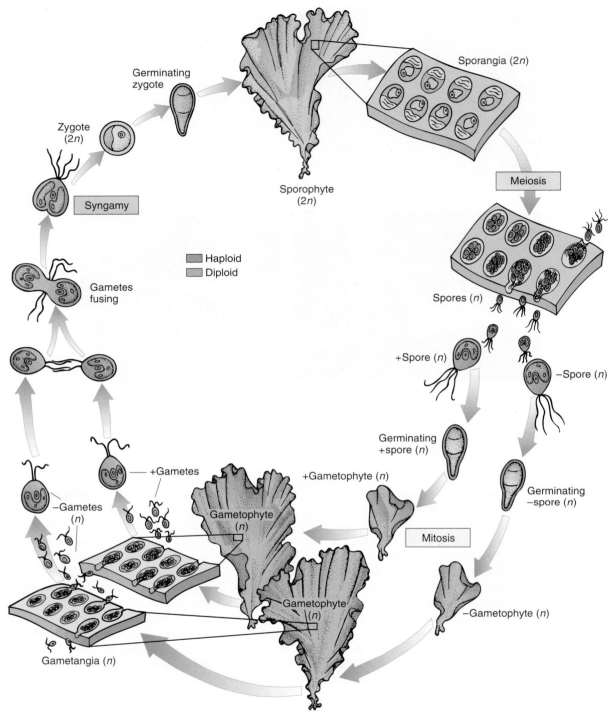

Figure 19-20 The life cycle of *Ulva*. Details are given in text. Two multicellular, isomorphic generations occur. The types of reproductive structures and the size of the nuclei are the only reliable criteria for distinguishing between sporophytes and gametophytes.

not have good artificial light or temperature control, and few pure chemicals were available for making culture solutions. Growing algae to study their life cycles was and is extremely difficult.

In 1938, P. Kornmann succeeded in obtaining zoospores from *Derbesia marina*, a branching, filamentous alga made up of giant coenocytic cells (Figure 19-21). The zoospores were carefully maintained, but instead of growing into another *Derbesia*-like individual as expected, they grew into individuals of a totally different genus, *Halicystis ovalis* (Figure 19-21). A member of *Hali-*

cystis looks nothing at all like *Derbesia*; instead, it is composed of a single large, spherical coenocytic cell attached to the substrate by a small holdfast. Almost its entire volume is one giant vacuole, with only a thin layer of protoplasm next to the wall. When mature, *Halicystis* individuals, which are the gametophyte stage, produce either male or female anisogametes that undergo syngamy and establish the zygote. After germination, the zygote grows into a *Derbesia* sporophyte. To classify different life stages as different genera is obviously not correct, and thus, the name *Halicystis ova-*

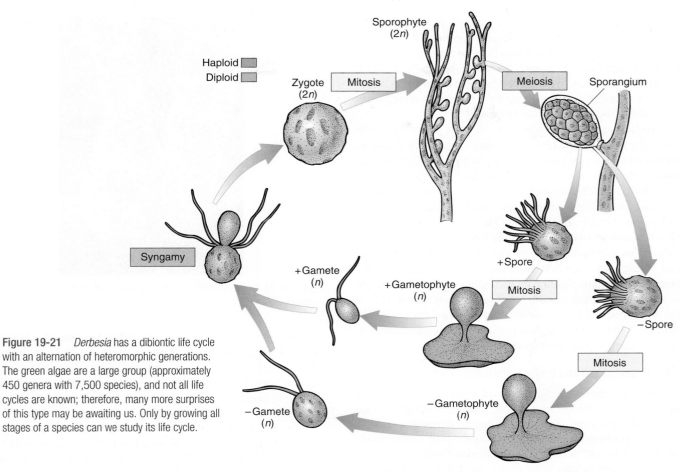

Figure 19-21 *Derbesia* has a dibiontic life cycle with an alternation of heteromorphic generations. The green algae are a large group (approximately 450 genera with 7,500 species), and not all life cycles are known; therefore, many more surprises of this type may be awaiting us. Only by growing all stages of a species can we study its life cycle.

lis has been eliminated. Flowering plants also have an alternation of heteromorphic generations, but the gametophytes grow inside the sporophytes; therefore, the question of which goes with which is easily solved.

Parenchymatous Species Several groups of green algae undergo a true parenchymatous growth, which is the basis of parenchyma in embryophytes. One group, in chlorophytes, divides by means of a phycoplast, which never occurs in plants, so this group is a remote possibility. The other group, **charophytes**, undergo cell division by means of a phragmoplast, just as plant cells do. In chlorophytes, the flagellar root apparatus (which anchors flagella to the cell) consists of four bands arranged in a cross; no true plant is known to have this type of root apparatus. In charophytes the flagellar root complex is similar to that of the motile cells of true plants: One major band of microtubules extends down into the cytoplasm from the basal body. Consequently, great attention is being given to charophytes.

An interesting example is *Chara* (Figure 19-22), which has a stem-like body divided into nodes and internodes, with whorls of branches arising at internodes. The body is several cells thick, composed of true parenchyma tissue derived from cell division in all three planes and originating from an apical meristem that contains a prominent apical cell. Although these features seem to correspond to those of flowering plants, virtually all of the resemblance

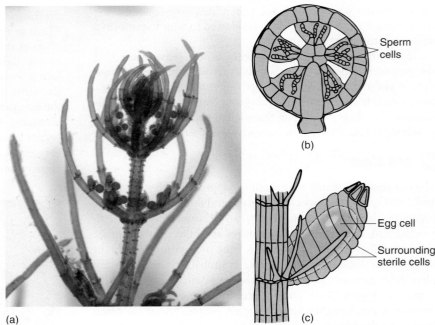

Figure 19-22 (a) An individual of *Chara*, with a parenchymatous body, microgametangia (spherical, reddish orange) and megagametangia (oval). (b) The microgametangium of *Chara* consists of an outer layer of sterile cells. Some inner cells mature into sperm cells, whereas others remain sterile and act as spacers. (c) The megagametangium is multicellular at maturity but not at initiation. The sterile cells around the egg are not sibling cells of the egg but rather are filaments that grow up from the cell below the egg. This is a multicellular gametangium, but it is filamentous, not parenchymatous.

is spurious because the earliest vascular land plants had no nodes, internodes, or branches. If *Chara* is the sister group to the embryophyte clade, the only features that may be homologous rather than analogous are the parenchymatous body and growth by an apical meristem. Simple parenchymatous bodies also occur in members of *Coleochaete,* another group of charophytes that are being studied as possible close relatives of land plants.

Reproduction in *Chara* is significant. Embryophytes have multicellular reproductive structures with sterile cells, and on this basis, *Chara* would have to be classified as a plant rather than an alga. Its sperm cells are produced in a truly multicellular gametangium whose outer cells are sterile; only the inner cells convert to sperm cells (Figure 19-22b). At maturity, the outer cells separate slightly and the motile sperm cells swim away. The egg is formed as the terminal cell of a short filament three cells long, but the subterminal cell subdivides, and those cells grow upward and surround the egg (Figure 19-22c). After fertilization, the sterile cells surrounding the fertilized egg deposit thickenings on their inner walls, those adjacent to the zygote. The resting structure thus consists not only of a thick-walled zygote but also protective sterile cells. When the resting cell germinates, it grows out as haploid filaments that soon establish an apical meristem and parenchymatic growth.

■ Green Algae and Embryophytes

Figure 19-3 shows charophytes and embryophytes as sister groups sharing a common ancestor. As such, they form a monophyletic clade, and some people have suggested that this clade be named the **streptophytes**, or even that the definition of embryophytes should be extended, that charophytes should be considered true plants, not algae. Most people hesitate to establish new formal classifications for clades and nodes; consequently, the charophyte + embryophyte clade is referred to with the informal name streptophyte.

Notice that the phylogenetic hypothesis depicted in Figure 19-3 also proposes a monophyletic clade that consists of red algae + green algae + embryophytes, and glaucophytes could be added as well. Names have been suggested for this group also, the **archaeplastids** (because this **clade** is based on the original plastid endosymbiosis) or **primoplantae**. So far, these names have not been used very often.

■ Red Algae

The **red algae** constitute a large group (approximately 400 genera and 3,900 species) of especially distinct and fascinating algae (see Figure 19-2h, **Figure 19-23**, and **Figure 19-24**). Numerous structural, biochemical, and reproductive features set them off from other algae as well as from embryophytes. One of the most important biochemical distinctions is that like cyanobacteria they contain phycobilin accessory pigments that are aggregated into phycobilisomes (see Figure 19-6). Their red color is due to the presence of phycoerythrin; however, they are often purple, brown, or black because of the additional presence of phycocyanin, just as in cyanobacteria. Carotenoid accessory pigments are also present, as is chlorophyll *a.* The actual quantities of each type of pigment vary with depth. Those algae that grow in bright surface waters have an array of pigments suitable for relatively intense light, whereas algae in deeper waters have a different complex of pigments better suited both to dim light and to the altered spectrum present as a result of the water's differential color absorption.

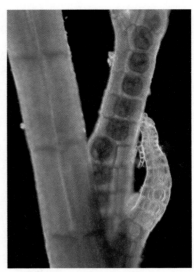

Figure 19-23   Many red algae have highly branched, filamentous bodies. *Polysiphonia* (×20).

Excess photosynthate is stored as **floridean starch**, a branched polymer of glucose somewhat similar to glycogen; it occurs as granules in the cytoplasm, never in the chloroplast. Other reserves occur that contain unusual sugars such as floridoside and isofloridoside, indicating that the carbohydrate metabolism of red algae has characteristic features not shared by plants.

In addition to a thin layer of cellulose, walls of red algae contain a thick layer of slimy mucilages called sulfated galactans. These are important commercially as thickening, suspending, or stabilizing agents in puddings, ice creams, cheeses, and salad dressings. The culture medium **agar** is also extracted from them. The complete wall—cellulose layer plus mucilage—is quite thick, often as thick as the protoplast is wide. Almost all red algae are multicellular, and thus, a large fraction of the individual's volume consists of this apoplastic space. As a result, all cells, even the most internal ones, may have relatively direct contact with the surrounding water. In the largest family of red algae, the Corallinaceae or coralline red algae, such large amounts of calcium carbonate are deposited into the walls that they become rocklike (Figure 19-24). It was not until 1837 that they were recognized as algae rather than corals.

Figure 19-24   Coralline red algae have walls so heavily impregnated with calcium carbonate that the body is hard and brittle.

Walls of red algae lack plasmodesmata of the type that occurs in plants, but they do have distinctive **pit connections**. During cell division, the new wall is formed as a ring that grows inward from the original cell wall; the growth of the septum stops before it is complete, leaving a hole in the center. Vesicles then deposit a material that precipitates, forming a lens-shaped plug. Whether these are a means of intercellular transport or communication is not known, but when two separate filaments of cells come into contact, a "secondary pit connection" can form between them.

Red algae are usually multicellular; only a few unicellular species (*Porphyridium, Rhodospora*) have ever been discovered. Most red algae tend to be rather conspicuous, often beautiful individuals, and are filamentous, membranous, or foliaceous; true parenchymatous bodies are reported to occur in only a few algae, such as *Bangia* and *Porphyra*. Typically, red algae are smaller and less complex than brown algae.

Despite the rather large size that bodies of red algae can attain, little differentiation or specialization occurs among the cells. The greatest differentiation typically involves only cell size and pigmentation. Outer cells are smaller and more heavily pigmented; inner cells are larger and have fewer, smaller chloroplasts. Most species grow attached by rhizoids to rocks, shells, algae, or sea grasses. Numerous red algae (more than 40 genera) are parasitic, usually on other red algae. The basal cells of these species penetrate into the host, forming secondary pit connections with host cells.

Life cycles of most red algae are poorly known, but the few that have been studied well are all extremely complex, almost all involving at least one multicellular stage but none having any motile cells. Flagella and centrioles do not occur in any stage of any species. Many variations occur, and there is no "typical" red alga life cycle; Figure 19-25 is a generalized life cycle. Gametophytes bear gametangia that produce nonmotile sperm cells called spermatia

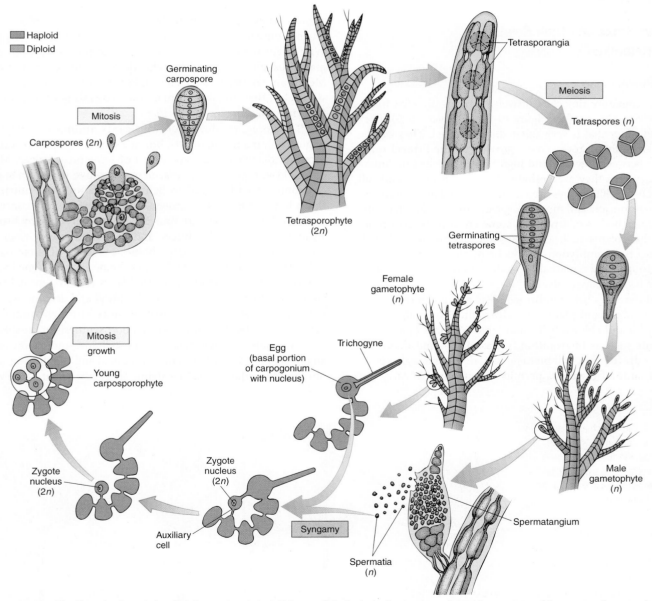

Figure 19-25 The life cycle of a red alga. Details are given in text. Note especially the lack of swimming cells and the presence of three generations, one haploid and two diploid.

and egg-like cells called carpogonia. Carpogonia are large cells with a long tubular extension that basically acts as a receptor for drifting spermatia. When a spermatium contacts the extension, plasmogamy occurs and the nucleus migrates to the carpogonium base, where karyogamy occurs. In any other group, this cell would be a zygote and would either grow or produce spores, but in many red algae, the fertilized carpogonium puts out another long filament that carries the diploid nucleus out of the carpogonium and deposits it into a totally different cell, an auxiliary cell, where mitosis begins and produces a mass of cells. This mass is a new generation, the carposporophyte, and it produces diploid carpospores by mitosis. The carposporophyte is a totally distinct generation that has no equivalent in any other algae or plants. Carpospores are released, float away and settle, and then germinate and grow into diploid individuals called tetrasporophytes, equivalent to a regular sporophyte. These have sporangia in which the cells undergo meiosis, producing haploid tetraspores that grow into gametophytes.

Figure 19-26 *Fucus* is a common brown alga you might see on any cool, rocky coast, just below the high tide level. It tolerates being exposed to air for several hours at low tide because its thick cell walls are extremely hydrophilic, slowing the evaporation of water.

Brown Algae and Their Relatives: The Heterokonts

Brown Algae

Brown algae are almost exclusively marine; only a few freshwater species are known (see Figure 19-2g and **Figure 19-26**). They prefer cold water that is very agitated and aerated. They can be found most easily on rocky coasts growing in the **littoral zone**, the region between low tide and high tide, also called the intertidal zone, where they are periodically exposed to air and full sunlight. Over 1,500 species are known, grouped into about 250 genera. Anatomically and morphologically, brown algae have the most complex bodies of any algae; some are considerably more complex than plants such as mosses and liverworts. Although very distinct from embryophytes biochemically and ecologically, the two groups have remarkable parallels in the types of bodies and life cycles that have evolved.

The differences between brown algae and green organisms (both green algae and plants) are clear cut. Brown algae have chlorophyll *a* and chlorophyll *c* and large amounts of a variety of pigments such as fucoxanthin, violaxanthin, and diatoxanthin (these are xanthophyll pigments). Carotenes are also present. Their suite of pigments permits brown algae to carry out photosynthesis at numerous levels in the ocean. Sunlight differs not only in intensity but also in quality at different depths: White light with a full spectrum occurs at the surface, but primarily just blue-green light reaches depths of 50 meters or more. Membranes of brown algal chloroplasts associate into grana-like stacks, but they are always small, each consisting of just two or three membranes (see Figure 19-8).

The storage product of brown algae is **laminarin** (a polymer of glucose), mannitol, or fats, but not starch. Laminarin constitutes up to 34% of the body weight of some brown algae. Most other algae, especially the marine species, live in such a stable environment with regard to light, temperature, and nutrients that photosynthesis and growth occur more or less continuously, so large reserves are unnecessary; however, many brown algae live in seasonal habitats, and they need energy reserves. For example, kelps have large bodies made up of three parts: holdfasts and stipes (stalks) that may be perennial (some live at least 17 years) and photosynthetic blades that are annual (**Figure 19-27**). When blades become moribund and decompose, the holdfasts and stalks must subsist on stored nutrients until the new blade can be formed and begin photosynthesis in the spring.

Cell walls of brown algae contain cellulose and alginic acid, an unusual polymer of D-mannuronic acid and L-guluronic acid not found in other algae. The alginic acid component of the wall

Figure 19-27 (a–d) In many kelps, the blade is annual but the stipe and holdfast are perennial. A new blade is formed each year; if the old blade has not been completely destroyed by wave action, it is sloughed off.

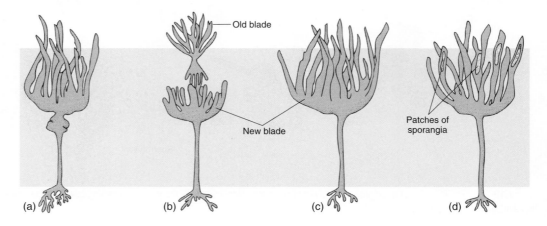

Old blade

New blade

Patches of sporangia

(a) (b) (c) (d)

is gummy or slimy and causes the filaments of cells to adhere into a compact body. It may have another unknown function, however, because as much as 24% of the body dry weight can be alginic acid in *Ascophyllum*.

A remarkable feature of brown algae is that they are all multicellular; no unicellular species is known to exist, and individuals of many species of kelps become huge and complex. One of the simplest brown algae is *Ectocarpus*. It has an alternation of isomorphic generations, both generations consisting of uniseriate branched filaments that arise from a prostrate branched system attached to rocks, shells, or other, larger algae. In the diploid sporophytes, some of the terminal cells of small lateral branches enlarge greatly and become sporangia. The nuclei divide repeatedly, and then 32 or 64 zoospores are released. The first nuclear division is meiotic, and the spores are therefore haploid. After swimming temporarily, zoospores settle down and grow into gametophytes that are almost identical to the sporophytes. The primary difference is that at the ends of the branches are multi-

cellular gametangia, not unicellular sporangia as in sporophytes. The gametes are anisogamous: Some settle and attract others by secreting a sex hormone ectocarpene. After fertilization, the zygote grows into a new sporophyte.

A more complex brown alga is *Fucus*, which is common on rocks in the intertidal zone (see Figure 19-26 and Figure 19-28). The diploid individuals are exposed at low tide, and their large dichotomously branched bodies (up to 2 m) are attached to rocks by holdfasts. The bodies are complex histologically with epidermis, cortex, and a central region. The ends of the branches are called **receptacles** and are swollen with large deposits of hydrophilic compounds. Scattered over the surface of the receptacles are minute openings that lead to small cavities, **conceptacles**; some conceptacle cells undergo meiosis, producing either large eggs or small sperms. At low tide, individuals are exposed to air, and the conceptacles contract, squeezing out gametes. When the tide comes in, gametes are washed free and fertilization occurs in the water. Fertilized eggs settle to the bottom and grow into new

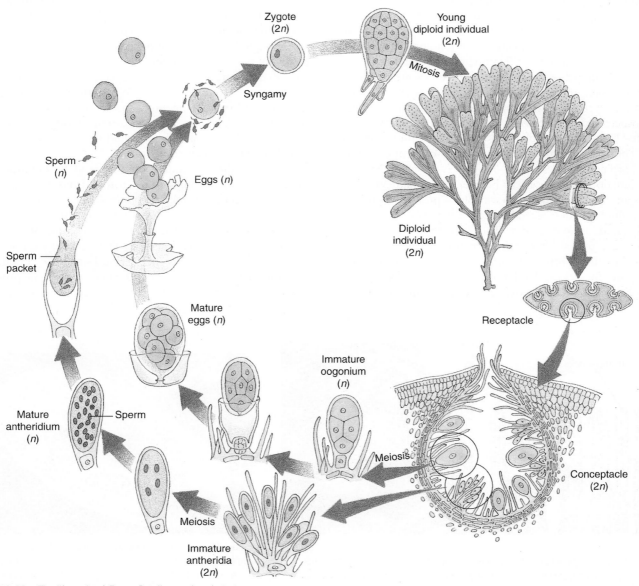

Figure 19-28 The life cycle of *Fucus*. Details are given in text.

diploid individuals. No free-living haploid generation occurs; *Fucus* is monobiontic.

Individuals of the kelp *Nereocystis* that measure up to 45 m—taller than most trees—are not uncommon, but they are not bulky, being less than 5 cm thick. Any alga that becomes as long as half a football field must have specialized regions to its body. The holdfasts of a kelp are located in deep, dark, poorly aerated water, whereas their blades exist in brightly lighted and well-aerated surface waters. To keep from being damaged by wave actions, the bodies must be firm, elastic, and thick—too thick for diffusion alone to mediate exchange of gases, nutrients, and wastes. Some species have an epidermis-like outer covering, called a meristoderm, which adds new layers to the surface. Interior to the meristoderm is a parenchymatous middle tissue that resembles plant cortex, and in the center is a cylinder of **trumpet cells** that resemble sieve tube members (**Figure 19-29**). Trumpet cells carry out long-distance transport of carbohydrates through the body. They are elongate, have large holes that occur in a sieve-like arrangement in the end walls, and the holes are lined with callose. If radioactive carbon dioxide is given to the blade, a short time later radioactive photosynthates can be detected in trumpet cells of the stipe. The flow rate can be as high as 65 to 78 cm/hr. Like sieve tube members, trumpet cells function only temporarily and then are replaced by new ones that differentiate from cortex cells.

Although trumpet cells and sieve tube members are extremely similar structurally and metabolically, they are not at all homologous. The most recent common ancestor shared by brown algae and angiosperms must have been a unicellular nonphotosynthetic protist that could not possibly have had either trumpet cells or phloem. This is an outstanding example of convergent evolution (a plesiomorphy) of tissues.

The junction between an air bladder (**pneumatocyst**) and a blade is an intercalary meristem capable of prolonged growth, which produces blades several meters long. Growth can be extremely rapid, up to 6 cm/day. Blades are photosynthetic and vegetative for a period after their formation by the intercalary meristem, but then portions become fertile and produce haploid zoospores by meiosis. Under inductive conditions, sporangia form either in patches or over the blade's entire surface. In *Macrocystis*, some blades remain vegetative and others are specialized and produce sporangia. *Macrocystis* releases as many as 76,000 spores/min/cm² of reproductive blade surface; the mean rate over long periods is 5,000 spores/min/cm². Spores grow into tiny, filamentous gametophytes that somewhat resemble small plants of *Ectocarpus*—an alternation of heteromorphic generations. Gametophytes produce oogametes, and after fertilization, a new sporophyte is formed.

■ Diatoms

This clade of heterokont algae contains about 200 genera and 5,000 species. Diatoms are easy to recognize because of their distinctive morphology. Each cell has a wall composed of two halves or frustules that fit together like a Petri dish and its lid (see Figure 19-2c and **Figure 19-30**). Each frustule is encrusted with silica, and when the diatom dies and the protoplasm degenerates, the siliceous frustules sink to the ocean floor and accumulate. Such deposits, known as **diatomaceous earth**, can become hundreds of meters thick and cover many square kilometers. Accumulations of this volume are possible because diatoms are the most abundant organisms in the oceans, and a single liter of seawater may easily contain more than 1 million individuals.

It might be thought that organisms consisting of a single cell could not possibly be very complex and certainly would not have enough structure to allow 5,000 species to be recognized, but frustules are extremely intricate. Where the two come together, they are held in place by girdle bands, and each species has its own characteristic complicated pattern of ridges, depressions, and pores. The cells are either round in face view (centric diatoms) or elongate (pennate diatoms).

When a diatom undergoes mitosis and cytokinesis, each progeny cell receives one of the frustules from the parental cell, using it as its outer frustule and synthesizing a new inner frustule. All cells that receive the inner frustule automatically mature into a cell that is smaller than the parent cell, and thus, the average cell size decreases with each cell cycle. This obviously cannot go on forever, and when a cell reaches a critical small size, sexual reproduction is triggered. The cells undergo meiosis, some producing 4, 8, or 16 sperm cells, others producing just one or two large egg

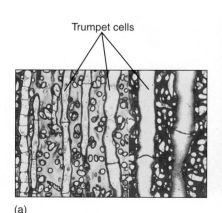

(a)

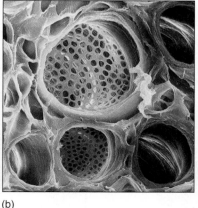

(b)

Figure 19-29 (a) A longitudinal section through the stipe of the kelp *Macrocystis*. In the center are numerous trumpet cells (×290). (b) A young trumpet cell, showing the enlarged end walls where holes have developed (×500).

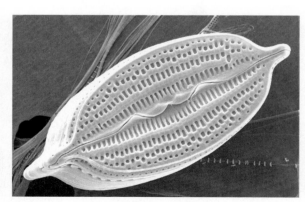

Figure 19-30 Pennate diatoms are elongated and bilaterally symmetrical. The numerous holes in their frustules allow nutrients to be taken up from seawater and wastes to be excreted (×4,500).

cells. After fertilization, the zygote grows into a large cell and then becomes dormant, surrounded by a wall different from that of a vegetative cell. The zygote finally "germinates," undergoing a round of nuclear and cell division, and reinitiating the process of decreasing cell size.

Yellow-Green Algae

Yellow-green algae occur mostly in fresh water, and formerly, many were thought to be green algae until chlorophyll *c* was discovered in them (see Figure 19-2e). They are somewhat diverse, some being unicellular, some filamentous, and some forming giant multinucleate cells. Filamentous forms, such as *Tribonema,* have walls like diatom frustules, each cell having two half-walls, but each of these frustule-like walls is firmly attached to the adjacent half-wall of the neighboring cell. Although daughter cells remain together after cell division, the organism can be called multicellular only in the broadest sense because no differentiation or specialization occurs in any part: The filament is really just a colony of individuals that happen to remain together.

Some yellow-green algae, such as *Vaucheria,* have an unusual body that consists of a long tubular coenocyte with a single large central vacuole, a thin peripheral layer of cytoplasm, and thousands of nuclei. During asexual reproduction, the cytoplasm at the tip of a tube is isolated by a transverse wall, and pairs of flagella form near each nucleus; then the old wall breaks open and the multiflagellate "cell" swims slowly away (Figure 19-31). It soon loses its flagella and begins growth at each end, forming a new tube. The greatest complexity is displayed during sexual reproduction. A mass of cytoplasm is walled off and pairs of flagella form, but then each nucleus organizes the protoplasm around it into an individual cell. When the original wall breaks down, numerous biflagellate sperm cells swim away. An egg cell is formed as a small multinucleate protrusion on a tube, but as the cross wall forms, all nuclei except one migrate out, leaving a single large uninucleate cell. After fertilization the zygote forms a thick wall and is abscised from the parent branch. After a period of dormancy, the zygote undergoes meiosis and germinates, forming a new tube filled with haploid nuclei.

Golden-Brown Algae

Golden-brown algae consist of about 70 genera and 325 species. They are single cells covered with numerous tiny siliceous scales that develop within special vesicles in the endoplasmic reticulum (see Figure 19-2d and Figure 19-32). Cells may be either uniflagellate or biflagellate; rarely they have no flagella, crawling by amoeboid motion instead. Although they are photoautotrophic because of their chloroplasts, they can also ingest bacteria by phagocytosis.

Golden-brown algae are particularly important because of their past biology. The living cells float in warm, sunny surface waters, like diatoms, but when they die, the lack of swimming motion allows their dense siliceous scales or frustules to sink and accumulate on the ocean floor as chalk (blackboard chalk used to be made from this). The remains of both diatoms and golden-brown algae, especially a group known as **coccolithophorids** (usually called coccoliths), are relatively inert and do not decompose. Drilling into sea floor sediments makes it possible to analyze the amount of diatom and coccolith remains and to estimate the rate at which they accumulated. It is possible to reconstruct the surface climate.

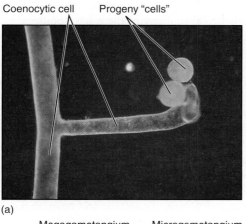

Coenocytic cell   Progeny "cells"

(a)

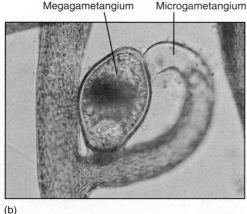

Megagametangium   Microgametangium

(b)

Figure 19-31 (a) *Vaucheria* occasionally walls off the tip of a coenocyte; then pairs of flagella form near each nucleus. The wall breaks open, and the "cell" swims away, settles, and grows into a new individual (×100). (b) During sexual reproduction, one segment of the coenocyte forms numerous biflagellate sperm cells, and another forms one large uninucleate egg. After the sperm cells swim away, all that is left of the gametangium is wall; unlike gametangia in true plants, no protective layer of sterile cells is present (×200).

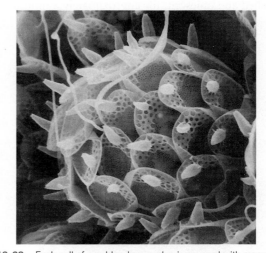

Figure 19-32 Each cell of a golden-brown alga is covered with many minute scales that are formed within vesicles of the endoplasmic reticulum. The pattern within each scale is constant for each species, but how it is controlled is unknown. *Synura uvella* (×5,000).

Accumulation was rapid when waters were warm and carbon dioxide was abundant for photosynthesis and slow when the climate was cool or if carbon dioxide was low. Shallow coastal seas tend to be warmer than the open oceans, and this also affects coccolith growth and reproduction. By also analyzing fossil animals, which are affected by temperature but not carbon dioxide, we can reconstruct past climates and ocean basin geology precisely.

Dinoflagellates

Dinoflagellates have many unusual characters (see Figure 19-2b and Figure 19-33). Their nuclear envelope and nucleolus persist throughout mitosis, and a typical spindle does not form. Instead, channels open in the nucleus and large bundles of microtubules pass through. Their chromosomes, which do not have centromeres and are permanently condensed, are attached to the nuclear envelope, whose expansion during the formation of the nuclear channels causes the chromosomes to split. There are no histones.

Dinoflagellates are almost exclusively motile and unicellular; only a few species such as *Gonyaulax monilata* form chains of similar cells with no differentiation. Their flagella have a strikingly characteristic arrangement in dinoflagellates: One long flagellum lies in a longitudinal groove with its distal end free and responsible for swimming. The other flagellum is flat and ribbon-like and lies completely within a transverse groove that encircles the cell. They differ from heterokonts by not having a tinsel flagellum. Most species are photosynthetic, but many are completely heterotrophic. Dinoflagellates have chlorophylls *a* and *c*, not *a* and *b*, and their carotenoids are quite unusual. In addition to β carotene, they also have several unique xanthophylls such as peridinin and dinoxanthin. Their reserve material is starch or oil. Dinoflagellates may lack a wall or have one consisting of cellulose plates whose arrangement and number are useful for identifying genera and species.

During dinoflagellate sexual reproduction, vegetative cells release small, naked cells that act as gametes. At the initiation of fusion, they become quiescent but then resume swimming, and the zygote may remain motile for 12 to 13 days. It then becomes thick walled and quiescent for approximately 2 months. During germination, it undergoes meiosis and a single haploid vegetative cell emerges—a new individual that grows and multiplies by mitosis.

Organisms with many relictual characters are often regarded as poorly adapted, perhaps surviving only in certain specialized, protected environments, but dinoflagellates are abundant throughout the oceans. Under conditions of ideal temperature and nutrients, population growth of some dinoflagellates, especially *Gonyaulax* and *Gymnodinium,* is explosive. Within a few days they become so numerous that their bodies actually color the water reddish brown—a "**red tide**" (Figure 19-34). The density of dinoflagellates can be as high as 30,000 cells per milliliter of sea water. Red tides are becoming increasingly frequent in the Gulf of Mexico, a phenomenon that is unexplained at present. This fascinating biological spectacle is also dangerous because these algae produce toxins, and their large concentrations kill fish and make other marine life poisonous to humans. The poisons of many species, *Gonyaulax catenella,* for instance, are potent neurotoxins, interfering with the movement of sodium ions across our nerve membranes.

Oomycetes

These protistans were thought to be fungi for many years. They have no chloroplasts, and thus, like fungi, are never photosynthetic. Also, many of them have bodies that consist of long, slender multinucleate, coenocytic tubes that have no cross walls, an organization typical of fungi; however, many of their characters are more like those of algae and plants but different from true fungi. For example, oomycetes have a cellulose wall but all true fungus cell walls are based on a polymer called chitin (as occurs in animals). Also, oomycetes have tinsel flagella, and their chromosomes are fairly large (chromosomes in fungi are extremely tiny). DNA sequences now strongly support inclusion of these groups in the heterokont protists.

Oomycetes are diverse in structure and nutrition. Lacking chloroplasts, they must be parasitic or saprophytic. Most, such as *Saprolegnia* and *Achlya,* are aquatic and have been called "water molds," but others live in dry habitats such as on and within leaves. Some consist of single cells that live as parasites inside

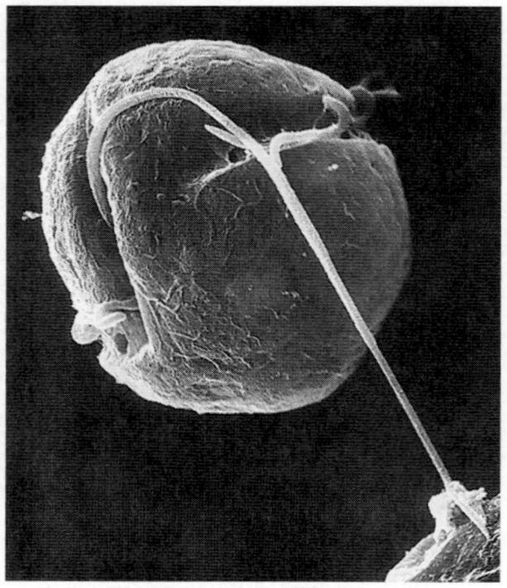

Figure 19-33   Dinoflagellates have two flagella, one of which lies in a groove that encircles the cell (×10,000).

Figure 19-34   The population density of dinoflagellates can rise so high that they color the water, producing a red tide. A few centimeters away, the density is so low that the water appears normal.

rotifers, minute aquatic animals. Many oomycetes, however, form extensive networks of slender tubular cells.

Sexual reproduction occurs in most oomycetes: A cell swells into a bulbous oogonium, which contains one or several diploid nuclei that undergo meiosis, each in its own mass of protoplasm (**Figure 19-35**). An adjacent cell elongates as an antheridium and contacts the oogonium; a hole forms, and microgametes pass directly into the oogonium, fertilizing the megagametes. The resulting oospores are diploid and are actually zygotes; each matures into a resting spore with a thick cellulose wall. These remain viable for many months even under dry, adverse conditions. Under favorable conditions, they germinate into diploid tubular cells.

Two oomycetes (*Plasmopara viticola* and *Phytophthora infestans*) are severe plant pathogens. The first attacks grapes and was introduced to France in the late 1870s on grafting stock imported from the United States. *Plasmopara viticola* caused tremendous damage throughout vineyards, except along roadsides where a mixture of lime and copper sulfate had been applied to the plants to keep passersby from stealing grapes. After it was recognized that this inhibits the pathogen without damaging the grape plants, it was used for treating entire vineyards.

Phytophthora infestans, an oomycete disease of potatoes, caused the great potato blight and famine in Ireland in 1846–1847. At that time, Ireland was overpopulated, and the potato was almost the only food for most people. The infection with *P. infestans* spread so rapidly and virulently that virtually the entire food crop for the country was destroyed; as many as 800,000 people starved to death the following winter, and more than 1 million emigrated, mostly to the United States. *Phytophthora infestans* is still a serious disease, and in cool, moist climates, it is controlled only by regular spraying with algicides.

Hypochytridiomycetes are a related group of nonphotosynthetic protistans that were formerly thought to be fungi.

Euglenoids

More than 800 species of **euglenoids** have been discovered, named, and placed into 36 genera (Figure 19-2a and **Figure 19-36**). The majority (25 genera) never have chloroplasts at all, but 11 do, derived through a secondary endosymbiosis with a green alga. Most euglenoids are unicellular, but in some species, a few cells remain together after cell division.

Green, photosynthetic euglenoids have chlorophylls *a* and *b* (just as do green algae), but carbohydrate is stored as the glucose polymer paramylon, not as starch. Euglenoid chloroplasts are extremely sensitive to antibiotics that interfere with 70S ribosomes, and exposure to streptomycin results in the loss of chloroplasts; colorless "bleached" individuals can be produced with this treatment. Many colorless individuals, both natural and bleached, take in food particles by phagocytosis (engulfing their food), and most absorb sugars and even proteins through their plasma membrane. Even chlorophyllous species grow faster in light if supplied with organic compounds, and all must obtain B vitamins from the environment.

Euglenoids swim actively with two flagella located at the cell's anterior end, one short and the other quite long. Euglenoids differ from other protists and plants in an important character: They never have a cellulose wall. Instead their surface contains a layer of elastic proteins that constitute a pellicle (also called a periplast), which is located under the plasma membrane and gives the cell some rigidity and shape.

Euglena is a representative example. Individuals are elongate cells that, because of their pellicle, have a somewhat ovoid

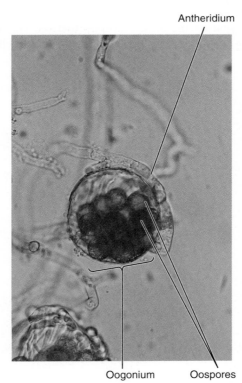

Figure 19-35 *Saprolegnia*: An antheridium has already contacted the oogonium and transferred microgametes. Oospores have formed. This situation is a form of oogamy (×3,500).

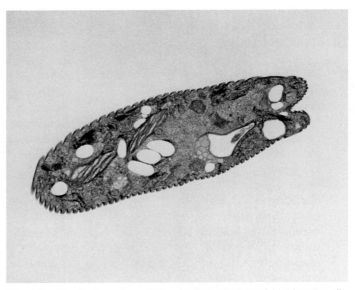

Figure 19-36 *Euglena*: The pellicle, now often called a periplast, is not a cell wall, but rather a thick layer of elastic proteins that accumulate just below the plasma membrane (×11,700).

shape, although they flex as they swim. At the anterior end is an invagination, the canal. The two flagella are attached at the base of the canal and extend upward, but only one is long enough to emerge from the canal and be useful for swimming. At the side of the canal, near the base of the flagella, is an eyespot composed of many small, orange droplets of carotenoids. There are numerous disk-shaped chloroplasts, each containing a region called a pyrenoid; in most algae, the pyrenoid is involved in polymerizing sugars into reserve polymers. That may also be true in *Euglena*, but the reserve paramylon accumulates outside the chloroplast, away from the pyrenoid.

Euglenoids have many features that must have undergone little change and thus probably are similar to those present in early eukaryotes. During mitosis, their nuclear envelope and nucleolus persist, and spindle microtubules form *within* the nucleus. Chromosomes are aligned lengthwise on the spindle, not crosswise as in other eukaryotes, and each chromatid appears to have its own minispindle consisting of four centromere microtubules and 8 to 12 longer microtubules. Cell division is by longitudinal cleavage, not by either a phragmoplast or a phycoplast, and after division is completed, chromosomes remain condensed. Neither meiosis nor sexual reproduction has been found in any euglenoid. Many features of normal mitosis and meiosis as occur in plants and animals apparently did not arise until after the euglenoid clade had become reproductively isolated from the rest of the early eukaryotes.

Summary

1. Several significant evolutionary divergences occurred in the early eukaryotes, long before the plant, animal, and fungus clades had originated. Euglenoids have no meiosis or sexual reproduction. Dinoflagellates lack histones and normal mitosis; several groups have surfaces that lack cellulose or have cellulose combined with chemicals not found in plants.

2. A primary endosymbiosis produced the chloroplasts of glaucophytes, red algae, green algae, and plants. Several events of secondary endosymbiosis of entire red or green algae produced chloroplasts of the other protists.

3. Algae are distinguished from plants (embryophytes) by having gametangia and sporangia in which fertile cells are not protected by a surrounding layer of sterile cells.

4. Cell division in unicellular algae is by membrane infurrowing and cell cleavage. In some green algae, it is by means of a phycoplast. In most multicellular algae, cell division occurs by phragmoplast and outward growth of the cell plate, but in red algae, walls grow inward from the existing side walls.

5. Charophytes have a flagellar root complex of one large band of microtubules, similar to that in plants. Other green algae have a root complex of four bands of microtubules, unlike that of plants.

6. Within green algae, several body types have evolved from unicellular forms: motile and nonmotile colonies, branched and unbranched filaments, coenocytes, membranous forms, and true parenchyma.

7. Several types of life cycle exist: dibiontic in which both a diploid and a haploid generation occurs; monobiontic in which only one generation is capable of mitosis.

8. Dibiontic life cycles may consist of an alternation of either isomorphic or heteromorphic generations. All plants have an alternation of heteromorphic generations.

9. Several evolutionary clades are evident in the green algae, but the charophyte clade is the one that most strongly resembles embryophytes.

10. Brown algae and their relatives make up the heterokont clade, sharing the synapomorphy (derived character) of having one tinsel flagellum and one whiplash flagellum.

Important Terms

agar	dinoflagellates	oomycetes
alternation of heteromorphic generations	embryophytes	parenchymatous body
	endosymbiont theory	phycoplast
alternation of isomorphic generations	euglenoids	pit connections
	filamentous body	pneumatocyst
anisogamy	floridean starch	primoplantae
archaeplastid clade	gametangia	receptacles
autogenous theory	golden-brown algae	red algae
brown algae	green algae	red tide
charophytes	heterokonts	sporangia
coccolithophorids	holdfast	streptophytes
coenocytic (siphonous) body	isogamous	trumpet cells
colony	laminarin	yellow-green algae
conceptacles	littoral zone	zoospores
conjugation	membranous body	
diatoms	monobiontic species	
dibiontic	oogamy	

Review Questions

1. All organisms are classified into three domains. Name the three and describe the organisms that are classified in each.

2. Reproductive structures are the critical factors in distinguishing algae from true plants of kingdom Plantae. What is one fundamental difference between reproductive structures in these two groups?

3. Which group of algae appears to be most closely related to the ancestors of true plants? Which features appear to be homologous?

4. Describe the endosymbiont theory of plastid origin. How many times might chloroplasts have arisen?

5. In contrast to the endosymbiont theory is the autogenous theory. Describe the basic idea of it. Do you think that the autogenous theory would predict that plastids and mitochondria would have DNA and 70S ribosomes?

6. Why do we think that cyanobacteria rather than purple or green bacteria gave rise to chloroplasts in the primary endosymbiosis? Which cyanobacteria are most similar to chloroplasts? Which algae were involved in the secondary symbioses?

7. Question 6 indicated that chloroplasts arose several distinct times in separate acts of endosymbiosis. If that is true, does the term "plastid" refer to a natural group of organelles that descended from a common ancestor? Would it be safer to use three separate names for these three separate structures?

8. List some of the unusual cyanobacterial features of red algal chloroplasts.

9. Describe cell division and mitosis in algae. How is the red alga method of cross-wall formation different from that of plants?

10. If after cell division the cells adhere to each other loosely, the resulting structure is a _____, not an individual organism. In such a structure, are there many types of different cells or are all cells more or less similar to each other?

11. Name and describe the six types of body construction that occur in green algae. What role does the middle lamella play in these? What is the importance of controlling the orientation of cell division? Which of these body types occurs in other groups of algae?

12. Describe monobiontic and dibiontic life cycles. Be careful to mention all possible types. What is the difference between a spore and a gamete? What is the difference between a spore and a zygote?

13. What is an alternation of isomorphic generations? What is an alternation of heteromorphic generations? If you have studied the life cycle of flowering plants, do flowering plants have an alternation of isomorphic or heteromorphic generations?

14. Describe a green alga with a life cycle involving alternation of heteromorphic generations. Why were the two generations originally named as separate species?

15. *Volvox* is a green alga that you might see in a biology or botany lab. Is it filamentous or membranous or colonial? What is its shape?

16. *Spirogyra* is another green alga you might see in a biology or botany lab. What is unusual about it? What gives it its name?

17. *Chara* is an unusual alga. What type of cell division do members of Charophyceae undergo—division with a phycoplast or phragmoplast? What is important about their flagella? Do true plants (especially those with xylem and phloem) have flagella? (Hint: think sperm cells in ferns and cycads.)

18. Red algae are unusual in many ways, but an especially significant trait is the types of accessory pigments in their chloroplasts. What pigments do they have? They are aggregated into a body called a _____.

19. Many red algae have cell walls so heavily impregnated with calcium carbonate that they are brittle. These are called the _____ red algae.

20. What are trumpet cells in brown algae? What type of cells do they resemble in true plants? Is their presence in brown algae a sign that brown algae are related to the ancestors of true plants, or is it just a remarkable case of convergent evolution?

21. If you have ever visited a rocky coast, you may have seen two brown algae—kelps and *Fucus* (these never occur on sandy beaches). Briefly describe the bodies of these two common seaweeds.

22. Dinoflagellates are abundant algae with many unusual features. The arrangement of their flagella is characteristic. Describe the flagella of a typical dinoflagellate.

23. Under ideal conditions, dinoflagellates can reproduce so rapidly that they actually give seawater a distinct color. What is this phenomenon called? Is it safe for fish and humans?

24. Diatoms have what chemical in their cell walls? When the cell dies and the protoplasm degenerates, the wall sinks to the ocean bottom and becomes part of a huge deposit known as _____. These deposits can be so extensive because diatoms are so abundant that just a single liter of sea water may easily contain more than _____ individuals.

25. Describe and compare the cell walls or cell coverings of dinoflagellates, euglenoids, diatoms, green algae, red algae, and true plants.

26. Describe features of oomycetes that caused people to believe they were fungi (these are believed to be homoplasies). Now describe feature that indicate they are heterokont algae (these are believed to be synapomorphies).

BotanyLinks

http://biology.jbpub.com/botany/5e

BotanyLinks contains a variety of resources and review material designed to assist in your study of botany.

Visit the Web Exercises area of BotanyLinks to complete this question:

1. Do you use or eat algae every day? Go to the BotanyLinks home page for information on this subject.

chapter 20

Nonvascular Plants: Mosses, Liverworts, and Hornworts

Concepts

In this chapter the origin and evolutionary diversification of true plants—embryophytes—are discussed. Several critical events occurred in the history of true plants: their origin from green algae, their adaptation to terrestrial habitats, the development of the vascular tissues xylem and phloem, and the origin of seeds, leaves, and woody growth. Plants are traditionally divided into those that (1) have neither vascular tissues nor seeds, the **nonvascular plants** (often called "bryophytes"); (2) have vascular tissue but not seeds, the **vascular cryptogams**; and (3) have both vascular tissue and seeds, **spermatophytes** (Figure 20-1 and Table 20-1). Nonvascular plants arose first, and some of their later members were the ancestors of the vascular plants. Many of the fundamental aspects of plant biology were established in the early land plants.

Chapter Opener Image: This rock is covered with a moss (*Grimmia laevigata*, the gray-green material) and lichens (the blue-gray material). Mosses typically grow in moist, shady habitats; however, there are also large numbers adapted to dry, sunny conditions as here at Inks Lake State Park in central Texas. This moss thrives side by side with cacti and yuccas in the blazing sun and with only sporadic, infrequent rains. But the important thing is that it does rain at some point, and there is dew or fog at other times…that is true of absolutely every place where plants grow: no plant ever grows where there is never liquid water at some time. Mosses always have small, simple bodies, and rather than being a handicap, this is an adaptation that allows them to live in small, rocky areas that have only a bit of occasional water, too little water to sustain a large tree. Mosses are confined to moist areas because their sperm cells must swim from one plant to another; the sperms do swim, but there is always liquid water at some time, so sexual reproduction is possible. It is important to learn the characteristic features and habitats of plants, but it is also important to be aware of alternative adaptations and their consequences.

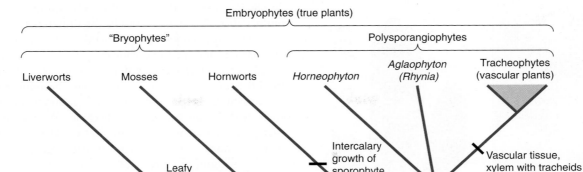

Figure 20-1   Recently proposed phylogeny of true plants. (MYBP = million years before present)

Beginning at some point earlier than 450 million years ago, certain green algae known as charophytes began to adapt to living on land. They evolved to survive the occasional drying of their streams, smaller lakes, and ocean-side mud flats. In environments where long, dry periods alternated with long, moist periods, the optimal survival strategy probably was formation of dormant, drought-resistant spores, but spore production requires major metabolic conversions; if the dry periods were brief, the ability to continue active metabolism by conserving water and avoiding desiccation would have been advantageous (Table 20-2). A large, compact, multicellular body would have had a low surface-to-volume ratio and would have automatically retained water better than a small unicellular or filamentous body. In addition, a water-proofing cuticle would have been selectively advantageous. Reproduction would have had to change to coordinate gamete production with periods of moisture because sperms had to swim. Also, gamete and spore mother cells would have needed

TABLE 20-1	Major Groups of Plants	
	Vascular Tissues	Seeds
Nonvascular Plants Mosses, liverworts, hornworts	absent	absent
Vascular plants without seeds Clubmosses, scouring rushes, ferns	present	absent
Seed Plants Cycads, conifers, angiosperms	present	present

TABLE 20-2	Means of Coping with Dry Periods
Not coping	This is the least expensive—the organism simply dies. Rapid extinction occurs unless the organism lives in a permanently moist environment such as an ocean, large river or lake, or rain forest.
Desiccation tolerance	Many mosses, liverworts, and lichens and some terrestrial algae have the ability to survive even when their body loses large amounts of water. The mechanism seems to be expensive because these organisms all grow slowly even in continually moist environments. Perhaps desiccation-tolerant membranes, proteins, and organelles have characteristics that do not permit rapid metabolism.
Formation of spores	Many organisms have a strategy in which most of the body is allowed to die during times of stress, while a few cells—spores—are made stress-tolerant. Fewer energy and nutrient resources are required to modify a few cells than the entire body; and the body, while alive, is capable of rapid metabolism. A disadvantage is that once the spores germinate, an entire new plant must be formed; if wet periods are brief, a new period of dryness may kill the sporlings before they are large enough to form new spores.
Desiccation avoidance	The organism avoids dry conditions by either retaining water within itself or tapping a safe, relatively permanent source of water. This method allows some metabolism at all times, and growth and reproduction can be somewhat independent of dry cycles. Such a survival strategy led to the vascular plants; a cuticle that retains water, a large body that has a low surface-to-volume ratio, and later, roots that penetrate deeper regions of soil that tend to be more permanently moist.

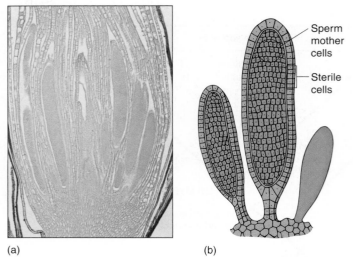

(a)   (b)

Figure 20-2 In true plants, gametangia and sporangia always have an outermost layer of cells that do not become gametes or spores. In these microgametangia of mosses, the sterile layer is only one cell thick, and each microgametangium produces numerous sperm cells (×25). Compare this with an algal gametangium.

protection from dryness by having a jacket of one or several layers of cells around them (**Figure 20-2** and **Figure 20-3**). Such reproductive modifications resulted in the grouping and protection of spore and gamete mother cells into the sporangia and gametangia characteristic of all embryophytes—more massive than those of algae and protected by a layer of sterile cells.

These simple modifications probably allowed the algal ancestors of plants not only to survive but also to be metabolically active during short dry periods; however, an automatic consequence of much greater advantage was that the algae were safe from predators while out of the water. Tremendous selective benefit resulted from mutations that enabled the algae to be active for even longer periods out of water. These mutations probably

involved increasing the size of the body and the impermeability of the cuticle, but there had to be the simultaneous evolution of stomatal pores and guard cells because a more protective cuticle also prevents the entry of carbon dioxide.

As a truly terrestrial existence became more successful, the mud flats and stream banks would have become crowded, and some plants must have grown over others and shaded them. With such shading, the environment became selective for mutations that produced an upright body that could grow into brighter light. The nonvascular plants mostly use turgid parenchyma cells and thus cannot grow very tall, but in the group of plants that would become vascular plants, the strengthening material that evolved was xylem, which is also good at conducting water. As phloem evolved, the basal part of the plant that remained in the shade could be nourished. Phloem permitted roots to extend deep into the dark soil, and xylem permitted transport upward of the water and nutrients those roots encountered. The presence of xylem, phloem, and roots had profound consequences: They freed plants from their muddy habitats and allowed them to grow anywhere moisture was available. The presence of the two vascular tissues, necessary for individuals even a few centimeters tall, suddenly gave plants the potential to become 100 meters tall, to branch, to form leaves, and to put sporangia high into the air, allowing the spores to be blown great distances, carrying the plant's genes over huge areas, and permitting the colonization of distant habitats.

Vascular tissue, especially phloem, also made feasible the evolution of truly heterotrophic tissues—roots, meristems, and organ primordia. Without phloem, each part of a plant can grow and develop only as rapidly as its photosynthesis permits, but phloem allows mobilization of sugars, minerals, and hormones throughout the entire body and their transfer to a shoot apical meristem or a group of sporangia, thus permitting a more vigorous, robust growth than could otherwise occur.

One obstacle to the total invasion of land was that plants were still reproductively amphibious. They had terrestrial bodies

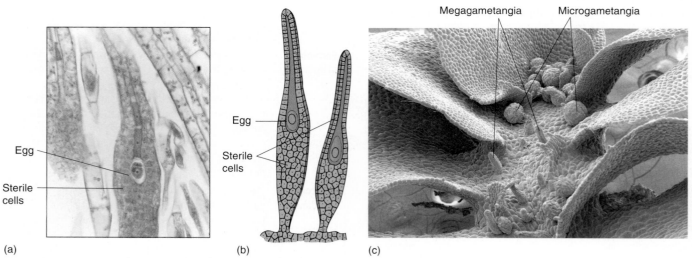

(a)   (b)   (c)

Figure 20-3 Megagametangia of true plants, such as these of a moss (a), also have a one-layered sterile jacket (b), but each produces only one egg, not four (×25). (c) This scanning electron micrograph shows the gametangia of a bryophyte (these are of a liverwort, but they are similar to those of mosses). The spheres are microgametangia in which sperm cells are produced, and the tubular structures are megagametangia, each with a single egg inside. The large flat sheets of cells are the leaves of this leafy liverwort (*Fossombronia*) (×50).

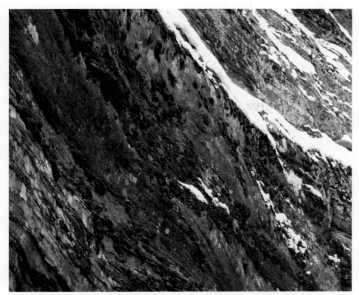

Figure 20-4 Zion National Park is a dry habitat, and this vertical rock wall is exceptionally dry; however, snow accumulates several times each winter and melts on warm days, and thus, there is sufficient liquid water during one or two months for mosses and liverworts to flourish. They remain dormant during the rest of the year.

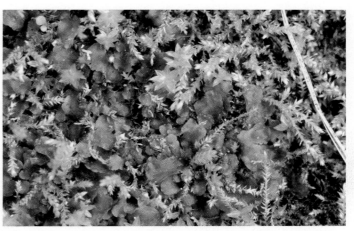

Figure 20-6 The tiny leafy shoots are gametophytes of a small moss. The numerous plate-like structures are gametophytes of a hornwort, *Phaeoceros.* Plants of both remain small and thin, never more than a few cells thick, and never elaborate.

but aquatic reproduction with swimming sperms. At some point in the life cycle, the environment had to be sufficiently wet that sperms could swim from one gametophyte to another. This requirement is a handicap, but not an insurmountable one: All plants need water to grow; no plant ever lives where there is never rain, fog, or dew (Figure 20-4).

Production of pollen and seeds eliminates the need for environmental water for reproduction. What changes must occur for seeds to evolve? Gametophytes must become so reduced that they mature completely within the walls of the spores (Figure 20-5). After this happened, retention of the megaspore and megagametophyte inside the parental sporophyte was feasible. The gametophytes functioned basically as tissues of the sporophyte and

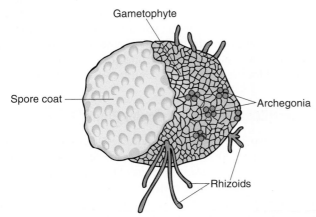

Figure 20-5 The megagametophyte of *Selaginella,* which has developed almost completely within the original wall of the megaspore. This is not part of a seed plant but is a necessary first step in seed evolution. Small megagametophytes can be protected and nurtured by their parent sporophyte. By doing so, the gametophyte generation benefits from all of those mutations that make the sporophyte adapted for life on land.

benefited from the land adaptations that had evolved; this is what occurs in ovules of seed plants. The microgametophyte, inside the microspore (known as pollen in seed plants), could be transferred by wind to the vicinity of the megasporangium, where it released sperms into fluids secreted by the megagametophyte. The sperms would swim in this tiny artificial pond and effect fertilization; only the last step of reproduction was still wholly aquatic. In the most derived seed plants, a pollen tube carries nonmotile sperms even further, to the egg itself.

Some gametophytes—for example, those of all nonvascular plants—do not need any protection by the parental sporophytes: In mosses, liverworts, and hornworts, gametophytes are the main photosynthetic phase of the life cycle (Figure 20-6), but even so, these gametophytes remain rather small and delicate, limiting the entire plant to environments mild enough for themselves (a plant can live and reproduce only in environments that are suitable to all essential aspects of its full life cycle). Plants should have been able to develop a life cycle in which gametophytes were just as large, tough, and woody as sporophytes, even having bark, leaves, roots, and so on, but this never happened. Some early land plants did have rather large gametophytes, but in all lines of evolution, gametophytes became simpler and smaller.

Characters of Nonvascular Plants

Most people are at least somewhat familiar with nonvascular plants because mosses are well known, and many people have heard of liverworts (Figure 20-7 and Figure 20-8). The third group, hornworts, is relatively unfamiliar, even to most botanists (Figure 20-6 and Figure 20-9). It is important to understand clearly what nonvascular plants are not: Spanish moss of the southeastern United States is the flowering plant *Tillandsia usneoides* of the pineapple family, not a true moss. Club mosses are lycophytes, not mosses, and the slimy, bright green "mosses" of ponds and slow-moving streams are green algae, usually *Spirogyra.* Several types of lichens, especially *Alectoria, Bryonia, Usnea,* and "reindeer moss," are frequently mistakenly thought to be mosses or liverworts.

Figure 20-7 Although mosses occur in hot, arid areas, they are most abundant in cool, moist regions. What you think of as the typical moss plant—green or gray-green and leafy—are masses of gametophytes; the slender stalk and capsules you may have noticed are sporophytes.

Figure 20-9 Hornworts such as this *Phaeoceros* are quite rare, and few people ever see them; they are easily confused with liverworts unless the tall "horns"—sporophytes—are present.

Technically, nonvascular plants are embryophytes that do not have vascular tissue. Being embryophytes, they have multicellular sporangia and gametangia: Reproductive cells are always surrounded by one or several layers of sterile cells (Figures 20-2 and 20-3). Their bodies are not composed of filaments as in many algae, but rather of true parenchyma derived by three-dimensional growth, usually from apical meristems: All mosses and many liverworts have leafy stems that look remarkably like small versions of flowering plants. Nonvascular plants are almost exclusively terrestrial and have a cuticle over much of their bodies, and many have stomata.

Like all plants, nonvascular plants have a life cycle with an alternation of heteromorphic generations; the sporophyte and gametophyte differ from each other structurally. Recall that sporophytes in flowering plants are the large plants with leaves and roots, whereas gametophytes are tiny and occur inside pollen grains and ovules. In nonvascular plants, the gametophyte is the larger, more prominent generation, and the sporophyte is

much smaller, more temporary, and often very inconspicuous. Any green moss plant that might be familiar to you is the gametophyte, the haploid phase, and it is photosynthetic, perennial and collects mineral nutrients; it typically can spread rapidly by asexual reproduction. The sporophytes it produces are usually very small. They absorb minerals only from the gametophytes and always remain attached to them. Sporophytes carry out so little photosynthesis that they could not support their own growth and sporogenesis. Unique sporophyte features in all species of nonvascular plants are that sporophytes are never independent of the gametophyte, never branch, and never have leaves. You will see sporophytes only if you examine mosses closely: They look like green or brown "hairs" standing up on the green gametophyte, but sporophytes are only present at certain times of the year.

Nonvascular plants can never grow to be really large, but being small and simple provides great selective advantage in certain habitats. The tiny parenchymatous bodies of mosses and liverworts permit them to thrive in microhabitats such as stone walls, fences, and bare rock, microhabitats that have too little water or soil for the larger, more complex vascular plants.

Classification of Nonvascular Plants

It is not known how closely related mosses, liverworts, and hornworts are (Figure 20-1). They have many features in common but also differ in significant respects. They are often treated as three distinct divisions: liverworts, division Hepatophyta; mosses, division Bryophyta; and hornworts, division Anthocerotophyta (Table 20-3). In the past, all three were often grouped together in division Bryophyta as three classes: mosses, Musci; liverworts, Hepaticae; and hornworts, Anthocerotae. Even today, they are informally referred to as "bryophytes." The three are treated here as divisions, and the generic term "bryophyte" is used only for mosses. All groups together are called "nonvascular plants."

Figure 20-8 This is *Marchantia,* the most abundant greenhouse liverwort; unfortunately, it is one of the least typical. As in mosses, the large green plant is the gametophyte, not the sporophyte. The cup-shaped structures contain clumps of cells that can be splashed out by rain and then grow into new plants.

TABLE 20-3 Alternative Methods of Classifying Nonvascular Plants*

A. Kingdom Plantae
  Division Bryophyta
  Division Hepatophyta
  Division Anthocerotophyta

B. Kingdom Plantae
  Division Bryophyta
    Class Musci
    Class Hepaticae
    Class Anthocerotae

*A assumes that they are less closely related than B does.

Division Bryophyta: Mosses

The Gametophyte Generation

Mosses are ubiquitous, occurring in all parts of the world and in almost every environment (Table 20-4). They are perennial and thrive in many places within cities (*Tortula* on walls and *Mnium* on soil).

Morphology The leafy stems, technically known as **gameto-phores**, of many moss plants grow close together, tightly appressed and forming dense mounds (*Grimmia, Pohlia*). In other species, particularly those of cool wet areas, the plants are more open and loose (*Anacolia, Climacium, Platygyrium*; see Figure 20-7). *Scouleria* gametophores grow as ribbons up to 15 cm long, submerged in rapidly flowing water. All moss stems have leaves (Figure 20-10), but because they are parts of a gametophyte, not a sporophyte, they are not homologous with those of vascular plants; that is, leaves of mosses did not evolve from the same structures as vascular plant stems and leaves did.

Moss gametophores grow from an apical meristem that contains a prominent apical cell (Figure 20-11). Derivative cells subdivide, producing the tissues of stem and leaves in rather precise arrangements. Leaves are aligned in three rows at least while young, and most have a midrib (costa). Leaves of almost all mosses are only one cell thick except at the midrib and along the margin (Figure 20-12a). In the family Polytrichaceae, common along roadsides in forested areas of the northern United States, genera such as *Polytrichum* and *Atrichum* have thin lamellae on the leaf upper surface, greatly increasing the photosynthetic tissue (Figure 20-12b). Cuticle occurs only on the upper surface of most moss leaves, the underside being uncutinized and capable of absorbing water directly from rain, dew, and fog. The lack of a cuticle means that mosses have little protection against desiccation; when the microhabitat dries, so do the leaves. No stomata

TABLE 20-4 Classification of Division Bryophyta

Class Sphagnopsida	*Sphagnum*
Class Andreaeopsida	*Andreaea, Neuroloma*
Class Bryopsida	*Atrichum, Bryum, Buxbaumia, Dicranum, Fissidens, Funaria, Grimmia, Mnium, Physcomitrium, Polytrichum*

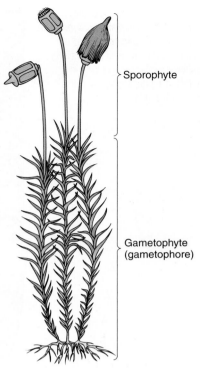

Figure 20-10 The individual shoots of a moss superficially resemble those of a flowering plant, having stems, leaves, nodes and internodes, and even buds. None of these structures is homologous to the equivalent organ in flowering plants, but the same set of names is used for both groups. Dark green, haploid tissue; light green, diploid tissue.

occur on moss leaves; they would be useless because the leaves are unistratose.

Moss stems are always slender and have little tissue differentiation (Figure 20-13). The surface layer is at most only slightly different from the underlying layers and is not called epidermis. The stem tissues, all called cortex, may be uniform in all parts, or the outer cells may be slightly narrower with walls that are somewhat thickened. Inner cells are larger, more parenchymatous, and chlorophyllous. In a few species of mosses, stems have hairs, but stomata do not occur.

Water Transport In some mosses, primarily the family Polytrichaceae, the innermost cortex is composed of cells called **hydroids** that conduct water and dissolved minerals. They are elongated cells that lose their cytoplasm when mature (Figure 20-14); their end walls are partially digested, but they are not removed entirely. Each hydroid is aligned with those above and below it. Species that have hydroids typically also have **leptoids**, cells that resemble sieve cells (Figure 20-14b). They are elongate, have relatively prominent interconnections with adjacent cells, and lack nuclei at maturity, although they do retain their cytoplasm. Adjacent parenchyma cells are unusually cytoplasmic and rich in enzymes, just as are companion cells.

The majority of mosses lack hydroids and leptoids; water is conducted along the exterior of their stems by capillary action. Leaves and stems are so small they form spaces narrow enough to act as capillary channels and transport water. In some species, such as *Funaria*, leaves curl and become more closely appressed to the stem as they dry; when rain or dew occurs, the dried plant

Figure 20-11 (a and b) Cross-section of the shoot apex of the moss *Physcomitrium pyriforme*: The apical cell is in the center, and its most recent daughter cell is to the left (1). The next older cell (2) has subdivided, as have successively older derivatives (3, 4, and 5) (×2,400). (c) The apical cell is shaped like an inverted pyramid.

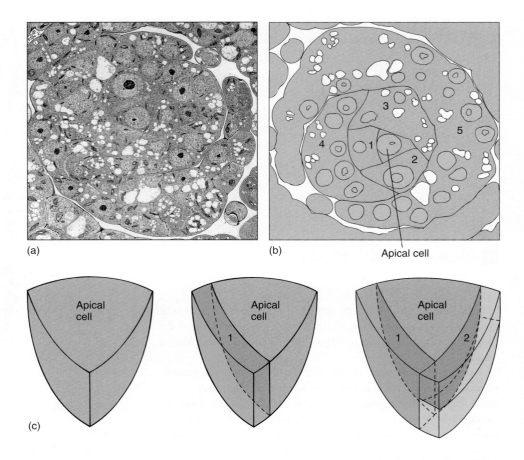

(a)

(b)

Apical cell

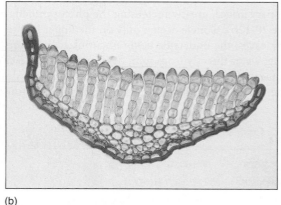

(c)

has even more capillary spaces than in the hydrated condition when the leaves are spread away from the stem. In mosses that lack leptoids, sugar is moved simply between parenchyma cells by slow transport.

At the base of the stem are **rhizoids**, small, multicellular trichome-like structures that penetrate the surface of the substrate. Rhizoids only anchor the stem; they do not absorb either water or minerals. They lack chloroplasts and have reddish walls.

Development Growth of the gametophore begins when a spore germinates and sends out a long, slender chlorophyllous cell.

This cell undergoes mitosis and produces a branched system of similar cells; the entire network is a **protonema** (plural, protonemata) (Figure 20-15 and Figure 20-16). A protonema superficially resembles a filamentous green alga but can be distinguished by its numerous small chloroplasts in each cell. Algal cells have only one or two large chloroplasts. Eventually, nodules of small cytoplasmic cells form on the protonema, organize an apical cell, and then grow upright as a stem with leaves—the gametophore. Protonemata are perennial and can grow extensively, producing many buds. The filamentous cells usually break when mosses

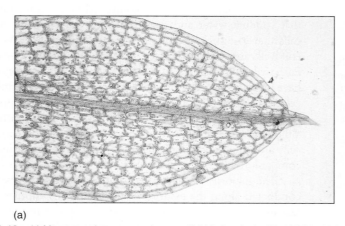

(a)

(b)

Figure 20-12 (a) Many moss leaves are only one cell thick, but the leaf "midrib" is thicker and is true parenchyma. (b) In the family Polytrichaceae, leaves bear long sheets of cells on their upper surface; this greatly increases the volume of photosynthetic tissue but does not decrease the high surface-to-volume ratio as it would if this were a solid tissue (×200).

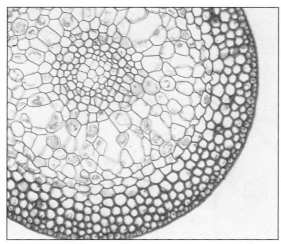

Figure 20-13 Although most moss stems do not have vascular tissues, they do support the shoot, and most have a layer or two of thick-walled cells (stained red here). Sugars and minerals must be transported from leaves to the shoot apex, gametangia, and sporophytes; therefore, living parenchyma cells are necessary (×40).

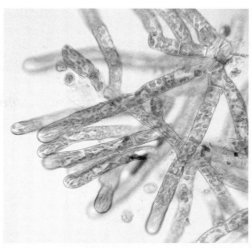

Figure 20-15 A moss spore has germinated and grown into these filaments of cells, which make up a protonema; later some cells will divide into apical meristems that will grow into gametophores. Although this looks like a green alga, algae do not have so many chloroplasts in each cell (×250).

are collected, and thus, a tuft of gametophores may appear to be independent plants when in fact they have all arisen from a single protonema.

Reproduction   The gametophore at some point produces gametangia. All mosses are oogamous; that is, every species has small biflagellate sperm cells and large nonmotile egg cells. Sperms are produced in microgametangia called **antheridia**, which consist of a short stalk, an outermost layer of sterile cells, and an inner

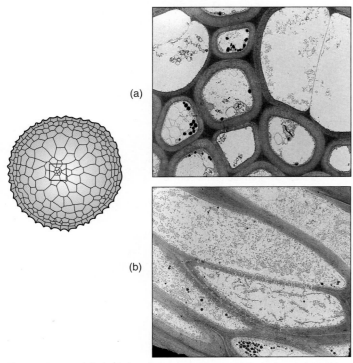

(a)

(b)

Figure 20-14 (a) Hydroids lose their cytoplasm at maturity, just as do tracheids and vessel elements. Hydroids, however, do not have specialized secondary walls. (b) Leptoids are analogous to sieve cells, phloem cells that are simpler than sieve tube members. All pores are small, and the cells remain alive at maturity.

mass of cells that differentiate into sperm cells (see Figure 20-2). Eggs occur in megagametangia called **archegonia**: Each is shaped like a vase with a long neck (see Figure 20-3). The neck is hollow at maturity, and the single egg is located at the base.

Antheridia and archegonia occur on the same gametophore in bisexual species (*Funaria, Pottia*), whereas other species have both male and female gametophores (*Barbula, Polytrichum, Rhacomitrium*). Depending on the species, gametangia may occur mixed with leaves along the gametophore stem or may be clustered at the stem tip, surrounded by a special cup-shaped whorl of leaves (**Figure 20-17**). Raindrops fall directly into the cup, splashing sperms out and carrying them as far as 50 cm. Without this mechanism, sperms must swim to archegonia (**Figure 20-18**).

When sperm cells are mature, the antheridium breaks open and liberates sperm cells either by contracting the sterile outer cells or by accumulating liquid below the sperms and pushing them out. Secretion of sucrose from the archegonia guides sperms as they swim toward the archegonia and then down the neck to the egg, where one sperm cell effects fertilization.

■ **The Sporophyte Generation**

In all embryophytes, the megagamete and subsequently the zygote are retained by the gametophyte; however, unlike those of mosses, megagametophytes of vascular plants are small and offer little nourishment or protection to the embryonic sporophytes. Megagametophytes of flowering plants typically have only six cells other than the egg, and both synergids and antipodals often degenerate just after the egg is fertilized (undergoes syngamy). All nutrition for the zygote is supplied by the grandparent generation, the sporophyte, by way of endosperm formation. In contrast, moss gametophytes are both large and photosynthetic, and they support the sporophyte throughout its entire life. The moss sporophyte is never an independent, free-living plant (Figure 20-16).

The zygote of a moss undergoes a transverse division, and the basal cell, located at the bottom of the archegonium, develops

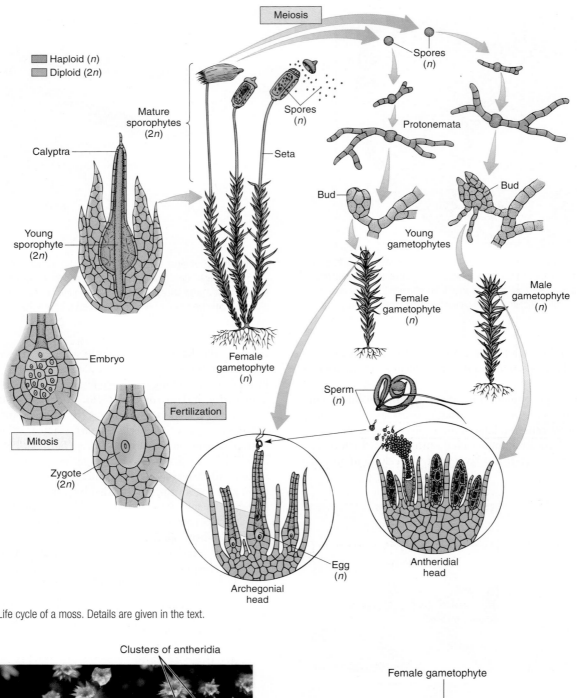

Figure 20-16 Life cycle of a moss. Details are given in the text.

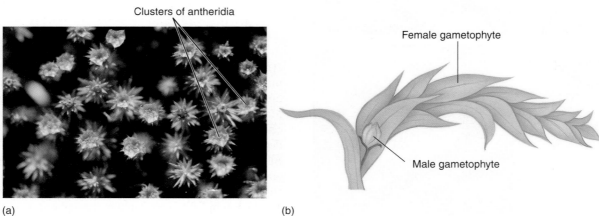

(a)

(b)

Figure 20-17 (a) In this moss, antheridia are clustered together at the gametophore apex; in other species, they might be located along the stem. (b) The phenomenon of dwarf males is common in many plants, animals, and algae. Each sperm cell is much smaller and less expensive to produce than each egg, but usually, many sperm cells are not carried to eggs; thus, a large number of sperm cells must be produced. If sperm delivery can be ensured, the number produced can be decreased, and even a small, dwarf male can produce adequate numbers. Delivery can be assured if the male gametophyte grows on the female as an epiphyte.

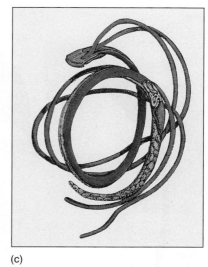

(a)   (b)   (c)

Figure 20-18 (a) The organelles of a young sperm cell have rather typical shapes (blue, nucleus; green, chloroplast; red, flagella and basal bodies; brown, mitochondrion; yellow, microtubules that anchor the flagella). (b and c) During differentiation, the organelles become modified, resulting in an elaborate spiral-shaped sperm cell at maturity (c). Cytosol and cell membrane are not shown.

into a small, bulbous tissue called the **foot**. The foot is the interface with the gametophore, from which it absorbs sugars, minerals, and water. Its cells are transfer cells in many species. The upper cell grows by cell division and expansion into a simple apical sporangium called the **capsule**, consisting of an outer layer of sterile cells and an inner column of sterile cells (the columella). A ring of sporogenous cells undergoes meiosis, producing haploid spores (Figure 20-19). Between the foot and the sporangium is a narrow stalk, the **seta** (plural, setae). All moss sporophytes have this basic, simple structure; none is ever branched or has leaves, bracts, or buds of any kind.

Although morphologically simple, the sporophyte is relatively complex structurally. It has a true epidermis with stomata at least on the base of the sporangium. Dehiscence of the sporangium is more elaborate than the opening of the gametangia: The apex of the sporangium differentiates as a caplike lid, the **operculum**, which separates from the rest of the sporangium as cells are torn apart. Cell breakage is elaborate and precise, resulting in one or two rows of beautiful, exquisitely complex teeth, called **peristome teeth** (Figure 20-19). The teeth respond to humidity, bending outward and opening the sporangium when the air is dry and bending inward and trapping the spores when the air is humid. Spores are released when they are light, dry, nonsticky, and easily carried by air currents. The apex of the sporangium in many species is covered by a **calyptra**, a layer of cells derived from the neck of the archegonium. As the embryo begins to grow, neck cells also proliferate. They keep pace with sporophyte growth at first but later grow more slowly and are torn away from the gametophore.

Virtually all mosses are homosporous: All spores are the same size and appear to be identical. A few species of *Macromitrium* and *Schlotheimia* produce two types of spores in each capsule. The larger spores develop into large gametophytes with archegonia, and the smaller spores grow into dwarf males that live epiphytically on the females (Figure 20-17b).

■ Metabolism and Ecology

The small size and lack of conducting tissues are two critical factors in the metabolism and ecology of mosses. Vascular plants tend to be large and their bulk protects them from short-term fluctuations in air humidity and moisture availability. Should the air become dry for several hours, most seed plant leaves and stems would not die, even if they were not receiving water from the roots by means of xylem. Very large plants, such as trees and succulents, can withstand dry conditions for weeks, months, or even years before losing a fatal amount of water. Leaves on moss gametophores, however, have only a thin, incomplete cuticle; if exposed to dry air for even a few minutes, the plants dry out. Without vascular tissue, stems and leaves can become desiccated, even while the rhizoids are in contact with moist soil or tree bark.

Several mechanisms compensate for the inability of mosses to retain water. Many species grow in permanently moist microhabitats such as rain forests, cloud forests, and the spray zones near waterfalls. Even habitats that are not constantly humid may have microhabitats that are continually moist: The foot of a rock cliff is usually damp because dew and mist concentrate there as runoff from the cliff. Shallow depressions in rock or soil retain moisture and are protected from drying winds. These sites are very small, but so are the mosses that live within them.

Other mosses are tolerant of desiccation. Drying does not damage them as it does most vascular plants and algae. Like lichens, many mosses can lose much of their water rather rapidly without dying or even being injured. As long as about 30% of their weight is water, they remain dormant but alive. If rain falls or dew forms, water is absorbed rapidly, and within a few minutes, respiration and photosynthesis are occurring at normal levels. In mild temperate regions, mosses may remain moist and metabolically active except for certain times during summer months. In deserts and dry range lands, on the other hand, mosses may be turgid primarily during winter. From spring to autumn, the

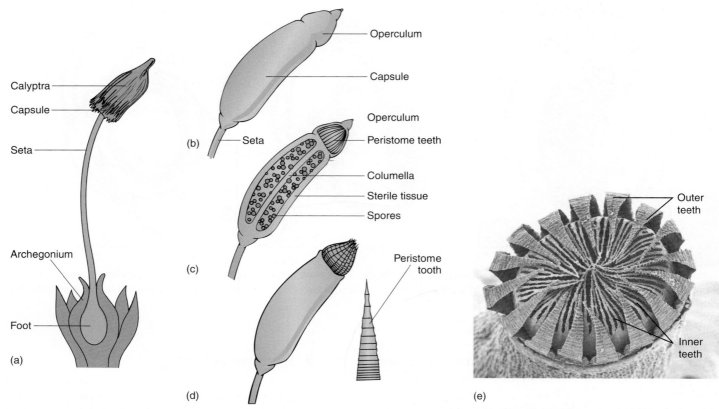

Figure 20-19 Aspects of a moss sporophyte. (a) The foot is embedded in the remnants of the archegonium. (b) External view of the capsule (sporangium) with the operculum still in place covering the teeth. (c) Median section view showing the columella and spores. (d) External view after the operculum has fallen off. Spores can be shed when the peristome teeth bend back. (e) Scanning electron micrograph view of peristome teeth.

mosses are dry and inactive every day, perhaps receiving enough dew on certain nights to be metabolically active for a few hours in the early morning.

Desiccated mosses are remarkably resistant to high or low temperature and to intense ultraviolet (UV) light. A common moss, *Tortula ruralis,* can be frozen safely in liquid nitrogen (−196°C) when dry and can also withstand brief periods at +100°C. Resistance to UV light is important for species that grow in full sun on the surface of rocks or at high altitudes on mountains, where the atmosphere is too thin to block UV radiation.

Many moss species thrive at low temperatures near or even below 0°C. When larger vascular plants die or abscise their leaves in winter, much more sunlight reaches the soil surface or tree trunks where mosses are growing. Snow or winter rains provide abundant moisture. Mosses that can carry out metabolism at cold winter temperatures can take advantage of the plentiful light and moisture.

Certain mosses can grow on hard, impervious surfaces because they have no roots that must penetrate the substrate. Consequently, mosses can be found on bare rock (*Andreaea, Hedwigia*), brick or mortar (*Tortula muralis*), and bark as well as soil. At the other extreme, *Sphagnum* mosses live on the surface of quiet water at the edges of lakes and ponds (**Figure 20-20**). Their leaves and stems are buoyant, and thick mats build up.

In almost all environments, mosses are important in the later establishment of other species, particularly vascular plants. Like lichens, they can colonize a bare surface and dissolve the rock with acids that leak from their cells. Leaves and stems of gametophores catch and hold dust particles, creating small pockets

of soil in which spores and seeds of vascular plants can become established. A similar process occurs even on the ground; in forests or grasslands, the moss layer on the soil may be thick enough to act as a moist, airy seed bed. It may hold water better than the soil and thus improve the microhabitat for seedlings.

Figure 20-20 *Sphagnum* moss grows at the edges of streams or ponds where there is little water movement. The plants become large and tangled, creating a mat that is solid enough to support the weight of an animal. These are known as quaking bogs because they quake and undulate with every step. Given enough time, seeds of vascular plants germinate, and leaves and debris blow onto the bog and decompose, slowly converting it to a rich soil.

Division Hepatophyta: Liverworts

Like mosses, **liverworts** are small plants that have an alternation of heteromorphic generations (Table 20-5). The smallest individuals of *Cephaloziella* are only 150 by 2,000 μm. Few species have plants that ever become large, the maximum being approximately 5 by 20 cm in *Monoclea*. Some species are leafy and greatly resemble mosses (Figure 20-21a); others form small, solid, ribbon-like gametophytes (see Figure 20-8 and Figure 20-21b). The sporophyte is even less conspicuous than in mosses and is also completely dependent on the gametophyte.

The Gametophyte Generation

Hepatic gametophytes are divided into two basic groups: **leafy liverworts** (orders Jungermanniales and Haplomitriales) and **thallose liverworts** (Figures 20-21 and Figure 20-22). In both, the gametophyte phase is initiated when spores germinate and establish a small, temporary protonematal phase. Liverwort protonemata are never as extensive or ramified or as long lived as those of mosses. Instead, after only a few cells are produced, an apical cell is established and growth of the gametophore begins.

The gametophore of leafy liverworts greatly resembles that of a moss—thin leaves on a slender stem; however, liverwort leaves typically have two rounded lobes with no midrib and no conducting tissue, whereas moss leaves are pointed and usually have a midrib. Liverwort leaves are arranged in three clearly defined rows, the leaves of two rows being much larger than those of the third. In prostrate liverworts, in which the shoot grows flat against the substrate, the underside row of leaves is reduced and may even be completely suppressed or replaced by rhizoids.

The gametophore stem grows by an apical cell with three, four, or five sides, one uppermost and the others embedded in the stem. Stem tissue is just simple parenchyma; no thick-walled cells or conducting cells occur in the majority of species. Virtually all liverworts have some or many cells that contain characteristic oil bodies.

Thallose liverworts show less resemblance to mosses. They are not leafy at all but rather flat and ribbon like or heart shaped

TABLE 20-5	Classification of Division Hepatophyta
Class Hepatopsida	
Order Marchantiales	*Conocephalum, Marchantia, Reboulia, Riccia, Ricciocarpus*
Order Sphaerocarpales	*Riella, Sphaerocarpos*
Order Monocleales	*Monoclea*
Order Metzgeriales	*Fossombronia, Pallavicinia, Pellia*
Order Jungermanniales	*Cephaloziella, Frullania, Jungermannia, Porella*
Order Haplomitriales	*Haplomitrium*

and bilaterally symmetrical. The body is sometimes referred to as a **thallus** (plural, thalli), a body without roots, stems, and leaves. Currently, "body" is much more commonly used than "thallus." Bodies of thallose liverworts are stratified and tend to be much thicker than those of leafy liverworts and mosses. The side next to the substrate bears unicellular rhizoids, and many cells contain large oil drops. Cells in the side away from the substrate have no oil but are rich in chlorophyll. The cells are loosely arranged as an aerenchyma with large air chambers that open to the exterior by means of large **air pores** (Figure 20-23). Air pores are not stomata—they have no guard cells and cannot be closed.

Some species are even simpler. *Sphaerocarpos texanus* is a small, thin ribbon a few cells thick at the center, but the rest of the body is only one cell thick. It has simple rhizoids on the bottom but no internal air chambers and no scales or other types of vegetative differentiation. These plants are basically as simple as the sea lettuce alga, *Ulva*, or the charophyte, *Coleochaete*. The liverworts *Pallavicinia* and *Pellia* are also this simple.

Liverwort gametophores may be either bisexual, producing both antheridia and archegonia, or unisexual, depending on the species. Leafy liverworts may bear their gametangia either mixed with regular leaves or positioned on specialized side branches and surrounded by modified leaves. In thallose

Scales   Groove with gametangia   Scales folded over desiccated but living body

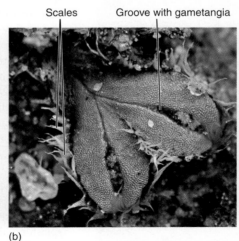

(a)   (b)   (c)

Figure 20-21 (a) Leafy liverworts such as this *Lophocolea* can be very easily confused with mosses. Their distinguishing features (not visible at this low magnification) are their typically lobed leaves, which grow from two apical points, and their basal pouch. Moss leaves are never like this. (b) Thallose liverworts have rather thick bodies and are not easily confused with mosses. This is a moist, active *Oxymitra incrassata*; notice its papery scales. White dots are air pores; clefts contain gametangia. When young, the gametophyte had one apical meristem and one cleft (upper left), and then the apex divided dichotomously so two apices and two clefts are now present. Each can divide again. (c) In dry seasons, plants of *Oxymitra* desiccate and fold inward, bringing the scales over their body, protecting themselves. They survive months of severe heat and drought.

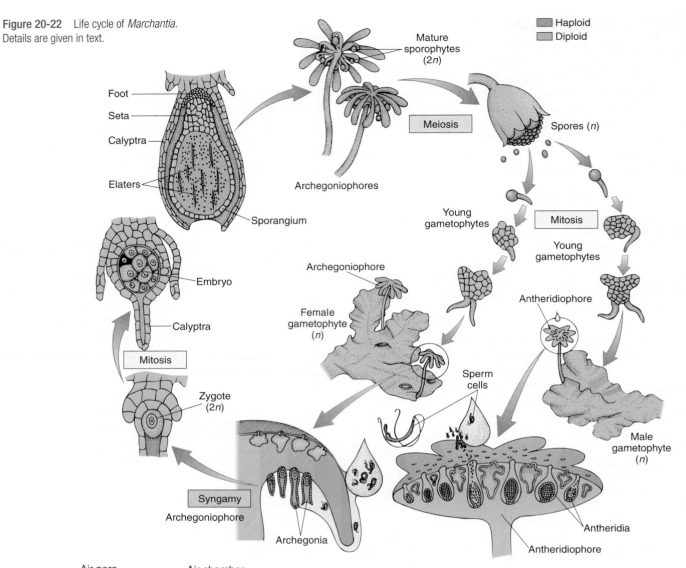

Figure 20-22 Life cycle of *Marchantia*. Details are given in text.

Haploid
Diploid

Foot
Seta
Calyptra
Elaters
Sporangium

Mature sporophytes (2n)

Meiosis

Spores (n)

Archegoniophores

Young gametophytes

Mitosis

Young gametophytes

Archegoniophore

Antheridiophore

Female gametophyte (n)

Embryo

Calyptra

Sperm cells

Mitosis

Zygote (2n)

Male gametophyte (n)

Syngamy

Archegoniophore

Archegonia

Antheridia

Antheridiophore

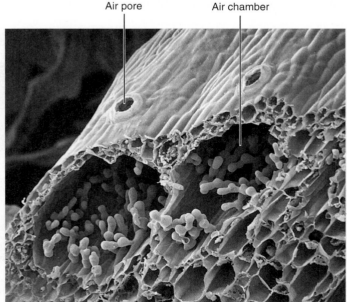

Air pore   Air chamber

Figure 20-23 Although gametophytes of thallose liverworts may become very thick (a few millimeters), they are never solid; rather, they have numerous chambers, and both carbon dioxide and water vapor can diffuse through them easily (×425).

species, antheridia and archegonia may be grouped together and surrounded by a tube of chlorophyllous cells. *Marchantia* is probably the most familiar thallose liverwort because it grows easily on moist soil in greenhouses. Its gametangial production is particularly elaborate. Male gametophores of *Marchantia* produce an umbrella-shaped outgrowth called an **antheridiophore** (**Figure 20-24**). It has a stalk several millimeters tall, and dozens of antheridia grow from its upper surface, each surrounded by a rim of sterile cells. **Archegoniophores** also are stalked, but their apex is a set of radiating fingers that project outward and droop downward; the underside has numerous archegonia (**Figure 20-25**). Archegoniophores and antheridiophores occur on separate plants in *Marchantia*; antheridia and archegonia are not produced by the same gametophyte.

If sperm cells are carried to the archegoniophore by raindrop splashing, they swim into the archegonium neck and fertilize the egg. The zygote is retained and grows into a small sporophyte (**Figure 20-26**). Surrounding gametophore tissue expands with it temporarily, forming a protective sheath, and the archegonial neck expands into a calyptra. In *Marchantia*, sporophytes are so small that each archegoniophore can support many of them.

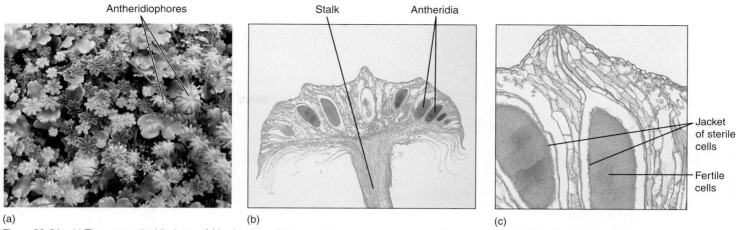

Figure 20-24 (a) These are antheridiophores of *Marchantia*; as their name indicates, these structures bear (-phore) antheridia. These are gametophytic tissues, and the antheridia are hidden within the upper surface (four archegoniophores are also present). (b) Low-magnification view of antheridiophore with antheridia (×8). (c) Like antheridia in mosses and all other embryophytes, those in liverworts consist of a jacket of sterile cells surrounding fertile cells that differentiate into sperms (×32).

The Sporophyte Generation

Little variability exists in the sporophytes of most liverworts, and their basic morphology is like that of mosses. In fundamental structure, however, the two groups are quite distinct. Most liverwort sporophytes have a foot, seta, and calyptra-covered sporangium, but the seta is extremely delicate, composed of clear, thin-walled cells that collapse quickly. Many members of Marchantiales have no seta at all. The sporangium is globose and while young is bright green and chlorophyllous, but it has no stomata. The outer layer of sterile cells is much thinner than in moss sporangia, often only one cell thick. The liverwort sporangium lacks a columella, the central mass of sterile cells found in mosses; dehiscence occurs by means of four longitudinal slits, not by an operculum, and there are no peristome teeth. Within the sporangium, some cells do not undergo meiosis but rather differentiate into **elaters**—single, elongate cells with spring-shaped walls. When the sporangium opens, the elaters uncoil, pushing the spores out (Figure 20-27). All liverworts are homosporous.

Strikingly simple sporophytes are produced in *Riccia* and *Ricciocarpus*. The zygote grows into a spherical mass within the archegonium of the gametophore body. No foot or seta is formed; instead, the inner cells of this mass undergo meiosis and produce spores. As the surrounding gametophore and sporophyte tissues age, die, and decay, the spores are finally liberated.

Figure 20-25 (a) Archegoniophores of *Marchantia* are easily recognizable because they have finger-like segments radiating from the stalk. Archegonia are located on the underside (×3). (b) Each archegonium consists of a long, tubular neck and a slightly swollen base, all only one cell thick. Each archegonium contains only one egg (×160).

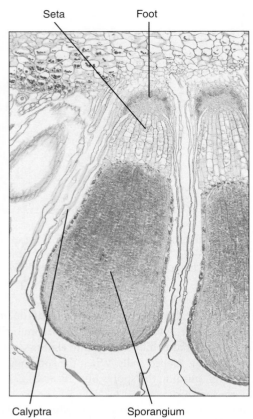

Figure 20-26 Liverwort sporophytes consist of foot, seta, and capsule (sporangium), but no elaborate set of teeth as in mosses. Instead, the apex breaks into several segments, all of which curl back and release the spores (×20).

Division Anthocerotophyta: Hornworts

Hornworts are a group of small, inconspicuous thalloid plants that grow on moist soil, hidden by grasses and other herbs (see Figures 20-6, 20-9, and 20-28 and Table 20-6). They rarely inhabit tree trunks or bare rock; therefore, one does not often encounter them unless looking for them specifically. Roadside cuts and stable eroded soil in shade are good sites for hornworts. Hornworts number approximately 100 to 150 named species in five or six genera, with *Anthoceros* and *Phaeoceros* being common examples. Many species names are probably

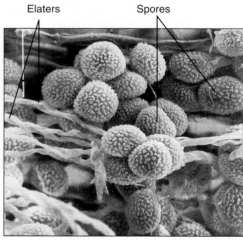

Figure 20-27 Spores and elaters of the liverwort *Haplometrium hookeri*.

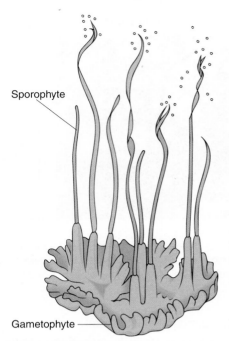

Figure 20-28 The "horns" of a hornwort are sporophytes that grow continuously from a basal meristem. The lower part of the sporophyte is surrounded by gametophyte tissue.

synonyms—that is, several names applied to the same species because many hornworts have such variable growth forms that individuals of the same species can differ greatly under varied conditions.

Hornworts superficially resemble thalloid liverworts, the gametophores of many species being ribbon shaped and thin, without a distinct stem or leaves; however, hornworts never contain oil bodies, whereas liverworts almost always do. As in all other embryophytes, an alternation of heteromorphic generations occurs, with the sporophyte depending on the larger, photosynthetically active gametophyte, but hornworts are quite distinct from all other embryophytes, including the other nonvascular plants. One of the most striking features is the presence of a single large chloroplast in each cell as opposed to the numerous small plastids present in all other nonalgal plants (Figure 20-29). The single chloroplast per cell is characteristic of many algae and is surprising in hornworts. Furthermore, hornwort chloroplasts have a pyrenoid, an algal feature absent in all other embryophytes. Hornwort chloroplasts probably represent the condition of the ancestors of all embryophytes. Unfortunately, we do not know what the chloroplasts of the earliest plants were like because they are not preserved in any known fossil.

The Gametophyte Generation

As few as three or four protonema cells are produced before the gametophore phase is established in most species. Gametophores are always thin, at least along the edges. Only in the center do they become more than four or five cells thick. They may be

TABLE 20-6	Classification of Division Anthocerotophyta
Class Anthocerotopsida	
Order Anthocerotales	*Anthoceros, Dendroceros, Megaceros, Notothylos, Phaeoceros*

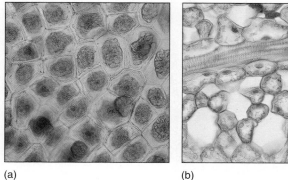

(a)   (b)

Figure 20-29 (a) Hornwort cells are unique among embryophytes: Each has just one plastid. Outside of hornworts, this feature is known only in algae. Although it may seem trivial, it must represent a significantly distinct cell/plastid relationship. The plastid's reproduction must be carefully coordinated with that of the cell, and a special mechanism is necessary to ensure that both progeny cells receive one each (×60). (b) Cells of privet leaf with many chloroplasts each (×60).

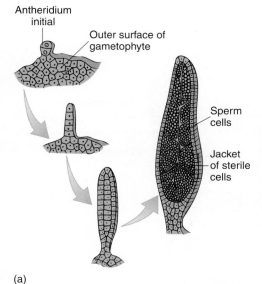

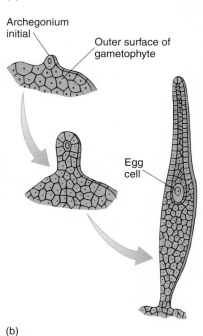

(a)

(b)

shaped like a ribbon or a heart, or they may grow outward irregularly, forming a disk. The upper surface in many is smooth, but in others thin, chlorophyllous lamellae grow upward. The gametophyte is parenchymatous, rather succulent but brittle. It does not tolerate drying; gametophores of hornworts typically live for less than 1 year in temperate climates. They act as winter annuals, appearing in the cool, moist autumn months, growing during winter, producing sporophytes in the spring, and dying before summer. In some species, they form oil-rich "tubers" as inner cells fill with oil and outer cells die; one thallus can produce several tubers.

Internally, hornwort gametophytes have numerous chambers. Young plants have mucilage chambers formed as cells break down and their contents are altered chemically into mucilage. These become invaded by *Nostoc* cyanobacteria. All hornworts form this symbiosis and presumably benefit by receiving nitrogen compounds from the *Nostoc*.

Gametangium development in hornworts is distinctive (**Figure 20-30**). Unlike all other embryophytes, the antheridial initials are not surface cells on the hornwort gametophyte. Instead, a special mucilage chamber forms near the upper surface; then cells lining the chamber grow into it and become antheridia. The chamber expands as the antheridia grow, and finally, its roof breaks; only then are antheridia exposed to the environment. As

Figure 20-30 (a and b) In mosses, the initial cell in the formation of either an antheridium or an archegonium lies on the surface of the gametophyte. Then cell divisions change this single cell into a mass that has sterile jacket cells and reproductive central cells. This pattern occurs in all mosses, liverworts, and vascular plants. The gametophytes of flowering plants are so reduced that the pattern is not obvious (c and d) The gametangia of hornworts are unique; the antheridia (c) do not form on the true surface of the gametophyte, and the egg (d) is not surrounded by discrete archegonial cells.

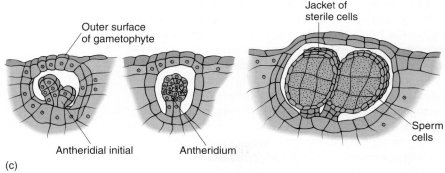

(c)

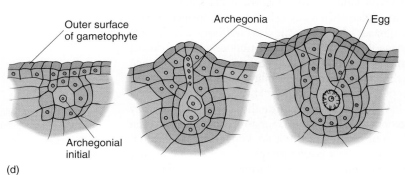

(d)

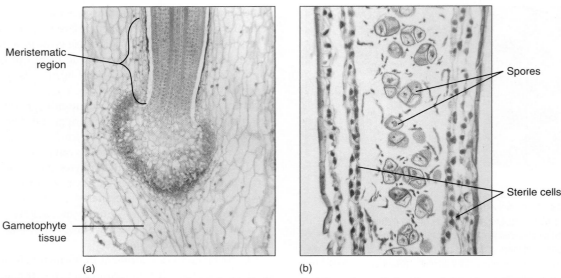

(a)   (b)

Figure 20-31 (a) The base of the sporophyte resembles a foot embedded in the gametophyte, and recently transfer cells have been discovered; therefore, active nutrient transport into the sporophyte must be occurring. Just above the foot is a meristematic region. (b) At a higher level, equivalent to Figure 20-32c, spores are mature.

sperm cells mature, sterile outer cells of antheridia transform their chloroplasts into chromoplasts, becoming orange or yellow.

Archegonia are formed from superficial cells, but archegonia do not completely surround the egg as do the flask-shaped archegonia of other embryophytes. Rather, the egg lies below a short neck and neck canal but is surrounded by vegetative thallus cells. Even the neck cells are not particularly distinct but rather are difficult to distinguish from ordinary thallus parenchyma cells.

After fertilization occurs, the zygote divides longitudinally; in mosses and liverworts, it divides transversely.

■ The Sporophyte Generation

Similarities between sporophytes of hornworts and those of mosses or liverworts are not easy to find. Hornworts have a foot embedded in gametophore tissue, but there is no seta or discrete sporangium (see Figure 20-28, Figure 20-31, and Figure 20-32). Instead, just above the foot is a meristem that continuously produces new sporangium tissues. As the newly formed cells are pushed upward, they grow, differentiate, mature, and die. They are simultaneously being replaced by more cells from the basal meristem. Consequently, the sporangium is a long, horn-like cylinder, typically 1 or 2 cm long in *Anthoceros* and *Phaeoceros,* but up to 12 cm in some species. At the tip, the sporangium is mature and open as a result of dehiscence along two linear apertures. The outer layer of sterile cells is thick, up to six cells deep, and chlorophyllous. It has stomata in *Anthoceros* and *Phaeoceros* but lacks them in *Notothylos, Dendroceros,* and *Megaceros.* Spores are green, golden yellow, brown, or black and in some species are multicellular when ready to be released. Hornworts have a columella as in mosses, but unlike the elaters of liverworts, those of hornworts (often called pseudoelaters) are multicellular and do not have spirally thickened walls. The basal meristem is active over a long period, depending on moisture availability and temperature, and large numbers of spores can be produced by each sporophyte. Several attempts have been made to remove the sporophyte surgically from the gametophyte and grow it in laboratory conditions, but even though it is chlorophyllous, it dies. It is not known whether death is caused by a lack of minerals, insufficient photosynthesis, or absence of growth factors, vitamins, or hormones from the gametophyte.

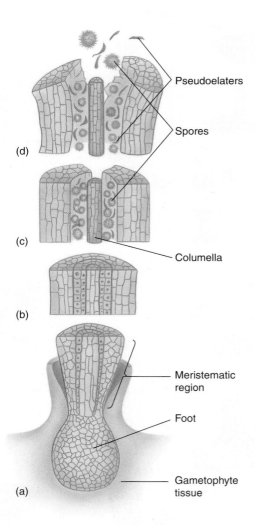

Figure 20-32 A longitudinal section through the sporophyte. At the base is a meristematic region (a); higher, above the basal meristem, sporocytes (spore mother cells) undergo meiosis (b). At higher levels the spores become mature (c) and are then released (d). This type of continuous meiosis is unknown in plants other than hornworts.

Summary

1. True plants first originated at least 450 million years ago, having evolved from green algal ancestors. Necessary modifications would have included methods of water conservation: low surface-to-volume ratio, compact body, cuticle, and coordination of reproduction with presence of water.

2. The presence of plants on land changed the land as a habitat; with crowding, competition for light would have become important and an upright habit selectively advantageous.

3. The evolution of vascular tissue had profound consequences: Relatively strong, vertical stems could be supported. Large plant bodies could be integrated, and parts could differentiate and specialize.

4. Nonvascular plants are classified here in three divisions, emphasizing their differences and possible separate evolutionary origins: Bryophyta (mosses), Hepatophyta (liverworts), and Anthocerotophyta (hornworts).

5. All three groups of nonvascular plants are embryophytes. They have gametangia and sporangia in which only the internal cells differentiate into gametes or spores; a jacket of sterile cells is always present.

6. In all three groups of vascular plants, the life cycle consists of an alternation of heteromorphic generations. The gametophyte is the larger and more persistent and photosynthetically active phase; the sporophyte depends almost entirely on the gametophyte for carbon, energy, and minerals.

7. The setae, stems, and leaf midribs of some mosses have cells that facilitate long-distance conduction: Leptoids transport sugars, and hydroids transport water. It is not known whether they are homologous or merely analogous to xylem and phloem.

8. Archegonia and antheridia may be borne on the same gametophore or on distinct, unisexual gametophores.

9. Neither the egg nor the zygote is released. The new sporophyte develops initially within the archegonium. When fully mature, it has a foot and sporangium; many have a seta.

10. A moss spore grows into a filamentous protonema that produces buds that develop into thick parenchymatous gametophores with stems, leaves, and rhizoids.

11. Liverworts may be either thallose or leafy. Liverworts contain oil droplets, have bilobed leaves, and have no thick leaf midrib, no columella in the sporangium, and no operculum or peristome teeth.

12. Hornwort sporophytes differ from those of all other plants: They have a foot and a basal meristem, and thus, sporogenous tissue is formed continuously.

13. Hornwort antheridia do not arise from surface cells, and the egg is not completely surrounded by distinct archegonial tissue.

Important Terms

air pores	elaters	peristome teeth
antheridiophore	foot	protonema
antheridium	gametophores	rhizoids
archegoniophores	hornworts	seta
archegonium	liverworts	thallus
calyptra	mosses	
capsule	operculum	

Review Questions

1. Some vascular plants produce seeds; others do not. Vascular plants that do not produce seeds are known as _____ _____, whereas vascular plants that do produce seeds are known as _____. Are there any plants that produce seeds but which do not have vascular tissue?

2. What are some of the modifications necessary if an alga is to become evolutionarily adapted to living on land? Is a single modification sufficient, or are several necessary?

3. Why would it be necessary for an evolutionary line to develop stomata and guard cells before it developed an extremely impervious cuticle? Why must vascular tissues precede the evolution of roots and active apical meristems?

4. The following organisms are often called mosses, but they are not actually closely related to mosses at all. What groups of plants do they actually belong to?
   a. Spanish moss
   b. Club mosses
   c. Slimy, bright green "mosses" of ponds and slow-moving streams
   d. "Reindeer moss"

5. What are the three groups of nonvascular plants? How would you determine whether an unknown specimen is a vascular plant?

6. The nonvascular plants of this chapter are believed to be true plants, just as ferns, conifers, and flowering plants are

true plants; however, there are two tissues that the nonvascular plants do not have. Which two tissues?

7. If the leptoids of mosses were found to contain a protein whose gene had the same nucleotide sequence as the gene that codes for P-protein, would that be significant evidence for either the homology or analogy of leptoids and phloem?

8. You will see sporophytes only if you examine mosses closely. They look like green or brown "_____" standing up on the green gametophyte, but sporophytes are (circle one: present almost all the time, only present at certain times of the year).

9. Do mosses have an alternation of isomorphic or heteromorphic generations? That is, can you easily tell a moss gametophyte from a moss sporophyte? When we look at leafy green moss plants, what are we seeing—the gametophyte or the sporophyte? In a flowering plant species, would the equivalent stage be the plant or the pollen grains and megagametophytes?

10. The leafy, green moss plants that are so familiar are gametophytes, haploid plants. This is very different from flowering plants and other seed plants. Does a leafy green moss plant grow from a spore or from a fertilized egg? Does the moss plant have both a paternal parent and a maternal parent?

11. Draw a single moss plant, similar to the one in Figure 20-10. Be certain to show the gametophyte and the sporophyte. Now draw one without the sporophyte, showing only the gametophyte. The sporophytes usually have only a very brief life, and after they shed their spores, the gametophytes let them die.

12. Draw and label the life cycle of a moss; be certain to show gametangia and sporangia. Which parts are haploid and which are diploid? Where and when does meiosis occur? Plasmogamy? Karyogamy?

13. In the majority of mosses, which lack hydroids and leptoids, water is conducted along the _____ of the plant by _____ action.

14. The leafy, green moss plants, being gametophytes, have gametangia, structures that produce gametes. What is the name of the gametangium that produces sperm cells? The gametangium that produces egg cells? Can one single moss gametophyte bear both of these? Do some species have plants that produce only sperm cells? Other plants that produce only egg cells?

15. The sporophyte of a moss usually has a stalk called a _____ and a simple apical sporangium called a _____.

16. Many people often think of mosses as plants adapted to rainy areas, areas that are usually wet. Are any mosses adapted to deserts? Can some mosses lose much of their water—the way a seed does before being planted—and still survive?

17. The liverwort *Marchantia* is one of the largest and most common. There is a good chance that you will find it if you search carefully in moist places (you may have to search many places). Is it a leafy liverwort or a thallose one? Describe its surface texture (see Figure 20-8). If you are very lucky, you may find it with archegoniophores and antheridiophores. What are these structures, and how would you recognize them if you saw them (what do they look like)?

18. Unlike *Marchantia,* some liverworts are as simple as a true plant can possibly get. Describe the body of *Sphaerocarpos texanus.* If you were shown one of these plants, how would you be certain it was not an alga (Hint: It would be almost impossible; there is only one way, not mentioned in the text. Algae tend to have only one chloroplast per cell; true plants—including *S. texanus*—always have many, except in the hornworts)? The point of this question is to have you think about how little difference there is between some algae and some true plants.

19. What are some of the ways in which liverworts differ from mosses? How do hornworts differ from both? Do the three have similar life cycles?

20. What are the "horns" of hornworts? What do they produce? They have a meristem. Where is it located?

21. An important consideration in the evolution of any organism is gene flow. What are some of the mechanisms by which genes move through the habitat in nonvascular plants? In a dense, cool forest, how strong are wind currents? Could they carry spores very far? What would you guess might be the maximum distance sperms can swim? How far can a raindrop splash a sperm or a spore?

BotanyLinks

http://biology.jbpub.com/botany/5e

BotanyLinks contains a variety of resources and review material designed to assist in your study of botany.

Visit the Web Exercises area of BotanyLinks to complete this question:

1. Do bryophytes affect your life in ways you do not realize, as is true for algae? Go to the BotanyLinks home page to begin researching this subject.

Vascular Plants Without Seeds

chapter 21

Concepts

What were the early vascular plants like? What features did they have, and what changes occurred as they diversified and evolved into the many types of plants alive today? Were they simple like some liverworts and hornworts and then gradually became more complex? Or were they completely different, with some giving rise to vascular plants by becoming more complex, whereas others gave rise to nonvascular plants by becoming simpler? Given all of the living plants present now plus the few fossils that we have from the time of the first embryophytes (at least 450 million years ago), what logical hypotheses can be proposed about the early embryophytes?

First, all extant plants (and all known fossil plants) have a **dibiontic life cycle**: Each species has a multicellular gametophyte and also a multicellular sporophyte. Some algae, especially *Coleochaete*, are **monobiontic**, having only one multicellular generation: A zygote undergoes only meiosis, producing more spores that will grow into new gametophytes. The zygote cannot undergo mitosis and cannot grow into a sporophyte. Thus, an important step in the evolution of embryophytes was the conversion of a monobiontic ancestor into dibiontic plants. A sporophyte generation had to come into existence. One hypothesis,

Chapter Opener Image: This is not a typical image of ferns. These are fern "fiddleheads" more technically called croziers, which are the developing, unopened leaves. While young like this, fiddleheads are edible and are cooked as a vegetable. Once mature, fern leaves are inedible. In addition to having leaves, ferns have stems, roots, xylem, and phloem. Like mosses, ferns shed their spores and never produce seeds, consequently they are known as the seedless vascular plants or vascular cryptogams ("crypto" means hidden, referring to the "hidden" spores, which were difficult to see before microscopes were invented: people thought ferns had invisible seeds).

Outline

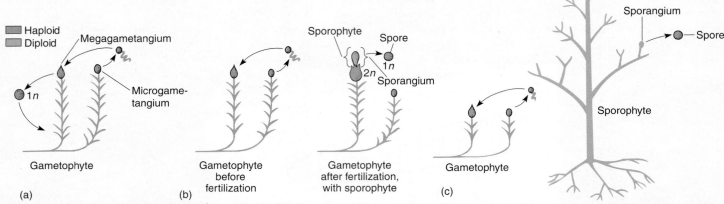

Figure 21-1 The interpolation theory. (a) The very earliest land plants were postulated as having no sporophyte; instead, the zygote "germinated" by meiosis, producing haploid spores that grew into new gametophytes. (b) At a later stage in evolution, the zygote would germinate mitotically and produce a simple sporophyte that in the early stages would have consisted of a sporangium and perhaps also a foot; this would have resembled a liverwort sporophyte. (c) With continued evolution, the sporophyte would have become progressively more elaborate while the gametophytes became simpler.

the **interpolation hypothesis**, postulates that a small sporophyte came into existence when a zygote germinated mitotically instead of meiotically (**Figure 21-1**). The sporophyte generation would have gradually evolved in complexity while the gametophyte generation remained small. A sporophyte generation was inserted (interpolated) into the monobiontic life cycle. In this hypothesis, nonvascular plants such as some liverworts might be intermediates in the progression from green algae to vascular plants. For example, the liverworts *Riccia* and *Ricciocarpus* have simple, almost

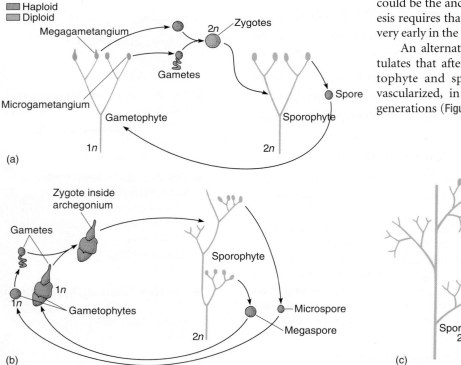

(a)

algae-like gametophytes, and their sporophytes consist of just a small globose sporangium with no foot or seta. The zygote undergoes several rounds of mitosis, then some cells undergo meiosis. It would take only a few evolutionary modifications to get such a simple sporophyte from an ancestor whose zygote merely underwent meiosis. If so, this sporophyte has been interpolated into the ancestor's monobiontic life cycle. From a *Riccia*-like condition, we could hypothesize that the sporophyte would evolve to be more elaborate, with a foot and seta, and then could later evolve to grow by an apical meristem, to branch, to live free of the gametophyte; it could be the ancestor of vascular plant sporophytes. This hypothesis requires that at least one of the nonvascular clades originated very early in the evolution of true plants.

An alternative hypothesis, the **transformation theory**, postulates that after the dibiontic life cycle originated, both gametophyte and sporophyte became larger, more complex, and vascularized, in a life cycle with an alternation of isomorphic generations (**Figure 21-2**). No living plants have gametophytes that

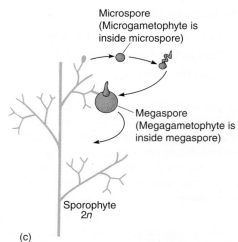

(c)

Figure 21-2 The transformation theory of the origin of the vascular plant life cycle postulates that in early land plants (a), gametophytes were upright and dichotomously branched, with epidermis, cuticle, and vascular tissue, just like sporophytes. (b) With time, sporophytes became larger and more complex, and gametophytes became simpler. In the species illustrated here, gametophytes have become so small that the microgametophyte develops within the spore wall and the megagametophyte protrudes from the spore only slightly. (c) With continued reduction, it is possible—but neither necessary nor inevitable—for the megaspore and its megagametophyte to be retained inside the megasporangium and remain on the parental sporophyte, an important step in the process of seed evolution.

look like sporophytes, but many algae do, and some fossil plants did: Some fossils of early vascular plants bore gametangia and grew among other similar plants that bore sporangia. The transformation theory postulates that these early ancestors diverged into two clades: (1) nonvascular plants in which sporophytes became much simpler and dependent on the gametophytes, and (2) the rest of the vascular plants, in which sporophytes became increasingly elaborate whereas gametophytes became very reduced.

Currently, one issue that is resolved is that all living and most fossil plants are dibiontic with an alternation of heteromorphic sporophytes and gametophytes. What we do not know is whether some liverworts or hornworts are relatively unchanged descendants of very early unvascularized ancestors or whether they originated later from more complex, vascularized ancestors and have since become simpler. This is discussed further after several early vascular plants have been described.

Early Vascular Plants

Rhyniophytes

The earliest fossils that definitely were vascular land plants belong to *Cooksonia*, a genus of extinct plants (Figure 21-3). These had upright stems that were simple, short cylinders (several centimeters long) with no leaves (they had "naked stems"). They had **equal dichotomous branching**, both branches being of equal size and vigor (Figure 21-4). Plants of *Cooksonia* had an epidermis with a cuticle, a cortex of parenchyma, and a simple bundle of xylem composed of tracheids with annular secondary walls. The ends of the branches were swollen and contained large, multicellular masses of sporogenous tissue surrounded by several layers of sterile cells. As in all plants, only the central cells were sporogenous, and the sporangium had to open to release the spores.

Plants of *Cooksonia* were homosporous; there were no separate microspores and megaspores. Fossils that have these general characters are called **rhyniophytes**.

Rhynia and *Aglaophyton* were other rhyniophytes, early vascular plants similar to *Cooksonia* (Figure 21-5). They had a prostrate rhizome, upright naked stems, and terminal sporangia. Stomata and guard cells occurred in the epidermis, and there was a layer that appears to have been a cuticle.

In the same rocks with *Rhynia* and *Aglaophyton* are fossils of similar plants. Two, *Lyonophyton* and *Sciadophyton,* were gametophytes, not sporophytes. The ends of their stems bore flattened cup-shaped areas that contained gametangia, both antheridia and archegonia, but not sporangia. These plants had upright, dichotomously branched stems, vascular tissue, stomata, and a cuticle. These were much larger and more complex than the gametophytes of any mosses, liverworts, or hornworts that are alive today. It is possible to suspect that because *Rhynia*-type sporophytes occur together with *Sciadophyton*-type gametophytes, they might have been alternate phases of the same species. If true, these plants had an alternation of isomorphic generations (Figure 21-2a), and later evolution into seed plants involved reduction of the gametophyte to just a few cells and elaboration of the sporophyte into a more complex plant. These may have given rise to the nonvascular plants by reduction of both gametophytes and sporophytes.

Another fossil from that same time, *Horneophyton*, had sporophytes with naked axes that branched dichotomously and were up to 20 cm tall (Figure 21-5c). They had stomata and terminal sporangia. Each sporangium had a short columella, and just as in hornworts, the sporogenous tissue surrounded the columella on all sides and the top. The stem base was swollen, similar to a hornwort foot, except that it bore rhizoids. The similarities between the hornworts *Anthoceros* and *Phaeoceros* and *Horneophyton* raise the possibility that vascular plants may have arisen from hornwort-like ancestors (Figure 21-5d, from left to right), but it is also possible to think about these transitions in reverse— that hornworts evolved from species like *Horneophyton* by means of an elaboration of the gametophyte and a reduction of the sporophyte (Figure 21-5d, from right to left).

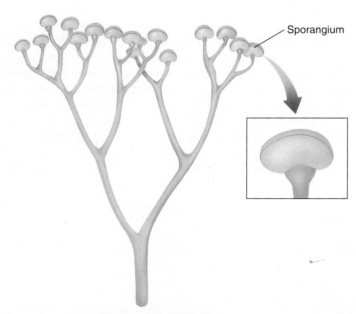

Figure 21-3 A reconstruction of *Cooksonia caledonica,* the earliest known plant that had xylem—tracheids with annular secondary walls. Important features are its sporangia at the ends of branches, the lack of leaves, and the dichotomous branching.

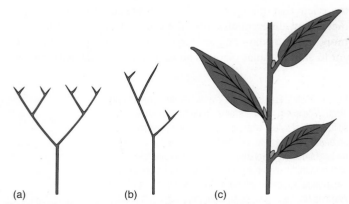

Figure 21-4 Branching patterns. (a) If a stem forks, resulting in two equal stems, it is a dichotomy or dichotomous branching. (b) It is pseudomonopodial branching if one stem is definitely larger and tends to form a trunk. (c) If one stem dominates the system absolutely, it is monopodial branching, as in most seed plants.

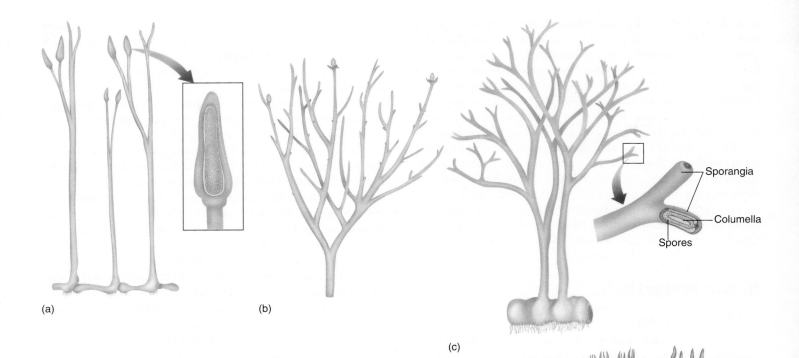

(a)

(b)

Sporangia

Columella

Spores

(c)

(d)

Figure 21-5 Reconstruction of *Aglaophyton major* (previously known as *Rhynia major*) (a) and *Rhynia gwynne-vaughanii* (b). They strongly resemble *Cooksonia*, and we know that they definitely had rhizomes, upright stems, and rhizoids. Inset shows the sporangium cut away, revealing spores. (c) Reconstruction of the extinct plant *Horneophyton* and its terminal sporangia. (d) At first, it was postulated that algae gave rise to true plants whose life cycle was dominated by the gametophyte and whose sporophyte was small and dependent, hemiparasitic on the gametophyte (as drawn on the left). It was postulated that this evolved into the vascular plants as the sporophyte became larger and more elaborate while the gametophyte became reduced (as drawn on the right). But it is possible that hornworts evolved by the reverse process, reading this series from right to left.

Xylem Structure of Early Vascular Plants   Early vascular plants had two types of xylem organization. In both, the center is a solid mass of xylem with no pith; this is a **protostele**. In an **endarch protostele**, protoxylem is located in the center and metaxylem differentiates on the outer edge of the xylem mass (**Figure 21-6a**). Protoxylem is the xylem that differentiates while cells are small and narrow, and metaxylem differentiates after the cells have expanded for a few more hours or days and are larger. The other type of stele present in early vascular plants is an **exarch protostele**, with metaxy-lem located in the center of the xylem mass and protoxylem on the edges as several groups next to the phloem (Figure 21-6b). Another type of stele, which did not evolve until later, is the **siphonostele**, one in which pith is present in the center, as occurs in the stems of ferns and **seed plants** (see Figure 21-29). Because xylem is often preserved well in fossils, its characteristics of exarch/endarch and protostele/siphonostele are usually available for study.

The xylem in many specimens of *Rhynia* and *Aglaophyton* is well preserved: It was a round cylinder without pith but with

Figure 21-6 Early vascular plants had two types of organization of xylem in their stems. In both, the center is a solid mass of xylem with no pith. This is a protostele. (a) In an endarch protostele, protoxylem is located in the center; metaxylem is on the periphery. Endarch protosteles occurred in the rhyniophytes. (b) In an exarch protostele, metaxylem is located interior to the xylem mass and protoxylem is on the edges, as several groups. Exarch protosteles are found in fossils of zosterophyllophyte stems.

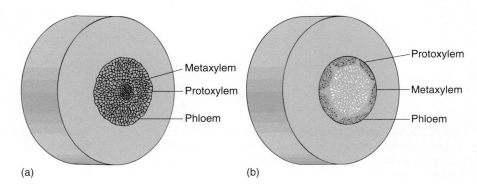

Metaxylem

Protoxylem

Phloem

Protoxylem

Metaxylem

Phloem

(a)

(b)

protoxylem in the center and metaxylem on the exterior—that is, an endarch protostele. All xylem cells were tracheids with annular thickenings. Around the xylem was a layer of phloem-like cells and then a parenchymatous cortex and epidermis.

Zosterophyllophytes

Another group of early vascular plants are the **zosterophyllophytes**, named after the principal genus *Zosterophyllum* (Figure 21-7a). They were small herbs without secondary growth. Many of their features were similar to those of rhyniophytes, but three characteristics make us think they were a distinct group: Their sporangia were lateral, not terminal; sporangia opened transversely along the top edge (Figure 21-7b), and their xylem was an exarch protostele, that is, protoxylem on the outer margin and metaxylem in the center (see Figure 21-6 and Table 21-1).

Although these distinctions may seem minor, we are certain that in all the more recent plants with large leaves (the euphyllophytes: ferns, conifers, flowering plants), sporangia are terminal rather than lateral; thus, plants such as *Rhynia* may have been the transitions between algae and later seed plants, but *Zosterophyllum* was not. Instead, some of the simplest vascular plants alive today, the lycophytes, have lateral sporangia, and they may represent a line of evolution based on *Zosterophyllum*-like ancestors (Figure 21-8).

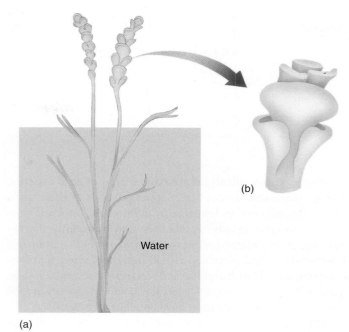

(a)

Figure 21-7 (a) *Zosterophyllum rehenanum* plants were quite similar to those of rhyniophytes, but the ends of fertile branches bore numerous lateral sporangia, not a single terminal one. Also, sporangia (b) opened by a suture that passed over the top of the sporangium, not up its side.

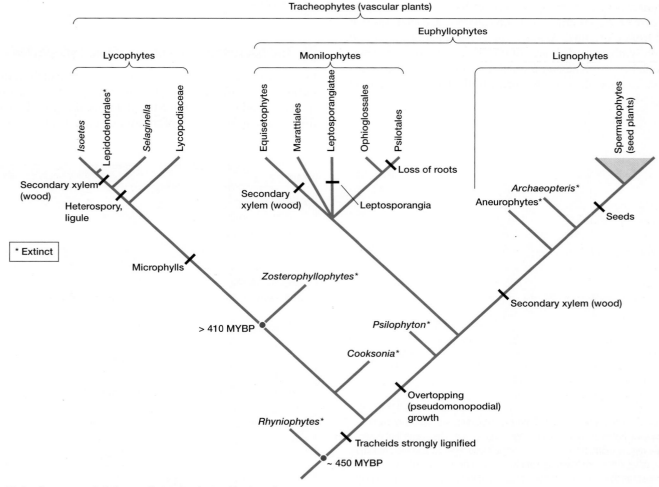

Figure 21-8 One proposed phylogeny of vascular plants without seeds.

TABLE 21-1	Characteristics of Rhyniophytes and Zosterophyllophytes	
	Rhyniophytes	Zosterophyllophytes
Sporangia		
location	terminal	lateral
dehiscence	along the side	across the top
Protoxylem	endarch	exarch

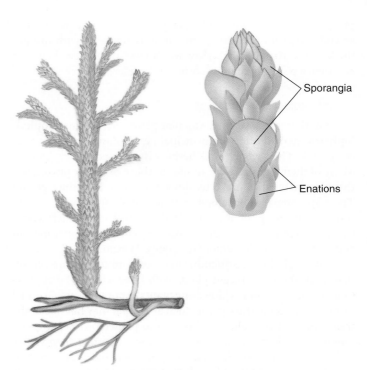

Figure 21-10 Reconstruction of *Asteroxylon,* an early lycophyte, showing its surface covered with enations, small flaps of photosynthetic tissue. In species without enations, stems were round and oriented vertically, not very good for harvesting sunlight.

Zosterophyllum plants (all are extinct) grew as small bunches, only approximately 15 cm high. Upper portions of their stems had cuticle, ordinary epidermal cells, and stomata, but lower portions did not, presumably because they grew in swampy, marshy areas. Stems of *Zosterophyllum* were naked (smooth), branched dichotomously, and contained a small amount of xylem that consisted of tracheids with annular and scalariform secondary walls. Sporangial walls were several layers thick, and all spores were the same size; therefore, they must have been homosporous.

Other genera of zosterophyllophytes show that significant morphological changes evolved quickly. In *Rebuchia,* sporangia occurred together on the ends of specialized branches (Figure 21-9). In *Crenaticaulis,* some branching was pseudomonopodial: Larger, trunk-like shoots bore smaller, shorter lateral shoots. These are the morphological bases for producing cones and trunks. Several zosterophyllophytes had a smooth surface (see Figure 21-7), but others had outgrowths called **enations** that ranged from quite small to long, thin scales. Enations increased the photosynthetic surface area of the plants, and in *Asteroxylon,* they contained stomata and a small trace of vascular tissue that ran from the stele through the cortex to the base of the enation (**Figure 21-10**).

The Microphyll Line of Evolution: Lycophytes

Lycophytes represent a distinct line of evolution out of the early land plants that resembled zosterophyllophytes. Lycophytes have lateral sporangia and exarch protosteles, and thus, they may have come from a *Zosterophyllum* type of ancestor (Figure 21-8 and Table 21-2).

Figure 21-9 This reconstruction of the zosterophyllophyte *Rebuchia ovata* shows several significant advances in body construction. The stems that bore sporangia were specialized, keeping sporangia close together and elevated into the wind; other stems branched repeatedly, forming a large photosynthetic surface. Some branches were larger than others, and thus, branching was no longer strictly equal dichotomies.

TABLE 21-2	Classification of Division Lycophyta	
Order Asteroxylales*		
Family Asteroxylaceae		*Asteroxylon*
Order Protolepidodendrales*		
Family Protolepidodendraceae		*Baragwanathia, Leclercqia, Protolepidodendron*
Order Lycopodiales		
Family Lycopodiaceae		*Lycopodium, Phylloglossum, Lycopodites**
Order Lepidodendrales*		
Family Lepidodendraceae		*Lepidodendron, Lepidophloios, Lepidostrobus, Sigillaria, Stigmaria*
Order Selaginellales		
Family Selaginellaceae		*Selaginella, Selaginellites**
Order Isoetales		
Family Isoetaceae		*Isoetes, Stylites, Isoetites**
Order Pleuromeiales*		
Family Pleuromeiaceae		*Pleuromeia*

*All species in this group are extinct.

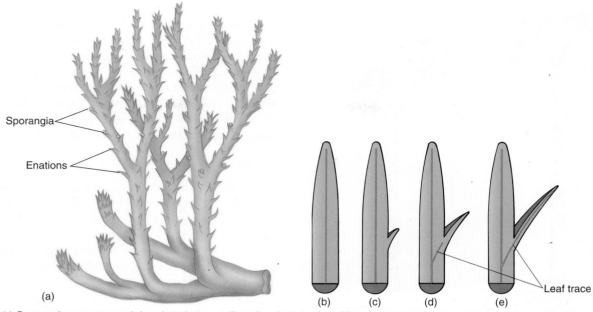

Sporangia

Enations

(a)

(b)　(c)　(d)　(e)

Leaf trace

Figure 21-11 (a) *Drepanophycus* was an early lycophyte that was still small and simple. (b–e) Microphylls in lycophytes are believed to have evolved as enations. Originally they were small, simple flaps of photosynthetic tissue (c). Later they became larger (d) and were vascularized (e). See Figure 21-20 for evolution of megaphylls.

■ Morphology

The earliest lycophytes were members of the genera *Drepanophycus* and *Baragwanathia* (Figure 21-11a). They were similar to their presumed ancestors, the zosterophyllophytes, with an important difference: Their enations were large, up to 4 cm long, and they contained a single well-developed trace of vascular tissue. Such enations must have increased photosynthesis, and they could be called leaves; however, "leaf" is an ambiguous term, and enations in the division Lycophyta are called **microphylls** for clarity (Figures 21-11b–e; the term "lycophyll" has been suggested

recently). "Micro-" refers to their evolution from small enations, not to their actual size. In some plants, they were up to 78 cm long. This is not the line of evolution that led to ferns and seed plants; microphylls are not the same as the leaves you are familiar with.

Another important advance was the evolution of true roots that allowed lycophyte sporophytes to anchor firmly, absorb efficiently, and thus to grow to tremendous size.

Many extinct lycophytes such as *Lepidodendron*, *Sigillaria*, and *Stigmaria* had a vascular cambium and secondary growth (Figures 21-12 to 21-14). Their wood looked remarkably like the

Figure 21-12 During the Carboniferous Period, Earth's vegetation was dominated by extensive forests of large lycophyte trees. None became truly massive like our redwoods, oaks, and elms, but the forests were both dense and vertically stratified. Flying insects were present, and early reptiles were just appearing; amphibians were the dominant land animals.

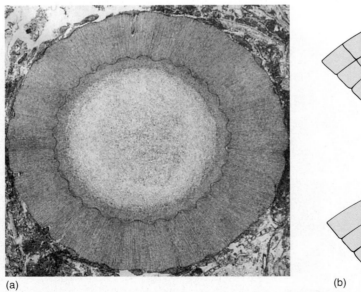

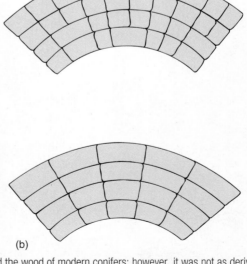

secondary xylem of pines and other living conifers, having a pith, rays, and elongate tracheids. However, the vascular cambium had one major flaw: Its cells apparently could not undergo radial longitudinal division; therefore, new fusiform initials could not be produced. As the wood grew to a larger circumference, cambial cells became increasingly wider tangentially (Figure 21-14b). No fossil has ever been found with wood more than about 10 cm thick. After that much secondary growth, the cambial cells may have stretched so much that they could no longer function.

■ Heterospory

In many extinct and extant lycophytes sporangia are clustered together in compact groups called **cones** or **strobili** (singular, strobilus), which protect them (Figure 21-15). Although many species remained homosporous, others became heterosporous, having microspores and megaspores that germinated to give rise to distinct microgametophytes and megagametophytes, respectively. Heterospory is a necessary precondition for the evolution of seeds (see Figure 21-2). In the lycophytes *Lepidostrobus* and *Lepidophloios*, the megaspore developed into a megagametophyte without enlarging: The megagametophyte existed completely within the wall of the original megaspore, which was up to 10 mm long. Furthermore, the megaspore in some species was retained within the sporophyll, protected by thick-walled cells of the sporangium. This is remarkably similar to ovules and seeds in seed plants, the most important difference being that in these lycophytes, the sporangium dehisced (much like modern anthers do), and the megaspore wall cracked, exposing the archegonia. Sperm cells could swim to the egg during fertilization; pollen tubes were unnecessary.

Figure 21-13 (a) Reconstruction of *Lepidodendron* showing a large tree with one distinct trunk (monopodial growth) and a well-developed root system. Leaves (microphylls) were sophisticated, and sporangia occurred in discrete strobili (cones). (b) *Sigillaria* was a large lycophyte tree in which the leaves usually were 1 m or more long. On the branch to the left, leaves have been left undrawn to reveal cones of sporangia.

Figure 21-14 (a) Wood of *Sigillaria* consisted of large tracheids and superficially resembled the wood of modern conifers; however, it was not as derived in its pitting and other features of its secondary wall, and it was formed by a vascular cambium that could not form new fusiform initials (×80). (b) In the vascular cambia of seed plants (top), as the cambium is pushed outward by accumulation of new secondary xylem, its circumference becomes greater, but fusiform initials divide by radial walls, creating new fusiform initials. An old cambium has many more cells than a young cambium. In arborescent lycophytes (bottom), cambial cells could not undergo this division; therefore, an old cambium had no more cells than a young one. Older cells became stretched circumferentially until they were no longer functional; secondary growth then ceased.

(a)

(b)

Cones

Figure 21-15 (a) *Lycopodium cernuum* is a common species that has extensive rhizomes, vertical chlorophyllous shoots, and sporangia clustered into cones. These shoots are leaning against and being supported by surrounding grasses and shrubs. (b) *Lycopodium obscurum*, with sporangia clustered into cones at the tips of branches. (c and d) *Lycopodium lucidulum* is one of several lycopod species in which sporangia are distributed among the leaves rather than in strobili.

(c)

(d)

Sporangia   Microphylls

Lycophytes are remarkable in that they represent an ancient line of evolution distinct from seed plants but having convergent evolution in several characters: leaves, roots, secondary growth, and almost seeds. In the Devonian and Carboniferous Periods (see diagram on the inside of the back cover), this group dominated the swampy areas of Earth with extensive forests of large trees (Figure 21-12), but most became extinct; currently, the entire division contains only five genera, *Lycopodium*, *Phylloglossum*, *Selaginella*, *Isoetes*, and *Stylites*.

■ Extant Genera

Lycopodium ("ground pine" or "club moss") is fairly common in forests from tropical regions to the arctic (Figure 21-15). All living species, approximately 200, are small herbs with prostrate rhizomes that have true roots and short upright branches. Microphylls are spirally arranged on their stems, and secondary growth never occurs. Sporangia may be arranged in cones or distributed along the shoots (*L. lucidulum*). All *Lycopodium* species are homosporous, a plesiomorphic (relictual) trait. Spores ger-

minate and grow into bisexual gametophytes that produce both antheridia and archegonia. In some species, gametophytes are green and photosynthetic; in others, they are subterranean and heterotrophic, nourished by fungi.

Selaginella is less common in temperate North America, and its plants are smaller and easily overlooked or mistaken to be mosses (Figure 21-16). Probably the best known species is the resurrection plant, *S. lepidophylla*, which curls up, turns brown, and appears dead upon drying but uncurls and regreens when moistened. Unlike *Lycopodium*, *Selaginella* has the apomorphic (derived) condition of being heterosporous, and the megagametophyte develops inside the megaspore wall (Figure 21-17a). The megaspore is not retained on the sporophyte, however, and is not seed like. Microgametophytes also develop within the spore wall and consist of a single vegetative cell and an antheridium (Figure 21-17b). Many flagellate sperms are produced and then released as the spore wall ruptures.

Selaginellas can be distinguished from lycopodiums by a small flap of tissue, the **ligule**, on the upper surface of *Selaginella* leaves. Although ligules are simple structures, and their adaptive

(a)

(b)

(c)

Figure 21-16 (a) Plants of *Selaginella willdenowii* are large and occur in moist habitats. (b) Plants of *S. lepidophylla* are also large, but occur in habitats that cycle between wet and severely dry. When dry, as shown here, the plant desiccates, curls up, and becomes dormant; the brown color is only on the underside of its leaves—the upper sides are green, even in this dry condition. If moistened with rain or dew, it quickly rehydrates, uncurls, and becomes active. (c) Plants of *S. wrightii* tend to be small, compact and easily mistaken for mosses. These are desert adapted and grow on bare rock in full sun throughout arid regions of the western United States.

advantages are unknown; they are ancient and can be used to distinguish early fossil selaginellas from fossil lycopods.

Isoetes is a genus of about 60 species of small, unusual plants called quillworts, which grow in wet, muddy areas that occasionally become dry (**Figure 21-18**). Their body consists of a small corm-like stem that has roots attached below and leaves above. *Isoetes* is heterosporous like *Selaginella,* and almost every leaf contains sporangia. Microphylls in this genus also have ligules. Weak cambial growth results in the production of additional cortex parenchyma to the exterior and a type of vascular tissue to the interior; the latter tissue is a mixture of tracheids, sieve elements, and parenchyma. *Stylites* plants are very similar to those of *Isoetes,* and several botanists have suggested that the two species of *Stylites* are really extreme forms of *Isoetes.*

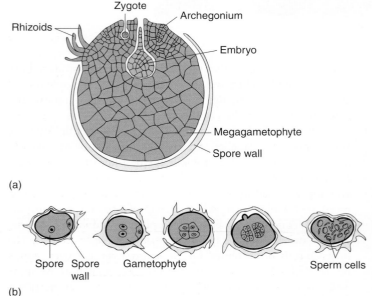
(a)

(b)

Figure 21-17 (a) The megagametophyte of some selaginellas develops almost entirely within the megaspore wall. The spore cracks open, exposing archegonia (megagametangia) and permitting fertilization by swimming sperm cells. (b) Microgametophytes of selaginellas develop within the microspore wall, then liberate motile sperms.

Figure 21-18 Although *Isoetes* is a small plant, its short basal corm-like stem has a small amount of secondary growth.

The anatomy and morphology preserved in fossil lycophytes indicate many instances of convergent evolution with seed plants. In both lines, elaborate, efficient leaves, wood, bark, and roots evolved. Sporangia became separated from nonreproductive organs and were grouped together into strobili. In both lycophytes and seed plants, heterospory evolved, as did endosporial development of the megagametophyte. In seed plants, megaspores were retained in the megasporangium and evolved into seeds; this nearly happened in certain extinct lycophytes. We cannot tell from the fossils whether convergent evolution of specialized metabolisms also occurred, but *Isoetes* does have CAM photosynthesis. It also has a unique means of obtaining its carbon dioxide—absorption by the roots from soil or mud. Its leaves have a thick cuticle but no stomata. We can only wonder what the metabolism of the extinct tree-like lycophytes was like. The extinction of so many of the lycophytes is unfortunate; they were complex, sophisticated plants that had many superb adaptations. They probably underwent interesting responses to changes in season, and they must have been able to resist numerous types of pathogenic bacteria, fungi, and insects. Those aspects of their biology will probably remain unknown to us forever.

The Megaphyll Line of Evolution: Euphyllophytes

Trimerophytes

Division Trimerophytophyta was proposed in 1968 for three genera of extinct plants, *Trimerophyton, Psilophyton,* and *Pertica.* Their fossils strongly resemble those of rhyniophytes, having terminal sporangia that dehisced laterally, homospory, dichotomous branching, and an endarch vascular cylinder of tracheids. Trimerophytes, however, are considered a distinct advancement out of rhyniophytes because of several special features. Most important is the trend of **overtopping**: Trimerophytes had an unequal branching in which one stem was more vigorous (see Figure 21-4b). In later species, the inequality was so pronounced that main stems and lateral stems can be identified easily. Finally, *Pertica* displays **pseudomonopodial branching**, that is, a single main trunk rather than a series of dichotomies (**Figure 21-19**). The plants have small lateral branches, some fertile and bearing sporangia and others sterile and acting as leaves.

Simultaneously, the positioning of branches became more regular and controlled. In rhyniophytes, the points of dichotomy were irregular and unpredictable, but in some species of trimerophytes, lateral branches were arranged in a regular spiral phyllotaxy. Other types of phyllotaxy that occurred were alternate, decussate (opposite leaves arranged in four rows), distichous (leaves in two rows), tetrastichous (alternate leaves, in four rows), and even whorled. Correlated with the evolution of a pseudomonopodial growth habit and the presence of numerous photosynthetic lateral branches was an increase in the vigor and robustness of the plants. Individuals of *Pertica quadrifaria* had stems 1.5 cm wide and about 1 m tall, whereas those of *P. dalhousii* are estimated to have been as much as 3 m tall, approximately the height of tall shrubs today. Although a dense stand of *P. dalhousii* would not be called a forest, there would have been vertical stratification of light, air movement, and humidity within the canopy.

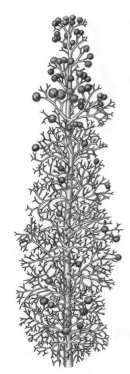

Figure 21-19 *Pertica quadrifaria* had one main trunk from which grew small branches; the smallest twigs still show dichotomous branching, but larger stems branched pseudomonopodially. This branch pattern also evolved separately in lycophytes, but here in the trimerophytes, it is the ancestor to the stem structure of seed plants. The globular structures are clusters of sporangia.

Trimerophytes became distinct from rhyniophytes during the Lower Devonian and existed until the Upper Devonian Period and then came to an end not by going extinct but rather by evolving into the ancestors of ferns and seed plants.

Origin of Megaphylls (Euphylls)

At least three distinct types of homoplasic (analogous) structures called leaves occur in plants: (1) leaves on gametophytes of nonvascular plants; (2) enations/microphylls of zosterophyllophytes and lycophytes; and (3) **megaphylls**, leaves that evolved from branch systems and are present in all seed plants, ferns, and equisetophytes. Megaphyll evolution is summarized by the **telome theory**. Imagine a plant like *Pertica* consisting of a main stem and smooth, cylindrical, dichotomously branching lateral stems (Figure 21-19). The ultimate twigs, those of the last dichotomy, are known as **telomes**. Now imagine that all subdivisions of a lateral branch become aligned in one plane (**planation**; Figure 21-20) and that parenchyma develops between telomes and even lower branches (**webbing**). This is not a leaf, but it has suitable characteristics to be the ancestor to leaves. Also, the leaves we see on trees and herbs of angiosperms are the result of about 300 million years of evolutionary refinement. If the branch system involved in this evolution produced sporangia, the resulting structure would not be just a leaf, but rather a **sporophyll**, a sporangium-bearing, leaf-like structure. For one reason or another, the plants with this organization outcompeted those with microphylls, the lycophytes. This is not to say that the megaphyll type of organization itself

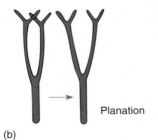
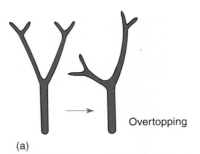
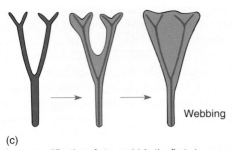

Overtopping   Planation   Webbing

(a)   (b)   (c)

Figure 21-20 The leaves of ferns, *Equisetum,* and seed plants are megaphylls (euphylls) that arose by evolutionary modification of stems. (a) In the first step, one branch (the main axis) overtops the other, which remains smaller and lateral. (b) Next, lateral branch systems restricted their branching to just one plane (planation); they stopped producing three-dimensional branch systems. (c) Finally, the spaces between close branches developed a thin sheet of chlorophyll-containing cells in a type of webbing.

was the main reason for their success, but today, megaphyllous plants are by far the more common. We believe that all megaphyllous plants form a monophyletic clade, now referred to as the **euphyllophytes**.

Monilophytes

Several current studies suggest that megaphyllous plants (**euphyllophytes**) are united by three synapomorphies: (1) their roots have exarch xylem, (2) they have megaphylls, and (3) they have a 30-kilobase inversion in the large single-copy region of their plastid DNA (**Box 21-1**). The euphyllophytes as presented in Figure 21-8 contains two sister clades, the **monilophytes** and the woody plants (**lignophytes**). Many monilophytes are plants we know as ferns (Leptosporangiatae, Marattiales, Ophioglossales); others have been called "fern allies" (equisetophytes, Psilotales). If we want to use the term "ferns" in the sense of a monophyletic, natural group, we must also refer to equisetophytes (horsetails) and Psilotales as ferns. This sounds strange to many experienced botanists because horsetails differ from all other ferns in many respects. Remember that Figure 21-8 is a model, a hypothesis. It would not be surprising to some of us if within a few years, new data would indicate that equisetophytes form a clade sister to ferns, but not part of ferns themselves. In this book, *Equisetum* and its clade will be treated as sister to the ferns.

Equisetophytes

Equisetophytes have been classified as division Arthrophyta (also called Sphenophyta). They consist of several genera of extinct plants and one genus, *Equisetum,* with 15 extant species known as **horsetails** or **scouring rushes** (Table 21-3). The living plants are all herbs without any secondary growth, and although certain species may attain a height of up to 10 m, they are usually less than 1 m tall (**Figure 21-21**). Their aerial stems have a characteristic jointed structure, with a whorl of fused leaves at the nodes. The leaves are small and have just a single trace of vascular tissue, but they are small megaphylls, not microphylls. If branches are

TABLE 21-3	Classification of Division Arthrophyta
Order Hyeniales*	
Family Hyeniaceae	*Hyenia*
Order Calamitales*	
Family Calamitaceae	*Arthropitys, Asterophyllites, Calamites*
Order Sphenophyllales*	
Family Sphenophyllaceae	*Sphenophyllum*
Order Equisetales	
Family Equisetaceae	*Equisetum, Equisetites**

*All species in this group are extinct.

Figure 21-21 *Equisetum* plants are usually small, having vertical shoots that arise from subterranean, highly branched rhizomes. In some species such as *E. hyemale* (a), each shoot is both photosynthetic and reproductive; in other species, such as *E. arvense* (b), some shoots are vegetative only (shown here), and separate reproductive shoots are produced at a different time of year. (c) This is *Equisetum giganteum*, in which the plants grow exceptionally large.

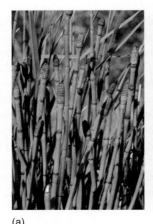

(a)   (b)   (c)

Botany and Beyond

Box 21-1 Molecular Studies of the Evolution of Early Land Plants

The evolutionary relationships of early land plants are still not understood well. Plants like mosses, liverworts, *Rhynia,* and lycopods are so simple that there are not many characters to compare. Furthermore, when two groups do resemble each other, we cannot always be certain whether the common characters are homologous (derived from the same ancestral character) or analogous (derived from different ancestral characters and resembling each other only as a result of convergent evolution).

Analysis of the arrangement of genes in chloroplasts has been used to investigate which early land plants are related. The circular DNA from chloroplasts has been isolated from many plants; then DNA restriction enzymes were used to analyze the order in which genes occur. Currently, many groups have been studied, and it turns out that mosses, liverworts, and lycophytes have one arrangement whereas euphyllophytes all have an alternative arrangement.

From numerous studies on many types of plants, it had been concluded that chloroplast genomes are quite stable and this type of rearrangement is rare. It seems unlikely that this type of rearrangement could occur twice, producing similar gene arrangements in three distinct groups. Thus, the presence of the rearrangement in all euphyllophytes is strong evidence that all constitute a monophyletic clade, that monilophytes, early woody plants, and modern seed plants all shared a common ancestor.

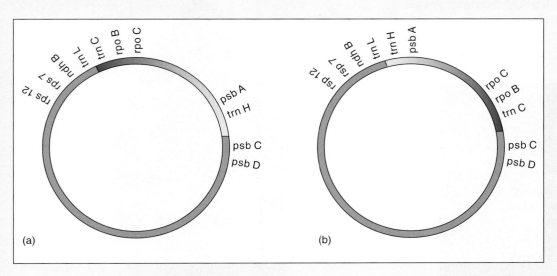

Figure B21-1   The order of genes on the circular DNA of plastids of liverworts, mosses, and lycophytes (a), and of seed plants (b). The section shown in blue is reversed in euphyllophytes.

present, they alternate with the leaves rather than being located in the leaf axils. The vertical, aerial stems arise from deep subterranean rhizomes; in harsh habitats, aerial shoots die during winter, but rhizomes persist and plants can spread vigorously. True roots are present, being produced at the rhizome's nodes.

Internal structure is as distinctive as external morphology (**Figure 21-22**). Stems have a pith, and thus, these are siphonosteles, not protosteles as in all other plants mentioned so far (compare Figure 21-22 with Figure 21-6). Protoxylem forms next to the pith, interior to the metaxylem, and thus, it is endarch, a rhyniophyte/trimerophyte trait distinctly different from the exarch protoxylem of zosterophyllophytes. Vessels are rare outside the flowering plants, occurring only in *Selaginella,* several ferns, three genera of gymnosperms, and *Equisetum.* Stem elongation causes some cells to be stretched and torn, forming canals. The cortex of *Equisetum* is composed of large cells, the outer

ones chlorophyllous. Air canals may be located in the outer cortex. The epidermis contains stomatal pores and guard cells, as well as such large amounts of silica that the stems are tough and rigid.

Reproductive structures in *Equisetum* are specialized; sporangia always occur in groups of five to ten located on an umbrellashaped **sporangiophore** (**Figure 21-23**). This has a short stalk and a flat, shield-shaped head from which the sporangia project, parallel to the sporangiophore stalk. Sporangiophores are always arranged in compact spirals forming a strobilus. *Equisetum* is homosporous; therefore, all plants have only one type of sporangiophore and strobilus, but in some species, the strobilus occurs at the tip of a green photosynthetic shoot, whereas in others it is borne on a special, colorless reproductive shoot.

After spores are released, they germinate on moist soil and develop into small (1 mm across) green gametophytes. They have

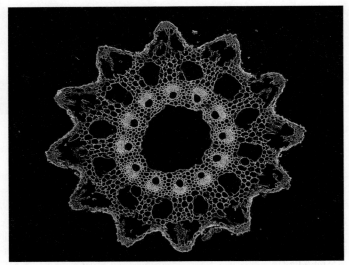

Figure 21-22 The stem anatomy of *Equisetum* is dominated by numerous canals. The center of the pith (this is a siphonostele) is torn apart, forming a central canal or pith canal, and around the remnants of the pith are carinal canals formed by breakdown of protoxylem. Just exterior to the carinal canals are metaxylem and phloem; vascular bundles are endarch. The cortex contains cortical canals that alternate with carinal canals, each cortical canal being located near a furrow in the stem surface. This highly characteristic structure allows us to identify fossils related to *Equisetum*.

Sporangiophores

Figure 21-23 Sporangia of *Equisetum* occur in groups of 5 to 10 clustered together on a sporangiophore. Sporangiophores are grouped into strobili.

no vascular tissue and no epidermis or stomata; the body is just a small mass of parenchyma. Gametophytes of *Equisetum* are either male or bisexual. Antheridia release numerous multiflagellate sperms that swim to the archegonia and eggs. The megagamete and zygote are never released, but rather are retained and nourished by the gametophyte.

The evolutionary line of arthrophytes can be traced back through the trimerophytes to the rhyniophytes. In two groups of early arthrophytes, the Sphenophyllales and the Calamitales, a vascular cambium produced secondary xylem. Presumably secondary phloem was produced also, but the preservation of the fossils is not good enough to be certain (**Figure 21-24**). These plants became large trees up to 30 cm in diameter and more than 20 m tall. They had true **monopodial growth**, a main trunk, lateral branches, true leaves, and true roots.

The equisetophyte vascular cambium evolved independently from that of the lycophytes, yet the two suffered from the same defect: The fusiform initials could not undergo radial longitudinal division to produce more fusiform initials. As the wood accumulated and pushed the cambium outward, the arthrophyte fusiform initials finally became too large to function, and secondary growth ceased.

Figure 21-24 *Calamites* was an obvious relative to living equisetums, having jointed stems with whorls of leaves or branches at each node. A plant like this would have been a small tree about 10 m tall.

Figure 21-25 Although *Equisetum* lacks many derived features, it is by no means primitive or poorly adapted. Here it is a successful weed, competing well with both a corn crop and its human farmers.

Although equisetophytes are an ancient clade with only a few surviving species, the plants are by no means primitive, poorly adapted, or rare. They occur worldwide in moist habitats and grow vigorously enough to be weeds in cultivated areas (Figure 21-25).

■ Ferns

Aside from the equisetophyte line of evolution out of the trimerophytes, probably two more led to the plants that dominate Earth at present: ferns (Table 21-4) and woody plants (lignophytes, almost all of which are seed plants).

Early ferns first appeared in the Devonian Period and then diversified greatly (Figures 21-26 and 21-27). Whereas equisetophytes only have 15 species of living plants, the rest of the monilophytes have more than 12,000 species, and almost all of those are what are called **leptosporangiate ferns** (Leptosporangiatae; these are the ferns that are familiar to you). The other three groups of monilophytes (Marattiales, Ophioglossales, and Psilotales) have less than 300 species of living plants altogether, and fewer people will be familiar with them.

Ferns can be found in almost any habitat. Moist, shady forests and lakesides are often considered "typical" fern habitats, but species of *Woodsia* and *Cheilanthes* occur in dry, hot deserts; *Salvinia* and *Azolla* grow floating on water; and *Ceratopteris* lives submerged below water (Figure 21-28). Other genera contain epiphytes (*Polypodium*) or vines (*Lygodium*). All ferns are perennial and herbaceous; none is woody, but some do achieve the size of small trees—the tree ferns *Angiopteris, Cyathea, Cnemidaria,* and others. Although called tree ferns, they never have secondary xylem.

The fern sporophyte consists of a single axis, either a vertical shoot or a horizontal rhizome, that bears both true roots and megaphyllous leaves. The vascular system of the stem is an endarch siphonostele, a derived trait also present in equisetophytes and

TABLE 21-4	Classification of Division Pteridophyta
Class Cladoxylopsida*	
Class Coenopteridopsida*	
Class Ophioglossopsida	
Order Ophioglossales	
Family Ophioglossaceae	*Ophioglossum, Botrychium*
Class Marattiopsida	
Order Marattiales	
Family Marattiaceae	*Marattia, Angiopteris, Psaronius**
Class Filicopsida	
Order Filicales	
Family Schizaeaceae	*Schizaea*
Family Gleicheniaceae	*Gleichenia*
Family Osmundaceae	*Osmunda*
Family Matoniaceae	*Matonia*
Family Polypodiaceae[†]	*Adiantum* (maidenhair fern), *Asplenium* (spleenwort), *Blechnum, Cheilanthes, Dryopteris* (shield fern), *Pellaea, Platycerium, Polypodium, Polystichum* (sword fern), *Pteridium* (bracken fern), *Woodsia*
Family Cyatheaceae	*Cnemidaria, Cyathea*
Family Hymenophyllaceae	*Trichomanes*
Order Marsileales	
Family Marsileaceae	*Marsilea, Regnellidium*
Order Salviniales	
Family Salviniaceae	*Azolla, Salvinia*

*All species in this group are extinct.
[†]This family contains almost all the living species of ferns.

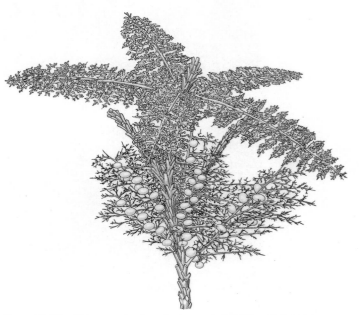

Figure 21-26 This reconstruction of a fern ancestor, *Rhacophyton,* shows spirally arranged branch systems that had evolved nearly to the point of being called leaves. Some are fertile with sporangia, and some are sterile; the former evolved into sporophylls, the latter into foliage leaves.

Figure 21-27 A later fern, *Phlebopteris,* from the Triassic Period. Even though the Triassic occurred from 230 to 190 million years ago, plants had evolved so many modern features that this fossil is not an ancestor to ferns in general, but to a particular family of living ferns, the Matoniaceae.

seed plants (**Figure 21-29**). At each node, a **leaf trace** diverges from the siphonostele, leaving a small segment of the vascular cylinder as just parenchyma; this region is a **leaf gap**. A vascular cambium has been reported to occur in one fern, *Botrychium.*

Leaves of ferns may be leathery or delicate, only one cell thick in filmy ferns, but many layers thick in most. They have an upper layer of palisade parenchyma and a lower region of spongy mesophyll. Leaves are small (*Trichomanes*) or up to several meters long (tree ferns), but they are almost always compound with a rachis and leaflets (**Figure 21-30**). Staghorn fern (*Platycerium*) and bird's-nest fern (*Asplenium*) have simple leaves. Fern leaf primordia have a distinct apical cell, unlike leaves of seed plants, and as the primordium grows, it curves inward, producing a tightly coiled young leaf commonly known as a fiddlehead (**Figure 21-31**). As the leaf expands, it uncoils and becomes flat. Fern leaves, especially larger ones, usually contain a considerable amount of vascular tissue. Leaf veins typically branch dichotomously.

In some species, for example *Blechnum spicant* and cinnamon fern (*Osmunda cinnamomea*), certain leaves serve only for photosynthesis, and others serve only as sporophylls; however, in most ferns, each leaf does both (**Figure 21-32**). On the underside of the leaf are **sori** (singular, **sorus**), clusters of sporangia where meiosis occurs (**Figure 21-33**). Most ferns are homosporous; only two groups of water ferns (Marsileaceae and Salviniaceae) are heterosporous.

(a)

(b)

(c)

(d)

Figure 21-28 (a) Many ferns grow in hot, dry climates with little rainfall. If transplanted to habitats often described as typical for ferns—cool, moist, and shady—these would die within days. *Azolla* (b) and *Salvinia* (c) do not look like ferns at first glance, but the structure of their leaves and sporangia shows they are ferns. (d) Several genera of ferns such as this *Cyathea* are referred to as tree ferns because they have very long vertical trunks and huge fronds; however, unlike trees, these never have secondary growth and are actually giant herbs.

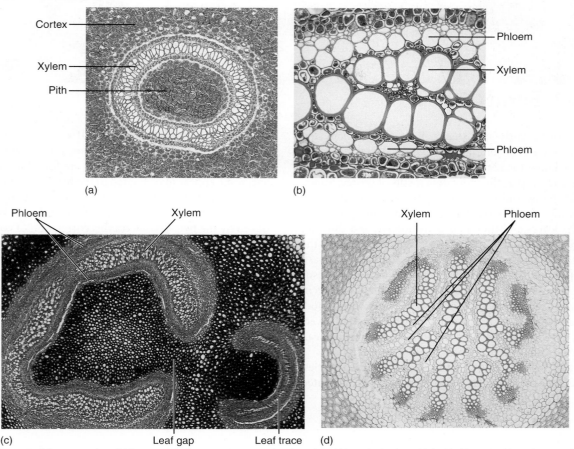

Figure 21-29 Anatomy of fern stems is variable—not only from species to species, but also within a single plant. (a) In this *Dennstaedtia* and many other ferns, xylem forms a complete ring in transverse section. (b) Phloem occurs on both the exterior and interior of the xylem in ferns; phloem interior to xylem is extremely rare in seed plants (×20). (c) In this *Adiantum,* the vascular tissue makes almost a complete cylinder. A leaf gap occurs where vascular tissue moves to the leaves. Leaf gaps are characteristic of ferns; they do not occur in the microphyll line of evolution (lycophytes). (d) Members of the microphyll line of evolution, such as this *Lycopodium annotinum,* do not have pith (×20).

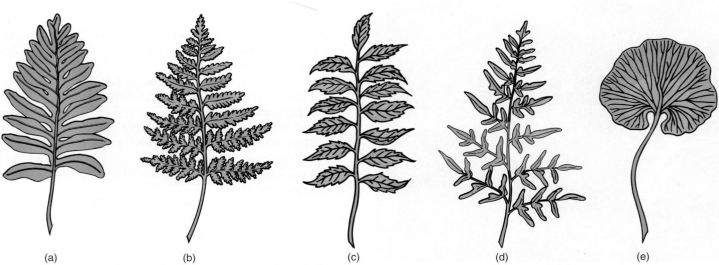

Figure 21-30 (a–e) Fern leaves (fronds) have a petiole and a lamina, the lamina being almost always pinnately compound. The midrib is a rachis, and the leaflets may also be compound.

Box 21-2 Form Genera

Although we can be fairly sure that a plant like *Sciadophyton* was the gametophyte of a plant like *Rhynia* or *Cooksonia,* we can probably never know for certain. These fossils are the separate pieces of a life cycle, just as the individual fossils of spores, leaves, fruits, flowers, stems, and roots are the pieces of plants. How do we know which fossil parts came from the same species? As a plant dies, its leaves and flowers usually abscise, and thus, we have numerous individual leaf and flower fossils mixed with the remains of bare stems. It is rare to find, at present or in the past, entire meadows or forests that consist of just a single species; usually many species grow mixed together. We find many types of fossil leaves mixed with many types of fossil stems, flowers, wood, and other plant parts. If we find a branch with leaves or flowers still attached, that is, if there is an organic connection, we can establish which organs are part of one plant.

Until we know which parts were portions of the same species, we must use **form genera**, which are created for types of isolated organs, tissues, spores, or pollen. If a fossil leaf is found which appears to be distinct from all others, it is named as a new form genus of leaf. As more fossil leaves are found in rocks of similar age, some appear to be related to the first and are assigned to that form genus. Mixed with the leaves may be twigs or large pieces of wood. We may be almost certain that the leaves have been produced by the twigs, based on this correlation of common occurrence at the same time and the same place; however, the leaves cannot be assigned to the form genus of the wood unless they are actually found still connected to a twig. Only when an organic connection has been established can two form genera be combined. For spores and pollen, no organic connection ever occurs, but occasionally, an unopened sporangium or anther is found; therefore, certain spores or pollen grains can be associated with those sporangia or anthers. Even when two form genera have been shown to be parts of the same species, it is still often much simpler to continue using the names of the form genera.

When fern spores germinate, they grow into small, simple heart-shaped or ribbon-shaped photosynthetic gametophytes with unicellular rhizoids on the lower surface but with no vascular tissue and no epidermis (**Figure 21-34**). Each usually bears

Figure 21-31 Fern leaves (but not equisetophyte leaves) have a highly characteristic development: Young leaves are tightly coiled and uncurl as they expand. This is true not only of the rachis, but of the leaflets as well. This is circinate vernation.

both antheridia and archegonia. Antheridia develop early and can be difficult to detect because they occur among rhizoids; archegonia develop later, close to the gametophyte apex. When the environment is sufficiently moist, antheridia release motile sperm cells that could easily be mistaken for unicellular green algae except for the absence of plastids. These swim to the archegonia and fertilize the egg. As in *Equisetum,* a fern zygote is retained by the gametophyte that nourishes it; however, the young sporophyte soon produces its first leaves and roots and becomes independent. Its continued growth eventually destroys the small gametophyte.

Eusporangia and Leptosporangia Ferns contain two types of sporangia that differ in fundamental aspects of their development. The **eusporangium** is initiated when several surface cells undergo periclinal divisions, resulting in a small multilayered plate of cells (**Figure 21-35**). The outer cells develop into the sporangium wall, and the inner cells proliferate into sporogenous tissue. This results in a relatively large sporangium with many spores. This is the fundamental type of sporangium, and it exists in virtually all plants other than Leptosporangiate; even the anthers and ovules of flowering plants develop this way.

Leptosporangia are initiated when a single surface cell divides periclinally and forms a small outward protrusion (**Figure 21-36**). This undergoes several more divisions, which result in a small set of sporogenous cells and a thin covering of sterile cells; only a few spores are produced. Having leptosporangia is a derived trait occurring only in the clade Leptosporangiatae.

Figure 21-32 (a) This cinnamon fern (*Osmunda cinnamomea*) contains both foliage leaves that never bear sporangia and highly distinct sporophylls that do bear sporangia. (b) Most ferns have only one type of leaf; the broad, thin shape of its leaflets is selectively advantageous for photosynthesis, and they bear clusters of sporangia (sori) on their underside. These sori are naked; they have no covering. (c) The sporangia of this maidenhair fern (*Adiantum*) occur near the margin of the leaflets and are protected by a flap of tissue called an indusium.

(a)

(b)

(c)

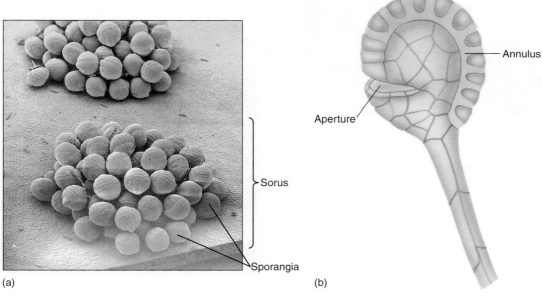

Figure 21-33 (a) A sorus contains many sporangia, each of which consists of a stalk and the actual sporangium body where meiosis occurs. (b) The sporangium opens when specialized cells of the annulus dehydrate, shrink, and suddenly crack open, expanding rapidly and throwing the spores (×40).

Sorus

Sporangia

Annulus

Aperture

(a)

(b)

Psilotum and *Tmesipteris* These two small genera contain the simplest of all living vascular plants (**Figure 21-37**). *Psilotum* is constructed very much like *Rhynia* and *Aglaophyton*, and until the mid 1970s, they were often placed together, even though *Rhynia* and *Aglaophyton* are extinct. Plants of *Psilotum* ("whisk ferns") are now known to be highly derived, highly simplified ferns. They are small plants with prostrate rhizomes and upright stems that branch dichotomously; they have an epidermis, cortex, and a simple vascular cylinder with no pith—a protostele (**Figure 21-38**). Xylem consists of annularly or helically thickened tracheids. *Psilotum* is unique among living vascular plants in that it has lost the capacity to make roots and leaves. Shoots do have occasional small, scale-like projections of tissues resembling enations.

Particularly unusual are the gametophytes, short, branched cylinders less than 2 mm in diameter. Their surface is covered

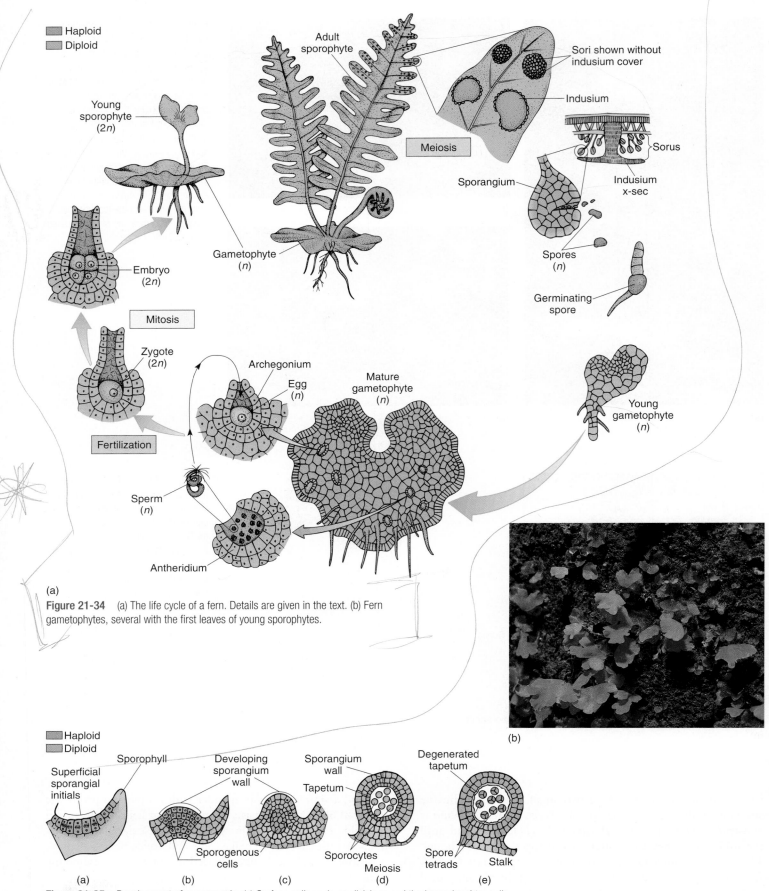

Figure 21-34 (a) The life cycle of a fern. Details are given in the text. (b) Fern gametophytes, several with the first leaves of young sporophytes.

Figure 21-35 Development of eusporangia. (a) Surface cells undergo divisions and the inner daughter cells become sporogenous. (b–e) Further divisions result in a large number of spores and a thick sporangium wall.

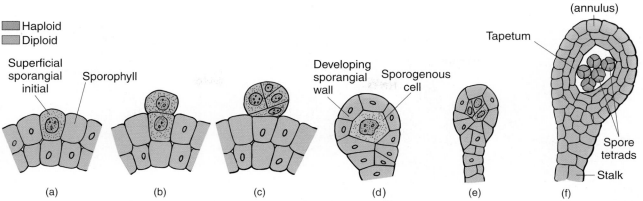

Figure 21-36 Development of leptosporangia. (a and b) The sporangium is initiated by division in just a single surface cell. (c–f) After a few cell divisions, the sporangium has a stalk, a thin wall, and only a few spores.

with elongate cells, rhizoids, that anchor and perhaps also absorb (**Figure 21-39**). Gametophytes have no chlorophyll but instead are heterotrophic, forming either a symbiotic or a parasitic association with soil fungi; fungi invade most cells of the gametophyte and provide it with sugars and minerals. *Psilotum* is the only species whose gametophytes contain vascular tissue: There is a small amount of vascular tissue similar to that of sporophytes. A central mass of tracheids is surrounded by phloem and an endodermis.

Psilotum occurs in tropical and subtropical regions. In the United States, it can be found in the Gulf Coast states from Florida to Texas, as well as in Hawaii. *Tmesipteris* is limited to Australasia, primarily Australia and other South Pacific islands.

The Term "Vascular Cryptogams"

Traditionally, the plants of this chapter have been referred to informally as **vascular cryptogams** (or often as "**ferns and fern allies**"). This name indicates that they have vascular tissue and that because they lack seeds their reproduction is hidden (crypto). Although these are two important features, they are shared ancestral features—symplesiomorphies—and thus do not indicate they are closely related. What is most striking about these plants are the many features they lack: seeds, flowers, fruits, and so on; however, we construct cladograms on the basis of shared derived features—synapomorphies—not on a lack of features, and consequently, we do not group vascular cryptogams together

Figure 21-37 The growth habit of *Psilotum* stems— dichotomous axes with eusporangia.

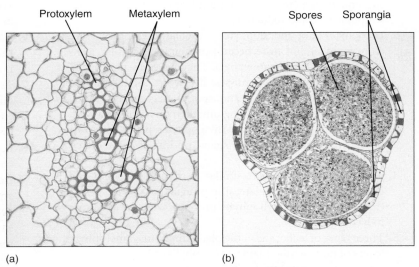

Figure 21-38 (a) Vascular structure of *Psilotum* is simple: a protostele and phloem surrounding the xylem. Protoxylem is exarch (to the exterior of the metaxylem), which is similar to the zosterophyllophytes rather than the rhyniophytes (×100). (b) Sporangia of *Psilotum* are actually considered to be three sporangia fused together; each is at the end of an extremely short branch, basically a three-branched shoot subtended by a scale (×8).

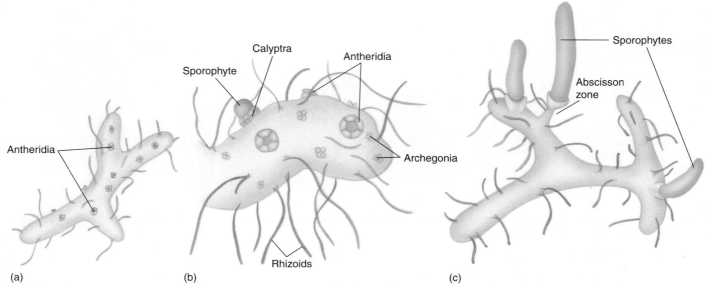

Figure 21-39 Gametophytes of *Psilotum* are small cylinders (a) that live heterotrophically, nourished by endophytic soil fungi. (b) Each gametophyte bears both antheridia and archegonia, and because the egg and zygote are retained in the archegonium, the young sporophyte initially is nourished by the parent gametophyte; during early development, there is proliferation of gametophyte tissue as a protective calyptra. (c) As the sporophyte grows and enlarges, it forms an abscission zone and detaches from the gametophyte, becoming independent.

formally. Also, if we tried to consider vascular cryptogams as a clade made up of several subclades (lycophytes and monilophytes for example), we would be creating a group that contains an ancestral group but not all its descendants: The ancestor of the monilophyte line is also the ancestor of lignophytes. To leave lignophytes out would be to create a paraphyletic group, which is not correct. Instead, when we use a term such as "vascular cryptogam," we are referring to a grade, to basically a level of evolutionary advancement. Similarly, the term "bryophyte" is often used for all the nonvascular plants, not to indicate that they are related, but that they are at a level of evolution that does not include the presence of vascular tissue.

Summary

1. Some early gametophytes were the most elaborate ever and subsequent gametophyte evolution from them has involved simplification and reduction.

2. Lycophytes and monilophytes (vascular cryptogams) are embryophytes, not algae, because of their multicellular, jacketed sporangia, and gametangia. They have vascular tissue, and thus, they are not mosses, liverworts, or hornworts. Although some have small endosporial megagametophytes, none is retained in the sporangium; therefore, none is a seed.

3. The earliest vascular plants are rhyniophytes, plants with simple dichotomously branched stems, no leaves, and endarch protoxylem.

4. Zosterophyllophytes were similar to rhyniophytes but had exarch protoxylem and lateral sporangia that opened across their tops. Zosterophyllophytes probably were the ancestors to lycophytes.

5. Three types of leaf occur in the plant kingdom: gametophyte leaves of nonvascular plants, microphylls of lycophytes, and megaphylls of all nonlycophyte vascular plants. Microphylls evolved from enations; megaphylls are modified telomes—branch systems.

6. Vascular cambia evolved at least three times; cells in the two that evolved in ancient lycophytes and equisetophytes could not divide with a radial longitudinal wall.

7. Trimerophytes were transitional between early vascular plants and euphyllophytes. Evolutionary advances in the trimerophytes were pseudomonopodial branching and the first steps in megaphyll evolution.

8. Equisetophytes contain only one small genus of living, herbaceous species and several extinct genera of large, complex, tree-like plants. All had a characteristic jointed structure and various systems of canals.

9. Ferns have many living species and tremendous diversity. No ferns have secondary growth or seeds, but all have megaphylls.

Important Terms

cones	megaphylls	siphonostele
enations	microphylls	sorus
endarch protostele	monilophytes	sporophyll
equal dichotomous branching	monopodial growth	strobili
euphyllophytes	overtopping	telome theory
exarch protostele	planation	vascular cryptogams
horsetails	protostele	webbing
leaf gap	rhyniophytes	zosterophyllophytes
leaf trace	scouring rushes	
lignophytes	seed plants	

Review Questions

1. The earliest fossils that definitely were vascular land plants belong to _____, a genus of extinct plants. They branched _____, both branches being of equal size and vigor.

2. Draw a plant of *Rhynia*. Be certain to include the reproductive parts. Now assume that it has a life cycle with an alternation of isomorphic generations and draw a complete life cycle.

3. If *Rhynia* or its contemporaries were the ancestors to the ferns, how did the gametophytes and sporophytes change during evolution?

4. In the vascular bundles of flowering plants, protoxylem is closest to the center of the stem, and metaxylem is farther out. Is this an endarch or an exarch arrangement? Seed plants always have just this one arrangement, but what about the early vascular plants. Were they endarch or exarch, or did both types occur originally?

5. The vascular bundles of flowering plants surround a pith, but the earliest vascular plants had no pith. A vascular system with a solid mass of xylem with no pith is called a _____. A stele that does have a pith is called a _____.

6. Rocks that contain fossils of *Rhynia* also often contain fossils of *Sciadophyton*. We know that *Rhynia* was a sporophyte and that *Sciadophyton* was a gametophyte. Because they grew together, we suspect that they may be the two generations of a single species. If so, did *Rhynia* have an alternation of isomorphic or heteromorphic generations? Also, if this is true, evolution has caused one generation to become more complex, the other to become less complex. Which is which? Explain.

7. There are two alternative hypotheses about the life cycle of the early vascular plants, the transformation hypothesis and the interpolation hypothesis. If *Rhynia* and *Sciadophyton* are the two generations of one species, which hypothesis would be favored? Briefly describe the other hypothesis.

8. What are the zosterophyllophytes? How did they differ from rhyniophytes? Why do we think they are related to lycophytes but not to ferns and seed plants?

9. The reconstruction of *Asteroxylon* in Figure 21-10 shows thin, leaf-like flaps of tissue on the plant's surface. What are these called? Did they ever have stomata in any of the zosterophyllophytes? Did they have vascular bundles? Did they ever become large (Hint: look at Figure 21-13)?

10. What are microphylls? Are they related to the enations of *Asteroxylon*?

11. The lycophytes once contained many species of large trees that formed extensive forests. Briefly describe plants of *Lepidodendron* and *Sigillaria*. Also describe their wood.

12. Name two genera of living lycophytes. What are their common names? About how big do they get? Are they leafy or do they have naked stems? In a plant identification book, they would probably be listed with ferns.

13. Trimerophytes were plants that probably evolved from rhyniophytes but with more derived features. In one feature, certain stems grow longer than others, and thus, rather than having dichotomous branching, they have _____ branching (displayed especially by *Pertica*). Simultaneously, the positioning of branches became more _____ and _____.

14. Describe the evolution of megaphylls. What are telomes?

15. Describe the trimerophytes. From what group did they evolve, and what lines of evolution did they produce? Even though all rhyniophytes and trimerophytes are now extinct, would you consider them unsuccessful?

16. What are the two common names of equisetums? What is the appearance of the plants? Their approximate height? Look at Figures 21-21 and 21-24. Equisetums have strobili (in plant identification books, these will be called cones) and canals. Briefly describe the canals and strobili.

17. Ferns first appeared in the Devonian Period. Look at the inside of the back cover. How long ago was the Devonian Period? Unlike all the groups mentioned so far in this chapter, most of this group is still extant, not extinct. About how many species of modern (i.e., not extinct) ferns are there?

18. Name two genera of ferns that are found in deserts. Name two that grow floating on water. One that lives underwater. There are ferns called "tree ferns." Do tree ferns have woody trunks with secondary xylem?

19. A very important feature of ferns is shown in Figure 21-29: Their shoot xylem is not solid as it was in the rhyniophytes. In the evolution of ferns and seed plants, there was the evolution of pith. A stele with a pith is called a _____ _____. Another significant feature of ferns is that the phloem occurs both to the outer side as well as the inner side of the xylem. Where does it occur in the vascular bundles of flowering plants?

20. You may have noticed that the undersides (but never the upper sides) of fern leaves have brown dots or brown streaks or brown patches. The brown dots are called _____ (singular, _____), clusters of _____ where meiosis occurs. If a leaf carries these structures, the leaf is called a _____.

21. When fern spores germinate, they grow into small, simple heart-shaped or ribbon-shaped _____ with unicellular _____ on the lower surface but with no vascular tissue and no _____.

22. Briefly describe eusporangia and leptosporangia. Which is the type that occurs in all other vascular plants? Which is the type that occurs in most ferns?

BotanyLinks

http://biology.jbpub.com/botany/5e
BotanyLinks contains a variety of resources and review material designed to assist in your study of botany.

Seed Plants I: Gymnosperms

Concepts

T he life cycle of vascular cryptogams is an alternation of independent, heteromorphic generations. A disadvantage of this life cycle is that the new sporophyte, while developing from the zygote, is temporarily dependent on a tiny gametophyte for its start in life. Consequently, many new sporophytes perish. All of the genes of each gametophyte and half the genes of each new sporophyte embryo are identical to those of the maternal sporophyte, and thus, any mutation that improves the survival of the gametophytes or embryos contributes to the reproductive success of the maternal sporophyte. It would be advantageous if the embryo could use the photosynthetic and absorptive capacity of the leaves and roots of

Chapter Opener Image: At this stage, these cones of a cedar tree (*Cedrus*) have been pollinated and seeds are developing inside but they are not quite mature enough to be released. These cones are the equivalent of unripe fruits in flowering plants. When these cones were pollinated, they were much smaller, about one centimeter long, and the scales were separated from each other so that wind-borne pollen could just settle into the ovules between the cone scales. When these cones are mature, they will be dry, tough and inedible, like the fruits of many angiosperms. Fir cones fall apart when they are mature, and the seeds simply fall to the ground; in other conifers, the cones expand and drop their seeds, and in still others they open only when heated by a fire. It is very rare for any gymnosperm cone to be fleshy and edible, but those of junipers (*Juniperus*) are. The clear material on these cones is resin, a secretion that repels any animal that would try to eat these cones and get to the seeds.

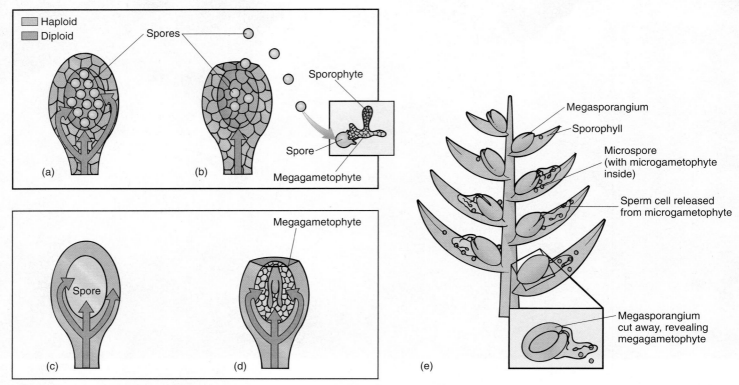

Figure 22-1 (a and b) Plants that release megaspores. Only limited amounts of nutrient can be placed in a spore, and after spore release (b), the sporophyte can do nothing more to nourish or protect the gametophytes or new sporophytes. (c and d) Plants that retain megaspores. Mutations that cause a reduction in the number of megaspores and their retention in the sporangium can be selectively advantageous. The sporophyte can nourish and protect its gametophytes and also the subsequent progeny sporophytes, at least for a while. In (d), the embryo is shown embedded in tissue of the megagametophyte; this occurs in gymnosperms—seed plants that are not flowering plants. In flowering plants, the tiny megagametophyte is quickly replaced by endosperm. (e) Megasporangia that are packed together in a cone composed of sporophylls automatically have a means of catching and retaining microspores. These germinate into microgametophytes with antheridia that release sperms within a millimeter or two of the megagametophyte and its archegonia.

the previous sporophyte. In order for this to happen, the megagametophyte and embryo must be retained inside the maternal sporophyte; this is accomplished by retaining the megaspore and allowing the megagametophyte to develop within the sporangium (Figures 22-1, a–d). Such an arrangement requires some alteration in the microgametophyte as well because retention of the megagametophyte changes its position. Free-living gametophytes develop on the soil, whereas retained ones develop high on the plant in strobili, a position that cannot be reached by swimming sperm cells produced by microgametophytes living at the soil surface. The modifications that overcame this problem are simple: The megasporophylls with the megagametophytes were arranged in upright cone-like structures, allowing microspores to be carried by wind and dropped into the megasporangiate cone (Figure 22-1e). There they would germinate into a microgametophyte, produce their antheridia and sperm cells, and carry out fertilization. Fertilization itself has changed little, with a moist sporophyll replacing moist soil. These changes produced the first seed plants, the division Pteridospermophyta or seed ferns (the "p" in "pterido" is silent: "terido sperm AH fit a") and the early members of division Coniferophyta, known as class Cordaitles. These two groups are now extinct, but their descendants still exist as the living seed plants. In several lines of evolution, those that produced the cycads and *Ginkgo,* this method of fertilization still exists, and beautiful swimming sperms are produced. In other lines, those leading to

conifers and to flowering plants, sperms have become nonmotile and are carried to the egg by growth of the microgametophyte as a pollen tube.

Evolution of seeds was preceded by evolution of a vascular cambium. Vascular cambia had arisen independently in some lycophytes and equisetophytes, but cells in this new cambium could undergo radial longitudinal divisions, thus allowing the cambium to grow in circumference as wood accumulated. In contrast to the rather slender lycophyte and equisetophyte trees, this new type of cambium could—and still does—produce trees as massive as giant redwoods, oaks, and hickories. We believe that this cambium arose just once, in one group of plants that then gave rise to a monophyletic group of woody plants, the **lignophytes** (Figure 22-2). Shortly afterward, seeds originated, establishing the seed plants, **spermatophytes.** The plants that existed after the origin of wood but before that of seeds were trees that reproduced with spores, the way ferns reproduce.

We are still not certain of the exact evolutionary relationships of all groups of extant seed plants and their extinct ancestors. Notice that in Figure 22-2, many groups emerge from the same point of the cladogram; this is an unresolved polychotomy, and it just means that we do not have enough information to know the order in which the various groups and their innovations arose. One problem is that we cannot extract and analyze DNA from such long-dead fossils.

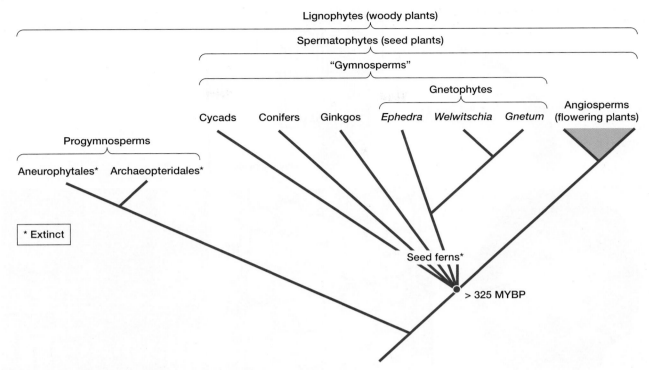

Figure 22-2 One recently proposed cladogram of seed plants.

As you read the following descriptions, keep in mind that certain features are often associated with certain others: There appear to be two major suites of characters. Some plants produce a small amount of very soft, spongy, parenchymatous wood (**manoxylic wood**, see Figure 22-26), have large, compound leaves and radially symmetrical seeds. This suite occurs in cycads and their relatives. The second suite consists of hard, strong wood with little parenchyma (**pycnoxylic wood**; Figure 22-3), small, simple leaves and flattened seeds (bidirectionally symmetrical). These occur in conifers and their relatives. The coniferophyte syndrome (and plants) probably evolved out of the cycadophyte syndrome, but as Figure 22-2 shows, we are not yet certain of this.

An old classification from the 1800s grouped all the seed plants together in a single division, Spermatophyta, with two classes, class Gymnospermae and class Angiospermae (Figure 22-2). The **gymnosperms** are those plants with "naked ovules," that is, ovules located on flat sporophylls, for example, pine cones. **Angiosperms** are the flowering plants, those with carpels, which are believed to be sporophylls that form a tube-like, closed structure; fruits are mature carpels. This classification emphasized both the close relationship of all seed plants and the idea that angiosperms originated from some group of gymnosperms, but if angiosperms evolved from some type of gymnosperm, then the group "gymnosperms" is not natural because it leaves out some of the descendants (angiosperms) of the ancestors. In cladistic terms, an incomplete group is a paraphyletic group. Consequently, class "Gymnospermae" should be abandoned.

The divisions of living seed plants commonly accepted now and used in this book are (1) division Cycadophyta, (2) division Coniferophyta, (3) division Ginkgophyta, (4) division Gnetophyta (these four are gymnosperms), and (5) division Magnoliophyta

(the flowering plants). Numerous extinct groups exist, the most significant for us being progymnosperms and seed ferns.

Division Progymnospermophyta: Progymnosperms

Trimerophytes were an important group of plants that evolved from rhyniophytes. One evolutionary trend initiated in trimerophytes was pseudomonopodial branching and the first steps in the evolution of megaphylls (euphylls). These evolutionary trends were continued in horsetails and ferns, which evolved out of trimerophytes.

A third group to evolve from trimerophytes was the now-extinct **progymnosperms** (Table 22-1) so named because some gave rise later to conifers, cycads, and the other gymnosperms. Like ferns and horsetails, progymnosperms also developed megaphyllous leaves; however, another feature was just as significant—the evolution of a vascular cambium with unlimited growth potential and capable of producing both secondary xylem and secondary phloem.

As early as 360 million years ago, the vascular cambium that evolved in progymnosperms was capable of undergoing radial longitudinal divisions, and thus, it could function indefinitely, producing large amounts of both secondary xylem and phloem (Figures 22-3 and Figure 22-4). The wood was almost indistinguishable from that of many living conifers. It contained elongate tracheids, most with circular bordered pits. It had little or no axial parenchyma. Rays were tall and uniseriate and consisted of procumbent ray tracheids. This wood must have been strong, effective at conduction, and capable of supporting a large mass of leaves and branches. At least some species, *Triloboxylon* and *Proteokalon*, for example, had a cork cambium that produced bark.

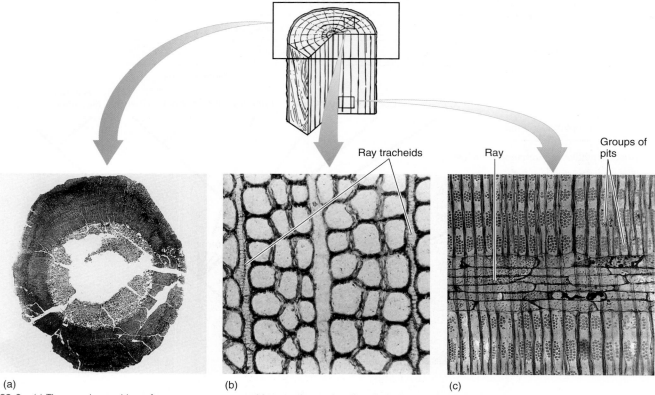

Figure 22-3 (a) The vascular cambium of progymnosperms was able to produce a virtually unlimited amount of both secondary xylem and secondary phloem, as in this *Callixylon brownii*. (b and c) Progymnosperm wood had large vertical tracheids with pitted walls and an abundance of rays with ray tracheids. Ray tracheids do not occur in flowering plants but are common in conifers as horizontal tracheids that permit rapid conduction horizontally in the sapwood. (b) *Callixylon erianum* (×700). (c) *C. newberryi* (×375).

Progymnosperms produced true woody trees: Trunks were up to 1.5 m in diameter and 12 m tall.

Although progymnosperm wood was similar to that of conifers, the two groups must be kept separate because progymnosperms did not have seeds or even ovule precursors. The species that arose later were heterosporous, but megaspores were shed through longitudinal slits in the sporangia; they were not retained in an indehiscent sporangium as occurs in seed plants. Although leaves and wood of progymnosperms were quite advanced, their reproduction was remarkably simple.

Aneurophytales

The order Aneurophytales contains the more relictual progymnosperms, such as *Aneurophyton, Protopteridium, Proteokalon, Tetraxylopteris, Triloboxylon,* and *Eospermatopteris* (**Figure 22-5**).

TABLE 22-1	Classification of Division Progymnospermophyta*
Order Aneurophytales	
Family Aneurophytaceae	*Aneurophyton, Eospermatopteris, Proteokalon, Protopteridium, Tetraxylopteris, Triloboxylon*
Order Archaeopteridales	
Family Archaeopteridaceae	*Archaeopteris, Callixylon*
*All species of this group are extinct.	

They varied in stature from shrubs (*Protopteridium, Tetraxylopteris*) to large trees, up to 12 m tall. They all had a vascular cambium and secondary growth, but the primary xylem of their stems was a protostele like that of rhyniophytes and trimerophytes. Aneurophytes further resembled trimerophytes in having little webbing between their ultimate branches; these could not yet be called leaves.

Archaeopteridales

A more derived progymnosperm was *Archaeopteris* in the order Archaeopteridales. These were trees up to 8.4 m tall with abundant wood and secondary phloem (Figure 22-4). Stems of *Archaeopteris* had a siphonostele, pith surrounded by a ring of primary xylem bundles, much like modern conifers and dicots. Although the "fronds" of archaeopterids resembled fern leaves, close examination reveals that they were actually planated branch systems and that the ultimate "leaflets" were really spirally arranged simple leaves. Webbing was only partial in *A. macilenta* (**Figure 22-6**), but it is complete in *A. halliana* and *A. obtusa*; these can be considered full-fledged megaphylls.

Reproduction in archaeopterids was heterosporous. Megaspores measured up to 300 μm in diameter and microspores only 30 μm in diameter, and each type was produced in its own distinctive sporangium—broad megasporangia and slender microsporangia. Sporangia were terminal on short branches mixed with sterile, leaf-like branch systems. Megaspores were released from the sporangia, not retained. Seeds were not produced.

Figure 22-4 Reconstruction of *Archaeopteris,* a small tree about 6 m tall. Although extinct now, *Archaeopteris* flourished about 360 million years ago across northern North America, Canada, and Europe.

Evolution of Seeds

Investigating the life cycle of extinct plants based on fossil material is difficult, but not impossible. In free-sporing species, spores can be identified with sporophytes if some spores were trapped in a sporangium attached to leaves or wood dur-

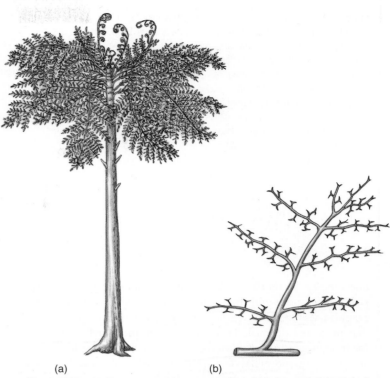

Figure 22-5 (a) *Eospermatopteris* was a member of the Aneurophytales; it had a trunk and what appear to be frond-like leaves, but these were just branch systems. Planation and webbing had not yet evolved. (b) A portion of the "leaf" of *Aneurophyton.*

ing fossilization. Unfortunately, spores cannot be identified with gametophytes except when the gametophyte is microscopic and develops within the spore wall. If megaspores are retained in the megasporangium, at least some of the investigation is easier.

Currently, the earliest known progymnosperm species with heterospory is *Chauleria* from the Middle Devonian Period, approximately 390 million years ago. In an Upper Devonian fossil, *Archaeosperma arnoldii,* the megasporangium produced only

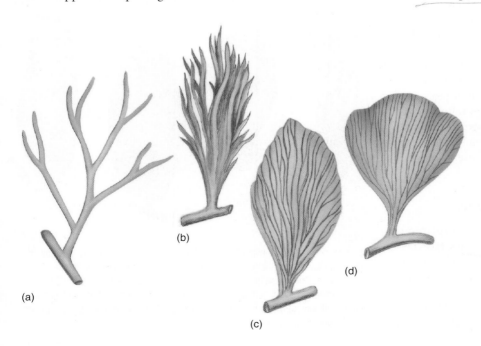

Figure 22-6 During the existence of the plants classified in the extinct genus *Archaeopteris,* the final transitions from telomes to leaves occurred. (a) *Archaeopteris fissilis.* (b) *A. macilenta.* (c) *A. halliana.* (d) *A. obtusa.* What appear to be veins in the leaves were actually the shoot axes of the telomes.

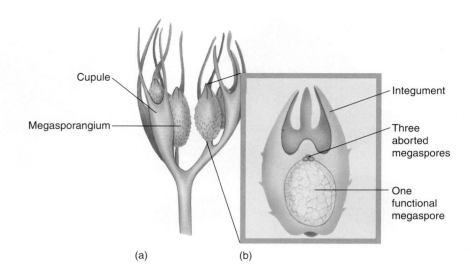

Figure 22-7 A reconstruction of the megasporangium (b) and adjacent telomes (a) of *Archaeosperma arnoldii*. Each sporangium contained only one large megaspore; the other three products of the spore mother cell meiosis apparently degenerated early. Tissue immediately around the megaspore was the megasporangium wall. Attached to it and extending upward as finger-like projections was a layer called an integument. Above the sporangium was a space surrounded by integument tissue, a calm, wind-free pollen chamber where pollen or spores settled. Around each megasporangium and integument was another set of partially fused telomes.

Labels on figure: Cupule; Megasporangium; Integument; Three aborted megaspores; One functional megaspore; (a); (b)

one megaspore mother cell, and this produced only one large, viable megaspore and three small, aborted cells. The megasporangium was surrounded by a layer of tissue, an **integument**, that projected upward. There was a large **micropyle**, a hole in the integument that permitted the sperm cells to swim to the egg after the megaspore had developed into a megagametophyte and had produced eggs (**Figure 22-7**). This is similar to angiosperm ovules, and some fossils help us understand the early stages. In *Genomosperma kidstoni*, the megasporangium was closely surrounded by sterile telomes (**Figure 22-8**). In *G. latens*, the telomes were fused at the base, and in *Eurystoma angulare*, they were similar to *Archaeosperma*, fused into one structure except at the tip. The sheath of sterile branches must have been important in trapping wind-blown microspores, but at first it would have allowed them to settle anywhere on the megasporangium. As they fused to each other and to the megasporangium, the space at the top of the megasporangium became the place where microspores settled, acting as a **pollen chamber** or holding area. As megasporangia evolved into ovules with integuments, other telomes on nearby branches became modified into cupules (Figure 22-7a), which may have later given rise to the carpel in angiosperms.

Simultaneously, microspores were evolving into pollen grains. Changes occurred in the nature of their wall, in their method of germination, and in the nature of the microgametophyte they produced. By the Carboniferous Period, just after the Upper Devonian Period (approximately 340 million years ago; see the timeline on the inside of the back cover), four types of pollen had become common. In some, internal structure is preserved well enough to see that these grains already had internal microgametophytes remarkably similar to those of modern gymnosperms.

Division Pteridospermophyta: Seed Ferns

Progymnosperms gave rise to another line of gymnospermous plants in addition to the conifers: the cycadophytes (see Figure 22-2). These are classified as three divisions: Pteridospermophyta (seed ferns, all extinct), Cycadophyta (cycads, extant), and Cycadeoidophyta (cycadeoids, all extinct).

The earliest **seed ferns** appeared in the Upper Devonian Period—others appeared later. Not all are closely related to each

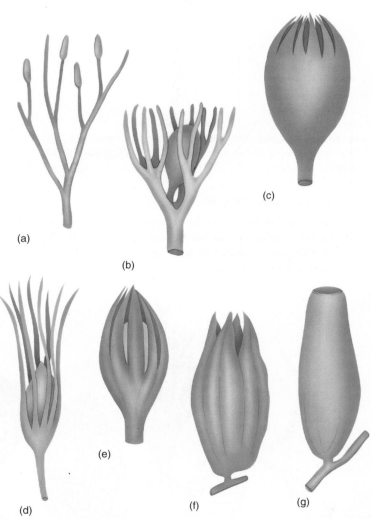

Figure 22-8 (a–c) A hypothetical evolution of an integument from telomes. In (a), sporangia are neither clustered nor particularly associated with sterile telomes. (b) One megasporangium is surrounded by sterile telomes. (c) The sterile telomes have fused at their bases. (d–g) Actual fossils that correspond to the hypothesis. (d) *Genomosperma kidstoni*. (e) *G. latens*. (f) *Eurystoma angulare*. (g) *Stamnostoma hyttonense*.

Labels on figure: (a); (b); (c); (d); (e); (f); (g)

other, they form a grade (a level of evolution) rather than a clade (all of the descendants of a common ancestor, regardless of their level of evolution): Seed ferns were any woody plant with fern-like foliage that bore seeds instead of sori (clusters of sporangia) on its leaves. Many resembled modern tree ferns (except they had wood), others were vines (Figure 22-9).

Pteridosperms are thought to have evolved from the Aneurophytales because the earliest seed ferns, such as *Stenomyelon*, had a three-ribbed protostele as in the aneurophytes *Triloboxylon* and *Aneurophyton*. In later species of *Stenomyelon, Calamopitys,* and *Lyginopteris,* most central cells of the stem differentiated as parenchyma, not tracheids. They had a ring of vascular bundles surrounding a pith, an arrangement that also occurs in the stems of all gymnosperms and angiosperms. Seed ferns had a long-lived vascular cambium that produced both xylem and phloem; this too is similar to gymnosperms and angiosperms. Although their wood was basically similar to that of their progymnosperm ancestors, there were interesting differences. Tracheids were much longer and wide enough that several rows of circular bordered pits could occur on each radial wall; conifer tracheids are so narrow that only one or two rows can fit on each radial wall. Rays in pteridosperm wood were many cells wide, not just one cell wide, and they were very tall, being large wedges of parenchyma: Seed fern wood was manoxylic, much softer and less dense than wood of conifers and progymnosperms, and manoxylic wood also occurs in cycads (Figure 22-10) and cycadeoids. Around the stem of pteridosperms was a thick cortex that contained distinctive radial plates of sclerenchyma just below the epidermis. The inner cortex contained secretory ducts. In older plants, a cork cambium and bark formed exterior to the secondary phloem; the cortex was shed with the first bark.

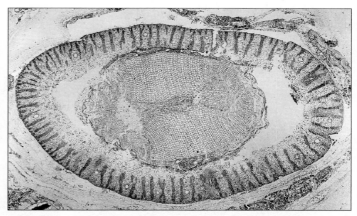

Figure 22-10 Transverse section of *Schopfiastrum* stem, showing the beginning of a pith, parenchymatous wood, and the distinctive cortex with plates of sclerenchyma (×10).

Leaves of seed ferns were similar to those of true ferns in overall organization—large, compound, and planar. Unlike ferns, however, the foliage leaves of seed ferns bore seeds (Figure 22-11). Within the seed fern ovule, the megasporangium (nucellus) was large, and bundles of vascular tissue ran into and through it. The integument was attached to the megasporangium only at the base and was vascularized. Seeds could be extremely large, up to 11 cm long and 6 cm in diameter in the now extinct *Pachytesta incrassata.*

■ Division Coniferophyta: Conifers

Because conifers are the most familiar living gymnosperms, they will be discussed first (Figure 22-12 and Table 22-2). They are the most diverse (approximately 50 genera and 550 species), and all are trees of moderate to gigantic size; the giant redwoods of California (*Sequoiadendron giganteum*) reach 90 m in height and 10 m in diameter (10 meters is the distance on a football field from the goal line to the 30 yard line, and this is trunk diameter).

Figure 22-9 Reconstruction of a swamp-forest during the Carboniferous Period. The large trees are related to *Lepidodendron* (*Sigillaria rugosa* on the left, *S. saulli* on the right), and the smaller fern-like plants are seed ferns (*Neuropteris decipiens*), not true ferns. Note the seeds at the ends of some fronds.

Figure 22-11 Seed ferns, such as this *Emplectopteris,* bore seeds along their leaves, not in cones. Otherwise, seed fern leaves were remarkably analogous (not homologous) to the leaves of true ferns in terms of venation and tissue structure.

(a)

(b)

(c)

(d)

Figure 22-12 Division Coniferophyta is large and diverse, containing many familiar plants. (a) Pine (*Pinus*). (b) European larch (*Larix decidua*). (c) Big tree redwood (*Sequoiadendron giganteum*). (d) Western juniper (*Juniperus occidentalis*).

TABLE 22-2   Classification of Division Coniferophyta

Order Cordaitales*	
Order Voltziales*	
Family Voltziaceae	*Lebachia*
Order Coniferales	
Family Pinaceae	*Abies* (firs), *Cedrus* (cedar), *Larix* (larch), *Picea* (spruce), *Pinus* (pine), *Pseudotsuga* (Douglas fir), *Tsuga* (hemlock)
Family Araucariaceae	*Agathis, Araucaria* (Norfolk Island pine and monkey puzzle)
Family Taxodiaceae	*Metasequoia* (dawn redwood), *Sequoia* (redwood), *Sequoiadendron* (giant sequoia), *Taxodium* (bald cypress)
Family Podocarpaceae	*Dacrydium, Podocarpus*
Family Cupressaceae	*Chamaecyparis, Cupressus* (cypress), *Juniperus* (juniper), *Libocedrus* (incense cedar), *Thuja* (arbor-vitae)
Family Cephalotaxaceae	*Cephalotaxus*
Order Taxales	
Family Taxaceae	*Taxus* (yew), *Torreya*

*All species of this group are extinct.

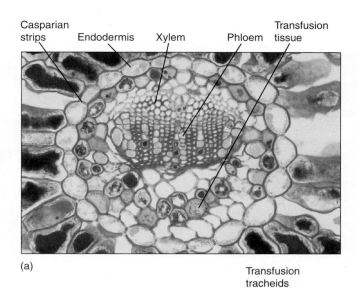

(a)

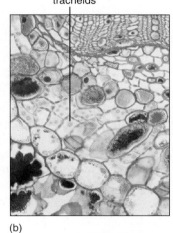

(b)

Figure 22-14   (a) The vascular bundle of this Douglas fir leaf has an endodermis, xylem, phloem, and transfusion tissue (×80). (b) High magnification of a pine leaf bundle showing secondary phloem and transfusion tracheids. Casparian strips are visible in the endodermis walls (×130).

Conifers are never vines, herbs, or annuals, and they never have bulbs or rhizomes. Conifer leaves are always simple needles or scales. Leaves of most conifers are perennial, persisting for many years (**Figure 22-13**); leaves of *Agathis* and *Araucaria* remain even on very old trunks.

The venation of conifer leaves is often simple, with just one or two long veins running down the center of a needle-shaped leaf or several parallel veins in scale-shaped leaves. Leaf veins have an endodermis and a tissue called transfusion tissue, consisting of transfusion parenchyma cells and transfusion tracheids (**Figure 22-14**). The latter are more or less cuboidal and have prominent circular bordered pits. Transfusion parenchyma is intermixed with the tracheids, the two cell types forming a complex three-dimensional pattern that facilitates transfer of materials between the ordinary vascular tissues and the mesophyll tissue outside the endodermis.

Just as in their progymnosperm ancestor *Archaeopteris,* wood of modern conifers lacks vessels, and their phloem lacks sieve tubes. Tracheids are so narrow that only one or two rows of circular bordered pits occur on their radial walls. All conifers have pollen cones and seed cones, most of which are woody, but in *Juniperus* and *Podocarpus,* seed cones superficially resemble the

(a)

(b)

(c)

Figure 22-13   Conifer leaves are never compound. They are always simple and leathery, needle like in some, and scale like in others. (a) In bristlecone fir (*Abies bracteata*), each leaf has two white bands where stomata are located. (b) Scale-like leaves of incense-cedar (*Calocedrus decurrens*). (c) Pines have two types of leaves—small, papery brown scales on the long shoots (visible on the branch) and the long needles borne by the short shoots located in the axils of the papery scale leaves. The long needles were produced last year.

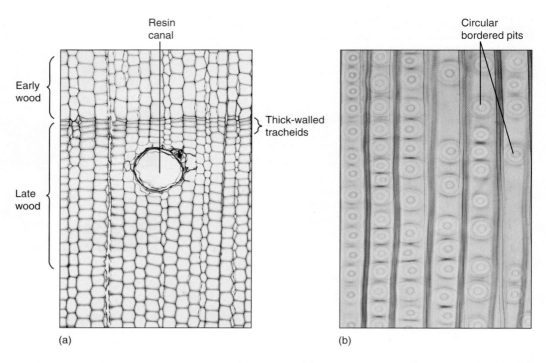

Figure 22-15 (a) Wood of pine, like that of all conifers, lacks vessels; it consists almost exclusively of tracheids. Growth rings are visible because late wood contains narrow, thick-walled tracheids, whereas early wood contains wide, thin-walled tracheids (transverse section, ×30). (b) Pine tracheids are narrow and their circular bordered pits are wide, and thus, only one row of pits fits on a given wall, unlike the dozens of small pits that cover each wall of an angiosperm vessel (×80).

Resin canal

Early wood

Late wood

Thick-walled tracheids

Circular bordered pits

(a)

(b)

fruits of flowering plants. The conifers are still a very successful group, forming extensive forests covering over 17 million km²; in many of these forests, flowering plants exist only as herbs, shrubs, or small trees growing in the conifers' shade.

The pines are good representatives for closer examination. The trees are monopodial, with one main trunk bearing many branches; the wood is composed exclusively of tracheids, but annual rings, spring wood, and summer wood are all visible because large-diameter tracheids are produced in the spring, followed by narrow-diameter tracheids in the summer (Figure 22-15). Rays are thin and tall and contain both ray parenchyma and ray tracheids. Resin canals, which produce the thick, sticky pitch, run vertically among the tracheids and horizontally in the rays. The wood has almost no axial parenchyma. Phloem contains sieve cells and albuminous cells and also has tall, narrow rays (Figure 22-16). A cork cambium produces a thick, tough bark that provides excellent protection even from forest fires.

Pines, like several other conifers, have two types of shoot, each with a characteristic type of leaf. Tiny papery leaves occur on **long shoots** and in their axils are **short shoots** that produce the familiar long needle leaves (see Figure 22-13c and Figure 22-17). The leaves have many xeromorphic characters: thick cuticle, sunken stomata, cylindrical shape.

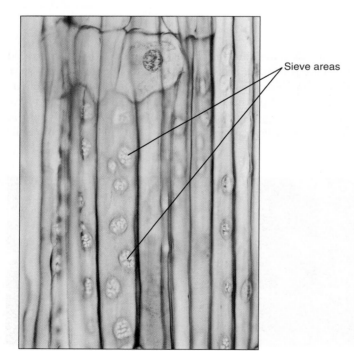

Sieve areas

Figure 22-16 Phloem of conifers contains sieve cells, not sieve tube members; sieve cells are long and narrow with sieve areas over much of their surface, but they never have horizontal cross walls (sieve plates) with enlarged sieve pores. The sieve areas in this pine are particularly abundant and easy to see (×25).

Long shoot   Short shoots   Leaves

Figure 22-17 In *Cedrus atlantica*, the nature of short shoots is obvious because they form more needles each year and so slowly grow into a visible shoot.

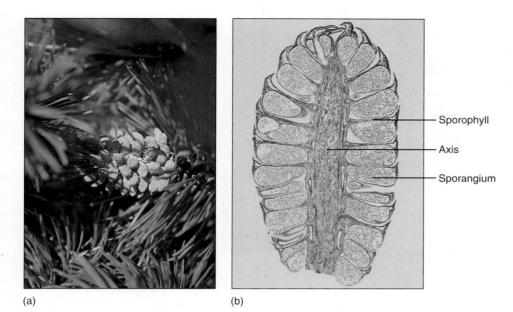

(a)

(b)

Figure 22-18   (a) Pollen cones typically occur in clusters near the ends of branches. As the microsporangia dehisce, pollen is liberated to the wind and blown away for distribution. Such a method of gene transfer is inefficient because so few pollen grains land on seed cones; it is successful primarily because conifers grow as dense forests, where each conifer is surrounded by hundreds of potential partners. (b) Pollen cones are simple cones; they have one single stem axis and bear microsporophylls. These are not microsporangia but rather leaf-like structures that bear microsporangia (×10).

— Sporophyll

— Axis

— Sporangium

Like all conifers, pines have both pollen cones and seed cones. Pollen cones are **simple cones** with a single short unbranched axis that bears microsporophylls (Figure 22-18). Microspore mother cells undergo meiosis and form microspores; then each of these develops endosporially into a small gametophyte with four cells, one of which is a generative cell as in flowering plants (Figure 22-19 and Figure 22-20). The gametophytes are shed from the tree as pollen and carried by wind; a small percentage land in seed cones, but the great majority land elsewhere and die.

Seed cones are more complex than pollen cones: They are **compound cones**, each consisting of a shoot with axillary buds. The short axis bears leaves called **cone bracts** rather than sporophylls (Figure 22-21). Each bract has an axillary bud that bears the megasporophylls. In some fossil conifers, the individual structures can still be seen, but in all modern conifers, extensive fusion has occurred: The axillary bud is microscopic, and its megasporophylls are fused laterally, forming an **ovuliferous scale** (Figure 22-22). In larch, fir, and Douglas fir (*Pseudotsuga*), the ovuliferous scale is

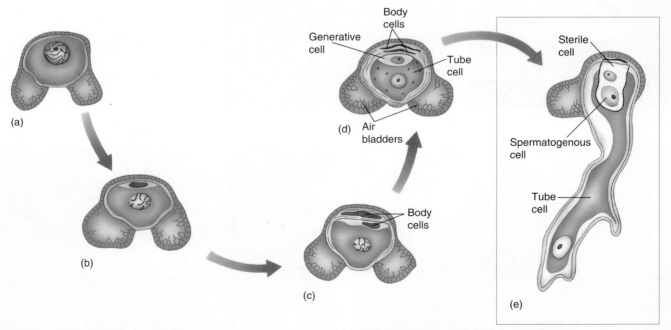

(a)

(b)

(c)

(d)

(e)

Body cells

Generative cell

Tube cell

Air bladders

Body cells

Sterile cell

Spermatogenous cell

Tube cell

Figure 22-19   (a and b) The microspore (pollen) of pine has one cell and two large air bladders that increase its buoyancy in air. The microspore develops into a small gametophyte in a process similar to that in pollen of flowering plants, except that a few more body cells are formed. First, two mitotic divisions produce two small body cells (c) that degenerate and a large cell that divides, resulting in a generative cell (d) and another body cell, called the tube cell (e). As in angiosperms, the body cell becomes the pollen tube, and the generative cell divides into two nonmotile sperm cells.

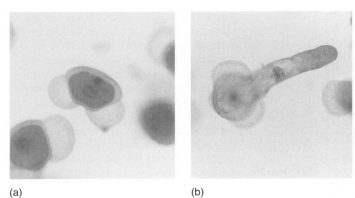

(a)   (b)

Figure 22-20 (a) The two air bladders of pine pollen are easily visible in this light micrograph. The addition of the hollow bladders adds virtually no weight to the pollen but increases its volume so that the pollen's density (weight per volume) is decreased and it does not sink so quickly in air (×200). (b) This pollen grain of pine has germinated and its tube cell is elongating (×200).

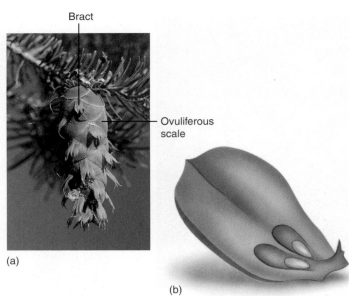

Bract

Ovuliferous scale

(a)

(b)

Figure 22-21 (a) The long three-pointed bracts of this Douglas fir (*Pseudotsuga*) cone are the true leaves of the cone axis. Each one contains an axillary bud whose leaves have fused together side by side into the flat, shield-like ovuliferous scale just behind each bract. These fused leaves that constitute the ovuliferous scale are actually megasporophylls. (b) If an ovuliferous scale were pulled from the cone and turned over, the two ovules would be visible. The scale obviously does not look like a set of leaves fused together, but look at Figure 22-6.

(a)   (b)

Figure 22-22 (a) In seed cones of pine, all parts are fused together; even the bract is fused to the ovuliferous scale. When morphologists first began working on pine cones without knowing about fossil structures, they were completely baffled and could not explain such a complex structure. (b) Larch (*Larix* species). (c) Monkey puzzle (*Araucaria araucana*). (d) *Taxus* has individual seeds, each surrounded by a red, fleshy aril (part of the seed coat); there are no ovulate cones in this genus.

Aril   Seed

(c)   (d)

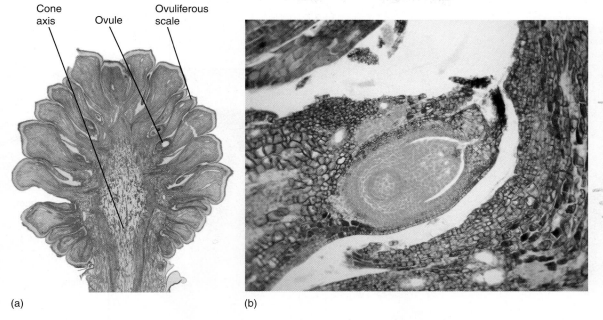

Cone axis   Ovule   Ovuliferous scale

(a)

(b)

Figure 22-23 This section of a pine cone was made just as the megaspore mother cells were about to begin meiosis. Each ovuliferous scale carries two ovules, each of which contains an integument and a nucellus (the actual megasporangium). As in flowering plants, each nucellus usually has only one megasporocyte (megaspore mother cell) and produces only one surviving megaspore. (a) ×2. (b) ×25.

still distinct and can be seen, but in most conifers, it is fused to the bract.

Inside each megasporangium, a single large megaspore mother cell undergoes meiosis, with three of the resulting cells degenerating and only one surviving as the megaspore (**Figure 22-23**). The mega-

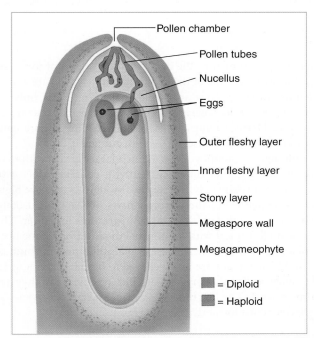

Pollen chamber

Pollen tubes

Nucellus

Eggs

Outer fleshy layer

Inner fleshy layer

Stony layer

Megaspore wall

Megagameophyte

■ = Diploid

■ = Haploid

Figure 22-24 Ovules of pine and other conifers are much larger than those of flowering plants. This cellular megagametophyte is not as large as a moss or liverwort gametophyte, but it is much more plantlike than the megagametophytes of angiosperms. Above the two eggs is the thick, indehiscent megasporangium (nucellus). Sperms cannot swim to the archegonia and eggs; they must be carried by a pollen tube that digests its way through the megasporangium. Between the megasporangium and the integuments is the pollen chamber.

sporangium does not dehisce; the megaspore is retained inside and grows into a large coenocytic (multinucleate) megagametophyte by free nuclear divisions and may have as many as 7,200 nuclei. It is much larger and more complex than the simple embryo sac of angiosperms. Development can take as long as a year, but finally walls form, converting the coenocyte into a cellular megagametophyte. Two or three archegonia form as sets of cells, each surrounding a large egg (**Figure 22-24**). Conifer eggs are gigantic cells loaded with carbohydrate and protein. The egg nucleus, although haploid, is swollen to a volume much larger than that typical for entire cells. It is probably filled with DNA synthetases and RNA polymerases, ready for extremely rapid activity once karyogamy with a sperm nucleus occurs.

Conifer pollen arrives before the egg is mature, and more than a year may pass between pollination and fertilization. A massive pollen tube slowly digests its way toward the megagametophyte as the egg forms. Because the megasporangium does not open, a passageway digested by a pollen tube is necessary. This differs greatly from the free-living gametophytes of mosses and ferns, which have exposed archegonia that sperm swim into. The two or three eggs in one megagametophyte can all be fertilized, but only one zygote develops into an embryo; the other one or two die. A zygote does not immediately form an embryo in conifers; instead, some of the first cells elongate as a **suspensor** that pushes the other cells deep into the megagametophyte (**Figure 22-25**). These other cells, called the **proembryo**, develop into the embryo. There is no double fertilization as occurs in angiosperms; rather, the female gametophyte continues to grow and acts as a haploid nutritive tissue similar to endosperm. The mature embryo has the same organization as an angiosperm embryo (radicle, hypocotyl, epicotyl, and cotyledons) but always has many cotyledons, not just one or two. The seed also resembles that of a flowering plant, but it is borne in a cone, not a fruit. In two conifers, the cones become fleshy, fruit-like, and brightly colored—red in *Podocarpus*, blue in *Juniperus*.

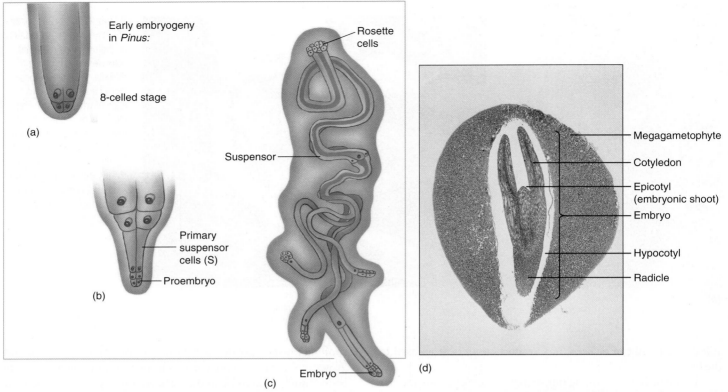

Figure 22-25 (a) Just as in angiosperms, the zygote of conifers produces a suspensor that thrusts the proembryo into the megagametophyte (b). (c) The suspensor can divide and form multiple embryos from each zygote, but only one embryo survives. (d) A conifer seed looks remarkably like the seed of a flowering plant. It has a seed coat (removed here) derived from the integument, an embryo with several cotyledons, and a nutritive tissue that is actually the megagametophyte, not endosperm. After fertilization, the parental sporophyte transports large amounts of nutrients, supplying everything needed for the embryos, which grow completely heterotrophically. It also fills all cells of the megagametophyte with carbohydrates, proteins, and mineral nutrients—this acts just like the endosperm of flowering plants and is often called endosperm (×25).

Division Cycadophyta: Cycads

Modern cycads are frequently confused with either ferns or young palm trees because they have stout trunks with pinnately compound leaves (Figure 22-26). Most cycads are short plants less than 1 or 2 m tall, but *Macrozamia* can reach heights of 18 m. The trunk is covered with bark and persistent leaf bases that remain on the plant even after the lamina and petiole have abscised. Internally, cycad stems are similar to those of seed ferns—a thick cortex containing secretory ducts surrounds a small amount of manoxylic (parenchymatous) wood. Tracheids are long and wide, and rays are massive. Even very old stems have only a small amount of wood; most support is provided by the tough leaf bases. A prominent pith contains secretory canals.

Unlike seed ferns, cycad foliage leaves do not bear ovules. Instead, cycads produce seed cones and pollen cones, each on separate plants; cycads are always dioecious. Pollen cones consist of spirally arranged shield-shaped microsporophylls that bear clusters of microsporangia (Figure 22-27). Upon germination, pollen grains produce a branched pollen tube and large, multiflagellated sperm cells. Seed cones are variable, with those of *Cycas revoluta* usually considered the most relictual (Figure 22-28). In this species, the seed cone is a large, loose aggregation of leaf-like, pinnately compound megasporophylls. Six to eight large ovules occur near the base, but the upper half of the megasporophyll is rather leaf like, similar to the ovule-bearing organs of some seed ferns. Megasporophylls of the cycad *Zamia floridana* are considered more derived. They are shield-shaped and bear only two ovules, and the entire cone is quite compact. Cycad ovules are like those of seed ferns, having a large, vascularized megasporangium and a loosely attached, vascularized integument.

Although Cycadophyta was a much larger group with many more species in earlier times, currently, it contains nine or ten genera and approximately 100 species. Modern cycads are highly prized ornamentals in the warmest parts of the United States; only a few can withstand freezing temperatures in winter. They are almost all tropical with an unusual distribution: Some occur in Cuba and Mexico, others in Australia, still others in southeast Asia or Africa. Cycads are now believed to have been widespread and to have occupied many habitats, but in most areas, they have become extinct. The few remaining, widely scattered species are the survivors of a formerly extensive group.

Division Cycadeoidophyta: Cycadeoids

The cycadeoids (all extinct) had vegetative features almost identical to those of cycads (Figure 22-29). The two groups differ only in subtle details of the differentiation of stomatal complexes and in leaf trace organization. On such characters alone, cycadeoids would never be considered distinct from cycads; however,

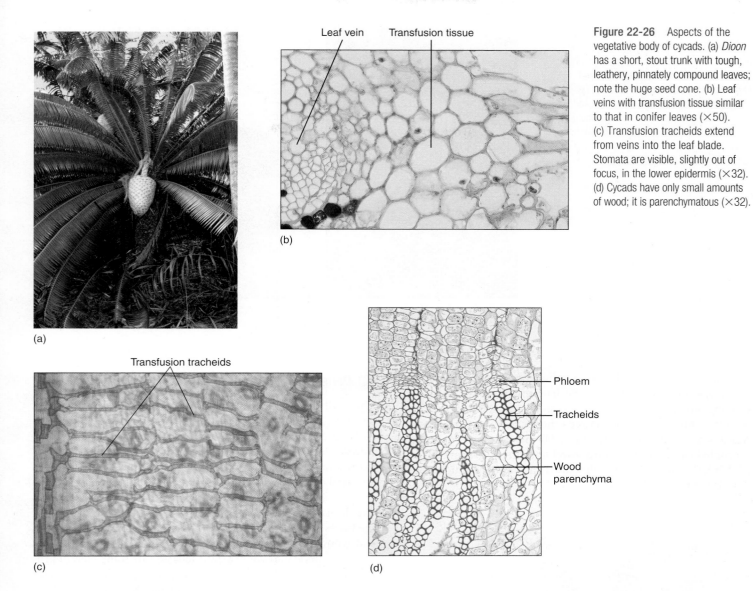

Figure 22-26 Aspects of the vegetative body of cycads. (a) *Dioon* has a short, stout trunk with tough, leathery, pinnately compound leaves; note the huge seed cone. (b) Leaf veins with transfusion tissue similar to that in conifer leaves (×50). (c) Transfusion tracheids extend from veins into the leaf blade. Stomata are visible, slightly out of focus, in the lower epidermis (×32). (d) Cycads have only small amounts of wood; it is parenchymatous (×32).

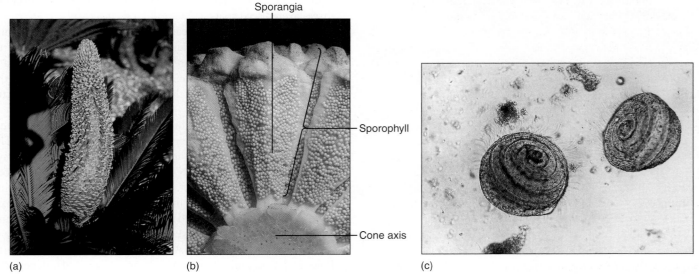

Figure 22-27   (a) Pollen cone, more than 30 cm long, in *Cycas circinnalis*. It is a simple cone, with one axis that bears microsporophylls. (b) Microsporophylls in cycads may bear many microsporangia. (c) Sperm cells of cycads have hundreds of flagella and must swim to the egg cell. *Zamia* (×165).

Figure 22-28 (a) Cycad seed cones are simple, with megasporophylls borne on the only axis. They have no bracts or axillary buds. (b) Cycad megasporophylls are not fused structures as are ovuliferous scales of conifers; instead, the megasporophyll is simple and strongly resembles a foliage leaf, at least in *Cycas circinnalis*. It is not difficult to see how this type of sporophyll could evolve from a seed fern sporophyll (see Figure 22-11).

(a)

Megasporophyll

Seed

(b)

individual cones of cycadeoids contained both microsporophylls and megasporophylls. Each ovule had a stalk, and the megasporangium was surrounded by an integument that extended out into a long micropyle. Between the ovules were thick, fleshy scales. Microsporophylls were located below the cluster of megasporophylls and curved upward, enveloping the megasporophylls. Each microsporophyll was cup shaped and contained numerous microsporangia.

Figure 22-29 Vegetatively, cycadeoids such as *Cycadeoidea* were similar to cycads. They had a broad cortex and a tough outer protective layer formed by persistent leaf bases.

Division Ginkgophyta: Maidenhair Tree

This division contains a single living species, *Ginkgo biloba* (Figure 22-30). It may seem unusual to erect an entire division for a single species, but *G. biloba* (the "maidenhair tree") is itself unusual. It looks very much like a large dicot tree with a stout trunk and many branches, but its wood is like that of conifers: It lacks vessels and axial parenchyma. It has "broad leaves," but they have dichotomously branched veins like seed ferns, not reticulate venation like dicots. Ginkgos have both short shoots, which bear most of the leaves, and long shoots. Reproduction in *Ginkgo* is dioecious and gymnospermous, but cones are not produced. Instead, ovules occur in pairs at the ends of a short stalk and are completely unprotected at maturity. Pollen is produced in an organ that resembles a catkin, having a stalk and several sporangiophores that each have two microsporangia. Like ovules of pteridosperms and cycads, the ovules of *Ginkgo* are large (1.5 to 2.0 m in diameter) and develop a three-layered seed coat.

A *Ginkgo* tree itself is beautiful and is a popular ornamental because the leaves turn brilliant yellow in autumn, but the microsporangiate ("male") trees are preferred; when the megasporangiate ("female") trees produce seeds, the outer fleshy layer of the seed emits butyric acid, which has a putrid odor that is difficult to tolerate.

The exact ancestors of ginkgos are not known, but they must have been one of the seed ferns or a closely related group. The cladogram of Figure 22-2 proposes that cycads and ginkgos have a common ancestor, but DNA analysis cannot include the seed ferns, all of which are extinct. Ginkgos became abundant during the Mesozoic Era, especially in the mid Jurassic Period (approximately 170 million years ago). Fossilized remains of leaves and wood can be found in almost all areas of the world, especially in high latitudes such as Alaska, Canada, and Siberia near the North Pole, and Patagonia, South Africa, and New Zealand near the South Pole. Ginkgos began to die out early in the Tertiary Period, but two species, *G. biloba* and *G. adiantoides,* persisted. The latter

(a)

(b)

(c)

(d)

Figure 22-30 (a) A tree of *Ginkgo biloba* could easily be mistaken for a dicot tree. (b) *Ginkgo* leaves are broad. They have dichotomous venation like leaves of cycads and seed ferns, not reticulate venation like dicots. (c) Microsporophylls occur in small, cone-like clusters (think of pine pollen cones [Figure 22-18], not pine seed cones [Figure 22-22]) mixed with foliage leaves (petioles are visible) on short shoots. (d) Ovules occur in pairs at the end of a stalk-like megasporophyll. The ovule itself, the integument, is exposed; no bract or ovuliferous scale protects it.

species became extinct during the Pliocene Epoch, approximately 10 to 12 million years ago.

Division Gnetophyta

Gnetophyta (pronounced as if spelled neat AH fit a) contains three groups of enigmatic plants: *Gnetum* with 30 species (Figure 22-31), *Ephedra* with about 40 species (Figure 22-32), and *Welwitschia mirabilis,* the only species in the genus (Figure 22-33).

Gnetums are mostly vines or small shrubs with broad leaves similar to those of dicots. They are native to southeast Asia, tropical Africa, and the Amazon Basin. Plants of *Ephedra* are tough shrubs and bushes that are very common in desert regions in northern Mexico and southwestern United States and dry mountains in South America. Their leaves are reduced and scale like. The few living plants of *Welwitschia* exist only in deserts of South Africa or in cultivation. They have a short, wide stem and only two leaves, but the leaves grow perennially from a basal meristem, becoming increasingly longer.

All three genera are unusual in being gymnosperms with vessels in their wood. This had been thought to show that they might be related to primitive angiosperms; however, their vessel elements evolved from tracheids with circular bordered pits, whereas those of angiosperms were derived from scalariform tracheids. Furthermore, angiosperms are thought to have evolved from vessel-less ancestors, the vessels evolving after flowers, not before them.

Unlike the pollen cones of all other gymnosperms, those of gnetophytes are compound and contain small bracts. Seed cones are also compound and contain extra layers of tissue around the ovules; the tissue is variously interpreted as an extra integument, bract, or sporophyll.

The few fossils of gnetophyte organs or tissues are only several million years old, too recent to be of much help in understanding the evolution and ancestry of the group. The pollen is distinctive, being spindle-shaped and having narrow ridges. It is easy to recognize, and fossil pollen of this type occurs as far back as the late Triassic Period, but pollen has not helped reveal their

(a)

(b)

Figure 22-31 *Gnetum.* (a) Plants of *Gnetum* strongly resemble dicots, having broad leaves and woody stems. (b) Ovules are not borne in cones.

Figure 22-32 *Ephedra.* (a) Plants of *Ephedra* often occur in dry areas and strongly resemble many types of desert-adapted dicots. Although their reproductive structures are gymnospermous, the microsporangiate cone (b) could be mistaken for a staminate imperfect flower. The naked ovules reveal that they are not angiosperms (c).

(a)

(b)

(c)

Figure 22-33 *Welwitschia mirabilis.* (a) Whole plant with torn leaves. (b) Microsporangiate strobili. (c) Megasporangiate strobilus.

(a)

(c)

(b)

origins either. Figure 22-2 includes them as one of several groups emerging from the same point. Certain aspects of their anatomy and reproduction have been interpreted as indicating that gnetophytes and flowering plants constitute two sister clades with a common ancestor. If so, the two would form the group "Division Anthophyta" or "**anthophytes**"; however, there are too few data to be confident of this, and currently, it is best to recall the principle of skepticism, realize that we do not yet know, and must continue searching for more data about their phylogenetic relationships. Remember that Figure 22-2 is a model, a hypothesis.

Summary

1. In a seed, the embryo draws its nutrition from the surrounding megagametophyte (or endosperm), which in turn is supplied with nutrients by the parental sporophyte. The sporangium and associated structures provide protection for the megaspore, megagametophyte, and embryo.

2. In all modern seed plants, the megasporangium (nucellus) does not open; therefore, a pollen tube must deliver the sperm cells to the archegonia. Sperm cells are still motile in many seed plants.

3. Progymnosperms arose from trimerophytes with the evolution of a vascular cambium of unlimited growth potential which produced solid wood with little parenchyma. The evolution of true seeds also began in the progymnosperms, first with retention of the megaspore and then with formation of an integument.

4. One order of progymnosperms, Archaeopteridales, gave rise to conifers; another progymnosperm order, Aneurophytales, evolved into pteridosperms, cycads, cycadeoids, and perhaps angiosperms.

5. Conifers constitute a diverse division of large trees with solid wood, simple pollen cones, and compound seed cones.

6. Pteridosperms (seed ferns) gave rise to cycads and cycadeoids. Important evolutionary changes were formation of pith and of ovules on more or less unspecialized foliage leaves. Their wood contained abundant axial parenchyma.

7. Cycads are modern gymnosperms with compound leaves, parenchymatous wood, and simple seed cones consisting of an axis and a rather leaflike megasporophyll.

8. *Ginkgo biloba* is the sole living species of its division. It resembles a dicot tree except that its leaves have dichotomous veins and it bears naked ovules, surrounded by neither a cone nor a carpel.

9. Reproduction in gnetophytes is gymnospermous, but plants of *Ephedra* and *Gnetum* resemble dicots. *Welwitschia* has only two leaves, which grow indeterminately from basal meristems.

Important Terms

angiosperms
anthophytes
aril
compound cones
cone bracts
gymnosperms
integument

lignophytes
long shoots
manoxylic wood
micropyle
ovuliferous scale
pollen chamber
proembryo

progymnosperms
pycnoxylic wood
seed ferns
short shoots
simple cones
spermatophytes
suspensor

Review Questions

1. What is the disadvantage of a life cycle in which alternate generations are completely independent of the preceding generation? Are there advantages to this type of life cycle?

2. What is a cone? How are cones of conifers similar to those of lycophytes and cycads? How do they differ?

3. The classification of the seed plants has varied. If a person groups all seed plants together in one large Division Spermatophyta, then they have two classes—the class _____ with naked seeds and the class _____, plants with carpels. On the other hand, many people (and this book) use four divisions, not one. List the three divisions of gymnosperms and the one division of flowering plants.

4. If angiosperms evolved from some type of gymnosperm, then the group "gymnosperms" is not natural. Why not? In cladistic terms, an incomplete group is a _____ group.

5. Describe progymnosperms. What were the significant evolutionary advances that characterized progymnosperms?

6. The section Evolution of Seeds describes the fossil seeds *Archaeosperma arnoldii*. Its megasporangium was surrounded by a layer of tissue, called an _____ that projected upward, and there was a large _____, a hole that permitted sperm cells to swim to the egg. If you have studied flowering plant reproduction, how many of these same features occur in flowering plants?

7. In progymnosperms, microgametophytes produced and released sperm cells that could swim—but not very far. What is a pollen chamber and how did it help fertilization?

8. Are any conifers herbs? Annuals? Vines? Do all conifers have needles like pines (Hint: Figure 22-13).

9. Describe a conifer pollen cone. Why is it a simple cone? Describe a conifer seed cone. Why is it a compound cone?

10. Seed cones of conifers are more complex than pollen cones. Seed cones are _____ cones, each consisting of a shoot with axillary buds. The main axis of the cone bears leaves called _____, rather than sporophylls, and each of these has an axillary bud that

bears the _____. The axillary bud is microscopic, and its _____ are fused laterally, forming an _____ scale.

11. Unlike pollination in flowering plants, conifer pollen arrives before the egg is mature, and more than a _____ may pass between pollination and fertilization. The pollen germinates, and a massive _____ _____ slowly digests its way toward the _____ as the egg forms.

12. In most conifers, as the seeds mature, the cone scales and ovuliferous scales become hard and tough—a typical pine cone (although firs have fir cones, not pine cones, and larches have larch cones, and so on); however, in _____, the cone becomes red and fleshy, and in _____, it becomes blue and fleshy.

13. Seed ferns are a group of extinct plants; they resembled tree ferns, but what did they have on their leaves instead of sori?

14. There is a good chance you have seen cycads. What do they look like? What two other types of plants can they be confused with (Hint: some are called "cardboard palms")? Do cycads ever have simple leaves?

15. Do any seed plants have flagella? Do their sperm cells have flagella? Look at Figure 22-27.

16. What is the one and only species in the Division Ginkgophyta? Being part of the "gymnosperms," does it have needle-shaped or scale-shaped leaves? The leaf venation is unusual because the leaves have dichotomously branched veins like _____ _____, not reticulate venation like _____.

17. What are the three groups classified in Gnetophyta?

18. Would it have been possible, on a theoretical basis, for an indehiscent megasporangium to evolve before integument-penetrating pollen tubes evolved? Why or why not? What problem would be involved?

BotanyLinks

http://biology.jbpub.com/botany/5e

BotanyLinks contains a variety of resources and review material designed to assist in your study of botany.

Visit the Web Exercises area of BotanyLinks to complete this question:

1. Many conifers are big trees that live for centuries. With all of the cutting of forests and loss of habitat, do any forests of old trees still exist? Go to the BotanyLinks home page for more information on this subject.

Seed Plants II: Angiosperms

chapter 23

Concepts

O f all the clades that have arisen since embryophytes became distinct from charophyte green algae, the flowering plant clade—the angiosperms—contains the greatest number living species—257,000. These are all classified together in a single division, the **Magnoliophyta**. It is common to refer to angiosperms as the most advanced group of plants, or the most derived; occasionally, they are called the peak of plant evolution, but these statements should be examined carefully. Certainly the group as a whole has a greater number of derived features than any other group, and especially important is the large number of derived characters that are unique to angiosperms and that distinguish them from other plant clades. The group is named for having flowers—a significant synapomorphy that at once unites all angiosperms and separates them from

Chapter Opener Image: The lily flower (*Lilium longiflorum*) shows two important ways in which angiosperms differ from gymnosperms. First, the pollen-producing organs (anthers) and the ovule-producing ones (carpels) occur together in most flowers whereas they always occur as separate cones in gymnosperms. Second, carpels are "closed," which means they have no opening that allows pollen to just fall into the carpels and onto the ovules; instead the pollen must germinate and send a pollen tube through the stigma and style, carrying the two sperm cells downward to the ovule and egg. This results in a prolonged period of contact during which the stigma and style might reject the pollen tube if it is not the proper type. In many angiosperm species, the carpels exert considerable control over which pollen tubes are allowed to grow and which are not.

all other clades. Other important derived characters are numerous methods of mutually beneficial interactions with animals that result in the plants being pollinated or having their seeds distributed while the animals received nectar or pollen or some other food. At the same time, angiosperms and animals have become more adept at fighting each other: As angiosperms evolved to provide reward for pollinators, natural selection also was giving them numerous new types of antiherbivore compounds, ranging from outright poisons (poison ivy, poison sumac, and deadly nightshade) to milder chemicals that deter feeding (chili peppers, mustard oil, and onion oils). Wood of almost all angiosperms is more complex than that of gymnosperms: It contains vessels, which provide a low-friction pathway of water conduction, and it contains axial wood parenchyma (parenchyma cells mixed among the vessels and tracheids), which makes wood a living tissue that protects itself from wood-boring insects and fungi and can influence the distribution of water as it is pulled upward through the plant.

Do these features, however, make angiosperms the most advanced of all groups? Keep in mind that the simple spore-based reproduction of ferns has allowed them to reproduce successfully, spread into many habitats, and diversify into numerous sizes and shapes. Conifers also are extremely successful; one clade in particular, Pinaceae, is a relative newcomer, having arisen within the conifer clade after the angiosperm clade had already made quite a few advancements. Pinaceae is diversifying and speciating rapidly, producing not only many new species but also new genera.

The point is that we must not think about evolution as a linear progression forward and upward, with each new group leaving the previous ones behind. Instead, think in terms of real plants and populations. Some population of a species encounters an isolation barrier that causes it to diversify until at some point some of its members cannot reproduce sexually with others, and at that point, two very similar but distinct species exist. Both are free to continue evolving. After the early embryophytes became distinct from charophytes, they diverged into two clades, the zosterophyllophyte/lycophyll clade and the rhyniophyte/euphyll clade: Both were extremely successful, undergoing large numbers of modifications

and diversifying greatly. After the early ferns originated, they too continued to diversify, with most variation occurring in habit, leaf size and shape, type of sporangium, type of indusium, and type of sporophyll. None evolved to have wood, flowers, or fruits, but they are not more primitive than the angiosperms; instead, different features have evolved in the two groups.

The earliest fossils clearly recognizable as parts of flowering plants are preserved pollen grains in rocks more than 130 million years old. We have not found any fossilized leaves or flowers from that time, and thus, we believe the earliest flowering plants must have lived in hilly or drier areas, areas where few sites were suitable for fossilizing plant parts. Large calm lakes with fine-grained sediments are best for fossilizing plants, and these are rare in dry areas or rocky hillsides. The first angiosperms probably were small trees or shrubs because plants of this type survive well in drier regions and because the most likely ancestors—other seed plants such as seed ferns or gymnosperms—were woody trees.

The evolutionary transformation of a gymnosperm into a flowering plant was not a simple process, and it involved many alterations. The most obvious was the conversion of gymnospermous sporophylls into stamens and carpels, resulting in the formation of flowers. Because gymnosperms have rather flat, leaf-like sporophylls, arranged in spirals around an axis, and microsporophylls never occur with megasporophylls except in cycadeoids, those flowering plants that have similar reproductive parts are believed to be the most relictual. For example, in a group known as the basal angiosperms, such as *Amborella*, Nymphaeales (water lilies), Austrobaileyales, and Magnoliales (*Magnolia*) have flat stamens without distinct filament and anther portions, arranged in spirals, and their sporogenous tissues (microspore mother cells) form relatively large, prominent internal masses (**Figure 23-1**).

In most **angiosperm carpels**, the edges of sporophyll primordia crowd against each other and grow shut, sometimes leaving a visible suture, sometimes closing so completely that no sign of a seam remains. This is called a **closed carpel**, and it develops into a fruit that encloses the embryos as they develop into seeds ("angio" means "clothed"). Carpels in the basal angiosperms are leaf like,

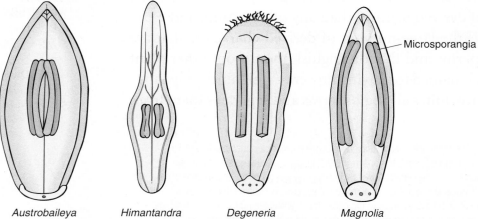

Austrobaileya *Himantandra* *Degeneria* *Magnolia*

Figure 23-1 Several groups of angiosperms have stamens with what are believed to be relictual features; these features have not undergone much evolutionary change, so they are still similar to the ancestral features. The stamens (microsporophylls) of these four living angiosperms are much fleshier, flatter, and more leaflike than most other stamens of living species. Also, the sporogenous tissue is rather massive, similar to that of cycads and cycadeoids. *Austrobaileya, Himantandra,* and *Degeneria* are unfamiliar to most students; they are plants of South Pacific islands.

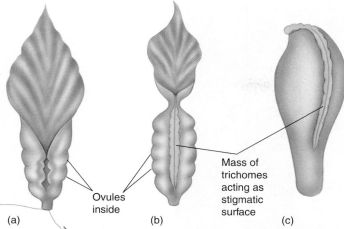

Ovules
inside

Mass of
trichomes
acting as
stigmatic
surface

(a)　　　　(b)　　　　(c)

Figure 23-2 Carpels are megasporophylls; therefore, we expect the most relictual living carpels to be rather leaflike, somewhat resembling megasporophylls of cycads. Several living angiosperms do have carpels resembling young leaves that have failed to open (a); the two halves of the blade are pressed together, a rather large amount of vascular tissue is present, and a stigma and style are absent. Hairs along the blade margin act like a stigma. If such a carpel produced ovules only at the base and if the upper part elongated, it would begin to resemble a more derived, modern carpel (c).

resembling young leaves whose blades have not yet opened. Instead of a stigma and style, the ovary edges have rows of secretory hairs that secrete a thick liquid that both seals the seam and functions as a stigmatic surface (**Figure 23-2**). Whereas ovules in gymnosperms are exposed on the megasporophyll surface, the folded nature of the megasporophyll in flowering plants offers ovules some protection, and probably much more important is the fact that the pollen tube (microgametophyte) must now interact with the stigmatic area in addition to the nucellus, allowing more opportunity for the sporophyte to "test" the pollen. Weak pollen or that of the wrong species can be inhibited from germinating, and the sporophyte permits only healthy pollen of the proper species to pass through the stigma and reach its ovules (**Figure 23-3**). Fewer ovules are wasted by exposure to improper or unhealthy pollen.

In the transition from gymnosperms to angiosperms, fertilization evolved such that the second sperm cell of the pollen tube fuses with the polar nuclei of the megagametophyte, producing the endosperm nucleus. This process of **double fertilization** is universal in flowering plants. Within the vegetative body, the major transitions were the evolution of vessel elements and sieve tubes. Leaves became broader and developed reticulate venation, and they became more polymorphic, adaptable to a variety of functions, not just photosynthesis.

Structural modifications are most easily seen and studied, but probably a more fundamental transition was the acquisition of developmental plasticity, the capacity to survive mutations that alter growth and development. Conifers lack developmental plasticity; for example, most people never have trouble recognizing an unfamiliar plant as being a conifer when they see one because all conifers greatly resemble each other. They are all large or giant trees with one or a few main stems and needle- or scale-shaped leaves. They are always green or brown, never brightly colored except for the red seeds of yew and the gray-blue berry-like cones of junipers. This uniformity is even more striking when one considers that gymnosperms dominated Earth for hundreds of millions of years and

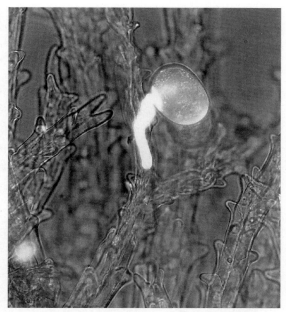

Figure 23-3 In many angiosperms, complex compatibility interactions occur between either the pollen and stigma or the pollen tube and stigma/style. In some cases, if the stigma and style detect an improper pollen tube, they deposit callose around it and inhibit its growth. The callose block is visible here as a white sheath (×60).

could have diversified into almost any habitat. On the other hand, angiosperms have tremendous developmental plasticity: They have diversified so greatly that not only hundreds of thousands of species, but also thousands of types, now exist. They range from gigantic (*Eucalyptus*, oaks, and elms) to tiny (mustards, certain daisies; **Figure 23-4a**); woody to herbaceous; perennial to biennial to annual; temperate to tropical; desert habitats to mesic to rain forest to aquatic; autotrophic to partially parasitic to fully parasitic; and epiphytic to subterranean (Figure 23-4b) to endophytic (living inside another plant, such as *Rafflesia* and *Tristerix*).

The evolutionary changes involved in the conversion of a gymnosperm line into an angiosperm line did not occur instantaneously, nor did they occur in all species. Double fertilization was probably the first transformation because it is universally present. It also occurs in some gnetophytes, and we are not certain whether it is a shared ancestral feature (a symplesiomorphy) indicating that flowering plants and gnetophytes belong to the same clade or whether it is independent, convergent evolution (homoplasy) in two unrelated groups. The ability to produce bisexual flowers must also have been one of the first transformations because all flowering plant clades—including basal angiosperms—have megasporophylls located above microsporophylls on a single axis. Although many angiosperms species produce either staminate or carpellate flowers, we are certain that in every case this has been a loss of parts from a bisexual ancestral flower. Vessel elements possibly evolved next because only one or two basal species lack them (*Amborella, Drimys*). This has been a conundrum for many years: Quite a few plants were considered to be basal angiosperms on the basis of having wood without vessels; it was thought that these were **primitively vesselless** (that they lacked vessels because their ancestors lacked them). Many other features, however, were inconsistent with the idea of these being a natural, early group, and the current hypothesis is that perhaps only *Amborella* is primitively

(a)

(b) Host plant   Tuber

(c)

(d)

Figure 23-4 (a) This dwarf daisy (*Monoptilon*) is by no means the smallest flowering plant. Many species of mustard, often the first flowers of spring, are also tiny. (b) *Ombrophytum subterraneum* is a parasite on roots of surrounding plants. The plant consists of an irregular, lumpy "tuber" and two inflorescences with tiny pink flowers. The root and stem base of the host are visible. (c) Many angiosperms, such as this *Hydrilla*, have adapted to living submerged in water, where they need little sclerenchyma and have no transpiration. (d) All the fruits and vegetables in this market were harvested from angiosperms. Almost all the plants we eat are angiosperms, and, in addition, most of our domesticated animals eat angiosperms such as hay, oats, and corn.

vesselless and that the others arose after vessels had originated but then these groups lost them. If so, they are **secondarily vesselless**, and their tracheid-based, gymnosperm-like wood is a derived feature that looks like a primitive one and misled us for years. Sieve tubes may have originated next; several species still have sieve cells in their phloem. All gymnosperms and seed ferns are or were woody plants and so are most of the basal angiosperms and eudicots; furthermore, wood in basal angiosperms has many gymnosperm-like features; therefore, we are confident that ancestral flowering plants were woody perennials. Many extant angiosperms are herbs, as are water lilies and peppers in the basal angiosperms; however, we are certain that this has resulted from the loss of the vascular cambium in these lines, and we are certain this has occurred repeatedly in many clades. The annual habit—the ability to germinate, grow, reproduce, and then die all within a single year—is a uniquely

angiosperm feature; no other group of seed plants can do this. Being an annual requires that plants produce flowers shortly after germination, during a period when most gymnosperm plants would still be producing a seedling body with juvenile features. Seedlings and adults often differ strongly in their body characters, and this ability of plants only a few months old to flower may be an important factor in angiosperms' developmental plasticity: Just because a plant has flowered does not mean that it has to die. After seedlings developed the capacity to flower, they could continue to live and grow large, retaining seedling features rather than developing adult ones.

Other derived features are the fusion of the carpels into a single structure (a **pistil**), fusion of petals into one structure (**sympetally**), and floral **zygomorphy**, that is, flowers that are bilaterally symmetrical, not radially symmetrical (Table 23-1).

TABLE 23-1 Relictual and Derived Features in Division Magnoliophyta

	Relictual (present in early angiosperms)	Derived (present only in some later, more modern angiosperms)
Habit		
Size	large bush, small tree?	large trees, herbs, bulbs, vines, many types
Leaf retention	evergreen	deciduous or leafless
Wood		
Vessels	none	present
Axial parenchyma	none or little	abundant; distribution is important
Rays	all narrow, tall	of several types in one plant: short/wide and narrow/tall
Flowers		
Presence of parts	complete flowers	incomplete/imperfect
Number of each type of part	many	few: sets of 3, 4, 5
Arrangement	spirals	whorls
Symmetry	radial	bilateral
Position of ovary	superior	inferior
Fusion of parts	none	much fusion
Pollination	wind? beetles?	many types
Fruit/seed dispersal	wind?	many types

Even more specialized characters such as succulence, parasitism, epiphytism, bulbs, corms, tubers, tendrils, and insect-trapping leaves originated only much later, after angiosperms had diversified into many clades; consequently, each of these innovations is present in only a few families or species. It should not be thought that the flowering plants are now "finished," that all their evolution has already happened. Some groups do seem to be changing very little now, but others, especially grasses, composites, bromeliads, and orchids, are still changing and evolving so rapidly that it is difficult to keep up with them. Similarly, we have no reason to expect that the flowering plants will always be the most derived group; lycophytes dominated Earth for millions of years, only to be displaced by gymnospermous seed plants, which in turn are now being overshadowed by flowering plants.

Changing Concepts About Early Angiosperms

Concepts about the nature of early angiosperms have changed as our knowledge of existing and fossil plants has become more complete. In the last century, wind-pollinated trees (alders, elms, oaks, and plane trees) were grouped together in a "subclass Hamamelidae" and were considered the most relictual living flowering plants (**Figure 23-5a**). This seemed reasonable because these species tend to be large trees with dense wood, their flowers are small and simple, usually without sepals and petals, and many gymnosperms have similar features. After hamamelid wood anatomy was studied, however, it became obvious that they could not be relictual because their wood contains vessels, fibers, and abundant parenchyma, features not found in gymnosperms. Furthermore, hamamelid flowers are simple through specialization and reduction: Wind-pollinated flowers do not need to attract pollinators, and thus, they do not need to be large or colorful—they do not need petals or sepals. Recent DNA studies indicate that this syndrome of large, wind-pollinated trees is a derived condition within angiosperms, and furthermore, it has evolved several times in various clades: This is an example of homoplasy, of convergent evolution, and the hamamelid taxa are now separated from each other and placed among the clades we believe to be their true relatives.

Approximately 100 years ago, C. E. Bessey developed the hypothesis of the **ranalean flower**, in which a *Magnolia*-type flower was thought to be relictual (Figure 23-5b). Such a flower is **generalized**; that is, it has all parts (sepals, petals, stamens, and carpels), and these are arranged spirally. Also, carpels occur in a superior position, above the other parts. It is easy to postulate the evolution of all of the various existing flower types from a ranalean, generalized ancestor. For example, monocot flowers often

Petal   Carpels   Stamens

(a)

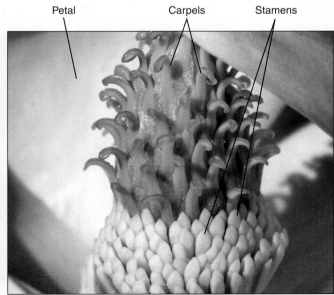

(b)

Figure 23-5 (a) Catkins of pecan (*Carya illinoensis*); the axis bears many flowers, all with stamens but not carpels. Other flowers on the same plant would be carpellate and give rise to pecan fruits after being pollinated. (b) A flower of *Magnolia* is considered the typical ranalean flower. All parts are rather massive and are not specialized for one particular type of pollinator. Very few flowers have so many carpels, almost always fewer than ten; in contrast, seed cones of conifers and cycads have large numbers of megasporophylls. Each carpel secretes a drop of liquid that catches pollen grains, much as conifer megasporangia do.

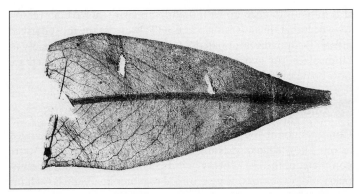

Figure 23-6 A leaf of *Magnoliaephyllum* showing strong similarities to the leaves of living magnolias; however, fine details visible by microscopy show that this is not identical to a true *Magnolia* leaf.

have three stamens and three carpels, whereas eudicot flowers usually have four or five of each; either condition could arise by a simple reduction of the large number present in a ranalean type flower. Also, the staminate or carpellate imperfect flowers of dioecious species could also arise easily by loss of parts.

Most botanists had long ago concluded that angiosperms are monophyletic. Features as complex as double fertilization, flowers, and developmental plasticity probably did not evolve more than once. Almost all recent DNA studies also indicate monophyly of angiosperms.

Currently, many paleobotanists and taxonomists believe that the transition from gymnosperm to angiosperm occurred during the Jurassic and Lower Cretaceous Periods of the Mesozoic Era. The earliest leaf fossils definitely considered to be those of angiosperms are from the Lower Cretaceous Period, approximately 130 million years ago. They represent both dicot and monocot leaves, for example *Magnoliaephyllum*, *Ficophyllum* (fig-like leaves), *Vitiphyllum* (grape-like leaves), and *Plantaginopsis* (simi-

lar to *Plantago*) (**Figure 23-6**). These fragments are rare in such old rocks, constituting only approximately 2% of all fossils present. In rocks about 50 million years more recent, from the Upper Cretaceous Period, many more types of angiosperm leaf fossils are present, outnumbering the leaf fossils of gymnosperms and ferns. These Upper Cretaceous leaf fossils originally were assigned to extant genera such as *Populus* (poplar), *Quercus* (oak), and *Magnolia* on the basis of overall shape and size, but careful examinations of fine venation, cuticle, and stomata often show that these fossils are not truly part of modern genera.

The original misidentification of these leaves contributed to the erroneous hypothesis that angiosperms must have originated in the Jurassic Period, become abundant enough to leave a few fossils in the Lower Cretaceous Period, but then suddenly, in the short span of 30 to 50 million years, rapidly diversified into most of the major extant families and even many of the modern genera. After this supposed burst, genera such as *Populus, Quercus,* and *Magnolia* would have had to remain relatively unchanged for tens of million years, up to the present. Such a burst of diversification followed by extreme stability did not seem plausible. With the discovery that the Upper Cretaceous and Tertiary leaf fossils only resemble leaves of modern genera but are not identical to them in subtle features, a much more gradual evolution can be assumed to have occurred. The early angiosperms were well established by the Lower and Mid Cretaceous Periods, but probably no modern genus was yet in existence, only precursors that greatly resembled modern genera. These would have evolved gradually into true poplars, oaks, magnolias, and so on.

The oldest wood that seems to be from an angiosperm comes from the Aptian Epoch (125 million years ago) of Japan; unequivocal angiosperm wood occurred approximately 120 million years ago. Flowers and fruits appear for the first time in the Lower Cretaceous Period, which must mean that their forerunners evolved in the Jurassic Period (**Figure 23-7**), about the same time that dinosaurs were flourishing (see the timeline on the inside cover on the

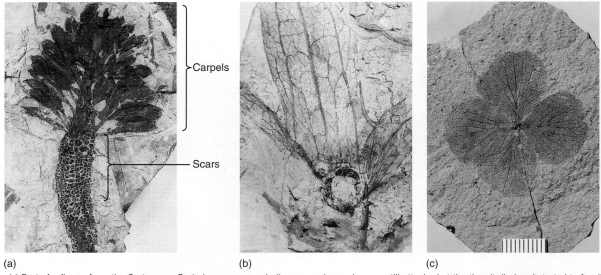

(a) (b) (c)

Figure 23-7 (a) Part of a flower from the Cretaceous Period; numerous spirally arranged carpels were still attached at the time it died and started to fossilize. The spiral arrangement is similar to that found in flowers of *Magnolia* and water lily. Below the carpels are spirally arranged scars where structures abscised; we do not know if they were sepals, petals, or stamens. (b) Part of a fossil fruit, *Paraoreomunnea puryearensis,* approximately 60 million years old. It was a winged fruit similar to those produced today by members of Juglandaceae, the family of walnuts and hickories. (c) Fossil of a *Hydrangea* flower, 44 million years old. This flower is much more modern and derived than that of (a), having fewer parts.

Figure 23-8 A reconstruction of a *Glossopteris,* a possible relative to the ancestors of flowering plants. The leaves had a prominent midrib and reticulate venation. The leaves were deciduous, and the wood had annual rings, indicating that it lived in a temperate climate.

back of the book). In the last several years, China has explored for fossils much more vigorously, and rich new fossil beds are yielding many new specimens of early angiosperms. Often the fossils consist only of abscised carpels or carpels attached to a receptacle. Unfortunately, sepals, petals, and stamens often abscise even while a flower is still on a plant, and as a result, they are rarely present with the carpels. Perianth parts and stamens of modern plants also tend to be delicate and to decay quickly rather than fossilize; the same may have been true of the early flowers. Pollen that could have come from either relictual dicots or monocots is found in Lower Cretaceous strata.

Much attention is now being given to gymnosperms of the Jurassic and Triassic Periods, and the focus is centering on cycadophytes and glossopterids (**Figure 23-8**). As more is learned about this group of gymnosperms, we realize how much the various groups had begun to develop angiosperm-like features.

As you read the following discussion about flowering plants, it will be helpful to study a timeline chart on the evolution of true plants.

Classification of Flowering Plants

Magnoliophyta is such a large group with so many families, genera, and species that it is rare for an individual taxonomist to attempt to study and classify the entire group. In the 1980s and 1990s, the most widely used monograph of the entire division was *An Integrated System of Classification of Flowering Plants* by Dr. Arthur Cronquist of the New York Botanical Garden. Since then, cladistic studies involving DNA, biochemistry, and anatomy often proposed phylogenies that agreed well with those of Dr. Cronquist for certain clades but disagreed for others. The areas where the two systems were inconsistent became the topics of more intense scrutiny of all characters, with subsequent adjustments in our concepts of phylogeny and classification. Currently, two widely used reference books are *Plant Systematics, A Phylogenetic Approach* (3rd edition) by W. S. Judd et al and *Plant Systematics* by Michael Simpson. Their approach is the basis for the treatment used here.

Soon after their origin, flowering plants began to follow two distinct lines of evolution, and currently, almost all angiosperms are classified as **monocots** or **eudicots** (**Figure 23-9**). No single

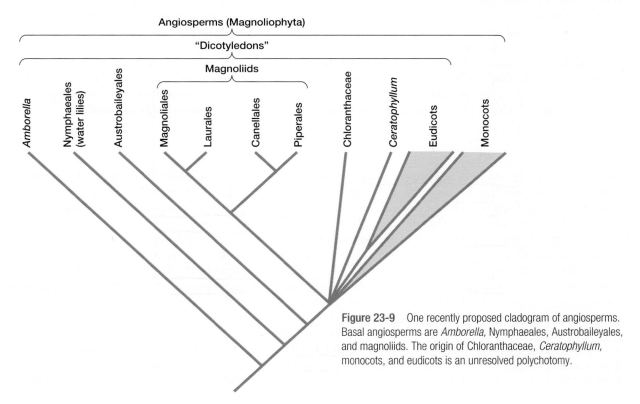

Figure 23-9 One recently proposed cladogram of angiosperms. Basal angiosperms are *Amborella,* Nymphaeales, Austrobaileyales, and magnoliids. The origin of Chloranthaceae, *Ceratophyllum,* monocots, and eudicots is an unresolved polychotomy.

Box 23-1 Maintaining Genetic Diversity

The flowering plants are particularly adept at evolving rapidly; in the short time they have existed, they have diversified greatly. The result is that hundreds of thousands of species exist, each differing from the others in features such as flowers, fruits, vegetative structure, and metabolism. Within this panoply of diverse plants, we and our ancestors have discovered a wealth of foods, medicines, and soul-soothing flowers.

Now, at the beginning of the 21st century, we have become aware of the horrific ecological destruction occurring on Earth, with vast areas of vegetation being cut or burned or flooded with artificial lakes behind hydroelectric dams. We have also begun to realize that we are causing the extinction of hundreds of species every year and, in the process, their genetic information is being lost. On a philosophical basis, we can debate whether one species, *Homo sapiens*, has the right to destroy so many other species, but on a more pragmatic basis, we should consider whether we are unwittingly destroying potential sources of new foods and medicines. Consider just two examples: A relatively unknown species, quinoa (*Chenopodium quinoa*), can be grown in relatively dry regions with poor soil, yet it produces an abundance of nutritious, protein-rich seeds. And *Pantadiplandra brazzeana* produces a small protein that is 2,000 times sweeter than sugar; this could possibly be used as a completely natural, safe, nutritious, low-calorie sweetener, decreasing the amount of sugar or artificial sweeteners that we consume.

How can botanists preserve the genetic resources of the plants, animals, and microbes that coexist with us? Several types of efforts are needed, and two particularly important ones are habitat preservation and germ plasm banks (germ plasm refers to the cells that produce gametes). Germ plasm banks were started several years ago as rather simple seed storage facilities, where bags of many types of seeds were maintained under cool, dry conditions. After several years, while most seeds were still healthy, they would be germinated, grown into mature plants, and then their fresh, healthy seeds would be collected and stored. After several years, the process would be repeated.

This is still done for many species, but now the seeds can be kept for extremely long periods, held at very low temperatures or even in liquid nitrogen. Furthermore, it is now possible to maintain some seeds as tissue or cell cultures, either grown slowly in culture tubes or kept as frozen cells. In the foreseeable future, it may be that some plants, or at least some of their genes, will be kept simply as DNA molecules in solution or in cloning vectors.

Currently, there are over 100 germ plasm banks throughout the world, preserving at least 3 million types of plant material. Examples are the Svalbard International Seedbank, built in an abandoned coal mine in the permafrost of northern Norway, and the National Seed Storage Laboratory (NSSL) located at Colorado State University and run by the United States. The NSSL was established in 1958 and has more than 232,000 seed types. At present, its size is being quadrupled, and its capacity will be more than 1 million seed samples. The Millennium Seed Bank Project at Kew Gardens in England reached its goal of safeguarding 10% of all the world's plant species in October, 2009.

The purpose of germ plasm banks is not merely to store and maintain plant genetic diversity but also to make it available. Botanists can request samples and use them for research and plant breeding programs. If some are found to have useful traits, they can be grown as crops. Of course, it is also hoped that in many cases we can use these plants to help restore or rebuild ecosystems that we have already damaged.

---

character always distinguishes a monocot from a eudicot, and some species would fool most botanists. In general, monocots, as their name implies, have only one cotyledon on each embryo, and other typical characters are the following: Their leaves usually have parallel veins because the leaves are elongate and strap shaped (grasses, lilies, and irises); vascular bundles are distributed throughout the stem, not restricted in one ring, and monocots never have ordinary secondary growth and wood (a few have anomalous secondary growth). Flowers of monocots have their parts arranged in groups or multiples of three: three sepals, three petals, three stamens, and three carpels (**Figure 23-10**). Eudicots are much more diverse and include a greater number of families, genera, and species. Eudicots have two cotyledons and reticulate venation in the leaves; vascular bundles occur in only one ring in the stem. They can be woody, herbaceous, or succulent or have any of many highly modified forms. Flower parts occur in sets of five most often and sometimes in sets of four, but they only rarely occur in sets of three.

A surprising result of cladistic studies is that the monocot/eudicot divergence did not occur right away, but instead, the early angiosperms diverged into several clades now called the **basal angiosperms**. These are not newly discovered species; it is just that we had classified them as monocots or dicots before, but many different DNA sequences indicate they had become reproductively separate from the other angiosperms very early. Of the basal angiosperms, *Amborella* and Austrobaileyales greatly resemble eudicots; however, they cannot be classified with the eudicot clade because the common ancestor that includes both them and

Figure 23-10 Lilies and their relatives are excellent for demonstrating the typical monocot flower: They have three of each type of appendage and all are large enough to be seen easily. The three carpels are fused together, but a transverse section would show three chambers. One petal and two stamens were removed to reveal the ovary.

(a)

(b)

Figure 23-11 (a) Flowers of water lilies have numerous sepals, stamens, and so on, with all parts arranged in spirals like the leaves of a rosette plant. This is believed to be a relictual feature similar to a gymnosperm cone. Monocot flowers would be derived from an ancestral flower like this by mutations that result in only three or six of each type of floral appendage. (b) *Amborella*.

the eudicots also includes monocots, and we cannot construct a group that includes some descendants of a common ancestor but excludes others. Similarly, Nymphaeales (water lilies) have many features of monocots, but classifying them as monocots would create a polyphyletic group.

Basal Angiosperms

The basal angiosperms contain the living descendants of several groups that originated while angiosperms were still a young clade. Their ancestors became reproductively isolated from the other early flowering plants before distinctive angiosperm traits had originated. It would be easy to assume that the descendants of the three earliest diverging groups would all display a large number of primitive traits and few derived ones: They "should" look primitive, and they "should" look like taxa that had just barely become distinct from some gymnosperm sister group. All organisms evolve, however, and so did the descendants of the basal angiosperms. The three groups of extant descendants of these clades—Amborellaceae, Nymphaeaceae, and Austrobaileyales—have not remained static evolutionarily and have not preserved all ancestral features intact. One group should be familiar to you: Nymphaeaceae are the water lilies (**Figure 23-11**). They are small, soft-bodied herbs with vascular bundles scattered like those in monocots, and they completely lack any wood; in most, their stems must be submerged underwater, and exposure to ordinary conditions on land would kill them. Their large colorful flowers are pollinated by animals. They have almost nothing in common with either gymnosperms or seed ferns. *Amborella trichopoda* (there is only one species in this clade) is very different, being small trees in forests of New Caledonia. Their wood contains tracheids but no vessels and little parenchyma. This species is dioecious, some plants having staminate flowers, others having carpellate flowers (and staminodes that indicate these evolved from bisexual ancestors). Both flowers are small with 5 to 11 tepals (not differentiated into sepals and petals). Staminate flowers have numerous stamens, and car-

pellate flowers have 5 or 6 carpels whose edges do not seal tightly but instead have interlocking, secretory hairs that act as a stigma. Austrobaileyales contains woody trees with bisexual flowers, with stamens and carpels being similar to those of *Amborella*.

Because water lilies, *Amborella,* and Austrobaileyales differ in so many features, they cannot all resemble the earliest angiosperms. Instead, one or the other—or all three—have undergone considerable evolutionary change since their clades originated. Water lilies have little in common with either gymnosperms or seed ferns; almost certainly many of their features have changed considerably. *Amborella* and Austrobaileyales appear to have changed little, but it is important to realize that they have not remained static. We cannot assume that every feature of these woody trees is like that of early angiosperms; instead, we assume

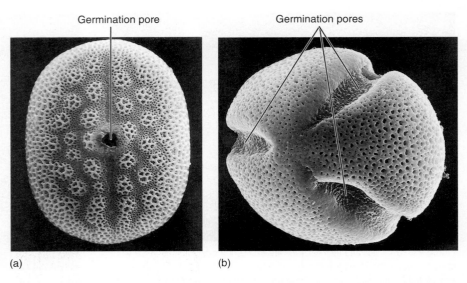

Germination pore   Germination pores

Figure 23-12 (a) All early fossil pollen has only one aperture, but almost all living eudicots with many derived features have three apertures; therefore, uniaperturate pollen in living angiosperms, such as this *Siphonoglossa*, is considered a relictual feature (×300). (b) Pollen of *Lophospermum* has three germination pores; the pores are located in grooves, something that does not occur in any early fossil pollen (×300).

(a)   (b)

TABLE 23-2	Representative Families of the Magnoliid Clade
Aristolochiaceae	*Asarum* (wild ginger), *Aristolochia*.
Lauraceae	*Cinnamomum* (*C. verum*, cinnamon; *C. comphora*, camphor), *Laurus* (*L. nobilis*, bay leaves), *Persea* (*P. americana*, avocado).
Magnoliaceae	*Liriodendron* (tulip tree), *Magnolia*.
Myristicaceae	*Myristica* (nutmeg, mace).
Piperaceae	*Piper* (*P. nigrum*, black pepper), *Peperomia* (ornamental).

that features present in them and gymnosperms are relictual features, plesiomorphies.

Ancestors of the magnoliid clade diverged from early angiosperms a bit later than *Amborella* and Austrobaileyales, but this conclusion is based mostly on DNA evidence—Morphologically, the magnoliids do not differ greatly from these two earlier clades (Table 23-2). For example, family Magnoliaceae contains trees with wood similar to that of gymnosperms in that it lacks vessels, fibers, and axial parenchyma; *Magnolia* flowers have numerous stamens and carpels arranged in spirals (see Figure 23-5b). Furthermore, carpels are not fused together into a pistil as occurs in almost all monocots and eudicots. Another important feature is that their pollen grains have only a single germination pore: They are **uniaperturate**, as are all other basal angiosperms and monocots. Eudicots have three germination pores (Figure 23-12). Other magnoliids are laurels and avocado (Laurales) and peppers and peperomias (Piperales).

Monocots

Currently, monocots are widely believed to have arisen from early angiosperms approximately 80 to 100 million, perhaps even 120 million years ago. Because all monocots lack ordinary secondary growth and wood, their ancestors were probably herbs with either no vascular cambium or little cambial activity. The gynoecia of many monocots are composed of several carpels (usually three) that are either free of each other or at most only slightly fused together (Figure 23-10). The perianth usually consists of three outer and three inner members; they often look so similar that rather than using the terms sepals and petals, the perianth members are just called **tepals**. Monocot tepals rarely fuse to each other to form a tube; instead, they remain free. These features must have been present in the group from which the monocots evolved. Water lilies (Nymphaeales) have similar features. An early set of angiosperms may have resembled Nymphaeales and may have given rise to both the water lilies and the monocots.

One theory about the parallel venation of monocot leaves postulates that the ancestors had broad leaves and lived as aquatic plants. The leaves evolved to a more reduced, simple type without a blade, a form more adaptive for a submerged leaf (Figures 23-13a and b). Some of these plants moved into drier habitats where their

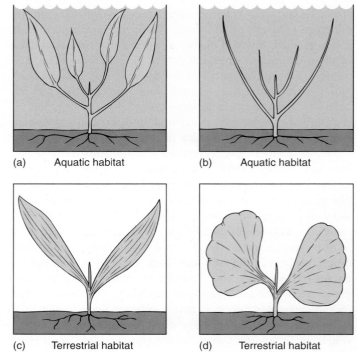

(a)   Aquatic habitat   (b)   Aquatic habitat

(c)   Terrestrial habitat   (d)   Terrestrial habitat

Figure 23-13 (a) Monocots are believed to have originated as semiaquatic plants that inhabited swamps and marshes. (b) The leaf may have consisted only of a leaf base and petiole, the entire blade having been lost evolutionarily. (c) A "pseudolamina" evolving from a petiole. (d) A "broad leaf" evolving from a petiole.

Figure 23-14 Some monocots, such as this banana (*Musa*), have broad leaves, but their lamina is not homologous to that of a dicot leaf. In this monocot broad leaf, all vascular bundles run parallel to each other from the midrib to the margin, whereas in a dicot lamina, they diverge and branch into a reticulate pattern.

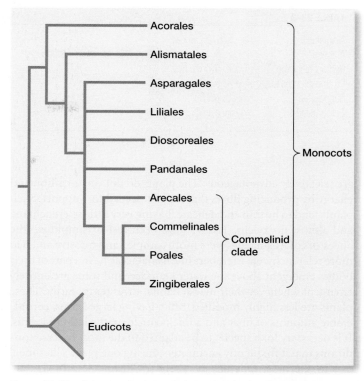

Figure 23-15 One recently proposed cladogram of monocots. There are many unresolved polychotomies.

leaves were not submerged and a broad or long lamina would be advantageous. Mutations that resulted in a basal meristem were selected, resulting in strap-shaped leaves (Figure 23-13c). In the evolutionary lines of "broadleaf" monocots such as palms, philodendrons, and dieffenbachias, a marginal type of meristem evolved; it was located at the end of the residual leaf, resulting in the formation of what appears to be a petiole and lamina (Figure 23-13d and Figure 23-14).

Early monocots diverged into a series of clades whose extant members are classified into approximately ten orders (Figure 23-15).

Alismatales Alismatales contains many aquatic herbs such as *Sagittaria* (arrowhead; Figure 23-16) and many aquarium plants such as *Hydrocharis, Najas,* and *Hydrilla* (Table 23-3; see Figure 23-4c).

These plants are most often found in swamps and marshes, partly or entirely submerged. Although many species retain a large number of plesiomorphic features, others have become highly modified in response to the unusual selection pressures associated with an aquatic habitat. Plants of Hydrocharitaceae, Najadaceae, Posidoniaceae, Ruppiaceae, and Zosteraceae grow completely submerged in seawater and are known as "sea grasses." They have no transpiration, and thus, mutations that result in the loss of stomata are not selectively disadvantageous. Air chambers make the plants buoyant; therefore, mutations that prevent the formation of lignified fibers

Figure 23-16 *Sagittaria,* a member of Alismatales, is dioecious; plants have either staminate (a) or carpellate (b) flowers.

(a)   (b)

TABLE 23-3	Representative Families of Alismatales
Alismataceae	*Sagittaria* (arrowhead).
Araceae	*Anthurium, Caladium, Dieffenbachia, Monstera, Philodendron.*
Hydrocharitaceae	*Elodea, Hydrilla, Najas.*
Potamogetonaceae.	*Potamogeton*
Ruppiaceae	(aquatic plants).
Zosteraceae	(aquatic plants).

are selectively advantageous. The plants do not waste carbon and energy by producing fibers that are not needed for support. Such plants tend to be thin and delicate, having very little sclerenchyma and almost no xylem. Tissues have been lost or simplified; the ancestors of Alismatales were more complex and massive and had more sclerenchyma. Members of Alismataceae have a part of their bodies emergent above the water's surface, and some are entirely emergent whenever their marsh habitat dries temporarily. These plants are less highly modified, often having large leaves, considerable amounts of fiber and a thick cuticle on the leaf epidermis. It is necessary for a species to be adapted to the most adverse conditions that it frequently encounters during one plant's lifetime.

In the most relictual members of Alismatales, flowers are large and showy, with three sepals and three petals, but in others, they are highly modified, especially in those species in which the flowers do not emerge from the water and pollination occurs below the surface. Such flowers often completely lack a perianth.

Family Araceae (aroids) contains numerous familiar houseplants: *Philodendron* (250 species), *Anthurium* (500 species), and *Arisaema* (Jack-in-the-pulpit; 100 species). *Dieffenbachia* (dumb cane) and 1,800 other species also belong here (**Figure 23-17**). This family is characterized by the evolution of a distinctive inflorescence: tiny flowers, either unisexual or perfect, embedded in a thick stalk, the spadix. Staminate flowers are located near the top of the spadix, carpellate flowers near the base. The spadix is surrounded by a large bract, the spathe.

Liliales In the past, Liliales was defined broadly as a large group with many highly derived families; it was basically defined as the "petaloid" monocots—those with large, colorful flowers. Recent DNA evidence indicates that Liliales is a smaller clade with approximately 11 families and 1,300 species. Many of its previous families are now grouped in Asparagales. Important features in Liliales are the presence of spots or lines on the petals and of rather ordinary nectaries formed at the bases of tepals or stamens; flowers in Asparagales typically have septal nectaries (discussed below). One of the largest families, Liliaceae, contains so many ornamental plants, mostly bulbs, that most people are familiar with them and think of them as the "typical" monocots (Figure 23-10). Liliaceae contains *Lilium* (lilies), *Tulipa* (tulips), *Calochortus* (mariposa lily), and *Fritillaria.* Colchicaceae contains *Gloriosa* (flame lily) and *Colchicum,* the source of the alkaloid colchicine that prevents microtubule polymerization.

Perhaps the most unusual member of Liliales is Smilacaceae, with the single genus *Smilax* (catbrier). Despite being closely related to lilies, these are tough, fibrous vines with leaves that have a petiole and broad lamina with reticulate venation—not lilylike at all. At first glance, it is easy to assume they are dicots.

Asparagales This is a large clade with many families, species, and types of biology. By examining this clade, we get a sense of evolution as diversification (Table 23-4). There are several morphologic and DNA synapomorphies that unite this group. Most form nectaries in an unusual way: Their carpels fuse side by side starting at their bases, the fused sides being called **septa**. They do not fuse all the way up to the style, however, and the open areas secrete nectar; these are **septal nectaries**. Beyond this, the group is extremely diverse in morphology (**Figure 23-18**), ranging from small, delicate bulbs like *Hyacinth,* chives, and onion (*Allium*) to vining epiphytes such as many orchids. Bulbs, rhizomes, and corms are common. Agaves and yuccas have giant, fibrous perennial

Figure 23-17 (a) This *Anthurium* (Araceae) is an aroid, one of the monocots that have a eudicot type of broad leaf; compare its reticulate venation with the parallel venation of banana (Figure 23-14). Many members of this family are adapted to conditions of low light and high humidity; most are members of tropical or subtropical habitats. (b) Araceae have a characteristic inflorescence with a sheath-like bract (a spathe) surrounding a thick column (a spadix). Tiny staminate flowers are embedded in the top of the spadix, carpellate ones in the base (hidden by the spathe here). *Peltandra virginica.*

(a)

Spathe
Staminate flowers
Spadix
Carpellate flowers

(b)

(a)

(b)

Figure 23-18 Members of Asparagales. (a) *Iris* flowers appear to have many unfamiliar parts, but the uppermost petaloid parts are actually stamens; the long dark regions are sporogenous tissues, the anthers. Some cultivated irises are sterile and do not have sporogenous tissue. In many angiosperms, structures that appear to be petals are really sterile stamens. (b) Like bamboos and palms, yuccas and agaves prove that herbs do not have to be soft and small. The leaves are long and narrowly triangular with parallel venation, and they may be thick at the base. They are arranged as a rosette around a short (agaves) or long (yuccas) stem. Most have a needle-sharp point and razor-like hooked spines along the edges of the leaf.

leaves. Iris has flattened sword shaped leaves, and many members such as *Narcissus* (daffodils) have delicate leaves that live only during spring and summer. Several members of Agavaceae (*Yucca brevifolia,* Joshua tree) and Ruscaceae (*Dracaena,* dragon tree) have anomalous cambia and grow to be large, highly branched trees. Asparagales has diversified biochemically as well, with different groups having distinctive chemical compounds, such as the sulfur compounds in onion and garlic or particular alkaloids in Amaryllidaceae. Flowers are large, colorful, and showy for the most part, and in Orchidaceae, numerous types of insect pollination have evolved, many accompanied by distinctive aromas.

The Orchidaceae is the largest and most diverse family of Asparagales. The most familiar, ornamental ones are epiphytic, but many are terrestrial and one is a subterranean parasite. Orchid flowers are highly modified from the ancestral conditions of angiosperms, being zygomorphic (bilaterally symmetrical) with complex shapes, colors, and fragrances that attract specific pollinators. Orchids are unusual in producing hundreds or thousands of tiny seeds in each fruit. The seeds are dustlike and so undeveloped at "germination" that they must form a symbiosis with certain fungi in order to survive long enough to form roots and leaves.

Dioscoreales   This small order has only one family, Dioscoreaceae, and is mentioned because it has a familiar, important food crop and unusual morphology. Yams are starchy "tubers" produced by several species of *Dioscorea* and are a major source of carbohydrates for many people of tropical areas. Also, dioscoreas have petiolate, broad leaves with reticulate venation and are easily mistaken for dicots.

■ Commelinoid Monocots

The following four orders of monocots are known as the **commelinoid monocots** (Figure 23-15) because they differ from the others in several unusual synapomorphies: They have unique types of epicuticular wax. Their walls have unusual types of hemicelluloses and ultraviolet-fluorescent compounds. Their pollen contains starch, as does their endosperm. Multiple studies of *rbcL, atpB,* and other DNA regions support this as a distinct clade.

Arecales   This order contains familiar plants, the palms, in family Arecaceae (an old name for the family is Palmae; **Figure 23-19**). There are about 3,500 species, all of which are easily recognizable by their solitary trunk (only a few species have branched trunks), which varies from 1.0 m wide in some and only 1.0 cm in others; all have scattered vascular bundles, which is typical of monocots. Leaves of palms always occur only near the shoot apex, never distributed along the length of the stem. In a few species, trunks are prostrate (palmetto palm, *Sabal*) or vines (*Daemonorops*). All

TABLE 23-4	Representative Families of Asparagales
Agavaceae	*Agave, Chlorophytum, Hosta, Yucca.*
Alliaceae	*Allium* (onion, garlic, shallots, chives, leeks).
Amaryllidaceae	*Amaryllis* (Cape belladonna), *Crinum* (crinum lily), *Haemanthus* (blood lily), *Hippeastrum* (amaryllis), *Hymenocallis* (spider lily), *Narcissus* (daffodil, jonquil).
Asparagaceae	*Asparagus.*
Asphodelaceae	*Aloe* (aloe vera), *Gasteria, Haworthia* (all with succulent leaves).
Hyacinthaceae	*Hyacinthus, Muscari, Scilla* (all ornamental bulbs).
Iridaceae	*Crocus, Gladiolus, Iris.*
Orchidaceae	*Dendrobium, Epidendrum, Maxillaria,* and many others.
Ruscaceae	*Dracaena* (dragon tree), *Liriope* (border grass), *Nolina, Sansevieria* (mother-in-law's tongue).

(a)

(b)

Figure 23-19 (a) Palm flowers usually occur in large inflorescences at the top of a plant. (b) Although palm flowers are small, they have typical monocot organization with their parts in sets of three. *Hyphaene coriacea.*

species have simple leaves that, after fully expanded, are torn by wind into either a pinnate pattern (feather palms) or palmate one (fan palms). Coconuts (*Cocos*) and dates (*Phoenix*) are two types of palm fruit, but many others exist. Palm flowers are seldom seen because they are usually tiny, approximately 5 mm across, and are formed only high up in the tree.

Poales This order contains the grass family Poaceae as well as several other familiar groups such as cattails (*Typha*), bromeliads, and sedges (Table 23-5).

Poaceae contains about 8,000 species and are much more than just the plants in the lawn (Figure 23-20). They also include most foods, such as wheat (*Triticum*), barley (*Hordeum*), oats (*Avena*), rye (*Secale*), corn (*Zea*), rice (*Oryza*), and sugar cane (*Saccharum*) as well as a major building material of the tropics, bamboo (the subfamily Bambusoideae). Humans began farming grass crops as early as 10,000 years ago, and approximately 50% of all calories consumed by people come from grass seeds. Also, our main sources of meat—cattle, pigs, and sheep—are raised on grassland and fed corn. Grasses are abundant in flat, open, dry regions in the central areas of all continents, such as the rangelands of the United States and Africa, the steppes of Russia and Ukraine, and the pampas of Argentina; this accounts for one quarter of all vegetated land on Earth. Some grasses grow as tight clumps, and others spread by rhizomes or runners; either way, their shoot apical meristems tend to be located near the ground, protected from fire and grazing animals. As a result, they survive both better than most other types of plants. Widespread grasslands are a new ecologic

development, however, arising less than 25 million years ago. During the Miocene epoch, extensive forests disappeared and were replaced by grasslands, making possible an increase in grazing animals and perhaps even in ancestral hominids switching from a life in trees to one on the ground in African rangelands.

All grasses are wind pollinated, so sepals and petals are of little importance and are reduced to bristle-like structures. The three carpels are fused together, but in most grasses, the three stigmas remain separate. An "ear" of corn is an inflorescence, and each "silk" is a compound style and stigma. Within the fused ovary of grasses there is just one ovule; once fertilized, it matures into a seed whose seed coat fuses firmly to the developing fruit wall. The objects we usually call "seeds" of corn, wheat, and oats are actually single-seeded dry fruits known as caryopses (singular, caryopsis). Nearly the entire bulk of the seed is endosperm; the embryo is small but well developed.

Figure 23-20 Grass flowers are very reduced, simplified, and wind pollinated. The usual two small, dry scales (called a lemma and a palea) are thought to be remnants of an ancestral bract and two fused tepals; the third tepal does not form. Three anthers with long filaments and three carpels fused together with long feathery styles and stigmas are usually present. Grass flowers occur grouped together in complex, compact inflorescences, and inflorescence characters are important for identifying grasses.

TABLE 23-5	Representative Families of Poales
Bromeliaceae	*Ananas comosus* (pineapple), *Tillandsia* (*T. usneoides*, Spanish moss, *T. recurvata*, ball moss).
Cyperaceae (sedges)	*Carex, Cyperus, Eleocharis.*
Juncaceae	*Juncus* (rushes).
Poaceae	*Avena sativa* (oats), *Hordeum vulgare* (barley), *Oryza sativa* (rice), *Saccharum officinale* (sugar cane), *Triticum aestivum* (wheat), *Zea mays* (corn).
Typhaceae	*Typha* (cattail).

(a)

Staminate flowers, withered

Fruits, produced by carpellate flowers

(b)

Figure 23-21 (a) Most sedges grow in wet, marshy areas. *Carex elata.* (b) Cattail, *Typha,* has hundreds of staminate flowers above and thousands of carpellate flowers below. As a cattail breaks apart, the tiny one-seeded fruits blow away; seedlings are miniscule, like short threads.

Closely related to grasses are sedges (Cyperaceae with 4,000 species; Figure 23-21) and rushes (Juncaceae). All three families contain plants of wet, marshy areas, with tiny, reduced flowers and small dry fruits. Sedges and rushes are often confused with grasses, but "Sedges have edges, rushes are round, and grasses are hollow from top to ground:" Sedge stems are triangular in cross-section. Those of rushes are round and solid, and those of grasses are round but hollow. This simple rhyme is mostly true, but there are exceptions.

Other members of Poales are cattails (Typhaceae) and Bromeliaceae (bromeliads). Cattails grow in ponds and marshy areas. Each plant spreads by thick, horizontal subterranean rhizomes, with some axillary buds growing out as more rhizomes, and other axillary buds growing upward and producing the familiar green leaves: What appears to be a cattail plant is really just a single branch. Their characteristic reproductive structures consist of a tip with many staminate flowers and thousands of carpellate flowers, although by the time we see them, the staminate flowers have usually withered and the carpellate flowers have matured into tiny fruits (Figure 23-21b).

Bromeliaceae contains some of the most beautiful tropical epiphytes, their large, brightly colored inflorescences being easily visible even in thick jungle vegetation (Figure 23-22). Epiphytic species extend as far northward as the subtropics, Spanish moss and ball moss (*Tillandsia*) occurring from Florida to Texas. Other species are terrestrial and usually xerophytic, such as *Puya* of Chile and Peru and pineapples (*Ananas*). These xerophytic terrestrial forms are believed to be relictual, and their capacity to withstand drought may have made it easier for bromeliads to adapt to life as epiphytes. Because bromeliads occur in only the Americas, it

is assumed that they evolved after South America separated from Africa approximately 80 million years ago. Had they evolved earlier, numerous species should occur in the African rain forests and coastal deserts.

Zingiberales This order contains some of the most familiar of all houseplants: *Maranta, Calathea,* canna lilies (*Canna*), and gingers (*Zingiber, Hedychium*), as well as some that are best known in the warmer southern states—banana (*Musa*) and bird-of-paradise (*Strelitzia*) (Table 23-6). Members of Zingiberales differ from most other monocots in that they tend to have large, showy flowers pollinated by insects, birds, or bats. Furthermore, many of their flowers have derived features; adjacent sepals are often fused to each other, forming a tube, and the same is often true of the petals; many are bilaterally symmetrical. The gynoecium typically consists of three carpels that have fused almost completely. Very often the gynoecium is inferior, located below the sepals and petals, a condition that is always considered quite derived, the result of considerable evolutionary modification.

TABLE 23-6	Representative Families of Zingiberales
Cannaceae	*Canna* (canna lily).
Marantaceae	*Maranta, Calathea.*
Musaceae	*Musa* (banana, plantain).
Strelitziaceae	*Strelitzia* (bird-of-paradise).
Zingiberaceae	*Elettaria* (cardamom), *Curcuma* (turmeric), *Nicolaia* (torch ginger), *Zingiber* (ginger).

(a)

(b)

(c)

Figure 23-22   (a) Many epiphytic bromeliads catch rainwater with a tank made by their overlapping leaves; they absorb water through their leaf bases rather than through their roots. Their flowers develop underwater. (b) Spanish moss, *Tillandsia usneoides,* is a bromeliad, not a moss. It grows attached to tree trunks as an epiphyte, not a parasite. This species grows in the southeastern United States, where humidity is high and rainfall is frequent. (c) Pineapple plants are terrestrial bromeliads; they grow with their roots in soil. The edible portion of pineapple is a multiple fruit, composed of an entire inflorescence.

The largest family, Zingiberaceae, has approximately 1,000 species. Almost all are tropical, and most are soft, non-leathery herbs. The leaves are broad and have a petiole (Figures 23-13 and 23-14). During the evolution of the Zingiberales, mutations that resulted in the petiole/leaf becoming flattened and broadened were selectively advantageous: The plants grow in the heavy shade of jungle understory where light interception is important.

Eudicots

The eudicots constitute a much larger group than the monocots and are divided into numerous clades, some of which have diversified into many families, others having fewer members (**Figure 23-23**). Virtually every type of organ, tissue, and metabolism has several or many variations, so they are more difficult to characterize than are the monocots. One especially distinctive feature is that their pollen grains have either three germination pores (**tricolpate**) or have some condition derived from the tricolpate mechanism (Figure 23-12). Furthermore, their flower parts are arranged in whorls, not spirals, and their stamens usually have a well-defined filament (stalk) and anther; they are not flat or leaflike at all. These characters are greatly

modified from those of sporophylls found in other vascular plants. Current cladograms of eudicots have a large number of orders, too many to be covered here. Those that have special importance and that contain plants that you might know are presented below.

Basal Eudicots

In older classifications in which all flowering plants were assigned to either Monocots or Dicots, many families did not seem to fit well in either group. Several, such as water lilies, are now assigned to the basal angiosperms; Ranunculales and several others are believed to be clades that diverged at early stages in eudicot evolution.

Ranunculaceae is interesting because its flowers have so little fusion of parts: Each flower usually has many stamens and carpels, all of which remain separate of the others. Stamens may remain free or may fuse together into a tube or even fuse to the carpels. Familiar members of Ranunculaceae that have numerous, free parts are buttercups, windflower (*Anemone*), and *Clematis* (**Figure 23-24**).

The poppy family, Papaveraceae (Figure 23-24), is well-known for its numerous ornamental species such as *Argemone* (prickly

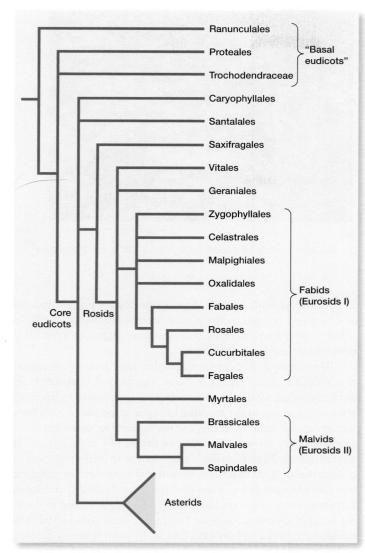

Figure 23-23 One recently proposed cladogram of eudicots. There are several unresolved polychotomies.

(a)

(b)

Figure 23-24 (a) Carpels in Ranunculales remain free of each other at maturity; they do not fuse during development. Consequently, each flower produces many separate fruits. *Clematis*. (b) Poppies (Papaveraceae) have numerous free stamens but fused carpels. They have no styles, and thus, the stigmatic surface is the top of the ovary.

poppy), *Eschscholzia* (California poppy), and *Papaver* (poppies). *Papaver somniferum* is opium poppy, the source of a milky latex harvested for opium; morphine, a strong analgesic (pain killer) that depresses the cerebral cortex, is extracted from opium, as is codeine. Morphologically, poppies are unusual in being either herbs or shrubs with a very soft wood.

Several hamamelid families are now believed to be basal eudicots rather than basal angiosperms. These are large trees with reduced wind-pollinated flowers, usually in dangling inflorescences containing many staminate flowers and just a few carpellate flowers. A familiar example is Platanaceae, sycamores (also called plane trees, *Platanus*): These large trees have small, inconspicuous clusters of flowers in the spring that mature into spherical clusters of hundreds of tiny dry fruits.

Caryophyllales DNA-based studies have combined a large number of families into a group called Caryophyllales (Table 23-7). Most of these families had long been known to be closely related and were also known as Caryophyllales; in new treatments, the old group is referred to as "core Caryophyllales." Examples of the

TABLE 23-7	Representative Families of Caryophyllales
Core Caryophyllales	
Aizoaceae	*Carpobrotus, Lampranthus, Mesembryanthemum* (all ornamentals).
Amaranthaceae	*Amaranthus* (a common weed in crops), *Beta* (*B. vulgaris*, beets), *Chenopodium* (a common weed in crops), *Spinacia* (spinach).
Cactaceae	*Ferocactus* (barrel cacti), *Mammillaria* (pincushion cacti), *Opuntia* (prickly pears and chollas), *Schlumbergera* (Christmas cacti).
Caryophyllaceae	*Dianthus* (carnations). *Gypsophila* (baby's breath).
Nyctaginaceae	*Bougainvillea, Mirabilis* (four o'clocks).
Phytolaccaceae	*Phytolacca* (pokeweed).
Portulacaceae	*Claytonia, Lewisia*.
Other Caryophyllales	
Droseraceae	*Dionaea* (Venus's flytrap), *Drosera* (sundew).
Nepenthaceae	*Nepenthes* (pitcher plant).
Polygonaceae	*Antigonon* (coral vine), *Fagopyrum* (buckwheat), *Rumex* (dock, a common weed).

(a)　　　　　　　　　　　　　　　　　　　　(b)

Figure 23-25　Caryophyllales: (a) *Echinocereus coccineus* (Cactaceae) and (b) *Carpobrotus* (Aizoaceae). The brilliant colors in the petals of both are due to betalain pigments.

core group are cacti (Cactaceae), iceplant (Aizoaceae), portulaca (Portulacaceae), bougainvillea and four-o'clocks (Nyctaginaceae), spinach, beets, and Russian thistle (Amaranthaceae), and carnations and chickweed (Caryophyllaceae) (**Figure 23-25**). The core Caryophyllales share many derived characters, but one is especially important. Whereas other flowering plants have **anthocyanin pigments** in their flowers, almost all Caryophyllales instead produce a group of water-soluble pigments called **betalains** (**Figure 23-26**). The ancestral group may have lacked petals in their flowers, a condition common in many Caryophyllales; although many appear to have petals, we believe that these are modified sepals in some families and modified stamens in others.

Another unifying character of Caryophyllales is that endosperm develops only a little and then fails to continue growing. Instead, nucellus cells proliferate and form a nutritive tissue called **perisperm**, which surrounds the developing embryo (**Figure 23-27**). Perisperm is usually absorbed almost completely by the time the seed is mature.

A third feature that unites this order and distinguishes it from all other taxa is the nature of sieve tube plastids. In Caryophylla-

les, phloem plastids contain deposits of fibrous protein located as a ring just interior to the plastid membrane (**Figure 23-28**). In other angiosperms, plastids contain either starch or crystalline protein with a different structure.

The Caryophyllales are postulated to have arisen from ancestors similar to Ranunculaceae. Most members of Caryophyllales are herbaceous, with either no wood, very little wood, or unusual, anomalous wood, so the ancestral group is suspected to have been herbaceous or shrubby, having lost the capacity for extended, massive secondary growth. The time of origin of Caryophyllales is approximately 70 to 80 million years ago, at an important time of continental drift. At that time, all land in the southern hemisphere was joined together in a giant continent now called Gondwana. Caryophyllales originated in southern Gondwana, and its earliest families spread across that continent. Then continental drift caused Gondwana to break up, first with South America moving westward and later with Australia, India, and Antarctica moving eastward. Families that had already come into existence were carried along with the moving land masses and are now present on

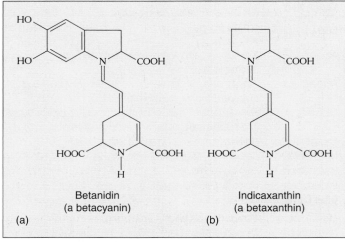

Betanidin
(a betacyanin)

Indicaxanthin
(a betaxanthin)

(a)　　　　　　　　　　　　　　(b)

Figure 23-26　Core Caryophyllales are characterized by the presence of betalain pigments in flowers and fruits. There are two basic types: (a) betacyanins (red to violet), (b) and betaxanthins (yellow).

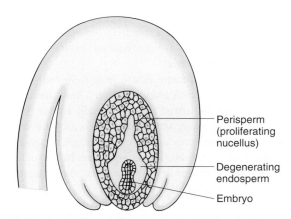

Perisperm (proliferating nucellus)

Degenerating endosperm

Embryo

Figure 23-27　In core Caryophyllales, endosperm develops for only a short while or not at all and is not sufficient to nourish the embryo. Instead, the nucellus (megasporangium) cells proliferate and act like endosperm, forming a tissue called perisperm.

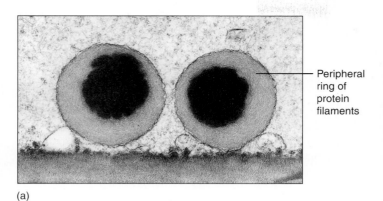

(a)

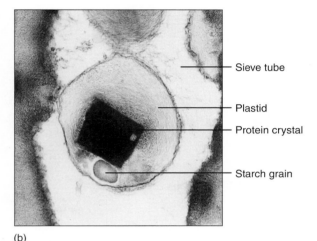

(b)

Figure 23-28 Plastids in sieve tube members are simple, but they may accumulate particles of starch or protein or both. The nature of the accumulated material is highly specific with regard to the family, and phloem plastid analysis is important in studying the evolution of flowering plants. Sieve tube plastids of *Monococcus* (a) lack a central cubic protein crystal, but those of *Macarthuria* have one (b). Both are members of Caryophyllales, and as a result, both have a peripheral ring of protein filaments (both × 40,000).

Figure 23-29 Mistletoe (*Phoradendron tomentosum*) is a hemiparasite; it carries out some photosynthesis. Its haustorial root system penetrates the host tree branch at the top of the photograph.

most or all the southern continents (except for Antarctica, whose southward drift doomed all the plants it carried). Families that originated after the breakup of Gondwana arose in individual continents and are now mostly confined to those. For example, cacti arose in South America and later spread to Central and North America, but its seeds could not cross either the Atlantic or Pacific Oceans; thus, cacti do not occur naturally on any other Gondwanan continent. Similarly, Didieriaceae arose in east Africa and Madagascar and occur nowhere else.

Several of the other families that are now placed in the expanded Caryophyllales are the buckwheats (Polygonaceae), pokeweeds (Plumbaginaceae), and two groups of carnivorous plants, Droseraceae (sundews and Venus's flytrap) and Nepenthaceae (pitcher plants).

Santalales This is a small order of highly modified plants, most of which are parasitic. The sandalwood family (Santalaceae) contains the large tree *Santalum* from which sandalwood incense is obtained. It appears to be an ordinary tree, but its roots make fine, almost imperceptible connections to roots of surrounding plants and parasitizes them. Also in this family are common mistletoes, *Viscum and Phoradendron*, which are used as

decorations at Christmas (**Figure 23-29**). These mistletoes have chlorophyll and are photosynthetic; therefore, they are just hemiparasitic, but some members are holoparasitic, having no chlorophyll at all. *Viscum minimum* lives entirely within its host plant, *Euphorbia*. Loranthaceae is a large family with about 900 species, all of which are parasitic, but none of which occurs in temperate North America.

◼ Rosid Clade

The remaining eudicots are members of two very large, very diverse clades, the rosids and asterids (see Figure 23-23). The **rosid clade** (named for the rose order Rosales) consists of many families that, taken as a whole, are so diverse with respect to vegetative body, flowers, chemistry, and ecology that it is difficult to see they are all related; however, some share enough characters with others to indicate a relationship, and those of the second group share different features with a third group and so on until a larger picture of phylogenetic relationships can be distinguished. DNA analysis has added a great deal of information to identifying patterns of relationships.

The rosids consist of several small orders and two large groups. One of the small orders has few members but great economic significance. Vitales contains Vitaceae, the grape family. *Vitis* is the genus of grapes that give us juice, raisins, and table grapes. *Vitis vinifera* is the wine grape, and despite the enormous variety of wines, virtually all are made from varieties of this one species. The other small order, Geraniales, contains the geranium family, Geraniaceae.

The two large clades of rosids are the **fabids** (also called **eurosids I**) and the **malvids** (**eurosids II**). These two clades together

TABLE 23-8 Representative Families of the Rosid Clade

Geraniaceae	*Erodium* (storksbill), *Geranium* (cranesbill), *Pelargonium* (geraniums used in gardens).
Vitaceae	*Parthenocissus* (Virginia creeper), *Vitis* (grapes).
Fabids (Eurosids I)	
Betulaceae	*Alnus* (alder), *Betula* (birch), *Corylus* (filbert, hazelnut).
Cannabaceae	*Cannabis* (marijuana, fibers), *Humulus* (hops, which provide the bitter taste to beer).
Clusiaceae	*Garcinia* (*G. mangostana*, mangosteen), *Mammea* (*M. americana*, mamey apple).
Cucurbitaceae	*Cucurbita* (gourds, pumpkin, squash), *Cucumis* (melons: cantaloupe, cucumber, honeydew, muskmelon), *Citrullus* (watermelon), *Luffa* (luffa sponge).
Euphorbiaceae	*Aleurites* (*A. fordii*, tung oil), *Euphorbia* (*E. millii*, crown-of-thorns, *E. pulcherrima*, poinsettia), *Hevea* (*H. brasiliensis*, rubber tree), *Sapium* (*S. sebiferum*, Chinese tallow tree).
Fabaceae (old name: Leguminosae)	*Albizia* (mimosa), *Arachis* (peanut), *Bauhinia* (orchid tree), *Glycine* (soybean), *Lens* (lentil), *Medicago* (alfalfa), *Melilotus* (sweet clover), *Phaseolus* (beans), *Pisum* (peas), *Trifolium* (clover), *Vicia* (vetch, beans), *Wisteria*.
Fagaceae	*Castanea* (chestnut), *Fagus* (beech), *Quercus* (oaks; bark of *Q. suber* is the source of cork).
Hypericaceae	*Hypericum* (St. John's wort).
Juglandaceae	*Carya* (*C. illinoensis*, pecan; *C. ovata*, shagbark hickory), *Juglans* (*J. regia*, walnut; *J. nigra*, black walnut).
Moraceae	*Artocarpus* (jackfruit, breadfruit), *Ficus* (figs), *Maclura* (Osage orange), *Morus* (mulberries).
Myrtaceae	*Eucalyptus*, *Pimenta* (*P. dioica*, allspice), *Syzygium* (*S. aromaticum*, cloves).
Onagraceae	*Fuchsia*, *Oenothera* (evening primrose).
Oxalidaceae	*Averrhoa* (*A. carambola*, star fruit), *Oxalis* (*Oxalis*, garden ornamental, *O. tuberosa*, oca, eaten in Andean South America).
Passifloraceae	*Passiflora* (passionflower).
Rhizophoraceae	*Bruguiera*, *Rhizophora* (both are mangroves that live in intertidal regions of warm coasts).
Rosaceae	rose family, see text for important examples.
Salicaceae	*Populus* (poplar, cottonwood, aspen), *Salix* (willow).
Violaceae	*Hekkingia* (although most of this family consists of small herbs, plants of *Hekkingia* grow to be shrubs or trees), *Viola* (violets, pansies).
Ulmaceae	*Ulmus* (elm).
Zygophyllaceae	*Guaiacum* (lignum vitae), *Larrea* (creosote bush).
Malvids (Eurosids II)	
Anacardiaceae	*Anacardium* (*A. occidentale*, cashew), *Mangifera* (mango), *Pistacia* (pistachio), *Rhus* (sumac), *Toxicodendron* (poison ivy, poison oak, poison sumac).
Brassicaceae	*Amoracia* (*A. rusticana*, horseradish), *Arabidopsis* (the complete genome of *A. thaliana* has been sequenced, and it is used extensively to study plant physiology and development), *Brassica* (*B. napus*, canola oil; *B. oleracea*: cabbage, broccoli, cauliflower, Brussels sprouts), *Raphanus* (*R. sativus*, radish), *Sinapis* (*S. alba*, mustard).
Burseraceae	*Boswella* (*B. carteri*, frankincense), *Commiphora* (*C. habessinica*, myrrh).
Malvaceae	*Althaea* (hollyhock), *Gossypium* (*G. hirsutum*, cotton), *Hibiscus*, *Malva* (mallow), *Ochroma* (balsa wood), *Theobroma* (*T. cacao*, chocolate), *Tilia* (linden).
Meliaceae	*Azadirachata* (*A. indica*, neem tree), *Melia azedarach* (chinaberry), *Swietenia* (mahogany).
Rutaceae	*Citrus* (citrus fruits: grapefruit, lemons, limes, oranges, tangerines), *Fortunella* (kumquat).
Sapindaceae	*Acer* (maples), *Aesculus* (horse chestnut); both are usually separated into their own families, Aceraceae and Hippocastanaceae.

contain more than 100 families, and it is difficult to give any universal characters (Table 23-8). As a group, rosids are more derived than basal tricolpates from which they arose and have derived characters that differ from those of Caryophyllales. Rosids are especially interesting in that none of them has any of the highly relictual features found in many basal angiosperms.

An important character in rosids is the presence of pinnately compound leaves. This is believed to have been the ancestral condition. Although some living species have simple leaves, these apparently have arisen from compound leaves, perhaps by suppression of all leaflets except one. Whereas simple leaves are an early, ancestral condition for the angiosperms as a whole, they are a later derived condition for rosids.

Although this clade is named for the order Rosales, roses should not be considered typical; they are just one group of many (Figure 23-30). Because the rosids consist of 14 large orders with over 50,000 species, only a few can be mentioned here. Five of the orders contain almost 75 percent of the species: Fabales (legumes), 19,000 species; Myrtales (*Eucalyptus* and evening primrose), 9,000; Malpi-ghiales (poinsettia), 1,600; Rosales (roses, elms, marijuana), 6,300; and Sapindales (maples, horse chestnuts, creosote bush, and the species whose resins are valued as frankincense—*Boswellia*—and myrrh—*Commiphora*), 5,800. Members of this subclass include roses, of course, and legumes (peas, beans, peanuts, *Mimosa*, redbud, and clover; Fabaceae = Leguminosae); *Fuchsia*, evening primrose (Onagraceae); dogwood (Cornaceae); the spurges that look like cacti and often have an extremely poisonous milky latex (Euphorbiaceae); maples (Aceraceae); and dill, celery, carrot, parsley, and hemlock (Apiaceae = Umbelliferae).

The rose family is important not only in an evolutionary sense but also economically. Rosaceae contains numerous ornamental genera, including *Rosa* (roses), *Crategus* (hawthorn), *Spiraea, Cotoneaster, Pyracantha, Photinia, Potentilla, Chaenomeles* (flowering quince), and *Sorbus* (mountain ash). The family also provides most of the fruits that can be grown in temperate climates: *Malus* (apple), *Pyrus* (almond, apricot, cherry, nectarine, peach, plum, and prune), *Eriobotrya* (loquat), *Fragaria* (strawberry), and *Rubus* (blackberry, loganberry, and raspberry).

Figure 23-30 Rosids. (a) *Hibiscus*. (b) St. John's wort (*Hypericum spathulatum*). (c) Immature gourd fruit (note the withered petals). (d) *Theobroma cacao,* the source of chocolate. Flowers and fruits are borne on the trunk and large branches, not on twigs. The seeds are harvested, fermented, roasted, and ground into cocoa powder. (e) Lupine (*Lupinus latifolius*). (f) Mexican plum (*Prunus mexicana*). (g) Strawberry (*Fragaria*).

Fabaceae (the legume family: beans, peas, lentils, and peanuts) is also a large family having bodies that vary from herbaceous annuals to shrubs, vines, and long-lived trees. It is important economically as a source of many foods, drugs, dyes, and woods, and many of its species are the dominant plants in arid areas and deserts. Many species have root nodules with symbiotic associations with nitrogen-fixing bacteria, and thus, they can grow in poor soils yet still produce protein-rich seeds. Legumes are a critically important source of proteins for people in arid areas.

■ Asterid Clade

The most derived large clade of eudicots is the **asterid clade** (Table 23-9), which contains plants such as sunflower, periwinkle, petunia, and morning glory. Asterids, being a sister clade of rosids,

originated perhaps as recently as 60 million years ago, and even its most basal members were much more highly derived than plants in the basal angiosperms. The majority of asterids can be easily distinguished from other angiosperms on the basis of three features: (1) They have sympetalous flowers (their petals are fused together into a tube); (2) they always have just a few stamens, not more than the number of petal lobes; and (3) stamens alternate with petals (**Figure 23-31**). Asterids exploit very specialized pollinators that recognize complex floral patterns, and such plants could not evolve before derived, sophisticated insects appeared.

Many chemical differences exist between this group and all others. It lacks many specialized chemicals found in other clades: It has no betalains (present in core Caryophyllales), benzyl-isoquinoline alkaloids (magnoliids), ellagic acid (**Figure 23-32**), or proanthocyanins (rosids). Instead, many asterids have **iridoid**

TABLE 23-9	Representative Families of the Asterid Clade
Basal Asterids	
Cornaceae	*Cornus* (dogwood), *Nyssa* (tupelo).
Ebenaceae	*Diospyros* (persimmon, ebony).
Ericaceae	*Arbutus* (madrone), *Erica* (heaths), *Gaultheria* (*G. procumbens*, oil of wintergreen), *Kalmia* (mountain laurel), *Rhododendron* (azaleas, rhododendrons), *Vaccinium* (cranberries, blueberries).
Hydrangeaceae	*Hydrangea*.
Polemoniaceae	*Phlox, Polemonium*.
Primulaceae	*Dodecatheon* (shooting star), *Primula* (primrose).
Sarraceniaceae	*Darlingtonia, Heliamphora, Sarracenia* (all are insectivorous pitcher plants).
Theaceae	*Camellia* (camellias, *C. sinensis*, tea).
Lamiids (Euasterids I)	
Acanthaceae	*Acanthus* (bear's breech), *Justicia* (shrimp plant), *Thunbergia* (blue-sky vine).
Apocynaceae	*Asclepias* (milkweed, butterfly weed), *Hoya* (wax plant), *Nerium* (oleander), *Plumeria* (frangipani), *Stapelia* (carrion flower), *Vinca* (periwinkle). Several species provide important medicines: *Catharanthus*, antileukemia; *Rauvolfia*, hypertension; *Strophanthus*, heart ailments.
Bignoniaceae	*Bignonia* (cross vine), *Campsis* (trumpet creeper), *Catalpa, Chilopsis* (desert willow), *Crescentia* (calabash tree).
Boraginaceae	*Heliotropium* (heliotrope), *Myosotis* (forget-me-not), *Phacelia*.
Convolvulaceae	*Convolvulus* (bindweed, morning glories), *Cuscuta* (a parasitic vine), *Ipomoea* (morning glories; *I. batatas*, sweet potato).
Gesneriaceae	*Saintpaulia* (African violet), *Sinningia* (gloxinia).
Lamiaceae (old name Labiatae)	*Calicarpa* (beautyberry), *Lavandula* (lavender), *Mentha* (peppermint, spearmint), *Nepeta* (catnip), *Ocimum* (basil), *Origanum* (oregano), *Rosmarinus* (rosemary), *Salvia* (sage), *Tectona* (teak), *Thymus* (thyme).
Lentibulariaceae (all are insectivorous herbs)	*Utricularia* (bladderwort), *Pinguicula* (butterwort).
Oleaceae	*Fraxinus* (ash), *Jasminum* (jasmine), *Ligustrum* (privet), *Olea* (*O. europaea,* olives used for food), *Osmanthus* (sweet olive, an ornamental), *Syringa* (lilac).
Orobanchaceae (all members are parasitic herbs)	*Castilleja* (indian paintbrush), *Conopholis, Epifagus* (this has the smallest plastid DNA circle known, with very few genes), *Orobanche, Striga* (a serious weed of cultivated crops).
Plantaginaceae	*Antirrhinum* (snapdragon), *Digitalis* (foxglove; also supplies digoxin, a drug that strengthens heart contractions), *Penstemon* (beardtongue).
Rubiaceae	*Cinchona* (quinine, used to treat malaria), *Coffea arabica* (coffee).
Scrophulariaceae	Many members of this family have been transferred either to Plantaginaceae or to Orobanchaceae.
Solanaceae	*Atropa* (belladonna), *Capsicum* (red, green, and cayenne peppers), *Datura* (jimsonweed), *Petunia, Solanum* (nightshades; *S. lycopersium*, tomato; *S. tuberosum*, potato). Almost all members of this family have extremely poisonous alkaloids.
Verbenaceae	*Lantana, Petraea* (queen's wreath), *Verbena*.
Campanulids (Euasterids II)	
Apiaceae (old name Umbelliferae)	*Anethum* (dill), *Apium* (celery), *Carum* (caraway), *Coriandrum* (coriander), *Cuminum* (cumin), *Daucus* (carrot), *Foeniculum* (fennel), *Petroselinum* (parsley), *Pimpinella* (anise).
Aquifoliaceae	*Ilex* (holly).
Araliaceae	*Hedera* (English ivy), *Schefflera* (umbrella tree).
Asteraceae (old name Compositae)	*Ambrosia* (ragweed, causes hayfever), *Artemesia* (wormwood, tarragon, sagebrush), *Chicorium* (chicory), *Cynara* (artichoke), *Dahlia, Helianthus* (sunflower), *Lactuca* (lettuce), *Leucanthemum* (chrysanthemum), *Taraxacum* (dandelion), *Zinnia*.
Campanulaceae	*Campanula* (bellflower), *Lobelia* (cardinal flower).
Caprifoliaceae	*Dipsacus* (teasel), *Lonicera* (honeysuckle).

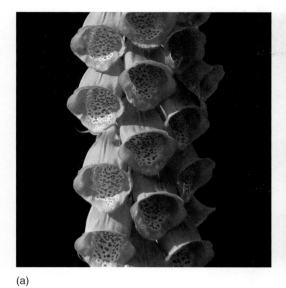
(a)

(b)

Figure 23-31 (a) In asterid flowers, such as *Digitalis purpurea,* all petals fuse to each other and form a tube; only certain pollinators can fit into the tube far enough to obtain nectar— and inadvertently carry out pollination. (b) Flowers of potato (*Solanum tuberosum*) also have fused petals, but they are so open that we can see that stamens alternate with petals.

compounds, which occur only rarely outside this group. Certain families produce numerous, very potent chemicals that deter animals or kill them outright. Apiaceae (celery, dill, and fennel) have secretory canals containing oils and resins, triterpenoid saponins, coumarins, falcarinone polyacetylenes, monoterpenes, and sesquiterpenes. The poison Socrates was given to kill himself came from *Conicum* (hemlock) in the Apiaceae. It is not important for you to learn this list or understand what these chemicals are, but rather to see that in addition to new morphological, anatomical, and reproductive characters originating evolutionarily, new defensive chemicals also arise.

Currently, asterids have the greatest number of species (approximately 60,000), but they are grouped into two small orders (Cornales, Ericales) and two groups of **orders**, **lamiids** (also called **euasterids I**) and **campanulids** (**euasterids II**) (**Figure 23-33**). One family (sunflowers, daisies; Asteraceae [old name is Compositae])

contains fully one third of all the species and is the largest family of eudicots. Examples of asterids are milkweeds (Asclepiadaceae); potato, tomato, red peppers, eggplant, tobacco, deadly nightshade, and petunia (Solanaceae); morning glory (Convolvulaceae); thyme, mints, and lavender (Lamiaceae); and Asteraceae with sunflowers, dandelions, lettuce, *Chrysanthemum*, ragweed, and thistle (**Figure 23-34**).

Many asterids are extremely important medicinally. Apocynaceae (oleander family) contains periwinkle, *Vinca,* from which are extracted vinblastine and vincristine, two of our most potent anticancer drugs. Rubiaceae contains, in addition to coffee (*Coffea*), *Cinchona,* from which we derive the antimalaria drug quinine. Another rubiaceous genus, *Cephaelis,* provides us with ipecac, an emetic used to induce vomiting in cases of oral poisoning. Heart disease is treated with cardiac glycosides extracted from *Digitalis* (Scrophulariaceae, the snapdragon family); these compounds

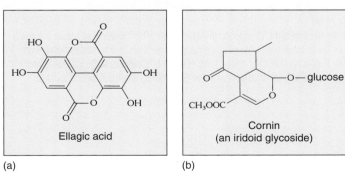

(a)   (b)

Figure 23-32 (a) Ellagic acid is common in several eudicot clades but absent from one of the most derived, the asterids. Its presence deters many insects from eating the plants, but by approximately 50 million years ago, many types of insects had evolved a tolerance to it. (b) Many asterids produce iridoid compounds such as this cornin. They are a relatively new class of chemical defense compounds, and thus, few insects tolerate them; iridoid compounds may be partly responsible for the success of the asterids by keeping herbivory to a minimum.

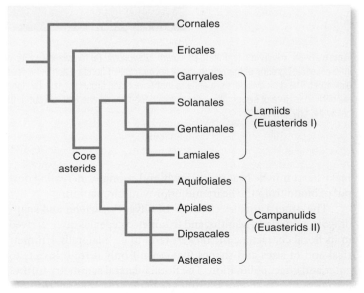

Figure 23-33 One recently proposed cladogram of asterids.

Figure 23-34 Asterids. (a) Potato (*Solanum tuberosum*). (b) Milkweed (*Asclepias viridis*). (c) Goatsbeard (*Tragopogon* in Asteraceae). The "flowers" of asters are really inflorescences, groups of many true, small flowers, called florets. Each outer "petal" is a ray floret (each with one greatly enlarged petal), and each central flower is a disk floret. The dark cylinders are anthers that have fused around styles. The innermost disk florets have not yet opened. (d) Flowers of *Cavendishia* (Ericaceae) show the derived feature of sympetally—petals fused together as one unit, except at their tips. (e) In *Hydrangea,* what appear to be large petals are really bracts located below the tiny flowers.

make heart muscle beat more slowly and strongly, increasing output of blood from the heart and improving circulation.

The order Lamiales (families of olive, *Penstemon* and snapdragon, gesneriads, acanths, and trumpet-creeper) is quite derived in its floral characters; the flowers tend to be bilaterally symmetrical and of sizes and shapes that permit only certain insects to enter and effect pollination. The floral bilateral symmetry further forces the insect to enter in one particular orientation. Pollinators capable of doing this tend to be sophisticated and able to rec-

ognize different flowers easily. They learn which flowers provide the greatest rewards and then search for other flowers of identical shape, color pattern, and fragrance, thus providing efficient pollination for the plant species.

The giant family Asteraceae (asters, daisies, and sunflowers) contains 1,100 genera and 20,000 species distributed worldwide in almost all habitats except dense, dark forests. They range from important food plants (*Lactuca*—lettuce) to ornamentals to weeds. The characteristic daisy/sunflower type of inflorescence makes

Box 23-2 Vegetarianism: Alternatives to Eating Meat

Our ancestors evolved to be omnivores. As a result, we modern humans can eat seeds, leaves, and fleshy roots because our flat molars allow us to grind and mash them before we swallow them, and our small intestine is so lengthy it ensures that plant matter remains inside us long enough to be digested. Similarly, we can eat meat because our incisor and canine teeth allow us to rip meat from bones (think of barbequed ribs, fried chicken, and pork chops) and our digestive system produces plenty of protein-digesting enzymes.

However, just because we have the ability to eat meat does not mean we must eat it. A large number of people have adopted some form of vegetarianism to reduce the amount of meat in their diet or eliminate it altogether. There are several types of **vegetarian diet**:

- *Ovo-lacto vegetarians* (often just called vegetarians) eat a diet based almost exclusively on plant products such as fruits, vegetables, greens, and seeds (in the form of foods like bread, pasta, and oatmeal) with some eggs and dairy products (milk, cheese, ice cream, and yogurt) (Figure B23-2). The concept here is that animals do not have to be killed to obtain eggs and milk. For some vegetarians the only acceptable animal-based food is eggs, for others only milk-based foods, and for others even fish and seafood is acceptable.
- *Vegans* avoid all foods obtained from animals, including meat, eggs, and all dairy products. Strict vegans also do not wear clothing made from animals, such as leather belts, shoes, and coats or anything made with fur or turtle shell.
- A *fruitarian* diet consists of only fruits, nuts, and seeds, items that can be gathered without harming a plant. Fruitarians will not eat roots such as carrots, beets, and cassava because those can be obtained only by killing the plants.
- *Flexitarians* have a flexible approach to meat and other animal products in their diet; they eat mostly plant foods and merely reduce the amount of meat they eat. For example, flexitarians might set the goal of having meat-free lunches or designating several days of the week as being meat-free. The advantage of a flexitarian approach is that a person can succumb to the occasional hamburger or steak without feeling they have failed and might as well give up.

There are several reasons for reducing or eliminating animal products in our diet and our daily lives. Many people do so for ethical reasons: They do not want to be responsible for animal suffering or death. There is a range of empathy here. Red meat is muscle tissue that can only be obtained by killing cattle. Cattle are mammals like we are, and they definitely feel pain and suffering. Obtaining milk from cows, goats, or other mammals (only mammals produce milk) does not involve killing them, so there is little or no suffering

Figure B23-2   This is an ovo-lacto vegetarian meal because it contains eggs and cheese in addition to plant-based foods. A dessert of pie and ice cream fits into such a diet.

in many dairy operations. Gathering eggs from chickens or other birds also does not involve harming them, although some poultry farms keep the animals crowded together in poor conditions and clip their beaks so they won't peck each other. Eating fish and seafood involves killing the animals, but many people feel little emotional connection to fish, shrimp, and clams.

When choosing your diet for ethical reasons, there are various issues to consider. Raising cattle in the United States and elsewhere is often cruel. Most cattle are raised on factory farms, crowded into large, dirty feedlots where the animals must stand in their own feces; they are given feed that includes antibiotics, growth supplements, and even the ground up debris from cattle that were slaughtered earlier. When it is time to kill them, the cattle are poked with prods that give them an electrical shock, forcing them into a line into the bright, noisy slaughterhouse (cattle are terrified by bright lights, noise, and having people moving around them, especially above them). Laws in many countries prohibit the slaughter of any animal that is not healthy enough to walk into the slaughterhouse, but that law is sometimes violated. PETA (People for the Ethical Treatment of Animals; http://www.peta.org/) has developed a set of guidelines for treating cattle more humanely, including giving them uncrowded pastures for grazing and safe food, and when they are to be killed the slaughterhouse should be quiet, dimly lighted, and with as few people as possible. The animals are still killed, of course, but at least their life and death were not miserable. For many people this humane treatment reduces their ethical objections and they will eat meat occasionally. To be certain the meat you eat has come from humanely treated animals, it is best to shop at reputable health food stores or organic food stores; don't be afraid to ask about the treatment of the animals.

Veal is especially objectionable because it is the flesh of calves that are killed while still so young they have never eaten anything other than their mother's milk. Often, the calves are kept in cages so small they cannot move because movement would cause their muscles to develop, making the meat less tender. The calves also may be kept in the dark to keep them quiet. It is difficult to justify treating animals like this just to have a particular kind of dinner.

Eggs and milk do not require the slaughter of animals, and therefore many people do not believe it is unethical to consume these products. Here too it is best if the chickens and cattle are treated humanely and are allowed access to open pastures or pens with fresh air and water. People with the strictest ethical standards try to avoid all animal products because they believe it is unfair for us to breed and maintain animals just for our own use. It is difficult, however, to eliminate the use of all animal products: Sugar is whitened by filtering it through the ash of animal bones, most cheese is made with rennet from calves' stomachs, clear juices (apple and grape) and alcoholic beverages are clarified with gelatin from animal hides and bones, and so on.

Other people avoid animal products for economic and ecological reasons. When we feed animals corn, soybeans, oats, and other feeds that we ourselves could eat, we are eating at a high trophic level. The plants are primary producers, the livestock animals are primary consumers, and we are secondary consumers. Typically, it takes 10 pounds of food at one trophic level to make 1 pound of food at the next level. Consequently, we need to feed cattle or chickens 10 pounds of oats just to get 1 pound of eggs or milk or meat. The other nine pounds are lost as carbon dioxide and methane (both greenhouse gasses), urine, feces, and waste. If we ate the plants ourselves, we would need to farm only one-tenth as much land as we use now, producing much less sewage. It is estimated that as much as 70% of all the wheat, corn, and other grains we cultivate are used to feed livestock. Also, huge areas of tropical rainforest have been and are being cut down and con-verted into pasture for cattle just to produce hamburger. Reducing the amount of animal products we eat is an important aspect of a philosophy of living more simply, of trying to minimize the impact each of us has on our environment.

An important consideration for anyone considering a vegetarian diet is whether it is healthful. The answer is easiest for a flexitarian diet: Virtually all studies show that diets low in meat and high in fruit and vegetables are much more healthful than ones with more meat. And it is now well established that even a strict vegan diet provides all the nutrients we need for a healthy life. We need to be especially careful to obtain adequate amounts of vitamin B_{12}, which is not found in any plant but which can be obtained as a supplement or by consuming yeast products. Many people believe that athletes and bodybuilders need a high protein intake and must have meat, but that has been shown repeatedly to not be true. Most of us Americans consume much more protein than we need, and the plants of a vegetarian or vegan diet provide adequate amounts for anyone. We must be careful to obtain adequate amounts of all eight essential amino acids. All plants, animals, fungi, and other organisms construct their proteins from the same 20 amino acids, and we humans can synthesize 12 of them within our own bodies, but our metabolism cannot make 8 of them. Dairy and egg foods supply adequate amounts, and so do the proteins of soy, amaranth, quinoa, and several other seed plants. The proteins of many food plants such as wheat, rice, beans, chickpeas, and others have adequate amounts of 18 or 19 amino acids but only sparse amounts of 1 or 2 others. We can avoid having a deficiency by eating what are known as complementary foods, foods in which one member has plenty of an amino acid that the other member lacks, and vice versa. For example, a combination of brown rice and beans, beans and corn, tofu (soy) and rice, or one of hummus and wheat pita bread provides adequate amounts of all 20 amino acids. Think about a typical Mexican meal that includes beans and corn tortillas; these two plant foods together provide all essential amino acids. The same is true of a typical American lunch: A peanut butter sandwich (peanuts and wheat) provides complementary proteins.

The take-home message here is that we should eat less meat and animal products. Both our bodies and our environment will be healthier.

them instantly recognizable. Members of Asteraceae have a wide range of unique chemical defenses against herbivores: sesquiter-pene lactones, monoterpenes, terpenoids, and latex canals that contain polyacetylene resins (**Figure 23-35**). The presence of these chemicals makes composites extremely resistant to animals that eat plants or lay eggs in them; it also causes them to be irritating to human skin, resulting in numerous cases of contact dermatitis. The Asteraceae is a young family, perhaps no more than 36 million years old, which probably originated in the Oligocene Epoch of the Tertiary Period.

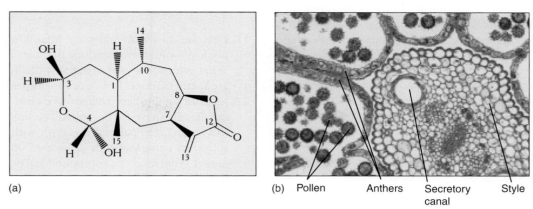

(a)

(b) Pollen   Anthers   Secretory   Style
                          canal

Figure 23-35 (a) Hymenoxon is a sesquiterpene lactone, a member of the chemical arsenal of asterids. This particular compound occurs in bitterweed (*Hymenoxys odorata*) of the southwestern United States; the plant is eaten by sheep only during droughts when no other plants are available. Sesquiterpene lactone causes hemorrhaging of all internal organs of the sheep. (b) Many members of Asteraceae contain secretory canals lined with cells that produce a variety of toxic compounds. This canal is in the style of sunflower (*Helianthus*).

Summary

1. All flowering plants, members of division Magnoliophyta, have closed carpels, long pollen tubes, double fertilization, and, with very few exceptions, sieve tubes and vessels.

2. Division Magnoliophyta is almost universally believed to be monophyletic; all of its evolutionary lines can be traced back to a single common ancestor. The earliest fossils of leaves and wood date to approximately 125 to 130 million years ago, and fossil pollen from earlier periods has been found.

3. The earliest angiosperms may have been weedy shrubs that grew in areas not conducive to fossilization.

4. Early characteristics of angiosperms are believed to be flowers with many sepals, petals, stamens, and carpels arranged in spirals; uniaperturate pollen; radially symmetrical flowers; insect pollination; and wood without vessels. Later, tracheid-like vessels with scalariform perforation plates arose. Early angiosperms are believed to have been perennial; none was annual or herbaceous.

5. Some major evolutionary modifications that have occurred in many groups are reduction in the number of each flower organ; imperfect flowers; bilateral symmetry of flowers; nectaries and fragrances that attract only specific pollinators; wind pollination; various seed or fruit distribution mechanisms; biennial or annual herbaceous habit; wood with shorter, wider vessel members; larger amounts of wood parenchyma; and numerous types of antipredator compounds—perhaps two or three generations of these.

6. Division Magnoliophyta contains basal angiosperms, monocots, and eudicots.

7. Monocots can generally be recognized by their long, strap-shaped leaves; flower parts in threes; parallel leaf venation; numerous vascular bundles in the stem, not arranged in a ring; and no ordinary secondary growth.

8. Eudicots can generally be recognized by broad, not strap-shaped leaves; reticulate leaf venation; flower parts in fives or fours, but not threes; pollen tricolpate or tricolpate-derived; vascular bundles in the stem arranged in one ring; and woody growth or an annual or biennial herbaceous body.

9. In the classification presented here, monocots contain several orders and the commelinoid clade, and eudicots contain basal eudicots, rosids (with fabids and malvids), and asterids (with lamiids and campanulids).

Important Terms

angiosperm carpels
anthocyanin pigments
asterid clade
basal angiosperms
betalain pigments
campanulid order
closed carpel
commelinoid monocots
Division Magnoliophyta
double fertilization

eudicots
fabids
generalized flower
iridoid compounds
lamiid order
malvids
monocots
perisperm
pistil
primitively vesselless

ranalean flower
rosid clade
secondarily vesselless
septal nectaries
sympetally
tricolpate pollen
uniaperturate pollen
zygomorphy

Review Questions

1. The earliest fossils clearly recognizable as parts of flowering plants are _____ _____ in rocks more than _____ million years old.

2. What transformations must have occurred if gymnosperms really were the ancestors of angiosperms? Which might have occurred earlier (are present in all relictual angiosperms) and which later (are absent from some of the most relictual angiosperms)?

3. If someone were to examine the stamens of most flowering plants, they would never automatically think of them as sporophylls because they have very little leaf-like structure, but look at the four stamens in Figure 23-1. These are from flowering plants. What leaf-like characters do they have? Are they flat and at least somewhat resemble the blade of a leaf? Do they have vascular bundles that branch, at least somewhat resembling the reticulate venation of a dicot leaf? Most dicot leaves have three vascular bundles entering their petioles. Which two of these stamens also have this leaf-like character? Which two have only one vascular bundle?

4. During fertilization in flowering plants, one sperm cell nucleus fuses with the egg nucleus, producing the diploid zygote. The other sperm nucleus fuses with the _____ _____, producing the triploid _____, which develops as temporary tissue that nourishes the embryo. This process is called _____ fertilization. What tissue nourishes the embryo in gymnosperm seeds?

5. Why were many modern genera of dicots initially thought to have come suddenly into existence about 130 million years ago? Why do we now think that those leaf fossils do not really belong to the modern genera they were named for?

6. What is meant by a closed carpel as opposed to an open cone scale?

7. Describe what the early angiosperms may have been like with regard to flowers, body, ecology, wood, and pollen. For each of the characters, what living angiosperms still have those characters? In what ways have those characters changed in certain lines of evolution (Hint: see Table 23-1)?

8. Concepts about the nature of the early angiosperms have changed (and may change again in the future). In the last century, members of the subclass Hamamelidae were considered the most relictual of the living flowering plants. What kind of pollination do these plants have—insect pollination, wind pollination, or bird pollination? Give the common names of four members of this group. Although these plants are woody trees like many conifers, their wood has three characters that wood of conifers does not have, so we no longer think that hamamelids are the most relictual flowering plants. What are the three characters?

9. About 100 years ago, C. E. Bessey developed the hypothesis of the "ranalean" flower. What are some of the characters of this type of flower? Is the ranalean flower wind pollinated or insect pollinated?

10. Do most botanists believe that the flowering plants are monophyletic, or do they believe that they are actually an unnatural group consisting of several lines of evolution that do not share a common ancestor?

11. Describe *Amborella* and water lilies. On the basis of morphology, would you conclude that they are closely related? What evidence is used to support the hypothesis that they are ancient lineages from basal angiosperms?

12. All plant families have a name that ends in "-aceae," and some families also have old names, still commonly used, without this ending. The palms are a family with two names. What is the new name that ends in "-aceae," and what is the older name? What kind of a compound leaf do the feather palms have? The fan palms? Why are palm flowers so seldom seen by most people?

13. Philodendron is a plant you may know. They and their close relatives are characterized by a distinctive inflorescence. It has tiny flowers embedded in a thick stalk called a _____, and it is surrounded by a large bract called a _____.

14. What is the name of the grass family? About how many species does it contain? Name seven genera (and give the common names) of grasses that are used for food. What kind of pollination do grasses have—animal or wind?

15. Zingiberales contains some of the most familiar of all houseplants. Name several of these.

16. Liliaceae contains many familiar plants. Name six. Which closely related family contains the irises? Which contains yams?

17. The Orchidaceae is the largest and most diverse family of monocots. What is the most common orchid habit (growth form)—being an epiphyte or being terrestrial? Are orchid flowers bilaterally symmetrical or radially symmetrical? Describe orchid seeds.

18. Certain eudicot clades contain plants with many relictual features. Name and describe several.

19. Core Caryophyllales contains many familiar plants. Can you list four members of this group? The group is unusual in having _____ pigments rather than anthocyanins in their flowers. Another feature of this group is that endosperm develops little or not at all, and instead, the seeds have _____, which develops through proliferation of cells of the _____.

20. Give the name of the families of each of the following:
    a. Camellias
    b. Chocolate
    c. Cotton
    d. Violets
    e. Pumpkins and watermelons
    f. Cranberries and blueberries

21. One important character in rosids is the presence of _____ _____ leaves. The subclass is named for the rose family, Rosaceae, and this one family has many genera that produce fruits that we like to eat. Name the genera of the following fruits, all in the rose family:
    a. Apples
    b. Almonds, apricots, cherries, peaches, and plums
    c. Strawberries
    d. Blackberries and raspberries

22. Textbooks often use cacti and euphorbias (spurges) to illustrate convergent evolution because many euphorbias have globular, succulent bodies that strongly resemble those of cacti, yet the two families are not closely related. What are the names of the euphorbia family and of the cactus family? Which major clade is each in?

23. Name the families that contain the following species:
    a. Sunflowers, dandelions, and daisies
    b. Milkweeds
    c. Potato, tomato, and tobacco
    d. Mints and lavender
    e. *Coffea* (coffee)

24. Think about the Coniferophyta and the Magnoliophyta of this chapter. Which has more families, genera, and species? Which has more diversity of body types? Which has species in the greatest number of habitats (deserts, forests, rain forests, underwater, grasslands, marshes, epiphytic habitats, and so on). Which group originated earlier in evolution, and which originated later? The older group has had more time to diversify. Is it the more diverse of the two?

BotanyLinks

http://biology.jbpub.com/botany/5e
BotanyLinks contains a variety of resources and review material designed to assist in your study of botany.

Visit the Web Exercises area of BotanyLinks to complete these questions:

1. Many plants provide us with medicinal chemicals. Go to the BotanyLinks home page to investigate which plants are useful, what chemicals they provide, and the illnesses that are treated.

2. Have you ever visited a botanical garden? Are there any close to where you live or are going to school? Go to the BotanyLinks home page to find a botanical garden located in your area. BotanyLinks includes a Directory of Organizations for this chapter.

Fungi

Concepts

At one time, fungi and bacteria were considered parts of the plant kingdom, primarily because they produce spores, have cell walls, and obviously are not animals. With recognition of the differences between prokaryotes and eukaryotes, bacteria were reclassified out of the plant kingdom, but fungi were left in. More recently, it has become clear that the organisms grouped together as fungi are definitely not plants. They lack plastids. Their walls do not contain cellulose. Their bodies are filamentous, not parenchymatous. The organization of large structures such as mushrooms and morels is completely different from that of plants with massive structures, and many biochemical pathways differ significantly.

Chapter Opener Image These button mushrooms (*Agaricus bisporus*) are being cultivated commercially and will be used as food. They are grown on a medium of wheat straw and usually horse or cattle manure that has been composted, which creates heat sterilizing the manure by killing most microbes in it. The manure and straw are mixed together and then either fungal spores or fungal cells are added and allowed to grow. If the manure were not sterile many different kinds of fungi and bacteria would grow faster than the mushrooms and ruin the crop: a mushroom farm is extremely clean. All fungi, including these mushrooms, use external digestion, excreting digestive enzymes that break down materials outside of the fungus' body. The fungus then absorbs the monomers. Fungi are extremely important in decomposing the bodies of plants, animals, and other organisms, thus, releasing the mineral nutrients in those bodies to the environment. If there were no decomposers like fungi and some bacteria, the world would be full of dead bodies that do not decay.

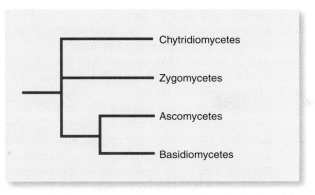

Figure 24-1 Cladogram representing one recent hypothesis about phylogeny of fungi. Deuteromycetes (imperfect fungi) are not on the cladogram; it is an artificial group.

Furthermore, the previous classification of fungi was not natural; it was not a monophyletic group. They were classified together only provisionally until we had enough information to understand their true evolutionary relationships. New data, especially DNA sequences, have clarified many relationships and have shown that many of the simpler fungi—those that do not make mushrooms or puffball-like reproductive structures—belong to several distinct clades and that their phylogeny is complex. Here, we focus on only a few of the more familiar and typical fungi.

Several characters were originally used to unite and define fungi: complete heterotrophy with no photosynthetic stages, formation of spores, the presence of chitin in their walls, and a lack of complex bodies with organs. These features, however, may result from convergent evolution as well as common ancestry. Unicellular fungi and those with flagella are grouped together as mastigomycetes, but many of these are thought to be closely related to heterokont algae and are not considered true fungi. Fungi known as zygomycetes, ascomycetes, basidiomycetes, and deuteromycetes are composed of filamentous individuals without motile stages (**Figure 24-1**). Many fungi still are poorly known, and the gaps in our knowledge are extensive.

In order to identify and classify a fungus, it is necessary to have the structures involved in sexual reproduction. A specimen cannot be classified definitively without these and is placed in the artificial group deuteromycetes. This taxonomic group serves basically as a "holding tank" until enough information can be obtained to classify its members correctly. All deuteromycetes are in reality members of some other fungal subdivision.

General Characteristics of Fungi

Fungi constitute a large group of organisms; although about 100,000 species have been named, 200,000 more species are estimated to remain undiscovered. Like insects and orchids, fungi are speciating more rapidly than they are being discovered; in the future, we may know more fungi, but we will know a smaller percentage of the total. By the way, the word *fungi* is pronounced as if spelled either funjee or funjai; the hard g (fun guy) is not used. Many names end in -mycetes, pronounced either as "my SEAT ease" (a more formal pronunciation) or with "MY seats" (more informal).

Nutrition

Modes of Nutrition A universal characteristic of fungi is that they are completely heterotrophic; no trace of photosynthesis occurs in any stage of any group. However, because they have walls, fungi cannot engulf food as animals do; instead, fungi must obtain soluble nutrients from the environment or from living, dying, or dead organisms. On this basis, fungi are subdivided into three types: (1) **biotrophs** (parasites), which draw nutrients slowly from living hosts, often without killing them; (2) **necrotrophs**, which attack living hosts so virulently that they kill the hosts and then absorb released nutrients; and (3) **saprotrophs**, fungi that attack organisms after they have died from other causes.

Many biotrophs secrete chemicals that cause the host cell membrane to become unusually permeable to sugars or amino acids; as they leak from the host cell, the fungus absorbs them. In sophisticated parasites, damage to the host cell is so slight the plant responds as though the fungus were a normal sink for metabolites, and extra sugars and amino acids are actually transported by the plant from other leaves to the site of infection, just as though the fungus were a developing fruit or other plant part. In biotrophic attacks, fungal cells may remain confined solely to intercellular spaces; in other species, the fungus creates a small hole in the plant cell wall and then inserts a small portion, the **haustorium**, of its filamentous cell through the hole (**Figure 24-2**). The haustorium is in close contact with the plant cell plasma membrane, which probably makes it easier to absorb nutrients.

Extracellular Digestion Fungi typically secrete digestive enzymes that attack host polymers, converting them to sugars, amino

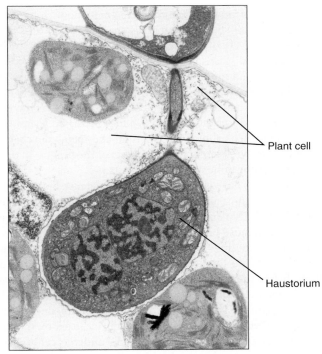

Plant cell

Haustorium

Figure 24-2 Some biotrophic fungi insert specialized portions, haustoria, through the walls of host cells. These never break the host cell's plasma membrane; therefore, technically they never enter the cell, and nutrients must still pass across the plasma membranes of host and fungus as well as the fungal cell wall (×5,000).

acids, and lipids that can be absorbed. Saprotrophs depend predominantly on this form of extracellular digestion, and many depolymerize and consume cellulose (brown rot fungi), hemicelluloses, and even lignin (white rot fungi). Many biotrophic and saprotrophic fungi can attack successfully only a few or just one host species or even just a single variety of one plant species. Many fungi are even tissue specific: Wilt-inducing fusariums must invade xylem and attack xylary middle lamellas; they cannot survive in cortex, pith, or phloem, even though those cells are rich in free monomers. Fungi transmitted by aphids and other phloem-sucking insects usually are able to attack only phloem.

Fungi, along with most bacteria, are agents of decay, rot, spoilage, and decomposition. As plants incorporate minerals and carbon into their molecules, these elements are locked away from other plants and are available only to the animals and biotrophs that consume the plants. Without microbial decay, dead plant and animal bodies would last almost forever, decomposing slowly, geochemically the way rocks decompose. Within a few years, all available minerals in the soil would be incorporated and unavailable, and new plant growth would cease. Decomposition mediated by fungi and bacteria is rapid, and within several months, the minerals are made available for new plant growth.

Toxins Necrotrophs secrete toxins that kill host cells; an example is the release of alternaric acid by *Alternaria solani.* Many toxins kill host cells by damaging their plasma membranes, causing nutrients to leak out rapidly and be readily available to the fungus. In many wilt diseases, the total mechanism that causes turgor loss is not known. In some *Fusarium* species, the fungus invades the xylem and then produces pectolytic enzymes that depolymerize the middle lamella, producing a slimy mucilage. This is drawn into tracheids and vessels by the water tension of transpiration, but the mucilage blocks pits and perforations, halting water conduction. "Soft rots" of fruits are also caused by fungi that attack the middle lamella.

Excessive levels of plant hormones are involved in many fungus-induced diseases. Gibberellic acid was discovered because rice plants in Japan suffer from lethal stem elongation induced by gibberellins secreted by the fungus *Gibberella fujikuroi.* Plant hormone may be produced and secreted by the fungus, or the fungus somehow induces the plant to produce increased levels of its own hormones. The abnormal structural and metabolic changes they cause may be essential to the survival of the fungus; in other cases, they may be nonessential side effects.

Some plants have elaborate antifungal defense mechanisms. One defense against an obligate biotroph is a **hypersensitive reaction**, in which plant cells die immediately upon contact with fungal cells. Because biotrophs can survive only in living hosts, the fungus cannot draw nutrients from these dead cells and must extend farther into the host. As it does so, however, more host cells die before the fungus can parasitize them. The fungus starves to death, and the plant has suffered the loss of just a few cells.

Some plants resist biotrophic and necrotrophic fungi by rapidly forming wound cork around any injury caused by the fungus. Cells adjacent to the site of invasion undergo cell division and then differentiate into cork with walls saturated with suberin and lignin; these are highly resistant to enzyme attack, and the fungus is trapped.

High levels of phenolic compounds in a plant vacuole may confer resistance; if the fungus' extracellular enzymes damage the cell membrane and vacuolar membrane, phenols flood out and denature all proteins, digestive enzymes included. With these enzymes inactivated, the fungus obtains no nutrition.

Many plants produce **phytoalexins**, lipid-like or phenolic compounds, in response to attack by fungi, bacteria, and even nematodes. During invasion, elicitors are released; these are usually large constituents of fungal wall that interact with the plant cell membrane, stimulating the cell to produce phytoalexins. In some cases, elicitors are extremely specific for either host or pathogen; a given fungus elicits a strong phytoalexin response in resistant plants but little or no response in susceptible ones. On the other hand, certain plants produce phytoalexins in response to some fungi but not others; the latter are pathogenic.

Phytoalexins damage fungi, bacteria, and nematodes in various ways. Some pathogens are highly sensitive to phytoalexins whereas others can inactivate them enzymatically. Certain virulent strains of fungi are even able to suppress the host's ability to produce phytoalexins in response to that fungus' elicitor. Very little is known about phytoalexins, but this is a fascinating area of research in co-evolution because it is essentially a gene-for-gene battle between two organisms.

Essential Nutrients Fungi require essential mineral elements, and their macroelement needs are almost identical to those of plants. Calcium, an essential macroelement for plants, either is a microelement for some fungi or is not needed at all, being replaced by strontium. The common essential microelements are iron, zinc, copper, manganese, molybdenum, and either calcium or strontium. Boron and cobalt are not required by many fungi, whereas other species need gallium or scandium.

Nitrogen metabolism is similar to that of plants; virtually all fungi absorb and assimilate ammonium, and many absorb and reduce either nitrate or nitrite if ammonium is not available. Some fungi cannot assimilate any form of inorganic nitrogen and must have amino acids to survive.

Many fungi synthesize all of their own compounds if supplied with minerals and a carbon/energy source such as sugar; other fungi must obtain certain vitamins from the environment. The most commonly required vitamin is thiamine, which many yeasts and filamentous fungi need. Biotin is required by many species, as is pyridoxine (B_6). Many yeasts are unable to synthesize their own pantothenic acid.

▪ Body

Cellular Organization The bodies of all fungi, except unicellular ones, are filamentous. Individual filaments are **hyphae** (singular, hypha), and they branch profusely, forming a network called a **mycelium** (plural, mycelia) (**Figure 24-3**). In zygomycetes, hyphae are long multinucleate cells known as **coenocytes**, but in ascomycetes and basidiomycetes, hyphae are septate and cellular.

Fungal cells are eukaryotic, having a membrane-bounded nucleus, a nucleolus, and a cytoplasm that contain membranous organelles. Mitochondria are abundant and similar to those in plants, but endoplasmic reticulum is sparse and smooth, rarely having ribosomes attached. Vacuoles are present but usually do not dominate the protoplasm except in older cells. Dictyosomes are rare throughout most of a hypha, and plastids of all types are absent. Vesicles are common and are involved in transporting

Hyphae

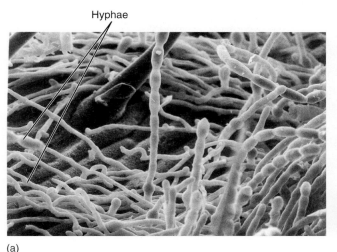

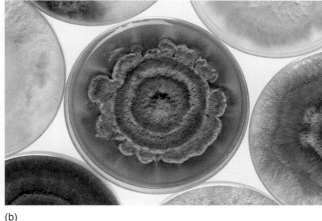

(a)   (b)

Figure 24-3 (a) All filamentous fungi are composed of hyphae, which are narrow, delicate, and usually transparent. Because hyphae are so long and slender (1.0 to 15 μm in diameter), they have a tremendous surface-to-volume ratio, ideal for absorbing nutrients (×1,000). (b) All hyphae of an individual constitute a mycelium; entire mycelia are virtually impossible to study in nature because they are so extensive, and hyphae are so delicate that they break in the attempt to free them from the substrate. On agar in culture dishes, mycelia tend to be very symmetrical, but they probably are more irregular in nature where small differences in the host or substrate affect growth, branching, and sporulation. Each circular colony grew from a single spore.

digestive enzymes across the plasma membrane to the exterior. Ribosomes are 80S, the eukaryotic type.

A hypha grows only at its tip, not throughout its length, just like roots, root hairs, rhizomes, and anything else that grows through a dense substrate. Most of the protoplasm and nuclei are aggregated in the most apical several millimeters of the hypha, and new wall materials are added at the apex. Walls of absorptive, vegetative hyphae are thin, less than 0.2 μm thick. The innermost wall layer is rich in **chitin**, a polymer arranged in microfibrils similar to cellulose and providing much of the strength of the wall. Exterior to the chitin layer is a predominantly proteinaceous layer, and outermost is a thick layer of polymerized sugars of complex structure.

Septa An important aspect of hyphal walls is the nature of the cross walls or **septa** (singular, septum). Ascomycetes and basidiomycetes regularly form septa as part of their normal development. In ascomycetes, the septum has a small, simple pore in its center that allows cytoplasmic continuity from cell to cell, and organelles migrate through it (**Figures 24-4a** and **c**). Basidiomycetes also have perforated septa, but the opening has a flange or collar and typically a hemispherical cap on each side (Figure 24-4b and d). When hyphae are cut or chewed open, the outward rush of cytoplasm causes septal pores to become plugged, and thus, the primary selective advantage of septa appears to be damage control rather than compartmentalization of hyphae into distinct cells.

Zygomycetes are filamentous, composed of hyphae, but these are usually nonseptate, each hypha being a long tubular cell with hundreds of nuclei. Certain slime molds are gigantic coenocytes with no walls of any type.

Nuclei and Mitosis Fungal nuclei are extremely small, less than 2 μm in diameter, whereas those of plants and animals are at least 5 to 8 μm in diameter. Chromatin in most species of fungi is very uniform, showing little differentiation into euchromatin and heterochromatin, and in some of the most important types,

such as yeast and *Neurospora,* chromosomes do not condense and become visible even during mitosis. In species that do have mitotic condensation, chromosomes are too small to be seen clearly or counted reliably by light microscopy. Consequently, we do not know how many chromosomes most fungi have; for those whose genetics have been studied extensively, we do know that they have fewer than 15 linkage groups.

The details of mitosis differ greatly, depending on the species; slime molds and nonfilamentous fungi have the greatest range of variation. During mitosis, the spindle forms inside the nucleus, which remains quite distinct because the nuclear envelope does not break down as it does in plants and animals (**Figure 24-5**). The spindle always contains just a few microtubules, and some studies have shown that only one microtubule attaches to each centromere face. Only species with flagella have centrioles; the rest lack these organelles completely. Chromosomes, if they condense, do not move to the center of the nucleus; there is no metaphase plate. Apparently, microtubules encounter the centromeres wherever they are, and during anaphase, one chromatid moves only a short distance, whereas the sister chromatid moves very far. In some species, microtubules form a dense column in the center of the nucleus, with chromosomes attached around the periphery.

During anaphase, the nucleus elongates and the nuclear envelope stretches. In some, especially yeasts, the nucleus becomes dumbbell shaped and then pinches in two. In others, the nucleus tears in the center as it stretches, and the two ends aggregate around the two sets of chromosomes. Parts of the original nuclear membrane may remain as fragments in the cytoplasm for quite some time.

In some species, the nucleolus is ejected from the nucleus during prophase, whereas in others, it remains intact and passes to one of the new nuclei. In others, it breaks down within the nucleus during division.

Fruiting Body Organization In ascomycetes and basidiomycetes, but not in the other fungi, some mycelia form a large, compact,

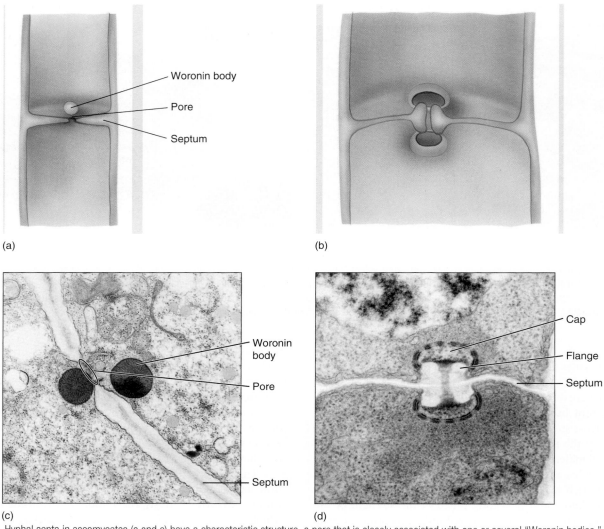

(a)

(b)

(c)

(d)

Figure 24-4   Hyphal septa in ascomycetes (a and c) have a characteristic structure, a pore that is closely associated with one or several "Woronin bodies." This septal structure is diagnostic for ascomycetes—all fungi with septa like this are assumed to be ascomycetes. The septum of a basidiomycete (b and d) is called a dolipore septum; it has a flange and a cap that is bounded by a layer of endoplasmic reticulum. Virtually all basidiomycetes have dolipore septa, except for the rusts and smuts (×20,000).

Figure 24-5   During fungal mitosis, a spindle forms inside the nuclear envelope, which remains intact during all or most of the division process. A second spindle may form outside as well (×19,000).

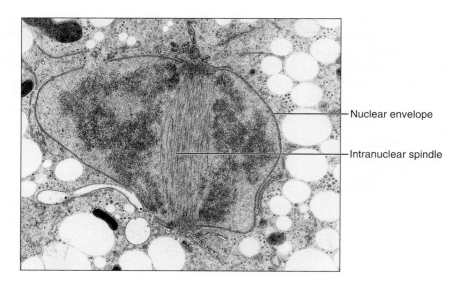

highly organized structure called a **fruiting body**, such as a morel, truffle, mushroom, bracket, or puffball, which is the principal means of producing spores sexually (Figure 24-6). Although these become quite large (mushrooms of *Termitomyces titanicus* can have a cap 1 meter in diameter), they are always composed of hyphae compacted together. Fungi never produce true three-dimensional parenchymatous tissues from an apical meristem, and perhaps as a consequence, they do not produce numerous distinct types of highly differentiated cells and tissues; however, some significant hyphal specialization can occur. In large structures, the outermost surface hyphae are often slender and have thick walls impregnated with pigments called melanins. Melanins are similar to lignins, being composed of phenolic compounds; they absorb ultraviolet light, protecting nuclei inside the body, and

they also deter insects from eating the fungus. Melanin may either be deposited on the surface of hyphae and spores in a cuticle-like layer or permeate the wall, binding to chitin microfibrils similarly to the way that lignin binds to cellulose in plant sclerenchyma.

Within large fruiting bodies, especially bracket fungi, are three types of hyphae: (1) generative hyphae, which are thin walled and produce spores; (2) skeletal hyphae, which are thick walled and unbranched; and (3) binding hyphae, which are also thick walled but are highly and irregularly branched (Figure 24-7). Binding hyphae adhere to the others and cement them into a solid structure; however, this complexity is exceptional; most fruiting bodies are much simpler. The equivalents of xylem, phloem, fibers, sclereids, collenchyma, and glandular cells are never formed in even the most complex fungus.

(a)

(b)

(c)

(d)

Figure 24-6   (a) Brackets are persistent, perennial fruiting bodies of the basidiomycete family Polyporaceae; these are *Trametes versicolor*. The underside of the bracket consists of hundreds of tubular pores lined with spore-producing hyphae. Because the pores are long and narrow, they must be perfectly vertical for the spores to fall out after they are released; brackets and other polypores are responsive gravitropically. (b) Puffballs are basidiomycete fruiting bodies in which the white outer covering breaks open, releasing millions of spores. (c) The fruiting body of a stinkhorn fungus (*Claphrus*) releases spores by decomposing. Its odor of rotting flesh attracts flies, which then disperse the spores. (d) Earthstars (*Geastrum*) are related to puffballs but have an outer protective layer that peels back, forming the star-like shape. The central spherical mass contains millions of spores that are released through the pore.

■ Spores

A universal character of fungi is their formation of spores, resistant resting stages that are the primary means of reproduction, dispersal, and survival. Spores are produced either asexually or sexually. The method of sexual spore formation differs in each clade and is discussed later. Asexual spores are described here.

In zygomycetes, asexual spores are typically **sporangiospores**—that is, spores that form inside the large swollen tip of a hypha. A large amount of cytoplasm and many nuclei aggregate at the tip. A septum seals the region off from the rest of the hypha, and then each nucleus organizes the cytoplasm around it into a spore (**Figure 24-8**). When the original hyphal wall breaks down, the spores are released. When sporangiospores germinate, they are nonmotile in most fungi.

In ascomycetes, basidiomycetes, and deuteromycetes, asexual spores are more often produced as **conidia** (singular, conidium), spores that do not form inside a sporangium. In the simplest type, the tip of a hypha forms septa, cutting off many uninucleate cells, each

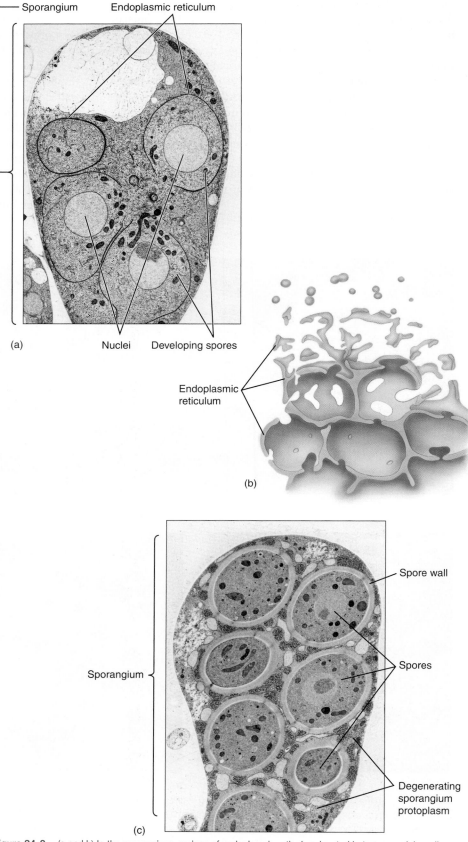

(a)

(b)

(c)

Figure 24-8 (a and b) In the sporangium, regions of endoplasmic reticulum located between nuclei swell and merge, gradually forming sheets. As this continues, cytoplasm is partitioned, and endoplasmic reticulum is modified into plasma membrane. (a) shows a complete cell; (b) shows just the endoplasmic reticulum. (c) The endoplasmic reticulum contents, which at first are amorphous, are converted to spore walls (a and c, ×1,400).

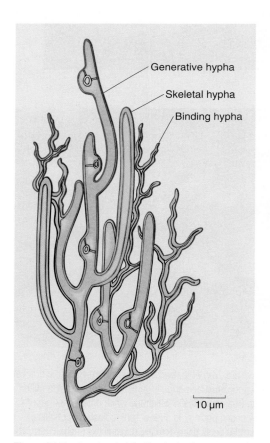

Figure 24-7 In a bracket fruiting body, three types of specialized hyphae occur.

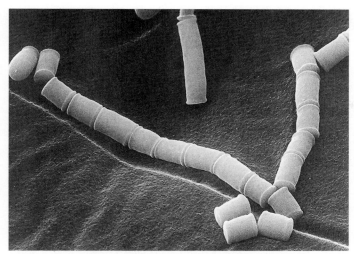

Figure 24-9 In the simplest method of conidium formation, nuclei space themselves at uniform intervals along the hypha; then septa form and partition this region into uninucleate spores, conidia. *Geotrichum candidum* (×3,250).

of which acts as an individual spore (**Figure 24-9**). In a more elaborate type of development, special flask-shaped cells push out a large bud of material, which forms a cross wall and becomes a conidium (**Figure 24-10**). Later, more material is pushed out from the basal cell, and a new conidium forms beneath the first, a process that is repeated until long chains of conidia have been formed. The conidium-producing structures, long, upright hyphae called conidiophores, elevate conidia somewhat, improving wind dispersal. When conidia germinate, they are never motile; they always grow out as a hypha that establishes a new mycelium.

Fungal spores are so small and light that even mild air currents lift them easily, and strong winds carry them great distances.

Figure 24-10 Much more abundant conidia are produced by conidiophores, as in *Aspergillus*. The tip of the hypha branches into flask-shaped "sterigmata," and these push out a bubble of wall, protoplasm, and a nucleus, which then mature into a conidium. The process is repeated indefinitely, producing new conidia under existing ones.

Spores have been collected in air samples taken at a 100-km altitude. The size and number of spores are inversely related: A given mycelium can either make a few large spores or many small ones (a puffball can release 1 billion spores). Larger spores have greater amounts of nutrients and on germination establish a somewhat extensive mycelium, at least part of which might encounter a suitable substrate or host. A small spore has few reserves; if it does not land in a favorable site, it has few nutrients that it can use to grow and explore for a better site. Of the multitudes of spores formed, at least some almost certainly land in good areas and establish a new individual.

Fungi are sensitive to fluctuating environmental conditions—winter cold and summer drought and heat. Many are biotrophic on ephemeral plants and animals that live only briefly, existing for several months merely as seeds, fertilized eggs, or larvae. Because these food sources are present episodically, it is selectively advantageous for fungi to have long-lived spores that survive times of environmental stress or food absence. The simplest type of super-resistant spore is a chlamydospore. It is formed when a mass of protoplasm, rich in reserves of oil or glycogen, accumulates within a short length of hypha and then deposits very thick melanized walls. The rest of the hypha dies as conditions become inhospitable, but the chlamydospore survives, usually at least a year, until favorable conditions or hosts return. Chlamydospores are especially common in soil fungi.

Sclerotia (singular, sclerotium) develop as a section of mycelium branches profusely and the hyphae attract each other, forming a compact aggregate. The outermost hyphae are swollen, and globose and have thick, melanized walls. The inner mass consists of large hyphae (filled with nutrients such as oil, glycogen, mannitol, and trehalose) and small, thin-walled hyphae (rich in cytoplasm and organelles). Rather than remaining distinct, the hyphae undergo numerous fusions with each other, forming a highly interconnected mass. Large amounts of mucilage are secreted around the hyphae, which seem to act as water-holding substances. Sclerotia are often formed by biotrophic fungi, and their germination frequently depends not merely on good conditions but on conditions favorable for the host as well. The sclerotia of *Sclerotium cepivorum* germinate only when a host root happens to grow next to it, and those of *Claviceps* germinate in the spring, just when the host grasses are starting to produce flowers.

■ Heterokaryosis and Parasexuality

In most fungi, sexual reproduction does not involve production of discrete unicellular gametes such as sperms and eggs. Instead, hyphae of one mycelium fuse with hyphae of a different mycelium if the two are compatible. The two hyphae are virtually identical morphologically, with no differentiation into male and female. They are distinct biochemically, however; we say that they are of two different **mating types**, and the designations + and − are used. In order for hyphae to fuse, they must be of different mating types: A + cannot fuse with another +. This requirement ensures that when a fusion does occur, the nuclei brought together are not identical, creating some genetic diversity (**Figure 24-11**).

Plasmogamy, the fusion of two hyphae, is usually not followed immediately by karyogamy; nuclear fusion is delayed slightly in zygomycetes and for a long time in ascomycetes and

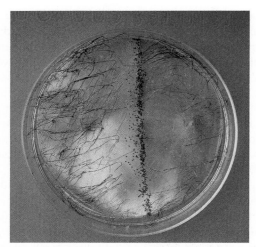

Figure 24-11 In fungi, plasmogamy is carried out by the mycelium, not by sperms and eggs. Hyphae fuse, undergoing plasmogamy; fruiting bodies such as mushrooms are only formed afterward. In this Petri dish, two compatible mycelia have been established by inoculating with + hyphae on one side and − hyphae on the other. As they grew together in the center of the dish, the two individuals merged at hundreds of points, each representing a distinct act of sexual reproduction. The dark structures are melanized spores formed after plasmogamy and karyogamy were completed.

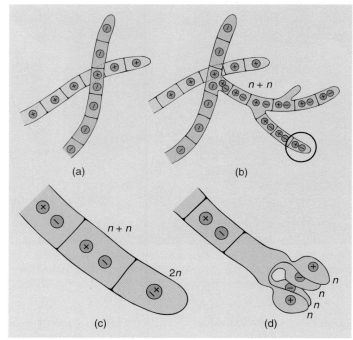

Figure 24-12 (a) Hyphae of two compatible individuals have encountered each other. Until this point, all nuclei in one hypha have been identical (either + or −), having been produced by mitosis of the original spore nucleus of each. Each mycelium is homokaryotic. (b) Plasmogamy has occurred; a hole is digested between two cells and the cytoplasm fuses, resulting in a binucleate cell. This heterokaryotic cell grows out and establishes a new hypha, each cell of which has two nuclei, one of each type. Each of the original homokaryotic hyphae continues to grow and may fuse again in the future, either with each other or with different mycelia. (c) Ultimately, in the tip cell of a heterokaryotic hypha, the nuclei undergo karyogamy to the diploid condition, as shown here. This is immediately followed by meiosis with crossing over. (d) Four meiospores are formed, two + and two −.

basidiomycetes. In these latter two groups, hyphae grow out that have a mixture of the two types of nuclei, + and −. This condition is termed **heterokaryosis**; a hypha before fusion, in which all of its nuclei are identical, is **homokaryotic** (Figures 24-12a and b). Being heterokaryotic is similar to being diploid: + nuclei may carry different alleles than − nuclei; thus, the hypha is heterozygous, and the effects of impaired alleles are masked. Heterokaryosis can be superior to diploidy, however, because the various hyphae of a single mycelium can each undergo fusion with different hyphae from numerous other mycelia; one mycelium then has several distinct heterokaryoses simultaneously.

Karyogamy does occur ultimately, in a special reproductive structure characteristic of each clade. Two nuclei of compatible mating types pair and fuse into a diploid nucleus (Figure 24-12c and d), followed immediately by meiosis, during which synapsis and crossing-over occur. Before karyogamy, the heterokaryotic mycelia of ascomycetes and basidiomycetes grow extensively, forming fruiting bodies; karyogamy occurs only in cells at the tip of the hyphae, but in an average-sized fruiting body, there can be millions or even billions of these (Figure 24-13). Each act of plasmogamy results in millions of karyogamies, resulting in millions of types of + and − spores, having undergone recombination of alleles. The spores blow away; each one that germinates grows into a new + or − mycelium.

In at least a few fungi, for example, *Aspergillus nidulans* and wheat rust fungus (*Puccinia graminis*), a **parasexual cycle** occurs in which compatible nuclei fuse prematurely, even though they are not part of a fruiting body. This happens only rarely, less than two nuclei in a thousand, but after they are diploid, meiosis may occur accompanied by crossing-over. The hypha returns to a heterokaryotic state with only haploid nuclei. If these nuclei become involved in the production of conidia or sporangiospores, they produce spores with a genetic makeup different from that of the original nuclei. For the many fungi that rarely or never undergo sexual reproduction (the deuteromycetes), the parasexual cycle is the only means of genetic recombination.

Metabolism

Although fungi are rather simple morphologically, they are diverse in their metabolism, ecology, and life cycles. Many are capable of either aerobic or anaerobic growth, switching to a fermentative respiration when oxygen is not available. This is the basis of the brewing and vinting (wine making) industries: Yeasts used for making beer and wine are ascomycetes. Similarly, yeast is used to leaven bread, that is, to make it rise by giving off carbon dioxide and thereby forming bubbles in dough. The characteristic flavors of cheeses come from the fungi used in making them: Camembert cheese contains *Penicillium camemberti*, and Roquefort cheese is made with *P. roqueforti*, and so on.

Fungi can attack many different substances or hosts. Although each species may be very specific in its nutritional requirements, so many fungi exist that almost any plant, animal, or prokaryote can be attacked, and fungi even attack other fungi. Common diseases are rusts and smuts, but damping off and heart rot are also fungal diseases. Fortunately for humans, most pathogenic fungi require warm, humid conditions; therefore, we are not often

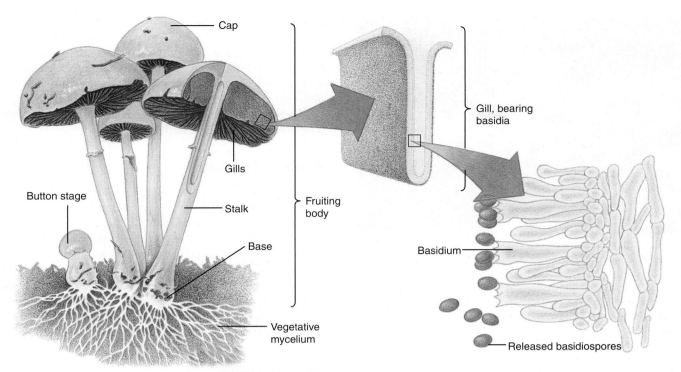

Figure 24-13 All hyphae that compose a mushroom are heterokaryotic, but only nuclei in cells at the tips of the hyphae that end in the gills undergo meiosis. These cells are the only ones that ever become truly diploid and then perform meiosis. All other cells remain heterokaryotic and die as the mushroom decomposes. In plants, too, only a few cells in anthers and ovules undergo meiosis.

attacked. The two most common fungal diseases for us, at least in temperate zones, are athlete's foot and ringworm, which is not a worm at all. Unfortunately, a common fungus that normally lives quietly within us, a species of *Candida,* has become a significant problem. When people are given drugs that impair the immune system in order to allow organ transplants or to treat cancer or if their immune system is damaged by AIDS, *Candida* becomes uncontrolled by the body and grows rapidly, becoming a serious disease instead of a relatively harmless biotroph.

The physiological diversity of fungi is also shown by the ability of some not only to survive but actually to thrive in harsh environments. As with bacteria, some are extreme thermophiles, growing best at temperatures up to 50°C but poorly at or below 20°C. *Mucor pusillus* and *Thermomyces lanuginosus* are two examples of thermophilic fungi; like most of the others, they tend to inhabit compost piles, wood chip piles, and garbage dumps, all special environments in which decomposition and bacterial respiration produce high temperatures in rich organic matter.

Psychrophilous fungi are those that grow best in cold conditions, in the range of −10°C to −15°C. *Sclerotinia borealis* grows most rapidly at 0°C in culture and shows some growth even at −7°C. Such fungi are inhabitants of extreme altitudes and latitudes; Arctic tundra contains many species of all classes of fungi. *Aureobasidium pullulans* and *Sporotrichum carnis* are just two of several fungi that grow on food and meat even while it is frozen; these can be stopped only by keeping the temperature below −15°C. Psychrophilous fungi called snow molds attack turf grasses during winter while they are covered with snow; the plants are healthy in autumn but diseased when snow melts in spring.

Xerophilous fungi grow on "dry" substrates, dryness being due not to lack of water but instead to a high concentration of solutes or other materials that bind water firmly. Examples of dry substrates are jams, jellies, dry grains, leather, dried fruit, and salted fish and meat. These fungi do not merely tolerate high solute content but actually require it: If the moisture content is raised, they do not grow as well. Xerophilous fungi draw water from their substrate by producing a high concentration of solutes, sugar alcohols such as mannitol, in their own cytoplasm. These reach such high concentrations that enzymes and membranes of the fungus are hydrated not just by water but also by the hydroxyl groups of the alcohols.

Slime Molds

Slime molds are distinct from true fungi. They are heterotrophic and form spores, but they lack walls and have a unique body organization. In true slime molds, the body is a large mass of protoplasm with a volume of several cubic centimeters containing thousands or millions of nuclei, all in the same cytoplasm and covered only by a plasma membrane (**Figure 24-14**). This mass of protoplasm, called a **plasmodium** (plural, plasmodia), is capable of migrating over a substrate, much like an amoeba, but is so large that it is easily visible to the naked eye. The plasmodium digests material from the substrate as it moves along; bacteria, yeasts, and decayed plant material are the most common nutrients and are engulfed just as bacteria are engulfed by an amoeba. Such consumption of particulate matter is not possible in true fungi because of their rigid cell walls.

Figure 24-14 This yellow slime mold is one large mass of protoplasm with no walls; the cytoplasm streams vigorously and surges forward—on a microscopic scale—causing the entire mass to roll over and engulf leaf litter, dead insects, and bacteria, parts of which are digested and absorbed. Slime molds are often inconspicuous on the decomposing leaves of rich soil, but they often appear right after bark mulch is spread on flower beds. This one has just begun to enter dormancy.

In response to environmental cues, a plasmodium aggregates its protoplasm, forming one or several mounds; it then extends upward, producing sporangia on stalks (**Figure 24-15**). Spores with true walls are formed and released; the spores are extremely resistant, surviving for many years even under adverse conditions. Meiosis occurs after the spores form; then three nuclei degenerate. Upon germination, the spores may release either an amoeboid or a flagellated cell, and the two forms are interconvertible. The cell grows and may undergo both nuclear and cellular division, proliferating into a population of haploid cells. Each cell contains mating type factors, and when two compatible cells meet, they fuse into one mass, mixing the two types of nuclei. Many cells may fuse, resulting in a multinucleate plasmodium. Karyogamy occurs shortly thereafter, and nuclei are then diploid. The plasmodium

continues to migrate and feed until induced to undergo another round of spore formation.

The phylogenetic relationships of slime molds to other organisms are not well known. Several distinct clades are called slime molds: cellular slime molds (acrasiomycetes), protostelids (protostelidiomycetes), and net slime molds (Labrinthulales). Many have received very little study, and not enough DNA sequences are available to construct a reliable cladogram.

Chytridiomycetes

Most chytridiomycetes have chitinous walls like other fungi, but they are distinct in having flagellated motile cells. They have a single posterior whiplash flagellum, one with a smooth surface, with the standard 9 + 2 arrangement of microtubules. All groups are primarily water molds, with most species living in streams, ponds, or lakes. Chytrids are either unicellular throughout their lives or form a small nonseptate mycelium, as in *Allomyces*. Chytrids live on various substrates, some being parasitic on other water molds and algae and others saprotrophic on dead insects and plant parts.

Zygomycetes

There are approximately 600 described species—most are terrestrial and live in decaying plant and animal matter in soil or forest litter. They are familiar as the mold on stale bread, *Rhizopus stolonifer*. Zygomycetes have simple mycelia composed of branched coenocytic hyphae; complex fruiting bodies are not formed. Haploid spores germinate by sending out a hypha that soon becomes multinucleate and branches profusely (**Figure 24-16**). Where hyphae touch the substrate, small projections, rhizoids, are sent down into the material, both anchoring the hyphae and absorbing nutrients by acting like roots. Above the rhizoids, more horizontal hyphae extend outward, continuing the expansion of the mycelium; other hyphae, sporangiophores, grow directly upward and form spores. Some spores are blown far, whereas others drop close to the

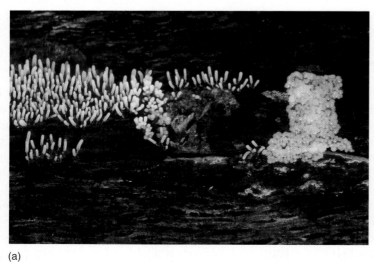

(a)

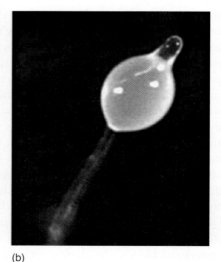

(b)

Figure 24-15 (a) Sporangia of *Stremonitis splendens:* Protoplasm aggregated into mounds and then formed a rigid structure which most of the protoplasm climbed. At the top, cell walls formed, spores differentiated, and nuclei underwent meiosis when the spores were approximately 18 to 30 hours old (×30). (b) Micrograph of a sporangium of *Dictyostelium discoideum;* the protoplasm has aggregated into spores (×550).

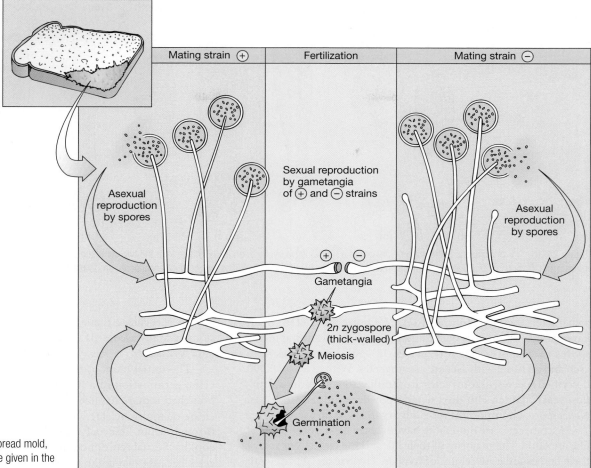

Mating strain (+) | Fertilization | Mating strain (−)

Asexual reproduction by spores

Sexual reproduction by gametangia of (+) and (−) strains

Asexual reproduction by spores

(+) (−)

Gametangia

2n zygospore (thick-walled)

Meiosis

Germination

Figure 24-16 Life cycle of bread mold, *Rhizopus stolonifer.* Details are given in the text.

sporangiophore. This is not a waste of spores because the long-distance spores may initiate new, distant colonies, and those that drop nearby reinfect the same substrate that the original mycelium is growing on, thus attacking and digesting it more quickly, before spores of other fungi land on it.

Sexual reproduction in zygomycetes occurs if hyphae of one individual come close to those of another of compatible mating strain. Each mycelium produces short, multinucleate branches that grow toward equivalent branches on the other mycelium (Figure 24-17). When the branches meet, the contacting walls break down, plasmogamy occurs, and the nuclei pair and fuse. The result is a large **zygosporangium** that has many diploid nuclei; it becomes dormant and inactive, often for months. When it germinates, nuclei undergo meiosis; a sporangiophore is formed (but not a mycelium), and the new, haploid spores blow away and continue the life cycle. No flagellated cells are produced at any point in zygomycetes.

Ascomycetes

When taxonomists study and classify a group of organisms, they usually hope to find a character that perfectly defines a natural group; that is, all members have the character, whereas all nonmembers lack it: a synapomorphy. Ascomycetes have such

Zygospores

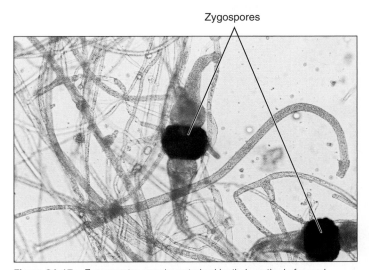

Figure 24-17 Zygomycetes are characterized by their method of sexual reproduction, the zygosporangium. After the two specialized hyphae meet, plasmogamy and the mingling of nuclei and cytoplasm occur, and nuclei pair and fuse in a common cytoplasm. This develops into a resting structure, often called a zygospore but more accurately considered a resistant sporangium. As it germinates, meiosis occurs, and a stalked sporangium is formed (×600).

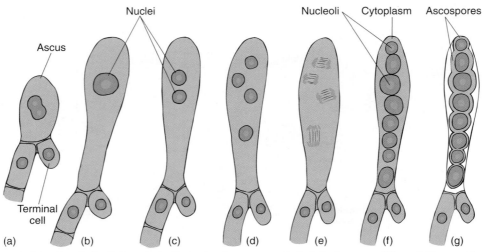

Figure 24-18 (a–g) An ascus is the subterminal cell of a reproductive hypha of an ascomycete; in it, the two compatible nuclei fuse and then immediately undergo meiosis and crossing over. Subsequent mitotic division may or may not occur, so either four or eight ascospores are formed.

a character, the **ascus** (plural, asci), a large sac-like cell in which karyogamy and meiosis occur and in which the resulting meiospores (**ascospores**) form (**Figure 24-18**). All ascomycetes have asci, and all fungi with asci are ascomycetes. With at least 30,000 described species, ascomycetes are familiar to you: morels, truffles, yeasts, Dutch elm disease, and powdery mildew on fruits. *Neurospora*, an important research organism, is an ascomycete.

Ascomycetes are thought to have evolved from a zygomycete-like ancestor. The majority of their life and the largest portion of their bodies are a mycelium of hyphae. Hyphae are septate with incomplete cross walls: They each contain a pore large enough for cytoplasm and nuclei to move from one "cell" to the next rather easily (see Figure 24-4). Ascomycetes form a rather organized, pseudoparenchymatous fruiting body, the **ascocarp**. Little differentiation of tissues occurs in an ascocarp—perhaps a firm covering layer and a region of asci (called the **hymenium;** Figure 24-19)

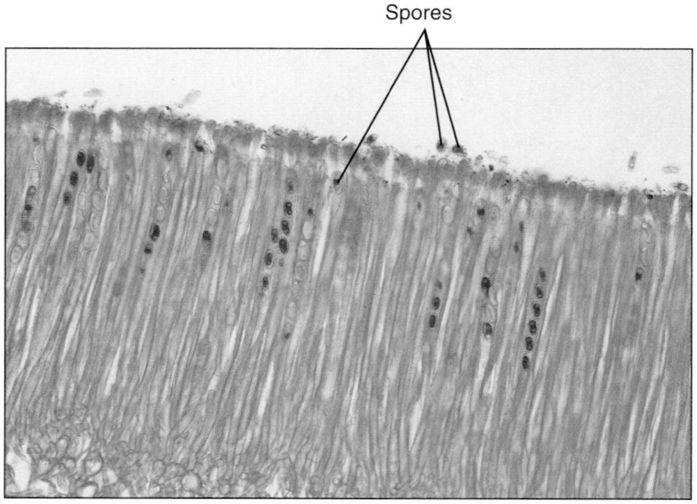

but no specialized tissues for conduction, support, or nutrient storage. Many ascocarps are ephemeral, decomposing a few days after formation.

The initial parts of the life cycle are like those in zygomycetes; germination of a haploid spore results in a mycelium of branched septate hyphae (**Figure 24-20**). Cells are either uninucleate or multinucleate, depending on the movement of nuclei through septa. Asexual reproduction occurs by formation of conidia. Sexual reproduction initially resembles that of zygomycetes because two compatible hyphae that happen to approach each other each produce special short, multinucleate branches. One is called an **ascogonium** and the other an **antheridium**; the ascogonium in a number of ascomycetes sends a small tube (**trichogyne**) to the antheridium and receives nuclei through the trichogyne (Figure 24-20). **The nuclei pair but do not fuse;** because the sets of genes are not in one nucleus, the cells are not considered diploid. Instead we say that they are dikaryotic. The ascogonium does not become dormant like a zygosporangium; instead, new hyphae sprout from it, forming a new mycelium. The nuclei migrate into these **ascogenous hyphae** in pairs, and because they undergo simultaneous mitosis, new pairs of nuclei are produced. Each cell therefore is heterokaryotic and contains two sets of genes.

Ascogenous hyphae, along with new branches of the original haploid hyphae, grow in an organized manner and produce the ascocarp. Ascogenous hyphae form the fertile layer, the hymenium, and the subterminal cell of each hypha forms an ascus. The ascus initial must contain one nucleus from each parent, and this is accomplished in the following manner: The tip bends over strongly, forming a hook (called a crozier) (Figure 24-18), and two complementary nuclei undergo simultaneous mitoses with parallel spindles. Two cell walls seal off one nucleus in a small, lateral tip cell and isolate two complementary nuclei in an ascus initial. The nuclei fuse and then undergo meiosis. Meiosis may be followed by mitosis, so the ascus contains either four or eight ascospores, or rarely 16, 32, 64. Immediately after nuclear division, all nuclei lie in a single mass of cytoplasm; this is organized

Figure 24-19 The hymenium of *Peziza,* an ascomycete, is composed of thousands, even millions, of asci. Because each ascus produces four or eight ascospores, the reproductive potential of each fruiting body is prodigious (×100).

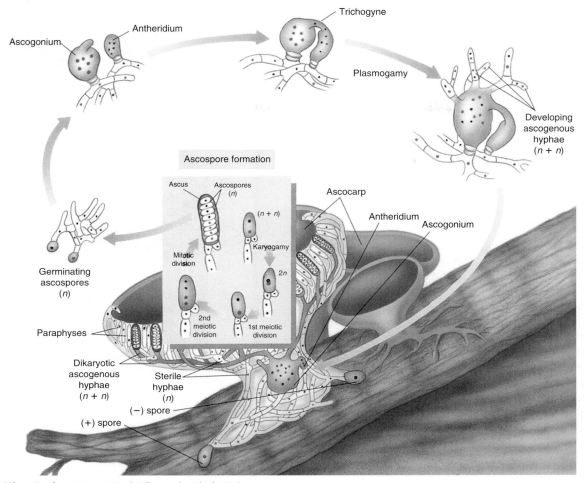

Figure 24-20   Life cycle of an ascomycete; details are given in the text.

into uninucleate spores, as described earlier for sporangiospores. Depending on the group, the ascus may develop a pore or a lid or may just disintegrate, liberating the ascospores, which blow away and initiate new haploid mycelia.

Ascocarps are of various types, all rather simple. The *cleistothecium* is spherical with no opening; sterile hyphae form the outermost layers, and asci and ascospores are located in the cen-

ter. Spores are released as the cleistothecium decays (**Figure 24-21**). A *perithecium* is a flask-shaped ascocarp that releases ascospores through a narrow opening. *Apothecia* are disk or saucer shaped, with the hymenium on the upper surface and sterile hyphae on the underside. Apothecia may be small, simple disks, as in *Humaria, Peziza,* and *Sarcoscypha,* or they may be large, as in morels (*Morchella*).

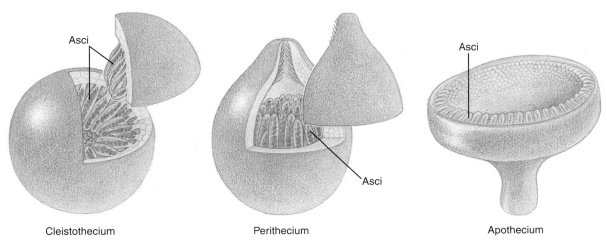

Figure 24-21   Types of ascocarps: cleistothecium, perithecium, and apothecium.

It is important to compare this life cycle to that in plants and animals: There are no sperms, eggs, or zygotes, but syngamy, meiosis, and crossing-over do occur. All of the essentials of sexual reproduction occur, but in a different manner. The same is true of basidiomycetes.

Although ascomycetes are generally more complicated than zygomycetes, some have become quite reduced. **Yeasts** are unicellular, without a mycelium, but individual cells still perform most of the functions of a large mycelium. They bud off new cells like conidia, or they fuse with other cells, acting like ascogonia and antheridia, developing into asci with karyogamy and meiosis taking place; ascospores are formed inside the original wall.

Basidiomycetes

Basidiomycetes (mushrooms, puffballs, and bracket fungi) are familiar fungi and, like ascomycetes, are clearly delimited by a synapomorphy, the **basidium** (plural, basidia), a club-shaped terminal cell within which karyogamy and meiosis occur and that produces **basidiospores** externally (see Figure 24-12 and **Figure 24-22**). Much of the basidiomycete life cycle is like that of the ascomycetes in that it begins with a haploid mycelium, also containing cross walls with pores (see Figure 24-4). When compatible hyphae encounter each other, no formation of ascogonia or antheridia occurs. Instead, with the exception of the rust fungi, regular hyphae contact each other and fuse (**Figure 24-23**). From this cell, a new dikaryotic mycelium is established, with each cell containing one of each type of nucleus.

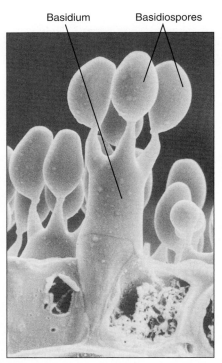

Figure 24-22 A basidium is the terminal cell of a reproductive hypha of a basidiomycete; in it, the two compatible nuclei fuse and then immediately undergo meiosis and crossing-over. Each basidium has four or eight projections (sterigmata); cytoplasm and a nucleus are pushed out through each, forming a basidiospore.

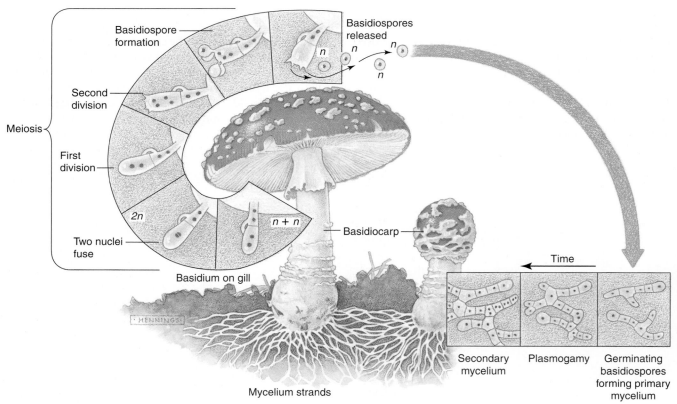

Figure 24-23 Life cycle of basidiomycetes; details are given in the text.

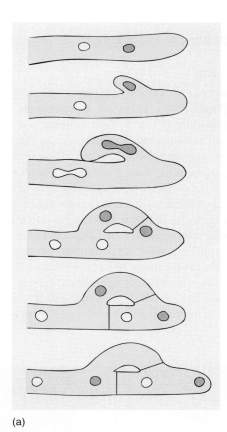

(a)

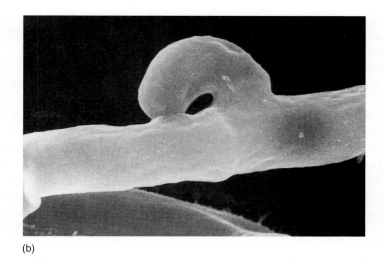

(b)

Figure 24-24 (a) Clamp connection in basidiomycetes is somewhat like ascus formation; backward growth occurs from the most apical portion of a hypha. This fuses with a lower portion of the same hypha and simultaneously two mitotic divisions of two nuclei occur, one of each parental type. Two new septa are formed, establishing a heterokaryotic tip cell. (b) Scanning electron micrograph (×20,000).

Ascomycetes maintain a dikaryotic condition in each ascogenous hypha cell by mere nuclear migration, but in basidiomycetes, the process is more carefully controlled. Each nuclear division is accompanied by the formation of a **clamp connection** (**Figure 24-24**), and each cell always has one nucleus from each parent. The dikaryotic mycelium grows in a diffuse manner, as do the initial haploid mycelia. When they receive the proper stimulus, certain regions undergo an organized growth to form the fruiting body, the **basidiocarp**, which may have generative, skeletal, and binding hyphae. Basidiocarps such as brackets are tough and

Figure 24-25 This bracket is a perennial basidiocarp; it persists from year to year, forming a fresh region of hymenium each year.

persistent, growing larger each year (**Figure 24-25**). Basidiocarps always contain an extensive surface area, most having numerous gills or pores, and the ends of the hyphae that terminate on these surfaces become basidia (see Figure 24-13). Karyogamy occurs in the terminal cell, followed by meiosis. Basidiospores do not form inside the basidium, however; instead four tiny projections form (**sterigmata**; singular, sterigma), and both cytoplasm and one nucleus are squeezed out through each sterigma. This protoplasm expands, forming a basidiospore on the tip of each sterigma. At maturity, the spores are released and blow away, establishing new mycelia.

Basidiomycetes are important as food; the mushrooms most commonly used in cooking are *Agaricus bisporus,* and in the United States, over 60,000 tons are eaten annually. Many mushrooms are dangerous. In many cases, mushroom identification is difficult, and poisonous species can greatly resemble edible ones.

Imperfect Fungi

For organisms that reproduce asexually, mutations that cause the loss of sexual reproduction are not immediately disadvantageous selectively. Certainly the new mutant line cannot evolve as rapidly without sexual reproduction, but it may be able to persist indefinitely using only asexual means of propagation. When this occurs in fungi, we say that the organism is **imperfect**, and it is assigned to the **deuteromycetes** (also called the fungi imperfecti). This is an artificial classification—we know that these are actually members of other groups, we just do not have enough information to classify each one properly. Currently, the group contains approximately 25,000 named species. In the past, these simple masses of hyphae were virtually impossible to classify because they have so little morphology to examine. However, electron microscopy has made it possible to examine the structure of their septa, so they can be assigned at least to zygomycetes, ascomycetes, or basidiomycetes; almost all are ascomycetes.

Many, perhaps most, deuteromycete species actually have sexual stages, but we have not found them. Think about how fungi are studied: A sample of soil or decomposing material is examined microscopically for spores and mycelia. Mycelia often look pretty much alike, but spores are typically quite distinctive. If a completely new spore is found, it is named as a new species, but one spore does not tell us much about the entire organism. It would be best to culture the spore along with another of compatible mating type, grow the mycelia, and then watch reproduction, both asexual and sexual. Many of the "imperfect" fungi probably are really the vegetative states of normal ascomycetes and basidiomycetes whose sexual stages have not yet been encountered. Whenever these stages are found, the species is reclassified out of the deuteromycetes.

Associations of Fungi With Other Organisms

Lichens

Lichens are an association of a fungus with an alga or a cyanobacterium (**Figure 24-26**). There are approximately 13,500 lichen "species"; because they are an association rather than a single organism, they are not considered true species. The fungi in most lichens are ascomycetes (only about a dozen are basidiomycetes), and the algae are most often green algae (especially the genera *Trebouxia* and *Trentopholia*), but cyanobacteria are also frequent. Lichens are commonly assumed to be a symbiotic relationship in which each organism benefits: The fungus (often referred to as the mycobiont) receives sugars, thiamine, and biotin from the alga (the phycobiont) or cyanobacterium, and the alga is protected from desiccation by the fungus. The body of many lichens is stratified with a thin, tough fungal upper layer that shields the autotroph, a middle layer in which most of the algal or cyanobacterial cells occur, and the lowest and thickest layer, composed of loosely

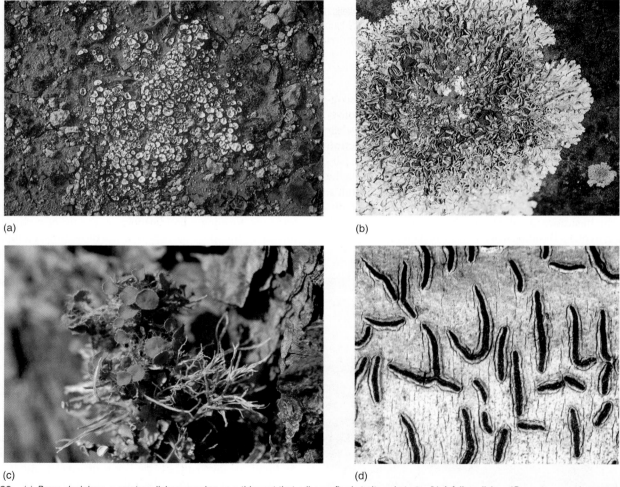

(a) (b)

(c) (d)

Figure 24-26 (a) *Psora decipiens,* a crustose lichen growing as a thin mat that adheres firmly to its substrate. (b) A foliose lichen (*Parmotrema* sp.) grows as a "leafy," open network of thin, flat sheets that project away from the substrate. (c) This is a fruticose or "shrubby" lichen (*Teloschistes chrysothalmus*); like a foliose lichen, it projects out from the substrate, but its body is composed of cylindrical elements. (d) The white tissue here is tree bark; a lichen is growing inside the bark, not on its surface. When the lichen must produce spores, it tears linear and Y-shaped openings called lirelli. The black color is due to thousands of microscopic fungus spores. Lirelli are visible to the naked eye, but you have to look closely. *Graphis scripta.*

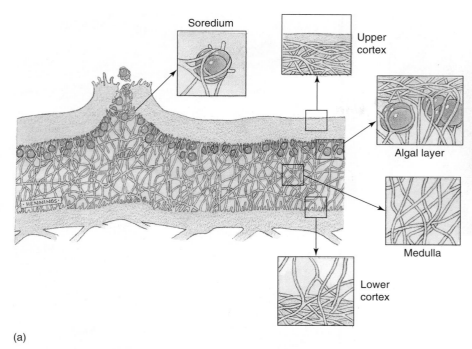

(b)

Figure 24-27 Diagram of the three-layered structure found in many lichens; different structures are also possible. This "body" has more complex differentiation and more tissue-like organization than most fungi or algae. (b) This lichen (*Ochrolechia africana*) has been sliced to show that the layers of algal cells (green) grow just below the upper surface of most lichens, in a position where they receive sunlight but are kept moist by a thin covering of fungal cells above them. Most of the fungal cells make up a spongy white mass between the algae and the substrate.

packed hyphae (**Figure 24-27**). This layer may be 0.5 mm thick and is involved in storing water and keeping the algae or cyanobacteria moist.

Lichens may actually represent specialized forms of parasitism: The autotrophs might not benefit from the association at all. In culture, they grow much more rapidly when free of the fungus than when combined with it in lichen form. Circumstantial evidence exists that the fungus interferes with the wall metabolism of the algae or cyanobacteria, such that sugars are secreted but polymerization is blocked. The fungus absorbs the unpolymerized sugars; presumably the autotroph secretes more sugar as a response to the continued thinness and weakness of the wall. In some, the fungus actually penetrates the alga, inserting haustoria into it.

Lichens are extraordinarily hardy, possibly because they can dry rapidly and enter a resistant state of suspended animation. With just a little moisture, as during a morning dew, lichens rapidly rehydrate and begin photosynthesis and growth. Many are metabolically active for only an hour or less each day; however, this is enough, and lichens grow on bare rock, on fence posts, in deserts, and in Antarctica (where approximately 350 species occur). Because their environments are typically extremely barren, lichens must be especially efficient at absorbing nutrients. This efficiency has become a problem for some because in many areas they absorb toxic pollutants from the air and die.

Although lichens are associations, in many ways they act remarkably like individuals. They undergo reproduction either by fragmenting or by producing **soredia** (singular, soredium), small masses of hyphae and autotrophic cells. Soredia are light enough to be distributed by wind, water, and small animals. The fungi produce spores, and the algae in some also produce reproductive cells. Whereas most lichen autotrophs can be grown in culture and their free-living form can be studied, only a few of the fungi can be grown successfully. Consequently, we do not know what the morphology

of the fungi would be without the autotroph. Many aspects of the form of the fungus have been found to be controlled by the autotroph. Some lichens, such as *Peltigera*, involve two autotrophs, one a green alga (*Coccomyxa*) and one a cyanobacterium (*Nostoc*). The two occur in separate areas of the lichen, and the fungal structure differs, depending on which alga is surrounded.

■ Fungus–Plant Associations

Mycorrhiza is the symbiotic association of roots and soil fungi. A different fungus–plant association occurs in orchids; orchid seeds are quite underdeveloped when mature, often having as few as ten cells, no root, no leaves, and no chlorophyll. They remain partly or entirely inactive until invaded by hyphae of soil fungi, usually the basidiomycete *Rhizoctonia*. After hyphae penetrate the base, the seed becomes active and starts to develop. The seedling enlarges and forms a bulb-like structure and then roots and leaves. In *Dactylorchis*, the first photosynthetic leaf is not produced until the second year after fungal invasion; in *Spiranthes spiralis*, it is not until the 11th year. Until then, the seedling is subterranean and heterotrophic. It actually appears as if the orchid is parasitizing the fungus because the fungi are capable of living freely without the orchid. In the association, the fungus degrades cellulose in leaf litter and converts it to sugar, which is then transported to the orchid. It is not known whether the fungus receives anything in return.

■ Fungi as Disease Agents of Plants

Mycorrhizal associations are beneficial to plants, as is the release of nitrogen by means of fungus-mediated decomposition; however, fungi mostly interact with living plants as disease organisms. Fungi parasitize plants, either weakening them and slowing their growth or killing them outright. Hundreds of fungus-caused plant diseases

Figure 24-28 An advanced case of brown rot of peach, caused by *Monilinia fructicola*.

Figure 24-29 Wheat heavily infected by rust, *Puccinia recondita*. The patches on the wheat leaf are fungal sporangia, in this case, uredinia.

are known and are a central concern of plant pathologists. Two particularly important diseases are discussed here.

Brown Rot of Stone Fruits

Stone fruits such as peaches, plums, and cherries are attacked by the ascomycete *Monilinia fructicola,* which causes brown rot (**Figure 24-28**). Ascospores are released from apothecia in early spring, just as peach trees are beginning to flower. Peaches produce flowers before their new leaves expand, and thus, the first wave of infection involves only flowers; these wither and the fungus produces masses of conidia. Conidia spread the infection to other flowers of the orchard and then attack leaves as they appear, causing leaf and twig blight. Any flowers that survive are able to produce fruit, which is resistant while young; however, as a fruit approaches maturity, it becomes susceptible to conidia from infected leaves. After conidia germinate, hyphae invade the fruit, dissolve the middle lamella, and cause tissues to become soft and brown. The fruit finally rots and shrivels, becoming a **mummy** that may remain on the tree. If the mummified fruit falls, conditions near the soil permit the fungus to form ascocarps and ascospores that initiate new infections the following spring. If the mummy remains on the tree, it does not undergo sexual reproduction, but it can produce copious amounts of conidia.

Rusts and Smuts

Many basidiomycete species cause serious plant diseases; the most important are rusts (order Uredinales) and smuts (order Ustilaginales). These are both in subclass Teliomycetidae, which is characterized by members that do not form basidiocarps. Rusts and smuts can attack hosts so virulently that they kill the plant; even in less severe infections so much damage is done that host growth and reproduction are adversely affected. In economic terms, rusts that attack cereals such as barley, corn, oats, rye, and wheat are most important, causing losses of millions of tons of grain annually (**Figure 24-29**). The most notorious species is the stem rust fungus *Puccinia graminis,* which occurs in many forms (each called a forma specialis), each adapted to a particular host: *P. graminis* forma specialis *tritici* attacks only

wheat (*Triticum*), *P. graminis* forma specialis *secali* attacks only rye (*Secale*), and so on. More than 250 different special forms and races of *P. graminis* have been currently identified; more are discovered every year.

The life cycle of *P. graminis* is complex but excellent for understanding many aspects of fungal biology. First, it is a **heteroecious** species; that is, it requires two different living hosts to complete its life cycle (**Figure 24-30**). Fungi that develop completely to maturity on a single host are **autoecious**. In the spring, basidiospores of *P. graminis* are released into the wind; in order to survive, they must land on young, tender leaves of barberry (*Berberis*). If the leaf is moist, the basidiospores germinate and produce a penetration peg that punctures the leaf cuticle; the hypha then grows into the leaf and parasitizes it. After the mycelium is large enough, it produces basket-shaped masses of hyphae, called spermogonia, on the upper side of the leaf. Within spermogonia are long receptive hyphae and small cells called spermatia. For development to progress, spermatia of one mating type must contact receptive hyphae of a compatible mating type; the two fuse, establishing a dikaryotic phase. After initiated, the dikaryotic condition spreads as the nucleus from the spermatium divides repeatedly, and the daughter nuclei enter adjacent cells of the hyphae by passing through the septal pores.

While this occurs, the haploid mycelium has begun producing another reproductive structure, the aecium, on the lower side of the leaf. This is sterile until nuclei from the spermatium finally arrive through the hyphae and convert its cells to the dikaryotic

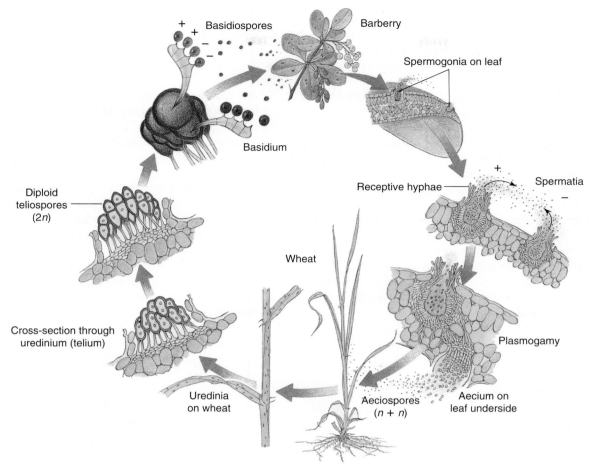

Figure 24-30 Life cycle of the rust *Puccinia graminis*. Details are given in the text.

condition. It then produces dikaryotic, binucleate aeciospores, which cannot germinate on barberry at all, even though they are produced on it. Instead, they must land on leaves of wheat.

Unlike basidiospores, aeciospores cannot penetrate the cuticle when they germinate; instead, they must find an open stoma. Once inside the wheat, hyphae spread extensively and then form a new type of spore, the binucleate, dikaryotic urediniospores in a third reproductive structure, the uredinium. Urediniospores are a means of rapid infection, either of the same plant or others. They are extremely effective and are known to blow hundreds of kilometers. Urediniospores are produced continually in massive numbers throughout summer.

In late summer or early autumn, the rust color changes to black as the uredinia stop producing urediniospores and begin producing yet another type of spore, dikaryotic teliospores. The uredinia are then called telia. In spring, the two nuclei in each cell of each teliospore fuse and then undergo meiosis; as the spore germinates, it produces four cells, each of which receives a nucleus and then produces a sterigma and a basidiospore, completing the life cycle.

This life cycle has the basic features of basidiomycetes, plus some additional ones. Spermatia are produced like other spores, but they do not germinate unless they are in contact with receptive hyphae, thus acting more like sex cells than spores. Aecia can be formed by haploid hyphae, but they are sterile unless they become dikaryotic. Their spores can germinate only on wheat, not on barberry. Obviously, aeciospores have all of the genetic programming to attack barberry because they are formed on this host, yet they themselves cannot live on it. Urediniospores, on the other hand, are formed on wheat and are specifically able to reinfect it.

Control of this disease in North America is almost entirely by means of breeding resistant strains. In eastern Europe, the disease is controlled by eliminating all barberry plants, but in North America, the fungus can spread solely by means of urediniospores. Mild winters in Mexico and the southwestern United States allow the fungus to survive by growing on winter wheat and then infecting more northerly fields during summer by means of wind-carried urediniospores. Winter wheat in Mexico is in turn infected in autumn by urediniospores blowing from Canadian fields. To fight the disease, infected fields are searched for resistant wheat plants; fortunately, alleles for resistance are dominant to those for susceptibility, so resistance can be bred quickly into new lines of wheat. A resistant strain is good for only a few years, however, before a mutation in the gigantic gene pool of *P. graminis* produces a new virulent pathological race.

Summary

1. Fungi are not considered plants because they lack plastids, they lack cellulose, their body construction is filamentous, and many of their metabolic pathways are not homologous with corresponding pathways in plants.

2. Fungi have three heterotrophic metabolisms: biotrophs, necrotrophs, and saprotrophs.

3. Fungi and bacteria break down plant and animal bodies, releasing their minerals and permitting nutrient recycling.

4. The fungal body is either a single cell or a mycelium of septate or nonseptate hyphae. Fungus walls are composed of chitin. In ascomycetes and basidiomycetes, large complex fruiting bodies are formed.

5. Conidia and sporangiospores are spores that are produced asexually.

6. Exchange of genetic material occurs by the fusion of compatible hyphae. Nuclear fusion is delayed, resulting in heterokaryotic hyphae. Shortly before formation of sexual spores, the nuclei fuse and then immediately undergo meiosis, with synapsis and crossing-over.

7. Slime molds consist of a plasmodium containing hundreds of thousands of nuclei but no individual cells.

8. Zygomycetes have mycelia of branched coenocytic hyphae; complex fruiting bodies are never formed. After specialized multinucleate hyphae grow together, plasmogamy and karyogamy result in a zygosporangium with many diploid nuclei.

9. Ascomycetes have sac-like asci where karyogamy and meiosis occur. Asci may be produced in ascocarps such as morels or truffles.

10. Basidiomycetes have club-like basidia where karyogamy and meiosis occur. Basidiomycetes produce basidiocarps such as mushrooms, puffballs, and brackets.

11. Deuteromycetes is an artificial group established for species whose sexual reproduction is unknown. Most have incomplete septa of the ascomycete type.

12. Lichens are associations of fungi and either algae or cyanobacteria. Most of the fungi are ascomycetes; a few are basidiomycetes.

Important Terms

ascocarp	fruiting body	mycelium
ascogenous hyphae	haustorium	necrotrophs
ascus	heterokaryosis	plasmodium
basidiocarp	homokaryotic	saprotrophs
basidium	hymenium	septum
biotrophs	hyphae	yeasts
chitin	imperfect fungi	zygosporangium
coenocytes	lichens	
conidium	mating types	

Review Questions

1. Why are fungi not considered plants? Why were they originally classified as plants?

2. How do biotrophs differ from necrotrophs and saprotrophs? Are any fungi autotrophs? Are any fungi lithotrophic like certain bacteria?

3. Many biotrophic and saprotrophic fungi are host specific. Two examples are mentioned in the text. What is attacked by wilt-inducing fusariums, and what is attacked by fungi that are transmitted by aphids?

4. Some plants have elaborate antifungal defense mechanisms. What is the name of the simple defense against an obligate biotroph? Very briefly, how does it work? Are the plant cells super-resistant to the fungus?

5. Fungi require essential mineral elements. Do they require any macroelements that are not essential for plants? Do fungi, in general, make most of their own organic compounds, or do they need to obtain many vitamins in their diets?

6. The bodies of all fungi, except unicellular ones, are filamentous, and individual filaments are called _____ (singular, _____). The network of branched filaments is called a _____ (plural, _____).

7. With an electron microscope, fungal cells can usually be distinguished immediately from plant cells because of the lack of _____ and because _____ are often not particularly abundant.

8. Whereas plant cell walls contain cellulose, those of most fungi contain the chemical _____.

9. Which fungi have no septa? Which have a small, simple pore? Which have a pore with a flange or collar and a hemispherical cap?

10. Are the nuclei of fungal cells typically larger or smaller than those of plants? What is an approximate diameter of fun-

gal nuclei? When fungi undergo mitosis, are their chromosomes easy to see and count?

11. In ascomycetes and basidiomycetes, but not in other fungi, some mycelia form a large, compact fruiting body. Give the common name of several fruiting bodies. How big is the cap of mushrooms of *Termitomyces titanicus*?

12. Can you think of any selective advantage of heterokaryosis as opposed to diploidy? Is a dikaryotic cell significantly different from a diploid one with respect to control of transcription, translation, and gene activation?

13. Look at Figure 24-8. It shows a process that never occurs in plants. In plants, walls form from cell plates and phragmoplasts. Figure 24-8a shows a multinucleate cell (a sporangium) forming walls around each nucleus. How is it doing that? How are the new walls being formed?

14. The size and number of spores are inversely related. What does that mean? That a fungus can make either many large spores or few small ones or what? What are some advantages of making large spores? What are some advantages of making small ones?

15. Very few fungi form sclerotia. What is a sclerotium?

16. Do most fungi produce sperms and eggs during sexual reproduction? How do nuclei of different mating types come together? Is there any such thing as a "male" fungus and a "female" fungus?

17. What are sporangiospores and conidia? What are meiospores? Name several types.

18. If a mycelium encounters another mycelium of a compatible mating type, the hyphae of the two can fuse and exchange nuclei. Do you think one mycelium might be fusing with many different mycelia at the same time? What would be the consequences with regard to the spores that it could later form?

19. What is the name of the fungus that gives Camembert cheese its flavor? What about the one for Roquefort cheese?

20. Psychrophilous fungi are those that grow best in cold. *Sclerotinia borealis* grows most rapidly at _____ °C and shows some growth even at _____ °C.

21. _____ fungi grow on "dry" substrates. Give several examples of dry substrates. Are they dry because they lack water?

22. Slime molds have a unique body organization, a large mass of protoplasm with thousands of nuclei. This body is called a _____.

23. What is the name (genus and species) of bread mold? What subdivision is it in? Describe their method of sexual reproduction.

24. What are an ascus and an ascospore? Which group of fungi is defined by their presence? Name several common examples.

25. Describe the life cycle of an ascomycete. Compare it with a life cycle of a flowering plant.

26. Many familiar fungi (or familiar fruiting bodies) occur in the basidiomycetes. Name several.

27. Describe a basidium and the formation of basidiospores.

28. Basidiocarps must be perfectly vertical if basidiospores are to fall out; basidiocarps detect gravity and respond gravitropically. Is this metabolism homologous or analogous to the corresponding metabolism in plants (do you think plants and basidiomycetes share a common ancestor that responded to gravity)?

29. What is an imperfect fungus? If you have already studied flowers, what is an imperfect flower? How has electron microscopy allowed us to assign some imperfect fungi to the zygomycetes, ascomycetes, or basidiomycetes?

30. A lichen is an association of a _____ with an _____ or a _____. Does the name "lichen" apply to a natural group of organisms, all descended from a common ancestor, or is it applied to many different things that are not necessarily related to each other?

BotanyLinks

http://biology.jbpub.com/botany/5e

BotanyLinks contains a variety of resources and review material designed to assist in your study of botany.

Visit the Web Exercises area of BotanyLinks to complete these questions:

1. Many mushrooms are delicious and deadly. Go to the BotanyLinks home page to learn how to cook with them and stay alive.

2. Foods provide nutrients for fungi and bacteria, and keeping our food safe from microbes is a constant battle. Go to the BotanyLinks home page for resources for investigating how this is done.

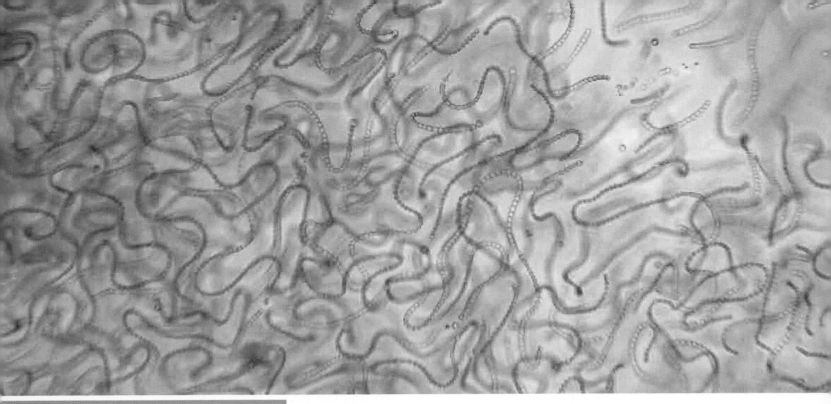

Prokaryotes: Domains Bacteria and Archaea

Concepts

Most prokaryotes are familiar to us as the well-known **bacteria**; a second group, **cyanobacteria**, is less familiar, although they are common in seawater, soils, and freshwater polluted with sewage. A third group contains many prokaryotes that live in unusual environments, such as hot pools and acidic habitats without oxygen; they are unique in numerous fundamental aspects and until recently were classified as a type of bacteria called archaebacteria. But now we realize that members of this third group differ from bacteria in so many features that they must be members of a separate clade, called **archaeans**, that has been distinct from both bacteria and eukaryotes for billions of years. All organisms are now classified into three **domains**, two of prokaryotes and one of eukaryotes: domain **Bacteria**, domain **Archaea**, and domain **Eukarya** (Figure 25-1).

Chapter Opener Image: Most bacteria are so small they appear as just tiny dots even with a high-power objective on a light microscope. These are a particular type of bacteria called cyanobacteria (this is a species of *Nostoc*) that have many important features. First, their cells are exceptionally large for those of a prokaryote, and these are easily visible with just a 10× objective, which was used for this micrograph. Second, cyanobacteria are found in almost any soil sample taken anywhere plants grow. Third, they take nitrogen (N_2) from air and convert it to amino acids (no eukaryote and few other bacteria can do that); consequently they are one of the most important sources of nitrogen-containing compounds for other organisms. Last but not least, chlorophyll and oxygenic photosynthesis originated in cyanobacteria, and certain cyanobacteria were the ancestors of chloroplasts in plants. To see the descendants of some of the first cyanobacteria, look at a soil sample as here or look at the chloroplasts in the leaf of any green plant.

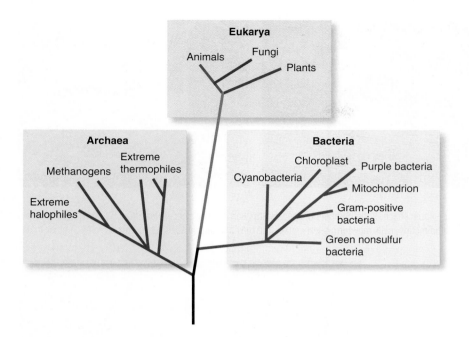

Figure 25-1 One current classification proposed for the evolution of all major groups of living organisms, based on rRNA analysis. Further analysis may suggest other changes; therefore, this is not to be regarded as absolute truth. Halophiles are "salt-loving" organisms. Thermophiles are "heat loving," and methanogens produce methane.

Although prokaryotes are not plants, they are included here because they allow us to perform comparative studies with plants. By comparing and contrasting plant biology and prokaryote biology, we can learn much about plants themselves. In many ways, all plants are so similar to each other that it can be impossible to know whether certain features are alike because they are related evolutionarily or because no alternatives are possible for that feature.

Consider photosynthesis. If photosynthesis in mosses, ferns, conifers, and flowering plants is compared, virtually no differences are found. All plants contain the same chlorophylls, accessory pigments, and electron carriers; all plant photosynthesis occurs in chloroplasts and always liberates oxygen. It would be reasonable to suspect that this particular photosynthetic mechanism is the only possible one; alternative types either must be theoretically impossible or selectively disadvantageous. In prokaryotes, however, we find not just one type of photosynthesis but actually five. Two types are similar to that which occurs in plants, but three are remarkably different, having different types of chlorophyll and electron carriers, and some produce sulfur, not oxygen, as a by-product. This places plant photosynthesis in a new perspective. Is one type most advantageous selectively under certain conditions? Could plant photosynthesis be improved if it used bacterial chlorophylls or bacterial electron carriers? If so, could the genes for the synthesis of those bacterial molecules be inserted into plants by genetic engineering?

Prokaryotes should also be studied because they are everywhere in the environment and play many ecological roles that affect plants. Some convert sulfur and atmospheric nitrogen (which plants cannot use) into sulfates and nitrates, compounds essential for plants. Prokaryotes, along with fungi, are agents of decay, attacking dead bodies enzymatically: Molecules containing nitrogen, sulfur, and other minerals are released and become available to living plants. Prokaryotes also cause numerous diseases. On the other hand, bacteria, especially actinobacteria (formerly called actinomycetes), provide many antibiotics, our most important medicines against bacterial disease.

Structure of the Prokaryotic Cell

Protoplasm

Cells of prokaryotes tend to be very small, usually only 1.0 to 5.0 μm in diameter, much smaller than any plant cell. Bacteria known as **mycoplasmas** are even smaller, some being spheres only 0.2 μm across, whereas certain cyanobacterial cells are up to 60 μm long. Almost all prokaryotes have walls that hold them into one of three basic forms: rods (**bacilli**), spheres (**cocci**), or long coils (**spirilla**) (Figure 25-2). Bacteria that lack walls are rare and have no permanent morphology, changing shape as conditions vary. Bacteria and archaeans are almost invariably unicellular and never have tissues or organs. Cyanobacterial cells have a mucilage sheath strong enough to hold long filaments of dozens of cells together, and the entire filament acts as a single individual (Figures 25-3 and 25-4). Within it, specialized cells, **heterocysts**, fix nitrogen and pass it to surrounding cells through microplasmodesmata, fine holes in their walls. If environmental cues stimulate dormancy, some cells form a thick wall and become **akinetes**, resistant spores; however, this seems to be the maximum amount of body-level differentiation in prokaryotes.

There is no nuclear envelope; instead, DNA occurs as one or several masses called **nucleoids** within the cytoplasm (Figure 25-5). DNA is in the form of closed circles, and no histones or nucleosomes are present; therefore, the **DNA** is said to be "**naked.**" In nonphotosynthetic prokaryotes, the rest of the cell is filled with cytoplasm, and in electron micrographs, it appears homogeneous, lacking mitochondria, plastids, dictyosomes, endoplasmic reticulum, and a central vacuole. Photosynthetic prokaryotes have an extensive invagination and folding of the plasma membrane, and this folded region is the site of photosynthesis. Ribosomes

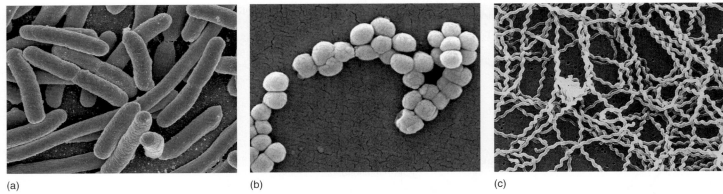

(a)   (b)   (c)

Figure 25-2   All bacteria and archaeans with walls have a constant, simple shape; large numbers of species are short rods (a), usually less than 5 μm long; many are spheres (b), and some have a spiral shape (c). In quiet habitats with little motion, cells may form short chains, but they are usually so weakly attached to each other that microscopic currents break the chains into individual cells.

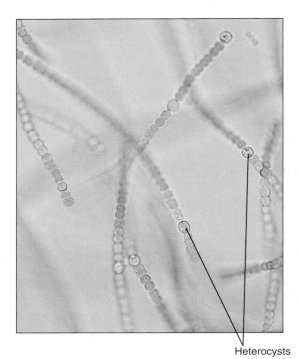

Heterocysts

Figure 25-3   Cyanobacteria, such as this *Nostoc,* are unique among prokaryotes in that some have truly multicellular bodies that may contain two types of cells. Here, the large round cells are heterocysts, specialized cells that fix nitrogen (N₂) into organic compounds. Once fixed, the nitrogen compounds are transported to surrounding cells through microplasmodesmata (×350).

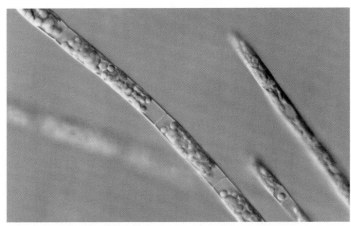

Figure 25-4   Long filaments of cyanobacteria may be broken into smaller pieces by disturbances in their medium, but some also have a self-controlled means of breaking up. Certain cells, "hormongonia," die, causing the filament to tear at this spot. This is not simply accidental death—the breaks occur with a regular spacing, producing fragments of a characteristic size. Lyngbya (×400).

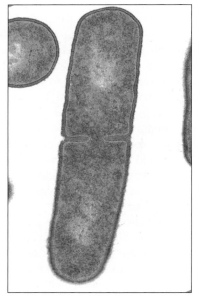

Figure 25-5   The light regions near each end of this dividing cell are the masses of DNA, the nucleoids. In many bacteria, each nucleoid contains many circles of DNA, not just two; there is no nuclear envelope. The number of circles and nucleoids may vary with age and vigor. The cell is dividing by inward growth of the wall.

are present and function as eukaryotic ribosomes, but the actual rRNA molecules differ somewhat, as do the proteins. They are called **70S ribosomes**. Many antibacterial drugs such as streptomycin, neomycin, and tetracycline interfere with 70S ribosomes but do not affect 80S ribosomes. Plastids and mitochondria of eukaryotes also have 70S ribosomes.

Photosynthetic prokaryotes (i.e., cyanobacteria, purple photosynthetic bacteria, and green photosynthetic bacteria) must be buoyant to stay in the upper lighted layers of ponds and lakes, the **photic zone**. Many of these organisms contain a gas vesicle that provides buoyancy. Whereas vesicles in eukaryotes are surrounded by a membrane of lipids and proteins, the gas vesicle is surrounded instead by a sheath of pure protein in a semicrystalline arrangement.

Cell Wall

Bacteria and cyanobacteria are divided into two groups: gram-positive and gram-negative, depending on their cell's reaction to the Gram's stain. The difference is related to wall structure, and in virtually all prokaryotes except archaeans, only two types of wall exist.

In **gram-positive cells**, the wall consists of a thick, 15 to 80 nm, layer of a polymer called peptidoglycan (**Figure 25-6**), which has a physical structure similar to that of a plant cell wall. One component is a long, strong polysaccharide made up of two sugars, N-acetylglucosamine and N-acetylmuramic acid, but just like cellulose, if molecules cannot be bound together laterally, they cannot form a strong wall. In bacteria, cross-linking is done by small peptide bridges, each containing only a few amino acids. The peptides react chemically with the sugars; therefore, tremendous cross-linking occurs, and the wall becomes one giant molecule. Eukaryotes never make N-acetylmuramic acid, nor do they have the unusual amino acid—diaminopimelic acid—that is part of the peptide cross-link; consequently, prokaryote wall metabolism has numerous steps that are exclusively prokaryotic. Certain antibiotics, such as penicillin, interfere only with those steps and so have no effect on eukaryotes.

In **gram-negative bacteria**, the peptidoglycan layer is present but thin, only about 10 nm, and a layer of lipopolysaccharide (LPS) lies exterior to it (Figure 25-6). The LPS layer is believed to be a lipid bilayer although the lipids are unusual chemically. The LPS layer does not contribute strength to the wall, but it may act instead as a selectively permeable barrier that keeps many large foreign molecules away from the plasma membrane. LPS is extremely toxic to animals; it is the reason the food-poisoning bacteria *Salmonella* and *Shigella* are so deadly to us.

Flagella

Many bacteria, but no cyanobacteria or archaeans, have flagella (**Figure 25-7**). These are extremely narrow tubes, approximately 20 nm across, made of just one type of protein, **flagellin**. The flagellum is hollow and is anchored to the cell by a set of rings, one attached in or near the cell membrane and another in the peptidoglycan layer; in gram-negative bacteria, a third set of rings lies near the LPS layer. Flagella are located at one or both ends of a cell (polar flagellation) or over the entire surface (peritrichous flagellation). They rotate continuously and permit the bacterium to swim rapidly, up to 20 times its length per second.

Motile bacteria usually are able to sense certain stimuli and then swim toward or away from them, for example, light (phototaxis) in photosynthetic bacteria, nutrients (chemotaxis) in others, and even magnetic fields in *Aquaspirillum magnetotacticum*.

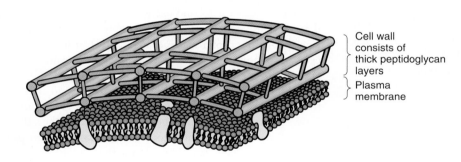

Cell wall consists of thick peptidoglycan layers

Plasma membrane

Gram-positive wall

Figure 25-6 Gram-positive and gram-negative bacteria both have a peptidoglycan layer, but gram-negative bacteria have an additional, external LPS layer that prevents many large molecules from reaching either the peptidoglycan layer or the cell membrane.

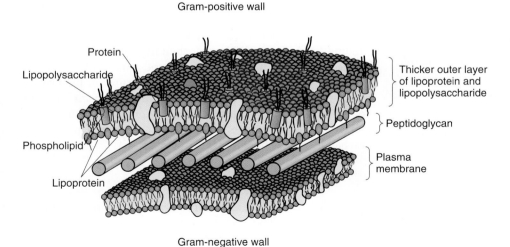

Protein

Lipopolysaccharide

Phospholipid

Lipoprotein

Thicker outer layer of lipoprotein and lipopolysaccharide

Peptidoglycan

Plasma membrane

Gram-negative wall

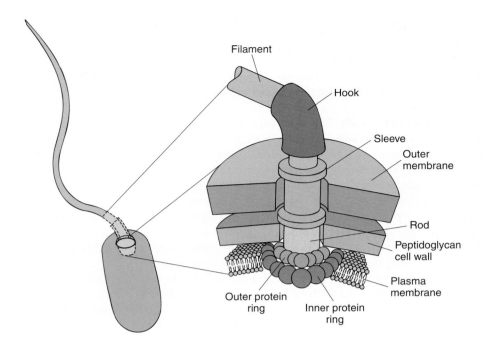

Figure 25-7 Diagram of a bacterial flagellum.

Filament

Hook

Sleeve

Outer membrane

Rod

Peptidoglycan cell wall

Plasma membrane

Outer protein ring

Inner protein ring

Cell Division and Reproduction

Cell Division

Nothing equivalent to eukaryotic nuclear mitosis, meiosis, or cell plate formation occurs during division of prokaryotic cells. Instead, most aspects are similar to replication of plastids and mitochondria: Cell division occurs by **binary fission**, in which the cell simply pinches in two (Figure 25-5). As the cell grows, its circles of DNA are duplicated, with multiple replication start sites and bidirectional replication. DNA replication is not closely coordinated with cell growth, and prokaryotic cells commonly have 10 to 20 (or more) circles of DNA, just as plastids and mitochondria do. All circles are attached to the plasma membrane and are pulled away from each other as the cell wall and membrane grow between them. When a critical size has been reached, the wall grows inward between two nucleoids, dividing the cell. After the new wall is complete, the cells separate. In flagellated bacteria, new flagella arise by an unknown process. Plastids and mitochondria never have walls or flagella. Many bacteria, *Escherichia coli,* for example, grow rapidly under ideal conditions, dividing every 20 minutes, but many archaeans grow extremely slowly in nature.

Exchange of Genetic Material

Genetic material can be exchanged between individual bacteria, similar to sexual reproduction in eukaryotes; however, only a few genes are exchanged at a time, and exchange occurs only rarely.

Transformation occurs when some bacteria die and their DNA breaks into rather large pieces, each with one or several genes. Other bacteria occasionally absorb one of these pieces; once inside the cell, the foreign genes are usually digested, but they may be accidentally inserted into the cell's DNA circle, thereby automatically becoming part of the cell's genome. As the cell duplicates its own DNA, the foreign genes are also duplicated and passed on to the progeny.

Transduction is similar to transformation but occurs when a virus invades a cell and reproduces. Small pieces of bacterial DNA can be accidentally incorporated into the virus; when the virus attacks a new bacterium, the DNA from the first bacterium is released into the second. This is not a particularly efficient method because the second bacterium may be killed by the virus; if it survives, its enzymes will probably depolymerize the foreign DNA rather than incorporate it. Most ordinary soils, however, have more than 50,000 bacteria per gram, and tremendous amounts of genetic exchange occur.

Conjugation involves specific structures and metabolisms that are the result of natural selection favoring genetic exchange. If two compatible bacteria come close to one another, the donor extends a narrow proteinaceous tube, a **conjugation pilus**, to the recipient (Figure 25-8). The donor then duplicates one of its DNA circles, one copy of which moves through the pilus into the recipient.

Plastids do not appear to have any regular genetic exchange among themselves, nor do mitochondria. Genome sequencing projects, however, have discovered plastid genes in mitochondria and vice versa. Perhaps there is a rare transformation-like process as individual organelles breakdown in long-lived plant cells; bits of DNA might be picked up by healthy organelles before cytosolic enzymes degrade them. Despite extensive searches, genetic exchange has ever been found in cyanobacteria.

Metabolism

Many aspects of plant metabolism center on sources of energy, carbon, and the flow of electrons that carry energy. Prokaryotes have numerous alternate methods for each of these processes: several types of photosynthesis, multiple types of respiration, and numerous sources of energy other than sunlight or sugars. Some bacteria, using only a few essential elements plus a simple sugar, build all of their own organic molecules just as plants do;

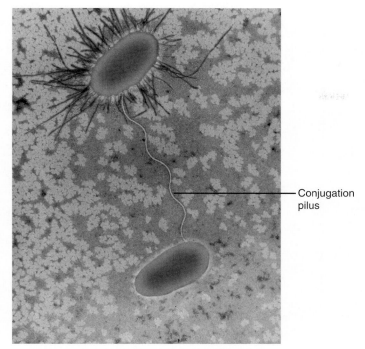

Conjugation
pilus

Figure 25-8 A conjugation pilus is an extremely narrow, proteinaceous tubule that grows from donor to recipient. It is so fragile that the connection is broken before much DNA passes from donor to recipient, even in gentle laboratory conditions.

they synthesize all their own vitamins, lipids, amino acids, and so on. Interestingly, in plant cells, many amino acids and lipids are synthesized only in plastids, not in cytosol. The eukaryotic portion of plants (their nuclei and cytoplasm) is dependent on their prokaryote-derived organelles (plastids and mitochondria), not only for photosynthesis and aerobic respiration, but also for numerous synthetic pathways.

Because bacteria and archaeans differ so greatly in the conditions to which each is adapted, it is extremely difficult to grow most of them in culture, and the metabolism and genetics of the great majority have not been studied at all. The metabolisms that are known are sufficiently out of the ordinary to amaze us, but more importantly, they teach us a great deal about the principles involved in the metabolism of life, not just of prokaryotes but of plants, fungi, and animals as well.

Classification of Prokaryotes

In plants and animals, the most important and most widely used characters for classification are rather easy to study. Features of anatomy and morphology, such as flower structure or the shapes of bones, teeth, and internal organs, offer abundant information for the analysis of evolutionary relationships. Unfortunately, prokaryotes are almost all unicellular, so they have very little anatomy and virtually no morphology. Consequently, their systematics and classification have been based almost entirely on their metabolism, wall chemistry, ability to carry out photosynthesis and use various substrates, and sensitivity to oxygen.

When a character is based on a complex metabolism that is the result of proper integration of numerous components, we are probably safe in assuming that the character is stable and conservative. There is only a slight possibility that it has arisen independently in two or more distinct lines. An example is oxygenic photosynthesis of cyanobacteria: chlorophyll *a*, all of the electron carriers of photosystem two (PS II), and the metabolic mechanism that ensures their proper placement in the correct membranes. We are confident that this suite of characters arose only once. Cyanobacteria are a natural, monophyletic group.

Some features are more difficult to assess. Mycoplasmas, the wall-less bacteria, share many unusual features: (1) They are always parasitic within plant or animal cells. (2) They have no wall. (3) They are very small and have a simple metabolism, and (4) they have little DNA, only enough for approximately 650 average-sized genes. This striking suite of characters is uniformly present in all mycoplasmas; however, many other parasites also have much simpler structures and metabolisms than their free-living relatives. Whereas it may take millions of years to build up a pathway evolutionarily, the pathway can be lost in a few generations in a parasite. Although they share many dramatic features, mycoplasmas actually could be an artificial group, composed of the advanced members of several different lines of convergent evolution, each of which began with a more complex ancestor.

Nucleotide sequencing is now the most powerful and reliable technique for analyzing prokaryote evolution. The most commonly sequenced nucleic acid is 16S ribosomal RNA, the prokaryote equivalent of 18S rRNA in eukaryotes. This type of RNA is especially valuable in studying the phylogeny of major groups for two reasons. First, it is always present in all organisms because ribosomes are invariably present. Second, 16S rRNA is extremely conservative; almost any mutation appears to be deleterious, so it evolves extremely slowly. Even if two groups diverged from a common ancestor hundreds of millions of years ago, large portions of their 16S rRNA are still identical (**Figure 25-9**).

Currently, many species have had at least some of their DNA sequenced, but they are still just a tiny fraction of the thousands of species of prokaryotes. Most classification of prokaryotes is still based on metabolism. We know that the current system contains many errors, but we do not know what changes to make to create a more natural system of classification. Research is progressing rapidly, with great confidence that a more accurate, natural classification will soon be developed (Figure 25-1).

Domain Archaea

Archaeans are distinct metabolically, but they resemble other prokaryotes structurally. They lack a true nucleus, mitochondria, plastids, and all other membrane-bounded organelles (**Figure 25-10**). They fall within the size range of other prokaryotes, and they have walls and 70S ribosomes. In electron micrographs, they resemble cells of bacteria. Metabolically, however, they are very different from bacteria: Their walls contain protein, glycoprotein, or polysaccharide, but they lack a true peptidoglycan component; therefore, they are immune to most antibiotics that interfere with bacterial wall synthesis. Their membranes have unusual lipids in which the fatty acids are attached to glycerol by ether linkages, not the esters that occur in all other organisms. Furthermore, some lipids are diglycerols; that is, each end of the

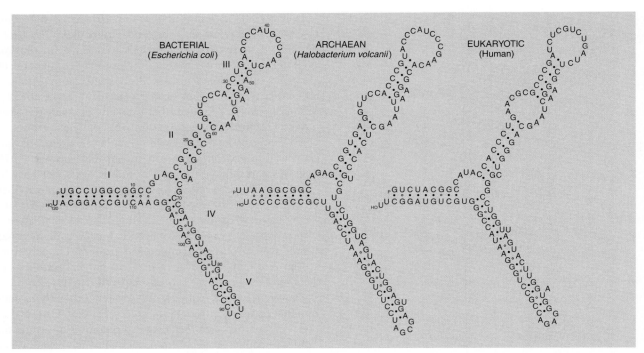

Figure 25-9 5S rRNA can be compared in all living groups, even those as diverse as archaeans, bacteria, and humans. Several nucleic acids may share the same nucleotide base at a particular position, whereas others have a different base; this may represent a homologous sequence descended from the same mutant DNA molecule, or they could be analogous, having descended from two identical but independent mutations. We can calculate the probability that any particular base will mutate and the probability that any two nucleic acids will show the same base-sequence changes purely by chance. Comparative statistics allow us to determine the degree to which two nucleic acids are related; that is, it gives an estimation of how long ago the two shared a common ancestor.

fatty acid is linked to a glycerol (**Figure 25-11**). Many drugs that inhibit ribosomes and protein synthesis in bacteria have no effect on archaeans. The overall metabolism of archaeans is exotic in that most thrive in certain rare environments: Some are methane producers; others are extremely halophilic (salt loving), and some are thermoacidophilic. Because of their unusual metabolic requirements, archaeans are now found in exotic environments: salt pools (*Halobacterium*), sulfur-rich hot acid pools (*Sulfolobus*)

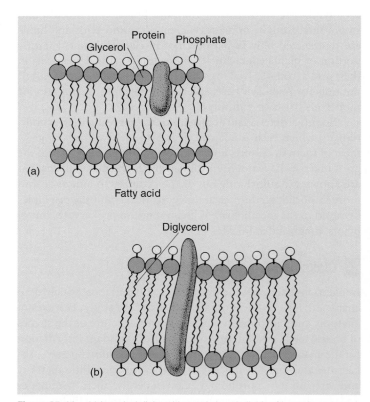

Figure 25-11 (a) In typical diglycerides and phospholipids of bacteria and eukaryotes, the fatty acids extend to one side of the glycerol, resulting in a bipolar molecule, and two layers are needed to form a stable membrane that is completely immersed in water. (b) A diglycerol molecule of archaens is polar at both ends and hydrophobic along the interior; a single layer can form a stable membrane.

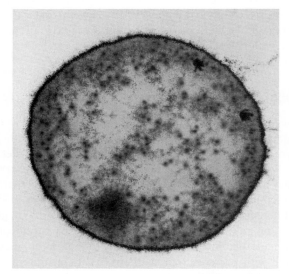

Figure 25-10 Archaeans such as this *Methanococcus voltal* have the same structure as bacteria when viewed with an electron microscope; they have no membrane-bounded organelles (×115,000).

Box 25-1 The "Misclassification" of the Blue-Green Algae

Until the 1970s, cyanobacteria were considered to be either primitive eukaryotic (blue-green) algae or perhaps a transition between prokaryotes and eukaryotes. Currently, there is no doubt that they are completely prokaryotic: They do not have a true nucleus. They have 70S ribosomes. They lack membrane-bounded organelles, and their walls are similar to those of gram-negative bacteria, completely lacking cellulose. It is easy to regard the misclassification of cyanobacteria as a remarkable mistake and to wonder how scientists could have misunderstood their nature. It is important to understand how this happened.

In comparing an electron micrograph of a bacterial or cyanobacterial cell with one of a plant or animal cell, it is impossible to confuse either of the prokaryotic cells with the eukaryotic ones; however, a comparison of a cyanobacterial cell with one from certain types of algae or fungi might cause trouble: The cytological differences between eukaryotes and prokaryotes are not always easy to see. Furthermore, whereas chloroplasts and mitochondria are relatively easy to see in the cells of angiosperms, even using a light microscope, they are much more difficult to see in the small cells of many true algae, especially if the algae have thick walls or refractile grains of starch or oil. Many aspects of cell structure that we take for granted were not well understood for many algae even into the 1970s.

Furthermore, only late in the 1960s did evidence begin to accumulate in support of the endosymbiont theory of eukaryotic cell origin. Before that time, it was generally believed that prokaryotes had gradually evolved into eukaryotes as their nucleoids and membrane systems became more complex and their cells became larger. Under that hypothesis, cyanobacteria appeared to have many of the characters of an intermediate stage between prokaryotes and eukaryotes. Of course, the oxygen-producing, chlorophyll *a*-mediated photosynthesis appeared to be a strong piece of evidence. Another supporting factor is that many cyanobacteria are considered to have true bodies. Some are composed of long, wide filaments much more massive than those of other prokaryotes and analogous to the bodies of many eukaryotic algae, especially green algae (see Figure 25-3). These bodies can also undergo some differentiation: Many cyanobacteria form heterocysts capable of fixing nitrogen. Thus, their bodies consist of at least two distinct types of cells, often arranged in a predictable pattern. Under certain conditions, the cells that do not fix nitrogen can be converted to akinetes. When this happens, differentiation begins in those cells located farthest from heterocysts; this requires a level of whole-body integration that is certainly quite simple but is just as certainly far more sophisticated than anything that occurs in other bacteria.

We can learn much about science and ourselves as scientists as we look at why cyanobacteria were misclassified for so long. Although they were studied by many very intelligent and reasonable scientists, three factors led to the error: (1) Critically important data, such as the prokaryotic ultrastructure and prokaryotic DNA organization, were not known. (2) Many of their characters are analogous to similar structures in eukaryotic algae but were misinterpreted as homologous. (3) Those scientists were working with a theory no longer considered to be valid, and this affected their evaluation of the evidence.

It is sobering to realize that these three factors may be operating in any given sector of our knowledge at present. In very few areas do we know absolutely everything about a subject; some data are always missing. How do we know whether they are trivial data or something that could completely outweigh everything else? For any structure or metabolism or behavior we study, we can usually find similar traits in some other organism. Are the two homologous (related) or analogous (convergent evolution)? How can we be certain? Regardless of what we study, we are working within the framework of numerous large, multifaceted theories. It is always possible that people working in a rather unrelated field will make a basic discovery that will change the theories and theoretical frameworks, thus requiring us to re-evaluate our findings. When we look at scientists of the past and are amused by their mistakes, we should consider what the scientists of the future will think about us.

(a)

(b)

Figure 25-12 (a) Hot pools such as this one at Yellowstone National Park are required for thermophilic archaeans. Some archaeans are the most thermophilic organisms known, living in water that is 100°C, the boiling point. These are not just thermotolerant organisms (tolerating heat); they are thermophiles—they require heat and die without such hot conditions. (b) In pools such as these at Mammoth Hot Springs, the top pool has the hottest water and the most thermophilic archaeans, which are brown here. In the outflow streams where the water is cooler, other species that are less thermophilic (green here) can be found. The highest temperatures tolerated by a few plants and insects is 50°C and by vertebrates 38°C.

(Figure 25-12), and anaerobic, H_2S-rich environments such as deep sea sediments, bogs, and sewage treatment plants (methanogenic archaeans).

Domain Bacteria

Domain Bacteria contains many more species than does Domain Archaea and has greater metabolic diversity (see Figure 25-1). The groups described below were chosen because they make nitrogen and sulfur available, decompose cellulose, or are plant pathogens. Most are probably not natural, monophyletic groups.

Gliding Bacteria

Gliding bacteria move only when in contact with a solid surface; they have no flagella. Many resemble cyanobacteria in consisting of long filaments of short disk-like cells. An example is *Beggiatoa,* which inhabits areas rich in H_2S, such as water polluted with sewage, decaying beds of seaweed, and deep mud layers of lakes. *Beggiatoa* oxidizes H_2S first to sulfur (S^0) and then to SO_4^{2-}. This can be important in detoxifying H_2S, which is poisonous to most organisms, and *Beggiatoa* forms associations with roots of cattails, rice, and other plants that live in stagnant water. *Beggiatoa* appears to benefit in that it receives either nutrients or perhaps the enzyme catalase from the roots while protecting them from H_2S.

Cytophaga is a genus of gliding bacteria that digest cellulose and chitin. The cells must actually bind to cellulose because the cellulases (cellulose-degrading enzymes) are part of the cell envelope and are not released. *Cytophaga* is easily cultured by putting soil onto moist filter paper: *Cytophaga* cells form yellow or orange colonies that are transparent because the underlying filter paper cellulose has been digested. Cellulose-digesting, gliding bacteria are important in humus formation and soil ecology. Many other genera of bacteria attack wood, some by tunneling through both the primary and secondary walls (Figure 25-13).

Nitrogen-Metabolizing Bacteria

Bacteria in this group are of special importance because they affect the amount of nitrogen available to all other life forms.

Nitrogen-fixing bacteria convert atmospheric N_2 into nitrites, nitrates, and ammonium. No eukaryote or archaean can do this, and thus, without nitrogen-fixing bacteria, life on Earth would be

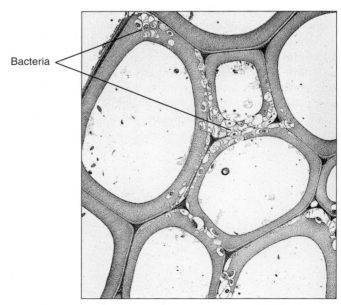
Bacteria

Figure 25-13 Wood-digesting bacteria are tunneling through all parts of the vessels in this wood (×4,000).

extremely limited. *Azotobacter* and *Derxia* are free-living examples of nitrogen-fixing bacteria, and *Rhizobium* and *Azospirillum* form symbiotic associations with plants.

Nitrifying bacteria, such as *Nitrosomonas* and *Nitrobacter,* oxidize ammonia in the environment to nitrate or nitrite. This affects plants because if a root takes up ammonia—the reduced form of nitrogen—then it can be used directly to synthesize amino acids, nucleotides, and so on. If, however, a plant takes up nitrate or nitrite—oxidized forms of nitrogen—then the plant must use reducing power and energy to reduce the nitrogen to the ammonium level before it can be used.

Denitrifying bacteria (*Hyphomicrobium*) convert nitrogen compounds to N_2 gas, making nitrogen completely unavailable to most other organisms and limiting plant growth.

■ Mycoplasmas

Mycoplasmas (*Mycoplasma* and *Acholeplasma*) are bacteria that completely lack a cell wall; they are variable in shape, being filamentous, branched, coccoid, and rod shaped. They are the smallest living cells known, coccoid forms being only 0.2 to 0.3 μm in diameter, and they pass through extremely fine filters capable of stopping all other cells. This presents an easy method for isolating mycoplasmas: Pass the material through a fine filter and then incubate it on medium that contains an antiwall antibiotic such as penicillin.

Many mycoplasmas are pathogenic in animals or plants (**Figure 25-14**). They may cause severe disease, such as citrus stubborn disease caused by *Spiroplasma,* or they may live in plants without producing any obvious disease symptoms. Electron microscopy has shown that many types of diseased plants contain mycoplasma-like inclusions. Very few have actually been isolated, cultured, and identified, so they usually are referred to as mycoplasma-like organisms rather than as mycoplasmas. Mycoplasmas may be a much larger and more damaging group of bacteria than is now realized.

■ Cyanobacteria

Until the 1970s, cyanobacteria (Figures 25-3 and 25-4) were thought to be closely related to true algae and were called blue-green algae (Figure 25-1; Box 25-1). Although some have cells as small as those typical of other bacteria (0.5 to 1.0 μm in diameter), most are larger, and some of the very largest cells of all prokaryotes are those of the cyanobacterium *Oscillatoria princeps,* up to 60 μm in diameter. Whereas some are unicellular (*Chroococcus*), most occur in large groups. These can be simple colonies held together by a mucilaginous matrix (*Microcystis*), or the cells may be more firmly attached and form long filaments that are either unbranched (*Oscillatoria* and *Lyngbya*) or branched (*Hapalosiphon*). The walls of cyanobacteria are like those of other gram-negative bacteria, but they also produce an extra, external layer of mucilage that binds cells and filaments together into a small, loose aggregation. In some genera, such as *Nostoc,* filaments are bound firmly and result in large spherical structures several centimeters across. In other cases, filaments adhere in large, flat sheets (see opening figure of this chapter). Almost any sample of soil will contain thousands of cyanobacteria large enough to see easily with a light microscope, and many filaments are visible with a dissecting microscope. Many lichens contain cyanobacteria as the photosynthetic partner.

Cyanobacteria are of special interest because they seem to be most closely related to the organisms that might have given rise to chloroplasts: They all have chlorophyll *a* and PS II and produce oxygen. One group in particular, **prochlorophytes**, appears especially interesting because they have both chlorophyll *a* and chlorophyll *b* and completely lack phycobilin pigments.

The ability of many cyanobacteria to fix nitrogen is important ecologically because these are widespread and often more active than other nitrogen-fixing bacteria. Typically, nitrogen fixation occurs in heterocysts that develop from vegetative cells (see Figure 25-3). If ammonium is present in the environment, heterocysts do not form; but when available nitrogen is depleted, heterocysts develop, fix nitrogen, and pass it to surrounding vegetative cells. Most species are free living, but others form symbiotic associations with plants: *Anabaena* grows with the water fern *Azolla* and the roots of many cycads; *Nostoc* lives within the body of some nonvascular plants such as *Blasia* and *Anthoceros.*

Cyanobacteria inhabit numerous extremely harsh environments, withstanding intense insolation and periods of almost complete desiccation. *Synechococcus* withstands temperatures of 73°C and lives in hot pools such as those in Yellowstone National Park. In such environments, cyanobacteria often extract lime from the water and secrete it around their own filaments, creating large, rock-like masses. In waters that are not quite so heated, such as those in shallow pools in hot, dry regions, cyanobacterial growths can become gigantic and so heavily encrusted with calcium carbonate that they appear to be stones. These deposits

Figure 25-14 Mycoplasmas cause numerous diseases in plants, such as this yellow blight in palm. Other diseases are aster yellows, elm phloem necrosis, pear decline, and corn stunt disease.

are called **stromatolites** (Figure 25-15) and are currently being formed in only a few localities, but thousands of them are known as fossils. The most ancient stromatolites, 3.2 billion years old, were formed just as oxygen was beginning to accumulate in the atmosphere; they may have been built by the very first organisms capable of the type of oxygenic photosynthesis that occurs in plants today.

Cyanobacteria are abundant in both fresh- and saltwater. They are normally a rather inconspicuous component of marine ecosystems, but when conditions are optimal, they grow rapidly, forming **blooms**—thick mats of filaments. They become especially conspicuous then because many form gas vacuoles, float and accumulate on the water's surface, accentuating the bloom. Such blooms have deleterious consequences because they are followed by the death of many of the cells that then sink and decompose. Decomposition consumes oxygen, leaving the water almost completely anaerobic and thus killing many fish.

Figure 25-15　Fossil stromatolites from the Bitter Springs rock formation in central Australia are known to be approximately 1 billion years old. Their cells strongly resemble those of cyanobacteria in living stromatolites still occurring at Shark Bay in Australia.

Summary

1. All organisms are grouped into three domains: Domain Bacteria, Domain Archaea, and Domain Eukarya.

2. Cells of prokaryotes differ significantly from those of eukaryotes, lacking membrane-bounded organelles and microtubules; walls contain polymers of amino sugars cross-linked by peptides.

3. Multicellular bodies are rare. The largest occur in cyanobacteria and contain only two or three cell types. Differentiated tissues and body-level integration have not evolved.

4. Prokaryotes have no histones; their DNA occurs as naked, closed circles in the cytoplasm. Regular, self-controlled exchange of DNA—conjugation—is known in only a few species.

5. Bacterial flagella are composed of flagellin, not tubulin, and are constructed as just a single tubule, not as a set of microtubules.

6. Prokaryotes carry out the same fundamental types of metabolism that occur in plants, but in many, the processes are simpler and help us understand how the more complex eukaryotic processes may have begun and evolved.

7. Almost all features used for the identification of bacteria and archaeans are chemical or metabolic. DNA and RNA sequence analyses have become the most important tools for analyzing the phylogeny of prokaryotes; their genomes are small enough that complete restriction maps are easy to prepare, and complete sequences of many species have been obtained.

8. Despite their structural similarity, archaeans share little nucleotide sequence homology with bacteria but have several features in common with eukaryotes.

Important Terms

akinetes	cocci	nucleoids
Archaea	conjugation	photic zone
archaeans	cyanobacteria	prochlorophytes
bacilli	Eukarya	70S ribosomes
Bacteria	flagellin	spirilla
bacteria	heterocysts	stromatolites
binary fission	mycoplasmas	transduction
blooms	naked DNA	transformation

Review Questions

1. What are the names of the two domains of organisms that do not have nuclei?

2. Considering that prokaryotes are not plants, what are some reasons they are included in botany textbooks? How many types of photosynthesis occur in true plants? How many in prokaryotes?

3. What does the name "prokaryote" mean? What are some differences in protoplasm organization between prokaryotes and eukaryotes?

4. What are the approximate dimensions of the cells of prokaryotes? Are they larger or smaller than plant cells? Mitochondria are about 1 μm in diameter and about 5 μm in length. Chloroplasts are approximately 4 to 6 μm in diameter. Are prokaryotes larger or smaller than these organelles?

5. Bacteria that have walls have one of three shapes. What are these three shapes?

6. Cells of cyanobacteria are often held together in long chains, and some of the cells in the chains are called heterocysts. What special metabolism do heterocysts carry out?

7. The ribosomes in the cytoplasm of plants—actually of all eukaryotes—are called 80S ribosomes. What are the ribosomes of prokaryotes called? What type of ribosomes are found in the plastids of plants and the mitochondria of all eukaryotes?

8. What are gram-negative and gram-positive bacteria? How does the structure of the wall affect the cell's sensitivity to drugs?

9. Many bacteria have flagella. Do any cyanobacteria have flagella? Is the structure of a prokaryote flagellum similar or dissimilar to those of eukaryotes?

10. Exchange of genetic material does occur in some bacteria. It appears to be similar to _____ _____, but the great majority of species (circle one: have, have not) been studied.

11. Describe each of the following: binary fission, transformation, transduction, and conjugation. Why is it difficult to do genetic studies with many species of prokaryotes?

12. Because bacteria differ so greatly in the conditions to which each is adapted, it is (circle one: very easy, extremely difficult) to grow most of them in culture. Consequently, the metabolism and genetics of the great majority of species (circle one: have, have not) been studied.

13. There are thousands of types of prokaryotes, but only three simple forms; therefore, thousands of different bacteria look exactly alike. We can use morphology to analyze the phylogeny of plants and animals, but for prokaryotes, it is necessary to use a molecule. Which molecules are most often used?

14. For many years, the prokaryotes have been classified as one kingdom, kingdom _____, with two divisions, division _____ ("_____-_____ algae") and division _____. (The classification of these may change, perhaps dramatically, several times in the next few years.)

15. What are some of the habitats inhabited by archaeans? Considering some of the chemicals that archaeans use in their metabolism, do you think these organisms might ultimately provide genes for genetically engineering new types of prokaryotes that can clean up toxic wastes?

16. Figure 25-1 lists archaeans as containing some organisms that are "Extreme thermophiles" and others that are "Extreme halophiles." What do these terms mean?

17. Figure 25-13 shows an electron micrograph of tree wood. What are all of the strange markings in the cell walls?

18. What is a mycoplasma? Do they tend to be beneficial or harmful to plants?

19. One group of cyanobacteria, called the prochlorophytes, is being studied as possible relatives to the prokaryotes that gave rise to chloroplasts by endosymbiosis. Which pigments do prochlorophytes have that are especially important?

20. In certain environments, cyanobacteria extract lime from water and create large, rock-like masses. What is the name of these rock-like structures? What is the age of the most ancient of these? What chemical was just starting to build up in Earth's atmosphere when these earliest fossil structures were formed?

21. Many bacteria are important components of soil and play an important part in the cycling of minerals. What effects do the following bacteria have on the plant life in an area: *Beggiatoa, Cytophaga, Azotobacter, Rhizobium, Nitrosomonas, Nitrobacter,* and *Hyphomicrobium*?

BotanyLinks

http://biology.jbpub.com/botany/5e

BotanyLinks contains a variety of resources and review material designed to assist in your study of botany.

Visit the Web Exercises area of BotanyLinks to complete this question:

1. New diseases of plants and animals appear from time to time (AIDS and Ebola virus in animals). How do plant scientists investigate new diseases? Go to the BotanyLinks home page for information on this subject.

PART 4

Ecology

The focus of this final part is the plant in its environment—plant biology in the broadest scope. An important aspect of the biology of a plant is to view it as a member of a population; population biology differs from the biology of individual plants. A dense population of grasses will attract the attention of grazers such as deer, and many plants will be eaten; however, if a grass plant is a member of a sparse, widely scattered population, grazers may overlook it, and it might not be harmed. The plants we are considering may be absolutely identical, but their fates will differ because of the nature of the populations of which they are a part.

Every plant is also affected by all other aspects of its immediate environment, which in turn is part of a larger environment—a continent or island or perhaps the water of a lake or ocean. Climate must also be considered. None of these factors is uniform in either time or space; there are differences in soil, rainfall patterns, disturbances, and day length. On a larger scale, periods of global warming alternate with ice ages while the continents drift across the surface of the planet. In addition, each organism is affected by at least some of the other organisms around it, and as they undergo their own evolution, their impacts on each other are altered. Earth's surface is composed of a rich, changing mosaic of diverse habitats. Within this milieu, evolution by natural selection has produced hundreds of thousands of species of prokaryotes, protists, algae, fungi, animals, and plants. Their various adaptations for particular aspects of the environment have determined where they will survive and where they will be outcompeted by organisms that are better adapted.

Part Opener Image: We people are the species that has the greatest impact on the ecology of all parts of Earth. Although this farming area is beautiful, almost all native plants and animals were cleared away to make room for fields, houses, and roads. An important aspect of ecological studies at present is to search for ways for us to do less harm to the environment, to preserve natural areas, and to minimize our negative impact on the other creatures that share Earth with us.

Chapter Outline

611

Populations and Ecosystems

Concepts

Ecology is the study of organisms in relationship to all aspects of their surroundings. Throughout this book, emphasis has been placed on the importance of analyzing structure, metabolism, and diversity in terms of adaptation and fitness. Mutations result in new alleles that alter the phenotype; then natural selection eliminates those that are less well adapted but retains those that increase fitness. Most factors have been discussed individually—for example, the effect of trichomes in deterring insects from chewing leaves and the ability of carotenoids to protect chlorophyll from excess sunlight. But most factors that affect a plant's health and survival are present together in the habitat, acting on the plant simultaneously. We must

Chapter Opener Image: Populations are rarely stable; change is much more common. These two photos show dry grassland in West Texas that experienced a wildfire, the Rock House Fire. It started on April 9, 2011, west of the small town of Marfa, and strong winds spread it rapidly northward for almost a month before it could be contained. There had been good rains the previous year, so grasses had grown large and thick and provided abundant fuel for the fire (left). Consequently, the fire was unusually hot and burned many trees, cacti, agaves, yuccas, and other native plants (right). If the previous year had been dry, there would have been less grass and the fire would have either been cooler, or it might not have been able to spread and many more plants would have survived. If the native mammals—buffalo— were still here, they would have eaten the grass, removing fuel. The buffalo, however, was hunted almost to extinction in the mid 1800s so that trains could run without hitting the animals on the tracks. Many plants survived the fire; most of the yuccas (right) are alive and should recover in 2012, if there are adequate rains. The ecology of an area is a complex interplay of many factors, some of which change from year to year.

Figure 26-1 The numerous individuals of one species constitute a population. They depend on each other for gametes during sexual reproduction and also support a healthy population of pollinators and seed dispersers. Because there are several species, the photograph shows a community of several populations. When the soil and atmospheric components are added, an ecosystem results.

try to understand the plant in relation to its entire habitat, to all components of its surroundings.

An individual plant never exists in isolation in a habitat; instead, there are other individuals of the same species and together they constitute a **population** (Figure 26-1). Individuals of plant populations often do not interact strongly, but the biology of the population is more than just the sum of all the biologies of individuals. In a species that consists of either dioecious or self-sterile individuals, a population can carry out successful sexual reproduction, but an individual cannot. If pollinated by animals, a single individual usually cannot produce enough nectar or pollen to keep even one pollinator alive, but a population of plants can sustain a population of pollinators. On the negative side, a population may be dense enough that spores from a pathogenic fungus are ensured of landing on at least one susceptible individual and then spreading to others, whereas if an individual could exist in isolation far removed

from others, it might be safe from pathogens, predators, and natural catastrophes such as fires (Figure 26-2).

A population also does not exist in isolation, but rather coexists with numerous populations of other plant species as well as populations of animals, fungi, protists, and prokaryotes. All of the populations together constitute a **community**, which when considered along with the physical, nonliving environment is an **ecosystem**. These contribute additional levels of interactions and complexity and make it more difficult to be precisely certain of the effects of factors on individuals. The presence of trichomes may deter leaf-eating insects, but those insects might have frightened away more damaging egg-laying insects. Besides, leaf-eating insects might have had a beneficial role by dropping nutrient-rich fecal pellets to the soil where they might have encouraged growth of mycorrhizal fungi. The structure, metabolism, and diversity of plants cannot be fully understood without understanding the ecosystem.

Plants in Relationship to Their Habitats

The **habitat** is the set of conditions in which an organism completes its life cycle. For migratory animals, the winter area, summer area, and migration routes are all habitat components. No plant is migratory, but portions of plants are spores, pollen, fruits, seeds, and vegetative propagules.

How much of the surroundings should be considered part of the habitat is debated. Many factors do not appear to affect certain plants at all. Small herbs on the forest floor do not seem to influence large trees; their presence or absence has little effect on the mineral nutrition of the trees or on their pollination (Figure 26-3).

Figure 26-3 The redwoods here completely dominate the habitat, each tree containing many times the bulk and volume of any single herb. They provide filtered light and protection from wind for the understory plants, which are incapable of tolerating full sunlight. Do the herbs affect the trees in any way? Yes, the redwood seeds germinate only after a low, cool fire is fueled by the understory herbs and shrubs.

Figure 26-2 It is more economical to cut trees growing in dense populations. Identical trees in sparse populations will survive.

An experiment might remove all of the small annuals and then examine whether the trees are affected; however, it would be difficult to measure the growth of entire trees, especially their roots, and it might take years to see any effect on the trees' metabolism. The small herbs may be important; perhaps they harbor spiders that catch insects that would otherwise kill the trees' seedlings. If so, then removing the herbs might cause an increase in the insect population and decreased survival of tree seedlings; with fewer seedlings of this species, it might be possible for the seedlings of a different species to survive better. After many years, the forest composition would be changed as a result of the removal of the herbs.

On the other hand, we do know that many components impact others directly. Pollinators are critically important aspects of the habitat for the plant species they pollinate, and any disease organisms or predators that prey on those pollinators are also important to the plant. Aspects of the habitat that definitely affect a plant constitute its **operational habitat**, whereas all components, whether with known effect or not, are its habitat.

For example, consider the redwood forests in California. It was discovered that there were no redwood seedlings in the national parks, and redwood seeds were not germinating. After a recent forest fire, redwood seeds sprouted and grew vigorously. The fire not only stimulated the seeds to germinate, but by burning the understory plants, released minerals that increased soil fertility. We now realize that shrubs and herbs are vital to the success of the giant redwoods: They fuel quick, cool fires that are too low to damage the large redwoods but are necessary for seed germination. The policy of preventing forest fires—a natural factor of this ecosystem—was harming the redwoods.

Habitat components are of two types, **abiotic** and **biotic**. Abiotic components are nonliving and are physical phenomena: climate, soil, latitude, altitude, and disturbances such as fires, floods, and avalanches. Biotic components are living factors: the plant itself, other plant species, and species of animals, fungi, protists, and prokaryotes.

■ Abiotic Components of the Habitat

Climate Climate is critically important to all organisms; most species are restricted to certain regions primarily because they cannot live in climatic conditions outside those regions. Climate itself has many components—temperature, rainfall, relative humidity, and winds being just a few.

The average temperature of a habitat is not as important as its extremes: the lowest winter temperature and highest summer temperature. Many species of bromeliads, aroids, and orchids are restricted to the tropics because those are frost-free habitats. Rain forests along the west coast of California, Oregon, and Washington receive adequate rain, but freezing winters prevent them from being suitable habitats for most tropical species. On the other hand, many temperate trees must have a winter dormancy period accompanied by weeks of subfreezing temperature in order to be vernalized and bloom. If cultivated in areas with warm winters, the plants grow and survive well but do not reproduce. The highest or lowest temperatures during a plant's lifetime are important. If a tree species must be ten years old before it can flower but killing temperatures occur every eight or nine years, the species cannot survive there.

The growing season of an area is often determined by the date of the last severe, killing frost in the spring and the first killing frost of autumn. The length of the growing season must be adequate for sufficient photosynthesis, growth, development, and reproduction; if not, a reproductive population cannot survive even if it can tolerate the temperature extremes of winter and summer.

Moisture occurs as rain or snow or as hail that supplies water but also damages leaves, buds, flowers, and animals. Habitats range from extremely dry (deserts) through progressively moister all the way to marshes, lakes, and rivers that are virtually all water. Just as with temperature, the total amount of precipitation may not be as important as seasonal extremes or the timing of precipitation. A constant drizzle that occurs almost year round supports certain types of ecosystems, whereas the same amount of rain, distributed as just winter snowfall and occasional summer thunderstorms, results in a different type of ecosystem.

Numerous metabolic processes respond proportionally to abiotic factors. Once there is sufficient moisture for marginal survival, increased amounts of water produce increased growth and reproduction. There is usually an upper limit: With too much water, roots drown for lack of soil oxygen (Figure 26-4a). Between the low and high extremes is the **tolerance range** of the organism. Ranges vary greatly from species to species (Figure 26-4b). Some are extremely broad. Most temperate plants across the northern United States and Canada, especially in the midwest, tolerate summer highs of over 100°F and winter lows below −40°F. Plants of southern Florida, Puerto Rico, and Hawaii tend to have much narrower tolerance ranges, being killed by both cool temperatures and hot ones.

Soil Factors Soils are formed by breakdown of rock. Initially, the resulting soil is thin and virtually identical to the parent rock in its chemical composition; consequently, young soils are variable in the amounts of macronutrients and micronutrients they have available. Because nitrogen is not a significant component of any type of rock, all young soils are deficient in it.

The first plants that invade a new soil, called **pioneers**, must tolerate severe conditions. The soil is sandy, with relatively large particle size, and most minerals are still locked in the rock matrix. The soil has little water-holding capacity, and the first plants have no neighbors to help moderate wind, provide transpired humidity, or otherwise temper the environment (Figure 26-5). Pioneer plants often are associated with nitrogen-fixing prokaryotes; many lichens contain cyanobacteria, and some angiosperms have root nodules. As pioneers live on the soil, they change it significantly; carbon dioxide from root respiration produces carbonic acid and accelerates chemical weathering. Dead plant parts such as leaves, fruits, roots, and bark become substrates for soil organisms, and their decay contributes humus, greatly increasing the soil's water-holding capacity. Roots may penetrate to the bedrock, entering larger cracks and then expanding and breaking the rock physically.

After many years, a thick soil may result that has a distinct soil profile with three layers or horizons (Figure 26-6). The **A horizon** is uppermost and is sometimes called the zone of leaching; it consists of litter and debris, and as this breaks down, rainwater washes nutrients from it downward into the next layer, the **B horizon**, or zone of deposition. The B horizon is the area where materials from the A horizon accumulate. It is rich in nutrients and contains both humus and clay. Below is the **C horizon**, composed mostly of parent rock and rock fragments.

Whereas young soils differ because of the chemical nature of their parental rock, older, more mature soils are less diverse. As the rock weathers, essential elements are absorbed by roots and become trapped in the plant body; nonessential elements are leached away.

(a)

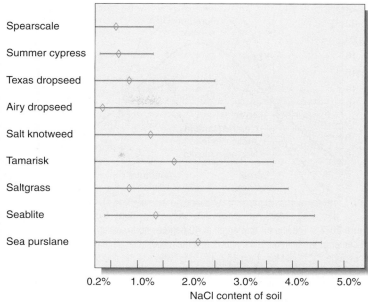

Spearscale	
Summer cypress	
Texas dropseed	
Airy dropseed	
Salt knotweed	
Tamarisk	
Saltgrass	
Seablite	
Sea purslane	

0.2% 1.0% 2.0% 3.0% 4.0% 5.0%

NaCl content of soil

(b)

Figure 26-4 (a) Swampy bayous of southern Louisiana contain so much water that most plants cannot grow; their roots drown. Bald cypress (*Taxodium*) is well adapted, however; its roots form "knees" filled with aerenchyma that allow atmospheric oxygen to diffuse into the submerged roots. (b) Plants vary in their ability to tolerate salty soil. These species occur around salt flats in Oklahoma and Kansas; sea purslane and seablite survive even in the presence of strong salt concentrations, whereas others tolerate less. By being tolerant of soil salt, these plants can occupy habitats on the edges of salt flats, which are free of other plants. Not shown are the many species that have virtually no salt tolerance. Diamond marks indicate the optimal concentrations.

As the plants or their parts die, they decay slowly, releasing the essential elements that re-enter the soil. There they are taken up again. As a result, essential elements cycle repeatedly, alternating between organisms and the A and B horizons, whereas other elements are gradually washed downward into the water table and are removed by underground water flow.

Latitude and Altitude Latitude contributes many factors to the abiotic environment. At the equator, all days are 12 hours long; no seasonal variation occurs, and plants cannot measure season by photoperiod. At progressively higher latitudes to either the north or south, summer days become progressively longer, as do winter nights (**Figure 26-7**). Above the Arctic and Antarctic Circles, mid-summer days are 24 hours long, as are mid-winter nights. At

Figure 26-5 This "soil"—a recent lava flow—consists mostly of rather large rock fragments with little water-holding capacity and very few dissolved essential elements. The plants growing here, pioneers, not only tolerate these conditions but actually change them. The acids they release as they decompose greatly accelerate chemical weathering and soil formation.

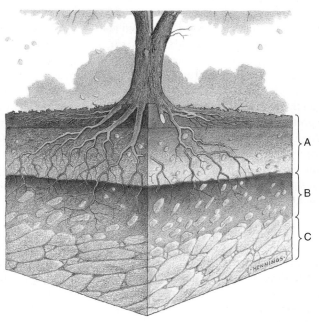

Figure 26-6 Most soils show three horizontal layers, or horizons. The relative thicknesses of the A horizon (zone of leaching) and B horizon (zone of deposition) depend on many factors, the abundance of vegetation and humus being especially important.

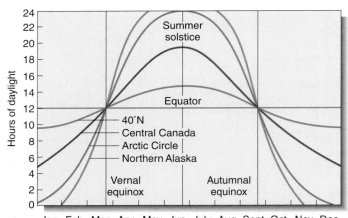

Figure 26-7 The line for day length at 40 degrees north, which corresponds to the center of the United States, shows that the longest day (summer solstice, about June 21) is just less than 15 hours long, whereas the shortest day (winter solstice, about December 21) is only about 10 hours long. Closer to the equator, the difference between the longest and shortest days is less, and near the equator the difference is too little for plants and animals to be able to use it as a seasonal indicator.

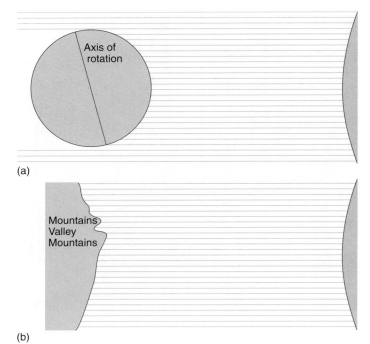

Figure 26-8 (a) During winter in the Northern Hemisphere, the North Pole is pointed away from the sun, and the amount of energy received per square meter is at a minimum. The South Pole points toward the sun, and the Southern Hemisphere receives more direct light, resulting in a large amount of energy per square meter. Six months later the positions are reversed. (b) The variability of energy received per square meter of surface is affected by topography. Away from the equator, the side of a mountain or valley that faces the equator also faces the sun and receives more energy. The sides that face the poles are cooler. Very often, the two sides of a valley running east to west have different vegetation.

intermediate and higher latitudes, day length is an excellent indicator of season, and some species are sensitive to photoperiod.

The amount of light energy that strikes a given area of Earth's surface each year also varies with latitude; in equatorial regions, the sun is always nearly overhead, and each square meter of surface receives a maximum amount of radiation (**Figure 26-8**). At higher latitudes, the sun is only rarely overhead, usually only near midsummer. At other times, when the sun is low, light strikes the Earth obliquely, and less energy is received per square meter; even at noon on winter days the sun is low in the sky and days are not bright. At high latitudes, temperatures fluctuate greatly on both a daily and seasonal basis. Soil formation is slow in the cold latitudes, and often, what soil has formed is blown away by strong winds. In much of Alaska and Canada, soil was scraped away during the ice ages 18,000 years ago.

Regions of high altitudes on mountain tops are similar to those at high latitude. There are high winds and poor soil; much or all the year is cold, and the growing season is short. Water may be present mostly as snow and ice, and thus, physiological drought is frequent. The areas may have varying day lengths, depending on their latitude. An additional stress present in high-altitude habitats is intense ultraviolet light. High altitudes are above much of Earth's atmosphere and thus are not fully shielded by ozone, oxygen, carbon dioxide, and water vapor.

Disturbance Disturbances are phenomena such as fires, landslides, snow avalanches, and floods; they produce a significant, often radical change in an ecosystem quickly (**Figure 26-9**). Disturbances affect the biotic factors directly, often completely eliminating many or all individuals from an area and also altering the soil, but they have little or no impact on other abiotic factors such as climate, latitude, or altitude. The elimination of large numbers of individuals by a disturbance alters species relationships in the ecosystem. Man-made disturbances have been caused by insecticides, herbicides, hunting, and habitat destruction (**Figure 26-10**).

Fire is a natural, common component of many dry ecosystems. With little moisture, fallen leaves and twigs decay so slowly that a thick layer of debris builds up. Living plants tend to have waxy cuticles and water-proofing resins that make them especially flammable. Lightning storms often occur without rainfall, starting fires that burn rapidly and cause great destruction. Many plants and animals of such ecosystems have become fire resistant as a result of natural selection caused by frequent fires. The bark of certain species of pine trees is so thick that a rapidly moving, moderate fire does no damage to the vascular cambium and other living tissues; the lower portions of the trunk have no branches because of self-pruning, so flames cannot reach high enough to ignite needles; only herbs and small shrubs are burned. Furthermore, seed cones of lodgepole pine and jack pine open only after being exposed to the heat of fire; this adaptation results in the release of seeds after many competing plants and predatory animals have been killed and the forest floor is open and sunny. Also, the soil is enriched by minerals in the ash, making it an ideal site for pine seedlings. However, if fires do not occur frequently enough, usually because firefighters put them out, understory shrubs and small trees grow tall and large; so much brush and dead wood accumulate that when fire does occur it is extremely hot. As it burns to the top of shrubs and understory trees, flames may reach the lowest branches of the pines, igniting the crowns (**Figure 26-11**). Once

(a)

(b)

(c)

(d)

Figure 26-9 (a) Fires kill not only plants but also pollinators, herbivores, and pathogenic fungi and bacteria. Fire also releases minerals back to the soil and decreases shading. A quick, cool fire does not damage rhizomes, tubers, seeds, or trees with thick bark, but an intense, hot fire can kill most organisms, sterilizing the soil. (b) Within a few months after a moderate fire, mosses, grasses, and wildflowers have begun regrowth. (c and d) The eruption of Mt. St. Helens radically altered the surrounding ecosystems, but there too plants, animals, and other organisms are recovering.

this occurs, the fire can spread rapidly through the canopy of trees, killing them.

Many grasses have adaptations that permit them to benefit from fires. Many prairie grasses in the midwest and the saw grasses of the Florida everglades grow in dense clumps with their shoot tips and leaf primordia at or below ground level, protected from fire by soil and the living bases of leaves. Leaf tips may be dead and dry, and when fire occurs, the dead portions burn, releasing their minerals, but the bases of the plants are unharmed.

Figure 26-10 Humans destroy habitats on a massive scale; hydroelectric dams permanently flood valleys upstream and eliminate the flooding that had been a natural part of the downstream habitat. Spawning grounds for fish are also destroyed. A few small dams are now being removed in an attempt to restore river ecology.

Figure 26-11 (a) If fire occurs frequently, understory shrubs are burned back before they become tall, and there is always a large space between them and the lowest branches of the dominant trees. Fire cannot get high enough to ignite the tree canopy. (b) If fire occurs infrequently, understory shrubs become tall, reaching the lowest branches of the dominant trees. Fire then burns upward from the shrubs into the canopy, igniting highly flammable needles, twigs, and cones. Even if the trees are not killed outright, their shoot meristems are destroyed, and no further growth is possible. (c and d) In most deserts, native grasses and herbs are so sparse that fire cannot occur, but introduced exotic herbs and grasses occasionally form a population dense enough to support fire. This has occurred in Arizona, putting cacti and yuccas at risk.

(a)

(b)

(c)

(d)

Annuals and short-lived plants do not survive fires, but their seeds, buried underground, do. Bulbs, rhizomes, tubers, or corms easily survive small fires.

Biotic Components of the Habitat

The Plant Itself An individual itself, just by being in a habitat, modifies the habitat and is a part of it. Habitat modification may be beneficial, detrimental, or neutral to the continued success of that species in its own habitat. In beech/oak forests of the northern United States, the trees modify the habitat by producing a dense canopy that results in a heavily shaded forest floor (Figure 26-12a). With such low light levels, few seedlings grow well, but two that do are those of oaks and beeches. As a result, mature trees create a habitat that suits their seedlings and aids their own successful reproduction.

Figure 26-12 (a) In this forest, the two dominant species, beeches and oaks, alter the habitat so that it is suitable for their own seedlings. (b) Although much more open and sunny than a beech/oak forest, a pine forest is still too shaded for pine seedlings. Pines seem to alter the habitat adversely for their own long-term survival, but the needles they drop are highly flammable and are the main cause of frequent fires that kill oak seedlings but not pine seedlings.

(a)

(b)

Figure 26-13 As a glacier retreats, the rubble and sand left behind is extremely poor soil, and of course, the climate is usually harsh; however, several species of pioneers can grow here, and their activity enriches the soil, permitting invasion by less hardy species.

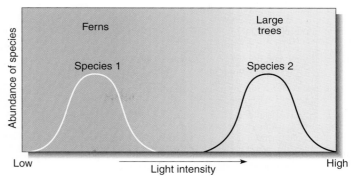

Figure 26-14 Both ferns and large trees need light, but many fern species cannot tolerate high light intensity and must have partial shading. Large trees often must have intense light; they do poorly if shaded. Growing together, they do not compete for sunlight; each uses a portion of the resource the other does not use. If either were removed, the other would not grow better.

Pine forests are more open, but still the forest floor is shaded by mature pines. Pine seedlings, however, need full sunlight and do not grow well below the canopy of older pines. Seedlings of other species flourish in these conditions and crowd out the few pine seedlings that may occur. Pine trees modify their habitat adversely for their continued success; only disturbances can create the open habitats needed for pine seedlings (Figure 26-12b).

As glaciers retreat, they leave behind moraines—great mounds of rubble, sand, and boulders. The soil is poor, with no humus and few available nutrients, but pioneer species such as alder, *Dryas* (in the rose family), willow, and fireweed colonize recently exposed moraines (Figure 26-13). They survive in the open conditions and tolerate low levels of nutrients; alders and *Dryas* have root nodules containing symbiotic nitrogen-fixing bacteria that supply nitrogen. Within a few years, decay of their leaves and bark has enriched the soil sufficiently that Sitka spruce and western hemlock become established. After this, however, the spruce/hemlock forest creates too much shade and eliminates the pioneers that have altered the habitat to their own detriment by enriching it.

Other Plant Species When several individuals, of either just one or several species, occur together, the possibility for interaction is created. If the interaction is basically beneficial for both organisms, it is described as **mutualism**, but if it is disadvantageous, it is **competition**. Competition is a situation in which two populations do not grow as well together as they do separately because they use the same limited supply of resources. Many plants are believed to compete with others for light, soil nutrients, water, and the attention of pollinators and seed dispersers, among other things. If a single plant were allowed to grow by itself, in many cases, it might grow more rapidly, become larger, and produce many more gametes than it would if other plants were nearby. Roots of other plants might grow among its roots and remove water and nutrients. The competitors might grow taller than it does and intercept sunlight by putting their leaves above its leaves. Their flowers, even if they did not produce more nectar, might still distract its pollinators such that its pollen would be carried to the stigmas of the wrong species and its own stigmas would receive foreign pollen.

The role of competition has been extensively debated. One theory postulates that the result of competition is **competitive exclusion**: Whichever species is less adapted is excluded from the ecosystem by superior competitors. The species that get sunlight and other resources win; those that do not lose and are eliminated. If this is true, then very little competition occurs in a typical ecosystem; each species is assumed to be adapted to a particular set of conditions, a **niche**, that no other species is adapted to use as efficiently. For example, some species are adapted to full sunlight and others to partial shade (Figure 26-14). In an ecosystem, the former must be a canopy tree and the latter an understory species. This theory of little competition predicts that if certain species are removed, the others do not benefit from the unused resources because they are not adapted for them. This is sometimes found to be the case.

The concept of niche is difficult to define exactly; basically, it refers to the set of aspects of the habitat that directly affect a species. For example, a particular marsh species occupies a particular semiaquatic niche defined by a range of soil moisture or flooding, a range of seasonal rainfall, a range of temperature, a paucity of root grazing caused by lack of swimming herbivores, and the presence of appropriate pollinators. Another species of marsh plant may grow in the same general geographical area, even the same marsh, but occupy a different niche because it may have a different pollinator or may grow in areas of the marsh that are slightly more acidic than the microhabitats occupied by the first species. As long as even one factor differs, they occupy different niches.

As an alternative to the theory of competitive exclusion, a second theory postulates that species overlap in their tolerance ranges (Figure 26-15a), and when grown together, each has exclusive use of the portion of the range not used by the other. In the overlap zone where the habitat is suitable for both species, usage is determined by competition: The one that is more adapted occupies the overlap zone exclusively. The weaker species occupies only part of its potential niche. The two can coexist, but if either is removed, the other then has its full range of resources and grows better. Examples of this type of competition have been found to occur.

A long-term result of competition should be species modification by natural selection. If two species compete with partial tolerance-range overlap, then mutations are beneficial if they

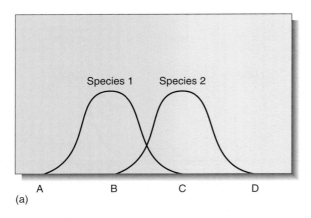

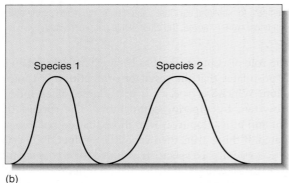

Figure 26-15 (a) Two species compete if they can use the same resource. Here, species 1 would have exclusive use of the resource under conditions A to B, whereas species 2 would have exclusive use between C and D. With conditions between B and C, they would compete. If the resource were water availability, species 1 might grow in drier areas and species 2 in wet areas, and both could grow in moist areas. (b) Over time, mutations that improve the ability of a species to compete may be selected. In this case, species 1 evolves such that it is specialized in dry habitats and now cannot live in moist conditions. This may increase its total fitness because moist environments have more insects and fungi that require expensive defense mechanisms, mechanisms that it no longer needs.

allow each to use more efficiently its exclusive portion of the range. Through time, the species diverge until competition is reduced (Figure 26-15b).

The geographic ranges of most populations are extensive enough that they contain a diversity of biotic and abiotic factors: Each may include hills, valleys, and plains; rocky soil and rich soil; and grasslands and open woodlands. Because of this diversity, a population may be competing for sunlight in one part of its geographic range but competing for water in another and for other factors in still another. Ecosystem diversity causes different subpopulations to specialize for certain features, especially if gene flow is not rapid and thorough. This could be the beginning of divergent speciation, of course, but long before the various subpopulations could be called subspecies, they would be considered **ecotypes**, each specialized in response to particular ecosystem factors at its locality. The various ecotypes resemble each other so strongly that they clearly belong to the same species but have enough differences for ecologists and taxonomists to suspect that separate ecotypes exist. Experiments are necessary to prove that

natural selection is in fact altering the genome because phenotypic differences can result simply from the growing conditions themselves. For example, beans grown in rich soil with adequate water are strikingly different from those grown in poor soil with little water and other stresses. To test whether ecotypes really exist, **transplant experiments** are performed: Plants from each site are transplanted to the alternate site, and plants from both sites may be grown together in a **common garden** at an intermediate site. If the transplanted individuals take on the phenotype of the naturally occurring plants at that site, then genetic factors were not important, just stresses; however, if the transplanted individuals retain their original phenotype or if they die, then genetic divergence had begun.

Also, success in competition depends on factors other than the ones involved in the competition. Species A may grow rapidly and shade species B when water is abundant, but species B may be more drought tolerant and can outgrow and outcompete species A when water is scarce. Water availability may vary over the geographic range such that A dominates in wet areas and B in dry areas. Also, because rainfall varies from year to year, species A may be more common some years and species B more abundant in other years.

Animal ecology includes a variety of negative interactions in addition to competition; predation—the killing of prey—and parasitism are two examples. Insectivorous plants do trap and kill insects, but no plant kills and digests other plants in a predator–prey relationship. Parasitic plants are not uncommon. Mistletoes are familiar examples, and hundreds of species of parasitic plants in at least 12 families are known.

Numerous animals, fungi, protists, and prokaryotes are saprotrophs, living on dead organisms; vultures, maggots, wood-rotting fungi, and decay bacteria are examples. No plant species of any type is known to attack and digest dead plant or animal material; roots do absorb the minerals, nitrates, and sulfates released from decomposing humus, but the roots do not participate in the enzymatic attack.

Organisms Other Than Plants Animals, fungi, and prokaryotes are obviously important biotic aspects of a plant's habitat. Interrelationships between plants and these other organisms can be either beneficial or detrimental for one or both partners.

Plants and animals have many relationships that are examples of mutualism, in which both species benefit. Most instances of pollination are mutualistic—the animal receives nectar or a portion of the pollen and the plant benefits from pollen transfer. Seed dispersal by fruit-eating animals (**frugivores**) also benefits both species as long as the animal does not chew the seeds and digest the embryos.

A famous example of mutualism is the association of certain ants and acacias in tropical Central America. Acacias are small trees that have enlarged, hollow thorns at their leaf base. The thorns are used as ready-made, stout, waterproof homes by ants (**Figure 26-16**). Because of the large number of thorns on a single plant, the ants benefit from having abundant housing, but in addition, acacias produce nectar, and their leaflet tips are modified into golden yellow food bodies (called Beltian bodies) that are filled with glycogen. The ants receive both housing and food. The plant benefits because this species of ant is aggressive; they

 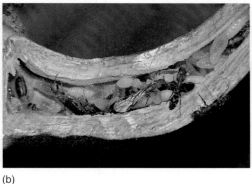 

(a) (b) (c)

Figure 26-16 (a) Ant *Acacia* plant. (b) These thorns on an ant *Acacia* are swollen and hollow and serve as excellent nests for ants. They form as part of the *Acacia*'s own normal development; the presence of ants does not induce their formation. (c) The leaf tips of ant acacias develop as Beltian bodies (yellow), food bodies rich in glycogen. As long as an *Acacia* is occupied by ants, all other vegetation is kept cleared away, but if the ants are killed with insecticide, the acacias are overgrown quickly.

patrol the plant and attack any animal that touches it, even large mammals. If the leaves of another plant come close to an *Acacia,* the ants destroy the leaf, keeping the *Acacia* unshaded. As a result, the plant is free of pathogenic fungi and insects, and it grows in full sunlight. This mutualism is obligate; the plant cannot survive in nature without the ants, and the ants are not found away from the plant. A similar mutualism occurs between ants and *Hydno-phytum* in Southeast Asia.

Commensal relationships, in which one species benefits and the other is unaffected, are also common between plants and animals. When birds build nests in trees, the birds benefit, and the tree is (usually) unharmed. When sticky fruits or seeds, such as cocklebur, stick to an animal's fur or feathers and then are dispersed, the plant benefits and the animal is unharmed. One-sided negative relationships also occur: Animals trample and kill small plants without being affected. Competition, in which both parties are adversely affected, probably is not common between plants and animals.

Predation is a relationship in which one species benefits and the other is harmed (**Figure 26-17**); the species that benefits seeks out the other and uses it specifically for food or some other form of resource. Animals that eat plants are **herbivores** and the process is **herbivory**, but it is often more precisely delimited as **browsing** (eating twigs and leaves of shrubs—deer, giraffes) or **grazing** (eating herbs—sheep, cattle). Insects also lay eggs in plants, and then their larvae feed on plant tissues as they tunnel through them. Aphids and spittlebugs suck sap rather than chew tissues. Insects and birds both harvest leaves and twigs for nest construction.

Many interrelationships between plants and fungi or bacteria are harmful to the plant, but the fungi and bacteria are described as being **pathogenic** rather than predatory. Either may cause mild disease or be so virulent that they kill the plants quickly. A large percentage of the microbes are saprotrophs, living on dead plant tissues such as leaves, logs, fallen fruit, and sloughed bark. This benefits the fungi and bacteria without harming the living plants, and it can actually help them. This process speeds up the release of mineral nutrients, especially nitrogen compounds, enriching the soil. A case of plants attacking and parasitizing fungi may be known. Orchid seeds are tiny and lack chlorophyll; they remain

moribund until invaded by soil fungi, then they turn green and grow well. It had been assumed that the seedling received nutrients and perhaps growth factors from the fungus and in return provided it with carbohydrate; however, tests have not revealed any benefit to the fungus, just to the plant.

Examples of plant–fungus mutualism are well known and recently have been shown to be of much greater importance than ever before suspected, as in the case of mycorrhizae. The mycorrhizal fungus transports phosphate into the plant and receives carbohydrate; both benefit. Many plant species grow only poorly in nature if soil fungi are killed with a fungicide, and it has been postulated that one reason the plains of the American midwest lack trees is because the type of mycorrhizae beneficial to trees cannot compete well with the mycorrhizae beneficial to grasses. A complex relationship has been discovered with Indian pipe (an achlorophyllous parasitic plant) and a mycorrhizal fungus: Indian pipe parasitizes the fungus, drawing nutrients from it, and the fungus in turn obtains carbohydrates from its other mycorrhizal partners that are chlorophyllous and photosynthetic. Indian pipe

Figure 26-17 Buffalo and plants are in a predator/prey relationship in which the buffalo benefits but the plants do not. The buffalo are herbivores, more specifically they are grazers, and they not only eat the plants but also trample them with their hooves and destroy them when they wallow to rid themselves of fleas and ticks.

Plants and People

Box 26-1 Niches in the Jet Age

A niche is defined as a set of conditions in which a particular species can thrive, outcompeting other organisms. Cacti occupy desert niches in North and South America, and water hyacinths proliferate in the waterways of the Amazon rain forest, filling the niche consisting of calm water surfaces in warm areas. But the desert niches occur in many geographical regions other than the western hemisphere, and tropical rivers flow quietly all around the world. Why do cacti and water hyacinths grow only in restricted areas? Basically, it is simply because they have not been able to spread to the other areas because their natural range—the site where they originated evolutionarily—is surrounded by vast regions of inhospitable barriers. Under natural conditions, cacti could establish themselves in the desert niches of Africa, the Middle East, China, and Australia only if migratory birds happened to carry seeds from one continent to another. Such long-distance dispersal is extremely rare.

Perhaps we should say that it was extremely rare. Humans have now far surpassed migratory birds as agents of long-distance dispersal. Planes and ships connect all parts of the world, causing an interchange of plant and animal species on an unprecedented scale. Some of the transport is accidental as seeds are caught in clothing of tourists or are attached to hides or other material being shipped. Zebra mussels came to the United States in the bilge water of ships, and insects are carried in fruit and produce and even in the stagnant water trapped in old tires being imported for recycling. Of course, even microbes and viruses are transported over long distances, as is the case with HIV, the virus that causes AIDS. Other transport is intentional, as living plants or animals are imported specifically for cultivation as crops or ornamentals. Carp were imported into our rivers as a food source; killer bees were brought to Argentina from Africa for research and then escaped from a broken laboratory cage. The vine kudzu was planted in the southern states as a ground cover to control erosion along road cuts and on canal banks, and water hyacinth apparently was introduced into our waterways simply because it is pretty.

All of these organisms, and many more, share a common feature—they have thrived in the new habitats in which we have placed them. All have found conditions that permit their rapid growth and reproduction. Because we did not bring along their natural predators or pathogens, they tend to be free of the organisms that could limit their expansion. They have all undergone population explosions and are proliferating rapidly. Zebra mussels are clogging waterways through the American northeast and are spreading down the Mississippi River; killer bees have spread throughout South America, Mexico, and Texas and into New Mexico in just ten years. In such cases, it turns out that these organisms are more highly adapted to the conditions here than are our native species. Consequently, the exotic, introduced species outcompete the indigenous species. Carp have crowded out many types of American fish, and kudzu covers thousands of square kilometers of forests, killing the trees by shading them and cutting off their sunlight.

No insecticides or herbicides are specific enough to control only the introduced species, and trying to eliminate plants, insects, and fish by hand is almost impossible. Most control efforts center on searching the home habitat to find predators or disease organisms and then introducing those into the new location, but because the original problem was caused by the introduction of an exotic organism, there is, of course, considerable reluctance to bring in another exotic organism. An ideal solution would be to find a pathogen that preys only on the problem species and on no other. Ideally, then, when the original exotic species is eliminated, the second will also die out for lack of food. Caution must be used because the pathogen may be able to attack native species that are related to the exotic species, or a mutation may occur in the pathogen that allows it to attack native species as well.

Already, the transport of species to new habitats where they can proliferate and outcompete native species has been a serious problem. It can only become worse as our jet-age global movement brings about increased travel and transport of material.

Figure B26-1   Kudzu (*Pueraria lobata*), which was introduced from Japan, has no pathogens or pests here in the United States. It grows rapidly, covering rocks, trees, telephone poles, and even buildings. Trees covered by kudzu die from a lack of sunlight.

basically parasitizes other plants, using the mycorrhizal fungus as a bridge.

The nonplant organisms add a great deal of complexity to a plant's habitat, and numerous types of interrelationships are possible. Only those relationships involving plants have been mentioned here, but the animals are competing with each other, as are the fungi and prokaryotes. Also, interactions occur between animals, fungi, protists, and prokaryotes. The ecosystem is extremely complex, and it is virtually impossible to predict how disruption of one part might affect other parts. Although the operational habitat may be simple, the real habitat contains so many factors linked to so many other factors that we must be careful in our treatment of ecosystems.

The Structure of Populations

Populations can be thought of as having many types of structure; their distribution through the habitat is an important one, as is the age structure of the individuals.

Geographic Distribution

Boundaries of the Geographic Range The ability of a plant species to spread throughout a geographic area is a result of its adaptations to the abiotic and biotic components of that area. Although most habitat components act on the plant simultaneously and most should be considered important, at any given time and locality, one factor alone determines the health of the plant. This factor, whatever it may be, is the **limiting factor**. As described for photosynthesis, at a medium level of carbon dioxide, increasing the amount of light causes an increase in photosynthesis, whereas increasing the concentration of carbon dioxide does not; light is the limiting factor. As light intensity is increased, however, a point is reached at which brighter light does not cause more rapid photosynthesis. Then, an increase in the level of carbon dioxide does result in greater photosynthesis, and carbon dioxide becomes the limiting factor.

The concept of limiting factors applies to all aspects of a plant's interaction with its habitat. In areas of high rainfall, more water probably does not result in better plant growth, but plants growing in the shady part of that ecosystem might benefit from more light, whereas those growing in sunny spots might grow faster if more nitrogen compounds were available in the soil. Still other plants might not respond to extra light or nitrogen but would benefit from decreased herbivory. If each plant received extra amounts of the factor that had been limiting it, its growth would increase until some other factor became limiting. If the shaded plant received extra light, its growth rate would increase to the point where nitrogen perhaps became limiting. If a plant is placed in an optimal environment and given adequate amounts of nutrients, light, water, and freedom from pathogens, growth and reproduction increase greatly, but not infinitely; at some point, the plant's innate capacity becomes the limiting factor. Crops on irrigated farms with weed control and pesticides are an example (Figure 26-18).

Any factor of the ecosystem can act as a limiting factor. Water is important to many species; most cannot live in desert regions because of lack of water, and most cannot live in marshes because of excess water. Extreme temperature inhibits plant growth in

Figure 26-18   On modern American farms, plants are given optimal amounts of fertilizer and water; pesticides and insecticides keep pathogens under control, and a variety of weed-control measures eliminate competition from other species. If the seeds are not planted too close together, each plant has adequate room and grows as rapidly as possible. It is limited only by its own innate capacity for growth.

many regions; even if given adequate water, some plants cannot conduct it as rapidly as it would be transpired at high temperatures. For other species, high temperatures apparently cause enzyme systems to lose synchronization, and metabolism does not function correctly. A lack of warmth in winter is a limiting factor that keeps many species restricted to the tropics; temperate rain forests have adequate moisture and even richer soil than do the natural habitats of most tropical species, but the temperate rain forest habitat has freezing winter temperatures.

Biotic factors are also critical; many desert plants grow much more rapidly if given more water than occurs in their habitat, but their ranges do not extend into moist regions: They cannot compete well against the plants that are already there and well-adapted to moist conditions. Another important consideration is that plant species that rely on animals for pollination or seed dispersal cannot reproduce where their animal partners do not exist; the geographic range of these plants may be set by the limiting factors of the animals.

Soil factors often produce abrupt boundaries for the geographic ranges of populations. Both mineral composition and soil texture are important (Figure 26-19). Soils derived from limestone, sandstone, or serpentine often have characteristic species growing on them. Beaches with loose, porous, sandy soils have species distinct from those on nearby soils that are more compact and contain more humus. The limiting factor for a particular species may be the same factor over its entire geographic range, but often it varies from area to area.

Local Geographic Distribution In addition to large-scale geographic distribution of the entire population, small-scale, local distribution of individuals with respect to each other is also important. Individuals have one of three types of local distribution: random, clumped, or uniform. The term **random distribution** is used whenever there is no obvious, identifiable pattern to the position

Figure 26-19 (a) The distribution of these plants is easy to understand. Most of this granite outcrop is so smooth that all seeds are washed off by rain—nothing can grow on it. Where it has cracked, however, soil accumulates, seeds germinate, and plants thrive. (b) Soil in the lowest part of this valley floor is too wet for the trees, and rocky soil on the slopes is too dry.

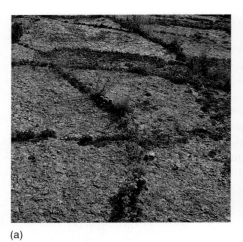

(a)

(b)

of individuals (**Figure 26-20a**). A random pattern has no predictive value; knowing the position of one plant does not let you estimate the position of another plant. In most habitats, many individuals seem to be distributed at random, but that may simply be due to the presence of many small-scale patterns that are difficult to detect or one large pattern that is too complex to see.

Clumped distributions are those in which the spacing between plants is either small or large, but rarely average (Figure 26-20b). This can result from many factors. Seeds of a plant often fall near the plant, not at uniform or random distances from it. If a bird or other animal eats many fruits and seeds, it will probably "deposit" them all together in a lump of organic fertilizer.

Uniform distributions are the types that occur in orchards and tree plantations; all individuals are evenly spaced from their neighbors (Figure 26-20c). In natural populations, uniform distributions are not extremely common; those that do occur are thought to result from intraspecies competition. The roots of one individual may establish a zone that prevents the germination or growth of others. Zones can also be established, at least theoretically, by the release from the plant of chemicals that inhibit other plants. Such chemicals are called **allelochemics** and the inhibition is **allelopathy**. One example may be the purple salvias of California (*Salvia leucophylla*); they grow in a relatively uniform spacing with a zone of bare soil surrounding each shrub (**Figure 26-21**). Several chemicals, particularly terpenes, are given off from these plants and have been shown to inhibit growth of plants in the laboratory; it is known also that these do accumulate in the soil near *Salvia*. It is not certain that these actually are allelochemics, however, because some experiments have shown that rabbits, mice,

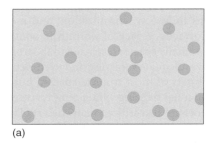

(a)

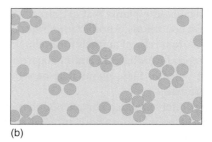

(b)

(c)

Figure 26-20 Types of distribution on a small scale. (a) This appears to be a random distribution, but examination of a larger area might have revealed a pattern. (b) Clumped distribution. (c) Uniform distribution.

Figure 26-21 Each bush of *Salvia leucophylla* is surrounded by a zone devoid of vegetation. Allelochemics given off by the salvias penetrate the top layers of soil and inhibit seed germination. If a seedling can get its root through the top zone, the plant may grow well, but herbivory by rabbits and mice that live in the protection of the salvias then becomes a problem.

and birds that live in the salvia shrubs are mostly responsible for the bare patches; if a wire cage is placed over a bare patch to exclude animals, plants do grow there. Presumably the bare patches are narrow because it is risky for the animals to venture too far from the cover of the salvia. It may be possible that terpenes are allelochemics under certain circumstances but that typically animal herbivory prevents other plants from growing near salvia.

Age Distribution: Demography

The manner in which a population responds to various factors in its habitat is affected partly by its **age distribution**, its **demography**—the relative proportions of young, middle-aged, and old individuals. Analysis of age distribution has been applied mostly to animal populations and may be difficult to apply to plants, but the fundamental aspects are important and easy to understand. Imagine a species in which a pair of individuals produces four offspring by the time they die; the four offspring also double their numbers, and thus, there are eight after they die. Future generations would contain 16, 32, 64, 128 individuals, and so on. This population is undergoing an exponential rate of increase (**Figure 26-22a**), and there are always greater numbers of young individuals than old ones. This is important in determining whether most members are very young and highly susceptible, moderately young and vigorous, older and well established, or very old and senescent.

Two factors affect the possible rate of population growth: generation time and intrinsic rate of natural increase. **Generation time**, the length of time from the birth of one individual until the birth of its first offspring, affects the rapidity of population growth: Annuals have a generation time of 1 year or less and can increase rapidly, whereas most conifers and angiosperm trees must be several years old before they produce their first seeds. For easier comparisons, we often measure population increase in terms of generations, not years.

The second factor, **intrinsic rate of natural increase** or **biotic potential**, is the number of offspring produced by an individual that actually live long enough to reproduce under ideal conditions. Even with optimal conditions, a large percentage of seeds do not germinate, and many seedlings die before they are old enough to reproduce; therefore, the biotic potential does not equal the number of seeds produced. For many species, biotic potential is a large number, represented in population equations as r. Plants that produce a large number of healthy, viable seeds over their lifetimes have a large r, a large biotic potential, and their populations can potentially increase greatly each generation. A species that produces fewer seeds than another species can reproduce faster than the second if it has a very short generation time, however. Mustard plants are small, live for only 1 year, and produce just a few seeds each, whereas oaks are large trees, each of which produces thousands of seeds in its lifetime. A mustard population, however, can grow more rapidly than can an oak population.

The biotic potential is measured under ideal conditions, but such conditions do not often occur in nature. Furthermore, after they do occur, even in a laboratory experiment, the very existence of the plants finally disrupts those ideal conditions. After the population becomes large, the plants must compete for water, nutrients, and space. The number of individuals in each population that can live in a particular ecosystem is limited; that number is the **carrying capacity** and it is symbolized by K. Theoretically, a population increases until the number of its individuals (N) becomes close to K; at that time, crowding and competition result in poorer growth, lower reproduction, and decreased chances that seedlings will be in suitable sites (Figure 26-22b). Birth rate (germination) decreases and death rate increases. These factors continue as the number of individuals continues to approach K; when N and K are equal, population growth stops because death rate equals birth rate.

Many factors cause death rate to increase and birth rate to decrease as population size approaches carrying capacity (N approaches K). A large, densely crowded population is an ideal target for herbivores; most of the progeny of one pair of insects are likely to find suitable plants wherever they go in a dense plant population, so most survive. The large plant population is ideal for the insects; therefore, the insects' population growth rate can increase toward their own biotic potential. The same is true of pathogenic fungi and bacteria. Even without considering predators or pathogens, large populations of plants may alter the environment physically, making it less ideal, as mentioned for the shading of pine seedlings by mature pine trees. A habitat filled with herbivores or pathogens or shade has a lower carrying capacity than the original habitat had.

On the basis of theory alone, we would expect that after a species invades a new habitat, it would undergo exponential growth as in Figure 26-22b, first increasing rapidly, later more slowly, and finally remaining stable with numbers that neither increase nor decrease. In reality, many factors prevent real populations from

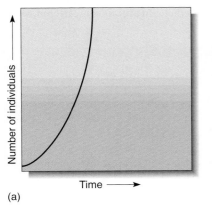

(a)

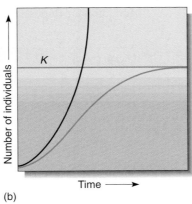

(b)

Figure 26-22 (a) Curve representing the number of individuals in a population that is growing exponentially. For this to occur indefinitely, growth must be controlled only by the organism's innate capacity for growth and the number of individuals present. In real situations, habitat limitations cause growth to be somewhat slower than the theoretical maximum. (b) Graph showing the more realistic situation in which the carrying capacity, K, is included. Instead of increasing infinitely (black curve), the growth rate (purple curve) decreases as the population size approaches the carrying capacity of the habitat. When population size equals carrying capacity, population increase stops; the death rate equals the birth (germination) rate.

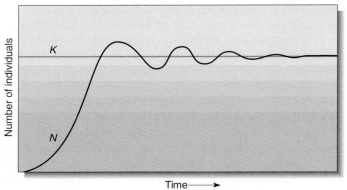

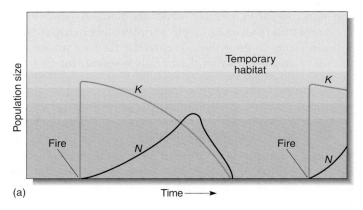

(a)

Figure 26-23 If a population increases slowly toward the carrying capacity of its environment, it may slow at *K*, as shown in Figure 26-22b. But often growth is too rapid and a large number of extra individuals temporarily survive; they then die off because of limited resources and the population falls below the carrying capacity.

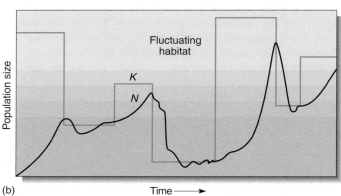

(b)

Figure 26-24 (a) Most habitats have no carrying capacity (*K* = 0) for species that are not a natural component of the habitat's ecosystem, but a disturbance such as a fire, flood, landslide, or disease outbreak may suddenly make it suitable for new species. Clearing an area for construction suddenly increases its carrying capacity for many weeds. They had been present in very low numbers because they compete poorly with the normal vegetation present before construction. The carrying capacity for the weed species soon begins to fall as other plants invade and the natural vegetation returns to areas not covered with asphalt or houses. (b) A site's carrying capacity may vary less drastically as the populations of competitors, pests, and pathogens rise and fall in their own cycles. If most of the herbivores are suddenly killed by a bacterial disease outbreak, the carrying capacity for the plants suddenly increases. This lasts only until new herbivores move in or some other pest responds to the increased number of uneaten plants.

acting like ideal ones. As the population approaches *K*, it may not slow sufficiently and may overshoot the carrying capacity; there would be too many individuals temporarily, followed by a die-off that drops the number far below the carrying capacity (**Figure 26-23**). With other species, as *N* approaches *K*, an explosive increase in the populations of many types of pests may occur, and the numbers of aphids, mites, caterpillars, and fungi may increase. The plant's pests have overshot their own *K*, and they kill so many plants that the plant population falls far below its *K*, perhaps almost to zero. From there, it may then increase again, undergoing major cycles. Alternatively, as with pines, as the population increases, they change the environment to one that favors other species; their own reproduction falls to zero, and as the adults die, the entire population is lost.

■ *r*- and *K*-Selection

As a population increases, theoretically it goes through a young phase in which numbers of individuals are low and resources are plentiful. Population growth is limited by the species' own biotic potential, *r*. Later, conditions are crowded, resources are scarcer, and population growth is governed by the carrying capacity of the ecosystem, *K*. The lifetime of any single individual is typically much shorter than the time required for a population to pass through this full development. Therefore, which is more advantageous to the species, to become adapted to *r* conditions or to *K* conditions? The two are very different and require distinct, often mutually exclusive adaptations.

r-Selection A disturbance usually produces *r* conditions. A fire or flood destroys many individuals in the area, and resources are plentiful for the few that remain, whether they are seeds, survivors, or immigrants carried in by wind, animals, or the flood itself. Pioneers that produce many seeds quickly have an advantage in that most seeds find suitable sites; if seeds form too slowly, the sites may be filled by seeds of other invaders (**Figure 26-24**). Because population density is low, the spread of predators and pathogens is slow, and thus, the threat from them is not great; having antiherbivore and antifungal defenses is not so important

because neither of those two agents is the limiting factor. Actually, the biotic potential is the limiting factor, and as a result, mutations that increase *r* are selectively advantageous (**Table 26-1**). Plants grow quickly, have few defensive compounds or structures, flower quickly, and produce many small seeds. Because most disturbances are impossible to predict and are widely scattered, seeds

TABLE 26-1	Selectively Advantageous Traits in *r*-Selected and *K*-Selected Species
Adaptive in *r*-Sites	Adaptive in *K*-Sites
annual	perennial
early maturity	late maturity
many small seeds	fewer, larger seeds
few mechanical or chemical defenses	many defenses

Figure 26-25 Most plants of disturbed habitats, such as these coastal dune plants, are *r*-selected species: Their wind-blown seeds land, germinate and grow quickly, and then produce many more seeds. They can grow so rapidly and reproduce so abundantly because almost no energy or mineral resources are spent on antiherbivore defenses or drought adaptations. Because of periodic storms, freshly exposed, highly disturbed sand dunes are always ready for invasion.

also must be adapted for widespread dispersal. *r*-Selected species typically are annuals or small shrubby perennials because the disturbed habitat gradually changes back into a crowded one that is no longer suitable for the pioneer *r* species. As more species of plants, animals, and fungi re-establish themselves in the area, the *r*-selected species are at a disadvantage; they have few defenses against predators and are too short to compete for sunlight. Only another disturbance can save them at this site; usually their population numbers fall to zero or close to it. The species itself survives because many seeds have emigrated to other sites, at least a few of which are appropriately disturbed areas (**Figure 26-25**).

Some types of disturbances are predictable: killing temperatures in temperate winters and lethal hot/dry conditions in desert summers. *r*-Selected species are ideally suited for these environments. In spring, the habitat becomes suitable and may be almost devoid of plants. Small annuals grow and reproduce quickly, and population growth is extremely rapid until the habitat becomes disturbed by winter or summer climate.

K-Selection Conditions in a crowded habitat, where a population is close to its carrying capacity, select for phenotypes very different from those that are beneficial in a disturbed habitat (Table 26-1). In a disturbed region, virtually every spot is a suitable site for seed germination and growth, but in a *K* habitat, almost every possible site is filled. After an individual dies, its site may become occupied by the seeds or rhizomes of a different species. It is advantageous to live for a long time, holding on to a site. To survive as a long-lived perennial is difficult, however, and large amounts of carbon and energy must be diverted into antipredator defenses. These resources are then not available for growth or reproduction, both of which are much slower than in an *r*-selected species. Many long-lived conifers such as redwoods, Douglas firs, and bristle-cone pines are good examples of *K*-selected species. *K*-selected species also face intense competition from other plant species, and therefore, adaptations that increase the ability to use scarce resources are beneficial. Examples of such adaptation may

be the capacity to use low amounts of light or soil strata that are poor in nutrients.

Species that are *r*-selected can occur next to ones that are *K*-selected. Avalanches in dense forests open up small sites suitable for *r*-selected species. The same is true for hurricanes, fires, and floods. The floor of a deciduous forest is temporarily an *r* site during springtime, between the time when temperatures become warm enough for germination and growth and the time when the canopy trees put out their new leaves. For several weeks, the forest floor is sunny, warm, rich in nutrients, and temporarily uncrowded. Small *r*-selected plants that grow and reproduce quickly can complete their life cycles before the larger trees come out of dormancy and block the sunlight.

The Structure of Ecosystems

Many concepts can be considered in the structure of an ecosystem. Different botanists think about ecosystems in ways that are influenced by their own interests. Four of the most commonly mentioned structures are the physiognomic structure, temporal structure, species diversity, and trophic levels.

Physiognomic Structure

The physical size and shape of the organisms and their distribution in relation to each other and to the physical environment constitute the **physiognomic structure**. Trees, shrubs, and herbs are the three most useful categories, but in addition, a system of **life forms** was defined by C. Raunkiaer in 1934 (**Table 26-2**). The criterion for classification was the means by which the plant survives stressful seasons, such as by placing buds below ground (geophytes: bulbs, rhizomes) or winterizing aerial buds (phanerophytes: trees, vines). Regions of the world that have similar climatic conditions have similar physiognomic structures unless the soil is particularly unusual (**Figure 26-26**).

An almost infinite number of combinations of life forms, vertical structures, and characteristics such as abundance of broadleaf plants, conifers, and sclerophyllous plants are possible (**Figure 26-27**). An almost infinite number of types of ecosystems might be expected but, actually, only a few basic types exist. Although differences exist between various types of forests or grasslands or marshes, each category is easily recognized as a common type of ecosystem.

TABLE 26-2	Life Forms of Raunkiaer
Life Form	Means of Surviving Stress
Therophytes	Annual life span; survive stress as seeds.
Geophytes	Buds are underground on rhizomes, bulbs, corms.
Hemicryptophytes	Buds are located at surface of soil, protected by leaf and stem bases: many grasses and rosette plants.
Chamaephytes	Buds are located above ground, but low enough to not be exposed to strong winds: small shrubs.
Phanerophytes	Buds are located high, on shoots at least 25 to 30 cm above ground: trees and large shrubs.

(a)

(b)

Figure 26-26 (a) This desert scrub vegetation near Tucson, Arizona, with small trees, large bushes, and saguaro cactus is characteristic of much of the southwestern United States and northern Mexico, where summers are hot with occasional rainstorms and winters are cool and moist but freezes are not severe. (b) The northern part of the central valley in Chile, just north of Santiago, has climatic conditions similar to those in the southwestern United States, and the vegetation of the two regions resembles each other in habit (life forms), distribution, and other features. Only by looking carefully does one notice that the species are very different.

■ Temporal Structure

The changes that an ecosystem undergoes with time constitute its **temporal structure**; the time span can be as short as a day or can encompass seasons or decades. For animals, a daily cycle can be especially obvious, with some animals active at night (nocturnal) and others during the day (diurnal). Many plants also have daily rhythms of flower opening and closing.

Plants change dramatically with the season, as do other organisms. Spring is typically a time of renewed activity, produc-

tion of flowers and new leaves. This is not simultaneous for all species; often the understory plants become active earliest, benefiting from the open canopy. Wind-pollinated flowers are usually produced before leaves expand and block the wind. Leafing out and flowering must be coordinated not only with the end of low-temperature stress conditions but also with the habits of pollinators and herbivores. It does a plant no good to produce flowers when temperatures are still too low for its pollinators. Through late spring and early summer, various species flower at distinct

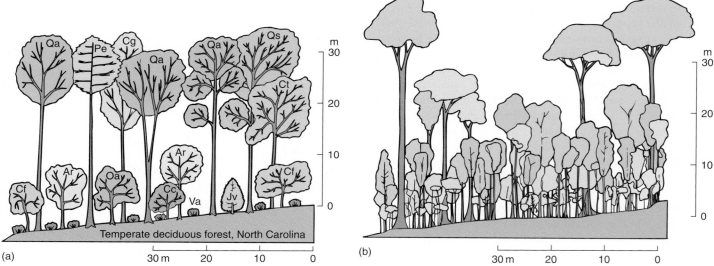

(a) Temperate deciduous forest, North Carolina

(b)

Figure 26-27 The presence of trees, shrubs, and herbs (phanerophytes, chamaephytes, and therophytes) gives an ecosystem vertical structure. The trees form the canopy. The shrubs, short trees, and saplings of tall trees form a middle level understory or subcanopy. The herbs and seedlings are the ground-level plants. Vines can occur in all three levels. (a) A temperate forest. Ar, *Acer rubrum* (red maple); Cc, *Cercis canadensis* (redbud); Cf, *Cornus florida* (dogwood); Cg, *Carya glabra* (pignut hickory); Ct, *Carya tomentosa* (mockernut hickory); Jv, *Juniperus virginiana* (red cedar); Oa, *Oxydendrum arboreum* (sourwood); Pe, *Pinus echinata* (shortleaf pine); Qa, *Quercus alba* (white oak); Qs, *Quercus stellata* (post oak); and Va, *Viburnum affine* (arrow-wood). (b) A tropical rain forest has a much more complex structure.

times, controlled by plant maturity, photoperiod, or adequate rainfall. In most ecosystems, there is no time during the summer when nothing is in bloom. The continual presence of some species in flower provides nectar and pollen for insects throughout the season.

Species that bloom early in spring probably form fruits and mature seeds during summer; species that flower later release their seeds in late summer or autumn. After seeds are dispersed, some remain dormant until the following spring. Others germinate and grow into a low rosette that survives the winter, even growing slightly on warm winter days. When spring arrives, the seedling is already well rooted and can begin to grow quickly while the seeds of competitors are just starting to germinate.

Late summer and autumn bring changes that depend on the ecosystem; in the northern United States, herbs die while shrubs and trees develop resting buds. The entire plant enters light dormancy; then the first cool days initiate deep dormancy. Leaves and fruits are abscised, removing the last sources of food for most animals. In the southern and southwestern United States, cooler autumn weather is often more welcome than the first warm days of spring because the summer is so much more severe than the winter. Most gardening is done in autumn rather than spring, and fall wildflowers are abundant and dramatic. The growing season extends at least to December for shrubs and many herbs, and the small rosette plants may never become truly dormant.

In tropical ecosystems, winter and summer do not exist, but an alternation of dry and wet seasons governs ecosystem change. Coastal marsh, wetland, and reef ecosystems may be strongly affected by seasonality; rainy seasons dilute the salt water, whereas in dry seasons rivers deliver less freshwater and mineral-rich silt.

Over long periods—many years—most ecosystems undergo gradual, often dramatic changes. This process of succession is discussed in the next chapter.

Species Composition

Species composition refers to the number and diversity of species that coexist in an ecosystem, and it depends on whether the climate is mild or stressful, the soil is rich or poor, and the species' tolerance ranges are broad or narrow (Figures 26-28 and 26-29). Stressful climates with poor soils support a low number of species because so few species are adapted to such conditions (Figure 26-30). On the other hand, mild climates and rich soils support an abundance of species because most plants have tolerance ranges that include such climatic and soil conditions (Figure 26-31). Competition is intense, but apparently natural selection has resulted in habitat partitioning, with each species occupying a narrow portion of the various resource gradients

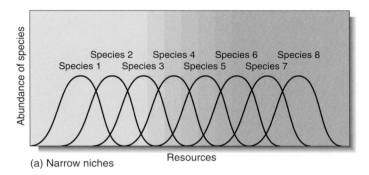

(a) Narrow niches

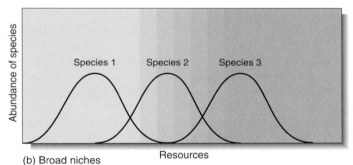

(b) Broad niches

Figure 26-29 Two ecosystems with similar ranges of resources can differ in the number of species they contain. One may be occupied by a large number of very specialized species, each adapted to only a narrow range of the resource (a), whereas the other may be occupied by a few species of generalists that grow well under a variety of conditions and exclude most competitors (b).

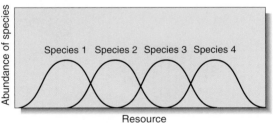

(a) Narrow range of resource

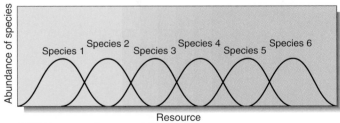

(b) Wide range of resource

Figure 26-28 In the ecosystem represented by (a), the range of the resource available is narrow compared with that of the ecosystem in (b); in general, it is able to support fewer species. For example, the resource may be water, with the ecosystem in (b) having a variety of areas that range from dry to moist to lakes or streams, whereas the ecosystem in (a) is just marshy or just desert.

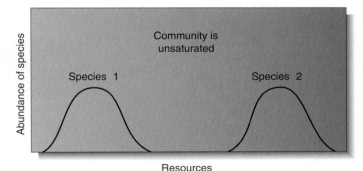

Figure 26-30 Adapting to certain harsh conditions is difficult, and only a few species may have succeeded. Consequently, harsh habitats may have many unused resources.

Figure 26-31 This climate is nonstressful, the soil is rich, and no toxins are present. Many species live here, and many more could if protected from competition. Species diversity here is high.

(Figure 26-29). The presence of a large number of species actually creates more niches that can be filled by new species; the presence of trees makes it possible for epiphytes and parasites to occur in the ecosystem.

Trophic Levels

Trophic levels are basically feeding levels. Each ecosystem contains some members, autotrophs, that bring energy into the system. Photosynthesis is by far the dominant method, accounting for virtually all energy input. Green vascular plants are most important, but algae and cyanobacteria carry out about one third of all photosynthesis worldwide, and lichens and nonvascular plants are important in cold, high latitudes and altitudes. Chemosynthetic bacteria bring chemical energy into ecosystems, and although this is minor now, it was the only method before photosynthesis evolved.

Autotrophs are known as **primary producers** for obvious reasons, and they are the first step of any food web. They are the energy and nutrient supply (food) for the herbivores, which constitute the **primary consumers** (sometimes called secondary producers). Herbivores are preyed on by carnivores, the **secondary consumers**. Omnivores exist at both trophic levels. **Decomposers** such as fungi and bacteria break down the remains of all types of organisms, even those of other decomposers (**Figure 26-32**).

As plants photosynthesize, energy and carbon compounds enter the ecosystem. As plants are eaten, their energy and carbon compounds move to the herbivore trophic level, then to the carnivore trophic level, and finally on to the decomposers. This is referred to as the **energy flow** and the **carbon flow** of the ecosystem. At each step, much of the food is used in respiration, resulting in the production of ATP, heat, carbon dioxide, and water. The carbon dioxide is released back to the atmosphere, where it can be used in photosynthesis again; the energy temporarily contributes to warming the planet, but it is ultimately radiated into space. The portion of the food that is not respired is available for growth and reproduction. As a very rough approximation, approximately 90% of an animal's food is respired, and 10% is retained as growth or gametes; therefore, when herbivores eat 1000 kg of grass, 900 kg are respired, and 100 kg become herbivore tissues. Similarly, a carnivore that eats the 100 kg of herbivore would only gain 10 kg and after dying would support just 1 kg of decomposers.

Along with the flow of energy and carbon, minerals flow through the ecosystem. Plants absorb minerals from soil and incorporate them into their bodies as amino acids, ATP, coenzymes, and so on, and these are consumed by herbivores. Many of these compounds are broken down by animals, and the minerals are lost as waste. After organisms die, the minerals that remain in

Figure 26-32 Energy and materials flow through the ecosystem as a network rather than in a straight line. Each level contributes to the decomposers (they even decompose other decomposers), and omnivores eat both plants and other animals. The primary producers never draw energy from the other stages, but many depend on animals and decomposers to return essential elements to the soil. At each step, respiration returns most of the carbon back to the air, and most of the energy is liberated as heat.

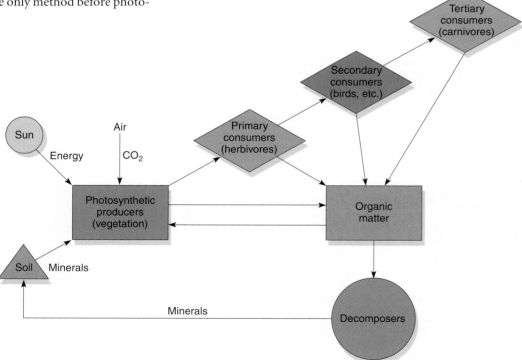

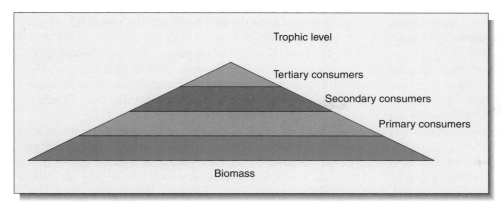

Figure 26-33 A pyramid such as this can represent the number of organisms at each level, the weight of all the individuals at each level, the respiration at each level, and so on. Such analysis can be difficult to do, but these relationships are extremely important. If a national park is established to preserve a species of wolf or mountain lion, it must be known how extensive its food web is and how large of an area must be set aside to ensure an adequate number of primary consumers.

their bodies are released by decomposers and become part of the soil again. In addition, minerals can be completely lost from an ecosystem as they are carried away by rainwater and streams, especially if erosion is occurring or as cities dump mineral-rich human waste into rivers. Such minerals are carried to the oceans and become part of marine ecosystems. Many algae, protists, and marine animals have bodies with heavy, mineral-rich shells and bones; as those organisms die, their bodies sink to the depths of the ocean and are unavailable to living organisms of the land or ocean surface.

The movement of energy and biomass from one trophic level to the next is often represented as a pyramid of energy or a pyramid of biomass (**Figure 26-33**). Such pyramids are useful for illustrating these principles in textbooks, but in reality, it is extremely difficult to measure all of the biomass of the primary producers of even a simple ecosystem. It is almost impossible to measure it accurately in a forest ecosystem, and it is impossible to measure the biomass or energy content of the fungi and bacteria that are acting as decomposers. Pyramids of numbers are more easily obtained; often there are fewer individuals in each higher trophic level, but of course, millions of aphids can live on one tree.

Pyramids of biomass and numbers have become important as a means of illustrating the impact of the human introduction of herbicides, pesticides, and toxic wastes. Although some of these are biodegradable, many are not, and they persist unchanged and toxic in the environment. Imagine a toxic herbicide applied at the low rate of one part per million—1 g per 1,000 kg of pasture grass.

Herbivores will eat the 1,000 kg of plants, and the pyramid of biomass shows that they will respire 900 kg away; however, the 1 g of herbicide is not oxidized or excreted. Instead, it accumulates in the animals, now at a concentration of 1 gm per 100 kg. This is 10 times stronger than the application rate. As carnivores eat the herbivores, the concentration rises to 1 g per 10 kg, now 100 times stronger. Human metabolism is no different; these chemicals move through the food web to us as well. Toxic materials are not distributed uniformly within the bodies of animals and humans but become concentrated in our livers and kidneys as well as in fatty tissues such as our brain and spinal cord.

The role of decomposers in the ecosystem is vitally important; decomposers are so ubiquitous as well as microscopic that we take them for granted. If there were no decomposers in nature, plants could grow only as rapidly as weathering breaks down rock; as soon as minerals were released, they would be absorbed and locked into a plant and would then move through the food web. When the organisms died, the minerals would remain locked in the nondecomposing body. Sterilized (canned) food does not decay even though moist and kept at room temperature because no decomposers are present. Only by the physical and chemical weathering of rock and dead bodies could plant growth continue, but with decomposers present, minerals are recycled rapidly; decomposers release the minerals to the environment through their membranes, and their own bodies are so small and delicate that after death they are broken down immediately by other decomposers or by inorganic chemical weathering.

Summary

1. Ecology is the study of organisms in relationship to all aspects of their surroundings. Each plant is part of a population of individuals of the same species, and these co-exist with other species, forming a community. When climate and physical surroundings are also considered, the whole is an ecosystem.

2. The biology of a population is not simply the sum of all the individuals because plants interact positively and negatively with each other.

3. Habitat components may be biotic or abiotic. Abiotic factors include climate, soil, latitude, altitude, and disturbances. Biotic factors include all living organisms and the results of their activities.

4. Disturbances are disruptions of portions of an ecosystem; although they may cause great destruction, they may also be necessary for maintaining the ecosystem such that there is not a succession, a long-term change in species present.

5. By existing in an ecosystem, a plant affects the ecosystem; the plant may produce changes that are detrimental for the survival of itself or its progeny, or it may alter the ecosystem such that it is more suitable for its progeny.

6. The amount of competition varies from species to species and from one ecosystem to another. In general, where climates are mild and soils rich, species diversity is great, and competition is most important in excluding species from

the ecosystem. In harsher conditions, fewer species exist, and the conditions themselves limit the number of species that exist successfully.

7. With a large geographic range, a species may face different types of competition or stress in different parts of its range.

8. Plants interact with animals, fungi, protists, and prokaryotes in numerous ways, but most interactions basically can be divided into mutualistic (beneficial to both) or predatory (detrimental to one, beneficial to the other).

9. Populations have at least two basic types of structure: (1) their geographic range, which is set by limiting factors, and their local distribution (random, clumped, or uni-

form), and (2) the distribution of the ages of all the individuals (demography).

10. The biotic potential of a species (r) is a measure of its reproductive success: the mean number of offspring per parent that survive to reproductive maturity. An r-selected species is one that is adapted to sites temporarily free of competitors and rich in resources.

11. The carrying capacity (K) of an ecosystem is a measure of the number of individuals of a particular species that can be sustained by the ecosystem. A K-selected species is one that is adapted to stable, relatively unchanging sites where competition may be very important.

Important Terms

biotic potential
carrying capacity
commensal relationships
community
competition
demography
ecosystem
ecotypes

habitat
herbivory
K-selection
life forms
limiting factor
mutualism
niche
pathogenic

pioneers
population
predation
primary consumers
primary producers
r-selection
tolerance range
trophic levels

Review Questions

1. Ecology is the study of organisms in relation to _____ aspects of their surroundings.

2. Define each of the following: population, community, habitat, and ecosystem. Why is the biology of a population different from the biology of an individual?

3. How much of a plant's environment should be considered its habitat? Would the habitat of a wind-pollinated, wind-dispersed species be simpler than one with animal-mediated pollination and seed dispersal? Why?

4. Consider the redwood forests of California. We know that the shade of the big trees affects the shrubs and herbs, but do the small plants affect the trees? What about the seeds of the redwood trees? What role did fire play? If a new disease were to kill off all the shrubs and herbs but not directly harm the trees, would the trees ultimately be impacted by the disease?

5. Habitat components are of two types, abiotic and biotic. List several components of each type.

6. Why are climate extremes often more important than average climatic conditions? What is the "growing season" of a plant, and what are some conditions that affect its length?

7. How are soils formed? Through the breakdown of what material? Why are new soils not suitable for most plant species?

8. What is a pioneer species? Are they more likely to be r- or K-selected? Why?

9. Why do parts of Alaska and Canada have such thin soil today? What happened to the soil that used to be there? Approximately how many years ago did this region lose its soil?

10. Fire is a critically important, natural disturbance in many ecosystems. Describe some plant adaptations that allow plants to survive fires.

11. If we could eliminate all other organisms so that a certain region contained only the soil and the plant itself, would this be a stable habitat? Does the plant itself change its habitat? Explain.

12. When several individuals occur together, the possibility for interaction is created. If the interaction is basically beneficial for both organisms, it is called _____, but if it is disadvantageous, it is _____. Do the two organisms have to be the same species or different species, or does it matter?

13. What is competition between plants? What is competitive exclusion? How does competition affect natural selection and evolution?

14. What is a niche? Is it a physical, tangible space or object? If a giant redwood grows side by side with a small fern, are they in the same niche? If not, how does the niche of one differ from that of the other?

15. What is a transplant experiment? What is a common garden? What is the point of using these two types of experiment?

16. When birds build nests in trees, the birds benefit, but the trees are usually unharmed. What kind of a relationship is this? Give two more examples of relationships like this.

17. Although most habitat components act on a plant simultaneously, at any given time and locality, one factor alone determines the health of the plant. This is called the _____ factor.

18. Plants often grow in clumped distributions. Give one cause of this.

19. What is the term that describes the length of time from the birth of one individual until the birth of its first offspring? Why do we not use this term for the time between the individual's birth and its own death?

20. Describe biotic potential. Describe carrying capacity. How do these two factors affect population growth and the ultimate stable size of a population?

21. Consider the population growth curve of Figure 26-22b. Would a young population (left side of curve) have a greater proportion of young individuals than an old population (right side of curve where it has reached K)? Why? Does it matter whether a population has a greater proportion of young, middle-aged, or old individuals? The answer is obvious for humans, but does it matter for plants?

22. Autotrophs are known as _____ _____, and they are the first step of any food web. Herbivores are _____ _____, and carnivores are _____ _____. Fungi and prokaryotes break down the remains of all types of organisms and are known as _____.

23. As plants photosynthesize, _____ and _____ compounds enter the ecosystem. How do these move to the next trophic level and then to the next? How are both lost from the ecosystem?

24. Look at the pyramid of biomass in Figure 26-33 and the text on the same page. If a toxic herbicide is applied at a low rate of one part per million, what happens to the concentration of the herbicide at each higher level of the pyramid? Does this affect our food? Does it affect us?

BotanyLinks

http://biology.jbpub.com/botany/5e

BotanyLinks contains a variety of resources and review material designed to assist in your study of botany.

chapter 27

Biomes

Concepts

Earth's land surface is covered almost entirely by **biomes**, extensive groupings of many ecosystems characterized by the distinctive aspects of the dominant plants. For example, some of the biomes of North America are the temperate deciduous forests, subalpine and montane coniferous forests, grasslands, and deserts. Plant life is absent only in the harshest deserts (the Atacama in Chile and Peru, the Sahara in Africa, the Gobi in China) and in the land regions covered permanently by ice (most of Antarctica and the tops of high mountains; **Figure 27-1**). In all other areas, the rock and soil are at least temporarily free of ice and have some liquid water; primary producers—plants, algae, and cyanobacteria—carry out photosynthesis and support food webs of consumers and decomposers.

Chapter Opener Image: The Chugach Mountains near Anchorage, Alaska, are located so far north that they receive little heat from sunlight even in the short summer, but they are very cold during the long winter. These conditions create two polar biomes, the tundra biome farther north and boreal (northern) coniferous forest biome seen here below snowline on these mountains. It was not always this way. About 55 million years ago, there was so much more carbon dioxide in the atmosphere that Earth was very warm and this region was inhabited by tropical forests, similar to what we see in the Amazon today (except with different species—the modern species had not yet evolved). Then about 30 million years ago the Earth began to cool and an ice age began: we are still in that ice age. There are warm periods (called interglacials) and cold periods (glacials) in an ice age. During the glacials, snow and ice were so deep that these mountains tops would not have been visible. At present we are in an interglacial, so it is warm enough for these trees and other plants to live here, establishing this biome.

Figure 27-1 Antarctica is one of the few places on Earth where plant life simply cannot exist; in the area shown, even if plants were introduced, they would die.

Biomes vary from extremely simple, as in **tundra**, to more complex grasslands, temperate forests, and tropical rain forests. Biome complexity and physiognomy are most strongly influenced by two abiotic factors: climate and soil. A particular type of biome, such as grassland or temperate deciduous forest, may occur in various regions of Earth because the same set of climatic and soil factors occurs in various regions. At all sites, the physiognomy—the appearance—of a biome is similar, but often the actual species present differ considerably from one area to another. For example, temperate grasslands are easily recognizable in the central plains of the United States, the steppes of Russia, the pampas of Argentina, and the veldt of Africa (**Figure 27-2**); all are dominated by grasses and large mammals and are devoid of trees except along rivers. Species that become adapted to the grassland niche of one continent do not occur on other continents because no birds, mammals, winds, or other means carry seeds such long distances. Despite the lack of gene flow between widely separated units of a biome, the physiognomic similarity persists as a result of convergent and parallel evolution: Climate, soil, and other habitat factors in each area select for similar phenotypes. If portions of a biome are not too widely separated, they will actually have some species in common.

The geographic locations of the Earth's biomes are determined by many factors, but two of overriding importance are (1) world climate and (2) positions of the continents. If either or both of these factors were different, the locations of the biomes would be different. It is important to realize that both do change: Climate undergoes cycles of cooling and warming, and continents

(a)

(b)

Figure 27-2 Under the proper conditions of climate and soil, grasslands almost invariably develop, as opposed to a rain forest or desert. Wherever those conditions exist, a specific type of vegetation can be expected. Even though the actual species present in one differ from those in the other, the biomes are still recognizable as grasslands because of convergent evolution due to similar environmental conditions. Grasslands in (a) Masai Mara, Kenya Africa; (b) Gillette, Wyoming; and (c) Argentina.

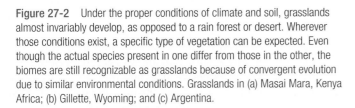

(c)

shift position because of **continental drift**. In the past, conditions on Earth's surface were quite different and so were the distributions of plants and animals. Similarly, the conditions will be altered in the future, and organisms will be affected by that as well.

World Climate

Earth's climatic conditions are the result of its tilted axis of rotation and the presence of the atmosphere and oceans.

Effects of Earth's Tilt

Imagine a planet whose axis of rotation is exactly perpendicular to the plane of its orbit. At the equator, the sun would rise exactly in the east every day of the year; it would pass directly overhead and set in the west. There would be no change, no seasons. At all sites not on the equator, 70 degrees N for instance, the sun would never be overhead; every day it would rise in the southeast, pass low in the sky, and set in the southwest. Maximum heating would be at the equator; all other regions would always receive only oblique lighting and would be much cooler. If there were no atmosphere or oceans, heat could not be transferred from the equator to the poles, and there would be a tremendous temperature gradient between those regions.

Earth's axis of rotation is tilted 23.5 degrees away from perpendicular to the orbital plane. At summer solstice, June 21 or 22, the North Pole points as directly toward the sun as possible, and in the Northern Hemisphere, the sun has reached its highest point in the sky. The sun appears to be overhead at noon for those people located at 23.5 degrees N latitude, the Tropic of Cancer, which runs just south of California, Texas, and Florida. The days are their longest, and summer has officially begun. North of the Arctic Circle, the sun is visible even at midnight.

As Earth continues its orbit, the axis of rotation points less toward the sun, which appears to rise and set more to the south and is lower in the southern sky at noon. By September 23, autumnal equinox, Earth has made one-quarter orbit. The sun is directly over the equator, and days are exactly 12 hours long: Autumn begins. After another 3 months, winter solstice, December 21 or 22, the South Pole points as directly as possible toward the sun; the sun is directly over the Tropic of Capricorn, 23.5 degrees S latitude. Summer begins in the Southern Hemisphere, but in the north, winter begins and days are at their shortest; above the Arctic Circle, the sun never rises. With another one-quarter orbit, the sun moves back toward the equator, arriving there at March 21, the vernal equinox.

The most intense solar heating moves seasonally northward, then southward. All parts of the planet experience seasonality, although in the region between 23.5 degrees N and 23.5 degrees S, the tropics, not as much change occurs between summer and winter as in the temperate regions outside this band.

Atmospheric Distribution of Heat

The atmosphere and oceans, being fluids, develop convection currents and massive flows when heated in one area and cooled in another. They distribute heat from the tropics to the temperate regions and all the way to the poles.

Much of the tropical zone is occupied by the Pacific, Atlantic, and Indian Oceans (**Figure 27-3**). Throughout the year, the water and air receive solar heat, causing tremendous amounts of evaporation. The air warms, expands, then rises high into the atmosphere. As it flows upward, the surrounding air pressure decreases, and the moist rising air expands even more. This expansion causes the air to cool, decreasing its ability to hold moisture. Water vapor condenses into rain and falls back to the surface in torrential storms, producing tropical rain forests in Central America, northern South America, central Africa, and Southeast Asia.

After rising, the air is pushed northward and southward by the continued rising of more tropical air below it (**Figure 27-4**). While spreading at high altitude, air radiates heat to space, cooling even more. By the time it reaches about 30 degrees N or S latitude (the horse latitudes), it has cooled, contracted, and become dense enough to sink. It encounters greater atmospheric pressure at lower altitudes and is compressed and heated. As it warms, its water-holding capacity increases. Because its capacity to hold water is greater but its water content is the same, the air becomes drier. But it had already lost much of its moisture while initially rising, so it is now extremely dry. Land areas below this descending air contain the world's hot, dry desert biomes (**Figure 27-3**).

Once back at Earth's surface, part of the air spreads toward the equator and part flows toward the poles. Earth's rotation causes the air moving toward the equator to be deflected westward. It moves as a northeast trade wind in the Northern Hemisphere and a southeast trade wind in the Southern Hemisphere. Air spreading toward the poles from the horse latitudes is deflected eastward and blows as a prevailing westerly.

The actual area of descending dry air varies with the season, being farther north when the sun is near the Tropic of Cancer (Northern Hemisphere summer) and farther south when the sun is near the Tropic of Capricorn. The United States, during its winter, is located entirely within the influence of the prevailing westerlies; winter weather comes from the Pacific and Arctic Oceans, moving eastward across the continent. During summer in the United States, the northeast trade winds have moved northward far enough to influence the states along the Gulf of Mexico; northeast trade winds bring summer storms westward from the Atlantic onto the east coast and gulf, supplying summer rains.

Continental Climate The size of a land mass influences the weather it receives. Larger islands such as the Hawaiian Islands, Guam, and Puerto Rico have mountains that force air to rise as it blows across them. Rising air cools, and rain forms. Low-lying, small islands such as the Florida Keys and the smallest islands in the Bahamas are too flat to affect the air; these **desert islands** are extremely dry, often with no fresh water streams or lakes (**Figure 27-5**).

Continents also cause air to rise, cool, and drop precipitation. If the topography is fairly flat, air rises gradually, and rains are distributed over an extensive area. Summer storms from the Atlantic and Gulf of Mexico move through the eastern United States, the Mississippi Valley, and the plains states; the land rises so gradually that rainfall covers half a continent.

If topography is mountainous, as on the United States' west coast, rain is dropped in a narrow area. The prevailing westerlies bring moist air from the Pacific Ocean onto land over California,

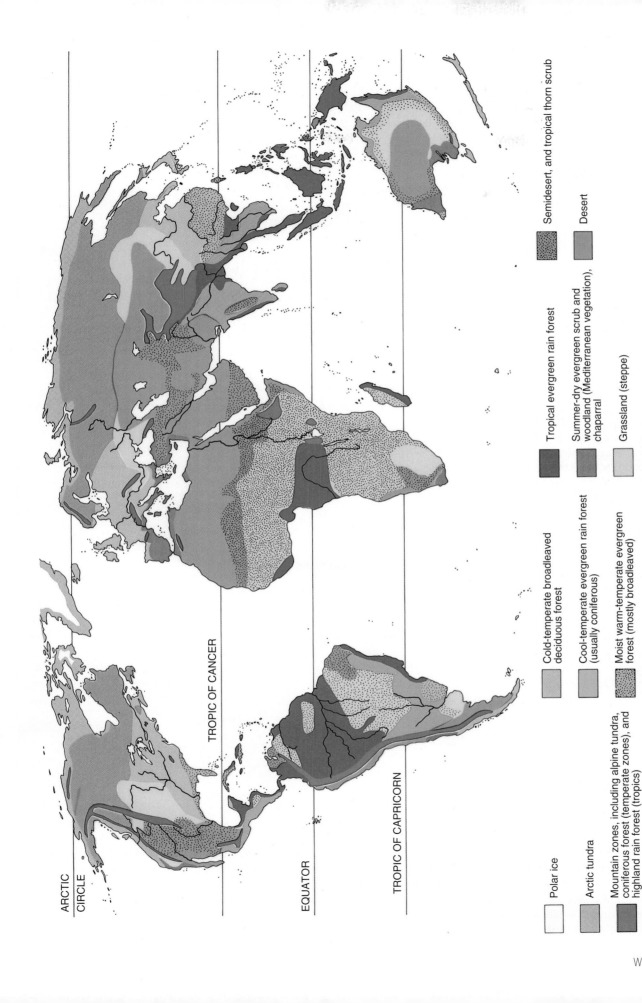

Figure 27-3 Most land on Earth is distributed rather far north at present, and the tropical zone, between the Tropics of Cancer and Capricorn, is mostly water. The greatest amount of solar energy falls on water, causing great evaporation and humidifying the atmosphere but leading to relatively little temperature change. If more land were in the tropics, the soil would heat more, which in turn would heat the air, but the oceans would stay cool and evaporate little moisture to the atmosphere; all regions would receive much less rain.

Legend:

- Polar ice
- Arctic tundra
- Mountain zones, including alpine tundra, coniferous forest (temperate zones), and highland rain forest (tropics)
- Taiga (coniferous forest belt)
- Cold-temperate broadleaved deciduous forest
- Cool-temperate evergreen rain forest (usually coniferous)
- Moist warm-temperate evergreen forest (mostly broadleaved)
- Tropical seasonal (monsoon) forest (usually at least partly dry-deciduous)
- Tropical evergreen rain forest
- Summer-dry evergreen scrub and woodland (Mediterranean vegetation), chaparral
- Grassland (steppe)
- Savanna, tropical woodlands, and thorn forests
- Semidesert, and tropical thorn scrub
- Desert

ARCTIC CIRCLE

TROPIC OF CANCER

EQUATOR

TROPIC OF CAPRICORN

Figure 27-4 Air rises above the equator and then spreads northward and southward at high altitudes. By the time it has reached the tropics, it is cool enough to contract, become dense, and sink. At the surface, some flows back toward the equator as trade winds, and some continues toward the poles as prevailing westerlies.

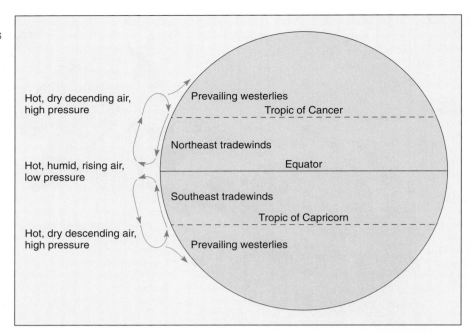

Figure 27-5 (a) When a mountain or other large land mass forces air to rise, the air cools below its dew point and drops rain or snow. On the lee side, descent of the air compresses and heats it, raising its ability to hold water. Rather than bringing rain to the area, it may actually dry out the soil. (b) Air blows undisturbed across low islands; therefore, they receive little rain even though the air may be extremely humid.

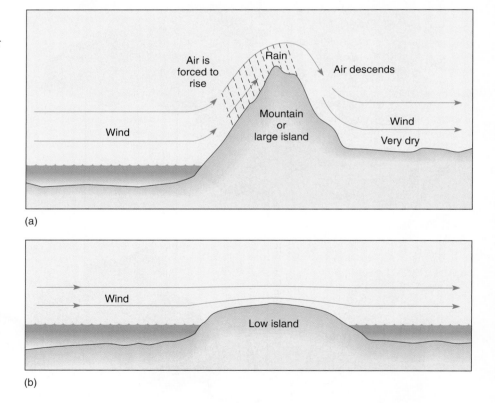

(a)

(b)

Oregon, Washington, Mexico, and western Canada. Air is immediately forced to rise by coastal mountain ranges; strong rains fall on the western slopes. After the air crosses the summit, however, it descends and warms, and rains cease. The eastern slopes are much drier than the western ones. The decreased rain on the landward side of mountains is called a **rain shadow**. As air continues east-

ward, it encounters the Sierra Nevadas and the Rocky Mountains and is forced upward again, and more rain falls. Descending east of the Rockies, rains cease, and the air moves across the central plains and only rarely provides precipitation.

During our summer, tropical storms sweep west through the Caribbean and then turn to the north or northeast and bring moisture inland as far west as central Texas and north toward the Great Lakes. Fortunately, no southern coastal mountain range blocks this movement or else the central and eastern United States would be very dry.

The size of a land mass also affects its temperature fluctuation. On a large island, moist air is always rising over the mountains, and clouds are frequent. Temperatures have a narrow range. Along the coasts of a continent, cloudy, rainy weather is frequent, and temperature fluctuation is mild. Farther inland, air is dry and clear; a lack of clouds exposes the land to full daytime insolation, and heating is rapid and extreme. At night, clear skies allow the land to radiate infrared energy into space, with none reflected by clouds; cooling is also rapid and extreme.

■ Oceanic Distribution of Heat

Air circulation patterns drive water in the Pacific Ocean and Atlantic Ocean basins in four giant circular currents, clockwise in the Northern Hemisphere and counterclockwise south of the equator. Ocean currents distribute heat from the tropics to the poles, lessening the temperature gradient that would otherwise exist. Also, as warm tropical surface water moves to higher latitudes, large amounts of water evaporate into the temperate prevailing westerlies, giving them more humidity than they would have if the oceans did not circulate.

As trade winds blow across the tropics from east to west, they push surface waters into equatorial currents. During the weeks that a particular mass of water flows along the equator as part of the Atlantic Equatorial currents, it absorbs huge amounts of energy and warms significantly. At the western side of the ocean basin, it is deflected northward by the tip of Brazil, and then part is deflected by Florida and enters the Gulf of Mexico. The water's warmth permits high evaporation into the air and keeps much of the gulf coast humid. The rest of the current moves along the east coast as the Gulf Stream; because this is a latitude that is dominated by westerly winds, the warm current does not keep the land as warm and wet as one would expect. Near New Jersey and New York, the current turns eastward toward Europe; at the turning point, cold polar water moves south along the east coast of Canada and the northeastern United States.

The Pacific Ocean has a similar pattern with westerly equatorial currents. The Philippines and Indonesia act as a barricade, deflecting water north and south. Much northern water heads northward then eastward as the Kuro Shio current toward Canada and Washington. Once at the west coast of North America, some water turns northward toward Alaska as the Alaska current, but the bulk turns south as the California current. The westerly winds absorb huge amounts of moisture as they blow for thousands of miles across the warm Pacific waters, and even though much falls as rain over the ocean, a large amount remains in the air and keeps the coasts wet.

Continental Drift

■ Present Position of the World's Continents

One of the most important factors in determining the climate of a region is that region's latitude. Near the equator in the tropical zone, climate is warm and humid, but a region located either to the north or south is exposed to cooler, drier conditions (Figure 27-3).

If the continents occupied different positions, they could cause the entire Earth to have an altered climate. The presence of the large continent of Antarctica at the South Pole allows huge amounts of fresh water to accumulate there as snow fields and glacial ice, lowering sea levels and exposing more land surface. It also increases ocean salinity by trapping fresh water for thousands of years. If Central America did not exist, water could circulate between the Pacific and Atlantic Oceans, resulting in new oceanic circulation patterns that would affect heat distribution to the poles.

If a continent is small and flat, moist oceanic air can blow across it and bring precipitation to all parts. If it is large and flat, the central regions are too far from the oceans to receive much moisture, most of it having fallen as rain or snow closer to the coastline. The center of the giant continent of Eurasia suffers this fate. Mountains create rain shadows if located on the side of the continent close to the source of wind. The United States would be moister if its highest mountains were in the east or if it were located farther south where the northeast trade winds could bring moisture in across the low Appalachians and Adirondacks.

■ Past Positions of the World's Continents

Cambrian Period During the Cambrian Period, while all life was still aquatic and nothing lived on land, several separate continents were distributed in a vast ocean. Eurasia consisted of either one continent or two that were already close and moving together; the collision of western Europe with eastern Europe and Asia caused the formation of the Ural Mountains. This ancient land mass, whether one or two continents, was also farther south than at present, located in a warmer climate. In the Southern Hemisphere was a giant continent called **Gondwanaland**; it was composed of South America, Africa, India, Australia, and Antarctica. Gondwanaland was constructed such that South America and Africa were attached, and Brazil and northwest Africa were located at the South Pole. The rest of the continent extended northward toward the tropics, but on the opposite side of the world from North America and Eurasia.

Middle and Late Paleozoic Era At about the time life was beginning to move onto land and rhyniophytes were evolving, the continents drifted together (Figure 27-6a). First, during the Silurian Period, North America collided with Eurasia, forming the Appalachian Mountains, which have almost entirely eroded away in the intervening 400 million years. This new continent is known as **Laurasia**. Next, Gondwanaland moved north and ran into Laurasia, pushing up the Alps. Virtually all land on the entire globe was located together as one supercontinent, **Pangaea**. The Laurasian portion was still in the tropics, and much of Gondwanaland was in the southern temperate and polar regions.

Pangaea, which existed for millions of years, through the end of the Paleozoic Era, had a diverse climate. The Laurasian portion was warm, moist, and tropical, whereas southern Gondwanaland was frozen and heavily glaciated. Large regions were swampy lowlands. Initially, Pangaea must have been relatively

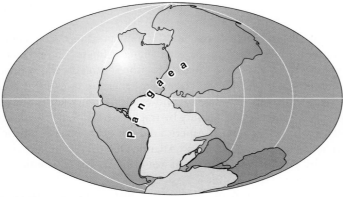

(a) 225 million years ago

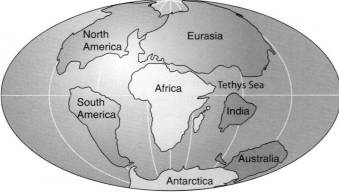

(b) 65 million years ago

Figure 27-6 The continents have changed not only position, but size, shape, and profile as certain areas rise into mountain chains and other areas are covered by ocean when the sea level rises because of polar ice cap melting. Oceanic circulation changes as continents deflect or channel current movement, and this affects both transfer of moisture to the atmosphere and precipitation patterns.

flat, but the collisions caused the formation of extensive mountain ranges in its center: the Appalachians and the Urals, as just mentioned. These must have caused rain shadows in the interior of Pangaea, and its enormous physical dimensions would have caused the central regions to be dry anyway. It was in the diverse conditions of Pangaea that most major groups of land plants arose: Aneurophytales, Trimerophytophyta, and Archaeopteridales appeared during the Devonian Period as Pangaea finished forming, and the diversification of horsetails, lycopods, Medullosales, and Cordaitales occurred during the Carboniferous Period.

Mesozoic Era In the early Mesozoic Era (the Triassic Period), just after the Permian Period, the climate worldwide may have warmed. There is no evidence of glaciation anywhere, although polar regions did remain cool; away from the poles, the climate was warm and equable. Many new plant groups evolved at this time: Caytoniales, Cycadeoidophyta, Cycadophyta, and Ginkgophyta. Ferns increased in number, but many of the first major groups started to disappear. As the Triassic Period continued, the climate became more arid; the Triassic Period is the driest of all geological times. By the time of the Jurassic Period,

Pangaea had many contrasting climates: the North American and Gondwanaland sections were arid, but the western Eurasia region was less so. Humid zones occurred in Australia and India, as well as in central and eastern Eurasia.

In the Jurassic Period, Pangaea began to break up; the North American segment moved northwestward. After separation from Eurasia occurred, the north Atlantic Ocean formed, producing a maritime influence on the two new coasts. Eurasia also began moving northward, separating from the southern portion of Pangaea. This resulted in the formation of a waterway, called the **Tethys Sea**, between the northern and southern continents. This area had alternated between marshy swampland and a shallow coastal sea; however, with continental separation, the Tethys Sea became deep, and oceanic circulation between southern and northern continents could occur on a large scale. The northward movement of the North American and Eurasian continents brought them out of the tropics and into the temperate zones, their northern regions extending into the north polar zones. Conifers evolved and spread throughout the regions that were becoming cooler and moister.

At approximately the same time, the Antarctic–Australian segment separated from the rest of Gondwanaland, and later Australia broke away and moved northward into the south temperate zone (Figure 27-6). India separated from Africa and moved rapidly over a long distance. At the time of separation, India was located at the edge of the south temperate zone (about 30 degrees S latitude) and had a flora adapted to such a climate. It migrated rapidly, in just a few tens of millions of years, into the tropics, across the equator, and into the north temperate zone. The rapid fluctuation of climate in less than 70 million years was too much of a disturbance, and massive extinctions occurred. After it arrived at the Eurasian continent, the two collided, forming the Himalayan Mountains. As the two continents neared each other, plants could invade the foreign territory, some migrating northward onto the mainland, others spreading south from Eurasia onto India. This was to the detriment of India; its flora and fauna had not had time to become well adapted to the conditions of the north temperate zone, whereas the invaders from the mainland were well adapted.

In the Cretaceous Period, approximately 110 million years ago, South America separated from Africa, forming the southern Atlantic Ocean and allowing more humid oceanic air to reach west Africa. Until the breakup of Pangaea, Africa had the misfortune of being the dry center of the supercontinent. Its climate was so arid and severe that it had few species of plants or animals. With the formation of the Atlantic Ocean, Africa's climate became much more humid and conducive to plant growth, but even today, the continent as a whole has many fewer indigenous species than one would expect if it had had a mild climate for a longer time.

Most continental pieces remained close together long enough for many newly evolved groups to migrate to most of the continents. The breakup of Pangaea occurred just as the flowering plants were becoming established, however, and as the North and South Atlantic Oceans widened, gene flow between the continents was reduced, and each developed unique floras. For example, the central deserts of Gondwanaland would have been ideal for cacti, but no cacti occur naturally in Africa, only in the Americas. The cactus family did not evolve until after the South Atlantic was wide enough to prevent the dispersal of cactus seeds from South America to Africa.

The latest major event in continental drift was (is) the collision of South America and North America. The two are coming together, and as always happens when continents collide, a mountain range forms. This collision started so recently, about 5 to 13 million years ago, that the mountains are short and are mostly still submerged below sea level. Only their tops protrude and are known as Central America and the Caribbean Islands. The formation of Central America created a continuous land bridge between the two continents, allowing plant species to be interchanged.

The Current World Biomes

As a result of continental drift, the United States is currently located almost exclusively in the north temperate zone and receives much of its weather from the Pacific and Arctic Oceans in winter and from the Pacific Ocean and tropical Atlantic Ocean in summer. Much of Alaska is situated in the polar zone, whereas Hawaii, Puerto Rico, and Guam lie in tropical waters.

Moist Temperate Biomes

Temperate Rain Forests The northwest coast of the United States is formed by a series of mountain ranges that force westerly winds from the Pacific Ocean to rise as soon as they come ashore. Rain on the western side of the Olympic Mountains in Washington is often above 300 cm/yr. Rains are reliably present through autumn, winter, and spring, with only a brief period of summer dryness. Winters are mild, and only rarely is there frost; summers are warm but not hot. These conditions extend from northern California to Alaska, but they end at the summit of the Coastal Range, approximately 200 km from the coast.

Plant life in the **temperate rain forest** biome is dominated by giant long-lived conifers. Coastal redwoods of California are up to 100 m tall, and Douglas fir, western hemlock, and western red cedar form a canopy that reaches 60 to 70 meters. Forests can attain old age with little disturbance: Hurricanes and tornadoes do not occur, and fires are rare. The stability of the forest and the daily fog result in a rich growth of epiphytic mosses, liverworts, and ferns; the ground is covered with shrubs, herbs, and ferns (**Figure 27-7**). Temperate rain forests, containing totally different species, also occur in southwest Chile, which has the same climate.

Drier Montane and Subalpine Forests As the westerly winds continue inland, they are drier and shed less rain on the Cascade Mountains and the Sierra Nevadas and then even less on the Rocky Mountains; each range feels the rain shadow effect of the preceding mountains. **Montane forests** occur at the bases of these mountains, subalpine forests at higher elevations (**Figure 27-8**). The first inland range in California is the Sierra Nevada; its lower elevations hold a montane forest of ponderosa pine, and some oaks from the valley floor may extend some distance up the slopes. At higher elevations is a mixed conifer forest, one containing many species of conifers: ponderosa pine, Douglas fir, white fir, incense cedar, and sugar pine. The most famous residents are

Figure 27-7 The Olympic National Park in Washington State encompasses a superb temperate rain forest, one of the moist temperate biomes. Abundant rain occurs almost every day. Temperatures are never extremely hot or cold, and disturbances are rare or nonexistent. The ecosystems in this biome are very stable.

the giant sequoia, the largest organisms in the world, much more massive than whales, being 80 m tall, up to 10 m in circumference, and weighing over 400 metric tons when dry. This species has been greatly harmed by lumbering and is currently found in only 75 groves. At the higher elevations are subalpine forests of

Figure 27-8 This is a drier montane forest (moist temperate biome), which may become warm in July and August and have several weeks without much precipitation. The forest is open and sunny and contains few epiphytes.

Figure 27-9 High elevations have increased precipitation and winds and decreased temperature and growing season. The montane forest gives way to alpine meadows. Many areas are flat enough that small bogs and marshes form, being rich in sedges.

Figure 27-10 Much of the dryness of the Rocky Mountains is due to edaphic (soil) conditions as well as to climate. Because these mountains are still very young and steep, soil formation is slow and erosion is rapid. Thin, rocky soil is not effective at holding any moisture that does fall, and in summer, plants often have little soil moisture available.

lodgepole pine, whitebark pine, and mountain hemlock. These form open stands and give way to alpine meadows at their upper boundaries (**Figure 27-9**).

The Rocky Mountains are our most massive mountain range, extending in a broad band from Alaska south through Canada, Idaho, Montana, Wyoming, Utah, Colorado, Arizona, New Mexico, and Mexico. Being the third range from the coast, it is the driest, receiving as little as 40 cm of rain per year; the southern Rockies are especially dry and warm. The subalpine forest contains Engelmann spruce over the entire length of the range, but in the north, other trees are alpine larch and whitebark pine; in the south, varieties of bristlecone pine are part of the subalpine forest. The montane forests typically contain Douglas fir in their higher elevations, ponderosa pine in lower ones. The montane forests on the Rocky Mountains are frequently subjected to fire, as often as every 5 years under natural conditions. Ponderosa pine is well adapted to fire and survives it easily. Other plants are killed, and fires create open grassy areas around the pines.

In drier montane and subalpine forests, soil is shallow, rocky, and very well drained; it tends to be acidic. Water stress in summer is not uncommon, due both to sparse rain and to rapid runoff through porous soils on steep slopes (**Figure 27-10**). Much of the available moisture comes as the melting of winter snow.

In the eastern United States, the Adirondacks and Appalachians are tall enough to support montane and subalpine forests. In the Adirondacks, a spruce/fir subalpine forest extends down to 760 m and contains balsam fir (**Figure 27-11**). In the warmer southern climate, the lower limit for subalpine forest is higher: 990 m in the Appalachians and 1,400 m in the Smoky Mountains. Above these elevations are stands of red and black spruce and Fraser's fir. Below the subalpine conifer forests are montane forests composed of hardwood, not conifers; these are described in the next section.

Temperate Deciduous Forests The climate that produces the **temperate deciduous forest** biome is one with cold winters and warm but not hot summers and relatively high precipitation in all seasons. Whereas the drier montane and subalpine forests of the Rocky Mountains have streams with running water restricted mostly to springtime when snowpack is melting, the northeastern temperate deciduous forest has streams that flow year round. Much of the precipitation in the northeastern region is derived from summer weather systems that move northward out of the central Atlantic Ocean or from winter storms that blow southward from Arctic seas.

Figure 27-11 The Appalachian Mountains are ancient and have been eroded extensively so that they are low and gently sloped and have a thick, rich soil. The higher elevations support subalpine forests of conifers, and the lower elevations have montane forests of hardwoods (dicots).

The temperate deciduous forest in the United States occupies lower, warmer regions, whereas higher, cooler elevations support the subalpine vegetation of the Appalachians and Adirondacks (Figure 27-12). Dominant trees vary geographically, but tall, broadleaf deciduous trees such as maple and oak are frequent everywhere, maples being more common in the north and oak in the south. Intermixed and forming a subcanopy are dogwood, hop hornbeam, and blue beech. The shrub layer is sparse because of heavy shade provided by broadleaf dominants, but witch hazel, spicebush, and gooseberry are common. The ground layer is covered with r-selected herbs that thrive during spring just before trees leaf out. Such a spring sunny period does not occur in evergreen forests.

The foliage of these angiosperm trees contains fewer defensive chemicals than do the needles and scale leaves of a conifer. There may be 3 to 4 metric tons of broadleaf foliage per hectare in a temperate deciduous forest, and as much as 5% is consumed by herbivores; the rest is abscised in autumn and decays quickly in the humid conditions produced by frequent rain. A thick layer of litter does not accumulate.

The forest is not uniform across its entire breadth. The geographic extent of temperate deciduous forest biome contains numerous soil types, various altitudes, and gradients of temperature and precipitation. As many as nine subdivisions have been recognized, such as oak/hickory forest (Illinois, Missouri, Arkansas), oak/chestnut forest (Pennsylvania, the Virginias), oak/pine forest (east Texas, northwest Louisiana, and northern parts of Mississippi, Alabama, Georgia, and the Carolinas), beech/maple forest (Michigan, Indiana, Ohio), maple/basswood forest (Wisconsin, Minnesota), and hemlock/white pine forest (northern Wisconsin to New York).

Forests in these areas have a rich species diversity, and their considerable vertical structure provides a diversity of habitats for animals and fungi. Much animal life is located above or below ground, not at its surface. Most birds make their nests on branches or in holes in the trunk. Small mammals may burrow, but many are

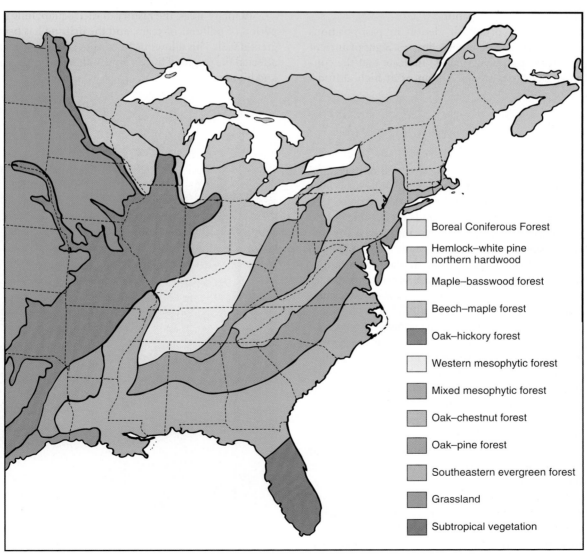

Boreal Coniferous Forest

Hemlock–white pine northern hardwood

Maple–basswood forest

Beech–maple forest

Oak–hickory forest

Western mesophytic forest

Mixed mesophytic forest

Oak–chestnut forest

Oak–pine forest

Southeastern evergreen forest

Grassland

Subtropical vegetation

Figure 27-12 The temperate deciduous forest (moist temperate biome) is an extensive and complex biome in the eastern United States consisting of many subdivisions. This is an area of low, nonmountainous topography with cold, wet winters and warm, rainy summers. Soils are deep.

Botany and Beyond

Box 27-1 Measuring Ancient Continental Positions and Climates

Continents move across Earth's surface, propelled by circulation patterns in the mantle. It is possible to discover where the continents have been located in the past by employing several techniques; the most important is studying paleomagnetism. Many rocks contain magnetic minerals, and when these rocks are fluid, as in lava or magma, the minerals become aligned with Earth's magnetic field. After rock solidifies, its magnetic alignment is fixed. If the magnetic orientation of a rock does not point to the north magnetic pole, we know that there has been a shift in the rock's position since it was formed. Studying the magnetic orientation of extensive blocks of rock of known age reveals the position of the rock at the time of formation.

Another technique important for identifying past positions of continents is the analysis of their past climates. Signs of ancient glaciation (characteristic scouring marks) indicate that the continent must have been near a polar region or at high altitude.

The mineral bauxite forms only under moist, warm conditions; therefore, its presence is an excellent indicator of tropical location in the past. Mineral deposits called evaporites are formed by evaporation of water from small, shallow seas located on a continent near a coast, like the Dead Sea today. They indicate that temperatures were high and precipitation low. Carbonic acid in seawater reacts with calcium to form carbonate, a temperature-dependent process; carbonates are found only in warm tropical waters. The presence of fossilized coral also indicates warm, shallow seas, whereas coal fields are derived from the fossilization of plants that lived in warm, rainy conditions.

In many areas, the history of each square mile is known for a period of millions of years, and the drifting has been mapped in great detail. This allows us to consider how continental drift has affected the origin of life on land and the evolution of ecosystems and biomes.

---

arboreal. As in the coniferous forest at higher elevations here and in the inland western mountains, strong climatic differences exist between summer and winter, accented by the deciduous nature of the broadleaf forest. Autumn leaf fall ends the feeding period for aerial insect leaf eaters but initiates the season for decomposers and insect herbivores of the litter zone. Many birds migrate to wintering grounds, but other species may migrate from farther north. Disturbance is not common; fires are rare and hurricanes that enter the regions usually lose destructive power quickly and change into widespread weaker storms (Figure 27-13).

Southeastern Evergreen Forests This biome shows the powerful effect that disturbance and soil type exert. The southeastern **evergreen forest** biome occurs at the southern edge of the oak/pine component of the temperate deciduous forest, along the northern portions of the Gulf States, the top of Florida, and the coast of the Carolinas. Its climate is similar to that of the inland region except that winters are warmer; frosts occur but the ground does not freeze. Precipitation is higher than in inland temperate deciduous forest areas, but the soil is so sandy and porous that rainwater percolates downward rapidly and runs off into streams. Shortly after a rain, the soil is dry. The region has a definite dry aspect to it.

In addition to rapid drainage, the biome is shaped by frequent fires; lightning initiates fires that burn rapidly through the litter of fallen pine needles that decompose only slowly (Figure 27-14). As a result of fire disturbance and soil conditions, the forest consists almost purely of fire-adapted longleaf pine with

Figure 27-13 Hurricanes move westward in the Atlantic but turn north-north-eastward as they approach land. Many enter land between Texas and Florida and then move up the Mississippi River Valley or just east of it. The winds lose their hurricane force immediately and change into a giant storm system that supplies summer rainfall.

Figure 27-14 Not all fires are the raging infernos you see on TV news. Low, cool fires burn frequently through southeastern evergreen forests, one of the moist, temperate biomes. This kills nonadapted species but leaves pines and cabbage palms: Without fires, seedlings of broadleaf trees would soon overtop the pines and make the forest floor too shaded for pine seedlings.

Figure 27-15 Many groups are undertaking efforts to re-establish natural prairies. Native grasses are planted and nonnative species weeded out. The animal communities must be re-established also to truly recreate the biome and make it self-perpetuating. Extensive acreage would be required to supply enough primary consumers to support the secondary consumers.

occasional oaks. The fact that fire rather than climate is the primary biome determinant has been proven by fire prevention: Repeatedly suppressing fire lets broadleaf seedlings survive and quickly changes the forest from pines to oaks, hickories, beeches, and evergreen magnolias.

Dry Temperate Biomes

Grasslands The entire central plains of North America, extending from the Texas coast to and beyond the Canadian border is—or more accurately, was—**grassland**, often referred to as prairie (see Figure 27-2). This part of North America has no mountains, being too far from the various continental collision zones; it is remarkably flat, with at most low, rolling hills. It is characterized climatically as drier than the forests discussed so far, being located in rain shadows of the prevailing westerlies but out of reach of many Atlantic weather systems. Rain is only about 85 cm/yr. Seasons vary from bitterly cold winters, especially in the north, to very hot summers. Climate and vegetation factors have produced some of the richest soil in existence anywhere. Rainfall is sufficient to promote reasonably rapid weathering of rock, but it is not so great as to leach away valuable elements. Grasses produce abundant foliage that, rather than abscising and falling to the soil, is eaten by herbivores, often large mammals such as cattle or, in the past, buffalo. Vegetable matter is returned to the soil as manure that decomposes rapidly, enriching the soil and increasing its water-holding capacity. Because of the lack of trees, except along rivers, no treetop habitats are available for animals such as squirrels. All small mammals burrow, and birds nest on the ground.

Because the soil is so rich, almost all of the grasslands have been converted to farms; virtually nothing remains of the original biome. Efforts are being made to re-establish grassland prairie by removing all domesticated plants and seeding in grasses and other species known to have occurred originally, such as composites, mints, and legumes (**Figure 27-15**). It is still too early to tell

how successful these reconstruction efforts will be, but many fear that the plots are too small (a few hectares) and should instead be extensive enough to support herds of buffalo if the full ecosystem is to be restored.

It is not known for certain what the critical factors were that caused this region to be grassland as opposed to forest. The buffalo may have been important; they roamed in unimaginable numbers, and tree seedlings would have been poorly adapted to survive their grazing (**Figure 27-16**). Also, Indians set prairies afire periodically, which caused grasses to grow especially luxuriantly the following year, providing better feed for buffalo; seedlings of tree species were killed by the fire. A recent hypothesis is that grasses support a type of mycorrhizal fungus that outcompetes and eliminates the type of fungus necessary for forest trees.

Extensive grasslands occur between the Cascade Mountains and Rocky Mountains, the result of the Cascade's rain shadow. These extend from northern Washington south through Oregon and Nevada. Much of them remain, but they are grazed; large parts have been cleared and used for irrigated farms.

Shrublands and Woodlands A **woodland** is similar to a forest except that trees are widely spaced and do not form a closed canopy. If grass grows between the trees, the biome is known as a **savanna** instead (**Figure 27-17**). **Shrublands** are similar except that trees are replaced by shrubs. Woodlands and shrublands often occur as the transition between a moist forest and a dry grassland or between grassland and desert. Soils often have a high clay content.

The **chaparral** in California is a well-known shrubland (**Figure 27-18a**). Its climate consists of a rainy, mild winter followed by a dry, hot summer; drought occurs every year. Much of the California chaparral is dominated by short shrubs, 1 to 3 m tall, but at higher elevations, manzanita, buckthorn, and scrub oak occur. Many plants have dimorphic root systems: Some roots spread extensively just below the soil surface, but a tap root system reaches great depths. The two together allow plants to gather

Figure 27-16 Large mammalian herbivores such as buffalo, deer, moose, and elk are important factors in their ecosystems. Grasses have basal meristems on their leaves and the shoot apical meristems are low, and thus, grasses can be grazed without being killed. Grazing of shrubs and tree saplings destroys the leaves and the shoot apical meristems; therefore, grazing tends to maintain a prairie as a grassland, free of trees. If too many cattle are fenced into an area, however, they overgraze it, killing even the grasses by eating leaves so quickly that no photosynthesis can occur. In this overgrazed pasture, the oaks and cacti have a selective advantage because their tough, bitter leaves or spines deter grazing.

Figure 27-17 If trees are widely spaced and grass grows between them, it is a woodland biome, a type of dry temperate biome. This is a transition biome, representing the interface between a forest and a grassland, but it is not just the mixing of the two. Many birds nest in the trees but feed in the grasslands; they cannot live in pure forest or pure grassland.

(a)

(b)

Figure 27-18 (a) This is California chaparral, one of the most famous shrubland biomes; such shrubland extends from California across the southwestern United States and northern Mexico into west Texas. Fire is an important disturbance that maintains this biome; it is being invaded by the large mammal *Homo sapiens,* which is protected by fire insurance policies and government disaster aid policies. (b) The resinous leaves and accumulated dry litter cause frequent fires to be inevitable. (c) Seedlings of *r*-selected species flourish in the spaces opened by the burning of the dominant shrubs. These spots remain open only temporarily because the shrubs are fire adapted and recover after a few years.

(c)

water from the deep, constant water table and from brief rains that penetrate only a few centimeters into the soil.

Fires occur frequently in California chaparral and are always in the news because of the houses they destroy. Low rainfall allows dry litter to accumulate without decomposing, and dead shrubs persist upright as dry sticks. An area typically burns every 30 to 40 years, with fires most frequent in summer. Winter rains cause flooding, erosion, and mudslides after a fire because no vegetation remains to hold soil in place. Although the shrubs and trees are fire adapted and resprout quickly, the main new growth is by annual and perennial herbs. These are present before the fire as seeds, bulbs, rhizomes, or other protected structures, and after the burn, they grow vigorously, free of shading by charred shrubs; a release of minerals from the ash also enriches the soil (Figure 27-18c). A few years after fire, larger shrubs dominate the biome again, and herbs are suppressed, perhaps by allelopathy, perhaps by recovery of the herbivore population.

Farther east, drier climates and higher elevations result in pinyon/juniper woodland instead of chaparral shrubland. Rainfall is only 25 to 50 cm, and soils are rocky, shallow, and infertile. The vegetation is a savanna of pinyon pine—small, slow-growing trees with short needles—and juniper trees that may have the stature of large shrubs (Figure 27-19). In Arizona and New Mexico, oaks may be important. Trees are widely spaced, and between them is a grassy vegetation; the species are bunch grasses that grow in clumps, not the mat-forming species of the central plains. Between the bunch grasses is bare soil. Sagebrush and bitterbush occur in the northern parts of the biome.

Desert The driest regions of temperate areas are occupied by **deserts**, where rainfall is less than 25 cm/yr. Deserts are either cold or hot, based on their winter temperatures (Figures 27-20 and 27-21). A hot desert has warm winter temperatures. In the United States, this climate occurs in rocky, mountainous areas of southern California, Arizona, New Mexico, and west Texas, but in other parts of the world, it may occur on flat, sandy plains, as in the Sahara. Three separate and highly distinct deserts actually

Figure 27-20 The Great Basin Desert is a cold desert (in the winter only); it extends from near Las Vegas in the south well into Washington, encompassing much of Nevada, eastern Oregon, and western Idaho. It has two dominant plants, sagebrush (*Artemesia tridentata*) and a bunch grass (*Bromus tectorum*). Even the northern parts have cacti (*Pediocactus* and prickly pear), although they tend to be inconspicuous.

occur in the southwestern United States and northern Mexico: the Chihuahuan Desert in west Texas, New Mexico and Mexico; the Sonoran Desert in Arizona and northwestern Mexico; and the Mohave Desert in southeastern California, southern Nevada, and northwestern Arizona.

Desert soils are rocky and thin; what little soil occurs may blow away, leaving nothing but pebbles. Brief, intense thundershowers wash soil out of mountains and deposit it in large alluvial fans at valley entrances; fans often have the deepest soil and their own distinct vegetation. The most abundant plants in our hot deserts are creosote bush, bur sage, agaves, and prickly pear. Most perennial plants have one or several defenses against herbivores—chemicals and spines being the most common. Joshua tree, an arborescent lily, grows in the Mohave desert, and numerous other needle-leaf yuccas and agaves are abundant; cacti are ubiquitous. Deserts are highly patchy ecosystems, and slight variations in soil type, drainage, elevation, or covering vegetation can cause abrupt changes in vegetation. In valleys or mountains, slopes that face the equator intercept light almost perpendicularly and so are much warmer and drier than those that face away from the equator and are lighted obliquely. The hotter side has the richer xerophyte vegetation.

Alpine Tundra The biome located above the highest point at which trees survive on a mountain, the timberline, is **alpine tundra** (Figure 27-22). In the equatorial region, elevations as high as 4,500 m can support tree growth, but in the cooler regions at 40 degrees N (approximately the middle of the United States), elevations as low as 3,500 m are too severe for trees. Alpine tundra is cold much of the year, with a short growing season limited by a late snow melt in spring and early snowfall in autumn. Soils are thin and have undergone little chemical weathering. Summer days can be surprisingly warm and clear, and many plants flourish in the brief summer. Nights are generally cold in all seasons, and severe, violent weather can occur at any time. The dominant forms of plant life are grasses, sedges, and herbs such as saxifrages,

Figure 27-19 East from the California chaparral is the drier pinyon/juniper woodland, shown here in southern Colorado and northern New Mexico. Soils are rocky, thin, and poor, and in many areas, chaparral grades into a desert or desert–grassland.

(a)

(b)

Figure 27-21 (a) For most North Americans, "desert" means Arizona and saguaro cactus (*Carnegiea*). (b) Streams of the western deserts have water only after a thunderstorm. At that time, the water flows as a flash flood, moving at tremendous speed and carrying not just silt but entire boulders. Their banks contain trees such as cottonwood or palo verde (*Cercidium*), whose deep roots tap moisture that remains in the stream bed after the surface water has run off.

buttercups, and composites. Dwarf plants growing with densely packed stems and leaves are common and are known as cushion plants. Much of the alpine tundra land occurs as flat meadows with shallow alpine marshes.

It is not known for certain why trees do not grow above the tree line. Those near the highest elevation are short and typically have branches mostly on the side away from the wind; at a slightly higher elevation, trees are extremely misshapen and gnarled, known by the German word krummholz; the forest is called an elfin forest. It is believed that blowing snow and ice abrade the trees' surfaces, permitting desiccation and death.

In North America, most alpine tundra biomes occur on the tall mountains in the west, but Mt. Washington in New Hampshire and Whiteface Mountain in New York are high enough to have regions of alpine tundra.

▪ Polar Biomes

Arctic Tundra   Polar regions contain few significantly dry areas (**Figure 27-23**). Precipitation, usually snow, may be low, but evaporation is also low—therefore, moisture persists. In the extreme northern latitudes, the ground freezes to great depths during winter, and summer is too cool to melt anything more than the top few centimeters. Below this, soil is permanently frozen and is known as **permafrost**.

Like alpine tundra, **arctic tundra** has a short growing season of 3 months or less, and temperatures are cool, averaging only about 10°C. Freezing temperatures can occur on any day of the year. Arctic soils have a high clay content and are poor in nitrogen because nitrogen-fixing microbes are sparse. Permafrost prevents drainage when the soil surface melts in summer, and thus, soils are waterlogged and marshy. Bogs, ponds, and shallow lakes are common.

Arctic tundra vegetation contains even more grasses and sedges than does alpine tundra, as well as many more mosses and lichens. Almost nothing is taller than 20 cm, even the dwarf willows and birches that occur. Many plants have underground storage tubers, bulbs, or succulent roots; more than 80% of a plant's biomass may be underground even during summer.

Figure 27-22 Plants in the Alpine tundra biome face short, cold growing seasons, and often snow is never gone from the shaded areas below cliffs. Soils are thin and ultraviolet light is intense.

Figure 27-23 Arctic tundra, one of the polar biomes, is marshy and wet when not frozen solid; permafrost prevents water from seeping into the soil, and the area is so flat that runoff is slow. This area was on the trailing edge during the formation of Pangaea and has not collided with any other tectonic plate; therefore, it has not undergone mountain building. Antarctic tundra is located on the Andes Mountains where drainage is excellent.

Boreal Coniferous Forests Just south of arctic tundra is a broad band of forest, the **boreal coniferous forest** (Figure 27-24). Boreal means northern, just as austral means southern (Australia). This forest occurs completely across Alaska and Canada and throughout northern Eurasia in Scandinavia and Russia. The Russian name for this biome is **taiga**, a term frequently used in the West. The boreal forest is an ancient biome, and its formation was strongly influenced by the diversification of division Coniferophyta just as the North American and Eurasian plates were breaking away from Pangaea and their northern parts were leaving the tropical zone and entering the north temperate, subarctic zone.

Boreal forest is almost exclusively coniferous; conifers appear to be adapted to this climate because they are evergreen and capable of photosynthesizing immediately whenever a sunny day occurs. Deciduous angiosperms would be limited to the short growing season, less than 4 months long. Conifers have drooping branches that shed snow easily; without this architecture, snow loads can easily break off limbs. Although the Coniferophyta contains many species, the boreal forest is not rich in diversity; it is not unusual for only two species to dominate thousands of square miles completely. Black spruce and white cedar may dominate western areas, whereas white spruce and balsam fir cover much of the eastern area. Shrubs are not abundant but include blueberries, blackberries, and gooseberries. Herbs are also sparse. Not much disturbance occurs in the boreal forest; fire may occur in the south and insect plagues in northern parts. This biome does contain many large mammals such as moose, caribou, deer, grizzly bear, and timber wolves. Boreal forest is continuous with and grades into subalpine and montane forests that occur farther south. Many species may occur in both biomes.

■ Tropical Biomes

Most tropical biomes are characterized by a lack of freezing temperatures. On high mountains in the tropics, cool or cold nights and winters occur, but only at very high elevations is frost encountered. Under conditions of high rainfall, tropical rain

(a)

(b)

Figure 27-24 (a) Within any square kilometer of boreal forest there may be thousands of individuals of the same species; therefore, the wind pollination of conifers is highly efficient. (b) Conifers dominate the boreal forest; flowering plants tend to be understory shrubs and herbs.

Figure 27-25 Unlike temperate rain forests, tropical rain forests (one of the tropical biomes) are never exposed to freezing conditions. Photosynthesis can occur throughout the year. Because leaves are not deciduous, their useful lifetime is not limited to just several summer months, but rather they can be effective for several years (Panama).

Figure 27-26 In tropical rain forests, many trees undergo massive flowering, involving not only simultaneous opening of all flowers on one tree but also the flowering of all trees of a particular species within an area. In some plants, this is triggered by changes in air temperature or humidity associated with a rainstorm at the end of the dry season.

forests develop, but the drier areas contain tropical grasslands and savannas.

Tropical Rainforests **Tropical rain forests** occur close to the equator; Hawaii, Puerto Rico, and Guam have extensive rain forests (**Figure 27-25**). Precipitation is high, typically over 200 cm/yr and often as much as 1,000 cm/yr (10 m—over 30 feet—of rain). Rains typically occur every day. The morning may be cool and fresh, but clouds develop rapidly, and rain almost invariably falls by noon. After a rain, there are large clouds in a clear sky and bright sunlight; the temperature quickly rises, and the relative humidity of the air is close to 100%.

High temperatures and moisture cause much more rapid soil transformation here than in other biomes. Many elements are leached from the soil, leaving behind just a matrix of aluminum and iron oxides. Humus decays rapidly, and there is little development of soil horizons. An extensive system of roots and mycorrhizae catch and recycle minerals as litter decays. Almost all available essential elements exist in the organisms, not in the soil.

The dominant trees are angiosperms; virtually no conifers occur naturally in tropical rain forests. The canopy is 30 to 40 m above ground level, but numerous large trees emerge above all others. A subcanopy may occur at about 10 to 25 m. Some trees have massive, gigantic trunks, but most are slender. In an undisturbed area where the canopy has remained intact for years, ground vegetation is minimal, and it is easy to walk through the forest. Localized disturbances happen when a large tree dies and falls, creating a gap in the canopy. Suddenly, light is available on the forest floor, and herbs and shrubs proliferate; however, a tree soon grows up and fills the canopy, blocking light.

Virtually all small shrubs and herbs occur as epiphytes, located high in the canopy, nearer the light. Orchids, bromeliads, aroids, and cacti are common. Numerous vines are anchored in the soil, but their stems grow to the canopy and then branch and leaf out profusely. Leaves and roots can be separated by a narrow stem 50 m long.

The tropical rain forest is synonymous with species diversity. A single hectare (a square 100 m on each side) may have over 40 species of trees, often up to 100, and a single tree may harbor thousands of species of insects, fungi, and epiphytic plants. With such diversity, each hectare may have only one or two individuals of a particular species, especially of trees. No dominants occur. With such low population density for most species, wind pollination is not successful. All plants must be pollinated by animals, and being noticed by a pollinator can be difficult; many subcanopy species have brilliantly colored flowers that are easy to see in the dim light, and scents tend to be so strong that insects and birds find the flowers quickly. Large trees can undergo a massive flowering that no pollinator could possibly ignore (**Figure 27-26**).

Tropical Grasslands and Savanna Those areas of the tropics with lower rainfall develop as thorn-scrub (low, spiny trees and shrubs; **Figure 27-27**), savannahs or grasslands (no trees; **Figure 27-28**). The best known savannas are in Africa, but they also occur in Brazil (the cerrados), Venezuela (the llanos), and Australia. Under natural conditions, undisturbed by humans, the vegetation consists mostly of bunch grasses up to 1.5 to 2 m tall. Most of the trees are open, flat-topped, widely scattered legumes. The South American savannas do not have many large grazers, but those of Africa are famous—zebras, wildebeests, and giraffes. An unexpected but very important grazer is the termite; termites are abundant and build giant nests of soil particles and plant debris. They bring in large amounts of plant material, digest it, and add their fecal material to the termite mound. Colonies are so abundant and the mounds so large that termites are a major link in nutrient cycling.

Figure 27-27   Along coastlines in tropical areas, the low-lying, flat terrain receives relatively little rainfall, and the sandy soil allows what little rain there is to seep away quickly. Conditions are desert like, except for very high humidity. There may be a coastal thorn-scrub biome, so named because the trees are short and most bear spines. Even the vine here is a cactus (*Acanthocereus*).

Figure 27-28   Tropical grassland is a tropical biome in which rainfall is not abundant enough to support forest. There is no cold winter, but rather three seasons: warm and wet, cool and dry, hot and dry.

Summary

1. Biomes are groupings of ecosystems; although controlled by diverse factors such as climate and soil, many biomes are distinctive enough to be easily recognizable.

2. Earth's climatic patterns result from several factors, including tilt of the axis of rotation, circulation of the atmosphere and oceans, and positions, shapes, and sizes of the land masses.

3. The warmest and most humid zone worldwide is located between the Tropics of Cancer and Capricorn. Temperate zones are farther from the equator and are cooler and drier; polar regions are cold.

4. The climate of a particular land mass is determined by its latitude, its size, the warmth of adjacent ocean currents, humidity and temperature of winds, and rain shadows produced by mountains.

5. The nature of many types of mineral deposits is affected by climatic conditions of heat and humidity; the minerals act as a record of past climatic conditions. Certain fossil life forms (coal and coral) are also good indicators. Paleomagnetism shows the past orientation and position of the continents.

6. In the late Paleozoic Era, just as land plants were appearing, all continents came together, forming Pangaea; during the Jurassic Period, Pangaea fragmented into the present continents. As each drifted, many moved into new latitudes, and their climates changed dramatically, affecting their life forms. Altered ocean circulation permitted moist air to cover more of the land surface.

7. Moist temperate biomes have high rainfall, and winters are cool or cold. Several biomes are temperate rain forest, montane and subalpine forest, temperate deciduous forest, and southeastern evergreen forest.

8. Dry temperate biomes have cold winters and moderate rainfall: grasslands, shrublands, woodlands, desert, and alpine tundra.

9. Polar biomes have low light intensity, long summer days, long winter nights, and permafrost. In arctic tundra, much of the biomass is located below ground as bulbs and tubers. Boreal forest (taiga) is an extensive circumpolar forest dominated by conifers.

10. Tropical biomes are influenced strongly by the complete absence of freezing temperatures; where rainfall is high, tropical rain forest develops. If rain is less abundant, tropical grasslands exist.

Important Terms

alpine tundra	grassland	taiga
biomes	Laurasia	temperate deciduous forest
boreal coniferous forest	montane forests	temperate rain forest
chaparral	Pangaea	Tethys Sea
continental drift	permafrost	tropical rain forests
deserts	rain shadow	tundra
evergreen forest	savanna	woodland
Gondwanaland	shrublands	

Review Questions

1. Earth's land surface is covered by biomes, extensive groupings of many _____. Are these characterized by distinctive aspects of both plants and animals or by just one of the two?

2. Biomes vary from extremely simple to more complex. Name a simple biome. Name two complex ones. Biome complexity is most strongly influenced by two abiotic factors: _____ and _____.

3. Why are the Florida Keys desert islands? Why are the Hawaiian Islands and Puerto Rico not desert islands?

4. Why does the west coast of the United States and Canada receive so much more rain than the central plains? What is the source of rain that prevents the central plains from being a complete desert—the Pacific Ocean, the Atlantic Ocean, the Gulf of Mexico?

5. What is the latitude where you live? Which coast is closer to your location? Does it influence your climate significantly?

6. What is the precipitation pattern in your area—mostly in summer or winter? What is typically the longest time between rains—10 days, a month? Does this dry period occur in summer or winter?

7. Are the soils in your area thin, rocky, and immature or thick and rich?

8. In the late Paleozoic Era, at about the time life was beginning to move onto land, all of the continents drifted together. What is the name we use for that supercontinent? What is the name for the northern part of it? What is the name for the southern part? Later, the continent broke up, and as the northern part separated from the southern part, a waterway was formed. It is called the _____ Sea (it no longer exists). When North America drifted to the northwest and broke away from Europe, another waterway was formed. It is called the _____ _____ Ocean (this still exists).

9. The northwestern United States is a temperate rain forest biome, and the western side of the Olympic Mountains often receives more than _____ cm of rain per year. There are 2.5 cm in every inch. How many inches of rain is this?

10. The northeastern United States is dominated by temperate deciduous forests. What does "temperate" mean? What is a "deciduous forest"? Describe the climate that produces the temperate deciduous forest. What are its winters, summers, and rainfall like?

11. The entire central plains of North America is—or more accurately, was—grassland, often referred to as prairie. Why does the sentence say "was"? What has happened to the grassland of the central plains? What is the soil like—thick and rich, or thin and poor?

12. Chaparral in California is a famous shrubland. What is a frequent disturbance in chaparral, one that is in the news every year? How are native chaparral plants adapted to this disturbance?

13. What is your image of a desert—dry and lifeless? Look at Figure 27-21 of a desert in Arizona. Do you think this area looks like this all year long? Look at the various life forms in Table 27-1 below. Which life form is represented by the poppies (the plants with yellow flowers)? Which life form is the saguaro cactus (pronounced sa wah row)?

14. What is the boreal coniferous forest? Where is it located? The boreal forest is almost exclusively coniferous, whereas just to the south are the temperate deciduous forests. Why is the more northern region dominated by conifers?

15. There is no huge, extensive austral coniferous forest. Why not (Hint: look at Figure 27-3; is there any land at the corresponding latitude in the Southern Hemisphere)?

16. Of the biomes discussed in this chapter, in which do you live? What are the dominant plants in your biome? What are the dominant animals other than humans? What are the most significant causes of ecological alteration in your area—urban development, logging, agriculture? What have been some of the changes caused by this?

17. Name all of the state and national parks or wilderness areas located within one day's drive of your home. How many of these have you visited? Of all the biomes mentioned in this chapter, how many have you traveled through? Did you notice the changes in vegetation?

BotanyLinks

http://biology.jbpub.com/botany/5e

BotanyLinks contains a variety of resources and review material designed to assist in your study of botany.

Visit the Web Exercises area of BotanyLinks to complete this question:

1. What resources are available for studying the world's biomes and biodiversity? Go to the BotanyLinks home page to research this subject.

TABLE 27-1	Life Forms of Raunkiaer
Life Form	Means of Surviving Stress
Therophytes	Annual life span; survive stress as seeds.
Geophytes	Buds are underground on rhizomes, bulbs, corms.
Hemicryptophytes	Buds are located at surface of soil, protected by leaf and stem bases: many grasses and rosette plants.
Chamaephytes	Buds are located above ground, but low enough to not be exposed to strong winds: small shrubs.
Phanerophytes	Buds are located high, on shoots at least 25 to 30 cm above ground: trees and large shrubs.

Numbers after definitions are the chapters where the principal discussions occur. Italicized terms are defined elsewhere in the Glossary.

A

A channel The groove in the ribosome small subunit in which the free amino acid-carrying tRNA occurs. Alternative: *P channel*. 15

A horizon The uppermost soil layer, the zone of leaching. 26

abiotic Refers to things that are not and never have been alive. Compare: *biotic*. 26

abscisic acid A hormone involved in resistance to stress conditions, stomatal closure, and other processes. 14

abscission zone The region at the base of an organ, such as a leaf or fruit, in which cells die and tear, permitting the organ to fall cleanly away from the stem with a minimum of damage. 6

absorption spectrum A graph of the relative ability of a pigment to absorb different wavelengths of light. Compare: *action spectrum*. 10

accessory fruit A fruit that contains nonovarian tissue. Synonym: *false fruit*. Alternative: *true fruit*. 9

accessory pigment A pigment that has an absorption spectrum different from that of chlorophyll *a* and that transfers its absorbed energy to chlorophyll *a*. 10

acid rain Rain that has become acidic due to air pollution; it can damage plant cuticle as well as speed the leaching of minerals from soil. 13

actinomorphic Synonym for regular flower; radially symmetrical. 9

action spectrum A graph of the relative rates of reaction of a process as influenced by different wavelengths of light. Compare: *absorption spectrum*. 10

activation energy The energy needed to overcome the electrical repulsion between two molecules, such that they can react chemically. 2

active transport The forced pumping of molecules from one side of a membrane to the other by means of molecular pumps located in the membrane. 3, 12

adaptive radiation Divergent evolution in which a species rapidly diverges into many new species. 17

adenosine triphosphate (ATP) A cofactor that contributes either energy or a phosphate group or both to a reaction; as it does so, it loses either one or two phosphate groups, becoming either ADP or AMP. 2, 10, 11

adult plant A plant that is mature enough to flower. Alternative: *juvenile plant*. 14

adventitious Refers to an organ that forms in an unusual place; refers primarily to roots that form on leaves, nodes, or cuttings rather than on another root. 7

agamospermy A set of methods of asexual reproduction that involve cells of the ovule and result in seeds and fruit. 9

aggregate fruit A fruit that develops from the crowding together of several separate carpels of one flower. Alternatives: *simple fruit* and *multiple fruit*. 9

albuminous cell In gymnosperm phloem, a nurse cell connected to and controlling an enucleate sieve cell. Compare: *companion cell*. 5

albuminous seed A seed that contains large amounts of endosperm. Alternative: *exalbuminous seed*. 9

all-or-none response A situation in which an organism either responds to a stimulus or does not respond; the level of response is not correlated with the level of stimulus. Alternative: *dosage-dependent response*. 14

alleles Versions of a gene that differ from each other in their nucleotide sequences. 16

allelochemic See *allelopathy*.

allelopathy The inhibition of germination or growth of one species by chemicals (allelochemics) given off by another species. 26

allopatric speciation Speciation that occurs when two or more populations of one species are physically separated such that they cannot interbreed. Alternative: *sympatric speciation*. 17

alternation of generations A type of plant life cycle in which a diploid spore-forming plant gives rise to haploid gamete-forming plants, which in turn give rise to more diploid spore-forming plants. The generations may be similar morphologically (isomorphic) or dissimilar (heteromorphic). 9, 19–22

amino acid A small molecule containing an amino group and a carboxyl group; the monomers of proteins. 2

amino acid attachment site In transfer RNA, the 3′ end where the amino acid is carried. 15

amino sugar A sugar that contains an amino group; a component of the chitin of fungal cell walls. 2, 24

amylase An enzyme that digests amylose (starch). 2

amylopectin A simple, branched polysaccharide containing only glucose residues. Much of starch is amylopectin. 2

amyloplast See *plastid*.

amylose A simple unbranched polysaccharide containing only glucose residues. Much of starch is amylose. 2

anabolism Metabolism in which large molecules are constructed from small ones. Alternative: *catabolism*. 10

analogous features Features that resemble each other but are not based on homologous genes, those related by descent from common ancestral genes. Alternative: *homologous features*. 18

anaphase The third phase of mitosis; at the metaphase-anaphase transition, centromeres divide and the two chromatids of a chromosome become independent chromosomes. During anaphase, the two are pulled to opposite poles of the spindle by spindle microtubules. 4

anaphase I The third phase of meiosis I, similar to anaphase of mitosis except that at the metaphase I-anaphase I transition, no division of centromeres occurs. Instead, one homolog is pulled away from the other in each pair, thus reducing the number of sets of chromosomes in each daughter nucleus to the haploid condition from the diploid. 4

anaphase II A phase of meiosis II, similar to anaphase of mitosis. 4

ancestral group The group of organisms that were the ancestors of another group. This can refer to the ancestral group that gave rise to just one species or to a set of species (such as the ancestral group of a genus or family). 18

androecium (pl.: androecia) A collective term referring to all the stamens of one flower. 9

angiosperm Informal term for flowering plants, members of division Magnoliophyta; their seeds develop within a fruit. Also called anthophytes. 5, 22, 23

angiospermous sporophyll The sporophyll of a flowering plant, the carpel that encloses the ovule. 23

anion An ion carrying a negative charge. 2

anisogamy A type of sexual reproduction in which the two gametes are only slightly different; usually one is larger and both are motile. Alternatives: *isogamy* and *oogamy*. 19

annual plant A plant that completes its life cycle in one year or less. Compare: *biennial* and *perennial*. 5, 9

annual ring In secondary xylem, the set of wood, usually early wood and late wood, produced in one year. 8

annular thickening A pattern of secondary wall deposition in tracheids and vessel elements; the wall occurs as separate rings. 5

anomalous secondary growth Any form of secondary growth that does not conform to that typically occurring in gymnosperms and dicots. 8

anoxygenic photosynthesis Bacterial photosynthesis that does not use water for an electron donor and does not release oxygen as a waste product. 10

anther The portion of a stamen that contains sporogenous tissue which produces microspores (pollen). 9, 23

antheridiophore In liverworts, an umbrella-shaped outgrowth of the gametophyte, bearing antheridia. 20

antheridium (pl.: antheridia) A small structure that produces sperm cells, or the equivalent of sperm cells in ascomycete fungi. 24, 20–23

anthropomorphism Attributing to plants certain capacities that only humans have, such as decision-making. 1

anticlinal wall A wall perpendicular to a nearby surface, especially the outer surface of the plant. Alternative: *periclinal wall*. 8

anticodon In transfer RNA, the nucleotide triplet complementary to the codon of mRNA. 15

antipodal cell One of several (usually three) cells in the angiosperm megagametophyte, located opposite the egg cell and the synergids. 9

apical dominance The suppression of axillary buds by the growing, active apical bud of a shoot. 14

apomorphy A feature present in one (autapomorphy) or several (synapomorphy) derived members of a group, but which is not present in the ancestral members. 18

apoplast The intercellular spaces and cell walls of a plant; all the volume of a plant that is not occupied by protoplasm (the symplast). 3, 5, 7

aquaporin A membrane protein that permits water to cross the membrane rapidly. 3

archaebacteria Prokaryotes that have unusual types of metabolism and membrane lipids, perhaps representing extremely archaic forms of life. 25

archegoniophore In liverworts, an outgrowth of the gametophyte, bearing archegonia and having a stalk with radiating fingers of tissue. 20

archegonium (pl.: archegonia) Any structure in true plants (not algae) that produces an egg; the megagametangium of true plants. 20, 21

aril A thick fleshy envelope around some seeds. 22

artificial selection The process in which humans purposefully alter the gene pool of a species by selective breeding. 17

artificial system of classification A classification not based on evolutionary, phylogenetic relationships but on other characters. Alternative: *natural system of classification*. 18

ascocarp Fruiting body of ascomycete fungi. 24

ascogenous hyphae See *ascogonium*.

ascogonium (pl.: ascogonia) In ascomycete fungi, a small tubular branch that contacts an antheridium and receives nuclei from it; it then produces heterokaryotic ascogenous hyphae, which contribute to the ascocarp. 24

ascospores Spores produced by meiosis in asci. 24

ascus (pl.: asci) In ascomycete fungi, the swollen end of a hypha where karyogamy and meiosis occur. 24

atactostele The vascular system of monocots, a set of bundles not restricted to forming one ring. Compare: *eustele* and *protostele*. 5

autotroph An organism that synthesizes its own organic compounds, using only carbon dioxide and mineral nutrients. Compare: *heterotroph*. 10

auxiliary cell In red algae, a cell that receives the diploid nucleus from the fertilized carpogonium. 19

auxins Hormones involved in cell elongation, apical dominance, and rooting, among other processes. 14

axial bud A bud located in a leaf axil, just above the attachment point of a leaf. May be either a leaf bud or a floral bud. Alternative: *terminal bud*. 5

axial tissue In a woody stem or root, the tissue derived from fusiform cambium cells. Alternative: *ray*. 8

axoneme In flagella, the set of 9 + 2 microtubules. 19

B

B horizon In soil, the zone of deposition, which receives leached minerals from the A horizon above it. 26

bacteriochlorophyll Light-harvesting pigment involved in bacterial photosynthesis. 10

bacteriophage A virus that attacks bacteria. 15

bar A measure of pressure and of water potential; equals 0.987 atmospheres or 1.02 kg/cm^2. 12

basal angiosperms The several clades of angiosperms that arose before the rest of the angiosperms diverged into the monocots and eudicots. 18, 23

basal body See *centrioles*.

basidiocarp In basidiomycete fungi, a fruiting body such as a mushroom or puffball. 24

basidiospores Spores produced by meiosis in basidia. 24

basidium (pl.: basidia) In basidiomycete fungi, the cell in which karyogamy and meiosis occur and which then produces basidiospores. 24

betalains Water-soluble pigments characteristic of the flowering plant class Caryophyllales. 23

biennial plant A plant that requires 2 years to complete its life cycle, with cold winter temperature in the first year being necessary for reproduction. 5, 14, 23

biliprotein A protein in cyanobacteria and red algae which associates with phycobilin, forming a phycobilisome. 19, 25

binary fission Cell division in prokaryotes; a term occasionally used for cell division in years. 25

binomial classification A system of providing scientific names to organisms, each name consisting of the genus name and the species epithet. 18

biome An extensive grouping of ecosystems, characterized by the distinctive aspects of dominant plants. 27

biotic Refers to living things. Compare: *abiotic*. 26

biotic potential The intrinsic rate of natural increase, the number of offspring produced by an individual which live long enough to reproduce. Symbol: *r*. 26

biotroph A fungus that slowly draws nutrients from living hosts, often without killing them. Synonym: *parasite*. 24

biparental inheritance Inheritance of genes from two parents, the most common case for nuclear genes. Alternative: *uniparental inheritance*. 16

bivalent chromosome During meiosis, homologous chromosomes pair with each other; the pairs are bivalents. 4

blind pit See *pit*.

bloom (algal) A sudden increase in the numbers of algae when environmental conditions become particularly favorable for growth. 25

blue-light responses Plant developmental responses triggered by blue-light; the photoreceptor is a small protein called phototropin. 14

bordered pit In a xylem-conducting cell, a pit in the secondary wall having a thickened rim (border). Alternative: *simple pit*. 5

bract A small, often thickened and protective leaflike structure; bracts usually protect developing inflorescences. 9

broadleaf plant Informal term for any member of the flowering plant class Magnoliopsida: a dicot. Alternative: *monocot*. 5, 23

browser An herbivore that eats twigs and leaves of shrubs. Compare: *grazer*. 26

bryophyte A term without uniform definition; used by some to refer only to mosses, by others to refer to mosses, liverworts, and hornworts. 20

bud scale A small, specialized leaf, usually waxy or corky, that protects an unopened bud. 5, 6

bulb A short, subterranean, vertical stem that has fleshy, scalelike leaves. Example: onion. 5

bulbil A small bulblike axillary plantlet that serves as a means of vegetative reproduction. 9

bundle sheath A set of cells, which may be parenchyma, collenchyma, or sclerenchyma, that encases some or all of the vascular bundles of a leaf. 6, 10

bundle sheath extension A set of cells, usually fibers, that extends from the bundle sheath to the upper or lower (or both) epidermis of a leaf. 6

buttress root A root that grows asymmetrically such that it becomes very tall and thin, extending up the trunk as much as a meter or more and giving the trunk lateral stability. 7

C

C_3 cycle Synonym for Calvin/Benson cycle, one type of stroma reaction. 10

C_4 metabolism (C_4 photosynthesis) A set of metabolic reactions in which carbon dioxide is fixed temporarily into organic acids that are transported to bundle sheaths, where they release the carbon dioxide and C_3 photosynthesis occurs. 10

C horizon The deepest soil layer, composed of parental rocks and rock fragments. 26

callose A long-chain carbohydrate polymer that seals certain regions, e.g., damaged sieve elements (12) or growing pollen tubes (9).

Calvin/Benson cycle Synonym for C_3 cycle, one type of stroma reaction. 10

calyptra (pl.: calyptras) In nonvascular plants, a small sheath of cells, derived from the archegonium, which covers the top of the capsule. 20

calyx (pl.: calyces) A collective term for all the sepals of one flower. 9

capsule In mosses and liverworts, the sporophyte generation. 20

carbohydrates Organic compounds composed of carbon backbones with hydrogens and oxygens attached in a ratio of about 2:1; sugars, starch, and cellulose are examples. 2

carbon fixation Photosynthetic conversion of carbon dioxide into an organic molecule, with carbon being reduced in the process. 10

carotenoid A class of lipid-soluble accessory pigments in chloroplasts and chromoplasts. 10

carpel Organ of a flower that contains ovules and is involved in the production of megaspores, seeds, and fruits. See *gynoecium*. 9, 23

carpogonium (pl.: carpogonia) In red algae, an egglike cell that fuses with a spermatium. 19

carpospores In red algae, diploid spores produced by the carposporophyte that grow into tetrasporophytes. 19

carposporophyte In red algae, the mass of diploid cells that arise after the auxiliary cell receives the diploid nucleus. 19

carrying capacity The number of individuals of a population that can live in a particular ecosystem. 26

caryopsis (pl.: caryopses) A single-seeded dry fruit that is fused to the enclosed seed; found in grasses and often mistaken for a seed rather than a fruit and a seed. 23

Casparian strip (band) A layer of impermeable lignin and suberin in the walls of endodermal cells, preventing diffusion of material through that portion of the wall. 7

catabolism Metabolism in which large molecules are broken down into smaller ones. Alternative: *anabolism*. 10

catalyst A material that reduces the activation energy of a reaction, permitting it to occur more rapidly at a lower temperature. 2

cation A positively charged ion. 2

cation exchange In soil, the release of an essential element cation from a soil particle and its replacement by a proton. 13

cavitation In xylem, the breaking of a water column when tension overcomes the cohesive nature of water; an embolism forms. 12

cDNA Complementary DNA. 15

cell cycle arrest When cells stop dividing, they undergo cell cycle arrest. The cells may become dormant temporarily (dormant buds, seeds) or they may differentiate and mature. 4

cell plate During cell division, the new cell wall forms inside a large vesicle surrounded by phragmoplast microtubules. The wall, vesicle, and phragmoplast together constitute the cell plate. 4

cellulose A polysaccharide composed only of glucose residues linked by beta-1, 4-glycosidic bonds; it is the major strengthening component of plant cell walls. 2, 3, 4

centimorgan (cM) Synonym for *map unit*. 16

central cell In the megagametophyte in a flower's ovule, the cell that contains two nuclei (usually) and develops into endosperm after fertilization. 9

centrifugal growth Growth outward from a common point. 19

centrioles In animals and some fungi and algae, organelles that act as basal bodies for organizing the microtubules of flagella. 3

centromere Region of a chromosome that holds the two chromatids together prior to anaphase of mitosis or anaphase II of meiosis. Spindle microtubules attach to centromeres and move the chromosomes during division. 4

CF_0-CF_1 complex Part of ATP synthase. CF_0 is a proton channel; CF_1 is a set of enzymes that phosphorylates ADP to ATP. 10

charophytes This clade of green algae is the sister group to the true plants (embryophytes). 19

chemical messenger A chemical that, by its presence, carries information from one area to another. 14

chemiosmotic phosphorylation The synthesis of ATP from ADP and phosphate using the energy of an osmotic gradient and a gradient of electrical charge; occurs in chloroplasts and mitochondria. 10

chemoautotroph A bacterium that has the ability to obtain energy from chemical reactions (without photosynthesis) and obtains its carbon from carbon dioxide. Compare: *photoautotroph*. 25

chemosynthetic origin of life Theory that life began through a series of chemical reactions on primitive Earth when conditions were quite different than they are today. 17

chiasma (pl.: chiasmata) During diplotene of prophase I of meiosis, as homologous chromosomes begin to move away from each other, they are held together by chiasmata, thought to be tangles in the chromosomes. During diakinesis, the chiasmata slide to the ends of the chromosomes and disappear. 4

chitin A polymer in fungal cell walls, composed in part of amino sugars. 2, 24

chlamydospore In fungi, a segment of hyphae rich in oil or glycogen and having thick, melanized walls; an extremely resistant spore. 24

chlorophyll Pigment involved in capturing the light energy that drives photosynthesis; found in plants, algae, and cyanobacteria. See *bacteriochlorophyll*. 3, 10

chloroplast See *plastids*.

chlorosis A common symptom of mineral deficiency, a yellowing of leaves due to lack of chlorophyll. 13

chromatid A portion of a chromosome consisting of one DNA double helix and its histones. Before S phase, each chromosome consists of just one chromatid, but after DNA replication in S phase, each chromosome consists of two chromatids. 4

chromatin The complex formed when histone proteins bind to DNA. 3, 15, 16

chromoplast See *plastids*.

chromosome Each nuclear DNA double helix is complexed with histone into a chromosome, which consists structurally of one (pre-S phase) or two (post-S phase) chromatids plus a centromere (4). The circles of DNA found in prokaryotes (25), plastids, and mitochondria are occasionally called chromosomes.

chrysolaminarin Reserve polysaccharide found in algae of division Chrysophyta (diatoms and their relatives). 19

cilium (pl.: cilia) Similar to a flagellum, only shorter. 3, 19

circadian rhythm An endogenous rhythm whose period is approximately 24 hours long. 14

circinate vernation Refers to the development of a fern leaf in which it must uncoil as it expands. 21

circular bordered pit A pit that is circular in cross-section and has a thickened rim (border) that slightly overarches the pit chamber. See *pit*. 5

cis-acting factor Portions of a strand of DNA that affect the activity of a gene on the same DNA strand. Alternative: trans-*acting factor*. 15

cis-position In a carbon-carbon double bond, the situation in which two groups of interest lie on the same side of the bond. Alternative: trans-*position*. 2

cis-unsaturated fatty acid An unsaturated fatty acid (it has one or more carbon–carbon double bonds), and the two hydrogens at the double bond are on the same side of the molecule. Compare with *trans*-unsaturated fatty acid. 2

citric acid cycle Metabolic pathway in which acetyl-CoA is oxidized to carbon dioxide as reduced electron carriers are generated. Synonyms: *Krebs cycle* and *tricarboxylic acid cycle*. 11

cladogram A branching diagram showing the phylogenetic relationships of several or many taxa. 18

cladophyll A flattened stem that resembles a leaf. 5

clamp connection In basidiomycete fungi, the connection between two adjacent cells in hyphae; clamp connections are part of a mechanism that ensures heterokaryosis. 24

cleavage furrow In the cell division of some algae, the inward furrowing of the cell membrane. 19

climacteric fruits Fruits that undergo a sudden burst of metabolism and ripening (the climacteric) as the last step of maturation. Alternative: *nonclimacteric fruits*. 14

clumped distribution A distribution of plants in space such that they occur in groups. Alternatives: *uniform* and *random distribution*. 26

codon In mRNA, a set of three nucleotides that specifies an amino acid to be incorporated into a protein. 15

coenocyte A cell, usually large, that has many nuclei, up to several thousand. 4, 19, 24

coenzyme Synonym for *cofactor*.

coenzyme A A carrier molecule, able to pick up an acetyl group, becoming acetyl-CoA. 2, 11

coevolution A type of evolution in which two species become increasingly adapted to each other, resulting in a highly specific interaction. 9

cofactor A small molecule essential to the activity of an enzyme. 2

cohesion-tension hypothesis Hypothesis that as water is pulled upward by transpiration, its molecules cohere sufficiently to withstand the tension. 12

coleoptile The outermost sheathing leaf of a grass seedling, providing protection for the shoot within. 14

collateral vascular bundle A vascular bundle that consists of both xylem and phloem. 5

collenchyma Collenchyma cells have only primary walls, but these are thickened at the corners of the cell and thin elsewhere. The walls are plastically deformable; if stretched to a new shape, they retain that shape. Alternatives: *parenchyma* and *sclerenchyma*. 5

colony A group of cells all derived from one recent mother cell and held together by an extracellular matrix, but not closely adhering to each other and not integrated as a single individual. 14, 19

commensal relationship An interaction of two species in which one benefits and the other is unaffected. 26

common ancestor During divergent evolution, a common ancestor gives rise to two or more groups. In a cladogram, it is the node that gives rise to the various taxa being considered. 18

community All the populations of a region. 26

companion cell In the phloem of angiosperms, a nurse cell that is connected to, and is a sister cell to, an enucleate sieve tube member. Alternative: *albuminous cell*. 5

compartmentalization The formation of numerous compartments, usually surrounded by semipermeable membranes, such that each compartment has a distinct metabolism. 3

compatibility barrier Chemical interactions that prevent the fertilization of a gamete by an inappropriate gamete. 9

competition An interaction of two species which is disadvantageous to one or both. Compare: *mutualism*. 26

competitive exclusion The inability of a species to grow in part of its range due to competition from another species more adapted to that part of the range. 26

competitive inhibitor A small molecule that inhibits an enzyme by binding to the active site. Alternative: *noncompetitive inhibitor*. 2

complementary DNA DNA synthesized with reverse transcription; the DNA is complementary to the RNA substrate and similar to the gene that coded for the RNA. 15

complete dominance A situation in which the presence of one allele completely masks the presence of the homologous allele. Alternative: *incomplete dominance*. 16

complete flower A flower having sepals, petals, stamens, and carpels. Alternative: *incomplete flower*. 9

compound cone A cone with several lateral axes attached to a main axis; conifer seed cones are compound. Compare: *simple cone*. 22

compound leaf A leaf in which the blade consists of several separate parts (leaflets), all attached to a common petiole. **palmately compound** All leaflets are attached to the same point. **pinnately compound** Leaflets are attached to the rachis, an extension of the petiole. Alternative: *simple leaf*. 6

compression wood See *reaction wood*.

conceptacle In the brown alga *Fucus*, a small cavity in which sperms and eggs are produced. 19

cone A compact collection of reproductive structures on a short axis. Synonym: *strobilus*. 21–22

cone scale In conifer seed cones, a scale that bears seeds; it is a flattened shoot with fused sporophylls. Synonym: *ovuliferous scale*. 22

conidiophore In fungi, a hypha that gives rise to conidia. 24

conidium (pl.: conidia) In fungi, spores formed by the segmentation of the end of a hypha. 24

conjugation A method of genetic exchange occurring in bacteria and certain green algae; DNA is passed through a tube that joins two adjacent cells. 19, 25

conjugation pilus (pl.: pili) In bacteria, the narrow tube through which DNA passes during conjugation. 25

conjunctive tissue In monocots, the pithlike region in which vascular bundles are located in stems and roots. 5

continental climate A climate characterized by dry air and great changes of temperature from summer to winter. 27

convergent evolution Evolution of two phenotypically distinct species, organs or metabolisms such that they strongly resemble each other, usually because they are responding to similar selection pressures. Compare: *parallel evolution* and *divergent speciation*. 17

cork cambium (pl.: cambia) A layer of cells that produces the cork cells of bark. Also called phellogen. 8

cork cell A cell in bark that has walls encrusted with suberin; cork prevents loss of water through the bark and prevents entry of pathogens. 8

corm A subterranean, vertical stem that is thick and fleshy and has only thin papery leaves. Example: gladiolus. 5

corolla A collective term for all the petals of a single flower. 9

cortex In stems and roots, the primary tissue located between the epidermis and the phloem. 5

cotyledon In embryos of seed plants, the rather leaflike structures involved in either nutrient storage (most dicots and gymnosperms) or nutrient transfer from the endosperm (most monocots). 9, 22

covalent bond A chemical bond in which electrons are shared between two atoms. 2

cpDNA Plastid DNA; it is in the form of closed circles, without histones. Usually there are many circles per plastid. See *mtDNA*. 3, 19

Crassulacean acid metabolism (CAM) A metabolism in which carbon dioxide is absorbed at night and fixed temporarily into organic acids. During daytime, the acids break down, carbon dioxide is released, and C_3 photosynthesis occurs. 10

crista (pl.: cristae) One of the tubular or vesicular folds of the mitochondrial inner membranes. 3, 11

critical night length The length of darkness that must be exceeded by short-day plants, or not exceeded by long-day plants, for flowering to be initiated. 14

cross-fertilization The fertilization of a gamete by a genetically distinct gamete derived from a different parent: Alternative: *self-fertilization;* see *cross-pollination*. 16

cross-field pitting In secondary xylem, pit-pairs formed between ray cells and axial tracheids or vessels. 8

crossing over During prophase I of meiosis, after homologous chromosomes have paired and a synaptonemal complex has formed, the DNA of the homologs breaks and the end of each homolog is attached to the other homolog, resulting in two new chromosomes. 4, 16

cross-pollination The pollination of a flower by pollen from a completely different plant. Alternative: *self-pollination*. 9

cuticle A layer of cutin on epidermal cells; the cuticle reduces water loss but also unavoidably restricts the entry of carbon dioxide. 5

cutin A polymer of fatty acids that is water impermeable; it forms a layer (cuticle) on the epidermis. 2, 5

cyanide-resistant respiration Synonym for *thermogenic respiration*. 11

cyanophycin granule A storage particle of nitrogen compounds, found in some cyanobacteria. 25

cyclic electron transport The flow of electrons from P700 back to plastoquinone in photosynthesis, such that there is proton pumping but no synthesis of NADPH. Alternative: *noncyclic electron transport*. 10

cytochromes Small electron carriers that contain iron. 10, 11

cytokinesis Division of the protoplasm of a cell, as opposed to nuclear division, karyokinesis. 4

cytokinins A class of hormones involved in cell division, apical dominance, and embryo development among other things. 14

cytoplasm Protoplasm consists of nucleus, vacuoles, and cytoplasm. 3

cytosol Synonym for *hyaloplasm*.

D

dark reactions Synonym for *stroma reactions*.

day-neutral plant A plant that is induced to flower by factors other than night length. 14

degenerate code In the genetic code, many amino acids are coded by several codons, not just one each. 15

dehydration reaction A chemical reaction in which a proton is lost from one reactant and a hydroxyl from the other, creating a water molecule. 2

deletion mutation A mutation involving loss of DNA. 16

demography The study of the age distribution of the individuals of a population. 26

denitrification The conversion, by microbes, of nitrate to nitrogen gas, thus making nitrogen no longer available to plants. 25

deoxyribonucleic acid (DNA) The information molecule in nuclei, plastids, mitochondria, and prokaryotes. 2, 14, 15, 16

deoxyribose A five-carbon sugar occurring in DNA. 2

derived features Features present in modern organisms but not in their ancestors. Alternative: *relictual features*. 1, 18

determinate growth Growth that stops at a genetically predetermined size. Typical of leaves and flowers but not of whole shoots and roots. Alternative: *indeterminate growth*. 5

diakinesis see *prophase I*.

diatom Common name for algae of the class Bacillariophyceae; they have two silica shells that fit together like the parts of a Petri dish. 19

diatomaceous earth Oceanic sediments formed by accumulation of the silica shells of diatoms. 19

diatoxanthin A xanthophyll pigment found in brown algae. 19

dibiontic A life cycle with an alternation of generations. Alternative: *monobiontic*. 19

dichotomous branching A forking that results in two nearly equal branches (or leaf veins or secretory ducts). Alternatives: *monopodial* and *sympodial branching*. 19–20, 25

dicot Informal term for any member of the flowering plant class Magnoliopsida, a broadleaf plant. Alternative: *monocot*. 5

dicotyledon Synonym for *dicot*.

dictyosome A stack of thin vesicles held together in a flat or cup-shaped array; dictyosomes receive vesicles from endoplasmic reticulum along their forming face, then modify the material in the vesicle lumen or synthesize new material. Vesicles swell and are released from the maturing face. See *Golgi apparatus*. 3

differentially (selectively) permeable membrane A membrane that permits the passage of certain types of particles and inhibits the passage of others. 12

diffuse porous wood Wood in which the vessels of late wood are about as numerous and as wide as those of early wood. Alternative: *ring porous wood*. 8

diffusion The random motion of particles from regions of higher concentration to regions of lower concentration. 12

dihybrid cross A cross in which two characters are considered simultaneously. 16

dikaryotic The condition of having two genetically different nuclei present in one cell, especially in ascomycete and basidiomycete fungi. Alternative: *monokaryotic*. 24

dimorphic Having two distinct morphologies, such as in plants with a juvenile and an adult form. 5, 9

dinoflagellate Common name for algae of division Pyrrhophyta. Most have two flagella, each of which lies in a groove. 19

dioecy The condition in which a species has two types of sporophyte—one with stamens and one with carpels. See *monoecy*. 9

diploid Refers to two full sets of chromosomes in each nucleus, as typically found in sporophytes and zygotes. See *haploid*. 4, 16

diplotene See *prophase I*.

disaccharide A small carbohydrate composed of just two simple sugar residues. 2

distal Refers to the position of a structure distant from a point of reference; relative to a stem, a leaf blade is distal to the petiole. Alternative: *proximal*. 5

diurnal Daytime; a diurnal plant opens its flowers at sunrise. Alternative: *Nocturnal*. 14

divergent speciation The evolution of part of a species into a new species, with the remainder of the species continuing as the original species or evolving into a third, new species. Alternative: *phyletic speciation*. 17

DNA cloning Producing large numbers of identical copies of DNA, usually by inserting it into bacteria and allowing the bacteria to multiply. 15

DNA denaturation Synonym for *DNA melting*.

DNA hybridization The slow cooling of a mixture of short DNA molecules such that complementary strands encounter each other and hydrogen bond into double helices. 15

DNA ligase An enzyme that can attach two strands of DNA to each other, repairing nicks and linking Okazaki fragments into complete molecules of DNA. 15

DNA melting Heating a DNA double helix gently until the hydrogen bonds are broken and one molecule separates from the complementary molecule. Synonym: DNA denaturation. 15

DNA microarray A microscope slide containing numerous dots, each dot being a fragment of DNA from a genome being studied. 15

DNases (DNAases) Enzymes that digest DNA. 15

domain All living organisms are now classified as belonging to one of three large clades called domains. Two domains (Bacteria and Archaea) contain prokaryotes; the third domain (Eukarya) contains eukaryotes. 1, 18, 25

dominant trait A trait whose phenotype completely masks that of the alternative (recessive) allele in the heterozygous condition. 16

dosage-dependent response A situation in which the amount of response is correlated with the amount of stimulus. Alternative: *all-or-none response*. 14

double fertilization The process unique to angiosperms in which one sperm fertilizes the egg (forming a zygote) and the other sperm fertilizes the polar nuclei (forming the primary endosperm nucleus). 9

duplication division Synonym for *mitosis*.

E

early wood Synonym for *spring wood*.

ecosystem The set of physical nonliving environmental factors of a region, plus the communities of organisms of that region. 26

ecotypes Races of a species that are each adapted to particular environmental factors in certain parts of the species range. 26

edaphic Refers to the soil: soil factors are edaphic factors. 26

egg apparatus A name for the egg cell and the one or two adjacent synergids in an angiosperm megagametophyte. 9

elater In the sporangia of liverworts and horsetails, small twisted cells that push the spores out of the sporangium. 20–21

elasticity A property of sclerenchyma walls; if stretched to a new size or shape, they return to their original size and shape once the deforming force is removed. Alternative: *plasticity*. 5

electron carrier A cofactor that carries electrons between reactions. Examples: nicotinamide adenine dinucleotide (phosphate), flavin adenine dinucleotide. 2

electron transport chain A series of electron carriers that transfer electrons from a donor, which becomes oxidized, to a receptor, which becomes reduced. 10, 11, 13, 25

electronegativity A measure of an atom's or a molecule's tendency to give off or take on electrons. 2

elicitor Something that provokes a response; in pathogenic fungi, a component of the fungal cell wall that provokes the plant to produce phytoalexins. 24

Embden-Meyerhoff pathway Synonym for *glycolysis*.

embolism A packet of water vapor formed in xylem when the water column cavitates (breaks). Often called an air bubble. 12

embryo sac A common synonym for the megagametophyte of flowering plants. 9

embryophyte A rarely used term for all plants that are not algae—that is, all that have multicellular reproductive structures with sterile tissue. 19

enation A small, projecting flap of tissue, thought to have been the ancestor of leaves in the lycophytes. 23

end-product inhibition A type of regulatory mechanism in which an enzyme at the front of a metabolic pathway is inhibited by the end product. Synonym: *feedback inhibition*. 2

endergonic reaction A reaction that absorbs energy. 2

endocarp The innermost layer of the fruit wall, the pericarp. See *exocarp* and *mesocarp*. 9

endocytosis A process of absorbing material into a cell by forming an invagination in the plasma membrane, then pinching it shut and forming a vesicle. See *exocytosis*. 3

endodermis A sheath of cells surrounding the vascular tissue of roots (and occasionally horizontal stems); their Casparian strips prevent uncontrolled diffusion between root cortex and root vascular tissue by means of walls and intercellular spaces. 6

endogenous rhythm A rhythm generated entirely within an organism whose periodicity is not maintained by an external rhythm. 14

endomembrane system The membranous organelles of a cell. 3

endophyte The portion of a parasitic plant that grows entirely within the body of the host plant. 11

endoplasmic reticulum (ER) A system of narrow tubes and sheets of membrane that form a network throughout the cytoplasm. If ribosomes are attached, it is rough ER (RER) and is involved in protein synthesis. If no ribosomes are attached, it is smooth ER (SER) and is involved in lipid synthesis. 3

endoreduplication The repeated synthesis of all nuclear DNA without its partitioning into separate nuclei by division. See *gene amplification*. 4

endosperm The tissue, usually polyploid, which is formed during double fertilization only in angiosperms and which nourishes the developing embryo and seedling. 9

endosymbiont theory The theory that postulates that plastids and mitochondria arose as prokaryotes that were living symbiotically within an early eukaryotic cell. 19

endosymbiosis (primary and secondary) In the endosymbiont theory, a single primary act of endosymbiosis occurred when a cyanobacterium was engulfed by an early eukaryote, initiating chloroplasts. Several acts of secondary endosymbiosis have occurred as other eukaryotes engulfed entire cells of green or red algae, establishing new groups of photosynthetic organisms, such as brown algae, dinoflagellates, and euglenoids. 19

endothermic reaction A reaction that absorbs heat. 2

enhancer elements Regions of DNA upstream from the structural region of a gene, which increase the ability of RNA polymerases to transcribe the gene. 15

entrainment The resetting of an endogenous rhythm by an exogenous stimulus. 14

entropy A measure of disorder in a system. 10

epicotyl In the embryo of a seed, the embryonic shoot, located above the cotyledons. Synonym: *plumule*. 9

epidermis The outermost layer of the plant primary body, covering leaves, flower parts, young stems, and roots. 5

epigynous See *inferior ovary*.

epiphyte A plant that grows on another plant, either attached to it or climbing over it, but not parasitizing it. 5, 26

epistasis The control of the expression of one gene by another, distinct gene, often involving the genes for the various enzymes catalyzing a single metabolic pathway. 16

essential element An element required for normal growth and reproduction. Major (macro) essential elements are needed in relatively large concentration; minor (micro, trace) essential elements are needed only in low concentrations. 13

essential organs In flowers, the spore-producing organs—stamens and carpels. Alternative: *nonessential organs*. 9

ethylene A hormone involved in fruit ripening, the initiation of aerenchyma in submerged roots and stems, and other aspects of development. 14

eudicot This is the clade of angiosperms that contains most species formerly known as dicots. They have broad leaves and pollen with more than one germination pore. 18, 23

eukaryotes Organisms that have true nuclei and membrane-bounded organelles. Eukaryotes are plants, animals, fungi, and protists, but not bacteria. Alternative: *prokaryotes*. 3, 25

euphyllophytes Plants that have megaphylls (synonym: *euphyll*), the horsetails, ferns, and seed plants. 21

eustele The vascular system of the stems of seed plants, composed of bundles around a pith. Compare: *protostele, siphonostele,* and *atactostele*. 21, 22

eutrophication The process that occurs as rivers and lakes receive too many mineral nutrients (usually as pollution from fertilized fields); they develop an overabundance of algae, much of which dies and decays, depleting the oxygen in the water and harming fish. 13

evolution The change of nucleotide sequences in a species' DNA through natural selection, genetic drift, or accident. 17

exalbuminous seed A seed with little or no endosperm at maturity. Alternative: *albuminous seed*. 9

excited state electron An electron that has absorbed a quantum and moved to a higher orbital; it has more energy than when it is in its ground state. 10

exergonic reaction A reaction that releases energy. 2

exocarp Outermost layer of the fruit wall; in fleshy fruits, the rind or peel. See *mesocarp* and *endocarp*. 9

exocytosis The transfer of vesicle or vacuole-carried material to the outside of a cell by fusion of the plasma membrane with the membrane of the vesicle or vacuole. See *endocytosis*. 3

exons Those portions of the structural region of a gene whose information is actually translated into protein. Alternative: *intron*. 15

exothermic reaction A reaction that releases heat. 2

extrinsic protein Synonym for *peripheral protein*. Alternative: *intrinsic protein*. 3

eyespot Synomym for *stigma* in phototactic algae.

F

facilitated diffusion The diffusion of hydrophilic (polar or charged) molecules through a membrane by means of channels composed of hydrophilic, membrane-spanning proteins. 3

facultative aerobe (facultative anaerobe) An organism that can use oxygen if present or survive without it. 11, 25

false fruit See *accessory fruit*.

fascicular cambium The portion of the vascular cambium that develops within a vascular bundle. Alternative: *interfascicular cambium*. 8

fatty acid Long chains of carbon with only hydrogens as side groups and a carboxyl group at one end. If all carbon-carbon bonds are single, the fatty acid is saturated; if it contains any double bonds, it is unsaturated. 2, 3, 11

feedback inhibition Synonym for *end-product inhibition*.

fermentation Synonym for *anaerobic respiration*. See *respiration*.

ferredoxin An iron-containing proteinaceous electron carrier in photosynthesis. 10

fertilization Fusion of two gametes (or their equivalents in some fungi and algae). Synonym: *syngamy*. 9, 19–20

fiber A sclerenchyma cell that is long and tapered and has pointed ends; provides a tissue with strength and flexibility. Compare: *sclereid*. 5, 8

field capacity The amount of water held by a soil after drainage due to gravity has been completed. 13

filament The stalk of a stamen, it elevates the anther. 9, 23

first filial generation (F₁) The progeny of an experimental cross. 16

first-order reaction A reaction involving only a single molecule, which breaks apart during the reaction. 2

flagellin Protein monomer that makes up the flagella of prokaryotes. 25

flagellum (pl.: flagella) Like a cilium but longer, an organelle of locomotion. Both flagella and cilia have a 9 + 2 arrangement of microtubules. 3, 20, 24

floridean starch A storage polysaccharide found in red algae. 19

fluid mosaic Biological membranes are two-dimensional fluids in which various types of lipids and intrinsic proteins can diffuse laterally. 3

fluorescence The spontaneous emission of a quantum by an excited electron, which allows the electron to return to its ground state. 10

form genus A genus based on a character that is not a reliable indicator of evolutionary relationships. Unrelated species are grouped together. 18

founder The first individual(s) that establishes a population in a new habitat; especially important in adaptive radiation. 17

frameshift error During the translation of mRNA, a misalignment of the ribosome such that the triplets it reads are not true codons. 15

freely permeable membrane A membrane that allows everything to pass through. See *impermeable* and *semipermeable membrane*. 3

fret A set of thylakoid membranes that connect grana in chloroplasts. 10

frond Nontechnical term for the leaf of a fern. 21

frugivore A fruit-eating animal, important in the dissemination of seeds. 9

fruit In angiosperms, the structure that forms from carpels and associated tissues after fertilization. 9

fruiting body In fungi, the spore-producing and spore-disseminating structures such as a mushroom or puffball. 24

fucoxanthin Xanthophyll pigment found in brown algae. 19

funiculus (pl.: funiculi) The stalk of an ovule. 9, 23

fusiform initials In a vascular cambium, the long cells with tapered ends that give rise to axial cells of the secondary xylem and secondary phloem. Alternative: *ray initials*. 8

G

G₁ Part of interphase of the cell cycle, G₁ (gap 1) is the interval between cell division and the synthesis of DNA in the nucleus. G₁ is often the longest phase, during which the nucleus actively directs cytoplasmic metabolism. 4

G₂ Part of interphase of the cell cycle between the synthesis of DNA and the beginning of nuclear division. 4

gametangium (pl.: gametangia) Any structure that produces gametes. 19–23

gamete A haploid sex cell, such as an egg or sperm. **megagamete** A large, immobile gamete; an egg. **microgamete** A small, often mobile gamete; a sperm. 9, 19–23

gametophore The leafy stem of a moss gametophyte. 20

gametophyte A haploid plant that produces gametes. Alternative: *sporophyte*. **megagametophyte** A gametophyte that produces megagametes (eggs) only. **microgametophyte** A gametophyte that produces microgametes (sperms) only. 9, 19–23

gas vesicle In certain prokaryotes, a conical vesicle of gas, surrounded by a layer of protein; provides buoyancy. 25

gene In DNA, a sequence of nucleotides which contains information necessary for the metabolism and structure of an organism. 15

gene amplification The repeated synthesis of the DNA of just one or a few genes, not the entire genome. See *endoreduplication*. 4, 15

gene family A set of several copies of an ancestral gene, all located within a single haploid genome. The various copies arise through gene duplication and then may evolve independently. 16

gene flow The movement of alleles within a population by the movement of pollen or seeds. 17

gene pool The total population of all the alleles in all the sex cells of the individuals of a population. 17

generative cell In the pollen grains of seed plants, the cell that gives rise directly to the sperm cells. Alternative: *vegetative cell*. 9, 22

genetic code The set of nucleotide triplets in DNA that code for amino acids to be inserted during protein synthesis. 15

genetic drift In a small gene pool, the alteration in allele frequencies mostly by accidents rather than by natural selection. 17

genetically modified (GM) plant A plant whose genotype and phenotype have been altered using recombinant DNA techniques. This term is not used for hybrids that have been produced in crosses using selective breeding. 15

genome All the alleles of an organism. 4

genotype The set of alleles present in an organism's genome. Alternative: *phenotype*. 15

genus (pl.: genera) A group of species closely related by descent from a common ancestor. 18

gibberellic acid A natural gibberellin. 14

gibberellins A class of hormones involved in stem elongation, seed germination, and other processes. 14

gluconeogenesis Formation of glucose from 3-phosphoglyceraldehyde. 10, 11

glucose One of the most abundant simple sugars, a six-carbon monosaccharide. Component of starch, cellulose, and many metabolic pathways. 2, 10, 11

glycocalyx (pl.: glycocalyces) A secretion of mucilaginous material that surrounds many prokaryotes. 25

glycolipid Lipid molecules with sugars attached. See *glycoprotein*. 3

glycolysis The metabolic pathway by which glucose is broken down to pyruvic acid. Synonym: *Embden-Meyerhoff pathway*. 11

glycoprotein A protein with sugars attached; often the sugars occur in short chains less than ten sugars long. See *glycolipid*. 3

glyoxysome See *microbody*.

Golgi apparatus A collection of interconnected dictyosomes, as many as several thousand, the entire set forming a cup-shaped apparatus. Often treated as a synonym for dictyosome; Golgi apparatuses are rare in plants, common in animals. Synonym: *Golgi body*. 3

Golgi body See *Golgi apparatus*.

Gondwanaland Southern portion of the ancient continent of Pangaea. Compare: *Laurasia*. 27

granum (pl.: grana) A set of flat vesicles in chloroplasts, involved in chemiosmotic phosphorylation. 3, 11

grazer An herbivore that eats low herbs such as grasses. Compare: *browser*. 26

ground meristem A term refering to any expanse of meristematic tissue that produces a somewhat uniform mature tissue. 5

ground state electron An electron in its most stable orbital, when it contains the least amount of energy. Alternative: *excited state electron*. 10

growth ring Synonym for *annual ring* in wood. "Growth ring" is preferred because rings are occasionally not strictly annual (sometimes two are produced in one year, and none might be produced in adverse years). 8

guard cells A pair of epidermal cells capable of adjusting their size and shape, causing the stomatal pore to open when they swell and close when they shrink. 5

gymnosperm Common name for plants with naked seeds; all seed plants that are not angiosperms. Examples: conifers and cycads. 22

gynoecium (pl.: gynoecia) A collective term referring to all the carpels of a flower. 9

H

habit The characteristic shape or appearance of the individuals of a species. 5

habitat The set of conditions in which an organism completes its life cycle. 26

hairpin loop A small kink in the end of a messenger RNA molecule that is being transcribed; the kink appears involved in signaling that transcription should stop. 15

half-inferior ovary An ovary that is partly inferior, such that the sepals, petals, and stamens appear attached to its side; those appendages are perigynous. Alternatives: *superior* and *inferior ovary*. 9

haploid Refers to one full set of chromosomes per nucleus. Gametes, spores, and gametophytes typically have haploid nuclei. See *diploid*. 4, 16, 19

hardwood A term applied to both dicot trees and shrubs and to their wood, because in general dicot wood contains fibers. Alternative: *softwood*. 8

haustorium (pl.: haustoria) The structure by which a parasite enters and draws nutrients from a plant; in fungi, it is a hypha; in mistletoes and similar parasites, it is a modified root. 7, 24

heartwood The colored, aromatic wood in the center of a trunk or branch; all the wood parenchyma cells have died and no water conduction is occurring. Alternative: *sapwood*. 8

helical thickening A pattern of secondary wall deposition in tracheids and vessel elements; the wall occurs as one or two helical bands. 5

hemicelluloses A set of cell wall polysaccharides that crosslink cellulose molecules in plant cell walls. 3

hemiparasite A parasite that draws water, minerals, and perhaps some organic material from its host but also carries out photosynthesis. Compare: *holoparasite*. 26

herb A plant that consists only of primary tissues; lacking wood. 5

herbivore An animal that eats plants. Compare *browser* and *grazer*. 26

heteroblasty The phenomenon in which an individual plant produces several different types of leaves. 6

heterocyst In cyanobacteria, specialized cells in which nitrogen fixation occurs. 25

heterokaryosis In fungi, a condition in which each cell has two or more nuclei of different mating types. Alternative: *homokaryosis*. 24

heterokont A clade of organisms with the synapomorphy of having two different types of flagella: one whiplash, one tinsel. 19

heteromorphic generations A dibiontic life cycle in which sporophyte and gametophyte are easily distinguishable morphologically. Alternative: *isomorphic generations*. 19

heterospory A condition in which the life cycle of a plant contains two types of spores, microspores and megaspores. Alternative: *homospory*. 9, 19–23

heterotroph An organism that obtains its carbon from organic molecules, not from carbon dioxide. Compare: *autotroph*. 10

heterozygote A diploid organism with two different alleles for a particular gene. Alternative: *homozygote*. 16

hexose monophosphate shunt Synonym for *pentose phosphate pathway*.

hilum (pl.: hila) Scar produced when a seed breaks from the funiculus. 9

histones A set of basic nuclear proteins that complex together and with DNA, first forming nucleosomes and then complexing further into chromosomes. 3, 4, 15, 16

hnRNA Heterogeneous nuclear RNA. 15

holdfast The portion of a nonmotile, attached alga that holds the alga to its substrate. 19

holoparasite A parasite that draws all its water, minerals, and organic material from its host and is unable to carry out photosynthesis. Compare: *hemiparasite*. 26

homeotic mutation A mutation that causes organs to be produced in unusual places on a body or in altered numbers. Flowers that have excess numbers of petals are the result of homeotic mutations. 14

homokaryosis In fungi, a condition in which all nuclei in the mycelium are identical genetically. Alternative: *heterokaryosis*. 24

homologous chromosomes In a diploid nucleus, each type of chromosome is present as a pair, one inherited paternally, the other maternally. 4

homologous features Features that are the phenotypic expression of homologous genes, those related by descent from common ancestral genes. Alternative: *analogous features*. 18

homospory A condition in which the life cycle of a plant contains only one type of spore. Alternative: *heterospory*. 9, 19–23

homozygote A diploid organism with two identical alleles for a particular gene. Alternative: *heterozygote*. 16

hormone A chemical that is produced by one part of a plant, often in response to a stimulus, and then is transported to other parts and induces responses in appropriate sites. 14, 15

hyaloplasm The liquid substance of protoplasm, excluding all the organelles such as nuclei, plastids, ribosomes: Synonym: *cytosol*. 3

hybrid sterility A postzygotic isolation mechanism in which a hybrid zygote can grow into an adult but cannot form fertile gametes. 17

hydrogen bonding The weak attraction between polar molecules. 2, 12

hydroids In the stems of some mosses, cells that resemble tracheids and are involved in water conduction. 20

hydrolysis The breaking of a chemical bond by adding water to it, a proton being added to one product and a hydroxyl to the other. 2

hydrophilic "Water loving"-refers to compounds that are relatively soluble in water and other polar solvents and insoluble in lipids and other nonpolar solvents. Opposite: *hydrophobic*. 2, 3

hydrophobic "Water fearing"-refers to compounds that are relatively insoluble in water and other polar solvents and soluble in lipids and other nonpolar solvents. Opposite: *hydrophilic*. 2, 3

hydroponics The growing of plants in a water solution, without the use of soil. 13

hymenium (pl.: hymenia) In a fungus fruiting body, the layer of fertile cells that produce spores. 24

hypersensitive reaction An extreme susceptibility to a fungus, such that a plant cell dies before nutrients can be drawn from it. 24

hyphae (sing.: hypha) The long, narrow filaments, either coenocytic or cellular, that constitute the body of a fungus. 24

hypocotyl The portion of an embryo axis located between the cotyledons and the radicle. 9

hypogynous See *superior ovary*. 9

hypothesis (pl.: hypotheses) A model of a phenomenon constructed from observations of the phenomenon. It must make testable predictions about the outcome of future observations or experiments. 1

I

immobile essential element An element that cannot be removed from mature tissues; if a plant becomes deficient in an immobile element, young tissues show symptoms even though older tissues may have extra. Alternative: *mobile essential element*. 13

imperfect flower A flower lacking either stamens or carpels or both. Alternative: *perfect flower*. 9

imperfect fungi Fungi that do not undergo sexual reproduction or whose sexual reproduction has never been observed. 24

impermeable A barrier that allows nothing to pass through. See *freely permeable* and *semipermeable membrane*. 3

incipient plasmolysis As a plant cell is losing water and shrinking, the point at which the protoplast just begins no longer to exert pressure against the wall. 12

included phloem In certain types of anomalous secondary growth, patches of secondary phloem may be located within the secondary xylem. 8

incomplete dominance A situation in which the phenotypes of both alleles of a heterozygote are expressed. Alternative: *complete dominance*. 16

incomplete flower A flower that is missing one or more of the four basic appendages (sepals, petals, stamens, carpels, or any combination). Alternative: *complete flower*. 9

indehiscent Remaining closed at maturity, not opening; true of many fruits and the megasporangia of seed plants. 9, 22, 23

independent assortment In a double heterozygote, the distribution of one allele of one gene during meiosis is not linked to the distribution of either allele of the homologous gene. Alternative: *linked genes*. 16

indeterminate growth Growth not limited by a plant's own genetic development program: most trees have indeterminate growth. Alternative: *determinate growth*. 5

indole acetic acid (IAA) An auxin. 14

infection thread During the invasion of roots by nitrogen-fixing bacteria, the bacteria are encased in an invagination of the plant cell wall, the infection thread. 7

inferior ovary An ovary located below the sepals, petals, and stamens; those appendages are epigynous. Alternatives: *superior* and *half-inferior ovary*. 9

inflorescence A discrete group of flowers. 9

inner bark The innermost, living layer of bark, located between the vascular cambium and the innermost cork cambium. Alternative: *outer bark*. 8

insertion mutation A mutation involving insertion of the new DNA into a sequence of pre-existing DNA. 16

insertion sequence A small transposable element that contains only the genes coding for the enzymes necessary for the element's excision and insertion. See *transposon*. 16

integument In flowers, the covering layer over the nucellus of an ovule. Usually two integuments (inner and outer) are present. 9, 23

intercalary meristem See *meristem*.

interfascicular cambium The portion of the vascular cambium that develops from parenchyma cells located between vascular bundles. Alternative: *fascicular cambium*. 8

interkinesis (pl.: interkineses) The portion of meiosis that occurs between telophase I and prophase II. There is no duplication of DNA. 4

internode Portion of a stem where there are no leaves; portion between nodes. 5

interphase The portion of the cell cycle that is not cell division; interphase consists of G_1, S, and G_2. 4

interpolation theory The theory that vascular plants arose from monobiontic algae and a sporophyte generation was gradually interpolated into the life cycle. Compare: *transformation theory*. 21

intrinsic protein A protein that is an integral part of a membrane, deeply embedded in it; it cannot be washed out of the membrane easily. Alternative: *peripheral protein*. 3

intron A portion of the structural region of a gene whose information is not translated into protein; instead, the intron-transcribed portions of the hnRNA are removed and digested. Alternative: *exon*. 15

inversion mutation A mutation in which a portion of the DNA double helix is excised, turned end for end, and ligated into place with reverse order. 16

ionic bond Chemical bond between two molecules, one of which is negatively charged, the other positively charged. 2

isogamy Sexual reproduction in which all gametes are structurally identical; there are no sperms and eggs. Alternatives: *anisogamy* and *oogamy*. 9, 19–21

isolation mechanism A structure or metabolism that inhibits the movement of substances from one region to another. 12

isomorphic generations A dibiontic life cycle in which sporophytes and gametophytes are almost indistinguishable. Alternative: *heteromorphic generations*. 19

isopentenyl adenosine A natural cytokinin. 14

isotype specimen A specimen obtained from the same plant or clone as the type specimen. 18

J

juvenile plant A plant that is too immature to flower, even if otherwise appropriate stimuli are present. Alternative: *adult plant*. 14

K

K Symbol for carrying capacity. 26

karyogamy Fusion of the nuclei of two gametes after protoplasmic fusion (plasmogamy). See *syngamy*. 9, 24

karyokinesis (pl.: karyokineses) Division of a nucleus, as opposed to cell division, cytokinesis. The two types of karyokinesis are mitosis and meiosis. 4, 16

kinetin An artificial cytokinin. 14

kinetochore In a chromosome, the kinetochore is the point at which spindle microtubules attach to the centromere. 4

Krebs cycle Synonym for *citric acid cycle* and *tricarboxylic acid cycle*.

L

labyrinthine wall Synonym for *transfer wall*.

lamina (pl.: laminae or laminas) The broad, expanded part of a leaf; not the petiole. 6

laminarin A reserve polysaccharide found in brown algae. 19

late wood Synonym for *summer wood*.

lateral meristem A name describing the position of the vascular cambium and cork cambium. 8

lateral veins The major, large vascular bundles of a leaf, which are attached to the midrib or the petiole. Larger than minor veins. 6

Laurasia Northern portion of the ancient continent of Pangaea. Compare: *Gondwanaland*. 27

leaf axil Portion of a node immediately above the attachment point of a leaf. 5

leaf gap In fern vascular tissue, an area above a leaf trace where there is no conducting tissue. 21

leaf primordium (pl.: primordia) An extremely early stage in leaf development, when the leaf exists only as a pronounced bulging of the shoot apical meristem. 6

leaf scar The region on a stem where a leaf was attached prior to abscission. 6

leaf sheath Almost exclusively in monocots, the basal portion of a leaf, wrapped around the stem above a node. 6

leaf trace A vascular bundle that extends from the stem vascular bundles through the cortex and enters a leaf. 6

lenticel In bark, a region of cork cells with intercellular spaces, permitting diffusion of oxygen into inner tissues. 8

leptoids In the stems of some mosses, elongate cells that resemble sieve cells and are involved in carbohydrate conduction. 20

leptotene See *prophase* I.

lethal allele An allele whose expression results in death. 16

leucoplast See *plastid*.

life forms A classification of the ways plants are adapted morphologically for surviving stressful seasons. 26

light compensation point The level of illumination at which photosynthetic fixation of carbon dioxide just matches respiratory loss. 10

light-dependent reactions In photosynthesis, the set of reactions directly driven by light. Alternative: *stroma reactions*. 10

lignin A complex compound that impregnates most secondary cell walls, making them stronger, more waterproof, and resistant to attack by fungi, bacteria, and animals. 3, 8, 11

lignophyte Synonym for a plant that develops wood. This is used as an informal name for the clade that contains the woody plants. 22

ligule In selaginellas, a small flap of tissue on the upper surface of a leaf. 21

limiting factor The growth of any organism depends on the availability of many factors such as water, light, and various nutrients; the rate of the organism's growth is limited by whichever factor is in short supply. 13

linkage group A set of genes that do not undergo independent assortment, being part of one chromosome. 16

linked genes Genes located close to each other on a chromosome undergo crossing over only rarely, so they are linked. Alternative: *independent assortment*. 16

lipid body A spherical droplet of oil or other lipid, common in the cells of many seeds. 3

lipids A class of compounds that are hydrophobic and water insoluble. Examples: fats, oils, and waxes. 2

lipopolysaccharide layer In gram-negative bacteria, a layer of sugar-bearing lipids located just exterior to the peptidoglycan layer. 25

lithotroph An organism (all are prokaryotes) that obtains energy by oxidizing inorganic mineral elements. 11

littoral zone On a sea coast, the region between low tide and high tide; a common habitat for brown algae. 19

locule The cavity within a structure such as a sporangium, gametangium, or carpel. 23

long-day plant A plant that is induced to flower by nights shorter than the critical night length. 14

lumen The interior of any structure such as a vesicle, vacuole, oil chamber, or resin duct. 3

lysis The bursting of an animal cell or a plant protoplast due to excessive absorption of water. Do not confuse with plasmolysis. 12

lysosome An organelle that contains digestive enzymes and is involved in recycling the components of worn-out organelles; in plants, the central vacuole acts as the lysosome. 3

M

macronutrient Synonym for major *essential element*.

mannitol A reserve polysaccharide in brown algae. 19

manoxylic wood Wood that contains significant amounts of axial parenchyma, such as that of cycads. Compare: *pycnoxylic wood*. 22

map unit A measure of the separation of genes on a chromosome; one map unit equals a 1% probability that crossing over will occur between them. Synonym: *centimorgan (cM)*. 16

maternal inheritance Uniparental inheritance for genes contributed by the megagamete but not by the microgamete. 16

mating types Biochemical distinctions such that gametes of identical mating type cannot fuse; only those of compatible mating types can undergo syngamy. 24

matric potential A component of water potential; a measure of the effect of a matrix on a substance's ability to absorb or release water. 12

matrix The liquid within a mitochondrion. 3, 11

megagametangium (pl.: megagametangia) A structure that produces megagametes (eggs). 19

megagamete In an oogamous species, the larger, immobile gamete. Synonym: *egg*. 9, 19–23

megagametophyte See *gametophyte*.

megapascal (MPa) A unit for quantifying pressure; 1 MPa is about 10 bars or 10 atmospheres of pressure. 12

megaphyll A leaf that has evolved from a branch system. Present in ferns and all seed plants. Alternative: *microphyll*. 21–23

megasporangium (pl.: megasporangia) A structure that produces megaspores. 9, 19–23

megaspore A large spore that grows into a megagametophyte that produces egg cells. Alternative: *microspore*. 9, 19–23

megaspore mother cell Synonym for *megasporocyte*.

megasporocyte In a heterosporous species, a cell that undergoes meiosis, resulting in the production of a megaspore. Synonym: *megaspore mother cell*. 9, 21, 22

meiosis Reduction division, a process in which nuclear chromosomes are duplicated once but divided twice, such that the resulting nuclei each have only one half as many chromosomes as the mother cell. 4, 16

meiosis I The first division of meiosis, during which the chromosome number per nucleus is reduced. Synapsis and crossing over occur during meiosis I. 4

meiosis II The second division of meiosis, during which centromeres divide and the two chromatids of one chromosome become independent chromosomes. 4

meristem A group of cells specialized for the production of new cells. **apical meristem** Located at the farthest point of the tissue or organ produced. **basal meristem** Located at the base. **intercalary meristem** Located between the apex and the base. **lateral meristem** Located along the side. 5

meristoderm In brown algae, the outer layer of the body, which is both meristematic and photosynthetic. 19

mesocarp The middle layer of the fruit wall. See *exocarp* and *endocarp*. 9

mesokaryotes A term proposed for dinoflagellate algae, because their nuclear and cytoplasmic organization appears to be intermediate between those of prokaryotes and eukaryotes. 19

mesophyll All tissues of a leaf except the epidermis. 6

metaphase The second phase of mitosis during which chromosomes move to the center of the spindle, the metaphase plate. 4

metaphase I The second phase of meiosis I, similar to metaphase of mitosis except that homologous pairs of chromosomes are involved. 4

metaphase II A phase of meiosis II, similar to metaphase of mitosis. 4

metaphase plate The region in the center of the spindle where chromosomes become aligned during nuclear division. 4

metaphloem The part of the primary phloem that differentiates late, after adjacent cells have completed their elongation. Alternative: *protophloem*. 5

metaxylem The part of the primary xylem that differentiates late, after adjacent cells have completed their elongation. Alternative: *protoxylem*. 5

MPa Megapascal, a unit of pressure measurement equivalent to about 70 pounds per square inch. 12

microbody A class of two types of small, vesicle-like organelles. Peroxisomes are involved in photorespiration, the detoxification of harmful products of photosynthesis. Glyoxysomes are involved in respiring stored fatty acids. 3, 10, 11

microfibril As adjacent molecules of cellulose are synthesized, they crystalize into a microfibril, which may be 10 to 25 nm wide. 3

microfilaments A structural element composed of actin and believed to be involved in the movement of organelles other than flagella, cilia, or chromosomes. 3

microgametangium (pl.: microgametangia) A structure that produces microgametes (sperm cells). 19

microgametophyte See *gametophyte*.

micronutrient Synonym for *minor essential element*.

microphyll The type of leaf that evolved from an enation; present in lycophytes. Alternative: *megaphyll*. 21, 22

microplasmodesmata (sing.: microplasmodesma) In cyanobacteria, small holes in the walls of heterocysts, connecting the heterocysts to adjacent cells. 25

micropyle In an ovule, the small apical opening created where the integuments do not meet; the pollen tube enters through the micropyle. 9, 23

microRNA general term for any of several types of RNA that are extremely short (a few dozen nucleotides or less) and appear to be involved in controlling gene expression. 15

microsporangium (pl.: microsporangia) A structure that produces microspores. 9, 19–23

microspore A small spore that grows into a microgametophyte that produces sperm cells. Alternative: *megaspore*. 23

microspore mother cell Synonym for *microsporocyte*.

microsporocyte In a heterosporous species, a cell that undergoes meiosis, resulting in the production of four microspores. Synonym: *microspore mother cells*. 9

microtubules A skeletal element in eukaryotic cells, composed of alpha and beta tubulin. Microtubules constitute the mitotic spindle, phragmoplast, and axial component of flagella. 3

middle lamella (pl.: lamellae or lamellas) The layer of adhesive pectin substances that acts as the glue that holds the cells of a multicellular plant together. 3

midrib The large, central vascular bundle of a leaf. Alternative: *lateral* and *minor veins*. 6

minor essential element See *essential element*.

minor veins The smallest veins of a leaf, branching off lateral veins. There is no criterion that strictly distinguishes between minor veins and lateral veins. 6

mitochondrion (pl.: mitochondria) The eukaryotic organelle involved in aerobic respiration, particularly the citric acid cycle and respiratory electron transport. 3, 11

mitosis (pl. mitoses) Duplication division—a type of nuclear division (karyokinesis) in which nuclear chromosomes are first duplicated, then divided in half, one daughter nucleus receiving one set, the other daughter nucleus receiving the other set. Alternative: *meiosis*. 4, 16, 19

mobile essential element An element that can be removed from mature tissues and transported to young or newly formed tissues. Alternative: *immobile essential element*. 13

molecular pump An integral membrane protein that forces molecules from one side of a membrane to the other, using energy in the process called active transport. 3, 12

monilophytes The clade containing ferns in a broad sense (*Psilotum, Equisetum*, Leptosporangiatae, Marattiales, and Ophioglossales). 21

monobiontic A life cycle with only one free-living generation; there is no alternation of generations. Alternative: *dibiontic*. 19

monocot Informal term for any member of the flowering plant class Liliopsida. Examples, lily, iris, palm, agave. Alternative: *dicot*. 5

monocotyledon Synonym for *monocot*.

monoecy The condition in which a species has imperfect flowers (some staminate, others carpellate), but both are located on the same sporophyte. See *dioecy*. 9

monohybrid cross A cross in which only a single trait is analyzed, disregarding all other traits. 16

monokaryotic In fungi, having one type of nucleus per cell. Alternative: *dikaryotic*. 24

monomer The subunit of a polymer. 2

monopodial branching Branching in which one shoot is dominant and forms a distinct trunk, all other shoots being significantly different from the trunk. Alternatives: *sympodial* and *dichotomous branching*. 21–23

monosaccharide Synonym for *simple sugar*; the monomer of polysaccharides. 2

morphogenic (morphogenetic) response A response in which the quality of the plant changes, such as conversion from a vegetative state to a floral state. 14

motor cells Cells that swell and shrink in plant organs capable of repeated, reversible movement, such as insect traps and petioles of leaves that undergo sleep movements. 12

mtDNA Mitochondrial DNA: it is in the form of closed circles, without histones. See *cpDNA*. 3

mucigel The mucilaginous, slimy material secreted by root caps and root hairs. 7

multiple fruit A fruit formed by the crowding together of the individual fruits of an entire inflorescence. Alternatives: *simple fruit* and *aggregate fruit*. 9

murein Synonym for *peptidoglycan*. 25

mutagen Any chemical or physical force that causes a change in the sequence of nucleotides in DNA. 16

mutation Any change in the sequence of DNA. 16

mutualism An interaction of two species in which both species benefit. Compare: *competition*. 26

mycelium (pl.: mycelia) The diffuse mass of hyphae which constitutes the vegetative body of a fungus. 20

mycorrhizae (sing.: mycorrhiza) Fungi that form a symbiotic relationship with roots, usually of benefit to plants because they provide phosphorus. **ectomycorrhizae** A type in which the fungi invade only the outermost cells of the root. **endomycorrhizae** A type in which the fungi invade all cells of the root cortex. 7, 13

N

nastic response A nongrowth response that is stereotyped and not oriented with regard to the stimulus. 14

natural selection The preferential survival, in natural conditions, of those individuals whose alleles cause them to be more adapted than other individuals with different alleles. 17

natural system of classification A classification based on evolutionary, phylogenetic relationships. Alternative: *artificial system of classification*. 18

necrotroph A fungus similar to a parasite which attacks its host virulently, killing it and absorbing the released nutrients. 24

nectary A gland that secretes a sugary solution that typically attracts pollinators. 9, 23

negative feedback system An enzyme system in which the product inhibits the action of one or several of the enzymes that produce it. 14

netted venation Synonym for *reticulate venation*, typically found in leaves of dicots. 6

niche The set of conditions exploited best by one species. 26

nitrification The conversion, by microbes, of ammonia to nitrate. 25

nitrogen assimilation The incorporation of ammonium into organic compounds within an organism. 13

nitrogen fixation The conversion of atmospheric nitrogen into any compound that can be used by plants, typically either nitrate or ammonium. 13

nitrogenase The enzyme responsible for nitrogen fixation. 13

nocturnal Nighttime; a nocturnal plant opens its flowers at dusk. Alternative: *diurnal*. 13

node Point on a stem where a leaf is attached. 5

nodule In roots of plants that form symbiotic associations with nitrogen-fixing bacteria, regions of the root swell, forming nodules whose cells contain the bacteria. 7

nonclimacteric fruits Fruits that ripen slowly and steadily, without a sudden burst of metabolism (the climacteric) at the end. Alternative: *climacteric fruits*. 14

noncompetitive inhibitor A small molecule that inhibits an enzyme by attaching to some site other than the active site. Alternative: *competitive inhibitor*. 2

noncyclic electron transport The flow of electrons from water to NADPH during the light-dependent reactions of photosynthesis. Alternative: *cyclic electron transport*. 10, 25

nondisjunction During meiosis II, the nonseparation of the two chromatids of a chromosome, such that one daughter cell receives both while the other daughter cell receives none. 16

nonessential organs In flowers, the sepals and petals. Alternative: *essential organs*. 9

nonpolar molecule A molecule that does not carry even a partial charge anywhere. 2

nonstoried cambium A cambium in which the fusiform initials are not aligned horizontally. Alternative: *storied cambium*. 8

nonvascular plant A plant that lacks vascular tissue: mosses, liverworts, and hornworts. 20

nucellus (pl.: nucelli) The megasporangium of an ovule. 9, 23

nuclear envelope A set of two membranes, the inner and the outer nuclear envelopes, that surround the nucleus. 3

nuclear pores Structures in the nuclear envelope that are involved in transport of material between nucleus and cytoplasm. 3

nucleic acid A polymer of nucleotides. 2, 15, 16

nucleoid In prokaryotes, the portion of the protoplasm where DNA circles are concentrated. 25

nucleolus (pl.: nucleoli) Organelles located within the nucleus, nucleoli are areas where ribosomal RNAs are synthesized and assembled into ribosomal subunits. 3, 15

nucleoplasm The substance located in the nucleus-DNA, histones, RNA, enzymes, nucleic acids, water. 3

nucleosome A nuclear particle composed of histones with DNA wrapped around them; a basic aspect of chromosome structure. 15

nucleotide The monomer of nucleic acids; each nucleotide consists of a nitrogenous base, a sugar, and a phosphate group. 2, 15, 16

nucleus (pl.: nuclei) In eukaryotic cells, the organelle that contains DNA and is involved in inheritance, metabolism control and ribosome synthesis. 3

O

obligate aerobe An organism that must have oxygen to survive. Synonym: strict aerobe. Alternative: *obligate anaerobe*. 11, 25

obligate anaerobe An organism that is killed by exposure to oxygen. Synonym: strict anaerobe. Alternative: *obligate aerobe*. 11, 25

Okazaki fragments During DNA replication, the set of short fragments that grow discontinuously along one strand of DNA and must be ligated into one continuous DNA molecule. 16

oligosaccharide A compound made up of a few simple sugars (monosaccharides). 2

ontogeny Synonym for development, morphogenesis. 14, 15

oogamy A type of sexual reproduction in which the two gametes are distinctly different structurally; one is a microgamete (sperm), and the other is a megagamete (egg). Alternatives: *anisogamy* and *isogamy*. 9, 19–23

oogonium Synonym for *megagametangium* in certain organisms. 19

open reading frame (ORF) A sequence of DNA that has many components of a gene (promotor, transcription start and end sites, exon/intron boundaries) but which is not actually known to act as a gene. 15

operational habitat Those aspects of a habitat that definitely affect the organism being considered. 26

operculum (pl.: opercula) In mosses, the lidlike top of a sporangium. 20

organ A structure composed of a variety of tissues; seed plants are considered to have only three organs: roots, stems, and leaves. 5

organelles The "little organs" of a cell, such as nuclei, plastids, mitochondria, and ribosomes. Many are membrane-bounded compartments, others are nonmembrane structures composed of protein or protein and RNA. 3

ortholog Genes in different species that evolved from the same ancestral gene. 15

osmosis Diffusion through a membrane. 12

osmotic potential A component of water potential; a measure of the effect of solute particles on a substance's ability to absorb or release water. 12

outcrossing Synonym for *cross fertilization*. 16

outer bark The outermost dead layers of bark, from the surface to the innermost cork cambium. Alternative: *inner bark*. 8

ovary In a flower, the base of the carpel; the region that contains ovules and will develop into a fruit. 9, 23

overtopping In the evolution of unequal branching, the ability of one shoot to grow for a longer time than the other shoot that resulted from the branching. 21, 22

ovule The structure in a carpel that contains the megasporangium and will develop into a seed. 9, 23

ovuliferous scale The scale that bears the ovule in gymnosperm seed cones. Synonym: *cone scale*. 22

oxidation state A measure of the number of electrons added to or removed from a molecule during an oxidation-reduction reaction. 10, 11

oxidative phosphorylation The formation of ATP from ADP and phosphate, powered by energy released through respiration. 10, 11

oxidize To raise the oxidation state of a molecule by removing an electron from it. 10, 11

oxidizing agent An electron carrier that is not carrying electrons. Alternative: *reducing agent*. 10, 11

P

P680 The reaction center of photosystem II. 10

P700 The reaction center of photosystem I. 10

P_{fr} See *phytocrome*.

P_r See *phytochrome*.

P channel The groove in the ribosome small subunit in which the nascent protein-carrying tRNA occurs. Alternative: *A channel*. 15

pachytene See *prophase I*.

palisade parenchyma Any part of leaf mesophyll in which the cells are elongate and aligned parallel to each other. 6

palmately compound See *compound leaf*.

Pangaea The ancient supercontinent composed of all the world's land, it existed in the late Paleozoic Era and consisted of Laurasia (north) and Gondwanaland (south). 27

parallel evolution The evolution of similar homologous features in two or more groups due to similar selective pressures, the features being derived from a common ancestral feature. Compare: *convergent evolution*. 17

parallel venation Almost exclusively in monocot leaves, a pattern in which all veins run approximately parallel to each other, either from the base of the leaf to its tip or from the midrib to the margin. Alternative: *reticulate venation*. 6

paralog Genes within single species that evolved from the same ancestral gene. 15

paramylon A storage polysaccharide in euglenoid algae. 19

paraphyletic group A clade that contains an ancestral taxon and several but not all its descendants. 18

parasexual cycle In fungi, a condition in which compatible nuclei of a heterokaryotic mycelium fuse, then undergo meiosis and crossing over, even though they are not in a sporangium. 24

parasite See *biotroph*.

parenchyma Cells with only thin primary walls; all other features are highly variable from type to type. Alternatives: *collenchyma* and *sclerenchyma*. 5

parental type chromosome A chromosome that, after meiosis, has not undergone crossing over. Alternative: *recombinant chromosome*. 16

parsimony The concept of minimum complexity; the simplest hypothesis that explains several observations is the most parsimonious. In cladistics, a cladogram with the least number of steps is the most parsimonious. 18

passage cell A cell in the endodermis that has only Casparian strips whereas all surrounding endodermis cells have thickened waterproof walls. 7

pectic substances A set of polysaccharides that constitute the middle lamella and act as the glue that holds together the cells of multicellular plants. 3

pedicel The stalk of an individual flower. Compare: *peduncle*. 9

peduncle The stalk of an inflorescense, a group of flowers. Compare: *pedicel*. 9

pellicle In euglenoid algae, a layer of elastic proteins on the cell surface. 19

pentose phosphate pathway A type of respiration in which glucose is converted either to ribose or erythrose. Synonyms: *hexose monophosphate shunt* and *phosphogluconate pathway*. 11

peptide bond The chemical bond that holds the amino acid residues together in a protein. 2

peptidoglycan The polymer that constitutes the main source of strength in the wall of bacteria. 25

perennial plant A plant that lives for more than 2 years. Compare: *annual* and *biennial plant*. 5, 9

perfect flower A flower that has both stamens and carpels. Alternative: *imperfect flower*. 9

perforation In a vessel element, the hole(s) where both primary and secondary walls are missing. Alternative: *pit*. 5

pericarp Technical term for the fruit wall, composed of one or more of the following: exocarp, mesocarp, endocarp. 9

periclinal wall A wall that is parallel to a nearby surface, especially the outer surface of the plant. Alternative: *anticlinal wall*. 8

pericycle An irregular band of cells in the root, located between the endodermis and the vascular tissue. 7

periderm Technical term for bark; it consists of cork, cork cambium, and any enclosed tissues such as secondary phloem. 8

perigynous See *half-inferior ovary*.

peripheral protein A membrane protein that is only weakly associated with the surface of the membrane. Alternative: *intrinsic protein*. 3

perisperm A nutritive tissue in seeds of the dicot order Caryophyllales, formed as nucellus cells proliferate. 9, 23

peristome teeth In a moss capsule, the one or two sets of teeth-like structures around the mouth of the sporangium. 20

permeable membrane A membrane through which materials can pass. Alternative: *impermeable membrane*. Compare: *differentially permeable membrane*. 2, 12

peroxisome See *microbody*.

petals The appendages, usually colored, on a flower, most often involved in attracting pollinators. See also *corolla*. 9, 23

petiole The stalk of a leaf. 6

petiolule The stalk that attaches a leaflet to the rachis of a compound leaf. 6

phage Synonym for *bacteriophage*.

phellem Technical term for cork. 8

phelloderm Parenchyma cells produced to the inside by the cork cambium; usually only a layer or two are formed, and phelloderm is not present in all species. 8

phellogen Synonym for *cork cambium*.

phenotype The physical, observable characteristics of an organism. Alternative: *genotype*. 16, 17

phloem The portion of vascular tissues involved in conducting sugars and other organic compounds, along with some water and minerals. Alternative: *xylem*. 5, 8

phosphogluconate pathway Synonym for *pentose phosphate pathway*.

phospholipid A type of lipid containing two fatty acids and a phosphate group bound to glycerol. 2

phosphorylation The attaching of a phosphate group to a substrate. See *chemiosmotic phosphorylation*, *photophosphorylation*, and *substrate level phosphorylation*. 2, 10, 11

photic zone In aquatic environments, upper regions that are illuminated sufficiently to allow photosynthesis. 25

photoautotroph An organism that obtains its energy through photosynthesis and its carbon from dioxide. Compare: *chemoautotroph*. 10, 11

photoperiod In reference to cycles of light and darkness, the length of time that uninterrupted light is present. 14

photophosphorylation The formation of ATP from ADP and phosphate by means of light energy; a part of photosynthesis. 10, 25

photorespiration The oxidation of phosphoglycolate produced when RuBP carboxylase adds oxygen, not carbon dioxide, to RuBP. 10

photosynthetic unit A cluster of photosynthetic pigments and electron carriers embedded in the chloroplast membrane. 10

photosystem I The pigments and electron carriers that transfer electrons from P700 to NADPH. 10

photosystem II The pigments and electron carriers that transfer electrons from water to P700 in photosystem I. 10

phototroph An organism that obtains its energy through photosynthesis. 25

phragmoplast During cell division, the phragmoplast is a set of short microtubules oriented parallel to the spindle microtubules; it catches dictyosome vesicles and guides them to the site where the new cell wall (cell plate) is forming. See *phycoplast*. 4, 19

phragmosome In cell division, a set of microtubules, actin filaments, and cytoplasm that is involved in dividing a large vacuole and creating a cytoplasmic bridge through which the phragmoplast and cell plate can grow. 4

phycobilins The accessory pigments of cyanobacteria and red algae. Phycocyanin absorbs blue light, and phycoerythrin absorbs red light. 19, 25

phycobilisome In cyanobacteria and red algae, a particle involved in photosynthesis and composed of phycobilins and biliproteins. 19, 25

phycocyanin See *phycobilins*.

phycoerythrin See *phycobilins*.

phycoplast In the cell division of some algae, a set of microtubules oriented parallel to the plane of the new cell wall and involved in wall formation. Compare: *phragmoplast*. 19

phyletic speciation The evolution of one species into a new species, such that the original species no longer exists. Alternative: *divergent speciation*. 17

phyllode Synonym for *cladophyll*.

phyllotaxy The arrangement of leaves and axillary buds on a stem. 5

phylogenetic relationships The evolutionary relationships that result as one taxon evolves into others that then evolve into still more. 18

phytoalexins Lipid-like or phenolic compounds produced by plants in response to attacks by fungi. 24

phytochrome A pigment involved in many aspects of morphogenesis in which the stimulus is red light or the length of a dark period. P_{fr} absorbs far-red light, P_r absorbs red light. 14

phytoferritin A protein molecule that binds and stores iron; mostly found in plastids. 3

pinna (pl.: pinnas or pinnae) Technical name for a leaflet of a fern. 21

pinnately compound See *compound leaf*.

pioneers The first plants to inhabit an area that previously had no life. 26

pit In a sclerenchyma cell, an area where there is no secondary wall over the primary wall and material can pass into or out of the cell.
 blind pit A pit that does not meet another pit in the adjacent cell.
 pit-pair A set of aligned pits in adjacent sclerenchyma cells. 5

pit connection In red algae, a large hole in the wall between two cells. 19

pit membrane The set of two primary walls and middle lamella that occurs between the two pits of a pit-pair. 5

pit plug In red algae, material that fills the hole (pit connection) between two cells. 19

pith The region of parenchyma located in the center of most shoots and some roots, surrounded by vascular bundles. 5

placenta (pl.: placentas or placentae) Tissue in the ovary of a carpel to which the ovules are attached. 9

planation In the telome theory of the origin of megaphylls, the concept that all branching occurred in one plane, resulting in a flat system. 21

plant growth substance Term used for any hormone-like compound, whether natural or artificial. 14

plasma membrane The semipermeable membrane that surrounds the protoplasm of a cell. Synonym: *plasmalemma*. 3

plasmalemma Synonym for *plasma membrane*.

plasmid A small circle of DNA occurring in some bacteria and acting like a bacterial chromosome. 15

plasmodesma (pl.: plasmodesmata) A narrow hole in a primary wall, containing some cytoplasm, plasma membrane, and a desmotubule; a means of communication between cells. See *symplast*. 3, 5

plasmodium (pl.: plasmodia) The body of a slime mold, a large mass of protoplasm with hundreds or thousands of nuclei. 24

plasmogamy The fusion of the cytoplasm of two gametes during sexual reproduction. See *karyogamy*. 9, 24

plasmolysis The shrinking of a cell due to loss of water. Do not confuse with lysis. 12

plasticity A property of collenchyma walls; once stretched to a new shape or size, usually by growth, the wall retains that new shape or size. Alternative: *elasticity*. 5

plastids A family of organelles within plant cells only. **proplastids** Young plastids common in meristematic cells. **chloroplasts** Chlorophyll-rich plastids that carry out photosynthesis. **amyloplasts** Plastids that store starch. **chromoplasts** Plastids that contain red or yellow pigments, located in flowers and fruits. **leucoplasts** Colorless plastids. 3

plastochron The length of time required for an apical meristem to make the cells of one node and internode. 5

plastocyanin A copper-containing electron carrier. 10

plastoglobulus (pl.: plastoglobuli) A droplet of lipid located within a plastid. 3

plastoquinone A class of lipid-soluble electron carriers. 10

pleiotropic effects The multiple phenotypic expressions of a single allele whose activity affects various aspects of metabolism. 16

pleomorphic bacteria Bacteria that lack cell walls and thus do not have a constant shape. 25

plumule Synonym for *epicotyl*.

plurilocular gametangium (pl.: gametangia) In brown algae, a multicellular structure in which each cell produces a gamete. Alternative: *unilocular sporangium*. 19

pneumatocyst Synonym for air bladder, a swollen, hollow structure in brown algae; it increases buoyancy. 19

poikilohydry The ability of some simple plants and lichens to allow their bodies to become extremely dehydrated without dying, then being capable of rehydrating and becoming metabolically active quickly. 12

point mutation A mutation affecting only a single nucleotide. 16

polar molecule A molecule that has a partial positive charge at one site and a partial negative charge at another site. 2

polar nuclei (sing.: nucleus) The two nuclei of the central cell of the megagametophyte in a flowering plant; after fertilization, they become the endosperm nucleus. 9

polar transport Transport in one direction based on an organ's structure, regardless of its spatial orientation. 14

polarity In development and morphogenesis, having two different ends, usually a shoot/root polarity, or a petiole/leaf tip polarity. 14

pollen In seed plants, the microspores and microgametophytes. 9, 21–23

pollen chamber In gymnosperms, a cavity just above the nucellus in the ovule, the site where pollen accumulates and germinates. 22

pollen tube After landing on a compatible stigma or gymnosperm megasporophyll, a pollen grain germinates with a tubelike process that carries the sperm cells to the vicinity of the egg cell. 9, 22

polymer A large compound composed of a number of subunits, monomers. 2

polymerase chain reaction A method of copying minute quantities of DNA using bacterial enzymes. 15

polyploid Refers to a nucleus that contains three or more sets of chromosomes. 16

polysaccharide A compound made up of many simple sugars (monosaccharides). 2

polysome The complex formed when numerous ribosomes bind to the same molecule of messenger RNA. 3, 15

population All the individuals of a species that live in a particular area at the same time and can interact with each other. 17, 26

positive feedback system An enzyme system in which the product stimulates the action of one or several of the enzymes that produce it. The ripening of climacteric fruits is an example. 14

postzygotic isolation mechanisms Phenomena that prevent successful interbreeding of two populations but that act after fertilization; the two sets of chromosomes are incompatible and cannot produce a fertile adult. Alternative: *prezygotic isolation mechanism.* 17

P-protein Phloem-protein, or fibrillar protein that plugs sieve pores and prevents leakage if sieve elements are damaged. 12

predation A relationship in which one species benefits and the other is harmed; not often used in botany. 26

preprophase band Prior to cell division, the preprophase band is a set of microtubules that encircles the cell just interior to the cell membrane and located at the site where the new wall will attach to the pre-existing wall. 4

presentation time The length of time a stimulus must be present in order for an organism to perceive it. 14

pressure flow hypothesis The hypothesis that flow in phloem is due to active loading in sources and active unloading in sinks. 12

pressure potential A component of water potential; a measure of the effect of pressure or tension on a substance's ability to absorb or release water. 12

prezygotic isolation mechanism Phenomena that prevent successful interbreeding of two populations but that act so early that fertilization is not possible. Alternative: *postzygotic isolation mechanism.* 17

primary cell wall A cell wall present on all plant cells except some sperm cells; it is formed during cell division and is usually thin, but some may be thick. See *secondary cell wall* and *collenchyma.* 3

primary consumer In ecology, a synonym for *herbivore.* 26

primary endosperm nucleus The nucleus formed by the fusion of one sperm nucleus and two polar nuclei. 9

primary growth The production of new cells by shoot and root apical meristems and leaf primordia. Alternative: *secondary (woody) growth.* 5, 8

primary phloem The phloem of the primary body; it differentiates from cells derived from apical meristems or forms in leaves, flowers, and fruits. Compare: *secondary phloem.* 5, 8

primary pit field An area of a primary cell wall that is especially thin and contains numerous plasmodesmata. See *plasmodesma.* 3, 5

primary plant body The herbaceous body produced by apical meristem (roots, stems, leaves, flowers, and fruits). Alternative: *secondary (woody) plant body.* 5, 8

primary producer In ecology, synonym for *autotroph.* 26

primary structure of a protein The order of the sequence of amino acids in the protein. 2, 15, 16

primary tissues The tissues derived more or less directly from an apical meristem or leaf primordium; the tissues of the primary plant body. Alternative: *secondary tissues.* 2, 5

primary xylem The xylem of the primary body; the xylem produced from cells derived from apical meristems or formed in leaves, flowers, and fruits. Compare: *secondary xylem.* 5, 8

primer RNA During DNA replication, a short piece of RNA that is synthesized against open DNA and from which DNA polymerase can begin building a new molecule of DNA. 16

primordium (pl.: primordia) A small mass of cells that grows into an organ such as a leaf or petal. 5

procambium (pl.: procambia) Tissue that matures into primary xylem and primary phloem. 5

prochlorophytes A group of prokaryotes that have both chlorophyll *a* and *b*; believed to be closely related to the ancestors of plastids in algae and plants. 19, 25

procumbent cell In rays in secondary xylem, cells that are longer radially than they are tall; they typically have little or no cross-field pitting. Alternative: *upright cell.* 8

proembryo An early stage of embryo development, usually considered to encompass the stages between the zygote and the initiation of the cotyledon primordia. 9

progymnosperms A group of extinct plants believed to have been the ancestors of gymnosperms. 2

prokaryotes Organisms that have no true nucleus or membrane-bounded organelles. Prokaryotes are eubacteria, cyanobacteria, and archaebacteria. Alternative: *eukaryotes.* 3, 25

promoter region That portion of a gene in which control molecules and RNA polymerases bind during gene activation and transcription. Alternative: *structural region.* 15

propagules Parts of a plant involved in reproduction and dissemination. Examples are seeds, bulbs, plant pieces that can form roots and grow into a new plant. 9, 26

prophase The initial phase of mitosis during which the nucleolus and nuclear membrane break down, chromosomes begin to condense, and the spindle begins to form. 4

prophase I The first phase of meiosis, similar to prophase of mitosis, with the following additional processes: **leptotene** Initiation of chromosome condensation. **zygotene** The pairing of homologous chromosomes (synapsis). **pachytene** Formation of the synaptonemal complex. **diplotene** Homolog separation and chiasmata become visible. **diakinesis** Complete separation of homologs and terminalization of chiasmata. 4

proplastid See *plastids.*

protease An enzyme that digests protein. 2

protoderm A term that refers to any immature epidermal cell. 5

protonema (pl.: protonemata) In nonvascular plants, the mass of alga-like cells that grow from the spore during germination. 20

protophloem The part of the primary phloem that differentiates early, while adjacent cells are still elongating. Alternative: *metaphloem.* 5

protoplasm All the substance of a cell, usually considered not to include the cell wall. The protoplasm of a single cell is a protoplast. See *cytoplasm* and *hyaloplasm.* 3

protostele A vascular cylinder that has no pith; common in roots and early vascular plants. Alternative: *siphonostele.* 21–23

protoxylem The part of the primary xylem that differentiates early, while adjacent cells are still elongating. Alternative: *metaxylem.* 5, 21–23

provascular tissue Cells in the primary plant body that later differentiate into xylem, phloem, or vascular cambium. Synonym: *procambium.* 5

proximal Refers to the position of a structure near a point of reference; relative to a stem, a petiole is proximal to a leaf blade. Alternative: *distal.* 5

pseudomonopodial branching A type of sympodial branching that strongly resembles monopodial branching, having what appears to be one main shoot. 21

pulvinus (pl.: pulvini) A jointlike region of a petiole where motor cells are located and flexion occurs during nastic responses. 14

purine One of the two types of nitrogenous bases occurring in nucleotides; purines have two ring structures. Compare: *pyrimidine*. 2

pycnoxylic wood Wood with little or no axial parenchyma, such as that of gymnosperms and progymnosperms. Compare: *manoxylic wood*. 22

pyrenoid In many algae and hornworts, a region of the choroplast believed to be involved in polymerization of polysaccharides. 19, 20

pyrimidine One of the two types of nitrogenous bases occurring in nucleotides; pyrimidines have only a single ring structure. Compare: *purine*. 2

Q

quantitative trait locus (QTL) analysis Most phenotype characters are the result of interactions of many genes (loci); QTL analysis involves numerous crosses to determine just how much various genes contribute to a particular phenotype. 16

quantum (pl.: quanta) A particle of electromagnetic energy. Synonym: *photon*. 10

quaternary structure The association of several proteins into a large structure such as a microtubule or synthetic complex. 2

quiescent center Portion of the root apical meristem in which cell division does not occur. 7

R

r Symbol for biotic potential.

rachis (pl.: rachises) The extension of the petiole in a compound leaf; all leaflets are attached to the rachis. 6

radial system (of wood) The set of rays within wood. 8

radially symmetrical Divisible into two equal halves by any median longitudinal section. Synonym: *actinomorphic*. 9

radicle The main root of a seed; it is the direct continuation of the embryonic stem. 7, 9

rain shadow The diminished amount of rainfall on the leeward side of a mountain compared to the side that faces an ocean. 27

random distribution In ecology, the distribution of individuals in the habitat with no obvious, identifiable pattern. Compare: *clumped* and *uniform distribution*. 26

raphe In seeds, a ridge caused by the fusion of the funiculus to the side of the ovule. In diatoms, a groove in the shell. 19

raphide A long, narrow, needle-like crystal, occurring in clusters in specialized cells. 3

ray In secondary xylem and phloem, a radial series of cells produced by ray initials. Alternative: *axial tissue*. 8

ray initials In a vascular cambium, the short cells that give rise to the rays of the secondary xylem and phloem. Alternative: *fusiform initials*. 8

ray tracheid Horizontal tracheids in the secondary xylem rays of gymnosperms. 8

reaction center A special chlorophyll *a* molecule actually involved in the transfer of electrons in photosynthesis. 10

reaction wood Wood formed in response to mechanical stress. **tension wood** The reaction wood of dicots, formed on the upper side of a branch. **compression wood** The reaction wood of gymnosperms, formed on the lower side of a branch. 8

reannealing Synonym for DNA hybridization by slowly cooling a mixture of single-stranded DNA molecules. 15

receptacle The stem (axis) of a flower, to which all the other parts are attached. 8 In the brown alga *Fucus*, the ends of the branches where conceptacles are located. 19

recessive trait A trait whose phenotype is completely masked by that of the alternative (dominant) allele in the heterozygous condition. 16

recombinant chromosome A chromosome that results from crossing-over in meiosis, being composed of parts of the paternal and maternal homologs. Alternative: *parental type chromosome*. 16

recombinant DNA DNA constructed from pieces of DNA from several sources, either through crossing-over in meiosis or laboratory manipulation. 15

red tide A bloom of algae in which the cells become so numerous as to give the water a reddish tint. 19

redox potential The tendency of a molecule to accept or donate electrons during a chemical reaction. 10

reduce To lower the oxidation state of a molecule by adding an electron to it. 10, 11

reducing agent An electron carrier that is carrying electrons. Alternative: *oxidizing agent*. 10, 11

reducing power The ability of an electron carrier to force electrons onto another compound. 10, 11

reduction division Synonym for *meiosis*.

regular flower A radially symmetrical flower. Alternative: *zygomorphic flower*. 9

relictual feature A feature that occurs in a modern organism and was inherited relatively unchanged from an ancient ancestor. Alternative: *derived feature*. 1, 18

replication fork In DNA replication, the point at which the double helix opens and formation of new DNA occurs. 16

replicon During DNA replication, a short segment of DNA that has opened and where replication is occurring. 16

reproductive barrier Any physical or metabolic phenomenon that prevents two members of a species from interbreeding. 17

reproductive isolation The inability of some members of a species to interbreed with other members through either biotic or abiotic reproductive barriers. 17

respiration The breakdown of molecules such that part of their energy is used to make ATP. If oxygen is required as an electron acceptor, the process is aerobic respiration; if not, it is anaerobic respiration (fermentation). 11, 25

respiratory quotient An indicator of the type of substrate being respired, RQ (carbon dioxide liberated)/(oxygen consumed). 11

resting phase Old synonym for *interphase*.

restriction endonucleases Enzymes that recognize specific sites in DNA double helices, then cut the two strands in complementary sites. 15

restriction map A map of a DNA molecule made by exposing it to restriction endonucleases, thus showing the number of cleavage sites and the number of bases between sites. 15

reticulate thickening A pattern of secondary wall deposition in tracheids and vessel elements; the wall is deposited as a series of intersecting bands that have a netlike appearance. 5

reticulate venation A netlike pattern of veins in a leaf, found primarily in leaves of dicots (broadleaf plants). Synonym: *netted venation*. Alternative: *parallel venation*. 6

retrovirus A virus whose genetic material is single-stranded RNA; the most common type of plant virus. 15

reverse transcriptase An enzyme that catalyzes the synthesis of DNA, using RNA as a template. 15

rhizoids In certain fungi, algae, and nonvascular plants, cells or parts of cells that project into the substrate and anchor the organism. 19–20, 24

rhizome A fleshy, horizontal, subterranean stem involved in allowing the plant to migrate laterally. Example: bamboo, iris. 5

rhyniophytes The common name for *Rhynia* and its close relatives, the earliest vascular plans. 21

ribonucleic acid (RNA) A polymer of ribose-containing nucleotides; there are three classes: messenger RNA, transfer RNA, and ribosomal RNA. 2, 15, 16

ribophorin An integral membrane protein that binds ribosomes to the membrane, usually in rough endoplasmic reticulum. 3

ribose A five-carbon sugar occurring in ribonucleic acid, among other things. 2

ribosome An organelle responsible for protein synthesis; ribosomes consist of a large subunit and a small subunit, both made of proteins and ribosomal RNA (rRNA). 3, 15

ring porous wood Wood in which the early wood has more numerous and larger vessels than the late wood. Alternative: *diffuse porous wood*. 8

RNA polymerase I The enzyme that transcribes ribosomal genes into ribosomal RNA. 15

RNA polymerase II The enzyme responsible for transcribing genes into messenger RNA. 15

RNA polymerase III The enzyme responsible for transcribing the 5S RNA gene into 5S RNA. 15

root cap A layer of parenchyma cells that cover and protect the root apex. 7

root hair zone Region of a root tip, just proximal to the zone of elongation, where epidermal cells grow out as root hairs. 7

root pressure Water in root xylem is under pressure due to the active accumulation of salts by the endodermis and the accompanying influx of water. 7, 12

rough ER (RER) See *endoplasmic reticulum*.

rubisco Synonym for *RuBP carboxylase*.

RuBP carboxylase The enzyme in photosynthesis that carboxylates RuBP, thus bringing carbon into the plant's metabolism. Synonym: *rubisco*. 10

S

S phase The synthesis phase of the cell cycle, during which nuclear DNA is replicated (synthesized). 4, 16

saprobe Synonym for *saprotroph*.

saprotroph A fungus that attacks an organism that has died from other causes. 24

sapwood The light-colored, light-scented outermost wood of a trunk or branch; conduction is still occurring and many wood parenchyma cells are alive. Alternative: *heartwood*. 8

scalariform thickening A pattern of secondary wall deposition in tracheids and vessel elements; the wall is interrupted by broad, short pits that cause the wall to have a ladder-like appearance. 5

scientific method A means of analyzing the physical universe. Observations are used as the basis for constructing a hypothesis that predicts the outcome of future observations or experiments. Anything that can never be verified cannot be accepted as part of a scientific hypothesis. 1

scientific name The binomial name of a species, consisting of the genus name and the species epithet. 18

sclereid A sclerenchyma cell that is rather cubical, not long like a fiber. Masses of sclereids provide a tissue with strength and rigidity. Alternative: *fiber*. 5

sclerenchyma Sclerenchyma cells have both a primary wall and an elastic secondary wall; if stretched to a new size or shape, the wall returns to its original size and shape after the deforming force is removed. Alternatives: *parenchyma* and *collenchyma*. 5

sclerotium (pl.: sclerotia) In fungi, a mass of tightly adhering, resistant hyphae able to survive harsh conditions for years. 24

scutellum (pl.: scutella) In grass seeds, the single cotyledon, which is shield-shaped and digests and absorbs the endosperm during germination. 9

secondary cell wall A cell wall present only in certain cells (sclerenchyma) and formed only after cell division has been completed. When present, the secondary wall is located interior to the primary wall and is typically impregnated with lignin. See *primary cell wall, sclerenchyma,* and *xylem*. 3, 5

second-order reaction A reaction involving the collision of two particles. 2

secondary consumer In ecology, a synonym for carnivores and omnivores. 26

secondary (woody) growth Growth that occurs by means of either the vascular cambium or the cork cambium. It results in wood and bark, the secondary tissues. Alternative: *primary growth*. 5, 8

secondary phloem Phloem derived from the vascular cambium. Compare: *primary phloem*. 5, 8

secondary (woody) plant body The wood and bark produced by the vascular cambium and cork cambium. Alternative: *primary plant body*. 5, 8

secondary structure Short sequences of regular helix or regular pleating in a protein. 2

secondary tissues The tissues of the secondary plant body-those produced by the vascular cambium and the cork cambium. Alternative: *primary tissues*. 8

secondary xylem Xylem derived from the vascular cambium. Compare: *primary xylem*. 5, 8

seed coat The protective layer on a seed; the seed coat develops from one or both integuments. Synonym: *testa*. 9

selectively permeable membrane Synonym for *differentially permeable membrane*.

self-assembly The automatic assembly of a larger structure solely due to interaction of charges and hydrophobic/hydrophilic regions on the molecules. 2

selfing Pollinating a plant's stigma with pollen from the same plant or a plant of identical genotype. 16

self-pollination The pollination of a flower by pollen from the same flower or another flower on the same plant. Alternative: *cross-pollination*. 9

semiconservative replication Refers to the fact that during DNA replication, one new molecule is paired with one original molecule such that in every new chromosome, half the DNA is conserved from the pre-existing chromosome. 16

semipermeable membrane A membrane that is relatively permeable to some substances and relatively impermeable to others. Synonym: *differentially permeable*. See *impermeable* and *freely permeable*. 3, 12

sepal In flowers, the outermost of the fundamental appendages, most often providing protection of the flower during its development. See *calyx*. 9

septum (pl.: septa) Synonym for cross wall, especially with regard to fungi and algae. 24

sessile Refers to an organ that has no stalk but rather is attached directly to the stem or other underlying organ. For leaves, alternative is *petiolate*. 6

seta (pl.: setae) In mosses and liverworts the stalk of the capsule, located between the foot and the sporangium. 20

sexual reproduction Reproduction in which genomes of two individuals are brought together in one nucleus followed by meiosis with crossing-over. 9, 16, 24

short-day plant A plant that is induced to flower by nights longer than the critical night length. 14

sieve area In phloem, an area on a sieve element wall in which numerous sieve pores occur. 5

sieve cell The phloem conducting cell in nonangiosperms; sieve cells are long and tapered with small sieve areas over much of their surfaces. Alternative: *sieve tube member*. 5

sieve element Refers to either or both types of phloem-conducting cells: sieve cells and sieve tube members. 5

sieve plate In phloem, the end walls of sieve tube members, bearing one or several large sieve areas with large sieve pores. 5

sieve pore In sieve elements, the holes (enlarged plasmodesmata) in the primary walls; sieve pores permit movement of phloem sap from one sieve element to another. 5

sieve tube In the phloem of angiosperms, a column of sieve tube members interconnected by large sieve areas and sieve pores. 5

simple cone A cone with just one axis, bearing only sporophylls. Most cones are simple cones. Alternative: *compound cone*. 22

simple fruit A fruit that develops from a single carpel or the fused carpels of a single flower. Alternatives: *aggregate fruit* and *multiple fruit*. 9

simple leaf A leaf in which the blade consists of just one part. Alternative: *compound leaf*. 6

simple pit In sclerenchyma cells, a pit with no border. Alternative: *bordered pit*. 6

simple sugar A sugar that is not composed of smaller sugar molecules, the monomer for polysaccharides. Synonym: *monosaccharide*. 2

sink In phloem transport, any organ or tissue that receives material transported by the phloem. Alternative: *source*. 12

siphonostele A vascular cylinder that contains pith; common in stems but absent in early vascular plants. Compare: *protostele*. 21–23

siphonous In algae, a synonym for coenocyte-a long, tubular cell with many nuclei and few or no cross walls. 19

skepticism The concept of systematic doubt or suspended belief; keeping an open mind so that even if numerous observations support a hypothesis, new evidence is still considered as it becomes available. 1

smooth ER (SER) See *endoplasmic reticulum*.

softwood A term applied to both gymnosperms and their wood, because few gymnosperms have any fibers in their wood. Alternative: *hardwood*. 8

somatic mutation A mutation in a cell that is not a gamete and does not give rise to gametes. 16

sorus (pl.: sori) In ferns, a cluster of sporangia on the underside of leaves. 21

source In phloem transport, any organ or tissue that supplies material to be transported. Alternative: *sink*. 12

speciation The conversion of one species or population of a species into a new species. 17

species A set of individuals that are closely related by descent from a common ancestor and can reproduce with each other but not with members of any other species. 18

spermatium (pl.: spermatia) Generally, a synonym for sperm cell, but in botany it usually refers only to the nonmobile sperms of red algae. 19

spermatophytes Plants that produce seeds. Alternative: *vascular cryptogams*. 21

spherosome See *lipid body*.

spindle The framework of microtubules that pulls the chromosomes from the center of the cell to the poles during nuclear division. 4

split genome virus A type of virus in which the genome occurs as two or more separate double helices of DNA, each packaged in separate viral particles. 15

spongy mesophyll Any part of leaf mesophyll in which the cells are not aligned parallel to each other and are separated by large intercellular spaces. 6

sporangiophore In the arthrophytes (sphenophytes), a stalked, umbrella-like structure that bears sporangia. 21

sporangiospores In fungi, spores that form inside the swollen tip of a hypha. 24

sporangium (pl.: sporangia) A structure that produces spores. 9, 19–24

spore A single cell that is a means of asexual reproduction; it can grow into a new organism but cannot fuse like a gamete. 9, 19–25

sporophyll A leaf that bears sporangia. 21–23

sporophyte A diploid plant that produces spores. Alternative: *gametophyte*. 9, 19–23

spring wood In secondary xylem, the wood formed early in the season, usually with an abundance of vessels in angiosperms or with wide tracheids in gymnosperms. Also called early wood. Alternative: *summer wood*. 8

stamens The organs of a flower involved in producing microspores (pollen). See also *androecium, anther*, and *filament*. 9, 23

start codon In messenger RNA, a codon (set of three nucleotides) that indicates the beginning of information for protein synthesis. 15

statocytes Cells within the root cap that detect the direction of gravity. 14

statolith A type of starch grain that is so dense it sinks to the bottom of a cell's cytoplasm, indicating the direction of gravity. 14

stele The set of vascular tissues in a root or stem (but not in a leaf). 5, 21, 22

sterigma (pl.: sterigmata) In basidiomycete fungi, the narrow tube that connects basidiospores to basidia. 24

stigma (pl.: stigmas) In the carpel of a flower, the receptive tissue to which pollen adheres. 9, 23 In algae, an eyespot, a set of pigment droplets involved in detecting light direction. 19

stipe The stalk of certain organisms-fern leaves, kelps, mushrooms. 19, 21, 24

stipules Small flaps of tissue located at the base of a leaf, near its attachment to the stem. Stipules may range from quite leaflike to small and inconspicuous. 6

STM/CC complex In many species, sieve tube members and guard cells are believed to function together in phloem loading and conduction, making it more accurate to speak of both rather than of just the sieve tube members. 12

stolon An aerial stem with elongate internodes; it establishes plantlets periodically when it contacts soil. Example: strawberry. 5

stoma (pl.: stomata) A word sometimes used to mean "stomatal pore," the intercellular space between guard cells through which carbon dioxide and water are exchanged, and sometimes used to mean "stomatal complex," the stomatal pore plus guard cells plus associated cells. 5

stomatal pore The intercellular space between two guard cells; carbon dioxide is absorbed through the pore and water is lost. 5

stop codon In messenger RNA, a codon (set of three nucleotides) that indicates the end of information for protein synthesis. 15

storied cambium A vascular cambium in which fusiform initials are aligned horizontally. Alternative: *nonstoried cambium*. 8

streptophyte An informal name for the clade that contains charophytes, embryophytes, and their most recent common ancestor. 19

strict aerobe Synonym for *obligate aerobe*.

strict anaerobe Synonym for *obligate anaerobe*.

strobilus (pl.: strobili) Synonym for *cone*.

stroma reactions In plant photosynthesis, the set of reactions that occur in the stroma and are not directly powered by light. Synonym: dark reactions. Alternative: *light-dependent reactions*. 10

stromatolite Large stonelike growths of cyanobacteria, formed in shallow, warm sea water. Some are 2.7 billion years old. 25

structural region The portion of a gene consisting of nucleotide triplets that specify which amino acids are to be incorporated into protein. Alternative: *promotor region*. 15

style In the carpel, the tissue that elevates the stigma above the ovary. 9, 23

subapical meristem The region of a shoot or root just proximal to the apical meristem. 5

suberin Lipid material that causes the hydrophobic properties of cork cell walls and the Casparian strip of the endodermis. 7, 8

substrate The reactant acted upon by an enzyme. 2

substrate level phosphorylation The formation of ATP from ADP by having a phosphate group transferred to it from a substrate molecule. 10, 11

substrate specificity The ability of an enzyme to distinguish one substrate from similar substrates. 2

summer wood In secondary xylem, the wood formed late in the season, usually with few or no vessels in angiosperms, or with narrow tracheids in gymnosperms. Also called late wood. Alternative: *spring wood*. 8

superior ovary An ovary located above the sepals, petals, and stamens; those appendages are hypogenous. Alternatives: *inferior* and *half-inferior ovary*. 9

suspensor In seed plant embryos, the stalk of cells that pushes the embryo into the endosperm. 9

symbiotic relationship A relationship in which two or more organisms live closely together. 19

sympatric speciation Speciation that occurs within a limited geographic range; populations are separated by biotic reproductive barriers, not by physical differences. Alternative: *allopatric speciation*. 17

sympetally The condition of having the petals of a flower fused together into a tube. 23

symplast The protoplasm of all the cells in a plant are interconnected by plasmodesmata; the entire mass is the symplast or symplasm. See *apoplast*. 3, 5, 7

sympodial branching A branching pattern in which what appears to be one main shoot (the trunk) is actually a series of lateral branches, each of which displaces the apex of the shoot that bears it. Alternatives: *monopodial* and *pseudomonopodial branching*. 21–23

synapsis (pl.: synapses) The pairing of homologous chromosomes during zygotene of prophase I of meiosis. Synapsis precedes crossing-over. 4

synaptonemal complex In prophase I of meiosis, after homologous chromosomes have paired (undergone synapsis), a protein complex, the synoptonemal complex, holds them together. It is composed of a central element connected by fine fibers to the lateral elements that actually connect to DNA. 4

synergid In the egg apparatus of an angiosperm megagametophyte, there is an egg and one or two adjacent cells, synergids; the pollen tube enters one of the synergids. 9

syngamy The fusion of a sperm and an egg. 9

T

TATA box In the promotor region of many genes, a region rich in thymine- and adenine-containing nucleotides, believed important for RNA polymerase binding. 15

taxis A response in which a cell swims toward or away from a stimulus. 14

taxon A term that refers to any taxonomic group such as species, genus, family, and so on. 18

teleology The interpretation that objects and processes have a purpose, such as, "The purpose of most leaves is photosynthesis," in contrast to a statement of fact, such as, "Most leaves carry out photosynthesis." 1, 18

telome In a plant with dichotomous branching, the last two twigs produced by the last bifurcation. 21

telome theory The theory that leaves (megaphylls) of arthrophytes, ferns, and seed plants evolved from branch systems (telomes) by overtopping, planation, and webbing. 21

telophase The fourth and last phase of mitosis, during which the chromosomes decondense, the nucleolus and nuclear envelope reform, the spindle depolymerizes, and the phragmoplast appears. 4

telophase I The fourth phase of meiosis I, similar to telophase of mitosis. However, in many organisms, telophase I and prophase II are often shortened or eliminated, and full nuclei are not formed between meiosis I and II. 4

telophase II A phase of meiosis II, similar to telophase of mitosis. 4

temperate virus A virus whose genome has been incorporated into the host's genome, being replicated simultaneously with host DNA; the virus produces few or no symptoms. 15

tendril An organ that attaches a vine to a support by wrapping around it. It may be a modified leaf, leaflet, or shoot. Example: grape. 5, 6

tension wood See *reaction wood*.

tepal Refers to members of a perianth when it is not certain if they are really sepals or petals. 9

terminal bud A bud located at the extreme apex of a shoot; usually present only in winter as a dormant bud. Alternative: *axillary bud*. 5

tertiary structure of a protein The overall three-dimensional shape of an entire protein molecule. 2

test cross A cross involving one parent known to be homozygous recessive for the trait being considered. 16

testa (pl.: testas) Synonym for *seed coat*.

tetrads During meiosis I, after homologs have paired and condensed sufficiently, the four chromatids are visible as a tetrad. 4

tetraploid Refers to four full sets of chromosomes within a single nucleus. 4, 16

tetraspores In red algae, the spores produced by meiosis in the tetrasporophyte. 19

tetrasporophyte In red algae, the diploid generation equivalent to a sporophyte generation in other algae. 19

thallophyte An old, rarely used term to distinguish organisms that are not embryophytes; algae and fungi. 19, 24

thallus In plants, algae, and fungi, a simple body that lacks vascular tissues and the complex organs of vascular plants. 20

theory After a hypothesis has been confirmed by numerous observations or experiments, it is considered to be a theory. 1

thermogenic respiration Respiration in which electron transport is uncoupled from ATP synthesis, so heat is generated. Synonym: *cyanide-resistant respiration*. 11

thigmotropic response A tropic response with touch as the stimulus. 14

threshold The level or intensity of stimulus that must be present during the presentation time in order for an organism to perceive it. 14

thylakoids The photosynthetic membranes of chloroplasts. 3, 10

ti plasmid A plasmid from the bacterium *Agrobacterium tumefaciens;* this is a commonly used vector for recombinant DNA studies in plants. 15

tinsel flagellum A flagellum covered with many minute hairlike projections. Alternative: *whiplash flagellum*. 19, 24

tolerance range The range of environmental conditions in which an organism can live and reproduce. 26

tonoplast The vacuolar membrane. 3

trace element See *essential element*.

tracheary element A term refering to either or both types of xylem-conducting cell: tracheids and vessel elements. 5

tracheid A xylem-conducting cell; tracheids tend to be long and tapered, and they never have a perforation—a complete hole in the primary wall—as vessel elements do. 5, 12, 23

trans-acting factor Molecules that affect the activity of a gene but which are not part of the DNA strand that contains the gene; trans-acting factors must diffuse to the DNA from some other location in a cell. Alternative: cis-*acting factor*. 15

transamination The transferral of an amino group from one molecule to another; important in the synthesis of amino acids. 13

transcription The "reading" of DNA by RNA polymerase with the simultaneous production of RNA. 15

transduction In a receptive tissue after the perception of a stimulus, transduction is the change that allows the tissue to communicate that the stimulus has occurred. 14 In bacteria, a method of genetic exchange occurring when a bacterium incorporates DNA carried in by a virus. 25

transfer cells Cells involved in rapid short-distance transfer of material; they have transfer (labyrinthine) walls. 12

transfer wall In transfer cells, walls whose inner surface is highly convoluted, thus increasing the surface area of the plasma membrane and the number of molecular pumps present. Synonym: *labyrinthine wall*. 12

transformation In bacteria, a method of genetic exchange occurring when a bacterium incorporates a piece of DNA from the environment. 25

transformation theory The theory that vascular plants arose from algae that had an alternation of isomorphic generations, each of which was gradually transformed into the types of sporophytes and gametophytes present today. Compare: *interpolation theory*. 21

translation In protein synthesis, the utilization of mRNA to guide the incorporation of amino acids into protein. 15

translocation Long-distance transport of water and nutrients by xylem and phloem. 12

transpiration Loss of water vapor through the epidermis. **transcuticular transpiration** Loss through the cuticle. **transstomatal transpiration** Loss through stomata. 12

transposable element A region of DNA that codes for enzymes that catalyze the release of the element and its insertion into a different site in the DNA. See also *transposon* and *insertion sequence*. 16

***trans*-position** In a carbon-carbon double bond, the situation in which two groups of interest lie on opposite sides of the bond. Alternative: *cis-position*. 2

transposon A large transposable element that carries, in addition to the insertion sequence, other genes that code for proteins not directly associated with transposition. 16

***trans*-unsaturated fatty acid,** usually just called a ***trans*-fat** An unsaturated fatty acid (it has one or more carbon–carbon double bonds) and the two hydrogens at the double bond are on opposite sides of the molecule. *Trans*-unsaturated fatty acids are synthetic, not natural, and they are not healthful. Compare with *cis*-unsaturated fatty acid. 2

tricarboxylic acid cycle Synonym for *citric acid cycle* and *Krebs cycle*. 11

trichogyne In ascomycete fungi, a narrow tube that transfers nuclei from an antheridium to an ascogonium. 24

trichome A plant hair; often restricted to structures that contain only cells derived from the epidermis. 5

tricolpate pollen Pollen that has three germination pores as opposed to just one; tricolpate pollen is a synapomorphy that unites eudicots. 23

triglyceride A type of lipid consisting of three fatty acids bound to one molecule of glycerol. 2

triploid A nucleus that contains three sets of chromosomes. 16

tropic response A growth response oriented with regard to the stimulus. Synonym: *tropism*. 14

true fruit A fruit that developed only from carpel tissue, not containing any other tissue. Alternative: *accessory fruit*. 9

trumpet hyphae In some brown algae, the phloem-like cells that conduct photosynthate through the alga. 19

tuber A short, fleshy, horizontal stem, involved in storing nutrients but not in migrating laterally. Example: potato. 5

tubulin See *microtubules*.

turgid Filled with water to such a degree that the surface of the cell or plant is firm. 12

turgor pressure The pressure with which a protoplast presses against the cell wall when a cell is turgid. 12

tylosis (pl.: tyloses) After a vessel stops conducting because of cavitation, adjacent cells may push cytoplasm into the vessel through pits, plugging the vessel. 8

type specimen A single specimen that is the absolute standard for the species and its scientific name. 18

U

uniaperturate pollen Pollen that has only a single germination pore. This is an ancestral condition (a symplesiomorphy) for the angiosperms. 23

uniform distribution In ecology, the distribution of individuals in the habitat such that they are evenly spaced, as in an orchard, rather than being clumped or occurring at random. 26

unilocular sporangium In brown algae, a multinucleate, coenocytic sporangium where meiosis occurs. Alternative: *plurilocular gametangia*. 19

uniparental inheritance Inheritance of genes from just one parent, the most common case for plastid and mitochondrial genes; sperm cells typically do not contribute these organelles to the zygote. Alternative: *biparental inheritance*. Synonym: *maternal inheritance*. 16

uniseriate Consisting of just one row; often used for things that consist of one layer. 5

upright cell In rays in secondary xylem, cells that are taller than long; they typically have extensive cross-field pitting. Alternative: *procumbent cell*. 8

V

vacuole A membrane-bounded (tonoplast) space larger than a vesicle which stores material, either dissolved in water or as a crystalline or flocculent mass. 3

vacuole membrane Synonym for *tonoplast*.

valence electrons The electrons that actually participate during a chemical reaction. 2

variegation A pattern of spots, stripes, or patches in leaves or other organs, caused by plastid mutations. 16

vascular bundle A column of vascular tissue, typically both xylem and phloem together, but in leaves sometimes consisting of only one or the other. 5

vascular cambium The meristem that produces secondary vascular tissues—secondary xylem (wood) and secondary phloem (inner bark). 8

vascular cryptogams The vascular plants that do not produce seeds, such as lycopods, horsetails, and ferns. Alternative: *spermatophytes*. 21

vector Pieces of DNA that are used to insert experimental, recombinant DNA into bacteria or eukaryotes. 15

vegetarian diet A diet centered mostly or exclusively on foods derived from plants (vegetables, fruits, grains, sugars) and excluding or minimizing any food derived from an animal. Some vegetarian diets exclude meat but include eggs or milk (an ovo-lacto-vegetarian diet); others avoid all animal products (a strict vegetarian diet). 23

vegetative Refers to phenomena or parts of a plant not involved in sexual reproduction. 9

vegetative cell In the pollen grain of seed plants, the cell or cells that do not give rise to the sperm cells; the cell that is not the generative cell. 9, 22

venation The pattern of veins in a tissue or organ. 5, 6

vernalization The cold treatment necessary for biennials to initiate flowering. 14

vesicle A small space enclosed by a single membrane. Vesicles are similar to vacuoles, but smaller, generally being unresolvable by light microscopy. 3

vessel In xylem, a column of vessel elements interconnected by pairs of perforations. 5

vessel element (vessel member) A xylem conducting cell that has one or two perforations—a complete large hole in the primary wall that permits water to flow easily from one vessel member to another. 5, 23

vesselless (primitively, secondarily) A plant that lacks vessels is vesselless. If it is part of a clade that has never had vessels, it is primitively vesselless, but if one of its ancestors had vessels but has since lost them, it is secondarily vesselless. 23

violaxanthin A xanthophyll pigment in brown algae. 19

W

water potential The chemical potential of water; a measure of the ability of a substance to absorb or release water relative to another substance. Components: osmotic potential, pressure potential, matric potential. 12

wax An extremely hydrophobic polymer of long-chain fatty acids; wax contributes to the water-retaining capacity of the epidermis. 2

webbing In the telome theory of the origin of megaphylls, the concept that the lamina originated by the production of parenchyma cells between the telomes. 21

whiplash flagellum A flagellum that has a smooth surface. Alternative: *tinsel flagellum*. 19, 24

whorl A set of leaves or flower parts, all attached to the stem or receptacle at the same level. 5, 6

wood Secondary xylem. 8

woody plant A plant that undergoes secondary growth by means of a vascular cambium which produces secondary xylem (wood) and secondary phloem. Alternative is an herbaceous plant. 5, 8

X

xerophyte A plant adapted to desert conditions. 5, 6, 27

xylem The water- and mineral-conducting portion of vascular tissues, containing either tracheids or vessel elements or both; parenchyma, fibers, and sclereids are also frequent components of xylem. Alternative: *phloem*. 5

Z

zeatin A natural cytokinin. 14

zone of elongation Region of a root tip, just proximal to the root apical meristem, where cells undergo pronounced elongation. 7

zoospore A spore capable of swimming. 19

zygomorphic flower A bilaterally symmetrical flower. Alternative: *regular flower* (actinomorphic). 9, 23

zygosporangium In zygomycete fungi, a large multinucleate sporangium formed by fusion of two compatible hyphae. 24

zygote The diploid cell formed as the result of the fusion of two gametes. 4, 9, 19, 20

zygotene See *prophase I*.

INDEX

Italicized numbers indicate a term found in a figure or table

A

A channel, 371
A horizon, 614, *615*
ABA (abscisic acid), *343,* 345, *345*
Abaxial side, 128
ABC model of flower organization, 356–358, *357, 358*
Abiological reproductive barriers, 426–427, *427*
Abiotic components, of habitat, 614–618, *615–618*
ABPI (auxin binding protein 1), 348
Abscisic acid (ABA), *343,* 345, *345*
Abscission zone, 133, *134*
Absorption spectrum, 240, *240*
Abutilon, 69
Acacia (genus)
    A. aphylla, 444
    A. drepanolobium, 444
    ants and, 620–621, *621*
    phenotypic changes, 443, *444*
Acanthocereus, 651
Acceptor molecules, 245
Accessory fruits, 222, *223*
Accessory pigments, 241, *241*
Accidents, gene pool change and, 419–420, *420*
Acer saccharum (sugar maple), *116*
Acetabularia, 468
Acetaldehyde, 264
Acetyl, 267
Acetyl CoA, 267, *269*
Acid rain, 323, *323*
Acidity
    definition of, 20
    of soil, 322, *322*
Acids, *20,* 20–21
Actin, 65
Actinomorphic (regular) flowers, 219
Action spectra, 240–241
Activated electron, 240
Activated state, 240
Activation energy, 22, *22,* 24
Active transport
    ATP in, 289
    definition of, 49, *50,* 289
    in roots, *49*
    of sugars, 300, *301,* 302
Adansonia grandidieri (baobab tree), 304
Adaptive radiation, 428–429
Adaptive responses, to scarcity of water, 338
Adenine, *30*
Adenosine, *33*
Adenosine diphosphate (ADP)
    phosphorylation of, 233, *233*
    structure of, 33, *33*
Adenosine monophosphate (AMP), 33, *33*
Adenosine triphosphate (ATP)
    in active transport, 289
    dephosphorylation of, 33, *33, 34*
    structure of, 33, *33*
    synthesis, 233, *233,* 243–244, *243–245*
Adenosine triphosphate synthetase (ATP synthetase), 244, *244*
Adhesive property, of water, 303, *303*
Adiantum, 519
ADP. *See* Adenosine diphosphate
Adult phase, 352, *352*

Advanced features (apomorphic features; apomorphies), 11, 444, *450*
Adventitious roots, 154, *163*
Aeonium, 128
Aeonium tabuliforme, 129
Aerial roots
    of orchids, 163, *163*
    of screw pine, 155
Aerobic respiration
    citric acid cycle, 265, 267–268, *268, 269*
    definition of, 261
    glycolysis, 265
    mitochondrial electron transport chain, 268, *270, 271*
    NADH shuffle, 268, 270–271, *272*
    oxygen in, 264–265
    total energy yield, 279, *279*
Aesculus arguta (buckeye), *141*
Aesculus glabra (Ohio buckeye), *184*
Agar, 474
Agaricus bisporus (button mushroom), 576
Agave
    A. americana or giant agave, 96, *97, 284*
    A. parryi, 127
    fiber cells, *102*
    leaves of, *127,* 128
Age distribution, *625,* 625–626, *626*
Aggregate fruits, 222
Aggregation, chemosynthesis and, 433
Aglaophyton, 521
Aglaophyton major, 506
Agriculture, 2, 42
Agrobacterium tumefasciens, 382, *383*
Air, water availability in, 294–295
Air pores, 495, *496*
Airplane plant (*Chlorophytum*), *106*
Aizoaceae (*Carpobrotus*), *564*
Akinetes, 599
Alaska, O'Malley Summit, *229*
Albuminous cells, 117
Albuminous seed, 214, *215*
Alcoholic beverages, fermentation of, 280, 282–284, *284*
Alfalfa (*Medicago sativa*), *326*
Algae. *See also* Green algae
    blue-green, 605
    brown, *4, 461,* 476–478, *476–478*
    cell division in, 87–88, *88, 89*
    cladogram, *462*
    development, *346,* 346–347, *347*
    dinoflagellate, *461, 480,* 480–481, *481*
    divisions of, *466*
    euglenoid, *461*
    flagella, 465, *465*
    global warming and, 14, *14*
    importance of, 460
    marine, 294
    number in soil, *314*
    plants and, 3, 93
    red, *461,* 474–476, *475*
    unicellular, 346
    Valonia, 87
    vs. embryophytes, 462, *462*
    yellow-green, *461, 479, 479*
Alismatales, *557,* 557–558, *558*

Alkaloids, *17,* 131
Allele frequency, 417–418, *418*
Alleles
    definition of, 390
    lethal, 410
Allelochemics, 624, *624*
Allelopathy, 624
Allium cepa (onion)
    cell cycle phase duration, *75*
    DNA, *393*
    stomata in epidermis, *135*
Allopatric speciation (geographic speciation), 426–427, *427*
All-or-none response, 341–342
Alnus rugosa (speckled alder), *116*
Aloe, 96, *97*
Alpha helix, 28–29, *29*
α-Ketoglutarate, 267–268, *269*
Alpha-1,4-glycosidic bond, 26–27, *27*
Alpha-tubulin, 36, 64, *64*
Alpine tundra, 647–648, *648*
Alternation of generations
    definition of, 204
    heteromorphic, 468, *469*
    isomorphic, 468, *469*
Altitude, 615–616, *616*
American elm (*Ulmus americana*), *116*
Amino acids
    activation of, 370
    attachment site, 370
    structure of, 27, *28*
    synthesis of, 56
Amino groups, 325
Amino sugars, 27
Amorphophallus, 442
AMP (adenosine monophosphate), 33, *33*
Amylopectin, 26, 27, *27*
Amylose, 26, 27, *27*
Amyoplasts, starch storage in, *56,* 56–57
Anabaena, 326
Anabolic metabolism, 246–248, *247*
Anabolic reactions, 246
Anabolism, 246
Anaerobic respiration (fermentation)
    of alcoholic beverages, 280, 282–284, *284*
    definition of, 261
    grain-based, 280, 282–284, *284*
    pathways in, *262,* 263–264, *263, 264, 267*
    total energy yield, 276, 279, *279*
    wine-making yeast, *265*
Anaphase
    I, of meiosis, *84,* 85
    process of, 81
    in root tips, *80*
Ancestral condition, 444
Ancestral groups, 452
Ancient Greeks, 327, 440
Androecium, 206
Aneurophytales, 530, *531*
Aneurophyton, 531
Angiosperm carpels, 548
Angiosperms, 547–575. *See also* Basal angiosperms; Eudicots; Monocots
    bodies of, *346,* 346–347, *347*
    cladogram, *553*

evolution of, 453, *453*, 598, *599*
  nitrogen and, 313
  nitrogen fixation and, *325*, *326*
  photosynthesis in, *255*, 255–256, *256*
Cyanogenic glycosides, 36
Cyathea, *518*
Cycadeoidea, *542*
Cycadeoidophyta, 540, *541*, 542, *542*
Cycadeoids, 540, *541*, 542, *542*
Cycadophyta, 540, *541*
Cycads, 47, 96, *97*, 540, *541*, *542*
Cycas circinnalis, *541*, *542*
Cyclic electron transport, 244, *245*
Cymbidium hybridum, *77*
Cynodon (Bermuda grass), *301*
Cysteine, 27, *28*, 29
Cytochrome b6/f complex, 242, *244*
Cytochrome oxidase, 268
Cytochromes
  a, 268
  a₃, 268
  b, 268
  c, 268
  c₁, 268
  as electron carriers, 235, *236*
Cytokinesis
  in algae, 88, *88*, 89
  definition of, 77
  stages of, 81, *81*, 83, *83*
  types of, 465
Cytokinins, *343*, 344, *344*
Cytophaga, 606, *606*
Cytoplasm, *51*, 55
Cytosine, *30*
Cytosol (hyaloplasm)
  components of, 63, *63*
  volume in plant cell, *54*

D

Dactylorchis, 593
Daisies (asters), 96, *97*
Darwin, Charles, 400, 418, 441
Datura (jimsonweed), 36
Daucus carota (carrot), *75*, 322
Daughter cells, 45, 72
Day-neutral plants, 353, *353*
Dead cells, 47
Deadly nightshade *(Atropa)*, 36
Decomposers, 630, *630*
Decussate phyllotaxy, 104, *105*
Degeneracy, DNA protection and, 366
Degeneria, 548
Dehisce, 208
Dehiscent fruits, 223
Dehydration reaction, *26*, *33*
Deletions, 393–394
Demography, *625*, 625–636, *626*
Denatured protein, 29
Dendrobium, *143*
Dendrochronology, *188*, 188–189
Denitrification, 328
Denitrifying bacteria, 328
Deoxyribonucleic acid. *See* DNA
Deoxyribonucleotides, 29
Deoxyribose, 29
Derbesia marina, *472*, *473*
Derived features, 11
Desert islands, 636
Desert plants, *308*
Desertification, 8

Deserts
  ecosystem characteristics of, 647, *647*, *648*
  plant biology of, 304–305
  soils in, 316, *316*
  as wasteland, 299, *299*
Desiccation avoidance, *485*
Desiccation tolerance, *485*
Desmotubule, *68*
Determinate growth, 123
Determinate inflorescence, 221, *222*
Determinate organogenesis, 123
Deuterium, *18*
Deuterium oxide, 81
Deuteromycetes, 591–592
Diakinesis, 85
Diatomaceous earth, *478*, 478–479
Diatoms, *478*, 478–479
Dibiontic life cycle, 503
Dibiontic species, 468, *469*, 471
Dicots, *184*, *451. See also* Basal angiosperms
Dictyosome vesicles, *83*
Dictyosomes
  creation of, *59*, 59–60, *60*
  volume in plant cell, *54*
Dictyostelium discoideum, *586*
Differential activation of genes, 363, *364*
Differential gene expression, 333
Differential growth, 339
Differentially permeable membranes (selectively
      permeable membranes), 49, *49*, *50*, 289
Differentiation
  of cells, 356–357, *357*
  definition of, 333–334
  of guard cells, 334
  of stem, 118, 120, *121*, 122, *122*, 124
  of vascular tissue, 349
Diffuse porous wood, *182*, 184
Diffusion, 289
Digitalis, 36
Digitalis purpurea, *569*
Digitaria (crab grass), *301*
Digitoxin, 36
Digoxin, 36
Dihybrid crosses
  crossing over, 405
  definition of, 403
  independent assortment, 404, *404*, 405
  linkage, 405, *406*, *407*, 407–408
Dihydroxyacetone phosphate, 248
Dinoflagellate algae, *461*, *480*, 480–481, *481*
Dinteranthus, *144*
Dioecious species, 218
Dioecy, 218
Dionaea muscipula (Venus' flytrap)
  leaves of, 148, *148*
  motor cells of, 298
  perception mechanism, 337, 340–341, *341*
Dioon, *541*
Dioscorea (yams), 304
Dioscoreales, 559
1,3-Diphosphoglycerate, 246
Diploid, 84
Diplotene, 85
Disaccharide, 26
Distichous phyllotaxy, 104, *105*
Disturbances, man-made and natural, 616–618,
      617, *618*
Diurnal species, 340
Divergent speciation, *425*, *425*, 426–429, *427*, *428*
Division of labor, 44–45, *45*, 107, *107*

Division phase, of cell cycle, 77–86, *78*, *80*–86
Divisions, in classification, *439*, 442. *See also specific
      classification divisions*
DNA (deoxyribonucleic acid)
  in cell nucleus, 10
  of eukaryotic cells, 463
  fragments, identifying, 376–379
  in gene transcription, *368*, 368–369, *369*
  in haploid set of chromosomes, *393*
  during leaf growth, *90*, 90
  ligated, *393*
  mitochondrial, 55
  naked, 599, *600*
  nuclear, 51
  in prokaryotic cells, 89, *89*, 463
  replication, *392*, 392–393, *393*
  during S phase, *76*, 76–77
  storage of, 364–365
  structure of, 29, *30*, 363
DNA cloning, 379–380, *380*, *382*
DNA denaturation, 375, *375*
DNA hybridization, 375–376
DNA ligase, 375
DNA melting, 375, *375*
DNA microarray, 379, *379*
DNA polymerase, 392
DNA repair processes, 396–397
DNA sequence analysis, 364
DNA sequencing
  for phylogenetic studies, 444, *445*
  technique, 380–381, *381*, *382*
DNases (DNAases), 365
Domains, 10, 46, 598
Dominant allele, 401
Dormancy
  germination induction, 340, *340*
  purpose of, 106
  signal for, 335, *336*
Dorsal surface, 128
Dosage-dependent responses, 342
Double bonds
  carbon-carbon, *21*, 21–22
  in unsaturated fatty acids, 31
Double fertilization, 213, 549
"Double" flowers, 356, 362
Douglas fir *(Pseudotsuga)*, 535, *538*
Down syndrome, 411
Dracaena, 145
Drepanophycus, *509*
Drosophyllum (insectivorous plant), 60
Drought, *420*, 422
Drought avoidance, 304
Drought tolerance, 304
Dry fruit, 222–223, *223*, 224
Dry habitat, coping with, *485*, 485–485, *487*
Dunalliela (alga), 65
Duplication division (mitosis), 77–81, *78*, *80*
Dwarf daisy *(Monoptilon)*, *550*
Dynein, 64, *65*

E

Early wood, 182
Earth
  atmospheric distribution of heat, 636, *638*,
      638–639, *639*
  axis of rotation tilt, 636
  climate, 42
  conditions before origin of life, *431*, 431–432
  history of, 460, *461*
Earthstars *(Geastrum)*, *581*

G

G₁ (gap 1), *74*, 74–75, *75*
G₂ phase (gap 2 phase), 77
Galen, 4–5
Galileo, *419*
Gameophores, 489, *489, 490*
Gametangia, 468, *486*
Gametes (sex cells), 83, 203
Gametophyte generation
    Hepatophyta, 495–496, *495–498*
    hornworts, *498*, 498–500, *499*
Gametophytes
    development, *213*
    hornwort, *487*
    megagametophytes, 211–212, *212, 213*
    microgametophytes, 211, *211–213*
    types of, 204, *204*
Gardenia, 322
Gazania splendens, 11
Geastrum (earthstars), *581*
Gel electrophoresis, 376, *377*
Gene amplification, 77, *77*
Gene family, 410–411, *411*
Gene flow, 426, *426*
Gene pool, change factors for, 419–423, *420–423*
Genealogy, 452
Gene-environment interaction, 9
Generalized flower, 551
Generation time, 625
Generative cell, 211
Genes
    analysis of, 375–381, *375–382*
    differential activation of, 363
    human, 407
    information in, 7
    protecting, *364*, 364–365, *365*
    on same chromosome, linkage of, 405, *406, 407,*
        407–408
    on separate chromosomes, 404, *404, 405*
    structure of, 366–368, *367*
    transcription of, *368*, 368–369, *369*
Genetic code, 365–366, *366*
Genetic diversity, 422, 554
Genetic drift, *428*, 428–429, 430
Genetic engineering
    definition of, 364
    evolution and, 373
    methods, 381–382, *382, 383*
Genetic information, storing, 364–369, *364–369*
Genetic mutations. *See* Mutations
Genetically diverse species, 200
Genetically modified plants (GM), 373
Genetics
    definition of, 390
    dihybrid crosses, 403–405, *404–407*, 407–408
    DNA replication, *392*, 392–393, *393*
    of haploid plants, 412, *412*
    monohybrid crosses, 397, *398, 399*, 399–403,
        401–403
Genome, 75
Genome sequencing, 381, *382*
Genomosperma kidstoni, 532
Genotypes, 390, 391, *391*
Genus (genera), 439, *439*
Geographic distribution, of populations, 623–625,
    623–625
Geographic range, boundaries of, 623, *623*
Geographic speciation (allopatric speciation),
    426–427, *427*
Geometric increase, 82, *82*

Geophytes, *627*
Geranium
    bark, *73*
    parenchyma cells of, *99*
    primary body of, *94*
Germ plasma banks, 554
Germanium phaeum, 77
Germination
    induction, 340, *340*
    temperature and, 359
Germination pores, 556, *556*
Giant cells, 47
Gibberellic acid, 578
Gibberellin, *343, 344*, 344–345, *345*
Ginkgo biloba (Maidenhair tree), 52, 542–543, *543*
Glaciers, 619, *619*
Gladiolus, 153, 164
Gladioluses, 106
Glandular cells, 98
Gleditsia triacanthos (honey locust), *184*
Gliding bacteria, 606, *606*
Global warming (global climate change)
    algae and, 14, *14*
    causes of, 8
    description of, 1–2
    environmental stimuli and, 359
    photosynthesis and, 238–239
    vineyards and, 283
Glossopteris, 420, 553
Gluconeogenesis, *247*, 247–248
Glucose
    energy recovery from, 259, *260*
    polymerization of, 26–27, *27*
    structure of, 25–26, *26*
    transport, *288*
Glucose residue, 27
Glucose-6-phosphate, 248
Glycolipids, 48
Glycolysis
    aerobic respiration and, 265
    anaerobic respiration and, 261, *262*, 264
Glycoproteins, 48
Glyoxysomes, 63
Glyphosphate, 373
GM (genetically modified) plants, 373
Gnetophyta (division), *543*, 543–544, *544*
Gnetum, 543, 543
Goatsbeard (*Tragopogon*), *570*
Golden-brown algae, 479–480, 479f
Golgi body (Golgi apparatus), 60
Gondwanaland, 639
Gonium, 346, 470
Gonyaulax monilata, 480
Grade classification, 454
Gram-negative cells, 601, *601*
Gram-positive cells, 601, *601*
Grana, 243, *243*
Granum (grana), 56, *56*
Grapes (*Vitis*)
    V. riparia, 140, *140*
    vines, 106
Graphis scripta, 592
Graptopetalum, 357
Grasses
    Bermuda or *Cynodon*, 301
    bunch or *Bromus tectorum, 647*
    crab or *Digitaria*, 301
    growth of, 560, *560*
    identification of, 96
    Paspalum, 301

Grasslands
    characteristics of, 645, *645, 646*
    tropical, 650, *651*
Gravity, direction of, 336, *337*
Gravity-sensing systems, 340, *341*
Grazing, 621
Greek philosophers, 4
Green algae
    body construction, *467*, 467–468, *468*
    characteristics of, *461*, 466–467
    Cladophora, 88
    classification of, 466
    coenocytic or siphonous, 467, *468*, 471–473, *473*
    embryophytes, 462, *462*, 474
    filamentous bodies, 346, 467, *467*, 471, *471*
    life cycles, 468, *469*, 470, *470*
    parenchymatous species, 467, *467*, 473–474, *474*
    representative genera, 470–473, *470–474*
    vs. brown algae, 476
Green bacteria, 14
Greenbrier (*Smilax*), *159*
Greenhouse effect, 238
Greenhouse gas, 238
Grimmia laevigata (moss), *484*
Ground meristem, 122, *122*
Ground pine (*Lycopodium*), 511
Ground state, 240
Growth. *See also* Secondary growth
    anomalous forms in woody plants, 191–193,
        192–194, 193
    cancerous *vs.* controlled, 79
    as characteristic of living organisms, 13
    controlled *vs.* cancerous, 79
    determinate, 123
    differential, 339
    direction of gravity and, 336, *337*
    establishment, 193
    exponential of population, 625, *625*
    indeterminate, 123
    of leaves, 90, *90*, 123
    monopodial, 516
    plant *vs.* human, 123
    primary, 124, 193, *194*
    secondary in monocots, 193, *193, 194*
    of stem, 118, 120, *121*, 122, *122*, 124
    of wood, 195–196
Growth phase, of cell cycle, 74–77, *74–77*
Growth rates, 82, *82*
Growth rings
    analysis of, *188*, 188–189
    cellular arrangements in, 182, *182*, 184
Guanine, *30*
Guanosine triphosphate, 233
Guard cells
    differentiation of, 334
    morphogenesis, 334
    short-distance intercellular transport and,
        298, *298*
    stomatal pore and, 110, *110*
    water transport control, 310
Gymnocalycium (cactus)
    evolution of, *424*
    G. horstii, 449
    plastid mutations, *409*
Gymnosperms, 527–546
    classification, 529, *529*
    conifers, 533, *534–540*, 535–537, *539*
    cycads, 540, *541, 542*
    Gnetophyta, *543*, 543–544, *544*
    maidenhair tree, 542–543, *543*

sporophyte generation, 491, 493, *494*
structures of, *3*
true, 96
water transport, 489–490, *491*
Mother cell, 45
Motile colonial green algae, 467, *467, 470,* 470–471
Motor cells, 298, *300*
Mucigel, 155
Mucilage, *58, 60*
Multicellular organisms
cellular interactions in, 68
division of labor in, 44–45, *45*
Multiple fruits, 222
Mummy, 594
Muscarinic receptor, 36
Mushrooms, 3, *4*
Mutualism, 619, 620
Mutations
causes, 393–396, *394, 395, 428*
deleterious, 417
DNA repair processes, 396–397
effects, 396
evolution and, 7, *9*
gene pool change and, 419
homeotic, 357–358, *358*
phenotypic change and, 443
plant diversity and, 11
selective advantage and, 93–94, 423
somatic, 396
ultraviolet light and, 108
Mutualism, plant-fungus, 621, 623
Mycelium, 578, *579*
Mycoplasmas, 599, 607, *607*
Mycorrhiza (mycorrhizae)
ectomycorrhizal relationship, 164
endomycorrhizal association, 164, *165*
phosphorus absorption and, 323–324
Myriophyllum heterophyllum, 130

N

NAD⁺. *See* Nicotinamide adenine dinucleotide
NADH
generation by respiration, *260,* 260–261
in heat-generating respiration, 271–273, *273*
production, 267, *268*
production during glycolysis, 261, *262,*
263–264, 264
NADH shuffle (malate-aspartate), 268, 270–271, *272*
NADP⁺ (nicotinamide adenine dinucleotide
phosphate), 34–35, 234, *235*
Naked DNA, 599
Natural disasters, gene pool changes and,
419–420, *420*
Natural disturbances of habitat, 616–618, *617, 618*
Natural selection
death rates and, 430
definition of, 9
elimination of synthetic pathways, 23
factors not part of, 423
gene pool change and, 421–423, *422, 423*
multiple pressures, 423–424
situations without, 423
speciation and, 425–429, *425–429*
Natural system of classification, 439
Necrosis, 318
Necrotrophs, 577, 578
Nectar guides, *207*
Negative feedback loop, 355
Negative gravitropism, 338–339, *339*

Negative tropism, 338
Nepenthes (Pitcher plant), 147–148, *148*
Nerium oleander (oleander), *135*
Nereocystis (kelp), 478
Neuropteris decipiens, 533
Neutrons, 18
Niacin deficiency, 23
Niches, 619, 622, *622*
Nicotiana (tobacco)
historical aspects, *441*
N. silvestris, 355
N. tabacum, 131, *131, 153*
nicotine-like alkaloids in, 36
plasmodesmata, *69*
Nicotinamide adenine dinucleotide (NAD⁺)
reduction during glycolysis, 261, *262,* 263–264, *264*
structure of, *34*
Nicotinamide adenine dinucleotide phosphate
(NADP⁺), 34–35, 234, *235*
Nicotine, 131
Nicotine-like alkaloids, 36
Nitrifying bacteria, 328
Nitrification, 328
Nitrogen
from animals, 328–329, *329*
assimilation of, 325–326, *326*
fungi and, 578
metabolism, processes in, 324–326, *325, 326*
natural sources, 327, *327*
oxidation states, 324, *324*
prokaryotes and, 328
reduction, 325, *326*
Nitrogen fixation, 164, *165,* 324–325, *325, 326*
Nitrogenase, 324
Nitrogen-metabolizing bacteria, 606–607
Nitrous acid, mutations and, 393, *394*
Nocturnal species, 340
Nodes, 103, 444
Nolina parryi, 308
Nomenclature, 439
Noncompetitive inhibitor, 37
Noncyclic electron transport, 244, *245*
Nondisjunction, 411
Nonessential organs, 218
Nonmotile colonies, of green algae, 467, *467*
Nonpolar molecule, 21
Nonstoried cambium, 176
Nonvascular plants. *See* Bryophyta (bryophytes)
Norway spruce (*Picea abies*), 303
Nostoc, 325, 326, 499, 598, *600*
Nucellus, of ovule, 208
Nuclear envelope, 51, *51, 88*
Nuclear genes, 445
Nuclear pores, 51, *51*
Nuclei. *See* Nucleus (nuclei)
Nucleic acid hybridization, 375, 375–376
Nucleic acids, 29–30, *30*
Nucleoids, 599, *600*
Nucleolus (nucleoli), *44,* 53, *53, 58*
Nucleoplasm, 53
Nucleosome, 365
Nucleotide sequencing analysis, of prokaryote
evolution, 603, *604*
Nucleotides, 29, *30,* 363, *364*
Nucleus (nuclei)
algal, division of, 87–88, *88*
function of, 51
fungal, 579, *580*
structure of, *44,* 51, *51, 53,* 53–54, *58*

unusual, 47
volume in plant cell, 53, *54,* 61
Nutrition, fungal
essential, 578
extracellular digestion, 577–578
toxins, 578
Nymphaeaceae, 555, *555*

O

Oak (*Quercus*)
flower, *358*
leaves, *130*
Q. nigra, 170
vessels, *423*
Oat (*Avena*), 69
Obligate aerobes, 261, *261*
Obligate anaerobes, 261, *261*
Observations, 12
Oceanic distribution of heat, 639
Ochrolechia africana, 593
Ocotillos (*Fouquieria splendens*), 299
Oenothera (evening primrose), *3*
Ohio buckeye (*Aesculus glabra*), 184
Oil bodies, 66, *66*
Oils, definition of, 31
Okazaki fragments, 392, *393*
Oleander (*Nerium oleander*), 135
Oligosaccharides, 26, *27*
Ombrophytum subterraneum, 550
Onion (*Allium*)
A. cepa, 75, 135, *393*
bulbs, 106, *106*
pH and, *322*
plasmodesmata, *69*
Oogamy, 468
Oomycetes, 480–481, *481*
Open-reading frame (ORF), 381
Operational habitat, 614
Operculum, 493
Opuntia (cactus)
chollas, *202*
evolutionary diversification of, 446
O. polyacantha or prickly pear, *94, 394*
shoots of, *288*
Orange dodder (*Cuscuta*), 231, *277*
Orange tree, *97*
Orchidaceae, 559
Orchids
aerial roots of, 163, *163*
pistils, 209
structure of, 94–95, *95*
Order, 439, 442
Oreocereus (cactus)
O. celsianus, 449
O. trollii, 438
ORF (open-reading frame), 381
Organelles. *See also specific organelles*
definition of, 43
division of, *89,* 89–90
volumes in plant cells, 53–54, *54*
Organic molecules
functional groups, 24, *25*
polymeric construction and, 24–25
Organisms. *See also specific organisms*
ability to synthesize organic compounds, 23
Organization
chemosynthesis and, 433
of living organisms, 13
Organs, plant. *See* Leaves; Roots; Stems

Secondary phloem, 185, *185*, *186*
Secondary plant body, 98, *98*
Secondary structure, 28–29, *29*
Secondary tissues, 170
Secondary vascular bundles, 193
Secondary xylem, in wood, 176, *176–182*, 180, 182, *184*, 184–185
Second-order reactions, 22, *22*, 24, 35
Sedges
　growth of, 561, *561*
　identification of, 96, *97*
　roots of, 76
Sedum spectabilis (stonecrop), *135*
Seed coat (testa), 215, *216*
Seed cones
　cycad, *542*
　larch, *538*
Seed dispersal, for phyletic speciation, 426, *426*
Seed ferns, 532–533, *533*
Seed plants, *3*, 506, *529*. See also specific seed plants
Seedlings, leaf shapes of, 132, *132*
Seeds
　angiosperm, *205*
　civilization and, *210*, 210–211
　deposited, *224*, 225
　development of, *213–216*, 214–215
　dispersal, 222, *222*
　evolution of, 528, 531–532, *532*
　minerals storage in, 329
　oil in, 160–161
　plant adaptation and, 11
　production, 487
　sexual reproduction and, 201–203
Seepweed (*Sueda*), 316
Selaginella
　megagametophyte, *487*
　poikilohydry, 305
　S. lepidophylla or resurrection plant, 305, *511–512*, *512*
　S. peruviana, *97*
　S. willdenowii, *512*
Selectively permeable membranes (differentially permeable membranes), 49, *49*, *50*, 289
Self-assembly, 29
Selfing (self-fertilization), *399*, 399–400
Self-pollination, 217
Semiconservative replication, 392
Semideserts, 299, *299*
Sempervivum, *135*
Sempervivum arachnoideum, *254*
Senecio rotundifolia, 144, *144*
Sensitive plant (*Mimosa pudica*), 298
Sensory systems, 336
Septal nectaries, 558
Sepals, *203*, *206*, *207*
Septum (septa)
　asparagales, 558
　fungal, 579, *579*, *580*
Sequence of deoxyribonucleotides, 30
Sequoiadendron giganteum (redwoods), 533, *534*
SER (smooth endoplasmic reticulum), 59
Serpentine soil, 316
Sessile leaf, 128
Seta (setae), 493
Sex cells (gametes), 83
Sexual reproduction
　advantages of, *201*, 201–202
　disadvantages of, 202
　fertilization, 213, *213*

gametophytes and, *211*, 211–212, *212*
genetics and, 390–392, *391*, *392*
of green algae, 468
oomycetes, 481
plant life cycle and, 203–204, *203–205*
Sheathing leaf base, 129, *129*
Shepherd's purse (*Capsella*), 87, 214
Shoots
　long, 536, *536*
　nutrient storage in, 106
　short, 536, *536*
　vs. stems, 103
Short-day plants, 353–354, *353–356*
Shrublands, 645
Sieve areas, 116, *117*
Sieve cells, 116, *117*, 118, 536
Sieve elements
　in active transport of sugars, 300, *301*
　types of, 116–118, *117*
Sieve plates, 116, *117*, 118
Sieve pore, 116, *117*
Sieve tube, 116–118
Sieve tube members, 116–117, *117*, *118*
Sieve tube plastids, 564, *565*
Sigillaria, *510*
Sigillaria rugosa, *533*
Sigillaria saulli, *533*
Silt, 320
Silverswords (*Argyroxiphium*), 428
Simple cones, 537, *537*
Simple fruits, 222
Simple leaves, 130, *130*
Simple sugars (monosaccharides), *25*, 25–26, *26*
Single bonds, 21, *21*
Sinks, 301, *301*
Siphonoglossa, 556
Siphonostele, 506
Siphonous body, 467
Sister chromatids, *80*
Skepticism, 5
Skunk cabbage (*Symplocarpus foetidus*), 272, *273*
Sleep movements, 340
Slime molds, 585–586, *586*
Smilax (greenbrier), *159*
Smooth endoplasmic reticulum (SER), 59
Smuts, 594
Snapdragon (*Antirrhinum majus*), 358
Sodium, reaction with chlorine, 18–19, *19*
Softwoods, 176
Soil
　boundaries of geographic range as, 623, *624*
　endodermis, selective absorption of substances and, 323, *324*
　fallow, 313
　as habitat component, 614–615, *615*
　mineral availability and, *319–322*, 319–324, *324*
　mineral nutrition and, 313–314
　organisms in, 314, *314*
　particle size, 320, *320*
　pH, 322, *322*
　sandy, 320, *320*, *321*
　spaces in, 155, 156f
　texture, 304
　types, water potentials of, 307, *307*
　water potential of, 291, *291*
　wet *vs.* dry, 303, *303*
Solanum tuberosum (potato)
　chromosomes, *75*
　classification of, 569, *570*

flowers of, *569*
soil pH and, *322*
tubers, 93, *106*
Somatic mutations, 396
Sophora (mescal bean), 36
Soredia, 593
Sori (sorus), 518, *521*
Sotols (*Dasylirion leiophyllum*), *420*
Sources, 302, *302*
Soybeans, 322
Space exploration, life on other planets and, 12, 13
Spadix, 221–222
Spanish moss (*Tillandsia usneoides*), *562*
Spearmint, *131*
Speciation, 425–429, *425–429*
Species, 439, *439*
Species composition, of ecosystems, *629*, 629–630, *630*
Species epithet, 442
Specific mass transfer, 300, *301*
Speckled alder (*Alnus rugosa*), *116*
Speculative philosophy, 4
Sperm cells
　fertilization of, 203, *204*, 213, *213*
　as haploid cells, *83*, 84
　in pollen tube, 211
Spermatophytes, 484, 528, *529*
Sphaerocarpos, 119
Sphagnum moss, *494*
Spike, 221, *222*
Spinach
　leaves, development of plastid genome in, 90, *90*
　pH and, *322*
Spindle microtubules, *80*, 81
Spindles, 79–80, 88
Spines, 146–147, *147*
Spirilla, 599, *600*
Spirits, fermentation of, 283, *284*
Spirogyra, 471, *471*
Split genome viruses, 384
Spongy mesophyll, 136, *136*, *137*, *139*
Sporangia
　fungal, 586
　of green algae, 468
　in true plants, 486
Sporangiophores, 515–516, *516*
Sporangiospores, 582, *582*
Spores
　formation of, 485
　fungal, 582, 582–583, *583*
　in plant life cycle, 203
Sporophylls, 513, *517*
Sporophyte generation (sporophyte phase)
　Bryophyta, 491, 493, *494*
　definition of, 203
　Hepatophyta, 497, *498*
　hornworts, 500, *500*
Sporophytes, 203, *494*
Squash mosaic virus, 385
St. John's wort (*Hypericum spathulatum*), *567*
Stachys (Lamb's ears), *135*
Stachys recta (mint family), *303*
Stamens
　structure of, *203*, *204*, 206, 208, *208*
　style maturation types, 217, *217*
Stamnostoma hyttonense, 532
Starch
　polymerization of, 26–27, *27*
　storage in amyloplasts, *56*, 56–57

Note: Unless otherwise indicated, all photographs and illustrations are under copyright of Jones & Bartlett Learning or have been kindly provided by the author.

Title page photo: Courtesy of Kate Rolland.

Section 3 opener: Courtesy of Chris McCoy.

Chapter 1

1.5a Courtesy of Chris McCoy. **1.5b** Courtesy of Chris McCoy. **1.8a** Courtesy of R. Fulginiti, University of Texas, Austin. **1.8b** Courtesy of R. Fulginiti, University of Texas, Austin.

Chapter 3

3.3 Courtesy of R. Fulginiti, University of Texas, Austin. **3.6a** © Don W. Fawcett/Photo Researchers, Inc. **3.13** © Dr Kari Lounatmaa/Photo Researchers, Inc. **3.17** © Keith R. Porter/Photo Researchers, Inc. **3.19a** Courtesy of Dr. Constantin Craciun, Babes-Bolyai University. **3.22b** Courtesy of R. Fulginiti, University of Texas, Austin. **3.23a** Courtesy of H. Mollenhauer, Texas A&M University. **3.23b** Courtesy of H. Mollenhauer, Texas A&M University. **3.26** © E.H. Newcomb & S. E. Frederick/Biological Photo Service. **3.27a** © Ablestock/Alamy Image. **3.29a** © Don W. Fawcett/Photo Researchers, Inc. **3.30** Courtesy of Louisa Howard, Dartmouth College, Electron Microscope Facility. **3.33** Courtesy of G. Montenegro, Universidad Católica, Santiago, Chile. **3.35a** Courtesy of Y. Sano, Laboratory of Woody Plant Biology, Hokkaido University. **3.37a, left** Courtesy of W. W. Thomson and M. Lazzard, University of California, Riverside. **3.37b** Courtesy of A. Koller, University of Georgia and T. Rost, Uiversity of California, Davis.

Chapter 4

4.10 © Biophoto Associates/Photo Researchers, Inc. **4.13** Photo by James Richardson. **4.14b** © Pr. G. Gimènez-Martìn/Photo Researchers, Inc. **4.14c** © Pr. G. Gimènez-Martìn/Photo Researchers, Inc. **4.14d** © Pr. G. Gimènez-Martìn/Photo Researchers, Inc. **4.14e** © Pr. G. Gimènez-Martìn/Photo Researchers, Inc. **4.17a** Courtesy of Dr. Kevin C. Vaughn, Southern Weed Science Research Unit/ARS/USD. **4.17b** Courtesy of Dr. Kevin C. Vaughn, Southern Weed Science Research Unit/ARS/USDA. **4.23b** Courtesy of J. W. LaClaire II, University of Texas, Austin. **4.23c** Courtesy of J. W. LaClaire II, University of Texas, Austin. **4.25a** Courtesy of J. L. Scott and K. W. Bullock, College of William and Mary. **4.25b** Courtesy of J. Dodge, University of London. **4.27** Courtesy of C. Pueschell, State University of New York, Binghamton. **4.29** Courtesy of P. Beech, Deakin University, Australia.

Chapter 5

5.5 Courtesy of Leo Martin, Phoenix, Arizona. **5.7a** Courtesy of W. W. Thomson and R. Balsamo, University of California, Riverside. **5.7b** Courtesy of M. J. Talbot and C. E. Offler, University of Newcastle, Australia. **5.23** Courtesy of Urs Eggli, Municipal Succulent Collection, Zurich. **5.24b** Courtesy of Urs Eggli, Municipal Succulent Collection, Zurich. **5.24c** Courtesy of Urs Eggli, Municipal Succulent Collection, Zurich. **5.28a** Courtesy of N.C. Brown Center for Ultrastructure Studies, SUNY–Environmental Science and Forestry, Syracuse, NY 13210. **5.28b** Courtesy of N.C. Brown Center for Ultrastructure Studies, SUNY-Environmental Science and Forestry, Syracuse, NY 13210. **5.28d** Courtesy of D. Romanovicz and J. D. Mauseth, University of Texas, Austin. **5.33** Courtesy of E. Schneider, University of Minnesota and S. Carlquist, Santa Barabara Botanic Gardens. **5.35** Courtesy of E. Schneider and S. Carlquist, Santa Barabara Botanic Gardens. **5.38b** Courtesy of Robert Warmbrodt, National Agricultural Library, Agricultural Research Service, USDA. **5.38c** Courtesy of Robert Warmbrodt, National Agricultural Library, Agricultural Research Service, USDA.

Chapter 6

6.19bc Courtesy of L. G. Clark, Iowa State University, and X. Londono, Columbian Bamboo Society. **6.21** Courtesy of G. Montenegro, Universidad Católica, Santiago, Chile. **6.27ab** Courtesy of N. R. Lersten and J. D. Curtis, Iowa State University. **6.29** Courtesy of Christian R. Lacroix and Usher Posluszny, University of Guelph. **6.30a** Courtesy of Christian R. Lacroix and Usher Posluszny, University of Guelph. **6.30b** Courtesy of Christian R. Lacroix. **6.31abc** Courtesy of Nancy Dengler, University of Toronto. **6.34ab** Courtesy of B. K. Kirchoff, University of North Carolina, and R. Rutishauser, Institute of Systematic Botany, Zurich. **6.45c** © Ariel Bravy/ShutterStock, Inc. Box **6.1d** © AlenKadr/ShutterStock, Inc.

Chapter 7

7.11 Courtesy of Lewis J. Feldman, University of California, Berkeley. **7.20** Courtesy of Judy Jernstedt, University of California, Davis. **7.21cd** Courtesy of R. L. Peterson and D. Hern, University of Guelph. **7.22a** © Nigel Cattlin/Alamy. **7.22b** Courtesy of E. H. Newcomb, University of Wisconsin Department of Botany. **7.23** Courtesy of C. L. Calvin, Rancho Santa Ana Botanical Garden.

Chapter 8

8.4ab Courtesy of R. Evert, University of Wisconsin. **8.9** Courtesy of D. A. Eggert and D. D. Gaunt, University of Illinois and American Journal of Botany, Botanical Society of America. **8.11** Courtesy of N. C. Brown Center for Ultrastructure Studies, SUNY–Environmental Science and Forestry, Syracuse, NY. **8.12e** Courtesy of Yuzou Sano, Laboratory of Woody Plant Biology, Hokkaido University.

Chapter 9

Opener Courtesy of Chris McCoy. **9.2c** Courtesy of Dr. Michael Grant, University of Colorado. **9.11a** Courtesy of Alan Prather, Michigan State University. **9.11bc** Courtesy of Beryl Simpson, University of Texas. **9.13** Courtesy of S. Muccifora. **9.14ac** Courtesy of Allison Snow, Ohio State University. **9.19d** © AgStock Images, Inc./Alamy. **9.24** Courtesy of J. B. and M. E. Nasralah, Cornell University. **9.32a** © Tomo Jesenicnik/ShutterStock, Inc. **9.32b** © Martin Maun/ShutterStock, Inc.

Chapter 10

10.21 © stevehullphotography/ShutterStock, Inc. **10.27a** Courtesy of R. Muhaidat and R. Sage, University of Toronto. Box **10.1a** © Mikhail Olykainen/ShutterStock, Inc. Box **10.1b** © Kent Knudson/PhotoLink/Photodisc/Getty Images.

Chapter 11

11.08c © Shi Yali/ShutterStock, Inc. **11.16** © FK/Alamy Images.

Chapter 13

13.1 Courtesy of Johannes A. Postma, MSc, Penn State University. **13.2** © Holt Studios International Ltd/Alamy Images. **13.4a** Courtesy of Matthew Godfrey. **13.4b** Courtesy of Dr. H. J. Arnott, University of Texas, Arlington. **13.5** Courtesy of CIAT from the files of Howard F. Schwartz, Colorado State University, Bugwood.org. **13.6a** © blickwinkel/Alamy Images. **13.6b** © Holt Studios International Ltd/Alamy Images. Box **6.2** © JoLin/ShutterStock, Inc.

Chapter 14

14.5ab Courtesy of Judith Verbeke. **14.9** © David Scheuber/ShutterStock, Inc. **14.11ab** Micrographs courtesy of Randy Moore. **14.15a** Courtesy of Sima Tokajian, PhD, Lebanese American University. **14.15b** Courtesy of Dennis Gray, University of Florida. **14.18** © WILDLIFE GmbH/Alamy. **14.34b** Courtesy of Chris McCoy. **14.35b** Courtesy of Chris McCoy. Box **14.1b** © Marevision/age fotostock

Chapter 15

Opener Courtesy of Chris McCoy. **15.8a** © Dr. Donald Fawcett, O. L. Miller, B. R. Beatty/Visuals Unlimited, Inc. **15.19** Courtesy of Dr. Jason Kang/National Cancer Institute. **15.25b** Courtesy of M. Bonness and T. J. Mabry, University of Texas, Austin. **15.25c** Courtesy of Dennis Gray, University of Florida. **15.26** Courtesy of Elliot W. Kitajima, PhD, University of São Paulo, Brazil. **15.27a** Courtesy of G. Stubbs, Vanderbilt University. K. Namba, Osaka University and D. Caspar, Florida State University. **15.29ab** Courtesy of Dr. Thomas Isakeit, Texas A&M University.

Chapter 16

Opener Courtesy of Chris McCoy. **16.1** © Hemera/Thinksock. **16.15** Courtesy of the National Seed Storage Laboratory, Fort Collins, Colorado/USDA ARS.

Chapter 17

17.2 Courtesy of NASA. **17.3** © Image Asset Managemen/age fotostock. **17.5** Courtesy of Dr. Thomas N. Taylor, University of Kansas. **17.6a** © Emilia Kun/ShutterStock, Inc. **17.6b** © Jerry Horbert/ShutterStock, Inc. **17.6c** © Uschi Hering/ShutterStock, Inc. **17.6d** © Anne Kitzman/ShutterStock, Inc. **17.7** © Charles A.

Blakeslee/age fotostock. **17.8** © Mike Boyatt/age fotostock. **17.17** Photo by Forest & Kim Starr. **17.18b** Courtesy of G. Ardisson, Université de Nice. **17.20** Courtesy of NASA/JPL/Space Science Institute. **17.21** © Reuters/Edgar Montana/Landov. **17.22b** Courtesy UC San Diego. **17.24** Courtesy of Doctor William Schopf, Professor of Paleobiology & Director of IGPP CSEOL, University of California.

Chapter 18
Box 18.1a © Visual Arts Library (London)/Alamy Images. **Box 18.1b** © National Library of Medicine.

Chapter 19
Opener © Claude Huot/ShutterStock, Inc. **19.2a** © Dr. Peter Siver/Visuals Unlimited/Alamy Images. **19.2b** Courtesy of Dr. José Luis Iriarte M., Universidad Austral de Chile. **19.2c** Courtesy of D. Romanovicz, Institute for Cellular and Molecular Biology, University of Texas, Austin. **19.2d** © Andrew Syred/Photo Researchers, Inc. **19.2e** © Michael Abbey/Photo Researchers, Inc. **19.2h** © Marevision/age fotostock. **19.6** Courtesy of E. Gantt. **19.13** © Phototake/Alamy Images. **19.17** Courtesy of Dr. Ursula Goodenough, Washington University in St. Louis. **19.18** Courtesy of Great Lakes, EPA. **19.19a** © Michael Abbey/Photo Researchers, Inc. **19.19b** © blickwinkel/Alamy Images. **19.19c** © M. I. Walker/Phto Researchers, Inc. **19.23** © FLPA/Alamy Images. **19.24** © Jeffrey L. Rotman/Corbis. **19.29a** Courtesy of M. L. Shih, J.-Y. Floch, and L. M. Srivastava, Simon Fraser University. **19.29b** Courtesy of M. L. Shih, J.-Y. Floch, and L. M. Srivastava, Simon Fraser University. **19.30** Courtesy of Elizabeth Ruck, University of Texas, Austin. **19.31a** © Tom Adams/Visuals Unlimited, Inc. **19.31b** © Dr. Robert Calentine/Visuals Unlimited, Inc. **19.32** Courtesy of the Provasoli-Guillard National Center for Marine Algae and Microbiota (NCMA). **19.33** © Photo Researchers, Inc. **19.34** Courtesy of P. Alejandro Díaz. **19.35** Photo by James Richardson. **19.36** © Phototake/Alamy Images.

Chapter 20
20.8 © Robert and Linda Mitchell. **20.11a** Courtesy of R. Fulginiti, University of Texas, Austin. **20.14a** Courtesy of Dan Scheirer, Duke University. **20.14b** Courtesy of Dan Scheirer, Duke University. **20.17a** © Oldrich Simek/age fotostock. **20.18a** Courtesy of Karen Renzaglia and Fred Alsop III. **20.18b** Courtesy of Karen Renzaglia and Fred Alsop III. **20.18c** Courtesy of Karen Renzaglia and Fred Alsop III. **20.19e** © Jessica M. Budke, University of Connecticut. **20.20** © Maslov Dmitry/ShutterStock, Inc. **20.21a** Courtesy of Dr. Andrew Spink (www.andrewspink.nl). **20.23** © Biophoto Associates/Science Source/Photo Researchers, Inc. **20.24a** Courtesy of Chrissen Gemmil, Centre of Biodiversity and Ecology Research, University of Waikato. **20.25a** © P00004 Verbiesen Henk/age fotostock. **20.27** Courtesy of Sharon Bartholomew-Began, West Chester University.

Chapter 21
Opener Courtesy of Tommy R. Navarre. **21.12** © Field Museum of Natural History, #CSGEO75400c. **21.14a** Courtesy of T. Delevoryas, University of Texas. **21.15b** © Michael P. Gadomski/Photo Researchers, Inc. **21.15c** Photo by James Richardson. **21.15d** Photo by James Richardson. **21.18** Courtesy of U. S. Fish and Wildlife Services. **21.21c** Courtesy of Chad Husby/Montgomery Botanical Center. **21.22** © Jim Solliday/Biological Service. **21.27** Courtesy of T. Delevoryas, University of Texas and R. C. Hope. **21.28c** © Eduardo M. Rivero/age fotostock. **21.28d** Courtesy of Chad Husby/Montgomery Botanical Center. **21.31** Courtesy of Dr. Shu-Chuan Hsiao, National Chunghsing University. **21.32a** Courtesy of Tommy R. Navarre. **21.33a** © Biophoto Associates/Photo Researchers, Inc. **21.34b** Courtesy of Dr. Shu-Chuan Hsiao, National Chunghsing University.

Chapter 22
22.3abc Courtesy of C. B. Beck, University of Michigan. **22.9** © Field Museum of Natural History, #CSGEO75400c. **22.10** Courtesy of B. M. Stidd and T. L. Phillips, Western Illinois University and American Journal of Botany, Botanical Society of America. **22.18b** © Robert and Linda Mitchell. **22.23a** © Thomas Mounsey/ShutterStock, Inc. **22.23b** © Jubal Harshaw/ShutterStock, Inc. **22.25d** © Phototake/Alamy Images. **22.27c** Courtesy of Dr. Knut Norstog. Used with permission by Paul

Norstog. **22.31a** Courtesy of Shigenobu Aoki. **22.31b** © Robert and Linda Mitchell. **22.32b** Courtesy of the Dale A. Zimmerman Herbarium, Western New Mexico University. **22.32c** © Dan Cheatham/Photo Researchers, Inc. **22.33a** © mdd/ShutterStock, Inc. **22.33c** © Tier und Naturfotografie/age fotostock.

Chapter 23
23.3 Courtesy of Dr. H. L. Mogensen. **23.6** Courtesy of D. Dilcher, Florida Museum of Natural History, University of Florida. **23.7ab** Courtesy of D. Dilcher, Florida Museum of Natural History, University of Florida. **23.7c** Courtesy of S. R. Manchester, Florida Museum of Natural History. **23.11b** Photo by Thomas J Lemieux, University of Colorado. **23.12a** Courtesy of R. A. Hilsenbeck, Sul Ross State University. **23.12b** Courtesy of W. J. Elisens, University of Oklahoma. **23.18b** Courtesy of David Bogler, Missouri Botanical Garden. **23.21a** © Geoff Kidd/age fotostock. **23.22c** © iofoto/ShutterStock, Inc. **23.28ab** Courtesy of H. Dietmar Behnke, University of Heidelberg. **23.30d** © Dr. Morley Read/ShutterStock, Inc.

Chapter 24
24.2 Courtesy of C. W. Mims, J. Taylor, and E. A. Richardson, University of Georgia. **24.3a** Courtesy of Dr. Garry Cole, University of Texas, San Antonio. **24.3b** Courtesy of Scott Bauer/USDA. **24.4c** Courtesy of J. W. Kimbrough, University of Florida. **24.4d** Courtesy of C. W. Mims, University of Georgia. **24.5** Courtesy of E. C. Swann and C. W. Mims, University of Georgia. **24.6a** Courtesy of C. W. Mims, University of Georgia. **24.6b** © Smolych Iryna/ShutterStock, Inc. **24.6c** © P00017 d'Hendecourt A/age fotostock. **24.8a** Courtesy of C. W. Mims, E. A. Richardson, and J. W. Kimbrough, University of Georgia. **24.8c** Courtesy of C. W. Mims, University of Georgia. **24.9** Courtesy of Dr. Garry Cole, University of Texas, San Antonio. **24.10** © Phototake/Alamy Images. **24.11** Photo by Dennis Drenner. **24.15a** © Daniel L. Geiger/SNAP/Alamy Images. **24.15b** Courtesy of Dr. Catherine Pears, Oxford University. **24.17** Photo by James Richardson . **24.22** © Biophoto Associates/Photo Researchers, Inc. **24.24b** © S. L. Flegler/Visual Unlimited. **24.25** Courtesy of C. W. Mims, University of Georgia. **24.28** Photo by Clemson University–USDA Cooperative Extension Slide Series, Bugwood.org. **24.29** © Damian Herde/ShutterStock, Inc.

Chapter 25
25.2a Courtesy of Rocky Mountain Laboratories, NIAID, NIH. **25.2b** Courtesy of Janice Carr/CDC. **25.2c** Courtesy of Rob Weyant and Janice Carr/CDC. **25.4** © Dr. Peter Siver/Visuals Unlimited/Alamy Images. **25.5** Courtesy of P. Beech, Deakin University, Australia. **25.8** © Phototake/Alamy Images. **25.10** © Dr. M. Rohde, BBF/Photo Researchers, Inc. **25.13** Courtesy of A. P. Singh, CSIRO and SCION, Rotorua, New Zealand. T. Nilsson and G.F. Daniel, Swedish University of Agricultural Sciences, Uppsala, Sweden. **25.14** © luchschen/ShutterStock, Inc. **25.15** © Chung Ooi Tan/ShutterStock, Inc.

Chapter 26
26.9a © Stephanie Swartz/ShutterStock, Inc. **26.9c** Courtesy of D.R. Mullineaux/Cascades Volcano Observatory/USGS. **26.9d** Courtesy of Lyn Topinka/Cascades Volcano Observatory/USGS. **26.11cd** Courtesy of Jan Emming. **26.12a** © Aleksander Bolbot/ShutterStock, Inc. **26.12b** © Janno/ShutterStock, Inc. **26.16a** © Marco Alegria/ShutterStock, Inc. **26.16b** © Robert and Linda Mitchell. **26.17** © Robynrg/ShutterStock, Inc. **26.21** Courtesy of Roger del Moral, University of Washington. **Box 26.1** © Danny E. Hooks/ShutterStock, Inc.

Chapter 27
27.1 © steve estvanik/ShutterStock, Inc. **27.2a** © Steffen Foerster Photography/ShutterStock, Inc. **27.7** © Natalia Bratslavsky/ShutterStock, Inc. **27.9** © Lauren Hamilton/ShutterStock, Inc. **27.11** © Alex Neauville/ShutterStock, Inc. **27.13** © Zbynek Burival/ShutterStock, Inc. **27.14** © Anna Galejeva/ShutterStock, Inc. **27.15** Courtesy of NOAA Restoration Center. **27.17** © Terry W. Ryder/ShutterStock, Inc. **27.18a** © Eugene Buchko/ShutterStock, Inc. **27.18b** © Peter Weber/ShutterStock, Inc. **27.18c** Courtesy of Jan L. Beyers, USDA Forest Service. **27.21b** © Josef F. Stuefer/ShutterStock, Inc. **27.22** © Jeffrey M. Frank/ShutterStock, Inc. **27.23** © Zastavkin/ShutterStock, Inc. **27.26** © Dr. Morley Read/ShutterStock, Inc. **27.28** Photo by Forest & Kim Starr.